Claude Cohen-Tannoudji
Bernard Diu
Franck Laloë

# QUANTUM MECHANICS

## Volume I

Translated from the French by **Susan Reid Hemley, Nicole Ostrowsky, Dan Ostrowsky**

WILEY-VCH Verlag GmbH & Co. KGaA

**HERMANN**
**Publishers in arts and science**  **Paris**

Published by Hermann and by John Wiley & Sons. Inc.



*Mécanique quantique* was originally published in French by Hermann in 1973; second revised and enlarged edition 1977.

***Library of Congress Cataloging in Publication Data :***

Cohen-Tannoudji, C.
Quantum mechanics.

Translation of Mecanique quantique.
"A Wiley-Interscience publication."
Includes index.

1. Quantum theory. I. Diu, Bernard, joint author. II. Laloë, Franck, joint author. III. Title.

QC174.12.C6313 530.1'2 76-5874
ISBN 978-0-471-16433-3 (Volume 1)
ISBN 978-0-471-16435-7 (Volume 2)
ISBN 978-0-471-56952-7 (Set)

# Directions for Use

This book is made up of chapters and their complements :

– *The chapters* contain the fundamental concepts. Except for a few additions and variations, they correspond to a course given in the last year of a typical undergraduate physics program.

These fourteen chapters are *complete in themselves* and can be studied independently of the complements.

– *The complements* follow the appropriate chapter. They are listed at the end of each chapter in a "*reader's guide*" which discusses the difficulty and importance of every one of them. Each is labelled by a letter followed by a subscript which gives the number of the corresponding chapter (for example, the complements of chapter V are, in order, $A_V$, $B_V$, $C_V$...). They can be recognized immediately by the symbol ● which appears at the top of each of their pages.

The complements vary : some are intended to expand the treatment of the corresponding chapter or to provide more detailed discussion of certain points; others describe concrete examples or introduce various physical concepts. One of the complements (usually the last one) is a collection of exercises.

The *difficulty* of the complements varies. Some are very simple examples or extensions of the chapter, while others are more difficult (some are at graduate level); in any case, the reader should have studied the material in the chapter before using the complements.

*The student should not try to study all the complements of a chapter at once.* In accordance with his aims and interests, he should choose a small number of them (two or three, for example), plus a few exercises. The other complements can be left for later study.

Some passages within the book have been set in small type and these can be omitted on a first reading.

# Table

## VOLUME I

## Complements of chapter IV



## Complements of chapter V

## Complements of chapter VI

## Complements of chapter VII



## VOLUME II



## Complements of chapter VIII

## Complements of chapter XI



## Chapter XII An application of perturbation theory : the fine and hyperfine structure of the hydrogen atom . . . 1209



## Complements of chapter XII

## Complements of chapter XIII

## Complements of chapter XIV

# Quantum mechanics

# QUANTUM MECHANICS

# Introduction

## Structure and level of this text

It is hardly necessary to emphasize the importance of quantum mechanics in modern physics and chemistry. Current university programs naturally reflect this importance. In French universities, for example, an essentially qualitative introduction to fundamental quantum mechanical ideas is given in the second year. In the final year of the undergraduate physics program, basic quantum mechanics and its most important applications are studied in detail.

This book is the direct result of several years of teaching quantum mechanics in the final year of the undergraduate program, first in two parallel courses at the Faculté des Sciences in Paris and then at the Universités Paris VI and Paris VII. We felt it to be important to mark a clear separation, in the structure of this book, between the two different but complementary aspects (lectures and recitations) of the courses given during this time. This is why we have divided this text into two distinct parts (see the "Directions for Use" at the beginning of the book). On the one hand, the chapters are based on the lectures given in the two courses, which we compared, discussed and expanded before writing the final version. On the other hand, the "complements" grew out of the recitations, exercises and problems given to the students, and reports that some of them prepared. Ideas also came from other courses given under other circumstances or at other levels (particularly in the graduate programs). As we pointed out in the "Directions for Use", the chapters as a whole constitute, more or less, a course we would envisage teaching to fourth-year college students or those whose level is equivalent. However, the complements are not intended to be treated in a single year. The reader, teacher or student, must choose between them in accordance with his interests, tastes and goals.

Throughout the writing of this book, our constant concern has been to address ourselves to students majoring in physics, like those we have taught over the past several years. Except in a few complements, we have not overstepped those limits. In addition, we have endeavored to take into account what we have seen of students' difficulties in understanding and assimilating quantum mechanics, as well as their questions. We hope, of course, that this book will also be of use to other readers such as graduate students, beginning research workers and secondary school teachers.

The reader is not required to be familiar with quantum physics : few of our students were. However, we do think that the quantum mechanics course we propose (see "General approach", below) should be supplemented by a more descriptive and more experimentally oriented course, in atomic physics for example.

## General approach

We feel that familiarity with quantum mechanics can best be acquired by using it to solve specific problems. We therefore introduce the postulates of quantum mechanics very early (in chapter III), so as to be able to apply them in the rest of the book. Our teaching experience has shown it to be preferable to introduce all the postulates together in the beginning rather than presenting them in several stages. Similarly, we have chosen to use state spaces and Dirac notation from the very beginning. This avoids the useless repetition which results from presenting the more general bra-ket formalism only after having developed wave mechanics uniquely in terms of wave functions. In addition, a belated change in the notation runs the risk of confusing the student, and casting doubts on concepts which he has only just acquired and not yet completely assimilated.

After a chapter of qualitative introduction to quantum mechanical ideas, which uses simple optical analogies to familiarize the reader with these new concepts, we present, in a systematic fashion, the mathematical tools (chapter II) and postulates of quantum mechanics as well as a discussion of their physical content (chapter III). This enables the reader, from the beginning, to have an overall view of the physical consequences of the new postulates. Starting with the complements of chapter III we take up applications, beginning with the simplest ones (two-level systems, the harmonic oscillator, etc.) and becoming gradually more complicated (the hydrogen atom, approximation methods, etc.). Our intention is to provide illustrations of quantum mechanics by taking many examples from different fields such as atomic physics, molecular physics and solid state physics. In these examples we concentrate on the quantum mechanical aspect of the phenomena, often neglecting specific details which are treated in more specialized texts. Whenever possible, the quantum mechanical results are compared with the classical ones in order to help the reader develop his intuition concerning quantum mechanical effects.

This essentially deductive viewpoint has led us to avoid stressing the historical introduction of quantum mechanical ideas, that is, the presentation and discussion of experimental facts which force us to reject the classical ideas. We have thus had to forego the inductive approach, which is nevertheless needed if physics is to be faithfully portrayed as a science in continual evolution, provoked by constant confrontation with experimental facts. Such an approach seems to us to be better suited to an atomic physics text or an introductory quantum physics course on a more elementary level.

Similarly, we have deliberately avoided any discussion of the philisophical implications of quantum mechanics and of alternative interpretations that have been proposed. Such discussions, while very interesting (see section 5 of the bibliography), seem to us to belong on another level. We feel that these questions can be fruitfully considered only after one has mastered the "orthodox" quantum theory whose impressive successes in all fields of physics and chemistry compelled its acceptance.

*Acknowledgements*

The teaching experiences out of which this text grew were group efforts, pursued over several years. We want to thank all the members of the various groups and particularly, Jacques Dupont-Roc and Serge Haroche, for their friendly collaboration, for the fruitful discussions we have had in our weekly meetings and for the ideas for problems and exercises which they have suggested. Without their enthusiasm and their valuable help, we would never have been able to undertake and carry out the writing of this book.

Nor can we forget what we owe to the physicists who introduced us to research, Alfred Kastler and Jean Brossel for two of us and Maurice Lévy for the third. It was in the context of their laboratories that we discovered the beauty and power of quantum mechanics. Neither have we forgotten the importance to us of the modern physics taught at the C.E.A. by Albert Messiah, Claude Bloch and Anatole Abragam, at a time when graduate studies were not yet incorporated into French university programs.

We wish to express our gratitude to Ms. Aucher, Baudrit, Boy, Brodschi, Emo, Heyvaerts, Lemirre, Touzeau for preparation of the manuscript.

# Foreword

This book is essentially a translation of the French edition which appeared at the end of 1973.

The text has undergone a certain number of modifications. The most important one is an addition of a detailed bibliography, with suggestions concerning its use appearing at the end of each chapter or complements.

This book was originally conceived for French students finishing their undergraduate studies or beginning their research work. It seems to us however that the structure of this book (separation into chapters and complements – see the "Directions for use") should make it suitable for other groups of readers. For example, for an undergraduate elementary Quantum Mechanics course, we would recommend using the most important chapters with their simplest complements. For a more advanced course, one could add the remaining chapters and use more difficult complements. Finally, it is hoped that some of the more advanced complements will help students in the transition from a regular Quantum Mechanics course to current research topics in various fields of Physics.

We wish to thank Nicole and Dan Ostrowsky, as well as Susan Hemley, for the care and enthusiasm which they brought to this translation. Their remarks often led to an improvement of the original text. In addition, we are grateful to Mrs. Audoin and Mrs. Mathieu for their aid in organizing the bibliography.

C. Cohen-Tannoudji
B. Diu
F. Laloë

CHAPTER I

# Waves and particles. Introduction to the fundamental ideas of quantum mechanics

## OUTLINE OF CHAPTER I

In the present state of scientific knowledge, quantum mechanics plays a fundamental role in the description and understanding of natural phenomena. In fact, phenomena which occur on a very small (atomic or subatomic) scale cannot be explained outside the framework of quantum physics. For example, the existence and the properties of atoms, the chemical bond and the propagation of an electron in a crystal cannot be understood in terms of classical mechanics. Even when we are concerned only with macroscopic physical objects (that is, whose dimensions are comparable to those encountered in everyday life), it is necessary, in principle, to begin by studying the behavior of their various constituent atoms, ions, electrons, in order to arrive at a complete scientific description. There are many phenomena which reveal, on a macroscopic scale, the quantum behaviour of nature. It is in this sense that it can be said that quantum mechanics is the basis of our present understanding of all natural phenomena, including those traditionally treated in chemistry, biology, etc...

From a historical point of view, quantum ideas contributed to a remarkable unification of the concepts of fundamental physics by treating material particles and radiation on the same footing. At the end of the nineteenth century, people distinguished between two entities in physical phenomena : matter and radiation. Completely different laws were used for each one. To predict the motion of material bodies, the laws of *Newtonian mechanics* (*cf.* appendix III) were utilized. Their success, though of long standing, was none the less impressive. With regard to radiation, the *theory of electromagnetism*, thanks to the introduction of Maxwell's equations, had produced a unified interpretation of a set of phenomena which had previously been considered as belonging to different domains : electricity, magnetism and optics. In particular, the electromagnetic theory of radiation had been spectacularly confirmed experimentally by the discovery of Hertzian waves. Finally, *interactions between radiation and matter* were well explained by the Lorentz force. This set of laws had brought physics to a point which could be considered satisfactory, in view of the experimental data at the time.

However, at the beginning of the twentieth century, physics was to be marked by the profound upheaval that led to the introduction of relativistic mechanics and quantum mechanics. The relativistic "revolution" and the quantum "revolution" were, to a large extent, independent, since they challenged classical physics on different points. Classical laws cease to be valid for material bodies travelling at very high speeds, comparable to that of light (relativistic domain). In addition, they are also found to be wanting on an atomic or subatomic scale (quantum domain). However, it is important to note that classical physics, in both cases, can be seen as an approximation of the new theories, an approximation which is valid for most phenomena on an everyday scale. For example, Newtonian mechanics enables us to predict correctly the motion of a solid body, providing it is non-relativistic (speeds much smaller than that of light) and macroscopic (dimensions much greater than atomic ones). Nevertheless, from a fundamental point of view, quantum theory remains indispensable. It is the only theory which enables us to understand the very existence of a solid body and the values of the macroscopic parameters (density, specific heat, elasticity, etc...) associated with it. At the present time, we do not yet have at our disposal a fully satisfactory theory unifying quantum and relativistic mechanics since difficulties have arisen in this domain. However, most atomic and molecular phenomena are well explained by the *non-relativistic quantum mechanics* that we intend to examine here.

This chapter is an introduction to quantum ideas and "vocabulary". *No attempt is made here to be rigorous or complete.* The essential goal is to awaken the curiosity of the reader. Phenomena will be described which unsettle ideas as firmly anchored in our intuition as the concept of a trajectory. We want to render the quantum theory "plausible" for the reader by showing simply and qualitatively how it enables us to solve the problems which are encountered on an atomic scale. We shall later return to the various ideas introduced in this chapter and go into further detail, either from the point of view of the mathematical formalism (chap. II) or from the physical point of view (chap. III).

In the first section (§A), we introduce the basic quantum ideas (wave-particle duality, the measurement process), relying on well-known optical experiments. Then we show (§ B) how these ideas can be extended to material particles (wave function, Schrödinger equation). We next study in more detail the characteristics of the "wave packet" associated with a particle, and we introduce the Heisenberg uncertainty relations (§ C). Finally, we discuss some simple cases of typical quantum effects (§ D).

## A. ELECTROMAGNETIC WAVES AND PHOTONS

### 1. Light quanta and the Planck-Einstein relations

Newton considered light to be a beam of particles, able, for example, to bounce back upon reflection from a mirror. During the first half of the nineteenth century, the wavelike nature of light was demonstrated (interference, diffraction). This later enabled optics to be integrated into electromagnetic theory. In this framework, the speed of light, $c$, is related to electric and magnetic constants and light polarization phenomena can be interpreted as manifestations of the vectorial character of the electric field.

However, the study of *blackbody* radiation, which electromagnetic theory could not explain, led Planck to suggest the hypothesis of the *quantization of energy* (1900): for an electromagnetic wave of frequency $\nu$, the only possible energies are integral multiples of the quantum $h\nu$, where $h$ is a new fundamental constant. Generalizing this hypothesis, Einstein proposed a return to the particle theory (1905): light consists of a beam of *photons*, each possessing an energy $h\nu$. Einstein showed how the introduction of photons made it possible to understand, in a very simple way, certain as yet unexplained characteristics of the photoelectric effect. Twenty years had to elapse before the photon was actually shown to exist, as a distinct entity, by the Compton effect (1924).

These results lead to the following conclusion: the interaction of an electromagnetic wave with matter occurs by means of *elementary indivisible processes*, in which the radiation appears to be composed of particles, the photons. Particle parameters (the energy $E$ and the momentum $\mathbf{p}$ of a photon) and wave parameters (the angular frequency $\omega = 2\pi\nu$ and the wave vector $\mathbf{k}$, where $|\mathbf{k}| = 2\pi/\lambda$, with $\nu$ the frequency and $\lambda$ the wavelength) are linked by the fundamental relations:

$$\boxed{\begin{aligned} E &= h\nu = \hbar\omega \\ \mathbf{p} &= \hbar\mathbf{k} \end{aligned}} \qquad \text{(Planck-Einstein relations)} \qquad \text{(A-1)}$$

where $\hbar = h/2\pi$ is defined in terms of the Planck constant $h$:

$$h \simeq 6.62 \; 10^{-34} \text{ Joule} \times \text{second} \qquad \text{(A-2)}$$

During each elementary process, energy and total momentum must be conserved.

## 2. Wave-particle duality

Thus we have returned to a particle conception of light. Does this mean that we must abandon the wave theory? Certainly not. We shall see that typical wave phenomena such as interference and diffraction could not be explained in a purely particle framework. Analyzing Young's well-known double-slit experiment will lead us to the following conclusion : a complete interpretation of the phenomena can be obtained only by conserving *both* the wave aspect and the particle aspect of light (although they seem *a priori* irreconcilable). We shall then show how this paradox can be resolved by the introduction of the fundamental quantum concepts.

### a. ANALYSIS OF YOUNG'S DOUBLE-SLIT EXPERIMENT

The device used in this experiment is shown schematically in figure 1. The monochromatic light emitted by the source $\mathcal{S}$ falls on an opaque screen $\mathcal{P}$ pierced by two narrow slits $F_1$ and $F_2$, which illuminate the observation screen $\mathcal{E}$ (a photographic plate, for example). If we block $F_2$, we obtain on $\mathcal{E}$ a light intensity distribution $I_1(x)$ which is the diffraction pattern of $F_1$. In the same way, when $F_1$ is obstructed, the diffraction pattern of $F_2$ is described by $I_2(x)$. When the two slits $F_1$ and $F_2$ are open at the same time, we observe a system of interference fringes on the screen. In particular, we note that the corresponding intensity $I(x)$ is not the sum of the intensities produced by $F_1$ and $F_2$ separately:

$$I(x) \neq I_1(x) + I_2(x) \qquad \text{(A-3)}$$

How could one conceive of explaining, in terms of a particle theory (seen, in the preceding section, to be necessary), the experimental results just described ? The existence of a diffraction pattern when only one of the two slits is open could, for example, be explained as being due to photon collisions with the edges of the slit. Such an explanation would, of course, have to be developed more precisely, and a more detailed study would show it to be insufficient. Instead, let us concentrate on the interference phenomenon. We could attempt to explain it by an interaction between the photons which pass through the slit $F_1$ and those which pass through the slit $F_2$. Such an explanation would lead to the following prediction : if the intensity of the source $\mathscr{S}$ (the number of photons emitted per second) is diminished until the photons strike the screen practically one by one, the interaction between the photons must diminish and, eventually, vanish. The interference fringes should therefore disappear.

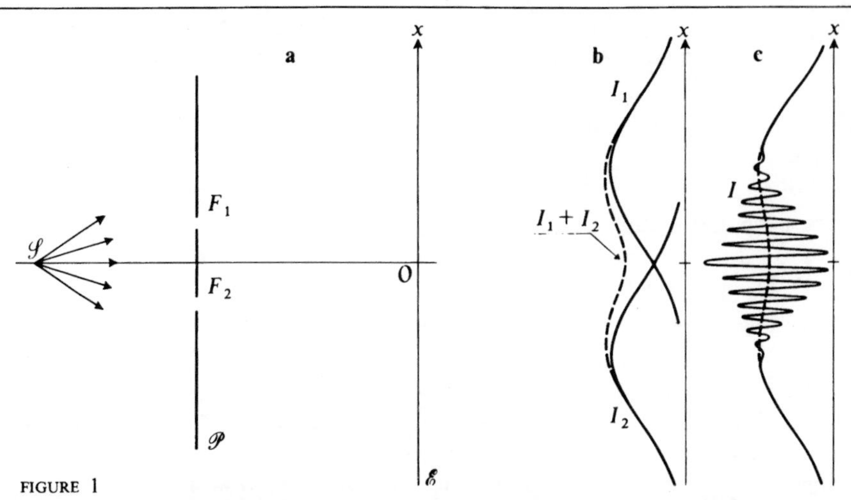


FIGURE 1

**Diagram of Young's double-slit light interference experiment (fig. a). Each of the slits $F_1$ and $F_2$ produces a diffraction pattern on the screen $\mathscr{E}$. The corresponding intensities are $I_1(x)$ and $I_2(x)$ (solid lines in figure b). When the two slits $F_1$ and $F_2$ are open simultaneously, the intensity $I(x)$ observed on the screen is not the sum $I_1(x) + I_2(x)$ (dashed lines in figures b and c), but shows oscillations due to the interference between the electric fields radiated by $F_1$ and $F_2$ (solid line in figure c).**

Before we indicate the answer given by experiment, recall that the wave theory provides a completely natural interpretation of the fringes. The light intensity at a point of the screen $\mathscr{E}$ is proportional to the square of the amplitude of the electric field at this point. If $E_1(x)$ and $E_2(x)$ represent, in complex notation, the electric fields produced at $x$ by slits $F_1$ and $F_2$ respectively (the slits behave like secondary sources), the total resultant field at this point when $F_1$ and $F_2$ are both open is * :

$$E(x) = E_1(x) + E_2(x) \tag{A-4}$$

Using complex notation, we then have :

$$I(x) \propto |E(x)|^2 = |E_1(x) + E_2(x)|^2 \tag{A-5}$$

Since the intensities $I_1(x)$ and $I_2(x)$ are proportional, respectively, to $|E_1(x)|^2$ and $|E_2(x)|^2$, formula (A-5) shows that $I(x)$ differs from $I_1(x) + I_2(x)$ by an interference term which depends on the phase difference between $E_1$ and $E_2$ and whose presence explains the fringes. The wave theory thus predicts that diminishing the intensity of the source $\mathscr{S}$ will simply cause the fringes to diminish in intensity but not to vanish.

* Since the experiment studied here is performed with unpolarized light, the vectorial character of the electric field does not play an essential role. For the sake of simplicity, we ignore it in this paragraph.

What actually happens when $\mathscr{S}$ emits photons practically one by one? *Neither the predictions of the wave theory nor those of the particle theory are verified.* In fact :

($i$) If we cover the screen $\mathscr{E}$ with a photographic plate and increase the exposure time so as to capture a large number of photons on each photograph, we observe when we develop them that *the fringes have not disappeared. Therefore, the purely corpuscular interpretation*, according to which the fringes are due to an interaction between photons, *must be rejected.*

($ii$) On the other hand, we can expose the photographic plate during a time so short that it can only receive a few photons. We then observe that each photon produces a *localized impact* on $\mathscr{E}$ and not a very weak interference pattern. *Therefore, the purely wave interpretation must also be rejected.*

In reality, as more and more photons strike the photographic plate, the following phenomenon occurs. Their individual impacts seem to be distributed in a *random manner*, and only when a great number of them have reached $\mathscr{E}$ does the distribution of the impacts begin to have a continuous aspect. The density of the impacts at each point of $\mathscr{E}$ corresponds to the interference fringes : maximum on a bright fringe and zero on a dark fringe. It can thus be said that the photons, as they arrive, build up the interference pattern.

The result of this experiment therefore leads, apparently, to a paradox. Within the framework of the particle theory, for example, it can be expressed in the following way. Since photon-photon interactions are excluded, each photon must be considered separately. But then it is not clear why the phenomena should change drastically according to whether only one slit or both slits are open. For a photon passing through one of the slits, why should the fact that the other is open or closed have such a critical importance?

Before we discuss this problem, note that in the preceding experiment we did not seek to determine through which slit each photon passed before it reached the screen. In order to obtain this information, we can imagine placing detectors (photomultipliers) behind $F_1$ and $F_2$. It will then be observed that, if the photons arrive one by one, each one passes through a well-determined slit (a signal is recorded either by the detector placed behind $F_1$ or by the one covering $F_2$ but not by both at once). But, obviously, the photons detected in this way are absorbed and do not reach the screen. Remove the photomultiplier which blocks $F_1$, for example. The one which remains behind $F_2$ tells us that, out of a large number of photons, about half pass through $F_2$. We conclude that the others (which can continue as far as the screen) pass through $F_1$. But the pattern that they gradually construct on the screen is not an interference pattern, since $F_2$ is blocked. It is only the diffraction pattern of $F_1$.

#### b. QUANTUM UNIFICATION OF THE TWO ASPECTS OF LIGHT

The preceding analysis shows that it is impossible to explain all the phenomena observed if only one of the two aspects of light, wave or particle, is considered. Now these two aspects seem to be mutually exclusive. To overcome this difficulty, it thus becomes indispensable to reconsider in a critical way the concepts of classical physics. We must accept the possibility that these concepts, although our everyday experience leads us to consider them well-founded, may not be valid in the new ("microscopic") domain which we are entering. For example, an essential characteristic of this new domain appeared when we placed counters behind Young's slits: *when one performs a measurement on a microscopic system, one disturbs it in a fundamental fashion.* This is a new property since, in the macroscopic domain, we always have the possibility of conceiving measurement devices whose influence on the system is practically as weak as one might wish. This critical revision of classical physics is imposed by experiment and must of course be guided by experiment.

Let us reconsider the "paradox" stated above concerning the photon which passes through one slit but behaves differently depending on whether the other slit is open or closed. We saw that if we try to detect the photons when they cross the slits, we prevent them from reaching the screen. More generally, a detailed experimental analysis shows that *it is impossible to observe the interference pattern and to know at the same time through which slit each photon has passed* (*cf.* complement $D_I$). Thus it is necessary, in order to resolve the paradox, to give up the idea that a photon inevitably passes through a particular slit. We are then led to question the concept, which is a fundamental one of classical physics, of a particle's trajectory.

Moreover, as the photons arrive one by one, their impacts on the screen gradually build up the interference pattern. This implies that, for a particular photon, we are not certain in advance where it will strike the screen. Now these photons are all emitted under the same conditions. Thus another classical idea has been destroyed: that the initial conditions completely determine the subsequent motion of a particle. We can only say, when a photon is emitted, that the probability of its striking the screen at $x$ is proportional to the intensity $I(x)$ calculated using wave theory, that is, to $|E(x)|^2$.

After many tentative efforts that we shall not describe here, the concept of *wave-particle duality* was formulated. We can summarize it schematically as follows★ :

(*i*) The particle and wave aspects of light are inseparable. *Light behaves simultaneously like a wave and like a flux of particles, the wave enabling us to calculate the probability of the manifestation of a particle.*

(*ii*) Predictions about the behavior of a photon can only be probabilistic.

(*iii*) The information about a photon at time $t$ is given by the wave $E(\mathbf{r}, t)$, which is a solution of Maxwell's equations. We say that this wave characterizes the state of the photons at time $t$. $E(\mathbf{r}, t)$ is interpreted as the *probability amplitude* of a photon appearing, at time $t$, at the point $\mathbf{r}$. This means that the corresponding probability is proportional to $|E(\mathbf{r}, t)|^2$.

COMMENTS:

(*i*) Since Maxwell's equations are linear and homogeneous, we can use a *superposition principle* : if $E_1$ and $E_2$ are two solutions of these equations, then $E = \lambda_1 E_1 + \lambda_2 E_2$, where $\lambda_1$ and $\lambda_2$ are constants, is also a solution.

★ It is worth noting that this interpretation of physical phenomena, generally considered to be "orthodox" at the present time, is still being contested today by certain physicists.

It is this superposition principle which explains wave phenomena in classical optics (interference, diffraction). In quantum physics, the interpretation of $E(\mathbf{r}, t)$ as a probability amplitude is thus essential to the persistence of such phenomena.

(*ii*) The theory merely allows one to calculate the probability of the occurence of a given event. Experimental verifications must thus be founded on the repetition of a large number of identical experiments. In the above experiment, a large number of photons, all produced in the same way, are emitted successively and build up the interference pattern, which is the manifestation of the calculated probabilities.

(*iii*) We are talking here about "the photon state" so as to be able to develop in § B an analogy between $E(\mathbf{r}, t)$ and the wave function $\psi(\mathbf{r}, t)$ which characterizes the quantum state of a material particle. This "optical analogy" is very fruitful. In particular, as we shall see in § D, it allows us to understand, simply and without recourse to calculation, various quantum properties of material particles. However, we should not push it too far and let it lead us to believe that it is rigorously correct to consider $E(\mathbf{r}, t)$ as characterizing the quantum state of a photon.

Furthermore, we shall see that the fact that $\psi(\mathbf{r}, t)$ is complex is essential in quantum mechanics, while the complex notation $E(\mathbf{r}, t)$ is used in optics purely for convenience (only its real part has a physical meaning). The precise definition of the (complex) quantum state of radiation can only be given in the framework of quantum electrodynamics, a theory which is simultaneously quantum mechanical and relativistic. We shall not consider these problems here (we shall touch on them in complement $K_V$).

## 3. The principle of spectral decomposition

Armed with the ideas introduced in § 2, we are now going to discuss another simple optical experiment, whose subject is the polarization of light. This will permit us to introduce the fundamental concepts which concern the measurement of physical quantities.

The experiment consists of directing a polarized plane monochromatic light wave onto an analyzer $A$. $Oz$ designates the direction of propagation of this wave and $\mathbf{e}_p$, the unit vector describing its polarization (*cf.* fig. 2). The analyzer $A$ transmits light polarized parallel to $Ox$ and absorbs light polarized parallel to $Oy$.

The classical description of this experiment (a description which is valid for a sufficiently intense light beam) is the following. The polarized plane wave is characterized by an electric field of the form:

$$\mathbf{E}(\mathbf{r}, t) = E_0 \mathbf{e}_p \, e^{i(kz - \omega t)} \tag{A-6}$$

where $E_0$ is a constant. The light intensity $I$ is proportional to $|E_0|^2$. After its passage through the analyzer $A$, the plane wave is polarized along $Ox$:

$$\mathbf{E}'(\mathbf{r}, t) = E'_0 \mathbf{e}_x \, e^{i(kz - \omega t)} \tag{A-7}$$

and its intensity $I'$, proportional to $|E'_0|^2$, is given by *Malus' law*:

$$I' = I \cos^2 \theta \tag{A-8}$$

[$\mathbf{e}_x$ is the unit vector of the $Ox$ axis and $\theta$ is the angle between $\mathbf{e}_x$ and $\mathbf{e}_p$].

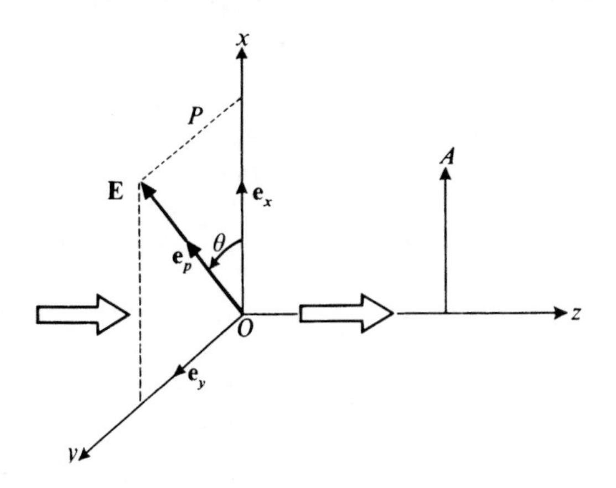


FIGURE 2

**A simple measurement experiment relating to the polarization of a light wave. A beam of light propagates along the direction *Oz* and crosses successively the polarizer *P* and the analyzer *A*. $\theta$ is the angle between *Ox* and the electric field of the wave transmitted by *P*. The vibrations transmitted by *A* are parallel to *Ox*.**

What will happen on the quantum level, that is, when $I$ is weak enough for the photons to reach the analyzer one by one? (We then place a photon detector behind this analyser.) First of all, the detector never registers a "fraction of a photon". Either the photon crosses the analyzer or it is entirely absorbed by it. Next (except in special cases that we shall examine in a moment), we cannot predict with certainty whether a given incident photon will pass or be absorbed. We can only know the corresponding probabilities. Finally, if we send out a large number $N$ of photons one after the other, the result will correspond to the classical law, in the sense that about $N \cos^2 \theta$ photons will be detected after the analyzer.

We shall retain the following ideas from this description :

(*i*) The measurement device (the analyzer, in this case) can give only certain privileged results, which we shall call *eigen* (or proper) *results*★. In the above experiment, there are only two possible results : the photon crosses the analyzer or it is stopped. One says that there is quantization of the result of the measurement, in contrast to the classical case [*cf.* formula (A-8)] where the transmitted intensity $I'$ can vary continuously, according to the value of $\theta$, between 0 and $I$.

★ The reason for this name will appear in chapter III.

(*ii*) To each of these eigen results corresponds an *eigenstate*. Here, the two eigenstates are characterized by:

$$\begin{aligned} &\mathbf{e}_p = \mathbf{e}_x \\ \text{or} \quad &\mathbf{e}_p = \mathbf{e}_y \end{aligned} \tag{A-9}$$

($\mathbf{e}_y$ is the unit vector of the $Oy$ axis). If $\mathbf{e}_p = \mathbf{e}_x$, we know with certainty that the photon will traverse the analyzer; if $\mathbf{e}_p = \mathbf{e}_y$, it will, on the contrary, definitely be stopped. The correspondence between eigen results and eigenstates is therefore the following. If the particle is, before the measurement, in one of the eigenstates, the result of this measurement is certain: it can only be the associated eigen result.

(*iii*) When the state before the measurement is arbitrary, only the probabilities of obtaining the different eigen results can be predicted. To find these probabilities, one decomposes the state of the particles into a linear combination of the various eigenstates. Here, for an arbitrary $\mathbf{e}_p$, we write:

$$\mathbf{e}_p = \mathbf{e}_x \cos\theta + \mathbf{e}_y \sin\theta \tag{A-10}$$

The probability of obtaining a given eigen result is then proportional to the square of the absolute value of the coefficient of the corresponding eigenstate. The proportionality factor is determined by the condition that the sum of all these probabilities must be equal to 1. We thus deduce from (A-10) that each photon has a probability $\cos^2\theta$ of traversing the analyzer and a probability $\sin^2\theta$ of being absorbed by it (we know that $\cos^2\theta + \sin^2\theta = 1$). This is indeed what was stated above. This rule is called in quantum mechanics the *principle of spectral decomposition*. Note that the decomposition to be performed depends on the type of measurement device being considered, since one must use the eigenstates which correspond to it: in formula (A-10), the choice of the axes $Ox$ and $Oy$ is fixed by the analyzer.

(*iv*) After passing through the analyzer, the light is completely polarized along $\mathbf{e}_x$. If we place, after the first analyzer $A$, a second analyzer $A'$, having the same axis, all the photons which traversed $A$ will also traverse $A'$. According to what we have just seen in point (*ii*), this means that, after they have crossed $A$, the state of the photons is the eigenstate characterized by $\mathbf{e}_x$. There has therefore been an abrupt change in the state of the particles. Before the measurement, this state was defined by a vector $\mathbf{E}(\mathbf{r}, t)$ which was collinear with $\mathbf{e}_p$. After the measurement, we possess an additional piece of information (the photon has passed) which is incorporated by describing the state by a different vector, which is now collinear with $\mathbf{e}_x$. This expresses the fact, already pointed out in § A-2, that *the measurement disturbs the microscopic system* (here, the photon) *in a fundamental fashion*.

COMMENT:

The certain prediction of the result when $\mathbf{e}_p = \mathbf{e}_x$ or $\mathbf{e}_p = \mathbf{e}_y$ is only a special case. The probability of one of the possible events is then indeed equal to 1. But, in order to verify this prediction, one must perform a large number of experiments. One must be sure that *all* the photons pass (or are stopped), since the fact that a particular photon crosses the analyzer (or is absorbed) is not characteristic of $\mathbf{e}_p = \mathbf{e}_x$ (or $\mathbf{e}_p = \mathbf{e}_y$).

## B. MATERIAL PARTICLES AND MATTER WAVES

### 1. The de Broglie relations

Parallel to the discovery of photons, the study of atomic emission and absorption spectra uncovered a fundamental fact, which classical physics was unable to explain : these spectra are composed of *narrow lines*. In other words, a given atom emits or absorbs only photons having well-determined frequencies (that is, energies). This fact can be interpreted very easily if one accepts that *the energy of the atom is quantized*, that is, it can take on only certain discrete values $E_i (i = 1, 2, ..., n, ...)$: the emission or absorption of a photon is then accompanied by a "jump" in the energy of the atom from one permitted value $E_i$ to another $E_j$. Conservation of energy implies that the photon has a frequency $\nu_{ij}$ such that:

$$h\nu_{ij} = |E_i - E_j| \tag{B-1}$$

Only frequencies which obey (B-1) can therefore be emitted or absorbed by the atom.

The existence of such discrete energy levels was confirmed independently by the Franck-Hertz experiment. Bohr interpreted this in terms of privileged electronic orbits and stated, with Sommerfeld, an empirical rule which permitted the calculation of these orbits for the case of the hydrogen atom. But the fundamental origin of these quantization rules remained mysterious.

In 1923, however, de Broglie put forth the following hypothesis : *material particles, just like photons, can have a wavelike aspect.* He then derived the Bohr-Sommerfeld quantization rules as a consequence of this hypothesis, the various permitted energy levels appearing as analogues of the normal modes of a vibrating string. Electron diffraction experiments (Davisson and Germer, 1927) strikingly confirmed the existence of a wavelike aspect of matter by showing that *interference patterns could be obtained with material particles such as electrons.*

One therefore associates with a material particle of energy $E$ and momentum $\mathbf{p}$, a wave whose angular frequency $\omega = 2\pi\nu$ and wave vector $\mathbf{k}$ are given by the same relations as for photons (*cf.* § A-1) :

$$\begin{cases} E = h\nu = \hbar\omega \\ \mathbf{p} = \hbar\mathbf{k} \end{cases} \tag{B-2}$$

In other words, the corresponding wavelength is :

$$\lambda = \frac{2\pi}{|\mathbf{k}|} = \frac{h}{|\mathbf{p}|} \quad \text{(de Broglie relation)} \tag{B-3}$$

COMMENT :

The very small value of the Planck constant $h$ explains why the wavelike nature of matter is very difficult to demonstrate on a macroscopic scale. Complement $A_I$ of this chapter discusses the orders of magnitude of the de Broglie wavelengths associated with various material particles.

## 2. Wave functions. Schrödinger equation

In accordance with de Broglie's hypothesis, we shall apply the ideas introduced in § A for the case of the photon to all material particles. Recalling the conclusions of this paragraph, we are led to the following formulation:

(*i*) For the classical concept of a trajectory, we must substitute the concept of a time-varying *state*. The quantum state of a particle such as the electron* is characterized by a *wave function* $\psi(\mathbf{r}, t)$, which contains all the information it is possible to obtain about the particle.

(*ii*) $\psi(\mathbf{r}, t)$ is interpreted as a *probability amplitude of the particle's presence*. Since the possible positions of the particle form a continuum, the probability $d\mathscr{P}(\mathbf{r}, t)$ of the particle being, at time $t$, in a volume element $d^3r = dx\,dy\,dz$ situated at the point $\mathbf{r}$ must be proportional to $d^3r$ and therefore infinitesimal. $|\psi(\mathbf{r}, t)|^2$ is then interpreted as the corresponding *probability density*, with:

$$d\mathscr{P}(\mathbf{r}, t) = C\,|\psi(\mathbf{r}, t)|^2\,d^3r \tag{B-4}$$

where $C$ is a normalization constant [see comment (*i*) at the end of § B-2].

(*iii*) *The principle of spectral decomposition* applies to the measurement of an arbitrary physical quantity:

– The result found must belong to a set of eigen results $\{ a \}$.

– With each eigenvalue $a$ is associated an eigenstate, that is, an eigenfunction $\psi_a(\mathbf{r})$. This function is such that, if $\psi(\mathbf{r}, t_0) = \psi_a(\mathbf{r})$ (where $t_0$ is the time at which the measurement is performed), the measurement will always yield $a$.

– For any $\psi(\mathbf{r}, t)$, the probability $\mathscr{P}_a$ of finding the eigenvalue $a$ for a measurement at time $t_0$ is found by decomposing $\psi(\mathbf{r}, t_0)$ in terms of the functions $\psi_a(\mathbf{r})$:

$$\psi(\mathbf{r}, t_0) = \sum_a c_a\,\psi_a(\mathbf{r}) \tag{B-5}$$

Then:

$$\mathscr{P}_a = \frac{|c_a|^2}{\sum_a |c_a|^2} \tag{B-6}$$

(the presence of the denominator insures that the total probability is equal to 1:

$$\left.\sum_a \mathscr{P}_a = 1\right).$$

– If the measurement indeed yields $a$, the wave function of the particle immediately after the measurement is:

$$\psi'(\mathbf{r}, t_0) = \psi_a(\mathbf{r}) \tag{B-7}$$

(*iv*) The equation describing the evolution of the function $\psi(\mathbf{r}, t)$ remains to be written. It is possible to introduce it in a very natural way, using the

* We shall not take into account here the existence of electron spin (*cf.* chap. IX).

Planck and de Broglie relations. Nevertheless, we have no intention of proving this fundamental equation, which is called the *Schrödinger equation*. We shall simply assume it. Later, we shall discuss some of its consequences (whose experimental verification will prove its validity). Besides, we shall consider this equation in much more detail in chapter III.

When the particle (of mass $m$) is subjected to the influence of a potential* $V(\mathbf{r}, t)$, the Schrödinger equation takes on the form :

$$i\hbar \frac{\partial}{\partial t} \psi(\mathbf{r}, t) = - \frac{\hbar^2}{2m} \Delta \psi(\mathbf{r}, t) + V(\mathbf{r}, t) \psi(\mathbf{r}, t) \tag{B-8}$$

where $\Delta$ is the Laplacian operator $\partial^2/\partial x^2 + \partial^2/\partial y^2 + \partial^2/\partial z^2$.
We notice immediately that this equation is linear and homogeneous in $\psi$. Consequently, for material particles, there exists a superposition principle which, combined with the interpretation of $\psi$ as a probability amplitude, is the source of wavelike effects. Note, moreover, that the differential equation (B-8) is first-order with respect to time. This condition is necessary if the state of the particle at a time $t_0$, characterized by $\psi(\mathbf{r}, t_0)$, is to determine its subsequent state.

Thus there exists a fundamental analogy between matter and radiation : in both cases, a correct description of the phenomena necessitates the introduction of quantum concepts, and, in particular, the idea of wave-particle duality.

COMMENTS :

(*i*) For a system composed of only one particle, the total probability of finding the particle anywhere in space, at time $t$, is equal to 1 :

$$\int \mathrm{d}\mathscr{P}(\mathbf{r}, t) = 1 \tag{B-9}$$

Since $\mathrm{d}\mathscr{P}(\mathbf{r}, t)$ is given by formula (B-4), we conclude that *the wave function $\psi(\mathbf{r}, t)$ must be square-integrable* :

$$\int |\psi(\mathbf{r}, t)|^2 \, \mathrm{d}^3 r \quad \text{is finite} \tag{B-10}$$

The normalization constant $C$ which appears in (B-4) is then given by the relation :

$$\frac{1}{C} = \int |\psi(\mathbf{r}, t)|^2 \, \mathrm{d}^3 r \tag{B-11}$$

(we shall later see that the form of the Schrödinger equation implies that $C$ is time-independent). One often uses wave functions which are normalized, such that :

$$\int |\psi(\mathbf{r}, t)|^2 \, \mathrm{d}^3 r = 1 \tag{B-12}$$

* $V(\mathbf{r}, t)$ designates a potential energy here. For example, it may be the product of an electric potential and the particle's charge. In quantum mechanics, $V(\mathbf{r}, t)$ is commonly called a potential.

The constant $C$ is then equal to 1.

(*ii*) Note the important difference between the concepts of classical states and quantum states. The classical state of a particle is determined at time $t$ by the specification of six parameters characterizing its position and its velocity at time $t$ : $x, y, z ; v_x, v_y, v_z$. The quantum state of a particle is determined by an *infinite number* of parameters : the values at the various points in space of the wave function $\psi(\mathbf{r}, t)$ which is associated with it. For the classical idea of a trajectory (the succession in time of the various states of the classical particle), we must substitute the idea of the propagation of the wave associated with the particle. Consider, for example, Young's double-slit experiment, previously described for the case of photons, but which in principle can also be performed with material particles such as electrons. When the interference pattern is observed, it makes no sense to ask through which slit each particle has passed, since the wave associated with it passed through both.

(*iii*) It is worth noting that, unlike photons, which can be emitted or absorbed during an experiment, material particles can neither be created nor destroyed. The electrons emitted by a heated filament for example already existed in the filament. In the same way, an electron absorbed by a counter does not disappear; it becomes part of an atom or an electric current. Actually, the theory of relativity shows that it is possible to create and annihilate material particles : for example, a photon having sufficient energy, passing near an atom, can materialize into an electron-positron pair. Inversely, the positron, when it collides with an electron, annihilates with it, emitting photons. However, we pointed out in the beginning of this chapter that we would limit ourselves here to the non-relativistic quantum domain, and we have indeed treated time and space coordinates asymmetrically. In the framework of non-relativistic quantum mechanics, material particles can neither be created nor annihilated. This conservation law, as we shall see, plays a role of primary importance. The need to abandon it is one of the important difficulties encountered when one tries to construct a relativistic quantum mechanics.

## C. QUANTUM DESCRIPTION OF A PARTICLE. WAVE PACKETS

In the preceding paragraph, we introduced the fundamental concepts which are necessary for the quantum description of a particle. In this paragraph, we are going to familiarize ourselves with these concepts and deduce from them several very important properties. Let us begin by studying a very simple special case, that of a free particle.

### 1. Free particle

Consider a particle whose potential energy is zero (or has a constant value) at every point in space. The particle is thus not subjected to any force; it is said to be free.

When $V(\mathbf{r}, t) = 0$, the Schrödinger equation becomes:

$$i\hbar \frac{\partial}{\partial t} \psi(\mathbf{r}, t) = -\frac{\hbar^2}{2m} \Delta \psi(\mathbf{r}, t) \tag{C-1}$$

This differential equation is obviously satisfied by solutions of the form:

$$\psi(\mathbf{r}, t) = A \, \mathrm{e}^{i(\mathbf{k}.\mathbf{r} - \omega t)} \tag{C-2}$$

(where $A$ is a constant), on the condition that $\mathbf{k}$ and $\omega$ satisfy the relation:

$$\omega = \frac{\hbar \mathbf{k}^2}{2m} \tag{C-3}$$

Observe that, according to the de Broglie relations [see (B-2)], condition (C-3) expresses the fact that the energy $E$ and the momentum $\mathbf{p}$ of a free particle satisfy the equation, which is well-known in classical mechanics:

$$E = \frac{\mathbf{p}^2}{2m} \tag{C-4}$$

We shall come back later (§ C-3) to the physical interpretation of a state of the form (C-2). We already see that, since

$$|\psi(\mathbf{r}, t)|^2 = |A|^2 \tag{C-5}$$

a plane wave of this type represents a particle whose probability of presence is uniform throughout all space (see comment below).

The principle of superposition tells us that every linear combination of plane waves satisfying (C-3) will also be a solution of equation (C-1). Such a superposition can be written:

$$\psi(\mathbf{r}, t) = \frac{1}{(2\pi)^{3/2}} \int g(\mathbf{k}) \, \mathrm{e}^{i[\mathbf{k}.\mathbf{r} - \omega(k)t]} \, \mathrm{d}^3 k \tag{C-6}$$

($d^3k$ represents, by definition, the infinitesimal volume element in $\mathbf{k}$-space: $\mathrm{d}k_x \mathrm{d}k_y \mathrm{d}k_z$). $g(\mathbf{k})$, which can be complex, must be sufficiently regular to allow differentiation inside the integral. It can be shown, moreover, that any square-integrable solution can be written in the form (C-6).

A wave function such as (C-6), a superposition of plane waves, is called a three-dimensional " wave packet". For the sake of simplicity, we shall often be led to study the case of a one-dimensional wave packet*, obtained from the superposition of plane waves all propagating parallel to $Ox$. The wave function then depends only on $x$ and $t$:

$$\psi(x, t) = \frac{1}{\sqrt{2\pi}} \int_{-\infty}^{+\infty} g(k) \, \mathrm{e}^{i[kx - \omega(k)t]} \, \mathrm{d}k \tag{C-7}$$

* A simple model of a two-dimensional wave packet is presented in complement $E_I$. Some general properties of three-dimensional wave packets are studied in complement $F_I$, which also shows how, in certain cases, a three-dimensional problem can be reduced to several one-dimensional problems.

In the following paragraph, we shall be interested in the form of the wave packet at a given instant. If we choose this instant as the time origin, the wave function is written:

$$\psi(x, 0) = \frac{1}{\sqrt{2\pi}} \int g(k)\, e^{ikx}\, dk \tag{C-8}$$

We see that $g(k)$ is simply the Fourier transform (*cf.* appendix I) of $\psi(x, 0)$ :

$$g(k) = \frac{1}{\sqrt{2\pi}} \int \psi(x, 0)\, e^{-ikx}\, dx \tag{C-9}$$

Consequently, the validity of formula (C-8) is not limited to the case of the free particle : whatever the potential, $\psi(x, 0)$ can always be written in this form. The consequences that we shall derive from this in §§ 2 and 3 below are thus perfectly general. It is not until § 4 that we shall return explicitly to the free particle.

COMMENT :

A plane wave of type (C-2), whose modulus is constant throughout all space [*cf.* (C-5)], is not square-integrable. Therefore, rigorously, it cannot represent a physical state of the particle (in the same way as, in optics, a plane monochromatic wave is not physically realizable). On the other hand, a superposition of plane waves like (C-7) can be square-integrable.

## 2. Form of the wave packet at a given time

The form of the wave packet is given by the $x$-dependence of $\psi(x, 0)$ defined by equation (C-8). Imagine that $|g(k)|$ has the shape depicted in figure 3; that is, it has a pronounced peak situated at $k = k_0$ and a width (defined, for example, at half its maximum value) of $\Delta k$.

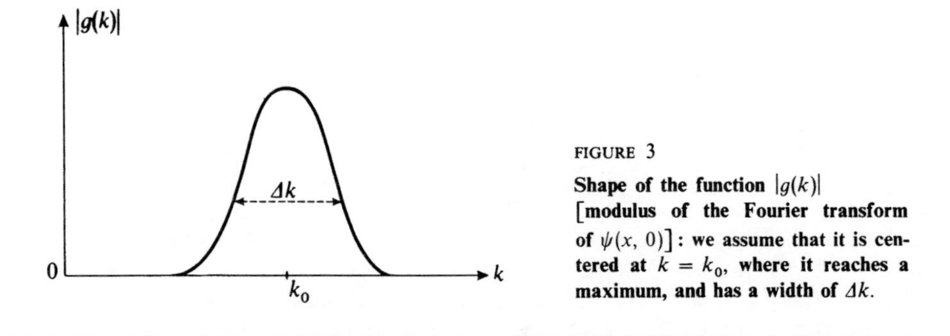


FIGURE 3

**Shape of the function $|g(k)|$ [modulus of the Fourier transform of $\psi(x, 0)$] : we assume that it is centered at $k = k_0$, where it reaches a maximum, and has a width of $\Delta k$.**

Let us begin by trying to understand qualitatively the behavior of $\psi(x, 0)$ through the study of a very simple special case. Let $\psi(x, 0)$, instead of being the superposition of an infinite number of plane waves $e^{ikx}$ as in formula (C-8), be the sum of only three plane waves. The wave vectors of these plane waves are $k_0$, $k_0 - \frac{\Delta k}{2}$, $k_0 + \frac{\Delta k}{2}$, and their amplitudes are proportional, respectively, to 1, 1/2 and 1/2. We then have :

$$\begin{aligned}\psi(x) &= \frac{g(k_0)}{\sqrt{2\pi}}\left[e^{ik_0x} + \frac{1}{2}e^{i\left(k_0 - \frac{\Delta k}{2}\right)x} + \frac{1}{2}e^{i\left(k_0 + \frac{\Delta k}{2}\right)x}\right] \\ &= \frac{g(k_0)}{\sqrt{2\pi}}\, e^{ik_0x}\left[1 + \cos\left(\frac{\Delta k}{2}x\right)\right]\end{aligned} \tag{C-10}$$

We see that $|\psi(x)|$ is maximum when $x = 0$. This result is due to the fact that, when $x$ takes on this value, the three waves are in phase and interfere constructively, as shown in figure 4. As one moves away from the value $x = 0$, the waves become more and more out of phase, and $|\psi(x)|$ decreases. The interference becomes completely destructive when the phase shift between $e^{ik_0x}$ and $e^{i(k_0 \pm \Delta k/2)x}$ is equal to $\pm\,\pi$ : $\psi(x)$ goes to zero when $x = \pm\frac{\Delta x}{2}$, $\Delta x$ being given by :

$$\Delta x\,.\,\Delta k = 4\pi \tag{C-11}$$

This formula shows that the smaller the width $\Delta k$ of the function $|g(k)|$, the larger the width $\Delta x$ of the function $|\psi(x)|$ (the distance between two zeros of $|\psi(x)|$).

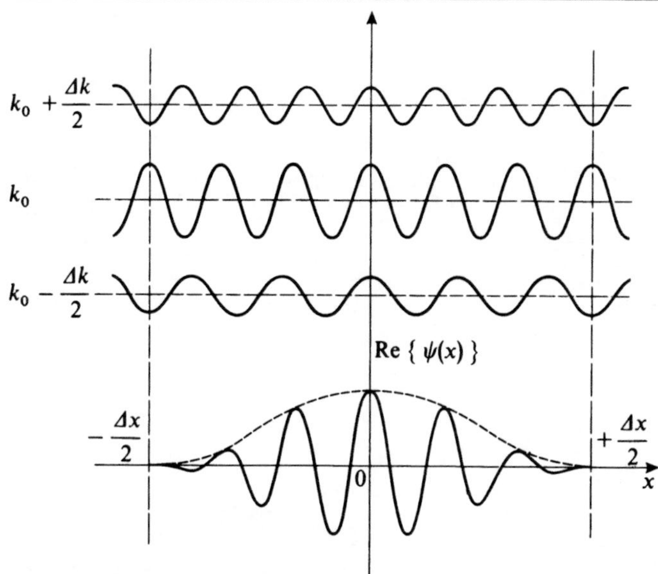


FIGURE 4

**The real parts of the three waves whose sum gives the function $\psi(x)$ of (C-10). At $x = 0$, the three waves are in phase and interfere constructively. As one moves away from $x = 0$, they go out of phase and interfere destructively for $x = \pm\,\Delta x/2$.**

**In the lower part of the figure, Re { $\psi(x)$ } is shown. The dashed-line curve corresponds to the function $\left[1 + \cos\left(\frac{\Delta k}{2}x\right)\right]$, which, according to (C-10), gives $|\psi(x)|$ (and therefore, the form of the wave packet).**

COMMENT:

Formula (C-10) shows that $|\psi(x)|$ is periodic in $x$ and therefore has a series of maxima and minima. This arises from the fact that $\psi(x)$ is the superposition of a finite number of waves (here, three). For a continuous superposition of an infinite number of waves, as in formula (C-8), such a phenomenon does not occur, and $|\psi(x, 0)|$ can have only one maximum.

Let us now return to the general wave packet of formula (C-8). Its form also results from an interference phenomenon: $|\psi(x, 0)|$ is maximum when the different plane waves interfere constructively.

Let $\alpha(k)$ be the argument of the function $g(k)$:

$$g(k) = |g(k)| \, e^{i\alpha(k)} \tag{C-12}$$

Assume that $\alpha(k)$ varies sufficiently smoothly within the interval $\left[k_0 - \frac{\Delta k}{2}, k_0 + \frac{\Delta k}{2}\right]$ where $|g(k)|$ is appreciable; then, when $\Delta k$ is sufficiently small, one can expand $\alpha(k)$ in the neighborhood of $k = k_0$:

$$\alpha(k) \simeq \alpha(k_0) + (k - k_0)[d\alpha/dk]_{k=k_0} \tag{C-13}$$

which enables us to rewrite (C-8) in the form:

$$\psi(x, 0) \simeq \frac{e^{i[k_0 x + \alpha(k_0)]}}{\sqrt{2\pi}} \int_{-\infty}^{+\infty} |g(k)| \, e^{i(k - k_0)(x - x_0)} \, dk \tag{C-14}$$

with:

$$x_0 = -[d\alpha/dk]_{k=k_0} \tag{C-15}$$

The form (C-14) is useful for studying the variations of $|\psi(x, 0)|$ in terms of $x$. When $|x - x_0|$ is large, the function of $k$ which is to be integrated oscillates a very large number of times within the interval $\Delta k$. We then see (*cf.* fig. 5-a, in which the real part of this function is depicted) that the contributions of the successive oscillations cancel each other out, and the integral over $k$ becomes negligible. In other words, when $x$ is fixed at a value far from $x_0$, the phases of the various waves which make up $\psi(x, 0)$ vary very rapidly in the domain $\Delta k$, and these waves destroy each other by interference. On the other hand, if $x \simeq x_0$, the function to be integrated over $k$ oscillates hardly at all (*cf.* fig. 5-b), and $|\psi(x, 0)|$ is maximum.

The position $x_M(0)$ of the center of the wave packet is therefore:

$$x_M(0) = x_0 = -[d\alpha/dk]_{k=k_0} \tag{C-16}$$

Actually the result (C-16) can be obtained very simply. An integral such as the one appearing in (C-8) will be maximum (in absolute value) when the waves having the largest amplitude (those with $k$ close to $k_0$) interfere constructively. This occurs when the $k$-dependent phases of these waves vary only slightly around $k = k_0$. To obtain the center of the wave packet, one then imposes (*stationary phase* condition) that the derivative with respect to $k$ of the phase is zero for $k = k_0$. In the particular case which we are studying, the phase of the wave corresponding to $k$ is $kx + \alpha(k)$. Therefore, $x_M(0)$ is that value of $x$ for which the derivative $x + d\alpha/dk$ is zero at $k = k_0$.

a

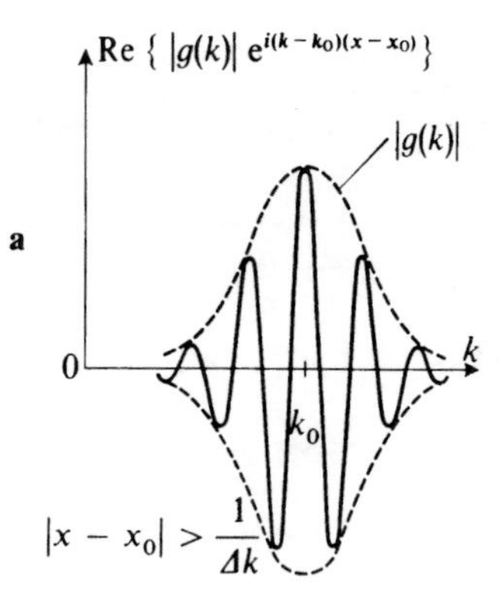


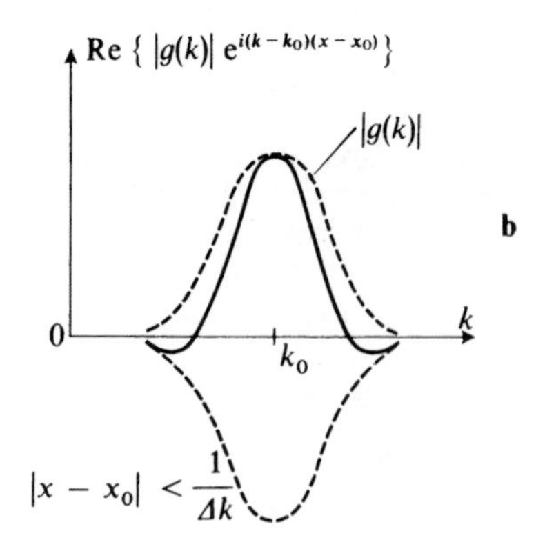


b

FIGURE 5

**Variations with respect to $k$ of the function to be integrated over $k$ in order to obtain $\psi(x, 0)$. In figure (a), $x$ is fixed at a value such that $|x - x_0| > 1/\Delta k$, and the function to be integrated oscillates several times within the interval $\Delta k$. In figure (b), $x$ is fixed such that $|x - x_0| < 1/\Delta k$, and the function to be integrated hardly oscillates, so that its integral over $k$ takes on a relatively large value. Consequently, the center of the wave packet [point where $|\psi(x, 0)|$ is maximum] is situated at $x = x_0$.**

---

When $x$ moves away from the value $x_0$, $|\psi(x, 0)|$ decreases. This decrease becomes appreciable if $e^{i(k-k_0)(x-x_0)}$ oscillates approximately once when $k$ traverses the domain $\Delta k$, that is, when:

$$\Delta k \,.\, (x - x_0) \simeq 1 \tag{C-17}$$

If $\Delta x$ is the approximate width of the wave packet, we therefore have :

$$\Delta k \,.\, \Delta x \gtrsim 1 \tag{C-18}$$

We are thus brought back to a classical relation between the widths of two functions which are Fourier transforms of each other. The important fact is that the product $\Delta x \,.\, \Delta k$ has a lower bound; the exact value of this bound clearly depends on the precise definition of the widths $\Delta x$ and $\Delta k$.

A wave packet such as (C-7) thus represents the state of a particle whose probability of presence, at the time $t = 0$, is practically zero outside an interval of approximate width $\Delta x$ centered at the value $x_0$.

COMMENT:

The preceding argument could lead one to believe that the product $\Delta x \,.\, \Delta k$ is always of the order of 1 [*cf.* (C-17)]. Let us stress the fact that this is a lower limit. Although it is impossible to construct wave packets for which the product $\Delta x \,.\, \Delta k$ is negligible compared to 1, it is perfectly possible to construct packets for which this product is as large as desired [see, for example, complement $G_I$, especially comment (*ii*) of §3-c]. This is why (C-18) is written in the form of an inequality.

### 3. Heisenberg uncertainty relation

In quantum mechanics, inequality (C-18) has extremely important physical consequences. We intend to discuss these now (we shall stay, for simplicity, within the framework of a one-dimensional model).

We have seen that a plane wave $e^{i(k_0 x - \omega_0 t)}$ corresponds to a constant probability density for the particle's presence along the $Ox$ axis, for all values of $t$. This result can be roughly expressed by saying that the corresponding value of $\Delta x$ is infinite. On the other hand, only one angular frequency $\omega_0$ and one wave vector $k_0$ are involved. According to the de Broglie relations, this means that the energy and momentum of the particle are well-defined : $E = \hbar\omega_0$ and $p = \hbar k_0$. Such a plane wave can, moreover, be considered to be a special case of (C-7), for which $g(k)$ is a « delta function » (appendix II):

$$g(k) = \delta(k - k_0) \tag{C-19}$$

The corresponding value of $\Delta k$ is then zero.

But this property can also be interpreted in the following manner, using the principle of spectral decomposition (*cf.* §§ A-3 and B-2). To say that a particle, described at $t = 0$ by the wave function $\psi(x, 0) = A\, e^{ikx}$, has a well-determined momentum, is to say that a measurement of the momentum at this time will definitely yield $p = \hbar k$. From this we deduce that $e^{ikx}$ *characterizes the eigenstate corresponding to* $p = \hbar k$. Since there exists a plane wave for every real value of $k$, the eigenvalues which one can expect to find in a measurement of the momentum on an arbitrary state include all real values. In this case, there is no quantization of the possible results : as in classical mechanics, all values of the momentum are allowed.

Now consider formula (C-8). In this formula, $\psi(x, 0)$ appears as a linear superposition of the momentum eigenfunctions in which the coefficient of $e^{ikx}$ is $g(k)$. We are thus led to interpret $|g(k)|^2$ (to within a constant factor) as the probability of finding $p = \hbar k$ if one measures, at $t = 0$, the momentum of a particle whose state is described by $\psi(x, t)$. In reality, the possible values of $p$, like those of $x$, form a continuous set, and $|g(k)|^2$ is proportional to a *probability density*: the probability $\overline{d\mathscr{P}}(k)$ of obtaining a value between $\hbar k$ and $\hbar(k + dk)$ is, to within a constant factor, $|g(k)|^2\, dk$. More precisely, if we rewrite formula (C-8) in the form:

$$\psi(x, 0) = \frac{1}{\sqrt{2\pi\hbar}} \int \bar{\psi}(p)\, e^{ipx/\hbar}\, dp \tag{C-20}$$

we know that $\bar{\psi}(p)$ and $\psi(x, 0)$ satisfy the Bessel-Parseval relation (appendix I):

$$\int_{-\infty}^{+\infty} |\psi(x, 0)|^2\, dx = \int_{-\infty}^{+\infty} |\bar{\psi}(p)|^2\, dp \tag{C-21}$$

If the common value of these integrals is $C$, $d\mathscr{P}(x) = \frac{1}{C} |\psi(x, 0)|^2\, dx$ is the probability of the particle being found, at $t = 0$, between $x$ and $x + dx$. In the same way :

$$\overline{d\mathscr{P}}(p) = \frac{1}{C} |\bar{\psi}(p)|^2\, dp \tag{C-22}$$

is the probability that the measurement of the momentum will yield a result included between $p$ and $p + \mathrm{d}p$ [relation (C-21) then insures that the total probability of finding any value is indeed equal to 1].

Now let us go back to the inequality (C-18). We can write it as :

$$\Delta x \,.\, \Delta p \gtrsim \hbar \tag{C-23}$$

($\Delta p = \hbar \Delta k$ is the width of the curve representing $|\overline{\psi}(p)|$). Consider a particle whose state is defined by the wave packet (C-20). We know that its position probability at $t = 0$, is appreciable only within a region of width $\Delta x$ about $x_0$ : its position is known within an uncertainty $\Delta x$. If one measures the momentum of this particle at the same time, one will find a value between $p_0 + \frac{\Delta p}{2}$ and $p_0 - \frac{\Delta p}{2}$, since $|\overline{\psi}(p)|^2$ is practically zero outside this interval : the uncertainty in the momentum is therefore $\Delta p$. The interpretation of relation (C-23) is then the following : it is impossible to define at a given time both the position of the particle and its momentum to an arbitrary degree of accuracy. When the lower limit imposed by (C-23) is reached, increasing the accuracy in the position (decreasing $\Delta x$) implies that the accuracy in the momentum diminishes ($\Delta p$ increases), and vice versa. This relation is called the *Heisenberg uncertainty relation.*

We know of nothing like this in classical mechanics. The limitation expressed by (C-23) arises from the fact that $h$ is not zero. It is the very small value of $h$ on the macroscopic scale which renders this limitation totally negligible in classical mechanics (an example is discussed in detail in complement $\mathrm{B_I}$).

COMMENT :

The inequality (C-18) with which we started is not an inherently quantum mechanical principle. It merely expresses a general property of Fourier transforms, numerous applications of which can be found in classical physics. For example, it is well known from electromagnetic theory that there exists no train of electromagnetic waves for which one can define the position and the wavelength with infinite accuracy at the same time. Quantum mechanics enters in when one associates a wave with a material particle and requires that the wavelength and the momentum satisfy de Broglie's relation.

## 4. Time evolution of a free wave packet

Until now, we have been concerned only with the form of a wave packet at a given instant; in this paragraph, we are going to study its time evolution. Let us return, therefore, to the case of a free particle whose state is described by the one-dimensional wave packet (C-7).

A given plane wave $\mathrm{e}^{i(kx - \omega t)}$ propagates along the $Ox$ axis with the velocity:

$$V_\varphi(k) = \frac{\omega}{k} \tag{C-24}$$

since it depends on $x$ and $t$ only through $\left(x - \frac{\omega}{k}t\right)$; $V_\varphi(k)$ is called the *phase velocity* of the plane wave.

We know that in the case of an electromagnetic wave propagating in a vacuum, $V_\varphi$ is independent of $k$ and equal to the speed of light $c$. All the waves which make up a wave packet move at the same velocity, so that the packet as a whole also moves with the same velocity, without changing in shape. On the other hand, we know that this is not true in a dispersive medium, where the phase velocity is given by:

$$V_\varphi(k) = \frac{c}{n(k)} \tag{C-25}$$

$n(k)$ being the index of the medium, which varies with the wavelength.

The case that we are considering here corresponds to a dispersive medium, since the phase velocity is equal to [*cf.* equation (C-3)]:

$$V_\varphi(k) = \frac{\hbar k}{2m} \tag{C-26}$$

We shall see that when the different waves thus have unequal phase velocities, the velocity of the maximum $x_M$ of the wave packet is not the average phase velocity $\frac{\omega_0}{k_0} = \frac{\hbar k_0}{2m}$, contrary to what one might expect.

As we did before, we shall begin by trying to understand qualitatively what happens, before taking a more general point of view. Therefore, let us return to the superposition of three waves considered in § C-2. For arbitrary $t$, $\psi(x, t)$ is given by:

$$\begin{aligned} \psi(x, t) &= \frac{g(k_0)}{\sqrt{2\pi}} \left\{ e^{i[k_0 x - \omega_0 t]} + \frac{1}{2} e^{i\left[\left(k_0 - \frac{\Delta k}{2}\right)x - \left(\omega_0 - \frac{\Delta\omega}{2}\right)t\right]} \right. \\ &\qquad \left. + \frac{1}{2} e^{i\left[\left(k_0 + \frac{\Delta k}{2}\right)x - \left(\omega_0 + \frac{\Delta\omega}{2}\right)t\right]} \right\} \\ &= \frac{g(k_0)}{\sqrt{2\pi}} e^{i(k_0 x - \omega_0 t)} \left[1 + \cos\left(\frac{\Delta k}{2}x - \frac{\Delta\omega}{2}t\right)\right] \end{aligned} \tag{C-27}$$

We see, therefore, that the maximum of $|\psi(x, t)|$, which was at $x = 0$ at $t = 0$, is now at the point:

$$x_M(t) = \frac{\Delta\omega}{\Delta k} t \tag{C-28}$$

and not at the point $x = \frac{\omega_0}{k_0} t$. The physical origin of this result appears in figure 6. Part *a*) of this figure represents the position at time $t = 0$ of three adjacent maxima (1), (2), (3), for the real parts of each of the three waves. Since the maxima denoted by the index (2) coincide at $x = 0$, there is constructive interference at this point, which thus corresponds to the position of the maximum of $|\psi(x, 0)|$. Since the phase velocity increases with $k$ [formula (C-26)], the maximum (3) of the wave $\left(k_0 + \frac{\Delta k}{2}\right)$ will gradually catch up with that of the wave $(k_0)$, which will in turn catch up with that of the wave $\left(k_0 - \frac{\Delta k}{2}\right)$. After a certain time, we shall thus have the situation shown in figure 6-*b*: it will be the maxima (3) which coincide and thus determine the position of the maximum $x_M(t)$ of $|\psi(x, t)|$. We clearly see in the figure that $x_M(t)$ is not equal to $\frac{\omega_0}{k_0} t$, and a simple calculation again yields (C-28).

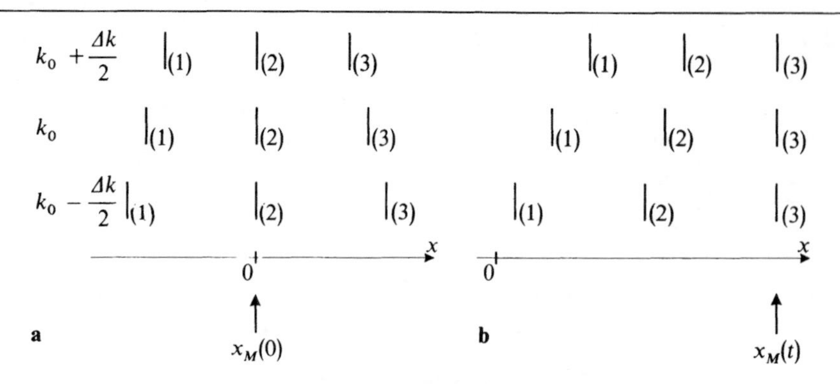


FIGURE 6

**Positions of the maxima of the three waves of figure 4 at time $t = 0$ (fig. a) and at a subsequent $t$ (fig. b). At time $t = 0$, it is the maxima (2), situated at $x = 0$, which interfere constructively: the position of the center of the wave packet is $x_M(0) = 0$. At time $t$, the three waves have advanced with different phase velocities $V_\varphi$. It is then the maxima (3) which interfere constructively and the center of the wave packet is situated at $x = x_M(t)$. We thus see that the velocity of the center of the wave packet (group velocity) is different from the phase velocities of the three waves.**

The shift of the center of the wave packet (C-7) can be found in an analogous fashion, by applying the "stationary phase" method. It can be seen from the form (C-7) of the free wave packet that, in order to go from $\psi(x, 0)$ to $\psi(x, t)$, all we need to do is change $g(k)$ to $g(k)\,e^{-i\omega(k)t}$. The reasoning of § C-2 thus remains valid, on the condition that we replace the argument $\alpha(k)$ of $g(k)$ by:

$$\alpha(k) - \omega(k)t \tag{C-29}$$

The condition (C-16) then gives:

$$x_M(t) = \left[\frac{d\omega}{dk}\right]_{k=k_0} t - \left[\frac{d\alpha}{dk}\right]_{k=k_0} \tag{C-30}$$

We are thus brought back to result (C-28) : the velocity of the maximum of the wave packet is:

$$V_G(k_0) = \left[\frac{d\omega}{dk}\right]_{k=k_0} \tag{C-31}$$

$V_G(k_0)$ is called the *group velocity* of the wave packet. With the dispersion relation given in (C-3), we obtain:

$$V_G(k_0) = \frac{\hbar k_0}{m} = 2V_\varphi(k_0) \tag{C-32}$$

This result is important, for it enables us to retrieve the classical description of the free particle, for the cases where this description is valid. For example, when one is dealing with a macroscopic particle (and the example of the dust particle discussed in complement $B_I$ shows how small it can be), the uncertainty relation does not introduce an observable limit on the accuracy with which its position and momentum are known. This means that we can construct, in order to describe such a particle in a quantum mechanical way, a wave packet whose characteristic widths $\Delta x$ and $\Delta p$ are negligible. We would then speak, in classical terms, of the position $x_M(t)$ and the momentum $p_0$ of the particle. But then its velocity must be $v = \frac{p_0}{m}$. This is indeed what is implied by formula (C-32), obtained in the quantum description: in the cases where $\Delta x$ and $\Delta p$ can both be made negligible, the maximum of the wave packet moves like a particle which obeys the laws of classical mechanics.

COMMENT:

We have stressed here the motion of the center of the free wave packet. It is also possible to study the way in which its form evolves in time. It is then easy to show that, if the width $\Delta p$ is a constant of the motion, $\Delta x$ varies over time and, for sufficiently long times, increases without limit (spreading of the wave packet). The discussion of this phenomenon is given in complement $G_I$, where the special case of a Gaussian wave packet is treated.

## D. PARTICLE IN A TIME-INDEPENDENT SCALAR POTENTIAL

We have seen, in § C, how the quantum mechanical description of a particle reduces to the classical description when Planck's constant $h$ can be considered to be negligible. In the classical approximation, the wavelike character does not appear because the wavelength $\lambda = \frac{h}{p}$ associated with the particle is much smaller than the characteristic lengths of its motion. This situation is analogous to the one encountered in optics. Geometrical optics, which ignores the wavelike properties of light, constitutes a good approximation when the corresponding wavelength can be neglected compared to the lengths with which one is concerned. Classical mechanics thus plays, with respect to quantum mechanics, the same role played by geometrical optics with respect to wave optics.

In this paragraph, we are going to be concerned with a particle in a time-independent potential. What we have just said implies that typically quantum effects (that is, those of wave origin) should arise when the potential varies appreciably over distances shorter than the wavelength, which cannot then be neglected. This is why we are going to study the behavior of a quantum particle placed in various "square potentials", that is, "step potentials", as shown in figure 7-a. Such a potential, which is discontinuous, clearly varies considerably over intervals of the order of the wavelength, however small it is: quantum effects must therefore always appear. Before beginning this investigation, we shall discuss some important properties of the Schrödinger equation when the potential is not time-dependent.

## 1. Separation of variables. Stationary states

The wave function of a particle whose potential energy $V(\mathbf{r})$ is not time-dependent must satisfy the Schrödinger equation :

$$i\hbar \frac{\partial}{\partial t} \psi(\mathbf{r}, t) = -\frac{\hbar^2}{2m} \Delta\psi(\mathbf{r}, t) + V(\mathbf{r})\, \psi(\mathbf{r}, t) \qquad \text{(D-1)}$$

### a. EXISTENCE OF STATIONARY STATES

Let us see if there exist solutions of this equation of the form:

$$\psi(\mathbf{r}, t) = \varphi(\mathbf{r})\, \chi(t) \qquad \text{(D-2)}$$

Substituting (D-2) into (D-1), we obtain :

$$i\hbar\, \varphi(\mathbf{r}) \frac{\mathrm{d}\chi(t)}{\mathrm{d}t} = \chi(t)\left[ -\frac{\hbar^2}{2m} \Delta\varphi(\mathbf{r}) \right] + \chi(t)\, V(\mathbf{r})\, \varphi(\mathbf{r}) \qquad \text{(D-3)}$$

If we divide both sides by the product $\varphi(\mathbf{r})\chi(t)$, we find :

$$\frac{i\hbar}{\chi(t)} \frac{\mathrm{d}\chi(t)}{\mathrm{d}t} = \frac{1}{\varphi(\mathbf{r})}\left[ -\frac{\hbar^2}{2m} \Delta\varphi(\mathbf{r}) \right] + V(\mathbf{r}) \qquad \text{(D-4)}$$

This equation equates a function of $t$ only (left-hand side) and a function of $\mathbf{r}$ only (right-hand side). This equality is only possible if each of these functions is in fact a constant, which we shall set equal to $\hbar\omega$, where $\omega$ has the dimensions of an angular frequency.

Setting the left-hand side equal to $\hbar\omega$, we obtain for $\chi(t)$ a differential equation which can easily be integrated to give:

$$\chi(t) = A\, \mathrm{e}^{-i\omega t} \qquad \text{(D-5)}$$

In the same way, $\varphi(\mathbf{r})$ must satisfy the equation :

$$-\frac{\hbar^2}{2m} \Delta\varphi(\mathbf{r}) + V(\mathbf{r})\, \varphi(\mathbf{r}) = \hbar\omega\, \varphi(\mathbf{r}) \qquad \text{(D-6)}$$

If we set $A = 1$ in equation (D-5) [which is possible if we incorporate, for example, the constant $A$ in $\varphi(\mathbf{r})$], we achieve the following result : the function

$$\psi(\mathbf{r}, t) = \varphi(\mathbf{r})\, \mathrm{e}^{-i\omega t} \qquad \text{(D-7)}$$

is a solution of the Schrödinger equation, on the condition that $\varphi(\mathbf{r})$ is a solution of (D-6). The time and space variables are said to have been separated.

A wave function of the form (D-7) is called a *stationary solution of the Schrödinger equation* : it leads to a time-independent probability density $|\psi(\mathbf{r}, t)|^2 = |\varphi(\mathbf{r})|^2$. In a stationary function, only one angular frequency $\omega$ appears; according to the Planck-Einstein relations, *a stationary state is a state with a well-defined energy* $E = \hbar\omega$ (energy eigenstate). In classical mechanics, when the potential energy is time-independent, the total energy is a constant of the motion; in quantum mechanics, there exist well-determined energy states.

Equation (D-6) can therefore be written:

$$\left[-\frac{\hbar^2}{2m}\,\Delta + V(\mathbf{r})\right]\varphi(\mathbf{r}) = E\,\varphi(\mathbf{r}) \tag{D-8}$$

or:

$$\boxed{H\,\varphi(\mathbf{r}) = E\,\varphi(\mathbf{r})} \tag{D-9}$$

where $H$ is the differential operator:

$$\boxed{H = -\frac{\hbar^2}{2m}\,\Delta + V(\mathbf{r})} \tag{D-10}$$

$H$ is a linear operator since, if $\lambda_1$ and $\lambda_2$ are constants, we have:

$$H[\lambda_1\varphi_1(\mathbf{r}) + \lambda_2\varphi_2(\mathbf{r})] = \lambda_1 H\,\varphi_1(\mathbf{r}) + \lambda_2 H\,\varphi_2(\mathbf{r}) \tag{D-11}$$

Equation (D-9) is thus the *eigenvalue equation* of the linear operator $H$: the application of $H$ to the « eigenfunction » $\varphi(\mathbf{r})$ yields the same function, multiplied by the corresponding « eigenvalue » $E$. *The allowed energies are therefore the eigenvalues of the operator H.* We shall see later that equation (D-9) has square-integrable solutions $\varphi(\mathbf{r})$ only for certain values of $E$ (*cf.* § D-2-c and § 2-c of complement $H_I$): this is the origin of *energy quantization.*

COMMENT:

Equation (D-8) [or (D-9)] is sometimes called the "time-independent Schrödinger equation", as opposed to the "time-dependent Schrödinger equation" (D-1). We stress their essential difference: equation (D-1) is a general equation which gives the evolution of the wave function, whatever the state of the particle; on the other hand, the eigenvalue equation (D-9) enables us to find, amongst all the possible states of the particle, those which are stationary.

### b. SUPERPOSITION OF STATIONARY STATES

In order to distinguish between the various possible values of the energy $E$ (and the corresponding eigenfunctions $\varphi(\mathbf{r})$), we label them with an index $n$. Thus we have:

$$H\varphi_n(\mathbf{r}) = E_n\varphi_n(\mathbf{r}) \tag{D-12}$$

and the stationary states of the particle have as wave functions:

$$\psi_n(\mathbf{r}, t) = \varphi_n(\mathbf{r})\,\mathrm{e}^{-iE_nt/\hbar} \tag{D-13}$$

$\psi_n(\mathbf{r}, t)$ is a solution of the Schrödinger equation (D-1). Since this equation is linear, it has a whole series of other solutions of the form

$$\psi(\mathbf{r}, t) = \sum_n c_n\,\varphi_n(\mathbf{r})\,\mathrm{e}^{-iE_nt/\hbar} \tag{D-14}$$

where the coefficients $c_n$ are arbitrary complex constants. In particular, we have :

$$\psi(\mathbf{r}, 0) = \sum_n c_n \varphi_n(\mathbf{r}) \tag{D-15}$$

Inversely, assume that we know $\psi(\mathbf{r}, 0)$, that is, the state of the particle at $t = 0$. We shall see later that any function $\psi(\mathbf{r}, 0)$ can always be decomposed in terms of eigenfunctions of $H$, as in (D-15). The coefficients $c_n$ are therefore determined by $\psi(\mathbf{r}, 0)$. The corresponding solution $\psi(\mathbf{r}, t)$ of the Schrödinger equation is then given by (D-14). All we need to do to obtain it is to multiply each term of (D-15) by the factor $e^{-iE_n t/\hbar}$, where $E_n$ is the eigenvalue associated with $\varphi_n(\mathbf{r})$. We stress the fact that these phase factors differ from one term to another. It is only in the case of stationary states that the $t$-dependence involves only one exponential [formula (D-13)].

## 2. One-dimensional "square" potentials. Qualitative study

We said at the beginning of § D that in order to display quantum effects we were going to consider potentials which varied considerably over small distances. We shall limit ourselves here to a qualitative study, so as to concentrate on the simple physical ideas. A more detailed study is presented in the complements of this chapter (complement $H_I$). To simplify the problem, we shall consider a one-dimensional model, in which the potential energy depends only on $x$ (the justification for such a model is given in complement $F_I$).

### a. PHYSICAL MEANING OF A SQUARE POTENTIAL

We shall consider a one-dimensional problem with a potential of the type shown in figure 7-a. The $Ox$ axis is divided into a certain number of constant-potential regions. At the border of two adjacent regions the potential makes an abrupt jump (discontinuity). Actually, such a function cannot really represent a physical potential, which must be continuous. We shall use it to represent schematically a potential energy $V(x)$ which actually has the shape shown in figure 7-b : there are no discontinuities, but $V(x)$ varies very rapidly in the neighborhood of certain values of $x$. When the intervals over which these variations occur are much smaller than all other distances involved in the problem (in particular, the wavelength associated with the particle), we can replace the true potential by the square potential of figure 7-a. This is an approximation, which would cease to be valid, for example, for a particle having too high an energy, whose wavelength would be very short.

The predictions of classical mechanics concerning the behavior of a particle in a potential such as that of figure 7 are easy to determine. For example, imagine that $V(x)$ is the gravitational potential energy. Figure 7-b then represents the real profile of the terrain on which the particle moves : the corresponding discontinuities are sharp slopes, separated by horizontal plateaus. Notice that, if we fix the total energy $E$ of the particle, the domains of the $Ox$ axis where $V > E$ are forbidden to it (its kinetic energy $E_k = E - V$ must be positive).

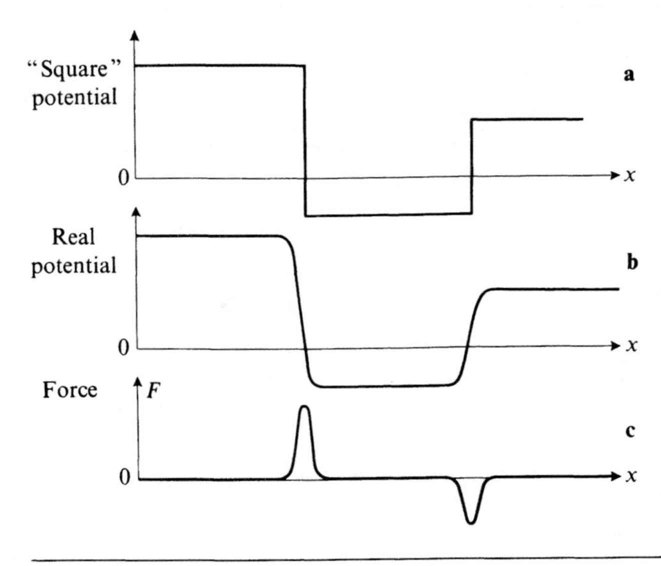


FIGURE 7

**Square potential (fig. a) which schematically represents a real potential (fig. b) for which the force has the shape shown in figure c.**

COMMENT:

The force exerted on the particle is $F(x) = -\dfrac{\mathrm{d}V(x)}{\mathrm{d}x}$. In figure 7-c, we have depicted this force, obtained from the potential $V(x)$ of figure 7-b. It can be seen that this particle, in all the regions where the potential is constant, is not subject to any force. Its velocity is then constant. It is only in the frontier zones between these plateaus that a force acts on the particle and, depending on the case, accelerates it or slows it down.

#### b. OPTICAL ANALOGY

We are going to consider the stationary states (§ D-1) of a particle in a one-dimensional "square" potential.

In a region where the potential has a constant value $V$, the eigenvalue equation (D-9) is written:

$$\left[ -\frac{\hbar^2}{2m}\frac{\mathrm{d}^2}{\mathrm{d}x^2} + V \right] \varphi(x) = E\varphi(x) \tag{D-16}$$

or:

$$\left[ \frac{\mathrm{d}^2}{\mathrm{d}x^2} + \frac{2m}{\hbar^2}(E - V) \right] \varphi(x) = 0 \tag{D-17}$$

Now, in optics, there exists a completely analogous equation. Consider a transparent medium whose index $n$ depends neither on $\mathbf{r}$ nor on time. In this medium, there can be electromagnetic waves whose electric field $\mathbf{E}(\mathbf{r}, t)$ is independent of $y$ and $z$ and has the form:

$$\mathbf{E}(\mathbf{r}, t) = \mathbf{e}E(x)\,\mathrm{e}^{-i\Omega t} \tag{D-18}$$

where **e** is a unit vector perpendicular to $Ox$. $E(x)$ must then satisfy:

$$\left[\frac{\mathrm{d}^2}{\mathrm{d}x^2} + \frac{n^2\Omega^2}{c^2}\right] E(x) = 0 \tag{D-19}$$

We see that equations (D-17) and (D-19) become identical if we set:

$$\frac{2m}{\hbar^2}(E - V) = \frac{n^2\Omega^2}{c^2} \tag{D-20}$$

Moreover, at a point $x$ where the potential energy $V$ [and, consequently, the index $n$ given by (D-20)] is discontinuous, the boundary conditions for $\varphi(x)$ and $E(x)$ are the same : these two functions, as well as their first derivatives, must remain continuous (*cf.* complement $H_I$, § 1-b). The structural analogy between the two equations (D-17) and (D-19) thus enables us to associate with a quantum mechanical problem, corresponding to the potential of figure 7-a, an optical problem : the propagation of an electromagnetic wave of angular frequency $\Omega$ in a medium whose index $n$ has discontinuities of the same type. According to (D-20), the relation between the optical and mechanical parameters is:

$$n(\Omega) = \frac{1}{\hbar\Omega}\sqrt{2mc^2(E - V)} \tag{D-21}$$

For the light wave, a region where $E > V$ corresponds to a transparent medium whose index is real. The wave is then of the form $e^{ikx}$.

What happens when $V > E$? Formula (D-20) gives a pure imaginary index. In (D-19), $n^2$ is negative and the solution is of the form $e^{-\rho x}$: it is the analogue of an "evanescent wave". Certain aspects of the situation recall the propagation of an electromagnetic wave in a metallic medium*.

Thus we can transpose the well-known results of wave optics to the problems which we are studying here. It is important, however, to realize that this is merely an analogy. The interpretation that we give for the wave function is fundamentally different from that which classical wave optics attributes to the electromagnetic wave.

#### c. EXAMPLES

##### α *Potential step and barrier*

Consider a particle of energy $E$ which, coming from the region of negative $x$, arrives at the potential "step" of height $V_0$ shown in figure 8.

If $E > V_0$, (the case in which the classical particle clears the potential step and continues towards the right with a smaller velocity), the optical analogy is the following : a light wave propagates from left to right in a medium of index $n_1$:

$$n_1 = \frac{c}{\hbar\Omega}\sqrt{2mE} \; ; \tag{D-22}$$

* This analogy should not be pushed too far, since the index $n$ of a metallic medium has both a real and a complex part (in a metal, an optical wave continues to oscillate as it damps out).

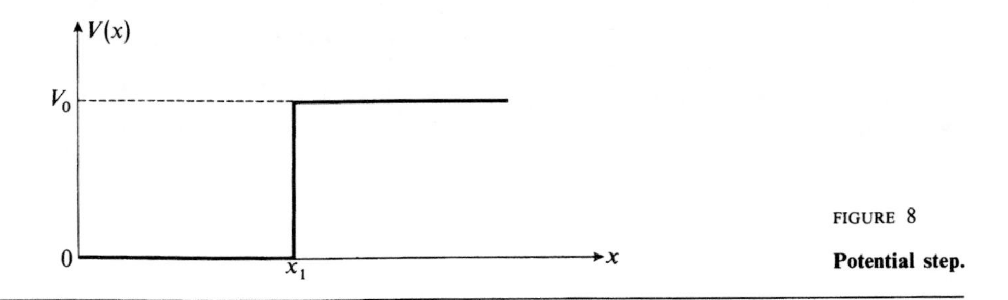


FIGURE 8

**Potential step.**

at $x = x_1$, there is a discontinuity, and the index, for $x > x_1$, is:

$$n_2 = \frac{c}{\hbar\Omega}\sqrt{2m(E - V_0)} \tag{D-23}$$

We know that the incident wave coming from the left splits into a reflected wave and a transmitted wave. Let us transpose this result to quantum mechanics: *the particle has a certain probability $\mathcal{P}$ of being reflected*, and only the probability $1 - \mathcal{P}$ of pursuing its course towards the right. This result is contrary to what is predicted by classical mechanics.

When $E < V_0$, the index $n_2$, which corresponds to the region $x > x_1$, becomes pure imaginary, and the incident light wave is totally reflected. The quantum prediction therefore coincides at this point with that of classical mechanics. Nevertheless, the existence, for $x > x_1$, of an evanescent wave, shows that the quantum particle has a non-zero probability of being found in this region.

The role of this evanescent wave is more striking in the case of a potential barrier (fig. 9). For $E < V_0$, a classical particle would always turn back. But, in the corresponding optical problem, we would have a layer of finite thickness, with an imaginary index, surrounded by a transparent medium. If this thickness is not much greater than the range $1/\rho$ of the evanescent wave, part of the incident wave is transmitted into the region $x > x_2$. Therefore, even for $E < V_0$, we find a *non-zero probability of the particle crossing the barrier*. This is called the "tunnel effect".

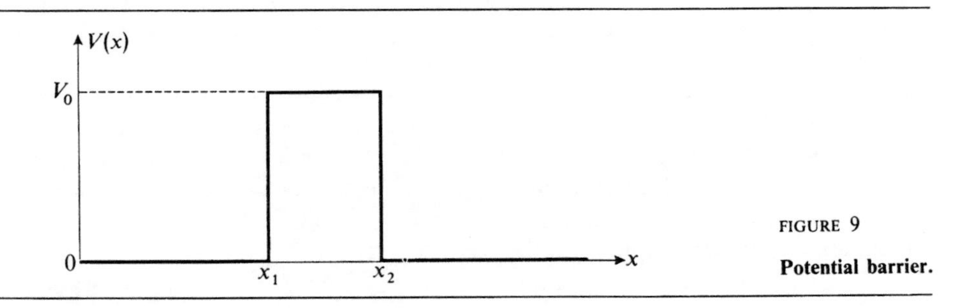


FIGURE 9

**Potential barrier.**

β *Potential well*

The function $V(x)$ now has the form shown in figure 10. The predictions of classical mechanics are the following: when the particle has a negative energy $E$ (but greater than $-V_0$), it can only oscillate between $x_1$ and $x_2$, with kinetic energy $E_k = E + V_0$; when the particle has a positive energy and arrives from the left, it undergoes an abrupt acceleration at $x_1$, then an equal deceleration at $x_2$, and then continues towards the right.

In the optical analogue of the case $-V_0 < E < 0$, the indices $n_1$ and $n_3$, which correspond to the regions $x < x_1$ and $x > x_2$, are imaginary, while the index $n_2$, which characterizes the interval $[x_1, x_2]$, is real. Thus we have the equivalent of

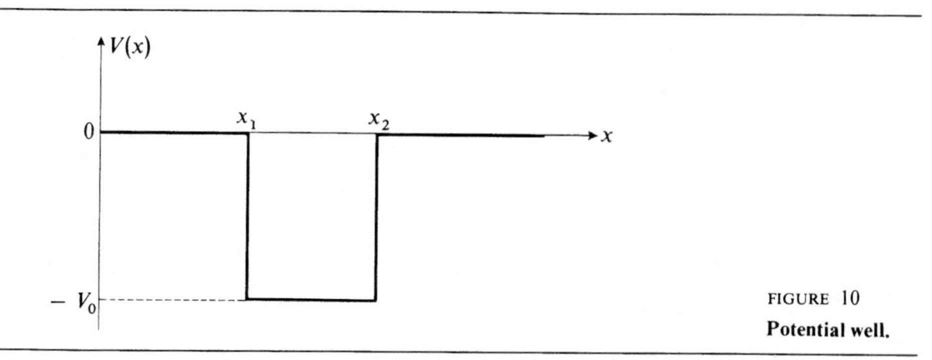


FIGURE 10
**Potential well.**

a layer of air, for example, between two reflecting media. The different waves reflected successively at $x_1$ and $x_2$ destroy each other through interference, except for certain well-determined frequencies ("normal modes") which allow stable stationary waves to be established. From the quantum point of view, this implies that *the negative energies are quantized*★, while, classically, all values included between $-V_0$ and 0 are possible.

For $E > 0$, the indices $n_1$, $n_2$ and $n_3$ are real:

$$n_1 = n_3 = \frac{c}{\Omega}\frac{1}{\hbar}\sqrt{2mE} \tag{D-24}$$

$$n_2 = \frac{c}{\Omega}\frac{1}{\hbar}\sqrt{2m(E + V_0)} \tag{D-25}$$

Since $n_2$ is greater than $n_1$ and $n_3$, the situation is analogous to that of a layer of glass in air. In order to obtain the reflected wave for $x < x_1$, or the transmitted wave in the region $x > x_2$, it is necessary to superpose an infinite number of waves which arise from successive reflections at $x_1$ and $x_2$ (multiple wave interferometer analogous to a Fabry-Pérot). We then find that, for certain incident frequencies, the wave is entirely transmitted. From the quantum point of view, the particle thus has, in general, a certain probability of being reflected. However, there exist energy values, called *resonance energies*, for which the probability of transmission is 1 and, consequently, the probability of reflection is 0.

These few examples show how much the predictions of quantum mechanics can differ from those of classical mechanics. They also clearly stress the primordial role of potential discontinuities (which represent, schematically, rapid variations).

★ The allowed energy values are not given by the well-known condition : $x_2 - x_1 = k\lambda_2/2$, for it is necessary to take into account the existence of evanescent waves, which introduce a phase shift upon reflection at $x = x_1$ and $x = x_2$ (*cf.* complement $H_I$, § 2-c).

CONCLUSION

In this chapter, we have introduced and discussed, in a qualitative and intuitive manner, certain fundamental ideas of quantum mechanics. We shall later return to these ideas (chap. III) so as to present them in a more precise and systematic way. Nevertheless, it is already clear that the quantum description of physical systems differs radically from the one given by classical mechanics (although the latter constitutes, in numerous cases, an excellent approximation). We have limited ourselves in this chapter to the case of physical systems composed of only one particle. The description of these systems at a given time is, in classical mechanics, founded on the specification of six parameters, which are the components of the position $\mathbf{r}(t)$ and the velocity $\mathbf{v}(t)$ of the particle. All the dynamical variables (energy, linear momentum, angular momentum) are determined by the specification of $\mathbf{r}(t)$ and $\mathbf{v}(t)$. Newton's laws enable us to calculate $\mathbf{r}(t)$ through the solution of second-order differential equations with respect to time. Consequently, they fix the values of $\mathbf{r}(t)$ and $\mathbf{v}(t)$ for any time $t$ when they are known for the initial time.

Quantum mechanics uses a more complicated description of phenomena. The dynamic state of a particle, at a given time, is characterized by a wave function. It no longer depends on only six parameters, but on an infinite number [the values of $\psi(\mathbf{r}, t)$ at all points $\mathbf{r}$ of space]. Moreover, the predictions of the measurement results are now only probabilistic (they yield only the probability of obtaining a given result in the measurement of a dynamical variable). The wave function is a solution of the Schrödinger equation, which enables us to calculate $\psi(\mathbf{r}, t)$ from $\psi(\mathbf{r}, 0)$. This equation implies a principle of superposition which leads to wave effects.

This upheaval in our conception of mechanics was imposed by experiment. The structure and behavior of matter on an atomic level are incomprehensible in the framework of classical mechanics. The theory has thereby lost some of its simplicity, but it has gained a great deal of unity, since matter and radiation are described in terms of the same general scheme (wave-particle duality). We stress the fact that this general scheme, although it runs counter to our ideas and habits drawn from the study of the macroscopic domain, is perfectly coherent. No one has ever succeeded in imagining an experiment which could violate the uncertainty principle (*cf.* complement $D_I$ of this chapter). In general, no observation has, to date, contradicted the fundamental principles of quantum mechanics. Nevertheless, at present, there is no global theory of relativistic and quantum phenomena, and nothing, of course, prevents the possibility of a new upheaval.

**References and suggestions for further reading:**

Description of physical phenomena which demonstrate the necessity of introducing quantum mechanical concepts: see the subsection "Introductory work – quantum physics" of section 1 of the bibliography; in particular, Wichmann (1.1) and Feynman III (1.2), chaps. 1 and 2.

History of the development of quantum mechanical concepts: references of section 4 of the bibliography, in particular, Jammer (4.8); also see references (5.11) and (5.12), which contain numerous references to the original articles.

Fundamental experiments: references to the original articles can be found in section 3 of the bibliography.

The problem of interpretation in quantum mechanics: section 5 of the bibliography; in particular, the "Resource Letter" (5.11), which contains many classified references.

Analogies and differences between matter waves and electromagnetic waves: Bohm (5.1), chap. 4; in particular, the table "Summary on Probabilities" at the end of the chapter.

See also the articles by Schrödinger (1.25), Gamow (1.26), Born and Biem (1.28), Scully and Sargent (1.30).

## COMPLEMENTS OF CHAPTER I

**$A_I$: ORDER OF MAGNITUDE OF THE WAVELENGTHS ASSOCIATED WITH MATERIAL PARTICLES**

**$B_I$: CONSTRAINTS IMPOSED BY THE UNCERTAINTY RELATIONS**

**$C_I$: THE UNCERTAINTY RELATIONS AND ATOMIC PARAMETERS**

$A_I$, $B_I$, $C_I$: very simple but fundamental reflections on the order of magnitude of quantum parameters

---

**$D_I$: AN EXPERIMENT ILLUSTRATING THE UNCERTAINTY RELATION**

$D_I$ : discussion of a simple thought experiment which attempts to invalidate the complementarity between the particle and wave aspects of light (easy, but could be reserved for subsequent study).

---

**$E_I$: A SIMPLE TREATMENT OF A TWO-DIMENSIONAL WAVE PACKET**

**$F_I$: THE RELATION BETWEEN ONE- AND THREE-DIMENSIONAL PROBLEMS**

**$G_I$: ONE-DIMENSIONAL GAUSSIAN WAVE PACKET : SPREADING OF THE WAVE PACKET**

$E_I$, $F_I$, $G_I$ : complements on wave packets (§ C of chapter I)

$E_I$ : reveals in a simple, qualitative way the relation which exists between the lateral extension of a two-dimensional wave packet and the angular dispersion of wave vectors (easy).

$F_I$: generalization to three dimensions of the results of § C of chapter I; shows how the study of a particle in three-dimensional space can, in certain cases, be reduced to one-dimensional problems (a little more difficult).

$G_I$: treats in detail a special case of wave packets for which one can calculate exactly the properties and the evolution (some difficulties in the calculation, but conceptually simple).

---

**$H_I$: STATIONARY STATES OF A PARTICLE IN ONE-DIMENSIONAL SQUARE POTENTIALS**

$H_I$: takes up in a more quantitative way the ideas of § D-2 of chapter I. Strongly recommended, since square potentials are often used to illustrate simply the implications of quantum mechanics (numerous complements and exercises proposed later in this book rely on the results of $H_I$).

---

**$J_I$: BEHAVIOR OF A WAVE PACKET AT A POTENTIAL STEP**

$J_I$: more precise study, for a special case, of the quantum behavior of a particle in a square potential. Since the particle is sufficiently well-localized in space (wave packet), one can follow its " motion " (average difficulty; important for the physical interpretation of the results).

**$K_I$ : EXERCISES**

## Complement $A_I$

## ORDER OF MAGNITUDE OF THE WAVELENGTHS ASSOCIATED WITH MATERIAL PARTICLES

De Broglie's relation:

$$\lambda = \frac{h}{p} \tag{1}$$

shows that, for a particle of mass $m$ and speed $v$, the smaller $m$ and $v$, the longer the corresponding wavelength.

To show that the wave properties of matter are impossible to detect in the macroscopic domain, take as an example a dust particle, of diameter 1 $\mu$ and mass $m \simeq 10^{-15}$ kg. Even for such a small mass and a speed of $v \simeq 1$ mm/s, formula (1) gives:

$$\lambda \simeq \frac{6.6 \times 10^{-34}}{10^{-15} \times 10^{-3}} \text{ meter} = 6.6 \times 10^{-16} \text{ meter} = 6.6 \times 10^{-6} \text{ Å} \tag{2}$$

Such a wavelength is completely negligible on the scale of the dust particle.

Consider, on the other hand, a thermal neutron, that is, a neutron ($m_n \simeq 1.67 \times 10^{-27}$ kg) with a speed $v$ corresponding to the average thermal energy at the (absolute) temperature $T$. $v$ is given by the relation:

$$\frac{1}{2} m_n v^2 = \frac{p^2}{2m_n} \simeq \frac{3}{2} kT \tag{3}$$

where $k$ is the Boltzmann constant ($k \simeq 1.38 \times 10^{-23}$ joule/degree). The wavelength which corresponds to such a speed is:

$$\lambda = \frac{h}{p} = \frac{h}{\sqrt{3m_n kT}} \tag{4}$$

For $T \simeq 300$ °K, we find:

$$\lambda \simeq 1.4 \text{ Å} \tag{5}$$

that is, a wavelength which is of the order of the distance between atoms in a crystal lattice. A beam of thermal neutrons falling on a crystal therefore gives rise to diffraction phenomena analogous to those observed with X-rays.

Let us now examine the order of magnitude of the de Broglie wavelengths associated with electrons ($m_e \simeq 0.9 \times 10^{-30}$ kg). If one accelerates an electron beam through a potential difference $V$ (expressed in volts), one gives the electrons a kinetic energy:

$$E = qV = 1.6 \times 10^{-19} \, V \text{ joule} \tag{6}$$

($q = 1.6 \times 10^{-19}$ coulomb is the electron charge.) Since $E = \frac{p^2}{2m_e}$, the associated wavelength is equal to:

$$\lambda = \frac{h}{p} = \frac{h}{\sqrt{2m_e E}} \tag{7}$$

that is, numerically:

$$\lambda = \frac{6.6 \times 10^{-34}}{\sqrt{2 \times 0.9 \times 10^{-30} \times 1.6 \times 10^{-19}\, V}} \text{ meter}$$

$$\simeq \frac{12.3}{\sqrt{V}} \text{ Å} \tag{8}$$

With potential differences of several hundreds of volts, one again obtains wavelengths comparable to those of X-rays, and electron diffraction phenomena can be observed with crystals or crystalline powders.

The large accelerators which are currently available are able to impart considerable energy to particles. This takes us out of the non-relativistic domain to which we have thus far confined ourselves. For example, electron beams are easily obtained for which the energy exceeds 1 GeV★ = $10^9$ eV (1 eV = 1 electron-volt = $1.6 \times 10^{-19}$ joule), while the electron rest mass is equal to $m_e c^2 \simeq 0.5 \times 10^6$ eV. This means that the corresponding speed is very close to the speed of light $c$. Consequently, the non-relativistic quantum mechanics which we are studying here does not apply. However, the relations:

$$E = h\nu \tag{9-a}$$

$$\lambda = \frac{h}{p} \tag{9-b}$$

remain valid in the relativistic domain. On the other hand, relation (7) must be modified since, relativistically, the energy $E$ of a particle of rest mass $m_0$ is no longer $p^2/2m_0$, but instead:

$$E = \sqrt{p^2c^2 + m_0^2c^4} \tag{10}$$

In the example considered above (an electron of energy 1 GeV), $m_e c^2$ is negligible compared to $E$, and we obtain:

$$\lambda \simeq \frac{hc}{E} = \frac{6.6 \times 10^{-34} \times 3 \times 10^8}{1.6 \times 10^{-10}} \text{ m} = 1.2 \times 10^{-15} \text{ m} = 1.2 \text{ fermi} \tag{11}$$

(1 fermi = $10^{-15}$ m). With electrons accelerated in this way, one can explore the structure of atomic nuclei and, in particular, the structure of the proton; nuclear dimensions are of the order of a fermi.

COMMENTS:

(*i*) We want to point out a common error in the calculation of the wavelength of a material particle of mass $m_0 \neq 0$, whose energy $E$ is known. This error consists of calculating the frequency $\nu$ using (9-a) and, then, by analogy with electromagnetic waves, of taking $c/\nu$ for the de Broglie

★ Translator's note : In the United States, this unit is sometimes written BeV.

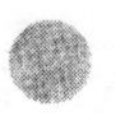



wavelength. Obviously, the correct reasoning consists of calculating, for example from (10) $\left(\text{or, in the non-relativistic domain, from the relation } E = \frac{p^2}{2m}\right)$ the momentum $p$ associated with the energy $E$ and then using (9-b) to find $\lambda$.

($ii$) According to (9-a), the frequency $\nu$ depends on the origin chosen for the energies. The same is true for the phase velocity $V_\varphi = \frac{\omega}{k} = \nu\lambda$. Note, on the other hand, that the group velocity $V_G = \frac{d\omega}{dk} = 2\pi \frac{d\nu}{dk}$ does not depend on the choice of the energy origin. This is important in the physical interpretation of $V_G$.

**References and suggestions for further reading:**

Wichmann (1.1), chap. 5; Eisberg and Resnick (1.3), § 3.1.

## Complement $B_I$

# CONSTRAINTS IMPOSED BY THE UNCERTAINTY RELATIONS



We saw in § C-3 of chapter I that the position and momentum of a particle cannot be simultaneously defined with arbitrary precision : the corresponding uncertainties $\Delta x$ and $\Delta p$ must satisfy the uncertainty relation:

$$\Delta x \, . \, \Delta p \gtrsim \hbar \tag{1}$$

Here we intend to evaluate numerically the importance of this constraint. We shall show that it is completely negligible in the macroscopic domain and that it becomes, on the other hand, crucial on the microscopic level.

## 1. Macroscopic system

Let us take up again the example of a dust particle (*cf.* complement $A_I$), whose diameter is on the order of 1 $\mu$ and whose mass $m \simeq 10^{-15}$ kg, having a speed $v = 10^{-3}$ m/sec. Its momentum is then equal to:

$$p = mv \simeq 10^{-18} \text{ joule sec/m} \tag{2}$$

If its position is measured to within 0.01 $\mu$, for example, the uncertainty $\Delta p$ in the momentum must satisfy:

$$\Delta p \simeq \frac{\hbar}{\Delta x} \simeq \frac{10^{-34}}{10^{-8}} = 10^{-26} \text{ joule sec/m} \tag{3}$$

Thus the uncertainty relation introduces practically no restrictions in this case since, in practice, a momentum measurement device is incapable of attaining the required relative accuracy of $10^{-8}$

In quantum terms, the dust particle is described by a wave packet whose group velocity is $v = 10^{-3}$ m/sec and whose average momentum is $p = 10^{-18}$ joule sec/m. But one can then choose such a small spatial extension $\Delta x$ and momentum dispersion $\Delta p$ that they are both totally negligible. The maximum of the wave packet then represents *the* position of the dust particle, and its motion is identical to that of the classical particle.

## 2. Microscopic system

Now let us consider an atomic electron. The Bohr model describes it as a classical particle. The allowed orbits are defined by quantization rules which

are assumed *a priori*: for example, the radius $r$ of a circular orbit and the momentum $p = mv$ of the electron travelling in it must satisfy:

$$pr = n\hbar \tag{4}$$

where $n$ is an integer.

For us to be able to speak in this way of an electron trajectory in classical terms, the uncertainties in its position and momentum must be negligible compared to $r$ and $p$ respectively:

$$\Delta x \ll r \tag{5-a}$$

$$\Delta p \ll p \tag{5-b}$$

which would mean that:

$$\frac{\Delta x}{r} \cdot \frac{\Delta p}{p} \ll 1 \tag{6}$$

Now the uncertainty relation imposes :

$$\frac{\Delta x}{r} \cdot \frac{\Delta p}{p} \gtrsim \frac{\hbar}{rp} \tag{7}$$

If we use formula (4) to replace $rp$ by $n\hbar$ on the right-hand side, this inequality can be written as:

$$\frac{\Delta x}{r} \cdot \frac{\Delta p}{p} \gtrsim \frac{1}{n} \tag{8}$$

We then see that (8) is incompatible with (6), unless $n \gg 1$. The uncertainty relation thus makes us reject the semi-classical picture of the Bohr orbits (see § C-2 of chapter VII).

**References and suggestions for further reading:**

Bohm (5.1), chap. 5, § 14.

## Complement $C_I$

# THE UNCERTAINTY RELATIONS AND ATOMIC PARAMETERS

The Bohr orbit has no physical reality when coupled with the uncertainty relations (*cf.* complement $B_I$). Later (chap. VII), we shall study the quantum theory of the hydrogen atom. We are going to show immediately, however, how the uncertainty relations enable one to understand the stability of atoms and even to derive simply the order of magnitude of the dimensions and the energy of the hydrogen atom in its ground state.

Let us consider, therefore, an electron in the coulomb field of a proton, which we shall assume to be stationary at the origin of the coordinate system. When the two particles are separated by a distance $r$, the potential energy of the electron is:

$$V(r) = -\frac{q^2}{4\pi\varepsilon_0}\frac{1}{r} \tag{1}$$

where $q$ is its charge (exactly opposite to that of the proton). We shall set:

$$\frac{q^2}{4\pi\varepsilon_0} = e^2 \tag{2}$$

Assume that the state of the electron is described by a spherically symmetric wave function whose spatial extent is characterized by $r_0$ (this means that the probability of presence is practically zero beyond $2r_0$ or $3r_0$). The potential energy corresponding to this state is then on the order of:

$$\overline{V} \simeq -\frac{e^2}{r_0} \tag{3}$$

For it to be as low as possible, it is necessary to take $r_0$ as small as possible. That is, the wave function must be as concentrated as possible about the proton.

But it is also necessary to take the kinetic energy into account. This is where the uncertainty principle comes in: if the electron is confined within a volume of linear dimension $r_0$, the uncertainty $\Delta p$ in its momentum is at least of the order of $\hbar/r_0$. In other words, even if the average momentum is zero, the kinetic energy $T$ associated with the state under consideration is not zero:

$$\overline{T} \gtrsim \overline{T}_{\text{min}} = \frac{1}{2m}(\Delta p)^2 \simeq \frac{\hbar^2}{2mr_0^2} \tag{4}$$

If we take $r_0$ smaller in order to decrease the potential energy, the minimum kinetic energy (4) increases.

The lowest total energy compatible with the uncertainty relation is thus the minimum of the function:

$$E_{\text{min}} = \overline{T}_{\text{min}} + \overline{V} = \frac{\hbar^2}{2mr_0^2} - \frac{e^2}{r_0} \tag{5}$$

This minimum is obtained for :

$$r_0 = a_0 = \frac{\hbar^2}{me^2} \tag{6}$$

and is equal to :

$$E_0 = -\frac{me^4}{2\hbar^2} \tag{7}$$

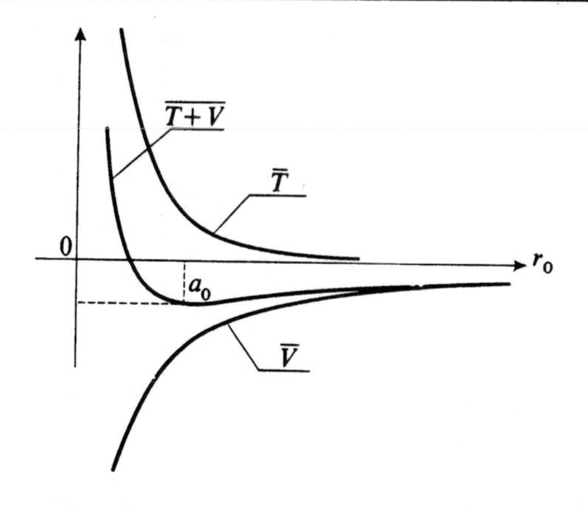


FIGURE 1

**Variation with respect to $r_0$ (extension of the wave function) of the potential energy $\overline{V}$, the kinetic energy $\overline{T}$, and the total energy $\overline{T} + \overline{V}$ of a hydrogen atom. The functions $\overline{T}$ and $\overline{V}$ vary inversely, so the total energy passes through a minimum value for some value of $\overline{T}$ and $\overline{V}$. The corresponding value $a_0$ of $r_0$ gives the order of magnitude of the hydrogen atom's size.**

Expression (6) is the one found in the Bohr model for the radius of the first orbit, and (7) gives correctly the energy of the ground state of the hydrogen atom (see chap. VII; the wave function of the ground state is indeed $e^{-r/a_0}$). Such quantitative agreement can only be accidental, since we have been reasoning on the basis of orders of magnitude. However, the preceding calculation reveals an important physical idea : because of the uncertainty relation, the smaller the extension of the wave function, the greater the kinetic energy of the electron. The ground state of the atom results from a compromise between the kinetic energy and the potential energy.

We stress the fact that this compromise, based on the uncertainty relation, is totally different from what would be expected in classical mechanics. If the electron moved in a classical circular orbit of radius $r_0$, its potential energy would be equal to:

$$V_{cl} = -\frac{e^2}{r_0} \tag{8}$$

The corresponding kinetic energy is obtained by equating the electrostatic force and the centrifugal force*:

$$\frac{e^2}{r_0^2} = m\frac{v^2}{r_0} \tag{9}$$

* In fact, the laws of classical electromagnetism indicate that an accelerated electron radiates, which already forbids the existence of stable orbits.

which gives:

$$T_{cl} = \frac{1}{2} m v^2 = \frac{1}{2} \frac{e^2}{r_0} \tag{10}$$

The total energy would then be equal to:

$$E_{cl} = T_{cl} + V_{cl} = -\frac{1}{2} \frac{e^2}{r_0} \tag{11}$$

The most favorable energetic situation would occur at $r_0 = 0$, which would give an infinite binding energy. Thus, we can say that it is the uncertainty relation which enables us to understand, as it were, the existence of atoms.

**References and suggestions for further reading:**

Feynman III (1.2), § 2-4. The same type of reasoning applied to molecules: Schiff (1.18), first section of § 49.

## Complement $D_I$

## AN EXPERIMENT ILLUSTRATING THE UNCERTAINTY RELATION

Young's double-slit experiment, which we analyzed in § A-2 of chapter I, led us to the following conclusions: both wave and particle aspects of light are needed to explain the observed phenomena; but they seem to be mutually exclusive, in the sense that *it is impossible to determine through which slit each photon has passed without destroying, by this very operation, the interference pattern.* The wave and particle aspects are sometimes said to be *complementary.*

We are going to consider Young's double-slit experiment again to demonstrate how complementarity and uncertainty relations are intimately related. To try to cast doubt on the uncertainty relation, one can imagine more subtle devices than the one of chapter I, which used photomultipliers placed behind the slits. We shall now analyze one of these devices.

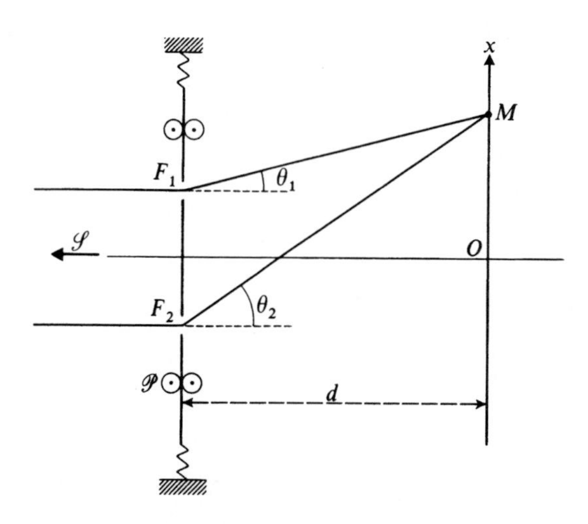


FIGURE 1

**Diagram of a device using a movable plate $\mathscr{P}$ whose momentum is measured before and after the passage of the photon to determine whether the photon passed through $F_1$ or through $F_2$ before arriving at point $M$ on the screen.**

Assume that the plate $\mathscr{P}$, in which the slits are pierced, is mounted so that it can move vertically in the same plane. Thus, it is possible to measure the vertical momentum transferred to it. Consider (fig. 1) a photon which strikes the observation screen $\mathscr{E}$ at point $M$ (for simplicity, we choose a source $\mathscr{S}$ at infinity). The momentum of this photon changes when it crosses $\mathscr{P}$. Conservation of momentum implies that the plate $\mathscr{P}$ absorbs the difference. But the momentum thus transferred to $\mathscr{P}$ depends on the path of the photon; depending on whether it passed through $F_1$ or $F_2$, the photon has a momentum of:

$$p_1 = -\frac{h\nu}{c}\sin\theta_1 \tag{1}$$

or:

$$p_2 = -\frac{h\nu}{c}\sin\theta_2 \tag{2}$$

$\left(\frac{h\nu}{c}\right.$ is the photon's momentum, $\theta_1$ and $\theta_2$ are the angles made by $F_1M$ and $F_2M$ with the incident direction.$\left.\right)$

We then allow the photons to arrive one by one and gradually construct the interference pattern on the screen $\mathscr{E}$. For each one, we determine through which slit it has passed by measuring the momentum acquired by the plate $\mathscr{P}$. It therefore seems that interference phenomena can still be observed on $\mathscr{E}$ although we know through which slit each photon has passed.

Actually, we shall see that the interference fringes are not visible with this device. The error in the preceding argument consists of assuming that only the photons have a quantum character. In reality, it must not be forgotten that quantum mechanics also applies to the plate $\mathscr{P}$ (macroscopic object). If we want to know through which hole a photon has passed, the uncertainty $\Delta p$ in the vertical momentum of $\mathscr{P}$ must be sufficiently small for us to be able to measure the difference between $p_1$ and $p_2$:

$$\Delta p \ll |p_2 - p_1| \tag{3}$$

But then the uncertainty relation implies that the position of $\mathscr{P}$ is only known to within $\Delta x$, with:

$$\Delta x \gtrsim \frac{h}{|p_2 - p_1|} \tag{4}$$

If we designate by $a$ the separation of the two slits and by $d$ the distance between the plate $\mathscr{P}$ and the screen $\mathscr{E}$, and if we assume that $\theta_1$ and $\theta_2$ are small ($d/a \gg 1$), we find (fig. 1):

$$\sin\theta_1 \simeq \theta_1 \simeq \frac{x - a/2}{d}$$

$$\sin\theta_2 \simeq \theta_2 \simeq \frac{x + a/2}{d} \tag{5}$$

($x$ denotes the position of the point of impact $M$ on $\mathscr{E}$). Formulas (1) and (2) then give:

$$|p_2 - p_1| \simeq \frac{h\nu}{c}|\theta_2 - \theta_1| \simeq \frac{h\,a}{\lambda\,d} \tag{6}$$

where $\lambda = \frac{c}{\nu}$ is the wavelength of light. Substituting this value into formula (4), we obtain:

$$\Delta x \gtrsim \frac{\lambda d}{a} \tag{7}$$

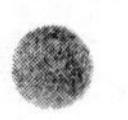



But $\dfrac{\lambda d}{a}$ is precisely the fringe separation we expect to find on $\mathscr{E}$. If the vertical position of the slits $F_1$ and $F_2$ is defined only to within an uncertainty greater than the fringe separation, it is impossible to observe the interference pattern.

The preceding discussion clearly shows that it is impossible to construct a quantum theory which is valid for light and not for material systems without running into serious contradictions. Thus, in the above example, if we could treat the plate $\mathscr{P}$ as a classical material system, we could invalidate the complementarity of the two aspects of light, and, consequently, the quantum theory of radiation. Inversely, a quantum theory of matter alone would come up against analogous difficulties. In order to obtain an overall coherence, we must apply quantum ideas to all physical systems.

**References and suggestions for further reading:**

Bohm (5.1), chaps. 5 and 6; Messiah (1.17), chap. IV § III; Schiff (1.18), § 4; Jammer (5.12), chaps. 4 and 5; also see reference (5.7).

## Complement $E_I$

# A SIMPLE TREATMENT OF A TWO-DIMENSIONAL WAVE PACKET



## 1. Introduction

In § C-2 of chapter I, we studied the shape of one-dimensional wave packets, obtained by superposing plane waves which all propagate in the same direction [formula (C-7)]. If this direction is that of the $Ox$ axis, the resulting function is independent of $y$ and $z$. It has a finite extension along $Ox$, but is not limited in the perpendicular directions : its value is the same at all points of a plane parallel to $yOz$.

We intend to examine here another simple type of wave packets : the plane waves which we are going to combine have coplanar wave vectors, which are (nearly) equal in magnitude but have slightly different directions. The goal is to show how the angular dispersion leads to a limitation of the wave packet in the directions perpendicular to the average wave vector.

We saw in § C-2 of chapter I how, by studying the superposition of three specific waves of the one-dimensional packet, one can understand the most important aspects of the phenomena. In particular, one can find the fundamental relation (C-18) of this chapter. We are going to limit ourselves here to a simplified model of this type. The generalization of the results which we are going to find can be carried out in the same way as in chapter I (see also complement $F_I$).

## 2. Angular dispersion and lateral dimensions

Consider three plane waves, whose wave vectors $\mathbf{k}_1$, $\mathbf{k}_2$ and $\mathbf{k}_3$ are shown in figure 1. All three are in the $xOy$ plane; $\mathbf{k}_1$ is directed along $Ox$; $\mathbf{k}_2$ and $\mathbf{k}_3$ are symmetric with respect to $\mathbf{k}_1$, the angle between each of them and $\mathbf{k}_1$ being $\Delta\theta$, which we assume to be small. Finally, the projections of $\mathbf{k}_1$, $\mathbf{k}_2$ and $\mathbf{k}_3$ on $Ox$ are equal:

$$k_{1x} \simeq k_{2x} = k_{3x} \simeq |\mathbf{k}_1| = k \tag{1}$$

The magnitudes of these three vectors differ only by terms which are second order in $\Delta\theta$, which we shall neglect. Their components along the $Oy$ axis are :

$$\begin{cases} k_{1y} = 0 \\ k_{2y} = -\,k_{3y} \simeq k\,\Delta\theta \end{cases} \tag{2}$$

We shall choose, as in § C-2 of chapter I, real amplitudes $g(\mathbf{k})$ which satisfy the relations:

$$g(\mathbf{k}_2) = g(\mathbf{k}_3) = \frac{1}{2} g(\mathbf{k}_1) \tag{3}$$

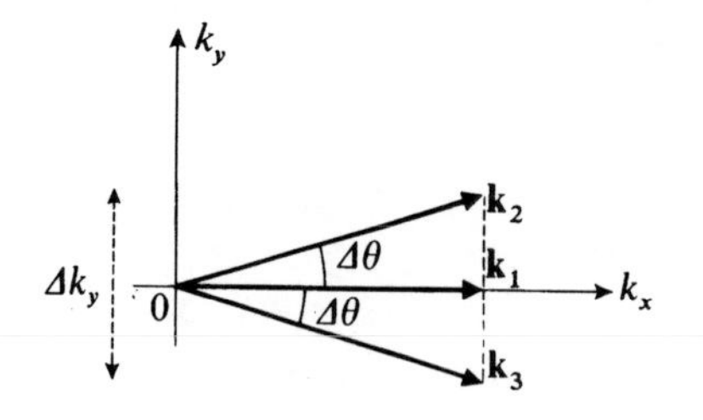


FIGURE 1

**The arrangement of the wave vectors $\mathbf{k}_1$, $\mathbf{k}_2$ and $\mathbf{k}_3$ associated with three plane waves which will be superposed to construct a two-dimensional wave packet.**

This model represents schematically a more complex situation, in which one would have a real wave packet, as in equation (C-6) of chapter I, with the following characteristics: all the wave vectors are perpendicular to $Oz$ and have the same projection on $Ox$ (only the component along $Oy$ varies); the function $|g(\mathbf{k})|$ has, with respect to this single variable $k_y$, the shape shown in figure 2; its width $\Delta k_y$ is related very simply to the angular dispersion $2\Delta\theta$:

$$\Delta k_y = 2k\Delta\theta \tag{4}$$

The superposition of the three waves defined above gives:

$$\begin{aligned}
\psi(x, y) &= \sum_{i=1}^{3} g(\mathbf{k}_i)\, e^{i\mathbf{k}_i \cdot \mathbf{r}} \\
&= g(\mathbf{k}_1)\left[ e^{ikx} + \frac{1}{2} e^{i(kx + k\,\Delta\theta\, y)} + \frac{1}{2} e^{i(kx - k\,\Delta\theta\, y)} \right] \\
&= g(\mathbf{k}_1)\, e^{ikx}[1 + \cos(k\,\Delta\theta\, y)]
\end{aligned} \tag{5}$$

(there is no $z$-dependence, which is why this is called a two-dimensional wave packet).

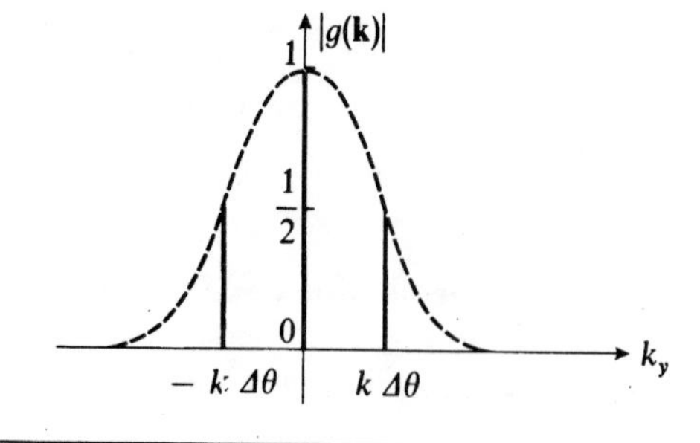


FIGURE 2

**The three values chosen for $k_y$ represent very schematically a peaked function $|g(\mathbf{k})|$ (dashed line).**

In order to understand what happens, we can use figure 3, where we represent, for each of the three components, the successive wave fronts corresponding to phase differences of $2\pi$. The function $|\psi(x, y)|$ has a maximum at $y = 0$ : the three waves interfere constructively on the $Ox$ axis. When we move away from this axis, $|\psi(x, y)|$ decreases (the phase shift between the components increases) and goes to zero at $y = \pm \frac{\Delta y}{2}$, where $\Delta y$ is given by:

$$\cos\left(k\Delta\theta\frac{\Delta y}{2}\right) = -1 \tag{6}$$

that is, for:

$$k\Delta\theta\,\Delta y = 2\pi \tag{7}$$

The phases of the $(\mathbf{k}_2)$ and $(\mathbf{k}_3)$ waves are then in opposition with that of the $(\mathbf{k}_1)$ wave (fig. 3). Using (4), we can rewrite (7) in a form which is analogous to that of relation (C-11) of chapter I:

$$\Delta y \,.\, \Delta k_y = 4\pi \tag{8}$$

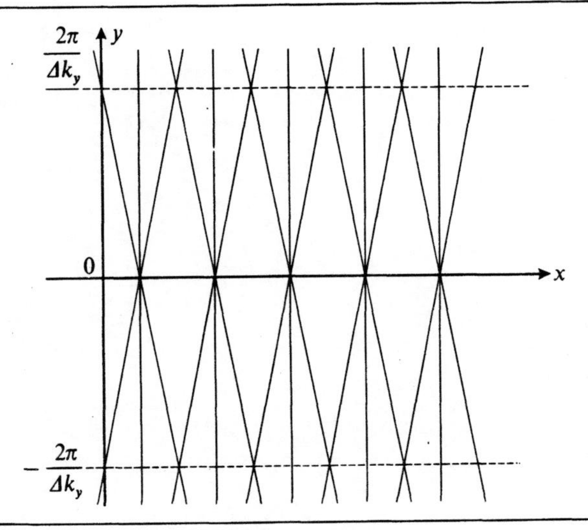


FIGURE 3

**Equal phase planes of the three waves associated with the three k vectors of figure 1: these waves are in phase at $y = 0$, but interfere destructively at $y = \pm 2\pi/\Delta k_y$.**

Thus an angular dispersion of the wave vectors limits the lateral dimensions of the wave packets. Quantitatively, this limitation has the form of an uncertainty relation [formulas (7) and (8)].

## 3. Discussion

Consider a plane wave with wave vector **k** propagating along $Ox$. Any attempt to limit its extension perpendicular to $Ox$ causes an angular dispersion to appear, that is, transforms it into a wave packet analogous to the ones we are studying here.

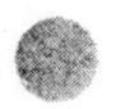


Assume, for example, that we place in the path of the plane wave a screen pierced by a slit of width $\Delta y$. This will give rise to a diffracted wave (*cf.* fig. 4). We know that the angular width of the diffraction pattern is given by:

$$2\Delta\theta \simeq 2\frac{\lambda}{\Delta y} \tag{9}$$

where $\lambda = \frac{2\pi}{|\mathbf{k}|}$ is the incident wavelength. This is indeed the same situation as above: formulas (7) and (9) are identical.

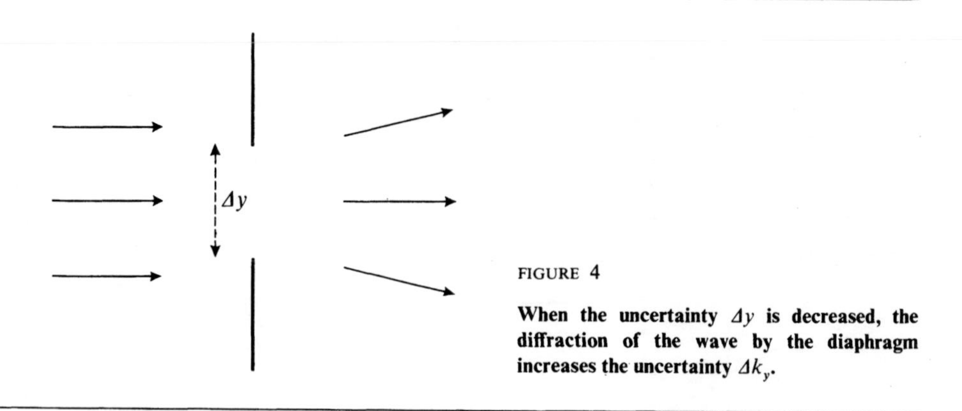


FIGURE 4

**When the uncertainty $\Delta y$ is decreased, the diffraction of the wave by the diaphragm increases the uncertainty $\Delta k_y$.**

# Complement $F_I$

## THE RELATIONSHIP BETWEEN ONE- AND THREE-DIMENSIONAL PROBLEMS



The space in which a classical or quantum particle moves is, of course, three-dimensional. This is why we wrote the Schrödinger equation (D-1) in chapter I for a wave function $\psi(\mathbf{r})$ which depends on the three components $x$, $y$, $z$ of $\mathbf{r}$. Nevertheless, we have repeatedly used in this chapter a one-dimensional model, in which only the $x$-variable is considered, without justifying this model in a very precise way. Therefore, this complement has two purposes: first (§ 1), to generalize to three dimensions the results given in § C of chapter I; then (§ 2), to show how one can, in certain cases, rigorously justify the one-dimensional model.

### 1. Three-dimensional wave packet

#### a. SIMPLE CASE

Let us begin by considering a very simple case, for which the following two hypotheses are satisfied:

– the wave packet is free $[V(\mathbf{r}) \equiv 0]$ and can therefore be written as in equation (C-6) of chapter I:

$$\psi(\mathbf{r}, t) = \frac{1}{(2\pi)^{3/2}} \int g(\mathbf{k})\, e^{i[\mathbf{k}.\mathbf{r} - \omega(\mathbf{k})t]}\, d^3k \tag{1}$$

– moreover, the function $g(\mathbf{k})$ is of the form:

$$g(\mathbf{k}) = g_1(k_x) \times g_2(k_y) \times g_3(k_z) \tag{2}$$

Recall the expression for $\omega(\mathbf{k})$ in terms of $\mathbf{k}$:

$$\omega(\mathbf{k}) = \frac{\hbar \mathbf{k}^2}{2m} = \frac{\hbar}{2m}(k_x^2 + k_y^2 + k_z^2) \tag{3}$$

Substitute (2) and (3) into (1). It is possible to separate the three integrations with respect to $k_x$, $k_y$ and $k_z$ to obtain:

$$\psi(\mathbf{r}, t) = \psi_1(x, t) \times \psi_2(y, t) \times \psi_3(z, t) \tag{4}$$

with:

$$\begin{cases} \psi_1(x, t) = \dfrac{1}{\sqrt{2\pi}} \displaystyle\int_{-\infty}^{+\infty} g_1(k_x)\, e^{i[k_x . x - \omega(k_x)t]}\, dk_x \\ \omega(k_x) = \dfrac{\hbar k_x^2}{2m} \end{cases} \tag{5}$$

and analogous expressions for $\psi_2(y, t)$ and $\psi_3(z, t)$.

$\psi_1(x, t)$ indeed has the form of a one-dimensional wave packet. In this particular case, $\psi(\mathbf{r}, t)$ is thus obtained simply by taking the product (4) of three one-dimensional wave packets, each of which evolves in a totally independent way.

#### b. GENERAL CASE

In the general case, where the potential $V(\mathbf{r})$ is arbitrary, formula (1) is not valid. It is then useful to introduce the three-dimensional Fourier transform $g(\mathbf{k}, t)$ of the function $\psi(\mathbf{r}, t)$ by writing:

$$\psi(\mathbf{r}, t) = \frac{1}{(2\pi)^{3/2}} \int g(\mathbf{k}, t)\, e^{i\mathbf{k}.\mathbf{r}}\, d^3k \tag{6}$$

A priori, the $t$-dependence of $g(\mathbf{k}, t)$, which brings in $V(\mathbf{r})$, is arbitrary. Moreover, there is no reason in general why we should be able to express $g(\mathbf{k}, t)$ in the form of a product, as in (2). In order to generalize the results of § C-2 of chapter I, we make the following hypothesis about its $\mathbf{k}$-dependence: $|g(\mathbf{k}, t)|$ is (at a given time $t$) a function which has a very pronounced peak for values of $\mathbf{k}$ close to $\mathbf{k}_0$ and takes on a negligible value when the tip of $\mathbf{k}$ leaves a domain $D_k$ centered at $\mathbf{k}_0$ and of dimensions $\Delta k_x$, $\Delta k_y$, $\Delta k_z$. As above, we set:

$$g(\mathbf{k}, t) = |g(\mathbf{k}, t)|\, e^{i\alpha(\mathbf{k}, t)} \tag{7}$$

so that the phase of the wave defined by the vector $\mathbf{k}$ can be written:

$$\xi(\mathbf{k}, \mathbf{r}, t) = \alpha(\mathbf{k}, t) + k_x . x + k_y . y + k_z . z \tag{8}$$

We can set forth an argument similar to that of § C-2 of chapter I. First of all, the wave packet attains a maximum when all the waves, for which the tip of $\mathbf{k}$ is in $D_k$, are practically in phase, that is, when $\xi$ varies very little within $D_k$. In general, $\xi(\mathbf{k}, \mathbf{r}, t)$ can be expanded about $\mathbf{k}_0$. Its variation between $\mathbf{k}_0$ and $\mathbf{k}$ is, to first order in $\delta\mathbf{k} = \mathbf{k} - \mathbf{k}_0$:

$$\delta\xi(\mathbf{k}, \mathbf{r}, t) \simeq \delta k_x \left[\frac{\partial}{\partial k_x} \xi(\mathbf{k}, \mathbf{r}, t)\right]_{\mathbf{k}=\mathbf{k}_0} + \delta k_y \left[\frac{\partial}{\partial k_y} \xi(\mathbf{k}, \mathbf{r}, t)\right]_{\mathbf{k}=\mathbf{k}_0} + \delta k_z \left[\frac{\partial}{\partial k_z} \xi(\mathbf{k}, \mathbf{r}, t)\right]_{\mathbf{k}=\mathbf{k}_0} \tag{9}$$

that is, more concisely*, using (8):

$$\delta\xi(\mathbf{k}, \mathbf{r}, t) \simeq \delta\mathbf{k} \,.\, [\nabla_{\mathbf{k}}\xi(\mathbf{k}, \mathbf{r}, t)]_{\mathbf{k}=\mathbf{k}_0}$$
$$\simeq \delta\mathbf{k} \,.\, [\mathbf{r} + [\nabla_{\mathbf{k}}\alpha(\mathbf{k}, t)]_{\mathbf{k}=\mathbf{k}_0}] \quad (10)$$

We see from (10) that the variation of $\xi(\mathbf{k}, \mathbf{r}, t)$ within the domain $D_k$ will be minimal for:

$$\mathbf{r} = \mathbf{r}_M(t) = -\, [\nabla_{\mathbf{k}}\alpha(\mathbf{k}, t)]_{\mathbf{k}=\mathbf{k}_0} \quad (11)$$

We have seen that, under these conditions, $|\psi(\mathbf{r}, t)|$ is maximum. Relation (11) therefore defines the position $\mathbf{r}_M(t)$ of the center of the wave packet and constitutes the generalization to three dimensions of equation (C-15) of chapter I.

In what domain $D_r$, centered at $\mathbf{r}_M$ and of dimensions $\Delta x$, $\Delta y$, $\Delta z$, does the wave packet (6) take on non-negligible values? $|\psi(\mathbf{r}, t)|$ becomes much smaller than $|\psi(\mathbf{r}_M, t)|$ when the various $\mathbf{k}$ waves destroy each other by interference, that is, when the variation of $\xi(\mathbf{k}, \mathbf{r}, t)$ within the domain $D_k$ is of the order of $2\pi$ (or roughly, of the order of 1 radian). Set $\delta\mathbf{r} = \mathbf{r} - \mathbf{r}_M$; if (11) is taken into account, relation (10) can be written:

$$\delta\xi(\mathbf{k}, \mathbf{r}, t) \simeq \delta\mathbf{k} \,.\, \delta\mathbf{r} \quad (12)$$

The condition $\delta\xi(\mathbf{k}, \mathbf{r}, t) \gtrsim 1$ immediately gives us the relations which exist between the dimensions of $D_r$ and those of $D_k$:

$$\begin{cases} \Delta x \,.\, \Delta k_x \gtrsim 1 \\ \Delta y \,.\, \Delta k_y \gtrsim 1 \\ \Delta z \,.\, \Delta k_z \gtrsim 1 \end{cases} \quad (13)$$

The Heisenberg uncertainty relations then follow directly from the relation $\mathbf{p} = \hbar\,\mathbf{k}$:

$$\begin{aligned} &\Delta x \,.\, \Delta p_x \gtrsim \hbar \\ &\Delta y \,.\, \Delta p_y \gtrsim \hbar \\ &\Delta z \,.\, \Delta p_z \gtrsim \hbar \end{aligned} \quad (14)$$

These inequalities constitute the generalization to three dimensions of (C-23) of chapter I.

Finally, note that the group velocity $\mathbf{V}_G$ of the wave packet can be obtained by differentiating (11) with respect to $t$:

$$\mathbf{V}_G = -\frac{\mathrm{d}}{\mathrm{d}t}[\nabla_{\mathbf{k}}\alpha(\mathbf{k}, t)]_{\mathbf{k}=\mathbf{k}_0} \quad (15)$$

In the special case of a free wave packet which does not, however, necessarily satisfy (2), we have:

$$\alpha(\mathbf{k}, t) = \alpha(\mathbf{k}, 0) - \omega(\mathbf{k})\,t \quad (16)$$

where $\omega(\mathbf{k})$ is given by (3). Formula (15) then gives:

$$\mathbf{V}_G = [\nabla_{\mathbf{k}}\omega(\mathbf{k})]_{\mathbf{k}=\mathbf{k}_0} = \frac{\hbar\mathbf{k}_0}{m} \quad (17)$$

which is the generalization of equation (C-31) of chapter I.

* The symbol $\nabla$ designates a "gradient" : by definition, $\nabla f(x, y, z)$ is the vector whose coordinates are $\partial f/\partial x$, $\partial f/\partial y$, $\partial f/\partial z$. The index $\mathbf{k}$ in $\nabla_{\mathbf{k}}$ means that, as in (9), the differentiations must be performed with respect to the variables $k_x$, $k_y$ and $k_z$.

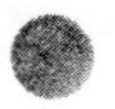



## 2. Justification of one-dimensional models

When the potential is time-independent, we saw in § D-1 of chapter I that it is possible to separate the time and space variables in the Schrödinger equation. This leads to the eigenvalue equation (D-8). We intend to show here how it is possible, in certain cases, to extend this method further and to separate as well the $x$, $y$, $z$ variables in (D-8).

Assume that the potential energy $V(\mathbf{r})$ can be written:

$$V(\mathbf{r}) = V(x, y, z) = V_1(x) + V_2(y) + V_3(z) \tag{18}$$

and let us see if there exist solutions of the eigenvalue equation of the form:

$$\varphi(x, y, z) = \varphi_1(x) \times \varphi_2(y) \times \varphi_3(z) \tag{19}$$

An argument analogous to the one set forth in chapter I (§ D-1-a) shows that this is possible if:

$$\left[ -\frac{\hbar^2}{2m}\frac{\mathrm{d}^2}{\mathrm{d}x^2} + V_1(x) \right] \varphi_1(x) = E_1\, \varphi_1(x) \tag{20}$$

and if we have two other similar equations where $x$ is replaced by $y$ (or $z$), $V_1$ by $V_2$ (or $V_3$), and $E_1$ by $E_2$ (or $E_3$). In addition, it is also necessary that the relation:

$$E = E_1 + E_2 + E_3 \tag{21}$$

be satisfied.

Equation (20) is of the same type as (D-8), but in one dimension. The $x$, $y$ and $z$ variables are separated★.

What happens, for example, if the potential energy $V(\mathbf{r})$ of a particle depends only on $x$? $V(\mathbf{r})$ can then be written in the form (18), where $V_1 = V$ and $V_2 = V_3 = 0$. Equations (20) in $y$ and $z$ correspond to the case already studied, in § C-1 of chapter I, of the free particle in one dimension; their solutions are plane waves $e^{ik_y \cdot y}$ and $e^{ik_z \cdot z}$. All that remains is to solve equation (20), which amounts to considering a problem in only one dimension; nevertheless, the total energy of the particle in three dimensions is now:

$$E = E_1 + \frac{\hbar^2}{2m}[k_y^2 + k_z^2] \tag{22}$$

The one-dimensional models studied in chapter I thus actually correspond to a particle in three dimensions moving in a potential $V(\mathbf{r})$ which depends only on $x$. The solutions $\varphi_2(y)$ and $\varphi_3(z)$ are then very simple and correspond to particles which are "free along $Oy$" or along $Oz$. This is why we have concentrated all our attention on the study of the $x$-equation.

★ It can be shown (*cf.* chap. II, § F-4-a-$\beta$) that, when $V(\mathbf{r})$ has the form (18), all the solutions of the eigenvalue equation (D-8) are linear combinations of those we find here.

## Complement $G_I$

# ONE-DIMENSIONAL GAUSSIAN WAVE PACKET: SPREADING OF THE WAVE PACKET



In this complement, we intend to study a particular (one-dimensional) free wave packet, for which the function $g(k)$ is gaussian. The reason why this example is interesting lies in the fact that the calculations can be carried out exactly and to the very end. Thus, we can first verify, in this special case, the various properties of wave packets which we pointed out in § C of chapter I. We shall then use these properties to study the variation in time of the width of this wave packet and to reveal the phenomenon of spreading over time.

## 1. Definition of a gaussian wave packet

Consider, in a one-dimensional model, a free particle $[V(x) \equiv 0]$ whose wave function at time $t = 0$ is:

$$\psi(x, 0) = \frac{\sqrt{a}}{(2\pi)^{3/4}} \int_{-\infty}^{+\infty} e^{-\frac{a^2}{4}(k-k_0)^2} e^{ikx} dk \tag{1}$$

This wave packet is obtained by superposing plane waves $e^{ikx}$ with the coefficients:

$$\frac{1}{\sqrt{2\pi}} g(k, 0) = \frac{\sqrt{a}}{(2\pi)^{3/4}} e^{-\frac{a^2}{4}(k-k_0)^2} \tag{2}$$

which correspond to a gaussian function centered at $k = k_0$ (and multiplied by a numerical coefficient which normalizes the wave function). This is why the wave packet (1) is called gaussian.

In the calculations which follow, we shall repeatedly come upon integrals of the type:

$$I(\alpha, \beta) = \int_{-\infty}^{+\infty} e^{-\alpha^2(\xi+\beta)^2} d\xi \tag{3}$$

where $\alpha$ and $\beta$ are complex numbers [for the integral (3) to converge, we must have Re $\alpha^2 > 0$]. The method of residues enables us to show that this integral does not depend on $\beta$:

$$I(\alpha, \beta) = I(\alpha, 0) \tag{4}$$

and that, when the condition $-\pi/4 < \text{Arg}\,\alpha < +\pi/4$ is fulfilled (which is always possible if Re $\alpha^2 > 0$), $I(\alpha, 0)$ is given by:

$$I(\alpha, 0) = \frac{1}{\alpha} I(1, 0) \tag{5}$$

Now all that remains is to evaluate $I(1, 0)$, which can be done classically, through a double integration in the $xOy$ plane and a change into polar coordinates:

$$I(1, 0) = \int_{-\infty}^{+\infty} e^{-\xi^2}\, d\xi = \sqrt{\pi} \tag{6}$$

Thus we have:

$$\int_{-\infty}^{+\infty} e^{-\alpha^2(\xi+\beta)^2}\, d\xi = \frac{\sqrt{\pi}}{\alpha} \tag{7}$$

with : $-\pi/4 < \text{Arg}\,\alpha < +\pi/4$.

Let us now calculate $\psi(x, 0)$. To do this, let us group, in the exponents of (1), the $k$-dependent terms into a perfect square, by writing them in the form:

$$-\frac{a^2}{4}(k - k_0)^2 + ikx = -\frac{a^2}{4}\left[k - k_0 - \frac{2ix}{a^2}\right]^2 + ik_0x - \frac{x^2}{a^2} \tag{8}$$

We can then use (7), which yields:

$$\psi(x, 0) = \left(\frac{2}{\pi a^2}\right)^{1/4} e^{ik_0x}\, e^{-x^2/a^2} \tag{9}$$

We find, as could be expected, that the Fourier transform of a gaussian function is also gaussian (*cf.* appendix I).

At time $t = 0$, the probability density of the particle is therefore given by:

$$|\psi(x, 0)|^2 = \sqrt{\frac{2}{\pi a^2}}\, e^{-2x^2/a^2} \tag{10}$$

The curve which represents $|\psi(x, 0)|^2$ is the familiar bell-shaped curve. The center of the wave packet [the maximum of $|\psi(x, 0)|^2$] is situated at the point $x = 0$. This is indeed what we could have found if we had applied the general formula (C-16) of chapter I since, in this particular case, the function $g(k)$ is real.

## 2. Calculation of $\Delta x$ and $\Delta p$; uncertainty relation

It is convenient, when one is studying a gaussian function $f(x) = e^{-x^2/b^2}$, to define its width $\Delta x$ precisely by :

$$\Delta x = \frac{b}{\sqrt{2}} \tag{11}$$

When $x$ varies from 0 to $\pm \Delta x$, $f(x)$ is reduced by a factor of $1/\sqrt{e}$. This definition, which is, of course, arbitrary, has the advantage of coinciding with that of the "root-mean-square deviation" of the $x$ variable (*cf.* chap. III, § C-5).

With this convention, we can calculate the width $\Delta x$ of the wave packet (10), which is equal to:

$$\Delta x = \frac{a}{2} \tag{12}$$

We can proceed in the same way to calculate the width $\Delta k$, since $|g(k, 0)|^2$ is also a gaussian function. This gives:

$$\Delta k = \frac{1}{a} \tag{13-a}$$

or:

$$\Delta p = \frac{\hbar}{a} \tag{13-b}$$

Thus we obtain:

$$\Delta x \,.\, \Delta p = \frac{\hbar}{2} \tag{14}$$

a result which is entirely compatible with Heisenberg's uncertainty relation.

## 3. Evolution of the wave packet

### a. CALCULATION OF $\psi(x, t)$

In order to calculate the wave function $\psi(x, t)$ at time $t$, all we need to do is use the general formula (C-6) of chapter I, which gives the wave function of a free particle; we obtain:

$$\psi(x, t) = \frac{\sqrt{a}}{(2\pi)^{3/4}} \int_{-\infty}^{+\infty} e^{-\frac{a^2}{4}(k-k_0)^2} \, e^{i[kx - \omega(k)t]} \, dk \tag{15}$$

with $\omega(k) = \dfrac{\hbar k^2}{2m}$ (dispersion relation for a free particle). We shall see that at time $t$, the wave packet still remains gaussian. Expression (15) can be transformed by grouping, as above, all the $k$-dependent terms in the exponents into a perfect square. We can then use (7), and we find:

$$\psi(x, t) = \left(\frac{2a^2}{\pi}\right)^{1/4} \frac{e^{i\varphi}}{\left(a^4 + \frac{4\hbar^2 t^2}{m^2}\right)^{1/4}} e^{ik_0 x} \exp\left\{-\frac{\left[x - \frac{\hbar k_0}{m}t\right]^2}{a^2 + \frac{2i\hbar t}{m}}\right\} \tag{16-a}$$

where $\varphi$ is real and independent of $x$:

$$\varphi = -\theta - \frac{\hbar k_0^2}{2m}t \quad \text{with} \quad \tan 2\theta = \frac{2\hbar t}{ma^2} \tag{16-b}$$

Let us calculate the probability density $|\psi(x, t)|^2$ of the particle at time $t$. We obtain:

$$|\psi(x, t)|^2 = \sqrt{\frac{2}{\pi a^2}} \frac{1}{\sqrt{1 + \frac{4\hbar^2 t^2}{m^2 a^4}}} \exp\left\{-\frac{2a^2\left(x - \frac{\hbar k_0}{m}t\right)^2}{a^4 + \frac{4\hbar^2 t^2}{m^2}}\right\} \tag{17}$$

Let us show that the norm of the wave packet, $\int_{-\infty}^{+\infty} |\psi(x, t)|^2 \, dx$, is not time-dependent (we shall see in chapter III that this property results from the fact that the Hamiltonian $H$ of the particle is Hermitian). We could, to this end, use (7) again in order to integrate expression (17) from $-\infty$ to $+\infty$. It is quicker to observe from expression (15) that the Fourier transform of $\psi(x, t)$ is given by:

$$g(k, t) = e^{-i\omega(k)t} g(k, 0) \tag{18}$$

$g(k, t)$ therefore obviously has the same norm as $g(k, 0)$. Now the Parseval-Plancherel equation tells us that $\psi(x, t)$ and $g(k, t)$ have the same norm, as do $\psi(x, 0)$ and $g(k, 0)$. From this we deduce that $\psi(x, t)$ has the same norm as $\psi(x, 0)$.

#### b. VELOCITY OF THE WAVE PACKET

We see in (17) that the probability density $|\psi(x, t)|^2$ is a gaussian function, centered at $x = V_0 t$, where the velocity $V_0$ is defined by:

$$V_0 = \frac{\hbar k_0}{m} \tag{19}$$

We could have expected this result, in view of the general expression (C-32) of chapter I, which gives the group velocity $V_G$.

#### c. SPREADING OF THE WAVE PACKET

Let us take up formula (17) again. The width $\Delta x(t)$ of the wave packet at time $t$, from definition (11), is equal to:

$$\Delta x(t) = \frac{a}{2}\sqrt{1 + \frac{4\hbar^2 t^2}{m^2 a^4}} \tag{20}$$

We see (*cf.* fig. 1) that the evolution of the wave packet is not confined to a simple displacement at a velocity $V_0$. The wave packet also undergoes a deformation. When $t$ increases from $-\infty$ to 0, the width of the wave packet decreases, reaching a minimum at $t = 0$. Then, as $t$ continues to increase, $\Delta x(t)$ grows without bound (spreading of the wave packet).

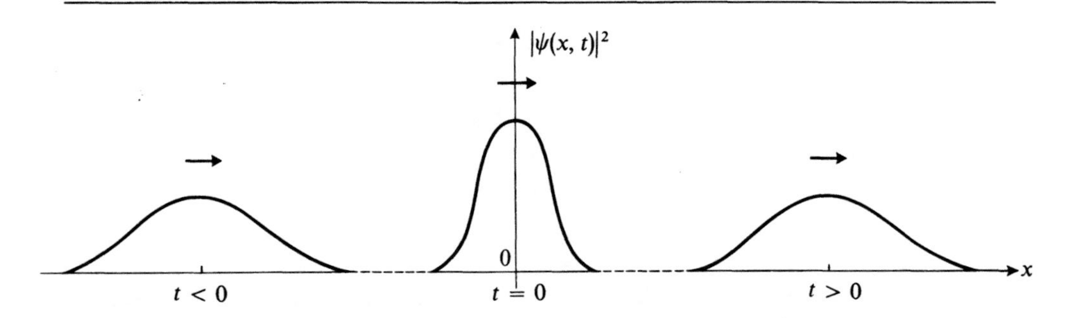


FIGURE 1

**For negative $t$, the gaussian wave packet decreases in width as it propagates. At time $t = 0$, it is a "minimum" wave packet : the product $\Delta x \,.\, \Delta p$ is equal to $\hbar/2$. Then, for $t > 0$, the wave packet spreads again as it propagates.**

It can be seen in (17) that the height of the wave packet also varies, but in opposition to the width, so the norm of $\psi(x, t)$ remains constant.

The properties of the function $g(k, t)$ are completely different. In fact [*cf.* formula (18)]:

$$|g(k, t)| = |g(k, 0)| \tag{21}$$

Therefore, the average momentum of the wave packet ($\hbar k_0$) and its momentum dispersion ($\hbar \Delta k$) do not vary in time. We shall see later (*cf.* chap. III) that this arises from the fact that the momentum is a constant of the motion for a free particle. Physically, it is clear that since the free particle encounters no obstacle, the momentum distribution cannot change.

The existence of a momentum dispersion $\Delta p = \hbar \Delta k = \hbar / a$ means that the velocity of the particle is only known to within $\Delta v = \dfrac{\Delta p}{m} = \dfrac{\hbar}{ma}$. Imagine a group of classical particles starting at time $t = 0$ from the point $x = 0$, with a velocity dispersion equal to $\Delta v$. At time $t$, the dispersion of their positions will be $\delta x_{cl} = \Delta v \, |t| = \dfrac{\hbar \, |t|}{ma}$; this dispersion increases linearly with $t$, as shown in figure 2. Let us draw on the same graph the curve which gives the evolution in time of $\Delta x(t)$; when $t$ becomes infinite, $\Delta x(t)$ practically coincides with $\delta x_{cl}$ [the branch of the hyperbola which represents $\Delta x(t)$ has for its asymptotes the straight lines which correspond to $\delta x_{cl}$]. Thus, we can say that, when $t$ is very large, there exists a quasi-classical interpretation of the width $\Delta x$. On the other hand, when $t$ approaches 0, $\Delta x(t)$ takes on values which differ more and more from $\delta x_{cl}$. The quantum particle must indeed constantly satisfy Heisenberg's uncertainty relation $\Delta x \,.\, \Delta p \geqslant \hbar/2$ which, since $\Delta p$ is fixed, imposes a lower limit on $\Delta x$. This corresponds to what can be seen in figure 2.

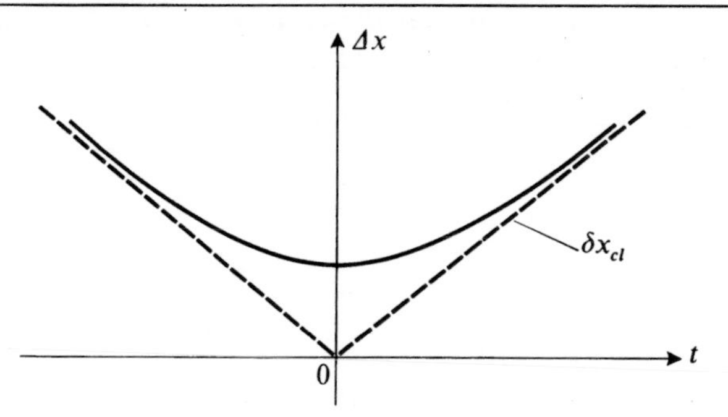


FIGURE 2

**Variation in time of the width $\Delta x$ of the wave packet of figure 1. For large $t$, $\Delta x$ approaches the dispersion $\delta x_{cl}$ of the positions of a group of classical particles which left $x = 0$ at time $t = 0$ with a velocity dispersion $\Delta p/m$.**

COMMENTS:

(*i*) The spreading of the packet of free waves is a general phenomenon which is not limited to the special case studied here. It can be shown that, for an arbitrary free wave packet, the variation in time of its width has the shape shown in figure 2 (*cf.* exercise 4 of complement $L_{III}$).

(*ii*) In chapter I, a simple argument led us in (C-17) to $\Delta x \,.\, \Delta k \simeq 1$, without making any particular hypothesis about $g(k)$, except for saying that $g(k)$ has a peak of width $\Delta k$ whose shape is that of figure 3 of chapter I (which is indeed the case in this complement). Then how did we obtain $\Delta x \,.\, \Delta k \gg 1$ (for example, for a gaussian wave packet when $t$ is large)?

Of course, this is only an apparent contradiction. In chapter I, in order to find $\Delta x \,.\, \Delta k \simeq 1$, we assumed in (C-13) that the argument $\alpha(k)$ of $g(k)$ could be approximated by a linear function in the domain $\Delta k$. Thus we implicitly assumed a supplementary hypothesis: that the nonlinear terms make a negligible contribution to the phase of $g(k)$ in the domain $\Delta k$. For example, for the terms which are of second order in $(k - k_0)$, it is necessary that:

$$\Delta k^2 \left[\frac{\mathrm{d}^2\alpha}{\mathrm{d}k^2}\right]_{k=k_0} \ll 2\pi \tag{22}$$

If, on the contrary, the phase $\alpha(k)$ cannot be approximated in the domain $\Delta k$ by a linear function with an error much smaller than $2\pi$, we find when we return to the argument of chapter I that the wave packet is larger than was predicted by (C-17).

In the case of the gaussian wave packet studied in this complement, we have $\Delta k \simeq \dfrac{1}{a}$ and $\alpha(k) = -\dfrac{\hbar k^2}{2m}t$. Consequently, condition (22) can be written $\left(\dfrac{1}{a}\right)^2 \dfrac{\hbar t}{m} \ll 2\pi$. Indeed, we can verify from (20) that, as long as this condition is fulfilled, the product $\Delta x \,.\, \Delta k$ is approximately equal to 1.

## Complement $H_I$

# STATIONARY STATES OF A PARTICLE IN ONE-DIMENSIONAL SQUARE POTENTIALS



We saw in chapter I (*cf.* § D-2) the interest in studying the motion of a particle in a "square potential" whose rapid spatial variations for certain values of $x$ introduce purely quantum effects. The shape of the wave functions associated with the stationary states of the particle was predicted by considering an optical analogy which enabled us to understand very simply how these new physical effects appear.

In this complement, we outline the quantitative calculation of the stationary states of the particle. We shall give the results of this calculation for a certain number of simple cases, and discuss their physical implications. We limit ourselves to one-dimensional models (*cf.* complement $F_I$).

## 1. Behavior of a stationary wave function $\varphi(x)$

### a. REGIONS OF CONSTANT POTENTIAL ENERGY

In the case of a square potential, $V(x)$ is a constant function $V(x) = V$ in certain regions of space. In such a region, equation (D-8) of chapter I can be written:

$$\frac{\mathrm{d}^2}{\mathrm{d}x^2}\varphi(x) + \frac{2m}{\hbar^2}(E - V)\,\varphi(x) = 0 \tag{1}$$

We shall distinguish between several cases:

(*i*) $E > V$

Let us introduce the positive constant $k$, defined by

$$E - V = \frac{\hbar^2 k^2}{2m} \tag{2}$$

The solution of equation (1) can then be written:

$$\varphi(x) = A\,\mathrm{e}^{ikx} + A'\,\mathrm{e}^{-ikx} \tag{3}$$

where $A$ and $A'$ are complex constants.

(*ii*) $E < V$

This condition corresponds to regions of space which would be forbidden to the particle by the laws of classical mechanics. In this case, we introduce the positive constant $\rho$ defined by:

$$V - E = \frac{\hbar^2 \rho^2}{2m} \tag{4}$$

and the solution of (1) can be written:

$$\varphi(x) = B\,\mathrm{e}^{\rho x} + B'\,\mathrm{e}^{-\rho x} \tag{5}$$

where $B$ and $B'$ are complex constants.

(*iii*) $E = V$. In this special case, $\varphi(x)$ is a linear function of $x$.

### b. BEHAVIOR OF $\varphi(x)$ AT A POTENTIAL ENERGY DISCONTINUITY

How does the wave function behave at a point $x = x_1$ where the potential $V(x)$ is discontinuous? It might be thought that at this point the wave function $\varphi(x)$ would behave strangely, becoming itself discontinuous, for example. The aim of this section is to show that this is not the case : $\varphi(x)$ and $\mathrm{d}\varphi/\mathrm{d}x$ are continuous, and it is only the second derivative $\mathrm{d}^2\varphi/\mathrm{d}x^2$ that is discontinuous at $x = x_1$.

Without giving a rigorous proof, let us try to understand this property. To do this, recall that a square potential must be considered (*cf.* chap. I, § D-2-a) as the limit, when $\varepsilon \longrightarrow 0$, of a potential $V_\varepsilon(x)$ equal to $V(x)$ outside the interval $[x_1 - \varepsilon, x_1 + \varepsilon]$, and varying continuously within this interval. Then consider the equation:

$$\frac{\mathrm{d}^2}{\mathrm{d}x^2}\varphi_\varepsilon(x) + \frac{2m}{\hbar^2}\left[E - V_\varepsilon(x)\right]\varphi_\varepsilon(x) = 0 \tag{6}$$

where $V_\varepsilon(x)$ is assumed to be bounded, independently of $\varepsilon$, within the interval $[x_1 - \varepsilon, x_1 + \varepsilon]$. Choose a solution $\varphi_\varepsilon(x)$ which, for $x < x_1 - \varepsilon$, coincides with a given solution of (1). The problem is to show that, when $\varepsilon \longrightarrow 0$, $\varphi_\varepsilon(x)$ tends towards a function $\varphi(x)$ which is continuous and differentiable at $x = x_1$. Let us grant that $\varphi_\varepsilon(x)$ remains bounded, whatever the value of $\varepsilon$, in the neighborhood of $x = x_1$*. Physically, this means that the probability density remains finite. Then, integrating (6) between $x_1 - \eta$ and $x_1 + \eta$, we obtain:

$$\frac{\mathrm{d}\varphi_\varepsilon}{\mathrm{d}x}(x_1 + \eta) - \frac{\mathrm{d}\varphi_\varepsilon}{\mathrm{d}x}(x_1 - \eta) = \frac{2m}{\hbar^2}\int_{x_1 - \eta}^{x_1 + \eta}\left[V_\varepsilon(x) - E\right]\varphi_\varepsilon(x)\,\mathrm{d}x \tag{7}$$

At the limit where $\varepsilon \longrightarrow 0$, the function to be integrated on the right-hand side of this expression remains bounded, owing to our previous assumption. Consequently, if $\eta$ tends towards zero, the integral also tends towards zero, and:

$$\frac{\mathrm{d}\varphi}{\mathrm{d}x}(x_1 + \eta) - \frac{\mathrm{d}\varphi}{\mathrm{d}x}(x_1 - \eta) \underset{\eta \to 0}{\longrightarrow} 0 \tag{8}$$

* This point could be proved mathematically from the properties of the differential equation (1).

Thus, at this limit, $d\varphi/dx$ is continuous at $x = x_1$, and so is $\varphi(x)$ (since it is the integral of a continuous function). On the other hand, $d^2\varphi/dx^2$ is discontinuous, and, as can be seen directly from (1), makes a jump at $x = x_1$ which is equal to $\frac{2m}{\hbar^2}\varphi(x_1)\sigma_V$ [where $\sigma_V$ represents the change in $V(x)$ at $x = x_1$].

COMMENT :

It is essential, in the preceding argument, that $V_\varepsilon(x)$ remain bounded. In certain exercises of complement $K_I$, for example, the case is considered for which $V(x) = \alpha\delta(x)$, an unbounded function whose integral remains finite. In this case, $\varphi(x)$ remains continuous, but $d\varphi/dx$ does not.

#### c. OUTLINE OF THE CALCULATION

The procedure for determining the stationary states in a "square potential" is therefore the following : in all regions where $V(x)$ is constant, write $\varphi(x)$ in whichever of the two forms (3) or (5) is applicable; then "match" these functions by requiring the continuity of $\varphi(x)$ and of $d\varphi/dx$ at the points where $V(x)$ is discontinuous.

## 2. Examination of certain simple cases

Let us now carry out the quantitative calculation of the stationary states, performed according to the method described above, for all the forms of $V(x)$ considered in § D-2-c of chapter I. Thus we shall verify that the form of the solutions is indeed the one predicted by the optical analogy.

#### a. POTENTIAL STEPS

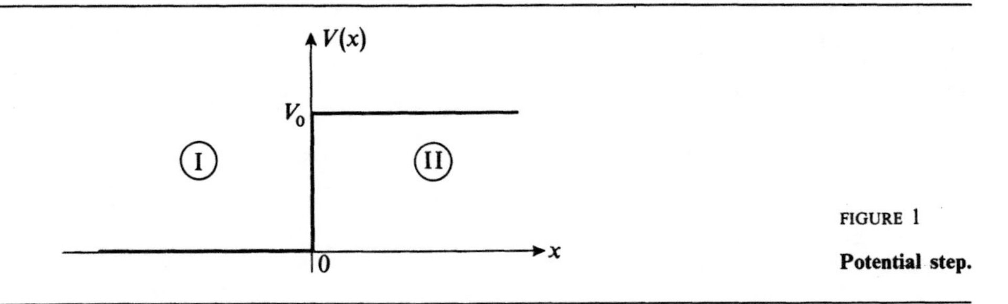


FIGURE 1

**Potential step.**

α. *Case where $E > V_0$; partial reflection*

Set:

$$\sqrt{\frac{2mE}{\hbar^2}} = k_1 \tag{9}$$

$$\sqrt{\frac{2m(E - V_0)}{\hbar^2}} = k_2 \tag{10}$$

The solution of (1) has the form (3) in the two regions I ($x < 0$) and II ($x > 0$):

$$\varphi_I(x) = A_1 \, e^{ik_1x} + A_1' \, e^{-ik_1x} \tag{11}$$

$$\varphi_{II}(x) = A_2 \, e^{ik_2x} + A_2' \, e^{-ik_2x} \tag{12}$$

Since equation (1) is homogeneous, the calculation method of § 1-c can only enable us to determine the ratios $A_1'/A_1$, $A_2/A_1$ and $A_2'/A_1$. In fact, the two matching conditions at $x = 0$ do not suffice for the determination of these three ratios. This is why we shall choose $A_2' = 0$, which amounts to limiting ourselves to the case of an incident particle coming from $x = -\infty$. The matching conditions then give:

$$\frac{A_1'}{A_1} = \frac{k_1 - k_2}{k_1 + k_2} \tag{13}$$

$$\frac{A_2}{A_1} = \frac{2k_1}{k_1 + k_2} \tag{14}$$

$\varphi_I(x)$ is the superposition of two waves. The first one (the term in $A_1$) corresponds to an incident particle, with momentum $p = \hbar k_1$, propagating from left to right. The second one (the term in $A_1'$) corresponds to a reflected particle, with momentum $-\hbar k_1$, propagating in the opposite direction. Since we have chosen $A_2' = 0$, $\varphi_{II}(x)$ consists of only one wave, which is associated with a transmitted particle. We shall see in chapter III (*cf.* § D-1-c-$\beta$) how it is possible, using the concept of a probability current, to define the transmission coefficient $T$ and the reflection coefficient $R$ of the potential step (see also § 2 of complement $B_{III}$). These coefficients give the probability for the particle, arriving from $x = -\infty$, to pass the potential step at $x = 0$ or to turn back. Thus we find:

$$R = \left| \frac{A_1'}{A_1} \right|^2 \tag{15}$$

and, for $T$*:

$$T = \frac{k_2}{k_1} \left| \frac{A_2}{A_1} \right|^2 \tag{16}$$

Taking (13) and (14) into account, we then have:

$$R = 1 - \frac{4k_1k_2}{(k_1 + k_2)^2} \tag{17}$$

$$T = \frac{4k_1k_2}{(k_1 + k_2)^2} \tag{18}$$

It is easy to verify that $R + T = 1$: it is certain that the particle will be either transmitted or reflected. Contrary to the predictions of classical mechanics, the incident particle has a non-zero probability of turning back. This point was explained in chapter I, using the optical analogy and considering the reflection

* The physical origin of the factor $k_2/k_1$ which appears in $T$ is discussed in § 2 of complement $J_I$.

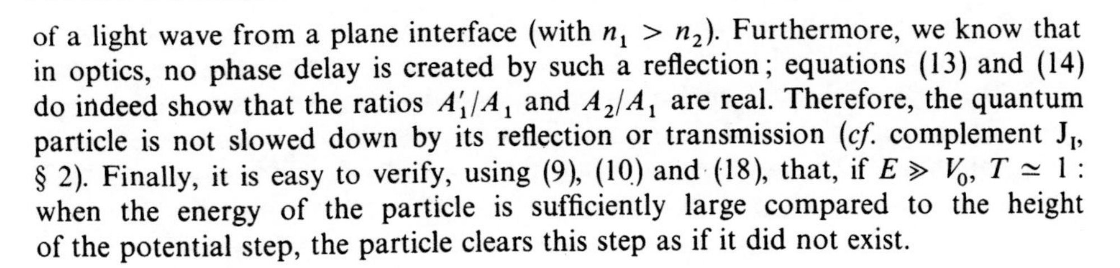

of a light wave from a plane interface (with $n_1 > n_2$). Furthermore, we know that in optics, no phase delay is created by such a reflection; equations (13) and (14) do indeed show that the ratios $A'_1/A_1$ and $A_2/A_1$ are real. Therefore, the quantum particle is not slowed down by its reflection or transmission (*cf.* complement $J_I$, § 2). Finally, it is easy to verify, using (9), (10) and (18), that, if $E \gg V_0$, $T \simeq 1$: when the energy of the particle is sufficiently large compared to the height of the potential step, the particle clears this step as if it did not exist.

β. *Case where $E < V_0$; total reflection*

We then replace (10) and (12) by:

$$\sqrt{\frac{2m(V_0 - E)}{\hbar^2}} = \rho_2 \tag{19}$$

$$\varphi_{II}(x) = B_2\, e^{\rho_2 x} + B'_2\, e^{-\rho_2 x} \tag{20}$$

For the solution to remain bounded when $x \longrightarrow +\infty$, it is necessary that:

$$B_2 = 0 \tag{21}$$

The matching conditions at $x = 0$ give in this case :

$$\frac{A'_1}{A_1} = \frac{k_1 - i\rho_2}{k_1 + i\rho_2} \tag{22}$$

$$\frac{B'_2}{A_1} = \frac{2k_1}{k_1 + i\rho_2} \tag{23}$$

The reflection coefficient $R$ is then equal to:

$$R = \left|\frac{A'_1}{A_1}\right|^2 = \left|\frac{k_1 - i\rho_2}{k_1 + i\rho_2}\right|^2 = 1 \tag{24}$$

As in classical mechanics, the particle is always reflected (total reflection). Nevertheless, there is an important difference, which has already been pointed out in chapter I: because of the existence of the evanescent wave $e^{-\rho_2 x}$, the particle has a non-zero probability of presence in the region of space which, classically, would be forbidden to it. This probability decreases exponentially with $x$ and becomes negligible when $x$ is greater than the "range" $1/\rho_2$ of the evanescent wave. Note also that the coefficient $A'_1/A_1$ is complex. A certain phase shift appears upon reflection, which, physically, is due to the fact that the particle is delayed when it penetrates the $x > 0$ region (*cf.* complement $J_I$, § 1 and also $B_{III}$, § 3). This phase shift is analogous to the one which appears when light is reflected from a metallic type of substance; however, there is no analogue in classical mechanics.

COMMENT:

When $V_0 \longrightarrow +\infty$, $\rho_2 \longrightarrow +\infty$, so that (22) and (23) yield:

$$\begin{cases} A'_1 \longrightarrow -A_1 \\ B'_2 \longrightarrow 0 \end{cases} \tag{25}$$

In the $x > 0$ region, the wave, whose range decreases without bound, tends towards zero. Since $(A_1 + A'_1) \longrightarrow 0$, the wave function $\varphi(x)$ goes to zero at $x = 0$, so that it remains continuous at this point. On the other hand, its derivative, which changes abruptly from the value $2ikA_1$ to zero, is no longer continuous. This is due to the fact that since the potential jump is infinite at $x = 0$, the integral of (7) no longer tends towards zero when $\eta$ tends towards 0.

### b. POTENTIAL BARRIERS

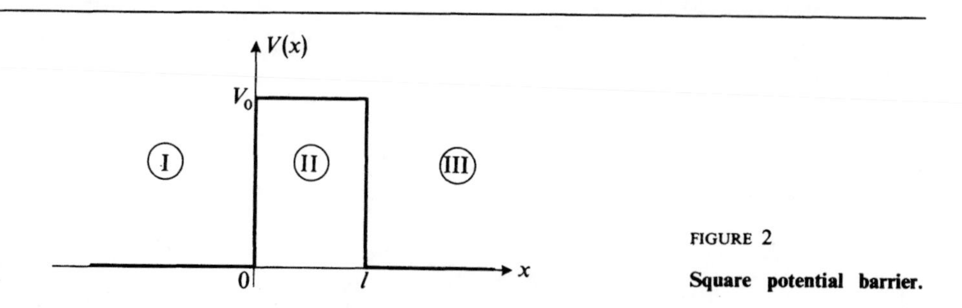


FIGURE 2

**Square potential barrier.**

#### α. *Case where $E > V_0$★; resonances*

Using notations (9) and (10), we find, in the three regions I ($x < 0$), II ($0 < x < l$) and III ($x > l$):

$$\varphi_{\text{I}}(x) = A_1\,e^{ik_1x} + A'_1\,e^{-ik_1x} \tag{26-a}$$

$$\varphi_{\text{II}}(x) = A_2\,e^{ik_2x} + A'_2\,e^{-ik_2x} \tag{26-b}$$

$$\varphi_{\text{III}}(x) = A_3\,e^{ik_1x} + A'_3\,e^{-ik_1x} \tag{26-c}$$

Let us choose, as above, $A'_3 = 0$ (incident particle coming from $x = -\infty$). The matching conditions at $x = l$ then give $A_2$ and $A'_2$ in terms of $A_3$, and those at $x = 0$ give $A_1$ and $A'_1$ in terms of $A_2$ and $A'_2$ (and, consequently, in terms of $A_3$). Thus we find:

$$\begin{aligned} A_1 &= \left[\cos k_2 l - i\frac{k_1^2 + k_2^2}{2k_1k_2}\sin k_2 l\right] e^{ik_1 l} A_3 \\ A'_1 &= i\frac{k_2^2 - k_1^2}{2k_1k_2}\sin k_2 l\, e^{ik_1 l} A_3 \end{aligned} \tag{27}$$

$A'_1/A_1$ and $A_3/A_1$ enable us to calculate the reflection coefficient $R$ and the transmission coefficient $T$ of the barrier, which here are equal to:

$$R = \left|\frac{A'_1}{A_1}\right|^2 = \frac{(k_1^2 - k_2^2)^2 \sin^2 k_2 l}{4k_1^2k_2^2 + (k_1^2 - k_2^2)^2 \sin^2 k_2 l} \tag{28-a}$$

★ $V_0$ can either be positive (the case of a potential barrier like the one shown in figure 2) or negative (a potential well).

$$T = \left| \frac{A_3}{A_1} \right|^2 = \frac{4k_1^2k_2^2}{4k_1^2k_2^2 + (k_1^2 - k_2^2)^2 \sin^2 k_2 l} \tag{28-b}$$

It is then easy to verify that $R + T = 1$. Taking (9) and (10) into account, we have :

$$T = \frac{4E(E - V_0)}{4E(E - V_0) + V_0^2 \sin^2 \left[\sqrt{2m(E - V_0)}\, l/\hbar\right]} \tag{29}$$

The variations with respect to $l$ of the transmission coefficient $T$ are shown in figure 3 (with $E$ and $V_0$ fixed): $T$ oscillates periodically between its minimum value, $\left[1 + \frac{V_0^2}{4E(E - V_0)}\right]^{-1}$, and its maximum value, which is 1. This function

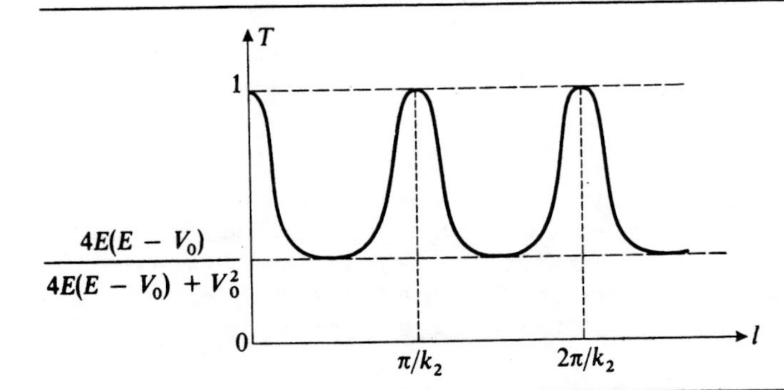


FIGURE 3

**Variations of the transmission coefficient $T$ of the barrier as a function of its width (the height $V_0$ of the barrier and the energy $E$ of the particle are fixed). Resonances appear each time that $l$ is an integral multiple of the half-wavelength $\pi/k_2$ in region II.**

is the analogue of the one which describes the transmission of a Fabry-Perot interferometer. As in optics, the resonances (obtained when $T = 1$, that is, when $k_2 l = n\pi$) correspond to the values of $l$ which are integral multiples of the half-wavelength of the particle in region II. When $E > V_0$, the reflection of the particle at each of the potential discontinuities occurs without a phase shift of the wave function (*cf.* § 2-a-α). This is why the resonance condition $k_2 l = n\pi$ corresponds to the values of $l$ for which a system of standing waves can exist in region II. On the other hand, far from the resonances, the various waves which are reflected at $x = 0$ and $x = l$ destroy each other by interference, so that the values of the wave function are small. A study of the propagation of a wave packet (analogous to the one in complement $J_I$) would show that, if the resonance condition is satisfied, the wave packet spends a relatively long time in region II. In quantum mechanics this phenomenon is called *resonance scattering*.

### β. *Case where $E < V_0$; tunnel effect*

We must now replace (26-b) by (20), $\rho_2$ still being given by (19). The matching conditions at $x = 0$ and $x = l$ enable us to calculate the transmission coefficient of the barrier. In fact, it is unnecessary to perform the calculations again: all we must do is replace, in the equations obtained in § α, $k_2$ by $-i\rho_2$. We then have :

$$T = \left| \frac{A_3}{A_1} \right|^2 = \frac{4E(V_0 - E)}{4E(V_0 - E) + V_0^2 \sinh^2 \left[\sqrt{2m(V_0 - E)}\, l/\hbar\right]} \tag{30}$$

with, of course, $R = 1 - T$. When $\rho_2 l \gg 1$, we have :

$$T \simeq \frac{16E(V_0 - E)}{V_0^2} e^{-2\rho_2 l} \tag{31}$$

We have already seen, in chapter I, why, contrary to the classical predictions, the particle has a non-zero probability of crossing the potential barrier. The wave function in region II is not zero, but has the behavior of an "evanescent wave" of range $1/\rho_2$. When $l \lesssim 1/\rho_2$, the particle has a considerable probability of crossing the barrier by the "tunnel effect". This effect has numerous physical applications : the inversion of the ammonia molecule (*cf.* complement $G_{IV}$), the tunnel diode, the Josephson effect, the $\alpha$-decay of certain nuclei, etc...

For an electron, the range of the evanescent wave is:

$$\left(\frac{1}{\rho_2}\right)_{el} \simeq \frac{1.96}{\sqrt{V_0 - E}} \text{ Å} \tag{32}$$

where $E$ and $V_0$ are expressed in electron-volts (this formula can be obtained immediately by replacing, in formula (8) of complement $A_I$, $\lambda = 2\pi/k$ by $2\pi/\rho_2$). Now consider an electron of energy 1 eV which encounters a barrier for which $V_0 = 2$ eV and $l = 1$ Å. The range of the evanescent wave is then 1.96 Å, that is, of the order of $l$ : the electron must then have a considerable probability of crossing the barrier. Indeed, formula (30) gives in this case:

$$T \simeq 0.78 \tag{33}$$

The quantum result is radically different from the classical result : the electron has approximately 8 chances out of 10 of crossing the barrier.

Let us now assume that the incident particle is a proton (whose mass is about 1 840 times that of the electron). The range $1/\rho_2$ then becomes:

$$\left(\frac{1}{\rho_2}\right)_{pr} \simeq \frac{1.96}{\sqrt{1\,840(V_0 - E)}} \text{ Å} \simeq \frac{4.6}{\sqrt{V_0 - E}} 10^{-2} \text{ Å} \tag{34}$$

If we retain the same values : $E = 1$ eV, $V_0 = 2$ eV, $l = 1$ Å, we find a range $1/\rho_2$ much smaller than $l$. Formula (31) then gives:

$$T \simeq 4 \times 10^{-19} \tag{35}$$

Under these conditions, the probability of the proton's crossing the potential barrier is negligible. This is all the more true if we apply (31) to macroscopic objects, for which we find such small probabilities that they cannot possibly play any role in physical phenomena.

### c. BOUND STATES ; SQUARE WELL POTENTIAL

#### α. *Well of finite depth*

We shall limit ourselves here to studying the case $-V_0 < E < 0$ (the case $E > 0$ was included in the calculations of the preceding section $b$-$\alpha$).

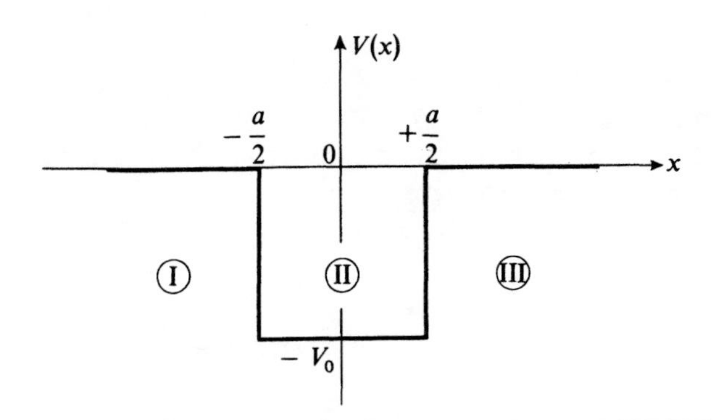


FIGURE 4

**Square well potential.**

In regions I $\left(x < -\frac{a}{2}\right)$, II $\left(-\frac{a}{2} \leqslant x \leqslant \frac{a}{2}\right)$, and III $\left(x > \frac{a}{2}\right)$, we have, respectively:

$$\varphi_{\rm I}(x) = B_1 \, e^{\rho x} + B'_1 \, e^{-\rho x} \tag{36-a}$$

$$\varphi_{\rm II}(x) = A_2 \, e^{ikx} + A'_2 \, e^{-ikx} \tag{36-b}$$

$$\varphi_{\rm III}(x) = B_3 \, e^{\rho x} + B'_3 \, e^{-\rho x} \tag{36-c}$$

with:

$$\rho = \sqrt{-\frac{2mE}{\hbar^2}} \tag{37}$$

$$k = \sqrt{\frac{2m(E + V_0)}{\hbar^2}} \tag{38}$$

Since $\varphi(x)$ must be bounded in region I, we must have:

$$B'_1 = 0 \tag{39}$$

The matching conditions at $x = -\frac{a}{2}$ then give:

$$A_2 = e^{(-\rho + ik)a/2} \, \frac{\rho + ik}{2ik} B_1$$

$$A'_2 = - \, e^{-(\rho + ik)a/2} \, \frac{\rho - ik}{2ik} B_1 \tag{40}$$

and those at $x = a/2$:

$$\frac{B_3}{B_1} = \frac{e^{-\rho a}}{4ik\rho} \left[(\rho + ik)^2 \, e^{ika} - (\rho - ik)^2 \, e^{-ika}\right]$$

$$\frac{B'_3}{B_1} = \frac{\rho^2 + k^2}{2k\rho} \sin ka \tag{41}$$

But $\varphi(x)$ must also be bounded in region III. Therefore, it is necessary that $B_3 = 0$, that is:

$$\left(\frac{\rho - ik}{\rho + ik}\right)^2 = e^{2ika} \tag{42}$$

Since $\rho$ and $k$ depend on $E$, equation (42) can only be satisfied for certain values of $E$. Imposing a bound on $\varphi(x)$ in all regions of space thus entails the quantization of energy. More precisely, two cases are possible:

($i$) if:

$$\frac{\rho - ik}{\rho + ik} = - e^{ika} \tag{43}$$

we have:

$$\frac{\rho}{k} = \tan\left(\frac{ka}{2}\right) \tag{44}$$

Set:

$$k_0 = \sqrt{\frac{2mV_0}{\hbar^2}} = \sqrt{k^2 + \rho^2} \tag{45}$$

We then obtain:

$$\frac{1}{\cos^2\left(\frac{ka}{2}\right)} = 1 + \tan^2\frac{ka}{2} = \frac{k^2 + \rho^2}{k^2} = \left(\frac{k_0}{k}\right)^2 \tag{46}$$

Equation (43) is thus equivalent to the system of equations:

$$\left\{\begin{array}{ll} \left|\cos\left(\frac{ka}{2}\right)\right| = \frac{k}{k_0} & \text{(47-a)} \\ \tan\left(\frac{ka}{2}\right) > 0 & \text{(47-b)} \end{array}\right.$$

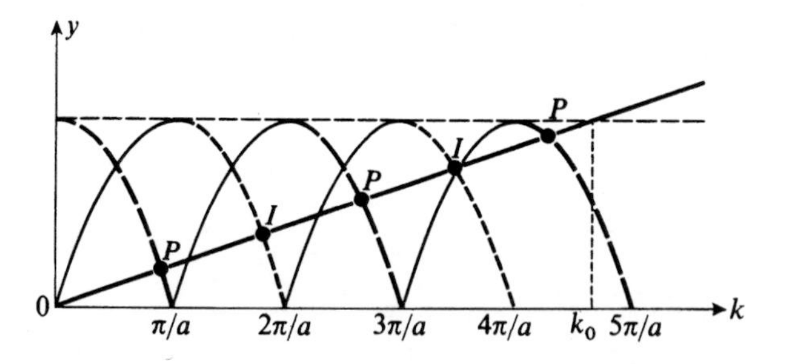


FIGURE 5

**Graphic solution of equation (42), giving the energies of the bound states of a particle in a square well potential. In the case shown in the figure, there exist five bound states, three even (associated with the points $P$ of the figure), and two odd (points $I$).**

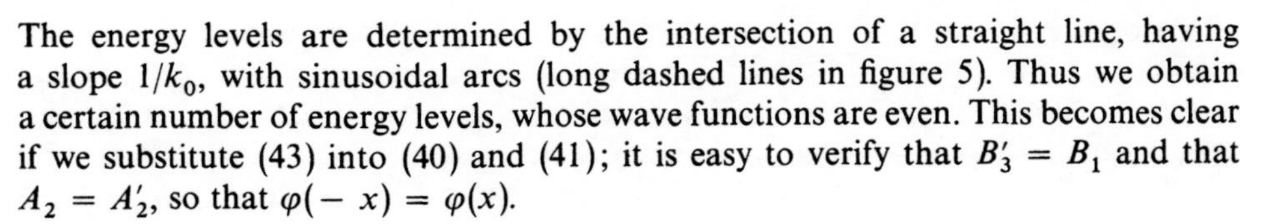

The energy levels are determined by the intersection of a straight line, having a slope $1/k_0$, with sinusoidal arcs (long dashed lines in figure 5). Thus we obtain a certain number of energy levels, whose wave functions are even. This becomes clear if we substitute (43) into (40) and (41); it is easy to verify that $B_3' = B_1$ and that $A_2 = A_2'$, so that $\varphi(-x) = \varphi(x)$.

(*ii*) if:

$$\frac{\rho - ik}{\rho + ik} = e^{ika} \tag{48}$$

a calculation of the same type leads to:

$$\begin{cases} \left|\sin\left(\dfrac{ka}{2}\right)\right| = \dfrac{k}{k_0} & \text{(49-a)} \\ \tan\left(\dfrac{ka}{2}\right) < 0 & \text{(49-b)} \end{cases}$$

The energy levels are then determined by the intersection of the same straight line as before with other sinusoidal arcs (*cf.* short dashed lines in figure 5). The levels thus obtained fall between those found in (*i*). It can easily be shown that the corresponding wave functions are odd.

COMMENT:

If $k_0 \leqslant \dfrac{\pi}{a}$, that is, if:

$$V_0 \leqslant V_1 = \frac{\pi^2\hbar^2}{2ma^2} \tag{50}$$

figure 5 shows that there exists only one bound state of the particle, and this state has an even wave function. Then, if $V_1 \leqslant V_0 < 4V_1$, a first odd level appears, and so on: when $V_0$ increases, there appear alternatively even and odd levels. If $V_0 \gg V_1$, the slope $1/k_0$ of the straight line of figure 5 is very small: for the lowest energy levels, we practically have:

$$k = \frac{n\pi}{a} \tag{51}$$

where $n$ is an integer, and consequently:

$$E = \frac{n^2\pi^2\hbar^2}{2ma^2} - V_0 \tag{52}$$

β. *Infinitely deep well*

Assume $V(x)$ to be zero for $0 < x < a$ and infinite everywhere else. Set:

$$k = \sqrt{\frac{2mE}{\hbar^2}} \tag{53}$$

According to the comment made at the end of § 2-a-β of this complement, $\varphi(x)$ must be zero outside the interval $[0, a]$, and continuous at $x = 0$, as well as at $x = a$.

Now for $0 \leqslant x \leqslant a$:

$$\varphi(x) = A\, e^{ikx} + A'\, e^{-ikx} \tag{54}$$

Since $\varphi(0) = 0$, it can be deduced that $A' = -A$, which leads to:

$$\varphi(x) = 2iA \sin kx \tag{55}$$

Moreover, $\varphi(a) = 0$, so that:

$$k = \frac{n\pi}{a} \tag{56}$$

where $n$ is an arbitrary positive integer. If we normalize function (55), taking (56) into account, we then obtain the stationary wave functions:

$$\varphi_n(x) = \sqrt{\frac{2}{a}} \sin\left(\frac{n\pi x}{a}\right) \tag{57}$$

with energies:

$$E_n = \frac{n^2\pi^2\hbar^2}{2ma^2} \tag{58}$$

The quantization of the energy levels is thus, in this case, particularly simple.

COMMENTS:

(*i*) Relation (56) simply expresses the fact that the stationary states are determined by the condition that the width $a$ of the well must contain an integral number of half-wavelengths, $\pi/k$. This is not the case when the well has a finite depth (*cf.* § α); the difference between the two cases arises from the phase shift of the wave function which occurs upon reflection from a potential step (*cf.* § 2-a-β).

(*ii*) It can easily be verified from (51) and (52) that, if the depth $V_0$ of a finite well approaches infinity, we find the energy levels of an infinite well.

**References and suggestions for further reading:**

Eisberg and Resnick (1.3), chap. 6; Ayant and Belorizky (1.10), chap. 4; Messiah (1.17), chap. III; Merzbacher (1.16), chap. 6; Valentin (16.1), annex V.

## Complement $J_I$

## BEHAVIOR OF A WAVE PACKET AT A POTENTIAL STEP



In complement $H_I$, we determined the stationary states of a particle in various "square" potentials. For certain cases (a step potential, for example), the stationary states obtained consist of unbounded plane waves (incident, reflected and transmitted). Of course, since they cannot be normalized, such wave functions cannot really represent a physical state of the particle. However, they can be linearly superposed to form normalizable wave packets. Moreover, since such a wave packet is expanded directly in terms of stationary wave functions, its time evolution is very simple to determine. All we need to do is multiply each of the coefficients of the expansion by an imaginary exponential $e^{-iEt/\hbar}$ with a well-defined frequency $\frac{E}{h}$ (chap. I, § D-1-b).

We intend, in this complement, to construct such wave packets and study their time evolution for the case where the potential presents a "step" of height $V_0$, as in figure 1 of complement $H_I$. In this way, we shall be able to describe precisely the quantum behavior of the particle when it arrives at the potential step by determining the motion and the deformation of its associated wave packet. This will also enable us to confirm various results obtained in $H_I$ through the study of the stationary states alone (reflection and transmission coefficients, slowing down upon reflection, etc...).

We shall set:

$$\sqrt{\frac{2mE}{\hbar^2}} = k$$

$$\sqrt{\frac{2mV_0}{\hbar^2}} = K_0 \tag{1}$$

and, as in complement $H_I$, we shall distinguish between two cases, corresponding to $k$ smaller or greater than $K_0$.

### 1. Total reflection: $E < V_0$

In this case, the stationary wave functions are given by formulas (11) and (20) of complement $H_I$ ($k_1$ will be simply called $k$ here), the coefficients $A_1$, $A'_1$, $B_2$ and $B'_2$ of these formulas being related by equations (21), (22) and (23) of $H_I$.

We are going to construct a wave packet from these stationary wave functions by linearly superposing them. We shall choose only values of $k$ less than $K_0$, so as to have the waves forming the packet undergo total reflection. In order to do

this, we shall choose a function $g(k)$ (which characterizes the wave packet) which is zero for $k > K_0$. We are going to concentrate our attention on the negative region of the $x$-axis, to the left of the potential barrier. In complement $H_I$, relation (22) shows that the coefficients $A_1$ and $A_1'$ of expression (11) for a stationary wave in this region have the same modulus. Therefore, we can set:

$$\frac{A_1'(k)}{A_1(k)} = e^{-2i\theta(k)} \tag{2}$$

with [*cf.* formula (19) of $H_I$]:

$$\tan\theta(k) = \frac{\sqrt{K_0^2 - k^2}}{k} \tag{3}$$

Finally, the wave packet which we are going to consider can be written, at time $t = 0$, for negative $x$:

$$\psi(x, 0) = \frac{1}{\sqrt{2\pi}} \int_0^{K_0} dk\, g(k) \left[ e^{ikx} + e^{-2i\theta(k)}\, e^{-ikx} \right] \tag{4}$$

As in § C of chapter I, we assume that $|g(k)|$ has a pronounced peak of width $\Delta k$ about the value $k = k_0 < K_0$.

In order to obtain the expression for the wave function $\psi(x, t)$ at any time $t$, we simply use the general relation (D-14) of chapter I:

$$\begin{aligned} \psi(x, t) = {} & \frac{1}{\sqrt{2\pi}} \int_0^{K_0} dk\, g(k)\, e^{i[kx - \omega(k)t]} \\ & + \frac{1}{\sqrt{2\pi}} \int_0^{K_0} dk\, g(k)\, e^{-i[kx + \omega(k)t + 2\theta(k)]} \end{aligned} \tag{5}$$

where $\omega(k) = \hbar k^2/2m$. By construction, *this expression is valid only for negative* $x$. Its first term represents the incident wave packet; its second term, the reflected packet. For simplicity, we shall assume $g(k)$ to be real. The stationary phase condition (*cf.* chap. I, § C-2) then enables us to calculate the position $x_i$ of the center of the incident wave packet: if, at $k = k_0$, we set the derivative with respect to $k$ of the argument of the first exponential equal to zero, we obtain:

$$x_i = t \left[ \frac{d\omega}{dk} \right]_{k = k_0} = \frac{\hbar k_0}{m}\, t \tag{6}$$

In the same way, the position $x_r$ of the center of the reflected packet is obtained by differentiating the argument of the second exponential. Differentiating equation (3), we find:

$$\begin{aligned} [1 + \tan^2\theta]\, d\theta &= \left[ 1 + \frac{K_0^2 - k^2}{k^2} \right] d\theta \\ &= -\frac{dk}{k^2} \sqrt{K_0^2 - k^2} - \frac{dk}{\sqrt{K_0^2 - k^2}} \end{aligned} \tag{7}$$

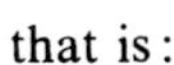

that is:

$$\frac{K_0^2}{k^2}\,\mathrm{d}\theta = -\frac{K_0^2}{k^2}\frac{1}{\sqrt{K_0^2 - k^2}}\,\mathrm{d}k \tag{8}$$

Thus we have:

$$x_r = -\left[t\frac{\mathrm{d}\omega}{\mathrm{d}k} + 2\frac{\mathrm{d}\theta}{\mathrm{d}k}\right]_{k=k_0} = -\frac{\hbar k_0}{m}t + \frac{2}{\sqrt{K_0^2 - k_0^2}} \tag{9}$$

Formulas (6) and (9) enable us to describe more precisely the motion of the particle, localized in a region of small width $\Delta x$ centered at $x_i$ or $x_r$.

First of all, let us consider what happens for negative $t$. The center $x_i$ of the incident wave packet propagates from left to right with a constant velocity $\hbar k_0/m$. On the other hand, we see from formula (9) that $x_r$ is positive, that is, situated outside the region $x < 0$ where expression (5) for the wave function is valid. This means that, for all negative values of $x$, the various waves of the second term of (5) interfere destructively: *for negative $t$, there is no reflected wave packet*, but only an incident wave packet like those we studied in § C of chapter I.

The center of the incident wave packet arrives at the barrier at time $t = 0$. During a certain interval of time around $t = 0$, the wave packet is localized in the region $x \simeq 0$ where the barrier is, and its form is relatively complicated. But, *when $t$ is sufficiently large*, we see from (6) and (9) that *it is the incident wave packet which has disappeared*, and we are left with only the reflected wave packet. It is now $x_i$ which is positive, while $x_r$ has become negative: the waves of the incident packet interfere destructively for all negative values of $x$, while those of the reflected packet interfere constructively for $x = x_r < 0$. The reflected wave packet propagates towards the left at a speed of $-\hbar k_0/m$, opposite to that of the incident packet, whose mirror image it is; its form is unchanged*. Moreover, formula (9) shows that the *reflection has introduced a delay $\tau$*, given by :

$$\tau = -2\left[\frac{\mathrm{d}\theta/\mathrm{d}k}{\mathrm{d}\omega/\mathrm{d}k}\right]_{k=k_0} = \frac{2m}{\hbar k_0\sqrt{K_0^2 - k_0^2}} \tag{10}$$

Contrary to what is predicted by classical mechanics, the particle is not instantaneously reflected. Note that the delay $\tau$ is related to the phase shift $2\theta(k)$ between the incident wave and the reflected wave for a given value of $k$. Nevertheless, it should be observed that the delay of the wave packet is not simply proportional to $\theta(k_0)$, as would be the case for an unbounded plane wave, but to the *derivative* $\mathrm{d}\theta/\mathrm{d}k$ evaluated at $k = k_0$. Physically, this delay is due to the fact that, for $t$ close to zero, the probability of presence of the particle in the region $x > 0$, which is classically forbidden, is not zero [evanescent wave, see comment (*i*) below]. It can be said, metaphorically, that the particle spends a time of the order of $\tau$ in this region before retracing its steps. Formula (10) shows that the closer the

* We assume $\Delta k$ to be small enough for the spreading of the wave packet to be negligible during the time interval considered.

average energy $\frac{\hbar^2 k_0^2}{2m}$ of the wave packet is to the height $V_0$ of the barrier, the longer the delay $\tau$.

COMMENTS:

(*i*) Here we have stressed the study of the wave packet for $x < 0$, but it is also possible to study what happens for $x > 0$. In this region, the wave packet can be written:

$$\psi(x, t) = \frac{1}{\sqrt{2\pi}} \int_0^{K_0} dk\, g(k) B_2'(k)\, e^{-\rho(k)x}\, e^{-i\omega(k)t} \tag{11}$$

where :

$$\rho(k) = \sqrt{K_0^2 - k^2} \tag{12}$$

$B_2'(k)$ is given by equation (23) of complement $H_I$ when we replace $A_1$ by 1, $k_1$ by $k$ and $\rho_2$ by $\rho$. An argument analogous to the one in § C-2 of chapter I then shows that the modulus $|\psi(x, t)|$ of expression (11) is maximum when the phase of the function to be integrated over $k$ is stationary. Now, according to expressions (22) and (23) of $H_I$, the argument of $B_2'$ is half that of $A_1'$, which, according to (2), is equal to $-2\theta(k)$. Consequently, if we expand $\omega(k)$ and $\theta(k)$ in the neighborhood of $k = k_0$, we obtain, for the phase of the function to be integrated over $k$ in (11):

$$\left\{ -\left[\frac{d\theta}{dk}\right]_{k=k_0} - \left[\frac{d\omega}{dk}\right]_{k=k_0} t \right\}(k - k_0) = -\frac{\hbar k_0}{m}(k - k_0)\left(t - \frac{\tau}{2}\right) \tag{13}$$

[we have used (10) and the fact that $g(k)$ is assumed real]. From this we can deduce that $|\psi(x, t)|$ is maximum in the $x > 0$ region for $t = \frac{\tau}{2}$★. The time at which the wave packet turns back is therefore $\tau/2$, which gives us the same delay $\tau$ upon reflection that we obtained above. We also see from expression (13) that, as soon as $\left|t - \frac{\tau}{2}\right|$ exceeds the time $\Delta t$ defined by:

$$\frac{hk_0}{m} \Delta k\, \Delta t \simeq 1 \tag{14}$$

where $\Delta k$ is the width of $g(k)$, the waves go out of phase and expression (11) for $|\psi(x, t)|$ becomes negligible. Thus, the wave packet as a whole remains in the $x > 0$ region during an interval of time $\Delta t$ of the order of:

$$\Delta t = \frac{1/\Delta k}{\hbar k_0/m} \tag{15}$$

which corresponds approximately to the time it takes, in the $x < 0$ region, to travel a a distance comparable to its width $1/\Delta k$.

(*ii*) Since $\Delta k$ is assumed to be much smaller than $k_0$ and $K_0$, the comparison of (10) and (15) shows that:

$$\Delta t \gg \tau \tag{16}$$

★ Note that the phase (13) does not depend on $x$, contrary to what we found in chapter I for a free wave packet. It follows that, in the $x > 0$ region, $|\psi(x, t)|$ does not have a pronounced peak which moves with respect to time.

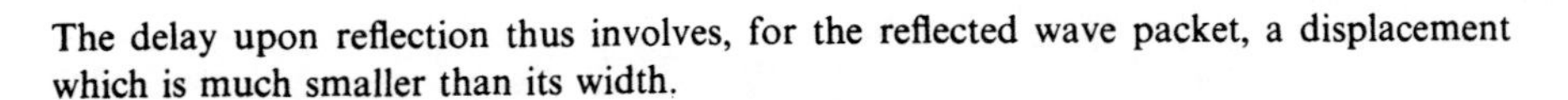


The delay upon reflection thus involves, for the reflected wave packet, a displacement which is much smaller than its width.

## 2. Partial reflection: $E > V_0$

This time, we shall consider a function $g(k)$ of width $\Delta k$, centered at a value $k = k_0 > K_0$, which is zero for $k < K_0$. The wave packet is formed in this case by superposing, with coefficients $g(k)$, the stationary wave functions whose expressions are given by formulas (11) and (12) of complement $H_I$. We shall choose $A'_2 = 0$ so as to have the particle being considered arrive at the barrier from the negative region of the $Ox$ axis, and we shall take $A_1 = 1$. The coefficients $A'_1(k)$ and $A_2(k)$ are obtained from formulas (13) and (14) of complement $H_I$ (in which $A_1$ is replaced by 1, $k_1$ by $k$, and $k_2$ by $\sqrt{k^2 - K_0^2}$). In order to describe the wave packet by a single expression, valid for all values of $x$, we can use the Heaviside "step function" $\theta(x)$ defined by:

$$\begin{aligned} \theta(x) &= 0 \qquad \text{if } x < 0 \\ \theta(x) &= 1 \qquad \text{if } x > 0 \end{aligned} \tag{17}$$

The wave packet under consideration can then be written:

$$\begin{aligned} \psi(x, t) = {} & \theta(-x)\frac{1}{\sqrt{2\pi}}\int_{K_0}^{+\infty} \mathrm{d}k\, g(k)\, \mathrm{e}^{i[kx - \omega(k)t]} \\ & + \theta(-x)\frac{1}{\sqrt{2\pi}}\int_{K_0}^{+\infty} \mathrm{d}k\, g(k)\, A'_1(k)\, \mathrm{e}^{-i[kx + \omega(k)t]} \\ & + \theta(x)\frac{1}{\sqrt{2\pi}}\int_{K_0}^{+\infty} \mathrm{d}k\, g(k)\, A_2(k)\, \mathrm{e}^{i[\sqrt{k^2 - K_0^2}\,x - \omega(k)t]} \end{aligned} \tag{18}$$

This time, we find three wave packets: incident, reflected and transmitted. As in § 1 above, the stationary phase condition gives the position of their respective centers $x_i$, $x_r$ and $x_t$. Since $A'_1(k)$ and $A_2(k)$ are real, we find:

$$x_i = \frac{\hbar k_0}{m} t \tag{19-a}$$

$$x_r = -\frac{\hbar k_0}{m} t \tag{19-b}$$

$$x_t = \frac{\hbar\sqrt{k_0^2 - K_0^2}}{m} t \tag{19-c}$$

A discussion analogous to that of (6) and (9) leads to the following conclusions: *for negative t, only the incident wave packet exists; for sufficiently large positive t, only the reflected and transmitted wave packets exist* (fig. 1). Note that there is no delay, either upon reflection or upon transmission [this is due to the fact that the coefficients $A'_1(k)$ and $A_2(k)$ are real].

The incident and reflected wave packets propagate with velocities of $\hbar k_0/m$ and $-\hbar k_0/m$ respectively. Let us assume $\Delta k$ to be sufficiently small that, within

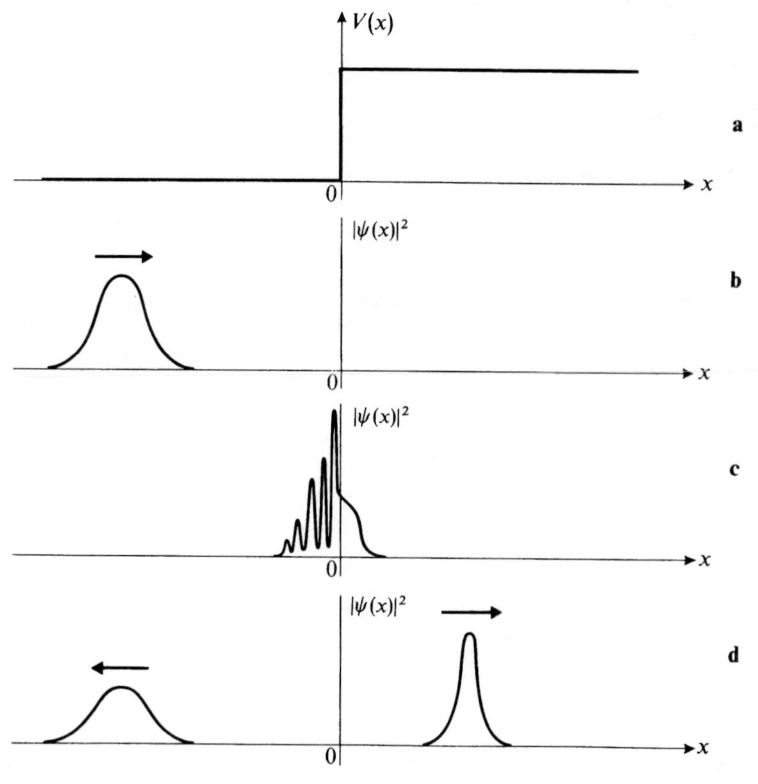


FIGURE 1

**Behavior of a wave packet at a potential step, in the case $E > V_0$. The potential is shown in figure a. In figure b, the wave packet is moving towards the step. Figure c shows the wave packet during the transitory period in which it splits in two. Interference between the incident and reflected waves are responsible for the oscillations of the wave packet in the $x < 0$ region. After a certain time (fig. d), we find two wave packets. The first one (the reflected wave packet) is returning towards the left; its amplitude is smaller than that of the incident wave packet, and its width is the same. The second one (the transmitted wave packet) propagates towards the right; its amplitude is slightly greater than that of the incident wave packet, but it is narrower.**

the interval $\left[k_0 - \frac{\Delta k}{2}, k_0 + \frac{\Delta k}{2}\right]$, we can neglect the variation of $A'_1(k)$ compared to that of $g(k)$. We can then, in the second term of (18), replace $A'_1(k)$ by $A'_1(k_0)$ and take it outside the integral. It is then easy to see that the reflected wave packet has the same form as the incident wave packet, its mirror image. Its amplitude is smaller, however, since, according to formula (13) of complement $H_I$, $A'_1(k_0)$ is less than 1. The reflection coefficient $R$ is, by definition, the ratio between the probabilities of finding the particle in the reflected wave packet and in the incident packet. Therefore, we have $R = |A'_1(k_0)|^2$, which indeed corresponds to equation (15) of complement $H_I$ [recall that we have chosen $A_1(k_0) = 1$].

The situation is different for the transmitted wave packet. We can still use the fact that $\Delta k$ is very small in order to simplify its expression: we replace $A_2(k)$

by $A_2(k_0)$, and $\sqrt{k^2 - K_0^2}$ by the approximation :

$$\sqrt{k^2 - K_0^2} \simeq \sqrt{k_0^2 - K_0^2} + (k - k_0)\left[\frac{\mathrm{d}\sqrt{k^2 - K_0^2}}{\mathrm{d}k}\right]_{k = k_0}$$

$$\simeq q_0 + (k - k_0)\frac{k_0}{q_0} \tag{20}$$

with :

$$q_0 = \sqrt{k_0^2 - K_0^2} \tag{21}$$

The transmitted wave packet can then be written :

$$\psi_t(x, t) \simeq A_2(k_0)\, \mathrm{e}^{iq_0x} \frac{1}{\sqrt{2\pi}} \int_{K_0}^{+\infty} \mathrm{d}k\, g(k)\, \mathrm{e}^{i\left[(k - k_0)\frac{k_0}{q_0}x - \omega(k)t\right]} \tag{22}$$

Let us compare this expression to the one for the incident wave packet :

$$\psi_i(x, t) = \mathrm{e}^{ik_0x} \frac{1}{\sqrt{2\pi}} \int_{K_0}^{+\infty} \mathrm{d}k\, g(k)\, \mathrm{e}^{i[(k - k_0)x - \omega(k)t]} \tag{23}$$

We see that :

$$|\psi_t(x, t)| \simeq A_2(k_0) \left| \psi_i\left(\frac{k_0}{q_0}x, t\right) \right| \tag{24}$$

The transmitted wave packet thus has a slightly greater amplitude than that of the incident packet : according to formula (14) of complement $\mathrm{H_I}$, $A_2(k_0)$ is greater than 1. However, its width is smaller, since, if $|\psi_i(x, t)|$ has a width $\Delta x$, formula (24) shows that the width of $|\psi_t(x, t)|$ is :

$$(\Delta x)_t = \frac{q_0}{k_0}\Delta x \tag{25}$$

The transmission coefficient (the ratio between the probabilities of finding the particle in the transmitted packet and in the incident packet) is seen to be the product of two factors :

$$T = \frac{q_0}{k_0}|A_2(k_0)|^2 \tag{26}$$

This indeed corresponds to formula (16) of complement $\mathrm{H_I}$, since $A_1(k_0) = 1$. Finally, note that, taking into account the contraction of the transmitted wave packet along the $Ox$ axis, we can find its velocity :

$$V_t = \frac{\hbar k_0}{m} \times \frac{q_0}{k_0} = \frac{\hbar q_0}{m} \tag{27}$$

**References and suggestions for further reading:**

Schiff (1.18), chap. 5, figs 16, 17, 18, 19; Eisberg and Resnick (1.3), § 6-3, fig. 6-8; also see reference (1.32).

## Complement $K_I$

### EXERCISES

**1.** A beam of neutrons of constant velocity, mass $M_n$ ($M_n \simeq 1.67 \times 10^{-27}$ kg) and energy $E$, is incident on a linear chain of atomic nuclei, arranged in a regular fashion as shown in the figure (these nuclei could be, for example, those of a long linear molecule). We call $l$ the distance between two consecutive nuclei, and $d$, their size ($d \ll l$). A neutron detector $D$ is placed far away, in a direction which makes an angle of $\theta$ with the direction of the incident neutrons.

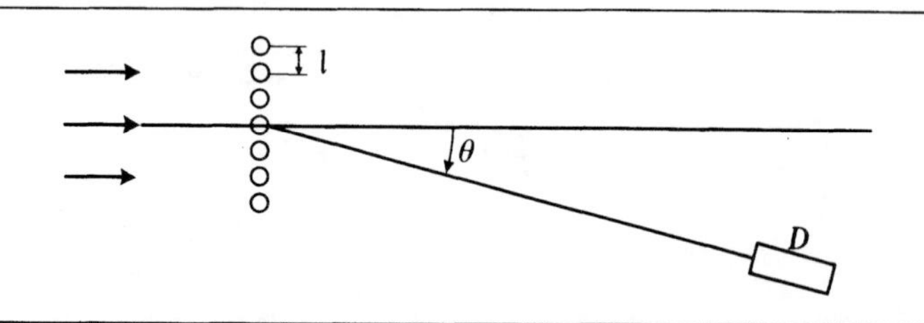


*a*) Describe qualitatively the phenomena observed at $D$ when the energy $E$ of the incident neutrons is varied.

*b*) The counting rate, as a function of $E$, presents a resonance about $E = E_1$. Knowing that there are no other resonances for $E < E_1$, show that one can determine $l$. Calculate $l$ for $\theta = 30°$ and $E_1 = 1.3 \times 10^{-20}$ joule.

*c*) At about what value of $E$ must we begin to take the finite size of the nuclei into account?

**2. Bound state of a particle in a "delta function potential"**

Consider a particle whose Hamiltonian $H$ [operator defined by formula (D-10) of chapter I] is:

$$H = -\frac{\hbar^2}{2m}\frac{d^2}{dx^2} - \alpha\,\delta(x)$$

where $\alpha$ is a positive constant whose dimensions are to be found.

*a*) Integrate the eigenvalue equation of $H$ between $-\varepsilon$ and $+\varepsilon$. Letting $\varepsilon$ approach 0, show that the derivative of the eigenfunction $\varphi(x)$ presents a discontinuity at $x = 0$ and determine it in terms of $\alpha$, $m$ and $\varphi(0)$.

*b*) Assume that the energy $E$ of the particle is negative (bound state). $\varphi(x)$ can then be written:

$$x < 0 \qquad \varphi(x) = A_1\, e^{\rho x} + A_1'\, e^{-\rho x}$$

$$x > 0 \qquad \varphi(x) = A_2\, e^{\rho x} + A_2'\, e^{-\rho x}$$

Express the constant $\rho$ in terms of $E$ and $m$. Using the results of the preceding question, calculate the matrix $M$ defined by:

$$\begin{pmatrix} A_2 \\ A'_2 \end{pmatrix} = M \begin{pmatrix} A_1 \\ A'_1 \end{pmatrix}$$

Then, using the condition that $\varphi(x)$ must be square-integrable, find the possible values of the energy. Calculate the corresponding normalized wave functions.

*c*) Trace these wave functions graphically. Give an order of magnitude for their width $\Delta x$.

*d*) What is the probability $\overline{\mathrm{d}\mathscr{P}(p)}$ that a measurement of the momentum of the particle in one of the normalized stationary states calculated above will give a result included between $p$ and $p + \mathrm{d}p$? For what value of $p$ is this probability maximum? In what domain, of dimension $\Delta p$, does it take on non-negligible values? Give an order of magnitude for the product $\Delta x \,.\, \Delta p$.

### 3. Transmission of a "delta function" potential barrier

Consider a particle placed in the same potential as in the preceding exercise. The particle is now propagating from left to right along the $Ox$ axis, with a positive energy $E$.

*a*) Show that a stationary state of the particle can be written:

$$\begin{cases} \text{if} \quad x < 0 \quad \varphi(x) = \mathrm{e}^{ikx} + A\,\mathrm{e}^{-ikx} \\ \text{if} \quad x > 0 \quad \varphi(x) = B\,\mathrm{e}^{ikx} \end{cases}$$

where $k$, $A$ and $B$ are constants which are to be calculated in terms of the energy $E$, of $m$ and of $\alpha$ (watch out for the discontinuity in $\dfrac{\mathrm{d}\varphi}{\mathrm{d}x}$ at $x = 0$).

*b*) Set $- E_L = - m\alpha^2/2\hbar^2$ (bound state energy of the particle). Calculate, in terms of the dimensionless parameter $E/E_L$, the reflection coefficient $R$ and the transmission coefficient $T$ of the barrier. Study their variations with respect to $E$; what happens when $E \longrightarrow \infty$? How can this be interpreted? Show that, if the expression of $T$ is extended for negative values of $E$, it diverges when $E \longrightarrow - E_L$, and discuss this result.

**4.** Return to exercise 2, using, this time, the Fourier transform.

*a*) Write the eigenvalue equation of $H$ and the Fourier transform of this equation. Deduce directly from this the expression for $\overline{\varphi}(p)$, the Fourier transform of $\varphi(x)$, in terms of $p$, $E$, $\alpha$ and $\varphi(0)$. Then show that only one value of $E$, a negative one, is possible. Only the bound state of the particle, and not the ones in which it propagates, is found by this method; why? Then calculate $\varphi(x)$ and show that one can find in this way all the results of exercise 2.

*b*) The average kinetic energy of the particle can be written (*cf.* chap. III):

$$E_k = \frac{1}{2m} \int_{-\infty}^{+\infty} p^2 \, |\overline{\varphi}(p)|^2 \, \mathrm{d}p$$

Show that, when $\overline{\varphi}(p)$ is a "sufficiently smooth" function, we also have:

$$E_k = -\frac{\hbar^2}{2m}\int_{-\infty}^{+\infty} \varphi^*(x)\ \frac{d^2\varphi}{dx^2}\ dx$$

These formulas enable us to obtain, in two different ways, the energy $E_k$ for a particle in the bound state calculated in *a*). What result is obtained? Note that, in this case, $\varphi(x)$ is not "regular" at $x = 0$, where its derivative is discontinuous. It is then necessary to differentiate $\varphi(x)$ in the sense of distributions, which introduces a contribution of the point $x = 0$ to the average value we are looking for. Interpret this contribution physically: consider a square well, centered at $x = 0$, whose width $a$ approaches 0 and whose depth $V_0$ approaches infinity (so that $aV_0 = \alpha$), and study the behavior of the wave function in this well.

## 5. Well consisting of two delta functions

Consider a particle of mass $m$ whose potential energy is

$$V(x) = -\alpha\delta(x) - \alpha\delta(x - l) \qquad \alpha > 0$$

where $l$ is a constant length.

*a*) Calculate the bound states of the particle, setting $E = -\frac{\hbar^2\rho^2}{2m}$. Show that the possible energies are given by the relation

$$e^{-\rho l} = \pm\left(1 - \frac{2\rho}{\mu}\right)$$

where $\mu$ is defined by $\mu = \frac{2m\alpha}{\hbar^2}$. Give a graphic solution of this equation.

(*i*) *Ground state*. Show that this state is even (invariant with respect to reflection about the point $x = l/2$), and that its energy $E_S$ is less than the energy $-E_L$ introduced in problem 3. Interpret this result physically. Represent graphically the corresponding wave function.

(*ii*) *Excited state*. Show that, when $l$ is greater than a value which you are to specify, there exists an odd excited state, of energy $E_A$ greater than $-E_L$. Find the corresponding wave function.

(*iii*) Explain how the preceding calculations enable us to construct a model which represents an ionized diatomic molecule ($H_2^+$, for example) whose nuclei are separated by a distance $l$. How do the energies of the two levels vary with respect to $l$? What happens at the limit where $l \longrightarrow 0$ and at the limit where $l \longrightarrow \infty$? If the repulsion of the two nuclei is taken into account, what is the total energy of the system? Show that the curve which gives the variation with respect to $l$ of the energies thus obtained enables us to predict in certain cases the existence of bound states of $H_2^+$, and to determine the value of $l$ at equilibrium. In this way we obtain a very elementary model of the chemical bond.

*b*) Calculate the reflection and transmission coefficients of the system of two delta function barriers. Study their variations with respect to $l$. Do the resonances thus obtained occur when $l$ is an integral multiple of the de Broglie wavelength of the particle? Why?

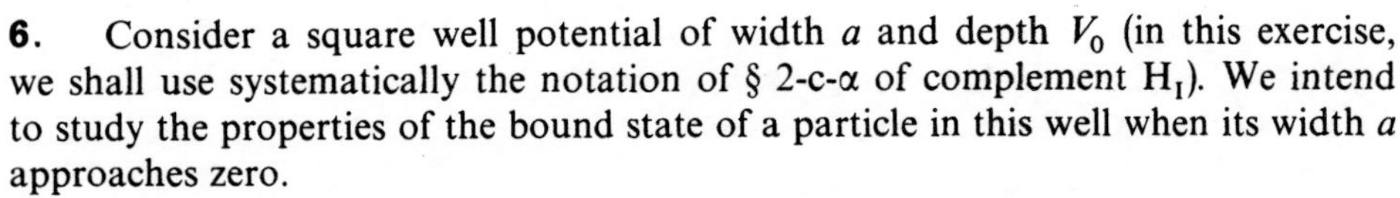

**6.** Consider a square well potential of width $a$ and depth $V_0$ (in this exercise, we shall use systematically the notation of § 2-c-α of complement $H_I$). We intend to study the properties of the bound state of a particle in this well when its width $a$ approaches zero.

*a*) Show that there indeed exists only one bound state and calculate its energy $E$ $\left(\text{we find } E \simeq -\dfrac{mV_0^2a^2}{2\hbar^2}\text{, that is, an energy which varies with the square of the area } aV_0 \text{ of the well}\right)$.

*b*) Show that $\rho \longrightarrow 0$ and that $A'_2 = A_2 \simeq B_1/2$. Deduce from this that, in the bound state, the probability of finding the particle outside the well approaches 1.

*c*) How can the preceding considerations be applied to a particle placed, as in exercise 2, in the potential $V(x) = -\alpha\delta(x)$?

**7.** Consider a particle placed in the potential

$$V(x) = 0 \qquad \text{if } x \geqslant a$$
$$V(x) = -V_0 \qquad \text{if } 0 \leqslant x < a,$$

with $V(x)$ infinite for negative $x$. Let $\varphi(x)$ be a wave function associated with a stationary state of the particle. Show that $\varphi(x)$ can be extended to give an odd wave function which corresponds to a stationary state for a square well of width $2a$ and depth $V_0$ (*cf.* complement $H_I$, § 2-c-α). Discuss, with respect to $a$ and $V_0$, the number of bound states of the particle. Is there always at least one such state, as for the symmetric square well?

**8.** Consider, in a two-dimensional problem, the oblique reflection of a particle from a potential step defined by:

$$V(x, y) = 0 \qquad \text{if } x < 0$$
$$V(x, y) = V_0 \qquad \text{if } x > 0$$

Study the motion of the center of the wave packet. In the case of total reflection, interpret physically the differences between the trajectory of this center and the classical trajectory (lateral shift upon reflection). Show that, when $V_0 \longrightarrow +\infty$, the quantum trajectory becomes asymptotic to the classical trajectory.

CHAPTER II

# The mathematical tools of quantum mechanics

OUTLINE OF CHAPTER II

This chapter is intended to be a general survey of the basic mathematical tools used in quantum mechanics. We shall give a simple condensed presentation aimed at facilitating the study of subsequent chapters for readers unfamiliar with these tools. We make no attempt to be mathematically complete and rigorous. We feel it preferable to limit ourselves to a practical point of view, uniting in a single chapter the various concepts which are useful in quantum mechanics. In particular, we wish to stress the convenience of the Dirac notation for carrying out the various calculations which we shall have to perform.

In this spirit, we shall try to simplify the discussion as much as possible. Neither the general definitions nor the rigorous proofs which would be required by a mathematician are to be found here. For example, we shall sometimes speak of infinite-dimensional spaces and reason as if they had a finite number of dimensions. Moreover, many terms (square-integrable functions, basis, etc...) will be employed with a meaning which, although commonly used in physics, is not exactly the one used in pure mathematics.

We begin in § A by studying the wave functions introduced in chapter I. We show that these wave functions belong to an abstract vector space, which we call the "wave function space $\mathscr{F}$". This study will be carried out in detail as it introduces some basic concepts of the mathematical formalism of quantum mechanics : scalar products, linear operators, basis, etc... Starting in § B, we shall develop a more general formalism, characterizing the state of a system by a "state vector" belonging to a vector space : the "state space $\mathscr{E}$". Dirac notation, which greatly simplifies calculations in this formalism, is introduced. § C is intended to study the idea of a representation. The reading of § D is particularly recommended to the reader who is unfamiliar with the diagonalization of an operator : this operation will be constantly useful to us in what follows. In § E, we treat two important examples of representations. In particular, we show how the wave functions studied in § A are the "components" of state vectors in a particular representation. Finally, we introduce in § F the concept of a tensor product. This concept will be illustrated more concretely by a simple example in complement $D_{IV}$.

## A. ONE-PARTICLE WAVE FUNCTION SPACE

The probabilistic interpretation of the wave function $\psi(\mathbf{r}, t)$ of a particle was given in the preceding chapter : $|\psi(\mathbf{r}, t)|^2 d^3r$ represents the probability of finding, at time $t$, the particle in a volume $d^3r = \mathrm{d}x\,\mathrm{d}y\,\mathrm{d}z$ about the point $\mathbf{r}$. The total probability of finding the particle somewhere in space is equal to 1, so we must have :

$$\int d^3r \, |\psi(\mathbf{r}, t)|^2 = 1 \tag{A-1}$$

where the integration extends over all space.

Thus, we are led to studying the set of square-integrable functions. These are functions for which the integral (A-1) converges. This set is called $L^2$ by mathematicians and it has the structure of a Hilbert space.

From a physical point of view, it is clear that the set $L^2$ is too wide in scope: given the meaning attributed to $|\psi(\mathbf{r}, t)|^2$, the wave functions which are actually used possess certain properties of regularity. We can only retain the functions $\psi(\mathbf{r}, t)$ which are everywhere defined, continuous, and infinitely differentiable (for example, to state that a function is really discontinuous at a given point in space has no physical meaning, since no experiment enables us to have access to real phenomena on a very small scale, say of $10^{-30}$ m). It is also possible to confine ourselves to wave functions which have a bounded domain (which makes it certain that the particle can be found within a finite region of space, for example inside the laboratory). We shall not try to give a precise, general list of these supplementary conditions: we shall call $\mathscr{F}$ the set of wave functions composed of sufficiently regular functions of $L^2$ ($\mathscr{F}$ is a subspace of $L^2$).

## 1. Structure of the wave function space $\mathscr{F}$

### a. $\mathscr{F}$ AS A VECTOR SPACE

It can easily be shown that $\mathscr{F}$ satisfies all the criteria of a vector space. As an example, we demonstrate that if $\psi_1(\mathbf{r})$ and $\psi_2(\mathbf{r}) \in \mathscr{F}$, then*:

$$\psi(\mathbf{r}) = \lambda_1\psi_1(\mathbf{r}) + \lambda_2\psi_2(\mathbf{r}) \in \mathscr{F} \tag{A-2}$$

where $\lambda_1$ and $\lambda_2$ are two arbitrary complex numbers.

In order to show that $\psi(\mathbf{r})$ is square-integrable, expand $|\psi(\mathbf{r})|^2$ :

$$|\psi(\mathbf{r})|^2 = |\lambda_1|^2 \, |\psi_1(\mathbf{r})|^2 + |\lambda_2|^2 \, |\psi_2(\mathbf{r})|^2 + \lambda_1^*\lambda_2\psi_1^*(\mathbf{r})\psi_2(\mathbf{r}) + \lambda_1\lambda_2^*\psi_1(\mathbf{r})\psi_2^*(\mathbf{r}) \tag{A-3}$$

The last two terms of (A-3) have the same modulus, which has as an upper limit:

$$|\lambda_1| \, |\lambda_2| \, [|\psi_1(\mathbf{r})|^2 + |\psi_2(\mathbf{r})|^2]$$

$|\psi(\mathbf{r})|^2$ is therefore smaller than a function whose integral converges, since $\psi_1$ and $\psi_2$ are square-integrable.

### b. THE SCALAR PRODUCT

#### α. *Definition*

With each pair of elements of $\mathscr{F}$, $\varphi(\mathbf{r})$ and $\psi(\mathbf{r})$, taken in this order, we associate a *complex number*, denoted by $(\varphi, \psi)$, which, by definition, is equal to:

$$\boxed{(\varphi, \psi) = \int d^3r \;\; \varphi^*(\mathbf{r}) \;\; \psi(\mathbf{r})} \tag{A-4}$$

$(\varphi, \psi)$ is the *scalar product of* $\psi(\mathbf{r})$ *by* $\varphi(\mathbf{r})$ [this integral always converges if $\varphi$ and $\psi$ belong to $\mathscr{F}$].

#### β. *Properties*

They follow from definition (A-4) :

* The symbol $\in$ signifies : "belongs to".

$$(\varphi, \psi) = (\psi, \varphi)^* \tag{A-5}$$

$$(\varphi, \lambda_1\psi_1 + \lambda_2\psi_2) = \lambda_1(\varphi, \psi_1) + \lambda_2(\varphi, \psi_2) \tag{A-6}$$

$$(\lambda_1\varphi_1 + \lambda_2\varphi_2, \psi) = \lambda_1^*(\varphi_1, \psi) + \lambda_2^*(\varphi_2, \psi) \tag{A-7}$$

The scalar product is *linear* with respect to the second function of the pair, *antilinear* with respect to the first one. If $(\varphi, \psi) = 0$, $\varphi(\mathbf{r})$ and $\psi(\mathbf{r})$ are said to be *orthogonal.*

$$(\psi, \psi) = \int d^3r\, |\psi(\mathbf{r})|^2 \tag{A-8}$$

is a *real, positive* number, which is *zero if and only if* $\psi(\mathbf{r}) \equiv 0$.

$\sqrt{(\psi, \psi)}$ is called the norm of $\psi(\mathbf{r})$ [it can easily be verified that this number has all the properties of a norm]. The scalar product chosen above thus permits the definition of a norm in $\mathscr{F}$.

Let us mention, finally, (*cf.* complement $A_{II}$) the *Schwarz inequality*:

$$|(\psi_1, \psi_2)| \leqslant \sqrt{(\psi_1, \psi_1)}\ \sqrt{(\psi_2, \psi_2)} \tag{A-9}$$

This becomes an equality if and only if the two functions $\psi_1$ and $\psi_2$ are proportional.

### c. LINEAR OPERATORS

#### α. *Definition*

A linear operator $A$ is, by definition, a mathematical entity which associates with every function $\psi(\mathbf{r}) \in \mathscr{F}$ another function $\psi'(\mathbf{r})$, the correspondence being linear :

$$\psi'(\mathbf{r}) = A\psi(\mathbf{r}) \tag{A-10-a}$$

$$A[\lambda_1\psi_1(\mathbf{r}) + \lambda_2\psi_2(\mathbf{r})] = \lambda_1 A\psi_1(\mathbf{r}) + \lambda_2 A\psi_2(\mathbf{r}) \tag{A-10-b}$$

Let us cite some simple examples of linear operators : the parity operator $\Pi$, whose definition is:

$$\Pi\psi(x, y, z) = \psi(-x, -y, -z) \tag{A-11}$$

the operator which performs a multiplication by $x$, which we shall call $X$, and which is defined by:

$$X\psi(x, y, z) = x\psi(x, y, z) \tag{A-12}$$

finally, the operator which differentiates with respect to $x$, which we shall call $D_x$, and whose definition is:

$$D_x\psi(x, y, z) = \frac{\partial\psi(x, y, z)}{\partial x} \tag{A-13}$$

[the two operators $X$ and $D_x$, acting on a function $\psi(\mathbf{r}) \in \mathscr{F}$, can transform it into a function which is no longer necessarily square-integrable].

β. *Product of operators*

Let $A$ and $B$ be two linear operators. Their product $AB$ is defined by:

$$\boxed{(AB)\psi(\mathbf{r}) = A[B\psi(\mathbf{r})]} \tag{A-14}$$

$B$ is first allowed to act on $\psi(\mathbf{r})$, which gives $\varphi(\mathbf{r}) = B\psi(\mathbf{r})$, then $A$ operates on the new function $\varphi(\mathbf{r})$.

In general, $AB \neq BA$. We call the *commutator* of $A$ and $B$ the operator written $[A, B]$ and defined by

$$\boxed{[A, B] = AB - BA} \tag{A-15}$$

We shall calculate, as an example, the commutator $[X, D_x]$. In order to do this, we shall take an arbitrary function $\psi(\mathbf{r})$:

$$\begin{aligned}[X, D_x]\ \psi(\mathbf{r}) &= \left(x\frac{\partial}{\partial x} - \frac{\partial}{\partial x}x\right)\psi(\mathbf{r}) \\ &= x\frac{\partial}{\partial x}\psi(\mathbf{r}) - \frac{\partial}{\partial x}[x\psi(\mathbf{r})] \\ &= x\frac{\partial}{\partial x}\psi(\mathbf{r}) - \psi(\mathbf{r}) - x\frac{\partial}{\partial x}\psi(\mathbf{r}) = -\psi(\mathbf{r})\end{aligned} \tag{A-16}$$

Since this is true for all $\psi(\mathbf{r})$, it can be deduced that:

$$[X, D_x] = -1 \tag{A-17}$$

## 2. Discrete orthonormal bases in $\mathscr{F}$ : $\{ u_i(\mathbf{r}) \}$

### a. DEFINITION

Consider a countable set of functions of $\mathscr{F}$, labeled by a discrete index $i$ ($i = 1, 2, ..., n, ...$):

$$u_1(\mathbf{r}) \in \mathscr{F}, \quad u_2(\mathbf{r}) \in \mathscr{F}, \quad ..., \quad u_i(\mathbf{r}) \in \mathscr{F}, \quad ...$$

– The set $\{u_i(\mathbf{r})\}$ is *orthonormal* if:

$$(u_i, u_j) = \int d^3r\ u_i^*(\mathbf{r})\ u_j(\mathbf{r}) = \delta_{ij} \tag{A-18}$$

where $\delta_{ij}$, the Kronecker delta function, is equal to 1 for $i = j$ and to 0 for $i \neq j$.

– It constitutes a *basis** if every function $\psi(\mathbf{r}) \in \mathscr{F}$ can be expanded in one and only one way in terms of the $u_i(\mathbf{r})$:

$$\boxed{\psi(\mathbf{r}) = \sum_i c_i\, u_i(\mathbf{r})} \tag{A-19}$$

* When the set $\{ u_i(\mathbf{r}) \}$ constitutes a basis, it is sometimes said to be a complete set of functions. It must be noted that the word complete is used with a meaning different from the one it usually has in mathematics.

#### b. COMPONENTS OF A WAVE FUNCTION IN THE $\{ u_i(\mathbf{r}) \}$ BASIS

Multiply the two sides of (A-19) by $u_j^*(\mathbf{r})$ and integrate over all space. From (A-6) and (A-18)* :

$$(u_j, \psi) = \left(u_j, \sum_i c_i\, u_i\right) = \sum_i c_i (u_j, u_i)$$

$$= \sum_i c_i\, \delta_{ij} = c_j \tag{A-20}$$

that is:

$$\boxed{c_i = (u_i, \psi) = \int d^3r\;\; u_i^*(\mathbf{r})\;\; \psi(\mathbf{r})} \tag{A-21}$$

The component $c_i$ of $\psi(\mathbf{r})$ on $u_i(\mathbf{r})$ is therefore equal to the scalar product of $\psi(\mathbf{r})$ by $u_i(\mathbf{r})$. Once the $\{ u_i(\mathbf{r}) \}$ basis has been chosen, it is equivalent to specify $\psi(\mathbf{r})$ or the set of its components $c_i$ with respect to the basis functions. The set of numbers $c_i$ is said to *represent* $\psi(\mathbf{r})$ in the $\{ u_i(\mathbf{r}) \}$ basis.

COMMENTS:

(*i*) Note the analogy with an orthonormal basis $\{ \mathbf{e}_1, \mathbf{e}_2, \mathbf{e}_3 \}$ of the ordinary three-dimensional space, $R^3$. The fact that $\mathbf{e}_1$, $\mathbf{e}_2$ and $\mathbf{e}_3$ are orthogonal and unitary can indeed be expressed by:

$$\mathbf{e}_i \,.\, \mathbf{e}_j = \delta_{ij} \qquad (i, j = 1, 2, 3) \tag{A-22}$$

Any vector $\mathbf{V}$ of $R^3$ can be expanded in this basis:

$$\mathbf{V} = \sum_{i=1}^{3} v_i\;\; \mathbf{e}_i \tag{A-23}$$

with

$$v_i = \mathbf{e}_i \,.\, \mathbf{V} \tag{A-24}$$

Formulas (A-18), (A-19) and (A-21) thus generalize, as it were, the well-known formulas, (A-22), (A-23) and (A-24). However, it must be noted that the $v_i$ are real numbers, while the $c_i$ are complex numbers.

(*ii*) The same function $\psi(\mathbf{r})$ obviously has different components in two different bases. We shall study the problem of a change in basis later.

(*iii*) We can also, in the $\{ u_i(\mathbf{r}) \}$ basis, represent a linear operator $A$ by a set of numbers which can be arranged in the form of a matrix. We shall take up this question again in § C, after we have introduced Dirac notation.

#### c. EXPRESSION FOR THE SCALAR PRODUCT IN TERMS OF THE COMPONENTS

Let $\varphi(\mathbf{r})$ and $\psi(\mathbf{r})$ be two wave functions which can be expanded as follows:

$$\varphi(\mathbf{r}) = \sum_i b_i\, u_i(\mathbf{r})$$

$$\psi(\mathbf{r}) = \sum_j c_j\, u_j(\mathbf{r}) \tag{A-25}$$

* To be completely rigorous, one should make certain that one can interchange $\sum_i$ and $\int d^3r$. We shall systematically ignore this kind of problem.

Their scalar product can be calculated by using (A-6), (A-7) and (A-18):

$$(\varphi, \psi) = \left( \sum_i b_i u_i , \sum_j c_j u_j \right) = \sum_{i,j} b_i^* c_j (u_i, u_j)$$
$$= \sum_{i,j} b_i^* c_j \, \delta_{ij}$$

that is:

$$\boxed{(\varphi, \psi) = \sum_i b_i^* c_i} \qquad \text{(A-26)}$$

In particular:

$$\boxed{(\psi, \psi) = \sum_i |c_i|^2} \qquad \text{(A-27)}$$

The scalar product of two wave functions (or the square of the norm of a wave function) can thus be very simply expressed in terms of the components of these functions in the $\{ u_i(\mathbf{r}) \}$ basis.

COMMENT:

Let $\mathbf{V}$ and $\mathbf{W}$ be two vectors of $R^3$, with components $v_i$ and $w_j$. The analytic expression of their scalar product is well-known:

$$\mathbf{V} \cdot \mathbf{W} = \sum_{i=1}^{3} v_i \, w_i \qquad \text{(A-28)}$$

Formula (A-26) can therefore be considered to be a generalization of (A-28).

**d. CLOSURE RELATION**

Relation (A-18), called the orthonormalization relation, expresses the fact that the functions of the set $\{ u_i(\mathbf{r}) \}$ are normalized to 1 and orthogonal with respect to each other. We are now going to establish another relation, called the closure relation, which expresses the fact that this set constitutes a basis.

If $\{ u_i(\mathbf{r}) \}$ is a basis of $\mathscr{F}$, there exists an expansion such as (A-19) for every function $\psi(\mathbf{r}) \in \mathscr{F}$. Substitute into (A-19) expression (A-21) for the various components $c_i$ [the name of the integration variable must be changed, since $\mathbf{r}$ already appears in (A-19)]:

$$\psi(\mathbf{r}) = \sum_i c_i \, u_i(\mathbf{r}) = \sum_i (u_i, \psi) \, u_i(\mathbf{r})$$
$$= \sum_i \left[ \int d^3r' \; u_i^*(\mathbf{r}') \; \psi(\mathbf{r}') \right] u_i(\mathbf{r}) \qquad \text{(A-29)}$$

Interchanging $\sum_i$ and $\int d^3r'$, we obtain:

$$\psi(\mathbf{r}) = \int d^3r' \; \psi(\mathbf{r}') \left[ \sum_i u_i(\mathbf{r}) \; u_i^*(\mathbf{r}') \right] \qquad \text{(A-30)}$$

$\sum u_i(\mathbf{r})u_i^*(\mathbf{r}')$is therefore a function $F(\mathbf{r}, \mathbf{r}')$ of $\mathbf{r}$ and of $\mathbf{r}'$ such that, for every function $\psi(\mathbf{r})$, we have:

$$\psi(\mathbf{r}) = \int d^3r' \; \psi(\mathbf{r}') \; F(\mathbf{r}, \mathbf{r}') \qquad \text{(A-31)}$$

Equation (A-31) is characteristic of the function $\delta(\mathbf{r} - \mathbf{r}')$ (*cf.* appendix II). From this it can be deduced that:

$$\boxed{\sum_i u_i(\mathbf{r}) \; u_i^*(\mathbf{r}') = \delta(\mathbf{r} - \mathbf{r}')} \qquad \text{(A-32)}$$

*Reciprocally*, if an orthonormal set $\{ u_i(\mathbf{r}) \}$ satisfies the closure relation (A-32), it constitutes a basis. Any function $\psi(\mathbf{r})$ can indeed be written in the form:

$$\psi(\mathbf{r}) = \int d^3r' \; \psi(\mathbf{r}') \; \delta(\mathbf{r} - \mathbf{r}') \qquad \text{(A-33)}$$

Substituting expression (A-32) for $\delta(\mathbf{r} - \mathbf{r}')$ into this expression, we obtain formula (A-30). To return to (A-29), all we must do is again interchange summation and integration. This equation then expresses the fact that $\psi(\mathbf{r})$ can always be expanded in terms of the $u_i(\mathbf{r})$ and gives the coefficients of this expansion.

COMMENT:

We shall re-examine the closure relation using the Dirac notation in §C, and we shall see that it can be given a simple geometric interpretation.

### 3. Introduction of "bases" not belonging to $\mathscr{F}$

The $\{ u_i(\mathbf{r}) \}$ bases studied above are composed of square-integrable functions. It can also be convenient to introduce "bases" of functions not belonging to either $\mathscr{F}$ or $L^2$, but in terms of which any wave function $\psi(\mathbf{r})$ can nevertheless be expanded. We are going to give examples of such bases and we shall show how it is possible to extend to them the important formulas which were established in the preceding section.

#### a. PLANE WAVES

For simplicity, we treat the one-dimensional case. We shall therefore study square-integrable functions $\psi(x)$ which depend only on the $x$ variable. In chapter I we saw the advantage of using the Fourier transform $\bar{\psi}(p)$ of $\psi(x)$:

$$\psi(x) = \frac{1}{\sqrt{2\pi\hbar}} \int_{-\infty}^{+\infty} \mathrm{d}p \; \bar{\psi}(p) \, \mathrm{e}^{ipx/\hbar} \qquad \text{(A-34-a)}$$

$$\bar{\psi}(p) = \frac{1}{\sqrt{2\pi\hbar}} \int_{-\infty}^{+\infty} \mathrm{d}x \; \psi(x) \, \mathrm{e}^{-ipx/\hbar} \qquad \text{(A-34-b)}$$

Consider the function $v_p(x)$, defined by:

$$\boxed{v_p(x) = \frac{1}{\sqrt{2\pi\hbar}} e^{ipx/\hbar}} \tag{A-35}$$

$v_p(x)$ is a plane wave, with the wave vector $p/\hbar$. The integral over the whole $x$ axis of $|v_p(x)|^2 = \frac{1}{2\pi\hbar}$ diverges. Therefore $v_p(x) \notin \mathscr{F}_x$. We shall designate by $\{ v_p(x) \}$ the set of all plane waves, that is, of all functions $v_p(x)$ corresponding to the various values of $p$. The number $p$, which varies continuously between $-\infty$ and $+\infty$, will be considered as a *continuous index* which permits us to label the various functions of the set $\{ v_p(x) \}$ [recall that the index $i$ used for the set $\{ u_i(\mathbf{r}) \}$ considered above was *discrete*].

Formulas (A-34) can be rewritten using (A-35):

$$\psi(x) = \int_{-\infty}^{+\infty} \mathrm{d}p \ \bar{\psi}(p) \ v_p(x) \tag{A-36}$$

$$\bar{\psi}(p) = (v_p, \psi) = \int_{-\infty}^{+\infty} \mathrm{d}x \ v_p^*(x) \ \psi(x) \tag{A-37}$$

These two formulas can be compared to (A-19) and (A-21). Relation (A-36) expresses the idea that every function $\psi(x) \in \mathscr{F}_x$ can be expanded in one and only way in terms of the $v_p(x)$, that is, the plane waves. Since the index $p$ varies continuously and not discretely, the summation $\sum_i$ appearing in (A-19) must be replaced by an integration over $p$. Relation (A-37), like (A-21), gives the component $\bar{\psi}(p)$ of $\psi(x)$ on $v_p(x)$ in the form of a scalar product* $(v_p, \psi)$. The set of these components, which correspond to the various possible values of $p$, constitutes a function of $p$, $\bar{\psi}(p)$, the Fourier transform of $\psi(x)$.

*Thus, $\bar{\psi}(p)$ is the analogue of $c_i$.* These two complex numbers, which depend either on $p$ or on $i$, represent the *components of the same function $\psi(x)$ in two different bases: $\{ v_p(x) \}$ and $\{ u_i(x) \}$*.

This point also appears clearly if we calculate the square of the norm of $\psi(x)$. According to Parseval's relation [app. I, formula (45)], we have:

$$\boxed{(\psi, \psi) = \int_{-\infty}^{+\infty} \mathrm{d}p \ |\bar{\psi}(p)|^2} \tag{A-38}$$

a formula which resembles (A-27), if we replace $c_i$ by $\bar{\psi}(p)$ and $\sum_i$ by $\int \mathrm{d}p$.

Let us show that the $v_p(x)$ satisfy a closure relation. Using the formula [*cf.* appendix II, equation (34)]:

$$\frac{1}{2\pi} \int_{-\infty}^{+\infty} \mathrm{d}k \ e^{iku} = \delta(u) \tag{A-39}$$

* We have only defined the scalar product for two square-integrable functions, but this definition can easily be extended to cases like this one, provided that the corresponding integral converges.

we find:

$$\boxed{\int_{-\infty}^{+\infty} \mathrm{d}p \; v_p(x) \; v_p^*(x') = \frac{1}{2\pi} \int \frac{\mathrm{d}p}{\hbar} \, \mathrm{e}^{i\frac{p}{\hbar}(x-x')} = \delta(x - x')} \tag{A-40}$$

This formula is the analogue of (A-32) with, again, the substitution of $\int \mathrm{d}p$ for $\sum_i$.

Finally, let us calculate the scalar product $(v_p, v_{p'})$ in order to see if there exists an equivalent of the orthonormalization relation. Again using (A-39), we obtain:

$$(v_p, v_{p'}) = \int_{-\infty}^{+\infty} \mathrm{d}x \; v_p^*(x) \; v_{p'}(x)$$

that is:

$$\boxed{(v_p, v_{p'}) = \frac{1}{2\pi} \int \frac{\mathrm{d}x}{\hbar} \, \mathrm{e}^{i\frac{x}{\hbar}(p'-p)} = \delta(p - p')} \tag{A-41}$$

Compare (A-41) and (A-18). Instead of having two discrete indices $i$ and $j$ and a Kronecker delta $\delta_{ij}$, we now have two continuous indices $p$ and $p'$ and a delta function of the difference between the indices, $\delta(p - p')$. Note that if we set $p = p'$, the scalar product $(v_p, v_p)$ diverges; again we see that $v_p(x) \notin \mathscr{F}_x$. Although this constitutes a misuse of the term, we shall call (A-41) an "orthonormalization" relation. It is also sometimes said that the $v_p(x)$ are "orthonormalized in the Dirac sense".

The generalization to three dimensions presents no difficulties. We consider the plane waves:

$$v_{\mathbf{p}}(\mathbf{r}) = \left(\frac{1}{2\pi\hbar}\right)^{3/2} \mathrm{e}^{i\mathbf{p}.\mathbf{r}/\hbar} \tag{A-42}$$

The functions of the $\{ v_{\mathbf{p}}(\mathbf{r}) \}$ basis now depend on the three continuous indices $p_x$, $p_y$, $p_z$, condensed into the notation $\mathbf{p}$. It is then easy to show that the following formulas are valid:

$$\psi(\mathbf{r}) = \int d^3p \; \bar{\psi}(\mathbf{p}) \; v_{\mathbf{p}}(\mathbf{r}) \tag{A-43}$$

$$\bar{\psi}(\mathbf{p}) = (v_{\mathbf{p}}, \psi) = \int d^3r \; v_{\mathbf{p}}^*(\mathbf{r}) \; \psi(\mathbf{r}) \tag{A-44}$$

$$(\varphi, \psi) = \int d^3p \; \bar{\varphi}^*(\mathbf{p}) \; \bar{\psi}(\mathbf{p}) \tag{A-45}$$

$$\int d^3p \; v_{\mathbf{p}}(\mathbf{r}) \; v_{\mathbf{p}}^*(\mathbf{r}') = \delta(\mathbf{r} - \mathbf{r}') \tag{A-46}$$

$$(v_{\mathbf{p}}, v_{\mathbf{p}'}) = \delta(\mathbf{p} - \mathbf{p}') \tag{A-47}$$

They represent the generalizations of (A-36), (A-37), (A-38), (A-40) and (A-41).

Thus the $v_{\mathbf{p}}(\mathbf{r})$ can be considered to constitute a "continuous basis". All the formulas established above for the discrete basis $\{ u_i(\mathbf{r}) \}$ can be extended to this continuous basis, using the correspondence rules summarized in table (II-1).

| | | |
|---|---|---|
| $i$ | $\longleftrightarrow$ | $\mathbf{p}$ |
| $\sum_i$ | $\longleftrightarrow$ | $\int d^3p$ |
| $\delta_{ij}$ | $\longleftrightarrow$ | $\delta(\mathbf{p} - \mathbf{p}')$ |

Table (II-1)

### b. "DELTA FUNCTIONS"

In the same way, let us introduce a set of functions of $\mathbf{r}$, $\{ \xi_{\mathbf{r}_0}(\mathbf{r}) \}$, labeled by the continuous index $\mathbf{r}_0$ (condensed notation for $x_0$, $y_0$, $z_0$) and defined by:

$$\boxed{\xi_{\mathbf{r}_0}(\mathbf{r}) = \delta(\mathbf{r} - \mathbf{r}_0)} \tag{A-48}$$

$\{ \xi_{\mathbf{r}_0}(\mathbf{r}) \}$ represents the set of delta functions centered at the various points $\mathbf{r}_0$ of space; $\xi_{\mathbf{r}_0}(\mathbf{r})$ is obviously not square-integrable : $\xi_{\mathbf{r}_0}(\mathbf{r}) \notin \mathscr{F}$.

Then consider the following relations, which are valid for every function $\psi(\mathbf{r})$ belonging to $\mathscr{F}$:

$$\psi(\mathbf{r}) = \int d^3r_0 \; \psi(\mathbf{r}_0) \; \delta(\mathbf{r} - \mathbf{r}_0) \tag{A-49}$$

$$\psi(\mathbf{r}_0) = \int d^3r \; \delta(\mathbf{r}_0 - \mathbf{r}) \; \psi(\mathbf{r}) \tag{A-50}$$

They can be rewritten, using (A-48), in the form:

$$\psi(\mathbf{r}) = \int d^3r_0 \; \psi(\mathbf{r}_0) \; \xi_{\mathbf{r}_0}(\mathbf{r}) \tag{A-51}$$

$$\psi(\mathbf{r}_0) = (\xi_{\mathbf{r}_0}, \psi) = \int d^3r \; \xi^*_{\mathbf{r}_0}(\mathbf{r}) \; \psi(\mathbf{r}) \tag{A-52}$$

(A-51) expresses the fact that every function $\psi(\mathbf{r}) \in \mathscr{F}$ can be expanded in one and only one way in terms of the $\xi_{\mathbf{r}_0}(\mathbf{r})$. (A-52) shows that the component of $\psi(\mathbf{r})$ on the function $\xi_{\mathbf{r}_0}(\mathbf{r})$ (we are dealing here with real basis functions) is precisely the value $\psi(\mathbf{r}_0)$ of $\psi(\mathbf{r})$ at the point $\mathbf{r}_0$. (A-51) and (A-52) are analogous to (A-19) and (A-21): we simply replace the discrete index $i$ by the continuous index $\mathbf{r}_0$, and $\sum_i$ by $\int d^3r_0$.

*$\psi(\mathbf{r}_0)$ is therefore the equivalent of $c_i$*: these two complex numbers, which depend either on $\mathbf{r}_0$ or on $i$, represent the components of the same function $\psi(\mathbf{r})$ in two different bases: $\{ \xi_{\mathbf{r}_0}(\mathbf{r}) \}$ and $\{ u_i(\mathbf{r}) \}$.

Formula (A-26) becomes here:

$$\boxed{(\varphi, \psi) = \int d^3r_0 \;\; \varphi^*(\mathbf{r}_0)\; \psi(\mathbf{r}_0)} \tag{A-53}$$

We see that the application of (A-26) to the case of the continuous basis $\{ \xi_{\mathbf{r}_0}(\mathbf{r}) \}$ results in the definition (A-4) of the scalar product.

Finally, note that the $\xi_{\mathbf{r}_0}(\mathbf{r})$ satisfy " orthonormalization " and closure relations of the same type as those for the $v_{\mathbf{p}}(\mathbf{r})$. Thus we have [formula (28) of appendix II]:

$$\boxed{\begin{aligned}\int d^3r_0 \;\; \xi_{\mathbf{r}_0}(\mathbf{r}) \; \xi^*_{\mathbf{r}_0}(\mathbf{r}') &= \int d^3r_0 \;\; \delta(\mathbf{r} - \mathbf{r}_0) \; \delta(\mathbf{r}' - \mathbf{r}_0) \\ &= \delta(\mathbf{r} - \mathbf{r}')\end{aligned}} \tag{A-54}$$

and:

$$\boxed{\begin{aligned}(\xi_{\mathbf{r}_0}, \xi_{\mathbf{r}'_0}) &= \int d^3r \;\; \delta(\mathbf{r} - \mathbf{r}_0) \;\; \delta(\mathbf{r} - \mathbf{r}'_0) \\ &= \delta(\mathbf{r}_0 - \mathbf{r}'_0)\end{aligned}} \tag{A-55}$$

All the formulas established for the discrete basis $\{ u_i(\mathbf{r}) \}$ can thus be generalized for the continuous basis $\{ \xi_{\mathbf{r}_0}(\mathbf{r}) \}$, using the correspondence rules summarized in table (II-2).

$$\boxed{\begin{aligned} i &\longleftrightarrow \mathbf{r}_0 \\ \sum_i &\longleftrightarrow \int d^3r_0 \\ \delta_{ij} &\longleftrightarrow \delta(\mathbf{r}_0 - \mathbf{r}'_0)\end{aligned}}$$

Table (II-2)

IMPORTANT COMMENT:

The usefulness of the continuous bases which we have just introduced is revealed more clearly in what follows. However, we must not lose sight of the following point : *a physical state must always correspond to a square-integrable wave function.* In no case can $v_{\mathbf{p}}(\mathbf{r})$ or $\xi_{\mathbf{r}_0}(\mathbf{r})$ represent the state of a particle. These functions are nothing more than intermediaries, very useful in calculations involving operations on the wave functions $\psi(\mathbf{r})$ which are used to describe a physical state.

An analogous situation is encountered in classical optics, where the plane monochromatic wave is a mathematically very useful, but physically unrealizable, idealization. Even the most selective filters always permit the passage of a frequency band $\Delta\nu$, which may be very small but is never exactly zero.

The same holds true for the functions $\xi_{\mathbf{r}_0}(\mathbf{r})$. We can imagine a square-integrable wave function, localized about $\mathbf{r}_0$, for example:

$$\xi^{(\varepsilon)}_{\mathbf{r}_0}(\mathbf{r}) = \delta^{(\varepsilon)}(\mathbf{r} - \mathbf{r}_0) = \delta^{(\varepsilon)}(x - x_0)\delta^{(\varepsilon)}(y - y_0)\delta^{(\varepsilon)}(z - z_0)$$

where the $\delta^{(\varepsilon)}$ are functions which have a peak of width $\varepsilon$ and amplitude $\frac{1}{\varepsilon}$, centered at $x_0$, $y_0$ or $z_0$, such that $\int_{-\infty}^{+\infty} \delta^{(\varepsilon)}(x - x_0)\, dx = 1$ (see § 1-b of appendix II for examples of such functions). When $\varepsilon \longrightarrow 0$, $\xi_{\mathbf{r}_0}^{(\varepsilon)}(\mathbf{r}) \longrightarrow \xi_{\mathbf{r}_0}(\mathbf{r})$, which is no longer square-integrable. But, in fact, it is impossible to have a physical state which corresponds to this limit : as localized as the physical state of a particle may be, $\varepsilon$ is never exactly zero.

### c. GENERALIZATION : CONTINUOUS "ORTHONORMAL" BASES

#### α. *Definition*

Generalizing the results obtained in the two preceding paragraphs, we shall call a *continuous "orthonormal" basis*, a set of functions of $\mathbf{r}$, $\{ w_\alpha(\mathbf{r}) \}$, labeled by a continuous index $\alpha$, which satisfy the two following relations, called *orthonormalization* and *closure* relations:

$$(w_\alpha, w_{\alpha'}) = \int d^3r \; w_\alpha^*(\mathbf{r}) \; w_{\alpha'}(\mathbf{r}) = \delta(\alpha - \alpha') \tag{A-56}$$

$$\int d\alpha \; w_\alpha(\mathbf{r}) \; w_\alpha^*(\mathbf{r}') = \delta(\mathbf{r} - \mathbf{r}') \tag{A-57}$$

COMMENTS:

(*i*) If $\alpha = \alpha'$, $(w_\alpha, w_\alpha)$ diverges. Therefore, $w_\alpha(\mathbf{r}) \notin \mathscr{F}$.

(*ii*) $\alpha$ can represent several indices, as is the case for $\mathbf{r}_0$ and $\mathbf{p}$ in the above examples.

(*iii*) It is possible to imagine a basis which includes both functions $u_i(\mathbf{r})$, labeled by a discrete index, and functions $w_\alpha(\mathbf{r})$, labeled by a continuous index. In this case, the set of $u_i(\mathbf{r})$ does not form a basis; the set of $w_\alpha(\mathbf{r})$ must be added to it.

Let us cite an example of this situation. Consider the case of the square well studied in § D-2-c of chapter I (see also complement $\text{H}_\text{I}$). As we shall see later, the set of stationary states of a particle in a time-independent potential constitutes a basis. For $E < 0$, we have discrete energy levels, to which correspond square-integrable wave functions labeled by a discrete index. But these are not the only possible stationary states. Equation (D-17) of chapter I is also satisfied, for all $E > 0$, by solutions which are bounded but which extend over all space and are thus not square-integrable.

In the case of a " mixed " (discrete and continuous) basis, $\{ u_i(\mathbf{r}), w_\alpha(\mathbf{r}) \}$, the orthonormalization relations are:

$$\begin{aligned} (u_i, u_j) &= \delta_{ij} \\ (w_\alpha, w_{\alpha'}) &= \delta(\alpha - \alpha') \\ (u_i, w_\alpha) &= 0 \end{aligned} \tag{A-58}$$

And the closure relation becomes:

$$\sum_i u_i(\mathbf{r}) \; u_i^*(\mathbf{r}') + \int d\alpha \; w_\alpha(\mathbf{r}) \; w_\alpha^*(\mathbf{r}') = \delta(\mathbf{r} - \mathbf{r}') \tag{A-59}$$

β. *Components of a wave function* $\psi(\mathbf{r})$

We can always write :

$$\psi(\mathbf{r}) = \int d^3r' \; \psi(\mathbf{r}') \; \delta(\mathbf{r} - \mathbf{r}') \tag{A-60}$$

Using the expression for $\delta(\mathbf{r} - \mathbf{r}')$ given by (A-57), and assuming that we can reverse the order of $\int d^3r'$ and $\int d\alpha$, we obtain :

$$\psi(\mathbf{r}) = \int d\alpha \left[ \int d^3r' \; w_\alpha^*(\mathbf{r}') \; \psi(\mathbf{r}') \right] w_\alpha(\mathbf{r})$$

that is :

$$\boxed{\psi(\mathbf{r}) = \int d\alpha \; c(\alpha) \; w_\alpha(\mathbf{r})} \tag{A-61}$$

with

$$\boxed{c(\alpha) = (w_\alpha, \psi) = \int d^3r' \; w_\alpha^*(\mathbf{r}') \; \psi(\mathbf{r}')} \tag{A-62}$$

(A-61) expresses the fact that every wave function $\psi(\mathbf{r})$ has a unique expansion in terms of the $w_\alpha(\mathbf{r})$. The component $c(\alpha)$ of $\psi(\mathbf{r})$ on $w_\alpha(\mathbf{r})$ is equal, according to (A-62), to the scalar product $(w_\alpha, \psi)$.

γ. *Expression for the scalar product and the norm in terms of the components*

Let $\varphi(\mathbf{r})$ and $\psi(\mathbf{r})$ be two square-integrable functions whose components in terms of the $w_\alpha(\mathbf{r})$ are known:

$$\varphi(\mathbf{r}) = \int d\alpha \; b(\alpha) \; w_\alpha(\mathbf{r}) \tag{A-63}$$

$$\psi(\mathbf{r}) = \int d\alpha' \; c(\alpha') \; w_{\alpha'}(\mathbf{r}) \tag{A-64}$$

Calculate their scalar product:

$$\begin{aligned} (\varphi, \psi) &= \int d^3r \; \varphi^*(\mathbf{r}) \; \psi(\mathbf{r}) \\ &= \int d\alpha \int d\alpha' \; b^*(\alpha) \; c(\alpha') \int d^3r \; w_\alpha^*(\mathbf{r}) \; w_{\alpha'}(\mathbf{r}) \end{aligned} \tag{A-65}$$

The last integral is given by (A-56):

$$(\varphi, \psi) = \int d\alpha \int d\alpha' \; b^*(\alpha) \; c(\alpha') \; \delta(\alpha - \alpha')$$

that is:

$$\boxed{(\varphi, \psi) = \int d\alpha \; b^*(\alpha) \; c(\alpha)} \tag{A-66}$$

In particular:

$$(\psi, \psi) = \int \mathrm{d}\alpha \, |c(\alpha)|^2 \tag{A-67}$$

All the formulas of § A-2 can thus be generalized, using the correspondence rules of table (II-3).

$$i \longleftrightarrow \alpha$$
$$\sum_i \longleftrightarrow \int \mathrm{d}\alpha$$
$$\delta_{ij} \longleftrightarrow \delta(\alpha - \alpha')$$

Table (II-3)

The most important formulas established in this section are assembled in table (II-4). In fact, it is not necessary to remember them in this form: we shall see that the introduction of Dirac notation enables us to rederive them very simply.

Table (II-4)

| | Discrete basis $\{ u_i(\mathbf{r}) \}$ | Continuous basis $\{ w_\alpha(\mathbf{r}) \}$ |
|---|---|---|
| Ortho-normalization relation | $(u_i, u_j) = \delta_{ij}$ | $(w_\alpha, w_{\alpha'}) = \delta(\alpha - \alpha')$ |
| Closure relation | $\sum_i u_i(\mathbf{r}) \, u_i^*(\mathbf{r}') = \delta(\mathbf{r} - \mathbf{r}')$ | $\int \mathrm{d}\alpha \; w_\alpha(\mathbf{r}) \, w_\alpha^*(\mathbf{r}') = \delta(\mathbf{r} - \mathbf{r}')$ |
| Expansion of a wave function $\psi(\mathbf{r})$ | $\psi(\mathbf{r}) = \sum_i c_i \, u_i(\mathbf{r})$ | $\psi(\mathbf{r}) = \int \mathrm{d}\alpha \; c(\alpha) \, w_\alpha(\mathbf{r})$ |
| Expression for the components of $\psi(\mathbf{r})$ | $c_i = (u_i, \psi) = \int d^3r \; u_i^*(\mathbf{r}) \, \psi(\mathbf{r})$ | $c(\alpha) = (w_\alpha, \psi) = \int d^3r \; w_\alpha^*(\mathbf{r}) \, \psi(\mathbf{r})$ |
| Scalar product | $(\varphi, \psi) = \sum_i b_i^* c_i$ | $(\varphi, \psi) = \int \mathrm{d}\alpha \; b^*(\alpha) \, c(\alpha)$ |
| Square of the norm | $(\psi, \psi) = \sum_i \lvert c_i \rvert^2$ | $(\psi, \psi) = \int \mathrm{d}\alpha \, \lvert c(\alpha) \rvert^2$ |

## B. STATE SPACE. DIRAC NOTATION

### 1. Introduction

In chapter I, we stated the following postulate : the quantum state of a particle is defined, at a given instant, by a wave function $\psi(\mathbf{r})$. The probabilistic interpretation of this wave function requires that it be square-integrable. This requirement led us to study the $\mathscr{F}$-space (§ A). We then found, in particular, that the same function $\psi(\mathbf{r})$ can be represented by several distinct sets of components, each one corresponding to the choice of a basis [table (II-5)]. This result can be interpreted in the following manner : $\{ c_i \}$, or $\bar{\psi}(\mathbf{p})$, or $c(\alpha)$, characterizes the state of a particle just as well as the wave function $\psi(\mathbf{r})$ [if the basis being used has been specified previously]. Furthermore, $\psi(\mathbf{r})$ itself appears, in table (II-5), on the same footing as $\{ c_i \}$, $\bar{\psi}(\mathbf{p})$ and $c(\alpha)$: the value $\psi(\mathbf{r}_0)$ which the wave function takes on at a point $\mathbf{r}_0$ of space can be considered as its component with respect to a specific function $\xi_{\mathbf{r}_0}(\mathbf{r})$ of a particular basis (the $\delta$ function basis).

| *Basis* | *Components of* $\psi(\mathbf{r})$ |
|---|---|
| $u_i(\mathbf{r})$ | $c_i$, $i = 1, 2, ..., n, ...$ |
| $v_{\mathbf{p}}(\mathbf{r})$ | $\bar{\psi}(\mathbf{p})$ |
| $\xi_{\mathbf{r}_0}(\mathbf{r})$ | $\psi(\mathbf{r}_0)$ |
| $w_\alpha(\mathbf{r})$ | $c(\alpha)$ |

Table (II-5)

We thus find ourselves in a situation which is analogous to the one encountered in ordinary space, $R^3$ : the position of a point in space can be described by a set of three numbers, which are its coordinates with respect to a system of axes defined in advance. If one changes axes, another set of coordinates corresponds to the same point. But the geometrical vector concept and vector calculation enable us to avoid referring to a system of axes; this considerably simplifies both formulas and reasoning.

We are going to use a similar approach here : each quantum state of a particle will be characterized by a *state vector*, belonging to an abstract space, $\mathscr{E}_{\mathbf{r}}$, called the *state space* of a particle. The fact that the space $\mathscr{F}$ is a subspace of $L^2$ means that $\mathscr{E}_{\mathbf{r}}$ is a subspace of a Hilbert space. We are going to define the notation and the rules of vector calculation in $\mathscr{E}_{\mathbf{r}}$.

Actually, the introduction of state vectors and the state space does more than merely simplify the formalism. It also permits a generalization of the formalism. Indeed, there exist physical systems whose quantum description cannot be given by a wave function : we shall see in chapters IV and IX that this is the case when the spin degrees of freedom are taken into account, even for a single particle. Consequently, the first postulate that we shall set forth in chapter III will be the following: *the quantum state of any physical system is characterized by a state vector, belonging to a space $\mathscr{E}$ which is the state space of the system.*

Therefore, in the rest of this chapter, we are going to develop a vector calculus in $\mathscr{E}$. The concepts which we are going to introduce and the results which we shall obtain are valid for whatever physical system we might consider. Nevertheless, to illustrate these concepts and results, we shall apply them to the simple case of a (spinless) particle, since this is the case we have previously considered.

We shall begin, in this paragraph, by defining the *Dirac notation*, which will prove to be very useful in the formal manipulations which we shall have to perform.

## 2. "Ket" vectors and "bra" vectors

### a. ELEMENTS OF $\mathscr{E}$: KETS

#### α. *Notation*

Any element, or vector, of $\mathscr{E}$-space is called a *ket vector*, or, more simply, a *ket*. It is represented by the symbol $| \ \rangle$, inside which is placed a distinctive sign which enables us to distinguish the corresponding ket from all others, for example: $| \psi \rangle$.

In particular, since the concept of a wave function is now familiar to us, we shall define the space $\mathscr{E}_{\mathbf{r}}$ of the states of a particle by associating with every square-integrable function $\psi(\mathbf{r})$ a ket vector $| \psi \rangle$ of $\mathscr{E}_{\mathbf{r}}$:

$$\psi(\mathbf{r}) \in \mathscr{F} \iff | \psi \rangle \in \mathscr{E}_{\mathbf{r}} \tag{B-1}$$

Afterwards, we shall transpose the different operations which we introduced for $\mathscr{F}$ into $\mathscr{E}_{\mathbf{r}}$. Although $\mathscr{F}$ and $\mathscr{E}_{\mathbf{r}}$ are isomorphic, we shall carefully distinguish between them in order to avoid confusion and to reserve the possibilities of generalization mentioned above in § B-1. We stress the fact that an $\mathbf{r}$-dependence no longer appears in $| \psi \rangle$; only the letter $\psi$ appears to remind us with which function it is associated. $\psi(\mathbf{r})$ will be interpreted (§ E) as the set of the components of the ket $| \psi \rangle$ in a particular basis, $\mathbf{r}$ playing the role of an index [*cf.* § A-3-b and table (II-5)]. Consequently, the procedure which we are adopting here consists in initially characterizing a vector by its components in a privileged coordinate system, which will later be treated on the same footing as all other coordinate systems.

We shall designate by $\mathscr{E}_x$ the state space of a (spinless) particle in only one dimension, that is, the abstract space constructed as in (B-1), but using wave functions which depend only on the $x$ variable.

#### β. *Scalar product*

With each pair of kets $| \varphi \rangle$ and $| \psi \rangle$, taken in this order, we associate a complex number, which is their scalar product, $(| \varphi \rangle, | \psi \rangle)$, and which satisfies the various properties described by equations (A-5), (A-6) and (A-7). We shall later rewrite these formulas in Dirac notation after we have introduced the concept of a "bra".

In $\mathscr{E}_{\mathbf{r}}$, the scalar product of two kets will coincide with the scalar product defined above for the associated wave functions.

#### b. ELEMENTS OF THE DUAL SPACE $\mathscr{E}^*$ OF $\mathscr{E}$: BRAS

##### α. *Definition of the dual space $\mathscr{E}^*$*

Recall, first of all, the definition of a *linear functional* defined on the kets $| \psi \rangle$ of $\mathscr{E}$. A linear functional $\chi$ is a linear operation which associates a complex number with every ket $| \psi \rangle$:

$$
\begin{aligned}
&| \psi \rangle \in \mathscr{E} \xrightarrow{\chi} \text{number } \chi(| \psi \rangle) \\
&\chi(\lambda_1 | \psi_1 \rangle + \lambda_2 | \psi_2 \rangle) = \lambda_1 \chi(| \psi_1 \rangle) + \lambda_2 \chi(| \psi_2 \rangle)
\end{aligned}
\tag{B-2}
$$

Linear functional and linear operator must not be confused. In both cases, one is dealing with linear operations, but the former associates each ket with a complex number, while the latter associates another ket.

It can be shown that the set of linear functionals defined on the kets $| \psi \rangle \in \mathscr{E}$ constitutes a vector space, which is called the *dual space* of $\mathscr{E}$ and which will be symbolized by $\mathscr{E}^*$.

##### β. *Bra notation for the vectors of $\mathscr{E}^*$*

Any element, or vector, of the space $\mathscr{E}^*$ is called a *bra vector*, or, more simply, a *bra*. It is symbolized by $\langle \, |$. For example, the bra $\langle \chi |$ designates the linear functional $\chi$ and we shall henceforth use the notation $\langle \chi | \psi \rangle$ to denote the *number* obtained by causing the linear functional $\langle \chi | \in \mathscr{E}^*$ to act on the ket $| \psi \rangle \in \mathscr{E}$:

$$
\boxed{\chi(| \psi \rangle) = \langle \chi | \psi \rangle}
\tag{B-3}
$$

The origin of this terminology is the word "bracket", used to denote the symbol $\langle | \rangle$. Hence the name "bra" for the left-hand side, and the name "ket" for the right-hand side of this symbol.

#### c. CORRESPONDENCE BETWEEN KETS AND BRAS

##### α. *To every ket corresponds a bra*

The existence of a scalar product in $\mathscr{E}$ will now enable us to show that we can associate, with every ket $| \varphi \rangle \in \mathscr{E}$, an element of $\mathscr{E}^*$, that is, a bra, which will be denoted by $\langle \varphi |$.

The ket $| \varphi \rangle$ does indeed enable us to define a linear functional: the one which associates (in a linear way), with each ket $| \psi \rangle \in \mathscr{E}$, a complex number which is equal to the scalar product $(| \varphi \rangle, | \psi \rangle)$ of $| \psi \rangle$ by $| \varphi \rangle$. Let $\langle \varphi |$ be this linear functional; it is thus defined by the relation:

$$
\boxed{\langle \varphi | \psi \rangle = (| \varphi \rangle, | \psi \rangle)}
\tag{B-4}
$$

β. *This correspondance is antilinear*

In the space $\mathscr{E}$, the scalar product is antilinear with respect to the first vector. In the notation of (B-4), this is expressed by:

$$\begin{aligned}(\lambda_1 | \varphi_1 \rangle + \lambda_2 | \varphi_2 \rangle, | \psi \rangle) &= \lambda_1^*(| \varphi_1 \rangle, | \psi \rangle) + \lambda_2^*(| \varphi_2 \rangle, | \psi \rangle) \\ &= \lambda_1^* \langle \varphi_1 | \psi \rangle + \lambda_2^* \langle \varphi_2 | \psi \rangle \\ &= (\lambda_1^* \langle \varphi_1 | + \lambda_2^* \langle \varphi_2 |) | \psi \rangle \qquad \text{(B-5)}\end{aligned}$$

It appears from (B-5) that the bra associated with the ket $\lambda_1 | \varphi_1 \rangle + \lambda_2 | \varphi_2 \rangle$ is the bra $\lambda_1^* \langle \varphi_1 | + \lambda_2^* \langle \varphi_2 |$:

$$\boxed{\lambda_1 | \varphi_1 \rangle + \lambda_2 | \varphi_2 \rangle \Longrightarrow \lambda_1^* \langle \varphi_1 | + \lambda_2^* \langle \varphi_2 |} \qquad \text{(B-6)}$$

The ket $\Longrightarrow$ bra correspondence is therefore antilinear.

COMMENT :

If $\lambda$ is a complex number, and $| \psi \rangle$ a ket, $\lambda | \psi \rangle$ is a ket ($\mathscr{E}$ is a vector space). We are sometimes led to write it as $| \lambda\psi \rangle$ :

$$| \lambda \psi \rangle = \lambda | \psi \rangle \qquad \text{(B-7)}$$

One must then be careful to remember that $\langle \lambda\psi |$ *represents the bra* associated with the ket $| \lambda\psi \rangle$. Since the correspondence between a bra and a ket is antilinear, we have :

$$\langle \lambda \psi | = \lambda^* \langle \psi | \qquad \text{(B-8)}$$

γ. *Dirac notation for the scalar product*

We now have at our disposal two distinct notations for designating the scalar product of $| \psi \rangle$ by $| \varphi \rangle$: $(| \varphi \rangle, | \psi \rangle)$ or $\langle \varphi | \psi \rangle$, $\langle \varphi |$ being the bra associated with the ket $| \varphi \rangle$. Henceforth we shall use only the (Dirac) notation : $\langle \varphi | \psi \rangle$. Table (II-6) summarizes, in Dirac notation, the properties of the scalar product, already given in § A-1-b.

$$\langle \varphi | \psi \rangle = \langle \psi | \varphi \rangle^* \qquad \text{(B-9)}$$

$$\langle \varphi | \lambda_1\psi_1 + \lambda_2\psi_2 \rangle = \lambda_1 \langle \varphi | \psi_1 \rangle + \lambda_2 \langle \varphi | \psi_2 \rangle \qquad \text{(B-10)}$$

$$\langle \lambda_1\varphi_1 + \lambda_2\varphi_2 | \psi \rangle = \lambda_1^* \langle \varphi_1 | \psi \rangle + \lambda_2^* \langle \varphi_2 | \psi \rangle \qquad \text{(B-11)}$$

$$\langle \psi | \psi \rangle \text{ real, positive; zero if and only if } | \psi \rangle = 0 \qquad \text{(B-12)}$$

δ. *Is there a ket to correspond to every bra ?*

Although to every ket there corresponds a bra, we shall see in two examples chosen in $\mathscr{F}$, that it is possible to find bras which have no corresponding kets. We shall later show why this difficulty does not hinder us in quantum mechanics.

(*i*) Counter-examples chosen in $\mathscr{F}$

For simplicity, we shall reason in one dimension.

Let $\xi_{x_0}^{(\varepsilon)}(x)$ be a sufficiently regular real function, such that $\int_{-\infty}^{+\infty} dx\, \xi_{x_0}^{(\varepsilon)}(x) = 1$, and having the form of a peak of width $\varepsilon$ and amplitude $1/\varepsilon$, centered at $x = x_0$ [see fig. 1; $\xi_{x_0}^{(\varepsilon)}(x)$ is, for example, one of the functions considered in § 1-b of appendix II]. If $\varepsilon \neq 0$, $\xi_{x_0}^{(\varepsilon)}(x) \in \mathscr{F}_x$ (the square of its norm is of the order of $1/\varepsilon$). Denote by $| \xi_{x_0}^{(\varepsilon)} \rangle$ the corresponding ket:

$$\xi_{x_0}^{(\varepsilon)}(x) \Longleftrightarrow | \xi_{x_0}^{(\varepsilon)} \rangle \tag{B-13}$$

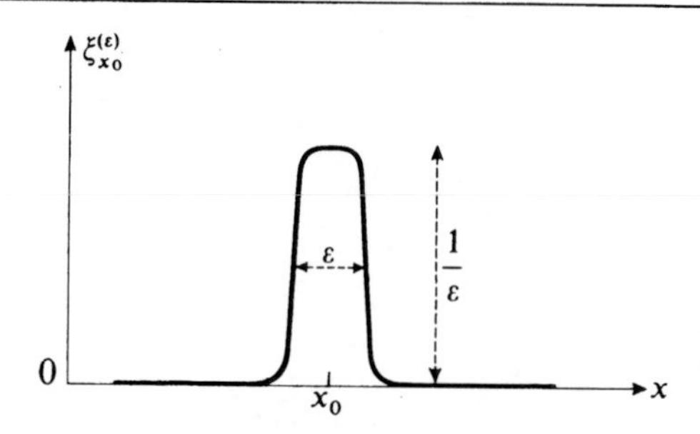


FIGURE 1

**$\xi_{x_0}^{(\varepsilon)}(x)$ is a function which has a peak at $x = x_0$ (of width $\varepsilon$ and amplitude $1/\varepsilon$), whose integral between $-\infty$ and $+\infty$ is equal to 1.**

If $\varepsilon \neq 0$, $| \xi_{x_0}^{(\varepsilon)} \rangle \in \mathscr{E}_x$. Let $\langle \xi_{x_0}^{(\varepsilon)} |$ be the bra associated with this ket; for every $| \psi \rangle \in \mathscr{E}_x$, we have:

$$\langle \xi_{x_0}^{(\varepsilon)} | \psi \rangle = (\xi_{x_0}^{(\varepsilon)}, \psi) = \int_{-\infty}^{+\infty} dx\ \xi_{x_0}^{(\varepsilon)}(x)\ \psi(x) \tag{B-14}$$

Now let $\varepsilon$ approach zero. On the one hand:

$$\lim_{\varepsilon \to 0} \xi_{x_0}^{(\varepsilon)}(x) = \xi_{x_0}(x) \notin \mathscr{F}_x \tag{B-15}$$

[the square of the norm of $\xi_{x_0}^{(\varepsilon)}(x)$, which is of the order of $1/\varepsilon$, diverges when $\varepsilon \longrightarrow 0$]; therefore:

$$\lim_{\varepsilon \to 0} | \xi_{x_0}^{(\varepsilon)} \rangle \notin \mathscr{E}_x \tag{B-16}$$

On the other hand, when $\varepsilon \longrightarrow 0$, integral (B-14) approaches a perfectly well-defined limit, $\psi(x_0)$ [since, for sufficiently small $\varepsilon$, $\psi(x)$ can be replaced in (B-14) by $\psi(x_0)$ and removed from the integral]. Consequently, $\langle \xi_{x_0}^{(\varepsilon)} |$ approaches a bra which we shall denote by $\langle \xi_{x_0} |$: $\langle \xi_{x_0} |$ is the linear functional which associates, with every ket $| \psi \rangle$ of $\mathscr{E}_x$, the value $\psi(x_0)$ taken on by the associated wave function at the point $x_0$:

$$\begin{gathered} \lim_{\varepsilon \to 0} \langle \xi_{x_0}^{(\varepsilon)} | = \langle \xi_{x_0} | \in \mathscr{E}_x^* \\ \text{If } | \psi \rangle \in \mathscr{E}_x,\ \langle \xi_{x_0} | \psi \rangle = \psi(x_0) \end{gathered} \tag{B-17}$$

Thus we see that *the bra $\langle \xi_{x_0} |$ exists, but no ket corresponds to it.*

In the same way, let us consider a plane wave which is truncated outside an interval of width $L$:

$$v_{p_0}^{(L)}(x) = \frac{1}{\sqrt{2\pi\hbar}} e^{ip_0x/\hbar} \quad \text{if } -\frac{L}{2} \leqslant x \leqslant +\frac{L}{2} \tag{B-18}$$

with the function $v_{p_0}^{(L)}(x)$ going rapidly to zero outside this interval (while remaining continuous and differentiable). We shall denote by $| v_{p_0}^{(L)} \rangle$ the ket associated with $v_{p_0}^{(L)}(x)$:

$$v_{p_0}^{(L)}(x) \in \mathscr{F}_x \Longleftrightarrow | v_{p_0}^{(L)} \rangle \in \mathscr{E}_x \tag{B-19}$$

The square of the norm of $v_{p_0}^{(L)}$, which is practically equal to $L/2\pi\hbar$, diverges if $L \longrightarrow \infty$. Therefore:

$$\operatorname*{Lim}_{L\to\infty} | v_{p_0}^{(L)} \rangle \notin \mathscr{E}_x \tag{B-20}$$

Now let us consider the bra $\langle v_{p_0}^{(L)} |$ associated with $| v_{p_0}^{(L)} \rangle$. For every $| \psi \rangle \in \mathscr{E}_x$, we have:

$$\langle v_{p_0}^{(L)} | \psi \rangle = (v_{p_0}^{(L)}, \psi) \simeq \frac{1}{\sqrt{2\pi\hbar}} \int_{-L/2}^{L/2} \mathrm{d}x \, \mathrm{e}^{-ip_0x/\hbar} \psi(x) \tag{B-21}$$

When $L \longrightarrow \infty$, $\langle v_{p_0}^{(L)} | \psi \rangle$ has a limit: the value $\bar{\psi}(p_0)$ of the Fourier transform $\bar{\psi}(p)$ of $\psi(x)$ for $p = p_0$. Therefore, when $L \longrightarrow \infty$, $\langle v_{p_0}^{(L)} |$ tends towards a perfectly, well-defined bra $\langle v_{p_0} |$:

$$\operatorname*{Lim}_{L\to\infty} \langle v_{p_0}^{(L)} | = \langle v_{p_0} | \in \mathscr{E}_x^*$$

$$\text{If } | \psi \rangle \in \mathscr{E}_x, \langle v_{p_0} | \psi \rangle = \bar{\psi}(p_0) \tag{B-22}$$

Here again, *no ket corresponds to the bra* $\langle v_{p_0} |$.

#### (*ii*) Physical resolution of the preceding difficulties

This dissymmetry of the correspondence between kets and bras is related, as the preceding examples show, to the existence of "continuous bases" for $\mathscr{F}_x$. Since the functions constituting these "bases" do not belong to $\mathscr{F}_x$, we cannot associate a ket of $\mathscr{E}_x$ with them. However, their scalar product with an arbitrary function of $\mathscr{F}_x$ is defined, and this permits us to associate with them a linear functional in $\mathscr{E}_x$, that is, a bra belonging to $\mathscr{E}_x^*$. The reason for using such "continuous bases" lies in their usefulness in certain practical calculations. The same reason (which will become more apparent in what follows) leads us here to reestablish the symmetry between kets and bras by introducing "generalized kets", defined using functions which are not square-integrable, but whose scalar product with every function of $\mathscr{F}_x$ exists. In what follows, we shall work with "kets" such as $| \xi_{x_0} \rangle$ or $| v_{p_0} \rangle$, associated with $\xi_{x_0}(x)$ or $v_{p_0}(x)$. It must not be forgotten that these generalized "kets" cannot, strictly speaking, represent physical states. They are merely intermediaries, useful in calculations involving certain operations which we shall have to perform on the true kets of the space $\mathscr{E}_x$, which actually characterize realizable quantum states.

This method poses a certain number of mathematical problems, which can be avoided by adopting the following physical point of view: $| \xi_{x_0} \rangle$ (or $| v_{p_0} \rangle$) actually denotes $| \xi_{x_0}^{(\varepsilon)} \rangle$ (or $| v_{p_0}^{(L)} \rangle$) where $\varepsilon$ is very small (or $L$ is very large) compared to all the other lengths in the problem we are considering. In all the intermediary calculations where $| \xi_{x_0}^{(\varepsilon)} \rangle$ (or $| v_{p_0}^{(L)} \rangle$) appears, the limit $\varepsilon = 0$ (or $L \longrightarrow \infty$) is never attained, so that one is always working in $\mathscr{E}_x$. The physical result obtained at the end of the calculation depends very little on the value of $\varepsilon$, as long as $\varepsilon$ is sufficiently small with respect to all the other lengths: it is then possible to neglect $\varepsilon$, that is, to set $\varepsilon = 0$, in the final result (the procedure to be used for $L$ is analogous).

The objection could be raised that, unlike $\{ \xi_{x_0}(x) \}$ and $\{ v_{p_0}(x) \}$, $\{ \xi_{x_0}^{(\varepsilon)}(x) \}$ and $\{ v_{p_0}^{(L)}(x) \}$ are not orthonormal bases, insofar as they do not rigorously satisfy the closure relation. In fact, they fulfill it approximately. For example, the expression $\int \mathrm{d}x_0 \, \xi_{x_0}^{(\varepsilon)}(x) \xi_{x_0}^{(\varepsilon)}(x')$ is a function of $(x - x')$ which can serve as an excellent approximation for $\delta(x - x')$. Its graphical representation is practically a triangle of base $2\varepsilon$ and height $\dfrac{1}{\varepsilon}$, centered at $x - x' = 0$ (appendix II, § 1-c-iv). If $\varepsilon$ is negligible compared to all the other lengths in the problem, the difference between this expression and $\delta(x - x')$ is physically inappreciable.

In general, the dual space $\mathscr{E}^*$ and the state space $\mathscr{E}$ are not isomorphic, except, of course, if $\mathscr{E}$ is finite-dimensional★: although to each ket $|\psi\rangle$ of $\mathscr{E}$ there corresponds a bra $\langle\psi|$ in $\mathscr{E}^*$, the converse is not true. Nevertheless, we shall agree to use, in addition to vectors belonging to $\mathscr{E}$ (whose norm is finite), *generalized kets* with infinite norms but whose scalar product with every ket of $\mathscr{E}$ is finite. Thus, to each bra $\langle\varphi|$ of $\mathscr{E}^*$, there will correspond a ket. But generalized kets do not represent physical states of the system.

## 3. Linear operators

### a. DEFINITIONS

These are the same as those of § A-1-c.

A linear operator $A$ associates with every ket $|\psi\rangle \in \mathscr{E}$ another ket $|\psi'\rangle \in \mathscr{E}$, the correspondence being linear:

$$|\psi'\rangle = A|\psi\rangle \tag{B-23}$$

$$A(\lambda_1|\psi_1\rangle + \lambda_2|\psi_2\rangle) = \lambda_1 A|\psi_1\rangle + \lambda_2 A|\psi_2\rangle \tag{B-24}$$

The product of two linear operators $A$ and $B$, written $AB$, is defined in the following way:

$$(AB)|\psi\rangle = A(B|\psi\rangle) \tag{B-25}$$

$B$ first acts on $|\psi\rangle$ to give the ket $B|\psi\rangle$; $A$ then acts on the ket $B|\psi\rangle$. In general, $AB \neq BA$. The commutator $[A, B]$ of $A$ and $B$ is, by definition:

$$[A, B] = AB - BA \tag{B-26}$$

Let $|\varphi\rangle$ and $|\psi\rangle$ be two kets. We call the *matrix element* of $A$ between $|\varphi\rangle$ and $|\psi\rangle$, the scalar product:

$$\langle\varphi|(A|\psi\rangle) \tag{B-27}$$

Consequently, this is *a number* which depends linearly on $|\psi\rangle$ and antilinearly on $|\varphi\rangle$.

### b. EXAMPLES OF LINEAR OPERATORS: PROJECTORS

#### α. *Important comment about Dirac notation*

We have begun to sense, in the preceding, the simplicity and convenience of the Dirac formalism. For example, $\langle\varphi|$ denotes a linear functional (a bra), and $\langle\psi_1|\psi_2\rangle$, the scalar product of two kets $|\psi_1\rangle$ and $|\psi_2\rangle$. The number associated by the linear functional $\langle\varphi|$ with an arbitrary ket $|\psi\rangle$ is then written simply by juxtaposing the symbols $\langle\varphi|$ and $|\psi\rangle$: $\langle\varphi|\psi\rangle$. This is the scalar product of $|\psi\rangle$ by the ket $|\varphi\rangle$ corresponding to $\langle\varphi|$ (which is why it is useful to have a one-to-one correspondence between kets and bras).

★ It is true that the Hilbert space $L^2$ and its dual space are isomorphic; however, we have taken for the wave function space $\mathscr{F}$ a subspace of $L^2$, which explains why $\mathscr{F}^*$ is "larger" than $\mathscr{F}$.

Now assume that we write $\langle \varphi |$ and $| \psi \rangle$ in the opposite order:

$$| \psi \rangle \langle \varphi | \tag{B-28}$$

We shall see that if we abide by the rule of juxtaposition of symbols, this expression represents an operator. Choose an arbitrary ket $| \chi \rangle$ and consider:

$$| \psi \rangle \langle \varphi | \chi \rangle \tag{B-29}$$

We already know that $\langle \varphi | \chi \rangle$ is a complex number; consequently, (B-29) is a ket, obtained by multiplying $| \psi \rangle$ by the scalar $\langle \varphi | \chi \rangle$. $| \psi \rangle \langle \varphi |$, applied to an arbitrary ket, gives another ket: it is an operator.

Thus we see that *the order of the symbols is of critical importance.* Only complex numbers can be moved about with impunity, because of the linearity of the space $\mathscr{E}$ and of the operators which we shall use. Indeed, if $\lambda$ is a number:

$$\begin{cases} | \psi \rangle \lambda = \lambda | \psi \rangle \\ \langle \psi | \lambda = \lambda \langle \psi | \\ A\lambda | \psi \rangle = \lambda A | \psi \rangle \qquad \text{(where } A \text{ is a linear operator)} \\ \langle \varphi | \lambda | \psi \rangle = \lambda \langle \varphi | \psi \rangle = \langle \varphi | \psi \rangle \lambda \end{cases} \tag{B-30}$$

But, for kets, bras and operators, the order must always be carefully respected in writing the formulas: this is the price that must be paid for the simplicity of the Dirac formalism.

β. *The projector $P_\psi$ onto a ket $| \psi \rangle$*

Let $| \psi \rangle$ be a ket which is normalized to one:

$$\langle \psi | \psi \rangle = 1 \tag{B-31}$$

Consider the operator $P_\psi$ defined by:

$$P_\psi = | \psi \rangle \langle \psi | \tag{B-32}$$

and apply it to an arbitrary ket $| \varphi \rangle$:

$$P_\psi | \varphi \rangle = | \psi \rangle \langle \psi | \varphi \rangle \tag{B-33}$$

$P_\psi$, acting on an arbitrary ket $| \varphi \rangle$, gives a ket proportional to $| \psi \rangle$. The coefficient of proportionality $\langle \psi | \varphi \rangle$ is the scalar product of $| \varphi \rangle$ by $| \psi \rangle$.

The "geometrical" significance of $P_\psi$ is therefore clear: it is the "orthogonal projection" operator onto the ket $| \psi \rangle$.

This interpretation is confirmed by the fact that $P_\psi^2 = P_\psi$ (projecting twice in succession onto a given vector is equivalent to projecting a single time). To see this, we write:

$$P_\psi^2 = P_\psi P_\psi = | \psi \rangle \langle \psi | \psi \rangle \langle \psi | \tag{B-34}$$

In this expression, $\langle \psi | \psi \rangle$ is a number, which is equal to 1 [formula (B-31)]. Therefore:

$$P_\psi^2 = | \psi \rangle \langle \psi | = P_\psi \tag{B-35}$$

γ. *Projector onto a subspace*

Let $| \varphi_1 \rangle, | \varphi_2 \rangle, ..., | \varphi_q \rangle$, be $q$ normalized vectors which are orthogonal to each other:

$$\langle \varphi_i | \varphi_j \rangle = \delta_{ij} \quad ; \quad i, j = 1, 2, ..., q \tag{B-36}$$

We denote by $\mathscr{E}_q$ the subspace of $\mathscr{E}$ spanned by these $q$ vectors.

Let $P_q$ be the linear operator defined by:

$$P_q = \sum_{i=1}^{q} | \varphi_i \rangle \langle \varphi_i | \tag{B-37}$$

Calculating $P_q^2$:

$$P_q^2 = \sum_{i=1}^{q} \sum_{j=1}^{q} | \varphi_i \rangle \langle \varphi_i | \varphi_j \rangle \langle \varphi_j | \tag{B-38}$$

we get, using (B-36):

$$P_q^2 = \sum_{i=1}^{q} \sum_{j=1}^{q} | \varphi_i \rangle \langle \varphi_j | \delta_{ij} = \sum_{i=1}^{q} | \varphi_i \rangle \langle \varphi_i | = P_q \tag{B-39}$$

$P_q$ is therefore a projector. It is easy to see that $P_q$ projects onto the subspace $\mathscr{E}_q$, since for any $| \psi \rangle \in \mathscr{E}$:

$$P_q | \psi \rangle = \sum_{i=1}^{q} | \varphi_i \rangle \langle \varphi_i | \psi \rangle \tag{B-40}$$

$P_q$ acting on $| \psi \rangle$ gives the linear superposition of the projections of $| \psi \rangle$ onto the various $| \varphi_i \rangle$, that is, the projection of $| \psi \rangle$ onto the subspace $\mathscr{E}_q$.

## 4. Hermitian conjugation

### a. ACTION OF A LINEAR OPERATOR ON A BRA

Until now, we have only defined the action of a linear operator $A$ on kets. We are now going to see that it is also possible to define the action of $A$ on bras.

Let $\langle \varphi |$ be a well-defined bra, and consider the set of all kets $| \psi \rangle$. With each of these kets can be associated the complex number $\langle \varphi | (A | \psi \rangle)$, already defined above as the matrix element of $A$ between $| \varphi \rangle$ and $| \psi \rangle$. Since $A$ is linear and the scalar product depends linearly on the ket, the number $\langle \varphi | (A | \psi \rangle)$ depends linearly on $| \psi \rangle$. Thus, for fixed $\langle \varphi |$ and $A$, we can associate with every ket $| \psi \rangle$ a number which depends linearly on $| \psi \rangle$. The specification of $\langle \varphi |$ and $A$ therefore defines a new linear functional on the kets of $\mathscr{E}$, that is, a new bra belonging to $\mathscr{E}^*$. We shall denote this new bra by $\langle \varphi | A$. The relation which defines $\langle \varphi | A$ can thus be written:

$$\boxed{(\langle \varphi | A) | \psi \rangle = \langle \varphi | (A | \psi \rangle)} \tag{B-41}$$

The operator $A$ associates with every bra $\langle \varphi |$ a new bra $\langle \varphi | A$. Let us show that the correspondence is linear. In order to do this, consider a linear combination of bras $\langle \varphi_1 |$ and $\langle \varphi_2 |$:

$$\langle \chi | = \lambda_1 \langle \varphi_1 | + \lambda_2 \langle \varphi_2 | \tag{B-42}$$

(which means that $\langle \chi | \psi \rangle = \lambda_1 \langle \varphi_1 | \psi \rangle + \lambda_2 \langle \varphi_2 | \psi \rangle$). From (B-41), we have:

$$\begin{aligned} (\langle \chi | A) | \psi \rangle &= \langle \chi | (A | \psi \rangle) \\ &= \lambda_1 \langle \varphi_1 | (A | \psi \rangle) + \lambda_2 \langle \varphi_2 | (A | \psi \rangle) \\ &= \lambda_1 (\langle \varphi_1 | A) | \psi \rangle + \lambda_2 (\langle \varphi_2 | A) | \psi \rangle \end{aligned} \tag{B-43}$$

Since $| \psi \rangle$ is arbitrary, it follows that:

$$\begin{aligned} \langle \chi | A &= (\lambda_1 \langle \varphi_1 | + \lambda_2 \langle \varphi_2 |) A \\ &= \lambda_1 \langle \varphi_1 | A + \lambda_2 \langle \varphi_2 | A \end{aligned} \tag{B-44}$$

Equation (B-41) therefore defines a linear operation on bras. The bra $\langle \varphi | A$ is the bra which results from the action of the linear operator $A$ on the bra $\langle \varphi |$.

COMMENTS:

(*i*) From definition (B-41) of $\langle \varphi | A$, we see that the place of the parenthesis in the symbol defining the matrix element of $A$ between $| \varphi \rangle$ and $| \psi \rangle$ is of no importance. Therefore, we shall henceforth designate this matrix element by the notation $\langle \varphi | A | \psi \rangle$:

$$\langle \varphi | A | \psi \rangle = (\langle \varphi | A) | \psi \rangle = \langle \varphi | (A | \psi \rangle) \tag{B-45}$$

(*ii*) The relative order of $\langle \varphi |$ and $A$ is very important in the notation $\langle \varphi | A$ (*cf.* § 3-b-α above). One must write $\langle \varphi | A$ and not $A \langle \varphi |$: $\langle \varphi | A$ acting on a ket $| \psi \rangle$ gives a number $\langle \varphi | A | \psi \rangle$; $\langle \varphi | A$ is therefore indeed a bra. On the other hand, $A \langle \varphi |$, acting on a ket $| \psi \rangle$, would give $A \langle \varphi | \psi \rangle$, that is, an operator (the operator $A$ multiplied by the number $\langle \varphi | \psi \rangle$). We have not defined any mathematical object of this sort: $A \langle \varphi |$ therefore has no meaning.

#### b. THE ADJOINT OPERATOR $A^\dagger$ OF A LINEAR OPERATOR $A$

We are now going to see that the correspondence between kets and bras, studied in § B-2-c, enables us to associate with every linear operator $A$ another linear operator $A^\dagger$, called the adjoint operator (or Hermitian conjugate) of $A$.

Let $| \psi \rangle$ then be an arbitrary ket of $\mathscr{E}$. The operator $A$ associates with it another ket $| \psi' \rangle = A | \psi \rangle$ of $\mathscr{E}$ (fig. 2).

To the ket $| \psi \rangle$ corresponds a bra $\langle \psi |$; in the same way, to $| \psi' \rangle$ corresponds $\langle \psi' |$. This correspondence between kets and bras thus permits us to define the action of the operator $A^\dagger$ on the bras: the operator $A^\dagger$ associates with the bra $\langle \psi |$ corresponding to the ket $| \psi \rangle$, the bra $\langle \psi' |$ corresponding to the ket $| \psi' \rangle = A | \psi \rangle$. We write: $\langle \psi' | = \langle \psi | A^\dagger$.

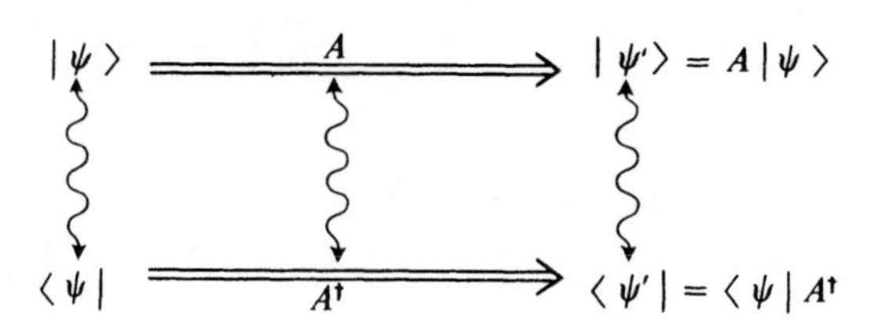


FIGURE 2

**Definition of the adjoint operator $A^\dagger$ of an operator $A$ using the correspondence between kets and bras.**

Let us show that the relation $\langle \psi' | = \langle \psi | A^\dagger$ is linear. We know that, to the bra $\lambda_1 \langle \psi_1 | + \lambda_2 \langle \psi_2 |$, corresponds the ket $\lambda_1^* | \psi_1 \rangle + \lambda_2^* | \psi_2 \rangle$ (the correspondence between a bra and a ket is antilinear). The operator $A$ transforms $\lambda_1^* | \psi_1 \rangle + \lambda_2^* | \psi_2 \rangle$ into $\lambda_1^* A | \psi_1 \rangle + \lambda_2^* A | \psi_2 \rangle = \lambda_1^* | \psi_1' \rangle + \lambda_2^* | \psi_2' \rangle$. Finally, to this ket corresponds the bra:

$$\lambda_1 \langle \psi_1' | + \lambda_2 \langle \psi_2' | = \lambda_1 \langle \psi_1 | A^\dagger + \lambda_2 \langle \psi_2 | A^\dagger.$$

From this we conclude that:

$$(\lambda_1 \langle \psi_1 | + \lambda_2 \langle \psi_2 |) A^\dagger = \lambda_1 \langle \psi_1 | A^\dagger + \lambda_2 \langle \psi_2 | A^\dagger \tag{B-46}$$

$A^\dagger$ is therefore a linear operator, defined by the formula:

$$\boxed{| \psi' \rangle = A | \psi \rangle \Longleftrightarrow \langle \psi' | = \langle \psi | A^\dagger} \tag{B-47}$$

From (B-47), it is easy to deduce another important relationship satisfied by the operator $A^\dagger$. Using the properties of the scalar product, one can always write:

$$\langle \psi' | \varphi \rangle = \langle \varphi | \psi' \rangle^* \tag{B-48}$$

where $|\varphi\rangle$ is an arbitrary ket of $\mathscr{E}$. Using expressions (B-47) for $|\psi'\rangle$ and $\langle \psi' |$, we obtain:

$$\boxed{\langle \psi | A^\dagger | \varphi \rangle = \langle \varphi | A | \psi \rangle^*} \tag{B-49}$$

a relation which is valid for all $|\varphi\rangle$ and $|\psi\rangle$.

COMMENT ABOUT NOTATION

We have already mentioned a notation which can lead to confusion: $|\lambda\psi\rangle$ and $\langle \lambda\psi |$, where $\lambda$ is a scalar [formulas (B-7) and (B-8)]. The same problem arises with the expressions $|A\psi\rangle$ and $\langle A\psi |$, where $A$ is a linear operator. $|A\psi\rangle$ is another way of designating the ket $A|\psi\rangle$:

$$| A \psi \rangle = A | \psi \rangle \tag{B-50}$$

$\langle A\psi |$ is the bra associated with the ket $|A\psi\rangle$. Using (B-50) and (B-47), we see that:

$$\langle A \psi | = \langle \psi | A^\dagger \tag{B-51}$$

When a linear operator $A$ is taken outside the bra symbol, it must be replaced by its adjoint $A^\dagger$ (and placed to the right of the bra).

#### c. CORRESPONDENCE BETWEEN AN OPERATOR AND ITS ADJOINT

By using (B-47) or (B-49), it is easy to show that

$$(A^\dagger)^\dagger = A \tag{B-52}$$

$$(\lambda A)^\dagger = \lambda^* A^\dagger \qquad \text{(where } \lambda \text{ is a number)} \tag{B-53}$$

$$(A + B)^\dagger = A^\dagger + B^\dagger \tag{B-54}$$

Now let us calculate $(AB)^\dagger$. To do this, consider the ket $| \varphi \rangle = AB | \psi \rangle$. Write it in the form $| \varphi \rangle = A | \chi \rangle$, setting $| \chi \rangle = B | \psi \rangle$. Then :

$$\langle \varphi | = \langle \psi | (AB)^\dagger = \langle \chi | A^\dagger = \langle \psi | B^\dagger A^\dagger$$

since $\langle \chi | = \langle \psi | B^\dagger$. From this, we deduce that :

$$\boxed{(AB)^\dagger = B^\dagger A^\dagger} \tag{B-55}$$

Note that *the order changes when one takes the adjoint of a product of operators.*

COMMENT :

Since $(A^\dagger)^\dagger = A$, we can write, using (B-51):

$$\langle A^\dagger \varphi | = \langle \varphi | (A^\dagger)^\dagger = \langle \varphi | A$$

Thus the left-hand side of (B-41) can be rewritten in the form $\langle A^\dagger \varphi | \psi \rangle$. In the same way, the right-hand side of this same equation can be put, with the notation of (B-50), into the form $\langle \varphi | A \psi \rangle$. From this results the following equation, sometimes used to define the adjoint operator $A^\dagger$ of $A$:

$$\langle A^\dagger \varphi | \psi \rangle = \langle \varphi | A \psi \rangle \tag{B-56}$$

#### d. HERMITIAN CONJUGATION IN DIRAC NOTATION

In the preceding section, we introduced the concept of an adjoint operator by using the correspondence between kets and bras. A ket $|\psi\rangle$ and its corresponding bra $\langle \psi |$ are said to be "Hermitian conjugates" of each other. The operation of Hermitian conjugation is represented by the wavy arrows in figure 2; we see that it associates $A^\dagger$ with $A$. This is the reason why $A^\dagger$ is also called the Hermitian conjugate operator of $A$.

The operation of Hermitian conjugation changes the order of the objects to which it is applied. Thus we see in figure 2 that $A|\psi\rangle$ becomes $\langle\psi|A^\dagger$. The ket $|\psi\rangle$ is changed into $\langle\psi|$, and $A$ into $A^\dagger$. Moreover, the order is reversed. In the same way, we saw in (B-55) that the Hermitian conjugate of a product of two operators is equal to the product of the Hermitian conjugates taken in the opposite order. Finally, let us show that :

$$(| u \rangle \langle v |)^\dagger = | v \rangle \langle u | \tag{B-57}$$

($|u\rangle$ is replaced by $\langle u|$, and $\langle v|$ by $|v\rangle$, and the order is changed). Applying relation (B-49) to the operator $|u\rangle\langle v|$, we find :

$$\langle \psi | (|u\rangle\langle v|)^\dagger | \varphi \rangle = [\langle \varphi | (|u\rangle\langle v|) | \psi \rangle]^* \tag{B-58}$$

Now, if we use property (B-9) of the scalar product:

$$[\langle \varphi | (|u\rangle\langle v|) | \psi \rangle]^* = \langle \varphi | u \rangle^* \langle v | \psi \rangle^* = \langle \psi | v \rangle \langle u | \varphi \rangle$$
$$= \langle \psi | (|v\rangle\langle u|) | \varphi \rangle \tag{B-59}$$

By comparing (B-58) and (B-59), we can derive (B-57).

The result of the operation of Hermitian conjugation on a constant remains to be found. We see from (B-6) and (B-53) that this operation simply transforms $\lambda$ into $\lambda^*$ (complex conjugation). This is in agreement with the fact that $\langle \varphi | \psi \rangle^* = \langle \psi | \varphi \rangle$.

Therefore, the Hermitian conjugate of a ket is a bra, and vice versa; that of an operator is its adjoint; that of a number, its complex conjugate. In Dirac notation, the operation of Hermitian conjugation is very simple to perform; it suffices to apply the following rule:

RULE

To obtain the Hermitian conjugate (or the adjoint) of any expression composed of constants, kets, bras and operators, one must :

- *Replace*
  - the constants by their complex conjugates
  - the kets by the bras associated with them
  - the bras by the kets associated with them
  - the operators by their adjoints
- *Reverse the order* of the factors (the position of the constants, nevertheless, is of no importance).

EXAMPLES

$\lambda \langle u | A | v \rangle | w \rangle \langle \psi |$ is an operator ($\lambda$ and $\langle u | A | v \rangle$ are numbers). The adjoint of this operator is obtained by using the preceding rule: $| \psi \rangle \langle w | \langle v | A^\dagger | u \rangle \lambda^*$, which can also be written $\lambda^* \langle v | A^\dagger | u \rangle | \psi \rangle \langle w |$, changing the position of the numbers $\lambda^*$ and $\langle v | A^\dagger | u \rangle$.

In the same way, $\lambda | u \rangle \langle v | w \rangle$ is a ket ($\lambda$ and $\langle v | w \rangle$ are constants). The conjugate bra is $\langle w | v \rangle \langle u | \lambda^*$, which can also be written $\lambda^* \langle w | v \rangle \langle u |$.

#### e. HERMITIAN OPERATORS

An operator $A$ is said to be Hermitian if it is equal to its adjoint, that is, if:

$$A = A^\dagger \tag{B-60}$$

Combining (B-60) and (B-49), we see that a Hermitian operator satisfies the relation:

$$\langle \psi | A | \varphi \rangle = \langle \varphi | A | \psi \rangle^* \tag{B-61}$$

which is valid for all $| \varphi \rangle$ and $| \psi \rangle$.

Finally, for a Hermitian operator, (B-56) becomes

$$\langle A \varphi | \psi \rangle = \langle \varphi | A \psi \rangle \tag{B-62}$$

We shall treat Hermitian operators in more detail later, when we consider the problem of eigenvalues and eigenvectors. Moreover, we shall see in chapter III that Hermitian operators play a fundamental role in quantum mechanics.

If formula (B-57) is applied to the case where $|u\rangle = |v\rangle = |\psi\rangle$, we see that the projector $P_\psi = |\psi\rangle\langle\psi|$ is Hermitian:

$$P_\psi^\dagger = |\psi\rangle\langle\psi| = P_\psi \tag{B-63}$$

COMMENT:

The product of two Hermitian operators $A$ and $B$ is Hermitian only if $[A, B] = 0$. Indeed, if $A = A^\dagger$ and $B = B^\dagger$, it can be shown using (B-55) that $(AB)^\dagger = B^\dagger A^\dagger = BA$, which is equal to $AB$ only if $[A, B] = 0$.

# C. REPRESENTATIONS IN STATE SPACE

## 1. Introduction

### a. DEFINITION OF A REPRESENTATION

Choosing a representation means choosing an orthonormal basis, either discrete or continuous, in the state space $\mathscr{E}$. Vectors and operators are then represented in this basis by *numbers*: components for the vectors, matrix elements for the operators. The vectorial calculus introduced in § B then becomes a matrix calculus with these numbers. The choice of a representation is, in theory, arbitrary. Actually, it obviously depends on the particular problem being studied: in each case, one chooses the representation which leads to the simplest calculations.

### b. AIM OF SECTION C

Using the Dirac notation, and for any arbitrary $\mathscr{E}$ space, we are going to treat again all the concepts introduced in §§ A-2 and A-3 for discrete and continuous bases of $\mathscr{F}$.

We shall write the two characteristic relations of a basis in Dirac notation: the orthonormalization and closure relations. Then we shall show how, using these two relations, it is possible to solve all specific problems involving a representation and the transformation from one representation to another.

## 2. Relations characteristic of an orthonormal basis

### a. ORTHONORMALIZATION RELATION

A set of kets, discrete ($\{|u_i\rangle\}$) or continuous ($\{|w_\alpha\rangle\}$), is said to be *orthonormal* if the kets of this set satisfy the orthonormalization relation:

$$\boxed{\langle u_i | u_j \rangle = \delta_{ij}} \tag{C-1}$$

or

$$\boxed{\langle w_\alpha \mid w_{\alpha'} \rangle = \delta(\alpha - \alpha')} \tag{C-2}$$

It can be seen that, for a continuous set, $\langle w_\alpha | w_\alpha \rangle$ does not exist : the $|w_\alpha \rangle$ have an infinite norm and therefore do not belong to $\mathscr{E}$. Nevertheless, the vectors of $\mathscr{E}$ can be expanded on the $|w_\alpha \rangle$. It is useful, consequently, to accept the $|w_\alpha \rangle$ as generalized kets (see the discussions in §§ A-3 and B-2-c).

#### b. CLOSURE RELATION

A discrete set, $\{ |u_i\rangle \}$, or a continuous one, $\{ |w_\alpha \rangle \}$, constitutes a *basis* if every ket $|\psi\rangle$ belonging to $\mathscr{E}$ has a unique expansion on the $|u_i\rangle$ or the $|w_\alpha\rangle$

$$\boxed{| \psi \rangle = \sum_i c_i \mid u_i \rangle} \tag{C-3}$$

$$\boxed{| \psi \rangle = \int \mathrm{d}\alpha \;\; c(\alpha) \mid w_\alpha \rangle} \tag{C-4}$$

Let us assume, moreover, that the basis is orthonormal. Then perform the scalar multiplication on both sides of (C-3) with $\langle u_j |$, and on both sides of (C-4) with $\langle w_{\alpha'} |$. We obtain, using (C-1) or (C-2), expressions for the components $c_j$ or $c(\alpha')$:

$$\langle u_j \mid \psi \rangle \; = c_j \tag{C-5}$$

$$\langle w_{\alpha'} \mid \psi \rangle = c(\alpha') \tag{C-6}$$

Then replace in (C-3) $c_i$ by $\langle u_i | \psi \rangle$, and in (C-4) $c(\alpha)$ by $\langle w_\alpha | \psi \rangle$:

$$\begin{aligned} | \psi \rangle &= \sum_i c_i \mid u_i \rangle = \sum_i \langle u_i \mid \psi \rangle \mid u_i \rangle \\ &= \sum_i \mid u_i \rangle \langle u_i \mid \psi \rangle = \left( \sum_i \mid u_i \rangle \langle u_i \mid \right) \mid \psi \rangle \end{aligned} \tag{C-7}$$

$$\begin{aligned} | \psi \rangle &= \int \mathrm{d}\alpha \;\; c(\alpha) \mid w_\alpha \rangle = \int \mathrm{d}\alpha \langle w_\alpha \mid \psi \rangle \mid w_\alpha \rangle \\ &= \int \mathrm{d}\alpha \mid w_\alpha \rangle \langle w_\alpha \mid \psi \rangle = \left( \int \mathrm{d}\alpha \mid w_\alpha \rangle \langle w_\alpha \mid \right) \mid \psi \rangle \end{aligned} \tag{C-8}$$

[since, in (C-7), we can place the number $\langle u_i | \psi \rangle$ after the ket $|u_i\rangle$; in the same way, in (C-8), we can place the number $\langle w_\alpha | \psi \rangle$ after the ket $| w_\alpha \rangle$].

Thus, we see two operators appear, $\sum_i | u_i \rangle \langle u_i |$ and $\int \mathrm{d}\alpha \, | w_\alpha \rangle \langle w_\alpha |$. They act on every ket $|\psi\rangle$ belonging to $\mathscr{E}$ to give the same ket $|\psi\rangle$. Since $|\psi\rangle$ is arbitrary, it follows that:

$$P_{\{u_i\}} = \sum_i | u_i \rangle \langle u_i | = \mathbb{1} \tag{C-9}$$

$$P_{\{w_\alpha\}} = \int \mathrm{d}\alpha \, | w_\alpha \rangle \langle w_\alpha | = \mathbb{1} \tag{C-10}$$

where $\mathbb{1}$ denotes the identity operator in $\mathscr{E}$. Relation (C-9), or (C-10), is called the *closure relation*. Conversely, let us show that relations (C-9) and (C-10) express the fact that the sets $\{ | u_i \rangle \}$ and $\{ | w_\alpha \rangle \}$ constitute bases. For every $| \psi \rangle$ belonging to $\mathscr{E}$, we can write:

$$\begin{aligned} | \psi \rangle = \mathbb{1} | \psi \rangle = P_{\{u_i\}} | \psi \rangle &= \sum_i | u_i \rangle \langle u_i | \psi \rangle \\ &= \sum_i c_i | u_i \rangle \end{aligned} \tag{C-11}$$

with:

$$c_i = \langle u_i | \psi \rangle \tag{C-12}$$

In the same way:

$$\begin{aligned} | \psi \rangle = \mathbb{1} | \psi \rangle = P_{\{w_\alpha\}} | \psi \rangle &= \int \mathrm{d}\alpha \, | w_\alpha \rangle \langle w_\alpha | \psi \rangle \\ &= \int \mathrm{d}\alpha \; c(\alpha) | w_\alpha \rangle \end{aligned} \tag{C-13}$$

with:

$$c(\alpha) = \langle w_\alpha | \psi \rangle \tag{C-14}$$

Thus, every ket $|\psi\rangle$ has a unique expansion on the $|u_i\rangle$ or on the $|w_\alpha\rangle$. Each of these two sets therefore forms a basis, a discrete one or a continuous one. We also see that relation (C-9), or (C-10), spares us the need of memorizing expressions (C-12) and (C-14) for the components $c_i$ and $c(\alpha)$.

COMMENTS:

(*i*) We shall see later (§ E) that, in the case of the $\mathscr{F}$-space, relations (A-32) and (A-57) can easily be deduced from (C-9) and (C-10).

(*ii*) Geometrical interpretation of the closure relation.

From the discussion of § B-3-b, we see that $\sum_i | u_i \rangle \langle u_i |$ is a projector: the projector onto the subspace $\mathscr{E}'$ spanned by $| u_1 \rangle, | u_2 \rangle, ... | u_i \rangle ...$ If the $| u_i \rangle$ form a basis, every ket of $\mathscr{E}$ can be expanded on the $| u_i \rangle$; the subspace $\mathscr{E}'$ is then identical with the $\mathscr{E}$-space itself. Consequently, it is reasonable for $\sum_i | u_i \rangle \langle u_i |$ to be equal to the identity operator : projecting onto $\mathscr{E}$ a ket which belongs to $\mathscr{E}$ does not modify this ket. The same argument can be applied to $\int d\alpha \, | w_\alpha \rangle \langle w_\alpha |$.

We can now find an equivalent of the closure relation for the three-dimensional space of ordinary geometry, $R^3$. If $\mathbf{e}_1$, $\mathbf{e}_2$, and $\mathbf{e}_3$ are three orthonormal vectors of this space, and $P_1$, $P_2$ and $P_3$ are the projectors onto these three vectors, the fact that $\{ \mathbf{e}_1, \mathbf{e}_2, \mathbf{e}_3 \}$ forms a basis in $R^3$ is expressed by the relation

$$P_1 + P_2 + P_3 = \mathbb{1} \tag{C-15}$$

On the other hand, $\{ \mathbf{e}_1, \mathbf{e}_2 \}$ constitutes an orthonormal set but not a basis of $R^3$. This is expressed by the fact that the projector $P_1 + P_2$ (which projects onto the plane spanned by $\mathbf{e}_1$ and $\mathbf{e}_2$) is not equal to $\mathbb{1}$; for example : $(P_1 + P_2)\mathbf{e}_3 = 0$.

Table (II-7) summarizes the only fundamental formulas which are required for any calculation to be performed in the $\{ |u_i\rangle \}$ or $\{ |w_\alpha\rangle \}$ representation.

| $\{ \| u_i \rangle \}$ representation | $\{ \| w_\alpha \rangle \}$ representation |
| --- | --- |
| $\langle u_i \| u_j \rangle = \delta_{ij}$ | $\langle w_\alpha \| w_{\alpha'} \rangle = \delta(\alpha - \alpha')$ |
| $P_{\{u_i\}} = \sum_i \| u_i \rangle \langle u_i \| = \mathbb{1}$ | $P_{\{w_\alpha\}} = \int d\alpha \, \| w_\alpha \rangle \langle w_\alpha \| = \mathbb{1}$ |

Table (II-7)

### 3. Representation of kets and bras

#### a. REPRESENTATION OF KETS

In the $\{ |u_i\rangle \}$ basis, the ket $|\psi\rangle$ is represented by the set of its components, that is, by the set of numbers $c_i = \langle u_i | \psi \rangle$. These numbers can be arranged vertically to form a one-column matrix (with, in general, a countable infinity of rows):

$$\begin{pmatrix} \langle u_1 | \psi \rangle \\ \langle u_2 | \psi \rangle \\ \vdots \\ \langle u_i | \psi \rangle \\ \vdots \end{pmatrix} \tag{C-16}$$

In the continuous $\{ | w_\alpha\rangle \}$ basis, the ket $|\psi\rangle$ is represented by a continuous infinity of numbers, $c(\alpha) = \langle w_\alpha | \psi \rangle$, that is, by a function of $\alpha$. It is then possible to draw a vertical axis, along which are placed the various possible values of $\alpha$. To each of these values corresponds a number, $\langle w_\alpha | \psi \rangle$:

$$\alpha \downarrow \begin{pmatrix} \vdots \\ \langle w_\alpha | \psi \rangle \\ \vdots \end{pmatrix} \tag{C-17}$$

#### b. REPRESENTATION OF BRAS

Let $\langle \varphi |$ be an arbitrary bra. In the $\{ | u_i \rangle \}$ basis, we can write:

$$\langle \varphi | = \langle \varphi | \mathbb{1} = \langle \varphi | P_{\{u_i\}} = \sum_i \langle \varphi | u_i \rangle \langle u_i | \tag{C-18}$$

$\langle \varphi |$ has a unique expansion on the bras $\langle u_i |$. The components of $\langle \varphi |$, $\langle \varphi | u_i \rangle$, are the complex conjugates of the components $b_i = \langle u_i | \varphi \rangle$ of the ket $| \varphi \rangle$ associated with $\langle \varphi |$.

In the same way, we obtain, in the $\{ | w_\alpha \rangle \}$ basis:

$$\langle \varphi | = \langle \varphi | \mathbb{1} = \langle \varphi | P_{\{w_\alpha\}} = \int \mathrm{d}\alpha \, \langle \varphi | w_\alpha \rangle \langle w_\alpha | \tag{C-19}$$

The components of $\langle \varphi |$, $\langle \varphi | w_\alpha \rangle$, are the complex conjugates of the components $b(\alpha) = \langle w_\alpha | \varphi \rangle$ of the ket $| \varphi \rangle$ associated with $\langle \varphi |$.

We have agreed to arrange the components of a ket vertically. Before describing how to arrange the components of a bra, let us show how the closure relation enables us to find simply the expression for the scalar product of two kets in terms of their components. We know that we can always place $\mathbb{1}$ between $\langle \varphi |$ and $| \psi \rangle$ in the expression for the scalar product:

$$\begin{aligned} \langle \varphi | \psi \rangle &= \langle \varphi | \mathbb{1} | \psi \rangle = \langle \varphi | P_{\{u_i\}} | \psi \rangle \\ &= \sum_i \langle \varphi | u_i \rangle \langle u_i | \psi \rangle = \sum_i b_i^* c_i \end{aligned} \tag{C-20}$$

In the same way:

$$\begin{aligned} \langle \varphi | \psi \rangle &= \langle \varphi | \mathbb{1} | \psi \rangle = \langle \varphi | P_{\{w_\alpha\}} | \psi \rangle \\ &= \int \mathrm{d}\alpha \, \langle \varphi | w_\alpha \rangle \langle w_\alpha | \psi \rangle = \int \mathrm{d}\alpha \; b^*(\alpha) \; c(\alpha) \end{aligned} \tag{C-21}$$

Let us arrange the components $\langle \varphi | u_i \rangle$ of the bra $\langle \varphi |$ horizontally, to form a row matrix (having one row and an infinite number of columns):

$$(\langle \varphi | u_1 \rangle \quad \langle \varphi | u_2 \rangle \; ...... \; \langle \varphi | u_i \rangle \; ......) \tag{C-22}$$

Using this convention, $\langle \varphi | \psi \rangle$ is the matrix product of the column matrix which represents $| \psi \rangle$ and the row matrix which represents $\langle \varphi |$. The result is a matrix having one row and one column, that is, a *number*.

In the $\{ | w_\alpha \rangle \}$ basis, $\langle \varphi |$ has a continuous infinity of components $\langle \varphi | w_\alpha \rangle$. The various values of $\alpha$ are placed along a horizontal axis. To each of these values corresponds a component $\langle \varphi | w_\alpha \rangle$ of $\langle \varphi |$:

$$\underset{\alpha}{\underrightarrow{(\ldots\ldots\ldots\ldots \langle \varphi | w_\alpha \rangle \ldots\ldots\ldots\ldots)}} \tag{C-23}$$

COMMENT:

In a given representation, the matrices which represent a ket $| \psi \rangle$ and the associated bra $\langle \psi |$ are Hermitian conjugates of each other (in the matrix sense): one passes from one matrix to the other by interchanging rows and columns and taking the complex conjugate of each element.

## 4. Representation of operators

### a. REPRESENTATION OF $A$ BY A "SQUARE" MATRIX

Given a linear operator $A$, we can, in a $\{ | u_i \rangle \}$ or $\{ | w_\alpha \rangle \}$ basis, associate with it a series of numbers defined by:

$$A_{ij} = \langle u_i | A | u_j \rangle \tag{C-24}$$

or

$$A(\alpha, \alpha') = \langle w_\alpha | A | w_{\alpha'} \rangle \tag{C-25}$$

These numbers depend on two indices and can therefore be arranged in a "square" matrix having a countable or continuous infinity of rows and columns. The usual convention is to have the first index fix the rows and the second, the columns. Thus, in the $\{ |u_i\rangle \}$ basis, the operator $A$ is represented by the matrix:

$$\begin{pmatrix} A_{11} & A_{12} \ldots & A_{1j} & \ldots \\ A_{21} & A_{22} \ldots & A_{2j} & \ldots \\ \vdots & \vdots & \vdots & \\ A_{i1} & A_{i2} \ldots & A_{ij} & \ldots \\ \vdots & \vdots & \vdots & \end{pmatrix} \tag{C-26}$$

We see that the $j$th column is made up of the components in the $\{ |u_i\rangle \}$ basis of the transform $A |u_j\rangle$ of the basis vector $|u_j\rangle$.

For a continuous basis, we draw two perpendicular axes. To a point which has for its abscissa $\alpha'$ and for its ordinate $\alpha$ there corresponds the number $A(\alpha, \alpha')$:

$$\alpha \downarrow \overset{\alpha' \rightarrow}{\begin{pmatrix} & \vdots & \\ \ldots\ldots\ldots & A(\alpha, \alpha') & \\ & & \end{pmatrix}} \tag{C-27}$$

Let us use the closure relation to calculate the matrix which represents the operator $AB$ in the $\{ |u_i\rangle \}$ basis:

$$\begin{aligned}\langle u_i | AB | u_j \rangle &= \langle u_i | A \mathbb{1} B | u_j \rangle \\ &= \langle u_i | A P_{\{u_k\}} B | u_j \rangle \\ &= \sum_k \langle u_i | A | u_k \rangle \langle u_k | B | u_j \rangle \end{aligned} \tag{C-28}$$

The convention chosen above for the arrangement of the elements $A_{ij}$ [or $A(\alpha, \alpha')$] is therefore consistent with the one relating to the product of two matrices: (C-28) expresses the fact that the matrix representing the operator $AB$ is the product of the matrices associated with $A$ and $B$.

#### b. MATRIX REPRESENTATION OF THE KET $|\psi'\rangle = A|\psi\rangle$

The problem is the following: knowing the components of $|\psi\rangle$ and the matrix elements of $A$ in a given representation, how can we calculate the components of $|\psi'\rangle = A|\psi\rangle$ in the same representation?

In the $\{ |u_i\rangle \}$ basis, the coordinate $c'_i$ of $|\psi'\rangle$ are given by:

$$c'_i = \langle u_i | \psi' \rangle = \langle u_i | A | \psi \rangle \tag{C-29}$$

If we simply insert the closure relation between $A$ and $|\psi\rangle$, we obtain:

$$\begin{aligned} c'_i &= \langle u_i | A \mathbb{1} | \psi \rangle = \langle u_i | A P_{\{u_j\}} | \psi \rangle \\ &= \sum_j \langle u_i | A | u_j \rangle \langle u_j | \psi \rangle \\ &= \sum_j A_{ij}\, c_j \end{aligned} \tag{C-30}$$

For the $\{ |w_\alpha\rangle \}$ basis, we obtain, in the same way:

$$\begin{aligned} c'(\alpha) &= \langle w_\alpha | \psi' \rangle = \langle w_\alpha | A | \psi \rangle \\ &= \int \mathrm{d}\alpha' \langle w_\alpha | A | w_{\alpha'} \rangle \langle w_{\alpha'} | \psi \rangle \\ &= \int \mathrm{d}\alpha' A(\alpha, \alpha')\; c(\alpha') \end{aligned} \tag{C-31}$$

The matrix expression for $|\psi'\rangle = A|\psi\rangle$ is therefore very simple. We see, for example from (C-30), that the column matrix representing $|\psi'\rangle$ is equal to the product of the column matrix representing $|\psi\rangle$ and the square matrix representing $A$:

$$\begin{pmatrix} c'_1 \\ c'_2 \\ \vdots \\ c'_i \\ \vdots \end{pmatrix} = \begin{pmatrix} A_{11} & A_{12} & \dots & A_{1j} & \dots \\ A_{21} & A_{22} & \dots & A_{2j} & \dots \\ \vdots & \vdots & & \vdots & \\ A_{i1} & A_{i2} & \dots & A_{ij} & \dots \\ \vdots & \vdots & & \vdots & \end{pmatrix} \begin{pmatrix} c_1 \\ c_2 \\ \vdots \\ c_j \\ \vdots \end{pmatrix} \tag{C-32}$$

### c. EXPRESSION FOR THE NUMBER $\langle \varphi | A | \psi \rangle$

By inserting the closure relation between $\langle \varphi |$ and $A$ and again between $A$ and $| \psi \rangle$, we obtain:

– for the $\{ | u_i \rangle \}$ basis:

$$\begin{aligned} \langle \varphi | A | \psi \rangle &= \langle \varphi | P_{\{u_i\}} A P_{\{u_j\}} | \psi \rangle \\ &= \sum_{i,j} \langle \varphi | u_i \rangle \langle u_i | A | u_j \rangle \langle u_j | \psi \rangle \\ &= \sum_{i,j} b_i^* \; A_{ij} \; c_j \end{aligned} \tag{C-33}$$

– for the $\{ | w_\alpha \rangle \}$ basis:

$$\begin{aligned} \langle \varphi | A | \psi \rangle &= \langle \varphi | P_{\{w_\alpha\}} A P_{\{w_{\alpha'}\}} | \psi \rangle \\ &= \iint \mathrm{d}\alpha \, \mathrm{d}\alpha' \, \langle \varphi | w_\alpha \rangle \langle w_\alpha | A | w_{\alpha'} \rangle \langle w_{\alpha'} | \psi \rangle \\ &= \iint \mathrm{d}\alpha \, \mathrm{d}\alpha' \; b^*(\alpha) \; A(\alpha, \alpha') \; c(\alpha') \end{aligned} \tag{C-34}$$

The interpretation of these formulas in the matrix formalism, is as follows: $\langle \varphi | A | \psi \rangle$ is a number, that is, a matrix with one row and one column, obtained by multiplying the column matrix representing $| \psi \rangle$ first by the square matrix representing $A$ and then by the row matrix representing $\langle \varphi |$. For example, in the $\{ | u_i \rangle \}$ basis:

$$\langle \varphi | A | \psi \rangle = (b_1^* \; b_2^* \dots b_i^* \dots) \begin{pmatrix} A_{11} & A_{12} & \dots & A_{1j} & \dots \\ A_{21} & A_{22} & \dots & A_{2j} & \dots \\ \vdots & \vdots & & \vdots & \\ A_{i1} & A_{i2} & \dots & A_{ij} & \dots \\ \vdots & \vdots & & & \end{pmatrix} \begin{pmatrix} c_1 \\ c_2 \\ \vdots \\ c_j \\ \vdots \end{pmatrix} \tag{C-35}$$

COMMENTS:

(*i*) It can be shown in the same way that the bra $\langle \varphi | A$ is represented by a row matrix, the product of the square matrix representing $A$ by the row matrix representing $\langle \varphi |$ [the first two matrices of the right-hand side of (C-35)]. Again we see the importance of the order of the symbols: the expression $A \langle \varphi |$ would lead to a matrix operation which is undefined (the product of a row matrix by a square matrix).

(*ii*) From a matrix point of view, equation (B-41) which defines $\langle \varphi | A$ merely expresses the associativity of the product of the three matrices which appear in (C-35).

(*iii*) Using the preceding conventions, we express $| \psi \rangle \langle \psi |$ by a square matrix:

$$\begin{pmatrix} c_1 \\ c_2 \\ \vdots \\ c_i \\ \vdots \end{pmatrix} (c_1^* \; c_2^* \ldots \; c_j^* \ldots) = \begin{pmatrix} c_1 c_1^* & c_1 c_2^* & \ldots & c_1 c_j^* & \ldots \\ c_2 c_1^* & c_2 c_2^* & \ldots & c_2 c_j^* & \ldots \\ \vdots & & & & \\ c_i c_1^* & c_i c_2^* & \ldots & c_i c_j^* & \ldots \\ \vdots & \vdots & & \vdots & \end{pmatrix} \tag{C-36}$$

This is indeed an operator, while $\langle \psi | \psi \rangle$, the product of a column matrix by a row matrix, is a number.

#### d. MATRIX REPRESENTATION OF THE ADJOINT $A^\dagger$ OF $A$

Using (B-49), we obtain easily:

$$(A^\dagger)_{ij} = \langle u_i | A^\dagger | u_j \rangle = \langle u_j | A | u_i \rangle^* = A_{ji}^* \tag{C-37}$$

or

$$A^\dagger(\alpha, \alpha') = \langle w_\alpha | A^\dagger | w_{\alpha'} \rangle = \langle w_{\alpha'} | A | w_\alpha \rangle^* = A^*(\alpha', \alpha) \tag{C-38}$$

Therefore, the matrices representing $A$ and $A^\dagger$ in a given representation are Hermitian conjugates of each other, in the matrix sense: *one passes from one to the other by interchanging rows and columns and then taking the complex conjugate.*

If $A$ is Hermitian, $A^\dagger = A$, and we can then replace $(A^\dagger)_{ij}$ by $A_{ij}$ in (C-37), and $A^\dagger(\alpha, \alpha')$ by $A(\alpha, \alpha')$ in (C-38):

$$A_{ij} = A_{ji}^* \tag{C-39}$$

$$A(\alpha, \alpha') = A^*(\alpha', \alpha) \tag{C-40}$$

A Hermitian operator is therefore represented by a Hermitian matrix, that is, one in which *any two elements which are symmetric with respect to the principal diagonal are complex conjugates of each other*. In particular, for $i = j$ or $\alpha = \alpha'$, (C-39) and (C-40) become:

$$A_{ii} = A_{ii}^* \tag{C-41}$$

$$A(\alpha, \alpha) = A^*(\alpha, \alpha) \tag{C-42}$$

*The diagonal elements of a Hermitian matrix are therefore always real numbers.*

## 5. Change of representations

### a. OUTLINE OF THE PROBLEM

In a given representation, a ket (or a bra, or an operator) is represented by a matrix. If we change representations, that is, bases, the same ket (or bra, or operator) will be represented by a different matrix. How are these two matrices related?

For the sake of simplicity, we shall assume here that we are going from one discrete orthonormal basis $\{ |u_i\rangle \}$ to another discrete orthonormal basis $\{ |t_k\rangle \}$. In § E, we shall study an example of changing from one continuous basis to another continuous basis.

The change of basis is defined by specifying the components $\langle u_i | t_k \rangle$ of each of the kets of the new basis in terms of each of the kets of the old one. We shall set:

$$S_{ik} = \langle u_i \mid t_k \rangle \tag{C-43}$$

$S$ is the matrix of the basis change (transformation matrix). If $S^\dagger$ denotes its Hermitian conjugate:

$$(S^\dagger)_{ki} = (S_{ik})^* = \langle t_k \mid u_i \rangle \tag{C-44}$$

The following calculations can be performed very easily, and without memorization, by using the two closure relations:

$$P_{\{u_i\}} = \sum_i | u_i \rangle \langle u_i | = \mathbb{1} \tag{C-45}$$

$$P_{\{t_k\}} = \sum_k | t_k \rangle \langle t_k | = \mathbb{1} \tag{C-46}$$

and the two orthonormalization relations:

$$\langle u_i \mid u_j \rangle = \delta_{ij} \tag{C-47}$$

$$\langle t_k \mid t_l \rangle = \delta_{kl} \tag{C-48}$$

COMMENT:

The transformation matrix, $S$, is unitary (complement $C_{II}$). That is, it satisfies:

$$S^\dagger S = SS^\dagger = I \tag{C-49}$$

where $I$ is the unit matrix. Indeed, we see that:

$$\begin{aligned}(S^\dagger S)_{kl} &= \sum_i S^\dagger_{ki} S_{il} = \sum_i \langle t_k | u_i \rangle \langle u_i | t_l \rangle \\ &= \langle t_k | t_l \rangle = \delta_{kl}\end{aligned} \tag{C-50}$$

In the same way:

$$\begin{aligned}(SS^\dagger)_{ij} &= \sum_k S_{ik} S^\dagger_{kj} = \sum_k \langle u_i | t_k \rangle \langle t_k | u_j \rangle \\ &= \langle u_i | u_j \rangle = \delta_{ij}\end{aligned} \tag{C-51}$$

#### b. TRANSFORMATION OF THE COMPONENTS OF A KET

To obtain the components $\langle t_k | \psi \rangle$ of a ket $| \psi \rangle$ in the new basis from its components $\langle u_i | \psi \rangle$ in the old basis, one simply inserts (C-45) between $\langle t_k |$ and $| \psi \rangle$:

$$\begin{aligned} \langle t_k | \psi \rangle &= \langle t_k | \mathbb{1} | \psi \rangle = \langle t_k | P_{\{u_i\}} | \psi \rangle \\ &= \sum_i \langle t_k | u_i \rangle \langle u_i | \psi \rangle \\ &= \sum_i S^\dagger_{ki} \langle u_i | \psi \rangle \end{aligned} \tag{C-52}$$

The inverse expressions can be derived in the same way, using (C-46):

$$\begin{aligned} \langle u_i | \psi \rangle &= \langle u_i | \mathbb{1} | \psi \rangle = \langle u_i | P_{\{t_k\}} | \psi \rangle \\ &= \sum_k \langle u_i | t_k \rangle \langle t_k | \psi \rangle \\ &= \sum_k S_{ik} \langle t_k | \psi \rangle \end{aligned} \tag{C-53}$$

#### c. TRANSFORMATION OF THE COMPONENTS OF A BRA

The principle of the calculation is exactly the same. For example:

$$\begin{aligned} \langle \psi | t_k \rangle &= \langle \psi | \mathbb{1} | t_k \rangle = \langle \psi | P_{\{u_i\}} | t_k \rangle \\ &= \sum_i \langle \psi | u_i \rangle \langle u_i | t_k \rangle \\ &= \sum_i \langle \psi | u_i \rangle S_{ik} \end{aligned} \tag{C-54}$$

#### d. TRANSFORMATION OF THE MATRIX ELEMENTS OF AN OPERATOR

If, in $\langle t_k | A | t_l \rangle$, we insert (C-45) between $\langle t_k |$ and $A$, and again between $A$ and $| t_l \rangle$, we obtain:

$$\begin{aligned} \langle t_k | A | t_l \rangle &= \langle t_k | P_{\{u_i\}} A \; P_{\{u_j\}} | t_l \rangle \\ &= \sum_{i,j} \langle t_k | u_i \rangle \langle u_i | A | u_j \rangle \langle u_j | t_l \rangle \end{aligned} \tag{C-55}$$

that is:

$$A_{kl} = \sum_{i,j} S^\dagger_{ki} \; A_{ij} \; S_{jl} \tag{C-56}$$

In the same way:

$$\begin{aligned} A_{ij} &= \langle u_i | A | u_j \rangle = \langle u_i | P_{\{t_k\}} A \; P_{\{t_l\}} | u_j \rangle \\ &= \sum_{k,l} \langle u_i | t_k \rangle \langle t_k | A | t_l \rangle \langle t_l | u_j \rangle \\ &= \sum_{k,l} S_{ik} \; A_{kl} \; S^\dagger_{lj} \end{aligned} \tag{C-57}$$

# D. EIGENVALUE EQUATIONS. OBSERVABLES

## 1. Eigenvalues and eigenvectors of an operator

### a. DEFINITIONS

$|\psi\rangle$ is said to be an eigenvector (or eigenket) of the linear operator $A$ if:

$$A|\psi\rangle = \lambda|\psi\rangle \tag{D-1}$$

where $\lambda$ is a complex number. We are going to study a certain number of properties of equation (D-1), *the eigenvalue equation of the linear operator A*. In general, this equation possesses solutions only when $\lambda$ takes on certain values, called *eigenvalues* of $A$. The set of the eigenvalues is called the *spectrum* of $A$.

Note that, if $|\psi\rangle$ is an eigenvector of $A$ with the eigenvalue $\lambda$, $\alpha|\psi\rangle$ (where $\alpha$ is an arbitrary complex number) is also an eigenvector of $A$ with the same eigenvalue:

$$A(\alpha|\psi\rangle) = \alpha A|\psi\rangle = \alpha\lambda|\psi\rangle = \lambda(\alpha|\psi\rangle) \tag{D-2}$$

To rid ourselves of this ambiguity, we could agree to normalize the eigenvectors to 1 :

$$\langle\psi|\psi\rangle = 1 \tag{D-3}$$

But this does not completely remove the ambiguity, since $e^{i\theta}|\psi\rangle$, where $\theta$ is an arbitrary real number, has the same norm as $|\psi\rangle$. We shall see below that, in quantum mechanics, the physical predictions obtained using $|\psi\rangle$ or $e^{i\theta}|\psi\rangle$ are the same.

The eigenvalue $\lambda$ is called *nondegenerate* (or *simple*) when its corresponding eigenvector is unique to within a constant factor, that is, when all its associated eigenkets are collinear. On the other hand, if there exist at least two linearly independent kets which are eigenvectors of $A$ with the same eigenvalue, this eigenvalue is said to be *degenerate*. Its *degree* (or *order*) *of degeneracy* is then the number of linearly independent eigenvectors which are associated with it (the degree of degeneracy of an eigenvalue can be finite or infinite). For example, if $\lambda$ is $g$-fold degenerate, there correspond to it $g$ independent kets $|\psi^i\rangle$ ($i = 1, 2, \ldots g$) such that :

$$A|\psi^i\rangle = \lambda|\psi^i\rangle \tag{D-4}$$

But then every ket $|\psi\rangle$ of the form:

$$|\psi\rangle = \sum_{i=1}^{g} c_i|\psi^i\rangle \tag{D-5}$$

is an eigenvector of $A$ with the eigenvalue $\lambda$, whatever the coefficients $c_i$, since :

$$A|\psi\rangle = \sum_{i=1}^{g} c_i\, A|\psi^i\rangle = \lambda\sum_{i=1}^{g} c_i|\psi^i\rangle = \lambda|\psi\rangle \tag{D-6}$$

Consequently, the set of eigenkets of $A$ associated with $\lambda$ constitutes a *g-dimensional vector space* (which can be infinite-dimensional), called the "*eigensubspace*" of the eigenvalue $\lambda$. In particular, it is equivalent to say that $\lambda$ is non-degenerate or to say that its degree of degeneracy is $g = 1$.

To illustrate these definitions, let us choose the example of a projector (§ B-3-b): $P_\psi = |\psi\rangle\langle\psi|$ (with $\langle\psi|\psi\rangle = 1$). Its eigenvalue equation is written:

$$P_\psi|\varphi\rangle = \lambda|\varphi\rangle$$

that is,

$$|\psi\rangle\langle\psi|\varphi\rangle = \lambda|\varphi\rangle \tag{D-7}$$

The ket on the left-hand side is always collinear with $|\psi\rangle$ or zero. Consequently, the eigenvectors of $P_\psi$ are : on the one hand, $|\psi\rangle$ itself, with an eigenvalue of $\lambda = 1$; on the other hand, all the kets $|\varphi\rangle$ orthogonal to $|\psi\rangle$, for which the associated eigenvalue is $\lambda = 0$. The spectrum of $P_\psi$ therefore includes only two values : 1 and 0. The first one is simple, the second, infinitely degenerate (if the state space considered is infinite-dimensional). The eigen-subspace associated with $\lambda = 0$ is the supplement* of $|\psi\rangle$ (see §D-2-c).

COMMENTS:

(*i*) Taking the Hermitian conjugate of both sides of equation (D-1), we obtain:

$$\langle\psi|A^\dagger = \lambda^*\langle\psi| \tag{D-8}$$

Therefore, if $|\psi\rangle$ is an eigenket of $A$ with an eigenvalue $\lambda$, it can also be said that $\langle\psi|$ is an eigenbra of $A^\dagger$ with an eigenvalue $\lambda^*$. However, let us stress the fact that, except in the case where $A$ is Hermitian (§ D-2-a), nothing can be said *a priori* about $\langle\psi|A$.

(*ii*) To be completely rigorous, one should solve the eigenvalue equation (D-1) in the space $\mathscr{E}$. That is, one should consider only those eigenvectors $|\psi\rangle$ which have a finite norm. In fact, we shall be obliged to use operators for which the eigenkets do not satisfy this condition (§ E). Therefore, we shall grant that vectors which are solutions of (D-1) can be "generalized kets".

### b. FINDING THE EIGENVALUES AND EIGENVECTORS OF AN OPERATOR

Given a linear operator $A$, how does one find all its eigenvalues and the corresponding eigenvectors? We are concerned with this question from a purely practical point of view. We shall consider the case where the state space is of finite dimension $N$, and we shall grant that the results can be generalized to an infinite-dimensional state space.

Let us choose a representation, for example, $\{|u_i\rangle\}$, and let us project the vector equation (D-1) onto the various orthonormal basis vectors $|u_i\rangle$:

$$\langle u_i|A|\psi\rangle = \lambda\langle u_i|\psi\rangle \tag{D-9}$$

Inserting the closure relation between $A$ and $|\psi\rangle$, we obtain:

$$\sum_j \langle u_i|A|u_j\rangle\langle u_j|\psi\rangle = \lambda\langle u_i|\psi\rangle \tag{D-10}$$

* In a vector space $\mathscr{E}$, two sub-spaces $\mathscr{E}_1$ and $\mathscr{E}_2$ are said to be supplementary if all kets $|\psi\rangle$ of $\mathscr{E}$ can be written $|\psi\rangle = |\psi_1\rangle + |\psi_2\rangle$ where $|\psi_1\rangle$ and $|\psi_2\rangle$ belong, respectively, to $\mathscr{E}_1$ and $\mathscr{E}_2$, and if $\mathscr{E}_1$ and $\mathscr{E}_2$ are disjoint (no common non-zero ket; the expansion $|\psi\rangle = |\psi_1\rangle + |\psi_2\rangle$ is then unique). Actually, there exists an infinity of sub-subspaces $\mathscr{E}_2$ supplementary to a given sub-space $\mathscr{E}_1$. One can fix $\mathscr{E}_2$ by forcing it to be orthogonal to $\mathscr{E}_1$. This shall be done throughout this book, even though the word "orthogonal" will not be explicitly written before supplement.

Example: In ordinary three-dimensional space, if $\mathscr{E}_1$ is a plane $P$, $\mathscr{E}_2$ can be any arbitrary straight line, not contained in $P$. The orthogonal supplement of $\mathscr{E}_1$ is the straight line passing through the origin and orthogonal to $P$.

With the usual notation :

$$\langle u_i \mid \psi \rangle = c_i$$
$$\langle u_i \mid A \mid u_j \rangle = A_{ij} \tag{D-11}$$

equations (D-10) can be written:

$$\sum_j A_{ij} \; c_j = \lambda c_i \tag{D-12}$$

or

$$\sum_j [A_{ij} - \lambda \delta_{ij}] c_j = 0 \tag{D-13}$$

(D-13) can be considered to be a system of equations where the unknowns are the $c_j$, the components of the eigenvector in the chosen representation. This system is linear and homogeneous.

α. *The characteristic equation*

The system (D-13) consists of $N$ equations ($i = 1, 2, ..., N$) in $N$ unknowns $c_j$ ($j = 1, 2, ..., N$). Since it is linear and homogeneous, it has a non-trivial solution (the trivial solution is the one for which all the $c_j$ are zero) if and only if the determinant of the coefficients is zero. This condition is written:

$$\boxed{\text{Det}\left[\mathscr{A} - \lambda I\right] = 0} \tag{D-14}$$

where $\mathscr{A}$ is the $N \times N$ matrix of elements $A_{ij}$ and $I$ is the unit matrix.

Equation (D-14), called the characteristic equation (or secular equation), enables us to determine all the eigenvalues of the operator $A$, that is, its spectrum. (D-14) can be written explicitly in the form:

$$\begin{vmatrix} A_{11} - \lambda & A_{12} & A_{13} \cdots & A_{1N} \\ A_{21} & A_{22} - \lambda & A_{23} \cdots & A_{2N} \\ \vdots & \vdots & \vdots & \vdots \\ A_{N1} & A_{N2} & A_{N3} \cdots & A_{NN} - \lambda \end{vmatrix} = 0 \tag{D.15}$$

This is an $N$th order equation in $\lambda$; consequently, it has $N$ roots, real or imaginary, distinct or identical. It is easy to show, by performing an arbitrary change of basis, that the characteristic equation is independent of the representation chosen. Therefore, *the eigenvalues of an operator are the roots of its characteristic equation.*

β. *Determination of the eigenvectors*

Now let us choose an eigenvalue $\lambda_0$, a solution of the characteristic equation (D-14), and let us look for the corresponding eigenvectors. We are going to distinguish between two cases:

($i$) First, assume that $\lambda_0$ is a simple root of the characteristic equation. We can then show that the system (D-13), when $\lambda = \lambda_0$, is comprised of $(N - 1)$ independent equations, the $N$th one following from the preceding ones and hence redundant. But we have $N$ unknowns; there is therefore an infinite number of solutions, but all the $c_j$ can be determined in a unique way in terms of one of them, say $c_1$. If we fix $c_1$, we obtain for the $(N - 1)$ other $c_j$ a system of $(N - 1)$ linear, inhomogeneous equations (the "right-hand side" of each equation is the term in $c_1$) with a non-zero determinant [the $(N - 1)$ equations are independent]. The solution of this system is of the form:

$$c_j = \alpha_j^0 \, c_1 \tag{D-16}$$

since the initial system (D-13) is linear and homogeneous. $\alpha_1^0$ is, of course, equal to 1 by definition, and the $(N - 1)$ coefficients $\alpha_j^0$ for $j \neq 1$ are determined from the matrix elements $A_{ij}$ and $\lambda_0$. The eigenvectors associated with $\lambda_0$ differ only by the value chosen for $c_1$. They are therefore all given by:

$$| \psi_0(c_1) \rangle = \sum_j \alpha_j^0 \, c_1 | u_j \rangle = c_1 | \psi_0 \rangle \tag{D-17}$$

with:

$$| \psi_0 \rangle = \sum_j \alpha_j^0 | u_j \rangle \tag{D-18}$$

Therefore, when $\lambda_0$ is a simple root of the characteristic equation, only one eigenvector corresponds to it (to within a constant factor): it is a non-degenerate eigenvalue.

(*ii*) When $\lambda_0$ is a multiple root of order $q > 1$ of the characteristic equation, there are two possibilities:

– in general, when $\lambda = \lambda_0$, the system (D-13) is still composed of $(N - 1)$ independent equations. Only one eigenvector then corresponds to the eigenvalue $\lambda_0$. The operator $A$ cannot be diagonalized in this case: the eigenvectors of $A$ are not sufficiently numerous for one to be able to construct with them alone a basis of the state space.

– nevertheless, when $\lambda = \lambda_0$, it may happen that the system (D-13) has only $(N - p)$ independent equations (where $p$ is a number greater than 1 but not larger than $q$). To the eigenvalue $\lambda_0$ there then corresponds an eigensubspace of dimension $p$, and $\lambda_0$ is a $p$-fold degenerate eigenvalue. Let us assume, for example, that, for $\lambda = \lambda_0$, (D-13) is composed of $(N - 2)$ linearly independent equations. These equations enable us to calculate the coefficients $c_j$ in terms of any two of them, for example $c_1$ and $c_2$:

$$c_j = \beta_j^0 c_1 + \gamma_j^0 c_2 \tag{D-19}$$

(obviously: $\beta_1^0 = \gamma_2^0 = 1 ; \gamma_1^0 = \beta_2^0 = 0$). All the eigenvectors associated with $\lambda_0$ are then of the form:

$$| \psi_0(c_1, c_2) \rangle = c_1 | \psi_0^1 \rangle + c_2 | \psi_0^2 \rangle \tag{D-20}$$

with:

$$| \psi_0^1 \rangle = \sum_j \beta_j^0 | u_j \rangle$$

$$| \psi_0^2 \rangle = \sum_j \gamma_j^0 \, | u_j \rangle \tag{D-21}$$

The vectors $| \psi_0(c_1, c_2) \rangle$ do indeed constitute a two-dimensional vector space, this being characteristic of a two-fold degenerate eigenvalue.

When an operator is Hermitian, it can be shown that the degree of degeneracy $p$ of an eigenvalue $\lambda$ is always equal to the multiplicity $q$ of the corresponding root in the characteristic equation. Since, in most cases, we shall be studying only Hermitian operators, we shall only need to know the multiplicity of each root of (D-14) to obtain immediately the dimension of the corresponding eigensubspace. Thus, in a space of finite dimension $N$, a Hermitian operator always has $N$ linearly independent eigenvectors (we shall see later that they can be chosen to be orthonormal): this operator can therefore be diagonalized (§ D-2-b).

## 2. Observables

### a. PROPERTIES OF THE EIGENVALUES AND EIGENVECTORS OF A HERMITIAN OPERATOR

We shall now consider the very important case in which the operator $A$ is Hermitian:

$$A^\dagger = A \tag{D-22}$$

(*i*) *The eigenvalues of a Hermitian operator are real*

Taking the scalar product of the eigenvalue equation (D-1) by $| \psi \rangle$, we obtain:

$$\langle \psi | A | \psi \rangle = \lambda \langle \psi | \psi \rangle \tag{D-23}$$

But $\langle \psi | A | \psi \rangle$ is a real number if $A$ is Hermitian, as we see from:

$$\langle \psi | A | \psi \rangle^* = \langle \psi | A^\dagger | \psi \rangle = \langle \psi | A | \psi \rangle \tag{D-24}$$

where the last equation follows from hypothesis (D-22). Since $\langle \psi | A | \psi \rangle$ and $\langle \psi | \psi \rangle$ are real, equation (D-23) implies that $\lambda$ must also be real.

If $A$ is Hermitian, we can, in (D-8), replace $A$ by $A^\dagger$ and $\lambda$ by $\lambda^*$, since we have just shown that $\lambda$ is real. Thus we obtain:

$$\langle \psi | A = \lambda \langle \psi | \tag{D-25}$$

which shows that $\langle \psi |$ is also an eigenbra of $A$ with the real eigenvalue $\lambda$. Therefore, whatever the ket $| \varphi \rangle$:

$$\langle \psi | A | \varphi \rangle = \lambda \langle \psi | \varphi \rangle \tag{D-26}$$

The Hermitian operator $A$ is said to act on the left in (D-26).

(*ii*) *Two eigenvectors of a Hermitian operator corresponding to two different eigenvalues are orthogonal.*

Consider two eigenvectors $| \psi \rangle$ and $| \varphi \rangle$ of the Hermitian operator $A$:

$$A | \psi \rangle = \lambda | \psi \rangle \tag{D-27-a}$$
$$A | \varphi \rangle = \mu | \varphi \rangle \tag{D-27-b}$$

Since $A$ is Hermitian, (D-27-b) can be written in the form:

$$\langle \varphi | A = \mu \langle \varphi | \tag{D-28}$$

Then multiply (D-27-a) by $\langle \varphi |$ on the left and (D-28) by $| \psi \rangle$ on the right:

$$\langle \varphi | A | \psi \rangle = \lambda \langle \varphi | \psi \rangle \tag{D-29-a}$$

$$\langle \varphi | A | \psi \rangle = \mu \langle \varphi | \psi \rangle \tag{D-29-b}$$

Subtracting (D-29-b) from (D-29-a), we find:

$$(\lambda - \mu) \langle \varphi | \psi \rangle = 0 \tag{D-30}$$

Consequently, if $(\lambda - \mu) \neq 0$, $| \varphi \rangle$ and $| \psi \rangle$ are orthogonal.

### b. DEFINITION OF AN OBSERVABLE

When $\mathscr{E}$ is finite-dimensional, we have seen (§ D-1-b) that it is always possible to form a basis with the eigenvectors of a Hermitian operator. When $\mathscr{E}$ is infinite-dimensional, this is no longer necessarily the case. This is why it is useful to introduce a new concept, that of an observable.

Consider a Hermitian operator $A$. For simplicity, we shall assume that the set of its eigenvalues forms a discrete spectrum $\{ a_n ; n = 1, 2, ... \}$, and we shall indicate later the modifications that must be made when all or part of this spectrum is continuous. The degree of degeneracy of the eigenvalue $a_n$ will be denoted by $g_n$ (if $g_n = 1$, $a_n$ is non-degenerate). We shall denote by $| \psi_n^i \rangle$ ($i = 1, 2, ..., g_n$) $g_n$ linearly independent vectors chosen in the eigensubspace $\mathscr{E}_n$ of $a_n$:

$$A | \psi_n^i \rangle = a_n | \psi_n^i \rangle ; \quad i = 1, 2, ..., g_n \tag{D-31}$$

We have just shown that every vector belonging to $\mathscr{E}_n$ is orthogonal to every vector of another subspace $\mathscr{E}_{n'}$, associated with $a_{n'} \neq a_n$; therefore:

$$\langle \psi_n^i | \psi_{n'}^j \rangle = 0 \quad \text{for } n \neq n' \text{ and arbitrary } i \text{ and } j \tag{D-32}$$

Inside each subspace $\mathscr{E}_n$, the $| \psi_n^i \rangle$ can always be chosen orthonormal, that is, such that:

$$\langle \psi_n^i | \psi_n^j \rangle = \delta_{ij} \tag{D-33}$$

If such a choice is made, the result is an *orthonormal system of eigenvectors of $A$*: the $| \psi_n^i \rangle$ satisfy the relations:

$$\boxed{\langle \psi_n^i | \psi_{n'}^{i'} \rangle = \delta_{nn'} \delta_{ii'}} \tag{D-34}$$

obtained by regrouping (D-32) and (D-33).

By definition, the Hermitian operator $A$ is an *observable* if this orthonormal system of vectors *forms a basis* in the state space. This can be expressed by the closure relation:

$$\boxed{\sum_{n=1}^{\infty} \sum_{i=1}^{g_n} | \psi_n^i \rangle \langle \psi_n^i | = \mathbb{1}} \tag{D-35}$$

COMMENTS:

(*i*) Since the $g_n$ vectors $| \psi_n^i \rangle$ ($i = 1, 2, ..., g_n$) which span the eigensubspace $\mathscr{E}_n$ of $a_n$ are orthonormal, the projector $P_n$ onto this subspace $\mathscr{E}_n$ can be written (*cf.* § B-3-b-γ):

$$P_n = \sum_{i=1}^{g_n} | \psi_n^i \rangle \langle \psi_n^i | \tag{D-36-a}$$

The observable $A$ is then given by:

$$A = \sum_n a_n \, P_n \tag{D-36-b}$$

(it is easy to verify that the action of both sides of this equation on all the kets $| \psi_n^i \rangle$ gives the same result).

(*ii*) Relation (D-35) can be generalized to include cases where the spectrum of eigenvalues is continuous by using the rules given in table (II-3). For example, consider a Hermitian operator whose spectrum is composed of a discrete part $\{ a_n$ (degree of degeneracy $g_n$) $\}$ and a continuous part $a(\nu)$ (assumed to be non-degenerate):

$$A \, | \psi_n^i \rangle = a_n \, | \psi_n^i \rangle ; \quad n = 1, 2, ... \quad i = 1, 2, ... \, g_n \tag{D-37-a}$$

$$A \, | \psi_\nu \rangle = a(\nu) \, | \psi_\nu \rangle ; \quad \nu_1 < \nu < \nu_2 \tag{D-37-b}$$

These vectors can always be chosen in such a way that they form an "orthonormal" system:

$$\begin{aligned} \langle \psi_n^i | \psi_{n'}^{i'} \rangle &= \delta_{nn'} \delta_{ii'} \\ \langle \psi_\nu | \psi_{\nu'} \rangle &= \delta(\nu - \nu') \\ \langle \psi_n^i | \psi_\nu \rangle &= 0 \end{aligned} \tag{D-38}$$

$A$ will be said to be an observable if this system forms a basis, that is, if:

$$\sum_n \sum_{i=1}^{g_n} | \psi_n^i \rangle \langle \psi_n^i | + \int_{\nu_1}^{\nu_2} \mathrm{d}\nu \, | \psi_\nu \rangle \langle \psi_\nu | = \mathbb{1} \tag{D-39}$$

## c. EXAMPLE: THE PROJECTOR $P_\psi$

Let us show that $P_\psi = | \psi \rangle \langle \psi |$ (with $\langle \psi | \psi \rangle = 1$) is an observable. We have already pointed out (§ B-4-e) that it is Hermitian, and that its eigenvalues are 1 and 0 (§ D-1-a); the first one is simple (associated eigenvector: $| \psi \rangle$), the second one is infinitely degenerate (associated eigenvectors: all kets orthogonal to $| \psi \rangle$).

Consider an arbitrary ket $| \varphi \rangle$ in the state space. It can always be written in the form:

$$| \varphi \rangle = P_\psi | \varphi \rangle + (\mathbb{1} - P_\psi) | \varphi \rangle \tag{D-40}$$

$P_\psi|\varphi\rangle$ is an eigenket of $P_\psi$ with the eigenvalue 1. Now, since $P_\psi^2 = P_\psi$:

$$P_\psi(P_\psi|\varphi\rangle) = P_\psi^2|\varphi\rangle = P_\psi|\varphi\rangle \tag{D-41}$$

$(\mathbb{1} - P_\psi)|\varphi\rangle$ is also an eigenket of $P_\psi$, but with the eigenvalue 0, as we see from:

$$P_\psi(\mathbb{1} - P_\psi)|\varphi\rangle = (P_\psi - P_\psi^2)|\varphi\rangle = 0. \tag{D-42}$$

Every ket $|\varphi\rangle$ can thus be expanded on these eigenkets of $P_\psi$; therefore, $P_\psi$ is an observable.

We shall see in § E-2 two other important examples of observables.

## 3. Sets of commuting observables

### a. IMPORTANT THEOREMS

#### α. Theorem I

*If two operators $A$ and $B$ commute, and if $|\psi\rangle$ is an eigenvector of $A$, $B|\psi\rangle$ is also an eigenvector of $A$, with the same eigenvalue.*

We know that, if $|\psi\rangle$ is an eigenvector of $A$, we have:

$$A|\psi\rangle = a|\psi\rangle \tag{D-43}$$

Applying $B$ to both sides of this equation, we obtain:

$$BA|\psi\rangle = aB|\psi\rangle \tag{D-44}$$

Since we assumed that $A$ and $B$ commute, we also have, replacing $BA$ on the left-hand side by $AB$:

$$A(B|\psi\rangle) = a(B|\psi\rangle) \tag{D-45}$$

This equation expresses the fact that $B|\psi\rangle$ is an eigenvector of $A$, with the eigenvalue $a$; the theorem is therefore proved.

Two cases may then arise:

($i$) If $a$ is a nondegenerate eigenvalue, all the eigenvectors associated with it are by definition colinear, and $B|\psi\rangle$ is necessarily proportional to $|\psi\rangle$. Therefore $|\psi\rangle$ is also an eigenvector of $B$.

($ii$) If $a$ is a degenerate eigenvalue, it can only be said that $B|\psi\rangle$ belongs to the eigensubspace $\mathscr{E}_a$ of $A$, corresponding to the eigenvalue $a$. Therefore, for any $|\psi\rangle \in \mathscr{E}_a$, we have:

$$B|\psi\rangle \in \mathscr{E}_a \tag{D-46}$$

$\mathscr{E}_a$ is said to be globally invariant (or stable) under the action of $B$. Theorem I can therefore be stated in another form:

*Theorem I':* If two operators $A$ and $B$ commute, every eigensubspace of $A$ is globally invariant under the action of $B$.

#### β. Theorem II

*If two observables $A$ and $B$ commute, and if $|\psi_1\rangle$ and $|\psi_2\rangle$ are two eigenvectors of $A$ with different eigenvalues, the matrix element $\langle\psi_1|B|\psi_2\rangle$ is zero.*

If $|\psi_1\rangle$ and $|\psi_2\rangle$ are eigenvectors of $A$, we can write:

$$A|\psi_1\rangle = a_1|\psi_1\rangle$$
$$A|\psi_2\rangle = a_2|\psi_2\rangle \tag{D-47}$$

According to theorem I, the fact that $A$ and $B$ commute means that $B|\psi_2\rangle$ is an eigenvector of $A$, with the eigenvalue $a_2$. $B|\psi_2\rangle$ is therefore (*cf.* § D-2-a) orthogonal to $|\psi_1\rangle$ (eigenvector of eigenvalue $a_1 \neq a_2$), which can be written:

$$\langle\psi_1|B|\psi_2\rangle = 0 \tag{D-48}$$

The theorem is therefore proved. Another proof can be given, which does not involve theorem I : since the operator $[A, B]$ is zero, we have:

$$\langle\psi_1|(AB - BA)|\psi_2\rangle = 0 \tag{D-49}$$

Using (D-47) and the Hermiticity of $A$ [*cf.* equation (D-25)], we obtain:

$$\langle\psi_1|AB|\psi_2\rangle = a_1\langle\psi_1|B|\psi_2\rangle$$
$$\langle\psi_1|BA|\psi_2\rangle = a_2\langle\psi_1|B|\psi_2\rangle \tag{D-50}$$

and (D-49) can be rewritten in the form:

$$(a_1 - a_2)\langle\psi_1|B|\psi_2\rangle = 0 \tag{D-51}$$

Since, by hypothesis, $(a_1 - a_2)$ is not zero, we can deduce (D-48) from this.

γ. *Theorem III (fundamental)*

*If two observables A and B commute, one can construct an orthonormal basis of the state space with eigenvectors common to A and B.*

Consider two commuting observables, $A$ and $B$. In order to simplify the notation, we shall assume that their spectra are entirely discrete. Since $A$ is an observable, there exists at least one orthonormal system of eigenvectors of $A$ which forms a basis in the state space. We shall denote these vectors by $|u_n^i\rangle$:

$$A|u_n^i\rangle = a_n|u_n^i\rangle; \quad n = 1, 2, \ldots$$
$$i = 1, 2, \ldots, g_n \tag{D-52}$$

$g_n$ is the degree of degeneracy of the eigenvalue $a_n$, that is, the dimension of the corresponding eigensubspace $\mathscr{E}_n$. We have:

$$\langle u_n^i|u_{n'}^{i'}\rangle = \delta_{nn'}\delta_{ii'} \tag{D-53}$$

What does the matrix look like which represents $B$ in the $\{|u_n^i\rangle\}$ basis? We know (*cf.* theorem II) that the matrix elements $\langle u_n^i|B|u_{n'}^{i'}\rangle$ are zero when $n \neq n'$ (on the other hand, we can say nothing *a priori* about what happens for $n = n'$ and $i \neq i'$). Let us arrange the basis vectors $|u_n^i\rangle$ in the order :

$$|u_1^1\rangle, |u_1^2\rangle, \ldots, |u_1^{g_1}\rangle \quad ; \quad |u_2^1\rangle, \ldots, |u_2^{g_2}\rangle \quad ; \quad |u_3^1\rangle, \ldots$$

We then obtain for $B$ a "block-diagonal" matrix, that is, of the form:

|  | $\mathscr{E}_1$ | $\mathscr{E}_2$ | $\mathscr{E}_3$ | ... |
|---|---|---|---|---|
| $\mathscr{E}_1$ |  | 0 | 0 | 0 |
| $\mathscr{E}_2$ | 0 |  | 0 | 0 |
| $\mathscr{E}_3$ | 0 | 0 |  | 0 |
| ⋮ | 0 | 0 | 0 |  |

(D-54)

(only the shaded parts contain non-zero matrix elements). The fact that the eigensubspaces $\mathscr{E}_n$ are globally invariant under the action of $B$ (*cf.* § α) is evident from this matrix.

Two cases can then arise:

(*i*) When $a_n$ is a nondegenerate eigenvalue of $A$, there exists only one eigenvector $|u_n\rangle$ of $A$, of eigenvalue $a_n$ (the index $i$ in $|u_n\rangle$ is then unnecessary): the dimension $g_n$ of $\mathscr{E}_n$ is then equal to 1. In the matrix (D-54), the corresponding "block" then reduces to a $1 \times 1$ matrix, that is, to a simple number. In the column associated with $|u_n\rangle$, all the other matrix elements are zero. This expresses the fact (*cf.* § α-*i*) that $|u_n\rangle$ is an eigenvector common to $A$ and $B$.

(*ii*) When $a_n$ is a degenerate eigenvalue of $A$ ($g_n > 1$), the "block" which represents $B$ in $\mathscr{E}_n$ is not, in general, diagonal : the $|u_n^i\rangle$ are not, in general, eigenvectors of $B$.

It can be seen, nevertheless, that, since the action of $A$ on each of the $g_n$ vectors $|u_n^i\rangle$ reduces to a simple multiplication by $a_n$, the matrix representing the restriction of $A$ to within $\mathscr{E}_n$ is equal to $a_n I$ (where $I$ is the $g_n \times g_n$ unit matrix). This expresses the fact that an arbitrary ket of $\mathscr{E}_n$ is an eigenvector of $A$ with the eigenvalue $a_n$. The choice, in $\mathscr{E}_n$ of a basis such as $\{ |u_n^i\rangle ; i = 1, 2, ..., g_n \}$ is therefore arbitrary. Whatever this basis, the matrix representing $A$ in $\mathscr{E}_n$ is always diagonal and equal to $a_n I$. We shall use this property to obtain a basis of $\mathscr{E}_n$ composed of vectors which are also eigenvectors of $B$.

The matrix representing $B$ in $\mathscr{E}_n$, when the basis chosen is

$$\{ |u_n^i\rangle ; i = 1, 2, ..., g_n \},$$

has for its elements:

$$\beta_{ij}^{(n)} = \langle u_n^i | B | u_n^j \rangle \tag{D-55}$$

This matrix is Hermitian ($\beta_{ji}^{(n)*} = \beta_{ij}^{(n)}$), since $B$ is a Hermitian operator. It is therefore diagonalizable, that is, *one can find in* $\mathscr{E}_n$ *a new basis* $\{ |v_n^i\rangle ; i = 1, 2, ..., g_n \}$ *in which* $B$ *is represented by a diagonal matrix*

$$\langle v_n^i | B | v_n^j \rangle = \beta_i^{(n)} \delta_{ij} \tag{D-56}$$

This means that the new basis vectors in $\mathscr{E}_n$ are eigenvectors of $B$:

$$B \,|\, v_n^i \rangle = \beta_i^{(n)} \,|\, v_n^i \rangle \tag{D-57}$$

As we saw above, these vectors are automatically eigenvectors of $A$ with an eigenvalue $a_n$ since they belong to $\mathscr{E}_n$. Let us stress the fact that *the eigenvectors of $A$ associated with degenerate eigenvalues are not necessarily eigenvectors of $B$*. What we have just shown is that *it is always possible to choose, in every eigensubspace of $A$, a basis of eigenvectors common to $A$ and $B$.*

If we perform this operation in all the subspaces $\mathscr{E}_n$, we obtain a basis of $\mathscr{E}$, formed by eigenvectors common to $A$ and $B$. The theorem is therefore proved.

COMMENTS:

(*i*) From now on, we shall denote by $|\, u_{n,p}^i \rangle$ the eigenvectors common to $A$ and $B$.

$$\begin{aligned} A \,|\, u_{n,p}^i \rangle &= a_n \,|\, u_{n,p}^i \rangle \\ B \,|\, u_{n,p}^i \rangle &= b_p \,|\, u_{n,p}^i \rangle \end{aligned} \tag{D-58}$$

The indices $n$ and $p$ which appear in $|\, u_{n,p}^i \rangle$ enable us to specify the eigenvalues $a_n$ and $b_p$ of $A$ and $B$. The additional index $i$ will eventually be used to distinguish between the different basis vectors which correspond to the same eigenvalues $a_n$ and $b_p$ (§ b below).

(*ii*) The converse of theorem III is very simple to prove: if there exists a basis of eigenvectors common to $A$ and $B$, these two observables commute. From (D-58), it is easy to deduce:

$$\begin{aligned} AB \,|\, u_{n,p}^i \rangle &= b_p \, A \,|\, u_{n,p}^i \rangle = b_p \, a_n \,|\, u_{n,p}^i \rangle \\ BA \,|\, u_{n,p}^i \rangle &= a_n \, B \,|\, u_{n,p}^i \rangle = a_n \, b_p \,|\, u_{n,p}^i \rangle \end{aligned} \tag{D-59}$$

and, subtracting these equations:

$$[A, B] \,|\, u_{n,p}^i \rangle = 0 \tag{D-60}$$

This relation is valid for all $i$, $n$ and $p$. Since, by hypothesis, the vectors $|\, u_{n,p}^i \rangle$ form a basis, (D-60) entails $[A, B] = 0$.

(*iii*) We shall occasionally solve the eigenvalue equation of an observable $C$ such that:

$$C = A + B \quad \text{with} \quad [A, B] = 0 \tag{D-61}$$

where $A$ and $B$ are also observables.

When one has found a basis $\{ \,|\, u_{n,p}^i \rangle \}$ of eigenvectors common to $A$ and $B$, the problem is solved, since we see immediately that $|\, u_{n,p}^i \rangle$ is also an eigenvector of $C$, with an eigenvalue $a_n + b_p$. The fact that $\{ \,|\, u_{n,p}^i \rangle \}$ constitutes a basis is obviously essential: this allows us, for example, to show simply that all the eigenvalues of $C$ are of the form $a_n + b_p$.

### b. COMPLETE* SETS OF COMMUTING OBSERVABLES (C.S.C.O.)**

Consider an observable $A$ and a basis of $\mathscr{E}$ composed of eigenvectors $|u_n^i\rangle$ of $A$. If none of the eigenvalues of $A$ is degenerate, the various basis vectors of $\mathscr{E}$ can be labelled by the eigenvalue $a_n$ (the index $i$ in $|u_n^i\rangle$ being in this case unnecessary). All the eigensubspaces $\mathscr{E}_n$ are then one-dimensional. Therefore, specifying the eigenvalue determines in a unique way the corresponding eigenvector (to within a constant factor). In other words, there exists only one basis of $\mathscr{E}$ formed by the eigenvectors of $A$ (we shall not consider here as distinct two bases whose vectors are proportional). It is then said that the observable $A$ constitutes, by itself, a C.S.C.O.

If, on the other hand, one or several eigenvalues of $A$ are degenerate, the situation is different. Specifying $a_n$ is no longer always sufficient to characterize a basis vector, since there correspond several independent vectors to degenerate eigenvalues. In this case, the basis of eigenvectors of $A$ is obviously not unique. One can choose any basis inside each of the eigensubspaces $\mathscr{E}_n$ of dimension greater than 1.

Let us then choose another observable $B$ which commutes with $A$, and let us construct an orthonormal basis of eigenvectors common to $A$ and $B$. By definition, $A$ and $B$ form a C.S.C.O. if this basis is unique (to within a phase factor for each of the basis vectors), that is, if, to each of the possible pairs of eigenvalues $\{ a_n, b_p \}$, there corresponds only one basis vector.

COMMENT:

In §a, we constructed a basis of eigenvectors common to $A$ and $B$ by solving the eigenvalue equation of $B$ inside each eigensubspace $\mathscr{E}_n$. For $A$ and $B$ to constitute a C.S.C.O., it is necessary and sufficient that, inside each of these subspaces, all the $g_n$ eigenvalues of $B$ be distinct. Since all the vectors of $\mathscr{E}_n$ correspond to the same eigenvalue $a_n$ of $A$, the $g_n$ vectors $|v_n^i\rangle$ can then be distinguished by the eigenvalue of $B$ which is associated with them. Note that it is not necessary that all the eigenvalues of $B$ be non-degenerate. Vectors $|v_n^i\rangle$ belonging to two distinct subspaces $\mathscr{E}_n$ can have the same eigenvalue for $B$. Moreover, if all the eigenvalues of $B$ were non-degenerate, $B$ alone would constitute a C.S.C.O.

If, for at least one of the possible pairs $\{ a_n, b_p \}$, there exist several independent vectors which are eigenvectors of $A$ and $B$ with these eigenvalues, the set $\{ A, B \}$ is not complete. Let us add to it, then, a third observable $C$, which commutes with both $A$ and $B$. We can then use the same argument as in § a above, generalizing it in the following way. When to a pair $\{ a_n, b_p \}$, there corresponds only one vector, this vector is necessarily an eigenvector of $C$. If there are several vectors, they form an eigensubspace $\mathscr{E}_{n,p}$, in which it is possible to choose a basis formed by vectors which are also eigenvectors of $C$. One can thus construct, in the state space, an orthonormal basis formed by eigenvectors common to $A$, $B$ and $C$. $A$, $B$ and $C$ form a C.S.C.O. if this basis is unique

* The word "complete" is used here in a sense which is totally unrelated to those referred to in the note of §A-2-a, p. 97. This use of the word "complete" is customary in quantum mechanics.

** To have a good understanding of the important concepts introduced in this section, the reader should apply them to a concrete example such as the one discussed in complement $H_{II}$ (solved exercises 11 and 12).

(to within multiplicative factors). Specifying a possible set of eigenvalues $\{ a_n, b_p, c_r \}$ of $A$, $B$, $C$ then characterizes only one of the vectors of this basis. If this is not the case, one adds to $A$, $B$, $C$ an observable $D$ which commutes with each of these three operators, and so on. In general, we are thus led to the following:

*By definition, a set of observables $A$, $B$, $C$... is called a complete set of commuting observables if*

(*i*) *all the observables $A$, $B$, $C$... commute by pairs,*

(*ii*) *specifying the eigenvalues of all the operators $A$, $B$, $C$... determines a unique* (to within a multiplicative factor) *common eigenvector.*

An equivalent way of saying this is the following:
*A set of observables $A$, $B$, $C$... is a complete set of commuting observables if there exists a unique orthonormal basis of common eigenvectors* (to within phase factors).

C.S.C.O.'s play an important role in quantum mechanics. We shall see numerous examples of them (see, in particular, § E-2-d).

COMMENTS:

(*i*) If $\{ A, B \}$ is a C.S.C.O., another C.S.C.O. can be obtained by adding to it any observable $C$, on the condition, of course, that it commutes with $A$ and $B$. However, it is generally understood that one is confined to "minimal" sets, that is, those which cease to be complete when any one of the observables is omitted.

(*ii*) Let $\{ A, B, C \}$ be a complete set of commuting observables. Since the specification of the eigenvalues $a_n$, $b_p$, $c_r$... determines a ket of the corresponding basis (to within a constant factor), this ket is sometimes denoted by $| a_n, b_p, c_r, ... \rangle$.

(*iii*) For a given physical system, there exist several complete sets of commuting observables. We shall see a particular example of this in § E-2-d.

## E. TWO IMPORTANT EXAMPLES OF REPRESENTATIONS AND OBSERVABLES

In this paragraph, we shall return to the $\mathscr{F}$-space of wave functions of a particle, or, more exactly, to the state space $\mathscr{E}_{\mathbf{r}}$ which is associated with it and which we shall define in the following way. Let there correspond to every wave function $\psi(\mathbf{r})$ a ket $| \psi \rangle$ belonging to $\mathscr{E}_{\mathbf{r}}$; this correspondence is linear. Moreover, the scalar product of two kets coincides with that of the functions which are associated with them:

$$\langle \varphi | \psi \rangle = \int d^3r \; \varphi^*(\mathbf{r}) \; \psi(\mathbf{r}) \tag{E-1}$$

$\mathscr{E}_{\mathbf{r}}$ is thus the state space of a (spinless) particle.

We are going to define and study, in this space, two representations and two operators which are particularly important. In chapter III we shall associate them with the position and the momentum of the particle under consideration. They will enable us, moreover, to apply and illustrate the concepts which we have introduced in the preceding sections.

## 1. The $\{ | \mathbf{r} \rangle \}$ and $\{ | \mathbf{p} \rangle \}$ representations

### a. DEFINITION

In §§ A-3-a and A-3-b, we introduced two particular "bases" of $\mathscr{F}$: $\{ \xi_{\mathbf{r}_0}(\mathbf{r}) \}$ and $\{ v_{\mathbf{p}_0}(\mathbf{r}) \}$. They are not composed of functions belonging to $\mathscr{F}$:

$$\xi_{\mathbf{r}_0}(\mathbf{r}) = \delta(\mathbf{r} - \mathbf{r}_0) \tag{E-2-a}$$

$$v_{\mathbf{p}_0}(\mathbf{r}) = (2\pi\hbar)^{-3/2} \, e^{\frac{i}{\hbar}\mathbf{p}_0 \cdot \mathbf{r}} \tag{E-2-b}$$

However, every sufficiently regular square-integrable function can be expanded in one or the other of these "bases".

This is why we shall remove the quotation marks and associate a ket with each of the functions of these bases (*cf.* § B-2-c). The ket associated with $\xi_{\mathbf{r}_0}(\mathbf{r})$ will be denoted simply by $| \mathbf{r}_0 \rangle$, and that associated with $v_{\mathbf{p}_0}(\mathbf{r})$, by $| \mathbf{p}_0 \rangle$:

$$\boxed{\begin{aligned} \xi_{\mathbf{r}_0}(\mathbf{r}) &\Longleftrightarrow | \mathbf{r}_0 \rangle \\ v_{\mathbf{p}_0}(\mathbf{r}) &\Longleftrightarrow | \mathbf{p}_0 \rangle \end{aligned}} \qquad \begin{aligned} &\text{(E-3-a)} \\ &\text{(E-3-b)} \end{aligned}$$

Using the bases $\{ \xi_{\mathbf{r}_0}(\mathbf{r}) \}$ and $\{ v_{\mathbf{p}_0}(\mathbf{r}) \}$ of $\mathscr{F}$, we thus define in $\mathscr{E}_{\mathbf{r}}$ two representations: the $\{ | \mathbf{r}_0 \rangle \}$ representation and the $\{ | \mathbf{p}_0 \rangle \}$ representation. A basis vector of the first one is characterized by three "continuous indices" $x_0$, $y_0$ and $z_0$, which are the coordinates of a point in three-dimensional space; for the second, the three indices are also the components of an ordinary vector.

### b. ORTHONORMALIZATION AND CLOSURE RELATIONS

Let us calculate $\langle \mathbf{r}_0 | \mathbf{r}'_0 \rangle$. Using the definition of the scalar product in $\mathscr{E}_{\mathbf{r}}$:

$$\langle \mathbf{r}_0 | \mathbf{r}'_0 \rangle = \int d^3r \;\, \xi^*_{\mathbf{r}_0}(\mathbf{r}) \; \xi_{\mathbf{r}'_0}(\mathbf{r}) = \delta(\mathbf{r}_0 - \mathbf{r}'_0) \tag{E-4-a}$$

where relation (A-55) has been used. In the same way:

$$\langle \mathbf{p}_0 | \mathbf{p}'_0 \rangle = \int d^3r \;\, v^*_{\mathbf{p}_0}(\mathbf{r}) \; v_{\mathbf{p}'_0}(\mathbf{r}) = \delta(\mathbf{p}_0 - \mathbf{p}'_0) \tag{E-4-b}$$

using (A-47). The bases which we have just defined are therefore orthonormal in the extended sense.

The fact that the set of the $| \mathbf{r}_0 \rangle$ or that of the $| \mathbf{p}_0 \rangle$ constitutes a basis in $\mathscr{E}_\mathbf{r}$ can be expressed by a closure relation in $\mathscr{E}_\mathbf{r}$. This is written in an analogous manner to (C-10), integrating here, however, over three indices instead of one.

We therefore have the fundamental relations:

| | |
|---|---|
| $\langle \mathbf{r}_0 \| \mathbf{r}'_0 \rangle = \delta(\mathbf{r}_0 - \mathbf{r}'_0)$ (a) | $\langle \mathbf{p}_0 \| \mathbf{p}'_0 \rangle = \delta(\mathbf{p}_0 - \mathbf{p}'_0)$ (c) |
| $\int d^3 r_0 \, \| \mathbf{r}_0 \rangle \langle \mathbf{r}_0 \| = \mathbb{1}$ (b) | $\int d^3 p_0 \, \| \mathbf{p}_0 \rangle \langle \mathbf{p}_0 \| = \mathbb{1}$ (d) |

(E-5)

c. COMPONENTS OF A KET

Consider an arbitrary ket $| \psi \rangle$, corresponding to the wave function $\psi(\mathbf{r})$. The preceding closure relations enable us to write it in either of these two forms:

$$| \psi \rangle = \int d^3 r_0 \, | \mathbf{r}_0 \rangle \langle \mathbf{r}_0 | \psi \rangle \tag{E-6-a}$$

$$| \psi \rangle = \int d^3 p_0 \, | \mathbf{p}_0 \rangle \langle \mathbf{p}_0 | \psi \rangle \tag{E-6-b}$$

The coefficients $\langle \mathbf{r}_0 | \psi \rangle$ and $\langle \mathbf{p}_0 | \psi \rangle$ can be calculated using the formulas:

$$\langle \mathbf{r}_0 | \psi \rangle = \int d^3 r \;\; \xi^*_{\mathbf{r}_0}(\mathbf{r}) \;\; \psi(\mathbf{r}) \tag{E-7-a}$$

$$\langle \mathbf{p}_0 | \psi \rangle = \int d^3 r \;\; v^*_{\mathbf{p}_0}(\mathbf{r}) \;\; \psi(\mathbf{r}) \tag{E-7-b}$$

We then find:

$$\boxed{\begin{aligned} \langle \mathbf{r}_0 | \psi \rangle &= \psi(\mathbf{r}_0) \\ \langle \mathbf{p}_0 | \psi \rangle &= \bar{\psi}(\mathbf{p}_0) \end{aligned}} \qquad \begin{aligned} &\text{(E-8-a)} \\ &\text{(E-8-b)} \end{aligned}$$

where $\bar{\psi}(\mathbf{p})$ is the Fourier transform of $\psi(\mathbf{r})$.

*The value* $\psi(\mathbf{r}_0)$ *of the wave function at the point* $\mathbf{r}_0$ *is thus shown to be the component of the ket* $| \psi \rangle$ *on the basis vector* $| \mathbf{r}_0 \rangle$ *of the* $\{ | \mathbf{r}_0 \rangle \}$ *representation*. The "wave function in momentum space" $\bar{\psi}(\mathbf{p})$ can be interpreted analogously. The possibility of characterizing $| \psi \rangle$ by $\psi(\mathbf{r})$ is thus simply a special case of the results of § C-3-a.

For example, for $| \psi \rangle = | \mathbf{p}_0 \rangle$, formula (E-8-a) gives:

$$\langle \mathbf{r}_0 | \mathbf{p}_0 \rangle = v_{\mathbf{p}_0}(\mathbf{r}_0) = (2\pi\hbar)^{-3/2} \, e^{\frac{i}{\hbar} \mathbf{p}_0 \cdot \mathbf{r}_0} \tag{E-9}$$

For $| \psi \rangle = | \mathbf{r}'_0 \rangle$, the result is indeed in agreement with the orthonormalization relation (E-5-a):

$$\langle \mathbf{r}_0 | \mathbf{r}'_0 \rangle = \xi_{\mathbf{r}'_0}(\mathbf{r}_0) = \delta(\mathbf{r}_0 - \mathbf{r}'_0) \quad \text{(E-10)}$$

Now that we have reinterpreted the wave function $\psi(\mathbf{r})$ and its Fourier transform $\bar{\psi}(\mathbf{p})$, we shall denote the basis vectors of the two representations we are studying here by $| \mathbf{r} \rangle$ and $| \mathbf{p} \rangle$, instead of $| \mathbf{r}_0 \rangle$ and $| \mathbf{p}_0 \rangle$. Formulas (E-8) can then be written:

$$\langle \mathbf{r} | \psi \rangle = \psi(\mathbf{r}) \quad \text{(E-8-a)}$$

$$\langle \mathbf{p} | \psi \rangle = \bar{\psi}(\mathbf{p}) \quad \text{(E-8-b)}$$

and the orthonormalization and closure relations (E-5) become:

$$\langle \mathbf{r} | \mathbf{r}' \rangle = \delta(\mathbf{r} - \mathbf{r}') \quad \text{(a)} \qquad \langle \mathbf{p} | \mathbf{p}' \rangle = \delta(\mathbf{p} - \mathbf{p}') \quad \text{(c)}$$

$$\int d^3r \, | \mathbf{r} \rangle \langle \mathbf{r} | = \mathbb{1} \quad \text{(b)} \qquad \int d^3p \, | \mathbf{p} \rangle \langle \mathbf{p} | = \mathbb{1} \quad \text{(d)} \qquad \text{(E-5)}$$

Of course, $\mathbf{r}$ and $\mathbf{p}$ are still considered to represent two sets of *continuous indices*, $\{ x, y, z \}$ and $\{ p_x, p_y, p_z \}$, which fix the basis kets of the $\{ | \mathbf{r} \rangle \}$ and $\{ | \mathbf{p} \rangle \}$ representations respectively.

Now let $\{ u_i(\mathbf{r}) \}$ be an orthonormal basis of $\mathscr{F}$. With each $u_i(\mathbf{r})$ is associated a ket $| u_i \rangle$ of $\mathscr{E}_\mathbf{r}$. The set $\{ | u_i \rangle \}$ forms an orthonormal basis in $\mathscr{E}_\mathbf{r}$; it therefore satisfies the closure relation:

$$\sum_i | u_i \rangle \langle u_i | = \mathbb{1} \quad \text{(E-11)}$$

Evaluate the matrix element of both sides of (E-11) between $| \mathbf{r} \rangle$ and $| \mathbf{r}' \rangle$:

$$\sum_i \langle \mathbf{r} | u_i \rangle \langle u_i | \mathbf{r}' \rangle = \langle \mathbf{r} | \mathbb{1} | \mathbf{r}' \rangle = \langle \mathbf{r} | \mathbf{r}' \rangle \quad \text{(E-12)}$$

According to (E-8-a) and (E-5-a), this relation can be written:

$$\sum_i u_i(\mathbf{r}) \; u_i^*(\mathbf{r}') = \delta(\mathbf{r} - \mathbf{r}') \quad \text{(E-13)}$$

The closure relation for the $\{ u_i(\mathbf{r}) \}$ [formula (A-32)] is therefore simply the expression in the $\{ | \mathbf{r} \rangle \}$ representation of the vectorial closure relation (E-11).

#### d. THE SCALAR PRODUCT OF TWO VECTORS

We have defined the scalar product of two kets of $\mathscr{E}_\mathbf{r}$ as being equal to that of the associated wave functions in $\mathscr{F}$ [equation (E-1)]. In light of the discussion in § c, this definition appears simply as a special case of formula (C-21). (E-1) can, in fact, be derived by inserting the closure relation (E-5-b) between $\langle \varphi |$ and $| \psi \rangle$:

$$\langle \varphi | \psi \rangle = \int d^3r \, \langle \varphi | \mathbf{r} \rangle \langle \mathbf{r} | \psi \rangle \quad \text{(E-14)}$$

and by interpreting the components $\langle \mathbf{r} | \psi \rangle$ and $\langle \mathbf{r} | \varphi \rangle$ as in (E-8-a).

If we place ourselves in the $\{ | \mathbf{p} \rangle \}$ representation, a well-known property of the Fourier transform is demonstrated (appendix I, § 2-c).

$$\begin{aligned} \langle \varphi | \psi \rangle &= \int d^3p \, \langle \varphi | \mathbf{p} \rangle \langle \mathbf{p} | \psi \rangle \\ &= \int d^3p \; \bar{\varphi}^*(\mathbf{p}) \; \bar{\psi}(\mathbf{p}) \end{aligned} \tag{E-15}$$

### e. CHANGING FROM THE $\{ | \mathbf{r} \rangle \}$ REPRESENTATION TO THE $\{ | \mathbf{p} \rangle \}$ REPRESENTATION

This is accomplished using the method indicated in § C-5, the only difference arising from the fact that we are dealing here with two continuous bases. Changing from one basis to the other brings in the numbers :

$$\langle \mathbf{r} | \mathbf{p} \rangle = \langle \mathbf{p} | \mathbf{r} \rangle^* = (2\pi\hbar)^{-3/2} \, e^{\frac{i}{\hbar}\mathbf{p}.\mathbf{r}} \tag{E-16}$$

A given ket $| \psi \rangle$ is represented by $\langle \mathbf{r} | \psi \rangle = \psi(\mathbf{r})$ in the $\{ | \mathbf{r} \rangle \}$ representation and by $\langle \mathbf{p} | \psi \rangle = \bar{\psi}(\mathbf{p})$ in the $\{ | \mathbf{p} \rangle \}$ representation. We already know [formula (E-7-b)] that $\psi(\mathbf{r})$ and $\bar{\psi}(\mathbf{p})$ are related by a Fourier transform. This is indeed what the formulas for the representation change yield:

$$\langle \mathbf{r} | \psi \rangle = \int d^3p \, \langle \mathbf{r} | \mathbf{p} \rangle \langle \mathbf{p} | \psi \rangle$$

that is:

$$\psi(\mathbf{r}) = (2\pi\hbar)^{-3/2} \int d^3p \; e^{\frac{i}{\hbar}\mathbf{p}.\mathbf{r}} \; \bar{\psi}(\mathbf{p}) \tag{E-17}$$

Inversely:

$$\langle \mathbf{p} | \psi \rangle = \int d^3r \, \langle \mathbf{p} | \mathbf{r} \rangle \langle \mathbf{r} | \psi \rangle$$

that is:

$$\bar{\psi}(\mathbf{p}) = (2\pi\hbar)^{-3/2} \int d^3r \; e^{-\frac{i}{\hbar}\mathbf{p}.\mathbf{r}} \; \psi(\mathbf{r}) \tag{E-18}$$

By applying the general formula (C-56), one can easily pass from the matrix elements $\langle \mathbf{r}' | A | \mathbf{r} \rangle = A(\mathbf{r}', \mathbf{r})$ of an operator $A$ in the $\{ | \mathbf{r} \rangle \}$ representation to the matrix elements $\langle \mathbf{p}' | A | \mathbf{p} \rangle = A(\mathbf{p}', \mathbf{p})$ of the same operator in the $\{ | \mathbf{p} \rangle \}$ representation:

$$A(\mathbf{p}', \mathbf{p}) = (2\pi\hbar)^{-3} \int d^3r \int d^3r' \; e^{\frac{i}{\hbar}(\mathbf{p}.\mathbf{r} - \mathbf{p}'.\mathbf{r}')} \; A(\mathbf{r}', \mathbf{r}) \tag{E-19}$$

An analogous formula enables one to calculate $A(\mathbf{r}', \mathbf{r})$ from $A(\mathbf{p}', \mathbf{p})$.

## 2. The R and P operators

### a. DEFINITION

Let $| \psi \rangle$ be an arbitrary ket of $\mathscr{E}_{\mathbf{r}}$ and let $\langle \mathbf{r} | \psi \rangle = \psi(\mathbf{r}) \equiv \psi(x, y, z)$ be the corresponding wave function. Using the definition of the operator $X$, the ket:

$$| \psi' \rangle = X | \psi \rangle \tag{E-20}$$

is represented, in the $\{ | \mathbf{r} \rangle \}$ basis, by the function $\langle \mathbf{r} | \psi' \rangle = \psi'(\mathbf{r}) \equiv \psi'(x, y, z)$ such that:

$$\psi'(x, y, z) = x \ \psi(x, y, z) \tag{E-21}$$

In the $\{ | \mathbf{r} \rangle \}$ representation, the $X$ operator therefore coincides with the operator which multiplies by $x$. Although we characterize $X$ by the way in which it transforms the wave functions, it is an operator which acts in the state space $\mathscr{E}_{\mathbf{r}}$. We can introduce two other operators, $Y$ and $Z$, in an analogous manner. Thus we define $X$, $Y$ and $Z$ by the formulas:

$$\langle \mathbf{r} | X | \psi \rangle = x \langle \mathbf{r} | \psi \rangle \tag{E-22-a}$$
$$\langle \mathbf{r} | Y | \psi \rangle = y \langle \mathbf{r} | \psi \rangle \tag{E-22-b}$$
$$\langle \mathbf{r} | Z | \psi \rangle = z \langle \mathbf{r} | \psi \rangle \tag{E-22-c}$$

where the numbers $x$, $y$, $z$ are precisely the three indices which label the ket $| \mathbf{r} \rangle$. $X$, $Y$ and $Z$ will be considered to be the "components" of a "vector operator" $\mathbf{R}$: for the moment, we shall treat this simply as a condensed notation, suggested by the fact that $x$, $y$, $z$ are the components of the ordinary vector $\mathbf{r}$.

Manipulation of the $X$, $Y$, $Z$ operators is particularly simple in the $\{ | \mathbf{r} \rangle \}$ representation. For example, in order to calculate the matrix element $\langle \varphi | X | \psi \rangle$, all we need to do is insert the closure relation (E-5-b) between $\langle \varphi |$ and $X$ and use definition (E-22):

$$\begin{aligned} \langle \varphi | X | \psi \rangle &= \int d^3r \langle \varphi | \mathbf{r} \rangle \langle \mathbf{r} | X | \psi \rangle \\ &= \int d^3r \ \varphi^*(\mathbf{r}) \ x \ \psi(\mathbf{r}) \end{aligned} \tag{E-23}$$

Similarly, we define the vector operator $\mathbf{P}$ by its components $P_x$, $P_y$, $P_z$, whose action, in the $\{ | \mathbf{p} \rangle \}$ representation, is given by:

$$\langle \mathbf{p} | P_x | \psi \rangle = p_x \langle \mathbf{p} | \psi \rangle \tag{E-24-a}$$
$$\langle \mathbf{p} | P_y | \psi \rangle = p_y \langle \mathbf{p} | \psi \rangle \tag{E-24-b}$$
$$\langle \mathbf{p} | P_z | \psi \rangle = p_z \langle \mathbf{p} | \psi \rangle \tag{E-24-c}$$

where $p_x$, $p_y$, $p_z$ are the three indices which appear in the ket $| \mathbf{p} \rangle$.

Let us ascertain how the $\mathbf{P}$ operator acts in the $\{ | \mathbf{r} \rangle \}$ representation.

To do so (*cf.* §C-5-d), we use the closure relation (E-5-d) and the transformation matrix (E-16) to obtain:

$$\langle \mathbf{r} | P_x | \psi \rangle = \int d^3p \langle \mathbf{r} | \mathbf{p} \rangle \langle \mathbf{p} | P_x | \psi \rangle$$
$$= (2\pi\hbar)^{-3/2} \int d^3p \; e^{\frac{i}{\hbar}\mathbf{p}.\mathbf{r}} \; p_x \; \bar{\psi}(\mathbf{p}) \tag{E-25}$$

We recognize in (E-25) the Fourier transform of $p_x\bar{\psi}(\mathbf{p})$, that is $\frac{\hbar}{i}\frac{\partial}{\partial x}\psi(\mathbf{r})$ [appendix I, relation (38-a)]. Therefore:

$$\langle \mathbf{r} | \mathbf{P} | \psi \rangle = \frac{\hbar}{i} \boldsymbol{\nabla} \langle \mathbf{r} | \psi \rangle \tag{E-26}$$

In the $\{ | \mathbf{r} \rangle \}$ representation, the **P** operator coincides with the differential operator $\frac{\hbar}{i}\boldsymbol{\nabla}$ applied to the wave functions. The calculation of a matrix element such as $\langle \varphi | P_x | \psi \rangle$ in the $\{ | \mathbf{r} \rangle \}$ representation is therefore performed in the following manner:

$$\langle \varphi | P_x | \psi \rangle = \int d^3r \langle \varphi | \mathbf{r} \rangle \langle \mathbf{r} | P_x | \psi \rangle$$
$$= \int d^3r \; \varphi^*(\mathbf{r}) \left[ \frac{\hbar}{i}\frac{\partial}{\partial x} \right] \psi(\mathbf{r}) \tag{E-27}$$

Placing ourselves in the $\{ | \mathbf{r} \rangle \}$ representation, we can also calculate the commutators between the $X$, $Y$, $Z$, $P_x$, $P_y$, $P_z$ operators. For example:

$$\langle \mathbf{r} | [X, P_x] | \psi \rangle = \langle \mathbf{r} | (XP_x - P_xX) | \psi \rangle$$
$$= x \langle \mathbf{r} | P_x | \psi \rangle - \frac{\hbar}{i}\frac{\partial}{\partial x} \langle \mathbf{r} | X | \psi \rangle$$
$$= \frac{\hbar}{i} x \frac{\partial}{\partial x} \langle \mathbf{r} | \psi \rangle - \frac{\hbar}{i}\frac{\partial}{\partial x} x \langle \mathbf{r} | \psi \rangle$$
$$= i\hbar \langle \mathbf{r} | \psi \rangle \tag{E-28}$$

This calculation is valid for all $| \psi \rangle$ and for any ket of the $| \mathbf{r} \rangle$ basis. Thus one finds★:

$$[X, P_x] = i\hbar \tag{E-29}$$

In the same way, we find all the other commutators between the components of **R** and those of **P**. The result can be written in the form:

$$\left.\begin{aligned} [R_i, R_j] &= 0 \\ [P_i, P_j] &= 0 \\ [R_i, P_j] &= i\hbar\delta_{ij} \end{aligned}\right\} \; i, j = 1, 2, 3 \tag{E-30}$$

★ The commutator $[X, P_x]$ is an operator, and it should, actually, be written $[X, P_x] = i\hbar\mathbb{1}$. However, we shall often replace the identity operator $\mathbb{1}$ by the number 1, except when it is important to make the distinction.

where $R_1$, $R_2$, $R_3$, and $P_1$, $P_2$, $P_3$ designate respectively $X$, $Y$, $Z$ and $P_x$, $P_y$, $P_z$. Formulas (E-30) are called *canonical commutation relations.*

#### b. R AND P ARE HERMITIAN

In order to show that $X$, for example, is a Hermitian operator, we can use formula (E-23):

$$\begin{aligned} \langle \varphi | X | \psi \rangle &= \int d^3r \; \varphi^*(\mathbf{r}) \; x \; \psi(\mathbf{r}) \\ &= \left[ \int d^3r \; \psi^*(\mathbf{r}) \; x \; \varphi(\mathbf{r}) \right]^* \\ &= \langle \psi | X | \varphi \rangle^* \end{aligned} \tag{E-31}$$

From §B-4-e, we know that equation (E-31) is characteristic of a Hermitian operator.

Similar proofs show that $Y$ and $Z$ are also Hermitian. For $P_x$, $P_y$ and $P_z$, the $\{ | \mathbf{p} \rangle \}$ representation can be used, and the calculations are then analogous to the preceding ones.

It is interesting to show that $\mathbf{P}$ is Hermitian by using equation (E-26), which gives its action in the $\{ | \mathbf{r} \rangle \}$ representation. Consider, for example, formula (E-27) and integrate it by parts:

$$\begin{aligned} \langle \varphi | P_x | \psi \rangle &= \frac{\hbar}{i} \int \mathrm{d}y \; \mathrm{d}z \int_{-\infty}^{+\infty} \mathrm{d}x \; \varphi^*(\mathbf{r}) \; \frac{\partial}{\partial x} \psi(\mathbf{r}) \\ &= \frac{\hbar}{i} \int \mathrm{d}y \; \mathrm{d}z \left\{ \Big[ \varphi^*(\mathbf{r}) \; \psi(\mathbf{r}) \Big]_{x=-\infty}^{x=+\infty} - \int_{-\infty}^{+\infty} \mathrm{d}x \; \psi(\mathbf{r}) \; \frac{\partial}{\partial x} \varphi^*(\mathbf{r}) \right\} \end{aligned} \tag{E-32}$$

Since the integral which yields the scalar product $\langle \varphi | \psi \rangle$ is convergent, $\varphi^*(\mathbf{r})\psi(\mathbf{r})$ approaches zero when $x \longrightarrow \pm \infty$. The first term on the right hand side of (E-32) is therefore equal to zero, and:

$$\begin{aligned} \langle \varphi | P_x | \psi \rangle &= -\frac{\hbar}{i} \int d^3r \; \psi(\mathbf{r}) \; \frac{\partial}{\partial x} \varphi^*(\mathbf{r}) \\ &= \left[ \frac{\hbar}{i} \int d^3r \; \psi^*(\mathbf{r}) \; \frac{\partial}{\partial x} \varphi(\mathbf{r}) \right]^* \\ &= \langle \psi | P_x | \varphi \rangle^* \end{aligned} \tag{E-33}$$

It can be seen that the presence of the imaginary number $i$ is essential. The differential operator $\frac{\partial}{\partial x}$, acting on the functions of $\mathscr{F}$, is not Hermitian, because of the sign change which is introduced by the integration by parts. However, $i\frac{\partial}{\partial x}$ is Hermitian, as is $\frac{\hbar}{i}\frac{\partial}{\partial x}$.

#### c. EIGENVECTORS OF R AND P

Consider the action of the $X$ operator on the ket $| \mathbf{r}_0 \rangle$; according to (E-22-a), we have:

$$\langle \mathbf{r} | X | \mathbf{r}_0 \rangle = x \langle \mathbf{r} | \mathbf{r}_0 \rangle = x \; \delta(\mathbf{r} - \mathbf{r}_0) = x_0 \; \delta(\mathbf{r} - \mathbf{r}_0) = x_0 \langle \mathbf{r} | \mathbf{r}_0 \rangle \tag{E-34}$$

This equation expresses the fact that the components, in the $\{|\mathbf{r}\rangle\}$ representation, of the ket $X\,|\,\mathbf{r}_0\,\rangle$ are equal to those of the ket $|\,\mathbf{r}_0\,\rangle$ multiplied by $x_0$. We therefore have:

$$X\,|\,\mathbf{r}_0\,\rangle = x_0\,|\,\mathbf{r}_0\,\rangle \tag{E-35}$$

An analogous argument shows that the kets $|\,\mathbf{r}_0\,\rangle$ are also eigenvectors of the $Y$ and $Z$ operators. Omitting the index zero which then becomes unnecessary, we can write:

$$\boxed{\begin{aligned} X\,|\,\mathbf{r}\,\rangle &= x\,|\,\mathbf{r}\,\rangle \\ Y\,|\,\mathbf{r}\,\rangle &= y\,|\,\mathbf{r}\,\rangle \\ Z\,|\,\mathbf{r}\,\rangle &= z\,|\,\mathbf{r}\,\rangle \end{aligned}} \tag{E-36}$$

The kets $|\,\mathbf{r}\,\rangle$ are therefore the eigenkets common to $X$, $Y$ and $Z$. Thus the notation $|\,\mathbf{r}\,\rangle$ which we chose above is justified : each eigenvector is labelled by a vector $\mathbf{r}$, whose components $x$, $y$, $z$ represent three continuous indices which correspond to the eigenvalues of $X$, $Y$, $Z$.

Similar arguments can be elaborated for the $\mathbf{P}$ operator, placing ourselves, this time, in the $\{\,|\,\mathbf{p}\,\rangle\,\}$ representation. We then obtain:

$$\boxed{\begin{aligned} P_x\,|\,\mathbf{p}\,\rangle &= p_x\,|\,\mathbf{p}\,\rangle \\ P_y\,|\,\mathbf{p}\,\rangle &= p_y\,|\,\mathbf{p}\,\rangle \\ P_z\,|\,\mathbf{p}\,\rangle &= p_z\,|\,\mathbf{p}\,\rangle \end{aligned}} \tag{E-37}$$

COMMENT:

This result can also be derived from equation (E-26), which gives the action of $\mathbf{P}$ in the $\{\,|\,\mathbf{r}\,\rangle\,\}$ representation. Using (E-9), we find:

$$\begin{aligned} \langle\,\mathbf{r}\,|\,P_x\,|\,\mathbf{p}\,\rangle &= \frac{\hbar}{i}\frac{\partial}{\partial x}\langle\,\mathbf{r}\,|\,\mathbf{p}\,\rangle = \frac{\hbar}{i}\frac{\partial}{\partial x}\;(2\pi\hbar)^{-3/2}\,e^{\frac{i}{\hbar}\mathbf{p}.\mathbf{r}} \\ &= p_x\;(2\pi\hbar)^{-3/2}\,e^{\frac{i}{\hbar}\mathbf{p}.\mathbf{r}} = p_x\,\langle\,\mathbf{r}\,|\,\mathbf{p}\,\rangle \end{aligned} \tag{E-38}$$

All the components of the ket $P_x\,|\,\mathbf{p}\,\rangle$ in the $\{\,|\,\mathbf{r}\,\rangle\,\}$ representation can be obtained by multiplying those of $|\,\mathbf{p}\,\rangle$ by the constant $p_x$: $|\,\mathbf{p}\,\rangle$ is an eigenket of $P_x$ with the eigenvalue $p_x$.

#### d. R AND P ARE OBSERVABLES

Relations (E-5-b) and (E-5-d) express the fact that the $\{\,|\,\mathbf{r}\,\rangle\,\}$ vectors and the $\{\,|\,\mathbf{p}\,\rangle\,\}$ vectors constitute bases in $\mathscr{E}_{\mathbf{r}}$. Therefore, $\mathbf{R}$ and $\mathbf{P}$ are observables.

Moreover, the specification of the three eigenvalues $x_0$, $y_0$, $z_0$ of $X$, $Y$, $Z$ uniquely determines the corresponding eigenvector $|\,\mathbf{r}_0\,\rangle$ : in the $\{\,|\,\mathbf{r}\,\rangle\,\}$ representation, its coordinates are $\delta(x - x_0)\delta(y - y_0)\delta(z - z_0)$. The set of the three operators $X$, $Y$, $Z$ therefore constitutes a C.S.C.O. in $\mathscr{E}_{\mathbf{r}}$.

It can be shown in the same way that the three components $P_x$, $P_y$, $P_z$ of **P** also constitute a C.S.C.O. in $\mathscr{E}_{\mathbf{r}}$.

Note that, in $\mathscr{E}_{\mathbf{r}}$, $X$ does not constitute a C.S.C.O. by itself. When the $x_0$ index is fixed, $y_0$ and $z_0$ can take on any real values. Thus, each eigenvalue $x_0$ is infinitely degenerate. On the other hand, in the state space $\mathscr{E}_x$ of a one-dimensional problem, $X$ constitutes a C.S.C.O. : the eigenvalue $x_0$ uniquely determines the corresponding eigenket $| x_0 \rangle$, its coordinates being $\delta(x - x_0)$ in the $\{ | x \rangle \}$ representation.

COMMENT:

We have found two C.S.C.O.'s in $\mathscr{E}_{\mathbf{r}}$, $\{ X, Y, Z \}$ and $\{ P_x, P_y, P_z \}$. We shall encounter others later. Consider, for example, the set $\{ X, P_y, P_z \}$: these three observables commute (equations (E-30)); moreover, if the three eigenvalues $x_0$, $p_{0y}$ and $p_{0z}$ are fixed, there corresponds to them only one ket, whose associated wave function is written:

$$\psi_{x_0, p_{0y}, p_{0z}}(x, y, z) = \delta(x - x_0) \frac{1}{2\pi\hbar} e^{\frac{i}{\hbar}(p_{0y} y + p_{0z} z)} \tag{E-39}$$

# F. TENSOR PRODUCT OF STATE SPACES*

## 1. Introduction

We introduced the state space of a physical system using the concept of a one-particle wave function. However, our reasoning has involved sometimes one- and sometimes three-dimensional wave functions. Now it is clear that the space of square-integrable functions is not the same for functions of one variable $\psi(x)$ as for functions of three variables $\psi(\mathbf{r})$: $\mathscr{E}_{\mathbf{r}}$ and $\mathscr{E}_x$ are therefore different spaces. Nevertheless, $\mathscr{E}_{\mathbf{r}}$ appears to be essentially a generalization of $\mathscr{E}_x$. Does there exist a more precise relation between these two spaces?

In this section, we are going to define and study the operation of taking the tensor product of vector spaces**, and apply it to state spaces. This will answer, in particular, the question we have just asked : $\mathscr{E}_{\mathbf{r}}$ can be constructed from $\mathscr{E}_x$ and two other spaces, $\mathscr{E}_y$ and $\mathscr{E}_z$, which are isomorphic to it (§ F-4-a below).

In the same way, we shall be concerned later (chapters IV and IX) with the existence, for certain particles, of an intrinsic angular momentum or spin. In addition to the external degrees of freedom (position, momentum), which are treated using the observables **R** and **P** defined in $\mathscr{E}_{\mathbf{r}}$, it will be necessary to take into account the internal degrees of freedom and to introduce spin observables which act in a spin state space $\mathscr{E}_s$. The state space $\mathscr{E}$ of a particle with spin will then be seen to be the tensor product of $\mathscr{E}_{\mathbf{r}}$ and $\mathscr{E}_s$.

* This section is not necessary for the understanding of chapter III. One can study it later when it becomes necessary to use tensor products (complement $D_{IV}$, or chapter IX).

** This operation is sometimes called the " Kronecker product ".

Finally, the concept of a tensor product of state spaces allows us to solve the following problem. Let $(S_1)$ and $(S_2)$ be two isolated physical systems (they are, for example, sufficiently far apart that their interactions are perfectly negligible). The state spaces which correspond to $(S_1)$ and $(S_2)$ are, respectively, $\mathscr{E}_1$ and $\mathscr{E}_2$. Now let us assume that we consider the set of these two systems to form one physical system $(S)$ (this becomes indispensable when they are close enough to interact). What is then the state space $\mathscr{E}$ of the global system $(S)$?

It can be seen from these examples how useful the definitions and results of this section are in quantum mechanics.

## 2. Definition and properties of the tensor product

Let $\mathscr{E}_1$ and $\mathscr{E}_2$ be two* spaces, of dimension $N_1$ and $N_2$ respectively ($N_1$ and $N_2$ can be finite or infinite). Vectors and operators of these spaces will be assigned an index, (1) or (2), depending on whether they belong to $\mathscr{E}_1$ or $\mathscr{E}_2$.

### a. TENSOR PRODUCT SPACE $\mathscr{E}$

#### α. *Definition*

By definition, the vector space $\mathscr{E}$ is called the *tensor product of $\mathscr{E}_1$ and $\mathscr{E}_2$*:

$$\mathscr{E} = \mathscr{E}_1 \otimes \mathscr{E}_2 \tag{F-1}$$

if there is associated with each pair of vectors, $|\varphi(1)\rangle$ belonging to $\mathscr{E}_1$ and $|\chi(2)\rangle$ belonging to $\mathscr{E}_2$, a vector of $\mathscr{E}$, denoted by**:

$$|\varphi(1)\rangle \otimes |\chi(2)\rangle \tag{F-2}$$

which is called the tensor product of $|\varphi(1)\rangle$ and $|\chi(2)\rangle$, this correspondence satisfying the following conditions:

(*i*) It is *linear* with respect to multiplication by complex numbers:

$$\begin{aligned} [\lambda\,|\varphi(1)\rangle] \otimes |\chi(2)\rangle &= \lambda[\,|\varphi(1)\rangle \otimes |\chi(2)\rangle] \\ |\varphi(1)\rangle \otimes [\mu\,|\chi(2)\rangle] &= \mu[\,|\varphi(1)\rangle \otimes |\chi(2)\rangle] \end{aligned} \tag{F-3}$$

(*ii*) It is distributive with respect to vector addition:

$$\begin{aligned} &|\varphi(1)\rangle \otimes [\,|\chi_1(2)\rangle + |\chi_2(2)\rangle] = |\varphi(1)\rangle \otimes |\chi_1(2)\rangle + |\varphi(1)\rangle \otimes |\chi_2(2)\rangle \\ &[\,|\varphi_1(1)\rangle + |\varphi_2(1)\rangle] \otimes |\chi(2)\rangle \\ &\qquad = |\varphi_1(1)\rangle \otimes |\chi(2)\rangle + |\varphi_2(1)\rangle \otimes |\chi(2)\rangle \end{aligned} \tag{F-4}$$

(*iii*) When a basis has been chosen in each of the spaces $\mathscr{E}_1$ and $\mathscr{E}_2$, $\{\,|u_i(1)\rangle\,\}$ for $\mathscr{E}_1$ and $\{\,|v_l(2)\rangle\,\}$ for $\mathscr{E}_2$, the set of vectors $|u_i(1)\rangle \otimes |v_l(2)\rangle$

* The following definitions can easily be extended to the tensor product of a finite number of spaces.

** This vector can be written either $|\varphi(1)\rangle \otimes |\chi(2)\rangle$ or $|\chi(2)\rangle \otimes |\varphi(1)\rangle$: the order of the two vectors is of no importance.

constitutes a basis in $\mathscr{E}$. If $N_1$ and $N_2$ are finite, the dimension of $\mathscr{E}$ is consequently $N_1 N_2$.

β. *Vectors of $\mathscr{E}$*

(*i*) Let us first consider a *tensor product vector*, $| \varphi(1) \rangle \otimes | \chi(2) \rangle$. Whatever $| \varphi(1) \rangle$ and $| \chi(2) \rangle$ may be, they can be expressed in the $\{ | u_i(1) \rangle \}$ and $\{ | v_l(2) \rangle \}$ bases respectively:

$$| \varphi(1) \rangle = \sum_i a_i | u_i(1) \rangle$$

$$| \chi(2) \rangle = \sum_l b_l | v_l(2) \rangle \tag{F-5}$$

Using the properties described in § α, the expansion of the vector $| \varphi(1) \rangle \otimes | \chi(2) \rangle$ in the $\{ | u_i(1) \rangle \otimes | v_l(2) \rangle \}$ basis can be written:

$$| \varphi(1) \rangle \otimes | \chi(2) \rangle = \sum_{i,l} a_i b_l | u_i(1) \rangle \otimes | v_l(2) \rangle \tag{F-6}$$

Therefore, *the components of a tensor product vector are the products of the components of the two vectors of the product.*

(*ii*) *There exist in $\mathscr{E}$ vectors which are not tensor products* of a vector of $\mathscr{E}_1$ by a vector of $\mathscr{E}_2$. Since $\{ | u_i(1) \rangle \otimes | v_l(2) \rangle \}$ constitutes by hypothesis a basis in $\mathscr{E}$, the most general vector of $\mathscr{E}$ is expressed by:

$$| \psi \rangle = \sum_{i,l} c_{i,l} | u_i(1) \rangle \otimes | v_l(2) \rangle \tag{F-7}$$

Given $N_1 N_2$ arbitrary complex numbers $c_{i,l}$, it is not always possible to put them in the form of products, $a_i b_l$, of $N_1$ numbers $a_i$ and $N_2$ numbers $b_l$. Therefore, in general, vectors $| \varphi(1) \rangle$ and $| \chi(2) \rangle$ of which $| \psi \rangle$ is the tensor product do not exist. However, *an arbitrary vector of $\mathscr{E}$ can always be decomposed into a linear combination of tensor product vectors*, as is shown by formula (F-7).

γ. *The scalar product in $\mathscr{E}$*

The existence of scalar products in $\mathscr{E}_1$ and $\mathscr{E}_2$ permits us to define one in $\mathscr{E}$ as well. We first define the scalar product of $| \varphi(1) \chi(2) \rangle = | \varphi(1) \rangle \otimes | \chi(2) \rangle$ by $| \varphi'(1) \chi'(2) \rangle = | \varphi'(1) \rangle \otimes | \chi'(2) \rangle$ by setting:

$$\langle \varphi'(1) \ \chi'(2) | \varphi(1) \ \chi(2) \rangle = \langle \varphi'(1) | \varphi(1) \rangle \langle \chi'(2) | \chi(2) \rangle \tag{F-8}$$

For two arbitrary vectors of $\mathscr{E}$, we simply use the fundamental properties of the scalar product [equations (B-9), (B-10) and (B-11)], since each of these vectors is a linear combination of tensor product vectors.

Notice, in particular, that the basis $\{ | u_i(1) \, v_l(2) \rangle = | u_i(1) \rangle \otimes | v_l(2) \rangle \}$ is orthonormal if each of the bases $\{ | u_i(1) \rangle \}$ and $\{ | v_l(2) \rangle \}$ is:

$$\begin{aligned} \langle u_{i'}(1) \ v_{l'}(2) | u_i(1) \ v_l(2) \rangle &= \langle u_{i'}(1) | u_i(1) \rangle \langle v_{l'}(2) | v_l(2) \rangle \\ &= \delta_{ii'} \delta_{ll'} \end{aligned} \tag{F-9}$$

### b. TENSOR PRODUCT OF OPERATORS

(*i*) First, consider a linear operator $A(1)$ defined in $\mathscr{E}_1$. We associate with it a linear operator $\tilde{A}(1)$ acting in $\mathscr{E}$, which we call the *extension of* $A(1)$ *in* $\mathscr{E}$, and which is characterized in the following way : when $\tilde{A}(1)$ is applied to a tensor product vector $| \varphi(1) \rangle \otimes | \chi(2) \rangle$, one obtains, by definition :

$$\tilde{A}(1)[| \varphi(1) \rangle \otimes | \chi(2) \rangle] = [A(1) | \varphi(1) \rangle] \otimes | \chi(2) \rangle \tag{F-10}$$

The hypothesis that $\tilde{A}(1)$ is linear is then sufficient for determining it completely. An arbitrary vector $| \psi \rangle$ of $\mathscr{E}$ can be written in the form (F-7). Definition (F-10) then gives the action of $\tilde{A}(1)$ on $| \psi \rangle$:

$$\tilde{A}(1) | \psi \rangle = \sum_{i,l} c_{i,l}[A(1) | u_i(1) \rangle] \otimes | v_l(2) \rangle \tag{F-11}$$

We obtain in an analogous manner the extension $\tilde{B}(2)$ of an operator $B(2)$ initially defined in $\mathscr{E}_2$.

(*ii*) Now let $A(1)$ and $B(2)$ be two linear operators acting respectively in $\mathscr{E}_1$ and $\mathscr{E}_2$. Their *tensor product* $A(1) \otimes B(2)$ is the linear operator in $\mathscr{E}$, defined by the following relation which describes its action on the tensor product vectors:

$$[A(1) \otimes B(2)] [| \varphi(1) \rangle \otimes | \chi(2) \rangle] = [A(1) | \varphi(1) \rangle] \otimes [B(2) | \chi(2) \rangle] \tag{F-12}$$

Here also, this definition is sufficient for characterizing $A(1) \otimes B(2)$.

COMMENTS:

(*i*) The extensions of operators are special cases of tensor products : if $\mathbb{1}(1)$ and $\mathbb{1}(2)$ are the identity operators in $\mathscr{E}_1$ and $\mathscr{E}_2$ respectively, $\tilde{A}(1)$ and $\tilde{B}(2)$ can be written:

$$\begin{aligned} \tilde{A}(1) &= A(1) \otimes \mathbb{1}(2) \\ \tilde{B}(2) &= \mathbb{1}(1) \otimes B(2) \end{aligned} \tag{F-13}$$

Inversely, the tensor product $A(1) \otimes B(2)$ coincides with the ordinary product of two operators $\tilde{A}(1)$ and $\tilde{B}(2)$ of $\mathscr{E}$:

$$A(1) \otimes B(2) = \tilde{A}(1)\tilde{B}(2) \tag{F-14}$$

(*ii*) It is easy to show that two operators such as $\tilde{A}(1)$ and $\tilde{B}(2)$ commute in $\mathscr{E}$:

$$[\tilde{A}(1), \tilde{B}(2)] = 0 \tag{F-15}$$

We must verify that $\tilde{A}(1)\tilde{B}(2)$ and $\tilde{B}(2)\tilde{A}(1)$ yield the same result when they act on an arbitrary vector of the $\{ | u_i(1) \rangle \otimes | v_l(2) \rangle \}$ basis:

$$\begin{aligned} \tilde{A}(1)\tilde{B}(2) | u_i(1) \rangle \otimes | v_l(2) \rangle &= \tilde{A}(1) \; | u_i(1) \rangle \otimes [B(2) | v_l(2) \rangle] \\ &= [A(1) | u_i(1) \rangle] \otimes [B(2) | v_l(2) \rangle] \end{aligned} \tag{F-16}$$

$$\begin{aligned} \tilde{B}(2)\tilde{A}(1) | u_i(1) \rangle \otimes | v_l(2) \rangle &= \tilde{B}(2) \; [A(1) | u_i(1) \rangle] \otimes | v_l(2) \rangle \\ &= [A(1) | u_i(1) \rangle] \otimes [B(2) | v_l(2) \rangle] \end{aligned} \tag{F-17}$$

(*iii*) The projector onto the tensor product vector $| \varphi(1)\chi(2) \rangle = | \varphi(1) \rangle \otimes | \chi(2) \rangle$, which is an operator acting in $\mathscr{E}$, is obtained by taking the tensor product of the projectors onto $| \varphi(1) \rangle$ and $| \chi(2) \rangle$:

$$| \varphi(1)\chi(2) \rangle \langle \varphi(1)\chi(2) | = | \varphi(1) \rangle \langle \varphi(1) | \otimes | \chi(2) \rangle \langle \chi(2) | \tag{F-18}$$

This relation follows immediately from the definition of the scalar product in $\mathscr{E}$.

(*iv*) Just as with vectors, there exist operators in $\mathscr{E}$ which are not tensor products of an operator of $\mathscr{E}_1$ and an operator of $\mathscr{E}_2$.

#### c. NOTATION

In quantum mechanics, the notation generally used is a simplified version of the one which we have defined here. This is the one we shall adopt, but it is important to interpret it correctly in the light of the preceding discussion.

First of all, the symbol $\otimes$ which indicates the tensor product is omitted, and the vectors or operators which are to be multiplied tensorially are simply juxtaposed:

$$|\varphi(1)\rangle|\chi(2)\rangle \quad \text{means} \quad |\varphi(1)\rangle \otimes |\chi(2)\rangle \tag{F-19}$$

$$A(1)B(2) \quad \text{means} \quad A(1) \otimes B(2) \tag{F-20}$$

Moreover, the extension in $\mathscr{E}$ of an operator of $\mathscr{E}_1$ or $\mathscr{E}_2$ is written in the same way as this operator itself:

$$A(1) \quad \text{means} \quad \tilde{A}(1) \quad \text{or} \quad A(1) \tag{F-21}$$

No confusion is possible in (F-19): until now we have never written two kets one after the other as we do here. Notice in particular that the expression $|\psi\rangle|\varphi\rangle$, where $|\psi\rangle$ and $|\varphi\rangle$ belong to the same space $\mathscr{E}$, is not defined in this space: it represents a vector of the space which is the tensor product of $\mathscr{E}$ by itself.

On the other hand, the notation in (F-20) and (F-21) is slightly ambiguous, especially in the latter, where two different operators are represented by the same symbol. However, it will be possible in practice to distinguish between them by the vector to which this symbol is applied : depending on whether it is a vector of $\mathscr{E}$ or of $\mathscr{E}_1$, we shall be dealing with $\tilde{A}(1)$, or with $A(1)$ in a strict sense. As for formula (F-20), it poses no problem when $\mathscr{E}_1$ and $\mathscr{E}_2$ are different, since we have, until now, defined only products of operators which act in the same space. Moreover, $A(1)$ $B(2)$ can be considered to be an ordinary product of operators of $\mathscr{E}$, if $A(1)$ and $B(2)$ are interpreted as designating, in fact, $\tilde{A}(1)$ and $\tilde{B}(2)$ [equation (F-14)].

## 3. Eigenvalue equations in the product space

The vectors of $\mathscr{E}$ which are tensor products of a vector of $\mathscr{E}_1$ and a vector of $\mathscr{E}_2$ play an important role in the discussion above. We shall see that this is also the case for the extensions of operators of $\mathscr{E}_1$ and $\mathscr{E}_2$.

#### a. EIGENVALUES AND EIGENVECTORS OF EXTENDED OPERATORS

α. *Eigenvalue equation of $A(1)$*

Consider an operator $A(1)$, for which we know, in $\mathscr{E}_1$, all the eigenstates and eigenvalues. We shall assume, for example, that the whole spectrum of $A(1)$ is discrete:

$$A(1)|\varphi_n^i(1)\rangle = a_n|\varphi_n^i(1)\rangle \quad ; \quad i = 1, 2, ..., g_n \tag{F-22}$$

We want to solve the eigenvalue equation of the extension of $A(1)$ in $\mathscr{E}$:

$$A(1)|\psi\rangle = \lambda|\psi\rangle; \quad |\psi\rangle \in \mathscr{E} \tag{F-23}$$

It can immediately be seen, from (F-10), that every vector of the form $|\varphi_n^i(1)\rangle|\chi(2)\rangle$ is an eigenvector of $A(1)$ with the eigenvalue $a_n$, whatever $|\chi(2)\rangle$ may be, since:

$$\begin{aligned} A(1)|\varphi_n^i(1)\rangle|\chi(2)\rangle &= [A(1)|\varphi_n^i(1)\rangle]|\chi(2)\rangle \\ &= a_n|\varphi_n^i(1)\rangle|\chi(2)\rangle \end{aligned} \tag{F-24}$$

Let us show that when $A(1)$ is an observable in $\mathscr{E}_1$, all the solutions of (F-23) can be obtained in this way. The set of the $|\varphi_n^i(1)\rangle$ then forms a basis in $\mathscr{E}_1$. Consequently, the orthonormal system of vectors $|\psi_n^{i,l}\rangle$ such that:

$$|\psi_n^{i,l}\rangle = |\varphi_n^i(1)\rangle|v_l(2)\rangle \tag{F-25}$$

where $\{|v_l(2)\rangle\}$ is a basis of $\mathscr{E}_2$, also forms a basis in $\mathscr{E}$. We therefore have an orthonormal basis constituted by the eigenvectors of $A(1)$ in $\mathscr{E}$, $\{|\psi_n^{i,l}\rangle\}$, which means that equation (F-23) is solved.

The following conclusions can be drawn :

– *If $A(1)$ is an observable in $\mathscr{E}_1$, it is also an observable in $\mathscr{E}$.* This results from the fact that the extension of $A(1)$ is Hermitian and from the fact that $\{|\psi_n^{i,l}\rangle\}$ constitutes a basis in $\mathscr{E}$.

– *The spectrum of $A(1)$ is the same in $\mathscr{E}$ as in $\mathscr{E}_1$* : the same eigenvalues $a_n$ appear in (F-22) and in (F-24).

– Nevertheless, an eigenvalue $a_n$ which is $g_n$-fold degenerate in $\mathscr{E}_1$ has, in $\mathscr{E}$, a degree of degeneracy $N_2 \times g_n$. We known that the eigensubspace associated with $a_n$ is spanned in $\mathscr{E}$ by the kets $|\psi_n^{i,l}\rangle = |\varphi_n^i(1)\rangle|v_l(2)\rangle$ with $n$ fixed and $i = 1, 2, ..., g_n$; $l = 1, 2, ..., N_2$. Therefore, even if $a_n$ is simple in $\mathscr{E}_1$, it is ($N_2$-fold) degenerate in $\mathscr{E}$.

The projector onto the eigensubspace corresponding to an eigenvalue $a_n$ is written, in $\mathscr{E}$ [*cf.* (F-18)]:

$$\begin{aligned} \sum_{i,l}|\psi_n^{i,l}\rangle\langle\psi_n^{i,l}| &= \sum_{i,l}|\varphi_n^i(1)\rangle\langle\varphi_n^i(1)| \otimes |v_l(2)\rangle\langle v_l(2)| \\ &= \sum_i |\varphi_n^i(1)\rangle\langle\varphi_n^i(1)| \otimes \mathbb{1}(2) \end{aligned} \tag{F-26}$$

using in $\mathscr{E}_2$ the closure relation relative to the $\{|v_l(2)\rangle\}$ basis. It is therefore the extension of the projector $P_n(1) = \sum_i |\varphi_n^i(1)\rangle\langle\varphi_n^i(1)|$ which is associated with $a_n$ in $\mathscr{E}_1$.

### β. *Eigenvalue equation of $A(1) + B(2)$*

We shall often need to solve, in a tensor product space such as $\mathscr{E}$, eigenvalue equations for operators of the form:

$$C = A(1) + B(2) \tag{F-27}$$

where $A(1)$ and $B(2)$ are observables whose eigenvalues and eigenvectors are known in $\mathscr{E}_1$ and $\mathscr{E}_2$ respectively:

$$A(1)|\varphi_n(1)\rangle = a_n|\varphi_n(1)\rangle$$
$$B(2)|\chi_p(2)\rangle = b_p|\chi_p(2)\rangle \quad \text{(F-28)}$$

[to simplify the notation, we assume the spectra of $A(1)$ and $B(2)$ to be discrete and non-degenerate in $\mathscr{E}_1$ and $\mathscr{E}_2$].

$A(1)$ and $B(2)$ commute [formulas (F-16) and (F-17)], and the $|\varphi_n(1)\rangle|\chi_p(2)\rangle$, which form a basis in $\mathscr{E}$, are eigenvectors common to $A(1)$ and $B(2)$:

$$A(1)|\varphi_n(1)\rangle|\chi_p(2)\rangle = a_n|\varphi_n(1)\rangle|\chi_p(2)\rangle$$
$$B(2)|\varphi_n(1)\rangle|\chi_p(2)\rangle = b_p|\varphi_n(1)\rangle|\chi_p(2)\rangle \quad \text{(F-29)}$$

They are also eigenvectors of $C$:

$$C|\varphi_n(1)\rangle|\chi_p(2)\rangle = (a_n + b_p)|\varphi_n(1)\rangle|\chi_p(2)\rangle \quad \text{(F-30)}$$

This gives us directly the solution of the eigenvalue equation of $C$.

Therefore : *the eigenvalues of $C = A(1) + B(2)$ are the sums of an eigenvalue of $A(1)$ and an eigenvalue of $B(2)$. One can find a basis of eigenvectors of $C$ which are tensor products of an eigenvector of $A(1)$ and an eigenvector of $B(2)$.*

COMMENT:

Equation (F-30) shows that the eigenvalues of $C$ are all of the form $c_{np} = a_n + b_p$. If two different pairs of values of $n$ and $p$ which give the same value for $c_{np}$ do not exist, $c_{np}$ is non-degenerate (recall that we have assumed $a_n$ and $b_p$ to be non-degenerate in $\mathscr{E}_1$ and $\mathscr{E}_2$ respectively). The corresponding eigenvector of $C$ is necessarily the tensor product $|\varphi_n(1)\rangle|\chi_p(2)\rangle$. If, on the other hand, the eigenvalue $c_{np}$ is, for example, two-fold degenerate (there exist $m$ and $q$ such that $c_{mq} = c_{np}$), all that can be asserted is that every eigenvector of $C$ corresponding to this eigenvalue is written:

$$\lambda|\varphi_n(1)\rangle|\chi_p(2)\rangle + \mu|\varphi_m(1)\rangle|\chi_q(2)\rangle \quad \text{(F-31)}$$

where $\lambda$ and $\mu$ are arbitrary complex numbers. In this case, therefore, there exist eigenvectors of $C$ which are not tensor products.

### b. COMPLETE SETS OF COMMUTING OBSERVABLES IN $\mathscr{E}$

We are finally going to show that if a C.S.C.O. has been chosen in both spaces $\mathscr{E}_1$ and $\mathscr{E}_2$, obtaining one in $\mathscr{E}$ is straightforward.

As an example, let us consider the case where $A(1)$ constitutes a C.S.C.O. by itself in $\mathscr{E}_1$, and the C.S.C.O. in $\mathscr{E}_2$ is composed of two observables, $B(2)$ and $C(2)$. This means (*cf.* § D-3-b) that all the eigenvalues $a_n$ of $A(1)$ are nondegenerate in $\mathscr{E}_1$:

$$A(1)|\varphi_n(1)\rangle = a_n|\varphi_n(1)\rangle \quad \text{(F-32)}$$

the ket $|\varphi_n(1)\rangle$ being unique to within a constant factor. On the other hand, in $\mathscr{E}_2$, some of the eigenvalues $b_p$ of $B(2)$ are degenerate, as are some of the eigenvalues $(c_r)$ of $C(2)$. Nevertheless, the basis of eigenvectors common to $B(2)$ and $C(2)$ is unique in $\mathscr{E}_2$, since there exists only one ket (to within a constant factor) which is an eigenvector of $B(2)$ and of $C(2)$ with the eigenvalues $b_p$ and $c_r$ fixed:

$$\begin{cases} B(2) \mid \chi_{pr}(2) \rangle = b_p \mid \chi_{pr}(2) \rangle \\ C(2) \mid \chi_{pr}(2) \rangle = c_r \mid \chi_{pr}(2) \rangle \\ \mid \chi_{pr}(2) \rangle \text{ unique to within a constant factor} \end{cases} \tag{F-33}$$

In $\mathscr{E}$, each of the eigenvalues $a_n$ is $N_2$-fold degenerate (*cf.* § F-3-a). Therefore, $A(1)$ no longer forms a C.S.C.O. by itself. Similarly, there exist $N_1$ linearly independent kets which are eigenvectors of $B(2)$ and $C(2)$ with the eigenvalues $b_p$ and $c_r$ respectively, and the set $\{ B(2), C(2) \}$ is not complete either. However, we saw in § F-3-a that the eigenvectors which are common to the three commuting observables $A(1)$, $B(2)$ and $C(2)$ are the $\mid \varphi_n(1)\chi_{pr}(2) \rangle = \mid \varphi_n(1) \rangle \mid \chi_{pr}(2) \rangle$:

$$\begin{aligned} A(1) \mid \varphi_n(1) \ \chi_{pr}(2) \rangle &= a_n \mid \varphi_n(1) \ \chi_{pr}(2) \rangle \\ B(2) \mid \varphi_n(1) \ \chi_{pr}(2) \rangle &= b_p \mid \varphi_n(1) \ \chi_{pr}(2) \rangle \\ C(2) \mid \varphi_n(1) \ \chi_{pr}(2) \rangle &= c_r \mid \varphi_n(1) \ \chi_{pr}(2) \rangle \end{aligned} \tag{F-34}$$

The system $\{ \mid \varphi_n(1)\chi_{pr}(2) \rangle \}$ constitutes a basis in $\mathscr{E}$, since this is the case for $\{ \mid \varphi_n(1) \rangle \}$ and $\{ \mid \chi_{pr}(2) \rangle \}$ in $\mathscr{E}_1$ and $\mathscr{E}_2$ respectively. Moreover, if a set of three eigenvalues $\{ a_n, b_p, c_r \}$ is chosen, only one vector $\mid \varphi_n(1)\chi_{pr}(2) \rangle$ corresponds to it. $A(1)$, $B(2)$ and $C(2)$ therefore constitute a C.S.C.O. in $\mathscr{E}$.

The preceding argument can be generalized without difficulty: *by joining two sets of commuting observables which are complete in $\mathscr{E}_1$ and $\mathscr{E}_2$ respectively, one obtains a complete set of commuting observables in $\mathscr{E}$.*

## 4. Applications

### a. ONE- AND THREE-DIMENSIONAL PARTICLE STATES

#### α. *State spaces*

Consider again, in the light of the preceding discussion, the problem posed in the introduction (§ F-1): how are $\mathscr{E}_x$ and $\mathscr{E}_{\mathbf{r}}$ related?

$\mathscr{E}_x$ is the state space of a one-dimensional particle, that is, the state space associated with the wave functions $\varphi(x)$. In $\mathscr{E}_x$, the observable $X$ which was studied in § E-2 constitutes a C.S.C.O. by itself (§ E-2-d); its eigenvectors are the basis kets of the $\{ \mid x \rangle \}$ representation. A vector $\mid \varphi \rangle$ of $\mathscr{E}_x$ is characterized, in this representation, by a wave function $\varphi(x) = \langle x \mid \varphi \rangle$; in particular, the basis ket $\mid x_0 \rangle$ corresponds to $\xi_{x_0}(x) = \delta(x - x_0)$.

In the same way, it is possible to introduce the spaces $\mathscr{E}_y$ and $\mathscr{E}_z$ associated with the wave functions $\chi(y)$ and $\omega(z)$. The observable $Y$ forms a C.S.C.O. in $\mathscr{E}_y$, as does $Z$ in $\mathscr{E}_z$. The corresponding eigenvectors are the basis kets of the $\{ \mid y \rangle \}$ and $\{ \mid z \rangle \}$ representations of $\mathscr{E}_y$ and $\mathscr{E}_z$ respectively. A vector $\mid \chi \rangle$ of $\mathscr{E}_y$ (or $\mid \omega \rangle$ of $\mathscr{E}_z$) is characterized in the $\{ \mid y \rangle \}$ (or $\{ \mid z \rangle \}$) representation by a function $\chi(y) = \langle y \mid \chi \rangle$ (or $\omega(z) = \langle z \mid \omega \rangle$). The function which corresponds to the basis ket $\mid y_0 \rangle$ (or $\mid z_0 \rangle$) is $\delta(y - y_0)$ (or $\delta(z - z_0)$).

Let us then form the tensor product:

$$\mathscr{E}_{xyz} = \mathscr{E}_x \otimes \mathscr{E}_y \otimes \mathscr{E}_z \tag{F-35}$$

We obtain a basis in $\mathscr{E}_{xyz}$ from the tensor product of the $\{ | x \rangle \}$, $\{ | y \rangle \}$ and $\{ | z \rangle \}$ bases. We shall denote it by $\{ | x, y, z \rangle \}$, with:

$$| x, y, z \rangle = | x \rangle | y \rangle | z \rangle \tag{F-36}$$

The basis kets are simultaneous eigenvectors of the $X$, $Y$ and $Z$ operators extended into $\mathscr{E}_{xyz}$:

$$\begin{aligned} X | x, y, z \rangle &= x | x, y, z \rangle \\ Y | x, y, z \rangle &= y | x, y, z \rangle \\ Z | x, y, z \rangle &= z | x, y, z \rangle \end{aligned} \tag{F-37}$$

Therefore, $\mathscr{E}_{xyz}$ coincides with $\mathscr{E}_{\mathbf{r}}$, the state space of a three-dimensional particle, and $| x, y, z \rangle$ with $| \mathbf{r} \rangle$:

$$| x, y, z \rangle \equiv | \mathbf{r} \rangle = | x \rangle | y \rangle | z \rangle \tag{F-38}$$

where $x$, $y$, $z$ are precisely the cartesian coordinates of $\mathbf{r}$.

There exist in $\mathscr{E}_{\mathbf{r}}$ kets $| \varphi \chi \omega \rangle = | \varphi \rangle | \chi \rangle | \omega \rangle$ which are the tensor products of three kets, one of $\mathscr{E}_x$, one of $\mathscr{E}_y$ and one of $\mathscr{E}_z$. Their components in the $\{ | \mathbf{r} \rangle \}$ representation are then [*cf.* formula (F-8)] :

$$\langle \mathbf{r} | \varphi \chi \omega \rangle = \langle x | \varphi \rangle \langle y | \chi \rangle \langle z | \omega \rangle \tag{F-39}$$

The associated wave functions are thus factorized : $\varphi(x) \chi(y) \omega(z)$. This is the case for the basis vectors themselves:

$$\langle \mathbf{r} | \mathbf{r}_0 \rangle = \delta(\mathbf{r} - \mathbf{r}_0) = \delta(x - x_0)\ \delta(y - y_0)\ \delta(z - z_0) \tag{F-40}$$

Note that the most general state of $\mathscr{E}_{\mathbf{r}}$ is not such a product. It is written :

$$| \psi \rangle = \int \mathrm{d}x\, \mathrm{d}y\, \mathrm{d}z\ \ \psi(x, y, z) | x, y, z \rangle \tag{F-41}$$

In $\psi(x, y, z) = \langle x, y, z | \psi \rangle$, the $x$-, $y$- and $z$-dependences cannot, in general, be factorized : each of the wave functions associated with the kets of $\mathscr{E}_{\mathbf{r}}$ is a wave function with three variables.

The results of § F-3 thus enable us to understand why $X$, which constitutes a C.S.C.O. by itself in $\mathscr{E}_x$, no longer has this property in $\mathscr{E}_{\mathbf{r}}$ (*cf.* § E-2-d): the eigenvalues of its extension in $\mathscr{E}_{\mathbf{r}}$ are the same as in $\mathscr{E}_x$, but they become infinitely degenerate because $\mathscr{E}_y$ and $\mathscr{E}_z$ are infinite-dimensional. Starting with a C.S.C.O. in $\mathscr{E}_x$, $\mathscr{E}_y$ and $\mathscr{E}_z$, we construct one for $\mathscr{E}_{\mathbf{r}}$ : $\{ X, Y, Z \}$, for example, but also $\{ P_x, Y, Z \}$ since $P_x$ forms a C.S.C.O. in $\mathscr{E}_x$, or $\{ P_x, P_y, Z \}$, etc...

β. *An important application*

Let us try to solve in $\mathscr{E}_{\mathbf{r}}$ the eigenvalue equation of an operator $H$ such that :

$$H = H_x + H_y + H_z \tag{F-42}$$

where $H_x$, $H_y$ and $H_z$ are the extensions of observables acting respectively in $\mathscr{E}_x$, $\mathscr{E}_y$ and $\mathscr{E}_z$. In practice, one recognizes that $H_x$, for example, is the extension of an observable of $\mathscr{E}_x$ because it is constructed using only the operators $X$ and $P_x$. Using the reasoning of § F-3-a-β, one first looks for the eigenvalues and eigenvectors of $H_x$ in $\mathscr{E}_x$, $H_y$ in $\mathscr{E}_y$ and $H_z$ in $\mathscr{E}_z$:

$$\begin{aligned} H_x \mid \varphi_n \rangle &= E_x^n \mid \varphi_n \rangle \\ H_y \mid \chi_p \rangle &= E_y^p \mid \chi_p \rangle \\ H_z \mid \omega_r \rangle &= E_z^r \mid \omega_r \rangle \end{aligned} \tag{F-43}$$

The eigenvalues of $H$ are then all of the form:

$$E^{n,p,r} = E_x^n + E_y^p + E_z^r \tag{F-44}$$

and an eigenvector corresponding to $E^{n,p,r}$ is the tensor product $\mid \varphi_n \rangle \mid \chi_p \rangle \mid \omega_r \rangle$; the wave function associated with this vector is the product:

$$\varphi_n(x)\, \chi_p(y)\, \omega_r(z) = \langle\, x \mid \varphi_n \rangle \langle\, y \mid \chi_p \rangle \langle\, z \mid \omega_r \rangle.$$

This is the type of situation that was considered in complement $F_{\text{I}}$ (§ 2) for the justification of the study of one-dimensional models. There, we were dealing with differential operators acting on wave functions:

$$H = -\frac{\hbar^2}{2m}\Delta + V(\mathbf{r}) \tag{F-45}$$

This equation can be decomposed as in (F-42) in the particular case where the potential can be written:

$$V(\mathbf{r}) = V_1(x) + V_2(y) + V_3(z) \tag{F-46}$$

#### b. STATES OF A TWO-PARTICLE SYSTEM

Consider a physical system which is made up of two (spinless) particles. We shall distinguish between them by numbering them (1) and (2). To describe the system quantum mechanically, we can generalize the concept of a wave function, introduced for the case of one particle. A state of the system can be characterized, at a given time, by a function of six spatial variables $\psi(\mathbf{r}_1, \mathbf{r}_2) = \psi(x_1, y_1, z_1; x_2, y_2, z_2)$. The probabilistic interpretation of such a two-particle wave function is the following: the probability $d\mathscr{P}(\mathbf{r}_1, \mathbf{r}_2)$, at the given time, of finding particle (1) in the volume $d^3r_1 = dx_1\, dy_1\, dz_1$ situated at the point $\mathbf{r}_1$, *and* particle (2) in the volume $d^3r_2 = dx_2\, dy_2\, dz_2$ about $\mathbf{r}_2$, is:

$$d\mathscr{P}(\mathbf{r}_1, \mathbf{r}_2) = C\, |\psi(\mathbf{r}_1, \mathbf{r}_2)|^2\, d^3r_1\, d^3r_2 \tag{F-47}$$

The normalization constant $C$ is obtained by imposing the condition that the total probability must be equal to 1 (conservation of the number of particles; *cf.* § B-2 of chapter I):

$$\frac{1}{C} = \int d^3r_1\, d^3r_2\, |\psi(\mathbf{r}_1, \mathbf{r}_2)|^2 \tag{F-48}$$

and the observables $X_1$, $Y_1$, $Z_1$ can be defined in $\mathscr{E}_{\mathbf{r}_1}$. Similarly, in the state space $\mathscr{E}_{\mathbf{r}_2}$ of particle (2), we introduce the $\{ | \mathbf{r}_2 \rangle \}$ representation and the observables $X_2$, $Y_2$, $Z_2$. Take the tensor product :

$$\mathscr{E}_{\mathbf{r}_1\mathbf{r}_2} = \mathscr{E}_{\mathbf{r}_1} \otimes \mathscr{E}_{\mathbf{r}_2} \tag{F-49}$$

The set of vectors:

$$| \mathbf{r}_1, \mathbf{r}_2 \rangle = | \mathbf{r}_1 \rangle | \mathbf{r}_2 \rangle \tag{F-50}$$

forms a basis in $\mathscr{E}_{\mathbf{r}_1\mathbf{r}_2}$. Consequently, every ket $| \psi \rangle$ of this space can be written :

$$| \psi \rangle = \int d^3r_1 \, d^3r_2 \;\; \psi(\mathbf{r}_1, \mathbf{r}_2) | \, \mathbf{r}_1, \mathbf{r}_2 \rangle \tag{F-51}$$

with

$$\psi(\mathbf{r}_1, \mathbf{r}_2) = \langle \, \mathbf{r}_1, \mathbf{r}_2 \, | \, \psi \, \rangle \tag{F-52}$$

Moreover, the square of the norm of $| \psi \rangle$ is equal to:

$$\langle \psi | \psi \rangle = \int d^3r_1 \;\; d^3r_2 \, | \, \psi(\mathbf{r}_1, \mathbf{r}_2)|^2 \tag{F-53}$$

For it to be finite, $\psi(\mathbf{r}_1, \mathbf{r}_2)$ must be square-integrable. Therefore, a wave function $\psi(\mathbf{r}_1, \mathbf{r}_2)$ is associated with each ket of $\mathscr{E}_{\mathbf{r}_1\mathbf{r}_2}$ : *the state space of a two-particle system is the tensor product of the spaces which correspond to each of the particles.* A C.S.C.O. is obtained in $\mathscr{E}_{\mathbf{r}_1\mathbf{r}_2}$ by joining, for example, $X_1, Y_1$, $Z_1$ and $X_2$, $Y_2$, $Z_2$.

Assume that the state of the system is described by a tensor product ket:

$$| \psi \rangle = | \psi_1 \rangle | \psi_2 \rangle \tag{F-54}$$

The corresponding wave function can then be factorized:

$$\psi(\mathbf{r}_1, \mathbf{r}_2) = \langle \, \mathbf{r}_1, \mathbf{r}_2 \, | \, \psi \, \rangle = \langle \, \mathbf{r}_1 \, | \, \psi_1 \, \rangle \langle \, \mathbf{r}_2 \, | \, \psi_2 \, \rangle = \psi_1(\mathbf{r}_1) \;\; \psi_2(\mathbf{r}_2) \tag{F-55}$$

In this case, one says that *there is no correlation* between the two particles. We shall analyze later (complement $D_{III}$) the physical consequences of such a situation.

The preceding can be generalized : when a physical system is composed of the union of two or several simpler systems, its state space is the tensor product of the spaces which correspond to each of the component systems.

**References and suggestions for further reading:**

Section 10 of the bibliography contains references to a certain number of mathematical texts, classified by subject. Under each heading, they are listed as much as possible in order of increasing difficulty. See also the quantum mechanics texts (sections 1 and 2 of the bibliography), which treat the mathematical problems at many different levels. They also contain other references.

For a very simple approach to the fundamental mathematical concepts needed to understand chapter II (vector spaces, operators, diagonalization of matrices, etc.), the reader can consult, for example : Arfken (10.4), chap. 4; Bak and Lichtenberg (10.3), chap. I; Bass (10.1), vol. I, chap. II to V. A more explicit application to quantum mechanics can be found in Jackson (10.5) (see, in particular, chap. 5), Butkov (10.8), chap. 10 (finite-dimensional linear spaces) and chap. 11 (infinite-dimensional vector spaces, spaces of functions). See also Meijer and Bauer (2.18), chap. 1, particularly the table at the end of this chapter.

## COMPLEMENTS OF CHAPTER II

| | |
|---|---|
| $A_{II}$: THE SCHWARZ INEQUALITY<br>$B_{II}$: REVIEW OF SOME USEFUL PROPERTIES OF LINEAR OPERATORS<br>$C_{II}$: UNITARY OPERATORS | $A_{II}$, $B_{II}$, $C_{II}$: review of some definitions and useful mathematical results (elementary level) intended for readers unfamiliar with these concepts; will serve as a reference later (especially $B_{II}$). |
| $D_{II}$: A MORE DETAILED STUDY OF THE $\{\lvert \mathbf{r} \rangle\}$ AND $\{\lvert \mathbf{p} \rangle\}$ REPRESENTATIONS<br>$E_{II}$: SOME GENERAL PROPERTIES OF TWO OBSERVABLES, $Q$ AND $P$, WHOSE COMMUTATOR IS EQUAL TO $i\hbar$ | $D_{II}$, $E_{II}$: complete § E of chapter II :<br>$D_{II}$: remains at the level of chapter II and can be read immediately after it.<br>$E_{II}$: adopts a more general and a slightly more formal point of view. Introduces, in particular, the translation operator. May be reserved for later study. |
| $F_{II}$: THE PARITY OPERATOR | $F_{II}$: discussion of the parity operator, particularly important in quantum mechanics; at the same time, a simple illustration of the concepts of chapter II; recommended for these two reasons. |
| $G_{II}$: AN APPLICATION OF THE PROPERTIES OF THE TENSOR PRODUCT : THE TWO-DIMENSIONAL INFINITE WELL | $G_{II}$: a simple application of the tensor product (§ F of chapter II); can be considered as a worked exercise. |
| $H_{II}$: EXERCISES | $H_{II}$: solutions are given for exercises 11 and 12; their aim is to familiarize the reader with the properties of commuting observables and the concept of a C.S.C.O. in a very simple special case. It is recommended that these exercises be done during the reading of § D-3 of chapter II. |

## Complément $A_{II}$

# THE SCHWARZ INEQUALITY

For any ket $| \psi \rangle$ belonging to the state space $\mathscr{E}$, we have:

$$\langle \psi | \psi \rangle \quad \text{real} \geqslant 0 \tag{1}$$

$\langle \psi | \psi \rangle$ being equal to zero only when $| \psi \rangle$ is the null vector [*cf.* equation (B-12) of chapter II]. Using inequality (1), we shall derive the Schwarz inequality, which states that, if $| \varphi_1 \rangle$ and $| \varphi_2 \rangle$ are any arbitrary vectors of $\mathscr{E}$, then:

$$\boxed{|\langle \varphi_1 | \varphi_2 \rangle|^2 \leqslant \langle \varphi_1 | \varphi_1 \rangle \langle \varphi_2 | \varphi_2 \rangle} \tag{2}$$

the equality being realized if and only if $| \varphi_1 \rangle$ and $| \varphi_2 \rangle$ are proportional.

Given $| \varphi_1 \rangle$ and $| \varphi_2 \rangle$, consider the ket $| \psi \rangle$ defined by:

$$| \psi \rangle = | \varphi_1 \rangle + \lambda | \varphi_2 \rangle \tag{3}$$

where $\lambda$ is an arbitrary parameter. Whatever $\lambda$ may be:

$$\langle \psi | \psi \rangle \\ = \langle \varphi_1 | \varphi_1 \rangle + \lambda \langle \varphi_1 | \varphi_2 \rangle + \lambda^* \langle \varphi_2 | \varphi_1 \rangle + \lambda\lambda^* \langle \varphi_2 | \varphi_2 \rangle \geqslant 0 \tag{4}$$

Let us chose for $\lambda$ the value:

$$\lambda = - \frac{\langle \varphi_2 | \varphi_1 \rangle}{\langle \varphi_2 | \varphi_2 \rangle} \tag{5}$$

In (4), the second and third terms of the right-hand side are then equal, and opposite in value to the fourth term, so that (4) reduces to:

$$\langle \varphi_1 | \varphi_1 \rangle - \frac{\langle \varphi_1 | \varphi_2 \rangle \langle \varphi_2 | \varphi_1 \rangle}{\langle \varphi_2 | \varphi_2 \rangle} \geqslant 0 \tag{6}$$

Since $\langle \varphi_2 | \varphi_2 \rangle$ is positive, we can multiply this inequality by $\langle \varphi_2 | \varphi_2 \rangle$, to obtain:

$$\langle \varphi_1 | \varphi_1 \rangle \langle \varphi_2 | \varphi_2 \rangle \geqslant \langle \varphi_1 | \varphi_2 \rangle \langle \varphi_2 | \varphi_1 \rangle \tag{7}$$

which is precisely (2). In (7), the equality can only be realized if $\langle \psi | \psi \rangle = 0$, that is, according to (3), if $| \varphi_1 \rangle = - \lambda | \varphi_2 \rangle$. The kets $| \varphi_1 \rangle$ and $| \varphi_2 \rangle$ are then proportional.

**References:**

Bass I (10.1), § 5-3; Arfken (10.4), § 9-4.

# Complement $B_{II}$

## REVIEW OF SOME USEFUL PROPERTIES OF LINEAR OPERATORS



The aim of this complement is to review a certain number of definitions and useful properties of linear operators.

### 1. Trace of an operator

#### a. DEFINITION

The trace of an operator $A$, written Tr $A$, is the sum of its diagonal matrix elements.

When a discrete orthonormal basis, $\{ | u_i \rangle \}$, is chosen for the space $\mathscr{E}$, one has, by definition:

$$\text{Tr}\, A = \sum_i \langle u_i | A | u_i \rangle \tag{1}$$

For the case of a continuous orthonormal basis $\{ | w_\alpha \rangle \}$, one has:

$$\text{Tr}\, A = \int \text{d}\alpha \, \langle w_\alpha | A | w_\alpha \rangle \tag{2}$$

When $\mathscr{E}$ is an infinite-dimensional space, the trace of the operator $A$ is defined only if expressions (1) and (2) converge.

**b. THE TRACE IS INVARIANT**

The sum of the diagonal elements of the matrix which represents an operator $A$ in an arbitrary basis does not depend on this basis.

Let us derive this property for the case of a change from one discrete orthonormal basis $\{ | u_i \rangle \}$ to another discrete orthonormal basis $\{ | t_k \rangle \}$. We have:

$$\sum_i \langle u_i | A | u_i \rangle = \sum_i \langle u_i | \left[ \sum_k | t_k \rangle \langle t_k | \right] A | u_i \rangle \tag{3}$$

(where we have used the closure relation for the $| t_k \rangle$ states). The right-hand side of (3) is equal to:

$$\sum_{i,k} \langle u_i | t_k \rangle \langle t_k | A | u_i \rangle = \sum_{i,k} \langle t_k | A | u_i \rangle \langle u_i | t_k \rangle \tag{4}$$

(since it is possible to change the order of two numbers in a product). We can then replace $\sum_i | u_i \rangle \langle u_i |$ in (4) by $\mathbb{1}$ (closure relation for the $| u_i \rangle$ states), and we obtain, finally:

$$\sum_i \langle u_i | A | u_i \rangle = \sum_k \langle t_k | A | t_k \rangle \tag{5}$$

We have therefore demonstrated the property of invariance for this case.

COMMENT:

If the operator $A$ is an observable, Tr $A$ can therefore be calculated in a basis of eigenvectors of $A$. The diagonal matrix elements are then the eigenvalues $a_n$ of $A$ (degree of degeneracy $g_n$) and the trace can be written:

$$\text{Tr}\, A = \sum_n g_n\, a_n \tag{6}$$

**c. IMPORTANT PROPERTIES**

$$\text{Tr}\, AB = \text{Tr}\, BA \tag{7a}$$

$$\text{Tr}\, ABC = \text{Tr}\, BCA = \text{Tr}\, CAB \tag{7b}$$

In general, the trace of the product of any number of operators is invariant when a cyclic permutation is performed on these operators.

Let us prove, for example, relation (7-a):

$$\begin{aligned} \text{Tr}\, AB &= \sum_i \langle u_i | AB | u_i \rangle = \sum_{i,j} \langle u_i | A | u_j \rangle \langle u_j | B | u_i \rangle \\ &= \sum_{i,j} \langle u_j | B | u_i \rangle \langle u_i | A | u_j \rangle = \sum_j \langle u_j | BA | u_j \rangle = \text{Tr}\, BA \end{aligned} \tag{8}$$

(twice using the closure relation on the $\{ | u_i \rangle \}$ basis). Relation (7-a) is thus proved; its generalization (7-b) presents no difficulty.

## 2. Commutator algebra

### a. DEFINITION

The commutator $[A, B]$ of two operators is, by definition:

$$[A, B] = AB - BA \tag{9}$$

### b. PROPERTIES

$$[A, B] = -[B, A] \tag{10}$$
$$[A, (B + C)] = [A, B] + [A, C] \tag{11}$$
$$[A, BC] = [A, B]C + B[A, C] \tag{12}$$
$$[A, [B, C]] + [B, [C, A]] + [C, [A, B]] = 0 \tag{13}$$
$$[A, B]^\dagger = [B^\dagger, A^\dagger] \tag{14}$$

The derivation of these properties is straightforward : it suffices to compare both sides of each equation after having written them out explicitly.

## 3. Restriction of an operator to a subspace

Let $P_q$ be the projector onto the $q$-dimensional subspace $\mathscr{E}_q$ spanned by the $q$ orthonormal vectors $| \varphi_i \rangle$:

$$P_q = \sum_{i=1}^{q} | \varphi_i \rangle \langle \varphi_i | \tag{15}$$

By definition, the restriction $\hat{A}_q$ of the operator $A$ to the subspace $\mathscr{E}_q$ is:

$$\hat{A}_q = P_q A P_q \tag{16}$$

If $| \psi \rangle$ is an arbitrary ket, it follows from this definition that:

$$\hat{A}_q | \psi \rangle = P_q A | \hat{\psi}_q \rangle \tag{17}$$

where:

$$| \hat{\psi}_q \rangle = P_q | \psi \rangle \tag{18}$$

is the orthogonal projection of $| \psi \rangle$ onto $\mathscr{E}_q$. Consequently, to make $\hat{A}_q$ act on an arbitrary ket $| \psi \rangle$, one begins by projecting this ket onto $\mathscr{E}_q$; then one lets the operator $A$ act on this projection, retaining only the projection in $\mathscr{E}_q$ of the resulting ket. The operator $\hat{A}_q$, which transforms any ket of $\mathscr{E}_q$ into a ket belonging to this same subspace, is therefore an operator whose action has been restricted to $\mathscr{E}_q$.

What can be said about the matrix which represents $\hat{A}_q$? Let us choose a basis $\{ | u_k \rangle \}$ whose first $q$ vectors belong to $\mathscr{E}_q$ (they are, for example, the $| \varphi_i \rangle$), the others belonging to the supplementary subspace. We have :

$$\langle u_i | \hat{A}_q | u_j \rangle = \langle u_i | P_q A P_q | u_j \rangle \tag{19}$$

that is:

$$\langle u_i | \hat{A}_q | u_j \rangle = \begin{cases} \langle u_i | A | u_j \rangle & \text{if } i, j \leqslant q \\ 0 & \text{if one of the two indices } i \text{ or } j \text{ is greater than } q \end{cases} \tag{20}$$

Therefore, the matrix which represents $\hat{A}_q$ is, as it were, "cut out" of the one which represents $A$. One retains only the matrix elements of $A$ associated with basis vectors $| u_i \rangle$ and $| u_j \rangle$, both belonging to $\mathscr{E}_q$, the other matrix elements being replaced by zeros.

## 4. Functions of operators

### a. DEFINITION; SIMPLE PROPERTIES

Consider an arbitrary linear operator $A$. It is not difficult to define the operator $A^n$: it is the operator which corresponds to $n$ successive applications of the operator $A$. The definition of the operator $A^{-1}$, the inverse of $A$, is also well known: $A^{-1}$ is the operator (if it exists) which satisfies the relations:

$$A^{-1}A = AA^{-1} = \mathbb{1} \tag{21}$$

How can we define, in a more general way, an arbitrary function of an operator? To do this, let us consider a function $F$ of a variable $z$. Assume that, in a certain domain, $F$ can be expanded in a power series in $z$:

$$F(z) = \sum_{n=0}^{\infty} f_n z^n \tag{22}$$

By definition, the corresponding function of the operator $A$ is the operator $F(A)$ defined by a series which has the same coefficients $f_n$:

$$F(A) = \sum_{n=0}^{\infty} f_n A^n \tag{23}$$

For example, the operator $e^A$ is defined by:

$$e^A = \sum_{n=0}^{\infty} \frac{A^n}{n!} = \mathbb{1} + A + A^2/2! + \ldots + A^n/n! + \ldots \tag{24}$$

We shall not consider the problems concerning the convergence of the series (23), which depends on the eigenvalues of $A$ and on the radius of convergence of the series (22).

Note that if $F(z)$ is a real function, the coefficients $f_n$ are real. If, moreover, $A$ is Hermitian, we see from (23) that $F(A)$ is Hermitian.

Let $| \varphi_a \rangle$ be an eigenvector of $A$ with eigenvalue $a$:

$$A | \varphi_a \rangle = a | \varphi_a \rangle \tag{25}$$

Applying the operator $n$ times in succession, we obtain:

$$A^n | \varphi_a \rangle = a^n | \varphi_a \rangle \tag{26}$$

Now let us apply series (23) to $| \varphi_a \rangle$; we obtain:

$$F(A) | \varphi_a \rangle = \sum_{n=0}^{\infty} f_n a^n | \varphi_a \rangle = F(a) | \varphi_a \rangle \tag{27}$$

This leads to the following rule : *when $| \varphi_a \rangle$ is an eigenvector of $A$ with the eigenvalue $a$, $| \varphi_a \rangle$ is also an eigenvector of $F(A)$, with the eigenvalue $F(a)$.*

This property leads to a second definition of a function of an operator. Let us consider a diagonalizable operator $A$ (this is always the case if $A$ is an observable), and let us choose a basis where the matrix associated with $A$ is actually diagonal (its elements are then the eigenvalues $a_i$ of $A$). $F(A)$ is, by definition, the operator which is represented, in this same basis, by the diagonal matrix whose elements are $F(a_i)$.

For example, if $\sigma_z$ is the matrix

$$\sigma_z = \begin{pmatrix} 1 & 0 \\ 0 & -1 \end{pmatrix} \tag{28}$$

it follows directly that:

$$e^{\sigma_z} = \begin{pmatrix} e & 0 \\ 0 & 1/e \end{pmatrix} \tag{29}$$

COMMENT:

Care must be taken, when functions of operators are used, with respect to the order of the operators. For example, the operators $e^A e^B$, $e^B e^A$, and $e^{A+B}$ are not, in general, equal when $A$ and $B$ are operators and not numbers. Consider :

$$e^A e^B = \sum_p \frac{A^p}{p!} \sum_q \frac{B^q}{q!} = \sum_{p,q} \frac{A^p}{p!}\frac{B^q}{q!} \tag{30}$$

$$e^B e^A = \sum_q \frac{B^q}{q!} \sum_p \frac{A^p}{p!} = \sum_{p,q} \frac{B^q A^p}{p!\,q!} \tag{31}$$

$$e^{A+B} = \sum_p \frac{(A+B)^p}{p!} \tag{32}$$

When $A$ and $B$ are arbitrary, the right-hand sides of (30), (31) and (32) have no reason to be equal (see exercise 7 of complement $H_{II}$). However, *when $A$ and $B$ commute*, we have :

$$[A, B] = 0 \Longrightarrow e^A e^B = e^B e^A = e^{A+B} \tag{33}$$

(a relation which is obvious, moreover, if the diagonal matrices which represent $e^A$ and $e^B$ are considered in a basis of eigenvectors common to $A$ and $B$).

#### b. AN IMPORTANT EXAMPLE: THE POTENTIAL OPERATOR

In one-dimensional problems, we shall often have to consider "potential" operators $V(X)$ (so called because they correspond to the classical potential energy $V(x)$ of a particle placed in a force field), where $V(X)$ is a function of the position operator $X$.

It follows from the preceding section that $V(X)$ has as eigenvectors the eigenvectors $| x \rangle$ of $X$, and we have simply:

$$V(X) | x \rangle = V(x) | x \rangle \tag{34}$$

The matrix elements of $V(X)$ in the $\{ | x \rangle \}$ representation are therefore:

$$\langle x | V(X) | x' \rangle = V(x)\delta(x - x') \tag{35}$$

Applying (34) and using the fact that $V(X)$ is Hermitian (the function $V(x)$ is real), we obtain :

$$\langle x | V(X) | \psi \rangle = V(x) \langle x | \psi \rangle = V(x)\psi(x) \tag{36}$$

This equation shows that in the $\{ | x \rangle \}$ representation, the action of the operator $V(X)$ is simply multiplication by $V(x)$.

The generalization of (34), (35) and (36) to three-dimensional problems can be performed without difficulty; in this case, we obtain:

$$V(\mathbf{R}) | \mathbf{r} \rangle = V(\mathbf{r}) | \mathbf{r} \rangle \tag{37}$$

$$\langle \mathbf{r} | V(\mathbf{R}) | \mathbf{r}' \rangle = V(\mathbf{r})\delta(\mathbf{r} - \mathbf{r}') \tag{38}$$

$$\langle \mathbf{r} | V(\mathbf{R}) | \psi \rangle = V(\mathbf{r})\psi(\mathbf{r}) \tag{39}$$

#### c. COMMUTATORS INVOLVING FUNCTIONS OF OPERATORS

Definition (23) shows that $A$ commutes with every function of $A$:

$$[A, F(A)] = 0 \tag{40}$$

Similarly, if $A$ and $B$ commute, so do $F(A)$ and $B$:

$$[B, A] = 0 \Longrightarrow [B, F(A)] = 0 \tag{41}$$

What will be the commutator of an operator with a function of another operator which does not commute with it? We shall restrict ourselves here to the case of the $X$ and $P$ operators, whose commutator is equal to :

$$[X, P] = i\hbar \tag{42}$$

Using relation (12), we can calculate:

$$[X, P^2] = [X, PP] = [X, P]P + P[X, P] = 2i\hbar\, P \tag{43}$$

More generally, let us show that:

$$[X, P^n] = i\hbar n P^{n-1} \tag{44}$$

If we assume that this equation is verified, we obtain:

$$\begin{aligned}[X, P^{n+1}] &= [X, PP^n] = [X, P]P^n + P[X, P^n] \\ &= i\hbar\, P^n + i\hbar n\, PP^{n-1} = i\hbar(n + 1)P^n\end{aligned} \tag{45}$$

Relation (44) is therefore established by recurrence.

Now let us calculate the commutator $[X, F(P)]$:

$$[X, F(P)] = \sum_n [X, f_n P^n] = \sum_n i\hbar n f_n P^{n-1} \tag{46}$$

If $F'(z)$ denotes the derivative of the function $F(z)$, we recognize in (46) the definition of the operator $F'(P)$. Therefore:

$$\boxed{[X, F(P)] = i\hbar\, F'(P)} \tag{47}$$

An analogous argument would have enabled us to obtain the symmetric relation:

$$\boxed{[P, G(X)] = -\, i\hbar\, G'(X)} \tag{48}$$

COMMENTS:

(*i*) The preceding argument is based on the fact that $F(P)$ (or $G(X)$) depends only on $P$ (or on $X$). It is more difficult to calculate a commutator such as $[X, \Phi(X, P)]$, where $\Phi(X, P)$ is an operator which depends on both $X$ and $P$: the difficulties arise from the fact that $X$ and $P$ do not commute.

(*ii*) Equations (47) and (48) can be generalized to the case of two operators $A$ and $B$ which both commute with their commutator. An argument modeled on the preceding one shows that, if we have:

$$[A, C] = [B, C] = 0 \tag{49}$$

with

$$C = [A, B] \tag{50}$$

then:

$$[A, F(B)] = [A, B]F'(B) \tag{51}$$

## 5. Differentiation of an operator

### a. DEFINITION

Let $A(t)$ be an operator which depends on an arbitrary variable $t$. By definition, the derivative $\frac{\mathrm{d}A}{\mathrm{d}t}$ of $A(t)$ with respect to $t$ is given by the limit (if it exists):

$$\frac{\mathrm{d}A}{\mathrm{d}t} = \lim_{\Delta t \to 0} \frac{A(t + \Delta t) - A(t)}{\Delta t} \tag{52}$$

The matrix elements of $A(t)$ in an arbitrary basis of $t$-independent vectors $| u_i \rangle$ are functions of $t$:

$$\langle u_i | A | u_j \rangle = A_{ij}(t) \tag{53}$$

Let us call $\left(\frac{\mathrm{d}A}{\mathrm{d}t}\right)_{ij} = \langle u_i \left| \frac{\mathrm{d}A}{\mathrm{d}t} \right| u_j \rangle$ the matrix elements of $\frac{\mathrm{d}A}{\mathrm{d}t}$. It is easy to verify the relation:

$$\left(\frac{\mathrm{d}A}{\mathrm{d}t}\right)_{ij} = \frac{\mathrm{d}}{\mathrm{d}t} A_{ij} \tag{54}$$

Thus we obtain a very simple rule: to obtain the matrix elements representing $\frac{\mathrm{d}A}{\mathrm{d}t}$, all we must do is take the matrix representing $A$ and differentiate each of its elements (without changing their places).

### b. DIFFERENTIATION RULES

They are analogous to the ones for ordinary functions:

$$\frac{\mathrm{d}}{\mathrm{d}t}(F + G) = \frac{\mathrm{d}F}{\mathrm{d}t} + \frac{\mathrm{d}G}{\mathrm{d}t} \tag{55}$$

$$\frac{\mathrm{d}}{\mathrm{d}t}(FG) = \frac{\mathrm{d}F}{\mathrm{d}t} G + F \frac{\mathrm{d}G}{\mathrm{d}t} \tag{56}$$

Nevertheless, care must be taken not to modify the order of the operators in formula (56).

Let us prove, for example, the second of these equations. The matrix elements of $FG$ are:

$$\langle u_i \mid FG \mid u_j \rangle = \sum_k \langle u_i \mid F \mid u_k \rangle \langle u_k \mid G \mid u_j \rangle \tag{57}$$

We have seen that the matrix elements of $\mathrm{d}(FG)/\mathrm{d}t$ are the derivatives with respect to $t$ of those of $(FG)$. Thus we have, differentiating the right-hand side of (57):

$$\begin{aligned}\langle u_i \mid \frac{\mathrm{d}}{\mathrm{d}t}(FG) \mid u_j \rangle &= \sum_k \left[ \langle u_i \mid \frac{\mathrm{d}F}{\mathrm{d}t} \mid u_k \rangle \langle u_k \mid G \mid u_j \rangle + \right. \\ &\qquad \left. + \langle u_i \mid F \mid u_k \rangle \langle u_k \mid \frac{\mathrm{d}G}{\mathrm{d}t} \mid u_j \rangle \right] \\ &= \langle u_i \mid \frac{\mathrm{d}F}{\mathrm{d}t} G + F \frac{\mathrm{d}G}{\mathrm{d}t} \mid u_j \rangle \end{aligned} \tag{58}$$

This equation is valid for any $i$ and $j$. Formula (56) is thus established.

### c. EXAMPLES

Let us calculate the derivative of the operator $e^{At}$. By definition, we have:

$$e^{At} = \sum_{n=0}^{\infty} \frac{(At)^n}{n!} \tag{59}$$

Differentiating the series term by term, we obtain:

$$\begin{aligned}\frac{\mathrm{d}}{\mathrm{d}t}e^{At} &= \sum_{n=0}^{\infty} n\frac{t^{n-1}A^n}{n!} \\ &= A\sum_{n=1}^{\infty}\frac{(At)^{n-1}}{(n-1)!} \\ &= \left[\sum_{n=1}^{\infty}\frac{(At)^{n-1}}{(n-1)!}\right]A\end{aligned} \tag{60}$$

We recognize inside the brackets the series which defines $e^{At}$ (taking as the summation index $p = n - 1$). The result is therefore:

$$\frac{\mathrm{d}}{\mathrm{d}t}e^{At} = A\,e^{At} = e^{At}A \tag{61}$$

In this simple case involving only one operator, it is unnecessary to pay attention to the order of the factors : $e^{At}$ and $A$ commute.

This is not the case if one is interested in differentiating an operator such as $e^{At}e^{Bt}$. Applying (56) and (61), we obtain :

$$\frac{\mathrm{d}}{\mathrm{d}t}(e^{At}e^{Bt}) = A\,e^{At}e^{Bt} + e^{At}B\,e^{Bt} \tag{62}$$

The right-hand side of this equation can be transformed into $e^{At}Ae^{Bt} + e^{At}Be^{Bt}$ or $e^{At}Ae^{Bt} + e^{At}e^{Bt}B$, for example. However, we can never obtain (unless, of course, $A$ and $B$ commute) an expression such as $(A + B)e^{At}e^{Bt}$. In this case, the order of the operators is therefore important.

COMMENT:

Even when the function involves only one operator, differentiation cannot always be performed according to the rules valid for ordinary functions. For example, when $A(t)$ has an arbitrary time-dependence, the derivative $\frac{\mathrm{d}}{\mathrm{d}t}e^{A(t)}$ *is generally not equal to* $\frac{\mathrm{d}A}{\mathrm{d}t}e^{A(t)}$. It can be seen by expanding $e^{A(t)}$ in a power series in $A(t)$ that $A(t)$ and $\frac{\mathrm{d}A}{\mathrm{d}t}$ must commute for this equality to hold.

#### d. AN APPLICATION: A USEFUL FORMULA

Consider two operators $A$ and $B$ which, by hypothesis, both commute with their commutator. In this case, we shall derive the relation:

$$e^{A}e^{B} = e^{A+B}\,e^{\frac{1}{2}[A,B]} \tag{63}$$

(Glauber's formula).

Let us define the operator $F(t)$, a function of the real variable $t$, by :

$$F(t) = e^{At}e^{Bt} \tag{64}$$

We have:

$$\frac{\mathrm{d}F}{\mathrm{d}t} = A e^{At} e^{Bt} + e^{At} B e^{Bt} = \left(A + e^{At} B e^{-At}\right) F(t) \tag{65}$$

Since $A$ and $B$ commute with their commutator, formula (51) can be applied in order to calculate :

$$\left[e^{At}, B\right] = t\left[A, B\right] e^{At} \tag{66}$$

Therefore:

$$e^{At} \; B = B \, e^{At} \; + t\left[A, B\right] e^{At} \tag{67}$$

Multiply both sides of this equation on the right by $e^{-At}$. Substituting the relation so obtained into (65), we obtain:

$$\frac{\mathrm{d}F}{\mathrm{d}t} = \left(A + B + t[A, B]\right) F(t) \tag{68}$$

The operators $A + B$ and $[A, B]$ commute by hypothesis. We can therefore integrate the differential equation (68) as if $A + B$ and $[A, B]$ were numbers. This yields:

$$F(t) = F(0) \, \mathrm{e}^{(A+B)t + \frac{1}{2}[A, B]t^2} \tag{69}$$

Setting $t = 0$, we see that $F(0) = \mathbb{1}$, and:

$$F(t) = \mathrm{e}^{(A+B)t + \frac{1}{2}[A, B]t^2} \tag{70}$$

Let us then set $t = 1$; we obtain equation (63), which is thus proven.

COMMENT:

When the operators $A$ and $B$ are arbitrary, equation (63) is not in general valid: it is necessary that both $A$ and $B$ commute with $[A, B]$. This condition may seem very restrictive. Actually, in quantum mechanics one often encounters operators whose commutator is a number: for example, $X$ and $P$, or the operators $a$ and $a^\dagger$ of the harmonic oscillator (*cf.* chap. V).

**References :**

See the subsections "General texts" and "Linear algebra – Hilbert spaces" of section 10 of the bibliography.

# Complement $C_{II}$

## UNITARY OPERATORS



### 1. General properties of unitary operators

#### a. DEFINITION; SIMPLE PROPERTIES

By definition, an operator $U$ is unitary if its inverse $U^{-1}$ is equal to its adjoint $U^\dagger$:

$$U^\dagger U = UU^\dagger = \mathbb{1} \tag{1}$$

Consider two arbitrary vectors $|\psi_1\rangle$ and $|\psi_2\rangle$ of $\mathscr{E}$, and their transforms $|\tilde{\psi}_1\rangle$ and $|\tilde{\psi}_2\rangle$ under the action of $U$:

$$\begin{aligned} |\tilde{\psi}_1\rangle &= U|\psi_1\rangle \\ |\tilde{\psi}_2\rangle &= U|\psi_2\rangle \end{aligned} \tag{2}$$

Let us calculate the scalar product $\langle\tilde{\psi}_1|\tilde{\psi}_2\rangle$; we obtain:

$$\langle\tilde{\psi}_1|\tilde{\psi}_2\rangle = \langle\psi_1|U^\dagger U|\psi_2\rangle = \langle\psi_1|\psi_2\rangle \tag{3}$$

The unitary transformation associated with the operator $U$ therefore conserves the scalar product (and, consequently, the norm) in $\mathscr{E}$. When $\mathscr{E}$ is finite-dimensional, moreover, this property is characteristic of a unitary operator.

COMMENTS:

(*i*) If $A$ is a Hermitian operator, the operator $T = e^{iA}$ is unitary, since:

$$T^\dagger = e^{-iA^\dagger} = e^{-iA} \tag{4}$$

and therefore:

$$\begin{aligned} T^\dagger T &= e^{-iA}e^{iA} = \mathbb{1} \\ TT^\dagger &= e^{iA}e^{-iA} = \mathbb{1} \end{aligned} \tag{5}$$

(obviously, $-iA$ commutes with $iA$).

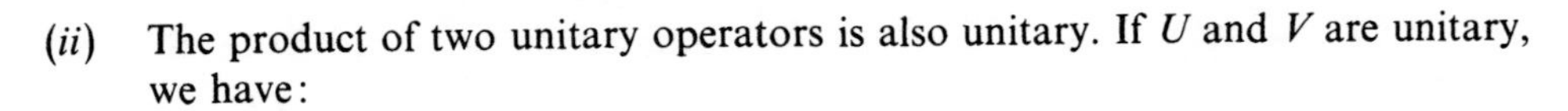

(*ii*) The product of two unitary operators is also unitary. If $U$ and $V$ are unitary, we have:

$$U^\dagger U = UU^\dagger = \mathbb{1}$$
$$V^\dagger V = VV^\dagger = \mathbb{1} \tag{6}$$

Let us now calculate:

$$(UV)^\dagger(UV) = V^\dagger U^\dagger UV = V^\dagger V = \mathbb{1}$$
$$(UV)(UV)^\dagger = UVV^\dagger U^\dagger = UU^\dagger = \mathbb{1} \tag{7}$$

These equations indeed show that the product operator $UV$ is unitary. This property, moreover, was foreseeable : when two transformations conserve the scalar product, so does the successive application of these two transformations.

(*iii*) In the ordinary three-dimensional space of real vectors, we are familiar with operators which conserve the norm and the scalar product: rotations, symmetry operations with respect to a point, to a plane, etc. In this case where the space is real, these operators are said to be orthogonal. Unitary operators constitute the generalization of orthogonal operators to complex spaces (with an arbitrary number of dimensions).

#### b. UNITARY OPERATORS AND CHANGE OF BASES

α. Let $\{ | v_i \rangle \}$ be an orthonormal basis of the state space $\mathscr{E}$, assumed to be discrete. Call $| \tilde{v}_i \rangle$ the transform of the vector $| v_i \rangle$ under the action of the operator $U$:

$$| \tilde{v}_i \rangle = U | v_i \rangle \tag{8}$$

Since the operator $U$ is unitary, we have:

$$\langle \tilde{v}_i | \tilde{v}_j \rangle = \langle v_i | v_j \rangle = \delta_{ij} \tag{9}$$

The $| \tilde{v}_i \rangle$ vectors are therefore orthonormal. Let us show that they constitute a basis of $\mathscr{E}$. To do so, consider an arbitrary vector $| \psi \rangle$ of $\mathscr{E}$. Since the set $\{ | v_i \rangle \}$ constitutes a basis, the vector $U^\dagger | \psi \rangle$ can be expanded on the $| v_i \rangle$:

$$U^\dagger | \psi \rangle = \sum_i c_i | v_i \rangle \tag{10}$$

Applying the operator $U$ to this equation, we obtain:

$$UU^\dagger | \psi \rangle = \sum_i c_i U | v_i \rangle \tag{11}$$

and, therefore:

$$| \psi \rangle = \sum_i c_i | \tilde{v}_i \rangle \tag{12}$$

This equation expresses the fact that any vector $| \psi \rangle$ can be expanded on the vectors $| \tilde{v}_i \rangle$, which therefore constitute a basis. Thus we can state the following result: a necessary condition for an operator $U$ to be unitary is that the vectors of an orthonormal basis of $\mathscr{E}$, transformed by $U$, constitute another orthonormal basis.

$\beta$. Conversely, let us show that this condition is sufficient. By hypothesis, we then have:

$$\begin{aligned} &| \tilde{v}_i \rangle = U | v_i \rangle \\ &\langle \tilde{v}_i | \tilde{v}_j \rangle = \delta_{ij} \\ &\sum_i | \tilde{v}_i \rangle \langle \tilde{v}_i | = \mathbb{1} \end{aligned} \tag{13}$$

and therefore:

$$\langle v_j | U^\dagger = \langle \tilde{v}_j | \tag{14}$$

Let us calculate:

$$\begin{aligned} U^\dagger U | v_i \rangle = U^\dagger | \tilde{v}_i \rangle &= \sum_j | v_j \rangle \langle v_j | U^\dagger | \tilde{v}_i \rangle \\ &= \sum_j | v_j \rangle \langle \tilde{v}_j | \tilde{v}_i \rangle = \sum_j | v_j \rangle \delta_{ij} \\ &= | v_i \rangle \end{aligned} \tag{15}$$

Relation (15), which is valid for all $i$, expresses the fact that the operator $U^\dagger U$ is the identity operator. Let us show, in the same way, that $UU^\dagger = \mathbb{1}$. To do this, consider the action of $U^\dagger$ on a vector $| v_i \rangle$:

$$\begin{aligned} U^\dagger | v_i \rangle &= \sum_j | v_j \rangle \langle v_j | U^\dagger | v_i \rangle \\ &= \sum_j | v_j \rangle \langle \tilde{v}_j | v_i \rangle \end{aligned} \tag{16}$$

We then have:

$$\begin{aligned} UU^\dagger | v_i \rangle &= \sum_j U | v_j \rangle \langle \tilde{v}_j | v_i \rangle \\ &= \sum_j | \tilde{v}_j \rangle \langle \tilde{v}_j | v_i \rangle \\ &= | v_i \rangle \end{aligned} \tag{17}$$

We deduce from this that $UU^\dagger = \mathbb{1}$: the operator $U$ is therefore unitary.

### c. UNITARY MATRICES

Let:

$$U_{ij} = \langle v_i | U | v_j \rangle \tag{18}$$

be the matrix elements of $U$. How can one see from the matrix representing $U$ if this operator is unitary?

Relation (1) gives us:

$$\langle v_i | U^\dagger U | v_j \rangle = \sum_k \langle v_i | U^\dagger | v_k \rangle \langle v_k | U | v_j \rangle \tag{19}$$

that is:

$$\sum_k U_{ki}^* U_{kj} = \delta_{ij} \tag{20}$$

When a matrix is unitary, the sum of the products of the elements of one column and the complex conjugates of the elements of another column is

- zero if the two columns are different,
- equal to 1 if they are not.

Let us cite some examples in which this rule can be easily verified.

EXAMPLES:

(*i*) The matrix which represents a rotation through an angle $\theta$ about $Oz$, in ordinary three-dimensional space:

$$R(\theta) = \begin{pmatrix} \cos\theta & -\sin\theta & 0 \\ \sin\theta & \cos\theta & 0 \\ 0 & 0 & 1 \end{pmatrix} \tag{21}$$

(*ii*) The rotation matrix in the state space of a spin $\frac{1}{2}$ particle (*cf.* chap. IX):

$$R^{(1/2)}(\alpha, \beta, \gamma) = \begin{pmatrix} e^{-\frac{i}{2}(\alpha+\gamma)} \cos\frac{\beta}{2} & -e^{\frac{i}{2}(\gamma-\alpha)} \sin\frac{\beta}{2} \\ e^{\frac{i}{2}(\alpha-\gamma)} \sin\frac{\beta}{2} & e^{\frac{i}{2}(\alpha+\gamma)} \cos\frac{\beta}{2} \end{pmatrix} \tag{22}$$

#### d. EIGENVALUES AND EIGENVECTORS OF A UNITARY OPERATOR

Let $|\psi_u\rangle$ be a normalized eigenvector of the unitary operator $U$ with eigenvalue $u$:

$$U|\psi_u\rangle = u|\psi_u\rangle \tag{23}$$

The square of the norm of the vector $U|\psi_u\rangle$ is:

$$\langle\psi_u|U^\dagger U|\psi_u\rangle = u^*u\langle\psi_u|\psi_u\rangle = u^*u \tag{24}$$

Since the unitary operator conserves the norm, we have, necessarily, $u^*u = 1$. The eigenvalues of a unitary operator must therefore be complex numbers of modulus 1:

$$u = e^{i\varphi_u} \quad \text{where } \varphi_u \text{ is real} \tag{25}$$

Consider two eigenvectors $|\psi_u\rangle$ and $|\psi_{u'}\rangle$ of $U$; we then have:

$$\begin{aligned} \langle\psi_u|\psi_{u'}\rangle &= \langle\psi_u|U^\dagger U|\psi_{u'}\rangle = u^*u'\langle\psi_u|\psi_{u'}\rangle \\ &= e^{i(\varphi_{u'}-\varphi_u)}\langle\psi_u|\psi_{u'}\rangle \end{aligned} \tag{26}$$

When the eigenvalues $u$ and $u'$ are different, we see from (26) that the scalar product $\langle\psi_u|\psi_{u'}\rangle$ is zero: two eigenvectors of a unitary operator corresponding to different eigenvalues are orthogonal.

## 2. Unitary transformations of operators

We saw in § 1-b that a unitary operator $U$ permits the construction, starting with one orthonormal basis $\{ | v_i \rangle \}$ of $\mathscr{E}$, of another one, $\{ | \tilde{v}_i \rangle \}$. In this section, we are going to define a transformation which acts, not on the vectors, but on the operators.

By definition, the transform $\tilde{A}$ of the operator $A$ will be the operator which, in the $\{ | \tilde{v}_i \rangle \}$ basis, has the same matrix elements as the operator $A$ in the $\{ | v_i \rangle \}$ basis:

$$\langle \tilde{v}_i | \tilde{A} | \tilde{v}_j \rangle = \langle v_i | A | v_j \rangle \tag{27}$$

Substituting (8) into this equation, we obtain:

$$\langle v_i | U^\dagger \tilde{A} U | v_j \rangle = \langle v_i | A | v_j \rangle \tag{28}$$

Since $i$ and $j$ are arbitrary, we have:

$$U^\dagger \tilde{A} U = A \tag{29}$$

or, multiplying this equation on the left by $U$ and on the right by $U^\dagger$:

$$\tilde{A} = UAU^\dagger \tag{30}$$

Equation (30) can be taken to be the definition of the transform $\tilde{A}$ of the operator $A$ by the unitary transformation $U$. In quantum mechanics, such transformations are often used: a first example is given in complement F of this chapter (§ 2-a).

How can the eigenvectors of $\tilde{A}$ be obtained from those of $A$? Let us consider an eigenvector $| \varphi_a \rangle$ of $A$, with an eigenvalue $a$:

$$A | \varphi_a \rangle = a | \varphi_a \rangle \tag{31}$$

Let $| \tilde{\varphi}_a \rangle$ be the transform of $| \varphi_a \rangle$ by the operator $U$: $| \tilde{\varphi}_a \rangle = U | \varphi_a \rangle$. We then have:

$$\begin{aligned} \tilde{A} | \tilde{\varphi}_a \rangle &= (UAU^\dagger) U | \varphi_a \rangle = UA(U^\dagger U) | \varphi_a \rangle \\ &= UA | \varphi_a \rangle = aU | \varphi_a \rangle \\ &= a | \tilde{\varphi}_a \rangle \end{aligned} \tag{32}$$

$| \tilde{\varphi}_a \rangle$ is therefore an eigenvector of $\tilde{A}$, with eigenvalue $a$. This can be generalized to the following rule: the eigenvectors of the transform $\tilde{A}$ of $A$ are the transforms $| \tilde{\varphi}_a \rangle$ of the eigenvectors $| \varphi_a \rangle$ of $A$; the eigenvalues are unchanged.

COMMENTS:

(*i*) The adjoint of the transform $\tilde{A}$ of $A$ by $U$ is the transform of $A^\dagger$ by $U$:

$$(\tilde{A})^\dagger = (UAU^\dagger)^\dagger = UA^\dagger U^\dagger = \widetilde{A^\dagger} \tag{33}$$

In particular, it follows from this equation that, if $A$ is Hermitian, $\tilde{A}$ is also.

(*ii*) Analogously, we have:

$$(\tilde{A})^2 = UAU^\dagger UAU^\dagger = UAAU^\dagger = \widetilde{A^2}$$

and, in general:

$$(\tilde{A})^n = \widetilde{A^n} \tag{34}$$

Using definition (23) of complement $B_{II}$, we can easily show that:

$$\widetilde{F(A)} = F(\tilde{A}) \tag{35}$$

where $F(A)$ is a function of the operator $A$.

## 3. The infinitesimal unitary operator

Let $U(\varepsilon)$ be a unitary operator which depends on an infinitely small real quantity $\varepsilon$; by hypothesis, $U(\varepsilon) \longrightarrow \mathbb{1}$ when $\varepsilon \longrightarrow 0$. Expand $U(\varepsilon)$ in a power series in $\varepsilon$:

$$U(\varepsilon) = \mathbb{1} + \varepsilon G + ... \tag{36}$$

We then have:

$$U^\dagger(\varepsilon) = \mathbb{1} + \varepsilon G^\dagger + ... \tag{37}$$

and:

$$U(\varepsilon)\, U^\dagger(\varepsilon) = U^\dagger(\varepsilon)\, U(\varepsilon) = \mathbb{1} + \varepsilon(G + G^\dagger) + ... \tag{38}$$

Since $U(\varepsilon)$ is unitary, the first-order terms in $\varepsilon$ on the right-hand side of (38) are zero; we therefore have:

$$G + G^\dagger = 0 \tag{39}$$

This equation expresses the fact that the operator $G$ is anti-Hermitian. It is convenient to set:

$$F = iG \tag{40}$$

so as to obtain the equation:

$$F - F^\dagger = 0 \tag{41}$$

which states that $F$ is Hermitian. An infinitesimal unitary operator can therefore be written in the form:

$$U(\varepsilon) = \mathbb{1} - i\varepsilon F \tag{42}$$

where $F$ is a Hermitian operator.

Substituting (42) into (30), we obtain:

$$\tilde{A} = (\mathbb{1} - i\varepsilon F)A(\mathbb{1} + i\varepsilon F^\dagger) = (\mathbb{1} - i\varepsilon F)A(\mathbb{1} + i\varepsilon F) \tag{43}$$

and, therefore:

$$\tilde{A} - A = -\, i\varepsilon[F, A] \tag{44}$$

The variation of the operator $A$ under the transformation $U$ is, to first order in $\varepsilon$, proportional to the commutator $[F, A]$.

# Complement $D_{II}$

## A MORE DETAILED STUDY OF THE $\{ | \mathbf{r} \rangle \}$ AND $\{ | \mathbf{p} \rangle \}$ REPRESENTATIONS



### 1. The $\{ | \mathbf{r} \rangle \}$ representation

#### a. THE R OPERATOR AND FUNCTIONS OF R

Let us calculate the matrix elements, in the $\{ | \mathbf{r} \rangle \}$ representation, of the $X$, $Y$, $Z$ operators. Using formula (E-36) of chapter II and the orthogonality relations of the kets $| \mathbf{r} \rangle$, we immediately obtain:

$$\begin{aligned} \langle \mathbf{r} | X | \mathbf{r}' \rangle &= x\delta(\mathbf{r} - \mathbf{r}') \\ \langle \mathbf{r} | Y | \mathbf{r}' \rangle &= y\delta(\mathbf{r} - \mathbf{r}') \\ \langle \mathbf{r} | Z | \mathbf{r}' \rangle &= z\,\delta(\mathbf{r} - \mathbf{r}') \end{aligned} \tag{1}$$

These three equations can be condensed into one:

$$\langle \mathbf{r} | \mathbf{R} | \mathbf{r}' \rangle = \mathbf{r}\,\delta(\mathbf{r} - \mathbf{r}') \tag{2}$$

The matrix elements, in the $\{ | \mathbf{r} \rangle \}$ representation, of a function $F(\mathbf{R})$ are also very simple [*cf.* equation (27) of complement $B_{II}$]:

$$\langle \mathbf{r} | F(\mathbf{R}) | \mathbf{r}' \rangle = F(\mathbf{r})\,\delta(\mathbf{r} - \mathbf{r}') \tag{3}$$

#### b. THE P OPERATOR AND FUNCTIONS OF P

Let us calculate the matrix element $\langle \mathbf{r} | P_x | \mathbf{r}' \rangle$:

$$\begin{aligned} \langle \mathbf{r} | P_x | \mathbf{r}' \rangle &= \int d^3p\, \langle \mathbf{r} | P_x | \mathbf{p} \rangle \langle \mathbf{p} | \mathbf{r}' \rangle \\ &= \int d^3p\, p_x \langle \mathbf{r} | \mathbf{p} \rangle \langle \mathbf{p} | \mathbf{r}' \rangle \\ &= (2\pi\hbar)^{-3} \int d^3p\, p_x\, e^{\frac{i}{\hbar}\mathbf{p}.(\mathbf{r} - \mathbf{r}')} \\ &= \left[ \frac{1}{2\pi\hbar} \int_{-\infty}^{+\infty} dp_x\, p_x e^{\frac{i}{\hbar}p_x(x - x')} \right] \times \end{aligned}$$

$$\times \left[ \frac{1}{2\pi\hbar} \int_{-\infty}^{+\infty} dp_y \, e^{\frac{i}{\hbar} p_y (y - y')} \right] \times$$

$$\times \left[ \frac{1}{2\pi\hbar} \int_{-\infty}^{+\infty} dp_z \, e^{\frac{i}{\hbar} p_z (z - z')} \right] \tag{4}$$

From this, it follows that, using the integral form of the "delta function" and its derivative [*cf.* appendix II, equations (34) and (53)]:

$$\langle \mathbf{r} \,|\, P_x \,|\, \mathbf{r}' \rangle = \frac{\hbar}{i} \delta'(x - x') \delta(y - y') \delta(z - z') \tag{5}$$

The matrix elements of the other components of **P** could be obtained in an analogous fashion.

Let us verify that the action of $P_x$ in the $\{ \,|\, \mathbf{r} \rangle \}$ representation can indeed be derived from formula (5). To do so, let us calculate:

$$\langle \mathbf{r} \,|\, P_x \,|\, \psi \rangle = \int d^3r' \, \langle \mathbf{r} \,|\, P_x \,|\, \mathbf{r}' \rangle \langle \mathbf{r}' \,|\, \psi \rangle \tag{6}$$

From (5):

$$\langle \mathbf{r} \,|\, P_x \,|\, \psi \rangle = \frac{\hbar}{i} \int \delta'(x - x') \, dx' \int \delta(y - y') \, dy' \int \delta(z - z') \psi(x', y', z') \, dz' \tag{7}$$

Using the relation

$$\int \delta'(-u) f(u) \, du = - \int \delta'(u) f(u) \, du = f'(0) \tag{8}$$

and taking $u = x' - x$, we obtain:

$$\langle \mathbf{r} \,|\, P_x \,|\, \psi \rangle = \frac{\hbar}{i} \frac{\partial}{\partial x} \psi(x, y, z) \tag{9}$$

which is indeed equation (E-26) of chapter II.

What is the value of the matrix element $\langle \mathbf{r} \,|\, G(\mathbf{P}) \,|\, \mathbf{r}' \rangle$ of a function $G(\mathbf{P})$ of the **P** operator? An analogous calculation gives us:

$$\begin{aligned} \langle \mathbf{r} \,|\, G(\mathbf{P}) \,|\, \mathbf{r}' \rangle &= \int d^3p \, \langle \mathbf{r} \,|\, G(\mathbf{P}) \,|\, \mathbf{p} \rangle \langle \mathbf{p} \,|\, \mathbf{r}' \rangle \\ &= (2\pi\hbar)^{-3} \int d^3p \, G(\mathbf{p}) \, e^{\frac{i}{\hbar} \mathbf{p}.(\mathbf{r} - \mathbf{r}')} \\ &= (2\pi\hbar)^{-3/2} \, \tilde{G}(\mathbf{r} - \mathbf{r}') \end{aligned} \tag{10}$$

where $\tilde{G}(\mathbf{r})$ is the inverse Fourier transform of the function $G(\mathbf{p})$:

$$\tilde{G}(\mathbf{r}) = (2\pi\hbar)^{-3/2} \int d^3p \, e^{\frac{i}{\hbar} \mathbf{p}.\mathbf{r}} G(\mathbf{p}) \tag{11}$$

### c. THE SCHRÖDINGER EQUATION IN THE $\{ | \mathbf{r} \rangle \}$ REPRESENTATION

In chapter III, we shall introduce the Schrödinger equation, which is of fundamental importance in quantum mechanics:

$$i\hbar \frac{\mathrm{d}}{\mathrm{d}t} | \psi(t) \rangle = H | \psi(t) \rangle \tag{12}$$

where $H$ is the Hamiltonian operator, which we shall define at that time. For a (spinless) particle in a scalar potential $V(\mathbf{r})$ [*cf.* equation (B-42) of chapter III]:

$$H = \frac{1}{2m} \mathbf{P}^2 + V(\mathbf{R}) \tag{13}$$

We are going to see how to write this equation in the $\{ | \mathbf{r} \rangle \}$ representation, that is, using the wave function $\psi(\mathbf{r}, t)$, defined by:

$$\psi(\mathbf{r}, t) = \langle \mathbf{r} | \psi(t) \rangle \tag{14}$$

Projecting (12) onto $| \mathbf{r} \rangle$, in the case where $H$ is given by formula (13), we obtain:

$$i\hbar \frac{\partial}{\partial t} \langle \mathbf{r} | \psi(t) \rangle = \frac{1}{2m} \langle \mathbf{r} | \mathbf{P}^2 | \psi(t) \rangle + \langle \mathbf{r} | V(\mathbf{R}) | \psi(t) \rangle \tag{15}$$

The quantities involved in this equation can be expressed in terms of $\psi(\mathbf{r}, t)$, since:

$$\frac{\partial}{\partial t} \langle \mathbf{r} | \psi(t) \rangle = \frac{\partial}{\partial t} \psi(\mathbf{r}, t) \tag{16}$$

$$\langle \mathbf{r} | V(\mathbf{R}) | \psi(t) \rangle = V(\mathbf{r}) \; \psi(\mathbf{r}, t) \tag{17}$$

The matrix element $\langle \mathbf{r} | \mathbf{P}^2 | \psi \rangle$ can be calculated by using the fact that in the $\{ | \mathbf{r} \rangle \}$ representation, $\mathbf{P}$ acts like $\frac{\hbar}{i} \boldsymbol{\nabla}$:

$$\begin{aligned} \langle \mathbf{r} | \mathbf{P}^2 | \psi(t) \rangle &= \langle \mathbf{r} | (P_x^2 + P_y^2 + P_z^2) | \psi(t) \rangle \\ &= -\hbar^2 \left( \frac{\partial^2}{\partial x^2} + \frac{\partial^2}{\partial y^2} + \frac{\partial^2}{\partial z^2} \right) \psi(x, y, z, t) \\ &= -\hbar^2 \Delta \psi(\mathbf{r}, t) \end{aligned} \tag{18}$$

The Schrödinger equation then becomes:

$$\boxed{i\hbar \frac{\partial}{\partial t} \psi(\mathbf{r}, t) = \left[ -\frac{\hbar^2}{2m} \Delta + V(\mathbf{r}) \right] \psi(\mathbf{r}, t)} \tag{19}$$

This is indeed the wave equation introduced in chapter I (§ B-2).

## 2. The $\{ \mid \mathbf{p} \rangle \}$ representation

### a. THE P OPERATOR AND FUNCTIONS OF P

We obtain without difficulty formulas analogous to (2) and (3):

$$\langle \mathbf{p} \mid \mathbf{P} \mid \mathbf{p}' \rangle = \mathbf{p}\,\delta(\mathbf{p} - \mathbf{p}') \tag{20}$$

$$\langle \mathbf{p} \mid G(\mathbf{P}) \mid \mathbf{p}' \rangle = G(\mathbf{p})\,\delta(\mathbf{p} - \mathbf{p}') \tag{21}$$

### b. THE R OPERATOR AND FUNCTIONS OF R

Arguments analogous to those of § 1 give us the formulas which correspond to (5) and (10):

$$\langle \mathbf{p} \mid X \mid \mathbf{p}' \rangle = i\hbar\,\delta'(p_x - p'_x)\,\delta(p_y - p'_y)\,\delta(p_z - p'_z) \tag{22}$$

and

$$\langle \mathbf{p} \mid F(\mathbf{R}) \mid \mathbf{p}' \rangle = (2\pi\hbar)^{-3/2}\,\overline{F}(\mathbf{p} - \mathbf{p}') \tag{23}$$

with

$$\overline{F}(\mathbf{p}) = (2\pi\hbar)^{-3/2} \int d^3r \; e^{-\frac{i}{\hbar}\mathbf{p}.\mathbf{r}}\, F(\mathbf{r}) \tag{24}$$

### c. THE SCHRÖDINGER EQUATION IN THE $\{ \mid \mathbf{p} \rangle \}$ REPRESENTATION

Let us introduce the "wave function in the $\{ \mid \mathbf{p} \rangle \}$ representation" by:

$$\overline{\psi}(\mathbf{p}, t) = \langle \mathbf{p} \mid \psi(t) \rangle \tag{25}$$

Using (12), we shall look for the equation which gives the time evolution of $\overline{\psi}(\mathbf{p}, t)$. Projecting (12) onto the ket $\mid \mathbf{p} \rangle$, we obtain:

$$i\hbar \frac{\partial}{\partial t} \langle \mathbf{p} \mid \psi(t) \rangle = \frac{1}{2m} \langle \mathbf{p} \mid \mathbf{P}^2 \mid \psi(t) \rangle + \langle \mathbf{p} \mid V(\mathbf{R}) \mid \psi(t) \rangle \tag{26}$$

Now we have:

$$\frac{\partial}{\partial t} \langle \mathbf{p} \mid \psi(t) \rangle = \frac{\partial}{\partial t} \overline{\psi}(\mathbf{p}, t) \tag{27}$$

$$\langle \mathbf{p} \mid \mathbf{P}^2 \mid \psi(t) \rangle = \mathbf{p}^2\, \overline{\psi}(\mathbf{p}, t) \tag{28}$$

The quantity which remains to be calculated is:

$$\langle \mathbf{p} \mid V(\mathbf{R}) \mid \psi(t) \rangle = \int d^3p' \, \langle \mathbf{p} \mid V(\mathbf{R}) \mid \mathbf{p}' \rangle \langle \mathbf{p}' \mid \psi(t) \rangle \tag{29}$$

Using (23), we find:

$$\langle \mathbf{p} \mid V(\mathbf{R}) \mid \psi(t) \rangle = (2\pi\hbar)^{-3/2} \int d^3p' \, \overline{V}(\mathbf{p} - \mathbf{p}')\,\overline{\psi}(\mathbf{p}', t) \tag{30}$$

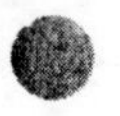



where $\overline{V}(\mathbf{p})$ is the Fourier transform of $V(\mathbf{r})$:

$$\overline{V}(\mathbf{p}) = (2\pi\hbar)^{-3/2} \int d^3r \; e^{-\frac{i}{\hbar}\mathbf{p}\cdot\mathbf{r}} \; V(\mathbf{r}) \tag{31}$$

The Schrödinger equation in the $\{ | \mathbf{p} \rangle \}$ representation is therefore written:

$$\boxed{i\hbar \frac{\partial}{\partial t} \overline{\psi}(\mathbf{p}, t) = \frac{\mathbf{p}^2}{2m} \overline{\psi}(\mathbf{p}, t) + (2\pi\hbar)^{-3/2} \int d^3p' \, \overline{V}(\mathbf{p} - \mathbf{p}') \overline{\psi}(\mathbf{p}', t)} \tag{32}$$

COMMENT:

Since $\overline{\psi}(\mathbf{p}, t)$ is the Fourier transform of $\psi(\mathbf{r}, t)$ [*cf.* formula (E-18) of chapter II], it would have been possible to find equation (32) by taking the Fourier transforms of both sides of equation (19).

## Complement $E_{II}$

# SOME GENERAL PROPERTIES OF TWO OBSERVABLES, $Q$ AND $P$, WHOSE COMMUTATOR IS EQUAL TO $i\hbar$



In quantum mechanics, one often encounters operators whose commutator is equal to $i\hbar$. This is the case, for example, when these two operators correspond to the two classical conjugate quantities $q_i$ and $p_i$ $\left(q_i\text{, the coordinate in a system of orthonormal axes, and the conjugate momentum } p_i = \frac{\partial \mathscr{L}}{\partial q_i}\right)$. In quantum mechanics, one associates with $q_i$ and $p_i$ the operators $Q_i$ and $P_i$ which satisfy the relation:

$$[Q_i, P_i] = i\hbar \tag{1}$$

In § E of chapter II, we encountered such operators : $X$ and $P_x$. In this complement, we shall take a more general point of view and show that it is possible to establish a whole series of important properties relative to two observables $P$ and $Q$ whose commutator is equal to $i\hbar$. All these properties are consequences of commutation relation (1).

## 1. The operator $S(\lambda)$: definition, properties

We shall consider two observables $P$ and $Q$, satisfying the relation:

$$[Q, P] = i\hbar \tag{2}$$

and we shall define the operator $S(\lambda)$, which depends on the real parameter $\lambda$, by:

$$S(\lambda) = e^{-i\lambda P/\hbar} \tag{3}$$

This operator is unitary; it is easy to verify the relations:

$$S^\dagger(\lambda) = S^{-1}(\lambda) = S(-\lambda) \tag{4}$$

Let us calculate the commutator $[Q, S(\lambda)]$. We can apply formula (51) of complement $B_{II}$, since $[Q, P] = i\hbar$ commutes with $Q$ and $P$:

$$[Q, S(\lambda)] = i\hbar \left(-\frac{i\lambda}{\hbar}\right) e^{-i\lambda P/\hbar} = \lambda S(\lambda) \tag{5}$$

This relation can also be written:

$$Q S(\lambda) = S(\lambda)[Q + \lambda] \tag{6}$$

Finally, note that:

$$S(\lambda) S(\mu) = S(\lambda + \mu) \tag{7}$$

## 2. Eigenvalues and eigenvectors of $Q$

### a. SPECTRUM OF $Q$

Assume that $Q$ has a non-zero eigenvector $| q \rangle$, with eigenvalue $q$:

$$Q | q \rangle = q | q \rangle \tag{8}$$

Apply equation (6) to the vector $| q \rangle$. This yields:

$$\begin{aligned} Q \; S(\lambda) | q \rangle &= S(\lambda)(Q + \lambda) | q \rangle \\ &= S(\lambda)(q + \lambda) | q \rangle = (q + \lambda) S(\lambda) | q \rangle \end{aligned} \tag{9}$$

This equation expresses the fact that $S(\lambda) | q \rangle$ is another non-zero eigenvector of $Q$, with an eigenvalue of $(q + \lambda)$ $\big(S(\lambda) | q \rangle$ is non-zero because $S(\lambda)$ is unitary$\big)$. Thus, starting with an eigenvector of $Q$, one can, by applying $S(\lambda)$, construct another eigenvector of $Q$, with any real eigenvalue ($\lambda$ can indeed take on any real value). The spectrum of $Q$ is therefore a continuous spectrum, composed of all possible values on the real axis*.

### b. DEGREE OF DEGENERACY

From now on, we shall assume, for simplicity, that the eigenvalue $q$ of $Q$ is non-degenerate (the results which we shall derive can be generalized to the case where $q$ is degenerate). Let us show that if $q$ is non-degenerate, all the other eigenvalues of $Q$ are also non-degenerate. Let us assume, for example, that the eigenvalue $q + \lambda$ is two-fold degenerate, and we shall show that we arrive at a contradiction. There would then exist two orthogonal eigenvectors, $| q + \lambda, \alpha \rangle$ and $| q + \lambda, \beta \rangle$, corresponding to the eigenvalue $q + \lambda$:

$$\langle q + \lambda, \beta | q + \lambda, \alpha \rangle = 0 \tag{10}$$

* This shows that in a space $\mathscr{E}$ of finite dimension $N$, there are no observables $Q$ and $P$, whose commutator is equal to $i\hbar$. The number of eigenvalues of $Q$ could not be simultaneously less than or equal to $N$ and infinite.

This result can be derived directly, moreover, by taking the trace of relation (2): $\text{Tr}\, QP - \text{Tr}\, PQ = \text{Tr}\, i\hbar$. When $N$ is finite, the traces on the left-hand side of this equation exist: they are finite and equal numbers [*cf.* complement $B_{II}$, formula (7-a)]. The equation becomes $0 = \text{Tr}\, i\hbar = Ni\hbar$, which is impossible.

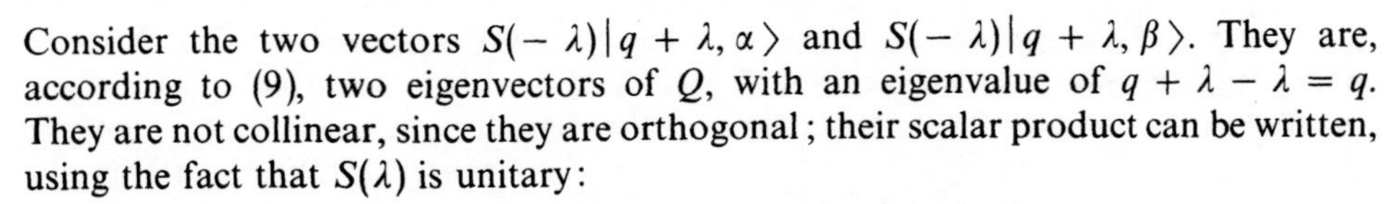

Consider the two vectors $S(-\lambda)|q+\lambda,\alpha\rangle$ and $S(-\lambda)|q+\lambda,\beta\rangle$. They are, according to (9), two eigenvectors of $Q$, with an eigenvalue of $q+\lambda-\lambda=q$. They are not collinear, since they are orthogonal; their scalar product can be written, using the fact that $S(\lambda)$ is unitary:

$$\langle q+\lambda,\beta\,|\,S^{\dagger}(-\lambda)S(-\lambda)\,|\,q+\lambda,\alpha\rangle = \langle q+\lambda,\beta\,|\,q+\lambda,\alpha\rangle = 0 \tag{11}$$

We reach the conclusion that $q$ is at least two-fold degenerate, which is contrary to the initial hypothesis. Consequently, all the eigenvalues of $Q$ must have the same degree of degeneracy.

### c. EIGENVECTORS

We shall fix the relative phases of the different eigenvectors of $Q$ with respect to the eigenvector $|0\rangle$, of eigenvalue 0, by setting:

$$|q\rangle = S(q)|0\rangle \tag{12}$$

Applying $S(\lambda)$ to both sides of (12) and using (7), we obtain:

$$S(\lambda)|q\rangle = S(\lambda)S(q)|0\rangle = S(\lambda+q)|0\rangle = |q+\lambda\rangle \tag{13}$$

The adjoint expression of (13) is written:

$$\langle q|S^{\dagger}(\lambda) = \langle q+\lambda| \tag{14}$$

or, using (4) and replacing $\lambda$ by$\gamma$ $-\lambda$:

$$\langle q|S(\lambda) = \langle q-\lambda| \tag{15}$$

## 3. The $\{\,|q\rangle\,\}$ representation

Since $Q$ is an observable, the set of its eigenvectors $\{\,|q\rangle\,\}$ constitutes a basis of $\mathscr{E}$. It is possible to characterize each ket by its "wave function in the $\{\,|q\rangle\,\}$ representation":

$$\psi(q) = \langle q|\psi\rangle \tag{16}$$

### a. THE ACTION OF $Q$ IN THE $\{\,|q\rangle\,\}$ REPRESENTATION

Let us calculate in the $\{\,|q\rangle\,\}$ representation the wave function associated with the ket $Q|\psi\rangle$. It is written:

$$\langle q|Q|\psi\rangle = q\langle q|\psi\rangle = q\psi(q) \tag{17}$$

[using (8) and the fact that $Q$ is Hermitian]. The action of $Q$ in the $\{\,|q\rangle\,\}$ representation is therefore simply a multiplication by $q$.

### b. THE ACTION OF $S(\lambda)$ IN THE $\{|q\rangle\}$ REPRESENTATION ; THE TRANSLATION OPERATOR

The wave function in the $\{|q\rangle\}$ representation associated with the ket $S(\lambda)|\psi\rangle$ is written [formula (15)]:

$$\langle q|S(\lambda)|\psi\rangle = \langle q - \lambda|\psi\rangle = \psi(q - \lambda) \tag{18}$$

The action of the operator $S(\lambda)$ in the $\{|q\rangle\}$ representation is therefore a translation of the wave function over a distance $\lambda$ parallel to the $q$-axis*. For this reason, $S(\lambda)$ is called the *translation operator.*

### c. THE ACTION OF P IN THE $\{|q\rangle\}$ REPRESENTATION

When $\varepsilon$ is an infinitely small quantity, we have:

$$S(-\varepsilon) = e^{i\varepsilon P/\hbar} = \mathbb{1} + i\frac{\varepsilon}{\hbar}P + O(\varepsilon^2) \tag{19}$$

Consequently :

$$\langle q|S(-\varepsilon)|\psi\rangle = \psi(q) + i\frac{\varepsilon}{\hbar}\langle q|P|\psi\rangle + O(\varepsilon^2) \tag{20}$$

On the other hand, equation (18) yields:

$$\langle q|S(-\varepsilon)|\psi\rangle = \psi(q + \varepsilon) \tag{21}$$

Comparison of (20) and (21) shows that:

$$\psi(q + \varepsilon) = \psi(q) + i\frac{\varepsilon}{\hbar}\langle q|P|\psi\rangle + O(\varepsilon^2) \tag{22}$$

It follows that:

$$\begin{aligned}\langle q|P|\psi\rangle &= \frac{\hbar}{i}\lim_{\varepsilon\to 0}\frac{\psi(q+\varepsilon) - \psi(q)}{\varepsilon} \\ &= \frac{\hbar}{i}\frac{\mathrm{d}}{\mathrm{d}q}\psi(q)\end{aligned} \tag{23}$$

The action of $P$ in the $\{|q\rangle\}$ representation is therefore that of $\frac{\hbar}{i}\frac{\mathrm{d}}{\mathrm{d}q}$. Equation (E-26) of chapter II is thus generalized.

## 4. The $\{|p\rangle\}$ representation. The symmetrical nature of the $P$ and $Q$ observables

Relation (23) enables us to obtain easily the wave function $v_p(q)$ associated, in the $\{|q\rangle\}$ representation, with the eigenvector $|p\rangle$ of $P$ with an eigenvalue of $p$ :

$$v_p(q) = \langle q|p\rangle = (2\pi\hbar)^{-1/2}\,e^{\frac{i}{\hbar}pq} \tag{24}$$

* The function $f(x - a)$ is the function which, at the point $x = x_0 + a$, takes on the value $f(x_0)$. It is therefore the function obtained from $f(x)$ by a translation of $+a$.

We can therefore write :

$$| p \rangle = (2\pi\hbar)^{-1/2} \int_{-\infty}^{+\infty} \mathrm{d}q \, \mathrm{e}^{\frac{i}{\hbar}pq} \, | q \rangle \tag{25}$$

A ket $| \psi \rangle$ can be defined by its "wave function in the $\{ \, | p \rangle \, \}$ representation":

$$\overline{\psi}(p) = \langle p \, | \, \psi \rangle \tag{26}$$

Using the adjoint relation of (25), we obtain:

$$\overline{\psi}(p) = (2\pi\hbar)^{-1/2} \int_{-\infty}^{+\infty} \mathrm{d}q \, \mathrm{e}^{-\frac{i}{\hbar}pq} \, \psi(q) \tag{27}$$

$\overline{\psi}(p)$ is therefore the Fourier transform of $\psi(q)$.

The action of the $P$ operator in the $\{ \, | p \rangle \, \}$ representation corresponds to a multiplication by $p$; that of the $Q$ operator corresponds, as can easily be shown using (27), to the operation $i\hbar \dfrac{\mathrm{d}}{\mathrm{d}p}$.

Thus we obtain symmetrical results in the $\{ \, | q \rangle \, \}$ and $\{ \, | p \rangle \, \}$ representations. This is not surprising : in our hypotheses, it is possible to exchange the $P$ and $Q$ operators, simply changing the sign of the commutator in relation (2). Instead of introducing the operator $S(\lambda)$, we could therefore have considered $T(\lambda')$ defined by:

$$T(\lambda') = \mathrm{e}^{i\lambda' Q/\hbar} \tag{28}$$

and we could have developed the same arguments, replacing $P$ by $Q$ and $i$ by $-i$ everywhere.

**References:**

Messiah (1.17), Vol. I, §VIII-6; Dirac (1.13), §25; Merzbacher (1.16), chap. 14, §7.

# Complement $F_{II}$

## THE PARITY OPERATOR



### 1. The parity operator

#### a. DEFINITION

Consider a physical system whose state space is $\mathscr{E}_{\mathbf{r}}$. The parity operator $\Pi$ is defined by its action on the basis vectors $|\mathbf{r}\rangle$ of $\mathscr{E}_{\mathbf{r}}$★:

$$\Pi\,|\,\mathbf{r}\,\rangle = |\,-\,\mathbf{r}\,\rangle \tag{1}$$

The matrix elements of $\Pi$ are therefore, in the $\{\,|\,\mathbf{r}\,\rangle\,\}$ representation:

$$\langle\,\mathbf{r}\,|\,\Pi\,|\,\mathbf{r}'\,\rangle = \langle\,\mathbf{r}\,|\,-\,\mathbf{r}'\,\rangle = \delta(\mathbf{r} + \mathbf{r}') \tag{2}$$

Consider an arbitrary vector $|\,\psi\,\rangle$ of $\mathscr{E}_{\mathbf{r}}$:

$$|\,\psi\,\rangle = \int d^3r\;\psi(\mathbf{r})\,|\,\mathbf{r}\,\rangle \tag{3}$$

If the variable change $\mathbf{r}' = -\,\mathbf{r}$ is performed, $|\,\psi\,\rangle$ can be written:

$$|\,\psi\,\rangle = \int d^3r'\;\psi(-\,\mathbf{r}')\,|\,-\,\mathbf{r}'\,\rangle \tag{4}$$

Now calculate $\Pi\,|\,\psi\,\rangle$; we obtain:

$$\Pi\,|\,\psi\,\rangle = \int d^3r'\;\psi(-\,\mathbf{r}')\,|\,\mathbf{r}'\,\rangle \tag{5}$$

★ Care must be taken not to confuse $|\,-\,\mathbf{r}_0\,\rangle$ and $-\,|\,\mathbf{r}_0\,\rangle$. The former is an eigenvector of $\mathbf{R}$, with eigenvalue $-\,\mathbf{r}_0$ and wavefunction $\xi_{-\mathbf{r}_0}(\mathbf{r}) = \delta(\mathbf{r} + \mathbf{r}_0)$. The latter is an eigenvector of $\mathbf{R}$ with eigenvalue $\mathbf{r}_0$ and wavefunction $-\,\xi_{\mathbf{r}_0}(\mathbf{r}) = -\,\delta(\mathbf{r} - \mathbf{r}_0)$.

Comparison of (3) and (5) shows that the action of $\Pi$ in the $\{ |\mathbf{r}\rangle \}$ representation is to change $\mathbf{r}$ to $-\mathbf{r}$:

$$\langle \mathbf{r} | \Pi | \psi \rangle = \psi(-\mathbf{r}) \tag{6}$$

Now let us consider a physical system $\mathscr{S}$ whose state vector is $|\psi\rangle$; $\Pi|\psi\rangle$ describes the physical system obtained from $\mathscr{S}$ by reflection through the origin of the axes.

#### b. SIMPLE PROPERTIES OF $\Pi$

The operator $\Pi^2$ is the identity operator. From (1) we have:

$$\Pi^2 |\mathbf{r}\rangle = \Pi(\Pi|\mathbf{r}\rangle) = \Pi|-\mathbf{r}\rangle = |\mathbf{r}\rangle \tag{7}$$

that is, since the kets $|\mathbf{r}\rangle$ form a basis of $\mathscr{E}_{\mathbf{r}}$:

$$\Pi^2 = \mathbb{1} \tag{8-a}$$

or:

$$\Pi = \Pi^{-1} \tag{8-b}$$

It is easy to show by recurrence that the operator $\Pi^n$ is

- equal to $\mathbb{1}$ when $n$ is even
- equal to $\Pi$ when $n$ is odd

We can rewrite (6) in the form:

$$\langle \mathbf{r} | \Pi | \psi \rangle = \langle -\mathbf{r} | \psi \rangle \tag{9}$$

Since this equation is valid for all $|\psi\rangle$, it can be deduced that:

$$\langle \mathbf{r} | \Pi = \langle -\mathbf{r} | \tag{10}$$

Moreover, the Hermitian conjugate expression of (1) is written:

$$\langle \mathbf{r} | \Pi^\dagger = \langle -\mathbf{r} | \tag{11}$$

Since the kets $|\mathbf{r}\rangle$ form a basis, it follows from (10) and (11) that $\Pi$ is Hermitian:

$$\Pi^\dagger = \Pi \tag{12}$$

Combining this equation with (8-b), we obtain:

$$\Pi^{-1} = \Pi^\dagger \tag{13}$$

$\Pi$ is therefore unitary as well.

#### c. EIGENSUBSPACES OF $\Pi$

Let $|\varphi_\pi\rangle$ be an eigenvector of $\Pi$, with an eigenvalue of $p_\pi$. Applying (8-a), we obtain:

$$|\varphi_\pi\rangle = \Pi^2 |\varphi_\pi\rangle = p^2 |\varphi_\pi\rangle \tag{14}$$

We therefore have $p_\pi^2 = 1$ : the eigenvalues of $\Pi$ are limited to 1 and $-1$. Since the space $\mathscr{E}_\mathbf{r}$ is infinite-dimensional, we immediately see that these eigenvalues are degenerate. An eigenvector of $\Pi$ with the eigenvalue $+1$ will be said to be even; an eigenvector with the eigenvalue $-1$, odd.

Consider the two operators $P_+$ and $P_-$ defined by:

$$P_+ = \frac{1}{2}(\mathbb{1} + \Pi)$$

$$P_- = \frac{1}{2}(\mathbb{1} - \Pi) \tag{15}$$

These operators are Hermitian; using (8-a), it is easy to show that:

$$\begin{aligned} P_+^2 &= P_+ \\ P_-^2 &= P_- \end{aligned} \tag{16}$$

$P_+$ and $P_-$ are thus the projectors onto two subspaces of $\mathscr{E}_\mathbf{r}$ which we shall call $\mathscr{E}_+$ and $\mathscr{E}_-$. Let us calculate the products $P_+P_-$ and $P_-P_+$ ; we obtain:

$$P_+P_- = \frac{1}{4}(\mathbb{1} + \Pi - \Pi - \Pi^2) = 0$$

$$P_-P_+ = \frac{1}{4}(\mathbb{1} - \Pi + \Pi - \Pi^2) = 0 \tag{17}$$

The two subspaces $\mathscr{E}_+$ and $\mathscr{E}_-$ are therefore orthogonal. Let us show that they are also supplementary. We see immediately from definition (15) that:

$$P_+ + P_- = \mathbb{1} \tag{18}$$

For all $|\psi\rangle \in \mathscr{E}$, we have, therefore:

$$|\psi\rangle = (P_+ + P_-)|\psi\rangle = |\psi_+\rangle + |\psi_-\rangle \tag{19}$$

with:

$$\begin{aligned} |\psi_+\rangle &= P_+|\psi\rangle \\ |\psi_-\rangle &= P_-|\psi\rangle \end{aligned} \tag{20}$$

Let us calculate the products $\Pi P_+$ and $\Pi P_-$ ; we obtain:

$$\Pi P_+ = \frac{1}{2}\Pi(\mathbb{1} + \Pi) = \frac{1}{2}(\Pi + \mathbb{1}) = P_+$$

$$\Pi P_- = \frac{1}{2}\Pi(\mathbb{1} - \Pi) = \frac{1}{2}(\Pi - \mathbb{1}) = -P_- \tag{21}$$

These equations enable us to show that the vectors $|\psi_+\rangle$ and $|\psi_-\rangle$ introduced in (20) are even and odd, respectively:

$$\begin{aligned} \Pi|\psi_+\rangle &= \Pi P_+|\psi\rangle = P_+|\psi\rangle = |\psi_+\rangle \\ \Pi|\psi_-\rangle &= \Pi P_-|\psi\rangle = -P_-|\psi\rangle = -|\psi_-\rangle \end{aligned} \tag{22}$$

The spaces $\mathscr{E}_+$ and $\mathscr{E}_-$ are therefore the eigensubspaces of $\Pi$, with the eigenvalues $+1$ and $-1$. In the $\{|\mathbf{r}\rangle\}$ representation, equations (22) can be written:

$$\begin{aligned} \langle \mathbf{r} | \psi_+ \rangle &= \psi_+(\mathbf{r}) = \langle \mathbf{r} | \Pi | \psi_+ \rangle = \psi_+(-\mathbf{r}) \\ \langle \mathbf{r} | \psi_- \rangle &= \psi_-(\mathbf{r}) = -\langle \mathbf{r} | \Pi | \psi_- \rangle = -\psi_-(-\mathbf{r}) \end{aligned} \tag{23}$$

The wave functions $\psi_+(\mathbf{r})$ and $\psi_-(\mathbf{r})$ are even and odd, respectively.

Relation (19) expresses the fact, therefore, that any ket $|\psi\rangle$ of $\mathscr{E}_\mathbf{r}$ can be decomposed into a sum of two eigenvectors of $\Pi$, $|\psi_+\rangle$ and $|\psi_-\rangle$, belonging respectively to the even subspace $\mathscr{E}_+$ and the odd subspace $\mathscr{E}_-$. Therefore, $\Pi$ is an observable.

## 2. Even and odd operators

### a. DEFINITIONS

In §2 of complement $C_{II}$, we defined the concept of a unitary transformation of operators. In the case of $\Pi$ [which is indeed unitary; see (13)], the transformed operator of an arbitrary operator $B$ is written:

$$\tilde{B} = \Pi B \Pi \tag{24}$$

and satisfies the relation [*cf.* equation (27) of complement $C_{II}$]:

$$\langle \mathbf{r} | \tilde{B} | \mathbf{r}' \rangle = \langle -\mathbf{r} | B | -\mathbf{r}' \rangle \tag{25}$$

The operator $\tilde{B}$ is said to be the parity transform of $B$.

In particular, if: $\tilde{B} = +B$ the operator $B$ is said to be even

if: $\breve{B} = -B$ the operator $B$ is said to be odd.

An even operator $B_+$ therefore satisfies:

$$B_+ = \Pi B_+ \Pi \tag{26}$$

or, multiplying this equation on the left by $\Pi$ and using (8-a):

$$\Pi B_+ = B_+ \Pi \tag{27}$$

$$[\Pi, B_+] = 0 \tag{28}$$

An even operator is therefore an operator which commutes with $\Pi$. It can be seen, similarly, that an odd operator $B_-$ is an operator which anticommutes with $\Pi$:

$$\Pi B_- + B_- \Pi = 0 \tag{29}$$

### b. SELECTION RULES

Let $B_+$ be an even operator. Let us calculate the matrix element $\langle \varphi | B_+ | \psi \rangle$; by hypothesis, we have:

$$\langle \varphi | B_+ | \psi \rangle = \langle \varphi | \Pi B_+ \Pi | \psi \rangle = \langle \varphi' | B_+ | \psi' \rangle \tag{30}$$

with:

$$\begin{aligned} |\varphi'\rangle &= \Pi |\varphi\rangle \\ |\psi'\rangle &= \Pi |\psi\rangle \end{aligned} \tag{31}$$

If one of the two kets, $|\varphi\rangle$ and $|\psi\rangle$, is even and the other odd ($|\varphi'\rangle = \pm|\varphi\rangle$, $|\psi'\rangle = \mp|\psi\rangle$), relation (30) yields:

$$\langle\varphi|B_+|\psi\rangle = -\langle\varphi|B_+|\psi\rangle = 0 \tag{32}$$

Hence the rule : the matrix elements of an even operator are zero between vectors of opposite parity.

If, now, $B_-$ is odd, relation (30) becomes:

$$\langle\varphi|B_-|\psi\rangle = -\langle\varphi'|B_-|\psi'\rangle \tag{33}$$

which is zero when $|\varphi\rangle$ and $|\psi\rangle$ are both either even or odd. Hence the rule : the matrix elements of an odd operator are zero between vectors of the same parity. In particular, the diagonal matrix element $\langle\psi|B_-|\psi\rangle$ (the mean value of $B_-$ in the state $|\psi\rangle$; *cf.* chapter III, §C-4) is zero if $|\psi\rangle$ has a definite parity.

### c. EXAMPLES

#### α. *The X, Y, Z operators*

In this case, we have:

$$\begin{aligned}\Pi X|\mathbf{r}\rangle &= \Pi X|x, y, z\rangle = x\Pi|x, y, z\rangle \\ &= x|-x, -y, -z\rangle = x|-\mathbf{r}\rangle\end{aligned} \tag{34}$$

and:

$$\begin{aligned}X\Pi|\mathbf{r}\rangle &= X|-\mathbf{r}\rangle = X|-x, -y, -z\rangle \\ &= -x|-x, -y, -z\rangle = -x|-\mathbf{r}\rangle\end{aligned} \tag{35}$$

Adding these two equations together, we obtain:

$$(\Pi X + X\Pi)|\mathbf{r}\rangle = 0 \tag{36}$$

or, since the vectors $|\mathbf{r}\rangle$ form a basis:

$$\Pi X + X\Pi = 0 \tag{37}$$

$X$ is therefore odd.

The proofs are the same for $Y$ and $Z$; $\mathbf{R}$ is therefore an odd operator.

#### β. *The $P_x$, $P_y$, $P_z$ operators*

Let us calculate the ket $\Pi|\mathbf{p}\rangle$; we obtain:

$$\begin{aligned}\Pi|\mathbf{p}\rangle &= (2\pi\hbar)^{-3/2}\int d^3r\, e^{i\mathbf{p}.\mathbf{r}/\hbar}\,\Pi|\mathbf{r}\rangle \\ &= (2\pi\hbar)^{-3/2}\int d^3r\, e^{i\mathbf{p}.\mathbf{r}/\hbar}\,|-\mathbf{r}\rangle \\ &= (2\pi\hbar)^{-3/2}\int d^3r'\, e^{-i\mathbf{p}.\mathbf{r}'/\hbar}\,|\mathbf{r}'\rangle \\ &= |-\mathbf{p}\rangle\end{aligned} \tag{38}$$

We then have, using an argument analogous to the one developed in $\alpha$:

$$\begin{aligned} \Pi P_x \mid \mathbf{p} \rangle &= p_x \mid -\mathbf{p} \rangle \\ P_x \Pi \mid \mathbf{p} \rangle &= -p_x \mid -\mathbf{p} \rangle \end{aligned} \tag{39}$$

and

$$\Pi P_x + P_x \Pi = 0 \tag{40}$$

The **P** operator is odd.

$\gamma$. *The parity operator*

$\Pi$ obviously commutes with itself; it is an even operator.

#### d. FUNCTIONS OF OPERATORS

Let $B_+$ be an even operator. Using relation (8-a), we obtain:

$$\Pi B_+^n \Pi = \underbrace{(\Pi B_+ \Pi)(\Pi B_+ \Pi) \ldots (\Pi B_+ \Pi)}_{n \text{ factors}} = B_+^n \tag{41}$$

An even operator raised to the $n$th power is even. Hence, any operator $F(B_+)$ is even.

Let $B_-$ be an odd operator; let us calculate the operator $\Pi B_-^n \Pi$:

$$\Pi B_-^n \Pi = \underbrace{(\Pi B_- \Pi)(\Pi B_- \Pi) \ldots (\Pi B_- \Pi)}_{n \text{ factors}} = (-1)^n (B_-)^n \tag{42}$$

An odd operator raised to the $n$th power is even if $n$ is even, odd if $n$ is odd. Consider an operator $F(B_-)$; this operator is even if the corresponding function $F(z)$ is even, odd if it is odd. In general, $F(B_-)$ has no definite parity.

### 3. Eigenstates of an even observable $B_+$

Let us consider an arbitrary even observable $B_+$ and an eigenvector $| \varphi_b \rangle$ of $B_+$ with an eigenvalue of $b$. Since $B_+$ is even, it commutes with $\Pi$. Applying the theorems of § D-3-a of chapter II, we obtain the following results:

$\alpha$. If $b$ is a non-degenerate eigenvalue, $| \varphi_b \rangle$ is necessarily an eigenvector of $\Pi$; it is therefore either an even or an odd vector. The mean value $\langle \varphi_b | B_- | \varphi_b \rangle$ of any odd observable $B_-$, such as **R**, **P**, etc..., is then zero.

$\beta$. If $b$ is a degenerate eigenvalue corresponding to the eigensubspace $\mathscr{E}_b$, the vectors of $\mathscr{E}_b$ do not all necessarily have a definite parity. $\Pi | \varphi_b \rangle$ may be a vector which is non-collinear with $| \varphi_b \rangle$; it is nevertheless a vector which has the same eigenvalue $b$. Moreover, it is possible to find a basis of eigenvectors common to $\Pi$ and $B_+$ in every subspace $\mathscr{E}_b$.

## 4. Application to an important special case

We shall often need to find the eigenstates of a Hamiltonian operator $H$, acting in $\mathscr{E}_{\mathbf{r}}$, of the form:

$$H = \frac{\mathbf{P}^2}{2m} + V(\mathbf{R}) \tag{43}$$

Since the $\mathbf{P}$ operator is odd, the $\mathbf{P}^2$ operator is even. When, in addition, the function $V(\mathbf{r})$ is even $\big(V(\mathbf{r}) = V(-\mathbf{r})\big)$, the operator $H$ is even. According to what we have just seen, it is then possible to look for the eigenstates of $H$ among the even or odd states. This often simplifies the calculation considerably.

We have already encountered a certain number of cases where the Hamiltonian $H$ is even : the square well, the infinite well (*cf.* complement $H_I$). We shall study others : the harmonic oscillator, the hydrogen atom, etc... It is easy to verify in all these special cases the properties which we have derived.

COMMENT :

If $H$ is even, and if one of its eigenstates $|\varphi_h\rangle$ which has no definite parity ($\Pi|\varphi_h\rangle$ non-collinear to $|\varphi_h\rangle$) has been found, it can be asserted that the corresponding eigenvalue is degenerate : since $\Pi$ commutes with $H$, $\Pi|\varphi_h\rangle$ is an eigenvector of $H$ with the same eigenvalue as $|\varphi_h\rangle$.

**References and suggestions for further reading :**

Schiff (1.18), §29; Roman (2.3), §5-3 d; Feynman I (6.3), chap. 52; Sakurai (2.7), chap. 3; articles by Morrison (2.28), Feinberg and Goldhaber (2.29), Wigner (2.30).

## Complement $G_{II}$

# AN APPLICATION OF THE PROPERTIES OF THE TENSOR PRODUCT: THE TWO-DIMENSIONAL INFINITE WELL



In complement $H_I$ (§ 2-c) we have already studied, in a one-dimensional problem, the stationary states of a particle placed in an infinite potential well. By using the concept of a tensor product (*cf.* chap. II, § F), we shall be able to generalize this discussion to the case of a two-dimensional infinite well (the introduction of a third dimension would not involve any additional theoretical difficulty).

## 1. Definition ; eigenstates

We shall consider a particle of mass $m$, restricted to a plane $xOy$, inside a "square box" of edge $a$ : its potential energy $V(x, y)$ becomes infinite when one of its coordinates $x$ or $y$ leaves the interval $[0, a]$:

$$V(x, y) = V_\infty(x) + V_\infty(y) \tag{1}$$

with:

$$\begin{aligned} V_\infty(u) &= 0 && \text{if} \quad 0 \leqslant u \leqslant a \\ &= +\infty && \text{if} \quad u < 0 \quad \text{or} \quad u > a \end{aligned} \tag{2}$$

The Hamiltonian of the quantum particle is then (chap. III, § B-5);

$$H = \frac{1}{2m}(P_x^2 + P_y^2) + V_\infty(X) + V_\infty(Y) \tag{3}$$

which can be written:

$$H = H_x + H_y \tag{4}$$

with:

$$\begin{aligned} H_x &= \frac{1}{2m} P_x^2 + V_\infty(X) \\ H_y &= \frac{1}{2m} P_y^2 + V_\infty(Y) \end{aligned} \tag{5}$$

We thus find ourselves in the important special case mentioned in chapter II (§ F-4-a-$\beta$), and we can consider the eigenstates of $H$ in the form:

$$| \Phi \rangle = | \varphi \rangle_x | \varphi \rangle_y \tag{6}$$

with:

$$H_x \mid \varphi \rangle_x = E_x \mid \varphi \rangle_x \; ; \; \mid \varphi \rangle_x \in \mathscr{E}_x$$
$$H_y \mid \varphi \rangle_y = E_y \mid \varphi \rangle_y \; ; \; \mid \varphi \rangle_y \in \mathscr{E}_y \tag{7}$$

We then have:

$$H \mid \Phi \rangle = E \mid \Phi \rangle$$

with

$$E = E_x + E_y \tag{8}$$

We have therefore reduced a two-dimensional problem to a one-dimensional problem, which, moreover, has already been solved (*cf.* complement $H_I$). Applying the results of this complement, and formulas (7) and (8), we therefore see that:

– the eigenvalues of $H$ are of the form:

$$E_{n,p} = \frac{1}{2ma^2}(n^2 + p^2)\,\pi^2\hbar^2 \tag{9}$$

where $n$ and $p$ are positive integers.

– to these energies correspond eigenstates $\mid \Phi_{n,p} \rangle$ which can be written in the form of tensor products:

$$\mid \Phi_{n,p} \rangle = \mid \varphi_n \rangle_x \mid \varphi_p \rangle_y \tag{10}$$

whose normalized wave function is:

$$\begin{aligned}\Phi_{n,p}(x, y) &= \varphi_n(x)\,\varphi_p(y) \\ &= \frac{2}{a} \sin \frac{n\pi x}{a} \sin \frac{p\pi y}{a}\end{aligned} \tag{11}$$

It is easy to verify that these wave functions vanish at the edges of the "square box" ($x$ or $y = 0$ or $a$), where the potential energy becomes infinite.

## 2. Study of the energy levels

### a. GROUND STATE

$n$ and $p$ are strictly positive integers*. The ground state is therefore obtained when $n = 1$, $p = 1$. Its energy is:

$$E_{1,1} = \frac{\pi^2\hbar^2}{ma^2} \tag{12}$$

This value is attained only for $n = p = 1$. The ground state is therefore not degenerate.

* The values $n = 0$ or $p = 0$, which give null wave functions (therefore impossible to normalize) are excluded.

### b. FIRST EXCITED STATES

The first excited state is obtained either for $n = 1$ and $p = 2$ or for $n = 2$ and $p = 1$. Its energy is:

$$E_{1,2} = E_{2,1} = \frac{5}{2}\frac{\pi^2\hbar^2}{ma^2} \tag{13}$$

This state is two-fold degenerate, since $| \Phi_{1,2} \rangle$ and $| \Phi_{2,1} \rangle$ are independent.

The second excited state corresponds to $n = p = 2$; it is not degenerate, and its energy is:

$$E_{2,2} = 4\frac{\pi^2\hbar^2}{ma^2} \tag{14}$$

The third excited state corresponds to $n = 1$, $p = 3$ and $n = 3$, $p = 1$, etc.

### c. SYSTEMATIC AND ACCIDENTAL DEGENERACIES

The general observation can be made that all levels for which $n \neq p$ are degenerate, since:

$$E_{n,p} = E_{p,n} \tag{15}$$

This degeneracy is related to a symmetry of the problem. The square well under consideration is symmetrical with respect to the first bisectrix of the $xOy$ plane. This is expressed by the fact that the Hamiltonian $H$ is invariant under the exchange:

$$\begin{aligned} X &\longleftrightarrow Y \\ P_x &\longleftrightarrow P_y \end{aligned} \tag{16}$$

(In the state space, an operator could be defined to correspond to a reflection about the first bisectrix. It could then be shown that, in the present case, this operator commutes with $H$). If an eigenstate of $H$ is known whose wave function is $\Phi(x, y)$, the state which corresponds to $\Phi'(x, y) = \Phi(y, x)$ is also an eigenstate of $H$ with the same eigenvalue. Consequently, if the function $\Phi(x, y)$ is not symmetric with respect to $x$ and $y$, the eigenvalue associated with it is necessarily degenerate. This is the origin of the degeneracy (15): for $n \neq p$, $\Phi_{n,p}(x, y)$ is not symmetrical with respect to $x$ and $y$ [formula (11)]. This interpretation is corroborated by the fact that if the symmetry is destroyed by choosing a well whose widths along $Ox$ and along $Oy$ are different (being equal to $a$ and $b$ respectively), the corresponding degeneracy disappears, and formula (9) becomes:

$$E_{n,p} = \frac{\pi^2\hbar^2}{2m}\left(\frac{n^2}{a^2} + \frac{p^2}{b^2}\right), \tag{17}$$

which implies:

$$E_{p,n} \neq E_{n,p} \tag{18}$$

Such degeneracies, whose origin lies in a symmetry of the problem are called *systematic degeneracies.*

COMMENT:

The other symmetries of the two-dimensional square well do not create systematic degeneracies because the eigenstates of $H$ are all invariant with respect to them. For example, for arbitrary $n$ and $p$, $\Phi_{n,p}(x, y)$ is simply multiplied by a phase factor if $x$ is replaced by $(a - x)$ and $y$ by $(a - y)$ (symmetry with respect to the center of the well).

Degeneracies may also arise which are not directly related to the symmetry of the problem. They are called *accidental degeneracies*. For example, in the case which we have discussed, it so happens that $E_{5,5} = E_{7,1}$ and $E_{7,4} = E_{8,1}$.

## Complement $H_{II}$

## EXERCISES

### Dirac notation. Commutators. Eigenvectors and eigenvalues

**1.** $|\varphi_n\rangle$ are the eigenstates of a Hermitian operator $H$ ($H$ is, for example, the Hamiltonian of an arbitrary physical system). Assume that the states $|\varphi_n\rangle$ form a discrete orthonormal basis. The operator $U(m, n)$ is defined by:

$$U(m, n) = |\varphi_m\rangle\langle\varphi_n|$$

*a.* Calculate the adjoint $U^\dagger(m, n)$ of $U(m, n)$.

*b.* Calculate the commutator $[H, U(m, n)]$.

*c.* Prove the relation:

$$U(m, n)U^\dagger(p, q) = \delta_{nq}U(m, p)$$

*d.* Calculate Tr $\{ U(m, n) \}$, the trace of the operator $U(m, n)$.

*e.* Let $A$ be an operator, with matrix elements $A_{mn} = \langle\varphi_m|A|\varphi_n\rangle$. Prove the relation:

$$A = \sum_{m,n} A_{mn}\, U(m, n)$$

*f.* Show that $A_{pq} = \text{Tr}\{ AU^\dagger(p, q) \}$.

**2.** In a two-dimensional vector space, consider the operator whose matrix, in an orthonormal basis $\{ |1\rangle, |2\rangle \}$, is written:

$$\sigma_y = \begin{pmatrix} 0 & -i \\ i & 0 \end{pmatrix}$$

*a.* Is $\sigma_y$ Hermitian? Calculate its eigenvalues and eigenvectors (giving their normalized expansion in terms of the $\{ |1\rangle, |2\rangle \}$ basis).

*b.* Calculate the matrices which represent the projectors onto these eigenvectors. Then verify that they satisfy the orthogonality and closure relations.

*c.* Same questions for the matrices:

$$M = \begin{pmatrix} 2 & i\sqrt{2} \\ -i\sqrt{2} & 3 \end{pmatrix}$$

and, in a three-dimensional space

$$L_y = \frac{\hbar}{2i} \begin{pmatrix} 0 & \sqrt{2} & 0 \\ -\sqrt{2} & 0 & \sqrt{2} \\ 0 & -\sqrt{2} & 0 \end{pmatrix}$$

**3.** The state space of a certain physical system is three-dimensional. Let $\{ | u_1 \rangle, | u_2 \rangle, | u_3 \rangle \}$ be an orthonormal basis of this space. The kets $| \psi_0 \rangle$ and $| \psi_1 \rangle$ are defined by:

$$| \psi_0 \rangle = \frac{1}{\sqrt{2}} | u_1 \rangle + \frac{i}{2} | u_2 \rangle + \frac{1}{2} | u_3 \rangle$$

$$| \psi_1 \rangle = \frac{1}{\sqrt{3}} | u_1 \rangle + \frac{i}{\sqrt{3}} | u_3 \rangle$$

*a.* Are these kets normalized?

*b.* Calculate the matrices $\rho_0$ and $\rho_1$ representing, in the $\{ | u_1 \rangle, | u_2 \rangle, | u_3 \rangle \}$ basis, the projection operators onto the state $| \psi_0 \rangle$ and onto the state $| \psi_1 \rangle$. Verify that these matrices are Hermitian.

**4.** Let $K$ be the operator defined by $K = | \varphi \rangle \langle \psi |$, where $| \varphi \rangle$ and $| \psi \rangle$ are two vectors of the state space.

*a.* Under what condition is $K$ Hermitian?

*b.* Calculate $K^2$. Under what condition is $K$ a projector?

*c.* Show that $K$ can always be written in the form $K = \lambda P_1 P_2$ where $\lambda$ is a constant to be calculated and $P_1$ and $P_2$ are projectors.

**5.** Let $P_1$ be the orthogonal projector onto the subspace $\mathscr{E}_1$, $P_2$ the orthogonal projector onto the subspace $\mathscr{E}_2$. Show that, for the product $P_1 P_2$ to be an orthogonal projector as well, it is necessary and sufficient that $P_1$ and $P_2$ commute. In this case, what is the subspace onto which $P_1 P_2$ projects?

**6.** The $\sigma_x$ matrix is defined by:

$$\sigma_x = \begin{pmatrix} 0 & 1 \\ 1 & 0 \end{pmatrix}$$

Prove the relation:

$$e^{i\alpha\sigma_x} = I \cos \alpha + i\sigma_x \sin \alpha$$

where $I$ is the $2 \times 2$ unit matrix.

**7.** Establish, for the $\sigma_y$ matrix given in exercise 2, a relation analogous to the one proved for $\sigma_x$ in the preceding exercise. Generalize for all matrices of the form:

$$\sigma_u = \lambda \sigma_x + \mu \sigma_y$$

with:

$$\lambda^2 + \mu^2 = 1$$

Calculate the matrices representing $e^{2i\sigma_x}$, $(e^{i\sigma_x})^2$ and $e^{i(\sigma_x + \sigma_y)}$. Is $e^{2i\sigma_x}$ equal to $(e^{i\sigma_x})^2$? $e^{i(\sigma_x + \sigma_y)}$ to $e^{i\sigma_x} e^{i\sigma_y}$?

**8.** Consider the Hamiltonian $H$ of a particle in a one-dimensional problem defined by:

$$H = \frac{1}{2m} P^2 + V(X)$$

where $X$ and $P$ are the operators defined in § E of chapter II and which satisfy the relation: $[X, P] = i\hbar$. The eigenvectors of $H$ are denoted by $| \varphi_n \rangle$: $H | \varphi_n \rangle = E_n | \varphi_n \rangle$, where $n$ is a discrete index.

*a.* Show that:

$$\langle \varphi_n | P | \varphi_{n'} \rangle = \alpha \langle \varphi_n | X | \varphi_{n'} \rangle$$

where $\alpha$ is a coefficient which depends on the difference between $E_n$ and $E_{n'}$. Calculate $\alpha$ (hint : consider the commutator $[X, H]$).

*b.* From this, deduce, using the closure relation, the equation:

$$\sum_{n'} (E_n - E_{n'})^2 |\langle \varphi_n | X | \varphi_{n'} \rangle|^2 = \frac{\hbar^2}{m^2} \langle \varphi_n | P^2 | \varphi_n \rangle$$

**9.** Let $H$ be the Hamiltonian operator of a physical system. Denote by $| \varphi_n \rangle$ the eigenvectors of $H$, with eigenvalues $E_n$:

$$H | \varphi_n \rangle = E_n | \varphi_n \rangle$$

*a.* For an arbitrary operator $A$, prove the relation:

$$\langle \varphi_n | [A, H] | \varphi_n \rangle = 0.$$

*b.* Consider a one-dimensional problem, where the physical system is a particle of mass $m$ and of potential energy $V(X)$. In this case, $H$ is written:

$$H = \frac{1}{2m} P^2 + V(X)$$

$\alpha$. In terms of $P$, $X$ and $V(X)$, find the commutators: $[H, P]$, $[H, X]$ and $[H, XP]$.

$\beta$. Show that the matrix element $\langle \varphi_n | P | \varphi_n \rangle$ (which we shall interpret in chapter III as the mean value of the momentum in the state $| \varphi_n \rangle$) is zero.

$\gamma$. Establish a relation between $E_k = \langle \varphi_n | \frac{P^2}{2m} | \varphi_n \rangle$ (the mean value of the kinetic energy in the state $| \varphi_n \rangle$) and $\langle \varphi_n | X \frac{dV}{dX} | \varphi_n \rangle$. Since the mean value of the potential energy in the state $| \varphi_n \rangle$ is $\langle \varphi_n | V(x) | \varphi_n \rangle$, how is it related to the mean value of the kinetic energy when:

$$V(X) = V_0 X^\lambda$$

$(\lambda = 2, 4, 6 \ldots; V_0 > 0)$?

**10** Using the relation $\langle x \mid p \rangle = (2\pi\hbar)^{-1/2} e^{ipx/\hbar}$, find the expressions $\langle x \mid XP \mid \psi \rangle$ and $\langle x \mid PX \mid \psi \rangle$ in terms of $\psi(x)$. Can these results be found directly by using the fact that in the $\{ \mid x \rangle \}$ representation, $P$ acts like $\frac{\hbar}{i}\frac{d}{dx}$?

### Sets of commuting observables and C.S.C.O.'S

**11.** Consider a physical system whose three-dimensional state space is spanned by the orthonormal basis formed by the three kets $\mid u_1 \rangle, \mid u_2 \rangle, \mid u_3 \rangle$. In the basis of these three vectors, taken in this order, the two operators $H$ and $B$ are defined by:

$$H = \hbar\omega_0 \begin{pmatrix} 1 & 0 & 0 \\ 0 & -1 & 0 \\ 0 & 0 & -1 \end{pmatrix} \qquad B = b \begin{pmatrix} 1 & 0 & 0 \\ 0 & 0 & 1 \\ 0 & 1 & 0 \end{pmatrix}$$

where $\omega_0$ and $b$ are real constants.

*a.* Are $H$ and $B$ Hermitian?

*b.* Show that $H$ and $B$ commute. Give a basis of eigenvectors common to $H$ and $B$.

*c.* Of the sets of operators: $\{ H \}, \{ B \}, \{ H, B \}, \{ H^2, B \}$, which form a C.S.C.O.?

**12.** In the same state space as that of the preceding exercise, consider two operators $L_z$ and $S$ defined by:

$$L_z \mid u_1 \rangle = \mid u_1 \rangle \qquad L_z \mid u_2 \rangle = 0 \qquad L_z \mid u_3 \rangle = - \mid u_3 \rangle$$
$$S \mid u_1 \rangle = \mid u_3 \rangle \qquad S \mid u_2 \rangle = \mid u_2 \rangle \qquad S \mid u_3 \rangle = \mid u_1 \rangle$$

*a.* Write the matrices which represent, in the $\{ \mid u_1 \rangle, \mid u_2 \rangle, \mid u_3 \rangle \}$ basis, the operators $L_z, L_z^2, S, S^2$. Are these operators observables?

*b.* Give the form of the most general matrix which represents an operator which commutes with $L_z$. Same question for $L_z^2$, then for $S^2$.

*c.* Do $L_z^2$ and $S$ form a C.S.C.O.? Give a basis of common eigenvectors.

**Solution of exercise 11**

*a.* $H$ and $B$ are Hermitian because the matrices which correspond to them are symmetric and real.

*b.* $| u_1 \rangle$ is an eigenvector common to $H$ and $B$. We therefore have, obviously, $HB | u_1 \rangle = BH | u_1 \rangle$. We see, then, that for $H$ and $B$ to commute, it is sufficient that the restrictions of these operators to the subspace $\mathscr{E}_2$, spanned by $| u_2 \rangle$ and $| u_3 \rangle$, commute. Now, in this subspace, the matrix representing $H$ is equal to $-\hbar\omega_0 I$ (where $I$ is the $2 \times 2$ unit matrix), which commutes with all $2 \times 2$ matrices. $H$ and $B$ therefore commute (this result could, of course, be obtained by calculating directly the matrices $HB$ and $BH$). The restriction of $B$ to $\mathscr{E}_2$ is written:

$$P_{\mathscr{E}_2} B P_{\mathscr{E}_2} = b\begin{pmatrix} 0 & 1 \\ 1 & 0 \end{pmatrix}$$

The normalized eigenvectors of this $2 \times 2$ matrix are easy to obtain; they are:

$$| p_2 \rangle = \frac{1}{\sqrt{2}}[ | u_2 \rangle + | u_3 \rangle ] \qquad \text{(eigenvalue } + b)$$

$$| p_3 \rangle = \frac{1}{\sqrt{2}}[ | u_2 \rangle - | u_3 \rangle ] \qquad \text{(eigenvalue } - b)$$

These vectors are automatically eigenvectors of $H$ since $\mathscr{E}_2$ is the eigensubspace of $H$ corresponding to the eigenvalue $-\hbar\omega_0$. To summarize, the eigenvectors common to $H$ and $B$ are given by:

| | eigenvalue of $H$ | eigenvalue of $B$ |
|---|---|---|
| $| p_1 \rangle = | u_1 \rangle$ | $\hbar\omega_0$ | $b$ |
| $| p_2 \rangle = \frac{1}{\sqrt{2}}[ | u_2 \rangle + | u_3 \rangle ]$ | $-\hbar\omega_0$ | $b$ |
| $| p_3 \rangle = \frac{1}{\sqrt{2}}[ | u_2 \rangle - | u_3 \rangle ]$ | $-\hbar\omega_0$ | $-b$ |

These vectors are the only (to within, of course, a phase factor) normalized eigenvectors common to $H$ and $B$.

*c.* It can be seen from the table that $H$ has a two-fold degenerate eigenvalue; it is therefore not a C.S.C.O. Similarly, $B$ also has a two-fold degenerate eigenvalue and is therefore not a C.S.C.O.: an eigenvector of $B$ with the eigenvalue $b$ can be $| p_1 \rangle$, or $| p_2 \rangle$, or $\frac{1}{\sqrt{3}} | u_1 \rangle + \frac{1}{\sqrt{3}} | u_2 \rangle + \frac{1}{\sqrt{3}} | u_3 \rangle$, for example. On the other hand, the set of the two operators $H$ and $B$ does constitute a C.S.C.O. We see from the above table that no two vectors $| p_j \rangle$ have the same eigenvalues for both $H$ and $B$. This is why, as has already been pointed out, the system of normalized eigenvectors common to $H$ and $B$ is unique (to within phase factors). Note that within the eigensubspace $\mathscr{E}_2$ of $H$ associated with the eigenvalue $-\hbar\omega_0$, the eigenvalues of $B$ are distinct ($b$ and $-b$). Similarly, in the

eigensubspace of $B$ spanned by $| p_1 \rangle$ and $| p_2 \rangle$, the eigenvalues of $H$ are distinct ($\hbar\omega_0$ and $-\hbar\omega_0$).

$H^2$ has for eigenvectors, with the eigenvalue $\hbar^2\omega_0^2$, $| p_1 \rangle$, $| p_2 \rangle$ and $| p_3 \rangle$. It is easy to see that $H^2$ and $B$ do not constitute a C.S.C.O., since two linearly independent eigenvectors $| p_1 \rangle$ and $| p_2 \rangle$ correspond to the pair of eigenvalues $\{ \hbar^2\omega_0^2, b \}$.

## Solution of exercise 12

*a.* Let us use the rule for constructing the matrix of an operator : "in the $n$th column of the matrix, write the components of the operator transform of the $n$th basis vector". We obtain easily :

$$L_z = \begin{pmatrix} 1 & 0 & 0 \\ 0 & 0 & 0 \\ 0 & 0 & -1 \end{pmatrix} \qquad S = \begin{pmatrix} 0 & 0 & 1 \\ 0 & 1 & 0 \\ 1 & 0 & 0 \end{pmatrix}$$

$$L_z^2 = \begin{pmatrix} 1 & 0 & 0 \\ 0 & 0 & 0 \\ 0 & 0 & 1 \end{pmatrix} \qquad S^2 = \begin{pmatrix} 1 & 0 & 0 \\ 0 & 1 & 0 \\ 0 & 0 & 1 \end{pmatrix}$$

These matrices are symmetric and real, and therefore Hermitian. Since the space is finite-dimensional, they can be diagonalized and therefore represent observables.

*b.* Let $M$ be an operator which commutes with $L_z$. $M$ cannot (*cf.* chap. II, §D-3-a) have any matrix elements between $| u_1 \rangle$ and $| u_2 \rangle$, or between $| u_2 \rangle$ and $| u_3 \rangle$, or between $| u_1 \rangle$ and $| u_3 \rangle$ (eigenvectors of $L_z$ with different eigenvalues). The matrix which represents $M$ is therefore necessarily diagonal, that is, of the form:

$$[M, L_z] = 0 \qquad \Longleftrightarrow \qquad M = \begin{pmatrix} m_{11} & 0 & 0 \\ 0 & m_{22} & 0 \\ 0 & 0 & m_{33} \end{pmatrix}$$

Let $N$ be an operator which commutes with $L_z^2$. The matrix representing $N$ can have elements between $| u_1 \rangle$ and $| u_3 \rangle$ (eigenvectors of $L_z^2$ with the same eigenvalue), but none between $| u_2 \rangle$ and $| u_1 \rangle$ or $| u_3 \rangle$. $N$ is therefore written:

$$[N, L_z^2] = 0 \qquad \Longleftrightarrow \qquad N = \begin{pmatrix} n_{11} & 0 & n_{13} \\ 0 & n_{22} & 0 \\ n_{31} & 0 & n_{33} \end{pmatrix}$$

It is therefore less restrictive to impose the condition that an operator commute with $L_z^2$ than with $L_z$ : $N$ is not necessarily a diagonal matrix. It can only be said that $N$ does not mix the vectors of the subspace $\mathscr{F}_2$ spanned by $| u_1 \rangle$ and $| u_3 \rangle$ with those of the one-dimensional subspace spanned by $| u_2 \rangle$. This property,

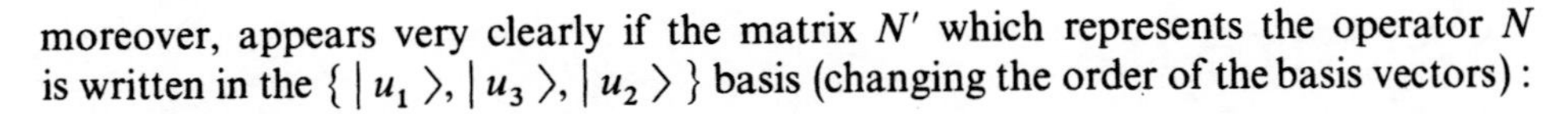



moreover, appears very clearly if the matrix $N'$ which represents the operator $N$ is written in the $\{ | u_1 \rangle, | u_3 \rangle, | u_2 \rangle \}$ basis (changing the order of the basis vectors):

$$N' = \begin{pmatrix} n_{11} & n_{13} & 0 \\ n_{31} & n_{33} & 0 \\ 0 & 0 & n_{22} \end{pmatrix}$$

Finally, since $S^2$ is the identity operator, any $3 \times 3$ matrix commutes with $S^2$, and its most general form is:

$$[P, S^2] = 0 \qquad \Longleftrightarrow \qquad P = \begin{pmatrix} p_{11} & p_{12} & p_{13} \\ p_{21} & p_{22} & p_{23} \\ p_{31} & p_{32} & p_{33} \end{pmatrix}$$

*c.* $| u_2 \rangle$ is an eigenvector common to $L_z^2$ and $S$. In the subspace $\mathscr{F}_2$ spanned by $| u_1 \rangle$ and $| u_3 \rangle$, $L_z^2$ and $S$ are written:

$$P_{\mathscr{F}_2} L_z^2 P_{\mathscr{F}_2} = \begin{pmatrix} 1 & 0 \\ 0 & 1 \end{pmatrix}$$

$$P_{\mathscr{F}_2} S P_{\mathscr{F}_2} = \begin{pmatrix} 0 & 1 \\ 1 & 0 \end{pmatrix}$$

The eigenvectors of the latter matrix are:

$$| q_2 \rangle = \frac{1}{\sqrt{2}} [ | u_1 \rangle + | u_3 \rangle ]$$

$$| q_3 \rangle = \frac{1}{\sqrt{2}} [ | u_1 \rangle - | u_3 \rangle ]$$

and the basis of eigenvectors common to $L_z^2$ and $S$ is:

| vector | eigenvalue of $L_z^2$ | eigenvalue of $S$ |
|---|---|---|
| $\| q_1 \rangle = \| u_2 \rangle$ | 0 | 1 |
| $\| q_2 \rangle = \frac{1}{\sqrt{2}} [ \| u_1 \rangle + \| u_3 \rangle ]$ | 1 | 1 |
| $\| q_3 \rangle = \frac{1}{\sqrt{2}} [ \| u_1 \rangle - \| u_3 \rangle ]$ | 1 | − 1 |

No two lines are alike in the table of eigenvalues of $L_z^2$ and $S$: these two operators therefore form a C.S.C.O. (this is not, however, the case for either one of them taken alone).

CHAPTER III

# The postulates of quantum mechanics

## OUTLINE OF CHAPTER III

## A. INTRODUCTION

In classical mechanics, the motion of any physical system is determined if the position $\mathbf{r}(x, y, z)$ and velocity $\mathbf{v}(\dot{x}, \dot{y}, \dot{z})$ of each of its points are known as a function of time. In general (appendix III), to describe such a system, one introduces generalized coordinates $q_i(t)$ $(i = 1, 2, ..., N)$, whose derivatives with respect to time, $\dot{q}_i(t)$, are the generalized velocities. Specifying the $q_i(t)$ and $\dot{q}_i(t)$ enables us to calculate, at any given instant, the position and velocity of any point of the system. Using the Lagrangian $\mathscr{L}(q_i, \dot{q}_i, t)$, one defines the conjugate momentum $p_i$ of each of the generalized coordinates $q_i$:

$$p_i = \frac{\partial \mathscr{L}}{\partial \dot{q}_i} \tag{A-1}$$

The $q_i(t)$ and $p_i(t)$ $(i = 1, 2, ..., N)$ are called the fundamental dynamical variables. All the physical quantities associated with the system (energy, angular momentum, etc.) can be expressed in terms of the fundamental dynamical variables. For example, the total energy of the system is given by the classical Hamiltonian $\mathscr{H}(q_i, p_i, t)$. The motion of the system can be studied by using either Lagrange's equations or the Hamilton-Jacobi canonical equations, which are written:

$$\frac{\mathrm{d}q_i}{\mathrm{d}t} = \frac{\partial \mathscr{H}}{\partial p_i} \tag{A-2-a}$$

$$\frac{\mathrm{d}p_i}{\mathrm{d}t} = -\frac{\partial \mathscr{H}}{\partial q_i} \tag{A-2-b}$$

In the special case of a system consisting of a single physical point of mass $m$, the $q_i$ are simply the three coordinates of this point, and the $\dot{q}_i$ are the components of its velocity $\mathbf{v}$. If the forces acting on this particle can be derived from a scalar potential $V(\mathbf{r}, t)$, the three conjugate momenta of its position $\mathbf{r}$ (that is, the components of its linear momentum $\mathbf{p}$) are equal to the components of its mechanical momentum $m\mathbf{v}$. The total energy is then written:

$$E = \frac{\mathbf{p}^2}{2m} + V(\mathbf{r}, t) \tag{A-3}$$

and the angular momentum with respect to the origin:

$$\mathscr{L} = \mathbf{r} \times \mathbf{p} \tag{A-4}$$

Since $\mathscr{H}(\mathbf{r}, \mathbf{p}, t) = \frac{\mathbf{p}^2}{2m} + V(\mathbf{r}, t)$, the Hamilton-Jacobi equations (A-2) here take on the well-known form:

$$\frac{\mathrm{d}\mathbf{r}}{\mathrm{d}t} = \frac{\mathbf{p}}{m} \tag{A-5-a}$$

$$\frac{\mathrm{d}\mathbf{p}}{\mathrm{d}t} = - \boldsymbol{\nabla} V \tag{A-5-b}$$

The classical description of a physical system can therefore be summarized as follows:

(*i*) The state of the system at a fixed time $t_0$ is defined by specifying $N$ generalized coordinates $q_i(t_0)$ and their $N$ conjugate momenta $p_i(t_0)$.

(*ii*) The value, at a given time, of the various physical quantities is completely determined when the state of the system at this time is known: knowing the state of the system, one can predict with certainty the result of any measurement performed at time $t_0$.

(*iii*) The time evolution of the state of the system is given by the Hamilton-Jacobi equations. Since these are first-order differential equations, their solution $\{ q_i(t), p_i(t) \}$ is unique if the value of these functions at a given time $t_0$ is fixed, $\{ q_i(t_0), p_i(t_0) \}$. The state of the system is known for all time if its initial state is known.

In this chapter, we shall study the postulates on which the quantum description of physical systems is based. We have already introduced them, in a qualitative and partial way, in chapter I. Here we shall discuss them explicitly, within the framework of the formalism developed in chapter II. These postulates will provide us with an answer to the following questions (which correspond to the three points enumerated above for the classical description):

(*i*) How is the state of a quantum system at a given time described mathematically?

(*ii*) Given this state, how can we predict the results of the measurement of various physical quantities?

(*iii*) How can the state of the system at an arbitrary time $t$ be found when the state at time $t_0$ is known?

We shall begin by stating the postulates of quantum mechanics (§ B). Then we shall analyze their physical content and discuss their consequences (§§ C, D, E).

## B. STATEMENT OF THE POSTULATES

### 1. Description of the state of a system

In chapter I, we introduced the concept of the quantum state of a particle. We first characterized this state at a given time by a square-integrable wave function. Then, in chapter II, we associated a ket of the state space $\mathscr{E}_{\mathbf{r}}$ with each wave function: choosing $| \psi \rangle$ belonging to $\mathscr{E}_{\mathbf{r}}$ is equivalent to choosing the corresponding function $\psi(\mathbf{r}) = \langle \mathbf{r} | \psi \rangle$. Therefore, the quantum state of a particle at a fixed time is characterized by a ket of the space $\mathscr{E}_{\mathbf{r}}$. In this form, the concept of a state can be generalized to any physical system.

*First Postulate:* At a fixed time $t_0$, the state of a physical system is defined by specifying a ket $| \psi(t_0) \rangle$ belonging to the state space $\mathscr{E}$.

It is important to note that, since $\mathscr{E}$ is a vector space, this first postulate implies a superposition principle : a linear combination of state vectors is a state vector. We shall discuss this fundamental point and its relations to the other postulates in §E.

## 2. Description of physical quantities

We have already used, in §D-1 of chapter I, a differential operator H related to the total energy of a particle in a scalar potential. This is simply a special case of the second postulate.

*Second Postulate:* Every measurable physical quantity $\mathscr{A}$ is described by an operator $A$ acting in $\mathscr{E}$; this operator is an observable.

COMMENTS:

(*i*) The fact that $A$ is an observable (*cf.* chap. II, §D-2) will be seen below (§3) to be essential.

(*ii*) Unlike classical mechanics (*cf.* §A), quantum mechanics describes in a fundamentally different manner the state of a system and the associated physical quantities : a state is represented by a vector, a physical quantity by an operator.

## 3. The measurement of physical quantities

### a. POSSIBLE RESULTS

The connection between the operator $H$ and the total energy of the particle appeared in §D-1 of chapter I in the following form: the only energies possible are the eigenvalues of the operator $H$. Here as well, this relation can be extended to all physical quantities.

*Third Postulate:* The only possible result of the measurement of a physical quantity $\mathscr{A}$ is one of the eigenvalues of the corresponding observable $A$.

COMMENTS:

(*i*) A measurement of $\mathscr{A}$ always gives a real value, since $A$ is by definition Hermitian.

(*ii*) If the spectrum of $A$ is discrete, the results that can be obtained by measuring $\mathscr{A}$ are quantized (§ C-2).

#### b. PRINCIPLE OF SPECTRAL DECOMPOSITION

We are going to generalize and discuss in more detail the conclusions of § A-3 of chapter I, where we analyzed a simple experiment performed on polarized photons.

Consider a system whose state is characterized, at a given time, by the ket $|\psi\rangle$, assumed to be normalized to 1:

$$\langle \psi | \psi \rangle = 1 \tag{B-1}$$

We want to predict the result of the measurement, at this time, of a physical quantity $\mathscr{A}$ associated with the observable $A$. This prediction, as we already know, is of a probabilistic sort. We are now going to give the rules which allow us to calculate the probability of obtaining any given eigenvalue of $A$.

##### α. *Case of a discrete spectrum*

First, let us assume that the spectrum of $A$ is entirely discrete. If all the eigenvalues $a_n$ of $A$ are non-degenerate, there is associated with each of them a unique (to within a constant factor) eigenvector $|u_n\rangle$:

$$A|u_n\rangle = a_n|u_n\rangle \tag{B-2}$$

Since $A$ is an observable, the set of the $|u_n\rangle$, which we shall take to be normalized, constitutes a basis in $\mathscr{E}$, and the state vector $|\psi\rangle$ can be written:

$$|\psi\rangle = \sum_n c_n |u_n\rangle \tag{B-3}$$

We postulate that the probability $\mathscr{P}(a_n)$ of finding $a_n$ when $\mathscr{A}$ is measured is:

$$\mathscr{P}(a_n) = |c_n|^2 = |\langle u_n|\psi\rangle|^2 \tag{B-4}$$

> *Fourth Postulate (case of a discrete non-degenerate spectrum):* When the physical quantity $\mathscr{A}$ is measured on a system in the *normalized* state $|\psi\rangle$, the probability $\mathscr{P}(a_n)$ of obtaining the *non-degenerate* eigenvalue $a_n$ of the corresponding observable $A$ is:
>
> $$\mathscr{P}(a_n) = |\langle u_n|\psi\rangle|^2$$
>
> where $|u_n\rangle$ is the normalized eigenvector of $A$ associated with the eigenvalue $a_n$.

If, now, some of the eigenvalues $a_n$ are degenerate, several orthonormalized eigenvectors $|u_n^i\rangle$ correspond to them:

$$A|u_n^i\rangle = a_n|u_n^i\rangle; \quad i = 1, 2, \ldots g_n \tag{B-5}$$

$|\psi\rangle$ can still be expanded in the orthonormal basis $\{|u_n^i\rangle\}$:

$$|\psi\rangle = \sum_n \sum_{i=1}^{g_n} c_n^i |u_n^i\rangle \tag{B-6}$$

In this case, the probability $\mathscr{P}(a_n)$ becomes:

$$\mathscr{P}(a_n) = \sum_{i=1}^{g_n} |c_n^i|^2 = \sum_{i=1}^{g_n} |\langle u_n^i | \psi \rangle|^2 \tag{B-7}$$

(B-4) is then seen to be a special case of (B-7), which can therefore be considered to be the general formula.

> *Fourth Postulate (case of a discrete spectrum):* When the physical quantity $\mathscr{A}$ is measured on a system in the *normalized* state $|\psi\rangle$, the probability $\mathscr{P}(a_n)$ of obtaining the eigenvalue $a_n$ of the corresponding observable $A$ is:
>
> $$\mathscr{P}(a_n) = \sum_{i=1}^{g_n} |\langle u_n^i | \psi \rangle|^2$$
>
> where $g_n$ is the degree of degeneracy of $a_n$ and $\{ | u_n^i \rangle \}$ $(i = 1, 2, ..., g_n)$ is an orthonormal set of vectors which forms a basis in the eigensubspace $\mathscr{E}_n$ associated with the eigenvalue $a_n$ of $A$.

For this postulate to make sense, it is obviously necessary that, if the eigenvalue $a_n$ is degenerate, the probability $\mathscr{P}(a_n)$ be independent of the choice of the $\{ |u_n^i\rangle \}$ basis in $\mathscr{E}_n$. To verify this, consider the vector:

$$|\psi_n\rangle = \sum_{i=1}^{g_n} c_n^i \, |u_n^i\rangle \tag{B-8}$$

where the coefficients $c_n^i$ are the same as those appearing in the expansion (B-6) of $|\psi\rangle$:

$$c_n^i = \langle u_n^i | \psi \rangle \tag{B-9}$$

$|\psi_n\rangle$ is the part of $|\psi\rangle$ which belongs to $\mathscr{E}_n$, that is, the projection of $|\psi\rangle$ onto $\mathscr{E}_n$. This is, moreover, what we find when we substitute (B-9) into (B-8):

$$\begin{aligned} |\psi_n\rangle &= \sum_{i=1}^{g_n} |u_n^i\rangle\langle u_n^i|\psi\rangle \\ &= P_n|\psi\rangle \end{aligned} \tag{B-10}$$

where:

$$P_n = \sum_{i=1}^{g_n} |u_n^i\rangle\langle u_n^i| \tag{B-11}$$

is the projector onto $\mathscr{E}_n$ (§ B-3-b of chapter II). Let us now calculate the square of the norm of $|\psi_n\rangle$. From (B-8):

$$\langle \psi_n | \psi_n \rangle = \sum_{i=1}^{g_n} |c_n^i|^2 \tag{B-12}$$

Therefore, $\mathscr{P}(a_n)$ *is the square of the norm of* $|\psi_n\rangle = P_n|\psi\rangle$, the projection of $|\psi\rangle$ onto $\mathscr{E}_n$. From this expression, it is clear that a change in the basis in $\mathscr{E}_n$ does not affect $\mathscr{P}(a_n)$. This probability is written:

$$\mathscr{P}(a_n) = \langle \psi | P_n^\dagger P_n | \psi \rangle \tag{B-13}$$

or, using the fact that $P_n$ is Hermitian ($P_n^\dagger = P_n$) and that it is a projector ($P_n^2 = P_n$):

$$\mathscr{P}(a_n) = \langle \psi | P_n | \psi \rangle \tag{B-14}$$

β. *Case of a continuous spectrum*

Now let us assume that the spectrum of $A$ is continuous and, for the sake of simplicity, non-degenerate. The system, orthonormal in the extended sense, of eigenvectors $| v_\alpha \rangle$ of $A$:

$$A | v_\alpha \rangle = \alpha | v_\alpha \rangle \tag{B-15}$$

forms a continuous basis in $\mathscr{E}$, in terms of which $| \psi \rangle$ can be expanded:

$$| \psi \rangle = \int \mathrm{d}\alpha \; c(\alpha) \, | v_\alpha \rangle \tag{B-16}$$

Since the possible results of a measurement of $\mathscr{A}$ form a continuous set, we must define a probability density, just as we did for the interpretation of the wave function of a particle (§ B-2 of chapter I). The probability $\mathrm{d}\mathscr{P}(\alpha)$ of obtaining a value included between $\alpha$ and $\alpha + \mathrm{d}\alpha$ is given by:

$$\mathrm{d}\mathscr{P}(\alpha) = \rho(\alpha)\, \mathrm{d}\alpha$$

with:

$$\rho(\alpha) = |c(\alpha)|^2 = |\langle v_\alpha | \psi \rangle|^2 \tag{B-17}$$

*Fourth Postulate (case of a continuous non-degenerate spectrum):* When the physical quantity $\mathscr{A}$ is measured on a system in the *normalized* state $| \psi \rangle$, the probability $\mathrm{d}\mathscr{P}(\alpha)$ of obtaining a result included between $\alpha$ and $\alpha + \mathrm{d}\alpha$ is equal to:

$$\mathrm{d}\mathscr{P}(\alpha) = |\langle v_\alpha | \psi \rangle|^2 \, \mathrm{d}\alpha$$

where $| v_\alpha \rangle$ is the eigenvector corresponding to the eigenvalue $\alpha$ of the observable $A$ associated with $\mathscr{A}$.

COMMENTS:

(*i*) It can be verified explicitly, in each of the cases considered above, that the total probability is equal to 1. For example, starting with formula (B-7), we find:

$$\sum_n \mathscr{P}(a_n) = \sum_n \sum_{i=1}^{g_n} |c_n^i|^2 = \langle \psi | \psi \rangle = 1 \tag{B-18}$$

since $| \psi \rangle$ is normalized. This last condition is therefore indispensable if the statements we have made are to be coherent. Nevertheless, it is not essential: if it is not fulfilled, it suffices to replace (B-7) and (B-17), respectively, by:

$$\mathscr{P}(a_n) = \frac{1}{\langle \psi | \psi \rangle} \sum_{i=1}^{g_n} |c_n^i|^2 \tag{B-19}$$

and:

$$\rho(\alpha) = \frac{1}{\langle \psi | \psi \rangle} |c(\alpha)|^2 \tag{B-20}$$

(*ii*) For the fourth postulate to be coherent, it is necessary for the operator $A$ associated with any physical quantity to be an observable: it must be possible to expand any state on the eigenvectors of $A$.

(*iii*) We have not given the fourth postulate in its most general form. Starting with the discussion of the cases we have envisaged, it is simple to extend the principle of spectral decomposition to any situation (continuous degenerate spectrum, partially continuous and partially discrete spectrum, etc...). In § E, and later in chapter IV, we shall apply this fourth postulate to a certain number of examples, pointing out certain implications of the superposition principle mentioned in § B-1.

γ. *An important consequence*

Consider two kets $|\psi\rangle$ and $|\psi'\rangle$ such that:

$$|\psi'\rangle = e^{i\theta} |\psi\rangle \tag{B-21}$$

where $\theta$ is a real number. If $|\psi\rangle$ is normalized, so is $|\psi'\rangle$:

$$\langle \psi' | \psi' \rangle = \langle \psi | e^{-i\theta} e^{i\theta} | \psi \rangle = \langle \psi | \psi \rangle \tag{B-22}$$

The probabilities predicted for an arbitrary measurement are the same for $|\psi\rangle$ and $|\psi'\rangle$ since, for any $|u_n^i\rangle$:

$$|\langle u_n^i | \psi' \rangle|^2 = |e^{i\theta} \langle u_n^i | \psi \rangle|^2 = |\langle u_n^i | \psi \rangle|^2 \tag{B-23}$$

Similarly, we can replace $|\psi\rangle$ by:

$$|\psi''\rangle = \alpha\, e^{i\theta} |\psi\rangle \tag{B-24}$$

without changing any of the physical results: there appear, in both the numerator and denominator of (B-19) and (B-20), factors of $|\alpha|^2$ which cancel. Therefore, *two proportional state vectors represent the same physical state.*

Care must be taken to interpret this result correctly. For example, let us assume that:

$$|\psi\rangle = \lambda_1 |\psi_1\rangle + \lambda_2 |\psi_2\rangle \tag{B-25}$$

where $\lambda_1$ and $\lambda_2$ are complex numbers. It is true that $e^{i\theta_1}|\psi_1\rangle$ represents, for all real $\theta_1$, the same physical state as $|\psi_1\rangle$, and $e^{i\theta_2}|\psi_2\rangle$ represents the same state as $|\psi_2\rangle$. But, in general:

$$|\varphi\rangle = \lambda_1 e^{i\theta_1} |\psi_1\rangle + \lambda_2 e^{i\theta_2} |\psi_2\rangle \tag{B-26}$$

*does not describe the same state as* $|\psi\rangle$ (we shall see in § E-1 that the *relative* phases of the expansion coefficients of the state vector play an important role). This is not true for the special case where $\theta_1 = \theta_2 + 2n\pi$, that is, where:

$$|\varphi\rangle = e^{i\theta_1}[\lambda_1 |\psi_1\rangle + \lambda_2 |\psi_2\rangle] = e^{i\theta_1} |\psi\rangle \tag{B-27}$$

In other words: *a global phase factor does not affect the physical predictions, but the relative phases of the coefficients of an expansion are significant.*

### c. REDUCTION OF THE WAVE PACKET

We have already introduced this concept in speaking of the measurement of the polarization of photons in the experiment described in § A-3 of chapter I. We are now going to generalize it, confining ourselves, nevertheless, to the case of a discrete spectrum (we shall take up the case of a continuous spectrum in § E).

Assume that we want to measure, at a given time, the physical quantity $\mathcal{A}$. If the ket $|\psi\rangle$, which represents the state of the system immediately before the measurement, is known, the fourth postulate allows us to predict the probabilities of obtaining the various possible results. But when the measurement is actually performed, it is obvious that only one of these possible results is obtained. Immediately after this measurement, we cannot speak of the "probability of having obtained" this or that value: we know which one was actually obtained. We therefore possess additional information, and it is understandable that the state of the system after the measurement, which must incorporate this information, should be different from $|\psi\rangle$.

Let us first consider the case where the measurement of $\mathcal{A}$ yields a simple eigenvalue $a_n$ of the observable $A$. We then postulate that the state of the system immediately after this measurement is the eigenvector $|u_n\rangle$ associated with $a_n$:

$$|\psi\rangle \overset{(a_n)}{\Longrightarrow} |u_n\rangle \tag{B-28}$$

COMMENTS:

(*i*) We have been speaking about states "immediately before" the measurement ($|\psi\rangle$) and "immediately after" ($|u_n\rangle$). The precise meaning of these expressions is the following: assume that the measurement takes place at the time $t_0 > 0$, and that we know the state $|\psi(0)\rangle$ of the system at the time $t = 0$. The sixth postulate (see § 4) describes how the system evolves

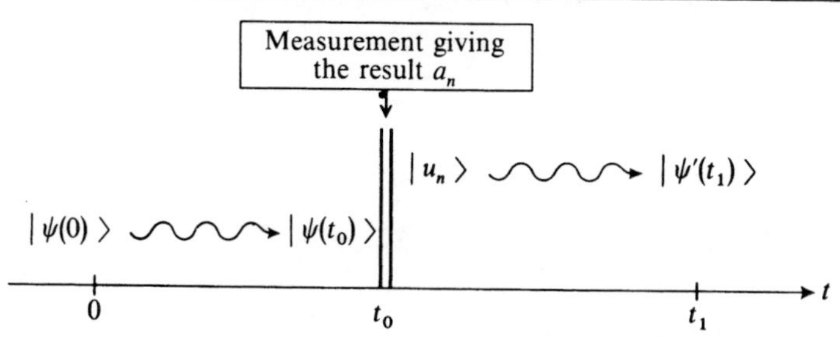


FIGURE 1

**When a measurement at time $t_0$ of the observable $A$ gives the result $a_n$, the state vector of the system undergoes an abrupt modification and becomes $|u_n\rangle$. This new initial state then evolves.**

over time, that is, enables us to calculate from $|\psi(0)\rangle$ the state $|\psi(t_0)\rangle$ "immediately before" the measurement. If the measurement has yielded the non-degenerate eigenvalue $a_n$, the state $|\psi'(t_1)\rangle$ at a time $t_1 > t_0$ must be calculated from $|\psi'(t_0)\rangle = |u_n\rangle$, the state "immediately after" the measurement, using the sixth postulate to determine the evolution of the state vector between the times $t_0$ and $t_1$ (fig. 1).

(*ii*) If we perform a second measurement of $\mathscr{A}$ immediately after the first one (that is, before the system has had time to evolve), we shall always find the same result $a_n$, since the state of the system immediately before the second measurement is $|u_n\rangle$, and no longer $|\psi\rangle$.

When the eigenvalue $a_n$ given by the measurement is degenerate, postulate (B-28) can be generalized as follows. If the expansion of the state $|\psi\rangle$ immediately before the measurement is written, with the same notation as in section b:

$$|\psi\rangle = \sum_n \sum_{i=1}^{g_n} c_n^i \, |u_n^i\rangle \tag{B-29}$$

the modification of the state vector due to the measurement is written:

$$|\psi\rangle \overset{(a_n)}{\Longrightarrow} \frac{1}{\sqrt{\sum_{i=1}^{g_n} |c_n^i|^2}} \sum_{i=1}^{g_n} c_n^i \, |u_n^i\rangle \tag{B-30}$$

$\sum_{i=1}^{g_n} c_n^i \, |u_n^i\rangle$ is the vector $|\psi_n\rangle$ defined above [formula (B-8)], that is, the projection of $|\psi\rangle$ onto the eigensubspace associated with $a_n$. In (B-30), we normalized this vector since it is always more convenient to use state vectors of norm 1 [comment (*i*) of §b above]. With the notation of (B-10) and (B-11), we can therefore write (B-30) in the form :

$$|\psi\rangle \overset{(a_n)}{\Longrightarrow} \frac{P_n|\psi\rangle}{\sqrt{\langle\psi|P_n|\psi\rangle}} \tag{B-31}$$

*Fifth Postulate:* If the measurement of the physical quantity $\mathscr{A}$ on the system in the state $|\psi\rangle$ gives the result $a_n$, the state of the system immediately after the measurement is the normalized projection, $\frac{P_n|\psi\rangle}{\sqrt{\langle\psi|P_n|\psi\rangle}}$, of $|\psi\rangle$ onto the eigensubspace associated with $a_n$.

The state of the system immediately after the measurement is therefore always an eigenvector of $A$ with the eigenvalue $a_n$. We stress the fact, however, that *it is not an arbitrary ket of the subspace* $\mathscr{E}_n$, but the part of $|\psi\rangle$ which belongs to $\mathscr{E}_n$ (suitably normalized, for convenience). In the light of § 3-b-γ above, equation (B-28) can be seen to be a special case of (B-30). When $g_n = 1$, the summation over $i$ disappears from (B-30), which becomes:

$$\frac{1}{|c_n|} c_n \, |u_n\rangle = e^{i \operatorname{Arg} c_n} \, |u_n\rangle \tag{B-32}$$

This ket indeed describes the same physical state as $| u_n \rangle$.

## 4. Time evolution of systems

We have already presented, in § B-2 of chapter I, the Schrödinger equation for one particle. Here we shall write it in the general case.

> *Sixth Postulate:* The time evolution of the state vector $|\psi(t)\rangle$ is governed by the Schrödinger equation:
>
> $$i\hbar \frac{\mathrm{d}}{\mathrm{d}t} | \psi(t) \rangle = H(t) | \psi(t) \rangle$$
>
> where $H(t)$ is the observable associated with the total energy of the system.

$H$ is called the *Hamiltonian operator* of the system, as it is obtained from the classical Hamiltonian (appendix III and §5 below).

## 5. Quantization rules

We are finally going to discuss how to construct, for a physical quantity $\mathscr{A}$ already defined in classical mechanics, the operator $A$ which describes it in quantum mechanics.

### a. STATEMENT

Let us first consider a system composed of a single particle, without spin, subject to a scalar potential. In this case:

> With the position $\mathbf{r}(x, y, z)$ of the particle is associated the observable $\mathbf{R}(X, Y, Z)$. With the momentum $\mathbf{p}(p_x, p_y, p_z)$ of the particle is associated the observable $\mathbf{P}(P_x, P_y, P_z)$.

Recall that the components of $\mathbf{R}$ and $\mathbf{P}$ satisfy the canonical commutation relations [chap. II, equations (E-30)]:

$$[R_i, R_j] = [P_i, P_j] = 0$$
$$[R_i, P_j] = i\hbar\delta_{ij} \tag{B-33}$$

Any physical quantity $\mathscr{A}$ related to this particle is expressed in terms of the fundamental dynamical variables $\mathbf{r}$ and $\mathbf{p}$: $\mathscr{A}(\mathbf{r}, \mathbf{p}, t)$. To obtain the corresponding observable $A$, one could simply replace, in the expression for $\mathscr{A}(\mathbf{r}, \mathbf{p}, t)$, the variables $\mathbf{r}$ and $\mathbf{p}$ by the observables $\mathbf{R}$ and $\mathbf{P}$*:

$$A(t) = \mathscr{A}(\mathbf{R}, \mathbf{P}, t) \tag{B-34}$$

* See, in complement $B_{II}$, the definition of a function of an operator.

However, this mode of action would be, in general, ambiguous. Assume, for example, that in $\mathscr{A}(\mathbf{r}, \mathbf{p}, t)$ there appears a term of the form :

$$\mathbf{r} \cdot \mathbf{p} = xp_x + yp_y + zp_z \tag{B-35}$$

In classical mechanics, the scalar product $\mathbf{r} \cdot \mathbf{p}$ is commutative, and one can just as well write :

$$\mathbf{p} \cdot \mathbf{r} = p_x x + p_y y + p_z z \tag{B-36}$$

But when $\mathbf{r}$ and $\mathbf{p}$ are replaced by the corresponding observables $\mathbf{R}$ and $\mathbf{P}$, the operators obtained from (B-35) and (B-36) are not identical [see relations (B-33)]:

$$\mathbf{R} \cdot \mathbf{P} \neq \mathbf{P} \cdot \mathbf{R} \tag{B-37}$$

Moreover, neither $\mathbf{R} \cdot \mathbf{P}$ nor $\mathbf{P} \cdot \mathbf{R}$ is Hermitian:

$$(\mathbf{R} \cdot \mathbf{P})^\dagger = (XP_x + YP_y + ZP_z)^\dagger = \mathbf{P} \cdot \mathbf{R} \tag{B-38}$$

To the preceding postulates, therefore, must be added a symmetrization rule. For example, the observable associated with $\mathbf{r} \cdot \mathbf{p}$ will be :

$$\frac{1}{2}(\mathbf{R} \cdot \mathbf{P} + \mathbf{P} \cdot \mathbf{R}) \tag{B-39}$$

which is indeed Hermitian. For an observable which is more complicated than $\mathbf{R} \cdot \mathbf{P}$, an analogous symmetrization is to be performed.

> The observable $A$ which describes a classically defined physical quantity $\mathscr{A}$ is obtained by replacing, in the suitably symmetrized expression for $\mathscr{A}$, $\mathbf{r}$ and $\mathbf{p}$ by the observables $\mathbf{R}$ and $\mathbf{P}$ respectively.

We shall see, however, that there exist quantum physical quantities which have no classical equivalent and which are therefore defined directly by the corresponding observables (this is the case, for example, for particle spin).

COMMENT:

The preceding rules, and commutation rules (B-33) in particular, are valid only in cartesian coordinates. It would be possible to generalize them to other coordinate systems; however, they would no longer have the same simple form as they do above.

#### b. IMPORTANT EXAMPLES

##### α. *The Hamiltonian of a particle in a scalar potential*

Consider a (spinless) particle of charge $q$ and mass $m$, placed in an electric field derived from a scalar potential $U(\mathbf{r})$. The potential energy of the particle is therefore $V(\mathbf{r}) = qU(\mathbf{r})$, and the corresponding classical Hamiltonian is written [appendix III, formula (29)]:

$$\mathscr{H}(\mathbf{r}, \mathbf{p}) = \frac{\mathbf{p}^2}{2m} + V(\mathbf{r}) \tag{B-40}$$

with:

$$\mathbf{p} = m\frac{\mathrm{d}\mathbf{r}}{\mathrm{d}t} = m\mathbf{v} \tag{B-41}$$

where $\mathbf{v}$ is the particle's velocity.

No difficulties are presented by the construction of the quantum operator $H$ which corresponds to $\mathscr{H}$. No symmetrization is necessary, since neither $\mathbf{P}^2 = P_x^2 + P_y^2 + P_z^2$ nor $V(\mathbf{R})$ involves products of noncommuting operators. We therefore have:

$$H = \frac{\mathbf{P}^2}{2m} + V(\mathbf{R}) \tag{B-42}$$

$V(\mathbf{R})$ is the operator obtained by replacing $\mathbf{r}$ by $\mathbf{R}$ in $V(\mathbf{r})$ (*cf.* complement $\mathrm{B_{II}}$, § 4).

In this particular case, the Schrödinger equation, given in the sixth postulate, becomes:

$$i\hbar\frac{\mathrm{d}}{\mathrm{d}t}|\,\psi(t)\,\rangle = \left[\frac{\mathbf{P}^2}{2m} + V(\mathbf{R})\right]|\,\psi(t)\,\rangle \tag{B-43}$$

β. *The Hamiltonian of a particle in a vector potential*

If the particle is now placed in an arbitrary electromagnetic field, the classical Hamiltonian becomes [appendix III, relation (66)]:

$$\mathscr{H}(\mathbf{r}, \mathbf{p}) = \frac{1}{2m}[\mathbf{p} - q\mathbf{A}(\mathbf{r}, t)]^2 + qU(\mathbf{r}, t) \tag{B-44}$$

where $U(\mathbf{r}, t)$ and $\mathbf{A}(\mathbf{r}, t)$ are the scalar and vector potentials which describe the electromagnetic field, and where $\mathbf{p}$ is given by:

$$\mathbf{p} = m\frac{\mathrm{d}\mathbf{r}}{\mathrm{d}t} + q\mathbf{A}(\mathbf{r}, t) = m\mathbf{v} + q\mathbf{A}(\mathbf{r}, t) \tag{B-45}$$

Once again, since $\mathbf{A}(\mathbf{r}, t)$ depends only on $\mathbf{r}$ and the parameter $t$ (and not on $\mathbf{p}$), construction of the corresponding quantum operator $\mathbf{A}(\mathbf{R}, t)$ presents no problem. The Hamiltonian operator $H$ is then given by:

$$H(t) = \frac{1}{2m}[\mathbf{P} - q\mathbf{A}(\mathbf{R}, t)]^2 + V(\mathbf{R}, t) \tag{B-46}$$

with

$$V(\mathbf{R}, t) = q\,U(\mathbf{R}, t) \tag{B-47}$$

and the Schrödinger equation is written:

$$i\hbar\frac{\mathrm{d}}{\mathrm{d}t}|\,\psi(t)\,\rangle = \left\{\frac{1}{2m}[\mathbf{P} - q\mathbf{A}(\mathbf{R}, t)]^2 + V(\mathbf{R}, t)\right\}|\,\psi(t)\,\rangle \tag{B-48}$$

COMMENT:

Care must be taken not to confuse **p** (the momentum of the particle, also called the conjugate momentum of **r**) with $m\mathbf{v}$ (the mechanical momentum of the particle): the difference between these two quantities appears clearly in (B-45). In quantum mechanics, there of course exists an operator associated with the velocity of the particle which is written here:

$$\mathcal{V} = \frac{1}{m}(\mathbf{P} - q\mathbf{A}) \tag{B-49}$$

$H$ is then given by:

$$H(t) = \frac{1}{2}m\mathcal{V}^2 + V(\mathbf{R}, t) \tag{B-50}$$

It is the sum of two terms, one corresponding to the kinetic energy and the other to the potential energy of the particle.

However, it is the conjugate momentum **p** and not the mechanical momentum $m\mathbf{v}$ which becomes in quantum mechanics the operator **P** which satisfies the canonical commutation relations (B-33).

# C. THE PHYSICAL INTERPRETATION OF THE POSTULATES CONCERNING OBSERVABLES AND THEIR MEASUREMENT

## 1. The quantization rules are consistent with the probabilistic interpretation of the wave function

It is natural to associate the observables **R** and **P**, whose action was defined in §E of chapter II, with the position and momentum of a particle. First of all, each of the observables $X$, $Y$, $Z$ and $P_x$, $P_y$, $P_z$ possesses a continuous spectrum, and experiments indeed show that all real values are possible for the six position and momentum variables. Moreover, we shall see that applying the fourth postulate to the case of these observables enables us to re-derive the probabilistic interpretation of the wave function as well as that of its Fourier transform (see §§ B-2 and C-3 of chapter I).

Let us consider, for simplicity, the one-dimensional problem. If the particle is in the normalized state $|\psi\rangle$, the probability that a measurement of its position will yield a result included between $x$ and $x + \mathrm{d}x$ is equal to [formula (B-17)]:

$$\mathrm{d}\mathcal{P}(x) = |\langle x | \psi \rangle|^2 \,\mathrm{d}x \tag{C-1}$$

where $|x\rangle$ is the eigenket of $X$ with the eigenvalue $x$. We again find that the square of the modulus of the wave function $\psi(x) = \langle x | \psi \rangle$ is the particle's position probability density. Now, to the eigenvector $|p\rangle$ of the observable $P$ corresponds the plane wave:

$$\langle x | p \rangle = \frac{1}{\sqrt{2\pi\hbar}} e^{\frac{ipx}{\hbar}} \tag{C-2}$$

and we have seen (§ C-3 of chapter I) that the de Broglie relations associate with this wave a well-defined momentum which is precisely $p$. In addition, the probability of finding, for a particle in the state $|\psi\rangle$, a momentum between $p$ and $p + \mathrm{d}p$ is:

$$\mathrm{d}\mathscr{P}(p) = |\langle p | \psi \rangle|^2 \, \mathrm{d}p = |\bar{\psi}(p)|^2 \, \mathrm{d}p \tag{C-3}$$

This is indeed what we found in § C-3 of chapter I.

## 2. Quantization of certain physical quantities

As we have already pointed out, the third postulate enables us to explain the quantization observed for certain quantities, such as the energy of atoms. But it does not imply that all quantities are quantized, since observables exist whose spectrum is continuous. The physical predictions based on the third postulate are therefore not at all obvious *a priori*. For example, when we study the hydrogen atom (chap. VII), we shall start from the total energy of the electron in the coulomb potential of the proton, from which we shall deduce the Hamiltonian operator. Solving its eigenvalue equation, we shall find that the bound states of the system can only correspond to certain discrete energies which we shall calculate. Thus we shall not only explain the quantization of the levels of the hydrogen atom, but also predict the possible energy values, which can be measured experimentally. We stress the fact that these results will be obtained using the same fundamental interaction law which is used in classical mechanics in the macroscopic domain.

## 3. The measurement process

The fourth and fifth postulates pose a certain number of fundamental problems which we shall not consider here. There is, in particular, the question of the "fundamental" perturbation involved in the observation of a quantum system (*cf.* chap. I, §§ A-2 and A-3). The origin of these problems lies in the fact that the system under study is treated independently from the measurement device, although their interaction is essential to the observation process. One should actually consider the system and the measurement device together as a whole. This raises delicate questions concerning the details of the measurement process.

We shall content ourselves with pointing out that the nondeterministic formulation of the fourth and fifth postulates is related to the problems that we have just mentioned. For example, the abrupt change from one state vector to another due to the measurement corresponds to the fundamental perturbation of which we have spoken. But it is impossible to predict what this perturbation will be, since it depends on the measurement result, which is not known with certainty in advance★.

We shall consider here only ideal measurements. To understand this concept, let us return, for example, to the experiment of §A-3 of chapter I on polarized photons. It is clear that when we grant that all photons polarized in a certain direction traverse the analyzer, we assume that the analyzer is perfect. In practice, obviously, it also absorbs some of the photons that it should let through. We shall therefore make the hypothesis, in the general case, that the measurement devices used are perfect : this amounts to assuming that the perturbation they provoke is due only to the quantum mechanical aspect of the measurement. Of course, the devices which can actually be constructed always present imperfections which affect the measurement and the system; but one can, in principle, constantly ameliorate them and thus approach the ideal limit defined by the postulates which we have stated.

## 4. Mean value of an observable in a given state

The predictions deduced from the fourth postulate are expressed in terms of probabilities. To verify them, it would be necessary to perform a large number of measurements under identical conditions. That is, one would have to measure the same quantity in a large number of systems which are all in the same quantum state. If these predictions are correct, the proportion of $N$ identical experiments resulting in a given event will approach, as $N \longrightarrow \infty$, the theoretically predicted probability $\mathscr{P}$ of this event. Such a verification can only be carried out in the limit where $N \longrightarrow \infty$; in practice, $N$ is of course finite, and statistical techniques must be used to interpret the results.

The mean value of the observable★★ $A$ in the state $|\psi\rangle$, which we shall denote by $\langle A \rangle_\psi$, or, more simply, by $\langle A \rangle$, is defined as the average of the results obtained when a large number $N$ of measurements of this observable are performed on systems which are all in the state $|\psi\rangle$. When $|\psi\rangle$ is given, the probabilities of finding all the possible results are known. The mean value $\langle A \rangle_\psi$ can therefore be predicted. We shall show that if $|\psi\rangle$ is normalized, $\langle A \rangle_\psi$ is given by the formula:

$$\boxed{\langle A \rangle_\psi = \langle \psi | A | \psi \rangle} \tag{C-4}$$

First consider the case where the entire spectrum of $A$ is discrete. Out of $N$ measurements of $A$ (the system being in the state $|\psi\rangle$ each time), the eigenvalue $a_n$ will be obtained $\mathscr{N}(a_n)$ times, with:

$$\frac{\mathscr{N}(a_n)}{N} \underset{N \to \infty}{\longrightarrow} \mathscr{P}(a_n) \tag{C-5}$$

★ Except, obviously, in the case where one is sure of the result that will be found (probability equal to 1 : the measurement does not modify the state of the system).

★★ We shall henceforth use the word "observable" to designate a physical quantity as well as the associated operator.

and:

$$\sum_n \mathcal{N}(a_n) = N \tag{C-6}$$

The mean value of the results of these $N$ experiments is the sum of the values found divided by $N$ (when $\mathcal{N}$ experiments have yielded the same result, this result will clearly appear $\mathcal{N}$ times in this sum). It is therefore equal to:

$$\frac{1}{N}\sum_n a_n \mathcal{N}(a_n) \tag{C-7}$$

Using (C-5), we see that when $N \longrightarrow \infty$, this mean value approaches:

$$\langle A \rangle_\psi = \sum_n a_n \mathcal{P}(a_n) \tag{C-8}$$

Now substitute into this formula expression (B-7) for $\mathcal{P}(a_n)$:

$$\langle A \rangle_\psi = \sum_n a_n \sum_{i=1}^{g_n} \langle \psi \mid u_n^i \rangle \langle u_n^i \mid \psi \rangle \tag{C-9}$$

Since:

$$A \mid u_n^i \rangle = a_n \mid u_n^i \rangle \tag{C-10}$$

(C-9) can be written in the form:

$$\begin{aligned} \langle A \rangle_\psi &= \sum_n \sum_{i=1}^{g_n} \langle \psi \mid A \mid u_n^i \rangle \langle u_n^i \mid \psi \rangle \\ &= \langle \psi \mid A \left[ \sum_n \sum_{i=1}^{g_n} \mid u_n^i \rangle \langle u_n^i \mid \right] \mid \psi \rangle \end{aligned} \tag{C-11}$$

Since the $\{ \mid u_n^i \rangle \}$ form an orthonormal basis of $\mathcal{E}$, the expression in brackets is equal to the identity operator (closure relation), and we obtain formula (C-4).

The argument is completely analogous for the case where the spectrum of $A$ is continuous (for simplicity, we shall continue to assume it to be non-degenerate). Consider $N$ identical experiments, and call $\mathrm{d}\mathcal{N}(\alpha)$ the number of experiments which have yielded a result included between $\alpha$ and $\alpha + \mathrm{d}\alpha$. We have, similarly:

$$\frac{\mathrm{d}\mathcal{N}(\alpha)}{N} \underset{N \to \infty}{\longrightarrow} \mathrm{d}\mathcal{P}(\alpha) \tag{C-12}$$

The mean value of the results obtained is $\frac{1}{N}\int \alpha \, \mathrm{d}\mathcal{N}(\alpha)$, which, when $N \longrightarrow \infty$, approaches:

$$\langle A \rangle_\psi = \int \alpha \, \mathrm{d}\mathcal{P}(\alpha) \tag{C-13}$$

Substitute into (C-13) the expression for $\mathrm{d}\mathcal{P}(\alpha)$ given by (B-17):

$$\langle A \rangle_\psi = \int \alpha \langle \psi \mid v_\alpha \rangle \langle v_\alpha \mid \psi \rangle \, \mathrm{d}\alpha \tag{C-14}$$

We can use the equation:

$$A \mid v_\alpha \rangle = \alpha \mid v_\alpha \rangle \tag{C-15}$$

to transform (C-14) into:

$$\langle A \rangle_\psi = \int \langle \psi | A | v_\alpha \rangle \langle v_\alpha | \psi \rangle \, \mathrm{d}\alpha$$

$$= \langle \psi | A \left[ \int \mathrm{d}\alpha \, | v_\alpha \rangle \langle v_\alpha | \right] | \psi \rangle \qquad \text{(C-16)}$$

Using the closure relation satisfied by the state $| v_\alpha \rangle$, we again find formula (C-4).

COMMENTS:

(*i*) $\langle A \rangle$, the average over a set of identical measurements, must not be confused with the time averages sometimes taken when dealing with time-dependent phenomena.

(*ii*) If the ket $| \psi \rangle$ representing the state of the system is not normalized, formula (C-4) becomes [*cf.* comment (*i*) of § B-3-b]:

$$\langle A \rangle_\psi = \frac{\langle \psi | A | \psi \rangle}{\langle \psi | \psi \rangle} \qquad \text{(C-17)}$$

(*iii*) In practice, to calculate $\langle A \rangle_\psi$ explicitly, one often places oneself in a particular representation. For example:

$$\langle X \rangle_\psi = \langle \psi | X | \psi \rangle$$

$$= \int d^3r \, \langle \psi | \mathbf{r} \rangle \langle \mathbf{r} | X | \psi \rangle$$

$$= \int d^3r \, \psi^*(\mathbf{r}) \, x \, \psi(\mathbf{r}) \qquad \text{(C-18)}$$

using the definition of the $X$ operator [*cf.* chap. II, relations (E-22)]. Similarly:

$$\langle P_x \rangle_\psi = \langle \psi | P_x | \psi \rangle$$

$$= \int d^3p \, \bar{\psi}^*(\mathbf{p}) \, p_x \, \bar{\psi}(\mathbf{p}) \qquad \text{(C-19)}$$

or, using the $\{ | \mathbf{r} \rangle \}$ representation:

$$\langle P_x \rangle_\psi = \int d^3r \, \langle \psi | \mathbf{r} \rangle \langle \mathbf{r} | P_x | \psi \rangle$$

$$= \int d^3r \, \psi^*(\mathbf{r}) \left[ \frac{\hbar}{i} \frac{\partial}{\partial x} \psi(\mathbf{r}) \right] \qquad \text{(C-20)}$$

since $\mathbf{P}$ is then represented by $\frac{\hbar}{i} \nabla$ [formula (E-26) of chapter II].

## 5. The root-mean-square deviation

$\langle A \rangle$ indicates the order of magnitude of the values of the observable $A$ when the system is in the state $|\psi\rangle$. However, this mean value does not give any idea of the dispersion of the results we expect when measuring $A$. Assume, for example, that the spectrum of $A$ is continuous and that, for a given state $|\psi\rangle$, the curve representing the variation with respect to $\alpha$ of the probability density $\rho(\alpha) = |\langle v_\alpha | \psi \rangle|^2$ has the shape shown in figure 2. For a system in the state $|\psi\rangle$, nearly all the values that can be found when $A$ is measured are included in an interval of width $\delta A$ containing $\langle A \rangle$, where the quantity $\delta A$ characterizes the width of the curve: the smaller $\delta A$, the more the measurement results are concentrated about $\langle A \rangle$.

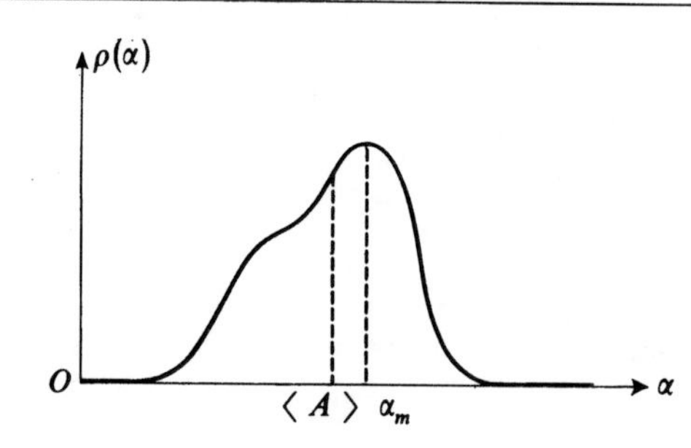


FIGURE 2

**Variation with respect to $\alpha$ of the probability density $\rho(\alpha)$. The mean value $\langle A \rangle$ is the abscissa of the center of gravity of the area under the curve (it does not necessarily coincide with the abscissa $\alpha_m$ of the maximum of the function).**

How can we define, in a general way, a quantity which characterizes the dispersion of the measurement results about $\langle A \rangle$? We might envisage the following method: for each measurement, take the difference between the value obtained and $\langle A \rangle$; then calculate the average of these deviations, dividing their sum by the number $N$ of experiments. It is easy to see, however, that the result obtained would be zero; we have, obviously:

$$\langle A - \langle A \rangle \rangle = \langle A \rangle - \langle A \rangle = 0 \tag{C-21}$$

By the very definition of $\langle A \rangle$, the average of the negative deviations balances exactly the average of the positive ones.

To avoid this compensation, it suffices to define $\Delta A$ such that $(\Delta A)^2$ is the mean of the squares of the deviations:

$$(\Delta A)^2 = \langle (A - \langle A \rangle)^2 \rangle \tag{C-22}$$

By definition, we therefore introduce the *root-mean-square deviation $\Delta A$* by setting:

$$\boxed{\Delta A = \sqrt{\langle (A - \langle A \rangle)^2 \rangle}} \tag{C-23}$$

Using the expression for the mean value given in (C-4), we then have:

$$\Delta A = \sqrt{\langle \psi | (A - \langle A \rangle)^2 | \psi \rangle} \tag{C-24}$$

This relation can also be written in a slightly different way:

$$\begin{aligned} \langle (A - \langle A \rangle)^2 \rangle &= \langle (A^2 - 2\langle A \rangle A + \langle A \rangle^2) \rangle \\ &= \langle A^2 \rangle - 2\langle A \rangle^2 + \langle A \rangle^2 \\ &= \langle A^2 \rangle - \langle A \rangle^2 \end{aligned} \tag{C-25}$$

The root-mean-square deviation $\Delta A$ is therefore also given by:

$$\Delta A = \sqrt{\langle A^2 \rangle - \langle A \rangle^2} \tag{C-26}$$

For example, in the case of the continuous spectrum of the observable $A$ considered above, $\Delta A$ is given by:

$$\begin{aligned} (\Delta A)^2 &= \int_{-\infty}^{+\infty} [\alpha - \langle A \rangle]^2 \rho(\alpha)\, \mathrm{d}\alpha \\ &= \int_{-\infty}^{+\infty} \alpha^2 \rho(\alpha)\, \mathrm{d}\alpha - \left[ \int_{-\infty}^{+\infty} \alpha\, \rho(\alpha)\, \mathrm{d}\alpha \right]^2 \end{aligned} \tag{C-27}$$

If definition (C-23) is applied to the observables **R** and **P**, it can be shown (complement $C_{III}$), using their commutation relations, that for any state $|\psi\rangle$, one has :

$$\begin{cases} \Delta X \,.\, \Delta P_x \geqslant \hbar/2 \\ \Delta Y \,.\, \Delta P_y \geqslant \hbar/2 \\ \Delta Z \,.\, \Delta P_z \geqslant \hbar/2 \end{cases} \tag{C-28}$$

In other words, we find the Heisenberg uncertainty relations again, but with a precise lower limit, which arises from the precise definition of the uncertainties.

## 6. Compatibility of observables

### a. COMPATIBILITY AND COMMUTATION RULES

Consider two observables $A$ and $B$ which commute:

$$[A, B] = 0 \tag{C-29}$$

We shall assume for simplicity that both of their spectra are discrete. According to the theorem proved in § D-3-a of chapter II, there exists a basis of the state space composed of eigenkets common to $A$ and $B$, which we shall denote by $|a_n, b_p, i\rangle$:

$$\begin{aligned} A\, |a_n, b_p, i\rangle &= a_n\, |a_n, b_p, i\rangle \\ B\, |a_n, b_p, i\rangle &= b_p\, |a_n, b_p, i\rangle \end{aligned} \tag{C-30}$$

(the index $i$ allows us to distinguish, if necessary, between the different vectors corresponding to one pair of eigenvalues). Therefore, for any $a_n$ and $b_p$ (chosen, respectively, in the spectra of $A$ and $B$), there exists at least one state $|a_n, b_p, i\rangle$ for which a measurement of $A$ will always give $a_n$ and a measurement of $B$ will always give $b_p$. Two such observables $A$ and $B$ which can be simultaneously determined are said to be compatible.

On the other hand, if $A$ and $B$ do not commute, a state cannot in general* be a simultaneous eigenvector of these two observables. They are said to be incompatible.

Let us examine more closely the measurement of two compatible observables on a system which is initially in an arbitrary (normalized) state $| \psi \rangle$. This state can always be written:

$$| \psi \rangle = \sum_{n,p,i} c_{n,p,i} \, | a_n, b_p, i \rangle \tag{C-31}$$

First assume that we measure $A$ and then, immediately afterwards, $B$ (before the system has had the time to evolve). Let us calculate the probability $\mathscr{P}(a_n, b_p)$ of obtaining $a_n$ in the first measurement and $b_p$ in the second one. We begin by measuring $A$ in the state $| \psi \rangle$; the probability of finding $a_n$ is therefore:

$$\mathscr{P}(a_n) = \sum_{p,i} |c_{n,p,i}|^2 \tag{C-32}$$

When we then measure $B$, the system is no longer in the state $| \psi \rangle$ but, if we have found $a_n$, in the state $| \psi'_n \rangle$:

$$| \psi'_n \rangle = \frac{1}{\sqrt{\sum_{p,i} |c_{n,p,i}|^2}} \sum_{p,i} c_{n,p,i} \, | a_n, b_p, i \rangle \tag{C-33}$$

The probability of obtaining $b_p$ when it is known that the first measurement has yielded $a_n$ is therefore equal to:

$$\mathscr{P}_{a_n}(b_p) = \frac{1}{\sum_{p,i} |c_{n,p,i}|^2} \sum_i |c_{n,p,i}|^2 \tag{C-34}$$

The probability $\mathscr{P}(a_n, b_p)$ sought corresponds to a "composite event" : to be in a favorable case, we must first find $a_n$ and then, having satisfied this first condition, find $b_p$. Therefore:

$$\mathscr{P}(a_n, b_p) = \mathscr{P}(a_n) \times \mathscr{P}_{a_n}(b_p) \tag{C-35}$$

Substituting into this formula expressions (C-32) and (C-34), we obtain:

$$\mathscr{P}(a_n, b_p) = \sum_i |c_{n,p,i}|^2 \tag{C-36}$$

Moreover, the state of the system becomes, immediately after the second measurement:

$$| \psi''_{n,p} \rangle = \frac{1}{\sqrt{\sum_i |c_{n,p,i}|^2}} \sum_i c_{n,p,i} \, | a_n, b_p, i \rangle \tag{C-37}$$

Therefore, if we decide to measure either $A$ or $B$ again, we are sure of the result ($a_n$ or $b_p$): $| \psi''_{n,p} \rangle$ is an eigenvector common to $A$ and $B$ with the eigenvalues $a_n$ and $b_p$ respectively.

Let us now return to the system in the state $| \psi \rangle$, and let us measure the two observables in the opposite order ($B$, then $A$). What is the probability $\mathscr{P}(b_p, a_n)$ of obtaining the same results as before? The reasoning is the same. We have here:

$$\mathscr{P}(b_p, a_n) = \mathscr{P}(b_p) \times \mathscr{P}_{b_p}(a_n) \tag{C-38}$$

* Some kets may be simultaneous eigenvectors of $A$ and $B$. But there would not be a sufficient number of them to form a basis, as would be the case if $A$ and $B$ commuted.

From (C-31), we see that:

$$\mathscr{P}(b_p) = \sum_{n,i} |c_{n,p,i}|^2 \tag{C-39}$$

and that, after a measurement of $B$ which yields $b_p$, the state of the system becomes:

$$|\varphi'_p\rangle = \frac{1}{\sqrt{\sum_{n,i} |c_{n,p,i}|^2}} \sum_{n,i} c_{n,p,i} \,|a_n, b_p, i\rangle \tag{C-40}$$

Therefore:

$$\mathscr{P}_{b_p}(a_n) = \frac{1}{\sum_{n,i} |c_{n,p,i}|^2} \sum_i |c_{n,p,i}|^2 \tag{C-41}$$

and finally:

$$\mathscr{P}(b_p, a_n) = \sum_i |c_{n,p,i}|^2 \tag{C-42}$$

If we have indeed found $b_p$ and then $a_n$, the system has gone into the state:

$$|\varphi''_{p,n}\rangle = \frac{1}{\sqrt{\sum_i |c_{n,p,i}|^2}} \sum_i c_{n,p,i} \,|a_n, b_p, i\rangle \tag{C-43}$$

When two observables are compatible, the physical predictions are the same, whatever the order of performing the two measurements (provided that the time interval which separates them is sufficiently small). The probabilities of obtaining either $a_n$ then $b_p$ or $b_p$ then $a_n$ are identical:

$$\mathscr{P}(a_n, b_p) = \mathscr{P}(b_p, a_n) = \sum_i |c_{n,p,i}|^2 = \sum_i |\langle a_n, b_p, i \,|\, \psi\rangle|^2 \tag{C-44}$$

Moreover, the state of the system immediately after the two measurements is in both cases (if the results are $a_n$ and $b_p$ for $A$ and $B$ respectively):

$$|\psi''_{n,p}\rangle = |\varphi''_{n,p}\rangle = \frac{1}{\sqrt{\sum_i |c_{n,p,i}|^2}} \sum_i c_{n,p,i} \,|a_n, b_p, i\rangle \tag{C-45}$$

New measurements of $A$ or $B$ will yield the same values again without fail.

The preceding discussion thus leads to the following result: when two observables $A$ and $B$ are compatible, the measurement of $B$ does not cause any loss of information previously obtained from a measurement of $A$ (and vice versa) but, on the contrary, adds to it. Moreover, the order of measuring the two observables $A$ and $B$ is of no importance. This last point, furthermore, enables us to envisage the simultaneous measurement of $A$ and $B$. The fourth and fifth postulates can be generalized to the case of such a simultaneous measurement, as can be seen from formulas (C-44) and (C-45). To the result $\{a_n, b_p\}$ correspond the orthonormal eigenvectors $|a_n, b_p, i\rangle$. From this, (C-44) and (C-45) can be seen to be applications of postulates (B-7) and (B-30).

On the other hand, if $A$ and $B$ do not commute, the preceding arguments are no longer valid. To understand this in a simple way, imagine that the state space $\mathscr{E}$ is replaced by the two-dimensional space of real vectors. The vectors $|u_1\rangle$

and $|u_2\rangle$ in figure 3 are eigenvectors of $A$ with eigenvalues $a_1$ and $a_2$ respectively; $|v_1\rangle$ and $|v_2\rangle$ are eigenvectors of $B$ with eigenvalues $b_1$ and $b_2$ respectively. Each of the two sets $\{|u_1\rangle, |u_2\rangle\}$ and $\{|v_1\rangle, |v_2\rangle\}$ forms an orthonormal basis in $\mathscr{E}$. We shall therefore represent them in figure 3 by two pairs of perpen-

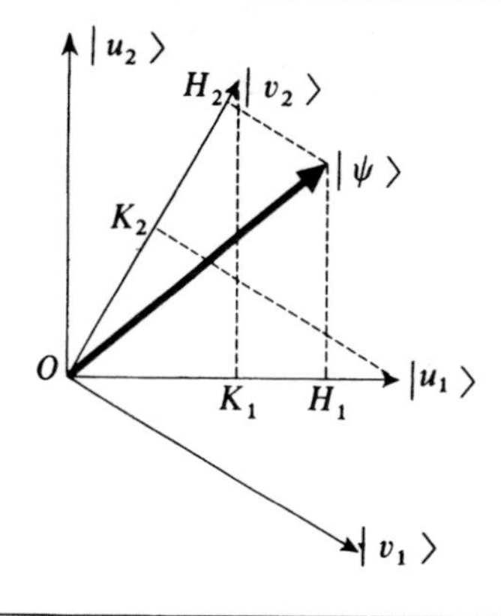


FIGURE 3

**Diagram associated with the successive measurement of two non-compatible observables $A$ and $B$. The state vector of the system is $|\psi\rangle$. The eigenvectors of $A$ are $|u_1\rangle$ and $|u_2\rangle$ (eigenvalues $a_1$ and $a_2$), which are different from those of $B$, $|v_1\rangle$ and $|v_2\rangle$ (eigenvalues $b_1$ and $b_2$).**

dicular unit vectors. The fact that $A$ and $B$ do not commute implies that these two pairs do not coincide. The physical system under study is initially in the normalized state $|\psi\rangle$, which is represented in the figure by an arbitrary unit vector. We measure $A$ and find, for example, $a_1$; the system goes into the state $|u_1\rangle$. We then measure $B$ and find, for example, $b_2$; the state of the system becomes $|v_2\rangle$:

$$|\psi\rangle \overset{(a_1)}{\Longrightarrow} |u_1\rangle \overset{(b_2)}{\Longrightarrow} |v_2\rangle \tag{C-46}$$

If, on the other hand, we perform the measurements in the opposite order, obtaining the same results:

$$|\psi\rangle \overset{(b_2)}{\Longrightarrow} |v_2\rangle \overset{(a_1)}{\Longrightarrow} |u_1\rangle \tag{C-47}$$

The final state of the system is not the same in both cases. We also see from figure 3 that:

$$\begin{aligned} \mathscr{P}(a_1, b_2) &= |OH_1|^2 \times |OK_2|^2 \\ \mathscr{P}(b_2, a_1) &= |OH_2|^2 \times |OK_1|^2 \end{aligned} \tag{C-48}$$

Although $|OK_1| = |OK_2|$, in general $|OH_1| \neq |OH_2|$ and:

$$\mathscr{P}(b_2, a_1) \neq \mathscr{P}(a_1, b_2) \tag{C-49}$$

Therefore: *two incompatible observables cannot be measured simultaneously*. It can be seen from (C-46) and (C-47) that *the second measurement causes the information supplied by the first one to be lost*. If, for example, after the sequence represented in (C-46), we measure $A$ again, we can no longer be sure of the result since $|v_2\rangle$ is not an eigenvector of $A$. All that was gained by the first measurement of $A$ is thus lost.

#### b. PREPARATION OF A STATE

Let us consider a physical system in the state $|\psi\rangle$ and measure the observable $A$ (whose spectrum we assume to be discrete).

If the measurement yields a non-degenerate eigenvalue $a_n$, the state of the system immediately after this measurement is the corresponding eigenvector $|u_n\rangle$. In this case, it suffices to know the result of this measurement to be abľe to determine unambiguously the state of the system after this measurement, as it does not depend on the initial ket. As we have already noted, at the end of § B-3-c, this is due to the fact that $\frac{c_n}{|c_n|}|u_n\rangle$ represents the same physical state as $|u_n\rangle$ itself.

The same does not hold true when the eigenvalue $a_n$ found in the measurement is degenerate. In:

$$|\psi'_n\rangle = \frac{1}{\sqrt{\sum_i |c_n^i|^2}} \sum_{i=1}^{g_n} c_n^i |u_n^i\rangle \tag{C-50}$$

the absolute values of the coefficients $c_n^i$ and their relative phases are significant (§ B-3-b-γ). Since the $c_n^i$ are fixed when the initial state $|\psi\rangle$ is specified, the state $|\psi'_n\rangle$ after the measurement therefore depends on $|\psi\rangle$.

However, we saw in § a that two compatible observables $A$ and $B$ can be measured simultaneously. If the result $(a_n, b_p)$ of this combined measurement corresponds to only one eigenvector $|a_n, b_p\rangle$ common to $A$ and $B$, there is no summation over $i$ in formula (C-37), which becomes:

$$|\psi''_{n,p}\rangle = \frac{c_{n,p}}{|c_{n,p}|} |a_n, b_p\rangle \tag{C-51}$$

This state is physically equivalent to $|a_n, b_p\rangle$. Again, specifying the result of the measurement uniquely determines the final state of the system, which is therefore independent of the initial ket $|\psi\rangle$.

If there are associated with $(a_n, b_p)$ several eigenvectors $|a_n, b_p, i\rangle$ of $A$ and $B$, we can go back to the first argument, measuring, at the same time as $A$ and $B$, a third observable $C$ which is compatible with both of them. We then arrive at the following conclusion: *for the state of the system after a measurement to be completely defined uniquely by the result obtained, this measurement must be made on a complete set of commuting observables* (§ D-3-b of chapter II). This is the property which justifies physically the introduction of the concept of a C.S.C.O.

The methods that can be used to *prepare a system in a well-defined quantum state* are analogous, in principle, to those used to obtain polarized light. When a polarizer is placed in the path of a light beam, the outgoing light is polarized along a direction which is characteristic of the polarizer and therefore independent of the state of polarization of the incoming light. Similarly, we can construct devices, intended to prepare a quantum system, in such a way that they only allow the passage of one state, corresponding to a particular eigenvalue of each of the observables of the complete set chosen. We shall study a concrete example of the preparation of a quantum system in chapter IV (§ B-1).

COMMENT:

The measurement of a C.S.C.O. enables us to prepare only one of the basis states associated with this C.S.C.O. However, it is obvious that changing the set of observables allows us to obtain other states of the system. We shall see explicitly in a concrete example, in §B-1 of chapter IV, that we can prepare in this way any state of the space $\mathscr{E}$.

# D. THE PHYSICAL IMPLICATIONS OF THE SCHRÖDINGER EQUATION

The Schrödinger equation plays a fundamental role in quantum mechanics since, according to the sixth postulate stated above, it is the equation which governs the time evolution of the physical system. In this section, we shall study in detail the most important properties of this equation.

## 1. General properties of the Schrödinger equation

### a. DETERMINISM IN THE EVOLUTION OF PHYSICAL SYSTEMS

The Schrödinger equation:

$$i\hbar \frac{\mathrm{d}}{\mathrm{d}t} | \psi(t) \rangle = H(t) | \psi(t) \rangle \tag{D-1}$$

is of first order in $t$. From this it follows that, given the initial state $| \psi(t_0) \rangle$, the state $| \psi(t) \rangle$ at any subsequent time $t$ is determined. There is no indeterminacy in the time evolution of a quantum system. Indeterminacy appears only when a physical quantity is measured, the state vector then undergoing an unpredictable modification (*cf.* fifth postulate). However, between two measurements, the state vector evolves in a perfectly deterministic way, in accordance with equation (D-1).

### b. THE SUPERPOSITION PRINCIPLE

Equation (D-1) is linear and homogeneous. It follows that its solutions are linearly superposable.

Let $| \psi_1(t) \rangle$ and $| \psi_2(t) \rangle$ be two solutions of (D-1). If the initial state of the system is $| \psi(t_0) \rangle = \lambda_1 | \psi_1(t_0) \rangle + \lambda_2 | \psi_2(t_0) \rangle$ (where $\lambda_1$ and $\lambda_2$ are two complex constants), to it corresponds, at time $t$, the state $| \psi(t) \rangle = \lambda_1 | \psi_1(t) \rangle + \lambda_2 | \psi_2(t) \rangle$. The correspondence between $| \psi(t_0) \rangle$ and $| \psi(t) \rangle$ is therefore linear. We shall later study (complement $F_{III}$) the properties of the linear operator $U(t, t_0)$ which transforms $| \psi(t_0) \rangle$ into $| \psi(t) \rangle$:

$$| \psi(t) \rangle = U(t, t_0) | \psi(t_0) \rangle \tag{D-2}$$

#### c. CONSERVATION OF PROBABILITY

##### α. *The norm of the state vector remains constant*

Since the Hamiltonian operator $H(t)$ which appears in (D-1) is Hermitian, the square of the norm of the state vector, $\langle \psi(t) | \psi(t) \rangle$, does not depend on $t$ as we shall now show :

$$\frac{\mathrm{d}}{\mathrm{d}t} \langle \psi(t) | \psi(t) \rangle = \left[ \frac{\mathrm{d}}{\mathrm{d}t} \langle \psi(t) | \right] | \psi(t) \rangle + \langle \psi(t) | \left[ \frac{\mathrm{d}}{\mathrm{d}t} | \psi(t) \rangle \right] \tag{D-3}$$

According to (D-1), we can write :

$$\frac{\mathrm{d}}{\mathrm{d}t} | \psi(t) \rangle = \frac{1}{i\hbar} H(t) | \psi(t) \rangle \tag{D-4}$$

Taking the Hermitian conjugates of both sides of (D-4), we find :

$$\frac{\mathrm{d}}{\mathrm{d}t} \langle \psi(t) | = -\frac{1}{i\hbar} \langle \psi(t) | H^{\dagger}(t) = -\frac{1}{i\hbar} \langle \psi(t) | H(t) \tag{D-5}$$

since $H(t)$ is Hermitian (it is an observable). Substituting (D-4) and (D-5) into (D-3), we obtain :

$$\begin{aligned} \frac{\mathrm{d}}{\mathrm{d}t} \langle \psi(t) | \psi(t) \rangle &= -\frac{1}{i\hbar} \langle \psi(t) | H(t) | \psi(t) \rangle + \frac{1}{i\hbar} \langle \psi(t) | H(t) | \psi(t) \rangle \\ &= 0 \end{aligned} \tag{D-6}$$

The property of the norm conservation is very useful in quantum mechanics. For example, it becomes indispensable when we interpret the square of the modulus $|\psi(\mathbf{r}, t)|^2$ of the wave function of a spinless particle as being the position probability density. The fact that the state $| \psi(t_0) \rangle$ of the particle is normalized at time $t_0$ is expressed by the relation :

$$\langle \psi(t_0) | \psi(t_0) \rangle = \int d^3r \, |\psi(\mathbf{r}, t_0)|^2 = 1 \tag{D-7}$$

where $\psi(\mathbf{r}, t_0) = \langle \mathbf{r} | \psi(t_0) \rangle$ is the wave function associated with $|\psi(t_0)\rangle$. Equation (D-7) means that the total probability of finding the particle in all space is equal to 1. The property of conservation of the norm which we have just proved is expressed by the equation :

$$\langle \psi(t) | \psi(t) \rangle = \int d^3r \, |\psi(\mathbf{r}, t)|^2 = \langle \psi(t_0) | \psi(t_0) \rangle = 1 \tag{D-8}$$

where $|\psi(t)\rangle$ is the solution of (D-1) which corresponds to the initial state $|\psi(t_0)\rangle$. In other words, time evolution does not modify the global probability of finding the particle in all space, which always remains equal to 1. Thus $|\psi(\mathbf{r}, t)|^2$ can be interpreted as a probability density.

### β. *Local conservation of probability. Probability densities and probability currents*

In this paragraph, we shall confine ourselves to the case of a physical system composed of *only one* (spinless) *particle.*

In this case, if $\psi(\mathbf{r}, t)$ is normalized,

$$\rho(\mathbf{r}, t) = |\psi(\mathbf{r}, t)|^2 \tag{D-9}$$

is a *probability density*: the probability $d\mathscr{P}(\mathbf{r}, t)$ of finding, at time $t$, the particle in an infinitesimal volume $d^3r$ located at the point $\mathbf{r}$ is equal to:

$$d\mathscr{P}(\mathbf{r}, t) = \rho(\mathbf{r}, t)d^3r \tag{D-10}$$

We have just shown that the integral of $\rho(\mathbf{r}, t)$ over all space remains constant for all time (and equal to 1 if $\psi$ is normalized). This does not mean that $\rho(\mathbf{r}, t)$ must be independent of $t$ at every point $\mathbf{r}$. The situation is analogous to the one encountered in electromagnetism. If, in an isolated physical system, there is a charge distributed in space with the volume density $\rho(\mathbf{r}, t)$, the total charge [the integral of $\rho(\mathbf{r}, t)$ over all space] is conserved over time. However, within the system, the spatial distribution of this charge may vary, giving rise to electric currents.

In fact, this analogy can be carried further. Global conservation of electrical charge is based on local conservation. If the charge $Q$ contained within a fixed volume $V$ varies over time, the closed surface $S$ which limits $V$ must be traversed by an electric current. More precisely, the variation $dQ$ during a time $dt$ of the charge contained within $V$ is equal to $-I\,dt$, where $I$ is the intensity of the current which traverses $S$, that is, the flux of the vector current density $\mathbf{J}(\mathbf{r}, t)$ leaving $S$. Using classical vector analysis, we can express local conservation of electrical charge in the form:

$$\frac{\partial}{\partial t}\rho(\mathbf{r}, t) + \operatorname{div}\mathbf{J}(\mathbf{r}, t) = 0 \tag{D-11}$$

We are going to show here that it is possible to find a vector $\mathbf{J}(\mathbf{r}, t)$, a *probability current*, which satisfies an equation identical to (D-11): there is then *local* conservation of probability. Thus, it is as if we were dealing with a "probability fluid" whose density and motion are described by $\rho(\mathbf{r}, t)$ and $\mathbf{J}(\mathbf{r}, t)$. If the probability of finding the particle in the (fixed) volume $d^3r$ about $\mathbf{r}$ varies over time, it means that the probability current has a non-zero flux accross the surface which limits this volume element.

First of all, let us assume that the particle under study is subject to a scalar potential $V(\mathbf{r}, t)$. Its Hamiltonian is then:

$$H = \frac{\mathbf{P}^2}{2m} + V(\mathbf{R}, t) \tag{D-12}$$

and the Schrödinger equation is written, in the $\{ |\mathbf{r}\rangle \}$ representation (see complement $D_{II}$):

$$i\hbar\frac{\partial}{\partial t}\psi(\mathbf{r}, t) = -\frac{\hbar^2}{2m}\Delta\psi(\mathbf{r}, t) + V(\mathbf{r}, t)\,\psi(\mathbf{r}, t) \tag{D-13}$$

$V(\mathbf{r}, t)$ must be real for $H$ to be Hermitian. The complex conjugate equation of (D-13) is therefore :

$$- i\hbar \frac{\partial}{\partial t} \psi^*(\mathbf{r}, t) = - \frac{\hbar^2}{2m} \Delta \psi^*(\mathbf{r}, t) + V(\mathbf{r}, t)\, \psi^*(\mathbf{r}, t) \tag{D-14}$$

Multiply both sides of (D-13) by $\psi^*(\mathbf{r}, t)$ and both sides of (D-14) by $- \psi(\mathbf{r}, t)$. Add the two equations thus obtained. It follows that :

$$i\hbar \frac{\partial}{\partial t} [\psi^*(\mathbf{r}, t)\, \psi(\mathbf{r}, t)] = - \frac{\hbar^2}{2m} [\psi^* \Delta \psi - \psi \Delta \psi^*] \tag{D-15}$$

that is:

$$\frac{\partial}{\partial t} \rho(\mathbf{r}, t) + \frac{\hbar}{2mi} [\psi^*(\mathbf{r}, t)\, \Delta \psi(\mathbf{r}, t) - \psi(\mathbf{r}, t)\, \Delta \psi^*(\mathbf{r}, t)] = 0 \tag{D-16}$$

If we set :

$$\begin{aligned} \mathbf{J}(\mathbf{r}, t) &= \frac{\hbar}{2mi} [\psi^* \nabla \psi - \psi \nabla \psi^*] \\ &= \frac{1}{m} \mathrm{Re}\left[ \psi^* \left( \frac{\hbar}{i} \nabla \psi \right) \right] \end{aligned} \tag{D-17}$$

equation (D-16) can be put into the form of (D-11) since :

$$\begin{aligned} \operatorname{div} \mathbf{J}(\mathbf{r}, t) &= \nabla \cdot \mathbf{J} \\ &= \frac{\hbar}{2mi} [(\nabla \psi^*) \cdot (\nabla \psi) + \psi^*(\nabla^2 \psi) - (\nabla \psi) \cdot (\nabla \psi^*) - \psi(\nabla^2 \psi^*)] \\ &= \frac{\hbar}{2mi} [\psi^* \Delta \psi - \psi \Delta \psi^*] \end{aligned} \tag{D-18}$$

We have therefore proved the equation of local conservation of probability and have found the expression for the probability current in terms of the normalized wave function $\psi(\mathbf{r}, t)$.

COMMENT :

The form of the probability current (D-17) can be interpreted as follows. $\mathbf{J}(\mathbf{r}, t)$ appears as the mean value, in the state $| \psi(t) \rangle$, of an operator $\mathbf{K}(\mathbf{r})$ given by :

$$\mathbf{K}(\mathbf{r}) = \frac{1}{2m} [\, | \mathbf{r} \rangle \langle \mathbf{r} | \mathbf{P} + \mathbf{P} | \mathbf{r} \rangle \langle \mathbf{r} | \,] \tag{D-19}$$

Now the mean value of the operator $| \mathbf{r} \rangle \langle \mathbf{r} |$ is $|\psi(\mathbf{r}, t)|^2$, that is, the probability density $\rho(\mathbf{r}, t)$, and $\frac{\mathbf{P}}{m}$ is the velocity operator $\mathscr{V}$. Therefore, $\mathbf{K}$ is the quantum operator constructed, with the help of an appropriate symmetrization, from the product of the probability density and the velocity of the particle. This indeed corresponds to the vector current density of a classical fluid (it is well known, for example, that the electrical current density associated with a fluid of charged particles is equal to the product of the charge volume density and the velocity of the particles).

If the particle is placed in an electromagnetic field described by the potentials $U(\mathbf{r}, t)$ and $\mathbf{A}(\mathbf{r}, t)$, we can use the preceding argument, starting with the Hamiltonian (B-46). We then find, in this case :

$$\mathbf{J}(\mathbf{r}, t) = \frac{1}{m} \text{Re}\left\{ \psi^* \left[ \frac{\hbar}{i} \nabla - q\mathbf{A} \right] \psi \right\} \tag{D-20}$$

We see that this expression can be obtained from (D-17), using the same rule as was used for the Hamiltonian: $\mathbf{P}$ is simply replaced by $\mathbf{P} - q\mathbf{A}$.

*Example of a plane wave.* Consider a wave function of the form :

$$\psi(\mathbf{r}, t) = A\, e^{i(\mathbf{k}.\mathbf{r} - \omega t)} \tag{D-21}$$

with : $\hbar\omega = \dfrac{\hbar^2 \mathbf{k}^2}{2m}$. The corresponding probability density :

$$\rho(\mathbf{r}, t) = |\psi(\mathbf{r}, t)|^2 = |A|^2 \tag{D-22}$$

is uniform throughout all space and does not depend on time. The calculation of $\mathbf{J}(\mathbf{r}, t)$ from (D-17) presents no difficulties and leads to:

$$\mathbf{J}(\mathbf{r}, t) = |A|^2 \frac{\hbar k}{m} = \rho(\mathbf{r}, t)\, \mathbf{v}_G \tag{D-23}$$

where $\mathbf{v}_G = \dfrac{\hbar \mathbf{k}}{m}$ is the group velocity associated with the momentum $\hbar\mathbf{k}$ (chap. I, §C-4). We see that the probability current is indeed equal to the product of the probability density and the group velocity of the particle. In this case, $\rho$ and $\mathbf{J}$ are time-independent: the flow of the probability fluid associated with a plane wave is in a *steady state* condition (since $\rho$ and $\mathbf{J}$ do not depend on $\mathbf{r}$ either, this state is also homogeneous and uniform).

#### d. EVOLUTION OF THE MEAN VALUE OF AN OBSERVABLE ; RELATIONSHIP WITH CLASSICAL MECHANICS

Let $A$ be an observable. If the state $|\psi(t)\rangle$ of the system is normalized (and we have just seen that this normalization is conserved for all $t$), the mean value of the observable $A$ at the instant $t$ is equal to★:

$$\langle A \rangle(t) = \langle \psi(t) | A | \psi(t) \rangle \tag{D-24}$$

We see that $\langle A \rangle(t)$ depends on $t$ through $|\psi(t)\rangle$ [and $\langle \psi(t)|$], which evolve over time according to the Schrödinger equation (D-4) [and (D-5)]. Moreover, the observable $A$ may depend explicitly on time, causing an additional variation of $\langle A \rangle(t)$ with respect to $t$.

We intend to study, in this section, the evolution of $\langle A \rangle(t)$ and to show how this enables us to relate classical mechanics to quantum mechanics.

★ The notation $\langle A \rangle(t)$ means that the mean value $\langle A \rangle$ of $A$ is a number which depends on $t$.

α. *General formula*

Differentiating (D-24) with respect to $t$, we obtain:

$$\frac{\mathrm{d}}{\mathrm{d}t}\langle \psi(t) | A(t) | \psi(t) \rangle = \left[\frac{\mathrm{d}}{\mathrm{d}t}\langle \psi(t) |\right] A(t) | \psi(t) \rangle + \langle \psi(t) | A(t) \left[\frac{\mathrm{d}}{\mathrm{d}t} | \psi(t) \rangle\right] + \langle \psi(t) | \frac{\partial A}{\partial t} | \psi(t) \rangle \tag{D-25}$$

Using (D-4) and (D-5) for $\frac{\mathrm{d}}{\mathrm{d}t}|\psi(t)\rangle$ and $\frac{\mathrm{d}}{\mathrm{d}t}\langle\psi(t)|$, we find:

$$\frac{\mathrm{d}}{\mathrm{d}t}\langle \psi(t) | A(t) | \psi(t) \rangle = \frac{1}{i\hbar}\langle \psi(t) | [A(t)\, H(t) - H(t)\, A(t)] | \psi(t) \rangle + \langle \psi(t) | \frac{\partial A}{\partial t} | \psi(t) \rangle \tag{D-26}$$

that is:

$$\boxed{\frac{\mathrm{d}}{\mathrm{d}t}\langle A \rangle = \frac{1}{i\hbar}\langle [A, H(t)] \rangle + \langle \frac{\partial A}{\partial t} \rangle} \tag{D-27}$$

COMMENT:

*The mean value* $\langle A \rangle$ *is a number* which depends only on $t$. It is essential to understand how this dependence arises. For example, consider the case of a spinless particle. Let $\mathscr{A}(\mathbf{r}, \mathbf{p}, t)$ be a classical quantity. In classical mechanics, $\mathbf{r}$ and $\mathbf{p}$ depend on time (they evolve according to the Hamilton equations), so that $\mathscr{A}(\mathbf{r}, \mathbf{p}, t)$ depends on $t$ explicitly, and implicitly through $\mathbf{r}$ and $\mathbf{p}$. To the classical quantity $\mathscr{A}(\mathbf{r}, \mathbf{p}, t)$ corresponds the Hermitian operator $A = \mathscr{A}(\mathbf{R}, \mathbf{P}, t)$, obtained by replacing, in $\mathscr{A}$, $\mathbf{r}$ and $\mathbf{p}$ by the operators $\mathbf{R}$ and $\mathbf{P}$ (quantization rules, see § B-5). The eigenstates and eigenvalues of $\mathbf{R}$ and $\mathbf{P}$ and, consequently, these observables themselves, no longer depend on $t$. *The time dependence of* $\mathbf{r}$ *and* $\mathbf{p}$, which characterizes the time evolution of the classical state, *no longer appears in* $\mathbf{R}$ *and* $\mathbf{P}$, *but in the quantum state vector* $|\psi(t)\rangle$, associated in the $\{ |\mathbf{r}\rangle \}$ representation with the wave function $\psi(\mathbf{r}, t) = \langle \mathbf{r} | \psi(t) \rangle$. In this representation, the mean value of $A$ is written:

$$\langle A \rangle = \int d^3r\, \psi^*(\mathbf{r}, t)\, \mathscr{A}\left(\mathbf{r}, \frac{\hbar}{i}\nabla, t\right) \psi(\mathbf{r}, t) \tag{D-28}$$

It is clear that integration over $\mathbf{r}$ leads to a number which only depends on $t$. With regard to classical mechanics, it is this number $\left[\text{and not the operator } \mathscr{A}(\mathbf{r}, \frac{\hbar}{i}\nabla, t)\right]$ which is to be compared with the value taken on by the classical quantity $\mathscr{A}(\mathbf{r}, \mathbf{p}, t)$ at time $t$ (*cf.* § γ below).

β. *Application to the observables* **R** *and* **P** *(Ehrenfest's theorem)*

Now let us apply the general formula (D-27) to the observables **R** and **P**. We shall consider, for simplicity, the case of a spinless particle in a scalar stationary potential $V(\mathbf{r})$. We then have:

$$H = \frac{\mathbf{P}^2}{2m} + V(\mathbf{R}) \tag{D-29}$$

so that we can write:

$$\frac{\mathrm{d}}{\mathrm{d}t}\langle \mathbf{R} \rangle = \frac{1}{i\hbar}\langle [\mathbf{R}, H] \rangle = \frac{1}{i\hbar}\left\langle \left[\mathbf{R}, \frac{\mathbf{P}^2}{2m}\right] \right\rangle \tag{D-30}$$

$$\frac{\mathrm{d}}{\mathrm{d}t}\langle \mathbf{P} \rangle = \frac{1}{i\hbar}\langle [\mathbf{P}, H] \rangle = \frac{1}{i\hbar}\langle [\mathbf{P}, V(\mathbf{R})] \rangle \tag{D-31}$$

The commutator which appears in (D-30) can easily be calculated from the canonical commutation relations; we obtain:

$$\left[\mathbf{R}, \frac{\mathbf{P}^2}{2m}\right] = \frac{i\hbar}{m}\mathbf{P} \tag{D-32}$$

For the one in formula (D-31), the following generalization of formula (B-33) must be used (*cf.* complement $\mathrm{B_{II}}$, formula (48)]:

$$[\mathbf{P}, V(\mathbf{R})] = - i\hbar \nabla V(\mathbf{R}) \tag{D-33}$$

where $\nabla V(\mathbf{R})$ denotes the set of three operators obtained by replacing **r** by **R** in the three components of the gradient of the function $V(\mathbf{r})$. Therefore:

$$\boxed{\begin{aligned} &\frac{\mathrm{d}}{\mathrm{d}t}\langle \mathbf{R} \rangle = \frac{1}{m}\langle \mathbf{P} \rangle && \text{(D-34)} \\ &\frac{\mathrm{d}}{\mathrm{d}t}\langle \mathbf{P} \rangle = - \langle \nabla V(\mathbf{R}) \rangle && \text{(D-35)} \end{aligned}}$$

These two equations express *Ehrenfest's theorem*. Their form recalls that of the classical Hamilton-Jacobi equations for a particle (appendix III, § 3):

$$\frac{\mathrm{d}}{\mathrm{d}t}\mathbf{r} = \frac{1}{m}\mathbf{p} \tag{D-36-a}$$

$$\frac{\mathrm{d}}{\mathrm{d}t}\mathbf{p} = - \nabla V(\mathbf{r}) \tag{D-36-b}$$

which reduce, in this simple case, to Newton's well-known equation:

$$\frac{\mathrm{d}\mathbf{p}}{\mathrm{d}t} = m\frac{\mathrm{d}^2\mathbf{r}}{\mathrm{d}t^2} = - \nabla V(\mathbf{r}) \tag{D-37}$$

γ. *Discussion of Ehrenfest's theorem; classical limit*

Let us analyze the physical meaning of Ehrenfest's theorem, that is, equations (D-34) and (D-35). We shall assume that the wave function $\psi(\mathbf{r}, t)$ which describes the state of the particle is a wave packet like the ones we studied in chapter I. $\langle \mathbf{R} \rangle$ then represents a set of three time-dependent numbers $\{ \langle X \rangle, \langle Y \rangle, \langle Z \rangle \}$. We shall call the point $\langle \mathbf{R} \rangle(t)$ the *center of the wave packet*★ at the instant $t$. The set of those points which correspond to the various values of $t$ constitutes the *trajectory followed by the center of the wave packet*. Recall, however, that one can never rigorously speak of the trajectory of the particle itself, whose state is described by the wave packet as a whole, which inevitably has a certain spatial extension. We see, nevertheless, that if this extension is much smaller than the other distances involved in the problem, we can approximate the wave packet by its center. In this limiting case, there is no appreciable difference between the quantum and classical descriptions of the particle.

It is therefore important to know the answer to the following question: does the motion of the center of the wave packet obey the laws of classical mechanics? This answer is supplied by Ehrenfest's theorem. Equation (D-34) expresses the fact that the velocity of the center of the wave packet is equal to the average momentum of this wave packet divided by $m$. Consequently, the left-hand side of (D-35) can be written $m \frac{\mathrm{d}^2}{\mathrm{d}t^2} \langle \mathbf{R} \rangle$, so that the answer to the preceding question will be affirmative if the right-hand side of (D-35) is equal to the classical force $\mathbf{F}_{cl}$ at the point where the center of the wave packet is situated:

$$\mathbf{F}_{cl} = [- \nabla V(\mathbf{r})]_{\mathbf{r}=\langle \mathbf{R} \rangle} \tag{D-38}$$

In fact, the right-hand side of (D-35) is equal to the average of the force over the whole wave packet, and, in general:

$$\langle \nabla V(\mathbf{R}) \rangle \neq [\nabla V(\mathbf{r})]_{\mathbf{r}=\langle \mathbf{R} \rangle} \tag{D-39}$$

(in other words, the mean value of a function is not equal to its value for the mean value of the variable). If we are rigorous, the answer to the question we asked is therefore negative.

COMMENT:

It is easy to convince ourselves of (D-39) if we consider a concrete example. Let us choose, for simplicity, a one-dimensional model, and assume that:

$$V(x) = \lambda x^n \tag{D-40}$$

where $\lambda$ is a real constant and $n$, a positive integer. From this we deduce the operator associated with $V(x)$:

$$V(X) = \lambda X^n \tag{D-41}$$

The left-hand side of (D-39) can be written $\left(\text{replacing } \nabla \text{ by } \frac{\mathrm{d}}{\mathrm{d}x}\right) \lambda n \langle X^{n-1} \rangle$. As for the right-hand side, it is equal to:

$$\left[\frac{\mathrm{d}V}{\mathrm{d}x}\right]_{x=\langle X \rangle} = [\lambda n \; x^{n-1}]_{x=\langle X \rangle} = \lambda n \langle X \rangle^{n-1} \tag{D-42}$$

★ The center and the maximum of a wave packet are, in general, distinct. They coincide, however, if the wave packet has a symmetrical shape (§C-5, fig. 2).

Now we know that in general $\langle X^{n-1} \rangle \neq \langle X \rangle^{n-1}$; for example, for $n = 3$, we have $\langle X^2 \rangle \neq \langle X \rangle^2$ (since the difference between these two quantities enters into the calculation of the root-mean-square deviation $\Delta X$).

Note however that for $n = 1$ or 2, $\langle X^{n-1} \rangle = \langle X \rangle^{n-1}$. The two sides of (D-39) are then equal. The same holds true, moreover, for $n = 0$, in which case both sides are equal to zero. For a free particle ($n = 0$), or a particle placed in a uniform force field ($n = 1$) or in a parabolic potential well ($n = 2$; the case of a harmonic oscillator), the motion of the center of the wave packet therefore rigorously obeys the laws of classical mechanics. We have already, moreover, established this result for the free particle ($n = 0$) (*cf.* chapter I, §C-4).

Although the two sides of (D-39) are not, in general, equal, there exist situations (called quasi-classical) where the difference between these two quantities is negligible: this is the case when the wave packet is sufficiently localized. To see this, let us write explicitly, in the $\{ |\mathbf{r}\rangle \}$ representation, the left-hand side of this equation:

$$\begin{aligned} \langle \nabla V(\mathbf{R}) \rangle &= \int d^3r\, \psi^*(\mathbf{r}, t)\, [\nabla V(\mathbf{r})]\, \psi(\mathbf{r}, t) \\ &= \int d^3r\, |\psi(\mathbf{r}, t)|^2\, \nabla V(\mathbf{r}) \end{aligned} \tag{D-43}$$

Let us assume the wave packet to be highly localized: more precisely, $|\psi(\mathbf{r}, t)|^2$ takes on non-negligible values only within a domain whose dimensions are much smaller than the distances over which $V(\mathbf{r})$ varies appreciably. Then, within this domain, centered about $\langle \mathbf{R} \rangle$, $\nabla V(\mathbf{r})$ is practically constant. Therefore, in (D-43), $\nabla V(\mathbf{r})$ can be replaced by its value for $\mathbf{r} = \langle \mathbf{R} \rangle$ and taken outside the integral, which is then equal to 1, since $\psi(\mathbf{r}, t)$ is normalized. Thus we find that for sufficiently localized wave packets:

$$\langle \nabla V(\mathbf{R}) \rangle \simeq [\nabla V(\mathbf{r})]_{\mathbf{r} = \langle \mathbf{R} \rangle} \tag{D-44}$$

In the macroscopic limit (where the de Broglie wavelengths are much smaller than the distances over which the potential varies*), wave packets can be made sufficiently small to satisfy (D-44) while retaining a good degree of definition for the momentum. The motion of the wave packet is then practically that of a classical particle of mass $m$ placed in the potential $V(\mathbf{r})$. The result that we have thus established is very important since it enables us to show that the equations of classical mechanics follow from the Schrödinger equation in certain limiting conditions satisfied, in particular, by most macroscopic systems.

## 2. The case of conservative systems

When the Hamiltonian of a physical system does not depend explicitly on time, the system is said to be *conservative*. In classical mechanics, the most important consequence of such a situation is the *conservation of energy* over time. It can

* See the order of magnitude of the de Broglie wavelengths associated with a macroscopic system in complement $A_I$.

also be said that the total energy of the system is a *constant of the motion*. We shall see in this section that in quantum mechanics as well, conservative systems possess important special properties in addition to the general properties of the preceding section.

### a. SOLUTION OF THE SCHRÖDINGER EQUATION

First, let us consider the eigenvalue equation for $H$:

$$H \mid \varphi_{n,\tau} \rangle = E_n \mid \varphi_{n,\tau} \rangle \tag{D-45}$$

For simplicity, we assume the spectrum of $H$ to be discrete. $\tau$ denotes the set of indices other than $n$ which are necessary for characterizing a unique vector $\mid \varphi_{n,\tau} \rangle$ (in general, these indices will fix the eigenvalues of operators which form a C.S.C.O. with $H$). Since, by hypothesis, $H$ does not depend explicitly on time, neither the eigenvalue $E_n$ nor the eigenket $\mid \varphi_{n,\tau} \rangle$ is $t$ dependent.

First, we are going to show that given the $E_n$ and the $\mid \varphi_{n,\tau} \rangle$, it is very simple to solve the Schrödinger equation, that is, to determine the time evolution of any state. Since the $\mid \varphi_{n,\tau} \rangle$ form a basis ($H$ is an observable), it is always possible, for every value of $t$, to expand any state $\mid \psi(t) \rangle$ of the system in terms of the $\mid \varphi_{n,\tau} \rangle$:

$$\mid \psi(t) \rangle = \sum_{n,\tau} c_{n,\tau}(t) \mid \varphi_{n,\tau} \rangle \tag{D-46}$$

with:

$$c_{n,\tau}(t) = \langle \varphi_{n,\tau} \mid \psi(t) \rangle \tag{D-47}$$

Since the $\mid \varphi_{n,\tau} \rangle$ do not depend on $t$, all the time dependence of $\mid \psi(t) \rangle$ is contained within the $c_{n,\tau}(t)$. To calculate the $c_{n,\tau}(t)$, let us project the Schrödinger equation onto each of the states $\mid \varphi_{n,\tau} \rangle$. This yields*:

$$i\hbar \frac{\mathrm{d}}{\mathrm{d}t} \langle \varphi_{n,\tau} \mid \psi(t) \rangle = \langle \varphi_{n,\tau} \mid H \mid \psi(t) \rangle \tag{D-48}$$

Since $H$ is Hermitian, it can be deduced from (D-45) that:

$$\langle \varphi_{n,\tau} \mid H = E_n \langle \varphi_{n,\tau} \mid \tag{D-49}$$

so that (D-48) can be written in the form:

$$i\hbar \frac{\mathrm{d}}{\mathrm{d}t} c_{n,\tau}(t) = E_n \, c_{n,\tau}(t) \tag{D-50}$$

This equation can be integrated directly to give:

$$c_{n,\tau}(t) = c_{n,\tau}(t_0) \, \mathrm{e}^{-iE_n(t-t_0)/\hbar} \tag{D-51}$$

* In (D-48), $\langle \varphi_{n,\tau} \mid$ can be placed to the right of $\frac{\mathrm{d}}{\mathrm{d}t}$, since $\langle \varphi_{n,\tau} \mid$ does not depend on $t$.

When $H$ does not depend explicitly on time, to find $|\psi(t)\rangle$, given $|\psi(t_0)\rangle$, proceed as follows :

(*i*) Expand $|\psi(t_0)\rangle$ in terms of a basis of eigenstates of $H$:

$$|\psi(t_0)\rangle = \sum_n \sum_\tau c_{n,\tau}(t_0) \, |\varphi_{n,\tau}\rangle \tag{D-52}$$

$c_{n,\tau}(t_0)$ is given by the usual formula:

$$c_{n,\tau}(t_0) = \langle \varphi_{n,\tau} | \psi(t_0) \rangle \tag{D-53}$$

(*ii*) Now, to obtain $|\psi(t)\rangle$ for arbitrary $t$, multiply each coefficient $c_{n,\tau}(t_0)$ of the expansion (D-52) by $e^{-iE_n(t-t_0)/\hbar}$, where $E_n$ is the eigenvalue of $H$ associated with the state $|\varphi_{n,\tau}\rangle$:

$$|\psi(t)\rangle = \sum_n \sum_\tau c_{n,\tau}(t_0) \, e^{-iE_n(t-t_0)/\hbar} \, |\varphi_{n,\tau}\rangle \tag{D-54}$$

The preceding argument can easily be generalized to the case where the spectrum of $H$ is continuous; formula (D-54) then becomes, with obvious notation:

$$|\psi(t)\rangle = \sum_\tau \int \mathrm{d}E \, c_\tau(E, t_0) \, e^{-iE(t-t_0)/\hbar} \, |\varphi_{E,\tau}\rangle \tag{D-55}$$

### b. STATIONARY STATES

An important special case is that in which $|\psi(t_0)\rangle$ is itself an eigenstate of $H$. Expansion (D-52) of $|\psi(t_0)\rangle$ then involves only eigenstates of $H$ with the same eigenvalue (for example, $E_n$):

$$|\psi(t_0)\rangle = \sum_\tau c_{n,\tau}(t_0) \, |\varphi_{n,\tau}\rangle \tag{D-56}$$

In formula (D-56), there is no summation over $n$, and the passage from $|\psi(t_0)\rangle$ to $|\psi(t)\rangle$ involves only one factor $e^{-iE_n(t-t_0)/\hbar}$, which can be taken outside the summation over $\tau$:

$$\begin{aligned} |\psi(t)\rangle &= \sum_\tau c_{n,\tau}(t_0) \, e^{-iE_n(t-t_0)/\hbar} \, |\varphi_{n,\tau}\rangle \\ &= e^{-iE_n(t-t_0)/\hbar} \sum_\tau c_{n,\tau}(t_0) \, |\varphi_{n,\tau}\rangle \\ &= e^{-iE_n(t-t_0)/\hbar} \, |\psi(t_0)\rangle \end{aligned} \tag{D-57}$$

$|\psi(t)\rangle$ and $|\psi(t_0)\rangle$ therefore differ only by the *global* phase factor $e^{-iE_n(t-t_0)/\hbar}$ These two states are physically indistinguishable (*cf.* discussion in §B-3-b-γ). From this we conclude that all the physical properties of a system which is in an eigenstate of $H$ do not vary over time; the eigenstates of $H$ are called, for this reason, *stationary states*.

It is also interesting to see how conservation of energy in a conservative system appears in quantum mechanics. Let us assume that, at time $t_0$, we measure the energy of such a system and we find, for example, $E_k$. Immediately after the measurement, the system is in an eigenstate of $H$, with an eigenvalue of $E_k$ (the postulate of the reduction of the wave packet). We have just seen that the eigenstates of $H$ are stationary states. Therefore, the state of the system will no longer evolve after the first measurement and will always remain an eigenstate of $H$ with an eigenvalue of $E_k$. It follows that a second measurement of the energy of the system, at any subsequent time $t$, will always yield the same result $E_k$ as the first one.

COMMENT:

One passes from (D-52) to (D-54) by multiplying each coefficient $c_{n,\tau}(t_0)$ of (D-52) by $e^{-iE_n(t-t_0)/\hbar}$. The fact that $e^{-iE_n(t-t_0)/\hbar}$ is a phase factor should not lead us to believe that $|\psi(t)\rangle$ and $|\psi(t_0)\rangle$ are always physically indistinguishable. Actually, expansion (D-52) involves, in general, *several eigenstates of H with different eigenvalues*. To these different possible values of $E_n$ correspond *different phase factors*. This modifies the *relative phases* of the expansion coefficients of the state vector and leads, consequently, to a state $|\psi(t)\rangle$ which is physically distinct from $|\psi(t_0)\rangle$.

Only in the case where only one value of $n$ enters into (D-52) [the case where $|\psi(t_0)\rangle$ is an eigenstate of $H$] is the time evolution described by a single phase factor, which is then a global one, of no physical importance. In other words, there is physical evolution over time only if the energy of the initial state is not known with certainty. We shall come back later to the relation between time evolution and energy uncertainty (*cf.* § D-2-e).

### c. CONSTANTS OF THE MOTION

By definition, a constant of the motion is an observable $A$ which does not depend explicitly on time and which commutes with $H$:

$$\begin{cases} \dfrac{\partial A}{\partial t} = 0 \\ [A, H] = 0 \end{cases} \tag{D-58}$$

For a conservative system, $H$ is therefore itself a constant of the motion.

Constants of the motion possess important properties which we are now going to derive.

($i$) If we substitute (D-58) into the general formula (D-27), we find:

$$\frac{d}{dt}\langle A \rangle = \frac{d}{dt}\langle \psi(t) | A | \psi(t) \rangle = 0 \tag{D-59}$$

Whatever the state $|\psi(t)\rangle$ of the physical system, *the mean value of A* in this state *does not evolve over time* (hence the term "constant of the motion").

(*ii*) Since $A$ and $H$ are two observables which commute, we can always find for them a system of common eigenvectors, which we shall denote by $\{ | \varphi_{n,p,\tau} \rangle \}$:

$$\begin{aligned} H \,|\, \varphi_{n,p,\tau} \rangle &= E_n \,|\, \varphi_{n,p,\tau} \rangle \\ A \,|\, \varphi_{n,p,\tau} \rangle &= a_p \,|\, \varphi_{n,p,\tau} \rangle \end{aligned} \tag{D-60}$$

We shall assume for simplicity that the spectra of $H$ and $A$ are discrete. The index $\tau$ fixes the eigenvalues of observables which form a C.S.C.O. with $H$ and $A$. Since the states $| \varphi_{n,p,\tau} \rangle$ are eigenstates of $H$, they are stationary states. If the system is in the state $| \varphi_{n,p,\tau} \rangle$ at the initial instant, it will therefore remain there indefinitely (to within a global phase factor). But the state $| \varphi_{n,p,\tau} \rangle$ is an eigenstate of $A$ as well. When $A$ is a constant of the motion, there therefore exist stationary states of the physical system (the states $| \varphi_{n,p,\tau} \rangle$) which always remain, for all $t$, eigenstates of $A$ with the same eigenvalue ($a_p$). The eigenvalues of $A$ are called, for this reason, *good quantum numbers*.

(*iii*) Finally, let us show that for an arbitrary state $| \psi(t) \rangle$, *the probability of finding the eigenvalue* $a_p$, when the constant of the motion $A$ is measured, *is not time-dependent*. $| \psi(t_0) \rangle$ can always be expanded on the $\{ | \varphi_{n,p,\tau} \rangle \}$ basis introduced above:

$$| \psi(t_0) \rangle = \sum_n \sum_p \sum_\tau c_{n,p,\tau}(t_0) \,|\, \varphi_{n,p,\tau} \rangle \tag{D-61}$$

From this we directly deduce :

$$| \psi(t) \rangle = \sum_n \sum_p \sum_\tau c_{n,p,\tau}(t) \,|\, \varphi_{n,p,\tau} \rangle \tag{D-62}$$

with:

$$c_{n,p,\tau}(t) = c_{n,p,\tau}(t_0) \, \mathrm{e}^{-iE_n(t-t_0)/\hbar} \tag{D-63}$$

According to the postulate of spectral decomposition, the probability $\mathscr{P}(a_p, t_0)$ of finding $a_p$ when $A$ is measured at time $t_0$, on the system in the state $| \psi(t_0) \rangle$, is equal to:

$$\mathscr{P}(a_p, t_o) = \sum_n \sum_\tau |c_{n,p,\tau}(t_0)|^2 \tag{D-64}$$

Similarly:

$$\mathscr{P}(a_p, t) = \sum_n \sum_\tau |c_{n,p,\tau}(t)|^2 \tag{D-65}$$

Now we see from (D-63) that $c_{n,p,\tau}(t)$ and $c_{n,p,\tau}(t_0)$ have the same modulus. Therefore, $\mathscr{P}(a_p, t) = \mathscr{P}(a_p, t_0)$, which proves the property stated above.

COMMENT:

If all but one of the probabilities $\mathscr{P}(a_p, t_0)$ are zero [leaving for example $\mathscr{P}(a_k, t_0)$ non-zero and, moreover, necessarily equal to 1], the physical system at time $t_0$ is in an eigenstate of $A$ with an eigenvalue of $a_k$. Since the $\mathscr{P}(a_p, t)$ do not depend on $t$, the state of the system at time $t$ remains an eigenstate of $A$ with an eigenvalue of $a_k$.

### d. BOHR FREQUENCIES OF A SYSTEM. SELECTION RULES

Let $B$ be an arbitrary observable of the system under consideration (it does not necessarily commute with $H$). Formula (D-27) enables us to calculate the derivative $\frac{\mathrm{d}}{\mathrm{d}t}\langle B \rangle$ of the mean value of $B$:

$$\frac{\mathrm{d}}{\mathrm{d}t}\langle B \rangle = \frac{1}{i\hbar}\langle [B, H] \rangle + \langle \frac{\partial B}{\partial t} \rangle \qquad \text{(D-66)}$$

For a conservative system, we know the general form (D-54) of $|\psi(t)\rangle$. Therefore, in this case, we can calculate explicitly $\langle \psi(t)|B|\psi(t)\rangle$ $\left(\text{and not merely } \frac{\mathrm{d}}{\mathrm{d}t}\langle B \rangle\right)$

The Hermitian conjugate expression of (D-54) is written (changing the summation indices):

$$\langle \psi(t) | = \sum_{n'}\sum_{\tau'} c^*_{n',\tau'}(t_0)\; \mathrm{e}^{iE_{n'}(t-t_0)/\hbar} \langle \varphi_{n',\tau'} | \qquad \text{(D-67)}$$

We can then, in $\langle \psi(t)|B|\psi(t)\rangle$, replace $|\psi(t)\rangle$ and $\langle \psi(t)|$ by expansions (D-54) and (D-67), respectively. Thus we obtain:

$$\begin{aligned}\langle \psi(t) | B | \psi(t) \rangle &= \langle B \rangle (t) \\ &= \sum_n\sum_\tau\sum_{n'}\sum_{\tau'} c^*_{n',\tau'}(t_0)\; c_{n,\tau}(t_0) \langle \varphi_{n',\tau'} | B | \varphi_{n,\tau} \rangle\, \mathrm{e}^{i(E_{n'}-E_n)(t-t_0)/\hbar}\end{aligned} \qquad \text{(D-68)}$$

From now on, we shall assume that $B$ does not depend explicitly on time: the matrix elements $\langle \varphi_{n',\tau'}|B|\varphi_{n,\tau}\rangle$ are therefore constant. Formula (D-68) then shows that the evolution of $\langle B \rangle(t)$ is described by a series of *oscillating terms*, whose frequencies $\frac{1}{2\pi}\frac{|E_{n'}-E_n|}{\hbar} = \left|\frac{E_{n'}-E_n}{h}\right| = \nu_{n'n}$ are characteristic of the system under consideration but independent of $B$ and of the initial state of the system. The frequencies $\nu_{n'n}$ are called the *Bohr frequencies* of the system. Thus, for an atom, the mean values of all the atomic quantities (electric and magnetic dipole moments, etc...) oscillate at the various Bohr frequencies of the atom. It is reasonable to imagine that only these frequencies can be radiated or absorbed by the atom. This remark allows us to understand intuitively the Bohr relation between the spectral frequencies emitted or absorbed and the differences in atomic energies.

It can also be seen from (D-68) that, while the frequencies involved in the motion of $\langle B \rangle(t)$ are independent of $B$, the same does not hold true for the respective *weights* of these frequencies in the variation of $\langle B \rangle$. The importance of each frequency $\nu_{n'n}$ depends on the matrix elements $\langle \varphi_{n',\tau'} | B | \varphi_{n,\tau} \rangle$. In particular, if these matrix elements are zero for certain values of $n$ and $n'$, the corresponding frequencies $\nu_{n'n}$ are absent from the expansion of $\langle B \rangle(t)$, whatever the initial state of the system. This is the origin of the *selection rules* which indicate what frequencies can be emitted or absorbed under given conditions. To establish these rules, one must study the non-diagonal matrix elements ($n \neq n'$) of the various atomic operators such as the electric and magnetic dipoles, etc...

Finally, the weights of the various Bohr frequencies also depend on the initial state, via $c^*_{n',\tau'}(t_0)c_{n,\tau}(t_0)$. In particular, if the initial state is a stationary state of energy $E_k$, the expansion of $| \psi(t_0) \rangle$ contains only one value of $n$ ($n = k$) and $c^*_{n',\tau'}(t_0)c_{n,\tau}(t_0)$ can be non-zero only for $n = n' = k$. In this case, $\langle B \rangle$ is not time-dependent.

COMMENT:

It can be directly verified, using (D-68), that the mean value of a constant of the motion is always time-independent. We see that if $B$ commutes with $H$, the matrix elements of $B$ are zero between two eigenstates of $H$ which correspond to different eigenvalues (*cf.* chap. II, § D-3-a). It follows that $\langle \varphi_{n',\tau'} | B | \varphi_{n,\tau} \rangle$ is zero for $n' \neq n$. The only terms of $B$ which are non-zero are thus constant.

### e. THE TIME-ENERGY UNCERTAINTY RELATION

We shall now see that for a conservative system, the greater the energy uncertainty, the more rapid the time evolution. More precisely, if $\Delta t$ is a time interval at the end of which the system has evolved to an appreciable extent, and if $\Delta E$ denotes the energy uncertainty, $\Delta t$ and $\Delta E$ satisfy the relation:

$$\boxed{\Delta t \,.\, \Delta E \gtrsim \hbar} \tag{D-69}$$

First, if the system is in an eigenstate of $H$, its energy is perfectly well-defined : $\Delta E = 0$. But we have seen that such a state is stationary; that is, it does not evolve. It can be said that, in this case, the evolution time $\Delta t$, is, in a sense, infinite [relation (D-69) indicates that when $\Delta E = 0$, $\Delta t$ must be infinite].

Now let us assume that $| \psi(t_0) \rangle$ is a linear superposition of two eigenstates of $H$, $| \varphi_1 \rangle$ and $| \varphi_2 \rangle$, with different eigenvalues $E_1$ and $E_2$:

$$| \psi(t_0) \rangle = c_1 | \varphi_1 \rangle + c_2 | \varphi_2 \rangle \tag{D-70}$$

Then:

$$| \psi(t) \rangle = c_1 \, e^{-iE_1(t-t_0)/\hbar} | \varphi_1 \rangle + c_2 \, e^{-iE_2(t-t_0)/\hbar} | \varphi_2 \rangle \tag{D-71}$$

If we measure the energy, we find either $E_1$ or $E_2$. The uncertainty of $E$ is therefore of the order of:

$$\Delta E \simeq |E_2 - E_1| \tag{D-72}$$

Now consider an arbitrary observable $B$ which does not commute with $H$. The probability of finding, in a measurement of $B$ at time $t$, the eigenvalue $b_m$ associated with the eigenvector $| u_m \rangle$ (we assume, for simplicity, $b_m$ to be non-degenerate) is given by:

$$\begin{aligned} \mathscr{P}(b_m, t) = |\langle u_m | \psi(t) \rangle|^2 &= |c_1|^2 \, |\langle u_m | \varphi_1 \rangle|^2 + |c_2|^2 \, |\langle u_m | \varphi_2 \rangle|^2 \\ &+ 2 \, \mathrm{Re} \, [c_2^* c_1 \, e^{i(E_2 - E_1)(t-t_0)/\hbar} \langle u_m | \varphi_2 \rangle^* \langle u_m | \varphi_1 \rangle] \end{aligned} \tag{D-73}$$

This equation shows that $\mathscr{P}(b_m, t)$ oscillates between two extreme values, with the Bohr frequency $\nu_{21} = \frac{|E_2 - E_1|}{h}$. The characteristic evolution time of the system is therefore:

$$\Delta t \simeq \frac{h}{|E_2 - E_1|} \tag{D-74}$$

and comparison with (D-72) shows that: $\Delta E \, . \, \Delta t \simeq h$.

Let us now assume that the spectrum of $H$ is continuous (and non-degenerate). The most general state $|\psi(t_0)\rangle$ can be written:

$$|\psi(t_0)\rangle = \int \mathrm{d}E \;\; c(E) \, |\varphi_E\rangle \tag{D-75}$$

where $|\varphi_E\rangle$ is the eigenstate of $H$ with the eigenvalue $E$. Let us assume that $|c(E)|^2$ has non-negligible values only in a domain of width $\Delta E$ about $E_0$ (fig. 4). $\Delta E$ then

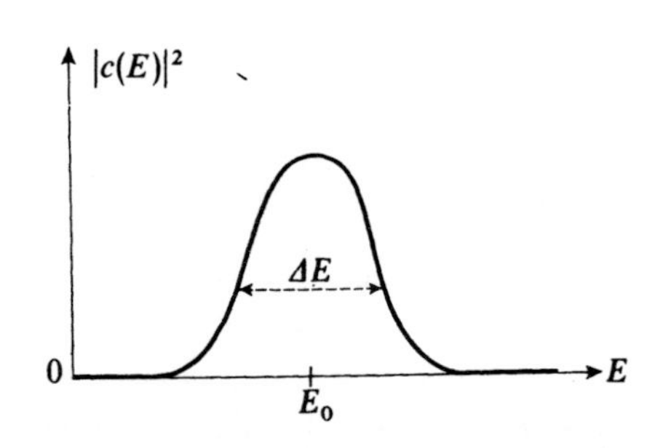


FIGURE 4

**By superposing the stationary states $|\varphi_E\rangle$ with the coefficients $c(E)$, we obtain a state $|\psi\rangle$ of the system where the energy is not perfectly well-defined. The corresponding uncertainty $\Delta E$ is given by the width of the curve which represents $|c(E)|^2$. According to the fourth uncertainty relation, the evolution of the state $|\psi(t)\rangle$ will be significant after a time $\Delta t$ such that $\Delta E \, . \, \Delta t \gtrsim \hbar$.**

represents the uncertainty of the energy of the system. $|\psi(t)\rangle$ is obtained by using (D-55):

$$|\psi(t)\rangle = \int \mathrm{d}E \;\; c(E) \, \mathrm{e}^{-iE(t-t_0)/\hbar} \, |\varphi_E\rangle \tag{D-76}$$

The quantity $\mathscr{P}(b_m, t)$ introduced above, which represents the probability of finding the eigenvalue $b_m$ when the observable $B$ is measured on the system in the state $|\psi(t)\rangle$, is here equal to:

$$\mathscr{P}(b_m, t) = |\langle u_m | \psi(t) \rangle|^2 = \left| \int \mathrm{d}E \;\; c(E) \mathrm{e}^{-iE(t-t_0)/\hbar} \langle u_m | \varphi_E \rangle \right|^2 \tag{D-77}$$

In general, $\langle u_m | \varphi_E \rangle$ does not vary rapidly with $E$ when $E$ varies about $E_0$. If $\Delta E$ is sufficiently small, the variation of $\langle u_m | \varphi_E \rangle$, in integral (D-77), can be neglected relative to that of $c(E)$. One can then replace $\langle u_m | \varphi_E \rangle$ by $\langle u_m | \varphi_{E_0} \rangle$ and take this quantity outside integral (D-77):

$$\mathscr{P}(b_m, t) \simeq |\langle u_m | \varphi_{E_0} \rangle|^2 \left| \int \mathrm{d}E \;\; c(E) \, \mathrm{e}^{-iE(t-t_0)/\hbar} \right|^2 \tag{D-78}$$

If this approximation is valid, we thus see that $\mathscr{P}(b_m, t)$ is, to within a coefficient, the square of the modulus of the Fourier transform of $c(E)$. According to the properties of the Fourier transform (*cf.* appendix I, §2-b), the width in $t$ of $\mathscr{P}(b_m, t)$, that is, $\Delta t$, is therefore related to the width $\Delta E$ of $|c(E)|^2$ by relation (D-69).

COMMENT:

(D-69) can be established directly for a free one-dimensional wave packet. One can associate with the momentum uncertainty $\Delta p$ of this wave packet an energy uncertainty $\Delta E = \frac{\mathrm{d}E}{\mathrm{d}p}\Delta p$. Since $E = \hbar\omega$ and $p = \hbar k$, we have $\frac{\mathrm{d}E}{\mathrm{d}p} = \frac{\mathrm{d}\omega}{\mathrm{d}k} = v_G$, where $v_G$ is the group velocity of the wave packet (chap. I, § C-4). Consequently:

$$\Delta E = v_G \, \Delta p \tag{D-79}$$

Now the characteristic evolution time $\Delta t$ is the time taken by this wave packet, travelling at the velocity $v_G$, to "pass" a point in space. If $\Delta x$ is the spatial extension of the wave packet, we therefore have:

$$\Delta t \simeq \frac{\Delta x}{v_G} \tag{D-80}$$

From this we deduce, combining (D-79) and (D-80):

$$\Delta E \, . \, \Delta t \simeq \Delta x \, . \, \Delta p \gtrsim \hbar \tag{D-81}$$

Relation (D-69) is often called the *fourth Heisenberg uncertainty relation*. It is clearly different, however, from the other three uncertainty relations which relate to the three components of **R** and **P** [formulas (14) of complement $F_I$]. In (D-69), only the energy is a physical quantity like **R** and **P**; $t$, on the other hand, *is a parameter*, with which no quantum mechanical operator is associated.

## E. THE SUPERPOSITION PRINCIPLE AND PHYSICAL PREDICTIONS

The physical meaning of the first postulate remains to be examined. According to this postulate, the states of a physical system belong to a vector space and are, consequently, linearly superposable.

One of the important consequences of the first postulate, when it is combined with the others, is the appearance of interference effects such as those which led us to wave-particle duality (chap. I). Our understanding of these phenomena is based on the concept of probability amplitudes, which we shall examine here with the aid of some simple examples.

## 1. Probability amplitudes and interference effects

### a. THE PHYSICAL MEANING OF A LINEAR SUPERPOSITION OF STATES

#### α. *The difference between a linear superposition and a statistical mixture*

Let $| \psi_1 \rangle$ and $| \psi_2 \rangle$ be two orthogonal normalized states:

$$\langle \psi_1 | \psi_1 \rangle = \langle \psi_2 | \psi_2 \rangle = 1$$
$$\langle \psi_1 | \psi_2 \rangle = 0 \tag{E-1}$$

($| \psi_1 \rangle$ and $| \psi_2 \rangle$ could be, for example, two eigenstates of the same observable $B$ associated with two different eigenvalues $b_1$ and $b_2$).

If the system is in the state $| \psi_1 \rangle$, we can calculate all the probabilities concerning the measurement results for a given observable $A$. For example, if $| u_n \rangle$ is the (normalized) eigenvector of $A$ which corresponds to the eigenvalue $a_n$ (assumed to be non-degenerate), the probability $\mathscr{P}_1(a_n)$ of finding $a_n$ when $A$ is measured on the system in the state $| \psi_1 \rangle$ is:

$$\mathscr{P}_1(a_n) = | \langle u_n | \psi_1 \rangle |^2 \tag{E-2}$$

An analogous quantity $\mathscr{P}_2(a_n)$ can be defined for the state $| \psi_2 \rangle$:

$$\mathscr{P}_2(a_n) = | \langle u_n | \psi_2 \rangle |^2 \tag{E-3}$$

Now consider a normalized state $| \psi \rangle$ which is a *linear superposition* of $| \psi_1 \rangle$ and $| \psi_2 \rangle$:

$$| \psi \rangle = \lambda_1 | \psi_1 \rangle + \lambda_2 | \psi_2 \rangle$$
$$|\lambda_1|^2 + |\lambda_2|^2 = 1 \tag{E-4}$$

It is often said that, when the system is in the state $| \psi \rangle$, one has a probability $|\lambda_1|^2$ of finding it in the state $| \psi_1 \rangle$ and a probability $|\lambda_2|^2$ of finding it in the state $| \psi_2 \rangle$. The exact meaning of this manner of speaking is the following: if $| \psi_1 \rangle$ and $| \psi_2 \rangle$ are two eigenvectors (here assumed to be normalized) of the observable $B$ corresponding to different eigenvalues $b_1$ and $b_2$, the probability of finding $b_1$ when $B$ is measured is $|\lambda_1|^2$ and that of finding $b_2$ is $|\lambda_2|^2$.

This could lead us to believe (wrongly, as we shall see), that a state such as (E-4) is a *statistical mixture* of the states $| \psi_1 \rangle$ and $| \psi_2 \rangle$ with the weights $|\lambda_1|^2$ and $|\lambda_2|^2$. In other words, if we consider a large number $N$ of identical systems, all in the state (E-4), we might imagine that this set of $N$ systems in the state $| \psi \rangle$ was completely equivalent to another set composed of $N|\lambda_1|^2$ systems in the state $| \psi_1 \rangle$ and $N|\lambda_2|^2$ systems in the state $| \psi_2 \rangle$. Such an interpretation of the state $| \psi \rangle$ is erroneous and leads to inaccurate physical predictions as we shall see.

Assume that we are actually trying to calculate the probability $\mathscr{P}(a_n)$ of finding the eigenvalue $a_n$ when the observable $A$ is measured on the system in the state $| \psi \rangle$ given by (E-4). If we interpret the state $| \psi \rangle$ as being a statistical mixture of the states $| \psi_1 \rangle$ and $| \psi_2 \rangle$ with the weights $|\lambda_1|^2$ and $|\lambda_2|^2$, then we can obtain $\mathscr{P}(a_n)$ by taking the weighted sum of the probabilities $\mathscr{P}_1(a_n)$ and $\mathscr{P}_2(a_n)$ calculated above [formulas (E-2) and (E-3)]:

$$\mathscr{P}(a_n) \stackrel{?}{=} |\lambda_1|^2 \mathscr{P}_1(a_n) + |\lambda_2|^2 \mathscr{P}_2(a_n) \tag{E-5}$$

Actually, the postulates of quantum mechanics unambiguously indicate how to calculate $\mathscr{P}(a_n)$. The correct expression for this probability is:

$$\mathscr{P}(a_n) = |\langle u_n | \psi \rangle|^2 \tag{E-6}$$

$\mathscr{P}(a_n)$ is therefore the square of the modulus of the *probability amplitude* $\langle u_n | \psi \rangle$. We see from (E-4) that this amplitude is the sum of two terms:

$$\langle u_n | \psi \rangle = \lambda_1 \langle u_n | \psi_1 \rangle + \lambda_2 \langle u_n | \psi_2 \rangle \tag{E-7}$$

Thus we obtain:

$$\begin{aligned}\mathscr{P}(a_n) &= |\lambda_1 \langle u_n | \psi_1 \rangle + \lambda_2 \langle u_n | \psi_2 \rangle|^2 \\ &= |\lambda_1|^2 |\langle u_n | \psi_1 \rangle|^2 + |\lambda_2|^2 |\langle u_n | \psi_2 \rangle|^2 \\ &+ 2 \,\text{Re}\, \{ \lambda_1 \lambda_2^* \langle u_n | \psi_1 \rangle \langle u_n | \psi_2 \rangle^* \}\end{aligned} \tag{E-8}$$

Taking (E-2) and (E-3) into account, we find that the correct expression for $\mathscr{P}(a_n)$ is therefore written:

$$\mathscr{P}(a_n) = |\lambda_1|^2 \mathscr{P}_1(a_n) + |\lambda_2|^2 \mathscr{P}_2(a_n) + 2 \,\text{Re}\, \{ \lambda_1 \lambda_2^* \langle u_n | \psi_1 \rangle \langle u_n | \psi_2 \rangle^* \} \tag{E-9}$$

This result is different from that of formula (E-5).

It is therefore wrong to consider $| \psi \rangle$ to be a statistical mixture of states. Such an interpretation eliminates all the *interference effects* contained in the double product of formula (E-9). We see that it is not only the moduli of $\lambda_1$ and $\lambda_2$ which play a role; the relative phase* of $\lambda_1$ and $\lambda_2$ is just as important, since it enters explicitly, through the intermediary of $\lambda_1 \lambda_2^*$, into the physical predictions.

β. *A concrete example*

Consider photons propagating along $Oz$ whose polarization state is represented by the unit vector (fig. 5):

$$\mathbf{e} = \frac{1}{\sqrt{2}} (\mathbf{e}_x + \mathbf{e}_y) \tag{E-10}$$

This state is a linear superposition of two orthogonal polarization states $\mathbf{e}_x$ and $\mathbf{e}_y$. It represents light which is linearly polarized at an angle of 45° with respect to $\mathbf{e}_x$ and $\mathbf{e}_y$. It would be absurd to assume that $N$ photons in the state $\mathbf{e}$ are equivalent to $N \times \left|\frac{1}{\sqrt{2}}\right|^2 = \frac{N}{2}$ photons in the state $\mathbf{e}_x$ and $N \times \left|\frac{1}{\sqrt{2}}\right|^2 = \frac{N}{2}$ photons in the state $\mathbf{e}_y$. If we place in the beam's trajectory an analyzer whose axis $\mathbf{e}'$ is perpendicular to $\mathbf{e}$, we know that *none* of the $N$ photons in the state $\mathbf{e}$ will pass through this analyzer. But, for the statistical mixture $\left\{\frac{N}{2} \text{ photons in the state } \mathbf{e}_x, \frac{N}{2} \text{ photons in the state } \mathbf{e}_y\right\}$, half the photons will pass through the analyzer.

* Multiplying $| \psi \rangle$ by a *global* phase factor $e^{i\theta}$ is equivalent to changing $\lambda_1$ and $\lambda_2$ to $\lambda_1 e^{i\theta}$ and $\lambda_2 e^{i\theta}$. It can be verified from (E-9) that such an operation does not modify the physical predictions, which depend only on $|\lambda_1|^2$, $|\lambda_2|^2$ and $\lambda_1 \lambda_2^*$.

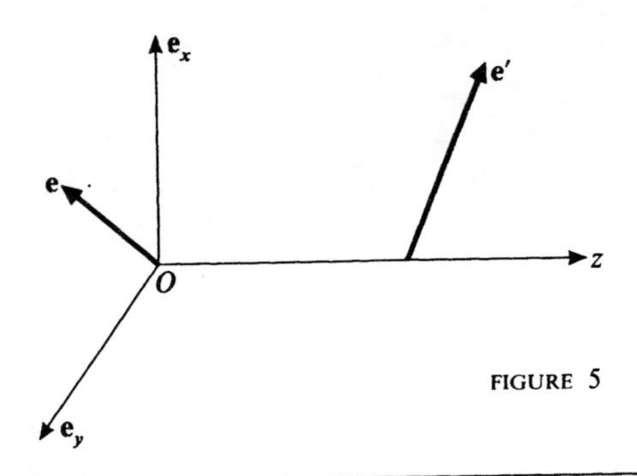


FIGURE 5

**A simple experiment which illustrates the difference between a linear superposition and a statistical mixture of states. If all the incident photons are in the polarization state**

$$\mathbf{e} = \frac{1}{\sqrt{2}}(\mathbf{e}_x + \mathbf{e}_y)$$

**none of them will pass through an analyzer whose axis $\mathbf{e}'$ is perpendicular to $\mathbf{e}$. If we had, on the contrary, a statistical mixture of photons polarized either along $\mathbf{e}_x$ or along $\mathbf{e}_y$ (in equal proportions; i.e., natural light), half of them would pass through the analyzer.**

In this concrete example, it is clear that a linear superposition such as (E-10), associated with *light polarized* at an angle of 45° with respect to $\mathbf{e}_x$ and $\mathbf{e}_y$, is physically different from a statistical mixture of equal proportions of the states $\mathbf{e}_x$ and $\mathbf{e}_y$ associated with *natural light* (an unpolarized beam).

We can also understand the importance of the relative phase of the expansion coefficients of the state vector, by considering the four states:

$$\mathbf{e}_1 = \frac{1}{\sqrt{2}}(\mathbf{e}_x + \mathbf{e}_y) \tag{E-11}$$

$$\mathbf{e}_2 = \frac{1}{\sqrt{2}}(\mathbf{e}_x - \mathbf{e}_y) \tag{E-12}$$

$$\mathbf{e}_3 = \frac{1}{\sqrt{2}}(\mathbf{e}_x + i\mathbf{e}_y) \tag{E-13}$$

$$\mathbf{e}_4 = \frac{1}{\sqrt{2}}(\mathbf{e}_x - i\mathbf{e}_y) \tag{E-14}$$

which differ only by the relative phase of the coefficients $\left(\text{this phase being equal to } 0, \pi, +\frac{\pi}{2} \text{ and } -\frac{\pi}{2}, \text{ respectively}\right)$. These four states are physically quite different: the first two represent light which is polarized linearly along the bisectors of $(\mathbf{e}_x, \mathbf{e}_y)$; the second two represent circularly polarized light (right and left respectively).

#### b. SUMMATION OVER THE INTERMEDIATE STATES

##### α. *Prediction of measurement results in two simple experiments*

(*i*) *Experiment 1*. Assume that the observable $A$ has been measured, at a given time, on a physical system, and that the non-degenerate eigenvalue $a$ has been found. If $| u_a \rangle$ is the eigenvector associated with $a$, the physical system, immediately after the measurement, is in the state $| u_a \rangle$.

Before the system has had time to evolve, we measure another observable $C$ which does not commute with $A$. Using the notation introduced in § C-6-a, we denote by $\mathscr{P}_a(c)$ the probability that this second measurement will yield the result $c$. Immediately before the measurement of $C$, the system is in the state $|u_a\rangle$. Therefore, if $|v_c\rangle$ is the eigenvector of $C$ associated with the eigenvalue $c$ (assumed to be non-degenerate), the postulates of quantum mechanics lead to:

$$\mathscr{P}_a(c) = |\langle v_c | u_a \rangle|^2 \tag{E-15}$$

(*ii*) *Experiment 2*. We now imagine *another* experiment, in which we measure successively and very rapidly three observables $A$, $B$, $C$, which do not commute with each other (the time separating two measurements is too short for the system to evolve). Denote by $\mathscr{P}_a(b, c)$ the probability, given that the result of the first measurement is $a$, that the results of the second and third will be $b$ and $c$ respectively. $\mathscr{P}_a(b, c)$ is equal to the product of $\mathscr{P}_a(b)$ (the probability that, the measurement of $A$ having yielded $a$, that of $B$ will yield $b$) and $\mathscr{P}_b(c)$ (the probability that, the measurement of $B$ having yielded $b$, that of $C$ will yield $c$):

$$\mathscr{P}_a(b, c) = \mathscr{P}_a(b)\,\mathscr{P}_b(c) \tag{E-16}$$

If all the eigenvalues of $B$ are assumed to be non-degenerate and if $|w_b\rangle$ denotes the corresponding eigenvectors, it follows that [using for $\mathscr{P}_a(b)$ and $\mathscr{P}_b(c)$ formulas analogous to (E-15)]:

$$\mathscr{P}_a(b, c) = |\langle v_c | w_b \rangle|^2 \, |\langle w_b | u_a \rangle|^2 \tag{E-17}$$

β. *The fundamental difference between these two experiments*

In both of these experiments, the state of the system after the measurement of the observable $A$ is $|u_a\rangle$ (the role of this measurement being to fix this initial state). It then becomes $|v_c\rangle$ after the last measurement, that of the observable $C$ (for this reason, $|v_c\rangle$ will be called the "final state"). It is possible in both cases to decompose the state of the system just before the measurement of $C$ in terms of the eigenvectors $|w_b\rangle$ of $B$, and to say that between the state $|u_a\rangle$ and the state $|v_c\rangle$, the system "can pass" through several different "intermediate states" $|w_b\rangle$. Each of these intermediate states defines a possible "path" between the initial state $|u_a\rangle$ and the final state $|v_c\rangle$ (fig. 6).

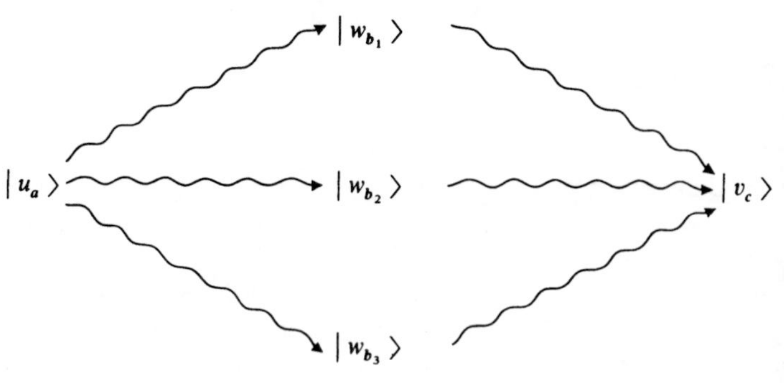


FIGURE 6

**Different possible "paths" for the state vector of the system when the system is allowed to evolve freely (without undergoing any measurement) between the initial state $|u_a\rangle$ and the final state $|v_c\rangle$. In this case, we must add together the probability amplitudes associated with these different paths, and not the probabilities.**

The difference between the two experiments described above is the following. In the first one, the path that the system has taken between the state $| u_a \rangle$ and the state $| v_c \rangle$ is not determined experimentally [we measure only the probability $\mathscr{P}_a(c)$ that, starting from $| u_a \rangle$, it ends up in $| v_c \rangle$]. On the other hand, in the second experiment, this path is determined, by measuring the observable $B$ [thus enabling us to obtain the probability $\mathscr{P}_a(b, c)$ that the system, starting from $| u_a \rangle$, passes through a given intermediate state $| w_b \rangle$ and ends up finally in $| v_c \rangle$].

We could then be tempted, in order to relate $\mathscr{P}_a(c)$ to $\mathscr{P}_a(b, c)$, to use the following argument : in experiment 1, the system is "free to pass" through all intermediate states $| w_b \rangle$; it would then seem that the global probability $\mathscr{P}_a(c)$ should be equal to the sum of all the probabilities $\mathscr{P}_a(b, c)$ associated with each of the possible "paths". Can we not then write:

$$\mathscr{P}_a(c) \overset{?}{=} \sum_b \mathscr{P}_a(b, c) \tag{E-18}$$

As we shall see, *this formula is wrong*. Let us go back to the exact formula (E-15) for $\mathscr{P}_a(c)$; this formula brings in the probability amplitude $\langle v_c | u_a \rangle$ which we can write, using the closure relation for the states $| w_b \rangle$:

$$\langle v_c | u_a \rangle = \sum_b \langle v_c | w_b \rangle \langle w_b | u_a \rangle \tag{E-19}$$

Substitute this expression into (E-15):

$$\begin{aligned} \mathscr{P}_a(c) &= \left| \sum_b \langle v_c | w_b \rangle \langle w_b | u_a \rangle \right|^2 \\ &= \sum_b |\langle v_c | w_b \rangle|^2 \, |\langle w_b | u_a \rangle|^2 + \\ &+ \sum_b \sum_{b' \neq b} \langle v_c | w_b \rangle \langle w_b | u_a \rangle \langle v_c | w_{b'} \rangle^* \langle w_{b'} | u_a \rangle^* \end{aligned} \tag{E-20}$$

Using (E-17), we therefore obtain:

$$\mathscr{P}_a(c) = \sum_b \mathscr{P}_a(b, c) + \sum_b \sum_{b' \neq b} \langle v_c | w_b \rangle \langle w_b | u_a \rangle \langle v_c | w_{b'} \rangle^* \langle w_{b'} | u_a \rangle^* \tag{E-21}$$

This equation enables us to understand why formula (E-18) is wrong: all the "cross terms" which appear in the square of the modulus of sum (E-19) are absent in (E-18). All the *interference effects between the different possible paths* are thus missing in (E-18).

If, therefore, we want to establish a relation between these two experiments, we see that it is necessary to reason in terms of probability amplitudes. *When the intermediate states of the system are not determined experimentally, it is the probability amplitudes, and not the probabilities, which must be summed.*

The error in the reasoning which led to the wrong relation (E-18) is obvious, moreover, if we remember the fifth postulate (reduction of the wave packet). In the second experiment, the measurement of the observable $B$ must, in fact, involve a perturbation of the system under study: during the measurement its state vector undergoes an abrupt change (projection onto one of the states $| w_b \rangle$). It is this unavoidable perturbation which is responsible for the disappearance of interference effects. In the first experiment, on the other hand, it is incorrect to say that the physical system "passes through one or another of the states $| w_b \rangle$"; it would be more accurate to say that it passes through all the states $| w_b \rangle$.

COMMENTS:

(*i*) The preceding discussion resembles in every respect that of § A-2-a of chapter I concerning Young's double-slit experiment. To determine the probability that a photon emitted by the source will arrive at a given point $M$ of the screen, one must first calculate the total electric field at $M$. In this problem, the electric field plays the role of a probability amplitude. When one is not trying to determine through which slit the photon passes, it is the electric fields radiated by the two slits, and not their intensities, that must be added together to obtain the total field at $M$ (whose square yields the desired probability). In other words, the field radiated by one of the slits at the point $M$ represents the amplitude for a photon, emitted by the source, to pass through this slit before arriving at $M$.

(*ii*) It is not necessary to retain the assumption that the measurements of $A$ and $C$ in experiment 1 and of $A$, $B$, $C$ in experiment 2 are performed very close together in time. If the system has had time to evolve between two of these measurements, we can use the Schrödinger equation to determine the modification of the state of the system due to this evolution [*cf.* complement $F_{III}$, comment (*ii*) of § 2].

### c. CONCLUSION: THE IMPORTANCE OF THE CONCEPT OF PROBABILITY AMPLITUDES

The two examples studied in §§ a and b demonstrate the importance of the concept of probability amplitudes. Formulas (E-5) and (E-18), as well as the arguments which lead to them, are incorrect since they represent an attempt to calculate a probability directly without first considering the corresponding probability amplitude. In both cases, the correct expression (E-8) or (E-20) has the form of a *square of a sum* (more precisely, the square of the modulus of this sum), while the incorrect formula (E-5) or (E-18) contains only a *sum of squares* (all the cross terms, responsible for interference effects, being omitted).

From the preceding discussion, we shall therefore retain the following ideas:

(*i*) The probabilistic predictions of quantum theory are always obtained by squaring the modulus of a *probability amplitude*.

(*ii*) When, in a particular experiment, no measurement is made at an intermediate stage, one must never reason in terms of the probabilities of the various results that could have been obtained in such a measurement, but rather in terms of their probability amplitudes.

(*iii*) The fact that the states of a physical system are linearly superposable means that a probability amplitude often presents the form of a sum of partial amplitudes. The corresponding probability is then equal to the square of the modulus of a sum of terms, and *the various partial amplitudes interfere with each other.*

## 2. Case in which several states can be associated with the same measurement result

In the preceding section, we stressed and illustrated the fact that, in certain cases, the probability of an event is given by the postulates of quantum mechanics in the form of a *square of a sum* of terms (more precisely, the square of the modulus of such a sum). Now the statement of the fourth postulate [formula (B-7)] involves a *sum of squares* (the sum of the squares of the moduli) when the measurement result whose probability is sought is associated with a degenerate eigenvalue. It is important to understand that these two rules are not contradictory but, on the contrary, complementary: each term of the sum of squares (B-7) can itself be the square of a sum. This is the first point on which we shall focus our attention in this section. Furthermore, this discussion will enable us to complete the statement of the postulates: we shall consider measurement devices whose accurary is limited (as is always, of course, the case) and see how to predict theoretically the possible results. Finally, we shall extend to the case of continuous spectra the fifth postulate of reduction of the wave packet.

### a. DEGENERATE EIGENVALUES

In the examples treated in § E-1, we always assumed that the results of the various measurements envisaged were simple, i.e. non-degenerate, eigenvalues of the corresponding observables. This hypothesis was intended to simplify these examples so that the origin of the interference effects appeared as clearly as possible.

Now let us consider a degenerate eigenvalue $a_n$ of an observable $A$. The eigenstates associated with $a_n$ form a vector subspace of dimension $g_n$, in which an orthonormal basis $\{ |u_n^i\rangle ; i = 1, 2, ..., g_n \}$ can be chosen.

The discussion of § C-6-b shows that knowing a measurement of $A$ has yielded $a_n$ is not sufficient to determine the state of the physical system after this measurement. We shall say that *several final states can be associated with the same result* $a_n$ : if the initial state (the state before the measurement) is given, the final state after the measurement is perfectly well-defined; but if the initial state is changed, the final state is, in general, different (for the same measurement result $a_n$). All final states associated with $a_n$ are linear combinations of the $g_n$ orthonormal vectors $|u_n^i\rangle$, with $i = 1, 2, ... g_n$.

Formula (B-7) indicates unambiguously how to find the probability $\mathscr{P}(a_n)$ that a measurement of $A$ on a system in the state $|\psi\rangle$ will yield the result $a_n$. One chooses an orthonormal basis, for example $\{ |u_n^i\rangle ; i = 1, 2, ..., g_n \}$, in the eigen-subspace which corresponds to $a_n$; one calculates the probability $|\langle u_n^i|\psi\rangle|^2$ of finding the system in each of the states of this basis; $\mathscr{P}(a_n)$ is then the sum of these $g_n$

probabilities. However, it must not be forgotten that each probability $|\langle u_n^i | \psi \rangle|^2$ can be the square of the modulus of a sum of terms. Consider, for example, the case envisaged in §E-1-a-α and assume now that the eigenvalue $a_n$ of the observable $A$, whose probability $\mathscr{P}(a_n)$ is to be calculated, is $g_n$-fold degenerate. Formula (E-6) is then replaced by:

$$\mathscr{P}(a_n) = \sum_{i=1}^{g_n} |\langle u_n^i | \psi \rangle|^2 \tag{E-22}$$

with:

$$\langle u_n^i | \psi \rangle = \lambda_1 \langle u_n^i | \psi_1 \rangle + \lambda_2 \langle u_n^i | \psi_2 \rangle \tag{E-23}$$

The discussion of §E-1-a-α remains valid for each of the terms of formula (E-22): $|\langle u_n^i | \psi \rangle|^2$, which is obtained from (E-23), is the square of a sum; $\mathscr{P}(a_n)$ is then the sum of these squares. §E-1-b can similarly be generalized to the case where the eigenvalues of the observables measured are degenerate.

Before summarizing the preceding discussions, we are going to study another important situation where several final states are associated with the same measurement result.

#### b. INSUFFICIENTLY SELECTIVE MEASUREMENT DEVICES

##### α. *Definition*

Assume that, in order to measure the observable $A$ in a given physical system, we have at our disposal a device which works in the following way:

(*i*) This device can give only two different answers*, which we shall denote, for convenience, by "yes" and "no".

(*ii*) If the system is in an eigenstate of $A$ whose eigenvalue is included in a given interval $\Delta$ of the real axis, the answer is always "yes"; this is also the case when the state of the system is any linear combination of eigenstates of $A$ associated with eigenvalues which are all included in $\Delta$.

(*iii*) If the state of the system is an eigenstate of $A$ whose eigenvalue falls outside $\Delta$, or any linear combination of such eigenstates, the answer is always "no"

$\Delta$ therefore characterizes the resolving power of the measurement device under consideration. If there exists only one eigenvalue $a_n$ of $A$ in the interval $\Delta$, the resolving power is infinite: when the system is in an arbitrary state, the probability $\mathscr{P}$(yes) of obtaining the answer "yes" is equal to the probability of finding $a_n$ in a measurement of $A$; the probability $\mathscr{P}$(no) of obtaining "no" is obviously equal to $1 - \mathscr{P}$(yes). If, on the other hand, $\Delta$ contains several eigenvalues of $A$, the device does not have a sufficient resolution to distinguish between these various eigenvalues: we shall say that it is *insufficiently selective*. We shall see how to calculate $\mathscr{P}$(yes) and $\mathscr{P}$(no) in this case.

To be able to study the perturbation created by such a measurement on the state of the system, we are going to add the following hypothesis: the device transmits without perturbation the eigenstates of $A$ associated with the eigenvalues of the interval $\Delta$ (as well as any linear combination of these eigenstates), while it "blocks" the eigenstates of $A$ associated with the eigenvalues outside $\Delta$ (as well as all their linear combinations). The device thus behaves like a perfect filter for all states associated with $\Delta$.

* The following arguments can easily be generalized to cases where the device can give several different answers having characteristics similar to those described in (*ii*) and (*iii*).

β. *Example*

Most measurement devices used in practice are insufficiently selective.

For example, to measure the abscissa of an electron propagating parallel to the $Oz$ axis, one can (fig. 7) place in the $xOy$ plane ($Oy$ is perpendicular to the plane of the figure) a plate with a slit whose axis is parallel to $Oy$, the abscissas of the edges being $x_1$ and $x_2$. It can then be seen that any wave packet which is entirely included between the $x = x_1$ and $x = x_2$ planes (a superposition of eigenstates of $X$ having eigenvalues $x$ contained within the interval $[x_1, x_2]$) will enter the region to the right of the slit ("yes" answer); in this case, it will not undergo any modification. On the other hand, any wave packet situated below the $x = x_1$ plane or above the $x = x_2$ plane will be blocked by the plate and will not pass to the right ("no" answer).

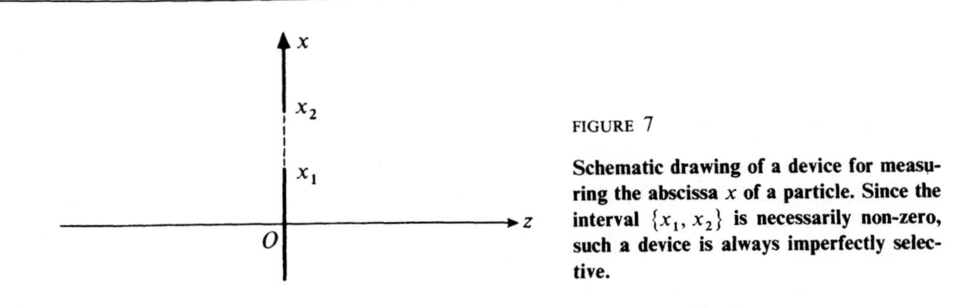


FIGURE 7

**Schematic drawing of a device for measuring the abscissa $x$ of a particle. Since the interval $\{x_1, x_2\}$ is necessarily non-zero, such a device is always imperfectly selective.**

γ. *Quantum description*

For such an insufficiently selective device, several final states are possible after a measurement which has yielded the answer yes, for example, the various eigenstates of $A$ which correspond to the eigenvalues of the interval $\Delta$.

The physical problem posed by such devices, and which we are now going to consider, consists of predicting the answer which will be obtained when a system in an arbitrary state enters the device. For example, for the apparatus of figure 7, what happens when we are dealing with a wave packet which is neither entirely contained between the $x = x_1$ and $x = x_2$ planes (in which case the answer is certainly yes) nor entirely situated outside this region (in which case the answer is certainly no)? We shall see that this is equivalent to measuring an observable whose spectrum is degenerate.

Consider the subspace $\mathscr{E}_\Delta$ spanned by all the eigenstates of $A$ whose eigenvalues $a_n$ are contained within the interval $\Delta$. The projector $P_\Delta$ onto this subspace is written (*cf.* § B-3-b-γ of chap. II):

$$P_\Delta = \sum_{a_n \in \Delta} \sum_{i=1}^{g_n} | u_n^i \rangle \langle u_n^i | \tag{E-24}$$

(the eigenvalues $a_n$ of the interval $\Delta$ can be degenerate, hence the additional index $i$; the vectors $| u_n^i \rangle$ are assumed to be orthonormal). $\mathscr{E}_\Delta$ is the subspace formed by all the possible states of the system after a measurement which has given the result yes.

Referring to the definition of the measurement device, we see that the response will certainly be yes for any state belonging to $\mathscr{E}_\Delta$, that is, for any eigenstate of $P_\Delta$ with the eigenvalue of + 1. The answer will certainly be no for any state belonging to the supplement of $\mathscr{E}_\Delta$, that is, for any eigenstates of $P_\Delta$ with the eigenvalue of 0. The yes and no answers which can be furnished by the measurement device therefore correspond to the eigenvalues + 1 and 0 of the observable $P_\Delta$ : it could be said that the device is actually measuring the observable $P_\Delta$ rather than $A$.

In the light of this interpretation, the case of an insufficiently selective measurement device can be treated in the framework of the postulates which we have stated. The probability $\mathscr{P}$(yes) of obtaining the answer yes is equal to the probability of finding the (degenerate) eigenvalue $+1$ of $P_\Delta$. Now an orthonormal basis in the corresponding eigensubspace is constituted by the set of states $| u_n^i \rangle$ which are eigenstates of $A$ with eigenvalues contained within the interval $\Delta$. Applying formula (B-7) to the eigenvalue $+1$ of the observable $P_\Delta$, we therefore obtain (for a system in the state $| \psi \rangle$):

$$\mathscr{P}(\text{yes}) = \sum_{a_n \in \Delta} \sum_{i=1}^{g_n} |\langle u_n^i | \psi \rangle|^2 \tag{E-25}$$

Since there are only two possible answers, we obviously have:

$$\mathscr{P}(\text{no}) = 1 - \mathscr{P}(\text{yes}) \tag{E-26}$$

The projector onto the eigensubspace associated with the eigenvalue $+1$ of the observable $P_\Delta$ is $P_\Delta$ itself; formula (B-14) therefore gives here:

$$\mathscr{P}(\text{yes}) = \langle \psi | P_\Delta | \psi \rangle \tag{E-27}$$

[this formula is equivalent to (E-25)].

Similarly, since the device does not perturb states belonging to $\mathscr{E}_\Delta$ and blocks those of the supplement of $\mathscr{E}_\Delta$, we find that the state of the system after a measurement which has given the result yes is:

$$| \psi' \rangle = \frac{1}{\sqrt{\displaystyle\sum_{a_n \in \Delta} \sum_{i=1}^{g_n} |\langle u_n^i | \psi \rangle|^2}} \sum_{a_n \in \Delta} \sum_{i=1}^{g_n} | u_n^i \rangle \langle u_n^i | \psi \rangle \tag{E-28}$$

that is:

$$| \psi' \rangle = \frac{1}{\sqrt{\langle \psi | P_\Delta | \psi \rangle}} P_\Delta | \psi \rangle \tag{E-29}$$

When $\Delta$ contains only one eigenvalue $a_n$, $P_\Delta$ reduces to $P_n$: formulas (B-14) and (B-31) are then seen to be special cases of formulas (E-27) and (E-29).

### c. RECAPITULATION: MUST ONE SUM THE AMPLITUDES OR THE PROBABILITIES ?

There are therefore cases (§ E-1) where, to calculate a probability, one takes the square of a sum, because several probability amplitudes must be added together. In other cases (§ E-2), one takes a sum of squares, because several probabilities must be added together. It is clearly important not to confuse these different cases and to know, in a given situation, if it is the probability amplitudes or the probabilities themselves which must be summed.

Young's double-slit experiment will again furnish us with a very convenient physical example which will enable us to illustrate and summarize the preceding discussions. Assume that we want to calculate the probability for a particular photon to strike the plate between two points $M_1$ and $M_2$ having abscissas of $x_1$ and $x_2$ (fig. 8). This probability is proportional to the total light intensity received by this portion of the plate. It is therefore a "sum of squares"; more precisely, it is the integral of the intensity $I(x)$ between $x_1$ and $x_2$. But each term $I(x)$ of this sum is obtained by squaring the electric field $\mathcal{E}(x)$ at $x$, which is equal to the sum of the electric fields $\mathcal{E}_A(x)$ and $\mathcal{E}_B(x)$ radiated at $M$ by the slits $A$ and $B$. $I(x)$ is therefore proportional to $|\mathcal{E}_A(x) + \mathcal{E}_B(x)|^2$, that is, to the square of a sum. $\mathcal{E}_A(x)$ and $\mathcal{E}_B(x)$ are the amplitudes associated with the two possible paths $SAM$ and $SBM$ which end at the same point $M$; they are added to obtain the amplitude at $M$ since one is not trying to determine through which slit the photon passes. Then, to calculate the total light intensity received by the interval $M_1M_2$, one adds the intensities which arrive at the various points of this interval.

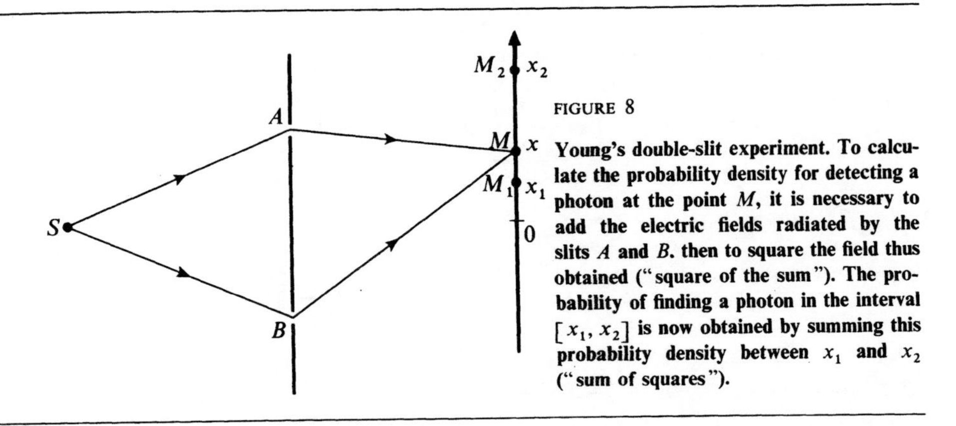


FIGURE 8

**Young's double-slit experiment. To calculate the probability density for detecting a photon at the point $M$, it is necessary to add the electric fields radiated by the slits $A$ and $B$. then to square the field thus obtained ("square of the sum"). The probability of finding a photon in the interval $[x_1, x_2]$ is now obtained by summing this probability density between $x_1$ and $x_2$ ("sum of squares").**

To sum up, the fundamental idea which must be retained from the discussions of this section can be expressed schematically in the following way:

| Add the amplitudes corresponding to the same final state, then the probabilities corresponding to orthogonal final states. |
|---|

### d. APPLICATION TO THE TREATMENT OF CONTINUOUS SPECTRA

When the observable we want to measure has a continuous spectrum, only insufficiently selective devices can be used: it is impossible to imagine a physical device which could isolate a single eigenvalue belonging to a continuous set. We shall see here how the study of § E-2-b enables us to be more precise and complete in our treatment of observables with continuous spectra.

#### α. *Example: measurement of the position of a particle*

Let $\psi(\mathbf{r}) = \langle \mathbf{r} | \psi \rangle$ be the wave function of a (spinless) particle. What is the probability of finding the abscissa of this particle within the interval $[x_1, x_2]$ of the $x$-axis, using, for example, a measurement device like the one in figure 7?

The subspace $\mathscr{E}_\Delta$ associated with this measurement result is the space spanned by the kets $| \mathbf{r} \rangle = | x, y, z \rangle$ which are such that: $x_1 \leqslant x \leqslant x_2$. Since these kets are orthonormal in the extended sense, application of the rule stated in § *c* above yields:

$$\begin{aligned}\mathscr{P}(x_1 \leqslant x \leqslant x_2) &= \int_{x_1}^{x_2} \mathrm{d}x \int_{-\infty}^{+\infty} \mathrm{d}y \int_{-\infty}^{+\infty} \mathrm{d}z \, |\langle x, y, z | \psi \rangle|^2 \\ &= \int_{x_1}^{x_2} \mathrm{d}x \int_{-\infty}^{+\infty} \mathrm{d}y \int_{-\infty}^{+\infty} \mathrm{d}z \, |\psi(\mathbf{r})|^2 \qquad \text{(E-30)}\end{aligned}$$

Formula (E-27) obviously leads to the same result, since the projector $P_\Delta$ is written here:

$$P_\Delta = \int_{x_1}^{x_2} \mathrm{d}x \int_{-\infty}^{+\infty} \mathrm{d}y \int_{-\infty}^{+\infty} \mathrm{d}z \, | \, x, y, z \, \rangle \langle \, x, y, z \, | \tag{E-31}$$

and we therefore have:

$$\begin{aligned} \mathscr{P}(x_1 \leqslant x \leqslant x_2) &= \langle \, \psi \, | \, P_\Delta \, | \, \psi \, \rangle \\ &= \int_{x_1}^{x_2} \mathrm{d}x \int_{-\infty}^{+\infty} \mathrm{d}y \int_{-\infty}^{+\infty} \mathrm{d}z \, \langle \, \psi \, | \, x, y, z \, \rangle \langle \, x, y, z \, | \, \psi \, \rangle \end{aligned} \tag{E-32}$$

To know the state $| \, \psi' \, \rangle$ of the particle after such a measurement, which has yielded the result yes, it suffices to apply formula (E-29):

$$\begin{aligned} | \, \psi' \, \rangle &= \frac{1}{N} P_\Delta \, | \, \psi \, \rangle \\ &= \frac{1}{N} \int_{x_1}^{x_2} \mathrm{d}x' \int_{-\infty}^{+\infty} \mathrm{d}y' \int_{-\infty}^{+\infty} \mathrm{d}z' \, | \, x', y', z' \, \rangle \langle \, x', y', z' \, | \, \psi \, \rangle \end{aligned} \tag{E-33}$$

where the normalization factor $N = \sqrt{\langle \, \psi \, | \, P \, | \, \psi \, \rangle}$ is known [formula (E-32)]. Let us calculate the wave function $\psi'(\mathbf{r}) = \langle \, \mathbf{r} \, | \, \psi' \, \rangle$ associated with the ket $| \, \psi' \, \rangle$:

$$\langle \, \mathbf{r} \, | \, \psi' \, \rangle = \frac{1}{N} \int_{x_1}^{x_2} \mathrm{d}x' \int_{-\infty}^{+\infty} \mathrm{d}y' \int_{-\infty}^{+\infty} \mathrm{d}z' \, \langle \, \mathbf{r} \, | \, \mathbf{r}' \, \rangle \, \psi(\mathbf{r}') \tag{E-34}$$

Now $\langle \, \mathbf{r} \, | \, \mathbf{r}' \, \rangle = \delta(\mathbf{r} - \mathbf{r}') = \delta(x - x') \, \delta(y - y') \, \delta(z - z')$. The integrations over $y'$ and $z'$ can therefore be performed immediately : they amount to replacing $y'$ and $z'$ by $y$ and $z$ in the function to be integrated. Equation (E-34) thus becomes :

$$\psi'(x, y, z) = \frac{1}{N} \int_{x_1}^{x_2} \mathrm{d}x' \, \delta(x - x') \, \psi(x', y, z) \tag{E-35}$$

If the point $x' = x$ is situated inside the interval of integration $[x_1, x_2]$, the result is the same as if we were integrating from $-\infty$ to $+\infty$:

$$\psi'(x, y, z) = \frac{1}{N} \psi(x, y, z) \qquad \text{for } x_1 \leqslant x \leqslant x_2 \tag{E-36}$$

On the other hand, if $x' = x$ falls outside the interval of integration, $\delta(x - x')$ is zero for all values of $x'$ included in this interval, and:

$$\psi'(x, y, z) = 0 \qquad \text{for } x > x_2 \quad \text{and} \quad x < x_1 \tag{E-37}$$

The part of $\psi(\mathbf{r})$ which corresponds to the interval accepted by the measurement device therefore persists, undeformed, immediately after the measurement [the factor $1/N$ simply assures that $\psi'(\mathbf{r})$ remain normalized]; the rest is suppressed by the measurement. The wave packet $\psi(\mathbf{r})$ representing the initial state of the particle is, as it were, "truncated" by the edges of the slit.

COMMENTS:

(*i*) This example clearly reveals the concrete meaning of the "reduction of the wave packet".

(*ii*) If a large number of particles, all in the same state $| \, \psi \, \rangle$, enter the device successively, the result will sometimes be yes and sometimes be no [with the probabilities $\mathscr{P}$(yes) and $\mathscr{P}(\text{no})$]. If the result is yes, the particle continues on its way, starting from the "truncated" state $| \psi' \rangle$; if the result is no, the particle is absorbed by the screen.

In the example we are considering here, the measurement device becomes all the more selective as $x_2 - x_1$ becomes smaller. We see, however, that it is impossible to make it perfectly selective because the spectrum of $X$ is continuous: however narrow the slit may be, the interval $[x_1, x_2]$ which it defines always contains an infinity of eigenvalues. Nevertheless, in the limiting case of a slit of an infinitely small width $\Delta x$, we find the equivalent of formula (B-17), which was the expression of the fourth postulate in the case of a continuous spectrum. Let us choose $x_1 = x_0 - \dfrac{\Delta x}{2}$ and $x_2 = x_0 + \dfrac{\Delta x}{2}$ (a slit of width $\Delta x$ centered at $x_0$), and assume that the wave function $\psi(\mathbf{r})$ varies very little within the interval $\Delta x$. Then, in (E-30), we can replace $|\psi(\mathbf{r})|^2$ by $|\psi(x_0, y, z)|^2$ and perform the integration over $x$:

$$\mathscr{P}\left(x_0 - \frac{\Delta x}{2} \leqslant x \leqslant x_0 + \frac{\Delta x}{2}\right) \simeq \Delta x \int_{-\infty}^{+\infty} \mathrm{d}y \int_{-\infty}^{+\infty} \mathrm{d}z \; |\psi(x_0, y, z)|^2 \qquad \text{(E-38)}$$

We indeed find a probability equal to the product of $\Delta x$ and a positive quantity which plays the role of a probability density at the point $x_0$. The difference with formula (B-17) lies in the fact that the latter applies to the case of a continuous but non-degenerate spectrum, while here the eigenvalues of $X$ are infinitely degenerate in $\mathscr{E}_{\mathbf{r}}$; this is the origin of the integrals over $y$ and $z$ which appear in (E-38) (summation over the indices associated with the degeneracy).

β. *Postulate of reduction of wave packets in the case of a continuous spectrum*

In § B-3-c, we confined ourselves, in the statement of the fifth postulate, to the case of a discrete spectrum. Formula (E-33) and its accompanying discussion enable us to understand the form assumed by this postulate when a continuous spectrum is considered: it is sufficient to apply the results of § E-2-b concerning insufficiently selective devices.

Let $A$ be an observable with a continuous spectrum (assumed, for simplicity, to be non-degenerate). The notation is the same as in § B-3-b-β.

If a measurement of $A$ on a system in the state $| \psi \rangle$ has yielded the result $\alpha_0$ to within $\Delta\alpha$, the state of the system immediately after this measurement is described by:

$$| \psi' \rangle = \frac{1}{\sqrt{\langle \psi | P_{\Delta\alpha}(\alpha_0) | \psi \rangle}} P_{\Delta\alpha}(\alpha_0) | \psi \rangle \qquad \text{(E-39)}$$

with:

$$P_{\Delta\alpha}(\alpha_0) = \int_{\alpha_0 - \frac{\Delta\alpha}{2}}^{\alpha_0 + \frac{\Delta\alpha}{2}} \mathrm{d}\alpha \, | v_\alpha \rangle \langle v_\alpha | \qquad \text{(E-40)}$$

Figures 9-a and 9-b illustrate this statement. If the function $\langle v_\alpha | \psi \rangle$ representing $| \psi \rangle$ in the $\{ | v_\alpha \rangle \}$ basis, has the form indicated in figure 9-a, the state of the system immediately after the measurement is represented, to within a normalization factor, by the function of figure 9-b [the calculation is analogous in all respects to the one which derives (E-36) and (E-37) from (E-33)].

We see that, even if $\Delta\alpha$ is very small, one can never actually prepare the system in the state $| v_{\alpha_0} \rangle$, which would be represented, in the $\{ | v_\alpha \rangle \}$ basis, by $\langle v_\alpha | v_{\alpha_0} \rangle = \delta(\alpha - \alpha_0)$. We can only obtain a narrow function centered at $\alpha_0$, since $\Delta\alpha$ is never exactly zero.

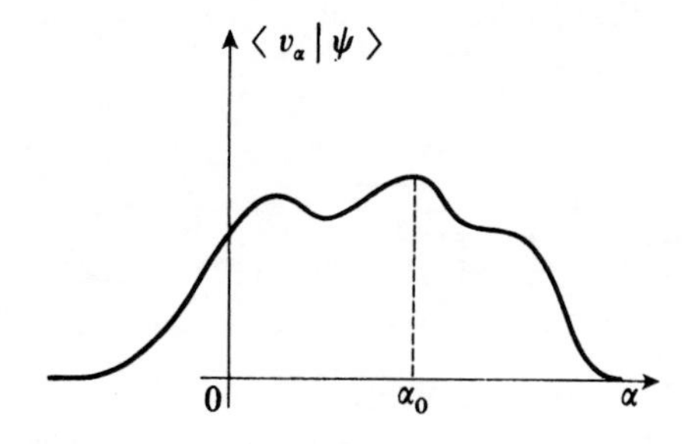


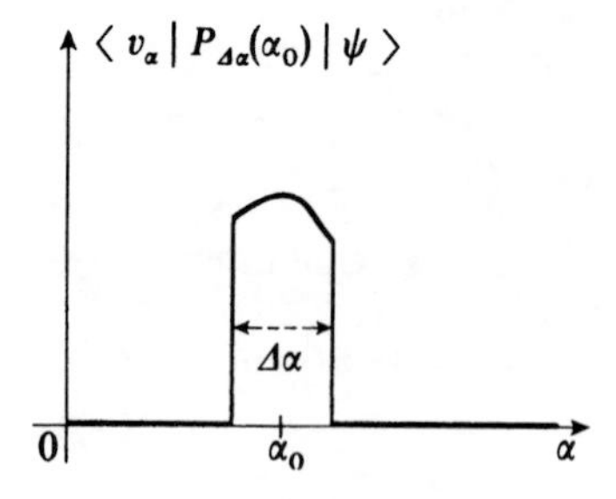


FIGURE 9

**Illustration of the postulate of reduction of wave packets in the case of a continuous spectrum: one measures the observable $A$, with eigenvectors $| v_\alpha \rangle$ and eigenvalues $\alpha$. The measurement device has a selectivity $\Delta\alpha$. If the value found is $\alpha_0$ to within $\Delta\alpha$, the effect of the measurement on the wave function $\langle v_\alpha | \psi \rangle$ is to "truncate" it about the value $\alpha_0$ (to normalize the new wave function, it is obviously necessary to multiply it by a factor larger than 1).**

**References and suggestions for further reading :**

Development of quantum mechanical concepts: references of section 4 of the bibliography, particularly Jammer (4.8).

Discussion and interpretation of the postulates: references of section 5 of the bibliography; Von Neumann (10.10), chaps. V and VI; Feynman III (1.2), §2.6, chap. 3 and §8.3.

Quantization rules using Poisson brackets: Dirac (1.13), §21; Schiff (1.18), §24.

Probability and statistics: see the corresponding subsection of section 10 of the bibliography.

COMPLEMENTS OF CHAPTER III

$A_{III}$: PARTICLE IN AN INFINITE POTENTIAL WELL

$B_{III}$: STUDY OF THE PROBABILITY CURRENT IN SOME SPECIAL CASES

$A_{III}$, $B_{III}$: direct applications of chapter III to simple cases. The accent is placed on the physical discussion of the results (elementary level).

$C_{III}$: ROOT-MEAN-SQUARE DEVIATIONS OF TWO CONJUGATE OBSERVABLES

$C_{III}$: a little more formal; general proof of the Heisenberg relations; may be skipped on a first reading.

$D_{III}$: MEASUREMENTS BEARING ON ONLY ONE PART OF A PHYSICAL SYSTEM

$D_{III}$: discussion of measurements bearing only on part of a system; a rather simple but somewhat formal application of chapter III; may be skipped on a first reading.

$E_{III}$: THE DENSITY OPERATOR

$F_{III}$: THE EVOLUTION OPERATOR

$G_{III}$: THE SCHRÖDINGER AND HEISENBERG PICTURES

$H_{III}$: GAUGE INVARIANCE

$J_{III}$: PROPAGATOR FOR THE SCHRÖDINGER EQUATION

$E_{III}$, $F_{III}$, $G_{III}$, $H_{III}$, $J_{III}$: complements which serve as an introduction to a more advanced quantum mechanics course. Aside from $F_{III}$, which is simple, they are on a higher level than the rest of this book, but they are comprehensible if chapter III has been read. To be reserved for subsequent study.

$E_{III}$: definition and properties of the density operator, which is used in the quantum mechanical description of systems whose state is imperfectly known (statistical mixture of states); fundamental tool of quantum statistical mechanics.

$F_{III}$: introduction of the evolution operator, which gives the quantum state of a system at an arbitrary instant $t$ in terms of its state at the instant $t_0$.

$G_{III}$: describes the evolution of a quantum system in a way which is different from, but equivalent to, that of chapter III. The time dependence now appears in the observables and not in the state of the system.

$H_{III}$: discussion of the quantum formalism in the case where the system is subject to an electromagnetic field. Although the description of the system involves the electromagnetic potentials, the physical properties depend only on the values of the electric and magnetic fields; they remain invariant when the potentials describing the electromagnetic field are changed.

$J_{III}$: an introduction to a different way of approaching quantum mechanics, based on a principle analogous to Huygens' principle in classical wave optics.

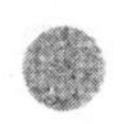



**$K_{III}$: UNSTABLE LEVELS; LIFETIMES**

$K_{III}$: simple introduction to the important physical concepts of instability and lifetimes; easy, but can be skipped on a first reading.

**$L_{III}$: EXERCISES**

**$M_{III}$: BOUND STATES OF A PARTICULE IN A "POTENTIAL WELL" OF ARBITRARY SHAPE**

**$N_{III}$: UNBOUND STATES OF A PARTICLE IN THE PRESENCE OF A POTENTIAL WELL OR BARRIER OF ARBITRARY SHAPE**

**$O_{III}$: QUANTUM PROPERTIES OF A PARTICLE IN A ONE-DIMENSIONAL PERIODIC STRUCTURE**

$M_{III}$, $N_{III}$, $O_{III}$: return to one-dimensional problems, considered from a more general point of view than in chapter I and its complements.

$M_{III}$: generalization to an arbitrary potential well of the principal results obtained in § 2-c of complement $H_I$; recommended, since easy and physically important.

$N_{III}$: study of unbound stationary states in an arbitrary potential; a little more formal; the definitions and results of this complement are necessary for complement $O_{III}$.

$O_{III}$: introduction of the concept (which is fundamental to solid state physics) of energy bands in a potential having a periodic structure (this concept will be treated differently in complement $F_{XI}$); rather difficult, can be reserved for later reading.

## Complement $A_{III}$

# PARTICLE IN AN INFINITE POTENTIAL WELL



In complement $H_I$ (§2-c-β), we studied the stationary states of a particle in a one-dimensional infinite potential well. Here we intend to re-examine this subject from a physical point of view. This will allow us to apply some of the postulates of chapter III to a concrete case. We shall be particularly interested in the results that can be obtained when the position or momentum of the particle is measured.

## 1. Distribution of the momentum values in a stationary state

### a. CALCULATION OF THE FUNCTION $\bar{\varphi}_n(p)$, OF $\langle P \rangle$ AND OF $\Delta P$

We have seen that the stationary states of the particle correspond to the energies*:

$$E_n = \frac{n^2\pi^2\hbar^2}{2ma^2} \tag{1}$$

and to the wave functions:

$$\varphi_n(x) = \sqrt{\frac{2}{a}} \sin\left(\frac{n\pi x}{a}\right) \tag{2}$$

(where $a$ is the width of the well and $n$ is any positive integer).

Consider a particle in the state $|\varphi_n\rangle$, with energy $E_n$. The probability of a measurement of the momentum $P$ of the particle yielding a result between $p$ and $p + \mathrm{d}p$ is:

$$\bar{\mathscr{P}}_n(p)\,\mathrm{d}p = |\bar{\varphi}_n(p)|^2\,\mathrm{d}p \tag{3}$$

with:

$$\bar{\varphi}_n(p) = \frac{1}{\sqrt{2\pi\hbar}} \int_0^a \sqrt{\frac{2}{a}} \sin\left(\frac{n\pi x}{a}\right) \mathrm{e}^{-ipx/\hbar}\,\mathrm{d}x \tag{4}$$

* We shall use the notation of complement $H_I$.

This integral is easy to calculate; it is equal to:

$$\overline{\varphi}_n(p) = \frac{1}{2i\sqrt{\pi\hbar a}} \int_0^a \left[ e^{i\left(\frac{n\pi}{a} - \frac{p}{\hbar}\right)x} - e^{-i\left(\frac{n\pi}{a} + \frac{p}{\hbar}\right)x} \right] dx$$

$$= \frac{1}{2i\sqrt{\pi\hbar a}} \left[ \frac{e^{i\left(\frac{n\pi}{a} - \frac{p}{\hbar}\right)a} - 1}{i\left(\frac{n\pi}{a} - \frac{p}{\hbar}\right)} - \frac{e^{-i\left(\frac{n\pi}{a} + \frac{p}{\hbar}\right)a} - 1}{-i\left(\frac{n\pi}{a} + \frac{p}{\hbar}\right)} \right] \tag{5}$$

that is:

$$\overline{\varphi}_n(p) = \frac{1}{2i}\sqrt{\frac{a}{\pi\hbar}}\, e^{i\left(\frac{n\pi}{2} - \frac{pa}{2\hbar}\right)} \left[ F\left(p - \frac{n\pi\hbar}{a}\right) + (-1)^{n+1} F\left(p + \frac{n\pi\hbar}{a}\right) \right] \tag{6}$$

with:

$$F(p) = \frac{\sin(pa/2\hbar)}{pa/2\hbar} \tag{7}$$

To within a proportionality factor, the function $\overline{\varphi}_n(p)$ is the sum (or the difference) of two "diffraction functions" $F\left(p \pm \frac{n\pi\hbar}{a}\right)$, centered at $p = \mp \frac{n\pi\hbar}{a}$. The "width" of these functions (the distance between the first two zeros, symmetrical with respect to the central value) does not depend on $n$ and is equal to $\frac{4\pi\hbar}{a}$. Their "amplitude" does not depend on $n$ either.

The function inside brackets in expression (6) is even if $n$ is odd, and odd if $n$ is even. The probability density $\overline{\mathscr{P}}_n(p)$ given in (3) is therefore an even function of $p$ in all cases, so that :

$$\langle P \rangle_n = \int_{-\infty}^{+\infty} \overline{\mathscr{P}}_n(p)\, p\, dp = 0 \tag{8}$$

The mean value of the momentum of the particle in the energy state $E_n$ is therefore zero.

Let us calculate, in the same way, the mean value $\langle P^2 \rangle_n$ of the square of the momentum. Using the fact that in the $\{ | x \rangle \}$ representation $P$ acts like $\frac{\hbar}{i}\frac{d}{dx}$, and performing an integration by parts, we obtain*:

$$\langle P^2 \rangle_n = \hbar^2 \int_0^a \left| \frac{d\varphi_n}{dx} \right|^2 dx$$

$$= \hbar^2 \int_0^a \frac{2}{a}\left(\frac{n\pi}{a}\right)^2 \cos^2\left(\frac{n\pi x}{a}\right) dx$$

$$= \left(\frac{n\pi\hbar}{a}\right)^2 \tag{9}$$

* Result (9) could also be derived from (6) by performing the integral $\langle P^2 \rangle_n = \int_{-\infty}^{+\infty} |\overline{\varphi}_n(p)|^2\, p^2\, dp$. This calculation, which presents no theoretical difficulties, is nevertheless not as direct as the one which is given here.

From (8) and (9), we get:

$$\Delta P_n = \sqrt{\langle P^2 \rangle_n - \langle P \rangle_n^2} = \frac{n\pi\hbar}{a} \tag{10}$$

The root-mean-square deviation therefore increases linearly with $n$.

### b. DISCUSSION

Let us trace, for different values of $n$, the curves which give the probability density $\overline{\mathscr{P}}_n(p)$. To do this, let us begin by studying the function inside brackets in expression (6). For the ground state ($n = 1$), it is the sum of two functions $F$, the centers of these two diffraction curves being separated by half their width (fig. 1-a). For the first excited level ($n = 2$), the distance between these centers is twice as large, and in this case, moreover, the difference of two functions $F$ must be taken (fig. 2-a). Finally, for an excited level corresponding to a large value of $n$, the centers of the two diffraction curves are separated by a distance much greater than their width.

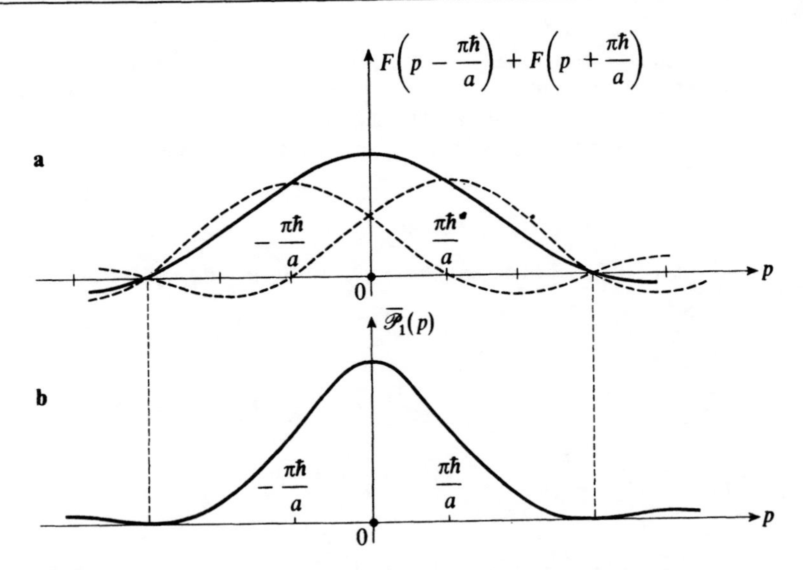


FIGURE 1

**The wave function $\overline{\varphi}_1(p)$, associated in the $\{ | p \rangle \}$ representation with the ground state of a particle in a infinite well, is obtained by adding two diffraction functions $F$ (curves in dashed lines in figure a). Since the centers of these two functions $F$ are separated by half their width, their sum has the shape represented by the solid-line curve in figure a. Squaring this sum, one obtains the probability density $\overline{\mathscr{P}}_1(p)$ associated with a measurement of the momentum of the particle (fig. b).**

Squaring these functions, one obtains the probability density $\overline{\mathscr{P}}_n(p)$ (*cf.* fig. 1-b and 2-b). Note that for large $n$ the interference term between $F\left(p - \frac{n\pi\hbar}{a}\right)$ and

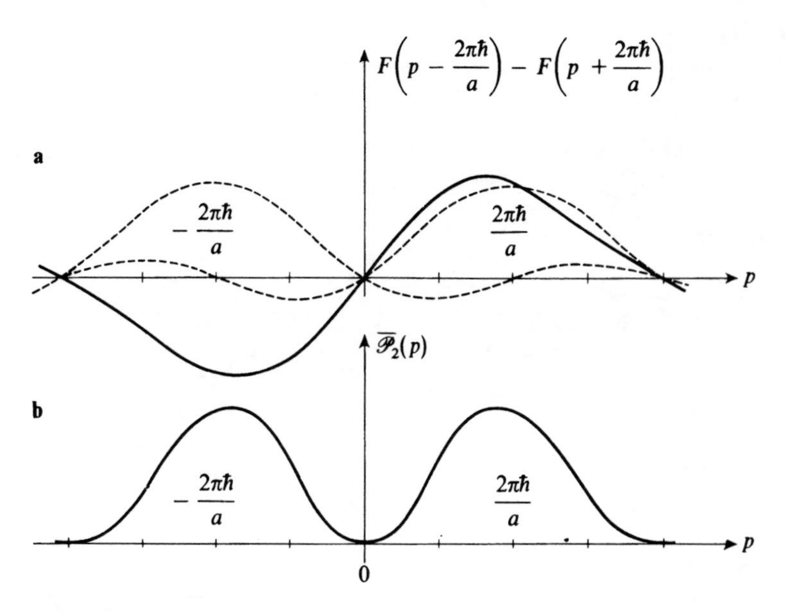


FIGURE 2

**For the first excited level, the function $\overline{\varphi}_2(p)$ is obtained by taking the difference between two functions $F$, which have the same width as in figure 1-a but are now more widely separated (dashed-line curve in figure a). The curve obtained is the solid line in figure a. The probability density $\overline{\mathscr{P}}_2(p)$ then has two maxima located in the neighborhood of $p = \pm 2\pi\hbar/a$ (fig. b).**

$F\left(p + \frac{n\pi\hbar}{a}\right)$ is negligible (because of the separation of the centers of the two curves):

$$\begin{aligned}\overline{\mathscr{P}}_n(p) &= \frac{a}{4\pi\hbar}\left[F\left(p - \frac{n\pi\hbar}{a}\right) + (-1)^{n+1}F\left(p + \frac{n\pi\hbar}{a}\right)\right]^2 \\ &\simeq \frac{a}{4\pi\hbar}\left[F^2\left(p - \frac{n\pi\hbar}{a}\right) + F^2\left(p + \frac{n\pi\hbar}{a}\right)\right] \qquad (11)\end{aligned}$$

The function $\overline{\mathscr{P}}_n(p)$ then has the shape shown in figure 3.

It can be seen that when $n$ is large, the probability density has two symmetrical peaks, of width $\frac{4\pi\hbar}{a}$, centered at $p = \pm\frac{n\pi\hbar}{a}$. It is then possible to predict with almost complete certainty the results of a measurement of the momentum of the particle in the state $|\,\varphi_n\,\rangle$: the value found will be nearly equal to $+\frac{n\pi\hbar}{a}$ or $-\frac{n\pi\hbar}{a}$, the relative accuracy* improving as $n$ increases (the two opposite

* The absolute accuracy is independent of $n$, since the width of the curves is always $\frac{4\pi\hbar}{a}$.

values $\pm \dfrac{n\pi\hbar}{a}$ being equally probable). This is simple to understand: for large $n$, the function $\varphi_n(x)$, which varies sinusoidally, performs numerous oscillations inside the well; it can then be considered to be practically the sum of two progressive waves corresponding to opposite momenta $p = \pm \dfrac{n\pi\hbar}{a}$.

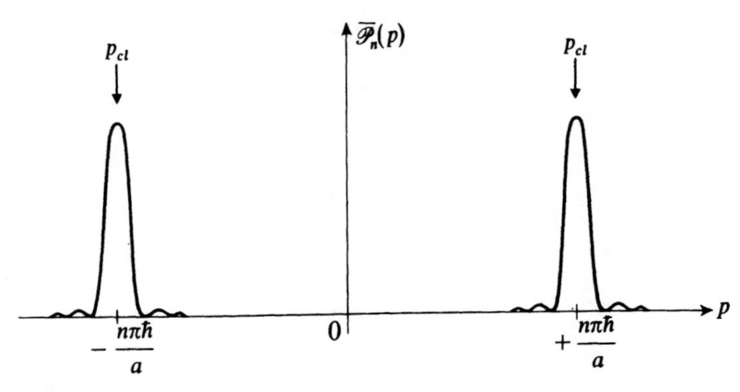


FIGURE 3

**When $n$ is large (a very excited level), the probability density has two pronounced peaks, centered at the values $p = \pm n\pi\hbar/a$, which are the momenta associated with the classical motion at the same energy.**

When $n$ decreases, the relative accuracy with which one can predict the possible values of the momentum diminishes. We see, for example, in figure 2-b, that when $n = 2$, the function $\overline{\mathscr{P}}_n(p)$ has two peaks whose widths are comparable to their distance from the origin. In this case, the wave function undergoes only one oscillation inside the well. It is not surprising that, for this sinusoid "truncated" at $x = 0$ and $x = a$, the wavelength (and therefore, the momentum of the particle) is poorly defined. Finally, for the ground state, the wave function is represented by half a sinusoidal arc: the relative values of the wavelength and momentum of the particle are then very poorly known (fig. 1-b).

COMMENTS:

(*i*) Let us calculate the momentum of a classical particle of energy $E_n$ given in (1); we have:

$$\frac{p_{cl}^2}{2m} = \frac{n^2\pi^2\hbar^2}{2ma^2} \tag{12}$$

that is:

$$p_{cl} = \pm \frac{n\pi\hbar}{a} \tag{13}$$

When $n$ is large, the two peaks of $\overline{\mathscr{P}}_n(p)$ therefore correspond to the classical values of the momentum.

(*ii*) We see that, for large $n$, although the absolute value of the momentum is well-defined, its sign is not. This is why $\Delta P_n$ is large: for probability distributions with two maxima like that of figure 3, the root-mean-square deviation reflects the distance between the two peaks; it is no longer related to their widths.

## 2. Evolution of the particle's wave function

Each of the states $| \varphi_n \rangle$, with its wave function $\varphi_n(x)$, describes a stationary state, which leads to time-independent physical predictions. Time evolution appears only when the state vector is a linear combination of several kets $| \varphi_n \rangle$. We shall consider here a very simple case, for which at time $t = 0$ the state vector $| \psi(0) \rangle$ is:

$$| \psi(0) \rangle = \frac{1}{\sqrt{2}} [| \varphi_1 \rangle + | \varphi_2 \rangle] \tag{14}$$

### a. WAVE FUNCTION AT THE INSTANT *t*

Apply formula (D-54) of chapter III; we immediately obtain:

$$| \psi(t) \rangle = \frac{1}{\sqrt{2}} \left[ e^{-i\frac{\pi^2\hbar}{2ma^2}t} | \varphi_1 \rangle + e^{-2i\frac{\pi^2\hbar}{ma^2}t} | \varphi_2 \rangle \right] \tag{15}$$

or, omitting a *global* phase factor of $| \psi(t) \rangle$:

$$| \psi(t) \rangle \propto \frac{1}{\sqrt{2}} [ | \varphi_1 \rangle + e^{-i\omega_{21}t} | \varphi_2 \rangle ] \tag{16}$$

with:

$$\omega_{21} = \frac{E_2 - E_1}{\hbar} = \frac{3\pi^2\hbar}{2ma^2} \tag{17}$$

### b. EVOLUTION OF THE SHAPE OF THE WAVE PACKET

The shape of the wave packet is given by the probability density:

$$|\psi(x, t)|^2 = \frac{1}{2} \varphi_1^2(x) + \frac{1}{2} \varphi_2^2(x) + \varphi_1(x) \varphi_2(x) \cos \omega_{21} t \tag{18}$$

We see that the time variation of the probability density is due to the interference term in $\varphi_1\varphi_2$. Only one Bohr frequency appears, $\nu_{21} = (E_2 - E_1)/h$, since the initial state (14) is composed only of the two states $| \varphi_1 \rangle$ and $| \varphi_2 \rangle$. The curves corresponding to the variation of the functions $\varphi_1^2$, $\varphi_2^2$ and $\varphi_1\varphi_2$ are traced in figures 4-a, b and c.

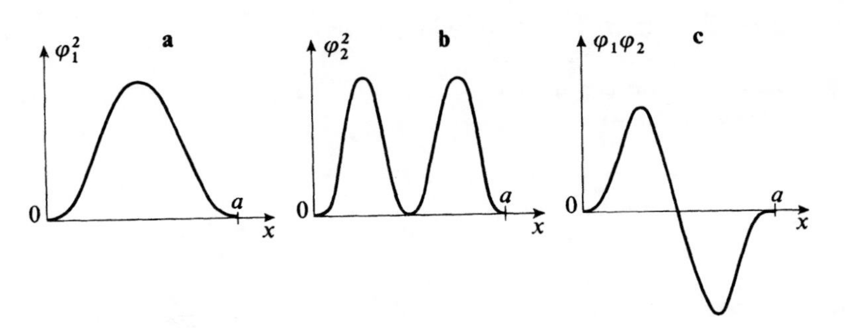


FIGURE 4

**Graphical representation of the functions $\varphi_1^2$ (the probability density of the particle in the ground state), $\varphi_2^2$ (the probability density of the particle in the first excited state) and $\varphi_1\varphi_2$ (the cross term responsible for the evolution of the shape of the wave packet).**

Using these figures and relation (18), it is not difficult to represent graphically the variation in time of the shape of the wave packet (*cf.* fig. 5): we see that the wave packet oscillates between the two walls of the well.

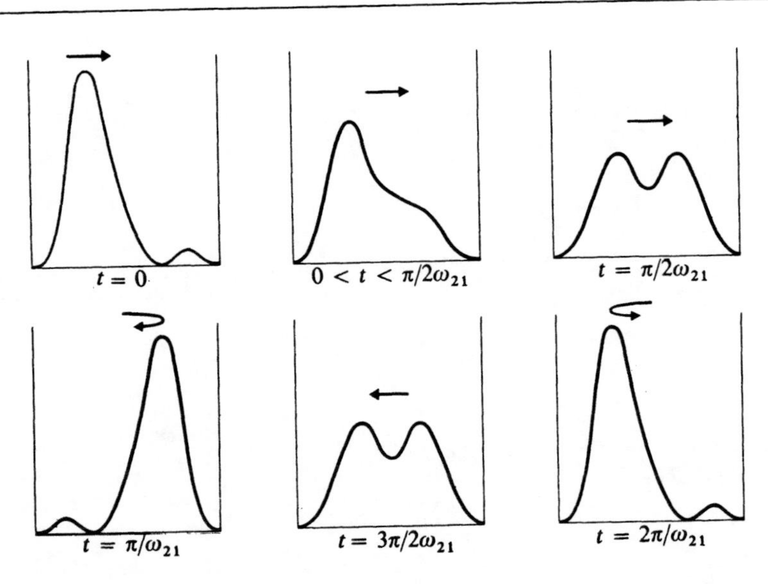


FIGURE 5

**Periodic motion of a wave packet obtained by superposing the ground state and the first excited state of a particle in an infinite well. The frequency of the motion is the Bohr frequency $\omega_{21}/2\pi$.**

### c. MOTION OF THE CENTER OF THE WAVE PACKET

Let us calculate the mean value $\langle X \rangle(t)$ of the position of the particle at time $t$. It is convenient to take:

$$X' = X - a/2 \tag{19}$$

since, by symmetry, the diagonal matrix elements of $X'$ are zero:

$$\begin{aligned}
\langle \varphi_1 | X' | \varphi_1 \rangle &\propto \int_0^a \left(x - \frac{a}{2}\right) \sin^2\left(\frac{\pi x}{a}\right) \mathrm{d}x = 0 \\
\langle \varphi_2 | X' | \varphi_2 \rangle &\propto \int_0^a \left(x - \frac{a}{2}\right) \sin^2\left(\frac{2\pi x}{a}\right) \mathrm{d}x = 0
\end{aligned} \tag{20}$$

We then have:

$$\langle X' \rangle(t) = \mathrm{Re}\,\{ \mathrm{e}^{-i\omega_{21}t} \langle \varphi_1 | X' | \varphi_2 \rangle \} \tag{21}$$

with:

$$\begin{aligned}
\langle \varphi_1 | X' | \varphi_2 \rangle &= \langle \varphi_1 | X | \varphi_2 \rangle - \frac{a}{2} \langle \varphi_1 | \varphi_2 \rangle \\
&= \frac{2}{a} \int_0^a x \sin\frac{\pi x}{a} \sin\frac{2\pi x}{a} \mathrm{d}x \\
&= -\frac{16a}{9\pi^2}
\end{aligned} \tag{22}$$

Therefore:

$$\langle X \rangle(t) = \frac{a}{2} - \frac{16a}{9\pi^2} \cos \omega_{21} t \tag{23}$$

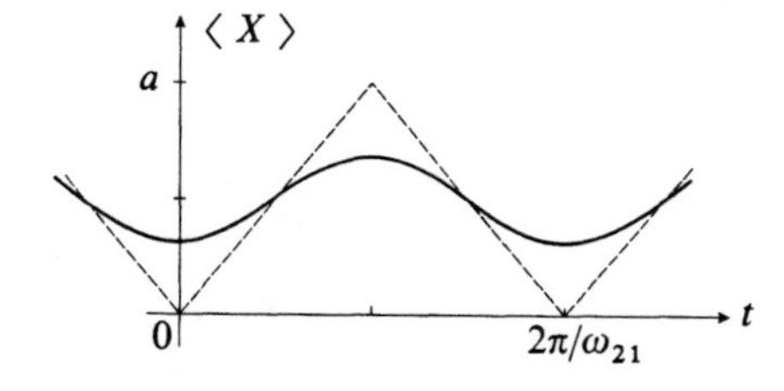


FIGURE 6

**Time variation of the mean value $\langle X \rangle$ corresponding to the wave packet of figure 5. The dashed line represents the position of a classical particle moving with the same period. Quantum mechanics predicts that the center of the wave packet will turn back before reaching the wall, as explained by the action of the potential on the "edges" of the wave packet.**

The variation of $\langle X \rangle(t)$ is represented in figure 6. In dashed lines, the variation of the position of a classical particle has been traced, for a particle moving to and fro in the well with an angular frequency of $\omega_{21}$ (since it is not subjected to any force except at the walls, its position varies linearly with $t$ between 0 and $a$ during each half-period).

We immediately notice a very clear difference between these two types of motion, classical and quantum mechanical. The center of the quantum wave packet, instead of turning back at the walls of the well, executes a movement of smaller amplitude and retraces its steps before reaching the regions where the potential is not zero. We see again here a result of §D-2 of chapter I: since the potential varies infinitely quickly at $x = 0$ and $x = a$, its variation within a domain of the order of the dimension of the wave packet is not negligible, and the motion of the center of the wave packet does not obey the laws of classical mechanics (see also chapter III, § D-1-d-$\gamma$). The physical explanation of this phenomenon is the following: before the center of the wave packet has touched the wall, the action of the potential on the "edges" of this packet is sufficient to make it turn back.

COMMENT:

The mean value of the energy of the particle in the state $|\psi(t)\rangle$ calculated in (15) is easy to obtain:

$$\langle H \rangle = \frac{1}{2}E_1 + \frac{1}{2}E_2 = \frac{5}{2}E_1 \tag{24}$$

as is:

$$\langle H^2 \rangle = \frac{1}{2}E_1^2 + \frac{1}{2}E_2^2 = \frac{17}{2}E_1^2 \tag{25}$$

which gives:

$$\Delta H = \frac{3}{2}E_1 \tag{26}$$

Note in particular that $\langle H \rangle$, $\langle H^2 \rangle$ and $\Delta H$ are not time-dependent; since $H$ is a constant of the motion, this could have been foreseen. In addition, we see from the preceding discussion that the wave packet evolves appreciably over a time of the order of:

$$\Delta t \simeq \frac{1}{\omega_{21}} \tag{27}$$

Using (26) and (27), we find:

$$\Delta H \, . \, \Delta t \simeq \frac{3}{2}E_1 \times \frac{\hbar}{3E_1} = \frac{\hbar}{2} \tag{28}$$

We again find the time-energy uncertainty relation.

### 3. Perturbation created by a position measurement

Consider a particle in the state $| \varphi_1 \rangle$. Assume that the position of the particle is measured at time $t = 0$, with the result $x = a/2$. What are the probabilities of the different results that can be obtained in a measurement of the energy, performed immediately after this first measurement?

One must beware of the following false argument: after the measurement, the particle is in the eigenstate of $X$ corresponding to the result found, and its wave function is therefore proportional to $\delta(x - a/2)$; if a measurement of the energy is then performed, the various values $E_n$ can be found, with probabilities proportional to:

$$\left| \int_0^a \mathrm{d}x \, \delta\left(x - \frac{a}{2}\right) \varphi_n^*(x) \right|^2 = \left| \varphi_n\left(\frac{a}{2}\right) \right|^2 = \begin{cases} 2/a & \text{if } n \text{ is odd} \\ 0 & \text{if } n \text{ is even} \end{cases} \tag{29}$$

Using this incorrect argument, one would find the probabilities of all values of $E_n$ corresponding to odd $n$ to be equal. This is absurd, since the sum of these probabilities would then be infinite.

This error results from the fact that we have not taken the norm of the wave function into account. To apply the fourth postulate of chapter III correctly, it is necessary to write the wave function as normalized just after the first measurement. However it is not possible to normalize the function $\delta(x - a/2)$*. The problem posed above must be stated more precisely.

As we saw in § E-2-b of chapter III, an experiment in which the measurement of an observable with a continuous spectrum is performed never yields any result with complete accuracy. For the case with which we are concerned, we can only say that:

$$\frac{a}{2} - \frac{\varepsilon}{2} \leqslant x \leqslant \frac{a}{2} + \frac{\varepsilon}{2} \tag{30}$$

where $\varepsilon$ depends on the measurement device used but is never zero.

If we assume $\varepsilon$ to be much smaller than the extension of the wave function before the measurement (here $a$), the wave function after the measurement will be practically $\sqrt{\varepsilon}\, \delta^{(\varepsilon)}\left(x - \frac{a}{2}\right)$ [$\delta^{(\varepsilon)}(x)$ is the null function everywhere except in the interval defined in (30), where it takes on the value $1/\varepsilon$; *cf.* appendix II, § 1-a]. This wave function is indeed normalized since:

$$\int \mathrm{d}x \left| \sqrt{\varepsilon}\, \delta^{(\varepsilon)}\left(x - \frac{a}{2}\right) \right|^2 = 1 \tag{31}$$

* We see concretely in this example that a $\delta$-function cannot represent a physically realizable state.

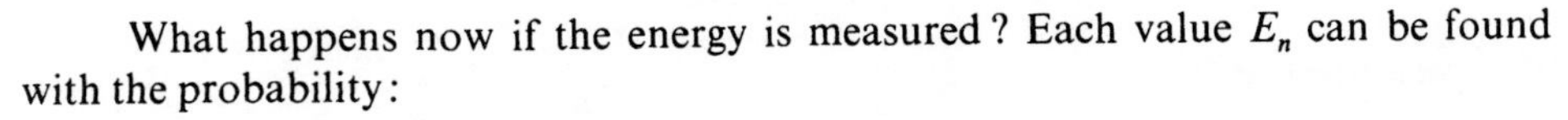



What happens now if the energy is measured? Each value $E_n$ can be found with the probability:

$$\mathscr{P}(E_n) = \left| \int \varphi_n^*(x) \sqrt{\varepsilon}\, \delta^{(\varepsilon)}\left(x - \frac{a}{2}\right) \mathrm{d}x \right|^2$$

$$= \begin{cases} \dfrac{8a}{\varepsilon}\left(\dfrac{1}{n\pi}\right)^2 \sin^2\left(\dfrac{n\pi\varepsilon}{2a}\right) & \text{if } n \text{ is odd} \\ 0 & \text{if } n \text{ is even} \end{cases} \tag{32}$$

The variation with respect to $n$ of $\mathscr{P}(E_n)$, for fixed $\varepsilon$ and odd $n$, is shown in figure 7. This figure shows that the probability $\mathscr{P}(E_n)$ becomes negligible when $n$ is much larger than $a/\varepsilon$. Therefore, however small $\varepsilon$ may be, the distribution of probabilities $\mathscr{P}(E_n)$ depends strongly on $\varepsilon$. This is why, in the first argument, where we set $\varepsilon = 0$ at the beginning, we could not obtain the correct result. We also see from the figure that the smaller $\varepsilon$ is, the more the curve extends towards large values of $n$. The interpretation of this result is the following: according to Heisenberg's uncertainty relations (*cf.* chap. I, § C-3), if one measures the position of the particle with great accuracy, one drastically changes its momentum. Thus kinetic energy is transferred to the particle, the amount increasing as $\varepsilon$ decreases.

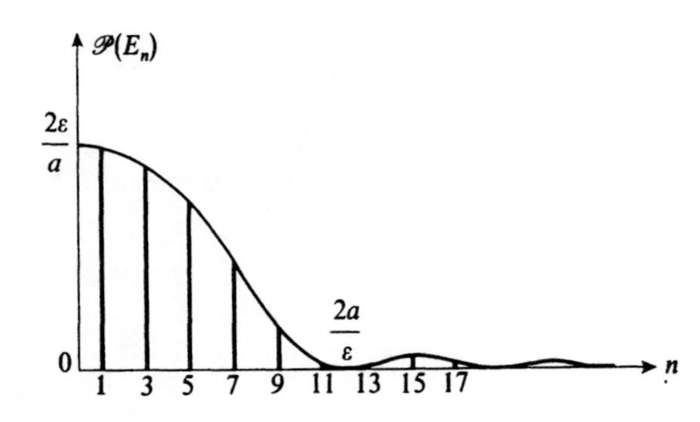


FIGURE 7

**Variation with $n$ of the probability $\mathscr{P}(E_n)$ of finding the energy $E_n$ after a measurement of the particle's position has yielded the result $a/2$ with an accuracy of $\varepsilon$ ($\varepsilon \ll a$). The smaller $\varepsilon$, the greater the probability of finding high energy values.**

## Complement $B_{III}$

## STUDY OF THE PROBABILITY CURRENT IN SOME SPECIAL CASES



The probability current associated with a particle having a wave function $\psi(\mathbf{r}, t)$ was defined in chapter III by the relation:

$$\mathbf{J}(\mathbf{r}, t) = \frac{\hbar}{2mi}\left[\psi^*(\mathbf{r}, t)\, \nabla\psi(\mathbf{r}, t) - \text{c.c.}\right] \tag{1}$$

(where c.c. is an abbreviation for complex conjugate). In this complement, we shall study this probability current in greater detail in some special cases: one- and two-dimensional "square" potentials.

### 1. Expression for the current in constant potential regions

Consider a one-dimensional problem, with a particle of energy $E$ placed in a constant potential $V_0$. In complement $H_I$, we distinguished between several cases.

(*i*) When $E > V_0$, the wave function is written:

$$\psi(x) = A\, e^{ikx} + A'\, e^{-ikx} \tag{2}$$

with:

$$E - V_0 = \frac{\hbar^2 k^2}{2m} \tag{3}$$

Substituting (2) into (1), we obtain:

$$J_x = \frac{\hbar k}{m}\left[\,|A|^2 - |A'|^2\,\right] \tag{4}$$

The interpretation of this result is simple: the wave function given in (2) corresponds to two plane waves of opposite momenta $p = \pm\,\hbar k$ with probability densities $|A|^2$ and $|A'|^2$.

(*ii*) When $E < V_0$, we have:

$$\psi(x) = B\, e^{\rho x} + B'\, e^{-\rho x} \tag{5}$$

with:

$$V_0 - E = \frac{\hbar^2 \rho^2}{2m} \tag{6}$$

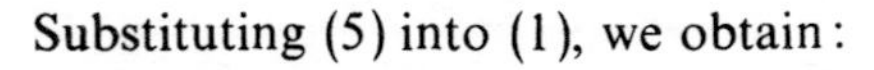

Substituting (5) into (1), we obtain:

$$J_x = \frac{\hbar \rho}{m} \left[ iB^* B' + \text{c.c.} \right] \tag{7}$$

In this case, we see that the two exponential waves must both have non-zero coefficients for the probability current to be non-zero.

## 2. Application to potential barrier problems

Let us apply these results to the potential barrier problems studied in complements $H_I$ and $J_I$. We shall therefore consider a particle of mass $m$ and energy $E$ propagating in the $Ox$ direction and arriving at $x = 0$ at a potential step of height $V_0$ (fig. 1).

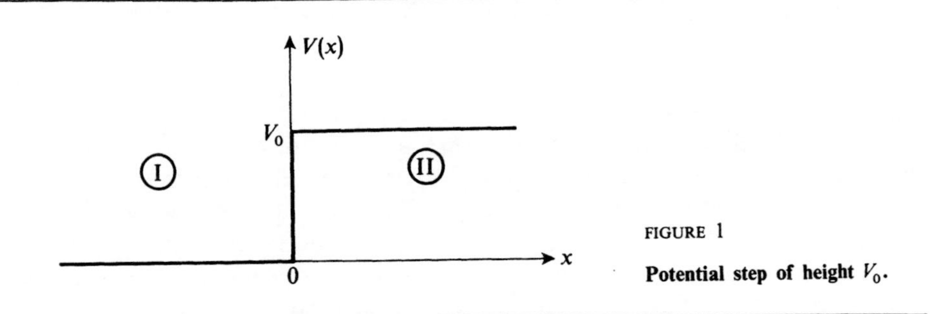


FIGURE 1

**Potential step of height $V_0$.**

### a. CASE WHERE $E > V_0$

Apply formula (4) to wave functions (11) and (12) of complement $H_I$, setting, as in that complement:

$$A_2' = 0 \tag{8}$$

In region I, the probability current is:

$$J_I = \frac{\hbar k_1}{m} \left[ |A_1|^2 - |A_1'|^2 \right] \tag{9}$$

and in region II:

$$J_{II} = \frac{\hbar k_2}{m} |A_2|^2 \tag{10}$$

$J_I$ is the difference of two terms, the first one corresponding to the incident current and the second, to the reflected current. The ratio of these two currents gives the reflection coefficient $R$ of the barrier:

$$R = \left| \frac{A_1'}{A_1} \right|^2 \tag{11}$$

which is precisely formula (15) of complement $H_I$.

Similarly, the transmission coefficient $T$ of the barrier is the ratio of the transmitted current $J_{II}$ to the incident current; we therefore have:

$$T = \frac{k_2}{k_1}\left|\frac{A_2}{A_1}\right|^2 \tag{12}$$

and we again find relation (16) of complement $H_I$.

#### b. CASE WHERE $E < V_0$

Since the expression for the wave function $\varphi_I(x)$ is the same as in § a, relation (9) is still valid. However, in region II, the wave function is:

$$\varphi_{II}(x) = B_2' \, e^{-\rho_2 x} \tag{13}$$

[since, in equation (20) of complement $H_I$, $B_2 = 0$]. Using (7), we thus obtain:

$$J_{II} = 0 \tag{14}$$

The transmitted flux is zero, as is consistent with relation (24) of $H_I$.

How should we interpret the fact that, in region II, the probability current is zero while the probability of finding the particle in this region is not? Let us refer to the results obtained in § 1 of complement $J_I$. We saw that part of the incident wave packet enters the classically forbidden region II and then turns back before setting out again in the negative $x$ direction (this incursion into region II being responsible for the delay upon reflection). In the steady state, we shall therefore have two probability currents in region II : a positive current corresponding to the entrance into this region of part of the incident wave packet; a negative current corresponding to the return towards region I of this part of the wave packet. These two currents are exactly equal, so the overall result is zero.

In the case of a one-dimensional problem, the structure of the probability current of the evanescent wave is therefore masked by the fact that the two opposite currents balance. This is why we are going to consider a two-dimensional problem, for the case of oblique reflection, so as to obtain a non-zero current and interpret its structure.

## 3. Probability current of incident and evanescent waves, in the case of reflection from a two-dimensional potential step

We shall consider the following two-dimensional problem : a particle of mass $m$, moving in the $xOy$ plane, has a potential energy $V(x, y)$ which is independent of $y$ and given by:

$$\begin{aligned} V(x, y) &= 0 \quad &\text{if} \quad x < 0 \\ V(x, y) &= V_0 \quad &\text{if} \quad x > 0 \end{aligned} \tag{15}$$

The present case corresponds to the one studied in §2 of complement $F_I$: the potential energy $V(x, y)$ is the sum of a function $V_1(x)$ (potential energy of a one-dimensional step) and a function $V_2(y)$, which is zero here. We can therefore look for a solution of the eigenvalue equation of the Hamiltonian in the form of a product:

$$\varphi(x, y) = \varphi_1(x)\,\varphi_2(y) \tag{16}$$

The functions $\varphi_1(x)$ and $\varphi_2(y)$ satisfy one-dimensional eigenvalue equations which correspond respectively to $V_1(x)$ and $V_2(y)$ and to energies $E_1$ and $E_2$ such that:

$$E_1 + E_2 = E \quad \text{(total energy of the particle)} \tag{17}$$

We shall assume $E_1 < V_0$: the equation which gives $\varphi_1(x)$ therefore corresponds to total reflection in a one-dimensional problem, and we can use formulas (11) and (20) of complement $H_I$. As for the function $\varphi_2(y)$, it can be obtained immediately since it corresponds to the case of a free particle ($V_2 = 0$): it is a plane wave. We therefore have, in region I ($x < 0$):

$$\varphi_I(x, y) = A\,e^{i(k_x x + k_y y)} + A'\,e^{i(-k_x x + k_y y)} \tag{18}$$

with:

$$k_x = \sqrt{\frac{2mE_1}{\hbar^2}} \qquad\qquad k_y = \sqrt{\frac{2mE_2}{\hbar^2}} \tag{19}$$

and, in region II ($x > 0$):

$$\varphi_{II}(x, y) = B\,e^{-\rho_x x}\,e^{ik_y y} \tag{20}$$

with:

$$\rho_x = \sqrt{\frac{2m(V_0 - E_1)}{\hbar^2}} \tag{21}$$

Equations (22) and (23) of $H_I$ give us the ratios $A'/A$ and $B/A$. Introducing the parameter $\theta$ defined by:

$$\tan\theta = \frac{\rho_x}{k_x} = \sqrt{\frac{V_0 - E_1}{E_1}}\,; \qquad\qquad 0 \leqslant \theta \leqslant \frac{\pi}{2} \tag{22}$$

we obtain:

$$\frac{A'}{A} = \frac{k_x - i\rho_x}{k_x + i\rho_x} = e^{-2i\theta} \tag{23}$$

and:

$$\frac{B}{A} = \frac{2k_x}{k_x + i\rho_x} = 2\cos\theta\,e^{-i\theta} \tag{24}$$

Let us apply relation (1), which defines the probability current. We obtain, in region I:

$$\mathbf{J}_I \begin{cases} (J_I)_x = \dfrac{\hbar k_x}{m}\left[\,|A|^2 - |A'|^2\right] = 0 \\ (J_I)_y = \dfrac{\hbar k_y}{m}\,|A\,\mathrm{e}^{ik_x x} + A'\,\mathrm{e}^{-ik_x x}|^2 \end{cases}$$

$$= \frac{\hbar k_y}{m}\,|A|^2\left[2 + 2\cos\,(2k_x x + 2\theta)\right] \tag{25}$$

and in region II:

$$\mathbf{J}_{II} \begin{cases} (J_{II})_x = 0 \\ (J_{II})_y = \dfrac{\hbar k_y}{m}\,|B|^2\,\mathrm{e}^{-2\rho_x x} = \dfrac{\hbar k_y}{m}\,4\,|A|^2\cos^2\theta\,\mathrm{e}^{-2\rho_x x} \end{cases} \tag{26}$$

In region I, only the $(J_I)_y$ component of the probability current is non-zero; this component is the sum of two terms:

– the term proportional to $2|A|^2$ which results from the sum of the currents of the incident and reflected waves (*cf.* fig. 2);

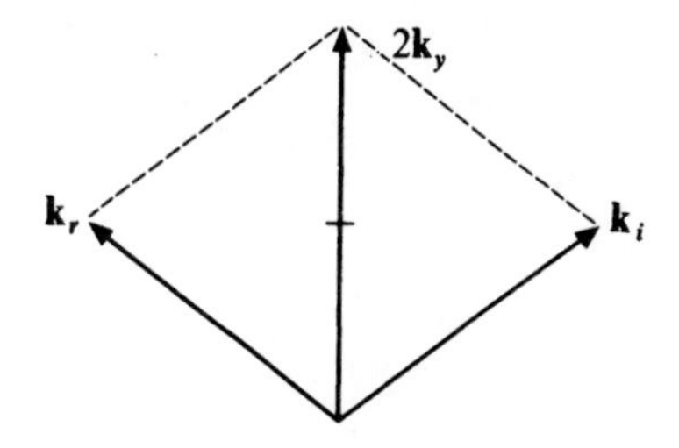


FIGURE 2

**The sum of the probability currents associated with the incident and reflected waves yields a probability current parallel to *Oy*.**

– the term containing $\cos\,(2k_x x + 2\theta)$, which represents an interference effect between the two waves; it is responsible for the oscillation of the probability current with respect to $x$ (*cf.* fig. 3).

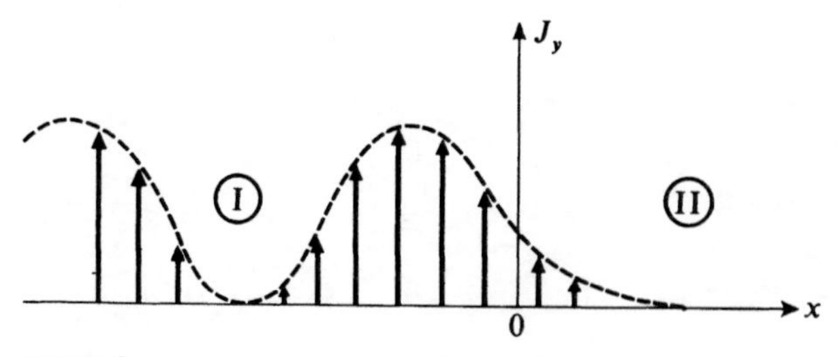


FIGURE 3

**Because of the interference between the incident and reflected waves, the probability current in region I is an oscillatory function of $x$; in region II, it decreases exponentially (evanescent wave).**

In region II, the probability current is again parallel to $Oy$. Its exponential decay corresponds to the decay of the evanescent wave. This probability current arises from the fact that the wave packets do enter the second region (*cf.* fig. 4) and, before turning back, propagate in the $Oy$ direction for a time of the order of the reflection delay $\tau$ [*cf.* complement $J_I$, equation (10)]. This penetration is also related to the lateral shift of the wave packet upon reflection (*cf.* fig. 4).

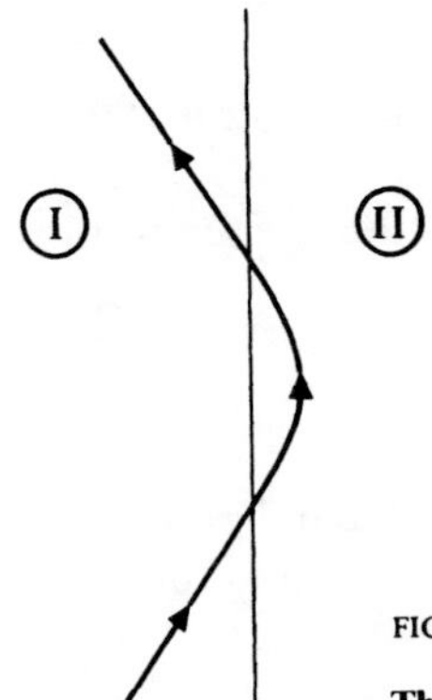


FIGURE 4

**The penetration of the particle into region II leads to a lateral shift upon reflection.**

# Complement $C_{III}$

## ROOT-MEAN-SQUARE DEVIATIONS OF TWO CONJUGATE OBSERVABLES



Two conjugate observables $P$ and $Q$ are two observables whose commutator $[Q, P]$ is equal to $i\hbar$. We shall show in this complement that the root-mean-square deviations (*cf.* §C-5 of chapter III) $\Delta P$ and $\Delta Q$, for any state vector of the system under study, satisfy the relation:

$$\Delta P \, . \, \Delta Q \geqslant \frac{\hbar}{2} \tag{1}$$

We shall then show that if the system is in a state where the product $\Delta P \, . \, \Delta Q$ is exactly equal to $\hbar/2$, the wave function associated with this state in the $\{ | q \rangle \}$ representation is a Gaussian wave packet (as is the wave function in the $\{ | p \rangle \}$ representation).

### 1. The uncertainty relation for *P* and *Q*

Consider the ket:

$$| \varphi \rangle = (Q + i\lambda P) | \psi \rangle \tag{2}$$

where $\lambda$ is an arbitrary real parameter. For all $\lambda$, the square of the norm $\langle \varphi | \varphi \rangle$ is positive. This is written:

$$\begin{aligned} \langle \varphi | \varphi \rangle &= \langle \psi | (Q - i\lambda P)(Q + i\lambda P) | \psi \rangle \\ &= \langle \psi | Q^2 | \psi \rangle + \langle \psi | (i\lambda QP - i\lambda PQ) | \psi \rangle + \langle \psi | \lambda^2 P^2 | \psi \rangle \\ &= \langle Q^2 \rangle + i\lambda \langle [Q, P] \rangle + \lambda^2 \langle P^2 \rangle \\ &= \langle Q^2 \rangle - \lambda\hbar + \lambda^2 \langle P^2 \rangle \geqslant 0 \end{aligned} \tag{3}$$

The discriminant of this expression, of second order in $\lambda$, is therefore negative or zero:

$$\hbar^2 - 4 \langle P^2 \rangle \langle Q^2 \rangle \leqslant 0 \tag{4}$$

and we have:

$$\langle P^2 \rangle \langle Q^2 \rangle \geqslant \frac{\hbar^2}{4} \tag{5}$$

Assuming $| \psi \rangle$ to be given, let us now introduce the two observables $Q'$ and $P'$ defined by:

$$\begin{aligned} P' &= P - \langle P \rangle = P - \langle \psi | P | \psi \rangle \\ Q' &= Q - \langle Q \rangle = Q - \langle \psi | Q | \psi \rangle \end{aligned} \tag{6}$$

$P'$ and $Q'$ are also conjugate observables, since we have:

$$[Q', P'] = [Q, P] = i\hbar \tag{7}$$

Result (5), obtained above for $P$ and $Q$, is therefore also valid for $P'$ and $Q'$:

$$\langle P'^2 \rangle \langle Q'^2 \rangle \geqslant \frac{\hbar^2}{4} \tag{8}$$

In addition, referring to definition (C-23) (chap. III) of the root-mean-square deviation and using (6), we see that:

$$\begin{aligned} \Delta P &= \sqrt{\langle P'^2 \rangle} \\ \Delta Q &= \sqrt{\langle Q'^2 \rangle} \end{aligned} \tag{9}$$

Relation (8) can therefore also be written:

$$\boxed{\Delta P \,.\, \Delta Q \geqslant \frac{\hbar}{2}} \tag{10}$$

Thus, if two observables are conjugate (as is the case when they correspond to a classical position $x_i$ and its conjugate momentum $p_i$), there exists an exact lower bound for the product $\Delta P \,.\, \Delta Q$. We thus generalize the Heisenberg uncertainty relation.

COMMENT:

This argument can easily be generalized to two arbitrary observables $A$ and $B$. One obtains:

$$\Delta A \,.\, \Delta B \geqslant \frac{1}{2} |\langle [A, B] \rangle| \tag{11}$$

## 2. The "minimum" wave packet

When the minimum value of the product $\Delta P \,.\, \Delta Q$ is attained:

$$\Delta P \,.\, \Delta Q = \frac{\hbar}{2} \tag{12}$$

the state vector $|\psi\rangle$ is said to correspond to a minimum wave packet for the observables $P$ and $Q$.

According to the preceding argument, relation (12) requires that the square of the norm of the ket:

$$|\varphi'\rangle = (Q' + i\lambda P')|\psi\rangle \tag{13}$$

be a second-order polynomial in $\lambda$ with a double root $\lambda_0$. When $\lambda = \lambda_0$, the ket $|\varphi'\rangle$ is therefore zero:

$$(Q' + i\lambda_0 P')|\psi\rangle = [Q - \langle Q \rangle + i\lambda_0(P - \langle P \rangle)]|\psi\rangle = 0 \tag{14}$$

On the other hand, if $\Delta P \,.\, \Delta Q > \hbar/2$, the polynomial which gives $\langle \varphi' \mid \varphi' \rangle$ can never be equal to zero (it is positive for all $\lambda$).

Therefore, the necessary and sufficient condition for the product $\Delta P \,.\, \Delta Q$ to take on its minimum value $\hbar/2$ is that the kets $(Q - \langle Q \rangle) \mid \psi \rangle$ and $(P - \langle P \rangle) \mid \psi \rangle$ be proportional. The proportionality coefficient $- i\lambda_0$ can easily be calculated. When $\Delta Q \,.\, \Delta P = \hbar/2$, the equation :

$$\langle \varphi' \mid \varphi' \rangle = \lambda^2(\Delta P)^2 - \lambda\hbar + (\Delta Q)^2 = 0 \tag{15}$$

has for its double root :

$$\lambda_0 = \frac{\hbar}{2(\Delta P)^2} = \frac{2(\Delta Q)^2}{\hbar} \tag{16}$$

Let us write relation (14) in the $\{ \mid q \rangle \}$ representation (for simplicity, we assume the eigenvalues $q$ of $Q$ to be non-degenerate). Using the fact (*cf.* complement $E_{II}$) that in this representation, $P$ acts like $\frac{\hbar}{i}\frac{d}{dq}$, we obtain:

$$\left[ q + \hbar\lambda_0 \frac{d}{dq} - \langle Q \rangle - i\lambda_0 \langle P \rangle \right] \psi(q) = 0 \tag{17}$$

with:

$$\psi(q) = \langle q \mid \psi \rangle \tag{18}$$

To integrate equation (17), it is convenient to introduce the function $\theta(q)$ defined by:

$$\psi(q) = e^{i\langle P \rangle q/\hbar}\, \theta(q - \langle Q \rangle) \tag{19}$$

Substituting (19) into (17), we thus obtain a more simple equation :

$$\left[ q + \lambda_0 \hbar \frac{d}{dq} \right] \theta(q) = 0 \tag{20}$$

whose solution is:

$$\theta(q) = C\, e^{-q^2/2\lambda_0\hbar} \tag{21}$$

(where $C$ is an arbitrary complex constant). Substituting (16) and (21) into (19), we obtain:

$$\psi(q) = C\, e^{i\langle P \rangle q/\hbar}\, e^{-\left[\frac{q - \langle Q \rangle}{2\,\Delta Q}\right]^2} \tag{22}$$

This function can be normalized by setting:

$$C = [2\pi(\Delta Q)^2]^{-1/4} \tag{23}$$

We thus arrive at the following conclusion: when the product $\Delta P \,.\, \Delta Q$ takes on its minimum value $\hbar/2$, the wave function in the $\{ \mid q \rangle \}$ representation is a Gaussian wave packet, obtained from the Gaussian function $\theta(q)$ by transformation (19) (which is equivalent to two changes of the origin, one on the $q$-axis and one on the $p$-axis).

COMMENT :

This argument in the $\{ | q \rangle \}$ representation can be repeated in the $\{ |p \rangle \}$ representation. One then finds that the wave function $\overline{\psi}(p)$ defined by:

$$\overline{\psi}(p) = \langle p | \psi \rangle = \frac{1}{\sqrt{2\pi\hbar}} \int_{-\infty}^{+\infty} \mathrm{d}q \; \mathrm{e}^{-ipq/\hbar} \; \psi(q) \tag{24}$$

is also a Gaussian function, given by:

$$\overline{\psi}(p) = \left[2\pi(\Delta P)^2\right]^{-1/4} \mathrm{e}^{-i\langle Q \rangle p/\hbar} \; \mathrm{e}^{-\left[\frac{p - \langle P \rangle}{2\Delta P}\right]^2} \tag{25}$$

up to a phase factor $\exp\left(i \langle Q \rangle \langle P \rangle / \hbar\right)$.

# Complement $D_{III}$

## MEASUREMENTS BEARING ON ONLY ONE PART OF A PHYSICAL SYSTEM



The concept of a tensor product, introduced in § F of chapter II, enabled us to see how to construct, starting with the state spaces of two subsystems, that of the global system obtained by considering them together. We intend to pursue this study here, using the postulates of chapter III to see what results can be obtained, when the state of the global system is known, from measurements bearing on only one subsystem.

### 1. Calculation of the physical predictions

Consider a system composed of two parts (1) and (2) (for example, a system of two electrons). If $\mathscr{E}(1)$ and $\mathscr{E}(2)$ are the state spaces of parts (1) and (2), the state space of the global system (1) + (2) is the tensor product $\mathscr{E}(1) \otimes \mathscr{E}(2)$. For example the state of a two-electron system is described by a wave function of six variables, $\psi(x_1, y_1, z_1 ; x_2, y_2, z_2)$, associated with a ket of $\mathscr{E}_{\mathbf{r}}(1) \otimes \mathscr{E}_{\mathbf{r}}(2)$ (*cf.* chap. II, § F-4-b).

It is possible to imagine measurements which bear on only one of the two parts [part (1), for example] of the global system. The observables $\tilde{A}(1)$ corresponding to these measurements are defined in $\mathscr{E}(1) \otimes \mathscr{E}(2)$ by extending the observables $A(1)$ acting only in $\mathscr{E}(1)$* (*cf.* chap. II, § F-2-b) :

$$A(1) \Longrightarrow \tilde{A}(1) = A(1) \otimes \mathbb{1}(2) \tag{1}$$

where $\mathbb{1}(2)$ is the identity operator in $\mathscr{E}(2)$.

The spectrum of $\tilde{A}(1)$ in $\mathscr{E}(1) \otimes \mathscr{E}(2)$ is the same as that of $A(1)$ in $\mathscr{E}(1)$. On the other hand, we have seen that all the eigenvalues of $\tilde{A}(1)$ are degenerate in $\mathscr{E}(1) \otimes \mathscr{E}(2)$, even if none of the eigenvalues of $A(1)$ is degenerate in $\mathscr{E}(1)$ [on the condition, of course, that the dimension of $\mathscr{E}(2)$ be greater than 1]. When a measurement is made on system (1) alone, the global system may therefore be in several different states after the measurement, whatever the result (the state after the measurement depends not only on the result but also on the state before the measurement). From a physical point of view, this multiplicity of states corresponds to the degrees of freedom of system (2), about which no information is sought in the measurement.

* For the sake of clarity, we shall adopt throughout this complement different notations for $A(1)$ and its extension $\tilde{A}(1)$.

Let $P_n(1)$ be the projector, in $\mathscr{E}(1)$, onto the eigensubspace related to the eigenvalue $a_n$ of $A(1)$:

$$P_n(1) = \sum_{i=1}^{g_n} | u_n^i(1) \rangle \langle u_n^i(1) | \tag{2}$$

where the kets $|u_n^i(1)\rangle$ are $g_n$ orthonormal eigenvectors associated with $a_n$. Let $\tilde{P}_n(1)$ be the projector, in $\mathscr{E}(1) \otimes \mathscr{E}(2)$, onto the eigensubspace related to the same eigenvalue $a_n$ of $\tilde{A}(1)$. $\tilde{P}_n(1)$ is obtained by extending $P_n(1)$ into $\mathscr{E}(1) \otimes \mathscr{E}(2)$:

$$\tilde{P}_n(1) = P_n(1) \otimes \mathbb{1}(2) \tag{3}$$

To write the identity operator $\mathbb{1}(2)$ of $\mathscr{E}(2)$, let us use the closure relation for an arbitrary orthonormal basis $\{ |v_k(2)\rangle \}$ of $\mathscr{E}(2)$:

$$\mathbb{1}(2) = \sum_k |v_k(2) \rangle \langle v_k(2) | \tag{4}$$

Substituting (4) into (3) and using (2), we obtain:

$$\tilde{P}_n(1) = \sum_{i=1}^{g_n} \sum_k | u_n^i(1)\, v_k(2) \rangle \langle u_n^i(1)\, v_k(2) | \tag{5}$$

Thus, knowing the state $|\psi\rangle$ of the global system (assumed to be normalized to 1), we can calculate the probability $\mathscr{P}^{(1)}(a_n)$ of finding the result $a_n$ in a measurement of $A(1)$ on part (1) of this system. Using general formula (B-14) of chapter III, which here gives:

$$\mathscr{P}^{(1)}(a_n) = \langle \psi | \tilde{P}_n(1) | \psi \rangle \tag{6}$$

we find:

$$\mathscr{P}^{(1)}(a_n) = \sum_{i=1}^{g_n} \sum_k |\langle u_n^i(1)\, v_k(2) | \psi \rangle|^2 \tag{7}$$

Similarly, the state $|\psi'\rangle$ of the system after the measurement can be calculated; according to formula (B-31) of chapter III, it is given by:

$$| \psi' \rangle = \frac{\tilde{P}_n(1) | \psi \rangle}{\sqrt{\langle \psi | \tilde{P}_n(1) | \psi \rangle}} \tag{8}$$

that is, using (5):

$$| \psi' \rangle = \frac{\sum_{i=1}^{g_n} \sum_k | u_n^i(1)\, v_k(2) \rangle \langle u_n^i(1)\, v_k(2) | \psi \rangle}{\sqrt{\sum_{i=1}^{g_n} \sum_k |\langle u_n^i(1)\, v_k(2) | \psi \rangle|^2}} \tag{9}$$

COMMENTS:

(*i*) The choice of an orthonormal basis, $\{ |v_k(2)\rangle \}$, in $\mathscr{E}(2)$ is arbitrary. We see from (3), (6) and (8) that the predictions concerning subsystem (1) do not depend on this choice. Physically, it is clear that if no measurement is performed on system (2), no state or set of states of this system can play a preferential role.

(*ii*) If the state $|\psi\rangle$ before the measurement is a tensor product :

$$|\psi\rangle = |\varphi(1)\rangle \otimes |\chi(2)\rangle \tag{10}$$

[where $|\varphi(1)\rangle$ and $|\chi(2)\rangle$ are two normalized states of $\mathscr{E}(1)$ and $\mathscr{E}(2)$], it is easy to see, using (3) and (8), that the state $|\psi'\rangle$ is also a tensor product :

$$|\psi'\rangle = |\varphi'(1)\rangle \otimes |\chi(2)\rangle \tag{11}$$

with :

$$|\varphi'(1)\rangle = \frac{P_n(1)\,|\,\varphi(1)\rangle}{\sqrt{\langle\,\varphi(1)\,|\,P_n(1)\,|\,\varphi(1)\,\rangle}} \tag{12}$$

The state of system (1) has therefore changed, but not that of system (2).

(*iii*) If the eigenvalue $a_n$ of $A(1)$ is non-degenerate in $\mathscr{E}(1)$ – or, more generally, if $A(1)$ actually represents a complete set of commuting observables of $\mathscr{E}(1)$ – the index $i$ is no longer necessary in formula (2) and those which follow. The state of the system after a measurement yielding the result $a_n$ can always be put in the form of a product of two vectors. This can be seen by writing relation (9) in the form :

$$|\psi'\rangle = |u_n(1)\rangle \otimes |\chi'(2)\rangle \tag{13}$$

where the normalized vector $|\chi'(2)\rangle$ of $\mathscr{E}(2)$ is given by :

$$|\chi'(2)\rangle = \frac{\sum_k |\,v_k(2)\,\rangle\langle\,u_n(1) v_k(2)\,|\,\psi\,\rangle}{\sqrt{\sum_k |\langle\,u_n(1)\,v_k(2)\,|\,\psi\,\rangle|^2}} \tag{14}$$

Therefore, whatever the state $|\psi\rangle$ of the global system before the measurement, the state of the system after a measurement bearing on part (1) alone is always a tensor product when this measurement is complete with respect to part (1) [although partial as regards the global system (1) + (2)].

## 2. The physical meaning of a tensor product state

To see what a product state represents physically, let us apply the results of the preceding paragraph to the particular case where the initial state of the global system is of the form (10). We immediately obtain, using (6) and (3):

$$\mathscr{P}^{(1)}(a_n) = \langle\,\varphi(1)\,\chi(2)\,|\,P_n(1) \otimes \mathbb{1}(2)\,|\,\varphi(1)\,\chi(2)\,\rangle \tag{15}$$

The very definition of the tensor product $P_n(1) \otimes \mathbb{1}(2)$ and the fact that $|\chi(2)\rangle$ is normalized then allow us to write :

$$\begin{aligned}\mathscr{P}^{(1)}(a_n) &= \langle\,\varphi(1)\,|\,P_n(1)\,|\,\varphi(1)\,\rangle\langle\,\chi(2)\,|\,\mathbb{1}(2)\,|\,\chi(2)\,\rangle \\ &= \langle\,\varphi(1)\,|\,P_n(1)\,|\,\varphi(1)\,\rangle\end{aligned} \tag{16}$$

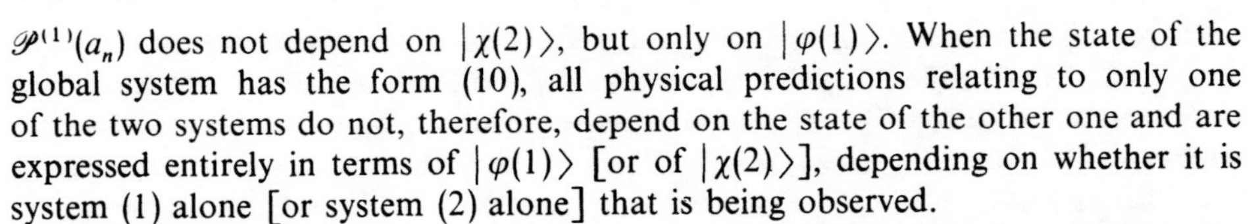

$\mathscr{P}^{(1)}(a_n)$ does not depend on $|\chi(2)\rangle$, but only on $|\varphi(1)\rangle$. When the state of the global system has the form (10), all physical predictions relating to only one of the two systems do not, therefore, depend on the state of the other one and are expressed entirely in terms of $|\varphi(1)\rangle$ [or of $|\chi(2)\rangle$], depending on whether it is system (1) alone [or system (2) alone] that is being observed.

A product state $|\varphi(1)\rangle \otimes |\chi(2)\rangle$ can therefore be considered to represent the simple juxtaposition of two systems, one in the state $|\varphi(1)\rangle$ and the other in the state $|\chi(2)\rangle$. In such a state, the two systems are also said to be *uncorrelated* (more precisely, the results of the two types of measurements, bearing either on one system or on the other, correspond to *independent random variables*). Such a situation is realized when the two systems have been separately prepared in the states $|\varphi(1)\rangle$ and $|\chi(2)\rangle$ and then united without interacting.

## 3. The physical meaning of a state which is not a tensor product

Now consider the case in which the state of the global system is not a product state, that is, where $|\psi\rangle$ cannot be written in the form $|\varphi(1)\rangle \otimes |\chi(2)\rangle$. The predictions of measurement results bearing on only one of the two systems can then no longer be expressed in terms of a ket $|\varphi(1)\rangle$ [or $|\chi(2)\rangle$] in which system (1) [or (2)] would be found. In this case, general formulas (6) and (7) must be used to find the probabilities of the various possible results. We assume here without proof that such a situation generally reflects the existence of *correlations* between systems (1) and (2). The results of measurements bearing on either system (1) or system (2) correspond to random variables which are not independent and can therefore be correlated. It can be shown, for example, that an interaction between the two systems transforms an initial state which is a product into one which is no longer a product: any interaction between two systems therefore introduces, in general, correlations between them.

When the state of the global system is not a product $|\varphi(1)\rangle \otimes |\chi(2)\rangle$, how can each partial system (1) or (2) be characterized, since the ket $|\varphi(1)\rangle$ or $|\chi(2)\rangle$ can no longer be associated with it? This question is very important since, in general, every physical system has interacted with others in the past (even if it is isolated at the instant when it is being studied). The state of the global system : { system (1) + systems (2) with which it has interacted in the past } is therefore not in general a product state, and it is not possible to associate a state vector $|\varphi(1)\rangle$ with system (1) alone. To resolve these difficulties, one must describe system (1), not by a state vector, but by an operator, called the density operator. Some indication of the corresponding formalism, fundamental to statistical quantum mechanics, is given in complement $E_{III}$ (§5-b).

However, system (1) can always be described by a state vector when a complete set of measurements has been performed on it. We have seen that whatever the state of the global system (1) + (2) before the measurement, a complete measurement on system (1) places the global system in a product state [*cf.* formulas (13) and (14)]. The vector associated with (1) is the unique eigenvector (to within a multiplicative factor) associated with the results of the complete set of measurements done on it. This set of measurements has therefore erased all correlations resulting from previous interactions between the two systems. If, at the moment of measurement, system (2) is already far away and no longer interacting with system (1), it can then be completely forgotten.

COMMENT:

It is easy to deduce from (14) that, when the state $| \psi \rangle$ before the measurement is not a product state, the state vector $| \chi'(2) \rangle$ associated with system (2) after the measurement depends on the result of the complete set of measurements made on system (1) [recall that this is not the case when $| \psi \rangle$ is a product state; *cf.* comment (*ii*) of § 1]. This result may at first seem surprising: the state of system (2) after a set of measurements performed on system (1) does depend on the result of these measurements, even if system (2), at the moment of measurement, is already very far away from system (1) and no longer interacting with it. To this "paradox", studied in detail by certain physicists, are attached the names of Einstein, Podolsky and Rosen.

**References and suggestions for further reading :**

The Einstein-Podolsky-Rosen paradox : see the subsection "Hidden variables and paradoxes" of section 5 of the bibliography; Bohm (5.1), §§22.15 to 22.19; d'Espagnat (5.3), chap. 7.

Photons produced in the decay of positronium : Feynman III (1.2), § 18.3; Dicke and Wittke (1.14), chap. 7.

## Complement $E_{III}$

# THE DENSITY OPERATOR



## 1. Outline of the problem

Until now, we have considered systems whose state is perfectly well known. We have shown how to study their time evolution and how to predict the results of various measurements performed on them. To determine the state of a system at a given instant, it suffices to perform on the system a set of measurements corresponding to a C.S.C.O. For example, in the experiment studied in § A-3 of chapter I, the polarization state of the photons is perfectly well known when the light beam has traversed the polarizer.

However, in practice, the state of the system is often not perfectly determined. This is true, for example, of the polarization state of photons coming from a source of natural (unpolarized) light, and also for the atoms of a beam emitted by a furnace at temperature $T$, where the atoms' kinetic energy is known only statistically. The problem posed by the quantum description of such systems is the following: how can we incorporate into the formalism the *incomplete information* we possess about the state of the system, so that our predictions make maximum use of this partial information? To do this, we shall introduce here a very useful mathematical tool, the density operator, which facilitates the simultaneous application of the postulates of quantum mechanics and the results of probability calculations.

## 2. The concept of a statistical mixture of states

When one has incomplete information about a system, one typically appeals to the concept of probability. For example, we know that a photon emitted by a source of natural light can have any polarization state with equal probability.

Similarly, a system in thermodynamic equilibrium at a temperature $T$ has a probability proportional to $e^{-E_n/kT}$ of being in a state of energy $E_n$.

More generally, the incomplete information one has about the system usually presents itself, in quantum mechanics, in the following way : the state of this system may be either the state $|\psi_1\rangle$ with a probability $p_1$ or the state $|\psi_2\rangle$ with a probability $p_2$, etc... Obviously :

$$p_1 + p_2 + ... = \sum_k p_k = 1 \tag{1}$$

We then say that we are dealing with a *statistical mixture* of states $|\psi_1\rangle, |\psi_2\rangle, ...$ with probabilities $p_1, p_2, ...$

Now let us see what happens to the predictions concerning the results of measurements performed on this system. If the state of the system were $|\psi_k\rangle$, we could use the postulates stated in chapter III to determine the probability of obtaining one or another measurement result. Since such a possibility (the state $|\psi_k\rangle$) has a probability of $p_k$, it is clear that the results obtained must be weighted by the $p_k$ and then summed over the various values of $k$, that is, over all the states of the statistical mixture.

COMMENTS :

(*i*) The various states $|\psi_1\rangle, |\psi_2\rangle, ...$ are not necessarily orthogonal. However, they can always be chosen normalized; in this complement, we shall assume that this is the case.

(*ii*) It must be noted that in the present case, the probabilities intervene on two different levels :

– first, in the initial information about the system (until now, we have not introduced probabilities at this stage : we considered the state vector to be perfectly well known, in which case all the probabilities $p_k$ are zero, except one, which is equal to 1);

– then, when the postulates concerning the measurement are applied (leading to probabilistic predictions, even if the initial state of the system is perfectly well known).

There are thus two totally different reasons necessitating the introduction of probabilities on these two levels : the incomplete nature of the initial information about the state of the system (such situations are also envisaged in classical statistical mechanics), and the (specifically quantum mechanical) uncertainty related to the measurement process.

(*iii*) A system described by a statistical mixture of states (with the probability $p_k$ of the state vector being $|\psi_k\rangle$) must not be confused with a system whose state $|\psi\rangle$ is a linear superposition of states★ :

$$|\psi\rangle = \sum_k c_k |\psi_k\rangle \tag{2}$$

It is often said in quantum mechanics, when the state vector is the ket $|\psi\rangle$ given in (2), that "the system has a probability $|c_k|^2$ of being in the state $|\psi_k\rangle$".

★ We assume, in this comment (*iii*), that the states $|\psi_k\rangle$ are orthonormal. This hypothesis is not essential but it simplifies the discussion.

If we want to be precise, this must be understood to mean that if we perform a set of measurements corresponding to a C.S.C.O. which has $|\psi_k\rangle$ as an eigenvector, the probability of finding the set of eigenvalues associated with $|\psi_k\rangle$ is $|c_k|^2$. But we have stressed, in § E-1 of chapter III, the fact that a system in the state $|\psi\rangle$ given in (2) *is not simply equivalent* to a system having the probability $|c_1|^2$ of being in the state $|\psi_1\rangle$, $|c_2|^2$ of being in the state $|\psi_2\rangle$, etc... In fact, for a linear combination of $|\psi_k\rangle$, there exist, in general, interference effects between these states (due to cross terms of the type $c_k c_{k'}^*$, obtained when the modulus of the probability amplitudes is squared) which are very important in quantum mechanics.

We therefore see that it is impossible, in general, to describe a statistical mixture by an "average state vector" which would be a superposition of the states $|\psi_k\rangle$. As we indicated earlier, when we take *a weighted sum of probabilities*, we can never obtain interference terms between the various states $|\psi_k\rangle$ of a statistical mixture.

## 3. The pure case. Introduction of the density operator

To study the behavior of a statistical mixture of states, we have envisaged one method: calculation of the physical predictions corresponding to a possible state $|\psi_k\rangle$; weighting the results so obtained by the probability $p_k$ associated with this state and summation over $k$. Although correct in principle, this method often leads to clumsy calculations. We have indicated [comment (*iii*)] that it is impossible to associate an "average state vector" with the system. Actually, it is an "average operator" and not an "average vector" which permits a simple description of the statistical mixture of states: the density operator.

Before studying this general case, we shall begin by again examining the simple case where the state of the system is perfectly known (all the probabilities $p_k$ are zero, except one). The system is then said to be in a *pure state*. We shall show that characterizing the system by its state vector is completely equivalent to characterizing it by a certain operator acting in the state space, the density operator. The usefulness of this operator will become apparent in § 4, where we shall show that nearly all the formulas involving this operator, and derived for the pure case, remain valid for the description of a statistical mixture of states.

### a. DESCRIPTION BY A STATE VECTOR

Consider a system whose state vector at the instant $t$ is:

$$|\psi(t)\rangle = \sum_n c_n(t) |u_n\rangle \tag{3}$$

where the $\{|u_n\rangle\}$ form an orthonormal basis of the state space, assumed to be discrete (extension to the case of a continuous basis presents no difficulties). The coefficients $c_n(t)$ satisfy the relation:

$$\sum_n |c_n(t)|^2 = 1 \tag{4}$$

which expresses the fact that $|\psi(t)\rangle$ is normalized.

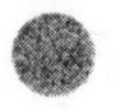



If $A$ is an observable, with matrix elements :

$$\langle u_n | A | u_p \rangle = A_{np} \tag{5}$$

the mean value of $A$ at the instant $t$ is :

$$\langle A \rangle (t) = \langle \psi(t) | A | \psi(t) \rangle = \sum_{n,p} c_n^*(t)\, c_p(t)\, A_{np} \tag{6}$$

Finally, the evolution of $| \psi(t) \rangle$ is described by the Schrödinger equation :

$$i\hbar \frac{\mathrm{d}}{\mathrm{d}t} | \psi(t) \rangle = H(t) | \psi(t) \rangle \tag{7}$$

where $H(t)$ is the Hamiltonian of the system.

#### b. DESCRIPTION BY A DENSITY OPERATOR

Relation (6) shows that the coefficients $c_n(t)$ enter into the mean values through quadratic expressions of the type $c_n^*(t)c_p(t)$. These are simply the matrix elements of the operator $| \psi(t) \rangle \langle \psi(t) |$, the projector onto the ket $| \psi(t) \rangle$ (*cf.* chap. II, §B-3-b), as can be seen from (3) :

$$\langle u_p | \psi(t) \rangle \langle \psi(t) | u_n \rangle = c_n^*(t) c_p(t) \tag{8}$$

It is therefore natural to introduce the density operator $\rho(t)$, defined by :

$$\rho(t) = | \psi(t) \rangle \langle \psi(t) | \tag{9}$$

The density operator is represented in the $\{ | u_n \rangle \}$ basis by a matrix called the *density matrix* whose elements are :

$$\rho_{pn}(t) = \langle u_p | \rho(t) | u_n \rangle = c_n^*(t) c_p(t) \tag{10}$$

We are going to show that the specification of $\rho(t)$ suffices to characterize the quantum state of the system; that is, it enables us to obtain all the physical predictions that can be calculated from $| \psi(t) \rangle$. To do this, let us write formulas (4), (6) and (7) in terms of the operator $\rho(t)$. According to (10), relation (4) indicates that the sum of the diagonal elements of the density matrix is equal to 1 :

$$\sum_n |c_n(t)|^2 = \sum_n \rho_{nn}(t) = \operatorname{Tr} \rho(t) = 1 \tag{11}$$

In addition, using (5) and (10), formula (6) becomes :

$$\begin{aligned} \langle A \rangle (t) &= \sum_{n,p} \langle u_p | \rho(t) | u_n \rangle \langle u_n | A | u_p \rangle \\ &= \sum_p \langle u_p | \rho(t) A | u_p \rangle \\ &= \operatorname{Tr} \{ \rho(t) A \} \end{aligned} \tag{12}$$

Finally, the time evolution of the operator $\rho(t)$ can be deduced from the Schrödinger equation (7):

$$\begin{aligned}\frac{\mathrm{d}}{\mathrm{d}t}\rho(t) &= \left(\frac{\mathrm{d}}{\mathrm{d}t}|\psi(t)\rangle\right)\langle\psi(t)| + |\psi(t)\rangle\left(\frac{\mathrm{d}}{\mathrm{d}t}\langle\psi(t)|\right)\\ &= \frac{1}{i\hbar}H(t)|\psi(t)\rangle\langle\psi(t)| + \frac{1}{(-i\hbar)}|\psi(t)\rangle\langle\psi(t)|H(t)\\ &= \frac{1}{i\hbar}[H(t), \rho(t)]\end{aligned} \tag{13}$$

Therefore, in terms of the density operator, conservation of probability is expressed by:

$$\mathrm{Tr}\,\rho(t) = 1 \tag{14}$$

The mean value of an observable $A$ is calculated using the formula:

$$\langle A\rangle(t) = \mathrm{Tr}\{A\rho(t)\} = \mathrm{Tr}\{\rho(t)A\} \tag{15}$$

and the time evolution obeys the equation:

$$i\hbar\frac{\mathrm{d}}{\mathrm{d}t}\rho(t) = [H(t), \rho(t)] \tag{16}$$

For completeness, we must also indicate how to calculate from $\rho(t)$ the probabilities $\mathscr{P}(a_n)$ of the various results $a_n$ which can be obtained in the measurement of an observable $A$ at time $t$. Actually, formula (15) can be used to do this. We know [see equation (B-14) of chapter III] that $\mathscr{P}(a_n)$ can be written as a mean value, that of the projector $P_n$ onto the eigensubspace associated with $a_n$:

$$\mathscr{P}(a_n) = \langle\psi(t)|P_n|\psi(t)\rangle \tag{17}$$

Using (15), we therefore obtain:

$$\mathscr{P}(a_n) = \mathrm{Tr}\{P_n\,\rho(t)\} \tag{18}$$

### c. PROPERTIES OF THE DENSITY OPERATOR IN A PURE CASE

In a pure case, a system can be described just as well by a density operator as by a state vector. However, the density operator presents a certain number of advantages.

First of all, we see from (9) that two state vectors $|\psi(t)\rangle$ and $e^{i\theta}|\psi(t)\rangle$ (where $\theta$ is a real number), which describe the same physical state, correspond to the same density operator. Using this operator therefore eliminates the drawbacks related to the existence of an arbitrary global phase factor for the state vector. Moreover, we see from (14), (15) and (18) that the formulas using the density operator are linear with respect to it, while expressions (6) and (17) are quadratic with respect to $|\psi(t)\rangle$. This is an important property which will be useful subsequently.

Finally, let us mention some properties of $\rho(t)$, which can be deduced directly from its definition (9):

$$\rho^{\dagger}(t) = \rho(t) \tag{19}$$

(the density operator is Hermitian)

$$\rho^2(t) = \rho(t) \tag{20}$$

$$\text{Tr}\, \rho^2(t) = 1 \tag{21}$$

These last two relations, which follow from the fact that $\rho(t)$ is a projector, are true only in a pure case. We shall see later that they are not valid for a statistical mixture of states.

## 4. A statistical mixture of states (non-pure case)

### a. DEFINITION OF THE DENSITY OPERATOR

Let us now return to the general case described in § 1, and consider a system for which (at a given instant) the various probabilities $p_1, p_2, \dots p_k, \dots$ are arbitrary, on the condition that they satisfy the relations:

$$\begin{cases} 0 \leqslant p_1, p_2, \dots, p_k, \dots \leqslant 1 \\ \sum\limits_k p_k = 1 \end{cases} \tag{22}$$

Under these conditions, how does one calculate the probability $\mathscr{P}(a_n)$ that a measurement of the observable $A$ will yield the result $a_n$?

Let:

$$\mathscr{P}_k(a_n) = \langle \psi_k \mid P_n \mid \psi_k \rangle \tag{23}$$

be the probability of finding $a_n$ if the state vector were $| \psi_k \rangle$. To obtain the desired probability $\mathscr{P}(a_n)$, one must, as we have already indicated, weight $\mathscr{P}_k(a_n)$ by $p_k$ and then sum over $k$:

$$\mathscr{P}(a_n) = \sum_k p_k\, \mathscr{P}_k(a_n) \tag{24}$$

Now, from (18), we have:

$$\mathscr{P}_k(a_n) = \text{Tr}\,\{ \rho_k\, P_n \} \tag{25}$$

where:

$$\rho_k = | \psi_k \rangle \langle \psi_k | \tag{26}$$

is the density operator corresponding to the state $| \psi_k \rangle$. Substituting (25) into (24), we have:

$$\begin{aligned} \mathscr{P}(a_n) &= \sum_k p_k\, \text{Tr}\,\{ \rho_k\, P_n \} \\ &= \text{Tr}\,\{ \sum_k p_k\, \rho_k\, P_n \} \\ &= \text{Tr}\,\{ \rho P_n \} \end{aligned} \tag{27}$$

where we have set:

$$\rho = \sum_k p_k \, \rho_k \tag{28}$$

We therefore see that the linearity of the formulas which use the density operator enables us to express all physical predictions in terms of $\rho$, the average of the density operators $\rho_k$; $\rho$ is, by definition, the density operator of the system.

### b. GENERAL PROPERTIES OF THE DENSITY OPERATOR

Since the coefficients $p_k$ are real, $\rho$ is obviously a Hermitian operator like each of the $\rho_k$.

Let us calculate the trace of $\rho$; it is equal to:

$$\text{Tr}\, \rho = \sum_k p_k \, \text{Tr}\, \rho_k \tag{29}$$

Now, as we saw in § 3-b, the trace of $\rho_k$ is always equal to 1; it follows that:

$$\text{Tr}\, \rho = \sum_k p_k = 1 \tag{30}$$

Relation (14) is therefore valid in the general case.

We have already given, in (27), the expression which enables us to calculate the probability $\mathscr{P}(a_n)$ in terms of $\rho$. Using this expression, we can easily generalize formula (15) to statistical mixtures:

$$\begin{aligned} \langle A \rangle &= \sum_n a_n \, \mathscr{P}(a_n) = \text{Tr}\, \{ \rho \sum_n a_n P_n \} \\ &= \text{Tr}\, \{ \rho A \} \end{aligned} \tag{31}$$

[we have used formula (D-36-b) of chapter II].

Now let us calculate the time evolution of the density operator. To do this, we shall assume that, unlike the state of the system, its Hamiltonian $H(t)$ is perfectly well known. One can then easily show that if the system at the initial time $t_0$ has the probability $p_k$ of being in the state $| \psi_k \rangle$, then, at a subsequent time $t$, it has the same probability $p_k$ of being in the state $| \psi_k(t) \rangle$ given by:

$$\begin{cases} i\hbar \dfrac{\text{d}}{\text{d}t} | \psi_k(t) \rangle = H(t) \, | \psi_k(t) \rangle \\ | \psi_k(t_0) \rangle = | \psi_k \rangle \end{cases} \tag{32}$$

The density operator at the instant $t$ will then be:

$$\rho(t) = \sum_k p_k \, \rho_k(t) \tag{33}$$

with:

$$\rho_k(t) = | \psi_k(t) \rangle \langle \psi_k(t) | \tag{34}$$

According to (16), $\rho_k(t)$ obeys the evolution equation:

$$i\hbar \frac{\text{d}}{\text{d}t} \rho_k(t) = [H(t), \rho_k(t)] \tag{35}$$

The linearity of formulas (33) and (35) with respect to $\rho_k(t)$ implies that :

$$i\hbar \frac{d}{dt} \rho(t) = [H(t), \rho(t)] \tag{36}$$

We can therefore generalize to a statistical mixture of states all the equations of § 3, with the exception of (20) and (21). We see that since $\rho$ is no longer a projector, we have, in general* :

$$\rho^2 \neq \rho \tag{37}$$

and, consequently :

$$\text{Tr}\, \rho^2 \leqslant 1 \tag{38}$$

Nevertheless, only one of the equations, (20) or (21), must be satisfied for us to be sure that we are dealing with a pure state.

Finally, we see from definition (28) that, for any ket $| u \rangle$, we have :

$$\begin{aligned} \langle u | \rho | u \rangle &= \sum_k p_k \langle u | \rho_k | u \rangle \\ &= \sum_k p_k |\langle u | \psi_k \rangle|^2 \end{aligned} \tag{39}$$

and consequently :

$$\langle u | \rho | u \rangle \geqslant 0 \tag{40}$$

$\rho$ is therefore a positive operator.

### c. POPULATIONS ; COHERENCES

What is the physical meaning of the matrix elements $\rho_{np}$ of $\rho$ in the $\{ | u_n \rangle \}$ basis ?

First, let us consider the diagonal element $\rho_{nn}$. According to (28), we have:

$$\rho_{nn} = \sum_k p_k [\rho_k]_{nn} \tag{41}$$

that is, using (26) and introducing the components :

$$c_n^{(k)} = \langle u_n | \psi_k \rangle \tag{42}$$

of $| \psi_k \rangle$ in the $\{ | u_n \rangle \}$ basis :

$$\rho_{nn} = \sum_k p_k |c_n^{(k)}|^2 \tag{43}$$

$|c_n^{(k)}|^2$ is a positive real number, whose physical interpretation is the following : if the state of the system is $| \psi_k \rangle$, it is the probability of finding, in a measurement, this system in the state $| u_n \rangle$. According to (41), if we take into account the

* Assume, for example, that the states $| \psi_k \rangle$ are orthogonal. In an orthonormal basis including the $| \psi_k \rangle$, $\rho$ is diagonal and its elements are the $p_k$. To obtain $\rho^2$, we simply replace $p_k$ by $p_k^2$. Relations (37) and (38) then follow from the fact that the $p_k$ are always less than 1 (except in the particular case where only one of them is non-zero : the pure case).

indeterminacy of the state before the measurement, $\rho_{nn}$ represents the average probability of finding the system in the state $| u_n \rangle$. For this reason, $\rho_{nn}$ is called the *population* of the state $| u_n \rangle$: if the same measurement is carried out $N$ times under the same initial conditions, where $N$ is a large number, $N\rho_{nn}$ systems will be found in the state $| u_n \rangle$. It is evident from (43) that $\rho_{nn}$ is a positive real number, equal to zero only if all the $|c_n^{(k)}|^2$ are zero.

A calculation analogous to the preceding one gives the following expression for the non-diagonal element $\rho_{np}$:

$$\rho_{np} = \sum_k p_k \, c_n^{(k)} c_p^{(k)*} \tag{44}$$

$c_n^{(k)} c_p^{(k)*}$ is a cross term, of the same type as those studied in § E-1 of chapter III. It expresses the interference effects between the states $|u_n\rangle$ and $|u_p\rangle$ which can appear when the state $|\psi_k\rangle$ is a coherent linear superposition of these states. According to (44), $\rho_{np}$ is the average of these cross terms, taken over all the possible states of the statistical mixture. In contrast to the populations, $\rho_{np}$ can be zero even if none of the products $c_n^{(k)} c_p^{(k)*}$ is: while $\rho_{nn}$ is a sum of real positive (or zero) numbers, $\rho_{np}$ is a sum of complex numbers. If $\rho_{np}$ is zero, this means that the average (44) has cancelled out any interference effects between $|u_n\rangle$ and $|u_p\rangle$. On the other hand, if $\rho_{np}$ is different from zero, a certain coherence subsists between these states. This is why the non-diagonal elements of $\rho$ are often called *coherences*.

COMMENTS:

(*i*) The distinction between populations and coherences obviously depends on the basis $\{ |u_n\rangle \}$ chosen in the state space. Since $\rho$ is Hermitian, it is always possible to find an orthonormal basis $\{ |\chi_l\rangle \}$ where $\rho$ is diagonal. $\rho$ can then be written:

$$\rho = \sum_l \pi_l \, |\chi_l\rangle \langle \chi_l| \tag{45}$$

Since $\rho$ is positive and $\text{Tr}\, \rho = 1$, we have:

$$\begin{cases} 0 \leqslant \pi_l \leqslant 1 \\ \sum_l \pi_l = 1 \end{cases} \tag{46}$$

$\rho$ can thus be considered to describe a statistical mixture of the states $|\chi_l\rangle$ with the probabilities $\pi_l$ (there are no coherences between the states $|\chi_l\rangle$).

(*ii*) If the kets $|u_n\rangle$ are eigenvectors of the Hamiltonian $H$, which is assumed to be time-independent:

$$H \, |u_n\rangle = E_n \, |u_n\rangle \tag{47}$$

we obtain directly from (36):

$$\begin{cases} i\hbar \dfrac{\text{d}}{\text{d}t} \rho_{nn}(t) = 0 \\[2ex] i\hbar \dfrac{\text{d}}{\text{d}t} \rho_{np}(t) = (E_n - E_p)\, \rho_{np} \end{cases} \tag{48}$$

that is:

$$\begin{cases} \rho_{nn}(t) = \text{constant} \\ \rho_{np}(t) = e^{\frac{i}{\hbar}(E_p - E_n)t} \rho_{np}(0) \end{cases} \tag{49}$$

The populations are constant, and the coherences oscillate at the Bohr frequencies of the system.

(*iii*) Using (40), one can prove the inequality:

$$\rho_{nn} \, \rho_{pp} \geqslant |\rho_{np}|^2 \tag{50}$$

It follows, for example, that $\rho$ can have coherences only between states whose populations are not zero.

## 5. Applications of the density operator

### a. SYSTEM IN THERMODYNAMIC EQUILIBRIUM

The first example we shall consider is borrowed from quantum statistical mechanics. Consider a system in thermodynamic equilibrium with a reservoir at the absolute temperature $T$. It can be shown that its density operator is then:

$$\rho = Z^{-1} \, e^{-H/kT} \tag{51}$$

where $H$ is the Hamiltonian operator of the system, $k$ is the Boltzmann constant, and $Z$ is a normalization coefficient chosen so as to make the trace of $\rho$ equal to 1:

$$Z = \text{Tr} \, \{ e^{-H/kT} \} \tag{52}$$

($Z$ is called the "partition function").

In the $\{ |u_n\rangle \}$ basis of eigenvectors of $H$, we have (*cf.* complement $B_{II}$, § 4-a):

$$\begin{aligned} \rho_{nn} &= Z^{-1} \langle u_n | e^{-H/kT} | u_n \rangle \\ &= Z^{-1} \, e^{-E_n/kT} \end{aligned} \tag{53}$$

and:

$$\begin{aligned} \rho_{np} &= Z^{-1} \langle u_n | e^{-H/kT} | u_p \rangle \\ &= Z^{-1} \, e^{-E_p/kT} \langle u_n | u_p \rangle \\ &= 0 \end{aligned} \tag{54}$$

At thermodynamic equilibrium, the populations of the stationary states are exponentially decreasing functions of the energy (the lower the temperature $T$, the more rapid the decrease), and the coherences between stationary states are zero.

### b. SEPARATE DESCRIPTION OF PART OF A PHYSICAL SYSTEM. CONCEPT OF A PARTIAL TRACE

We are now going to return to the problem mentioned in §3 of complement $D_{III}$. Let us consider two different systems (1) and (2) and the global system (1) + (2), whose state space is the tensor product:

$$\mathscr{E} = \mathscr{E}(1) \otimes \mathscr{E}(2) \tag{55}$$

Let $\{ | u_n(1) \rangle \}$ be a basis of $\mathscr{E}(1)$ and $\{ | v_p(2) \rangle \}$, a basis of $\mathscr{E}(2)$; the kets $| u_n(1) \rangle | v_p(2) \rangle$ form a basis of $\mathscr{E}$.

The density operator $\rho$ of the global system is an operator which acts in $\mathscr{E}$. We saw in chapter II (*cf.* § F-2-b) how to extend into $\mathscr{E}$ an operator which acts only in $\mathscr{E}(1)$ [or $\mathscr{E}(2)$]. We are going to show here how to perform the inverse operation: we shall construct from $\rho$ an operator $\rho(1)$ [or $\rho(2)$] acting only in $\mathscr{E}(1)$ [or $\mathscr{E}(2)$] which will enable us to make all the physical predictions about measurements bearing only on system (1) or system (2). This operation will be called a partial trace with respect to (2) [or (1)].

Let us introduce the operator $\rho(1)$ whose matrix elements are:

$$\langle u_n(1) | \rho(1) | u_{n'}(1) \rangle = \sum_p (\langle u_n(1) | \langle v_p(2) |) \rho (| u_{n'}(1) \rangle | v_p(2) \rangle) \tag{56}$$

By definition, $\rho(1)$ is obtained from $\rho$ by performing a partial trace on (2):

$$\rho(1) = \text{Tr}_2 \, \rho \tag{57}$$

Similarly, the operator:

$$\rho(2) = \text{Tr}_1 \, \rho \tag{58}$$

has matrix elements:

$$\langle v_p(2) | \rho(2) | v_{p'}(2) \rangle = \sum_n (\langle u_n(1) | \langle v_p(2) |) \rho (| u_n(1) \rangle | v_{p'}(2) \rangle) \tag{59}$$

It is clear why these operations are called partial traces. We know that the (total) trace of $\rho$ is:

$$\text{Tr} \, \rho = \sum_n \sum_p (\langle u_n(1) | \langle v_p(2) |) \rho (| u_n(1) \rangle | v_p(2) \rangle) \tag{60}$$

The difference between (60) and (56) [or (59)] is the following: for the partial traces, the indices $n$ and $n'$ (or $p$ and $p'$) are not required to be equal and the summation is performed only over $p$ (or $n$). We have, moreover:

$$\text{Tr} \, \rho = \text{Tr}_1(\text{Tr}_2 \, \rho) = \text{Tr}_2(\text{Tr}_1 \, \rho) \tag{61}$$

$\rho(1)$ and $\rho(2)$ are therefore, like $\rho$, operators whose trace is equal to 1. It can be verified from their definitions that they are Hermitian and, in general, that they satisfy all the properties of a density operator (*cf.* § 4-b).

Now let $A(1)$ be an observable acting in $\mathscr{E}(1)$ and $\tilde{A}(1) = A(1) \otimes \mathbb{1}(2)$, its extension in $\mathscr{E}$. We obtain, using (31), the definition of the trace, and the closure relation on the $\{ | u_n(1) \rangle | v_p(2) \rangle \}$ basis:

$$\begin{aligned}
\langle \tilde{A}(1) \rangle &= \text{Tr} \{ \rho \tilde{A}(1) \} \\
&= \sum_{n,p} \sum_{n',p'} (\langle u_n(1) | \langle v_p(2) |) \rho (| u_{n'}(1) \rangle | v_{p'}(2) \rangle) \\
&\qquad \times (\langle u_{n'}(1) | \langle v_{p'}(2) |) A(1) \otimes \mathbb{1}(2) (| u_n(1) \rangle | v_p(2) \rangle) \\
&= \sum_{n,p,n',p'} (\langle u_n(1) | \langle v_p(2) |) \rho (| u_{n'}(1) \rangle | v_{p'}(2) \rangle) \\
&\qquad \times \langle u_{n'}(1) | A(1) | u_n(1) \rangle \langle v_{p'}(2) | v_p(2) \rangle
\end{aligned} \tag{62}$$

Now:

$$\langle v_{p'}(2) \mid v_p(2) \rangle = \delta_{pp'} \tag{63}$$

We can therefore write (62) in the form:

$$\langle \tilde{A}(1) \rangle = \sum_{n,n'} \left[ \sum_p \langle u_n(1)\, v_p(2) \mid \rho \mid u_{n'}(1)\, v_p(2) \rangle \right] \langle u_{n'}(1) \mid A(1) \mid u_n(1) \rangle \tag{64}$$

Inside the brackets on the right-hand side of (64), we recognize the matrix element of $\rho(1)$ defined in (56). We therefore have:

$$\begin{aligned} \langle \tilde{A}(1) \rangle &= \sum_{n,n'} \langle u_n(1) \mid \rho(1) \mid u_{n'}(1) \rangle \langle u_{n'}(1) \mid A(1) \mid u_n(1) \rangle \\ &= \sum_n \langle u_n(1) \mid \rho(1)\, A(1) \mid u_n(1) \rangle \\ &= \text{Tr} \{ \rho(1)\, A(1) \} \end{aligned} \tag{65}$$

Let us compare this result with (31). We see that the partial trace $\rho(1)$ enables us to calculate all the mean values $\langle \tilde{A}(1) \rangle$ as if the system (1) were isolated and had $\rho(1)$ for a density operator. Making the same comment as for formula (17), we see that $\rho(1)$ also enables us to obtain the probabilities of all the results of measurements bearing on system (1) alone.

COMMENTS:

(*i*) We saw in complement $D_{III}$ that it is impossible to assign a state vector to system (1) [or (2)] when the state of the global system (1) + (2) is not a product state. We now see that the density operator is a much more simple tool than the state vector. In all cases (whether the global system is in a product state or not, whether it corresponds to a pure case or to a statistical mixture), one can always, thanks to the partial trace operation, assign a density operator to subsystem (1) [or (2)]. This permits us to calculate all the physical predictions about this subsystem.

(*ii*) Even if $\rho$ describes a pure state ($\text{Tr}\, \rho^2 = 1$), this is not in general true of the density operators $\rho(1)$ and $\rho(2)$ obtained from $\rho$ by a partial trace. It can be verified from (56) [or (59)] that $\text{Tr} \{ \rho^2(1) \}$ [or $\text{Tr} \{ \rho^2(2) \}$] is not generally equal to 1. This is another way of saying that it is not in general possible to assign a state vector to (1) [or (2)], except, of course, if the global system is in a product state.

(*iii*) If the global system is in a product state:

$$| \psi \rangle = | \varphi(1) \rangle \, | \chi(2) \rangle \tag{66}$$

we can verify directly that the corresponding density operator is written:

$$\rho = \sigma(1) \otimes \tau(2) \tag{67}$$

with:

$$\begin{aligned} \sigma(1) &= | \varphi(1) \rangle \langle \varphi(1) | \\ \tau(2) &= | \chi(2) \rangle \langle \chi(2) | \end{aligned} \tag{68}$$

More generally, we can envisage states of the global system for which the density operator $\rho$ can be factored as in (67) [$\sigma(1)$ and $\tau(2)$ can correspond to statistical mixtures as well as to pure cases]. The partial trace operation then yields:

$$\begin{aligned} \text{Tr}_2 \{ \sigma(1) \otimes \tau(2) \} &= \sigma(1) \\ \text{Tr}_1 \{ \sigma(1) \otimes \tau(2) \} &= \tau(2) \end{aligned} \tag{69}$$

An expression such as (67) therefore represents the simple juxtaposition of a system (1), described by the density operator $\sigma(1)$, and a system (2), described by the density operator $\tau(2)$.

(*iv*) Starting with an arbitrary density operator $\rho$ [which cannot be factored as in (67)], let us calculate $\rho(1) = \text{Tr}_2\, \rho$ and $\rho(2) = \text{Tr}_1\, \rho$. Then let us form the product:

$$\rho' = \rho(1) \otimes \rho(2) \tag{70}$$

Unlike the case envisaged in comment (*iii*), $\rho'$ is in general different from $\rho$. When the density operator cannot be factored as in (67), there is therefore a certain "correlation" between systems (1) and (2), which is no longer contained in the operator $\rho'$ of formula (70).

(*v*) If the evolution of the global system is described by equation (36), it is in general impossible to find a Hamiltonian operator relating to system (1) alone which would enable us to write an analogous equation for $\rho(1)$. While the definition, at any time, of $\rho(1)$ in terms of $\rho$ is simple, the evolution of $\rho(1)$ is much more difficult to describe.

**References and suggestions for further reading:**

Articles by Fano (2.31) and Ter Haar (2.32). Using the density operator to study relaxation phenomena : Abragam (14.1), chap. VIII; Slichter (14.2), chap. 5; Sargent, Scully and Lamb (15.5), chap. VII.

# Complement $F_{III}$

# THE EVOLUTION OPERATOR



In §D-1-b of chapter III, we saw that the transformation of $|\psi(t_0)\rangle$ (the state vector at the initial instant $t_0$) into $|\psi(t)\rangle$ (the state vector at an arbitrary instant) is linear. There therefore exists a linear operator $U(t, t_0)$ such that:

$$|\psi(t)\rangle = U(t, t_0)\,|\psi(t_0)\rangle \tag{1}$$

We intend to study here the principal properties of $U(t, t_0)$, which is, by definition, the evolution operator of the system.

## 1. General properties

Since the ket $|\psi(t_0)\rangle$ is arbitrary, it is clear from (1) that:

$$U(t_0, t_0) = \mathbb{1} \tag{2}$$

Also, substituting (1) into the Schrödinger equation, we obtain:

$$i\hbar\frac{\partial}{\partial t}\,U(t, t_0)\,|\psi(t_0)\rangle = H(t)\,U(t, t_0)\,|\psi(t_0)\rangle \tag{3}$$

from which, for the same reason as above:

$$i\hbar\frac{\partial}{\partial t}\,U(t, t_0) = H(t)\,U(t, t_0) \tag{4}$$

The first-order differential equation (4) completely defines $U(t, t_0)$, taking the initial condition (2) into account. Note, moreover, that (2) and (4) can be condensed into a single integral equation:

$$U(t, t_0) = \mathbb{1} - \frac{i}{\hbar}\int_{t_0}^{t} H(t')\,U(t', t_0)\,\mathrm{d}t' \tag{5}$$

Now let us consider the parameter $t_0$, which appears in $U(t, t_0)$ as a variable $t'$, just like $t$. We then write (1) in the form:

$$|\psi(t)\rangle = U(t, t')\,|\psi(t')\rangle \tag{6}$$

But $|\psi(t')\rangle$ can itself be obtained from a formula of the same type:

$$|\psi(t')\rangle = U(t', t'')\,|\psi(t'')\rangle \tag{7}$$

Substitute (7) into (6):

$$| \psi(t) \rangle = U(t, t') \, U(t', t'') \, | \psi(t'') \rangle \tag{8}$$

Since, moreover, $| \psi(t) \rangle = U(t, t'') | \psi(t'') \rangle$, we deduce ($| \psi(t'') \rangle$ being arbitrary):

$$U(t, t'') = U(t, t') U(t', t'') \tag{9}$$

It is easy to generalize this procedure and to obtain:

$$U(t_n, t_1) = U(t_n, t_{n-1}) \ldots U(t_3, t_2) U(t_2, t_1) \tag{10}$$

where $t_1, t_2, \ldots, t_n$ are arbitrary. If we assume that $t_1 < t_2 < t_3 < \ldots < t_n$, formula (10) is simple to interpret: to go from $t_1$ to $t_n$, the system progresses from $t_1$ to $t_2$, then from $t_2$ to $t_3$, ..., then, finally from $t_{n-1}$ to $t_n$.

Set $t'' = t$ in (9); taking (2) into consideration, we obtain:

$$\mathbb{1} = U(t, t') U(t', t) \tag{11}$$

or, interchanging the roles of $t$ and $t'$:

$$\mathbb{1} = U(t', t) U(t, t') \tag{12}$$

We therefore have:

$$U(t', t) = U^{-1}(t, t') \tag{13}$$

Now let us calculate the evolution operator between two instants separated by $\mathrm{d}t$. To do this, write the Schrödinger equation in the form:

$$\begin{aligned} \mathrm{d} | \psi(t) \rangle &= | \psi(t + \mathrm{d}t) \rangle - | \psi(t) \rangle \\ &= - \frac{i}{\hbar} H(t) \, | \psi(t) \rangle \, \mathrm{d}t \end{aligned} \tag{14}$$

that is:

$$| \psi(t + \mathrm{d}t) \rangle = \left[ \mathbb{1} - \frac{i}{\hbar} H(t) \, \mathrm{d}t \right] | \psi(t) \rangle \tag{15}$$

We then obtain, using the very definition of $U(t + \mathrm{d}t, t)$:

$$U(t + \mathrm{d}t, t) = \mathbb{1} - \frac{i}{\hbar} H(t) \, \mathrm{d}t \tag{16}$$

$U(t + \mathrm{d}t, t)$ is called the infinitesimal evolution operator. Since $H(t)$ is Hermitian, $U(t + \mathrm{d}t, t)$ is unitary (*cf.* complement $C_{II}$, § 3). It follows that $U(t, t')$ is also unitary since the interval $[t, t']$ can be divided into a very large number of infinitesimal intervals. Formula (10) then shows that $U(t, t')$ is a product of unitary operators; it is therefore a unitary operator. One can consequently write (13) in the form:

$$U^\dagger(t, t') = U^{-1}(t, t') = U(t', t) \tag{17}$$

It is not surprising that the transformation $U(t, t')$ is unitary, that is, that it conserves the norm of vectors on which it acts. We saw in chapter III (*cf.* §D-1-c) that the norm of the state vector does not change over time.

## 2. Case of conservative systems

When the operator $H$ does not depend on time, equation (4) can easily be integrated; taking the initial condition (2) into account, we obtain :

$$U(t, t_0) = e^{-iH(t-t_0)/\hbar} \tag{18}$$

One can verify directly from this formula all the properties of the evolution operator cited in § 1.

It is very simple to go from formula (D-52) to (D-54) of chapter III using (18). It suffices to apply the operator $U(t, t_0)$ to both sides of (D-52), noting that, since $|\varphi_{n,\tau}\rangle$ is an eigenvector of $H$ with the eigenvalue $E_n$:

$$\begin{aligned} U(t, t_0) | \varphi_{n,\tau} \rangle &= e^{-iH(t-t_0)/\hbar} | \varphi_{n,\tau} \rangle \\ &= e^{-iE_n(t-t_0)/\hbar} | \varphi_{n,\tau} \rangle \end{aligned} \tag{19}$$

COMMENTS:

(*i*) When $H$ is time-dependent, one might be tempted to believe, by analogy with formula (18), that the evolution operator is equal to the operator $V(t, t_0)$ defined by:

$$V(t, t_0) = e^{-\frac{i}{\hbar}\int_{t_0}^{t} H(t')dt'} \tag{20}$$

Actually, this is not true, since the derivative of an operator of the form $e^{F(t)}$ is not in general equal to $F'(t)\, e^{F(t)}$ (*cf.* complement $B_{II}$, §5-c):

$$i\hbar \frac{\partial}{\partial t} V(t, t_0) \neq H(t)V(t, t_0) \tag{21}$$

(*ii*) Let us again consider the experiments described in § E-1-b of chapter III. As we have already indicated [comment (*ii*) of §E-1-b-β], it is not necessary to assume that the measurements of the various observables $A$, $B$ and $C$ are made very close together in time. When the system has had the time to evolve between two successive measurements, the variations of the state vector can easily be taken into account by using the evolution operator. If $t_0$, $t_1$ and $t_2$ designate respectively the instants at which the measurements of $A$, $B$ and $C$ are performed, we then replace (E-15) by:

$$\mathscr{P}_a(c) = |\langle v_c | U(t_2, t_0) | u_a \rangle|^2 \tag{22}$$

and (E-17) by:

$$\mathscr{P}_a(b, c) = |\langle v_c | U(t_2, t_1) | w_b \rangle|^2 \, |\langle w_b | U(t_1, t_0) | u_a \rangle|^2 \tag{23}$$

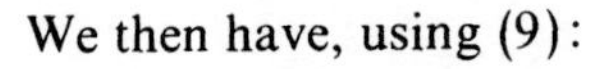



We then have, using (9):

$$\begin{aligned}\langle v_c \mid U(t_2, t_0) \mid u_a \rangle &= \langle v_c \mid U(t_2, t_1) U(t_1, t_0) \mid u_a \rangle \\ &= \sum_b \langle v_c \mid U(t_2, t_1) \mid w_b \rangle \langle w_b \mid U(t_1, t_0) \mid u_a \rangle \qquad (24)\end{aligned}$$

Substituting (24) into (22), we see, as in (E-21), that $\mathscr{P}_a(c)$ is not equal to $\sum_b \mathscr{P}_a(b, c)$.

**References and suggestions for further reading:**

The evolution operator is of fundamental importance in collision therory (see references for chapter VIII) and time-dependent perturbation theory (see references for chapter XIII).

## Complement $G_{III}$

## THE SCHRÖDINGER AND HEISENBERG PICTURES

In the formalism developed in chapter III, it is the time-independent operators which correspond to the observables of the system (*cf.* chap. III, § D-1-d). For example, the position, momentum and kinetic energy operators of a particle do not depend on time. The evolution of the system is entirely contained in that of the state vector $|\psi(t)\rangle$ [here written $|\psi_S(t)\rangle$, for reasons which will be evident later] and is obtained from the Schrödinger equation. This is why this approach is called the *Schrödinger picture*.

Nevertheless, we know that all the predictions of quantum mechanics (probabilities, mean values) are expressed in terms of scalar products of a bra and a ket or matrix elements of operators. Now, as we saw in complement $C_{II}$, these quantities are invariant when the same unitary transformation is performed on the kets and on the operators. This transformation can be chosen so as to make the transform of the ket $|\psi_S(t)\rangle$ a time-independent ket. Of course, the transforms of the observables cited above then depend on time. We thus obtain the *Heisenberg picture*.

To avoid confusion, in this complement, we shall systematically assign an index $S$ to the kets and operators of the Schrödinger picture and an index $H$ to those of the Heisenberg picture. Since the latter picture is used only in this complement, the index $S$ can be considered to be implicit in all the other complements and chapters.

The state vector $|\psi_S(t)\rangle$ at the instant $t$ is expressed in terms of $|\psi_S(t_0)\rangle$ by the relation:

$$|\psi_S(t)\rangle = U(t, t_0)|\psi_S(t_0)\rangle \tag{1}$$

where $U(t, t_0)$ is the evolution operator (*cf.* complement $F_{III}$). Since this operator is unitary, it is sufficient to perform the unitary transformation associated with the operator $U^\dagger(t, t_0)$ to obtain a constant transformed vector $|\psi_H\rangle$:

$$\begin{aligned}|\psi_H\rangle = U^\dagger(t, t_0)|\psi_S(t)\rangle &= U^\dagger(t, t_0)U(t, t_0)|\psi_S(t_0)\rangle \\ &= |\psi_S(t_0)\rangle\end{aligned} \tag{2}$$

In the Heisenberg picture, the state vector, which is constant, is therefore equal to $|\psi_S(t)\rangle$ at time $t_0$.

The transform $A_H(t)$ of an operator $A_S(t)$ is given by (complement $C_{II}$, §2):

$$A_H(t) = U^\dagger(t, t_0)A_S(t)U(t, t_0) \tag{3}$$

As we have already seen, $A_H(t)$ generally depends on time, even if $A_S$ does not.

Nevertheless, there exists an interesting special case in which, if $A_S$ is time-independent, the same is true of $A_H$: the case in which the system is conservative ($H_S$ does not depend on time) and $A_S$ commutes with $H_S$ ($A_S$ is then a constant of the motion; *cf.* chap. III, § D-2-c). In this case, we have:

$$U(t, t_0) = e^{-iH_s(t-t_0)/\hbar} \tag{4}$$

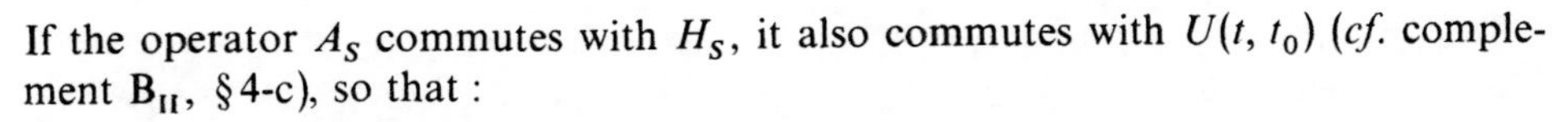


If the operator $A_S$ commutes with $H_S$, it also commutes with $U(t, t_0)$ (*cf.* complement $B_{II}$, §4-c), so that :

$$A_H(t) = U^\dagger(t, t_0)U(t, t_0)A_S = A_S \tag{5}$$

The operators $A_S$ and $A_H$ are therefore simply equal in this case (in particular, $H_S = H_H$, and the indices $S$ and $H$ are, in reality, unnecessary for the Hamiltonian). Since they are time-independent, we see that they indeed correspond to a constant of the motion.

When $A_S(t)$ is arbitrary, let us calculate the evolution of the operator $A_H(t)$. Using relation (4) of complement $F_{III}$, as well as its adjoint, we obtain :

$$\begin{aligned}\frac{\mathrm{d}}{\mathrm{d}t} A_H(t) = & -\frac{1}{i\hbar} U^\dagger(t, t_0)\, H_S(t)\, A_S(t)\, U(t, t_0) + U^\dagger(t, t_0) \frac{\mathrm{d}A_S(t)}{\mathrm{d}t} U(t, t_0) \\ & + \frac{1}{i\hbar} U^\dagger(t, t_0)\, A_S(t)\, H_S(t)\, U(t, t_0)\end{aligned} \tag{6}$$

In the first and last terms of this expression, let us insert between $A_S$ and $H_S$ the product $U(t, t_0)U^\dagger(t, t_0)$, which is equal to the identity operator [formula (17) of complement $F_{III}$]:

$$\begin{aligned}\frac{\mathrm{d}}{\mathrm{d}t} A_H(t) = & -\frac{1}{i\hbar} U^\dagger(t, t_0)\, H_S(t)\, U(t, t_0)\, U^\dagger(t, t_0)\, A_S(t)\, U(t, t_0) \\ & + U^\dagger(t, t_0) \frac{\mathrm{d}A_S(t)}{\mathrm{d}t} U(t, t_0) \\ & + \frac{1}{i\hbar} U^\dagger(t, t_0)\, A_S(t)\, U(t, t_0)\, U^\dagger(t, t_0)\, H_S(t)\, U(t, t_0)\end{aligned} \tag{7}$$

According to definition (3), we finally obtain:

$$i\hbar \frac{\mathrm{d}}{\mathrm{d}t} A_H(t) = [A_H(t), H_H(t)] + i\hbar \left( \frac{\mathrm{d}}{\mathrm{d}t} A_S(t) \right)_H \tag{8}$$

COMMENTS:

(*i*) Historically, the first picture was developed by Schrödinger, leading him to the equation which bears his name, and the second one, by Heisenberg, who calculated the evolution of matrices representing the various operators $A_H(t)$ (hence the name "matrix mechanics"). It was not until later that the equivalence of the two approaches was proved.

(*ii*) Using (8), one immediately obtains equation (D-27) of chapter III as we shall now show. In the Heisenberg picture, the evolution of the mean value

$$\langle A \rangle(t) = \langle \psi_S(t) | A_S(t) | \psi_S(t) \rangle$$

can be calculated, since:

$$\langle A \rangle(t) = \langle \psi_H | A_H(t) | \psi_H \rangle \tag{9}$$

On the right-hand side of (9), only $A_H(t)$ depends on time, so (D-27) can be obtained directly by differentiation. Note, nevertheless, that equation (8) is more general than (D-27) since, instead of expressing the equality of two mean values (that is, two matrix elements of operators), it expresses the equality of two operators.

(*iii*) When the system under consideration is composed of a particle of mass $m$ under the influence of a potential, equation (8) becomes very simple. We then have (confining ourselves to one dimension):

$$H_S(t) = \frac{P_S^2}{2m} + V(X_S, t) \tag{10}$$

and therefore [*cf.* formula (35) of complement $C_{II}$]:

$$H_H(t) = \frac{P_H^2}{2m} + V(X_H, t) \tag{11}$$

Substituting (11) into (8) and using the fact that $[X_H, P_H] = [X_S, P_S] = i\hbar$, we obtain, by an argument analogous to that of § D-1-d of chapter III :

$$\begin{aligned} \frac{\mathrm{d}}{\mathrm{d}t} X_H(t) &= \frac{1}{m} P_H(t) \\ \frac{\mathrm{d}}{\mathrm{d}t} P_H(t) &= -\frac{\partial V}{\partial X}(X_H, t) \end{aligned} \tag{12}$$

These equations generalize the Ehrenfest theorem [*cf.* chap. III, relations (D-34) and (D-35)]. They are similar to those which give the evolution of the classical quantities $x$ and $p$ [*cf.* chap. III, relations (D-36-a) and (D-36-b)]. An advantage of the Heisenberg picture is therefore that it leads to equations which are formally similar to those of classical mechanics.

**References and suggestions for further reading:**

The interaction picture: Messiah (1.17), chap. VIII, § 14; Schiff (1.18), § 24; Merzbacher (1.16), chap. 18, § 7.

## Complement $H_{III}$

## GAUGE INVARIANCE



### 1. Outline of the problem: scalar and vector potentials associated with an electromagnetic field; concept of a gauge

Consider an electromagnetic field, characterized by the values $\mathbf{E}(\mathbf{r}; t)$ of the electric field and $\mathbf{B}(\mathbf{r}; t)$ of the magnetic field at every instant and at all points in space: $\mathbf{E}(\mathbf{r}; t)$ and $\mathbf{B}(\mathbf{r}; t)$ are not independent since they satisfy Maxwell's equations. Instead of specifying these two vector fields, it is possible to introduce a scalar potential $U(\mathbf{r}; t)$ and a vector potential $\mathbf{A}(\mathbf{r}; t)$ such that:

$$\begin{cases} \mathbf{E}(\mathbf{r}; t) = -\nabla U(\mathbf{r}; t) - \dfrac{\partial}{\partial t}\mathbf{A}(\mathbf{r}; t) \\ \mathbf{B}(\mathbf{r}; t) = \nabla \times \mathbf{A}(\mathbf{r}; t) \end{cases} \tag{1}$$

It can be shown from Maxwell's equations (*cf.* appendix III, § 4-b-α) that there always exist functions $U(\mathbf{r}; t)$ and $\mathbf{A}(\mathbf{r}; t)$ which allow $\mathbf{E}(\mathbf{r}; t)$ and $\mathbf{B}(\mathbf{r}; t)$ to be expressed in the form (1). All electromagnetic fields can therefore be described by scalar and vector potentials. However, when $\mathbf{E}(\mathbf{r}; t)$ and $\mathbf{B}(\mathbf{r}; t)$ are given, $U(\mathbf{r}; t)$ and $\mathbf{A}(\mathbf{r}; t)$ are not uniquely determined. It can easily be verified that if we have a set of possible values for $U(\mathbf{r}; t)$ and $\mathbf{A}(\mathbf{r}; t)$, we obtain other potentials $U'(\mathbf{r}; t)$ and $\mathbf{A}'(\mathbf{r}; t)$ which describe the same electromagnetic field by the transformation:

$$\begin{aligned} U'(\mathbf{r}; t) &= U(\mathbf{r}; t) - \frac{\partial}{\partial t}\chi(\mathbf{r}; t) \\ \mathbf{A}'(\mathbf{r}; t) &= \mathbf{A}(\mathbf{r}; t) + \nabla\chi(\mathbf{r}; t) \end{aligned} \tag{2}$$

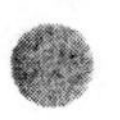



where $\chi(\mathbf{r}\,;t)$ is an arbitrary function of $\mathbf{r}$ and $t$. This can be seen by replacing $U(\mathbf{r}\,;t)$ by $U'(\mathbf{r}\,;t)$ and $\mathbf{A}(\mathbf{r}\,;t)$ by $\mathbf{A}'(\mathbf{r}\,;t)$ in (1) and verifying that $\mathbf{E}(\mathbf{r}\,;t)$ and $\mathbf{B}(\mathbf{r}\,;t)$ remain unchanged. Moreover, it can be shown that relations (2) give all the possible scalar and vector potentials associated with a given electromagnetic field.

When a particular set of potentials has been chosen to describe an electromagnetic field, a *choice of gauge* is said to have been made. As we have just said, an infinite number of different gauges can be used for the same field, characterized by $\mathbf{E}(\mathbf{r}\,;t)$ and $\mathbf{B}(\mathbf{r}\,;t)$. When one changes from one to another, one is said to perform a *gauge transformation.*

It often happens in physics that the equations of motion of a system involve, not the fields $\mathbf{E}(\mathbf{r}\,;t)$ and $\mathbf{B}(\mathbf{r}\,;t)$, but the potentials $U(\mathbf{r}\,;t)$ and $\mathbf{A}(\mathbf{r}\,;t)$. We saw an example of this in § B-5-b of chapter III, when we wrote the Schrödinger equation for a particle of charge $q$ in an electromagnetic field [*cf.* relation (B-48) of that chapter]. The following question can then be posed : do the physical results predicted by the theory depend only on the values of the fields $\mathbf{E}(\mathbf{r}\,;t)$ and $\mathbf{B}(\mathbf{r}\,;t)$ at all points in space, or do they also depend on the gauge used to write the equations? In the latter case, it would obviously be necessary, in order for the theory to make sense, to specify in which gauge the equations are valid.

The aim of this complement is to answer this question. We shall see that in classical mechanics (§ 2), as in quantum mechanics (§ 3), physical results are not modified when a gauge transformation is performed. The scalar and vector potentials can then be seen to be calculation tools; actually, all that counts are the values of the electric and magnetic fields at all points in space. We shall express this result by saying that classical and quantum mechanics possess the property of *gauge invariance.*

## 2. Gauge invariance in classical mechanics

### a. NEWTON'S EQUATIONS

In classical mechanics, the motion of a particle* of charge $q$ and mass $m$ placed in an electromagnetic field can be calculated from the force $\mathbf{f}$ exerted on it. This force is given by Lorentz' law:

$$\mathbf{f} = q[\mathbf{E}(\mathbf{r}\,;t) + \mathbf{v} \times \mathbf{B}(\mathbf{r}\,;t)] \tag{3}$$

where $\mathbf{v}$ is the velocity of the particle. To obtain the equations of motion which allow one to calculate the position $\mathbf{r}(t)$ of the particle at any instant $t$, one substitutes this equation into the fundamental dynamical equation (Newton's law) :

$$m\frac{\mathrm{d}^2}{\mathrm{d}t^2}\,\mathbf{r}(t) = \mathbf{f} \tag{4}$$

In this approach, only the values of the electric and magnetic fields enter into the calculation; therefore, the problem of gauge invariance does not arise.

* For simplicity, we shall assume in this complement that the system under study is composed of a single particle. Generalization to a more complex system formed by several particles placed in an electromagnetic field presents no difficulties.

### b. THE HAMILTONIAN FORMALISM

Instead of adopting the point of view of the preceding section, one can use other equations of motion, the Hamilton-Jacobi equations. It is not difficult to show (*cf.* appendix III) that the latter equations are completely equivalent to Newton's equations. However, since we used the Hamiltonian formalism in chapter III to quantize a physical system, it is useful to study how a gauge transformation appears in this formalism. Although the scalar and vector potentials do not enter into Newton's equations, they are indispensable for writing those of Hamilton. The property of gauge invariance is therefore less obvious for this second point of view.

#### α. *The dynamical variables of the system and their evolution*

To determine the motion of a particle subject to the Lorentz force written in (3), one can use the Lagrangian★:

$$\mathscr{L}(\mathbf{r}, \mathbf{v}\,; t) = \frac{1}{2} m\mathbf{v}^2 - q[U(\mathbf{r}\,; t) - \mathbf{v} \cdot \mathbf{A}(\mathbf{r}\,; t)] \tag{5}$$

This expression permits the calculation of the momentum $\mathbf{p}$, which is written:

$$\mathbf{p} = \nabla_{\mathbf{v}} \mathscr{L}(\mathbf{r}, \mathbf{v}\,; t) = m\mathbf{v} + q\mathbf{A}(\mathbf{r}\,; t) \tag{6}$$

It is then possible to introduce the classical Hamiltonian:

$$\mathscr{H}(\mathbf{r}, \mathbf{p}\,; t) = \frac{1}{2m}[\mathbf{p} - q\mathbf{A}(\mathbf{r}\,; t)]^2 + qU(\mathbf{r}\,; t) \tag{7}$$

In the Hamiltonian formalism, the state of the particle at a given time is defined by its position $\mathbf{r}$ and its momentum $\mathbf{p}$, which we shall call the fundamental dynamical variables, and no longer by its position and its velocity, as in §a above (and as in the Lagrange point of view). The momentum $\mathbf{p}$ (conjugate momentum of the position $\mathbf{r}$) must not be confused with the mechanical momentum $\boldsymbol{\pi}$:

$$\boldsymbol{\pi} = m\mathbf{v} \tag{8}$$

They are indeed different since, according to (6):

$$\boldsymbol{\pi} = \mathbf{p} - q\mathbf{A}(\mathbf{r}\,; t) \tag{9}$$

This relation allows us to calculate the mechanical momentum (and therefore the velocity) whenever the values of $\mathbf{r}$ and $\mathbf{p}$ are known. Similarly, all the other quantities associated with the particle (kinetic energy, angular momentum, etc...) are expressed in the Hamiltonian formalism as functions of the fundamental dynamical variables $\mathbf{r}$ and $\mathbf{p}$ (and, if necessary, of time).

The evolution of the system is governed by Hamilton's equations :

$$\begin{cases} \dfrac{\mathrm{d}}{\mathrm{d}t}\mathbf{r}(t) = \nabla_{\mathbf{p}} \mathscr{H}[\mathbf{r}(t), \mathbf{p}(t); t] \\[2ex] \dfrac{\mathrm{d}}{\mathrm{d}t}\mathbf{p}(t) = -\nabla_{\mathbf{r}} \mathscr{H}[\mathbf{r}(t), \mathbf{p}(t); t] \end{cases} \tag{10}$$

★ We state without proof a certain number of results of analytical mechanics which are established in appendix III.

where $\mathscr{H}$ is the function of $\mathbf{r}$ and $\mathbf{p}$ written in (7). These equations give the values, for all times, of the fundamental dynamical variables if they are known at the initial instant.

To write equations (10), it is necessary to choose a gauge $\mathscr{J}$, that is, a pair of potentials $\{ U(\mathbf{r}\,;t), \mathbf{A}(\mathbf{r}\,;t) \}$ describing the electromagnetic field. What happens if, instead of this gauge $\mathscr{J}$, we choose another one $\mathscr{J}'$, characterized by different potentials $U'(\mathbf{r}\,;t)$ and $\mathbf{A}'(\mathbf{r}\,;t)$, but describing the same fields $\mathbf{E}(\mathbf{r}\,;t)$ and $\mathbf{B}(\mathbf{r}\,;t)$? We shall label with a prime the values of the dynamical variables associated with the motion of the particle when the gauge chosen is $\mathscr{J}'$. As we pointed out in § a, Newton's equations indicate that the position $\mathbf{r}$ and the velocity $\mathbf{v}$ take on, at every instant, values independent of the gauge. Consequently, we have:

$$\begin{cases} \mathbf{r}'(t) = \mathbf{r}(t) & \text{(11-a)} \\ \boldsymbol{\pi}'(t) = \boldsymbol{\pi}(t) & \text{(11-b)} \end{cases}$$

Now, from (9):

$$\begin{aligned} \boldsymbol{\pi}(t) &= \mathbf{p}(t) - q\mathbf{A}[\mathbf{r}(t)\,;t] \\ \boldsymbol{\pi}'(t) &= \mathbf{p}'(t) - q\mathbf{A}'[\mathbf{r}'(t)\,;t] \end{aligned} \tag{12}$$

Therefore, the values $\mathbf{p}(t)$ and $\mathbf{p}'(t)$ of the momentum in the gauges $\mathscr{J}$ and $\mathscr{J}'$ are different; they must satisfy:

$$\mathbf{p}'(t) - q\mathbf{A}'[\mathbf{r}'(t)\,;t] = \mathbf{p}(t) - q\mathbf{A}[\mathbf{r}(t)\,;t] \tag{13}$$

If $\chi(\mathbf{r}\,;t)$ is the function appearing in formulas (2) which govern the gauge transformation from $\mathscr{J}$ to $\mathscr{J}'$, the values of the fundamental dynamical variables are transformed according to the formulas:

$$\begin{cases} \mathbf{r}'(t) = \mathbf{r}(t) & \text{(14-a)} \\ \mathbf{p}'(t) = \mathbf{p}(t) + q\boldsymbol{\nabla}\chi[\mathbf{r}(t)\,;t] & \text{(14-b)} \end{cases}$$

*In the Hamiltonian formalism, the value at each instant of the dynamical variables describing a given motion depends on the gauge chosen.* Moreover, such a result is not surprising since, in (7) and (10), the scalar and vector potentials appear explicitly in the equations of motion for the position and momentum.

### β. "*True physical quantities*" *and* "*non-physical quantities*"

#### (*i*) DEFINITIONS

We have just seen, in relations (14) for example, that it is possible to distinguish between two types of quantities associated with the particle : those which, like $\mathbf{r}$ or $\boldsymbol{\pi}$, have identical values at all times in two different gauges, and those which, like $\mathbf{p}$, have values which depend on the arbitrarily chosen gauge. We are thus led to the following general definition:

– A *true physical quantity* associated with the system under consideration is a quantity whose value at any time does not depend (for a given motion of the system) on the gauge used to describe the electromagnetic field.

– A *non-physical quantity*, on the other hand, is a quantity whose value is modified by a gauge transformation; thus, like the scalar and vector potentials, it is seen to be a calculation tool, rather than an actually observable quantity.

The problem then posed is the following: in the Hamiltonian formalism, all the quantities associated with the system appear in the form of functions of the fundamental dynamical variables $\mathbf{r}$ and $\mathbf{p}$; how can we know whether such a function corresponds to a true physical quantity or not?

### (*ii*) CHARACTERISTIC RELATION OF TRUE PHYSICAL QUANTITIES

Let us first assume that a quantity associated with the particle is described, in the gauge $\mathscr{J}$, by a function of $\mathbf{r}$ and $\mathbf{p}$ (which may depend on time) which we shall denote by $\mathscr{F}(\mathbf{r}, \mathbf{p}; t)$. If to this quantity corresponds, in another gauge $\mathscr{J}'$, the same function $\mathscr{F}(\mathbf{r}', \mathbf{p}'; t)$, the quantity is clearly not truly physical [except in the special case where the function $\mathscr{F}$ depends only on $\mathbf{r}$ and not on $\mathbf{p}$; see equations (14)]. Since the values of the momentum are different in the two gauges $\mathscr{J}$ and $\mathscr{J}'$, the same is obviously true for the values of the function $\mathscr{F}$.

To obtain the true physical quantities associated with the system, *we must therefore consider functions* $\mathscr{G}_{\mathscr{J}}(\mathbf{r}, \mathbf{p}; t)$ *whose form depends on the gauge chosen* (this is why we label these functions with an index $\mathscr{J}$). We have already seen an example of such a function: the mechanical momentum $\boldsymbol{\pi}$ is a function of $\mathbf{r}$ and $\mathbf{p}$ through the intermediate of the vector potential $\mathbf{A}$ [*cf.* (9)]. In this case, the function is different in the two gauges $\mathscr{J}$ and $\mathscr{J}'$; that is, it is of the form $\boldsymbol{\pi}_{\mathscr{J}}(\mathbf{r}, \mathbf{p}; t)$. The definition given in (*i*) thus implies that the function $\mathscr{G}_{\mathscr{J}}(\mathbf{r}, \mathbf{p}; t)$ describes a true physical quantity on the condition that:

$$\mathscr{G}_{\mathscr{J}}[\mathbf{r}(t), \mathbf{p}(t); t] = \mathscr{G}_{\mathscr{J}'}[\mathbf{r}'(t), \mathbf{p}'(t); t] \tag{15}$$

where $\mathbf{r}(t)$ and $\mathbf{p}(t)$ are the values taken on by the position and momentum in the gauge $\mathscr{J}$, and $\mathbf{r}'(t)$ and $\mathbf{p}'(t)$ are their values in the gauge $\mathscr{J}'$. If we substitute relations (14) into (15), we obtain:

$$\mathscr{G}_{\mathscr{J}}[\mathbf{r}(t), \mathbf{p}(t); t] = \mathscr{G}_{\mathscr{J}'}[\mathbf{r}(t), \mathbf{p}(t) + q\boldsymbol{\nabla}\chi(\mathbf{r}(t); t); t] \tag{16}$$

where this relation must be satisfied at every instant $t$ and for all possible motions of the system. Since, when $t$ is fixed, the values of the position and the momentum can be chosen independently, both sides of (16) must in fact be the same function of $\mathbf{r}$ and $\mathbf{p}$, which is written:

$$\mathscr{G}_{\mathscr{J}}[\mathbf{r}, \mathbf{p}; t] = \mathscr{G}_{\mathscr{J}'}[\mathbf{r}, \mathbf{p} + q\boldsymbol{\nabla}\chi(\mathbf{r}; t); t] \tag{17}$$

This relation is characteristic of the functions $\mathscr{G}_{\mathscr{J}}[\mathbf{r}, \mathbf{p}; t]$ associated with true physical quantities. Therefore, if *one considers the function* $\mathscr{G}_{\mathscr{J}'}[\mathbf{r}, \mathbf{p}; t]$ *for the gauge* $\mathscr{J}'$, *and if one replaces* $\mathbf{p}$ *by* $\mathbf{p} + q\boldsymbol{\nabla}\chi(\mathbf{r}; t)$ [where $\chi(\mathbf{r}; t)$ defines, according to (2), the gauge transformation from $\mathscr{J}$ to $\mathscr{J}'$], *one obtains a new function of* $\mathbf{r}$ *and* $\mathbf{p}$ *which must be identical to* $\mathscr{G}_{\mathscr{J}}[\mathbf{r}, \mathbf{p}; t]$. If this is not the case, the function considered corresponds to a quantity which is not truly physical.

*(iii)* EXAMPLES

Let us give some examples of functions $\mathcal{G}_{\mathcal{J}}\,[\mathbf{r},\ \mathbf{p};\ t]$ which describe true physical quantities. We have already encountered two : those corresponding to the position and to the mechanical momentum; the first is simply equal to $\mathbf{r}$ and the second to:

$$\boldsymbol{\pi}_{\mathcal{J}}(\mathbf{r}, \mathbf{p}\,; t) = \mathbf{p} - q\mathbf{A}(\mathbf{r}\,; t) \tag{18}$$

Since relations (11) express the fact that $\mathbf{r}$ and $\boldsymbol{\pi}$ are true physical quantities, we know *a priori* that relation (17) is satisfied by the corresponding functions.

However, let us verify this directly in order to familiarize ourselves with the use of this relation. As regards $\mathbf{r}$, we are dealing with a function which does not depend on $\mathbf{p}$ and whose form does not depend on the gauge*; this immediately implies (17). As regards $\boldsymbol{\pi}$, relation (18) yields:

$$\boldsymbol{\pi}_{\mathcal{J}'}(\mathbf{r}, \mathbf{p}\,; t) = \mathbf{p} - q\mathbf{A}'(\mathbf{r}\,; t) \tag{19}$$

Replace in this function $\mathbf{p}$ by $\mathbf{p} + q\boldsymbol{\nabla}\chi(\mathbf{r};\ t)$; we obtain the function:

$$\mathbf{p} + q\boldsymbol{\nabla}\chi(\mathbf{r}\,; t) - q\mathbf{A}'(\mathbf{r}\,; t) = \mathbf{p} - q\mathbf{A}(\mathbf{r}\,; t) \tag{20}$$

which is none other than $\boldsymbol{\pi}_{\mathcal{J}}(\mathbf{r};\ \mathbf{p};\ t)$; relation (17) is therefore satisfied.

Other true physical quantities are the kinetic energy:

$$\gamma_{\mathcal{J}}(\mathbf{r}, \mathbf{p}\,; t) = \frac{1}{2m}\left[\mathbf{p} - q\mathbf{A}(\mathbf{r}\,; t)\right]^2 \tag{21}$$

and the moment, with respect to the origin, of the mechanical momentum :

$$\boldsymbol{\lambda}_{\mathcal{J}}(\mathbf{r}, \mathbf{p}\,; t) = \mathbf{r} \times \left[\mathbf{p} - q\mathbf{A}(\mathbf{r}\,; t)\right] \tag{22}$$

In general, we see that whenever a function of $\mathbf{r}$ and $\mathbf{p}$ has the form :

$$\mathcal{G}_{\mathcal{J}}(\mathbf{r}, \mathbf{p}\,; t) = F\left[\mathbf{r}, \mathbf{p} - q\mathbf{A}(\mathbf{r}\,; t)\right] \tag{23}$$

(where $F$ is a function whose form is independent of the gauge $\mathcal{J}$ chosen), we obtain a true *physical quantity***. This result makes sense since (23) really expresses the fact that the values taken on by the quantity considered are obtained from those of $\mathbf{r}$ and $\boldsymbol{\pi}$, which we know to be gauge-invariant.

Let us also give some examples of functions describing quantities which are not true physical quantities. In addition to the momentum $\mathbf{p}$, we can cite the function :

$$\mathcal{E}(\mathbf{p}) = \frac{\mathbf{p}^2}{2m} \tag{24}$$

* It is not difficult to verify that, in general, any function $\mathcal{G}\,(\mathbf{r},\ t)$ which depends only on $\mathbf{r}$ (and, possibly, on the time), and whose form is the same in any gauge $\mathcal{J}$ chosen, describes a true physical quantity.

** One could also construct functions associated with true physical quantities in which the potentials are involved in a more complex way than in (23) (for example, the scalar product of the particle velocity and the electric field at the position of the particle).

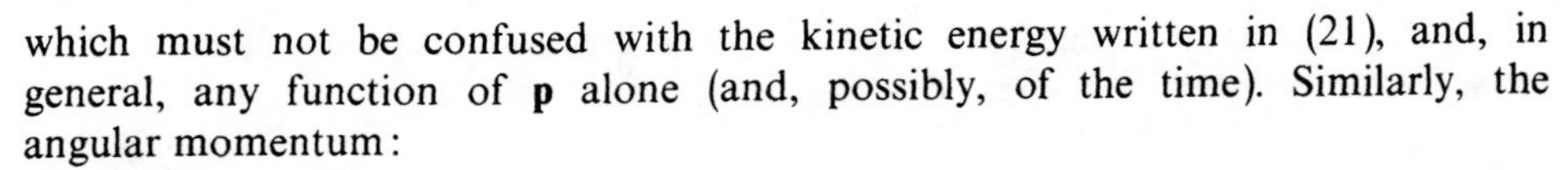


which must not be confused with the kinetic energy written in (21), and, in general, any function of $\mathbf{p}$ alone (and, possibly, of the time). Similarly, the angular momentum:

$$\mathscr{L}(\mathbf{r}, \mathbf{p}) = \mathbf{r} \times \mathbf{p} \tag{25}$$

cannot be considered to be a true physical quantity. Finally, let us cite the classical Hamiltonian, which, according to (7), is the sum of the kinetic energy $\gamma_{\mathscr{J}}(\mathbf{r}, \mathbf{p}; t)$, which is a true physical quantity, and the potential energy $qU$. Now, the latter [which should rigorously be written in the form of a gauge-dependent function $U_{\mathscr{J}}(\mathbf{r}; t)$] is not a true physical quantity since, at every point in space, its value changes when the gauge is changed.

## 3. Gauge invariance in quantum mechanics

In chapter III, we introduced the postulates of quantum mechanics by starting from the Hamiltonian formulation of classical mechanics. We are thus led to ask if the problem of gauge invariance, easily resolved in classical mechanics because of the existence of Newton's equations, is more complex in the framework of quantum mechanics. The following question then arises: are the postulates stated in chapter III valid for any arbitrarily chosen gauge $\mathscr{J}$ or only for a particular gauge?

In answering this question, we shall be guided by the results obtained in the preceding paragraph. Following the same type of reasoning, we shall see that there exists a close analogy between the consequences of a gauge transformation in the classical Hamiltonian formalism and in the quantum mechanical formalism. We shall thus establish the gauge invariance of quantum mechanics.

To do this, we shall begin (§a) by examining the results obtained when the quantization rules are applied in the same way in two different gauges. We shall then see (§b) that, like in classical mechanics, where the values of the dynamical variables generally change when the gauge is changed, *a given physical system must be characterized by a mathematical state vector* $|\psi\rangle$ *which depends on the gauge.* The transition from a state vector corresponding to one gauge $\mathscr{J}$ to that of another gauge $\mathscr{J}'$ is effected by a *unitary transformation.* The form of the Schrödinger equation, however, always remains the same (as do Hamilton's equations in classical mechanics). Finally, we shall examine the behavior, under a gauge transformation, of the observables associated with the system (§c). We shall then see that the simultaneous modification of the state vector and the observables is such that *the physical content of quantum mechanics does not depend on the gauge chosen.* Moreover, we shall demonstrate this by showing that the density and the probability current values are gauge invariant.

### a. QUANTIZATION RULES

The state space of a (spinless) particle is always $\mathscr{E}_{\mathbf{r}}$. However, we are clearly led by the results of §2 above to expect that the operator associated with a given quantity may be different in two different gauges. We shall therefore label these operators with an index $\mathscr{J}$.

The quantization rules associate, with the position **r** and the momentum **p** of the particle, operators **R** and **P** acting in $\mathscr{E}_{\mathbf{r}}$ such that:

$$[X, P_x] = [Y, P_y] = [Z, P_z] = i\hbar \tag{26}$$

(where all the other commutators between components of **R** and **P** are zero). In the $\{ | \mathbf{r} \rangle \}$ representation, the operator **R** acts like a multiplication by **r**, and **P** like the differential operator $\frac{\hbar}{i}\boldsymbol{\nabla}$. These rules are the same in all gauges. We can therefore write:

$$\begin{cases} \mathbf{R}_{\mathscr{J}'} = \mathbf{R}_{\mathscr{J}} & \text{(27-a)} \\ \mathbf{P}_{\mathscr{J}'} = \mathbf{P}_{\mathscr{J}} & \text{(27-b)} \end{cases}$$

In fact, these equations enable us to omit the index $\mathscr{J}$ for the observables **R** and **P**, and we shall henceforth do so.

The quantization of all other quantities associated with the particle is derived from this: in a given gauge, one takes the function of **r** and **p** which gives the classical quantity considered and (after having symmetrized, if necessary) replaces **r** by the operator **R** and **p** by **P**. We thus obtain the operator which, in the gauge chosen, describes this quantity. Consider some examples:

– The angular momentum operator, obtained from $\mathbf{r} \times \mathbf{p}$, is the same in all gauges:

$$\mathbf{L}_{\mathscr{J}'} = \mathbf{L}_{\mathscr{J}} \tag{28}$$

– The operator associated with the mechanical momentum, on the other hand, depends on the gauge chosen. In the gauge $\mathscr{J}$, it is given by:

$$\boldsymbol{\Pi}_{\mathscr{J}} = \mathbf{P} - q\mathbf{A}(\mathbf{R}\,; t) \tag{29}$$

If the gauge is changed, it becomes:

$$\boldsymbol{\Pi}_{\mathscr{J}'} = \mathbf{P} - q\mathbf{A}'(\mathbf{R}\,; t) \tag{30}$$

whose action in $\mathscr{E}_{\mathbf{r}}$ is different from that of $\boldsymbol{\Pi}_{\mathscr{J}}$:

$$\boldsymbol{\Pi}_{\mathscr{J}'} = \boldsymbol{\Pi}_{\mathscr{J}} - q\boldsymbol{\nabla}\chi(\mathbf{R}\,; t) \tag{31}$$

– Similarly, the operator*:

$$\boldsymbol{\Lambda}_{\mathscr{J}} = \mathbf{R} \times \boldsymbol{\Pi}_{\mathscr{J}} = \mathbf{R} \times [\mathbf{P} - q\mathbf{A}(\mathbf{R}\,; t)] \tag{32}$$

which describes the moment of the mechanical momentum, explicitly involves the vector potential chosen.

– Finally, the Hamiltonian operator is obtained from formula (7):

$$H_{\mathscr{J}} = \frac{1}{2m}[\mathbf{P} - q\mathbf{A}(\mathbf{R}\,; t)]^2 + qU(\mathbf{R}\,; t) \tag{33}$$

* It can be verified, by using the commutation relations of **R** and $\boldsymbol{\Pi}_{\mathscr{J}}$, that it is not necessary to symmetrize expression (32).

It is obvious that in another gauge, it becomes a different operator, since :

$$H_{\mathscr{J}'} = \frac{1}{2m}[\mathbf{P} - q\mathbf{A}'(\mathbf{R}\,;t)]^2 + qU'(\mathbf{R}\,;t) \neq H_{\mathscr{J}} \tag{34}$$

### b. UNITARY TRANSFORMATION OF THE STATE VECTOR ; FORM INVARIANCE OF THE SCHRÖDINGER EQUATION

#### α. *The unitary operator* $T_\chi(t)$

In classical mechanics, we denoted by $\{ \mathbf{r}(t), \mathbf{p}(t) \}$ and $\{ \mathbf{r}'(t), \mathbf{p}'(t) \}$ the values of the fundamental dynamical variables characterizing the state of the particle in two different gauges $\mathscr{J}$ and $\mathscr{J}'$. In quantum mechanics, we shall therefore denote by $| \psi(t) \rangle$ and $| \psi'(t) \rangle$ the state vectors relative to these two gauges, and the analogue of relations (14) is thus given by relations between mean values:

$$\langle \psi'(t) | \mathbf{R}_{\mathscr{J}'} | \psi'(t) \rangle = \langle \psi(t) | \mathbf{R}_{\mathscr{J}} | \psi(t) \rangle \tag{35-a}$$

$$\langle \psi'(t) | \mathbf{P}_{\mathscr{J}'} | \psi'(t) \rangle = \langle \psi(t) | \mathbf{P}_{\mathscr{J}} + q\boldsymbol{\nabla}\chi(\mathbf{R}\,;t) | \psi(t) \rangle \tag{35-b}$$

Using (27), we see immediately that this is possible only if $| \psi(t) \rangle$ and $| \psi'(t) \rangle$ are two different kets. We shall therefore seek a unitary transformation $T_\chi(t)$ which enables us to go from $| \psi(t) \rangle$ to $| \psi'(t) \rangle$:

$$| \psi'(t) \rangle = T_\chi(t) | \psi(t) \rangle \tag{36-a}$$

$$T_\chi^\dagger(t)\, T_\chi(t) = T_\chi(t)\, T_\chi^\dagger(t) = \mathbb{1} \tag{36-b}$$

Taking (27) into account, we see that equations (35) are satisfied for any $| \psi(t) \rangle$ on condition that:

$$\begin{cases} T_\chi^\dagger(t)\, \mathbf{R}\, T_\chi(t) = \mathbf{R} & (37\text{-a}) \\ T_\chi^\dagger(t)\, \mathbf{P}\, T_\chi(t) = \mathbf{P} + q\boldsymbol{\nabla}\chi(\mathbf{R}\,;t) & (37\text{-b}) \end{cases}$$

Multiplying (37-a) on the left by $T_\chi(t)$, we obtain:

$$\mathbf{R}\, T_\chi(t) = T_\chi(t)\, \mathbf{R} \tag{38}$$

The desired unitary operator commutes with the three components of $\mathbf{R}$; it can therefore be written in the form:

$$T_\chi(t) = e^{iF(\mathbf{R}\,;t)} \tag{39}$$

where $F(\mathbf{R}\,;t)$ is a Hermitian operator. Relation (48) of complement $B_{II}$ then allows us to write :

$$[\mathbf{P}, T_\chi(t)] = \hbar\boldsymbol{\nabla}\{ F(\mathbf{R}\,;t) \}\, T_\chi(t) \tag{40}$$

If we multiply this equation on the left by $T_\chi^\dagger(t)$ and substitute it into (37-b), we easily obtain the relation:

$$\hbar\boldsymbol{\nabla}\{ F(\mathbf{R}\,;t) \} = q\boldsymbol{\nabla}\chi(\mathbf{R}\,;t) \tag{41}$$

which is satisfied when:

$$F(\mathbf{R}\,;t) = F_0(t) + \frac{q}{\hbar}\chi(\mathbf{R}\,;t) \tag{42}$$

Omitting the coefficient $F_0(t)$, which corresponds, for the state vector $| \psi(t) \rangle$, to a global phase factor of no physical consequence, we obtain the operator $T_\chi(t)$ :

$$T_\chi(t) = e^{i\frac{q}{\hbar}\chi(\mathbf{R}\,;t)} \tag{43}$$

If, in (36-a), $T_\chi(t)$ is this operator, relations (35) are automatically satisfied.

COMMENTS:

(*i*) In the $\{ | \mathbf{r} \rangle \}$ representation, relations (36-a) and (43) imply that the wave functions $\psi(\mathbf{r}, t) = \langle \mathbf{r} | \psi(t) \rangle$ and $\psi'(\mathbf{r}, t) = \langle \mathbf{r} | \psi'(t) \rangle$ are related by:

$$\psi'(\mathbf{r}, t) = e^{i\frac{q}{\hbar}\chi(\mathbf{r},t)} \psi(\mathbf{r}, t) \tag{44}$$

*For the wave function, the gauge transformation corresponds to a phase change which varies from one point to another, and is not, therefore, a global phase factor.* The gauge invariance of physical predictions obtained by using the wave functions $\psi$ or $\psi'$, is therefore not obvious *a priori*.

(*ii*) If the system under study is composed of several particles having positions $\mathbf{r}_1, \mathbf{r}_2, ...$ and charges $q_1, q_2, ...$, (43) must be replaced by:

$$\begin{aligned} T_\chi(t) &= T_\chi^{(1)}(t)\, T_\chi^{(2)}(t) \ldots \\ &= e^{\frac{i}{\hbar}[q_1 \chi(\mathbf{R}_1, t) + q_2 \chi(\mathbf{R}_2, t) + \ldots]} \end{aligned} \tag{45}$$

β. *Time evolution of the state vector*

Now let us show that if the evolution of the ket $| \psi(t) \rangle$ obeys, in the gauge $\mathscr{J}$, the Schrödinger equation:

$$i\hbar \frac{\mathrm{d}}{\mathrm{d}t} | \psi(t) \rangle = H_{\mathscr{J}}(t) | \psi(t) \rangle \tag{46}$$

the state vector $| \psi'(t) \rangle$ given by (36) satisfies an equation of the same form in the gauge $\mathscr{J}'$:

$$i\hbar \frac{\mathrm{d}}{\mathrm{d}t} | \psi'(t) \rangle = H_{\mathscr{J}'}(t) | \psi'(t) \rangle \tag{47}$$

where $H_{\mathscr{J}'}(t)$ is given by (34).

To do this, let us calculate the left-hand side of (47); it is written:

$$\begin{aligned} i\hbar \frac{\mathrm{d}}{\mathrm{d}t} | \psi'(t) \rangle &= i\hbar \frac{\mathrm{d}}{\mathrm{d}t} \{ T_\chi(t) | \psi(t) \rangle \} \\ &= i\hbar \{ \frac{\mathrm{d}}{\mathrm{d}t} T_\chi(t) \} | \psi(t) \rangle + i\hbar T_\chi(t) \frac{\mathrm{d}}{\mathrm{d}t} | \psi(t) \rangle \end{aligned} \tag{48}$$

that is, according to (43) and (46)*:

$$\begin{aligned} i\hbar \frac{\mathrm{d}}{\mathrm{d}t} | \psi'(t) \rangle &= - q \{ \frac{\partial}{\partial t} \chi(\mathbf{R}; t) \} T_\chi(t) | \psi(t) \rangle + T_\chi(t) H_{\mathscr{J}}(t) | \psi(t) \rangle \\ &= \{ - q \frac{\partial}{\partial t} \chi(\mathbf{R}; t) + \tilde{H}_{\mathscr{J}}(t) \} | \psi'(t) \rangle \end{aligned} \tag{49}$$

* The function $\chi$ depends on $\mathbf{R}$ and not on $\mathbf{P}$; consequently $\chi(\mathbf{R}, t)$ commutes with $\frac{\partial}{\partial t} \chi(\mathbf{R}; t)$. This is why $T_\chi(t)$ can be differentiated as if $\chi(\mathbf{R}, t)$ were an ordinary function of the time and not an operator (*cf.* complement $B_{II}$, comment of §5-c).

where $\tilde{H}_{\mathcal{J}}(t)$ designates the transform of $H_{\mathcal{J}}(t)$ by the unitary operator $T_\chi(t)$:

$$\tilde{H}_{\mathcal{J}}(t) = T_\chi(t)\, H_{\mathcal{J}}(t)\, T_\chi^\dagger(t) \tag{50}$$

Equation (47) will therefore be satisfied if:

$$H_{\mathcal{J}'}(t) = \tilde{H}_{\mathcal{J}}(t) - q\frac{\partial}{\partial t}\chi(\mathbf{R}\,;t) \tag{51}$$

Now $\tilde{H}_{\mathcal{J}}(t)$ is given by:

$$\tilde{H}_{\mathcal{J}}(t) = \frac{1}{2m}[\tilde{\mathbf{P}} - q\mathbf{A}(\tilde{\mathbf{R}}\,;t)]^2 + qU(\tilde{\mathbf{R}}\,;t) \tag{52}$$

where $\tilde{\mathbf{R}}$ and $\tilde{\mathbf{P}}$ designate the transforms of $\mathbf{R}$ and $\mathbf{P}$ by the unitary operator $T_\chi(t)$. According to (37) :

$$\begin{cases} \tilde{\mathbf{R}} = T_\chi(t)\,\mathbf{R}\,T_\chi^\dagger(t) = \mathbf{R} & \text{(53-a)} \\ \tilde{\mathbf{P}} = T_\chi(t)\,\mathbf{P}\,T_\chi^\dagger(t) = \mathbf{P} - q\nabla\chi(\mathbf{R}\,;t) & \text{(53-b)} \end{cases}$$

These relations, substituted into (52), indicate that:

$$\tilde{H}_{\mathcal{J}}(t) = \frac{1}{2m}[\mathbf{P} - q\mathbf{A}(\mathbf{R}\,;t) - q\nabla\chi(\mathbf{R}\,;t)]^2 + qU(\mathbf{R}\,;t) \tag{54}$$

Using relations (2) to replace the potentials relative to the gauge $\mathcal{J}$ by those relative to $\mathcal{J}'$, we then obtain, using (34), relation (51). Therefore, the Schrödinger equation can be written in the same way in any gauge chosen.

### c. INVARIANCE OF PHYSICAL PREDICTIONS UNDER A GAUGE TRANSFORMATION

#### α. *Behavior of the observables*

Under the effect of the unitary transformation $T_\chi(t)$, any observable $K$ is transformed into $\tilde{K}$, with:

$$\tilde{K} = T_\chi(t)\,K\,T_\chi^\dagger(t) \tag{55}$$

We have already seen, in (53), that while $\tilde{\mathbf{R}}$ is simply equal to $\mathbf{R}$, $\tilde{\mathbf{P}}$ is not equal to $\mathbf{P}$. Similarly, $\tilde{\mathbf{\Pi}}_{\mathcal{J}}$ is different from $\mathbf{\Pi}_{\mathcal{J}}$ since:

$$\begin{aligned} \tilde{\mathbf{\Pi}}_{\mathcal{J}} &= \tilde{\mathbf{P}} - q\mathbf{A}(\tilde{\mathbf{R}}\,;t) \\ &= \mathbf{P} - q\nabla\chi(\mathbf{R}\,;t) - q\mathbf{A}(\mathbf{R}\,;t) \\ &= \mathbf{\Pi}_{\mathcal{J}} - q\nabla\chi(\mathbf{R}\,;t) \end{aligned} \tag{56}$$

Taking (27-a) and (31) into account, we see that relations (53-a) and (56) imply that the observables $\mathbf{R}$ and $\mathbf{\Pi}_{\mathcal{J}}$, associated with true physical quantities (position and mechanical momentum) are such that :

$$\begin{cases} \tilde{\mathbf{R}}_{\mathcal{J}} = \mathbf{R}_{\mathcal{J}'} \\ \tilde{\mathbf{\Pi}}_{\mathcal{J}} = \mathbf{\Pi}_{\mathcal{J}'} \end{cases} \tag{57}$$

On the other hand, the momentum $\mathbf{P}$ (which is not a true physical quantity) does not satisfy an analogous relation, since, from (27-b) and (53-b):

$$\tilde{\mathbf{P}}_{\mathcal{J}} \neq \mathbf{P}_{\mathcal{J}'} \tag{58}$$

We shall see that this result is a general one : *in quantum mechanics, for every true physical quantity, there is an operator* $G_{\mathscr{J}}(t)$ *which satisfies:*

$$\tilde{G}_{\mathscr{J}}(t) = G_{\mathscr{J}'}(t) \tag{59}$$

This relation is the quantum mechanical analogue of the classical relation (16). It shows that, except for the special case of **R** or a function of **R** alone, *the operator corresponding to a true physical quantity depends on the gauge* $\mathscr{J}$. We have already seen examples of this in (29) and (32).

To prove (59), one need only apply the quantization rules stated in chapter III to a function $\mathscr{G}_{\mathscr{J}}(\mathbf{r}, \mathbf{p}; t)$ and use relation (17), the characteristic relation for true physical classical quantities. We therefore replace **r** and **p** by the operators **R** and **P** and obtain (if necessary, after a symmetrization with respect to these operators) the operator $G_{\mathscr{J}}(t)$. If the form of the function $\mathscr{G}_{\mathscr{J}}$ depends on the gauge chosen, the operator $G_{\mathscr{J}}(t)$ also depends on $\mathscr{J}$. When the quantity associated with $\mathscr{G}_{\mathscr{J}}$ is a true physical quantity, we have, according to (17):

$$\mathscr{G}_{\mathscr{J}}[\mathbf{R}, \mathbf{P}; t] = \mathscr{G}_{\mathscr{J}'}[\mathbf{R}, \mathbf{P} + q\nabla\chi(\mathbf{R}; t); t] \tag{60}$$

Applying the unitary transformation $T_\chi(t)$ to this relation, we obtain:

$$\begin{aligned}\tilde{\mathscr{G}}_{\mathscr{J}}[\mathbf{R}, \mathbf{P}; t] &= \tilde{\mathscr{G}}_{\mathscr{J}'}[\mathbf{R}, \mathbf{P} + q\nabla\chi(\mathbf{R}; t); t] \\ &= \mathscr{G}_{\mathscr{J}'}[\tilde{\mathbf{R}}, \tilde{\mathbf{P}} + q\nabla\chi(\tilde{\mathbf{R}}; t); t]\end{aligned} \tag{61}$$

That is, taking (53) into account:

$$\tilde{\mathscr{G}}_{\mathscr{J}}[\mathbf{R}, \mathbf{P}; t] = \mathscr{G}_{\mathscr{J}'}[\mathbf{R}, \mathbf{P}; t] \tag{62}$$

After symmetrizing, if necessary, both sides of this relation, we indeed obtain (59).

Let us give some examples of true physical observables. In addition to **R** and $\mathbf{\Pi}_{\mathscr{J}}$, we can cite the moment $\mathbf{\Lambda}_{\mathscr{J}}$ of the mechanical momentum [*cf.* (32)], or the kinetic energy:

$$\Gamma_{\mathscr{J}} = \frac{\mathbf{\Pi}_{\mathscr{J}}^2}{2m} = \frac{1}{2m}[\mathbf{P} - q\mathbf{A}(\mathbf{R}; t)]^2 \tag{63}$$

On the other hand, **P** and **L** are not true physical quantities; neither is the Hamiltonian, since relation (51) implies in general that:

$$\tilde{H}_{\mathscr{J}}(t) \neq H_{\mathscr{J}'}(t) \tag{64}$$

COMMENT :

In classical mechanics, it is well known that the total energy of a particle moving in a time-independent electromagnetic field is a constant of the motion. It is indeed possible in this case to limit oneself to potentials which are also time-independent. We see from (51) that one then has:

$$\tilde{H}_{\mathscr{J}} = H_{\mathscr{J}'} \tag{65}$$

In this particular case, $H_{\mathscr{J}}$ is indeed a true physical observable which can therefore be interpreted to be the total energy of the particle.

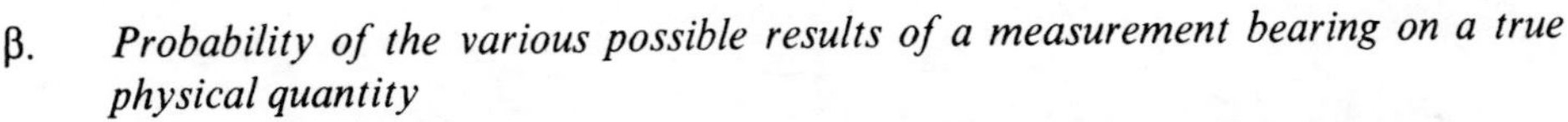


β. *Probability of the various possible results of a measurement bearing on a true physical quantity*

Assume that at time $t$ we want to measure a true physical quantity. In the gauge $\mathcal{J}$, the state of the system is described at this instant by the ket* $| \psi \rangle$, and the physical quantity, by the observable $G_{\mathcal{J}}$. Let $| \varphi_n \rangle$ be an eigenvector of $G_{\mathcal{J}}$, with the eigenvalue $g_n$ (assumed, for simplicity, to be non-degenerate):

$$G_{\mathcal{J}} | \varphi_n \rangle = g_n | \varphi_n \rangle \tag{66}$$

As calculated in the gauge $\mathcal{J}$ from the postulates of quantum mechanics, the probability of obtaining $g_n$ in the measurement envisaged is equal to:

$$\mathscr{P}_n = |\langle \varphi_n | \psi \rangle|^2 \tag{67}$$

What happens to this prediction when the gauge is changed? According to (59), the operator $G_{\mathcal{J}'}$ associated with the quantity under consideration in the new gauge $\mathcal{J}'$ can have the ket:

$$| \varphi'_n \rangle = T_\chi | \varphi_n \rangle \tag{68}$$

as an eigenvector, with the same eigenvalue $g_n$ as in (66). That is:

$$\begin{aligned} G_{\mathcal{J}'} | \varphi'_n \rangle &= T_\chi G_{\mathcal{J}} T_\chi^\dagger T_\chi | \varphi_n \rangle \\ &= T_\chi g_n | \varphi_n \rangle = g_n | \varphi'_n \rangle \end{aligned} \tag{69}$$

In the gauge $\mathcal{J}'$, $g_n$ still appears, therefore, as a possible measurement result. Moreover, calculation of the corresponding probability yields the same value as in the gauge $\mathcal{J}$, since, according to (36-a) and (68):

$$\langle \varphi'_n | \psi' \rangle = \langle \varphi_n | T_\chi^\dagger T_\chi | \psi \rangle = \langle \varphi_n | \psi \rangle \tag{70}$$

We have thus verified that the postulates of quantum mechanics lead to *gauge-invariant physical predictions*: the possible results of any measurement and the associated probabilities are invariant under a gauge transformation.

γ. *Probability density and current*

Let us calculate, from formulas (D-9) and (D-20) of chapter III, the probability density $\rho(\mathbf{r}, t)$ and probability current $\mathbf{J}(\mathbf{r}, t)$ in two different gauges $\mathcal{J}$ and $\mathcal{J}'$. For the first gauge, we have:

$$\rho(\mathbf{r}, t) = |\psi(\mathbf{r}, t)|^2 \tag{71}$$

and:

$$\mathbf{J}(\mathbf{r}, t) = \frac{1}{m} \operatorname{Re} \left\{ \psi^*(\mathbf{r}, t) \left[ \frac{\hbar}{i} \nabla - q\mathbf{A}(\mathbf{r}; t) \right] \psi(\mathbf{r}, t) \right\} \tag{72}$$

Relation (44) immediately shows that:

$$\rho'(\mathbf{r}, t) = |\psi'(\mathbf{r}, t)|^2 = \rho(\mathbf{r}, t) \tag{73}$$

* We do not indicate the time dependence because all the quantities must be evaluated at the time $t$ when we want to perform the measurement.

Moreover, it also implies that :

$$\mathbf{J}'(\mathbf{r}, t) = \frac{1}{m} \operatorname{Re}\left\{ e^{-i\frac{q}{\hbar}\chi(\mathbf{r};t)} \psi^*(\mathbf{r}, t) \left[ \frac{\hbar}{i} \boldsymbol{\nabla} - q\mathbf{A}'(\mathbf{r}; t) \right] e^{i\frac{q}{\hbar}\chi(\mathbf{r};t)} \psi(\mathbf{r}, t) \right\}$$
$$= \frac{1}{m} \operatorname{Re}\left\{ \psi^*(\mathbf{r}, t) \left[ \frac{\hbar}{i} \boldsymbol{\nabla} - q\mathbf{A}'(\mathbf{r}; t) + q\boldsymbol{\nabla}\chi(\mathbf{r}; t) \right] \psi(\mathbf{r}, t) \right\} \tag{74}$$

that is, taking (2) into account:

$$\mathbf{J}'(\mathbf{r}, t) = \mathbf{J}(\mathbf{r}, t) \tag{75}$$

The probability density and current are therefore invariant under a gauge transformation. This result could have been foreseen, moreover, from the conclusions of § β above, since [*cf.* relation (D-19) of chapter III] $\rho(\mathbf{r}, t)$ and $\mathbf{J}(\mathbf{r}, t)$ can be considered to be mean values of the operators $| \mathbf{r} \rangle \langle \mathbf{r} |$ and:

$$\mathbf{K}_{\mathscr{J}}(\mathbf{r}) = \frac{1}{2m} \{ | \mathbf{r} \rangle \langle \mathbf{r} | \mathbf{\Pi}_{\mathscr{J}} + \mathbf{\Pi}_{\mathscr{J}} | \mathbf{r} \rangle \langle \mathbf{r} | \} \tag{76}$$

It is not difficult to show that these two operators satisfy relation (59). They therefore describe true physical quantities whose mean values are gauge-invariant.

**References and suggestions for further reading:**

Messiah (1.17), chap. XXI, §§20 to 22; Sakurai (2.7), §8-1.

Gauge invariance, extended to other domains, has recently roused considerable interest in particle physics; see, for example, the article by Abers and Lee (16.35).

## Complement $J_{III}$

# PROPAGATOR FOR THE SCHRÖDINGER EQUATION



## 1. Introduction

Consider a particle described by the wave function $\psi(\mathbf{r}, t)$. The Schrödinger equation enables us to calculate $\frac{\partial}{\partial t}\psi(\mathbf{r}, t)$, that is, the rate of variation of $\psi(\mathbf{r}, t)$ with respect to $t$. It therefore gives the time evolution of the wave function $\psi(\mathbf{r}, t)$, using a differential point of view. One might wonder if it is possible to adopt a more global (but equivalent) point of view that would allow us to determine directly the value $\psi(\mathbf{r}_0, t)$ taken on by the wave function at a given point $\mathbf{r}_0$ and a given time $t$ from knowledge of the whole wave function $\psi(\mathbf{r}, t')$ at a previous time $t'$ (which is not necessarily infinitesimally close).

To consider this possibility, we can take our inspiration from another domain of physics, electromagnetism, where both points of view are possible. *Maxwell's equations* (the differential point of view) give the rates of variation of the various components of the electric and magnetic fields. *Huygens' principle* (the global point of view) permits the direct calculation, when a monochromatic field is known on a surface $\Sigma$, of the field at any point $M$: one sums the fields radiated at the point $M$ by *fictional secondary sources* $N_1$, $N_2$, $N_3$, ... situated on the surface $\Sigma$ and whose amplitude and phase are determined by the value of the field at $N_1$, $N_2$, $N_3$, ... (fig. 1).

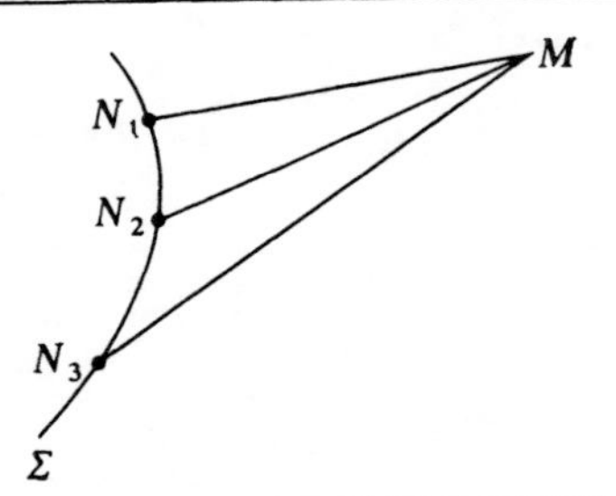


FIGURE 1

**In a diffraction experiment, Huygens' principle permits the calculation of the electric field at the point $M$, as a sum of fields radiated by secondary sources $N_1$, $N_2$, $N_3$, ... situated on a surface $\Sigma$.**

We intend to show in this complement that there exists an analogue in quantum mechanics of Huygens' principle. More precisely, we can write, for $t_2 > t_1$:

$$\psi(\mathbf{r}_2, t_2) = \int d^3r_1 \, K(\mathbf{r}_2, t_2; \mathbf{r}_1, t_1) \, \psi(\mathbf{r}_1, t_1) \tag{1}$$

a formula whose physical interpretation is the following: the probability amplitude of finding the particle at $\mathbf{r}_2$ at the instant $t_2$ is obtained by summing all the amplitudes "radiated" by the "secondary sources" $(\mathbf{r}_1, t)$, $(\mathbf{r}'_1, t_1)$... situated in space-time on the surface $t = t_1$, each of these sources contributing to a degree proportional to $\psi(\mathbf{r}_1, t_1)$, $\psi(\mathbf{r}'_1, t_1)$, ... (fig. 2). We shall prove the preceding formula, calculate $K$, called the *propagator* for the Schrödinger equation, and study its properties. We shall then indicate very qualitatively how it is possible to present all of quantum mechanics in terms of $K$ (the Lagrangian formulation of quantum mechanics; Feynman's point of view).

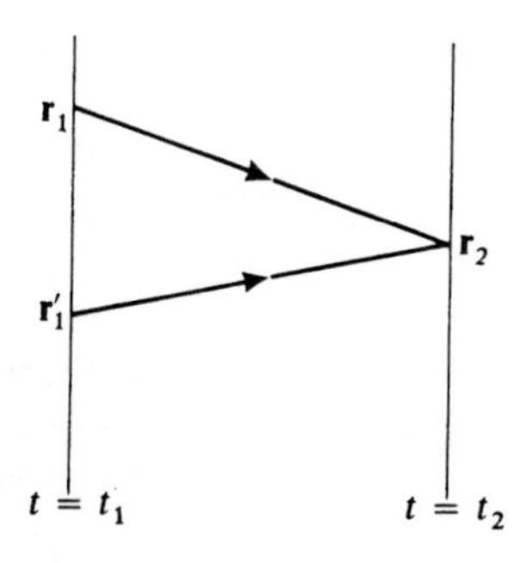


FIGURE 2

**The probability amplitude $\psi(\mathbf{r}_2, t_2)$ can be obtained by summing the contributions of the various amplitudes $\psi(\mathbf{r}_1, t_1)$, $\psi(\mathbf{r}'_1, t_1)$, etc... corresponding to one previous instant $t_1$. With each of the arrows of the figure is associated a "propagator" $K(\mathbf{r}_2, t_2; \mathbf{r}_1, t_1)$, $K(\mathbf{r}_2, t_2; \mathbf{r}'_1, t_1)$, etc...**

## 2. Existence and properties of a propagator $K(2, 1)$

### a. EXISTENCE OF A PROPAGATOR

The problem is to link directly the states of the system at two different times. This is possible if we use the evolution operator introduced in complement $F_{III}$, since we can write:

$$|\psi(t_2)\rangle = U(t_2, t_1) \, |\psi(t_1)\rangle \tag{2}$$

Given $|\psi(t_2)\rangle$, it is easy to find the wave function $\psi(\mathbf{r}_2, t_2)$:

$$\psi(\mathbf{r}_2, t_2) = \langle \mathbf{r}_2 | \psi(t_2) \rangle \tag{3}$$

Substituting (2) into (3) and inserting the closure relation:

$$\int d^3r_1 \, | \mathbf{r}_1 \rangle \langle \mathbf{r}_1 | = \mathbb{1} \tag{4}$$

between $U(t_2, t_1)$ and $| \psi(t_1) \rangle$, we obtain:

$$\psi(\mathbf{r}_2, t_2) = \int d^3r_1 \langle \mathbf{r}_2 | U(t_2, t_1) | \mathbf{r}_1 \rangle \langle \mathbf{r}_1 | \psi(t_1) \rangle$$

$$= \int d^3r_1 \langle \mathbf{r}_2 | U(t_2, t_1) | \mathbf{r}_1 \rangle \psi(\mathbf{r}_1, t_1) \tag{5}$$

The result is thus a formula identical to (1), on the condition that we set:

$$\langle \mathbf{r}_2 | U(t_2, t_1) | \mathbf{r}_1 \rangle = K(\mathbf{r}_2, t_2; \mathbf{r}_1, t_1)$$

In fact, since we want to use formulas of the type of (1) only for $t_2 > t_1$, we can set $K = 0$ for $t_2 < t_1$. The exact definition of $K$ then becomes:

$$K(\mathbf{r}_2, t_2; \mathbf{r}_1, t_1) = \langle \mathbf{r}_2 | U(t_2, t_1) | \mathbf{r}_1 \rangle \, \theta(t_2 - t_1) \tag{6}$$

where $\theta(t_2 - t_1)$ is the "step function":

$$\begin{aligned} \theta(t_2 - t_1) &= 1 \quad \text{if} \quad t_2 > t_1 \\ \theta(t_2 - t_1) &= 0 \quad \text{if} \quad t_2 < t_1 \end{aligned} \tag{7}$$

The introduction of $\theta(t_2 - t_1)$ is of both physical and mathematical interest. From the physical point of view, it is a simple way of compelling the secondary sources situated on the surface $t = t_1$ of figure 2 to "radiate" only towards the future. For this reason, $K(\mathbf{r}_2, t_2; \mathbf{r}_1, t_1)$ as defined by (6) is called the *retarded propagator*. From the mathematical point of view, we shall see later that $K(\mathbf{r}_2, t_2; \mathbf{r}_1, t_1)$, because of the factor $\theta(t_2 - t_1)$, obeys a partial differential equation whose right-hand side is a delta function, that is, the equation which defines a *Green's function*.

COMMENTS:

(*i*) Note, however, that equation (5) remains valid even if $t_2 < t_1$. It is possible, moreover, to introduce mathematically an "advanced" propagator which would be different from zero only for $t_2 < t_1$ and which would also obey the equation defining a Green's function. Since the physical meaning of such an advanced propagator is not obvious at this stage, we shall not study it here.

(*ii*) When no ambiguity is possible, we shall simply write $K(2, 1)$ for $K(\mathbf{r}_2, t_2; \mathbf{r}_1, t_1)$.

### b. PHYSICAL INTERPRETATION OF $K(2, 1)$

This interpretation follows very simply from definition (6): $K(2, 1)$ represents the probability amplitude that the particle, starting from the point $\mathbf{r}_1$ at time $t_1$, will arrive at the point $\mathbf{r}_2$ at a later time $t_2$. If we take as the initial state at time $t_1$ a state localized at the point $\mathbf{r}_1$:

$$| \psi(t_1) \rangle = | \mathbf{r}_1 \rangle \tag{8}$$

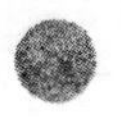



at time $t_2$, the state vector has become:

$$|\psi(t_2)\rangle = U(t_2, t_1)|\psi(t_1)\rangle = U(t_2, t_1)|\mathbf{r}_1\rangle \tag{9}$$

The probability amplitude of finding the particle at the point $\mathbf{r}_2$ at this time is then:

$$\langle \mathbf{r}_2 | \psi(t_2)\rangle = \langle \mathbf{r}_2 | U(t_2, t_1) | \mathbf{r}_1 \rangle \tag{10}$$

### c. EXPRESSION FOR $K(2, 1)$ IN TERMS OF THE EIGENSTATES OF $H$

Assume that the Hamiltonian $H$ does not depend explicitly on time, and call $|\varphi_n\rangle$ and $E_n$ its eigenstates and eigenvalues:

$$H|\varphi_n\rangle = E_n|\varphi_n\rangle \tag{11}$$

According to formula (18) of $F_{III}$, we have:

$$U(t_2, t_1) = e^{-iH(t_2 - t_1)/\hbar} \tag{12}$$

The closure relation:

$$\sum_n |\varphi_n\rangle\langle\varphi_n| = \mathbb{1} \tag{13}$$

enables us to write (12) in the form:

$$U(t_2, t_1) = e^{-iH(t_2 - t_1)/\hbar} \sum_n |\varphi_n\rangle\langle\varphi_n| \tag{14}$$

that is, taking (11) into account:

$$U(t_2, t_1) = \sum_n e^{-iE_n(t_2 - t_1)/\hbar} |\varphi_n\rangle\langle\varphi_n| \tag{15}$$

To calculate $K(2, 1)$, it then suffices to take the matrix element of both sides of (15) between $\langle \mathbf{r}_2|$ and $|\mathbf{r}_1\rangle$ and to multiply it by $\theta(t_2 - t_1)$. Since:

$$\langle \mathbf{r}_2 | \varphi_n \rangle = \varphi_n(\mathbf{r}_2) \tag{16}$$

$$\langle \varphi_n | \mathbf{r}_1 \rangle = \varphi_n^*(\mathbf{r}_1) \tag{17}$$

there results:

$$K(\mathbf{r}_2, t_2; \mathbf{r}_1, t_1) = \theta(t_2 - t_1) \sum_n \varphi_n^*(\mathbf{r}_1)\varphi_n(\mathbf{r}_2) e^{-iE_n(t_2 - t_1)/\hbar} \tag{18}$$

### d. EQUATION SATISFIED BY $K(2, 1)$

$\varphi_n(\mathbf{r}_2) e^{-iE_n t_2/\hbar}$ is a solution of the Schrödinger equation. We deduce from this that, in the $\{|\mathbf{r}\rangle\}$ representation:

$$\left[i\hbar\frac{\partial}{\partial t_2} - H\left(\mathbf{r}_2, \frac{\hbar}{i}\boldsymbol{\nabla}_2\right)\right]\varphi_n(\mathbf{r}_2) e^{-iE_n t_2/\hbar} = 0 \tag{19}$$

(where $\boldsymbol{\nabla}_2$ is a condensed notation which designates the three operators $\left.\frac{\partial}{\partial x_2}, \frac{\partial}{\partial y_2}, \frac{\partial}{\partial z_2}\right)$. Let us then apply, to both sides of equation (18), the operator :

$$i\hbar \frac{\partial}{\partial t_2} - H\left(\mathbf{r}_2, \frac{\hbar}{i}\nabla_2\right)$$

which acts only on the variables $\mathbf{r}_2$ and $t_2$. We know [*cf.* appendix II, relation (44)] that :

$$\frac{\partial}{\partial t_2}\theta(t_2 - t_1) = \delta(t_2 - t_1) \tag{20}$$

Consequently, using (19), we obtain :

$$\left[i\hbar \frac{\partial}{\partial t_2} - H\left(\mathbf{r}_2, \frac{\hbar}{i}\nabla_2\right)\right]K(\mathbf{r}_2, t_2 ; \mathbf{r}_1, t_1) = i\hbar\delta(t_2 - t_1)\sum_n \varphi_n^*(\mathbf{r}_1)\varphi_n(\mathbf{r}_2)\,\mathrm{e}^{-iE_n(t_2 - t_1)/\hbar} \tag{21}$$

Because of the presence of $\delta(t_2 - t_1)$, we can replace $t_2 - t_1$ by zero in the sum over $n$ which appears on the right-hand side of (21). This makes the exponential equal to 1. We are thus left with the quantity $\sum_n \varphi_n(\mathbf{r}_2)\varphi_n^*(\mathbf{r}_1)$, which, according to (13), (16) and (17), is equal to $\delta(\mathbf{r}_2 - \mathbf{r}_1)$ [taking the matrix element of (13) between $\langle \mathbf{r}_2 |$ and $| \mathbf{r}_1 \rangle$]. Finally, $K$ satisfies the equation :

$$\left[i\hbar \frac{\partial}{\partial t_2} - H\left(\mathbf{r}_2, \frac{\hbar}{i}\nabla_2\right)\right]K(\mathbf{r}_2, t_2 ; \mathbf{r}_1, t_1) = i\hbar\,\delta(t_2 - t_1)\delta(\mathbf{r}_2 - \mathbf{r}_1) \tag{22}$$

The solutions of equation (22), whose right-hand side is proportional to a four-dimensional "delta function", are called *Green's functions*. It can be shown that, to determine $K(2, 1)$ completely, it suffices to associate with (22) the boundary condition :

$$K(\mathbf{r}_2, t_2 ; \mathbf{r}_1, t_1) = 0 \qquad \text{if } t_2 < t_1 \tag{23}$$

Equations (22) and (23) have interesting implications, in particular with regard to perturbation theory, which we shall study in chapter XI.

## 3. Lagrangian formulation of quantum mechanics

### a. CONCEPT OF A SPACE-TIME PATH

Let us consider, in space-time, the two points $(\mathbf{r}_1, t_1)$ and $(\mathbf{r}_2, t_2)$ (*cf.* fig. 3 ; $t$ is plotted as the abscissa, and the ordinate axis represents the set of the three spatial axes). Choose $N$ intermediate times $t_{\alpha_i}$ ($i = 1, 2, ..., N$), evenly spaced between $t_1$ and $t_2$ :

$$t_1 < t_{\alpha_1} < t_{\alpha_2} < ... < t_{\alpha_{N-1}} < t_{\alpha_N} < t_2 \tag{24}$$

and, for each of them, a position $\mathbf{r}_{\alpha_i}$ in space. We can thus construct, when $N$ approaches infinity, a function $\mathbf{r}(t)$ (which we shall assume to be continuous) such that:

$$\mathbf{r}(t_1) = \mathbf{r}_1 \tag{25-a}$$

$$\mathbf{r}(t_2) = \mathbf{r}_2 \tag{25-b}$$

$\mathbf{r}(t)$ is said to define a *space-time path* between $(\mathbf{r}_1, t_1)$ and $(\mathbf{r}_2, t_2)$ : such a path might be thought of as the trajectory of a physical point leaving the point $\mathbf{r}_1$ at time $t_1$ and arriving at $\mathbf{r}_2$ at time $t_2$.

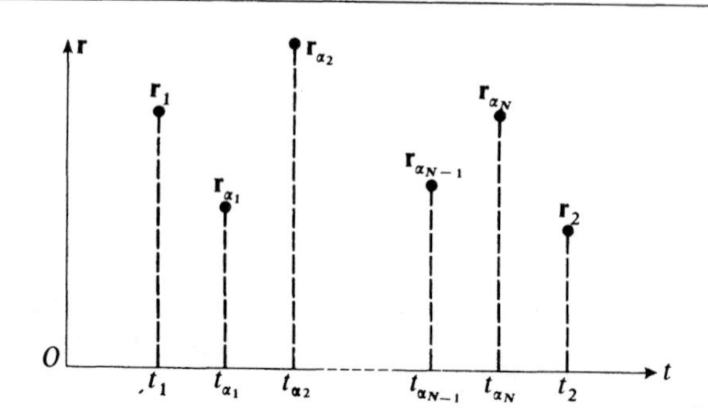


FIGURE 3

**Diagram associated with a "space-time path" : one picks $N$ intermediate times $t_{\alpha_i}(i = 1, 2, \ldots, N)$ evenly spaced between $t_1$ and $t_2$, and chooses for each of them a value of r.**

#### b. DECOMPOSITION OF $K(2, 1)$ INTO A SUM OF PARTIAL AMPLITUDES

Let us first return to the case where the number $N$ of intermediate times is finite. Formula (10) of complement $F_{III}$ enables us to write:

$$U(t_2, t_1) = U(t_2, t_{\alpha_N})\,U(t_{\alpha_N}, t_{\alpha_{N-1}}) \ldots U(t_{\alpha_2}, t_{\alpha_1})\,U(t_{\alpha_1}, t_1) \tag{26}$$

Let us take the matrix elements of both sides of (26) between $\langle \mathbf{r}_2 |$ and $| \mathbf{r}_1 \rangle$ and insert the closure relation relative to the $\{ | \mathbf{r} \rangle \}$ representation for each intermediate time $t_{\alpha_i}$. According to (6) and (24), we thus obtain:

$$K(2, 1) = \int d^3r_{\alpha_N} \int d^3r_{\alpha_{N-1}} \ldots \int d^3r_{\alpha_2} \int d^3r_{\alpha_1}\, K(2, \alpha_N)\, K(\alpha_N, \alpha_{N-1}) \ldots \times$$

$$\times\, K(\alpha_2, \alpha_1)\, K(\alpha_1, 1) \tag{27}$$

Now consider the product:

$$K(2, \alpha_N)\, K(\alpha_N, \alpha_{N-1}) \ldots K(\alpha_2, \alpha_1)\, K(\alpha_1, 1) \tag{28}$$

Generalizing the argument of § 2-b, we can interpret this term as being the probability amplitude for the particle, having left the point 1 $(\mathbf{r}_1, t_1)$, to arrive at the point 2 $(\mathbf{r}_2, t_2)$, having passed successively through all the points $\alpha_i(\mathbf{r}_{\alpha_i}, t_{\alpha_i})$ of figure 3. Note that, in formula (27), one is summing over all possible positions $\mathbf{r}_{\alpha_i}$ at each time $t_{\alpha_i}$.

We now let $N$ approach infinity*. A series of points $\alpha_i$ then defines a space-time path between 1 and 2, and the product (28) associated with it becomes the proba-

* In this treatment, we make no attempt to be mathematically rigorous.

bility amplitude which the particle has of following this path. Of course, the number of integrations in formula (27) becomes infinite. It is understandable, however, that the summation over the set of possible positions at each time should reduce to a summation over the set of possible paths. $K(2, 1)$ is thus seen to be a sum (in fact, an integral) which corresponds to the coherent superposition of the amplitudes associated with all possible space-time paths starting from 1 and ending at 2.

### c. FEYNMAN'S POSTULATES

The concepts of a propagator and a space-time path permit a new formulation of the postulate concerning the time evolution of physical systems. We shall outline here such a formulation for the case of a spinless particle.

We define $K(2, 1)$ directly as being the probability amplitude for the particle, starting from $\mathbf{r}_1$ at time $t_1$, to arrive at $\mathbf{r}_2$ at time $t_2$. We then postulate that:

(*i*) $K(2, 1)$ is the sum of an infinity of partial amplitudes, one for each of the space-time paths connecting $(\mathbf{r}_1, t_1)$ with $(\mathbf{r}_2, t_2)$.

(*ii*) The partial amplitude $K_\Gamma(2, 1)$ associated with one of these paths $\Gamma$ is determined in the following manner: let $S_\Gamma$ be the *classical action* calculated along $\Gamma$, that is:

$$S_\Gamma = \int_{(\Gamma)} \mathscr{L}(\mathbf{r}, \mathbf{p}, t)\, \mathrm{d}t \tag{29}$$

where $\mathscr{L}(\mathbf{r}, \mathbf{p}, t)$ is the Lagrangian of the particle (*cf.* appendix III). $K_\Gamma(2, 1)$ is then equal to:

$$K_\Gamma(2, 1) = N\, \mathrm{e}^{\frac{i}{\hbar}S_\Gamma} \tag{30}$$

where $N$ is a normalization constant (which can, moreover, be determined explicitly).

It can be shown that the Schrödinger equation follows as a consequence of these two postulates. Similarly, one can deduce the canonical commutation relation between the components of the observables $\mathbf{R}$ and $\mathbf{P}$. The two preceding postulates therefore permit a formulation of quantum mechanics which is different from that of chapter III, but equivalent.

### d. THE CLASSICAL LIMIT AND HAMILTON'S PRINCIPLE

The formulation which we have just evoked is particularly useful for discussing the relation between quantum and classical mechanics.

Consider a situation in which the actions $S_\Gamma$ are much larger than $\hbar$. In this case, the variation $\Delta S_\Gamma$ of the action between two different paths, even if its relative value is small $\left(\dfrac{\Delta S_\Gamma}{S_\Gamma} \ll 1\right)$, is usually much larger than $\hbar$. Consequently, the phase of $K_\Gamma(2, 1)$ varies rapidly, and the contributions to the global amplitude $K(2, 1)$ of most of the paths $\Gamma$ cancel out by interference. Let us assume, however, that there exists a path $\Gamma_0$ for which the *action is stationary* (that is, does not vary, to first order, when one goes from $\Gamma_0$ to another infinitesimally close path). The ampli-

tude $K_{\Gamma_0}(2, 1)$ then interferes constructively with those of the paths next to $\Gamma_0$, since, this time, their phases remain practically equal. Consequently, when the actions $S_\Gamma$ are much larger than $\hbar$, one is in a "quasi-classical" situation: to obtain $K(2, 1)$, one can ignore all the paths except $\Gamma_0$ and paths infinitely close to it; it can then be said that, between points 1 and 2, the particle follows the trajectory $\Gamma_0$. Now this is indeed the classical trajectory, defined by Hamilton's principle as being the path along which the action is minimal. Feynman's postulates therefore include, at the classical limit, Hamilton's principle of least action. They enable, us, moreover, to associate with it the following picture: it is the wave associated with the particle which, "exploring" the various possible paths, picks the one for which the action will be the smallest.

The Lagrangian formulation of quantum mechanics presents numerous other advantages, which we shall not examine in detail. Let us point out, for example, that it lends itself easily to relativistic generalization since one is already reasoning in space-time. Moreover, it can be applied to any classical system (not necessarily mechanical) governed by a variational principle (for example, a field).

However, it has a certain number of disadvantages on the mathematical level (summation over an infinite number of paths, the limit $N \longrightarrow \infty, ...$).

**References and suggestions for further reading:**

Feynman's original article (2.38); Feynman and Hibbs (2.25); Bjorken and Drell (2.6), chaps. 6 and 7.

# Complement $K_{III}$

# UNSTABLE STATES. LIFETIME



## 1. Introduction

Consider a conservative system (a system whose Hamiltonian $H$ is time-independent). Assume that at time $t = 0$ the state of the system is one of the eigenstates $| \varphi_n \rangle$ of the Hamiltonian, of energy $E_n$:

$$| \psi(0) \rangle = | \varphi_n \rangle \tag{1}$$

with:

$$H | \varphi_n \rangle = E_n | \varphi_n \rangle \tag{2}$$

In this case, the system remains indefinitely in the same state (a stationary state; § D-2-b of chapter III).

We shall study the hydrogen atom in chapter VII by solving the eigenvalue equation of its Hamiltonian, which is a time-independent operator. The states of the hydrogen atom (that is, the possible values of its energy) which we shall find are in very good agreement with the experimentally measured energies. However, it is known that most of these states are actually *unstable*: if, at the instant $t = 0$, the atom is in an excited state (an eigenstate $| \varphi_n \rangle$ corresponding to an energy $E_n$ greater than that of the ground state, which is the lowest energy state), it generally "falls back" into this ground state by emitting one or several photons. The state $| \varphi_n \rangle$ is not really, therefore, a stationary state in this case.

This problem arises from the fact that, in calculations of the type used in chapter VII, the system under study (the hydrogen atom) is treated as if it were totally isolated, while it is actually in constant interaction with the electromagnetic field. Although the evolution of the global system "atom + electromagnetic field" can be perfectly well described by a Hamiltonian, it is not rigorously possible to define a Hamiltonian for the hydrogen atom alone [*cf.* comment ($v$) of § 5-b of complement $E_{III}$]. However, since the coupling between the atom and the field happens to be weak (it can be shown that its "force" is characterized by the fine structure constant $\alpha \simeq \frac{1}{137}$, which we shall introduce in chapter VII), the approximation consisting of completely neglecting the existence of the electromagnetic field is very good, except, of course, if we are interested precisely in the instability of the states.

COMMENTS:

(*i*) If a strictly conservative, isolated system, at the initial instant, is in a state formed by a linear combination of several stationary states, it evolves over time and does not always remain in the same state. But its Hamiltonian is a constant of the motion, and consequently (*cf.* chapter III, § D-2-c), the probability of finding one energy value or another is independent of time, as is the mean value of the energy. On the other hand, in the case of an unstable state, one state is transformed irreversibly into another, with a loss of energy for the system: this energy is taken away by the photons emitted*.

(*ii*) The instability of the excited states of an atom is caused by the spontaneous emission of photons; the ground state is stable, since there exists no lower energy state. Recall, nevertheless, that atoms can also absorb light energy and so ascend to a higher energy level.

We intend to indicate here how to take the instability of a state into account phenomenologically. The description will not be rigorous, as we shall continue to consider the system as if it were isolated. We shall try to incorporate this instability as simply as possible into the quantum description of the system.

Complement $D_{XIII}$ presents a more rigorous treatment of this problem, justifying the phenomenological approach used here.

## 2. Definition of the lifetime

Experiments show that the instability of a state can often be characterized by just one parameter $\tau$, having the dimensions of a time, which is called the *lifetime* of the state. More precisely, if one prepares the system at time $t = 0$ in the unstable state $|\varphi_n\rangle$, one observes that the probability $\mathscr{P}(t)$ of its still being excited at a later time $t$ is equal to:

$$\mathscr{P}(t) = e^{-t/\tau} \tag{3}$$

This result can also be expressed in the following way. Consider a large number $\mathscr{N}$ of identical independent systems, all prepared at time $t = 0$ in the state $|\varphi_n\rangle$. At time $t$, there will remain $N(t) = \mathscr{N}\, e^{-t/\tau}$ in this state. Between times $t$ and $t + dt$, a certain number $dn(t)$ of systems leave the unstable state:

$$dn(t) = N(t) - N(t + dt) = -\frac{dN(t)}{dt}\,dt = N(t)\frac{dt}{\tau} \tag{4}$$

For each of the $N(t)$ systems which are still in the state $|\varphi_n\rangle$ at time $t$, a probability can therefore be defined:

$$d\varpi(t) = \frac{dn(t)}{N(t)} = \frac{dt}{\tau} \tag{5}$$

of their leaving this state during the time interval $dt$ following the instant $t$. We see that $d\varpi(t)$ is independent of $t$: the system is said to have a *probability per unit time* $\frac{1}{\tau}$ of leaving the unstable state.

* which, moreover, may also take away linear and angular momentum.

COMMENTS :

(*i*) Let us calculate the mean value of the time the system remains in the unstable state. It is equal to:

$$\int_0^{\infty} t\, e^{-t/\tau} \frac{dt}{\tau} = \tau \tag{6}$$

$\tau$ is therefore the mean time the system spends in the state $|\varphi_n\rangle$; this is why it is called the lifetime of this state.

For a stable state, $\mathscr{P}(t)$ is always equal to 1, and the lifetime $\tau$ is infinite.

(*ii*) A remarkable property of the lifetime $\tau$ is that it does not depend on the procedure used to prepare the system in the unstable state, that is, on its previous "history": the lifetime is a characteristic of the unstable state itself.

(*iii*) According to the time-energy uncertainty relation (§ D-2-e of chapter III), the time $\tau$ characteristic of the evolution of an unstable state is associated with an uncertainty in the energy $\Delta E$ given by:

$$\Delta E \simeq \frac{\hbar}{\tau} \tag{7}$$

One indeed finds that the energy of an unstable state cannot be determined with arbitrary accuracy, but only to within an uncertainty of the order of $\Delta E$. $\Delta E$ is called the *natural width* of this state. For the case of the hydrogen atom, the width of the various states is negligible compared to their separation. This explains why we can treat them, in a first approximation, as if they were stable.

## 3. Phenomenological description of the instability of a state

First let us consider a conservative system, prepared, at the initial time, in the eigenstate $|\varphi_n\rangle$ of the Hamiltonian $H$. According to the rule (D-54) of chapter III, the state vector, at time $t$, becomes:

$$|\psi(t)\rangle = e^{-iE_n t/\hbar} |\varphi_n\rangle \tag{8}$$

The probability $\mathscr{P}_n(t)$ of finding, in a measurement at time $t$, the system in the state $|\varphi_n\rangle$ is:

$$\mathscr{P}_n(t) = |e^{-iE_n t/\hbar}|^2 \tag{9}$$

Since the energy $E_n$ is real (since $H$ is an observable), this probability is constant and equal to 1: we again find that $|\varphi_n\rangle$ is a stationary state.

Let us examine what would happen if, in expression (9), we replaced the energy $E_n$ by the complex number:

$$E'_n = E_n - i\hbar\frac{\gamma_n}{2} \tag{10}$$

The probability $\mathscr{P}_n(t)$ then becomes:

$$\mathscr{P}'_n(t) = |e^{-i(E_n - i\hbar\frac{\gamma_n}{2})t/\hbar}|^2 = e^{-\gamma_n t} \tag{11}$$

In this case, the probability of finding the system in the state $|\varphi_n\rangle$ decreases exponentially with time, as in formula (3). Therefore, to take into account phenomenologically the instability of a state $|\varphi_n\rangle$ whose lifetime is $\tau_n$, it suffices to add, as in (10), an imaginary part to its energy, by setting:

$$\gamma_n = \frac{1}{\tau_n} \tag{12}$$

COMMENT:

When $E_n$ is replaced by $E'_n$, the norm of the state vector written in (8) becomes $e^{-\gamma_n t/2}$ and therefore varies with time. This result is not surprising. We saw in § D-1-c of chapter III that the conservation of the norm of the state vector arose from the Hermitian nature of the Hamiltonian operator; now, an operator whose eigenvalues are complex, as are the $E'_n$, cannot be Hermitian. Of course, as we pointed out in § 1, this is due to the fact that the system under study is part of a larger system (it is interacting with the electromagnetic field) and its evolution cannot be described rigorously by means of a Hamiltonian. It is already rather remarkable that its evolution can be simply explained by introducing a "Hamiltonian" with complex eigenvalues.

## Complement $L_{III}$

## EXERCISES

**1.** In a one-dimensional problem, consider a particle whose wave function is :

$$\psi(x) = N \frac{e^{ip_0 x/\hbar}}{\sqrt{x^2 + a^2}}$$

where $a$ and $p_0$ are real constants and $N$ is a normalization coefficient.

*a.* Determine $N$ so that $\psi(x)$ is normalized.

*b.* The position of the particle is measured. What is the probability of finding a result between $-\frac{a}{\sqrt{3}}$ and $+\frac{a}{\sqrt{3}}$?

*c.* Calculate the mean value of the momentum of a particle which has $\psi(x)$ for its wave function.

**2.** Consider, in a one-dimensional problem, a particle of mass $m$ whose wave function at time $t$ is $\psi(x, t)$.

*a.* At time $t$, the distance $d$ of this particle from the origin is measured. Write, as a function of $\psi(x, t)$, the probability $\mathscr{P}(d_0)$ of finding a result greater than a given length $d_0$. What are the limits of $\mathscr{P}(d_0)$ when $d_0 \longrightarrow 0$ and $d_0 \longrightarrow \infty$?

*b.* Instead of performing the measurement of question *a*, one measures the velocity $v$ of the particle at time $t$. Express, as a function of $\psi(x, t)$, the probability of finding a result greater than a given value $v_0$.

**3.** The wave function of a free particle, in a one-dimensional problem, is given at time $t = 0$ by :

$$\psi(x, 0) = N \int_{-\infty}^{+\infty} dk\, e^{-|k|/k_0}\, e^{ikx}$$

where $k_0$ and $N$ are constants.

*a.* What is the probability $\mathscr{P}(p_1, 0)$ that a measurement of the momentum, performed at time $t = 0$, will yield a result included between $-p_1$ and $+p_1$? Sketch the function $\mathscr{P}(p_1, 0)$.

*b.* What happens to this probability $\mathscr{P}(p_1, t)$ if the measurement is performed at time $t$? Interpret.

*c.* What is the form of the wave packet at time $t = 0$? Calculate for this time the product $\Delta X . \Delta P$; what is your conclusion? Describe qualitatively the subsequent evolution of the wave packet.

### 4. Spreading of a free wave packet

Consider a free particle.

*a.* Show, applying Ehrenfest's theorem, that $\langle X \rangle$ is a linear function of time, the mean value $\langle P \rangle$ remaining constant.

*b.* Write the equations of motion for the mean values $\langle X^2 \rangle$ and $\langle XP + PX \rangle$. Integrate these equations.

*c.* Show that, with a suitable choice of the time origin, the root-mean-square deviation $\Delta X$ is given by :

$$(\Delta X)^2 = \frac{1}{m^2} (\Delta P)_0^2 \, t^2 + (\Delta X)_0^2$$

where $(\Delta X)_0$ and $(\Delta P)_0$ are the root-mean-square deviations at the initial time.

How does the width of the wave packet vary as a function of time (see § 3-c of complement $G_I$)? Give a physical interpretation.

### 5. Particle subject to a constant force

In a one-dimensional problem, consider a particle of potential energy $V(X) = -fX$, where $f$ is a positive constant [$V(X)$ arises, for example, from a gravity field or a uniform electric field].

*a.* Write Ehrenfest's theorem for the mean values of the position $X$ and the momentum $P$ of the particle. Integrate these equations; compare with the classical motion.

*b.* Show that the root-mean-square deviation $\Delta P$ does not vary over time.

*c.* Write the Schrödinger equation in the $\{ | p \rangle \}$ representation. Deduce from it a relation between $\frac{\partial}{\partial t} | \langle p | \psi(t) \rangle |^2$ and $\frac{\partial}{\partial p} | \langle p | \psi(t) \rangle |^2$. Integrate the equation thus obtained; give a physical interpretation.

**6.** Consider the three-dimensional wave function

$$\psi(x, y, z) = N \, e^{-\left[\frac{|x|}{2a} + \frac{|y|}{2b} + \frac{|z|}{2c}\right]}$$

where $a$, $b$ and $c$ are three positive lengths.

*a.* Calculate the constant $N$ which normalizes $\psi$.

*b.* Calculate the probability that a measurement of $X$ will yield a result included between 0 and $a$.

*c.* Calculate the probability that simultaneous measurements of $Y$ and $Z$ will yield results included respectively between $-b$ and $+b$, and $-c$ and $+c$.

*d.* Calculate the probability that a measurement of the momentum will yield a result included in the element $dp_x dp_y dp_z$ centered at the point $p_x = p_y = 0$; $p_z = \hbar/c$.

**7.** Let $\psi(x, y, z) = \psi(\mathbf{r})$ be the normalized wave function of a particle. Express in terms of $\psi(\mathbf{r})$ the probability for:

*a.* a measurement of the abscissa $X$, to yield a result included between $x_1$ and $x_2$;

*b.* a measurement of the component $P_x$ of the momentum, to yield a result included between $p_1$ and $p_2$;

*c.* simultaneous measurements of $X$ and $P_z$, to yield:

$$x_1 \leqslant x \leqslant x_2$$
$$p_z \geqslant 0$$

*d.* simultaneous measurements of $P_x$, $P_y$, $P_z$, to yield:

$$p_1 \leqslant p_x \leqslant p_2$$
$$p_3 \leqslant p_y \leqslant p_4$$
$$p_5 \leqslant p_z \leqslant p_6$$

Show that this probability is equal to the result of $b$ when $p_3, p_5 \longrightarrow -\infty$; $p_4, p_6 \longrightarrow +\infty$;

*e.* a measurement of the component $U = \frac{1}{\sqrt{3}}(X + Y + Z)$ of the position, to yield a result included between $u_1$ and $u_2$.

**8.** Let $\mathbf{J}(\mathbf{r})$ be the probability current associated with a wave function $\psi(\mathbf{r})$ describing the state of a particle of mass $m$ [chap. III, relations (D-17) and (D-19)].

*a.* Show that:

$$m \int d^3r \; \mathbf{J}(\mathbf{r}) = \langle \mathbf{P} \rangle$$

where $\langle \mathbf{P} \rangle$ is the mean value of the momentum.

*b.* Consider the operator $\mathbf{L}$ (orbital angular momentum) defined by $\mathbf{L} = \mathbf{R} \times \mathbf{P}$. Are the three components of $\mathbf{L}$ Hermitian operators? Establish the relation:

$$m \int d^3r \, [\mathbf{r} \times \mathbf{J}(\mathbf{r})] = \langle \mathbf{L} \rangle.$$

**9.** One wants to show that the physical state of a (spinless) particle is completely defined by specifying the probability density $\rho(\mathbf{r}) = |\psi(\mathbf{r})|^2$ and the probability current $\mathbf{J}(\mathbf{r})$.

*a.* Assume the function $\psi(\mathbf{r})$ known and let $\xi(\mathbf{r})$ be its argument:

$$\psi(\mathbf{r}) = \sqrt{\rho(\mathbf{r})} \, e^{i\xi(\mathbf{r})}$$

Show that:

$$\mathbf{J}(\mathbf{r}) = \frac{\hbar}{m} \rho(\mathbf{r}) \, \nabla \xi(\mathbf{r})$$

Deduce that two wave functions leading to the same density $\rho(\mathbf{r})$ and current $\mathbf{J}(\mathbf{r})$ can differ only by a global phase factor.

*b.* Given arbitrary functions $\rho(\mathbf{r})$ and $\mathbf{J}(\mathbf{r})$, show that a quantum state $\psi(\mathbf{r})$ can be associated with them only if $\nabla \times \mathbf{v}(\mathbf{r}) = 0$, where $\mathbf{v}(\mathbf{r}) = \mathbf{J}(\mathbf{r})/\rho(\mathbf{r})$ is the velocity associated with the probability fluid.

*c.* Now assume that the particle is submitted to a magnetic field $\mathbf{B}(\mathbf{r}) = \nabla \times \mathbf{A}(\mathbf{r})$ [see chap. III, definition (D-20) of the probability current in this case]. Show that:

$$\mathbf{J} = \frac{\rho(\mathbf{r})}{m}\left[\hbar \nabla \xi(\mathbf{r}) - q\mathbf{A}(\mathbf{r})\right]$$

and:

$$\nabla \times \mathbf{v}(\mathbf{r}) = -\frac{q}{m}\mathbf{B}(\mathbf{r})$$

## 10. Virial theorem

*a.* In a one-dimensional problem, consider a particle with the Hamiltonian:

$$H = \frac{P^2}{2m} + V(X)$$

where:

$$V(X) = \lambda X^n$$

Calculate the commutator $[H, XP]$. If there exists one or several stationary states $|\varphi\rangle$ in the potential $V$, show that the mean values $\langle T\rangle$ and $\langle V\rangle$ of the kinetic and potential energies in these states satisfy the relation: $2\langle T\rangle = n\langle V\rangle$.

*b.* In a three-dimensional problem, $H$ is written:

$$H = \frac{\mathbf{P}^2}{2m} + V(\mathbf{R})$$

Calculate the commutator $[H, \mathbf{R}\,.\,\mathbf{P}]$. Assume that $V(\mathbf{R})$ is a homogeneous function of $n$th order in the variables $X$, $Y$, $Z$. What relation necessarily exists between the mean kinetic energy and the mean potential energy of the particle in a stationary state?

Apply this to a particle moving in the potential $V(r) = -e^2/r$ (hydrogen atom).

Recall that a homogeneous function $V$ of $n$th degree in the variables $x$, $y$ and $z$ by definition satifies the relation:

$$V(\alpha x, \alpha y, \alpha z) = \alpha^n V(x, y, z)$$

and satisfies Euler's identity:

$$x\frac{\partial V}{\partial x} + y\frac{\partial V}{\partial y} + z\frac{\partial V}{\partial z} = nV(x, y, z).$$

*c.* Consider a system of $N$ particles of positions $\mathbf{R}_i$ and momenta $\mathbf{P}_i$ ($i = 1, 2, \ldots N$). When their potential energy is a homogeneous ($n$th degree) function of the set of components $X_i$, $Y_i$, $Z_i$, can the results obtained above be genera-

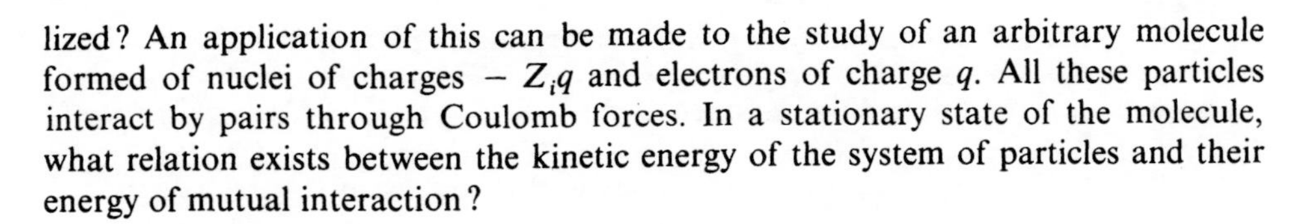

lized? An application of this can be made to the study of an arbitrary molecule formed of nuclei of charges $-Z_i q$ and electrons of charge $q$. All these particles interact by pairs through Coulomb forces. In a stationary state of the molecule, what relation exists between the kinetic energy of the system of particles and their energy of mutual interaction?

### 11. Two-particle wave function

In a one-dimensional problem, consider a system of two particles (1) and (2) with which is associated the wave function $\psi(x_1, x_2)$.

*a.* What is the probability of finding, in a measurement of the positions $X_1$ and $X_2$ of the two particles, a result such that:

$$x \leqslant x_1 \leqslant x + dx$$
$$\alpha \leqslant x_2 \leqslant \beta$$

*b.* What is the probability of finding particle (1) between $x$ and $x + dx$ [when no observations are made on particle (2)]?

*c.* Give the probability of finding at least one of the particles between $\alpha$ and $\beta$.

*d.* Give the probability of finding one and only one particle between $\alpha$ and $\beta$.

*e.* What is the probability of finding the momentum of particle (1) included between $p'$ and $p''$ and the position of particle (2) between $\alpha$ and $\beta$?

*f.* The momenta $P_1$ and $P_2$ of the two particles are measured; what is the probability of finding $p' \leqslant p_1 \leqslant p''$; $p''' \leqslant p_2 \leqslant p''''$?

*g.* The only quantity measured is the momentum $P_1$ of the first particle. Calculate, first from the results of *e* and then from those of *f*, the probability of finding this momentum included between $p'$ and $p''$. Compare the two results obtained.

*h.* The algebraic distance $X_1 - X_2$ between the two particles is measured; what is the probability of finding a result included between $-d$ and $+d$? What is the mean value of this distance?

### 12. Infinite one-dimensional well

Consider a particle of mass $m$ submitted to the potential:

$$V(x) = 0 \quad \text{if} \quad 0 \leqslant x \leqslant a.$$
$$V(x) = +\infty \quad \text{if} \quad x < 0 \quad \text{or} \quad x > a.$$

$|\varphi_n\rangle$ are the eigenstates of the Hamiltonian $H$ of the system, and their eigenvalues are $E_n = \dfrac{n^2\pi^2\hbar^2}{2ma^2}$ (*cf.* complement $H_I$). The state of the particle at the instant $t = 0$ is:

$$|\psi(0)\rangle = a_1|\varphi_1\rangle + a_2|\varphi_2\rangle + a_3|\varphi_3\rangle + a_4|\varphi_4\rangle$$

*a*. What is the probability, when the energy of the particle in the state $|\psi(0)\rangle$ is measured, of finding a value smaller than $\frac{3\pi^2\hbar^2}{ma^2}$?

*b*. What is the mean value and what is the root-mean-square deviation of the energy of the particle in the state $|\psi(0)\rangle$?

*c*. Calculate the state vector $|\psi(t)\rangle$ at the instant $t$. Do the results found in *a* and *b* at the instant $t = 0$ remain valid at an arbitrary time $t$?

*d*. When the energy is measured, the result $\frac{8\pi^2\hbar^2}{ma^2}$ is found. After the measurement, what is the state of the system? What is the result if the energy is measured again?

## 13. Infinite two-dimensional well (*cf.* complement $G_{II}$)

In a two-dimensional problem, consider a particle of mass $m$; its Hamiltonian $H$ is written:

$$H = H_x + H_y$$

with:

$$H_x = \frac{P_x^2}{2m} + V(X) \qquad H_y = \frac{P_y^2}{2m} + V(Y)$$

The potential energy $V(x)$ [or $V(y)$] is zero when $x$ (or $y$) is included in the interval $[0, a]$ and is infinite everywhere else.

*a*. Of the following sets of operators, which form a C.S.C.O.?

$$\{ H \}, \{ H_x \}, \{ H_x, H_y \}, \{ H, H_x \}$$

*b*. Consider a particle whose wave function is:

$$\psi(x, y) = N \cos\frac{\pi x}{a} \cos\frac{\pi y}{a} \sin\frac{2\pi x}{a} \sin\frac{2\pi y}{a}$$

when $0 \leqslant x \leqslant a$ and $0 \leqslant y \leqslant a$, and is zero everywhere else (where $N$ is a constant).

*α*. What is the mean value $\langle H \rangle$ of the energy of the particle? If the energy $H$ is measured, what results can be found, and with what probabilities?

*β*. The observable $H_x$ is measured; what results can be found, and with what probabilities? If this measurement yields the result $\frac{\pi^2\hbar^2}{2ma^2}$, what will be the results of a subsequent measurement of $H_y$, and with what probabilities?

*γ*. Instead of performing the preceding measurements, one now performs a simultaneous measurement of $H_x$ and $P_y$. What are the probabilities of finding :

$$E_x = \frac{9\pi^2\hbar^2}{2ma^2}$$

and :

$$p_0 \leqslant p_y \leqslant p_0 + \mathrm{d}p \ ?$$

**14.** Consider a physical system whose state space, which is three-dimensional, is spanned by the orthonormal basis formed by the three kets $|u_1\rangle$, $|u_2\rangle$, $|u_3\rangle$. In this basis, the Hamiltonian operator $H$ of the system and the two observables $A$ and $B$ are written:

$$H = \hbar\omega_0 \begin{pmatrix} 1 & 0 & 0 \\ 0 & 2 & 0 \\ 0 & 0 & 2 \end{pmatrix} \ ; \quad A = a \begin{pmatrix} 1 & 0 & 0 \\ 0 & 0 & 1 \\ 0 & 1 & 0 \end{pmatrix} \ ; \quad B = b \begin{pmatrix} 0 & 1 & 0 \\ 1 & 0 & 0 \\ 0 & 0 & 1 \end{pmatrix}$$

where $\omega_0$, $a$ and $b$ are positive real constants.

The physical system at time $t = 0$ is in the state:

$$|\psi(0)\rangle = \frac{1}{\sqrt{2}}|u_1\rangle + \frac{1}{2}|u_2\rangle + \frac{1}{2}|u_3\rangle$$

*a.* At time $t = 0$, the energy of the system is measured. What values can be found, and with what probabilities? Calculate, for the system in the state $|\psi(0)\rangle$, the mean value $\langle H \rangle$ and the root-mean-square deviation $\Delta H$.

*b.* Instead of measuring $H$ at time $t = 0$, one measures $A$; what results can be found, and with what probabilities? What is the state vector immediately after the measurement?

*c.* Calculate the state vector $|\psi(t)\rangle$ of the system at time $t$.

*d.* Calculate the mean values $\langle A \rangle(t)$ and $\langle B \rangle(t)$ of $A$ and $B$ at time $t$. What comments can be made?

*e.* What results are obtained if the observable $A$ is measured at time $t$? Same question for the observable $B$. Interpret.

### 15. Interaction picture

(It is recommended that complement $F_{III}$ and perhaps complement $G_{III}$ be read before this exercise is undertaken.)

Consider an arbitrary physical system. Denote its Hamiltonian by $H_0(t)$ and the corresponding evolution operator by $U_0(t, t')$:

$$\begin{cases} i\hbar \dfrac{\partial}{\partial t} U_0(t, t_0) = H_0(t)\, U_0(t, t_0) \\ U_0(t_0, t_0) = \mathbb{1} \end{cases}$$

Now assume that the system is perturbed in such a way that its Hamiltonian becomes:

$$H(t) = H_0(t) + W(t)$$

The state vector of the system in the "interaction picture", $|\psi_I(t)\rangle$, is defined from the state vector $|\psi_S(t)\rangle$ in the Schrödinger picture by:

$$|\psi_I(t)\rangle = U_0^\dagger(t, t_0)\,|\psi_S(t)\rangle$$

*a.* Show that the evolution of $|\psi_I(t)\rangle$ is given by:

$$i\hbar\frac{\mathrm{d}}{\mathrm{d}t}|\psi_I(t)\rangle = W_I(t)\,|\psi_I(t)\rangle$$

where $W_I(t)$ is the transform operator of $W(t)$ under the unitary transformation associated with $U_0^\dagger(t, t_0)$:

$$W_I(t) = U_0^\dagger(t, t_0)W(t)\,U_0(t, t_0)$$

Explain qualitatively why, when the perturbation $W(t)$ is much smaller than $H_0(t)$, the motion of the vector $|\psi_I(t)\rangle$ is much slower than that of $|\psi_S(t)\rangle$.

*b.* Show that the preceding differential equation is equivalent to the integral equation:

$$|\psi_I(t)\rangle = |\psi_I(t_0)\rangle + \frac{1}{i\hbar}\int_{t_0}^{t}\mathrm{d}t'\,W_I(t')\,|\psi_I(t')\rangle$$

where : $|\psi_I(t_0)\rangle = |\psi_S(t_0)\rangle$.

*c.* Solving this integral equation by iteration, show that the ket $|\psi_I(t)\rangle$ can be expanded in a power series in $W$ of the form:

$$|\psi_I(t)\rangle = \left\{\mathbb{1} + \frac{1}{i\hbar}\int_{t_0}^{t}\mathrm{d}t'W_I(t') + \frac{1}{(i\hbar)^2}\int_{t_0}^{t}\mathrm{d}t'W_I(t')\int_{t_0}^{t'}\mathrm{d}t''W_I(t'') + \ldots\right\}|\psi_I(t_0)\rangle$$

## 16. Correlations between two particles

(It is recommended that the complement $E_{III}$ be read in order to answer question *e* of this exercise.)

Consider a physical system formed by two particles (1) and (2), of the same mass $m$, which do not interact with each other and which are both placed in an infinite potential well of width $a$ (*cf.* complement $H_I$, § 2-c). Denote by $H(1)$ and $H(2)$ the Hamiltonians of each of the two particles and by $|\varphi_n(1)\rangle$ and $|\varphi_q(2)\rangle$ the corresponding eigenstates of the first and second particle, of energies $\dfrac{n^2\pi^2\hbar^2}{2ma^2}$ and $\dfrac{q^2\pi^2\hbar^2}{2ma^2}$. In the state space of the global system, the basis chosen is composed of the states $|\varphi_n\varphi_q\rangle$ defined by:

$$|\varphi_n\varphi_q\rangle = |\varphi_n(1)\rangle \otimes |\varphi_q(2)\rangle$$

*a.* What are the eigenstates and the eigenvalues of the operator $H = H(1) + H(2)$, the total Hamiltonian of the system? Give the degree of degeneracy of the two lowest energy levels.

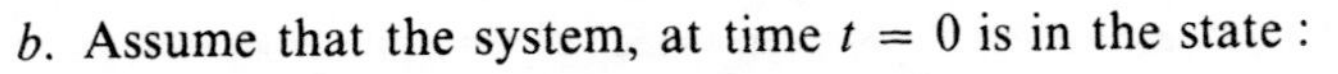

*b*. Assume that the system, at time $t = 0$ is in the state:

$$| \psi(0) \rangle = \frac{1}{\sqrt{6}} | \varphi_1 \varphi_1 \rangle + \frac{1}{\sqrt{3}} | \varphi_1 \varphi_2 \rangle + \frac{1}{\sqrt{6}} | \varphi_2 \varphi_1 \rangle + \frac{1}{\sqrt{3}} | \varphi_2 \varphi_2 \rangle$$

$\alpha$. What is the state of the system at time $t$?

$\beta$. The total energy $H$ is measured. What results can be found, and with what probabilities?

$\gamma$. Same questions if, instead of measuring $H$, one measures $H(1)$.

*c*. $\alpha$. Show that $| \psi(0) \rangle$ is a tensor product state. When the system is in this state, calculate the following mean values: $\langle H(1) \rangle$, $\langle H(2) \rangle$ and $\langle H(1)H(2) \rangle$. Compare $\langle H(1) \rangle \langle H(2) \rangle$ with $\langle H(1)H(2) \rangle$; how can this result be explained?

$\beta$. Show that the preceding results remain valid when the state of the system is the state $| \psi(t) \rangle$ calculated in *b*.

*d*. Now assume that the state $| \psi(0) \rangle$ is given by:

$$| \psi(0) \rangle = \frac{1}{\sqrt{5}} | \varphi_1 \varphi_1 \rangle + \sqrt{\frac{3}{5}} | \varphi_1 \varphi_2 \rangle + \frac{1}{\sqrt{5}} | \varphi_2 \varphi_1 \rangle$$

Show that $| \psi(0) \rangle$ cannot be put in the form of a tensor product. Answer for this case all the questions asked in *c*.

*e*. $\alpha$. Write the matrix, in the basis of the vectors $| \varphi_n \varphi_p \rangle$, which represents the density operator $\rho(0)$ corresponding to the ket $| \psi(0) \rangle$ given in *b*. What is the density matrix $\rho(t)$ at time $t$? Calculate, at the instant $t = 0$, the partial traces:

$$\rho(1) = \text{Tr}_2 \rho \quad \text{and} \quad \rho(2) = \text{Tr}_1 \rho$$

Do the density operators $\rho$, $\rho(1)$ and $\rho(2)$ describe pure states? Compare $\rho$ with $\rho(1) \otimes \rho(2)$; what is your interpretation?

$\beta$. Answer the same questions as in $\alpha$, but choosing for $| \psi(0) \rangle$ the ket given in *d*.

*The subject of the following exercises is the density operator: they therefore assume the concepts and results of complement* $E_{\text{III}}$ *to be known.*

**17.** Let $\rho$ be the density operator of an arbitrary system, where $| \chi_l \rangle$ and $\pi_l$ are the eigenvectors and eigenvalues of $\rho$. Write $\rho$ and $\rho^2$ in terms of the $| \chi_l \rangle$ and $\pi_l$. What do the matrices representing these two operators in the $\{ | \chi_l \rangle \}$ basis look like – first, in the case where $\rho$ describes a pure state and then, in the case of a statistical mixture of states? (Begin by showing that, in a pure case, $\rho$ has only one non-zero diagonal element, equal to 1, while for a statistical mixture, $\rho$ has several diagonal elements included between 0 and 1.) Show that $\rho$ corresponds to a pure case if and only if the trace of $\rho^2$ is equal to 1.

**18.** Consider a system whose density operator is $\rho(t)$, evolving under the influence of a Hamiltonian $H(t)$. Show that the trace of $\rho^2$ does not vary over time. Conclusion: can the system evolve so as to be successively in a pure state and a statistical mixture of states?

**19.** Let (1) + (2) be a global system, composed of two subsystems (1) and (2). $A$ and $B$ denote two operators acting in the state space $\mathscr{E}(1) \otimes \mathscr{E}(2)$. Show that the two partial traces $\text{Tr}_1 \{ AB \}$ and $\text{Tr}_1 \{ BA \}$ are equal when $A$ (or $B$) actually acts only in the space $\mathscr{E}(1)$, that is, when $A$ (or $B$) can be written:

$$A = A(1) \otimes \mathbb{1}(2) \qquad [\text{or } B = B(1) \otimes \mathbb{1}(2)].$$

Application: if the operator $H$, the Hamiltonian of the global system, is the sum of two operators which act, respectively, only in $\mathscr{E}(1)$ and only in $\mathscr{E}(2)$:

$$H = H(1) + H(2),$$

calculate the variation $\frac{\mathrm{d}}{\mathrm{d}t}\rho(1)$ of the reduced density operator $\rho(1)$. Give the physical interpretation of the result obtained.

**Exercise 5**

References: Flügge (1.24), §§40 and 41; Landau and Lifshitz (1.19), §22.

**Exercise 10**

References: Levine (12.3), chap. 14; Eyring et al (12.5), §18 b

**Exercise 15**

References: see references of complement $G_{III}$.

## RETURN TO ONE-DIMENSIONAL PROBLEMS

Now that we are more familiar with the mathematical formalism and the physical content of quantum mechanics, we can go into some of the results obtained in chapter I in more detail. In the three complements which follow, we shall study in a general way the quantum properties of a particle subject to a scalar potential* of arbitrary form, confining ourselves for simplicity to one-dimensional problems. We shall treat the bound stationary states of a particle, whose energies form a discrete spectrum (complement $M_{III}$), and then the unbound states corresponding to an energy continuum (complement $N_{III}$). In addition, we shall examine a special case which is very important because of its applications, particularly in solid state physics, that of a periodic potential (complement $O_{III}$).

## Complement $M_{III}$

## BOUND STATES OF A PARTICLE IN A "POTENTIAL WELL" OF ARBITRARY SHAPE



In complement $H_I$, we studied, for a special case (finite or infinite "square" well), the bound states of a particle in a potential well. We derived certain properties of these bound states: a discrete energy spectrum and a ground state energy greater than the classical minimum energy. These properties are, in fact, general, and have numerous physical consequences, as we shall show in this complement.

When the potential energy of a particle posesses a minimum (see figure 1-a), the particle is said to be placed in a "potential well"**. Before studying qualitatively the stationary states of a quantum particle in such a well, let us recall the corresponding motion of a classical particle. When its energy $E_{cl}$ takes on the minimum possible value $E_{cl} = - V_0$ (where $V_0$ is the depth of the well), the particle is motionless at the point $M_0$ whose abscissa is $x_0$. In the cas where $- V_0 < E_{cl} < 0$, the particle oscillates in the well, with an amplitude which increases with $E_{cl}$. Finally, when $E_{cl} > 0$, the particle does not remain in the well, but moves off towards infinity. The "bound states" of the classical particle therefore correspond to all negative energy values between $- V_0$ and 0.

* The effects of a vector potential **A** will be studied later, in particular in complement $E_{VI}$.

** The potential energy, of course, is only defined to within a constant. By convention, we set the potential equal to zero at infinity.

For a quantum particle, the situation is very different. Well-defined energy states $E$ are stationary states whose wave functions $\varphi(x)$ are solutions of the eigenvalue equation of the Hamiltonian $H$:

$$\left[ - \frac{\hbar^2}{2m} \frac{d^2}{dx^2} + V(x) \right] \varphi(x) = E \varphi(x) \tag{1}$$

Such a second-order differential equation has an infinite number of solutions, whatever the value chosen for $E$: if we pick arbitrary values of $\varphi(x)$ and its derivative at any given point, we can obtain $\varphi$ for any other value of $x$. Equation (1) alone cannot, therefore, restrict the possible energy values. However, we shall show here that if, in addition, we impose certain boundary conditions on $\varphi(x)$, only a certain number of values of $E$ remain possible (quantization of energy levels).

## 1. Quantization of the bound state energies

We shall call "bound states of the particle" states whose wave functions $\varphi(x)$ satisfy the eigenvalue equation (1) and are *square-integrable* [indispensable if $\varphi(x)$ is actually to describe the physical state of a particle]. These are therefore stationary states, for which the position probability density $|\varphi(x)|^2$ takes on non-negligible values only in a limited region of space [for $\int_{-\infty}^{+\infty} dx \, |\varphi(x)|^2$ to converge, $|\varphi(x)|^2$ must approach zero sufficiently rapidly when $x \longrightarrow \pm \infty$]. Bound states remind us of classical motion where the particle oscillates inside the well without ever being able to emerge (energy $E_{cl}$ negative, but greater than $- V_0$).

We shall see that in quantum mechanics, the fact that $\varphi(x)$ is required to be square-integrable implies that the possible energies form a discrete set of values which are also included between $- V_0$ and 0. To understand this, let us return to the potential shown in figure 1-a. For simplicity, we shall assume that $V(x)$ is identically equal to zero outside an interval $[x_1, x_2]$. If $x < x_1$ (region I), $V(x) = 0$, and the solution to equation (1) can immediately be written:

– if $E > 0$:

$$\varphi_I(x) = A \, e^{ikx} + A' \, e^{-ikx} \tag{2}$$

with:

$$k = \sqrt{\frac{2mE}{\hbar^2}} \tag{3}$$

– if $E < 0$:

$$\varphi_I(x) = B \, e^{\rho x} + B' \, e^{-\rho x} \tag{4}$$

with:

$$\rho = \sqrt{- \frac{2mE}{\hbar^2}} \tag{5}$$

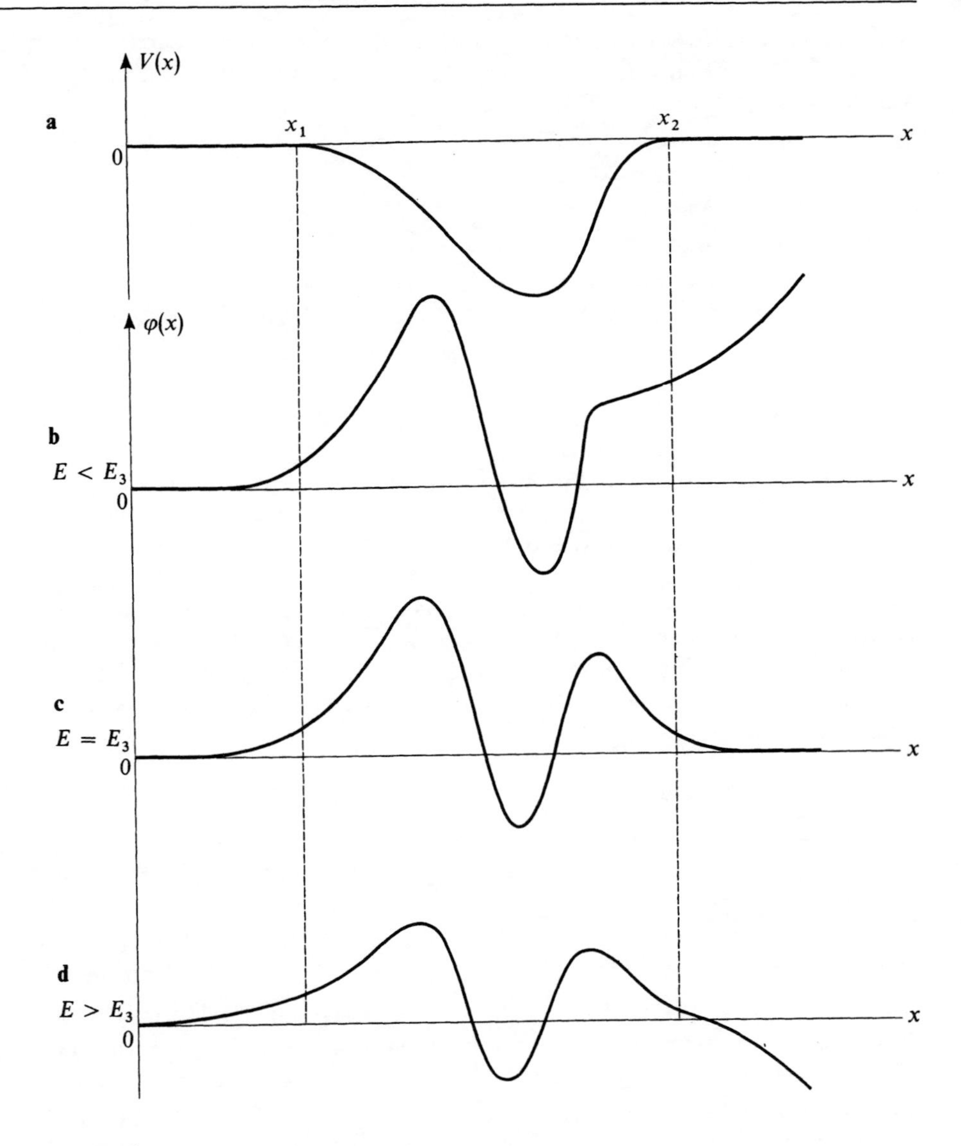


FIGURE 1

**Potential well of depth $V_0$ (fig. a), situated between the points $x = x_1$ and $x = x_2$. We choose a solution $\varphi(x)$ of the eigenvalue equation of $H$ which, for $x < x_1$, approaches zero exponentially when $x \longrightarrow -\infty$. We then extend this solution to the entire $x$-axis. For an arbitrary energy value $E$, $\varphi(x)$ diverges like $\tilde{B}(E)\, e^{\rho x}$ when $x \longrightarrow +\infty$: figure b represents the case where $\tilde{B}(E) > 0$; figure d, that where $\tilde{B}(E) < 0$. However, if the energy $E$ is chosen so as to make $\tilde{B}(E) = 0$, $\varphi(x)$ approaches zero exponentially when $x \longrightarrow +\infty$ (fig. c), and $\varphi(x)$ is square-integrable.**

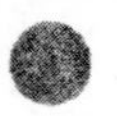



We are looking for a square-integrable solution; we must therefore eliminate the form (2) in which $\varphi_I(x)$ is a superposition of plane waves of constant modulus which cause the integral:

$$\int_{-\infty}^{x_1} dx \, |\varphi_I(x)|^2 \tag{6}$$

to diverge. Only possibility (4) remains, and we obtain our first result: *the bound states of the particle all have a negative energy*. In (4), we cannot retain the term in $e^{-\rho x}$, which diverges when $x \longrightarrow -\infty$. We are therefore left with:

$$\varphi_I(x) = e^{\rho x} \qquad \text{if} \qquad x < x_1 \tag{7}$$

[We have omitted the proportionality factor $B$ since the homogeneity of equation (1) allows us to define $\varphi(x)$ to within a multiplicative coefficient].

The value of $\varphi(x)$ in the interval $x_1 \leqslant x \leqslant x_2$ (region II) is obtained by extending $\varphi_I(x)$: we must look for the solution of equation (1) which is equal to $e^{\rho x_1}$ for $x = x_1$ and whose derivative at this point is equal to $\rho \, e^{\rho x_1}$. The function $\varphi_{II}(x)$ thus obtained depends on $\rho$ and, of course, on the exact expression for $V(x)$. Nevertheless, since (1) is a second-order differential equation, $\varphi_{II}(x)$ is determined uniquely by the preceding boundary conditions; it is, moreover, real (which enables us to trace curves such as those in figures 1-b, 1-c and 1-d).

All that now remains to be done is to obtain the solution when $x > x_2$ (region III); this solution can be written:

$$\varphi_{III}(x) = \tilde{B} \, e^{\rho x} + \tilde{B}' \, e^{-\rho x} \tag{8}$$

where $\tilde{B}$ and $\tilde{B}'$ are real constants determined by the two continuity conditions for $\varphi(x)$ and $d\varphi/dx$ at the point $x = x_2$. $\tilde{B}$ and $\tilde{B}'$ depend on $\rho$, as well as on the function $V(x)$.

We have therefore constructed a solution of equation (1), such as the one shown in figure 1-b. Is this solution square-integrable? We see from (8) that, in general, it is not, except when $\tilde{B}$ is zero (this special case is shown in figure 1-c). Now, for a given function $V(x)$, $\tilde{B}$ is a function of $E$ through the intermediary of $\rho$. The only values of $E$ for which a bound state exists are therefore solutions of the equation $\tilde{B}(E) = 0$. These solutions $E_1$, $E_2$, ... (*cf.* fig. 2) form a discrete spectrum which, of course, depends on the potential $V(x)$ chosen (we shall see in the following section that all the energies $E_i$ are greater than $-V_0$).

We thus arrive at the following result: *the bound state energy values possible for a particle placed in a potential well of arbitrary shape form a discrete set* (it is often said that the bound state energies are quantized). This result can be compared to the quantization of electromagnetic modes in a cavity. There is no analogue in classical mechanics, where, as we have seen, all energy values included between $-V_0$ and 0 are acceptable. In quantum mechanics, the lowest energy level $E_1$ is called the *ground state*, the energy level $E_2$ immediately above, the *first excited state*, the next energy level $E_3$, the second excited state, etc. The following schematic diagram is often associated with each of these states: inside the potential well representing $V(x)$, a horizontal line is drawn whose vertical position corresponds to the energy of the state and whose length gives

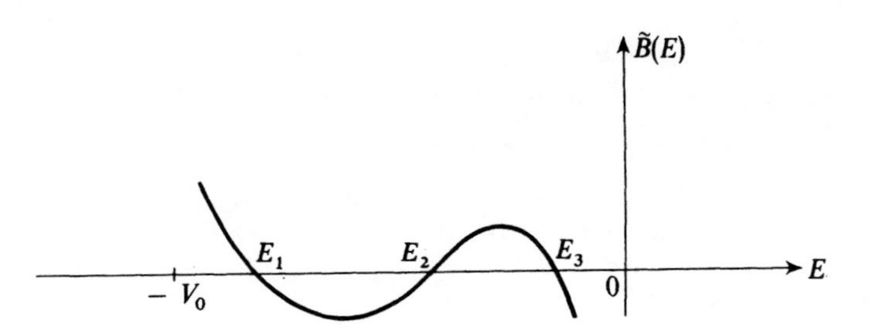


FIGURE 2

**Graphical representation of the function $\tilde{B}(E)$. The zeros of $\tilde{B}(E)$ give the values of $E$ for which $\varphi(x)$ is square-integrable (the situation in figure 1-c), that is, the energies $E_1$, $E_2$, $E_3$, ... of the bound states; all these energies are included between $-V_0$ and 0.**

an idea of the spatial extension of the wave function (this line actually covers the points of the axis which would be reached by a classical particle of the same energy). For the set of energy levels, we obtain a schematic diagram of the type shown of figure 3.

As we saw in chapter I, the phenomenon of energy quantization was one of the factors which led to the introduction of quantum mechanics. Discrete energy levels appear in a very large number of physical systems : atoms (*cf.* chap. VII, hydrogen atom), the harmonic oscillator (*cf.* chap. V), atomic nuclei, etc.

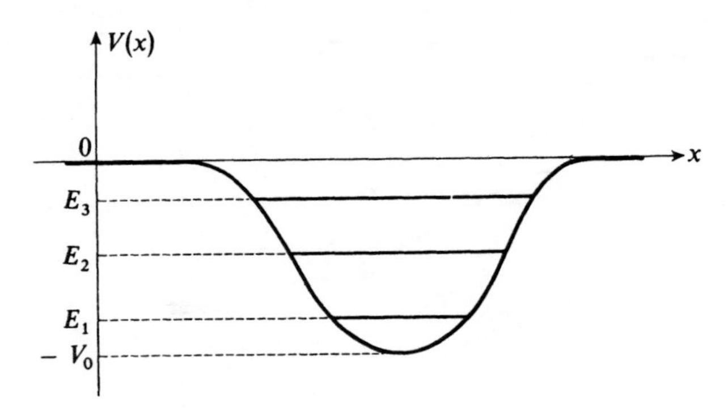


FIGURE 3

**Schematic representation of the bound states of a particle in a potential well. For each of these stationary states, one draws a horizontal line whose ordinate is equal to the energy of the corresponding level. The ends of this line are the points of intersection with the curve which represents the potential $V(x)$; that is, the line is confined to the region of classical motion for the same energy; this gives an idea of the extension of the wave function.**

## 2. Minimum value of the ground state energy

In this section, we shall show that the energies $E_1$, $E_2$, etc... are all greater than the minimum value $-V_0$ of the potential energy $V(x)$. We shall then see how this result can be easily understood using Heisenberg's uncertainty relation.

If $\varphi(x)$ is a solution of (1), we obtain, multiplying this equation by $\varphi^*(x)$ and integrating the relation thus obtained:

$$-\frac{\hbar^2}{2m}\int_{-\infty}^{+\infty} \mathrm{d}x\ \varphi^*(x)\frac{\mathrm{d}^2}{\mathrm{d}x^2}\varphi(x) + \int_{-\infty}^{+\infty} \mathrm{d}x\, V(x)\,|\,\varphi(x)|^2$$

$$= E\int_{-\infty}^{+\infty} \mathrm{d}x\ |\varphi(x)|^2 \tag{9}$$

For a bound state, the function $\varphi(x)$ can be normalized, and equation (9) can then be written simply:

$$E = \langle\, T \,\rangle + \langle\, V \,\rangle \tag{10}$$

with:

$$\langle\, T \,\rangle = -\frac{\hbar^2}{2m}\int_{-\infty}^{+\infty} \mathrm{d}x\ \varphi^*(x)\frac{\mathrm{d}^2}{\mathrm{d}x^2}\varphi(x) = \frac{\hbar^2}{2m}\int_{-\infty}^{+\infty} \mathrm{d}x \left|\frac{\mathrm{d}}{\mathrm{d}x}\varphi(x)\right|^2 \tag{11}$$

[where we have performed an integration by parts and used the fact that $\varphi(x)$ goes to zero when $|x| \longrightarrow \infty$] and:

$$\langle\, V \,\rangle = \int_{-\infty}^{+\infty} \mathrm{d}x\, V(x)\,|\varphi(x)|^2 \tag{12}$$

Relation (10) shows simply that $E$ is the sum of the mean value of the kinetic energy:

$$\langle\, T \,\rangle = \langle\, \varphi \,|\frac{P^2}{2m}|\, \varphi \,\rangle \tag{13}$$

and that of the potential energy:

$$\langle\, V \,\rangle = \langle\, \varphi \,|\, V(X) \,|\, \varphi \,\rangle \tag{14}$$

From relations (11) and (12), it follows immediately that:

$$\langle\, T \,\rangle > 0 \tag{15}$$

$$\langle\, V \,\rangle \geqslant \int_{-\infty}^{+\infty} \mathrm{d}x\ (-\,V_0)\,|\varphi(x)|^2 = -\,V_0 \tag{16}$$

Consequently:

$$E = \langle T \rangle + \langle V \rangle > \langle V \rangle \geqslant - V_0 \tag{17}$$

Since $E$ is negative, as we showed in § 1, we see that, *as in classical mechanics, the bound state energies are always between* $- V_0$ *and* 0.

There exists, nevertheless, an important difference between the classical and quantum situations: while, in classical mechanics, the particle can have an energy equal to $- V_0$ (case of a particle at rest at $M_0$) or slightly greater than $- V_0$ (case of small oscillations), the same is not true in quantum mechanics, where the lowest possible energy is the energy $E_1$ of the ground state, which is necessarily greater than $- V_0$ (*cf.* fig. 3). The Heisenberg uncertainty relations enable us to understand the physical origin of this result, as we shall now show.

If we try to construct a state of the particle for which the mean potential energy is as small as possible, we see from (12) that we must choose a wave function which is practically localized at the point $M_0$. The root-mean-square deviation $\Delta X$ is then very small, so $\Delta P$ is necessarily very large. Since:

$$\langle P^2 \rangle = (\Delta P)^2 + \langle P \rangle^2 \geqslant (\Delta P)^2 \tag{18}$$

the kinetic energy $\langle T \rangle = \langle P^2 \rangle / 2m$ is then also very large. Therefore, if the potential energy of the particle approaches its minimum, the kinetic energy increases without bound. The wave function of the ground state corresponds to a compromise, for which the sum of these two energies is a minimum. The ground state of the quantum particle is thus characterized by a wave function which has a certain spatial extension (*cf.* fig. 3), and its energy is necessarily greater than $- V_0$. Unlike the situation in classical mechanics, there exists no well-defined energy state in quantum mechanics where the particle is "at rest" at the bottom of the potential well.

COMMENT:

Since the energy of the bound states is included between $- V_0$ and 0, such states can exist only if the potential $V(x)$ takes on negative values in one or several regions of the $x$-axis. This is why we have chosen for this complement a potential "well" like the one shown in figure 1-a (while in the following complement, we shall not confine ourselves to the case of a potential well).

However, there is nothing to prevent $V(x)$ from being positive for certain values of $x$; for example, the "well" can be surrounded by potential "barriers" as is shown in figure 4 (we shall always assume the potential to be zero at infinity). In this case, certain classical motions of positive energy remain bounded, while in quantum mechanics, the same reasoning as above shows that the bound states always have an energy between $- V_0$ and 0. Physically, this difference arises from the fact that a potential barrier of finite height is never able to make a quantum particle turn back completely: the particle always has a non-zero probability of passing through by the tunnel effect.

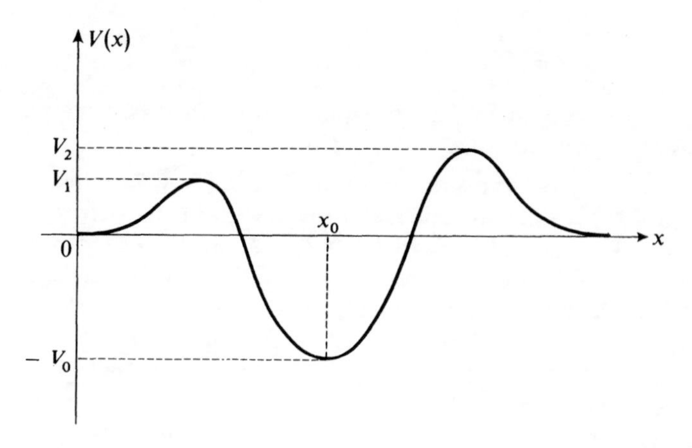


FIGURE 4

**Potential well of depth $-V_0$ situated between two potential barriers of height $V_1$ and $V_2$ (assuming, for example, $V_1 \leqslant V_2$). Classically, there exist particle states whose energy is between $-V_0$ and $V_1$ which remain confined between the two barriers. In quantum mechanics, a particle whose energy is between 0 and $V_1$ can penetrate the barrier by the tunnel effect; consequently, the bound states always have energies between $-V_0$ and 0.**

**References and suggestions for further reading:**

Feynman III (1.2), §16-6; Messiah (1.17), chap. III, §II; Ayant and Belorizky (1.10), chap. IV, §§1, 2, 3; Schiff (1.18), §8.

# Complement $N_{III}$

## UNBOUND STATES OF A PARTICLE IN THE PRESENCE OF A POTENTIAL WELL OR BARRIER OF ARBITRARY SHAPE



In complement $M_{III}$, we showed that bound states of a particle placed in a potential $V(x)$ have negative energies* and that they exist only if $V(x)$ is an attractive potential (a potential well which allows classical bounded motion). We had to reject positive energy values since they led to eigenfunctions $\varphi_k(x)$ of the Hamiltonian $H$ which, at infinity, behaved like superpositions of non-square-integrable exponentials $e^{\pm ikx}$. Nevertheless, we saw as early as chapter I, that, by superposing such functions linearly, one can construct square-integrable wave functions $\psi(x)$ (wave packets) which can therefore represent the physical state of a particle. It is clear that, since the states thus obtained involve several values of $k$ (that is, of the energy), they are no longer stationary states; the wave function $\psi(x)$ therefore evolves over time, propagating and becoming deformed. However, the fact that $\psi(x)$ is already expanded in terms of the eigenfunctions $\varphi_k(x)$ enables us to calculate this evolution very simply [as we did, for example, in complement $J_I$, where we used the properties of the $\varphi_k(x)$ to calculate the transmission and reflection coefficients of a potential barrier, the delay upon reflection, etc.]. This is why, despite the fact that each of the $\varphi_k(x)$ cannot alone represent a physical state, it is useful to study the positive energy eigenfunctions** of $H$, as we have already done, in complement $H_I$, for certain square potentials.

In this complement, we are going to study in a general way (confining ourselves, nevertheless, to one-dimensional problems) the effect of a potential $V(x)$ on the positive energy eigenfunctions $\varphi_k(x)$. We shall assume nothing about the shape of $V(x)$, which may present one or several barriers, wells, etc., except that $V(x)$ goes to zero outside a finite interval $[x_1, x_2]$ of the $x$-axis. We shall show that, in all cases, the effect of $V(x)$ on the functions $\varphi_k(x)$ can be described by a $2 \times 2$ matrix, $M(k)$, which possesses a certain number of general properties. We shall thus obtain various results which are independent of the shape of the potential $V(x)$ chosen. For example, we shall see that the transmission

* Recall that we chose the energy origin so as to make $V(x)$ zero at infinity.

** One might also consider studying the non-square-integrable negative energy eigenfunctions of $H$ (those whose energies do not belong to the discrete spectrum obtained in complement $M_{III}$). However, these functions diverge very rapidly (exponentially) at infinity, and one could not obtain square-integrable wave functions by linearly superposing them.

and reflection coefficients of a barrier (whether symmetrical or not) are the same for a particle coming from the left and for a particle of the same energy coming from the right. An additional aim of this complement $N_{III}$ is to serve as the point of departure for the calculations of complement $O_{III}$, in which we study the properties of a particle in a periodic potential $V(x)$.

## 1. Transmission matrix $M(k)$

### a. DEFINITION OF $M(k)$

In a one-dimensional problem, consider a potential $V(x)$ which is zero outside an interval $[x_1, x_2]$ of length $l$, but which varies in an arbitrary way inside this interval (fig. 1). We choose the $x$ origin to be in the middle of the interval $[x_1, x_2]$, so as to have $V(x)$ vary only for $|x| < l/2$. The equation satisfied by every wave function $\varphi(x)$ associated with a stationary state of energy $E$ is:

$$\left\{ \frac{d^2}{dx^2} + \frac{2m}{\hbar^2} [E - V(x)] \right\} \varphi(x) = 0 \tag{1}$$

In the rest of this complement, we shall choose, to characterize the energy, the parameter $k$ given by:

$$k = \sqrt{\frac{2mE}{\hbar^2}} \tag{2}$$

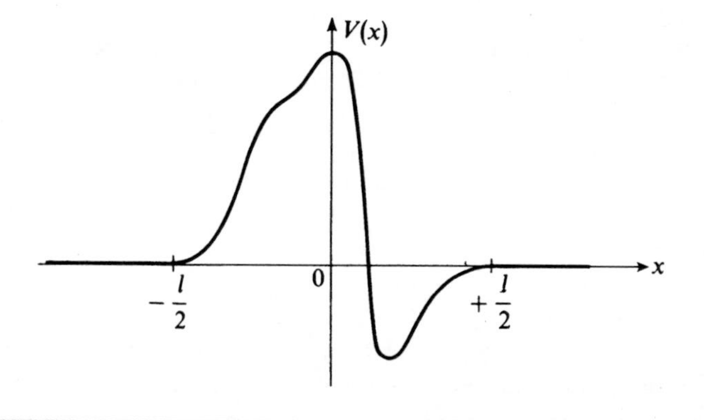


FIGURE 1

**The potential $V(x)$ under consideration varies in an arbitrary way within the interval $-l/2 \leqslant x \leqslant l/2$ and goes to zero outside this interval.**

In the region $x < -\frac{l}{2}$, the function $e^{ikx}$ satisfies equation (1); let us call $v_k(x)$ the solution of this equation that is identical to $e^{ikx}$ for $x < -\frac{l}{2}$.

When $x > +\frac{l}{2}$, $v_k(x)$ is necessarily a linear combination of two independent solutions $e^{ikx}$ and $e^{-ikx}$ of (1). This gives us:

$$\begin{cases} \text{if } x < -\frac{l}{2}: & v_k(x) = e^{ikx} \quad (3\text{-a}) \\ \text{if } x > +\frac{l}{2}: & v_k(x) = F(k)\, e^{ikx} + G(k)\, e^{-ikx} \quad (3\text{-b}) \end{cases}$$

where $F(k)$ and $G(k)$ are coefficients which depend on $k$, as well as on the shape of the potential under study. Similarly, we can introduce the solution $v'_k(x)$, which, for $x < -l/2$, is equal to $e^{-ikx}$.

$$\begin{cases} \text{if } x < -\frac{l}{2}: & v'_k(x) = e^{-ikx} \quad (4\text{-a}) \\ \text{if } x > +\frac{l}{2}: & v'_k(x) = F'(k)\, e^{ikx} + G'(k)\, e^{-ikx} \quad (4\text{-b}) \end{cases}$$

The most general solution $\varphi_k(x)$ of equation (1) (of second order in $x$), for a given value of $E$ (that is, of $k$), is a linear combination of $v_k$ and $v'_k$:

$$\varphi_k(x) = A\, v_k(x) + A' v'_k(x) \tag{5}$$

Relations (3-a) and (4-a) imply that:

$$\text{if } x < -\frac{l}{2}: \qquad \varphi_k(x) = A\, e^{ikx} + A'\, e^{-ikx} \tag{6-a}$$

while relations (3-b) and (4-b) yield:

$$\text{if } x > +\frac{l}{2}: \qquad \varphi_k(x) = \tilde{A}\, e^{ikx} + \tilde{A}'\, e^{-ikx} \tag{6-b}$$

with:

$$\begin{aligned} \tilde{A} &= F(k)\, A + F'(k)\, A' \\ \tilde{A}' &= G(k)\, A + G'(k)\, A' \end{aligned} \tag{7}$$

By definition, the matrix $M(k)$ is the $2 \times 2$ matrix:

$$M(k) = \begin{pmatrix} F(k) & F'(k) \\ G(k) & G'(k) \end{pmatrix} \tag{8}$$

which allows us to write relations (7) in the matrix form:

$$\begin{pmatrix} \tilde{A} \\ \tilde{A}' \end{pmatrix} = M(k) \begin{pmatrix} A \\ A' \end{pmatrix} \tag{9}$$

$M(k)$ therefore enables us to determine, given the behavior (6-a) of the wave function to the left of the potential, its behavior (6-b) to the right. We call $M(k)$ the "transmission matrix" of the potential.

COMMENT:

The current associated with a wave function $\varphi(x)$ is:

$$J(x) = \frac{\hbar}{2mi}\left[\varphi^*(x)\frac{d\varphi}{dx} - \varphi(x)\frac{d\varphi^*}{dx}\right] \tag{10}$$

Differentiating, we find:

$$\frac{d}{dx}J(x) = \frac{\hbar}{2mi}\left[\varphi^*(x)\frac{d^2\varphi}{dx^2} - \varphi(x)\frac{d^2\varphi^*}{dx^2}\right] \tag{11}$$

Taking (1) into account, we obtain:

$$\frac{d}{dx}J(x) = 0 \tag{12}$$

Therefore, the current $J(x)$ associated with a stationary state is the same at all points of the $x$-axis. Note, moreover, that (12) is simply the one-dimensional analogue of the relation:

$$\text{div}\ \mathbf{J}(\mathbf{r}) = 0 \tag{13}$$

which is valid, according to relation (D-11) of chapter III, for any stationary state of a particle moving in three-dimensional space. According to (12), the current $J_k(x)$ associated with $\varphi_k(x)$ can therefore be calculated for any $x$, choosing either the form (6-a) or the form (6-b) of $\varphi_k(x)$ :

$$J_k(x) = \frac{\hbar k}{m}[\,|A|^2 - |A'|^2\,] = \frac{\hbar k}{m}[\,|\tilde{A}|^2 - |\tilde{A}'|^2\,] \tag{14}$$

#### b. PROPERTIES OF $M(k)$

$\alpha$. It is easy to show, using the fact that the function $V(x)$ is real, that if $\varphi(x)$ is a solution of equation (1), $\varphi^*(x)$ is also. Now consider the function $v_k^*(x)$, which is a solution of (1); comparison of (3-a) and (4-a) shows that it is identical to $v_k'(x)$ when $x < -\frac{l}{2}$. We therefore have, for all $x$:

$$v_k^*(x) = v_k'(x) \tag{15}$$

Substituting relations (3-b) and (4-b) into this relation, we obtain:

$$F^*(k) = G'(k) \tag{16}$$

$$G^*(k) = F'(k) \tag{17}$$

It follows that the matrix $M(k)$ can be written in the simplified form:

$$M(k) = \begin{pmatrix} F(k) & G^*(k) \\ G(k) & F^*(k) \end{pmatrix} \tag{18}$$

$\beta$. We saw above [*cf.* (12)] that the probability current $J(x)$ does not depend on $x$ for a stationary state. We must therefore have [*cf.* (14)]:

$$|A|^2 - |A'|^2 = |\tilde{A}|^2 - |\tilde{A}'|^2 \tag{19}$$

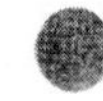

for any $A$ and $A'$. Now relations (9) and (18) yield:

$$\begin{aligned}|\tilde{A}|^2 - |\tilde{A}'|^2 &= [F(k)A + G^*(k)A'][F^*(k)A^* + G(k)A'^*] \\ &\quad - [G(k)A + F^*(k)A'][G^*(k)A^* + F(k)A'^*] \\ &= [|F(k)|^2 - |G(k)|^2][|A|^2 - |A'|^2]\end{aligned} \tag{20}$$

Condition (19) is therefore equivalent to:

$$|F(k)|^2 - |G(k)|^2 = \text{Det } M(k) = 1 \tag{21}$$

COMMENTS:

(*i*) We have made no particular assumptions about the shape of the potential. If it is even, that is, if $V(x) = V(-x)$, the matrix $M(k)$ possesses an additional property: it can be shown that $G(k)$ is a pure imaginary.

(*ii*) Relations (6) show that $A$ and $\tilde{A}'$ are the coefficients of "incoming" plane waves, i.e. waves associated with particles arriving respectively from $x = -\infty$ and $x = +\infty$ and moving towards the zone of influence of the potential (incident particles). On the other hand, $\tilde{A}$ and $A'$, are the coefficients corresponding to "outgoing" waves, associated with particles moving away from the potential (transmitted or reflected particles). It is useful to introduce the matrix $S$, which allows us to calculate the amplitude of the outgoing waves in terms of that of the incoming waves:

$$\begin{pmatrix} \tilde{A} \\ A' \end{pmatrix} = S(k) \begin{pmatrix} A \\ \tilde{A}' \end{pmatrix} \tag{22}$$

$S(k)$ can easily be expressed in terms of the elements of the matrix $M(k)$, as we now show. The relations:

$$\tilde{A} = F(k)\, A + G^*(k)\, A' \tag{23-a}$$

$$\tilde{A}' = G(k)\, A + F^*(k)\, A' \tag{23-b}$$

imply that:

$$A' = \frac{1}{F^*(k)} [\tilde{A}' - G(k)A] \tag{24}$$

Substituting this relation into (23-a), we obtain:

$$\tilde{A} = \frac{1}{F^*(k)} \left[\left(F(k)F^*(k) - G(k)G^*(k)\right)A + G^*(k)\tilde{A}'\right] \tag{25}$$

Taking (21) into account, we can then write the matrix $S(k)$:

$$S(k) = \frac{1}{F^*(k)} \begin{pmatrix} 1 & G^*(k) \\ -G(k) & 1 \end{pmatrix} \tag{26}$$

It is easy to verify, using (21) again, that:

$$S(k)\, S^\dagger(k) = S^\dagger(k)\, S(k) = 1 \tag{27}$$

$S(k)$ is therefore unitary. This matrix plays an important role in collision theory; we could have proved its unitary property from that of the evolution operator (*cf.* complement $F_{III}$), which simply expresses the conservation over time of the total probability of finding the particle somewhere on the $Ox$ axis (norm of the wave function).

## 2. Transmission and reflection coefficients

To calculate the reflection and transmission coefficients for a particle encountering the potential $V(x)$, one should (as in complement $J_I$) construct a wave packet with the eigenfunctions of $H$ which we have just studied. Consider, for example, an incident particle of energy $E_i$ coming from the left. The corresponding wave packet is obtained by superposing functions $\varphi_k(x)$, for which we set $\tilde{A}' = 0$, with coefficients given by a function $g(k)$ which has a marked peak in the neighborhood of $k = k_i = \sqrt{2mE_i/\hbar^2}$. We shall not go into these calculations in detail here; they are analogous in every way to those of complement $J_I$. They show that the reflection and transmission coefficients are equal, respectively, to $|A'(k_i)/A(k_i)|^2$ and $|\tilde{A}(k_i)/A(k_i)|^2$.

Since $\tilde{A}' = 0$, relations (22) and (26) yield:

$$\begin{aligned} \tilde{A}(k) &= \frac{1}{F^*(k)} A(k) \\ A'(k) &= -\frac{G(k)}{F^*(k)} A(k) \end{aligned} \tag{28}$$

The reflection and transmission coefficients are therefore equal to:

$$R_1(k_i) = \left| \frac{A'(k_i)}{A(k_i)} \right|^2 = \left| \frac{G(k_i)}{F(k_i)} \right|^2 \tag{29-a}$$

$$T_1(k_i) = \left| \frac{\tilde{A}(k_i)}{A(k_i)} \right|^2 = \frac{1}{|F(k_i)|^2} \tag{29-b}$$

[it is easy to verify that condition (21) insures that $R_1(k_i) + T_1(k_i) = 1$].

If we now consider a particle coming from the right, we must take $A = 0$, which gives:

$$\begin{aligned} \tilde{A}(k) &= \frac{G^*(k)}{F^*(k)} \tilde{A}'(k) \\ A'(k) &= \frac{1}{F^*(k)} \tilde{A}'(k) \end{aligned} \tag{30}$$

The transmission and reflection coefficients are now equal to:

$$T_2(k) = \left| \frac{A'(k)}{\tilde{A}'(k)} \right|^2 = \frac{1}{|F(k)|^2} \tag{31-a}$$

and:

$$R_2(k) = \left| \frac{\tilde{A}(k)}{\tilde{A}'(k)} \right|^2 = \left| \frac{G(k)}{F(k)} \right|^2 \tag{31-b}$$

Comparison of (29) and (31) shows that $T_1(k) = T_2(k)$ and that $R_1(k) = R_2(k)$: for a given energy, the transparency of a barrier (whether symmetrical or not) is therefore always the same for particles coming from the right and from the left.

In addition, from (21) we have:

$$|F(k)| \geqslant 1 \tag{32}$$

When the equality is realized, the reflection coefficient is zero and the transmission coefficient is equal to 1 (resonance). On the other hand, the inverse situation is not possible : since (21) imposes that $|F(k)| > |G(k)|$, one can never have $T = 0$ and $R = 1$ [except in the case where $F$ and $G$ tend simultaneously towards infinity]. Actually, such a situation can only occur for $k = 0$. To see this, divide the function $v_k(x)$ defined in (3) by $F(k)$. If $F(k)$ goes to infinity, the wave function will be identically zero on the left hand side, and hence necessarily, by extension, zero on the right hand side. However, this is impossible unless $k = 0$ and $F = -G$.

## 3. Example

Let us return to the square potentials studied in § 2-b of complement $H_I$: in the region $-l/2 < x < +l/2$, $V(x)$ is equal to a constant $V_0$* (see figure 2, where $V_0$ has been chosen to be positive).

First, let us assume that $E$ is smaller than $V_0$, and set:

$$\rho = \sqrt{2m(V_0 - E)/\hbar^2} \tag{33}$$

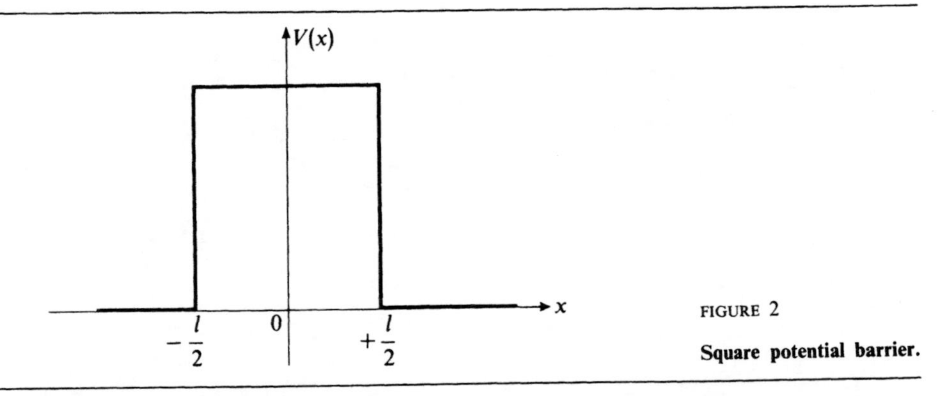


FIGURE 2

**Square potential barrier.**

An elementary calculation analogous to the one in complement $H_I$ yields:

$$M(k) = \begin{pmatrix} \left[\cosh \rho l + i\dfrac{k^2 - \rho^2}{2k\rho} \sinh \rho l\right] e^{-ikl} & -i\dfrac{k_0^2}{2k\rho} \sinh \rho l \\ i\dfrac{k_0^2}{2k\rho} \sinh \rho l & \left[\cosh \rho l - i\dfrac{k^2 - \rho^2}{2k\rho} \sinh \rho l\right] e^{ikl} \end{pmatrix} \tag{34}$$

* In fact, we are considering here a barrier which is displaced relative to that of complement $H_I$, since we are assuming it to be situated between $x = -l/2$ and $x = +l/2$, instead of between $x = 0$ and $x = l$.

with:

$$k_0 = \sqrt{\frac{2mV_0}{\hbar^2}} \tag{35}$$

($V_0$ is necessarily positive here, since we have assumed $E < V_0$).

If now we assume that $E > V_0$, we set:

$$k' = \sqrt{\frac{2m}{\hbar^2}(E - V_0)} \tag{36}$$

and:

$$k_0 = \sqrt{\varepsilon \frac{2mV_0}{\hbar^2}} \tag{37}$$

(where $\varepsilon = +1$ if $V_0 > 0$ and $-1$ if $V_0 < 0$). We thus obtain:

$$M(k) = \begin{pmatrix} \left[\cos k'l + i\dfrac{k^2 + k'^2}{2kk'}\sin k'l\right] e^{-ikl} & -i\varepsilon\dfrac{k_0^{\ 2}}{2kk'}\sin k'l \\ i\varepsilon\dfrac{k_0^{\ 2}}{2kk'}\sin k'l & \left[\cos k'l - i\dfrac{k^2 + k'^2}{2kk'}\sin k'l\right] e^{ikl} \end{pmatrix} \tag{38}$$

It is easy to verify that the matrices $M(k)$ written in (34) and (38) satisfy relations (16), (17) and (21).

**References and suggestions for further reading:**

Merzbacher (1.16), chap. 6, §§5, 6 and 8; see also the references of complement $M_{III}$.

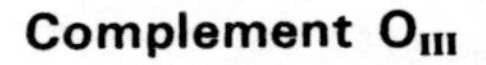

## Complement $O_{III}$

# QUANTUM PROPERTIES OF A PARTICLE IN A ONE-DIMENSIONAL PERIODIC STRUCTURE



In this complement, we are going to study the quantum properties of a particle placed in a potential $V(x)$ having a periodic structure. The functions $V(x)$ which we shall consider will not necessarily be periodic in the strict sense of the term; it suffices for them to have the shape of a periodic function in a finite region of the $x$-axis (fig. 1), that is, to be the result of juxtaposing $N$ times the same motif at regular intervals [$V(x)$ is truly periodic only in the limit $N \longrightarrow \infty$].

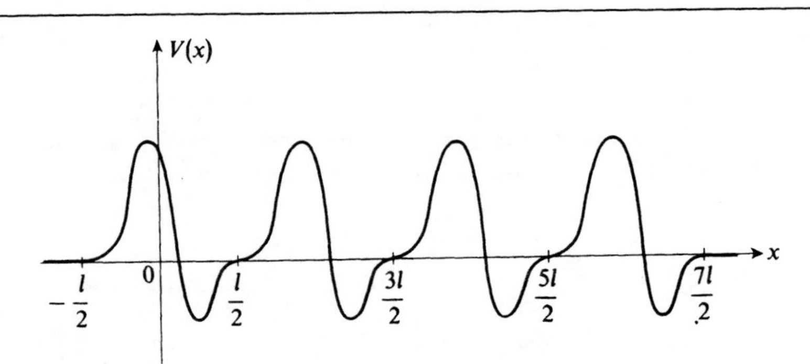


FIGURE 1

**Potential $V(x)$ having a periodic structure obtained by juxtaposing $N$ times the same motif ($N = 4$ in the figure).**

Such periodic structures are encountered, for example, in the study of a linear molecule formed by $N$ atoms (or groups of atoms) which are identical and equally spaced. They are also encountered in solid state physics, when one chooses a one-dimensional model in order to understand the disposition of the energy levels of an electron in a crystal. If $N$ is very large (as in the case of a linear macromolecule or a macroscopic crystal), the potential $V(x)$ is given in a wide region of space by a periodic function, and the properties of the particle can be expected to be practically the same as they would be if $V(x)$ were really periodic. However, from a physical point of view, the limit of infinite $N$ is never attained, and we shall be concerned here with the case where $N$ is arbitrary.

To study the effect of the potential $V(x)$ on an eigenfunction $\varphi(x)$ of the Hamiltonian $H$, of eigenvalue $E$, we shall introduce a $2 \times 2$ matrix, the iteration matrix $Q$, which depends on $E$. We shall show that the behavior of $\varphi(x)$ is totally different depending on whether the eigenvalues of the iteration matrix are real or imaginary. Since these eigenvalues depend on the energy $E$ chosen, we shall find it useful to distinguish between domains of energy corresponding to real eigenvalues and those which lead to imaginary eigenvalues. The concept of an *allowed or forbidden energy band* will thus be introduced.

COMMENTS:

(*i*) For the sake of convenience, we shall speak of a "potential barrier" to designate the motif which, repeated $N$ times, gives the potential $V(x)$ (fig. 1). However, this motif can also be a "potential well" or have an arbitrary shape.

(*ii*) Common usage in solid state physics reserves the letter $k$ to designate a parameter which is involved in the expression for stationary wave functions and which is not simply proportional to the square root of the energy. To conform to this usage, we shall henceforth use a notation slightly different from that of complement $N_{III}$; we shall replace $k$ by $\alpha$, setting:

$$\alpha = \sqrt{\frac{2mE}{\hbar^2}} \tag{1}$$

and we shall not introduce the letter $k$ until later (we shall see that $k$ is directly related to the eigenvalues of the matrix $Q$ when they are complex).

## 1. Passage though several successive identical potential barriers

Consider a potential $V(x)$ which is obtained by juxtaposing $N$ barriers as in figure 1: the first barrier is centered at $x = 0$, the second, at $x = l$, the third, at $x = 2l$, ..., the last at $x = (N - 1)l$. We intend to study the behavior, during passage through this set of barriers, of an eigenfunction $\varphi_\alpha(x)$ which is a solution of the eigenvalue equation of $H$:

$$\left\{\frac{d^2}{dx^2} + \frac{2m}{\hbar^2}[E - V(x)]\right\}\varphi_\alpha(x) = 0 \tag{2}$$

where $E$ and $\alpha$ are related by (1).

### a. NOTATION

To the left of the $N$ barriers, that is, for $x \leqslant -\frac{l}{2}$, $V(x)$ is zero, and the general solution of equation (2) is:

$$\text{if } x \leqslant -\frac{l}{2} : \varphi_\alpha(x) = A_0 \, e^{i\alpha x} + A'_0 \, e^{-i\alpha x} \tag{3-a}$$

Consider, as in § 1-a of complement $N_{III}$, the two functions $v_k(x)$ and $v'_k(x)$ which here become $v_\alpha(x)$ and $v'_\alpha(x)$. In the region of the first barrier, centered at $x = 0$, the general solution of (2) is written:

$$\text{if } -\frac{l}{2} \leqslant x \leqslant \frac{l}{2} : \varphi_\alpha(x) = A_1 \, v_\alpha(x) + A'_1 \, v'_\alpha(x) \tag{3-b}$$

Similarly, in the region of the second barrier, centered at $x = l$, we obtain:

$$\text{if } \frac{l}{2} \leqslant x \leqslant \frac{3l}{2} : \varphi_\alpha(x) = A_2 \, v_\alpha(x - l) + A'_2 \, v'_\alpha(x - l) \tag{3-c}$$

and, more generally, in the region of the $n$th barrier, centered at $x = (n - 1)l$:

$$\text{if } (n-1)l - \frac{l}{2} \leqslant x \leqslant (n-1)l + \frac{l}{2} :$$

$$\varphi_\alpha(x) = A_n \, v_\alpha[x - (n-1)l] + A'_n \, v'_\alpha[x - (n-1)l] \tag{3-d}$$

Finally, to the right of the $N$ barriers, that is, for $x \geqslant (N - 1)l + \frac{l}{2}$, $V(x)$ is again zero, and we have:

$$\text{if } x \geqslant (N-1)l + \frac{l}{2} : \varphi_\alpha(x) = C_0 \, e^{i\alpha[x-(N-1)l]} + C'_0 \, e^{-i\alpha[x-(N-1)l]} \tag{3-e}$$

We must now match these various expressions for $\varphi_\alpha(x)$ at $x = -\frac{l}{2}, +\frac{l}{2}, \ldots, (N-1)l + \frac{l}{2}$. This is what we shall do in the following section.

### b. MATCHING CONDITIONS

The functions $v_\alpha$ and $v'_\alpha$ depend on the form of the potential chosen. We shall show, however, that it is simple to calculate them, and their derivatives as well, at the two edges of each barrier, by using the results of complement $N_{III}$.

To do so, let us imagine that all but one of the barriers are removed, leaving, for example, the $n$th one, centered at $x = (n - 1)l$. Solution (3-d), always valid inside this barrier, must then be extended to the left and to the right by superposing plane waves. These waves are obtained by replacing, in formulas (6-a) and (6-b) of $N_{III}$, $x$ by $x - (n - 1)l$ and $k$ by $\alpha$, and adding an index $n$ to $A$, $A'$, $\tilde{A}$, $\tilde{A}'$. Thus we have, if the $n$th barrier is isolated:

for $x \leqslant (n - 1)l - \dfrac{l}{2}$:

$$A_n \, e^{i\alpha[x - (n-1)l]} + A'_n \, e^{-i\alpha[x - (n-1)l]} \tag{4}$$

for $x \geqslant (n - 1)l + \dfrac{l}{2}$:

$$\tilde{A}_n \, e^{i\alpha[x - (n-1)l]} + \tilde{A}'_n \, e^{-i\alpha[x - (n-1)l]} \tag{5}$$

with:

$$\begin{pmatrix} \tilde{A}_n \\ \tilde{A}'_n \end{pmatrix} = M(\alpha) \begin{pmatrix} A_n \\ A'_n \end{pmatrix} \tag{6}$$

where, with the change in notation taken into account, $M(\alpha)$ is the matrix $M(k)$ introduced in complement $N_{III}$. Consequently, at the left edge of the $n$th barrier, the function $\varphi_\alpha(x)$ defined in (3-d) has the same value and the same derivative as the superposition of plane waves (4). Similarly, at the right edge of this barrier, it has the same value and the same derivative as (5). These results enable us to write simply the matching conditions in the periodic structure.

Thus, at the left edge of the first barrier (that is, at $x = -l/2$), it is sufficient to note that (3-a) has the same value and the same derivative as $A_1 \, e^{i\alpha x} + A'_1 \, e^{-i\alpha x}$, which yields directly:

$$\begin{cases} A_0 = A_1 \\ A'_0 = A'_1 \end{cases} \tag{7}$$

(a result which was obvious from $N_{III}$).

At the right edge of the first barrier, which is the same as the left edge of the second one, we must write that $\tilde{A}_1 \, e^{i\alpha x} + \tilde{A}' \, e^{-i\alpha x}$ and $A_2 \, e^{i\alpha(x-l)} + A'_2 \, e^{-i\alpha(x-l)}$ have the same value and the same derivative, which yields:

$$\begin{cases} A_2 = \tilde{A}_1 \, e^{i\alpha l} \\ A'_2 = \tilde{A}'_1 \, e^{-i\alpha l} \end{cases} \tag{8}$$

Similarly, at the junction of the $n$th and $(n + 1)$th barriers $\left(x = nl - \dfrac{l}{2}\right)$, we obtain, setting equal the value and derivative of (5) and those of the expression obtained by replacing $n$ by $n + 1$ in (4):

$$\begin{cases} A_{n+1} = \tilde{A}_n \, e^{i\alpha l} \\ A'_{n+1} = \tilde{A}'_n \, e^{-i\alpha l} \end{cases} \tag{9}$$

Finally, at the right edge of the last barrier $\left(x = (N - 1)l + \dfrac{l}{2}\right)$, we must write that (3-e) has the same value and the same derivative as the expression obtained by replacing $n$ by $N$ in (5), which yields:

$$\begin{cases} C_0 = \tilde{A}_N \\ C'_0 = \tilde{A}'_N \end{cases} \tag{10}$$

### c. ITERATION MATRIX $Q(\alpha)$

Let us introduce the matrix $D(\alpha)$ defined by:

$$D(\alpha) = \begin{pmatrix} e^{i\alpha l} & 0 \\ 0 & e^{-i\alpha l} \end{pmatrix} \tag{11}$$

It enables us to write the matching condition (9) in the form:

$$\begin{pmatrix} A_{n+1} \\ A'_{n+1} \end{pmatrix} = D(\alpha) \begin{pmatrix} \tilde{A}_n \\ \tilde{A}'_n \end{pmatrix} \tag{12}$$

that is, taking (6) into account:

$$\begin{pmatrix} A_{n+1} \\ A'_{n+1} \end{pmatrix} = D(\alpha)\, M(\alpha) \begin{pmatrix} A_n \\ A'_n \end{pmatrix} \tag{13}$$

Iterating this equation and using (7), we then obtain:

$$\begin{aligned} \begin{pmatrix} A_{n+1} \\ A_{n+1} \end{pmatrix} &= [D(\alpha)\, M(\alpha)]^n \begin{pmatrix} A_1 \\ A'_1 \end{pmatrix} \\ &= [D(\alpha)\, M(\alpha)]^n \begin{pmatrix} A_0 \\ A'_0 \end{pmatrix} \end{aligned} \tag{14}$$

Finally, matching condition (10) can be transformed by using (6) and (14):

$$\begin{pmatrix} C_0 \\ C'_0 \end{pmatrix} = M(\alpha) \begin{pmatrix} A_N \\ A'_N \end{pmatrix} = M(\alpha) [D(\alpha)\, M(\alpha)]^{N-1} \begin{pmatrix} A_0 \\ A'_0 \end{pmatrix} \tag{15}$$

that is:

$$\begin{pmatrix} C_0 \\ C'_0 \end{pmatrix} = \underbrace{M(\alpha) D(\alpha) M(\alpha) D(\alpha) \ldots D(\alpha) M(\alpha)}_{N \text{ matrices } M(\alpha)} \begin{pmatrix} A_0 \\ A'_0 \end{pmatrix} \tag{16}$$

In this formula, which enables us to go from $\begin{pmatrix} A_0 \\ A'_0 \end{pmatrix}$ to $\begin{pmatrix} C_0 \\ C'_0 \end{pmatrix}$, a matrix $M(\alpha)$ is associated with each barrier, and a matrix $D(\alpha)$, with each interval between two successive barriers.

Relations (13) and (14) demonstrate the importance of the role played by the matrix:

$$Q(\alpha) = D(\alpha)\, M(\alpha) \tag{17}$$

which enters to the $n$th power when one goes from $\begin{pmatrix} A_1 \\ A'_1 \end{pmatrix}$ to $\begin{pmatrix} A_{n+1} \\ A'_{n+1} \end{pmatrix}$, that is, when one performs a translation through a distance $nl$ along the periodic structure. For this reason, we shall call $Q(\alpha)$ the "iteration matrix". Using formula (18) of complement $N_{III}$ and expression (11) for $D(\alpha)$, we obtain:

$$Q(\alpha) = \begin{pmatrix} e^{i\alpha l} F(\alpha) & e^{i\alpha l} G^*(\alpha) \\ e^{-i\alpha l} G(\alpha) & e^{-i\alpha l} F^*(\alpha) \end{pmatrix} \tag{18}$$

The calculation of $[Q(\alpha)]^n$ is facilitated if we change bases so as to make $Q(\alpha)$ diagonal; for this reason we shall study the eigenvalues of $Q(\alpha)$.

#### d. EIGENVALUES OF $Q(\alpha)$

Let $\lambda$ be an eigenvalue of $Q(\alpha)$. The characteristic equation of the matrix (18) is written:

$$[e^{i\alpha l}F(\alpha) - \lambda]\,[e^{-i\alpha l}F^*(\alpha) - \lambda] - |G(\alpha)|^2 = 0 \tag{19}$$

that is, taking into account relation (21) of complement $N_{III}$:

$$\lambda^2 - 2\lambda\, X(\alpha) + 1 = 0 \tag{20}$$

where $X(\alpha)$ is the real part of the complex number $e^{i\alpha l}F(\alpha)$:

$$X(\alpha) = \text{Re}\,[e^{i\alpha l}F(\alpha)] = \frac{1}{2}\,\text{Tr}\,Q(\alpha) \tag{21}$$

Recall [*cf.* complement $N_{III}$, relation (21)] that the modulus of $F(\alpha)$ is greater than 1; the same is therefore true of $e^{i\alpha l}F(\alpha)$.

The discriminant of the second-degree equation (20) is:

$$\Delta' = [X(\alpha)]^2 - 1 \tag{22}$$

Two cases may then arise:

(*i*) If the energy $E$ is such that:

$$|X(\alpha)| \leqslant 1 \tag{23}$$

(for example, if, in figure 2, $\alpha$ is between $\alpha_0$ and $\alpha_1$), one can set:

$$X(\alpha) = \cos\,[k(\alpha)l] \tag{24}$$

with:

$$0 \leqslant k(\alpha) \leqslant \frac{\pi}{l} \tag{25}$$

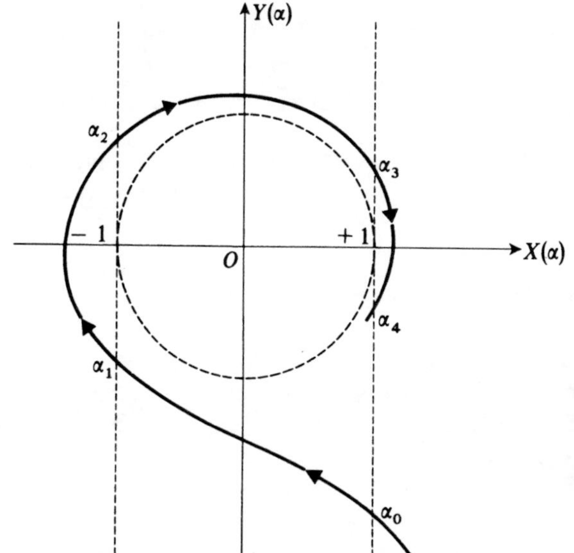


FIGURE 2

**Variation with respect to $\alpha$ of the complex number $e^{i\alpha l}F(\alpha) = X(\alpha) + iY(\alpha)$. Since $|F(\alpha)| > 1$, the curve obtained in the complex plane falls outside the circle centered at $O$ of unit radius. The following discussion shows that if $|X(\alpha)|$ is less than 1, that is, if the value of $\alpha$ chosen gives a point of the curve which is between the two vertical dashed lines of the figure, the corresponding energy falls in an "allowed band"; in the opposite case, it falls in a "forbidden band".**

A simple calculation then shows that the eigenvalues of $Q(\alpha)$ are given by:

$$\lambda = e^{\pm ik(\alpha)l} \tag{26}$$

There are therefore two eigenvalues, which are complex conjugates and whose modulus is equal to 1.

(*ii*) If, on the other hand, the energy $E$ gives a value of $\alpha$ such that:

$$|X(\alpha)| > 1 \tag{27}$$

(for example, if, in figure 2, $\alpha$ is between $\alpha_1$ and $\alpha_2$), one sets:

$$X(\alpha) = \varepsilon \cosh [\rho(\alpha)l] \tag{28}$$

with:

$$\rho(\alpha) \geqslant 0 \tag{29}$$

and $\varepsilon = +1$ if $X(\alpha)$ is positive, $\varepsilon = -1$ if $X(\alpha)$ is negative. We then find:

$$\lambda = \varepsilon\, e^{\pm \rho(\alpha)l} \tag{30}$$

In this case, both eigenvalues of $Q(\alpha)$ are real, and they are each other's inverse.

## 2. Discussion: the concept of an allowed or forbidden energy band

### a. BEHAVIOR OF THE WAVE FUNCTION $\varphi_\alpha(x)$

To apply (14), we begin by calculating the two column matrices $\Lambda_1(\alpha)$ and $\Lambda_2(\alpha)$ associated with the eigenvectors of $Q(\alpha)$ and corresponding respectively to the eigenvalues $\lambda_1$ and $\lambda_2$. We then decompose the matrix $\begin{pmatrix} A_1 \\ A'_1 \end{pmatrix}$ into the form:

$$\begin{pmatrix} A_1 \\ A'_1 \end{pmatrix} = c_1(\alpha)\,\Lambda_1(\alpha) + c_2(\alpha)\,\Lambda_2(\alpha) \tag{31}$$

which enables us to obtain directly:

$$\begin{pmatrix} A_n \\ A'_n \end{pmatrix} = \lambda_1^{n-1} c_1(\alpha)\,\Lambda_1(\alpha) + \lambda_2^{n-1} c_2(\alpha)\,\Lambda_2(\alpha) \tag{32}$$

It is clear from this expression that the behavior of the wave function is very different depending on whether $|X(\alpha)|$ is smaller or greater than 1 in the energy domain of the wave function. In the first case, formula (26) shows that the effect of traversing successive barriers is expressed in (32) by a phase shift in the components of the column matrix $\begin{pmatrix} A_n \\ A'_n \end{pmatrix}$ with respect to $\Lambda_1(\alpha)$ and $\Lambda_2(\alpha)$. The behavior of $\varphi_\alpha(x)$ here recalls that of a superposition of imaginary exponentials. On the other hand, if the energy is such that $|X(\alpha)| > 1$, formula (30) indicates that only one of the two eigenvalues (for example, $\lambda_1$) has a modulus greater than 1. For $n$ sufficiently large, we have, as a result:

$$\begin{pmatrix} A_n \\ A'_n \end{pmatrix} \simeq \varepsilon^{n-1} \, e^{(n-1)\rho(\alpha)l} \, c_1(\alpha) \, \Lambda_1(\alpha) \tag{33}$$

$A_n$ and $A'_n$ therefore increase exponentially with $n$ [except in the special case where $c_1(\alpha) = 0$]; the wave function $\varphi_\alpha(x)$ then increases in modulus as it traverses the successive potential barriers, and its behavior recalls that of a superposition of real exponentials.

### b. BRAGG REFLECTION; POSSIBLE ENERGIES FOR A PARTICLE IN A PERIODIC POTENTIAL

Depending on whether $\varphi_\alpha(x)$ behaves like a superposition of real or imaginary exponentials, the resulting phenomena can reasonably be expected to be very different.

Let us evaluate, for example, the transmission coefficient $T_N(\alpha)$ of the set of $N$ identical barriers. For these $N$ barriers, relation (15) shows that the matrix $M(\alpha)\ [Q(\alpha)]^{N-1}$ plays a role analogous to the one played by $M(\alpha)$ for a single barrier. Now, according to relation (29-b) of complement $N_{III}$, the transmission coefficient $T(\alpha)$ is expressed in terms of the element of this matrix which is placed in the first row and the first column [the inverse of $T_N(\alpha)$ is equal to the square of the modulus of this element]. What happens if the energy $E$ of the particle is chosen so as to make the eigenvalues of $Q(\alpha)$ real, that is, given by (30)? When $N$ becomes sufficiently large, the eigenvalue $\lambda_1 = \varepsilon\, e^{\rho(\alpha)l}$ becomes dominant, and the matrix $[Q(\alpha)]^{N-1}$ increases exponentially with $N$ [as can also be seen from relation (33)]. Consequently, the transmission coefficient decreases exponentially:

$$T_N(\alpha) \propto e^{-2N\rho(\alpha)l} \tag{34}$$

In this case, for large values of $N$, the set of $N$ potential barriers reflects the particle practically without fail. This is explained by the fact that the waves scattered by the different potential barriers interfere totally destructively for the transmitted wave, and constructively for the reflected wave. This phenomenon can therefore be likened to *Bragg reflection*. Note, moreover, that this destructive interference for the transmitted wave can be produced even if the energy $E$ is greater than the height of the barrier (a case where, in classical mechanics, the particle is transmitted).

Nevertheless, if the transmission coefficient of an isolated barrier is very close to 1, we have $|F(\alpha)| \simeq 1$ [for example, in figure 2, $|F(\alpha)| \longrightarrow 1$ if $\alpha$, that is, the energy, approaches infinity]. The point representing the complex number $e^{i\alpha l}F(\alpha)$ is then very close to the circle of unit radius centered at $O$. Figure 2 shows that the regions of the energy axis where $|X(\alpha)| > 1$, that is, where total reflection occurs, are very narrow and can practically be seen as isolated energy values. Physically, this is explained by the fact if the energy $E$ of the incident particle is much larger than the amplitude of variation of the potential $V(x)$, its momentum is well-defined, as is the associated wavelength. The Bragg condition $l = n\frac{\lambda}{2}$ (where $n$ is an integer) then gives well-defined energy values.

If, on the other hand, the energy $E$ of the particle falls in a domain where the eigenvalues are of modulus 1 as in (26), the elements of the matrix $[Q(\alpha)]^{N-1}$ no longer approach infinity when $N$ does. Under these conditions, the transmission coefficient $T_N(\alpha)$ does not approach zero when the number of barriers is increased. We are again dealing with a purely quantum mechanical phenomenon, related to the wavelike nature of the wave function, which enables it to propagate in the regular periodic potential structure without being exponentially attenuated. Note especially that the transmission coefficient $T_N(\alpha)$ is very different from the product of the individual transmission coefficients of the barriers taken separately (this product approaches zero when $N \longrightarrow \infty$ since all the factors are smaller than 1).

Another interesting problem, encountered particularly in solid state physics, is that of the quantization of energy levels for a particle placed in a series of identical and evenly spaced potential wells, that is, placed in a potential $V(x)$ having a periodic structure. This problem will be studied in detail in § 3; however, we can already guess the form of the spectrum of possible energies. If we assume that the energy of the particle is such that $|X(\alpha)| > 1$, equation (33) shows that the coefficients $A_n$ and $A'_n$ become infinite when $n \longrightarrow \infty$. It is clear that this possibility must be rejected, since it means that the wave function does not remain bounded. The corresponding energies are therefore forbidden; hence, the name of *forbidden bands* given to the energy domains for which $|X(\alpha)| > 1$. On the other hand, if the energy of the particle is such that $|X(\alpha)| < 1$, $A_n$ and $A'_n$ remain bounded when $n \longrightarrow \infty$; the corresponding regions of the energy axis are called *allowed bands*. To sum up, the energy spectrum is composed of finite intervals inside which all the energies are acceptable, separated by regions all of whose energies are forbidden.

## 3. Quantization of energy levels in a periodic potential; effect of boundary conditions

Consider a particle of mass $m$ placed in the potential $V(x)$ shown in figure 3. In the region $-\frac{l}{2} \leqslant x \leqslant Nl + \frac{l}{2}$, $V(x)$ has the form of a periodic function, composed of a series of $N + 1$ successive barriers of height $V_0$, centered at $x = 0, l, 2l, ..., Nl$. Outside this region, $V(x)$ undergoes arbitrary variations over distances comparable to $l$, then becomes equal to a positive constant value $V_e$. In what follows, the region $[0, Nl]$ will be called "inside the lattice" and the limiting regions $x \simeq -\frac{l}{2}$ and $x \simeq Nl + \frac{l}{2}$, "ends (or edges) of the lattice".

Physically, such a function $V(x)$ can represent the potential seen by an electron in a linear molecule or in a crystal (in a one-dimensional model). The potential wells situated at $x = \frac{l}{2}, \frac{3l}{2}, ...$ then correspond to the attraction of the electron by the various ions. Far from the crystal (or the molecule), the electron is not subject to any attractive forces, which is why $V(x)$ rapidly becomes constant outside the region $-\frac{l}{2} \leqslant x \leqslant Nl + \frac{l}{2}$.

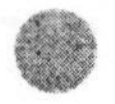



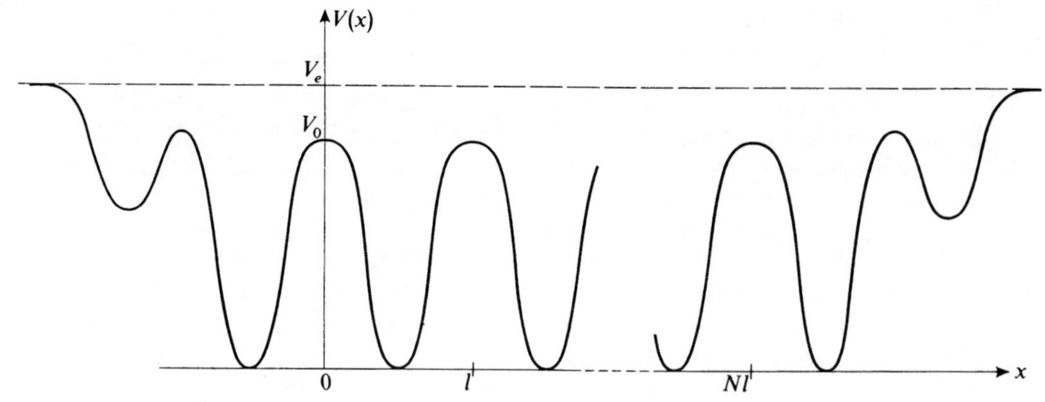


FIGURE 3

**Variation with respect to $x$ of the potential seen by an electron in a "one-dimensional crystal" and on its edges. Inside the crystal, the potential has a periodic structure; $V(x)$ is maximum between the ions (barriers at $x = 0, l, 2l, ...$) and minimum at the positions of the ions (wells at $x = l/2, 3l/2, ...$). On the edges of the crystal, $V(x)$ varies in a more or less complicated way over a distance comparable to $l$, then rapidly approaches a constant value $V_e$.**

The potential $V(x)$ that we have chosen fits perfectly into the framework of complement $M_{III}$ (apart from a change in the energy origin). We already know, therefore, that the bound states of the particle form a discrete spectrum of energies, all less than $V_e$. However, the potential $V(x)$ picked here also presents the remarkable peculiarity of having a periodic structure of the type of those considered in § 1 above; relying on the results of this section, we shall show that the conclusions of complement $M_{III}$ take on a special form in this case. For example, we stressed the fact in complement $M_{III}$ that it is the boundary conditions [$\varphi(x) \longrightarrow 0$ when $x \longrightarrow \pm \infty$] which introduce the quantization of the energy levels. The boundary conditions of the problem we are studying here, that is, the variation of the potential at the edges of the lattice, might thus be expected to play a critical role in determining the possible energies. Actually, this is not at all the case: we shall see that these energies depend practically only on the values of $V(x)$ in the region where it is periodic, and not on the edge effects (on condition, of course, that the number of potential wells is sufficiently large). In addition, we shall verify the result obtained intuitively in § 2-b, showing that most of the possible energies are grouped in allowed energy bands. Only a few stationary states, localized near the edges, depend on a critical manner on the variation of $V(x)$ in this region and can have an energy which falls in a forbidden band.

We shall therefore proceed essentially as in complement $M_{III}$, first examining precisely the conditions imposed on the wave function $\varphi_\alpha(x)$ of a stationary state.

### a. CONDITIONS IMPOSED ON THE WAVE FUNCTION

In the region where $V(x)$ is periodic, relation (3-d) gives the form of the wave function $\varphi_\alpha(x)$; the coefficients $A_n$ and $A_n'$ are determined from (32). To write (32) more explicitly, let us set

$$c_1(\alpha)A_1(\alpha) = \begin{pmatrix} f_1(\alpha) \\ f'_1(\alpha) \end{pmatrix}$$
$$c_2(\alpha)A_2(\alpha) = \begin{pmatrix} f_2(\alpha) \\ f'_2(\alpha) \end{pmatrix} \tag{35}$$

We then obtain:

$$\begin{aligned} A_n &= f_1(\alpha)\,\lambda_1^{n-1} + f_2(\alpha)\,\lambda_2^{n-1} \\ A'_n &= f'_1(\alpha)\,\lambda_1^{n-1} + f'_2(\alpha)\,\lambda_2^{n-1} \end{aligned} \tag{36}$$

Now let us examine the boundary conditions on the wave function $\varphi_\alpha(x)$. First of all, to the left, far from the lattice, $V(x)$ is equal to $V_e$ and $\varphi_\alpha(x)$ is written in the form:

$$\varphi_\alpha(x) = B\,e^{\mu(\alpha)x} \tag{37-a}$$

with:

$$\mu(\alpha) = \sqrt{\frac{2m}{\hbar^2}(V_e - E)} \tag{37-b}$$

(we eliminate the solution in $e^{-\mu(\alpha)x}$, which diverges when $x \longrightarrow -\infty$). The probability current associated with the function (37) is zero (*cf.* complement $B_{III}$, § 1). Now, for a stationary state, this current is independent of $x$ [*cf.* complement $N_{III}$, relation (12)]; it therefore remains zero at all $x$, even inside the lattice. According to relation (14) of complement $N_{III}$, the coefficients $A_n$ and $A'_n$ therefore necessarily have the same modulus. Thus, if we choose to express the boundary conditions on the left as relations between the coefficients $A_1$ and $A'_1$ [that is, by writing that the expression for $\varphi_\alpha(x)$ for $-\frac{l}{2} \leqslant x \leqslant \frac{l}{2}$ is the extension of the wave function (37)], we find a relation of the form:

$$\frac{A_1}{A'_1} = e^{i\chi(\alpha)} \tag{38-a}$$

$\chi(\alpha)$ is a real function of $\alpha$ (and therefore of the energy $E$) which depends on the precise behavior of $V(x)$ at the left-hand edge of the lattice [in what follows, we shall not need the exact expression for this function $\chi(\alpha)$; the essential point is that the boundary conditions on the left have the form (38-a)].

The same type of reasoning can obviously be applied on the right ($x \longrightarrow +\infty$), where the boundary conditions are written:

$$\frac{A_{N+1}}{A'_{N+1}} = e^{i\chi'(\alpha)} \tag{38-b}$$

where the real function $\chi'(\alpha)$ depends on the behavior of $V(x)$ on the right-hand edge of the lattice.

To sum up, we can say that the quantization of the energy levels can be obtained in the following manner:

– we start with two coefficients $A_1$ and $A'_1$ which satisfy (38-a); this insures that the function $\varphi_\alpha(x)$ will remain bounded when $x \longrightarrow -\infty$. Since $\varphi_\alpha(x)$ is defined to within a constant factor, we can choose, for example:

$$\begin{aligned} A_1 &= e^{i\chi(\alpha)/2} \\ A'_1 &= e^{-i\chi(\alpha)/2} \end{aligned} \tag{39}$$

– we then calculate, using (36), the coefficients $A_n$ and $A'_n$ so as to extend the wave function chosen throughout all the crystal. Note that the condition (39) implies that $\varphi_\alpha(x)$ is real (*cf.* complement $N_{III}$, § 1-b); calculation of $A_n$ and $A'_n$ must therefore yield:

$$A'_n = A_n^* \tag{40}$$

– finally, we write that the coefficients $A_{N+1}$ and $A'_{N+1}$ satisfy (38-b), a relation which insures that $\varphi_\alpha(x)$ will remain bounded when $x \longrightarrow +\infty$. In fact, relation (40) shows that the ratio $A_{N+1}/A'_{N+1}$ is automatically a complex number of unit modulus; condition (38-b) therefore amounts to an equality between the phases of two complex numbers. We thus obtain a real equation in $\alpha$, which has a certain number of real solutions giving the allowed energies.

We are going to apply this method, distinguishing between two cases: real eigenvalues of $Q(\alpha)$ [the case where $|X(\alpha)| > 1$] and imaginary ones [the case where $|X(\alpha)| < 1$].

## b. ALLOWED ENERGY BANDS: STATIONARY STATES OF THE PARTICLE INSIDE THE LATTICE

First assume that the energy $E$ is in a domain where $|X(\alpha)| < 1$.

### α. *Form of the quantization equation*

Taking (26) into account, relations (36) become:

$$\begin{cases} A_n = f_1(\alpha)\,\mathrm{e}^{i(n-1)k(\alpha)l} + f_2(\alpha)\,\mathrm{e}^{-i(n-1)k(\alpha)l} \\ A'_n = f'_1(\alpha)\,\mathrm{e}^{i(n-1)k(\alpha)l} + f'_2(\alpha)\,\mathrm{e}^{-i(n-1)k(\alpha)l} \end{cases} \tag{41}$$

Also, we have seen that the choice (39) of $A_1$ and $A'_1$ implies that $A'_n = A_n^*$ for all $n$. Now, it is easy to show that relations (41) yield two complex conjugate numbers only if:

$$\begin{aligned} f_1^*(\alpha) &= f'_2(\alpha) \\ f_2^*(\alpha) &= f'_1(\alpha) \end{aligned} \tag{42}$$

Condition (38-b) can then be written:

$$\frac{f_1(\alpha)\,\mathrm{e}^{2iNk(\alpha)l} + f_2(\alpha)}{f_2^*(\alpha)\,\mathrm{e}^{2iNk(\alpha)l} + f_1^*(\alpha)} = \mathrm{e}^{i\chi'(\alpha)} \tag{43}$$

This equation in $\alpha$ is the one which gives the quantization of the energy levels. To solve it, let us set:

$$\Theta(\alpha) = \mathrm{Arg}\left\{\frac{f_1^*(\alpha)\,\mathrm{e}^{i\chi'(\alpha)/2} - f_2(\alpha)\,\mathrm{e}^{-i\chi'(\alpha)/2}}{f_1(\alpha)\,\mathrm{e}^{-i\chi'(\alpha)/2} - f_2^*(\alpha)\,\mathrm{e}^{i\chi'(\alpha)/2}}\right\} \tag{44}$$

[$\Theta(\alpha)$ can, in principle, be calculated from $\chi(\alpha)$, $\chi'(\alpha)$ and the matrix $Q(\alpha)$]. Equation (43) can then be written simply:

$$\mathrm{e}^{2iNk(\alpha)l} = \mathrm{e}^{i\Theta(\alpha)} \tag{45}$$

The energy levels are therefore given by:

$$k(\alpha) = \frac{\Theta(\alpha)}{2Nl} + p\frac{\pi}{Nl} \tag{46}$$

with:

$$p = 0, 1, 2, ..., (N-1) \tag{47}$$

[the other values of $p$ must be excluded since condition (25) here forces $k(\alpha)$ to vary within an interval of width $\pi/l$]. We can already see that if $N$ is very large, we can write equation (46) in the simplified form:

$$k(\alpha) \simeq p\frac{\pi}{Nl} \tag{48}$$

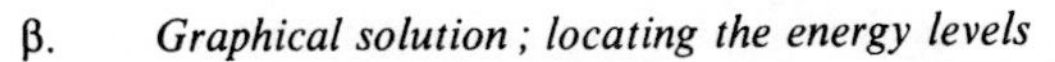

β. *Graphical solution ; locating the energy levels*

If we substitute definition (24) of $k(\alpha)$ into (46), we obtain an equation in $\alpha$ which gives the allowed energies. To solve it graphically, let us begin by tracing the curve which represents the function $X(\alpha) = \text{Re}\left[e^{i\alpha l}F(\alpha)\right]$. Because of the imaginary exponential $e^{i\alpha l}$, we expect this curve to have an oscillatory behavior, of the type of that shown in figure 4-a. Since $|F(\alpha)|$ is greater than 1 [*cf.* complement $N_{III}$, relation (32)], the amplitude of the oscillation is greater than 1, so the curve intersects the two straight lines $X(\alpha) = \pm 1$ at certain values $\alpha_0, \alpha_1, \alpha_2, ...$ of the variable $\alpha$. We then eliminate all regions of the $\alpha$-axis, bounded by these values, where the condition $|X(\alpha)| < 1$ is not satisfied. Using the set of arcs of curves thus obtained for $X(\alpha)$, we must represent the function:

$$k(\alpha) = \frac{1}{l}\,\text{Arc cos}\,X(\alpha) \tag{49}$$

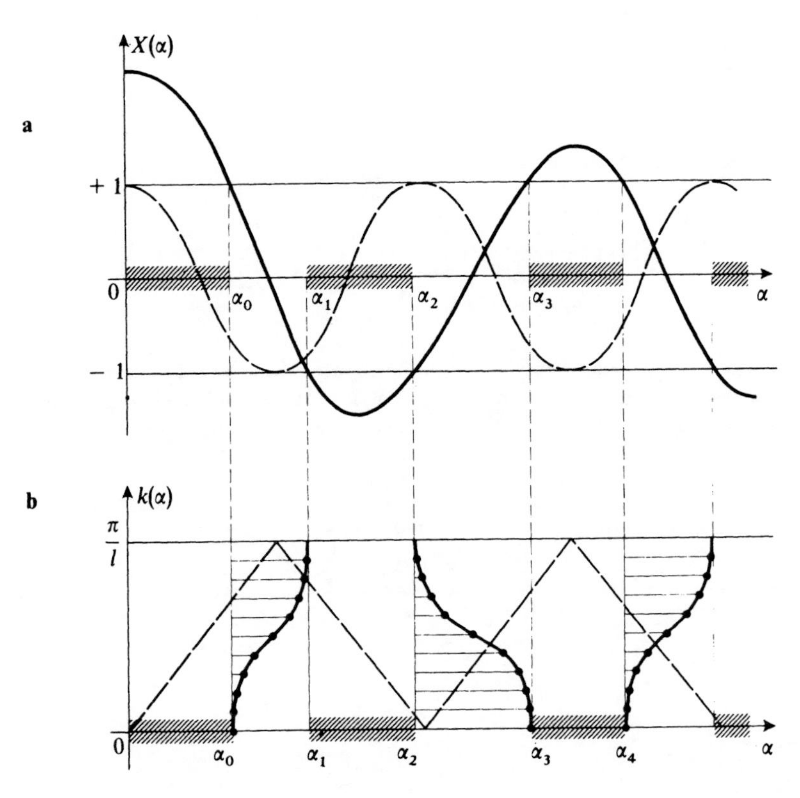


FIGURE 4

**Variation with respect to $\alpha$ of $X(\alpha) = \text{Re}[F(\alpha)\,e^{i\alpha l}]$ (see fig. 2) and of $k(\alpha) = \frac{1}{l}\,\text{Arc cos}\,[X(\alpha)]$. The values of $\alpha$ (that is, of the energy $E$) associated with stationary states are obtained (if $N \gg 1$) by cutting the curve which represents $k(\alpha)$ with the horizontal lines whose equations are $y = p\pi/Nl$ ($p = 0, 1, 2, ... N - 1$). The allowed bands are thus revealed. Each includes $N$ levels which are very close together (intervals $\alpha_0 \leqslant \alpha \leqslant \alpha_1$, etc.). The forbidden bands are represented by the shaded areas ($\alpha_1 < \alpha < \alpha_2$, etc...).**

**The dashed-line curves correspond to the special case where $V(x) = 0$ (a free particle).**

Taking into account the form of the Arc cosine function (*cf.* fig. 5), we are led to the curve whose shape is shown in figure 4-b. Equation (46) indicates that the energy levels correspond to the intersections of this curve with those which represent the functions $\frac{\Theta(\alpha)}{2Nl} + p\frac{\pi}{Nl}$, that is, if $N \gg 1$, with the horizontal lines whose equations are $y = p\frac{\pi}{Nl}$ (with $p = 0, 1, 2, ..., N-1$).

We thus obtain groups of $N$ levels, associated with equidistant values of $k(\alpha)$ and situated in the allowed bands defined by $\alpha_0 \leqslant \alpha \leqslant \alpha_1$, $\alpha_2 \leqslant \alpha \leqslant \alpha_3$, etc. Between these allowed bands are the forbidden bands (we shall examine their properties in § c).

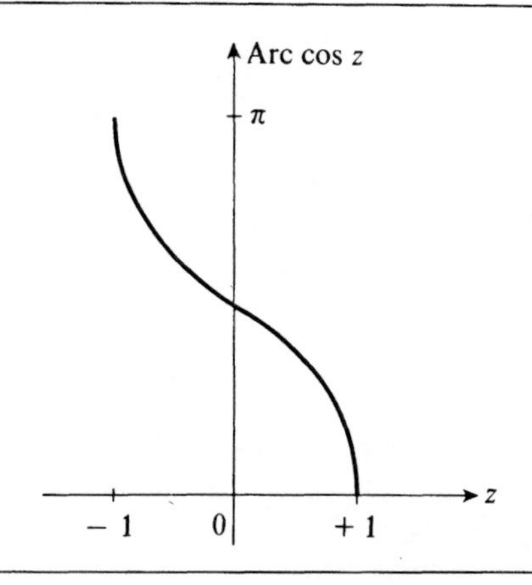


FIGURE 5

**The Arc cosine function.**

If we consider a particular allowed band, we can locate each level according to the value of $k(\alpha)$ which corresponds to it. This leads to choosing $k$ as the variable and considering $\alpha$ and, consequently, $E$ as functions $\alpha(k)$ and $E(k)$ of $k$. The variation of $\alpha$ with respect to $k$ is given directly by the curve of figure 4-b, so it suffices to evaluate the function $\frac{\hbar^2\alpha^2}{2m}$ to obtain the energy $E(k)$. The corresponding curve has the shape shown in figure 6.

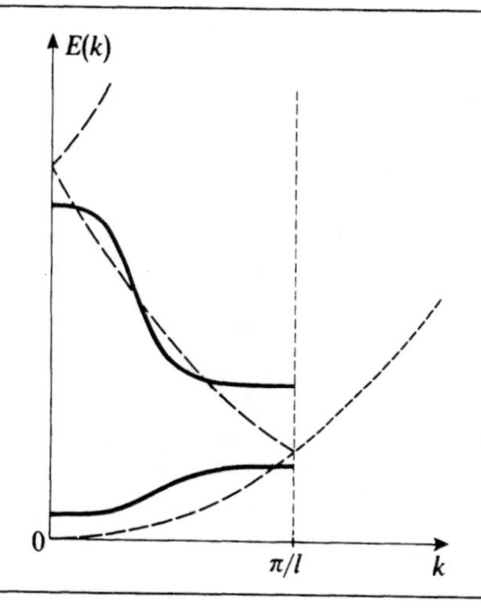


FIGURE 6

**Variation of the energy with respect to the parameter $k$. The solid lines correspond to the energies for the first two allowed bands (the values of $k$ which give the energy levels being equidistant inside the interval $0 \leqslant k \leqslant \pi/l$). The dashed lines correspond to the special case where the potential $V(x)$ is zero (a free particle); the allowed bands are contiguous, and there are no forbidden bands.**

COMMENT:

It is clear from figure 4-b that, to a given value of $k$, correspond several values of $\alpha$ and therefore of the energy; this is why several arcs appear in figure 6. Nevertheless, if, within a given allowed band, $X(\alpha)$ increases steadily from $-1$ to $+1$ (or decreases steadily from $+1$ to $-1$), only one energy level corresponds to each value of $k$ for this band, and *this band includes N energy levels.*

γ. *Discussion*

The preceding calculations show how, when we go from $N = 1$ to very high values of $N$, we move gradually from a set of discrete levels to allowed energy bands. Rigorously, these bands are formed by discrete levels, but their separation is so small for a macroscopic lattice that they practically constitute a continuum. When $k$ is taken as a parameter, the *density of states* (the number of possible energies per unit interval of $k$) *is constant* and equal to $Nl/\pi$. This property, which is very useful, explains why $k$ is generally chosen as the variable.

An important point appears in going from (46) to (48): when $N$ is large, the edge effects of the lattice, which enter only through the intermediary of the functions $\chi(\alpha)$, $\chi'(\alpha)$ and, in (46), $\Theta(\alpha)$, no longer play any role; only the form of the periodic potential inside the lattice is important in determining the possible energies.

It is interesting to consider the two following limiting cases:

(*i*) If $V(x) = 0$ (free particle), we have:

$$\begin{cases} F(\alpha) = 1 \\ X(\alpha) = \cos \alpha l \end{cases} \tag{50}$$

and we obtain:

$$\begin{aligned} &\text{if } 0 \leqslant \alpha \leqslant \frac{\pi}{l}: \quad k(\alpha) = \alpha \\ &\text{if } \frac{\pi}{l} \leqslant \alpha \leqslant \frac{2\pi}{l}: \quad k(\alpha) = \frac{2\pi}{l} - \alpha \\ &\text{etc...} \end{aligned} \tag{51}$$

(the corresponding broken line is shown in figure 4-b as a dashed line). Relation (50) shows that the condition $|X(\alpha)| \leqslant 1$ is always satisfied : as we know, forbidden bands do not exist for a free particle.

Figure 6 thus enables us to see the effect of the potential $V(x)$ on the curve $E(k)$. When forbidden bands appear, the curves representing the energy become deformed so as to have horizontal tangents for $k = 0$ and $k = \pi/l$ (edges of the band). Unlike what happens for a free particle, there exists a point of inflection for each band where the energy varies linearly with $k$.

(*ii*) If the transmission coefficient $T(\alpha)$ is practically zero, we have [*cf.* complement $N_{III}$, equations (29) and (21)]:

$$\begin{cases} |F(\alpha)| \gg 1 \\ |G(\alpha)| \gg 1 \end{cases} \tag{52}$$

In figure 2, the point representing the complex number $e^{i\alpha l}F(\alpha)$ is very far from the origin. We thus see in this figure that the regions of the $\alpha$-axis where $|X(\alpha)| < 1$ are extremely narrow. The allowed bands therefore shrink if the transmission coefficient of the elementary barrier decreases; in the limit of zero transmission, they reduce to individual levels in an isolated well. Inversely, as soon as the tunnel effect allows the particle to pass from one well to the next one, *each of the discrete levels of the well gives rise to an energy band, whose width increases as the transmission coefficient grows.* We shall return to this property in complement $F_{XI}$.

### c. FORBIDDEN BANDS: STATIONARY STATES LOCALIZED ON THE EDGES

#### α. *Form of the equations; energy levels*

Let us now assume that $\alpha$ belongs to a domain where $|X(\alpha)| > 1$. According to (30), relations (36) can then be written:

$$\begin{cases} A_n = \varepsilon^{n-1}[f_1(\alpha)\, e^{(n-1)\rho(\alpha)l} + f_2(\alpha)\, e^{-(n-1)\rho(\alpha)l}] \\ A'_n = \varepsilon^{n-1}[f'_1(\alpha)\, e^{(n-1)\rho(\alpha)l} + f'_2(\alpha)\, e^{-(n-1)\rho(\alpha)l}] \end{cases} \tag{53}$$

The fact that $A'_n = A^*_n$ for all $n$ means that we must have here:

$$\begin{cases} f'_1(\alpha) = f^*_1(\alpha) \\ f'_2(\alpha) = f^*_2(\alpha) \end{cases} \tag{54}$$

The quantization condition (38-b) then takes on the form:

$$\frac{A_{N+1}}{A'_{N+1}} = \frac{f_1(\alpha) + f_2(\alpha)\, e^{-2N\rho(\alpha)l}}{f^*_1(\alpha) + f^*_2(\alpha)\, e^{-2N\rho(\alpha)l}} = e^{i\chi'(\alpha)} \tag{55}$$

that is:

$$e^{-2N\rho(\alpha)l} = L(\alpha) \tag{56}$$

where the real function $L(\alpha)$ is defined by:

$$L(\alpha) = -\frac{f^*_1(\alpha)\, e^{i\chi'(\alpha)/2} - f_1(\alpha)\, e^{-i\chi'(\alpha)/2}}{f^*_2(\alpha)\, e^{i\chi'(\alpha)/2} - f_2(\alpha)\, e^{-i\chi'(\alpha)/2}} \tag{57}$$

Consider the case where $N \gg 1$; we then have $e^{-2N\rho(\alpha)l} \simeq 0$, and equation (56) reduces to:

$$L(\alpha) = 0 \tag{58}$$

The energy levels situated in the forbidden bands are therefore given by the zeros of the function $L(\alpha)$ (*cf.* fig. 7). $N$ enters neither into (57) nor into (58), so the number of these levels does not depend on $N$ (unlike the number of levels situated in an allowed band). Consequently, when $N \gg 1$, it can be said that practically all the levels are grouped in the allowed bands.

#### β. *Discussion*

The situation here is radically different from the one encountered in § b: the number $N$, that is, the length of the lattice, plays no role (provided, nevertheless, that it is sufficiently large); on the other hand, definition (57) of $L(\alpha)$ shows that the functions $\chi(\alpha)$ and $\chi'(\alpha)$ play an essential role in the problem. Since we already know that these functions depend on the behavior of $V(x)$ on the edges of the lattice, we expect to obtain states localized in these regions.

This is indeed the case. Equations (57) and (58) offer two possibilities:

($i$) if $f_1(\alpha) \neq 0$, the fact that $L(\alpha) = 0$ requires that:

$$\frac{f_1(\alpha)}{f^*_1(\alpha)} = \frac{f_1(\alpha)}{f'_1(\alpha)} = e^{i\chi'(\alpha)} \tag{59}$$

Let us return to definition (35) of $f_1(\alpha)$ and $f'_1(\alpha)$; we see that relation (59) shows that the wave function constructed from the first eigenvector of $Q(\alpha)$ satisfies the boundary conditions on the right. This is easy to understand: if we start at $x = 0$ with an arbitrary wave function which

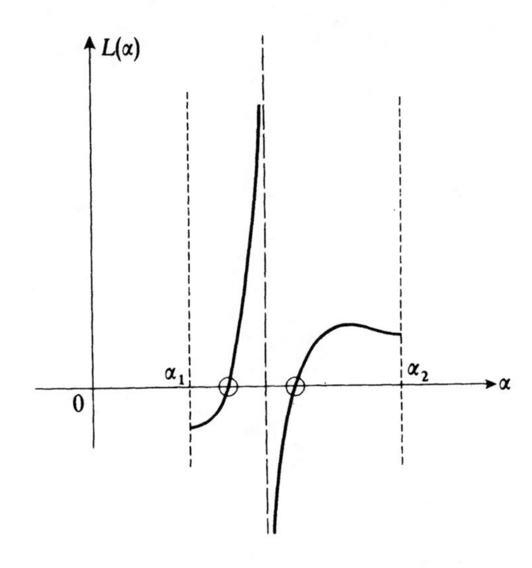


FIGURE 7

**Variation of $L(\alpha)$ with respect to $\alpha$ in a forbidden band. The zeros of $L(\alpha)$ give the stationary states which are localized on the edges of the lattice.**

satisfies the boundary conditions on the left, the matrix $\begin{pmatrix} A_1 \\ A'_1 \end{pmatrix}$ has components on the two eigenvectors of $Q(\alpha)$; the coefficients $A_{N+1}$ and $A'_{N+1}$ are then (if $N \gg 1$) essentially given by (33), which expresses the fact that the matrix $\begin{pmatrix} A_{N+1} \\ A'_{N+1} \end{pmatrix}$ is proportional to the column matrix of the first eigenvector of $Q(\alpha)$.

Note that since the eigenvalue $\lambda_1(\alpha)$ is greater than 1, the wave function grows exponentially when $x$ increases. The stationary state given by the first eigenvector of $Q(\alpha)$ is therefore localized at the right end of the lattice.

(*ii*) if $f_1(\alpha) = 0$, (54) gives $f'_1(\alpha) = 0$, and definitions (35) imply that $c_1(\alpha) = 0$: the corresponding stationary state is associated with the second eigenvector of $Q(\alpha)$. Aside from the fact that this state is localized at the left end of the lattice, the conclusions obtained in (*i*) remain valid.

**References and suggestions for further reading :**

Merzbacher (1.16), chap. 6, §7; Flügge (1.24), §§28 and 29; Landau and Lifshitz (1.19), §104; see also solid state physics texts (section 13 of the bibliography).

CHAPTER IV

# Application of the postulates to simple cases: spin $\frac{1}{2}$ and two-level systems

## OUTLINE OF CHAPTER IV

In this chapter, we intend to illustrate the postulates of quantum mechanics, which we stated and discussed in chapter III. We shall apply them to simple concrete cases, in which the dimension of the state space is finite (equal to two). The interest of these examples is not confined to their mathematical simplicity, which will allow a better understanding of the postulates and their consequences. It is also based on their physical importance: they exhibit typically quantum mechanical behavior which can be verified experimentally.

In §§A and B, we shall study the spin 1/2 case (which we shall take up again in more detail in chapter IX). First, we shall describe (§A-1) a fundamental experiment which revealed the quantization of a simple physical quantity, the angular momentum. We shall see that the component along *Oz* of the angular momentum (or magnetic moment) of a neutral paramagnetic atom can take on only certain values, which belong to a *discrete* set. Thus, for a silver atom in its ground state, there are only two possible values ($+\ \hbar/2$ and $-\ \hbar/2$) for the component $S_z$ of its angular momentum: a silver atom in the ground state is said to be a *spin* 1/2 *particle*. In §A-2, we indicate how quantum mechanics describes the "spin variables" of such a particle. In situations where one can dispense with a quantum treatment of the "external variables" **r** and **p**, the state of the particle ("spin state space") has only two dimensions. We shall then (§B) be able to illustrate and discuss the quantum mechanical postulates in this particularly simple case:we shall first see how to prepare silver atoms in any desired arbitrary spin state, in a real experiment. We shall then show how the measurement of the physical values of the spin on such silver atoms enables us to verify the quantum mechanical postulates experimentally. By integrating the corresponding Schrödinger equation, we shall study the evolution of a spin 1/2 particle in a uniform magnetic field (Larmor precession). Finally, in §C, we shall begin the study of *two-level systems*. Although these systems are not generally spin 1/2 particles, their study leads to calculations which are very similar to those developed in §§A and B. We shall treat in detail the effect of an external perturbation on the stationary states of a two-level system and use this very simple model to point out important physical effects.

## A. SPIN 1/2 PARTICLE: QUANTIZATION OF THE ANGULAR MOMENTUM

### 1. Experimental demonstration

First of all, we are going to describe and analyze the Stern-Gerlach experiment, which demonstrated the quantization of the components of an angular momentum (sometimes called "space quantization").

### a. THE STERN-GERLACH APPARATUS

The experiment consists of studying the deflection of a beam of neutral paramagnetic atoms (in this case, silver atoms) in a highly inhomogeneous magnetic field. The apparatus used is shown schematically in figure 1 ★.

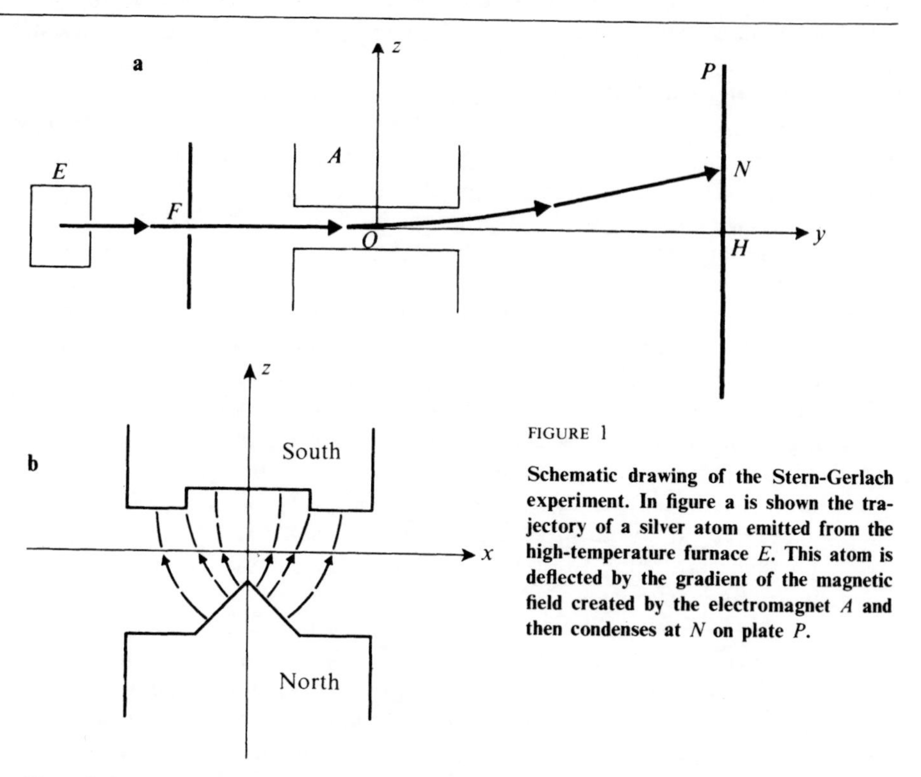


FIGURE 1

**Schematic drawing of the Stern-Gerlach experiment. In figure a is shown the trajectory of a silver atom emitted from the high-temperature furnace $E$. This atom is deflected by the gradient of the magnetic field created by the electromagnet $A$ and then condenses at $N$ on plate $P$.**

**Figure b shows a cross section in the $xOz$ plane of the electromagnet $A$; the lines of force of the magnetic field are shown in dashed lines. $B_z$ has been assumed to be positive and $\partial B_z/\partial z$, negative. Consequently, the trajectory of figure a corresponds to a negative component $\mathcal{M}_z$ of the magnetic moment, that is, to a positive component of $\mathcal{S}_z$ ($\gamma$ is negative for a silver atom).**

Silver atoms contained in a furnace $E$, which is heated to a high temperature, leave through a small opening and propagate in a straight line in the high vacuum which exists inside the whole apparatus. A collimating slit $F$ selects those atoms whose velocity is parallel to a particular direction that we shall choose for the $Oy$ axis. The atomic beam thus constructed traverses the gap of an electromagnet $A$ before condensing on a plate $P$.

Let us describe the characteristics of the magnetic field **B** produced by the electromagnet $A$. This magnetic field has a plane of symmetry (which we shall

★ We shall indicate here only the most important characteristics of this equipment. A more detailed description of the experimental technique can be found in a book on atomic physics.

designate by $yOz$) which contains the initial direction $Oy$ of the atomic beam. In the air-gap, it is the same at all points situated on any given line parallel to $Oy$ (the edges of the electromagnet are parallel to $Oy$, and we neglect edge effects). **B** has no component along $Oy$. Its largest component is along $Oz$; it varies strongly with $z$: in figure 1-b, the field lines are much closer together close to the north pole than close to the south pole of the magnet. Of course, since the magnetic field has a conserved flux (div $\mathbf{B} = 0$), it must also have a component along $Ox$ which varies with the distance $x$ from the plane of symmetry.

#### b. CLASSICAL CALCULATION OF THE DEFLECTION*

Note, first, that the silver atoms, being neutral, are not subject to the Lorentz force. On the other hand, they possess a permanent magnetic moment $\mathcal{M}$ (they are paramagnetic atoms); the resulting forces are derived from the potential energy:

$$W = - \mathcal{M} . \mathbf{B} \tag{A-1}$$

The existence, for an atom, of an electronic magnetic moment $\mathcal{M}$ and an angular momentum $\mathcal{S}$ is due to two causes: the motion of the electrons about the nucleus (the corresponding rotation of the charges being responsible for the appearance of an orbital magnetic moment) and the intrinsic angular momentum, or spin, (*cf.* chapter IX) of the electrons, with which is associated a spin magnetic moment. It can be shown (as we shall assume here without proof) that, for a given atomic level, $\mathcal{M}$ and $\mathcal{S}$ are proportional**:

$$\mathcal{M} = \gamma \mathcal{S} \tag{A-2}$$

The proportionality constant $\gamma$ is called the *gyromagnetic ratio* of the level under consideration.

Before the atoms traverse the electromagnet, the magnetic moments of the silver atoms which constitute the atomic beam are oriented randomly (isotropically). Let us study the action of the magnetic field on one of these atoms, whose magnetic moment $\mathcal{M}$ has a given direction at the entrance of the air-gap. From expression (A-1) for the potential energy, it is easy to deduce that the resultant of the forces exerted on the atom is:

$$\mathbf{F} = \nabla (\mathcal{M} . \mathbf{B}) \tag{A-3}$$

(this resultant would be equal to zero if the field **B** were uniform), and that their total moment relative to the position of the atom is:

$$\mathbf{\Gamma} = \mathcal{M} \times \mathbf{B} \tag{A-4}$$

* We shall confine ourselves here to outlining this calculation; for more details, we refer the reader to a book on atomic physics.

** In the case of silver atoms in the ground state (like those of the beam), the angular momentum $\mathcal{S}$ is simply equal to the spin of the outer electron, which is therefore solely responsible for the existence of the magnetic moment $\mathcal{M}$. This is because the outer electron has a zero orbital angular momentum, and the resultant orbital and spin angular momenta of the inner electrons are also zero. Moreover the experimental conditions realized in practice are such that effects linked to the spin of the nucleus are negligible. This is why the silver atom in the ground state, like the electron, has a spin 1/2.

The angular momentum theorem can be written:

$$\frac{d\mathscr{S}}{dt} = \mathbf{\Gamma} \tag{A-5}$$

that is:

$$\frac{d\mathscr{S}}{dt} = \gamma \, \mathscr{S} \times \mathbf{B} \tag{A-6}$$

The atom thus behaves like a gyroscope (fig. 2): $d\mathscr{S}/dt$ is perpendicular to $\mathscr{S}$, and the angular momentum turns about the magnetic field, the angle $\theta$ between $\mathscr{S}$ and $\mathbf{B}$ remaining constant. The rotational angular velocity is equal to the product of the gyromagnetic ratio $\gamma$ and the modulus of the magnetic field. The components of $\mathscr{M}$ which are perpendicular to the magnetic field therefore oscillate about zero, the component parallel to $\mathbf{B}$ remaining constant.

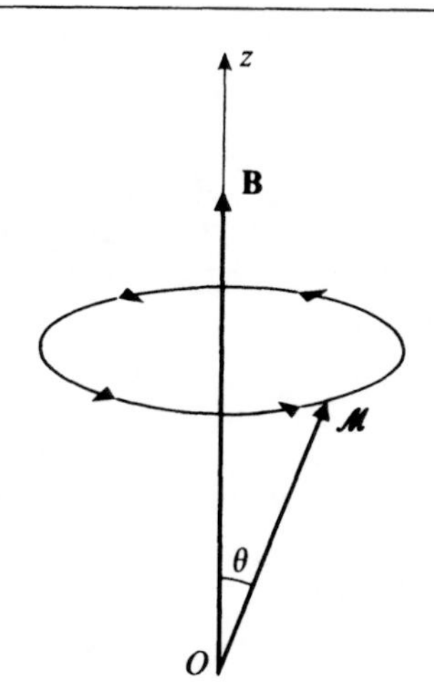


FIGURE 2

**The silver atom possesses a magnetic moment $\mathscr{M}$ and an angular momentum $\mathscr{S}$ which are proportional. Consequently, the effect of a uniform magnetic field B is to cause $\mathscr{M}$ to turn about B with a constant angular velocity (Larmor precession).**

To calculate the force $\mathbf{F}$ [formula (A-3)], we can, to a very good approximation, neglect in $W$ the terms proportional to $\mathscr{M}_x$ and $\mathscr{M}_y$ and take $\mathscr{M}_z$ to be constant. This is because the frequency of oscillation due to the rotation of $\mathscr{M}$ is so great that only the time-averaged values of $\mathscr{M}_x$ and $\mathscr{M}_y$ can play a role in $W$, and these are both zero. Consequently, it is as if the atom were submitted to the sole force:

$$\mathbf{F}' = \nabla(\mathscr{M}_z B_z) = \mathscr{M}_z \nabla B_z \tag{A-7}$$

In addition, the components of $\nabla B_z$ along $Ox$ and $Oy$ are zero: $\partial B_z/\partial y = 0$ because the magnetic field is independent of $y$ (§ a above), and $\partial B_z/\partial x = 0$ at all points of the plane of symmetry $yOz$. *The force on the atom is therefore parallel to Oz and proportional to $\mathscr{M}_z$*. Since it is this force which produces the deflection $HN$ of the atom (fig. 1), $HN$ is proportional to $\mathscr{M}_z$ (and hence, to $\mathscr{S}_z$). Consequently, measuring $HN$ is equivalent to measuring $\mathscr{M}_z$ or $\mathscr{S}_z$.

Since, at the entrance to the air-gap, the moments of the various atoms of the beam are distributed isotropically (all values of $\mathcal{M}_z$ included between $|\mathcal{M}|$ and $-|\mathcal{M}|$ are found), we expect the beam to form a single pattern, symmetrical with respect to $H$, on the plate $P$. The upper bound $N_1$ and the lower bound $N_2$ of this pattern correspond in principle to the maximum value $|\mathcal{M}|$ and minimum value $-|\mathcal{M}|$ of $\mathcal{M}_z$. In fact, the dispersion of the velocities and the finite width of the slit $F$ cause the atoms having a given value of $\mathcal{M}_z$ to condense, not at the same point, but in a spot centered about the deflection corresponding to the average velocity.

### c. RESULTS AND CONCLUSIONS

The results of the experiment (performed for the first time in 1922 by Stern and Gerlach) are in complete contradiction with the preceding predictions.

We do not observe a single spot centered at $H$, but *two spots* (fig. 3) centered at the points $N_1$ and $N_2$, symmetrical with respect to $H$ (the width of these two spots corresponds to the effect of the dispersion of the velocities and of the width of the slit $F$). The predictions of classical mechanics are therefore shown to be invalidated by the experiment.

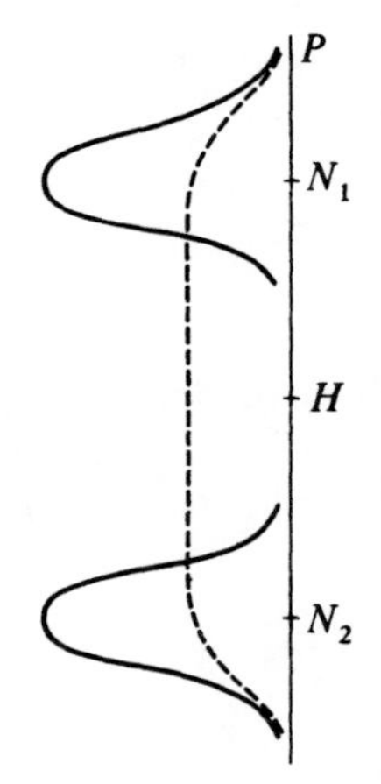


FIGURE 3

**Spots observed on the plate $P$ in the Stern-Gerlach experiment. The magnetic moments $\mathcal{M}$ of the atoms emitted from the furnace $E$ are distributed randomly in all directions of space, so classical mechanics predicts that a measurement of $\mathcal{M}_z$ can yield with equal probability all values included between $+|\mathcal{M}|$ and $-|\mathcal{M}|$. One should therefore observe only one spot (dashed lines in the figure). In reality, the result of the experiment is completely different: two spots, centered at $N_1$ and $N_2$, are observed. This means that a measurement of $\mathcal{M}_z$ can yield only two possible results (quantization of the measurement result).**

Now let us see how the preceding results can be interpreted. Of the physical quantities associated with a silver atom, some correspond to its external degrees of freedom (that is, are functions of its position $\mathbf{r}$ and its linear momentum $\mathbf{p}$), and others, to its internal degrees of freedom (also called spin degrees of freedom) $\mathcal{M}$ or $\mathcal{S}$.

Let us first show that, under these experimental conditions, it is not necessary to treat the external degrees of freedom quantum mechanically. To do this, we shall verify that it is possible, in order to describe the motion of the silver atoms, to construct wave packets whose width $\Delta z$ and momentum dispersion $\Delta p_z$ are completely negligible. $\Delta z$ and $\Delta p_z$ must satisfy the uncertainty relation:

$$\Delta z \,.\, \Delta p_z \gtrsim \hbar \qquad \text{(A-8)}$$

Numerically, the mass $M$ of a silver atom is equal to $1.8 \times 10^{-25}$ kg. $\Delta z$ and the velocity uncertainty $\Delta v_z = \Delta p_z/M$ must be such that:

$$\Delta z \,.\, \Delta v_z \gtrsim \frac{\hbar}{M} \simeq 10^{-9} \text{ M.K.S.A.} \tag{A-9}$$

Now what are the lengths and velocities involved in the problem? The width of the slit $F$ is equal to about 0.1 mm and the separation $N_1N_2$ of the two spots, several millimeters. The distance over which the magnetic field varies appreciably can be deduced from the values of the field in the middle of the air-gap ($B \simeq 10^4$ gauss) and its gradient $\left(\frac{\partial B}{\partial z} \simeq 10^5 \text{ gauss/cm}\right)$, which yields $B \Big/ \frac{\partial B}{\partial z} \simeq 1$ mm. In addition, the velocity of the silver atoms leaving a furnace at an absolute temperature of 1 000 °K is of the order of 500 m/s. However well-defined the beam is, the dispersion of the velocities along $Oz$ is not much less than several meters per second. It is then easy to find uncertainties $\Delta z$ and $\Delta v_z$, which, while satisfying (A-9), are negligible on the scale of the experiment being considered. As far as the external variables $\mathbf{r}$ and $\mathbf{p}$ of each atom are concerned, it is therefore not necessary to resort to quantum mechanics in this case. It is possible to reason in terms of quasi-pointlike wave packets moving along classical trajectories. Consequently, it is correct to claim that measurement of the deflection $HN$ constitutes a measurement of $\mathcal{M}_z$ or $\mathcal{S}_z$.

The results of the experiment thus lead us necessarily to the following conclusion: if we measure the component $\mathcal{S}_z$ of the intrinsic angular momentum of a silver atom in its ground state, we can find only one or the other of two values corresponding to the deflections $HN_1$ and $HN_2$. We are therefore obliged to reject the classical image of a vector $\boldsymbol{\mathcal{S}}$ whose angle $\theta$ with the magnetic field can take on any value: $\mathcal{S}_z$ is a quantized physical quantity whose discrete spectrum includes only two eigenvalues. When we study the quantum theory of angular momentum (chap. VI), we shall see that these eigenvalues are $+\hbar/2$ and $-\hbar/2$; we shall assume this here and say that the spin of the silver atom in its ground state is 1/2.

## 2. Theoretical description

We are now going to show how quantum mechanics describes the degrees of freedom of a silver atom, that is, of a spin 1/2 particle.

We do not yet possess all the necessary elements for the presentation of a deductive and rigorous theory of the spin 1/2 particle. Such a study will be developed in chapter IX, in the framework of the general theory of angular momentum. We shall therefore be forced here to assume without proof a small number of results which will be proved later, in chapter IX. Such a point of view is justified by the fact that the essential goal of the present chapter is to show the reader how to handle the quantum mechanical formalism in a simple and concrete case, and not to stress the angular momentum aspect of the spin 1/2. The idea is to give precise examples of kets and observables, to show how physical predictions can be extracted from them and how to distinguish clearly between the various stages of an experiment (preparation, evolution, measurement).

We saw in chapter III that with every measurable physical quantity must be associated, in quantum mechanics, an observable, that is, a Hermitian operator whose eigenvectors can form a basis in the state space. We must therefore define the state space and the observables corresponding to the components of $\mathcal{S}$ ($\mathcal{S}_x$, $\mathcal{S}_y$, $\mathcal{S}_z$ and, more generally, $\mathcal{S}_u = \mathcal{S} . \mathbf{u}$, where $\mathbf{u}$ is an arbitrary unit vector), which we know from § 1 to be measurable.

#### a. THE OBSERVABLE $S_z$ AND THE SPIN STATE SPACE

With $\mathcal{S}_z$ we must associate an observable $S_z$ which has, according to the results of the experiment described in § 1 above, two eigenvalues, $+ \hbar/2$ and $- \hbar/2$. We shall assume (see chap. IX) that these two eigenvalues are not degenerate, and we shall denote by $| + \rangle$ and $| - \rangle$ the corresponding orthonormal eigenvectors :

$$\begin{cases} S_z | + \rangle = + \dfrac{\hbar}{2} | + \rangle \\ S_z | - \rangle = - \dfrac{\hbar}{2} | - \rangle \end{cases} \tag{A-10}$$

with :

$$\begin{cases} \langle + | + \rangle = \langle - | - \rangle = 1 \\ \langle + | - \rangle = 0 \end{cases} \tag{A-11}$$

$S_z$ alone therefore forms a C.S.C.O., and the spin state space is the two-dimensional space $\mathscr{E}_S$ spanned by its eigenvectors $| + \rangle$ and $| - \rangle$. The fact that these eigenvectors constitute a basis of $\mathscr{E}_S$ is expressed by the closure relation :

$$| + \rangle \langle + | \ + \ | - \rangle \langle - | = \mathbb{1} \tag{A-12}$$

The most general (normalized) vector of $\mathscr{E}_S$ is a linear superposition of $| + \rangle$ and $| - \rangle$:

$$| \psi \rangle = \alpha | + \rangle + \beta | - \rangle \tag{A-13}$$

with:

$$|\alpha|^2 + |\beta|^2 = 1 \tag{A-14}$$

In the $\{ | + \rangle, | - \rangle \}$ basis, the matrix representing $S_z$ is obviously diagonal and is written:

$$(S_z) = \frac{\hbar}{2} \begin{pmatrix} 1 & 0 \\ 0 & -1 \end{pmatrix} \tag{A-15}$$

#### b. THE OTHER SPIN OBSERVABLES

With the $\mathcal{S}_x$ and $\mathcal{S}_y$ components of $\mathcal{S}$ will be associated the observables $S_x$ and $S_y$. The operators $S_x$ and $S_y$ must be represented in the $\{ | + \rangle, | - \rangle \}$ basis by $2 \times 2$ Hermitian matrices.

We shall see in chapter VI that in quantum mechanics, the three components of an angular momentum do not commute with each other but satisfy well-defined commutation relations. This will enable us to show that, in the case of a spin 1/2, with which we are concerned here, the matrices representing $S_x$ and $S_y$ in the basis of the eigenvectors $|+\rangle$ and $|-\rangle$ of $S_z$ are the following:

$$(S_x) = \frac{\hbar}{2}\begin{pmatrix} 0 & 1 \\ 1 & 0 \end{pmatrix} \tag{A-16}$$

$$(S_y) = \frac{\hbar}{2}\begin{pmatrix} 0 & -i \\ i & 0 \end{pmatrix} \tag{A-17}$$

For the moment, we shall assume this result.

As for the $\mathscr{S}_u$ component of $\boldsymbol{\mathscr{S}}$ along the unit vector **u**, characterized by the polar angles $\theta$ and $\varphi$ (fig. 4), it is written:

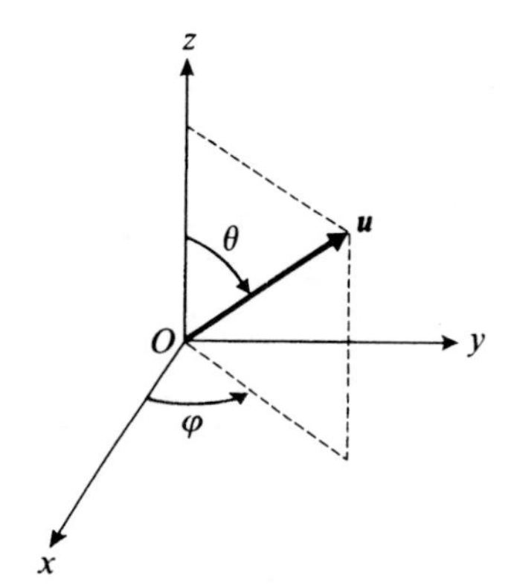


FIGURE 4

**Definition of the polar angles $\theta$ and $\varphi$ characterizing a unit vector u.**

$$\mathscr{S}_u = \boldsymbol{\mathscr{S}} . \mathbf{u} = \mathscr{S}_x \sin\theta\cos\varphi + \mathscr{S}_y \sin\theta\sin\varphi + \mathscr{S}_z \cos\theta \tag{A-18}$$

Using (A-15), (A-16) and (A-17), we easily find the matrix which represents the corresponding observable $S_u = \mathbf{S} . \mathbf{u}$ in the $\{ |+\rangle, |-\rangle \}$ basis:

$$\begin{aligned}(S_u) &= (S_x)\sin\theta\cos\varphi + (S_y)\sin\theta\sin\varphi + (S_z)\cos\theta \\ &= \frac{\hbar}{2}\begin{pmatrix} \cos\theta & \sin\theta\, e^{-i\varphi} \\ \sin\theta\, e^{i\varphi} & -\cos\theta \end{pmatrix}\end{aligned} \tag{A-19}$$

In what follows, we shall need to know the eigenvalues and eigenvectors of the observables $S_x$, $S_y$ and $S_u$. The calculations which enable us to obtain them from the matrices (A-16), (A-17) and (A-19) do not present any difficulty. We shall simply present the results here.

The $S_x$, $S_y$ and $S_u$ operators have the same eigenvalues, $+\hbar/2$ and $-\hbar/2$, as $S_z$. This result could have been expected, since it is always possible to rotate the Stern-Gerlach apparatus as a whole so as to make the axis defined by the magnetic field parallel either to $Ox$, to $Oy$, or to **u**. Since all directions of space have the same properties, the phenomena observed on the plate of the apparatus must be unchanged under such rotations: the measurement of $\mathcal{S}_x$, $\mathcal{S}_y$ or $\mathcal{S}_u$ can therefore yield only one of two results : $+\hbar/2$ or $-\hbar/2$.

As for the eigenvectors of $S_x$, $S_y$ and $S_u$, we shall denote them respectively by $|\pm\rangle_x$, $|\pm\rangle_y$ and $|\pm\rangle_u$ (the sign in the ket is that of the corresponding eigenvalue). Their expansions on the basis of eigenvectors $|\pm\rangle$ of $S_z$ is written:

$$|\pm\rangle_x = \frac{1}{\sqrt{2}}[|+\rangle \pm |-\rangle] \tag{A-20}$$

$$|\pm\rangle_y = \frac{1}{\sqrt{2}}[|+\rangle \pm i|-\rangle] \tag{A-21}$$

$$\begin{cases} |+\rangle_u = \cos\frac{\theta}{2}\,e^{-i\varphi/2}\,|+\rangle + \sin\frac{\theta}{2}\,e^{i\varphi/2}\,|-\rangle & \text{(A-22-a)} \\ |-\rangle_u = -\sin\frac{\theta}{2}\,e^{-i\varphi/2}\,|+\rangle + \cos\frac{\theta}{2}\,e^{i\varphi/2}\,|-\rangle & \text{(A-22-b)} \end{cases}$$

## B. ILLUSTRATION OF THE POSTULATES IN THE CASE OF A SPIN 1/2

Using the formalism which we have just described, we are now going to apply the postulates of quantum mechanics to a certain number of experiments on silver atoms which can actually be performed with the Stern-Gerlach apparatus. We shall thus be able to discuss the consequences of these postulates in a concrete case.

### 1. Actual preparation of the various spin states

In order to make predictions about the result of a measurement, we must know the state of the system (here, the spin of a silver atom) immediately before the measurement. We are going to see how to prepare a beam of silver atoms so that they are all in a given spin state.

#### a. PREPARATION OF THE STATES $|+\rangle$ AND $|-\rangle$

Let us assume that we pierce a hole in the plate $P$ of the apparatus represented in figure 1-a, at the position of the spot centered at $N_1$ (fig. 3). The atoms which are deflected downward continue to condense about $N_2$, while some of those which are deflected upward pass through the plate $P$ (fig. 5). Each of the atoms of the beam which propagates to the right of the plate is a physical system on which we have just performed a measurement of the observable $S_z$, the result being $+\hbar/2$. According to the fifth postulate of chapter III, this atom is in the eigenstate which corresponds to this result, that is, in the state $|+\rangle$ (since $S_z$ alone constitutes a C.S.C.O., the measurement result suffices to determine the state of the system after this measurement). The device in figure 5 thus produces a beam of atoms which are all in the spin state $|+\rangle$. This device acts like an "atomic polarizer" since it acts the same way on atoms as an ordinary polarizer does on photons.

Of course, if we pierced the plate around $N_2$ and not around $N_1$, we would obtain a beam all of whose atoms would be in the spin state $|-\rangle$.

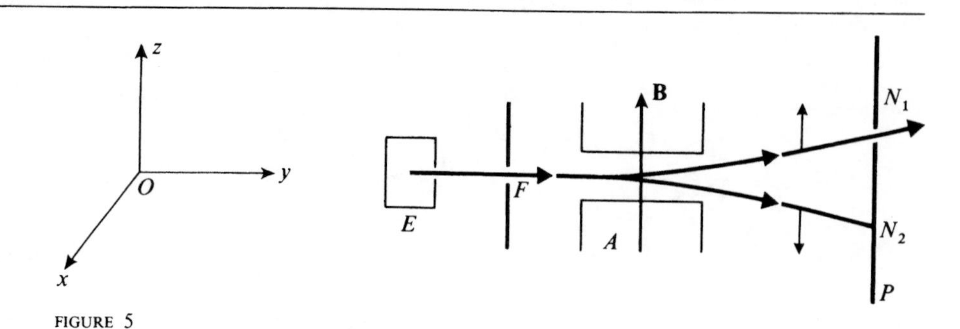


FIGURE 5

**When we pierce a hole in the plate $P$ at the position of the spot $N_1$, the atoms which pass through this hole are all in the spin state $|+\rangle$. The Stern-Gerlach apparatus is then acting like a polarizer.**

b. PREPARATION OF THE STATES $|\pm\rangle_x$, $|\pm\rangle_y$, $|\pm\rangle_u$

The observable $S_x$ also constitutes a C.S.C.O. since none of its eigenvalues is degenerate. To prepare one of its eigenstates, we must simply select, after a measurement of $S_x$, the atoms for which this measurement has yielded the corresponding eigenvalue. In practice, if we rotate the apparatus of figure 5 through an angle of $+\pi/2$ about $Oy$, we obtain a beam of atoms whose spin state is $|+\rangle_x$ (fig. 6).

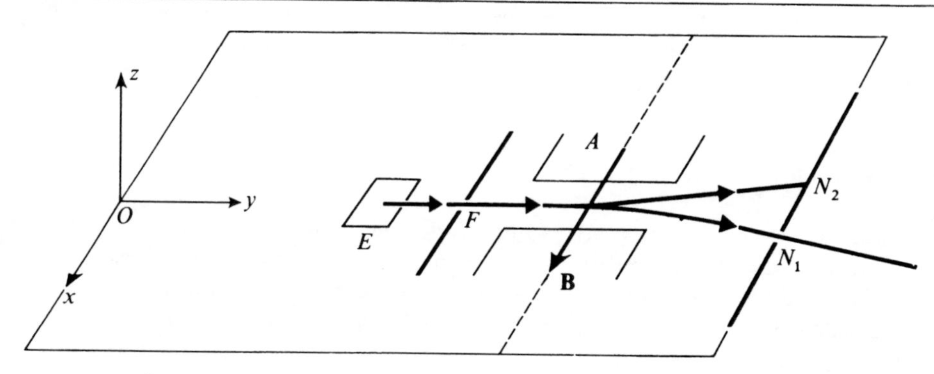


FIGURE 6

**When the apparatus of figure 5 is rotated through 90° about $Oy$, we obtain a polarizer which prepares atoms in the spin state $|+\rangle_x$.**

This method can immediately be generalized: by placing the Stern-Gerlach apparatus so that the axis of the magnetic field is parallel to an arbitrary unit vector **u**, and piercing the plate either at $N_1$ or at $N_2$, we can prepare silver atoms in the spin state $| + \rangle_u$ or $| - \rangle_u$*.

### c. PREPARATION OF THE MOST GENERAL STATE

We indicated above that the most general (normalized) ket of the spin state space is of the form:

$$|\psi\rangle = \alpha\,| + \rangle + \beta\,| - \rangle \tag{B-1}$$

with:

$$|\alpha|^2 + |\beta|^2 = 1 \tag{B-2}$$

Is it possible to prepare atoms whose spin state is described by the corresponding ket $|\psi\rangle$?

We are going to show that there exists, for all $|\psi\rangle$, a unit vector **u** such that $|\psi\rangle$ is collinear with the ket $| + \rangle_u$. We therefore choose two complex numbers $\alpha$ and $\beta$ which satisfy relation (B-2) but which are arbitrary in every other respect. Taking (B-2) into account, we find that there necessarily exists an angle $\theta$ such that:

$$\begin{cases} \cos\dfrac{\theta}{2} = |\alpha| \\[2ex] \sin\dfrac{\theta}{2} = |\beta| \end{cases} \tag{B-3}$$

If, in addition, we impose:

$$0 \leqslant \theta \leqslant \pi \tag{B-4}$$

the equation $\tan\dfrac{\theta}{2} = \left|\dfrac{\beta}{\alpha}\right|$ determines $\theta$ uniquely. We already know that only the difference of the phases of $\alpha$ and $\beta$ enters into the physical predictions. Let us therefore set:

$$\varphi = \text{Arg}\,\beta - \text{Arg}\,\alpha \tag{B-5}$$

$$\chi = \text{Arg}\,\beta + \text{Arg}\,\alpha \tag{B-6}$$

We thus have:

$$\begin{aligned} \text{Arg}\,\beta &= \frac{1}{2}\chi + \frac{1}{2}\varphi \\ \text{Arg}\,\alpha &= \frac{1}{2}\chi - \frac{1}{2}\varphi \end{aligned} \tag{B-7}$$

* The direction of the atomic beam is no longer necessarily $Oy$, but this is not important in what concerns us here.

With this notation, the ket $|\psi\rangle$ can be written:

$$|\psi\rangle = e^{i\frac{\chi}{2}}\left[\cos\frac{\theta}{2}e^{-i\frac{\varphi}{2}}|+\rangle + \sin\frac{\theta}{2}e^{i\frac{\varphi}{2}}|-\rangle\right] \tag{B-8}$$

If we compare this expression with formula (A-22-a), we see that $|\psi\rangle$ differs from the ket $|+\rangle_u$ (which corresponds to the unit vector **u** characterized by $\theta$ and $\varphi$) only by the phase factor $e^{i\chi/2}$, which has no physical significance.

Consequently, to prepare silver atoms in the state $|\psi\rangle$, it suffices to place the Stern-Gerlach apparatus (with its plate pierced at $N_1$) so that its axis is directed along the vector **u** whose polar angles are determined from $\alpha$ and $\beta$ by (B-3) and (B-5).

## 2. Spin measurements

We saw in § A that a Stern-Gerlach apparatus enables us to *measure* the component of the angular momentum $\mathcal{S}$ of silver atoms along a given axis. We have just pointed out, in § B-1, that an apparatus of the same type can be used to *prepare* an atomic beam in a given spin state. Consequently, if we place two Stern-Gerlach magnets one after the other, we can verify experimentally the predictions of the postulates. The first apparatus acts like a "polarizer": the beam which comes out of it is composed of a large number of silver atoms all in the same spin state. This beam then enters the second apparatus, which is used to measure a specified component of the angular momentum $\mathcal{S}$: this is, as it were, the "analyzer" (note the analogy with the optical experiment described in § A-3 of chapter I). We shall assume in this section that the spin state of the atoms of the beam does not evolve between the time they leave the "polarizer" and the time they enter the "analyzer", that is, between the preparation and the measurement. It would be easy to forgo this hypothesis, by using the Schrödinger equation to determine the spin evolution between the moment of preparation and the moment of measurement.

### a. FIRST EXPERIMENT

Let us choose the axes of the two apparatuses parallel to $Oz$ (fig. 7). The first one prepares the atoms in the state $|+\rangle$ and the second one measures $\mathcal{S}_z$. What is observed on the plate of the second apparatus?

Since the state of the system under study is an eigenstate of the observable $S_z$ which we want to measure, the postulates indicate that the measurement result is *certain*: we find, without fail, the corresponding eigenvalue $(+\hbar/2)$. Consequently, all the atoms of the beam must condense into a single spot on the plate of the second apparatus, at the position corresponding to $+\hbar/2$.

This is indeed what is observed experimentally: all the atoms strike the second plate in the vicinity of $N_1$, none hitting near $N_2$.

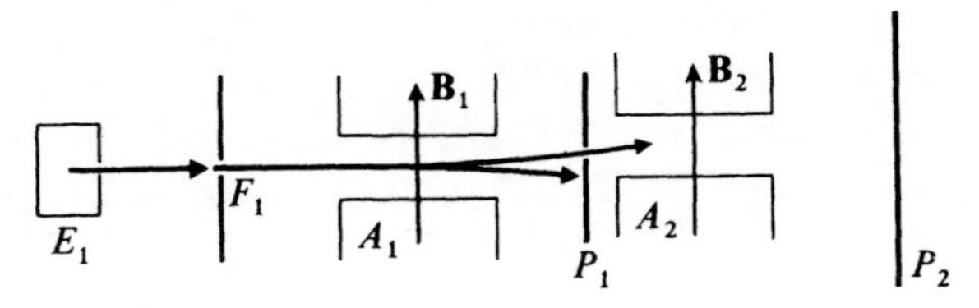


FIGURE 7

**The first apparatus (a "source" composed of the furnace $E_1$ and the slit $F_1$, plus a "polarizer" formed by the magnet $A_1$ and the pierced plate $P_1$) prepares the atoms in the state $|+\rangle$. The second one (an "analyzer" composed of the magnet $A_2$ and the plate $P_2$) measures the component $\mathcal{S}_z$. The result obtained is certain $(+\hbar/2)$.**

### b. SECOND EXPERIMENT

Now let us place the axis of the first apparatus along the unit vector **u**, with polar angles $\theta$, $\varphi = 0$ (**u** is therefore contained in the $xOz$ plane). The axis of the second apparatus remains parallel to $Oz$ (fig. 8). According to (A-22-a), the spin state of the atoms when they leave the "polarizer" is:

$$|\psi\rangle = \cos\frac{\theta}{2}|+\rangle + \sin\frac{\theta}{2}|-\rangle \tag{B-9}$$

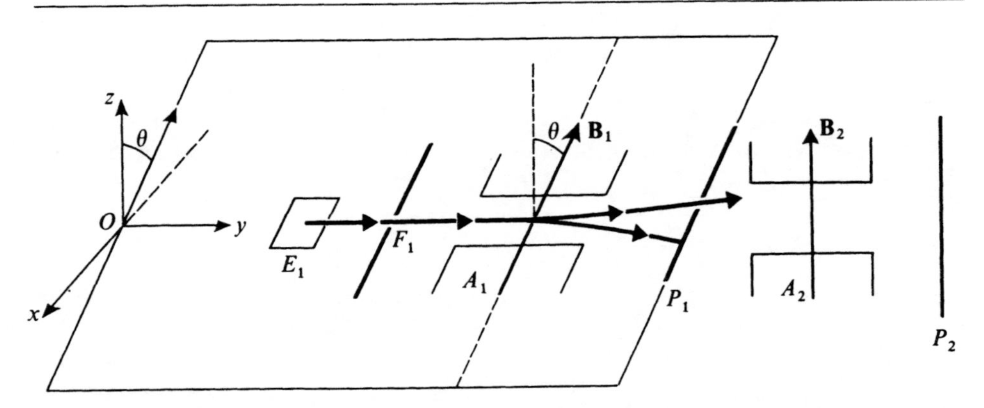


FIGURE 8

**The first apparatus prepares the spins in the state $|+\rangle_u$ (u is the unit vector of the $xOz$ plane which makes an angle $\theta$ with $Oz$). The second one measures the $\mathcal{S}_z$ component. The possible results are $+\hbar/2$ (probability $\cos^2\theta/2$) and $-\hbar/2$ (probability $\sin^2\theta/2$).**

The "analyzer" measures $\mathcal{S}_z$ on these atoms. What are the results?

This time, we find that certain atoms condense at $N_1$ and others, at $N_2$, although they have all been prepared in the same way: there is an indeterminacy in the behavior of each of the atoms taken individually. The postulate of spectral decomposition merely enables us to predict the probability of each atom's appearance at $N_1$ or $N_2$. Since (B-9) gives the expansion of the spin state of an atom in terms of the eigenstates of the observable being measured, we can calculate directly that these probabilities are, respectively, $\cos^2 \theta/2$ and $\sin^2 \theta/2$. Thus, when enough atoms have condensed on the plate, we observe that the intensity of the spots at $N_1$ and $N_2$ corresponds to numbers of atoms which are proportional, respectively, to $\cos^2 \theta/2$ and $\sin^2 \theta/2$.

COMMENT:

For any value of the angle $\theta$ (except exactly 0 or $\pi$), it is therefore always possible to find the two results $+\hbar/2$ and $-\hbar/2$ in a measurement of $S_z$. This prediction may seem to a certain extent paradoxical. For example, if $\theta$ is very small, the spin at the exit of the first apparatus points in a direction which is practically $Oz$, and yet one can find $-\hbar/2$ as well as $+\hbar/2$ in a measurement of $S_z$ $\left(\text{while, in classical mechanics, the result would be } \frac{\hbar}{2}\cos\theta \simeq \frac{\hbar}{2}\right)$. Nevertheless, the smaller $\theta$, the smaller the probability of finding $-\hbar/2$. Moreover, we shall see later [formula (B-11)] that the mean value of the results which would be obtained in a large number of identical experiments is $\langle S_z \rangle = \frac{\hbar}{2}\cos\theta$, which corresponds to the classical result.

### c. THIRD EXPERIMENT

Let us take a "polarizer" positioned as in §2-b, so as to prepare atoms in the state (B-9), and let us rotate the "analyzer" until its axis is directed along $Ox$, so that it measures the $\mathcal{S}_x$ component of the angular momentum.

To calculate the predictions of the postulates in this case, we must expand the state (B-9) in terms of the eigenstates of the observable $S_x$ [formula (A-20)]. We easily find:

$$\begin{cases} {}_x\langle + \mid \psi \rangle = \dfrac{1}{\sqrt{2}}\left(\cos\dfrac{\theta}{2} + \sin\dfrac{\theta}{2}\right) = \cos\left(\dfrac{\pi}{4} - \dfrac{\theta}{2}\right) \\ {}_x\langle - \mid \psi \rangle = \dfrac{1}{\sqrt{2}}\left(\cos\dfrac{\theta}{2} - \sin\dfrac{\theta}{2}\right) = \sin\left(\dfrac{\pi}{4} - \dfrac{\theta}{2}\right) \end{cases} \tag{B-10}$$

The probability of finding the eigenvalue $+\hbar/2$ of $S_x$ is therefore $\cos^2\left(\frac{\pi}{4} - \frac{\theta}{2}\right)$ and that of finding $-\hbar/2$, $\sin^2\left(\frac{\pi}{4} - \frac{\theta}{2}\right)$.

It is possible to verify these predictions by measuring the intensity of the two spots on the plate situated at the exit of the second Stern-Gerlach apparatus.

COMMENT:

The fact that it is $\left(\frac{\pi}{4} - \frac{\theta}{2}\right)$ that enters in here is not at all surprising: in § 2-b, the angle between the axes of the two apparatuses was $\theta$; it became $\left(\frac{\pi}{2} - \theta\right)$ after rotation of the second apparatus.

#### d. MEAN VALUES

In the situation of § 2-b, we find experimentally that, of a great number $\mathcal{N}$ of atoms, $\mathcal{N} \cos^2 \theta/2$ arrive at $N_1$ and $\mathcal{N} \sin^2 \theta/2$, at $N_2$. The measurement of $\mathcal{S}_z$ therefore yields $+ \hbar/2$ for each of the first group and $- \hbar/2$ for each of the second. If we calculate the mean value of these results, we obtain:

$$
\begin{aligned}
\langle S_z \rangle &= \frac{1}{\mathcal{N}}\left[\frac{\hbar}{2} \times \mathcal{N} \cos^2\frac{\theta}{2} - \frac{\hbar}{2} \times \mathcal{N} \sin^2\frac{\theta}{2}\right] \\
&= \frac{\hbar}{2}\cos\theta
\end{aligned}
\tag{B-11}
$$

It is easy to verify from formulas (B-9) and (A-10) that this is indeed the value of the matrix element $\langle \psi | S_z | \psi \rangle$.

Similarly, the average of the measurement results obtained in the experiment of § 2-c is equal to :

$$
\begin{aligned}
\langle S_x \rangle &= \frac{1}{\mathcal{N}}\left[\frac{\hbar}{2} \times \mathcal{N} \cos^2\left(\frac{\pi}{4} - \frac{\theta}{2}\right) - \frac{\hbar}{2} \times \mathcal{N} \sin^2\left(\frac{\pi}{4} - \frac{\theta}{2}\right)\right] \\
&= \frac{\hbar}{2}\sin\theta
\end{aligned}
\tag{B-12}
$$

To calculate the matrix element $\langle \psi | S_x | \psi \rangle$, we can use the matrix (A-16) which represents $S_x$ in the $\{ | + \rangle, | - \rangle \}$ basis. In this same basis, the ket $| \psi \rangle$ is represented by the column vector $\begin{pmatrix} \cos\theta/2 \\ \sin\theta/2 \end{pmatrix}$, and the bra $\langle \psi |$ by the corresponding row vector. We therefore have:

$$
\begin{aligned}
\langle \psi | S_x | \psi \rangle &= \frac{\hbar}{2}(\cos\theta/2 \quad \sin\theta/2)\begin{pmatrix} 0 & 1 \\ 1 & 0 \end{pmatrix}\begin{pmatrix} \cos\theta/2 \\ \sin\theta/2 \end{pmatrix} \\
&= \frac{\hbar}{2}\sin\theta
\end{aligned}
\tag{B-13}
$$

The mean value of $\mathcal{S}_x$ is indeed equal to the matrix element, in the state $| \psi \rangle$, of the associated observable $S_x$.

It is interesting to note that if we were dealing with a classical angular momentum of modulus $\hbar/2$ directed along the axis of the "polarizer", its components along $Ox$ and $Oz$ would be precisely $\frac{\hbar}{2}\sin\theta$ and $\frac{\hbar}{2}\cos\theta$. More generally, if we calculate [using the same technique as in (B-13)] the mean values of $S_x$, $S_y$ and $S_z$ in the state $| + \rangle_u$ [formula (A-22-a)], we find:

$$
\begin{aligned}
{}_u\langle + | S_x | + \rangle_u &= \frac{\hbar}{2}\sin\theta\cos\varphi \\
{}_u\langle + | S_y | + \rangle_u &= \frac{\hbar}{2}\sin\theta\sin\varphi \\
{}_u\langle + | S_z | + \rangle_u &= \frac{\hbar}{2}\cos\theta
\end{aligned}
\tag{B-14}
$$

These mean values are equal to the components of a classical angular momentum of modulus $\hbar/2$ oriented along the vector **u** whose polar angles are $\theta$ and $\varphi$. Therefore, we can also establish here a relation between classical mechanics and quantum mechanics through the mean values. However, we must not lose sight of the fact that *a measurement of* $\mathscr{S}_x$, *for example, on a given atom will never yield* $\frac{\hbar}{2} \sin\theta \cos\varphi$: the only results which can be found are $+\hbar/2$ and $-\hbar/2$. Only in taking the average of values obtained in a large number of identical measurements (same state of the system, here $| + \rangle_u$, and same observable measured, here $S_x$) do we obtain $\frac{\hbar}{2} \sin\theta \cos\varphi$.

COMMENT:

It is useful to consider again at this stage the problem of external degrees of freedom (position, momentum).

When a silver atom enters the second Stern-Gerlach apparatus in the spin state $| \psi \rangle$ given by (B-9), we have just seen that it is impossible to predict with certainty whether it will condense at $N_1$ or $N_2$. It seems difficult to reconcile this indeterminacy with the idea of a perfectly well-determined classical trajectory, given the initial state of the system.

In fact, this is not a real paradox. To say that the external degrees of freedom can be treated classically means only that it is possible to form wave packets which are much smaller than all the dimensions of the problem. It does not necessarily mean, as we shall see, that the particle itself follows a classical trajectory.

Let us first consider a silver atom which enters the apparatus in the initial spin state $| + \rangle$. The wave function which describes the external degrees of freedom of this particle is a wave packet whose spread is very small and whose center follows the

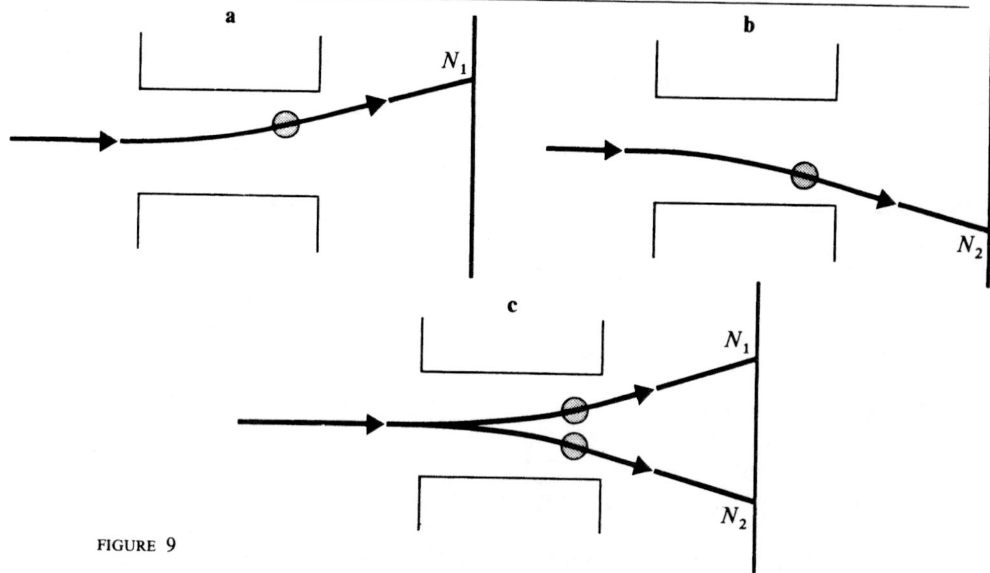


FIGURE 9

**When the spin is in the state $| + \rangle$ (fig. a) or $| - \rangle$ (fig. b), the center of the wave packet follows a well-defined trajectory which can be calculated classically. When the spin state is a linear superposition of $| + \rangle$ and $| - \rangle$, the wave packet splits into two parts and it is no longer possible to say that the atom follows a classical trajectory (despite the fact that the spread of each of the packets is much smaller than the characteristic dimensions of the problem).**

classical trajectory of figure 9-a. Similarly, if the silver atom enters with the spin state $| - \rangle$, the center of the wave packet associated with it follows the classical trajectory of figure 9-b.

If we now consider an atom which enters with the spin state $| \psi \rangle$ of formula (B-9), the corresponding initial state is a perfectly well-defined linear superposition of the two preceding initial states. Since the Schrödinger equation is linear, the wave function of the particle at a subsequent instant (fig. 9-c) is a linear superposition of the two wave packets of figures 9-a and 9-b. The particle therefore has a certain probability amplitude of being in one or the other of these two wave packets. We see that it does not follow a classical trajectory at all, unlike what happens to the centers of the two wave packets. Upon arrival on the screen, the wave function has non-zero values in two different regions, each very localized, around the points $N_1$ and $N_2$. The particle can therefore appear either near $N_1$ or near $N_2$, and we cannot predict with certainty at which of these two points it will arrive.

Note that the two wave packets of figure 9-c do not represent two different particles; they represent only one particle, whose wave function has two parts, each of which is very localized about a different point. The two wave packets, moreover, have a well-defined phase relation because they arise from the same initial wave packet, split into two under the influence of the gradient of **B**. We could recombine them to form one wave packet again by removing the screen (that is, by not performing the measurement) and by submitting them to a new field gradient, whose sign would be the opposite of the first one.

## 3. Evolution of a spin 1/2 particle in a uniform magnetic field

### a. THE INTERACTION HAMILTONIAN AND THE SCHRÖDINGER EQUATION

Consider a silver atom in a *uniform* magnetic field $\mathbf{B}_0$, and choose the $Oz$ axis along $\mathbf{B}_0$. The classical potential energy of the magnetic moment $\mathcal{M} = \gamma \mathcal{S}$ of this atom is then:

$$W = -\mathcal{M} \cdot \mathbf{B}_0 = -\mathcal{M}_z B_0 = -\gamma B_0 \mathcal{S}_z \qquad \text{(B-15)}$$

where $B_0$ is the modulus of the magnetic field. Let us set:

$$\omega_0 = -\gamma B_0 \qquad \text{(B-16)}$$

It is easy to see that $\omega_0$ has the dimensions of the inverse of a time, that is, of an angular velocity.

Since we are quantizing only the internal degrees of freedom of the particle, $\mathcal{S}_z$ must be replaced by the operator $S_z$, and the classical energy (B-15) becomes an operator: it is the Hamiltonian $H$ which describes the evolution of the spin of the atom in the field $\mathbf{B}_0$:

$$H = \omega_0 S_z \qquad \text{(B-17)}$$

Since this operator is time-independent, solving the corresponding Schrödinger equation amounts to solving the eigenvalue equation of $H$. We immediately see that the eigenvectors of $H$ are those of $S_z$:

$$\begin{aligned} H | + \rangle &= +\frac{\hbar\omega_0}{2} | + \rangle \\ H | - \rangle &= -\frac{\hbar\omega_0}{2} | - \rangle \end{aligned} \qquad \text{(B-18)}$$

There are therefore two energy levels, $E_+ = +\hbar\omega_0/2$ and $E_- = -\hbar\omega_0/2$ (fig. 10). Their separation $\hbar\omega_0$ is proportional to the magnetic field; they define a single "Bohr frequency":

$$\nu_{+-} = \frac{1}{h}(E_+ - E_-) = \frac{\omega_0}{2\pi} \tag{B-19}$$

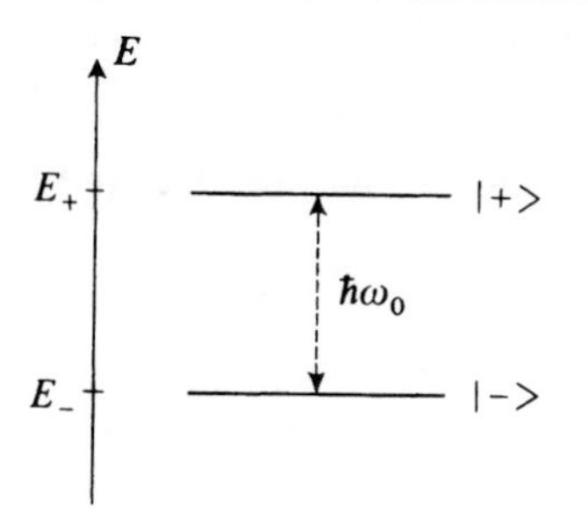


FIGURE 10

**Energy levels of a spin 1/2, of gyromagnetic ratio $\gamma$, placed in a magnetic field $\mathbf{B}_0$ parallel to $Oz$; $\omega_0$ is defined by $\omega_0 = -\gamma B_0$.**

COMMENTS:

(*i*) If the field $\mathbf{B}_0$ is parallel to the unit vector $\mathbf{u}$ whose polar angles are $\theta$ and $\varphi$, relation (B-17) must be replaced by:

$$H = \omega_0 S_u \tag{B-20}$$

where $S_u = \mathbf{S} . \mathbf{u}$ is the component of $\mathbf{S}$ along $\mathbf{u}$.

(*ii*) For the silver atom, $\gamma$ is negative; $\omega_0$ is therefore positive, according to (B-16). This explains the arrangement of the levels in figure 10.

### b. LARMOR PRECESSION

Let us assume that, at time $t = 0$, the spin is in the state:

$$|\psi(0)\rangle = \cos\frac{\theta}{2}\, e^{-i\varphi/2}\,|+\rangle + \sin\frac{\theta}{2}\, e^{i\varphi/2}\,|-\rangle \tag{B-21}$$

(we showed in § B-1-c that any state could be put in this form). To calculate the state $|\psi(t)\rangle$ at an arbitrary instant $t > 0$, we apply the rule (D-54) given in chapter III. In expression (B-21), $|\psi(0)\rangle$ is already expanded in terms of the eigenstates of the Hamiltonian, and we therefore obtain:

$$|\psi(t)\rangle = \cos\frac{\theta}{2}\, e^{-i\frac{\varphi}{2}}\, e^{-iE_+\frac{t}{\hbar}}\,|+\rangle + \sin\frac{\theta}{2}\, e^{i\frac{\varphi}{2}}\, e^{-iE_-\frac{t}{\hbar}}\,|-\rangle \tag{B-22}$$

or, using the values of $E_+$ and $E_-$:

$$|\psi(t)\rangle = \cos\frac{\theta}{2}\, e^{-i(\varphi+\omega_0 t)/2}\,|+\rangle + \sin\frac{\theta}{2}\, e^{i(\varphi+\omega_0 t)/2}\,|-\rangle \tag{B-23}$$

The presence of the magnetic field $\mathbf{B}_0$ therefore introduces a phase shift, proportional to the time, between the coefficients of the kets $|+\rangle$ and $|-\rangle$.

Comparing expression (B-23) for $|\psi(t)\rangle$ with that for the eigenket $|+\rangle_u$ of the observable $\mathbf{S}\,.\,\mathbf{u}$ [formula (A-22-a)], we see that the direction $\mathbf{u}(t)$ along which the spin component is $+\hbar/2$ with certainty is defined by the polar angles :

$$\begin{cases} \theta(t) = \theta \\ \varphi(t) = \varphi + \omega_0 t \end{cases} \tag{B-24}$$

The angle between $\mathbf{u}(t)$ and $Oz$ (the direction of the magnetic field $\mathbf{B}_0$) therefore remains constant, but $\mathbf{u}(t)$ revolves about $Oz$ at an angular velocity of $\omega_0$ (proportional to the magnetic field). Thus, we find in quantum mechanics the phenomenon which we described for a classical magnetic moment in § A-1-b, and which bears the name of *Larmor precession.*

From expression (B-17) for the Hamiltonian, it is obvious that the observable $S_z$ is a constant of the motion. It can be verified from (B-23) that the probabilities of obtaining $+\hbar/2$ or $-\hbar/2$ in a measurement of this observable are time-independent. Since the modulus of $e^{\pm i(\varphi + \omega_0 t)/2}$ is equal to 1, these probabilities are equal, respectively, to $\cos^2 \theta/2$ and $\sin^2 \theta/2$. The mean value of $S_z$ is also time-independent:

$$\langle \psi(t) | S_z | \psi(t) \rangle = \frac{\hbar}{2} \cos \theta \tag{B-25}$$

On the other hand, $S_x$ and $S_y$ do not commute with $H$ [it is easy to show this by using the matrices which represent $S_x$, $S_y$ and $S_z$ given in (A-15), (A-16) and (A-17)]. Thus, formulas (B-14) here become:

$$\begin{aligned} \langle \psi(t) | S_x | \psi(t) \rangle &= \frac{\hbar}{2} \sin \theta \cos (\varphi + \omega_0 t) \\ \langle \psi(t) | S_y | \psi(t) \rangle &= \frac{\hbar}{2} \sin \theta \sin (\varphi + \omega_0 t) \end{aligned} \tag{B-26}$$

In these expressions, we again find the single Bohr frequency $\omega_0/2\pi$ of the system. Moreover, the mean values of $S_x$, $S_y$ and $S_z$ behave like the components of a classical angular momentum of modulus $\hbar/2$ undergoing Larmor precession.

## C. GENERAL STUDY OF TWO-LEVEL SYSTEMS

The simplicity of the calculations presented in § B derives from the fact that the state space has only two dimensions.

There exist numerous other cases in physics which, to a first approximation, can be treated just as simply. Consider, for example, a physical system having two states whose energies are close together and very different from those of all other states of the system. Assume that we want to evaluate the effect of an external perturbation (or of internal interactions previously neglected) on these two states. When the intensity of the perturbation is sufficiently weak, it can be shown (*cf.* chap. XI) that its effect on the two states can be calculated, to a first approximation, by ignoring all the other energy levels of the system. All the calculations can then be perfomed in a two-dimensional subspace of the state space.

In this section, we shall study certain general properties of two-level systems (which are not necessarily spin 1/2 particles). Such a study is interesting because it enables us, using a mathematically simple model, to bring out some general and important physical ideas (quantum resonance, oscillation between two levels, etc...).

## 1. Outline of the problem

### a. NOTATION

Consider a physical system whose state space is two-dimensional (as we have already pointed out, this is usually an approximation: under certain conditions, we can confine ourselves to a two-dimensional subspace of the state space). For a basis, we choose the system of the two eigenstates $|\varphi_1\rangle$ and $|\varphi_2\rangle$ of the Hamiltonian $H_0$ whose eigenvalues are, respectively, $E_1$ and $E_2$:

$$\begin{aligned} H_0 |\varphi_1\rangle &= E_1 |\varphi_1\rangle \\ H_0 |\varphi_2\rangle &= E_2 |\varphi_2\rangle \end{aligned} \qquad \text{(C-1)}$$

This basis is orthonormal:

$$\langle \varphi_i | \varphi_j \rangle = \delta_{ij} \; ; \quad i, j = 1, 2 \qquad \text{(C-2)}$$

Assume that we want to take into account an external perturbation or interactions internal to the system, initially neglected in $H_0$. The Hamiltonian becomes:

$$H = H_0 + W \qquad \text{(C-3)}$$

The eigenstates and eigenvalues of $H$ will be denoted by $|\psi_\pm\rangle$ and $E_\pm$:

$$\begin{aligned} H |\psi_+\rangle &= E_+ |\psi_+\rangle \\ H |\psi_-\rangle &= E_- |\psi_-\rangle \end{aligned} \qquad \text{(C-4)}$$

$H_0$ is often called the unperturbed Hamiltonian and $W$, the perturbation or coupling. We shall assume here that $W$ is time-independent. In the $\{ |\varphi_1\rangle, |\varphi_2\rangle \}$ basis of eigenstates of $H_0$ (called unperturbed states), $W$ is represented by a Hermitian matrix:

$$(W) = \begin{pmatrix} W_{11} & W_{12} \\ W_{21} & W_{22} \end{pmatrix} \qquad \text{(C-5)}$$

$W_{11}$ and $W_{22}$ are real. Moreover:

$$W_{12} = W_{21}^* \qquad \text{(C-6)}$$

In the absence of coupling, $E_1$ and $E_2$ are the possible energies of the system, and the states $|\varphi_1\rangle$ and $|\varphi_2\rangle$ are stationary states (if the system is placed in one of these two states, it remains there indefinitely). The problem consists of evaluating the modifications which appear when the coupling $W$ is introduced.

#### b. CONSEQUENCES OF THE COUPLING

α. *$E_1$ and $E_2$ are no longer the possible energies of the system*

A measurement of the energy of the system can yield only one of the two eigenvalues $E_+$ and $E_-$ of $H$, which generally differ from $E_1$ and $E_2$.

The first problem which arises therefore consists of calculating $E_+$ and $E_-$ in terms of $E_1$, $E_2$ and the matrix elements $W_{ij}$ of $W$. This amounts to studying the effect of the coupling on the position of the energy levels.

β. *$| \varphi_1 \rangle$ and $| \varphi_2 \rangle$ are no longer stationary states*

Since $| \varphi_1 \rangle$ and $| \varphi_2 \rangle$ are not generally eigenstates of the total Hamiltonian $H$, they are no longer stationary states. If, for example, the system at time $t = 0$ is in the state $| \varphi_1 \rangle$, there is a certain probability $\mathscr{P}_{12}(t)$ of finding it in the state $| \varphi_2 \rangle$ at time $t$ : $W$ therefore induces transitions between the two unperturbed states. Hence the name "coupling" (between $| \varphi_1 \rangle$ and $| \varphi_2 \rangle$) given to $W$.

This dynamic aspect of the effect of $W$ constitutes the second problem with which we shall be concerned.

COMMENT:

In complement $C_{IV}$, the two problems we have just cited are considered by introducing the concept of a fictitious spin. It can indeed be shown that the Hamiltonian $H$ to be diagonalized has the same form as that of a spin 1/2 placed in a static magnetic field **B**, whose components $B_x$, $B_y$ and $B_z$ are simply expressed in terms of $E_1$, $E_2$ and the matrix elements $W_{ij}$. In other words, with every two-level system (not necessarily a spin 1/2), can be associated a spin 1/2 (called a fictitious spin) placed in a static field **B** and described by a Hamiltonian of identical form. All the results related to two-level systems which we are going to establish in this section can be interpreted in a simple geometric way in terms of magnetic moment, Larmor precession, and the various concepts introduced in §§ A and B of this chapter in connection with spin 1/2 particles. This geometrical interpretation is developed in complement $C_{IV}$.

### 2. Static aspect: effect of coupling on the stationary states of the system

#### a. EXPRESSIONS FOR THE EIGENSTATES AND EIGENVALUES OF $H$

In the $\{ | \varphi_1 \rangle, | \varphi_2 \rangle \}$ basis, the matrix representing $H$ is written:

$$(H) = \begin{pmatrix} E_1 + W_{11} & W_{12} \\ W_{21} & E_2 + W_{22} \end{pmatrix} \tag{C-7}$$

The diagonalization of matrix (C-7) presents no problems (it is performed in detail in complement $B_{IV}$). We find the eigenvalues:

$$E_+ = \frac{1}{2}(E_1 + W_{11} + E_2 + W_{22}) + \frac{1}{2}\sqrt{(E_1 + W_{11} - E_2 - W_{22})^2 + 4|W_{12}|^2}$$

$$E_- = \frac{1}{2}(E_1 + W_{11} + E_2 + W_{22}) - \frac{1}{2}\sqrt{(E_1 + W_{11} - E_2 - W_{22})^2 + 4|W_{12}|^2} \tag{C-8}$$

(it can be verified that if $W = 0$, $E_+$ and $E_-$ are identical to $E_1$ and $E_2$★). The eigenvectors associated with $E_+$ and $E_-$ are written :

$$| \psi_+ \rangle = \cos\frac{\theta}{2} e^{-i\varphi/2} | \varphi_1 \rangle + \sin\frac{\theta}{2} e^{i\varphi/2} | \varphi_2 \rangle \tag{C-9-a}$$

$$| \psi_- \rangle = - \sin\frac{\theta}{2} e^{-i\varphi/2} | \varphi_1 \rangle + \cos\frac{\theta}{2} e^{i\varphi/2} | \varphi_2 \rangle \tag{C-9-b}$$

where the angles $\theta$ and $\varphi$ are defined by:

$$\tan\theta = \frac{2|W_{12}|}{E_1 + W_{11} - E_2 - W_{22}} \quad \text{with} \quad 0 \leqslant \theta < \pi \tag{C-10}$$

$$W_{21} = |W_{21}| e^{i\varphi} \tag{C-11}$$

### b. DISCUSSION

α. *Graphical representation of the effect of coupling*

All the interesting effects which we shall discuss later arise from the fact that the perturbation $W$ possesses non-diagonal matrix elements $W_{12} = W_{21}^*$ (if $W_{12} = 0$, the eigenstates of $H$ are the same as those of $H_0$, the new eigenvalues being simply $E_1 + W_{11}$ and $E_2 + W_{22}$). To simplify the discussion, we shall therefore assume from now on that the matrix $(W)$ is purely non-diagonal, that is, that $W_{11} = W_{22} = 0$★★ Formulas (C-8) and (C-10) then become:

$$\begin{aligned} E_+ &= \frac{1}{2}(E_1 + E_2) + \frac{1}{2}\sqrt{(E_1 - E_2)^2 + 4|W_{12}|^2} \\ E_- &= \frac{1}{2}(E_1 + E_2) - \frac{1}{2}\sqrt{(E_1 - E_2)^2 + 4|W_{12}|^2} \end{aligned} \tag{C-12}$$

$$\tan\theta = \frac{2|W_{12}|}{E_1 - E_2} \qquad 0 \leqslant \theta < \pi \tag{C-13}$$

We now intend to study the effect of the coupling $W$ on the energies $E_+$ and $E_-$ in terms of the values of $E_1$ and $E_2$. Assume that $W_{12}$ is fixed and introduce the two parameters:

$$\begin{aligned} E_m &= \frac{1}{2}(E_1 + E_2) \\ \Delta &= \frac{1}{2}(E_1 - E_2) \end{aligned} \tag{C-14}$$

We see immediately from (C-12) that the variation of $E_+$ and $E_-$ with respect to $E_m$ is extremely simple : changing $E_m$ reduces to shifting the origin along the energy axis. Moreover, it can be verified from (C-9), (C-10) and (C-11) that the

★ If $E_1 > E_2$, $E_+$ approaches $E_1$ and $E_-$ approaches $E_2$ when $W$ approaches zero. On the other hand, if $E_1 < E_2$, $E_+$ approaches $E_2$ and $E_-$ approaches $E_1$.

★★ If $W_{11}$ and $W_{22}$ are non-zero, we simply set : $\tilde{E}_1 = E_1 + W_{11}$, $\tilde{E}_2 = E_2 + W_{22}$. All the results obtained in this section then remain valid if we replace $E_1$ and $E_2$ by $\tilde{E}_1$ and $\tilde{E}_2$.

vectors $|\psi_+\rangle$ and $|\psi_-\rangle$ do not depend on $E_m$. We are therefore concerned only with the influence of the parameter $\Delta$. Let us show on the same graph, in terms of $\Delta$, the four energies $E_1$, $E_2$, $E_+$ and $E_-$. We thus obtain for $E_1$ and $E_2$ two straight lines of slope $+1$ and $-1$ (shown in dashed lines in figure 11). Substituting (C-14) into (C-12), we find:

$$E_+ = E_m + \sqrt{\Delta^2 + |W_{12}|^2} \qquad \text{(C-15)}$$

$$E_- = E_m - \sqrt{\Delta^2 + |W_{12}|^2} \qquad \text{(C-16)}$$

When $\Delta$ varies, $E_+$ and $E_1$ describe the two branches of a hyperbola which is symmetrical with respect to the coordinate axes and whose asymptotes are the two straight lines associated with the unperturbed levels; the minimum separation between the two branches is $2|W_{12}|$ (solid lines in figure 11)*.

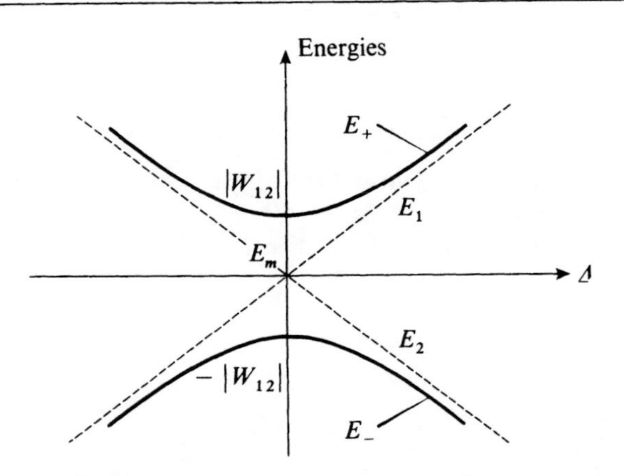


FIGURE 11

**Variation of the energies $E_+$ and $E_-$ with respect to the energy difference $\Delta = (E_1 - E_2)/2$. In the absence of coupling, the levels cross at the origin (dashed straight lines). Under the effect of the non-diagonal coupling $W$, the two perturbed levels " repel each other " and we obtain an " anti-crossing ": the curves giving $E_+$ and $E_-$ in terms of $\Delta$ are branches of a hyperbola (solid lines in the figure) whose asymptotes are the unperturbed levels.**

### β. *Effect of the coupling on the position of the energy levels*

In the absence of coupling, the energies $E_1$ and $E_2$ of the two levels "cross" at $\Delta = 0$. It is clear from figure 11 that under the effect of coupling, the two levels "repel each other"– that is, the energy values move further away from each other. The diagram in solid lines in figure 11 is often called, for this reason, an *anti-crossing* diagram.

* It is clear from figure 11 why, when $W \longrightarrow 0$:

$E_+ \longrightarrow E_1, E_- \longrightarrow E_2$ if $E_1 > E_2$
$E_+ \longrightarrow E_2, E_- \longrightarrow E_1$ if $E_1 < E_2$

Also, we see that, for any $\Delta$, we always have:

$$|E_+ - E_-| > |E_1 - E_2|$$

This is a result which appears rather often in other domains of physics (for example, in electrical circuit theory): *the coupling separates the normal frequencies.*

Near the asymptotes, that is, for $|\Delta| \gg |W_{12}|$, formulas (C-15) and (C-16) can be written in the form of a limited power series expansion in $\left|\frac{W_{12}}{\Delta}\right|$:

$$\begin{aligned} E_+ &= E_m + \Delta\left(1 + \frac{1}{2}\left|\frac{W_{12}}{\Delta}\right|^2 + \ldots\right) \\ E_- &= E_m - \Delta\left(1 + \frac{1}{2}\left|\frac{W_{12}}{\Delta}\right|^2 + \ldots\right) \end{aligned} \quad \text{(C-17)}$$

On the other hand, at the center of the hyperbola, for $E_2 = E_1$ ($\Delta = 0$), formulas (C-15) and (C-16) yield:

$$\begin{aligned} E_+ &= E_m + |W_{12}| \\ E_- &= E_m - |W_{12}| \end{aligned} \quad \text{(C-18)}$$

Therefore, *the effect of the coupling is much more important when the two unperturbed levels have the same energy.* The effect is then of first order, as can be seen from (C-18), while it is of second order when $\Delta \gg |W_{12}|$ [formulas (C-17)].

γ. *Effect of the coupling on the eigenstates*

When (C-14) is used, formula (C-13) becomes:

$$\tan\theta = \frac{|W_{12}|}{\Delta} \quad \text{(C-19)}$$

It follows that, when $\Delta \ll |W_{12}|$ (strong coupling), $\theta \simeq \pi/2$; on the other hand, when $\Delta \gg |W_{12}|$ (weak coupling), $\theta \simeq 0$ (assuming $\Delta \geqslant 0$).

At the center of the hyperbola, when $E_2 = E_1$ ($\Delta = 0$), we have:

$$\begin{aligned} |\psi_+\rangle &= \frac{1}{\sqrt{2}}\left[e^{-i\varphi/2}|\varphi_1\rangle + e^{i\varphi/2}|\varphi_2\rangle\right] \\ |\psi_-\rangle &= \frac{1}{\sqrt{2}}\left[-e^{-i\varphi/2}|\varphi_1\rangle + e^{i\varphi/2}|\varphi_2\rangle\right] \end{aligned} \quad \text{(C-20)}$$

while near the asymptotes (that is, for $\Delta \gg |W_{12}|$), we have, to first order in $|W_{12}|/\Delta$:

$$\begin{aligned} |\psi_+\rangle &= e^{-i\varphi/2}\left[|\varphi_1\rangle + e^{i\varphi}\frac{|W_{12}|}{2\Delta}|\varphi_2\rangle + \ldots\right] \\ |\psi_-\rangle &= e^{i\varphi/2}\left[|\varphi_2\rangle - e^{-i\varphi}\frac{|W_{12}|}{2\Delta}|\varphi_1\rangle + \ldots\right] \end{aligned} \quad \text{(C-21)}$$

In other words, for a weak coupling ($E_1 - E_2 \gg |W_{12}|$), the perturbed states differ very slightly from the unperturbed states. We see from (C-21) that to within a global phase factor $e^{-i\varphi/2}$, $|\psi_+\rangle$ is equal to the state $|\varphi_1\rangle$ slightly "contaminated" by a small contribution from the state $| \varphi_2 \rangle$. On the other hand, for a strong coupling ($E_1 - E_2 \ll |W_{12}|$), formulas (C-20) indicate that the states $| \psi_+ \rangle$ and $| \psi_- \rangle$ are very different from the states $| \varphi_1 \rangle$ and $| \varphi_2 \rangle$, since they are linear superpositions of them with coefficients of the same modulus.

Thus, like the energies, the eigenstates undergo significant modifications in the neighborhood of the point where the two unperturbed states cross.

#### c. IMPORTANT APPLICATION: THE PHENOMENA OF QUANTUM RESONANCE

When $E_1 = E_2 = E_m$, the corresponding energy of $H_0$ is two-fold degenerate. As we have just seen, the coupling $W_{12}$ lifts this degeneracy and, in particular, gives rise to a level whose energy is lowered by $|W_{12}|$. In other words, if the ground state of a physical system is two-fold degenerate (and sufficiently far from all the other levels), any (purely non-diagonal) coupling between the two corresponding states lowers the energy of the ground state of the system, which thus becomes more stable.

As a first example of this phenomenon, we shall cite the resonance stabilization of the benzene molecule $C_6H_6$. Experiments show that the six carbon atoms are situated at the vertices of a regular hexagon, and we would expect the ground state to include three double bonds between neighboring carbon atoms. Figures 12-a and 12-b represent two possible dispositions of these bonds. The nuclei are assumed here to be fixed because of their high masses. Thus, the electronic states $| \varphi_1 \rangle$ and $| \varphi_2 \rangle$, associated with figures 12-a and 12-b respectively, are different. If the structure of figure 12-a were the only one possible, the

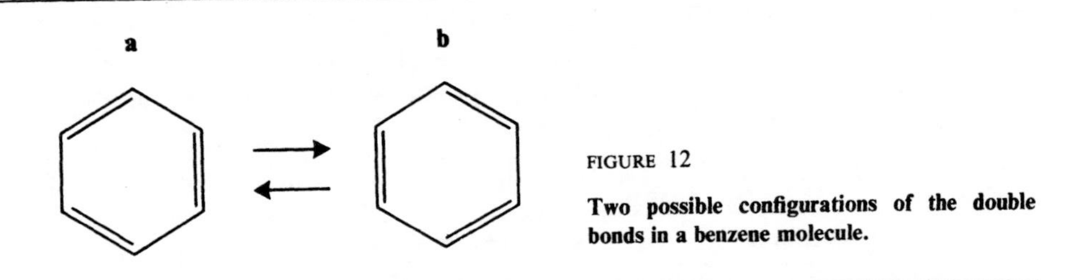


FIGURE 12

**Two possible configurations of the double bonds in a benzene molecule.**

ground state of the electronic system would have an energy of $E_m = \langle \varphi_1 | H | \varphi_1 \rangle$, where $H$ is the Hamiltonian of the electrons in the potential created by the nuclei. But the bonds can also be placed as shown in figure 12-b. By symmetry, we obviously have $\langle \varphi_2 | H | \varphi_2 \rangle = \langle \varphi_1 | H | \varphi_1 \rangle$, and we could conclude that the ground state of the molecule is doubly degenerate. However, the non-diagonal matrix element $\langle \varphi_2 | H | \varphi_1 \rangle$ of the Hamiltonian $H$ is not zero. This coupling between the states $| \varphi_1 \rangle$ and $| \varphi_2 \rangle$ gives rise to two distinct levels, one of which has an energy lower than $E_m$. The benzene molecule is therefore more stable than we would have expected. Moreover, in its true ground state, the configuration of the molecule cannot be represented either by figure 12-a or by figure 12-b : this state is a linear superposition of $| \varphi_1 \rangle$ and $| \varphi_2 \rangle$ [the coefficients of this superposition having, as in (C-20), the same modulus]. This is what is symbolized by the double arrow of figure 12, commonly used by chemists.

Another example is that of the (ionized) molecule $H_2^+$, composed of two protons $p_1$ and $p_2$ and one electron. The two protons, because of their large masses, can be considered to be fixed. Let us call $R$ the distance between them and $|\varphi_1\rangle$ and $|\varphi_2\rangle$, the states where the electron is localized around $p_1$ or around $p_2$, its wave function being that of the hydrogen atom it would form with $p_1$ or $p_2$ (fig. 13). As above, the diagonal elements $\langle\varphi_1|H|\varphi_1\rangle$ and

FIGURE 13

**In the $H_2^+$ ion, the electron could logically be localized either around the proton $p_1$ (fig. a) or around the proton $p_2$ (fig. b). In the ground state of the ion, the wave function of the electron is a linear superposition of the wave functions associated with figures a and b. Its probability of presence is symmetrical with respect to the plane bisecting $p_1p_2$.**

$\langle\varphi_2|H|\varphi_2\rangle$ of the Hamiltonian are equal because of symmetry; we shall denote them by $E_m(R)$. The two states $|\varphi_1\rangle$ and $|\varphi_2\rangle$ are not, however, stationary states, since the matrix element $\langle\varphi_1|H|\varphi_2\rangle$ is not zero. Here again, we obtain an energy level lower than $E_m(R)$ and, in the ground state, the wave function of the electron is a linear combination of those of figures 13-a and 13-b. The electron is thus no longer localized about one of the two protons alone, and it is this delocalization which, by lowering its potential energy, is responsible for the chemical bond*

## 3. Dynamical aspect: oscillation of the system between the two unperturbed states

### a. EVOLUTION OF THE STATE VECTOR

Let :

$$|\psi(t)\rangle = a_1(t)\,|\varphi_1\rangle + a_2(t)\,|\varphi_2\rangle \tag{C-22}$$

be the state vector of the system at the instant $t$. The evolution of $|\psi(t)\rangle$ in the presence of the coupling $W$ is given by the Schrödinger equation:

$$i\hbar\frac{\mathrm{d}}{\mathrm{d}t}|\psi(t)\rangle = (H_0 + W)\,|\psi(t)\rangle \tag{C-23}$$

* A more elaborate study of the ionized molecule $H_2^+$ will be presented in complement $G_{XI}$.

Let us project this equation onto the basis vectors $|\varphi_1\rangle$ and $|\varphi_2\rangle$. We obtain, using (C-5) (where we have set $W_{11} = W_{22} = 0$) and (C-22):

$$i\hbar\frac{\mathrm{d}}{\mathrm{d}t}a_1(t) = E_1\,a_1(t) + W_{12}\,a_2(t)$$
$$i\hbar\frac{\mathrm{d}}{\mathrm{d}t}a_2(t) = W_{21}\,a_1(t) + E_2\,a_2(t) \tag{C-24}$$

If $|W_{12}| \neq 0$, these equations constitute a linear system of homogeneous coupled differential equations. The classical method of solving such a system reduces, in fact, to the application of rule (D-54) of chapter III: look for the eigenvectors $|\psi_+\rangle$ (eigenvalue $E_+$) and $|\psi_-\rangle$ (eigenvalue $E_-$) of the operator $H = H_0 + W$ [whose matrix elements are the coefficients of equations (C-24)], and decompose $|\psi(0)\rangle$ in terms of $|\psi_+\rangle$ and $|\psi_-\rangle$:

$$|\psi(0)\rangle = \lambda\,|\psi_+\rangle + \mu\,|\psi_-\rangle \tag{C-25}$$

(where $\lambda$ and $\mu$ are fixed by the initial conditions). We then have:

$$|\psi(t)\rangle = \lambda\,\mathrm{e}^{-iE_+t/\hbar}\,|\psi_+\rangle + \mu\,\mathrm{e}^{-iE_-t/\hbar}\,|\psi_-\rangle \tag{C-26}$$

[which enables us to obtain $a_1(t)$ and $a_2(t)$ by projecting $|\psi(t)\rangle$ onto $|\varphi_1\rangle$ and $|\varphi_2\rangle$].

It can be shown that a system whose state vector is the vector $|\psi(t)\rangle$ given in (C-26) oscillates between the two unperturbed states $|\varphi_1\rangle$ and $|\varphi_2\rangle$. To see this, we shall assume that the system at time $t = 0$ is in the state $|\varphi_1\rangle$:

$$|\psi(0)\rangle = |\varphi_1\rangle \tag{C-27}$$

and calculate the probability $\mathscr{P}_{12}(t)$ of finding it in the state $|\varphi_2\rangle$ at time $t$.

### b. CALCULATION OF $\mathscr{P}_{12}(t)$: RABI'S FORMULA

As in (C-25), let us therefore expand the ket $|\psi(0)\rangle$ given in (C-27) on the $\{|\psi_+\rangle, |\psi_-\rangle\}$ basis. Inverting formulas (C-9), we obtain:

$$|\psi(0)\rangle = |\varphi_1\rangle = \mathrm{e}^{i\varphi/2}\left[\cos\frac{\theta}{2}\,|\psi_+\rangle - \sin\frac{\theta}{2}\,|\psi_-\rangle\right] \tag{C-28}$$

from which we deduce, using (C-26):

$$|\psi(t)\rangle = \mathrm{e}^{i\varphi/2}\left[\cos\frac{\theta}{2}\,\mathrm{e}^{-iE_+t/\hbar}\,|\psi_+\rangle - \sin\frac{\theta}{2}\,\mathrm{e}^{-iE_-t/\hbar}\,|\psi_-\rangle\right] \tag{C-29}$$

The probability amplitude of finding the system at time $t$ in the state $|\varphi_2\rangle$ is then written:

$$\langle\varphi_2|\psi(t)\rangle = \mathrm{e}^{i\varphi/2}\left[\cos\frac{\theta}{2}\,\mathrm{e}^{-iE_+t/\hbar}\,\langle\varphi_2|\psi_+\rangle - \sin\frac{\theta}{2}\,\mathrm{e}^{-iE_-t/\hbar}\,\langle\varphi_2|\psi_-\rangle\right]$$
$$= \mathrm{e}^{i\varphi}\sin\frac{\theta}{2}\cos\frac{\theta}{2}\left[\mathrm{e}^{-iE_+t/\hbar} - \mathrm{e}^{-iE_-t/\hbar}\right] \tag{C-30}$$

which enables us to calculate $\mathscr{P}_{12}(t) = |\langle \varphi_2 | \psi(t) \rangle|^2$. We thus find:

$$\mathscr{P}_{12}(t) = \frac{1}{2}\sin^2\theta\left[1 - \cos\left(\frac{E_+ - E_-}{\hbar}t\right)\right]$$
$$= \sin^2\theta \sin^2\left(\frac{E_+ - E_-}{2\hbar}t\right) \qquad \text{(C-31)}$$

or, using expressions (C-12) and (C-13):

$$\mathscr{P}_{12}(t) = \frac{4|W_{12}|^2}{4|W_{12}|^2 + (E_1 - E_2)^2}\sin^2\left[\sqrt{4|W_{12}|^2 + (E_1 - E_2)^2}\frac{t}{2\hbar}\right] \qquad \text{(C-32)}$$

Formula (C-32) is sometimes called Rabi's formula.

### c. DISCUSSION

We observe from (C-31) that $\mathscr{P}_{12}(t)$ oscillates over time with a frequency of $(E_+ - E_-)/h$, which is simply the unique Bohr frequency of the system. $\mathscr{P}_{12}(t)$ varies between zero and a maximum value which, according to (C-31), is equal to $\sin^2\theta$ and is attained for all values of $t$ such that $t = (2k + 1)\pi\hbar/(E_+ - E_-)$, with $k = 0, 1, 2, \ldots$ (fig. 14).

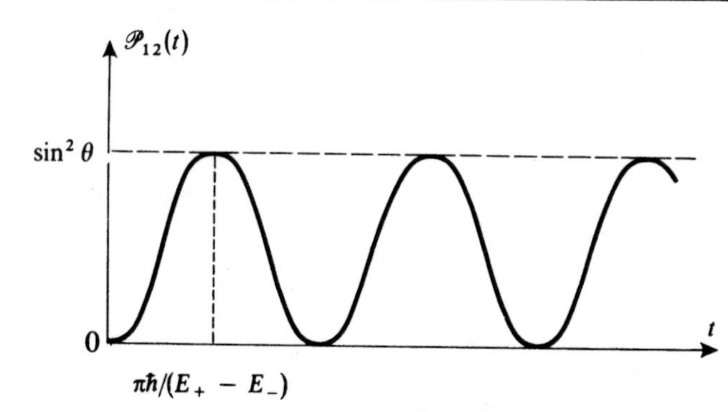


FIGURE 14

**Variation with respect to time of the probability $\mathscr{P}_{12}(t)$ of finding the system in the state $|\varphi_2\rangle$ when it was initially in the state $|\varphi_1\rangle$. When the states $|\varphi_1\rangle$ and $|\varphi_2\rangle$ have the same unperturbed energy, the probability $\mathscr{P}_{12}(t)$ can attain the value 1.**

The oscillation frequency $(E_+ - E_-)/h$, as well as the maximum value $\sin^2\theta$ of $\mathscr{P}_{12}(t)$, are functions of $|W_{12}|$ and $E_1 - E_2$, whose main features we are now going to describe.

When $E_1 = E_2$, $(E_+ - E_-)/h$ is equal to $2|W_{12}|/h$, and $\sin^2\theta$ takes on its greatest possible value, that is, 1 : at certain times, $t = (2k + 1)\pi\hbar/2|W_{12}|$, the system (which started from the state $|\varphi_1\rangle$) is in the state $|\varphi_2\rangle$. Therefore, any coupling between two states of equal energy causes the system to oscillate completely from one state to the other with a frequency proportional to the coupling*.

* The same phenomenon is found in other domains of physics. Consider, for example, two identical pendulums (1) and (2), suspended from the same support and having the same frequency. Let us assume that at time $t = 0$, pendulum (1) is set in motion. Because of the coupling assured by the common support, we know (*cf.* complement $H_V$) that after a certain time (the stronger the coupling, the shorter the time), we shall arrive at the complementary situation, in which only pendulum (2) is oscillating with the initial amplitude of pendulum (1), and so on.

When $E_1 - E_2$ increases, $(E_+ - E_-)/h$ does also, while $\sin^2 \theta$ decreases. For a weak coupling ($E_1 - E_2 \gg |W_{12}|$), $E_+ - E_-$ differs very little from $E_1 - E_2$, and $\sin^2 \theta$ becomes very small. This last result is not surprising since, in the case of a weak coupling, the state $|\varphi_1\rangle$ is very close to the stationary state $|\psi_+\rangle$ [*cf.* formulas (C-21)] : the system, having started in the state $|\varphi_1\rangle$, therefore evolves very little over time.

#### d. EXAMPLE OF OSCILLATION BETWEEN TWO STATES

Let us return to the example of the $H_2^+$ molecule. We shall assume that, at a certain time, the electron is localized about proton $p_1$: it is, for example, in the state shown in figure 13-a. According to the results of the preceding section, we know that it will oscillate between the two protons with a frequency equal to the Bohr frequency associated with the two stationary states $|\psi_+\rangle$ and $|\psi_-\rangle$ of the molecule. To this oscillation of the electron between the two states represented in 13-a and 13-b, corresponds an oscillation of the mean value of the electric dipole moment of the molecule (the dipole moment is non-zero when the electron is localized about one of the two protons and changes signs depending on whether the proton involved is $p_1$ or $p_2$). Thus we see concretely how, when the molecule is not in a stationary state, an oscillating electric dipole moment can appear. Such an oscillating dipole can exchange energy with an electromagnetic wave of the same frequency. Consequently, this frequency must appear in the absorption and emission spectrum of the $H_2^+$ ion.

Other examples of oscillations between two states are discussed in complements $F_{IV}$, $G_{IV}$ and $H_{IV}$.

**References and suggestions for further reading :**

The Stern-Gerlach experiment : original article (3.8); Cagnac and Pebay-Peyroula (11.2), chap. X; Eisberg and Resnick (1.3), §8-3; Bohm (5.1), §§22.5 and 22.6; Frisch (3.13).

Two-level systems: Feynman III (1.2), chaps. 6, 10 and 11.; Valentin (16.1), Annexe XII; Allen and Eberly (15.8), particularly chap. 3.

## COMPLEMENTS OF CHAPTER IV

| | |
|---|---|
| $A_{IV}$: THE PAULI MATRICES<br>$B_{IV}$: DIAGONALIZATION OF A 2 × 2 HERMITIAN MATRIX | $A_{IV}$, $B_{IV}$: technical study of 2 × 2 matrices; simple, and important for solving numerous quantum mechanical problems. |
| $C_{IV}$: FICTITIOUS SPIN 1/2 ASSOCIATED WITH A TWO-LEVEL SYSTEM | $C_{IV}$: establishes the close relation which exists between §§B and C of chapter IV; supplies a simple geometrical interpretation of the properties of two-level systems (easy, but not indispensable for what follows). |
| $D_{IV}$: SYSTEM OF TWO SPIN 1/2 PARTICLES | $D_{IV}$: simple illustration of the tensor product and the postulates of quantum mechanics (can be considered to be a set of worked exercises). |
| $E_{IV}$: SPIN 1/2 DENSITY MATRIX | $E_{IV}$: illustration, in the case of spin 1/2 particles, of concepts introduced in complement $E_{III}$. |
| $F_{IV}$: SPIN 1/2 PARTICLE IN A STATIC MAGNETIC FIELD AND A ROTATING FIELD: MAGNETIC RESONANCE | $F_{IV}$: study of a very important physical phenomenon with many applications: magnetic resonance; can be studied later. |
| $G_{IV}$: A SIMPLE MODEL OF THE AMMONIA MOLECULE | $G_{IV}$: example of a physical system whose study can be reduced, in a first approximation, to that of a two-level system; moderately difficult. |
| $H_{IV}$: COUPLING BETWEEN A STABLE STATE AND AN UNSTABLE STATE | $H_{IV}$: study of the influence of coupling between two levels with different lifetimes; easy, but requires the concepts introduced in complement $K_{III}$. |
| $J_{IV}$: EXERCISES | |

# Complement $A_{IV}$

# THE PAULI MATRICES



In § A-2 of chapter IV, we introduced the matrices which represent the three components $S_x$, $S_y$ and $S_z$ of a spin $\mathbf{S}$ in the $\{ | + \rangle, | - \rangle \}$ basis (eigenvectors of $S_z$). In quantum mechanics, it is often convenient to introduce the dimensionless operator $\boldsymbol{\sigma}$, proportional to $\mathbf{S}$ and given by:

$$\mathbf{S} = \frac{\hbar}{2} \boldsymbol{\sigma} \tag{1}$$

The matrices which represent the three components of $\boldsymbol{\sigma}$ in the $\{ | + \rangle, | - \rangle \}$ basis are called the "Pauli matrices".

## 1. Definition ; eigenvalues and eigenvectors

Let us go back to equations (A-15), (A-16) and (A-17) of chapter IV. Using (1), we see that the definition of the Pauli matrices is:

$$\sigma_x = \begin{pmatrix} 0 & 1 \\ 1 & 0 \end{pmatrix} \qquad \sigma_y = \begin{pmatrix} 0 & -i \\ i & 0 \end{pmatrix} \qquad \sigma_z = \begin{pmatrix} 1 & 0 \\ 0 & -1 \end{pmatrix} \tag{2}$$

These are Hermitian matrices, all three of which have the same characteristic equation:

$$\lambda^2 - 1 = 0 \tag{3}$$

The eigenvalues of $\sigma_x$, $\sigma_y$ and $\sigma_z$ are therefore:

$$\lambda = \pm 1 \tag{4}$$

which is consistent with the fact that those of $S_x$, $S_y$ and $S_z$ are $\pm \hbar/2$.

It is easy to obtain, from definition (2), the eigenvectors of $\sigma_x$, $\sigma_y$ and $\sigma_z$. They are the same, respectively, as those of $S_x$, $S_y$ and $S_z$, already introduced in § A-2 of chapter IV:

$$\begin{aligned} \sigma_x | \pm \rangle_x &= \pm | \pm \rangle_x \\ \sigma_y | \pm \rangle_y &= \pm | \pm \rangle_y \\ \sigma_z | \pm \rangle &= \pm | \pm \rangle \end{aligned} \tag{5}$$

with:

$$|\pm\rangle_x = \frac{1}{\sqrt{2}}[|+\rangle \pm |-\rangle]$$

$$|\pm\rangle_y = \frac{1}{\sqrt{2}}[|+\rangle \pm i|-\rangle] \tag{6}$$

## 2. Simple properties

It is easy to see from definition (2) that the Pauli matrices verify the relations:

$$\text{Det}(\sigma_j) = -1 \qquad j = x, y \quad \text{or} \quad z \tag{7}$$

$$\text{Tr}(\sigma_j) = 0 \tag{8}$$

$$\sigma_x^2 = \sigma_y^2 = \sigma_z^2 = I \qquad \text{(where } I \text{ is the } 2 \times 2 \text{ unit matrix)} \tag{9}$$

$$\sigma_x\sigma_y = -\sigma_y\sigma_x = i\sigma_z \tag{10}$$

as well as the equations which can be deduced from (10) by cyclic permutation of $x$, $y$ and $z$.

Equations (9) and (10) are sometimes condensed into the form:

$$\sigma_j \sigma_k = i \sum_l \varepsilon_{jkl} \sigma_l + \delta_{jk} I \tag{11}$$

where $\varepsilon_{jkl}$ is antisymmetric with respect to the interchange of any two of its indices. It is equal to:

$$\varepsilon_{jkl} = \begin{cases} 0 \text{ if the indices } j, k, l \text{ are not all different} \\ 1 \text{ if } j, k, l \text{ is an even permutation of } x, y, z \\ -1 \text{ if } j, k, l \text{ is an odd permutation of } x, y, z \end{cases} \tag{12}$$

From (10), we immediately conclude:

$$[\sigma_x, \sigma_y] = 2i\sigma_z \tag{13}$$

(and the relations obtained by cyclic permutation). This yields:

$$[S_x, S_y] = i\hbar S_z$$

$$[S_y, S_z] = i\hbar S_x \tag{14}$$

$$[S_z, S_x] = i\hbar S_y$$

We shall see later (*cf.* chap. VI) that equations (14) are characteristic of an angular momentum.

We also see from (10) that:

$$\sigma_x\sigma_y + \sigma_y\sigma_x = 0 \tag{15}$$

(the $\sigma_i$ matrices are said to anticommute with each other) and that, taking (9) into account:

$$\sigma_x\sigma_y\sigma_z = iI \tag{16}$$

Finally, let us mention an identity which is sometimes useful in quantum mechanics. If **A** and **B** denote two vectors whose components are numbers (or operators which commute with all operators acting in the two-dimensional spin state space):

$$(\boldsymbol{\sigma} \cdot \mathbf{A})(\boldsymbol{\sigma} \cdot \mathbf{B}) = \mathbf{A} \cdot \mathbf{B}\, I + i\boldsymbol{\sigma} \cdot (\mathbf{A} \times \mathbf{B}) \tag{17}$$

We see that, using formula (11) and the fact that **A** and **B** commute with $\boldsymbol{\sigma}$, we can write:

$$\begin{aligned}(\boldsymbol{\sigma} \cdot \mathbf{A})(\boldsymbol{\sigma} \cdot \mathbf{B}) &= \sum_{j,k} \sigma_j A_j \sigma_k B_k \\ &= \sum_{j,k} A_j B_k \left[ i \sum_l \varepsilon_{jkl} \sigma_l + \delta_{jk} I \right] \\ &= \sum_l i\sigma_l \left[ \sum_{j,k} \varepsilon_{jkl} A_j B_k \right] + \sum_j A_j B_j I \end{aligned} \tag{18}$$

In the second term, we recognize the scalar product $\mathbf{A} \cdot \mathbf{B}$. In addition, it is easy to see from (12) that $\sum_{j,k} \varepsilon_{jkl} A_j B_k$ is the $l$th component of the vector product $\mathbf{A} \times \mathbf{B}$. This proves (17). Note that if **A** and **B** do not commute, they must appear in the same order on both sides of the identity.

## 3. A convenient basis of the 2 × 2 matrix space

Consider an arbitrary 2 × 2 matrix:

$$M = \begin{pmatrix} m_{11} & m_{12} \\ m_{21} & m_{22} \end{pmatrix} \tag{19}$$

It can always be written as a linear combination of the four matrices:

$$I, \sigma_x, \sigma_y, \sigma_z \tag{20}$$

since, using (2), we can immediately verify that:

$$M = \frac{m_{11} + m_{22}}{2} I + \frac{m_{11} - m_{22}}{2} \sigma_z + \frac{m_{12} + m_{21}}{2} \sigma_x + i\, \frac{m_{12} - m_{21}}{2} \sigma_y \tag{21}$$

Therefore, any 2 × 2 matrix can be put in the form:

$$M = a_0 I + \mathbf{a} \cdot \boldsymbol{\sigma} \tag{22}$$

where the coefficients $a_0$, $a_x$, $a_y$ and $a_z$ are complex numbers.

Comparing (21) and (22), we see that $M$ is Hermitian if and only if the coefficients $a_0$ and $\mathbf{a}$ are real. These coefficients can be expressed formally in terms of the matrix $M$ in the following manner:

$$a_0 = \frac{1}{2} \operatorname{Tr}(M) \tag{23-a}$$

$$\mathbf{a} = \frac{1}{2} \operatorname{Tr}(M\boldsymbol{\sigma}) \tag{23-b}$$

These formulas can easily be proven from (8), (9) and (10).

# Complement $B_{IV}$
# DIAGONALIZATION OF A $2 \times 2$ HERMITIAN MATRIX



## 1. Introduction

In quantum mechanics, one must often diagonalize $2 \times 2$ matrices. When we need only the eigenvalues, it is very easy to solve the characteristic equation since it is of second degree. In principle, the calculation of the normalized eigenvectors is also extremely simple; however, if it is performed clumsily, it can lead to expressions which are unnecessarily complicated and difficult to handle.

The goal of this complement is to present a simple method of calculation which is applicable in all cases. After having changed the origin of the eigenvalues, we introduce the angles $\theta$ and $\varphi$, defined in terms of the matrix elements, which enable us to write the normalized eigenvectors in a simple easy-to-use form. The angles $\theta$ and $\varphi$ also have an interesting physical interpretation in the study of two-level systems, as we shall see in complement $C_{IV}$.

## 2. Changing the eigenvalue origin

Consider the Hermitian matrix:

$$(H) = \begin{pmatrix} H_{11} & H_{12} \\ H_{21} & H_{22} \end{pmatrix} \tag{1}$$

$H_{11}$ and $H_{22}$ are real. Moreover:

$$H_{12} = H_{21}^* \tag{2}$$

The matrix $(H)$ therefore represents, in an orthonormal basis, $\{ | \varphi_1 \rangle, | \varphi_2 \rangle \}$, a certain Hermitian operator $H$*.

* We use the letter $H$ because the Hermitian operator which we are trying to diagonalize is often a Hamiltonian. Nevertheless, the calculations presented in this complement can obviously be applied to any $2 \times 2$ Hermitian matrix.

Using the half-sum and half-difference of the diagonal elements $H_{11}$ and $H_{22}$, we can write $(H)$ in the following way:

$$(H) = \begin{pmatrix} \frac{1}{2}(H_{11} + H_{22}) & 0 \\ 0 & \frac{1}{2}(H_{11} + H_{22}) \end{pmatrix} + \begin{pmatrix} \frac{1}{2}(H_{11} - H_{22}) & H_{12} \\ H_{21} & -\frac{1}{2}(H_{11} - H_{22}) \end{pmatrix} \tag{3}$$

It follows that the operator $H$ itself can be decomposed into:

$$H = \frac{1}{2}(H_{11} + H_{22})\,\mathbb{1} + \frac{1}{2}(H_{11} - H_{22})\,K \tag{4}$$

where $\mathbb{1}$ is the identity operator and $K$ is the Hermitian operator represented in the $\{ | \varphi_1 \rangle, | \varphi_2 \rangle \}$ basis by the matrix:

$$(K) = \begin{pmatrix} 1 & \dfrac{2H_{12}}{H_{11} - H_{22}} \\ \dfrac{2H_{21}}{H_{11} - H_{22}} & -1 \end{pmatrix} \tag{5}$$

It is clear from (4) hat $H$ and $K$ have the same eigenvectors. Let $| \psi_\pm \rangle$ be these eigenvectors, and $E_\pm$ and $\kappa_\pm$, the corresponding eigenvalues for $H$ and $K$:

$$H | \psi_\pm \rangle = E_\pm | \psi_\pm \rangle \tag{6}$$

$$K | \psi_\pm \rangle = \kappa_\pm | \psi_\pm \rangle \tag{7}$$

From (4), we immediately conclude that:

$$E_\pm = \frac{1}{2}(H_{11} + H_{22}) + \frac{1}{2}(H_{11} - H_{22})\,\kappa_\pm \tag{8}$$

Finally, the first matrix appearing on the right-hand side of (3) plays a minor role : we could make it disappear by choosing for the eigenvalue origin $(H_{11} + H_{22})/2$*.

## 3. Calculation of the eigenvalues and eigenvectors

### a. ANGLES $\theta$ AND $\varphi$

Let $\theta$ and $\varphi$ be the angles defined in terms of the matrix elements $H_{ij}$ by :

$$\tan\theta = \frac{2\,|H_{21}|}{H_{11} - H_{22}} \qquad \text{with } 0 \leqslant \theta < \pi \tag{9}$$

$$H_{21} = |H_{21}|\,e^{i\varphi} \qquad \text{with } 0 \leqslant \varphi < 2\pi \tag{10}$$

* Furthermore, this new origin is the same, whatever the basis $\{ | \varphi_1 \rangle, | \varphi_2 \rangle \}$ initially chosen, since $H_{11} + H_{22} = \text{Tr}(H)$ is invariant under a change of orthonormal bases.

$\varphi$ is the argument of the complex number $H_{21}$. According to (2), we have $|H_{12}| = |H_{21}|$ and:

$$H_{12} = |H_{12}| \, e^{-i\varphi} \tag{11}$$

If we use (9), (10) and (11), the matrix $(K)$ becomes:

$$(K) = \begin{pmatrix} 1 & \tan\theta \, e^{-i\varphi} \\ \tan\theta \, e^{i\varphi} & -1 \end{pmatrix} \tag{12}$$

#### b. EIGENVALUES OF $K$

The characteristic equation of the matrix (12)

$$\text{Det}\left[(K) - \kappa I\right] = \kappa^2 - 1 - \tan^2\theta = 0 \tag{13}$$

directly yields the eigenvalues $\kappa_+$ and $\kappa_-$ of $(K)$:

$$\kappa_+ = +\frac{1}{\cos\theta} \tag{14-a}$$

$$\kappa_- = -\frac{1}{\cos\theta} \tag{14-b}$$

We see that they are indeed real (property of a Hermitian matrix, *cf.* § D-2-a of chapter II). If we want to express $1/\cos\theta$ in terms of $H_{ij}$, all we need to do is use (9) and notice that $\cos\theta$ and $\tan\theta$ have the same sign since $0 \leqslant \theta < \pi$:

$$\frac{1}{\cos\theta} = \frac{\sqrt{(H_{11} - H_{22})^2 + 4\,|H_{12}|^2}}{H_{11} - H_{22}} \tag{15}$$

#### c. EIGENVALUES OF $H$

Using (8), (14) and (15), we immediately obtain:

$$E_+ = \frac{1}{2}(H_{11} + H_{22}) + \frac{1}{2}\sqrt{(H_{11} - H_{22})^2 + 4\,|H_{12}|^2} \tag{16-a}$$

$$E_- = \frac{1}{2}(H_{11} + H_{22}) - \frac{1}{2}\sqrt{(H_{11} - H_{22})^2 + 4\,|H_{12}|^2} \tag{16-b}$$

COMMENTS:

(*i*) As we have already said, the eigenvalues (16) can easily be obtained from the characteristic equation of the matrix $(H)$. If we need only the eigenvalues of $(H)$, it is therefore not necessary to introduce the angles $\theta$ and $\varphi$ as we have done here. On the other hand, we shall see in the following section that this method is very useful when we need to use the normalized eigenvectors of $H$.

(*ii*) It can be verified immediately from formulas (16) that:

$$E_+ + E_- = H_{11} + H_{22} = \text{Tr}\,(H) \tag{17}$$

$$E_+ E_- = H_{11} H_{22} - |H_{12}|^2 = \text{Det}\,(H) \tag{18}$$

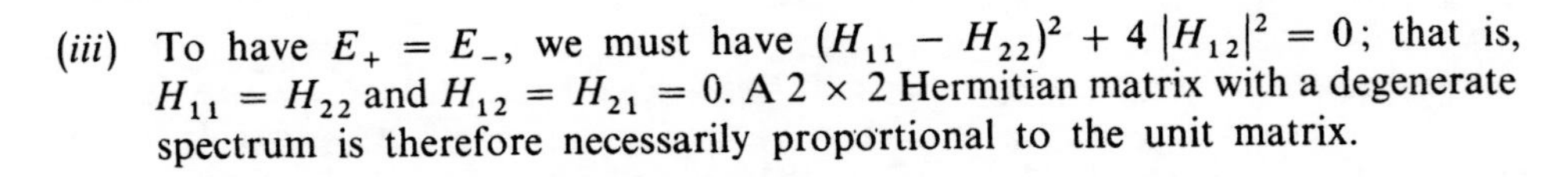

(*iii*) To have $E_+ = E_-$, we must have $(H_{11} - H_{22})^2 + 4\,|H_{12}|^2 = 0$; that is, $H_{11} = H_{22}$ and $H_{12} = H_{21} = 0$. A 2 × 2 Hermitian matrix with a degenerate spectrum is therefore necessarily proportional to the unit matrix.

### d. NORMALIZED EIGENVECTORS OF *H*

Let $a$ and $b$ be the components of $|\,\psi_+\rangle$ on $|\,\varphi_1\rangle$ and $|\,\varphi_2\rangle$. According to (7), (12) and (14-a), they must satisfy:

$$\begin{pmatrix} 1 & \tan\theta\, e^{-i\varphi} \\ \tan\theta\, e^{i\varphi} & -1 \end{pmatrix}\begin{pmatrix} a \\ b \end{pmatrix} = \frac{1}{\cos\theta}\begin{pmatrix} a \\ b \end{pmatrix} \tag{19}$$

which yields:

$$\left(1 - \frac{1}{\cos\theta}\right)a + \tan\theta\, e^{-i\varphi}\, b = 0 \tag{20}$$

that is:

$$-\left(\sin\frac{\theta}{2}\, e^{i\varphi/2}\right)a + \left(\cos\frac{\theta}{2}\, e^{-i\varphi/2}\right)b = 0 \tag{21}$$

The normalized eigenvector $|\,\psi_+\rangle$ can therefore be written:

$$|\,\psi_+\rangle = \cos\frac{\theta}{2}\, e^{-i\varphi/2}\,|\,\varphi_1\rangle + \sin\frac{\theta}{2}\, e^{i\varphi/2}\,|\,\varphi_2\rangle \tag{22}$$

An analogous calculation would yield:

$$|\,\psi_-\rangle = -\sin\frac{\theta}{2}\, e^{-i\varphi/2}\,|\,\varphi_1\rangle + \cos\frac{\theta}{2}\, e^{i\varphi/2}\,|\,\varphi_2\rangle \tag{23}$$

It can be verified that $|\,\psi_+\rangle$ and $|\,\psi_-\rangle$ are orthogonal.

COMMENT:

While the trigonometric functions of the angle $\theta$ can be expressed rather simply in terms of the matrix elements $H_{ij}$ [see, for example, formulas (9) and (15)], the same is not true of those of the angle $\theta/2$. Consequently, formulas (22) and (23) for the normalized eigenvectors $|\,\psi_+\rangle$ and $|\,\psi_-\rangle$ become complicated when $\cos\theta/2$ and $\sin\theta/2$ are replaced by their expressions in terms of $H_{ij}$; they are no longer very convenient. It is better to use expressions (22) and (23) directly, keeping the functions $\cos\theta/2$ and $\sin\theta/2$ during the entire calculation involving the normalized eigenvectors of $H$. Furthermore, the final result of the calculation often involves only functions of the angle $\theta$ (see, for example, the calculation of §C-3-b of chapter IV) and, consequently, can be expressed simply in terms of the $H_{ij}$. Expressions (22) and (23) thus enable us to carry out the intermediate calculations elegantly, avoiding unnecessarily complicated expressions. This is the advantage of the method presented in this complement. Another advantage concerns the physical interpretation and will be discussed in the next complement.

# Complement $C_{IV}$

## FICTITIOUS SPIN 1/2 ASSOCIATED WITH A TWO-LEVEL SYSTEM



### 1. Introduction

Consider a two-level system whose Hamiltonian $H$ is represented, in an orthonormal basis $\{ | \varphi_1 \rangle, | \varphi_2 \rangle \}$, by the Hermitian matrix $(H)$ [formula (1) of complement $B_{IV}$]★. If we choose $(H_{11} + H_{22})/2$ as the new energy origin, the matrix $(H)$ becomes:

$$(H) = \begin{pmatrix} \frac{1}{2}(H_{11} - H_{22}) & H_{12} \\ H_{21} & -\frac{1}{2}(H_{11} - H_{22}) \end{pmatrix} \tag{1}$$

Although the two-level system under consideration is not necessarily a spin 1/2, we can always associate with it a spin 1/2 whose Hamiltonian $\tilde{H}$ is represented by the same matrix $(H)$ in the $\{ | + \rangle, | - \rangle \}$ basis of eigenstates of the $S_z$ component of this spin. We shall see that $(H)$ can then be interpreted as describing the interaction of this "fictitious spin" with a static magnetic field $\mathbf{B}$, whose direction and modulus are very simply related to the parameters introduced in the preceding complement in the discussion of the diagonalization of $(H)$. Thus it is possible to give a simple physical meaning to these parameters.

Moreover, if the Hamiltonian $H$ is the sum $H = H_0 + W$ of two operators, we shall see that we can associate with $H$, $H_0$ and $W$ three magnetic fields, $\mathbf{B}$, $\mathbf{B}_0$ and $\mathbf{b}$, such that $\mathbf{B} = \mathbf{B}_0 + \mathbf{b}$. Introducing the coupling $W$ is equivalent, in terms of fictitious spin, to adding the field $\mathbf{b}$ to $\mathbf{B}_0$. We shall show that this point of view enables us to interpret very simply the different effects studied in § C of chapter IV.

### 2. Interpretation of the Hamiltonian in terms of fictitious spin

We saw in chapter IV that the Hamiltonian $\tilde{H}$ of the coupling between a spin 1/2 and a magnetic field $\mathbf{B}$, of components $B_x$, $B_y$, $B_z$, can be written :

$$\tilde{H} = -\gamma \, \mathbf{B}.\mathbf{S} = -\gamma(B_x S_x + B_y S_y + B_z S_z) \tag{2}$$

★ We are using the same notation as in complement $B_{IV}$ and chapter IV.

To calculate the matrix associated with this operator, we substitute into this relation the matrices associated with $S_x$, $S_y$, $S_z$ [chap. IV, relations (A-15), (A-16), (A-17)]. This immediately yields :

$$(\tilde{H}) = -\frac{\gamma\hbar}{2}\begin{pmatrix} B_z & B_x - iB_y \\ B_x + iB_y & -B_z \end{pmatrix} \tag{3}$$

Therefore, to make matrix (1) identical to $(\tilde{H})$, we must simply choose a "fictitious field" **B** defined by:

$$\begin{cases} B_x = -\dfrac{2}{\gamma\hbar}\,\text{Re}\,H_{12} \\ B_y = \dfrac{2}{\gamma\hbar}\,\text{Im}\,H_{12} \\ B_z = \dfrac{1}{\gamma\hbar}\,(H_{22} - H_{11}) \end{cases} \tag{4}$$

Note that the modulus $B_\perp$ of the projection $\mathbf{B}_\perp$ of **B** onto the $xOy$ plane is then equal to:

$$B_\perp = \frac{2}{\hbar}\left|\frac{H_{12}}{\gamma}\right| \tag{5}$$

According to formulas (9) and (10) of complement $B_{IV}$, the angles $\theta$ and $\varphi$ associated with the matrix $(H) = (\tilde{H})$ written in (3) are given by:

$$\begin{cases} \tan\theta = \dfrac{|\gamma B_\perp|}{-\gamma B_z} & 0 \leqslant \theta < \pi \\ -\gamma(B_x + iB_y) = |\gamma B_\perp|\, e^{i\varphi} & 0 \leqslant \varphi < 2\pi \end{cases} \tag{6}$$

The gyromagnetic ratio $\gamma$ is a simple calculation tool and can have an arbitrary value. If we agree to choose $\gamma$ negative, relations (6) show that the angles $\theta$ and $\varphi$ associated with the matrix $(H)$ are simply the polar angles of the direction of the field **B** (if we had chosen $\gamma$ positive, they would be those of the opposite direction).

Finally, we see that we can forget the two-level system with which we started and consider the matrix $(H)$ as representing, in the basis of the eigenstates $|+\rangle$ and $|-\rangle$ of $S_z$, the Hamiltonian $\tilde{H}$ of a spin 1/2 placed in a field **B** whose components are given by (4). $\tilde{H}$ can also be written:

$$\tilde{H} = \omega S_u \tag{7-a}$$

where $S_u$ is the operator $\mathbf{S}\cdot\mathbf{u}$ which describes the spin component along the direction **u**, whose polar angles are $\theta$ and $\varphi$, and $\omega$ is the Larmor angular velocity:

$$\omega = |\gamma|\,|\mathbf{B}| \tag{7-b}$$

The following table summarizes the various correspondences between the two-level system and the associated fictitious spin 1/2.

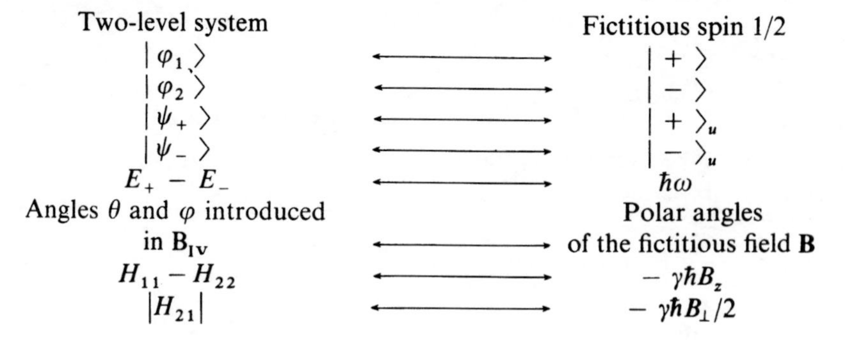

| Two-level system | | Fictitious spin 1/2 |
|---|---|---|
| $\lvert \varphi_1 \rangle$ | ⟷ | $\lvert + \rangle$ |
| $\lvert \varphi_2 \rangle$ | ⟷ | $\lvert - \rangle$ |
| $\lvert \psi_+ \rangle$ | ⟷ | $\lvert + \rangle_u$ |
| $\lvert \psi_- \rangle$ | ⟷ | $\lvert - \rangle_u$ |
| $E_+ - E_-$ | ⟷ | $\hbar\omega$ |
| Angles $\theta$ and $\varphi$ introduced in $B_{IV}$ | ⟷ | Polar angles of the fictitious field **B** |
| $H_{11} - H_{22}$ | ⟷ | $-\gamma\hbar B_z$ |
| $\lvert H_{21} \rvert$ | ⟷ | $-\gamma\hbar B_\perp/2$ |

## 3. Geometrical interpretation of the various effects discussed in § C of chapter IV

### a. FICTITIOUS MAGNETIC FIELDS ASSOCIATED WITH $H_0$, $W$ AND $H$

Assume, as in §C of chapter IV, that $H$ appears as the sum of two terms:

$$H = H_0 + W \tag{8}$$

In the $\{ \lvert \varphi_1 \rangle, \lvert \varphi_2 \rangle \}$ basis, the unperturbed Hamiltonian $H_0$ is represented by a diagonal matrix which, with a suitable choice of the energy origin, is written:

$$(H_0) = \begin{pmatrix} \dfrac{E_1 - E_2}{2} & 0 \\ 0 & -\dfrac{E_1 - E_2}{2} \end{pmatrix} \tag{9}$$

As far as the coupling $W$ is concerned, we assume, as in §C of chapter IV, that it is purely non-diagonal:

$$(W) = \begin{pmatrix} 0 & W_{12} \\ W_{21} & 0 \end{pmatrix} \tag{10}$$

The discussion of the preceding section then enables us to associate with $(H_0)$ and $(W)$ two fields $\mathbf{B}_0$ and $\mathbf{b}$ such that [*cf.* formulas (4) and (5)]:

$$\begin{cases} B_{0z} = \dfrac{E_2 - E_1}{\gamma\hbar} \\ B_{0\perp} = 0 \end{cases} \tag{11}$$

$$\begin{cases} b_z = 0 \\ b_\perp = \dfrac{2}{\hbar}\left| \dfrac{W_{12}}{\gamma} \right| \end{cases} \tag{12}$$

$\mathbf{B}_0$ is therefore parallel to $Oz$ and proportional to $(E_1 - E_2)/2$; $\mathbf{b}$ is perpendicular to $Oz$ and proportional to $|W_{12}|$. Since $(H) = (H_0) + (W)$, the field $\mathbf{B}$ associated with the total Hamiltonian is the vector sum of $\mathbf{B}_0$ and $\mathbf{b}$:

$$\mathbf{B} = \mathbf{B}_0 + \mathbf{b} \tag{13}$$

The three fields $\mathbf{B}_0$, $\mathbf{b}$ and $\mathbf{B}$ are shown in figure 1; the angle $\theta$ introduced in § C-2-a of chapter IV is the angle between $\mathbf{B}_0$ and $\mathbf{B}$, since $\mathbf{B}_0$ is parallel to $Oz$.

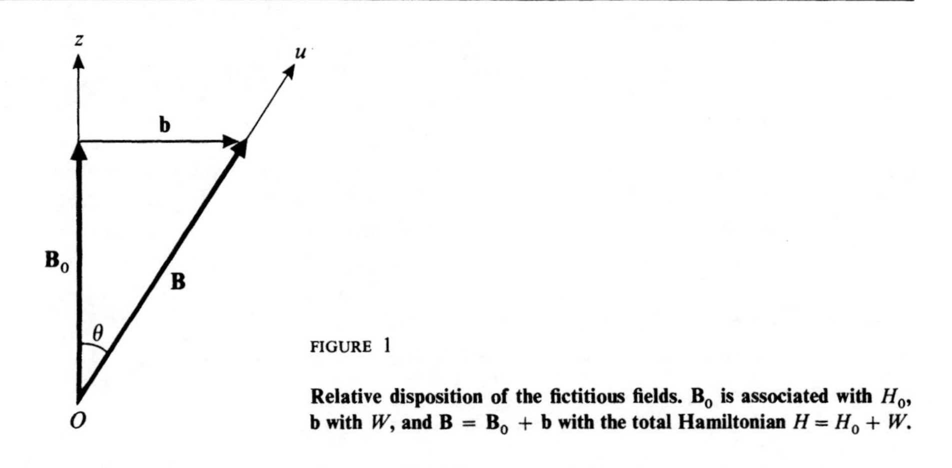


FIGURE 1

**Relative disposition of the fictitious fields. $\mathbf{B}_0$ is associated with $H_0$, b with $W$, and $\mathbf{B} = \mathbf{B}_0 + \mathbf{b}$ with the total Hamiltonian $H = H_0 + W$.**

The strong coupling condition introduced in § C-2 of chapter IV ($|W_{12}| \gg |E_1 - E_2|$) is equivalent to $|\mathbf{b}| \gg |\mathbf{B}_0|$ (fig. 2-a). The weak coupling condition ($|W_{12}| \ll |E_1 - E_2|$) is equivalent to $|\mathbf{b}| \ll |\mathbf{B}_0|$ (fig. 2-b).

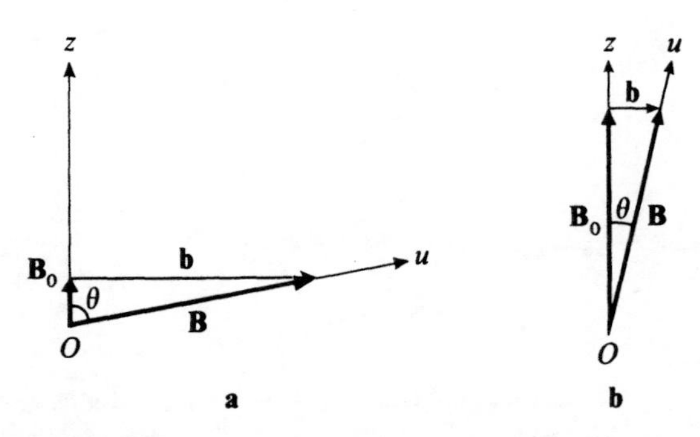


FIGURE 2

**Relative disposition of the fictitious fields $\mathbf{B}_0$, b and B in the case of strong coupling (fig. a) and weak coupling (fig. b).**

### b. EFFECT OF COUPLING ON THE EIGENVALUES AND EIGENVECTORS OF THE HAMILTONIAN

$E_1 - E_2$ and $E_+ - E_-$ correspond respectively to the Larmor angular velocities $\omega_0 = |\gamma|\,|\mathbf{B}_0|$ and $\omega = |\gamma|\,|\mathbf{B}|$ in the fields $\mathbf{B}_0$ and $\mathbf{B}$. We see in figure 1 that $\mathbf{B}_0$, $\mathbf{b}$ and $\mathbf{B}$ form a right triangle whose hypotenuse is $\mathbf{B}$; we therefore always have $|\mathbf{B}| \geqslant |\mathbf{B}_0|$, which again shows that $E_+ - E_-$ is always greater than $|E_1 - E_2|$.

For a weak coupling (fig. 2-b), the difference between $|\mathbf{B}|$ and $|\mathbf{B}_0|$ is very small in relative value, being of second order in $|\mathbf{b}|/|\mathbf{B}_0|$. From this we deduce immediately that $E_+ - E_-$ and $E_1 - E_2$ differ in relative value by terms of second order in $|W_{12}|/(E_1 - E_2)$. On the other hand, for a strong coupling (fig. 2-a), $|\mathbf{B}|$ is much larger than $|\mathbf{B}_0|$ and practically equal to $|\mathbf{b}|$; $E_+ - E_-$ is then much larger than $|E_1 - E_2|$ and practically proportional to $|W_{12}|$. We thus find again all the results of §C-2 of chapter IV.

As far as the effect of the coupling on the eigenvectors is concerned, it can also be understood very simply from figures 1 and 2. The eigenvectors of $H$ and $H_0$ are associated respectively with the eigenvectors of the components of $\mathbf{S}$ on the $Ou$ and $Oz$ axes. These two axes are practically parallel in the case of weak coupling (fig. 2-b) and perpendicular in the case of strong coupling (fig. 2-a). The eigenvectors of $S_u$ and $S_z$, and, consequently, those of $H$ and $H_0$, are very close in the first case and very different in the second one.

### c. GEOMETRICAL INTERPRETATION OF $\mathscr{P}_{12}(t)$

In terms of fictitious spin, the problem considered in §C-3 of chapter IV can be put in the following way: at time $t = 0$, the fictitious spin associated with the two-level system is in the eigenstate $|+\rangle$ of $S_z$; $\mathbf{b}$ is added to $\mathbf{B}_0$; what

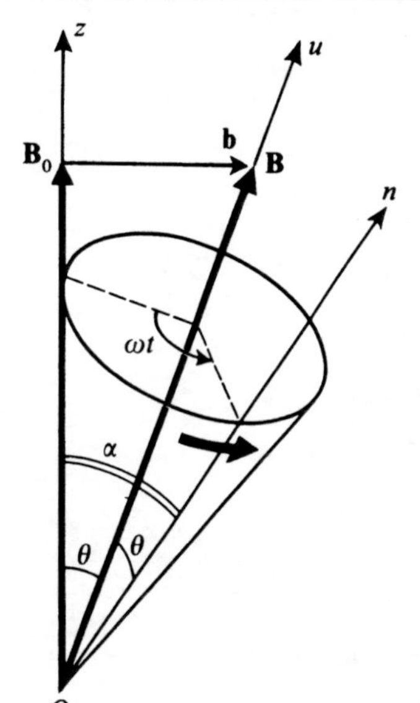


FIGURE 3

**Geometrical interpretation of Rabi's formula in terms of fictitious spin. Under the effect of the coupling (represented by b), the spin, initially oriented along $Oz$, precesses about B; consequently, the probability of finding $-\hbar/2$ in a measurement of its $S_z$ component on $Oz$ is an oscillating function of time.**

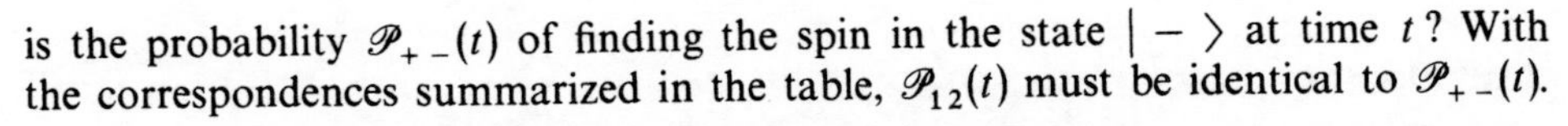



is the probability $\mathscr{P}_{+-}(t)$ of finding the spin in the state $| - \rangle$ at time $t$? With the correspondences summarized in the table, $\mathscr{P}_{12}(t)$ must be identical to $\mathscr{P}_{+-}(t)$.

The calculation of $\mathscr{P}_{+-}(t)$ is then very simple since the time evolution of the spin reduces to a Larmor precession about **B** (fig. 3). During this precession, the angle $\theta$ between the spin and the direction $Ou$ of **B** remains constant. At time $t$, the spin points in the direction $On$, making an angle $\alpha$ with $Oz$; the angle formed by the $(Oz, Ou)$ and $(Ou, On)$ planes is equal to $\omega t$. A classical formula of spherical trigonometry enables us to write:

$$\cos \alpha = \cos^2 \theta + \sin^2 \theta \cos \omega t \tag{14}$$

Now, when the spin points in a direction which makes an angle of $\alpha$ with $Oz$, the probability of finding it in the state $| - \rangle$ of $S_z$ is equal (*cf.* § B-2-b of chapter IV) to $\sin^2 \alpha/2 = (1 - \cos \alpha)/2$. From this we deduce, using (14), that:

$$\mathscr{P}_{+-}(t) = \sin^2 \frac{\alpha}{2} = \frac{1}{2} \sin^2 \theta (1 - \cos \omega t) \tag{15}$$

This result is identical, when we replace $\omega$ by $(E_+ - E_-)/\hbar$, to formula (C-31) of chapter IV (Rabi's formula). We have thus given this formula a purely geometrical interpretation.

**References and suggestions for further reading:**

Abragam (14.1), chap. II, §F; Sargent et al. (15.5), §7-5; Allen (15.7), chap. 2; see also the article by Feynman et al. (1.33).

# Complement $D_{IV}$

## SYSTEM OF TWO SPIN 1/2 PARTICLES



In this complement, we intend to use the formalism introduced in §A-2 of chapter IV to describe a system of two spin 1/2 particles. This case is hardly more complicated than that of a single spin 1/2 particle. Its interest, as far as the postulates are concerned, lies in the fact that none of the various spin observables alone constitutes a C.S.C.O. (while this is the case for one spin alone). Thus, we shall be able to consider measurements bearing either on one observable with a degenerate spectrum or simultaneously on two observables. In addition, this study provides a very simple illustration of the concept of a tensor product, introduced in §F of chapter II. We shall be concerned, as in chapter IV, only with the internal degrees of freedom (spin states), and we shall moreover assume that the two particles which constitute the system are not identical (systems of identical particles will be studied in a general way in chapter XIV).

## 1. Quantum mechanical description

We saw in chapter IV how to describe quantum mechanically the spin state of a spin 1/2 particle. Thus, all we need to do is apply the results of §F of chapter II in order to know how to describe systems of two spin 1/2 particles.

### a. STATE SPACE

We shall use the indices 1 and 2 to distinguish between the two particles. When particle (1) is alone, its spin state is defined by a ket which belongs to a two-dimensional state space $\mathscr{E}_S(1)$. Similarly, the spin states of particle (2) alone form a two-dimensional space $\mathscr{E}_S(2)$. We shall designate by $\mathbf{S}_1$ and $\mathbf{S}_2$ the spin observables of particles (1) and (2) respectively. In $\mathscr{E}_S(1)$ [or $\mathscr{E}_S(2)$], we choose as a basis the eigenkets of $S_{1z}$ (or $S_{2z}$), which we shall denote by $| 1 : + \rangle$ and $| 1 : - \rangle$ (or $| 2 : + \rangle$ and $| 2 : - \rangle$). The most general ket of $\mathscr{E}_S(1)$ can be written :

$$| \varphi(1) \rangle = \alpha_1 | 1 : + \rangle + \beta_1 | 1 : - \rangle \tag{1}$$

and that of $\mathscr{E}_S(2)$:

$$| \chi(2) \rangle = \alpha_2 \, | 2 : + \rangle + \beta_2 \, | 2 : - \rangle \tag{2}$$

($\alpha_1$, $\beta_1$, $\alpha_2$, $\beta_2$ are arbitrary complex numbers).

When we join the two particles to make a single system, the state space $\mathscr{E}_S$ of such a system is the tensor product of the two preceding spaces:

$$\mathscr{E}_S = \mathscr{E}_S(1) \otimes \mathscr{E}_S(2) \tag{3}$$

In the first place, this means that a basis of $\mathscr{E}_S$ can be obtained by multiplying tensorially the two bases defined above for $\mathscr{E}_S(1)$ and $\mathscr{E}_S(2)$. We shall use the following notation :

$$\begin{aligned} | + + \rangle &= | 1 : + \rangle | 2 : + \rangle \\ | + - \rangle &= | 1 : + \rangle | 2 : - \rangle \\ | - + \rangle &= | 1 : - \rangle | 2 : + \rangle \\ | - - \rangle &= | 1 : - \rangle | 2 : - \rangle \end{aligned} \tag{4}$$

In the state $| + - \rangle$, for example, the component along $Oz$ of the spin of particle (1) is $+ \hbar/2$, with absolute certainty; that of the spin of particle (2) is $- \hbar/2$, with absolute certainty. We shall agree here to denote by $\langle + - |$ the conjugate bra of the ket $| + - \rangle$; the order of the symbols is therefore the same in the ket and in the bra: the first symbol is always associated with particle (1) and the second, with particle (2).

The space $\mathscr{E}_S$ is therefore four-dimensional. Since the $\{ | 1 : \pm \rangle \}$ and $\{ | 2 : \pm \rangle \}$ bases are orthonormal in $\mathscr{E}_S(1)$ and $\mathscr{E}_S(2)$ respectively, the basis (4) is orthonormal in $\mathscr{E}_S$:

$$\langle \varepsilon_1 \varepsilon_2 | \varepsilon'_1 \varepsilon'_2 \rangle = \delta_{\varepsilon_1 \varepsilon'_1} \delta_{\varepsilon_2 \varepsilon'_2} \tag{5}$$

($\varepsilon_1$, $\varepsilon_2$, $\varepsilon'_1$, $\varepsilon'_2$ are to be replaced by $+$ or $-$ depending on the case; $\delta_{\varepsilon\varepsilon'}$ is equal to 1 if $\varepsilon$ and $\varepsilon'$ are identical and 0 if they are different). The system of vectors (4) also satisfies a closure relation in $\mathscr{E}_S$:

$$\begin{aligned} \sum_{\varepsilon_1 \varepsilon_2} | \varepsilon_1 \varepsilon_2 \rangle \langle \varepsilon_1 \varepsilon_2 | &= | + + \rangle \langle + + | + | + - \rangle \langle + - | + \\ & \quad | - + \rangle \langle - + | + | - - \rangle \langle - - | = \mathbb{1} \end{aligned} \tag{6}$$

#### b. COMPLETE SETS OF COMMUTING OBSERVABLES

We extend into $\mathscr{E}_S$ the observables $\mathbf{S}_1$ and $\mathbf{S}_2$ which were originally defined in $\mathscr{E}_S(1)$ and $\mathscr{E}_S(2)$ (as in chapter II, we shall continue to denote these extensions by $\mathbf{S}_1$ and $\mathbf{S}_2$). Their action on the kets of the basis (4) is simple: the components of $\mathbf{S}_1$, for example, act only on the part of the ket related to particle (1). In particular, the vectors of the basis (4) are simultaneous eigenvectors of $S_{1z}$ and $S_{2z}$:

$$\begin{aligned} S_{1z} | \varepsilon_1 \varepsilon_2 \rangle &= \frac{\hbar}{2} \varepsilon_1 | \varepsilon_1 \varepsilon_2 \rangle \\ S_{2z} | \varepsilon_1 \varepsilon_2 \rangle &= \frac{\hbar}{2} \varepsilon_2 | \varepsilon_1 \varepsilon_2 \rangle \end{aligned} \tag{7}$$

For the other components of $\mathbf{S}_1$ and $\mathbf{S}_2$, we apply the formulas given in §A-2 of chapter IV. For example, we know from relation (A-16) of chapter IV how $S_{1x}$ acts on the kets $| 1 : \pm \rangle$:

$$\begin{aligned} S_{1x} | 1 : + \rangle &= \frac{\hbar}{2} | 1 : - \rangle \\ S_{1x} | 1 : - \rangle &= \frac{\hbar}{2} | 1 : + \rangle \end{aligned} \tag{8}$$

From this we deduce the action of $S_{1x}$ on the kets (4):

$$\begin{aligned} S_{1x} | + + \rangle &= \frac{\hbar}{2} | - + \rangle \\ S_{1x} | + - \rangle &= \frac{\hbar}{2} | - - \rangle \\ S_{1x} | - + \rangle &= \frac{\hbar}{2} | + + \rangle \\ S_{1x} | - - \rangle &= \frac{\hbar}{2} | + - \rangle \end{aligned} \tag{9}$$

It is then easy to verify that, although the three components of $\mathbf{S}_1$ (or of $\mathbf{S}_2$) do not commute with each other, *any component of* $\mathbf{S}_1$ *commutes with any component of* $\mathbf{S}_2$.

In $\mathscr{E}_S(1)$, the observable $S_{1z}$ alone constituted a C.S.C.O., and the same was true of $S_{2z}$ in $\mathscr{E}_S(2)$. In $\mathscr{E}_S$, the eigenvalues of $S_{1z}$ and $S_{2z}$ remain $\pm \hbar/2$, but each of them is two-fold degenerate. To the eigenvalue $+ \hbar/2$ of $S_{1z}$, for example, correspond two orthogonal vectors, $| + + \rangle$ and $| + - \rangle$ [formulas (7)] and all their linear combinations. Therefore, in $\mathscr{E}_S$, neither $S_{1z}$ nor $S_{2z}$ (taken separately) constitutes a C.S.C.O. On the other hand, the set $\{ S_{1z}, S_{2z} \}$ is a C.S.C.O. in $\mathscr{E}_S$, as can be seen from formulas (7).

This is obviously not the only C.S.C.O. that can be constructed. For example, another one is $\{ S_{1z}, S_{2x} \}$. These two observables commute, as we noted above, and each of them constitutes a C.S.C.O. in the space in which it was initially defined. The eigenvectors which are common to $S_{1z}$ and $S_{2x}$ are obtained by taking the tensor product of their respective eigenvectors in $\mathscr{E}_S(1)$ and $\mathscr{E}_S(2)$. Using relation (A-20) of chapter IV, we find:

$$\begin{aligned} | 1 : + \rangle | 2 : + \rangle_x &= \frac{1}{\sqrt{2}} [ | + + \rangle + | + - \rangle ] \\ | 1 : + \rangle | 2 : - \rangle_x &= \frac{1}{\sqrt{2}} [ | + + \rangle - | + - \rangle ] \\ | 1 : - \rangle | 2 : + \rangle_x &= \frac{1}{\sqrt{2}} [ | - + \rangle + | - - \rangle ] \\ | 1 : - \rangle | 2 : - \rangle_x &= \frac{1}{\sqrt{2}} [ | - + \rangle - | - - \rangle ] \end{aligned} \tag{10}$$

#### c. THE MOST GENERAL STATE

The vectors (4) were obtained by multiplying tensorially a ket of $\mathscr{E}_S(1)$ and a ket of $\mathscr{E}_S(2)$. More generally, using an arbitrary ket of $\mathscr{E}_S(1)$ [such as (1)] and an arbitrary ket of $\mathscr{E}_S(2)$ [such as (2)], we can construct a ket of $\mathscr{E}_S$:

$$| \varphi(1) \rangle | \chi(2) \rangle \\ = \alpha_1\alpha_2 | + + \rangle + \alpha_1\beta_2 | + - \rangle + \alpha_2\beta_1 | - + \rangle + \beta_1\beta_2 | - - \rangle \quad (11)$$

The components of such a ket in the basis (4) are the products of the components of $| \varphi(1) \rangle$ and $| \chi(2) \rangle$ in the bases of $\mathscr{E}_S(1)$ and $\mathscr{E}_S(2)$ which were used to construct (4).

But *all the kets of* $\mathscr{E}_S$ *are not tensor products*. The most general ket of $\mathscr{E}_S$ is an arbitrary linear combination of the basis vectors:

$$| \psi \rangle = \alpha | + + \rangle + \beta | + - \rangle + \gamma | - + \rangle + \delta | - - \rangle \quad (12)$$

If we want to normalize $| \psi \rangle$, we must choose:

$$|\alpha|^2 + |\beta|^2 + |\gamma|^2 + |\delta|^2 = 1 \quad (13)$$

Given $| \psi \rangle$, it is not in general possible to find two kets $| \varphi(1) \rangle$ and $| \chi(2) \rangle$ of which it is the tensor product. For (12) to be of the form (11), we must have, in particular:

$$\frac{\alpha}{\beta} = \frac{\gamma}{\delta} \quad (14)$$

and this condition is not necessarily fulfilled.

## 2. Prediction of the measurement results

We are now going to envisage a certain number of measurements which can be performed on a system of two spin 1/2 particles and we shall calculate the predictions furnished by the postulates for each of them. We shall assume each time that the state of the system immediately before the measurement is described by the normalized ket (12).

#### a. MEASUREMENTS BEARING SIMULTANEOUSLY ON THE TWO SPINS

Since any component of $\mathbf{S}_1$ commutes with any component of $\mathbf{S}_2$, we can envisage measuring them simultaneously (chap. III, §C-6-a). To calculate the predictions related to such measurements, all we need to do is use the eigenvectors common to the two observables.

##### α. *First example*

First of all, let us assume that we are simultaneously measuring $S_{1z}$ and $S_{2z}$. What are the probabilities of the various results that can be obtained?

Since the set $\{ S_{1z}, S_{2z} \}$ is a C.S.C.O., there exists only one state associated with each measurement result. If the system is in the state (12) before the measurement, we can therefore find:

$$+\frac{\hbar}{2} \text{ for } S_{1z} \text{ and } +\frac{\hbar}{2} \text{ for } S_{2z}, \text{ with the probability } |\langle + + | \psi \rangle|^2 = |\alpha|^2$$

$$+\frac{\hbar}{2} \quad \text{''} \quad -\frac{\hbar}{2} \quad \text{''} \quad \text{''} \quad |\langle + - | \psi \rangle|^2 = |\beta|^2$$

$$-\frac{\hbar}{2} \quad \text{''} \quad +\frac{\hbar}{2} \quad \text{''} \quad \text{''} \quad |\langle - + | \psi \rangle|^2 = |\gamma|^2$$

$$-\frac{\hbar}{2} \quad \text{''} \quad -\frac{\hbar}{2} \quad \text{''} \quad \text{''} \quad |\langle - - | \psi \rangle|^2 = |\delta|^2 \quad (15)$$

β. *Second example*

We now measure $S_{1y}$ and $S_{2z}$. What is the probability of obtaining $+ \hbar/2$ for each of the two observables?

Here again, $\{ S_{1y}, S_{2z} \}$ constitutes a C.S.C.O. The eigenvector common to $S_{1y}$ and $S_{2z}$ which corresponds to the eigenvalues $+ \hbar/2$ and $+ \hbar/2$ is the tensor product of the vector $| 1 : + \rangle_y$ and the vector $| 2 : + \rangle$:

$$| 1 : + \rangle_y \, | 2 : + \rangle = \frac{1}{\sqrt{2}} \left[ | + + \rangle + i | - + \rangle \right] \tag{16}$$

Applying the fourth postulate of chapter III, we find that the probability we are looking for is:

$$\mathscr{P} = \left| \frac{1}{\sqrt{2}} \left[ \langle + + | - i \langle - + | \right] | \psi \rangle \right|^2$$
$$= \frac{1}{2} | \alpha - i\gamma |^2 \tag{17}$$

The result therefore appears in the form of a "square of a sum"*.

After the measurement, if we have actually found $+ \hbar/2$ for $S_{1y}$ and $+ \hbar/2$ for $S_{2z}$, the system is in the state (16).

**b. MEASUREMENTS BEARING ON ONE SPIN ALONE**

It is obviously possible to measure only one component of one of the two spins. In this case, since none of these components constitutes by itself a C.S.C.O., there exist several eigenvectors corresponding to the same measurement result, and the corresponding probability will be a "sum of squares".

* It must be remembered that the sign of $i$ changes when we go from (16) to the conjugate bra. If this were to be forgotten, the result obtained would be incorrect ($|\alpha + i\gamma|^2 \neq |\alpha - i\gamma|^2$ since $\alpha/\gamma$ is not in general real).

α. *First example*

We measure only $S_{1z}$. What results can be found, and with what probabilities?

The possible results are the eigenvalues $\pm \hbar/2$ of $S_{1z}$. Each of them is doubly degenerate. In the associated eigensubspace, we choose an orthonormal basis: we can, for example, take $\{ | + + \rangle, | + - \rangle \}$ for $+ \hbar/2$ and $\{ | - + \rangle, | - - \rangle \}$ for $- \hbar/2$. We then obtain:

$$\begin{aligned} \mathscr{P}\left(+\frac{\hbar}{2}\right) &= |\langle + + | \psi \rangle|^2 + |\langle + - | \psi \rangle|^2 \\ &= |\alpha|^2 + |\beta|^2 \\ \mathscr{P}\left(-\frac{\hbar}{2}\right) &= |\langle - + | \psi \rangle|^2 + |\langle - - | \psi \rangle|^2 \\ &= |\gamma|^2 + |\delta|^2 \end{aligned} \tag{18}$$

COMMENT:

Since we are not performing any measurement on the spin (2), the choice of the basis in $\mathscr{E}_S(2)$ is arbitrary. We can, for example, choose as a basis of the eigensubspace of $S_{1z}$ associated with the eigenvalue $+ \hbar/2$ the vectors:

$$| 1 : + \rangle | 2 : \pm \rangle_x = \frac{1}{\sqrt{2}}[| + + \rangle \pm | + - \rangle] \tag{19}$$

which again gives us :

$$\begin{aligned} \mathscr{P}\left(+\frac{\hbar}{2}\right) &= \frac{1}{2}|\alpha + \beta|^2 + \frac{1}{2}|\alpha - \beta|^2 \\ &= |\alpha|^2 + |\beta|^2 \end{aligned} \tag{20}$$

The general proof of the fact that the probability obtained is independent (in the case of a degenerate eigenvalue) of the choice of the basis in the corresponding eigensubspace was given in § B-3-b-α of chapter III.

β. *Second example*

Now it is $S_{2x}$ that we want to measure. What is the probability of obtaining $- \hbar/2$?

The eigensubspace associated with the eigenvalue $- \hbar/2$ of $S_{2x}$ is two-dimensional. We can choose for a basis in it:

$$\begin{aligned} | 1: + \rangle | 2: - \rangle_x &= \frac{1}{\sqrt{2}}[| + + \rangle - | + - \rangle] \\ | 1: - \rangle | 2: - \rangle_x &= \frac{1}{\sqrt{2}}[| - + \rangle - | - - \rangle] \end{aligned} \tag{21}$$

We then find:

$$\begin{aligned} \mathscr{P} &= \left| \frac{1}{\sqrt{2}}[\langle + + | - \langle + - |] | \psi \rangle \right|^2 + \left| \frac{1}{\sqrt{2}}[\langle - + | - \langle - - |] | \psi \rangle \right|^2 \\ &= \frac{1}{2}|\alpha - \beta|^2 + \frac{1}{2}|\gamma - \delta|^2 \end{aligned} \tag{22}$$

In this result, each of the terms of the "sum of squares" is itself the "square of a sum".

If the measurement actually yields $-\hbar/2$, the state $|\psi'\rangle$ of the system immediately after this measurement is the (normalized) projection of $|\psi\rangle$ onto the corresponding eigensubspace. We have just calculated the components of $|\psi\rangle$ on the basis vectors (21) of this subspace : they are equal, respectively, to $\frac{1}{\sqrt{2}}(\alpha - \beta)$ and $\frac{1}{\sqrt{2}}(\gamma - \delta)$. Consequently :

$$|\psi'\rangle = \frac{1}{\sqrt{\frac{1}{2}|\alpha - \beta|^2 + \frac{1}{2}|\gamma - \delta|^2}} \left[ \frac{1}{2}(\alpha - \beta)(|+ +\rangle - |+ -\rangle) + \frac{1}{2}(\gamma - \delta)(|- +\rangle - |- -\rangle) \right] \tag{23}$$

COMMENT :

We have considered, in this complement, only the components of $\mathbf{S}_1$ and $\mathbf{S}_2$ on the coordinate axes. It is obviously possible to measure their components $\mathbf{S}_1 \cdot \mathbf{u}$ and $\mathbf{S}_2 \cdot \mathbf{v}$ on arbitrary unit vectors $\mathbf{u}$ and $\mathbf{v}$. The reasoning is the same as above.

# Complement $E_{IV}$

## SPIN 1/2 DENSITY MATRIX



### 1. Introduction

The aim of this complement is to illustrate the general considerations developed in complement $E_{III}$, using a very simple physical system, that of a spin 1/2. We are going to study the density matrices which describe a spin 1/2 in a certain number of cases : perfectly polarized spin (pure case), unpolarized or partially polarized spin (statistical mixture). We shall thus be able to verify and interpret the general properties stated in complement $E_{III}$. In addition, we shall see that the expansion of the density matrix in terms of the Pauli matrices can be expressed very simply as a function of the mean values of the various spin components.

### 2. Density matrix of a perfectly polarized spin (pure case)

Consider a spin 1/2, coming out of an "atomic polarizer" of the type described in § B of chapter IV, which is in the eigenstate $| + \rangle_u$ (eigenvalue $+ \hbar/2$) of the $\mathbf{S} \cdot \mathbf{u}$ component of the spin (recall that the polar angles of the unit vector $\mathbf{u}$ are $\theta$ and $\varphi$). The spin state is then perfectly well-known and is written [*cf.* formula (A-22-a) of chapter IV] :

$$| \psi \rangle = \cos\frac{\theta}{2} e^{-i\varphi/2} | + \rangle + \sin\frac{\theta}{2} e^{i\varphi/2} | - \rangle \tag{1}$$

We saw in complement $E_{III}$ that, by definition, such a situation corresponds to a pure case. We shall say that the beam which leaves the "polarizer" is perfectly polarized. Recall also that, for each spin, the mean value $\langle \mathbf{S} \rangle$ is equal to $\frac{\hbar}{2}\mathbf{u}$ [chap. IV, relations (B-14)].

It is simple to write, in the $\{ | + \rangle, | - \rangle \}$ basis, the density matrix $\rho(\theta, \varphi)$ corresponding to the state (1). We write the matrix of the projector onto this state :

$$\rho(\theta, \varphi) = \begin{pmatrix} \cos^2\frac{\theta}{2} & \sin\frac{\theta}{2}\cos\frac{\theta}{2} e^{-i\varphi} \\ \sin\frac{\theta}{2}\cos\frac{\theta}{2} e^{i\varphi} & \sin^2\frac{\theta}{2} \end{pmatrix} \tag{2}$$

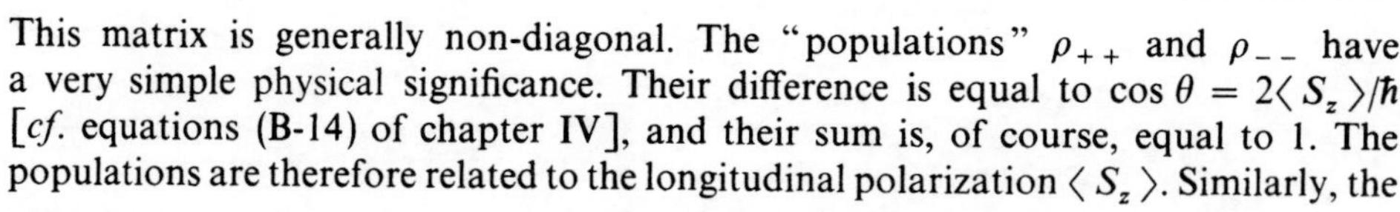

This matrix is generally non-diagonal. The "populations" $\rho_{++}$ and $\rho_{--}$ have a very simple physical significance. Their difference is equal to $\cos\theta = 2\langle S_z \rangle/\hbar$ [*cf.* equations (B-14) of chapter IV], and their sum is, of course, equal to 1. The populations are therefore related to the longitudinal polarization $\langle S_z \rangle$. Similarly, the modulus of the "coherences" $\rho_{+-}$ and $\rho_{-+}$ is $|\rho_{+-}| = |\rho_{-+}| = \frac{1}{2}\sin\theta = \frac{1}{\hbar}|\langle \mathbf{S}_\perp \rangle|$ (where $\langle \mathbf{S}_\perp \rangle$ is the projection of $\langle \mathbf{S} \rangle$ onto the $xOy$ plane). The argument of $\rho_{-+}$ is $\varphi$, that is, the angle between $\langle \mathbf{S}_\perp \rangle$ and $Ox$ : the coherences are therefore related to the transverse polarization $\langle \mathbf{S}_\perp \rangle$.

It can also be verified that:

$$[\rho(\theta, \varphi)]^2 = \rho(\theta, \varphi) \tag{3}$$

a relation characteristic of a pure state.

### 3. Example of a statistical mixture: unpolarized spin

Now let us consider the spin of a silver atom leaving a furnace, such as the one in figure 1 of chapter IV, and which has not passed through an "atomic polarizer" (the spin has not been prepared in a particular state). The only information we then possess about this spin is the following: it can point in any direction of space, and all directions are equally probable. With the notation of complement $E_{III}$, such a situation corresponds to a statistical mixture of the states $|+\rangle_u$ with equal weights. Formula (28) of complement $E_{III}$ defines the density matrix $\rho$ which corresponds to this case. Nevertheless, the discrete sum $\sum_k$ must here be replaced by an integral over all the possible directions:

$$\rho = \frac{1}{4\pi}\int \mathrm{d}\Omega\, \rho(\theta, \varphi) = \frac{1}{4\pi}\int_0^{2\pi} \mathrm{d}\varphi \int_0^{\pi} \sin\theta\, \mathrm{d}\theta\, \rho(\theta, \varphi) \tag{4}$$

(the factor $1/4\pi$ insures the normalization of the probabilities associated with the various directions). The integrals which give the matrix elements of $\rho$ are simple to calculate and lead to the following result:

$$\rho = \begin{pmatrix} 1/2 & 0 \\ 0 & 1/2 \end{pmatrix} \tag{5}$$

It is easy to deduce from (5) that $\rho^2 = \rho/2$, which shows that, in the case of a statistical mixture of states, $\rho^2$ is different from $\rho$.

In addition, if we calculate from (5) the mean values of $S_x$, $S_y$, $S_z$, we obtain :

$$\langle S_i \rangle = \mathrm{Tr}\,[\rho S_i] = \frac{1}{2}\mathrm{Tr}\, S_i = 0 \qquad i = x, y, z \tag{6}$$

We again find the fact that the spin is unpolarized : since all the directions are equivalent, the mean value of the spin is zero.

COMMENTS:

(*i*) It is clear from this example how the non-diagonal elements (coherences) of $\rho$ can disappear from the summation over the various states of the statistical mixture. As we saw in § 2, the coherences $\rho_{+-}$ and $\rho_{-+}$ are related to the transverse polarization $\langle \mathbf{S}_\perp \rangle$ of the spin. Upon summing the vectors $\langle \mathbf{S}_\perp \rangle$ corresponding to all (equiprobable) directions of the $xOy$ plane, we obviously find a null result.

(*ii*) The case of unpolarized spin is also very instructive, since it helps us to understand the impossibility of describing a statistical mixture by an "average state vector". Assume that we are trying to choose $\alpha$ and $\beta$ so that the vector:

$$| \psi \rangle = \alpha | + \rangle + \beta | - \rangle \tag{7}$$

with :

$$|\alpha|^2 + |\beta|^2 = 1 \tag{8}$$

represents an unpolarized spin, for which $\langle S_x \rangle$, $\langle S_y \rangle$ and $\langle S_z \rangle$ are zero. A simple calculation gives:

$$\begin{aligned} \langle S_x \rangle &= \frac{\hbar}{2}(\alpha^*\beta + \alpha\beta^*) \\ \langle S_y \rangle &= \frac{\hbar}{2i}(\alpha^*\beta - \alpha\beta^*) \\ \langle S_z \rangle &= \frac{\hbar}{2}(\alpha^*\alpha - \beta^*\beta) \end{aligned} \tag{9}$$

If we want to make $\langle S_x \rangle$ zero, we must choose $\alpha$ and $\beta$ so as to make $\alpha^*\beta$ a pure imaginary; similarly, $\alpha^*\beta$ must be real for $\langle S_y \rangle$ to be zero. We must therefore have $\alpha^*\beta = 0$; that is:

either $\alpha = 0$, which implies $|\beta| = 1$ and $\langle S_z \rangle = -\hbar/2$
or $\beta = 0$, which implies $|\alpha| = 1$ and $\langle S_z \rangle = \hbar/2$

Therefore, $\langle S_z \rangle$, $\langle S_x \rangle$ and $\langle S_y \rangle$ cannot all be zero at the same time; consequently, the state (7) cannot represent an unpolarized spin.

Furthermore, the discussion of § B-1-c of chapter IV shows that for any $\alpha$ and $\beta$ which satisfy (8), one can always associate with them two angles $\theta$ and $\varphi$ which fix a direction $\mathbf{u}$ such that $| \psi \rangle$ is an eigenvector of $\mathbf{S} \cdot \mathbf{u}$ with the eigenvalue $+ \hbar/2$. Thus we see directly that a state such as (7) always describes a spin which is perfectly polarized in a certain direction of space.

(*iii*) The density matrix (5) represents a statistical mixture of the various states $| + \rangle_u$, all the directions $\mathbf{u}$ being equiprobable (this is how we obtained it). We could, however, imagine other statistical mixtures which would lead to the same density matrix : for example, a statistical mixture of equal proportions of the states $| + \rangle$ and $| - \rangle$, or a statistical mixture of equal proportions of three states $| + \rangle_u$ such that the tips of the three corresponding vectors $\mathbf{u}$ are the vertices of an equilateral triangle centered at $O$. Thus we see that the same density matrix can be obtained in several different ways. In fact, since all the physical predictions depend only on the density matrix, it is impossible to distinguish physically between the various types of statistical mixtures which lead to the same density matrix. They must be considered to be different expressions for the same incomplete information which we possess about the system.

## 4. Spin 1/2 in thermodynamic equilibrium in a static field

Consider a spin 1/2 placed in a static field $\mathbf{B}_0$ parallel to $Oz$. We saw in § B-3-a of chapter IV that the stationary states of this spin are the states $|+\rangle$ and $|-\rangle$, of energies $+\hbar\omega_0/2$ and $-\hbar\omega_0/2$ (with $\omega_0 = -\gamma B_0$, where $\gamma$ is the gyromagnetic ratio of the spin). If we know only that the system is in thermodynamic equilibrium at the temperature $T$, we can assert that it has a probability $Z^{-1}\,\mathrm{e}^{-\hbar\omega_0/2kT}$ of being in the state $|+\rangle$ and $Z^{-1}\,\mathrm{e}^{+\hbar\omega_0/2kT}$ of being in the state $|-\rangle$, where $Z = \mathrm{e}^{-\hbar\omega_0/2kT} + \mathrm{e}^{+\hbar\omega_0/2kT}$ is a normalization factor ($Z$ is called the "partition function"). We have here another example of a statistical mixture, described by the density matrix:

$$\rho = Z^{-1}\begin{pmatrix} \mathrm{e}^{-\hbar\omega_0/2kT} & 0 \\ 0 & \mathrm{e}^{+\hbar\omega_0/2kT} \end{pmatrix} \tag{10}$$

Once more, it is easy to verify that $\rho^2 \neq \rho$. The non-diagonal elements are zero since all directions perpendicular to $\mathbf{B}_0$ (that is, to $Oz$) and fixed by the angle $\varphi$ are equivalent.

From (10), it is easy to calculate:

$$\begin{aligned} \langle S_x \rangle &= \mathrm{Tr}\,(\rho S_x) = 0 \\ \langle S_y \rangle &= \mathrm{Tr}\,(\rho S_y) = 0 \\ \langle S_z \rangle &= \mathrm{Tr}\,(\rho S_z) = -\frac{\hbar}{2}\,\tanh\left(\frac{\hbar\omega_0}{2kT}\right) \end{aligned} \tag{11}$$

We see that the spin acquires a polarization parallel to the field in which it is placed. The larger $\omega_0$ (that is, $B_0$) and the lower the temperature $T$, the greater the polarization. Since $|\tanh x| < 1$, this polarization is less than the value $\hbar/2$ which corresponds to a spin which is perfectly polarized along $Oz$. (10) can therefore be said to describe a spin which is "partially polarized" along $Oz$.

COMMENT:

The magnetization $\langle M_z \rangle$ is equal to $\gamma\langle S_z \rangle$. It is possible to calculate from (11) the paramagnetic susceptibility $\chi$ of the spin, defined by:

$$\langle M_z \rangle = \gamma \langle S_z \rangle = \chi B_0 \tag{12}$$

We find (Brillouin's formula):

$$\chi = \frac{\hbar\gamma}{2B_0}\,\tanh\left(\frac{\hbar\gamma B_0}{2kT}\right) \tag{13}$$

## 5. Expansion of the density matrix in terms of the Pauli matrices

We saw in complement $A_{IV}$ that the unit matrix $I$ and the Pauli matrices $\sigma_x$, $\sigma_y$ and $\sigma_z$ form a convenient basis for expanding a $2 \times 2$ matrix. We therefore set, for the density matrix $\rho$ of a spin 1/2 :

$$\rho = a_0 I + \mathbf{a} \cdot \boldsymbol{\sigma} \tag{14}$$

where the coefficients $a_i$ are given by [*cf.* complement $A_{IV}$, relations (23)]:

$$\begin{aligned} a_0 &= \frac{1}{2} \text{Tr}\, \rho \\ a_x &= \frac{1}{2} \text{Tr}\,(\rho \sigma_x) = \frac{1}{\hbar} \text{Tr}\,(\rho S_x) \\ a_y &= \frac{1}{2} \text{Tr}\,(\rho \sigma_y) = \frac{1}{\hbar} \text{Tr}\,(\rho S_y) \\ a_z &= \frac{1}{2} \text{Tr}\,(\rho \sigma_z) = \frac{1}{\hbar} \text{Tr}\,(\rho S_z) \end{aligned} \tag{15}$$

Thus we have:

$$\begin{aligned} a_0 &= \frac{1}{2} \\ \mathbf{a} &= \frac{1}{\hbar} \langle \mathbf{S} \rangle \end{aligned} \tag{16}$$

and $\rho$ can be written:

$$\rho = \frac{1}{2} I + \frac{1}{\hbar} \langle \mathbf{S} \rangle \cdot \boldsymbol{\sigma} \tag{17}$$

Therefore, the density matrix $\rho$ of a spin 1/2 can be expressed very simply in terms of the mean value $\langle \mathbf{S} \rangle$ of the spin.

COMMENT:

Let us square expression (17). We obtain, using identity (17) of complement $A_{IV}$ :

$$\rho^2 = \frac{1}{4} I + \frac{1}{\hbar^2} \langle \mathbf{S} \rangle^2 I + \frac{1}{\hbar} \langle \mathbf{S} \rangle \cdot \boldsymbol{\sigma} \tag{18}$$

The condition $\rho^2 = \rho$, characteristic of the pure case, is therefore equivalent, for a spin 1/2, to the condition:

$$\langle \mathbf{S} \rangle^2 = \frac{\hbar^2}{4} \tag{19}$$

This condition is obviously not satisfied for an unpolarized spin ($\langle \mathbf{S} \rangle$ is then zero) or for a spin in thermodynamic equilibrium (we saw in §4 that in this case $|\langle \mathbf{S} \rangle| < \hbar/2$). On the other hand, it can be verified, using formulas (B-14) of chapter IV, that, for a spin in the state $|\psi\rangle$ given in (1), $\langle \mathbf{S} \rangle^2$ is indeed equal to $\hbar^2/4$.

**References and suggestions for further reading:**

Abragam (14.1), chap. II, §C.

# Complement $F_{IV}$

## SPIN 1/2 PARTICLE IN A STATIC MAGNETIC FIELD AND A ROTATING FIELD: MAGNETIC RESONANCE



In chapter IV, we used quantum mechanics to study the evolution of a spin 1/2 in a static magnetic field. In this complement, we shall consider the case of a spin 1/2 simultaneously subjected to several magnetic fields, some of which can be time dependent, as is the case in magnetic resonance experiments. Before attacking this problem quantum mechanically, we shall briefly review several results obtained using classical mechanics.

### 1. Classical treatment; rotating reference frame

#### a. MOTION IN A STATIC FIELD; LARMOR PRECESSION

Consider a system of angular momentum $\mathbf{j}$ which possesses a magnetic moment $\mathbf{m} = \gamma\mathbf{j}$ collinear with $\mathbf{j}$ (the constant $\gamma$ is the gyromagnetic ratio of the system), placed in a static magnetic field $\mathbf{B}_0$, which exerts a torque $\boldsymbol{m} \times \mathbf{B}_0$ on the system. The classical equation of motion of $\mathbf{j}$ is:

$$\frac{\mathrm{d}\mathbf{j}}{\mathrm{d}t} = \mathbf{m} \times \mathbf{B}_0 \tag{1}$$

or:

$$\frac{\mathrm{d}}{\mathrm{d}t}\mathbf{m}(t) = \gamma\mathbf{m}(t) \times \mathbf{B}_0 \tag{2}$$

Performing a scalar multiplication of both sides of this equation by either $\mathbf{m}(t)$ or $\mathbf{B}_0$, we obtain:

$$\frac{\mathrm{d}}{\mathrm{d}t}[\mathbf{m}(t)]^2 = 0 \tag{3}$$

$$\frac{\mathrm{d}}{\mathrm{d}t}[\mathbf{m}(t) \cdot \mathbf{B}_0] = 0 \tag{4}$$

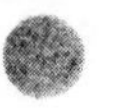


$\mathbf{m}(t)$ therefore evolves with a constant modulus, maintaining a constant angle with $\mathbf{B}_0$. If we project equation (2) onto the plane perpendicular to $\mathbf{B}_0$, we see that $\mathbf{m}(t)$ rotates about $\mathbf{B}_0$ (Larmor precession) with an angular velocity of $\omega_0 = -\gamma B_0$ (the rotation is clockwise if $\gamma$ is positive).

### b. INFLUENCE OF A ROTATING FIELD ; RESONANCE

Now assume that we add to the static field $\mathbf{B}_0$ a field $\mathbf{B}_1(t)$, perpendicular to $\mathbf{B}_0$, and which is of constant modulus and rotates about $\mathbf{B}_0$ with an angular velocity $\omega$ (*cf.* fig. 1). We set:

$$\begin{aligned} \omega_0 &= -\gamma B_0 \\ \omega_1 &= -\gamma B_1 \end{aligned} \tag{5}$$

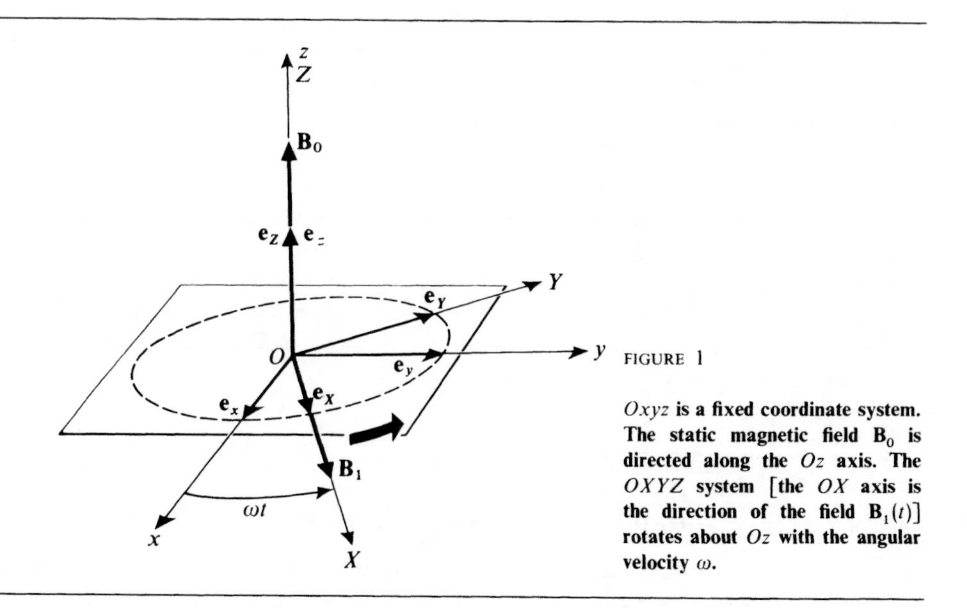


FIGURE 1

**$Oxyz$ is a fixed coordinate system. The static magnetic field $\mathbf{B}_0$ is directed along the $Oz$ axis. The $OXYZ$ system [the $OX$ axis is the direction of the field $\mathbf{B}_1(t)$] rotates about $Oz$ with the angular velocity $\omega$.**

We shall designate by $Oxyz$ (unit vectors $\mathbf{e}_x$, $\mathbf{e}_y$, $\mathbf{e}_z$) a fixed coordinate system, whose $Oz$ axis is the direction of the field $\mathbf{B}_0$, and by $OXYZ$ (unit vectors $\mathbf{e}_X$, $\mathbf{e}_Y$, $\mathbf{e}_Z$), the axes obtained from $Oxyz$ by rotation through an angle $\omega t$ about $Oz$ [$OX$ is the direction of the rotating field $\mathbf{B}_1(t)$]. The equation of motion of $\mathbf{m}(t)$ in the presence of the total field $\mathbf{B}(t) = \mathbf{B}_0 + \mathbf{B}_1(t)$ then becomes:

$$\frac{\mathrm{d}}{\mathrm{d}t}\mathbf{m}(t) = \gamma\, \mathbf{m}(t) \times [\mathbf{B}_0 + \mathbf{B}_1(t)] \tag{6}$$

To solve this equation, it is convenient to place ourselves, not in the absolute reference frame $Oxyz$, but in the rotating reference frame $OXYZ$, with respect to which the relative velocity of the vector $\mathbf{m}(t)$ is:

$$\left(\frac{\mathrm{d}\mathbf{m}}{\mathrm{d}t}\right)_{\mathrm{rel}} = \frac{\mathrm{d}\mathbf{m}}{\mathrm{d}t} - \omega\, \mathbf{e}_Z \times \mathbf{m}(t) \tag{7}$$

Let us set

$$\Delta\omega = \omega - \omega_0 \tag{8}$$

Substituting (6) into (7), we obtain:

$$\left(\frac{\mathrm{d}\mathbf{m}}{\mathrm{d}t}\right)_{\text{rel}} = \mathbf{m}(t) \times [\Delta\omega\, \mathbf{e}_Z - \omega_1 \mathbf{e}_X] \tag{9}$$

This equation is much simpler to solve than equation (6), since the coefficients of the right-hand side are now time-independent. Moreover, its form is analogous to that of (2) : the relative motion of the vector $\mathbf{m}(t)$ is therefore a rotation about the "effective field" $\mathbf{B}_{\text{eff}}$ (which is static with respect to the rotating reference frame), given by (*cf.* fig. 2):

$$\mathbf{B}_{\text{eff}} = \frac{1}{\gamma}[\Delta\omega\, \mathbf{e}_Z - \omega_1 \mathbf{e}_X] \tag{10}$$

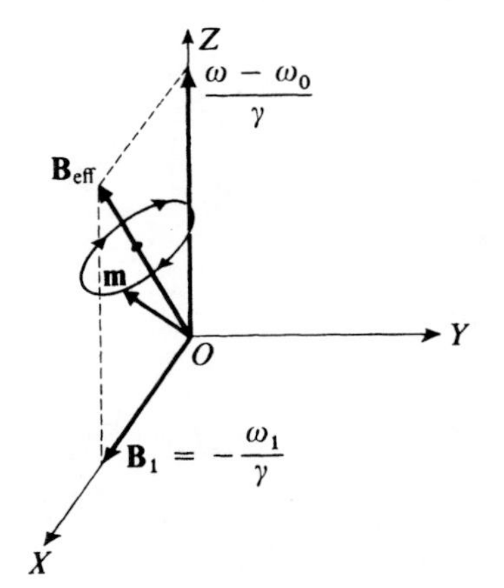


FIGURE 2

**In the rotating reference frame $OXYZ$, the effective field $\mathbf{B}_{\text{eff}}$ has a fixed direction, about which the magnetic moment $\mathbf{m}(t)$ rotates with a constant angular velocity (precession in the rotating reference frame).**

To obtain the absolute motion of $\mathbf{m}(t)$, we must combine this precession about $\mathbf{B}_{\text{eff}}$ with a rotation about $Oz$ of angular velocity $\omega$.

These first results already enable us to understand the essence of the magnetic resonance phenomenon. Let us consider a magnetic moment which, at time $t = 0$, is parallel to the field $\mathbf{B}_0$ (the case, for example, of a magnetic moment in thermodynamic equilibrium at very low temperatures : it is in the lowest energy state possible in the presence of the field $\mathbf{B}_0$). What happens when we apply a weak rotating field $\mathbf{B}_1(t)$? If the rotation frequency $\omega/2\pi$ of this field is very different from the natural frequency $\omega_0/2\pi$ (more precisely, if $\Delta\omega = \omega - \omega_0$ is much larger than $\omega_1$), the effective field is directed practically along $Oz$. The precession of $\mathbf{m}(t)$ about $\mathbf{B}_{\text{eff}}$ then has a very small amplitude and hardly modifies the direction of the magnetic moment. On the other hand, if the resonance condition $\omega \simeq \omega_0$ is satisfied ($\Delta\omega \ll \omega_1$), the angle between the field $\mathbf{B}_{\text{eff}}$ and $Oz$ is large. The precession of the magnetic moment then has a large amplitude and, at resonance ($\Delta\omega = 0$), the magnetic moment can even be completely flipped.

## 2. Quantum mechanical treatment

### a. THE SCHRÖDINGER EQUATION

Let $|+\rangle$ and $|-\rangle$ be two eigenvectors of the projection $S_z$ of the spin onto $Oz$, with respective eigenvalues $+\hbar/2$ and $-\hbar/2$. The state vector of the system can be written:

$$|\psi(t)\rangle = a_+(t)\,|+\rangle + a_-(t)\,|-\rangle \tag{11}$$

The Hamiltonian operator $H(t)$ of the system is*:

$$H(t) = -\,\mathbf{M}\,.\,\mathbf{B}(t) = -\,\gamma\,\mathbf{S}\,.\,[\mathbf{B}_0 + \mathbf{B}_1(t)] \tag{12}$$

that is, expanding the scalar product:

$$H(t) = \omega_0 S_z + \omega_1[\cos\omega t\; S_x + \sin\omega t\; S_y] \tag{13}$$

Using formulas (A-16) and (A-17) of chapter IV, we obtain the matrix which represents $H$ in the $\{\,|+\rangle, |-\rangle\,\}$ basis:

$$H = \frac{\hbar}{2}\begin{pmatrix} \omega_0 & \omega_1\,\mathrm{e}^{-i\omega t} \\ \omega_1\,\mathrm{e}^{i\omega t} & -\,\omega_0 \end{pmatrix} \tag{14}$$

Using (11) and (14), we can write the Schrödinger equation in the form:

$$\begin{cases} i\dfrac{\mathrm{d}}{\mathrm{d}t}a_+(t) = \dfrac{\omega_0}{2}\,a_+(t) + \dfrac{\omega_1}{2}\,\mathrm{e}^{-i\omega t}\,a_-(t) \\[2ex] i\dfrac{\mathrm{d}}{\mathrm{d}t}a_-(t) = \dfrac{\omega_1}{2}\,\mathrm{e}^{i\omega t}\,a_+(t) - \dfrac{\omega_0}{2}\,a_-(t) \end{cases} \tag{15}$$

### b. CHANGING TO THE ROTATING FRAME

Equations (15) constitute a linear homogeneous system with time-dependent coefficients. It is convenient to define new functions by setting:

$$\begin{aligned} b_+(t) &= \mathrm{e}^{i\omega t/2}\,a_+(t) \\ b_-(t) &= \mathrm{e}^{-i\omega t/2}\,a_-(t) \end{aligned} \tag{16}$$

Substituting (16) into (15), we obtain a system which has constant coefficients:

$$\begin{cases} i\dfrac{\mathrm{d}}{\mathrm{d}t}b_+(t) = -\,\dfrac{\Delta\omega}{2}\,b_+(t) + \dfrac{\omega_1}{2}\,b_-(t) \\[2ex] i\dfrac{\mathrm{d}}{\mathrm{d}t}b_-(t) = \dfrac{\omega_1}{2}\,b_+(t) + \dfrac{\Delta\omega}{2}\,b_-(t) \end{cases} \tag{17}$$

* In expression (12), $\mathbf{M}\,.\,\mathbf{B}(t)$ symbolizes the scalar product $M_xB_x(t) + M_yB_y(t) + M_zB_z(t)$, where $M_x$, $M_y$ and $M_z$ are operators (observables of the system under study), while $B_x(t)$, $B_y(t)$ and $B_z(t)$ are numbers (since we consider the magnetic field to be a classical quantity whose value is imposed by an external device which is independent of the system under study).

This system can also be written:

$$i\hbar \frac{\mathrm{d}}{\mathrm{d}t} | \tilde{\psi}(t) \rangle = \tilde{H} | \tilde{\psi}(t) \rangle \tag{18}$$

if we introduce the ket $| \tilde{\psi}(t) \rangle$ and the operator $\tilde{H}$ defined by:

$$| \tilde{\psi}(t) \rangle = b_+(t) | + \rangle + b_-(t) | - \rangle \tag{19}$$

$$\tilde{H} = \frac{\hbar}{2} \begin{pmatrix} -\Delta\omega & \omega_1 \\ \omega_1 & \Delta\omega \end{pmatrix} \tag{20}$$

Transformation (16) has led to equation (18), which is analogous to a Schrödinger equation in which the operator $\tilde{H}$, given in (20), plays the role of a time-independent Hamiltonian. $\tilde{H}$ describes the interaction of the spin with a *fixed* field, whose components are none other than those of the effective field introduced above in the $OXYZ$ frame [formula (10)]. We can therefore consider that the transformation (16) is the quantum mechanical equivalent of the change from the fixed $Oxyz$ frame to the rotating $OXYZ$ frame.

This result can be proved rigorously. According to (16), we can write:

$$| \tilde{\psi}(t) \rangle = R(t) | \psi(t) \rangle \tag{21}$$

where $R(t)$ is the unitary operator defined by:

$$R(t) = \mathrm{e}^{i\omega t S_z / \hbar} \tag{22}$$

We shall see later (*cf.* complement $\mathrm{B_{VI}}$) that $R(t)$ describes a rotation of the coordinate system through an angle $\omega t$ about $Oz$. (18) is therefore indeed the transformed Schrödinger equation in the rotating $OXYZ$ frame.

Equation (18) is very simple to solve. To determine $| \tilde{\psi}(t) \rangle$, given $| \tilde{\psi}(0) \rangle$, all we need to do is expand $| \tilde{\psi}(0) \rangle$ on the eigenvectors of $\tilde{H}$ (which can be calculated exactly) and then apply rule (D-54) of chapter III (which is possible since $\tilde{H}$ is not explicitly time-dependent). We then go from $| \tilde{\psi}(t) \rangle$ to $| \psi(t) \rangle$ by using formulas (16).

### c. TRANSITION PROBABILITY; RABI'S FORMULA

Consider a spin which, at time $t = 0$, is in the state $| + \rangle$:

$$| \psi(0) \rangle = | + \rangle \tag{23}$$

According to (16), this corresponds to:

$$| \tilde{\psi}(0) \rangle = | + \rangle \tag{24}$$

What is the probability $\mathscr{P}_{+-}(t)$ of finding this spin in the state $| - \rangle$ at time $t$? Since $a_-(t)$ and $b_-(t)$ have the same modulus, we can write:

$$\mathscr{P}_{+-}(t) = |\langle - | \psi(t) \rangle|^2 = |a_-(t)|^2 = |b_-(t)|^2 = |\langle - | \tilde{\psi}(t) \rangle|^2 \tag{25}$$

We must therefore calculate $|\langle - | \tilde{\psi}(t) \rangle|^2$, where $| \tilde{\psi}(t) \rangle$ is the solution of (18) which corresponds to the initial condition (24).

The problem we have just posed has already been solved, in § C-3-b of chapter IV. To use the calculations of that section, all we need to do is apply the following correspondences:

$$\begin{aligned}
| \varphi_1 \rangle &\longrightarrow | + \rangle \\
| \varphi_2 \rangle &\longrightarrow | - \rangle \\
E_1 &\longrightarrow -\frac{\hbar}{2} \Delta\omega \\
E_2 &\longrightarrow \frac{\hbar}{2} \Delta\omega \\
W_{12} &\longrightarrow \frac{\hbar}{2} \omega_1
\end{aligned} \tag{26}$$

Rabi's formula [equation (C-32) of chapter IV] then becomes:

$$\mathscr{P}_{+-}(t) = \frac{\omega_1^2}{\omega_1^2 + (\Delta\omega)^2} \sin^2 \left[ \sqrt{\omega_1^2 + (\Delta\omega)^2} \, \frac{t}{2} \right] \tag{27}$$

The probability $\mathscr{P}_{+-}(t)$ is, of course, zero at time $t = 0$ and then varies sinusoidally with respect to time between the values 0 and $\dfrac{\omega_1^2}{\omega_1^2 + (\Delta\omega)^2}$. Again, we have a resonance phenomenon. For $|\Delta\omega| \gg |\omega_1|$, $\mathscr{P}_{+-}(t)$ remains almost zero (*cf.* fig. 3-a); near resonance, the oscillation amplitude of $\mathscr{P}_{+-}(t)$ becomes large and, when the condition $\Delta\omega = 0$ is exactly satisfied, we have $\mathscr{P}_{+-}(t) = 1$ at times $t = \dfrac{(2n+1)\pi}{\omega_1}$ (*cf.* fig. 3-b).

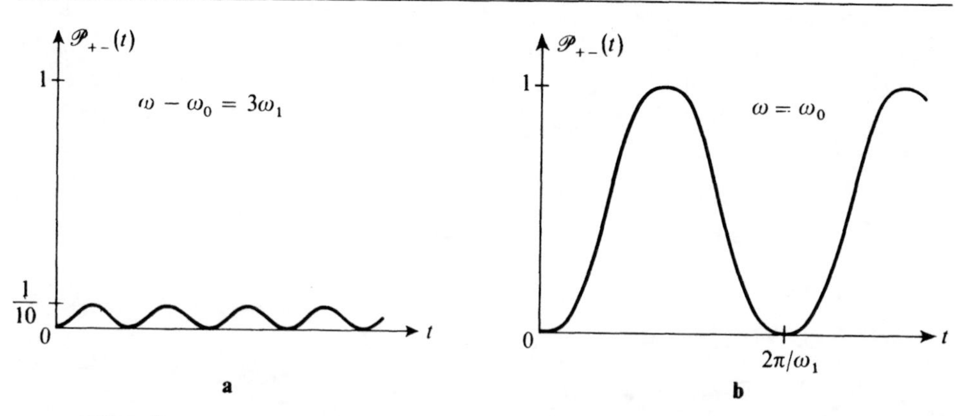


FIGURE 3

**Variation with respect to time of the transition probability between the states $| + \rangle$ and $| - \rangle$, under the effect of a rotating magnetic field $\mathbf{B}_1(t)$. Outside resonance (fig. a), this probability remains small; at resonance (fig. b), however small the field $\mathbf{B}_1$, there exist times when the transition probability is equal to 1.**

Thus we again find the result which we have already obtained classically : at resonance, a very weak rotating field is able to reverse the direction of the spin. Note, moreover, that the angular frequency of the oscillation of $\mathscr{P}_{+-}(t)$ is $\sqrt{\omega_1^2 + (\Delta\omega)^2} = |\gamma \mathbf{B}_{eff}|$. This oscillation corresponds, in the rotating frame, to the projection onto $OZ$ of the precession of the magnetic moment about the effective field, sometimes called "Rabi precession" [see also the calculation of $\mathscr{P}_{+-}(t)$ in complement $C_{IV}$, § 3-c].

#### d. CASE WHERE THE TWO LEVELS ARE UNSTABLE

We are now going to assume that the two states $| \pm \rangle$ correspond to two sublevels of an excited atomic level (whose angular momentum is assumed equal to 1/2). $n$ atoms are excited per unit time, all being raised to the state $| + \rangle$★. An atom decays, by spontaneous emission of radiation, with a probability per unit time of $1/\tau$, which is the same for the two sublevels $| \pm \rangle$. We know that, under these conditions, an atom which was excited at time $-t$ has a probability $e^{-t/\tau}$ of still being excited at time $t = 0$ (*cf.* complement $K_{III}$).

We assume that the experiment is performed in the steady state : in the presence of the fields $\mathbf{B}_0$ and $\mathbf{B}_1(t)$, the atoms are excited at a constant rate $n$ into the state $| + \rangle$. After a time much longer than the lifetime $\tau$, what is the number $N$ of atoms which decay per unit time from the state $| - \rangle$? If an atom is excited at time $-t$, the probability of finding it in the state $| - \rangle$ at $t = 0$ is $e^{-t/\tau}\mathscr{P}_{+-}(t)$, where $\mathscr{P}_{+-}(t)$ is given by relation (27). The total number of atoms in the state $| - \rangle$ is obtained by taking the sum of atoms excited at all previous times $-t$, that is, by calculating the integral:

$$\int_0^\infty e^{-t/\tau}\, \mathscr{P}_{+-}(t)\, n\, dt \tag{28}$$

This calculation presents no difficulties. Multiplying the number of atoms thus obtained by their probability $1/\tau$ of decay per unit time, we obtain:

$$N = \frac{n}{2} \frac{\omega_1^2}{(\Delta\omega)^2 + \omega_1^2 + (1/\tau)^2} \tag{29}$$

The variation of $N$ with respect to $\Delta\omega$ corresponds to a Lorentz curve whose half-width is:

$$L = \sqrt{\omega_1^2 + (1/\tau)^2} \tag{30}$$

In the experiment described above, let us measure, for various values of the magnetic field $B_0$ (that is, with $\omega$ assumed to be fixed, for various values of $\Delta\omega$), the number of atoms which decay from the level $| - \rangle$. According to (29), we obtain a resonance curve which has the shape shown in figure 4.

It is very interesting to obtain such a curve experimentally, since one can use it to determine several parameters:

– if we know $\omega$ and measure the value $B_0^m$ of the field $B_0$ which corresponds to the peak of the curve, we can deduce the value of the gyromagnetic ratio $\gamma$ through the relation $\gamma = -\omega/B_0^m$.

★ In practice, this excitation can be produced, for example, by placing the atoms in a light beam. When the incident photons are polarized, conservation of angular momentum, in certain cases, requires that the atoms which absorb them can attain only the state $| + \rangle$ (and not the state $| - \rangle$). Similarly, by detecting the polarization of the photons re-emitted by the atoms, one can know whether the atoms fall back into the ground state from the state $| + \rangle$ or the state $| - \rangle$.

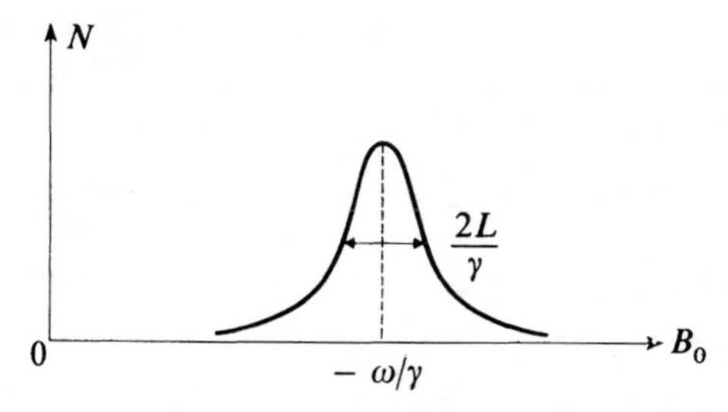


FIGURE 4

**Resonance curve. To observe a resonance phenomenon, we perform an experiment in which $n$ atoms are excited per unit time into the state $| + \rangle$. Under the effect of a field $\mathbf{B}_1(t)$, rotating at the frequency $\omega/2\pi$, the atoms undergo transitions towards the state $| - \rangle$. In the steady state, if we measure the number $N$ of atoms which decay per unit time from the state $| - \rangle$, we obtain a resonant variation when we scan the static field $B_0$ about the value $-\omega/\gamma$.**

– if we know $\gamma$, we can, by measuring the frequency $\omega/2\pi$ which corresponds to resonance, measure the static magnetic field $B_0$. Various magnetometers, often of very great precision, operate on this principle. In certain cases, one can derive interesting information from such a measurement of the field. If, for example, the spin being considered is that of a nucleus which belongs to a molecule or to a crystal lattice, one can find the local field seen by the nucleus, its variation with the site occupied, etc.

– if we trace the square $L^2$ of the half-width as a function of $\omega_1^2$, we obtain a straight line which, extrapolated to $\omega_1 = 0$, gives the lifetime $\tau$ of the excited level (*cf.* fig. 5).

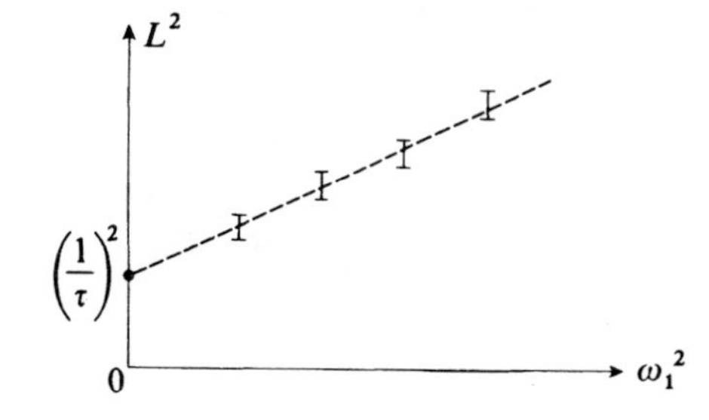


FIGURE 5

**The extrapolation to $\omega_1 = 0$ of the squared half-width $L$ of the resonance curve of figure 4 gives the lifetime of the level being studied.**

## 3. Relation between the classical treatment and the quantum mechanical treatment: evolution of $\langle \mathbf{M} \rangle$

The results obtained in §§1 and 2 are very similar, although we used classical mechanics in one case and quantum mechanics in the other. We are now going to show that this similarity is not accidental. It arises from the fact that the quantum mechanical evolution equations of the mean value of a magnetic moment placed in an arbitrary magnetic field are identical to the corresponding classical equations.

The mean value of the magnetic moment associated with a spin 1/2 is:

$$\langle \mathbf{M} \rangle(t) = \gamma \langle \mathbf{S} \rangle(t) \tag{31}$$

To calculate the evolution of $\langle \mathbf{M} \rangle(t)$, we use theorem (D-27) of chapter III:

$$i\hbar \frac{\mathrm{d}}{\mathrm{d}t} \langle \mathbf{M} \rangle(t) = \langle [\mathbf{M}, H(t)] \rangle \tag{32}$$

where $H(t)$ is the operator:

$$H(t) = -\mathbf{M} \cdot \mathbf{B}(t) \tag{33}$$

Let us calculate for example the commutator $[M_x, H(t)]$. Using the fact that the field components $B_y(t)$ and $B_z(t)$ are numbers (*cf.* note of §2-a), we find:

$$\begin{aligned}[M_x, H(t)] &= -\gamma^2[S_x, S_xB_x(t) + S_yB_y(t) + S_zB_z(t)] \\ &= -\gamma^2 B_y(t)[S_x, S_y] - \gamma^2 B_z(t)[S_x, S_z]\end{aligned} \tag{34}$$

Using relations (14) of complement $A_{IV}$, we obtain:

$$[M_x, H(t)] = i\hbar\gamma^2[B_z(t)S_y - B_y(t)S_z] \tag{35}$$

Substituting (35) into (32):

$$\frac{d}{dt}\langle M_x \rangle(t) = \gamma[B_z(t)\langle M_y \rangle(t) - B_y(t)\langle M_z \rangle(t)] \tag{36}$$

By cyclic permutation, we can calculate analogous expressions for the components on $Oy$ and $Oz$; the three equations obtained can be condensed into:

$$\frac{d}{dt}\langle \mathbf{M} \rangle(t) = \gamma \langle \mathbf{M} \rangle(t) \times \mathbf{B}(t) \tag{37}$$

Let us compare (37) with (6): the evolution of the mean value $\langle \mathbf{M} \rangle(t)$ obeys the classical equations exactly, whatever the time-dependence of the magnetic field $\mathbf{B}(t)$.

## 4. Bloch's equations

In practice, in a magnetic resonance experiment, it is not the magnetic moment of a single spin that is observed, but rather that of a great number of identical spins (as in the experiment described in §2-d above, where the number of atoms which decay from the state $|-\rangle$ is detected). Moreover, one is not concerned solely with the quantity $\mathscr{P}_{+-}(t)$, calculated above. One can also measure the global magnetization $\mathcal{M}$ of the sample under study: the sum of the mean values of the observable $\langle \mathbf{M} \rangle$ corresponding to each spin of the sample*. It is interesting, therefore, to obtain the equations of motion of $\mathcal{M}$, called the *Bloch equations*.

In order to understand the physical significance of the various terms appearing in these equations, we are going to derive them for a simple concrete case. The results obtained can be generalized to other more complicated situations.

### a. A CONCRETE EXAMPLE

Consider a beam of atoms issuing from an atomic polarizer of the type studied in §B-1-a of chapter IV. All the atoms of the beam** are in the spin state $|+\rangle$ and therefore have their magnetic moments parallel to $Oz$. They enter a cell $C$ through a small opening (fig. 6), rebound a certain number of times from the inside walls of the cell and, after a certain time, escape through the same opening.

* It is possible to detect, for example, the electromotive force emf induced in a coil by the variation of $\mathcal{M}$ with respect to time.

** For example, silver or hydrogen atoms in the ground state. For the sake of simplicity, all effects related to nuclear spin are neglected.

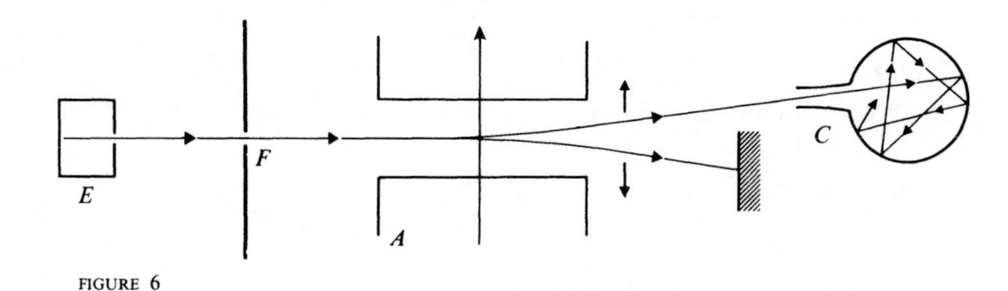


FIGURE 6

**Schematic drawing of an experimental device which supplies cell $C$ with atoms in the state $|+\rangle$.**

We shall denote by $n$ the number of polarized atoms entering the cell per unit time; $n$ is generally small and the atomic density inside the cell is low enough to allow atomic interactions to be neglected. Moreover, if the inside walls of the cell are suitably coated, collisions with the walls have little effect on the spin state of the atoms*. We shall assume that there is a probability per unit time $1/T_R$ for the elementary magnetization introduced into the cell by a polarized atom to disappear, either because of a depolarizing collision with the walls or simply because the atom has left the cell. $T_R$ is called the "relaxation time". The cell is placed in a magnetic field $\mathbf{B}(t)$ which may have a static component and a rotating component. The problem consists of finding the equation of motion of the global magnetization $\mathcal{M}(t)$ of the atoms which are inside the cell at time $t$. First, let us write the exact expression for $\mathcal{M}(t)$:

$$\mathcal{M}(t) = \sum_{i=1}^{\mathcal{N}} \langle \psi^{(i)}(t) | \mathbf{M} | \psi^{(i)}(t) \rangle = \sum_{i=1}^{\mathcal{N}} \mathcal{M}^{(i)}(t) \tag{38}$$

In (38), the sum is taken over the $\mathcal{N}$ spins which are already in the cell and which, at time $t$, have neither left nor undergone a depolarizing collision. $|\psi^{(i)}(t)\rangle$ is the state vector of such a spin $(i)$ at time $t$ [we are not counting, in (38), the spins which have undergone a depolarizing collision and have not yet left the cell, since their global contribution is zero: their spins point randomly in all directions].

Between times $t$ and $t + dt$, $\mathcal{M}(t)$ varies for three different reasons:

($i$) A certain proportion, $dt/T_R$, of the $\mathcal{N}$ spins undergo a depolarizing collision or leave the compartment; these spins disappear from the sum (38) and $\mathcal{M}(t)$ therefore decreases by:

$$d\mathcal{M}(t) = -\frac{dt}{T_R}\mathcal{M}(t) \tag{39}$$

($ii$) The other spins evolve freely in the field $\mathbf{B}(t)$. We saw in § 3 above that, for each of them, the evolution of the mean value of $\mathbf{M}$:

$$\mathcal{M}^{(i)}(t) = \langle \psi^{(i)}(t) |\mathbf{M}| \psi^{(i)}(t) \rangle$$

obeys the classical law:

$$d\mathcal{M}^{(i)}(t) = \gamma\mathcal{M}^{(i)}(t) \times \mathbf{B}(t)\, dt \tag{40}$$

Since the right-hand side of (40) is linear with respect to $\mathcal{M}^{(i)}(t)$, the contribution of these spins to the variation of $\mathcal{M}(t)$ is given by:

$$d\mathcal{M}(t) = \gamma\mathcal{M}(t) \times \mathbf{B}(t)\, dt \tag{41}$$

* For example, for hydrogen atoms bouncing off teflon walls, tens of thousands of collisions are required for the magnetic moment of the hydrogen atom to become disoriented.

(*iii*) Finally, a certain number, $n\,\mathrm{d}t$, of new spins have entered the cell. Each of them adds to the global magnetization a contribution $\boldsymbol{\mu}_0$, equal to the mean value of $\mathbf{M}$ in the state $|+\rangle$ $\left(\boldsymbol{\mu}_0 \text{ is parallel to } Oz \text{ and } |\boldsymbol{\mu}_0| = |\gamma|\frac{\hbar}{2}\right)$. $\mathcal{M}$ therefore increases by:

$$\mathrm{d}\mathcal{M}(t) = n\,\boldsymbol{\mu}_0\,\mathrm{d}t \tag{42}$$

The global variation of $\mathcal{M}$ is obtained by adding (39), (41) and (42). Dividing by $\mathrm{d}t$, we obtain the equation of motion of $\mathcal{M}(t)$ (Bloch equation):

$$\frac{\mathrm{d}}{\mathrm{d}t}\mathcal{M}(t) = n\,\boldsymbol{\mu}_0 - \frac{1}{T_R}\mathcal{M}(t) + \gamma\mathcal{M}(t)\times\mathbf{B}(t) \tag{43}$$

We have derived (43) in a specific case, making certain hypotheses. However, the main features of this equation remain valid for a great number of other experiments where the rate of variation of $\mathcal{M}(t)$ appears in the form of a sum of three terms:

– a source term (here $n\,\boldsymbol{\mu}_0$) which describes the preparation of the system. It would, in fact, be impossible to observe magnetic resonance without a preliminary polarization of the spins, which can be achieved through selection using a magnetic field gradient (as in the example studied here), a polarized optical excitation (as in the example studied in § 2-d above), cooling of the sample in a strong static field, etc.

– a damping term $\left(\text{here } -\frac{1}{T_R}\mathcal{M}(t)\right)$ which describes the disappearance or "relaxation" of the global magnetization under the effect of various processes : collisions, disappearance of atoms, change in atomic levels through spontaneous emission (as in the example studied in § 2-d), etc.

– a term which describes the precession of $\mathcal{M}(t)$ in the field $\mathbf{B}(t)$ [last term of (43)].

## b. SOLUTION IN THE CASE OF A ROTATING FIELD

When the field $\mathbf{B}(t)$ is the sum of a static field $\mathbf{B}_0$ and a rotating field $\mathbf{B}_1(t)$, such as those considered above, equations (43) can be solved exactly. As in §§ 1 and 2, one changes to the rotating frame $OXYZ$, with respect to which the relative variation of $\mathcal{M}(t)$ is:

$$\left(\frac{\mathrm{d}}{\mathrm{d}t}\mathcal{M}\right)_{\text{rel}} = n\,\boldsymbol{\mu}_0 - \frac{1}{T_R}\mathcal{M} + \gamma\mathcal{M}\times\mathbf{B}_{\text{eff}} \tag{44}$$

[where $\mathbf{B}_{\text{eff}}$ is defined by equation (10)].

Projecting this equation onto $OX$, $OY$ and $OZ$, we obtain a system of three linear differential equations with constant coefficients whose stationary solution (valid after a time much greater than $T_R$) is:

$$\begin{aligned}
(\mathcal{M}_X)_S &= -\,n\mu_0 T_R\,\frac{\omega_1\,\Delta\omega}{(\Delta\omega)^2 + \omega_1^2 + (1/T_R)^2}\\
(\mathcal{M}_Y)_S &= -\,n\mu_0\,\frac{\omega_1}{(\Delta\omega)^2 + \omega_1^2 + (1/T_R)^2}\\
(\mathcal{M}_Z)_S &= n\mu_0 T_R\left[1 - \frac{\omega_1^2}{(\Delta\omega)^2 + \omega_1^2 + (1/T_R)^2}\right]
\end{aligned} \tag{45}$$

The three components of the stationary magnetization $(\mathcal{M})_S$, when the field $B_0$ varies, have resonant variations about the value $B_0 = -\omega/\gamma$ (*cf.* fig. 7). $(\mathcal{M}_Y)_S$ and $(\mathcal{M}_Z)_S$ give absorption curves $\left(\text{Lorentz curves of width } \frac{2}{|\gamma|}\sqrt{\omega_1^2 + (1/T_R)^2}\right)$. $(\mathcal{M}_X)_S$ gives a dispersion curve (of the same width).

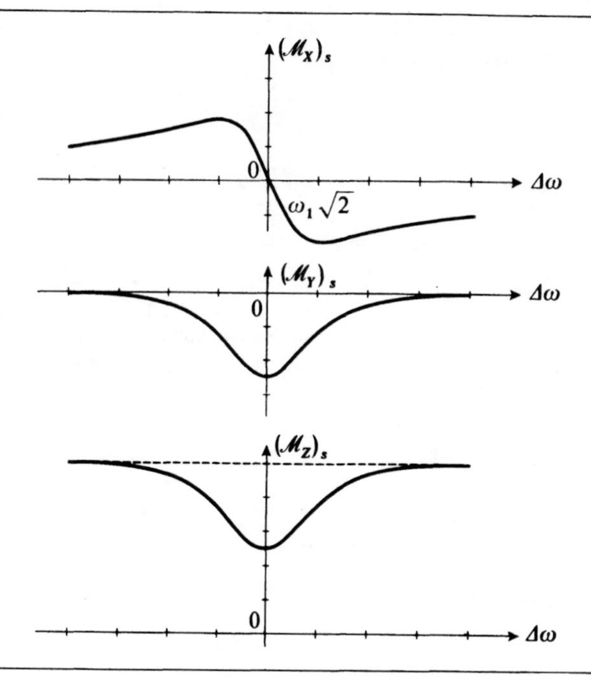


FIGURE 7

**Variation with respect to $\Delta\omega = \omega - \omega_0$ of the stationary values of the components of $\mathcal{M}$ in the rotating frame. One obtains a dispersion curve for $(\mathcal{M}_X)_S$ and absorption curves for $(\mathcal{M}_Y)_S$ and $(\mathcal{M}_Z)_S$. The three curves have the same width, $2\sqrt{\omega_1^2 + (1/T_R)^2}$, which increases with $\omega_1$. They have been drawn assuming that $\omega_1 = 1/T_R$ ("half-saturation").**

The comments made at the end of § 2-d above about the experimental interest of such curves can be repeated here.

**References and suggestions for further reading:**

Feynman II (7.2), chap. 35; Cagnac and Pebay-Peyroula (11.2), chaps. IX § 5, X § 5, XI §§ 2 to 5, XIX § 3; Kuhn (11.1), § VI, D.

See the references of section 14 of the bibliography, particularly Abragam (14.1) and Slichter (14.2).

## Complement $G_{IV}$

# A SIMPLE MODEL OF THE AMMONIA MOLECULE



## 1. Description of the model

In the ammonia molecule $NH_3$, the three hydrogen atoms form the base of a pyramid whose apex is the nitrogen atom (*cf.* fig. 1). We shall study this molecule by using a simplified model with the following features: the nitrogen atom, much heavier than its partners, is motionless; the hydrogen atoms form a rigid equilateral triangle whose axis always passes through the nitrogen atom. The potential energy of the system is thus a function of only

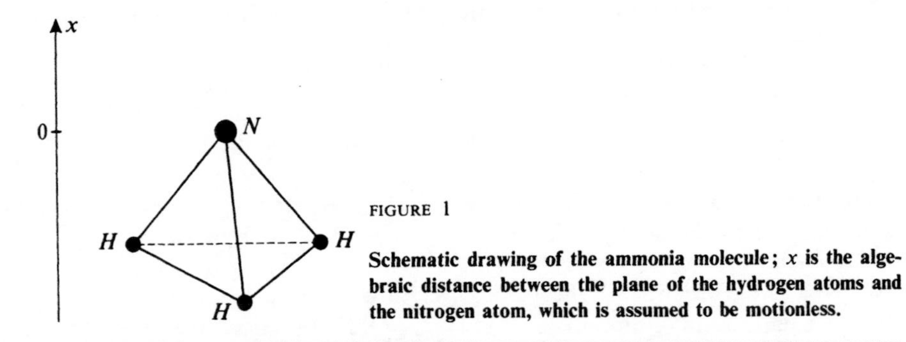


FIGURE 1

**Schematic drawing of the ammonia molecule; $x$ is the algebraic distance between the plane of the hydrogen atoms and the nitrogen atom, which is assumed to be motionless.**

one parameter, the (algebraic) distance $x$ between the nitrogen atom and the plane defined by the three hydrogen atoms*. The shape of this potential energy $V(x)$ is given by the solid-line curve in figure 2. The symmetry of the problem with respect to the $x = 0$ plane requires $V(x)$ to be an even function of $x$. The two minima of $V(x)$ correspond to two symmetrical configurations of the molecule in which, classically, it is stable; we shall choose the energy origin

* In this one-dimensional model, effects linked to the rotation of the molecule are obviously not taken into account.

such that its energy is then zero. The potential barrier at $x = 0$, of height $V_1$, expresses the fact that, if the nitrogen atom is in the plane of the hydrogen atoms, they repel it. Finally, the increase in $V(x)$ when $|x|$ is greater than $b$ corresponds to the chemical bonding force which insures the cohesion of the molecule.

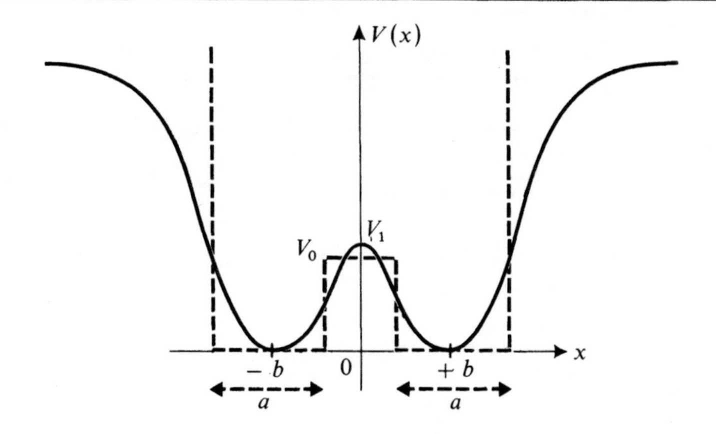


FIGURE 2

**Variation with respect to $x$ of the potential energy $V(x)$ of the molecule. $V(x)$ has two minima (classical equilibrium positions), separated by a potential barrier due to the repulsion for small $|x|$ between the nitrogen atom and the three hydrogen atoms. The "square potential" used to approximate $V(x)$ is shown in dashed lines.**

This model therefore reduces the problem to a one-dimensional one in which a fictitious particle of mass $m$ $\left(\text{it can be shown that the "reduced mass" } m \text{ of the system is equal to } \dfrac{3m_H\, m_N}{3m_H + m_N}\right)$ is under the influence of the potential $V(x)$. Under these conditions, what are the energy levels predicted by quantum mechanics? With respect to classical predictions, two major differences appear :

($i$) The Heisenberg uncertainty relation forbids the molecule to have an energy equal to the minimum of $V(x)$ ($V_{\text{min}} = 0$ in our case). We have already seen, in complements $C_I$ and $M_{III}$ why this energy must be greater than $V_{\text{min}}$.

($ii$) Classically, the potential barrier at $x = 0$ cannot be cleared by a particle whose energy is less than $V_1$ : the nitrogen atom thus always remains on the same side of the plane of the hydrogen atoms, and the molecule cannot invert itself. Quantum mechanically, such a particle can cross this barrier by the tunnel effect (*cf.* chap. I, § D-2-c) : the inversion of the molecule is therefore always possible. We are going to discuss the consequences of this effect.

We shall be concerned here only with a qualitative discussion of the physical phenomena and not with an exact quantitative calculation which would not have much significance in this approximate model. For example, we shall try to

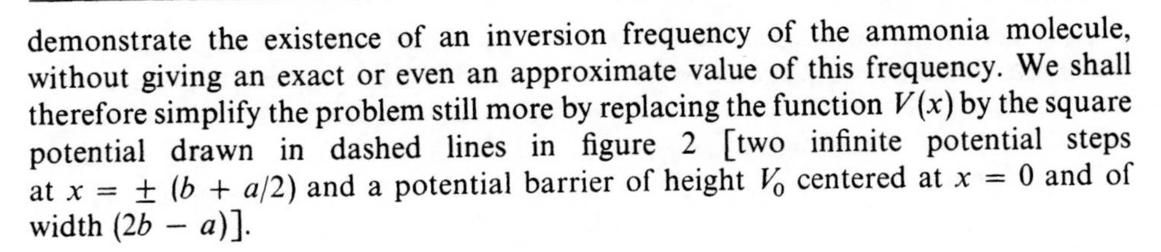

demonstrate the existence of an inversion frequency of the ammonia molecule, without giving an exact or even an approximate value of this frequency. We shall therefore simplify the problem still more by replacing the function $V(x)$ by the square potential drawn in dashed lines in figure 2 [two infinite potential steps at $x = \pm (b + a/2)$ and a potential barrier of height $V_0$ centered at $x = 0$ and of width $(2b - a)$].

## 2. Eigenfunctions and eigenvalues of the Hamiltonian

### a. INFINITE POTENTIAL BARRIER

Before calculating the eigenfunctions and eigenvalues of the Hamiltonian corresponding to the "square" potential of figure 2, we are going to assume, in this first stage, that the potential barrier $V_0$ is infinite (in which case, no tunnel effect is possible). This will lead us to a better understanding of the consequences of the tunnel effect across the finite potential barrier of figure 2. We shall therefore consider, first of all, a particle in a potential $\tilde{V}(x)$ composed of two infinite wells of width $a$ centered at $x = \pm b$ (fig. 3). If the particle is in one of these two wells, it obviously cannot go into the other one.

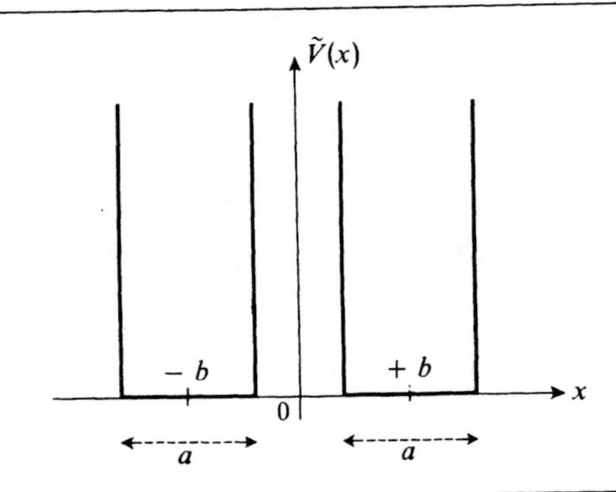


FIGURE 3

**When the height $V_0$ of the potential barrier of figure 2 is large, we have two practically infinite potential wells of width $a$ whose centers are separated by a distance of $2b$.**

Each of the two wells of figure 3 is similar to the one studied in complement $H_I$, in §2-c-$\beta$. We can therefore use the results obtained in this complement. The possible energies of the particle are:

$$E_n = \frac{\hbar^2 k_n^2}{2m} \tag{1}$$

with:

$$k_n = \frac{n\pi}{a} \tag{2}$$

(where $n$ is a positive integer). Each of the energy values is twofold degenerate, since two wave functions correspond to it:

$$\varphi_1^n(x) = \begin{cases} \sqrt{\dfrac{2}{a}} \sin\left[k_n\left(b + \dfrac{a}{2} - x\right)\right] & \text{if } b - \dfrac{a}{2} \leqslant x \leqslant b + \dfrac{a}{2} \\ 0 & \text{everywhere else} \end{cases}$$

$$\varphi_2^n(x) = \begin{cases} \sqrt{\dfrac{2}{a}} \sin\left[k_n\left(b + \dfrac{a}{2} + x\right)\right] & \text{if } b - \dfrac{a}{2} \leqslant -x \leqslant b + \dfrac{a}{2} \\ 0 & \text{everywhere else} \end{cases} \quad (3)$$

In the state $|\varphi_1^n\rangle$, the particle is in the infinite well on the right; in the state $|\varphi_2^n\rangle$, it is in the one on the left.

Figure 4 shows the first two energy levels of the molecule, which are two-fold degenerate. The Bohr frequency $(E_2 - E_1)/h$ associated with these two levels corresponds, as we saw in complement $A_{III}$ (§ 2-b), to the to-and-fro motion of the particle between the two sides of the well on the right (or on the left) when its state is a linear superposition of $|\varphi_1^1\rangle$ and $|\varphi_1^2\rangle$ (or of $|\varphi_2^1\rangle$ and $|\varphi_2^2\rangle$). Physically, such an oscillation represents a molecular vibration of the plane of the three hydrogen atoms about its stable equilibrium position, which corresponds to $x = +b$ (or $x = -b$). The frequency of this oscillation falls in the infrared part of the spectrum.

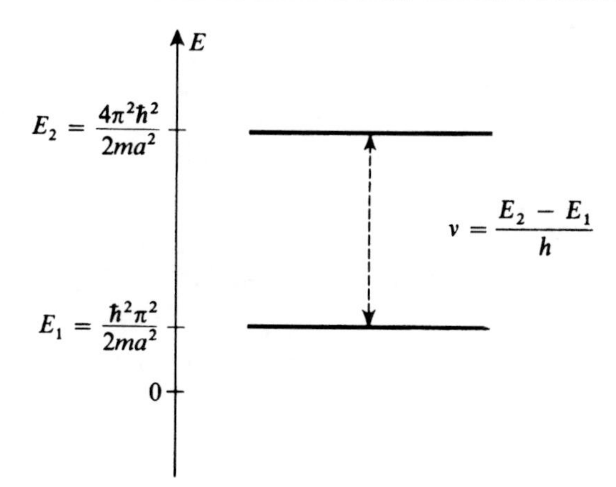


FIGURE 4

**First energy levels obtained in the potential wells of figure 3. The oscillation of the system in one of the two wells at the Bohr frequency $\nu = (E_2 - E_1)/h$ represents the vibration of the molecule about one of its two classical equilibrium positions.**

In the rest of the calculations, it is convenient to change bases, in each of the eigensubspaces of the Hamiltonian of the particle. Since the function $V(x)$ is even, this Hamiltonian $H$ commutes with the parity operator $\Pi$ (*cf.* complement $F_{II}$, §4). In this case, a basis of eigenvectors of $H$ can be found which are even or odd;

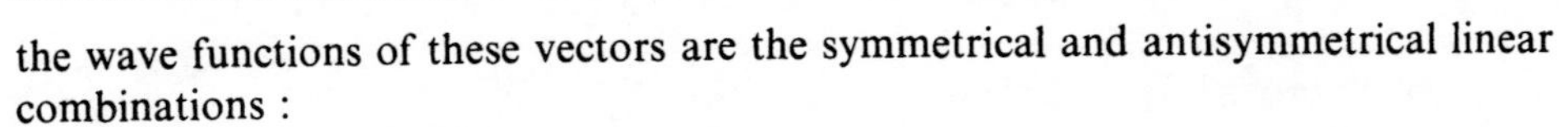



the wave functions of these vectors are the symmetrical and antisymmetrical linear combinations :

$$\begin{aligned} \varphi_s^n(x) &= \frac{1}{\sqrt{2}}\left[\, \varphi_1^n(x) + \varphi_2^n(x) \,\right] \\ \varphi_a^n(x) &= \frac{1}{\sqrt{2}}\left[\, \varphi_1^n(x) - \varphi_2^n(x) \,\right] \end{aligned} \tag{4}$$

In the states $|\, \varphi_s^n \rangle$ and $|\, \varphi_a^n \rangle$, the particle can be found in one or the other of the the two potential wells.

In what follows, we shall confine ourselves to the study of the ground state, for which the wave functions $\varphi_1^1(x)$, $\varphi_2^1(x)$, $\varphi_s^1(x)$ and $\varphi_a^1(x)$ are shown in figure 5.

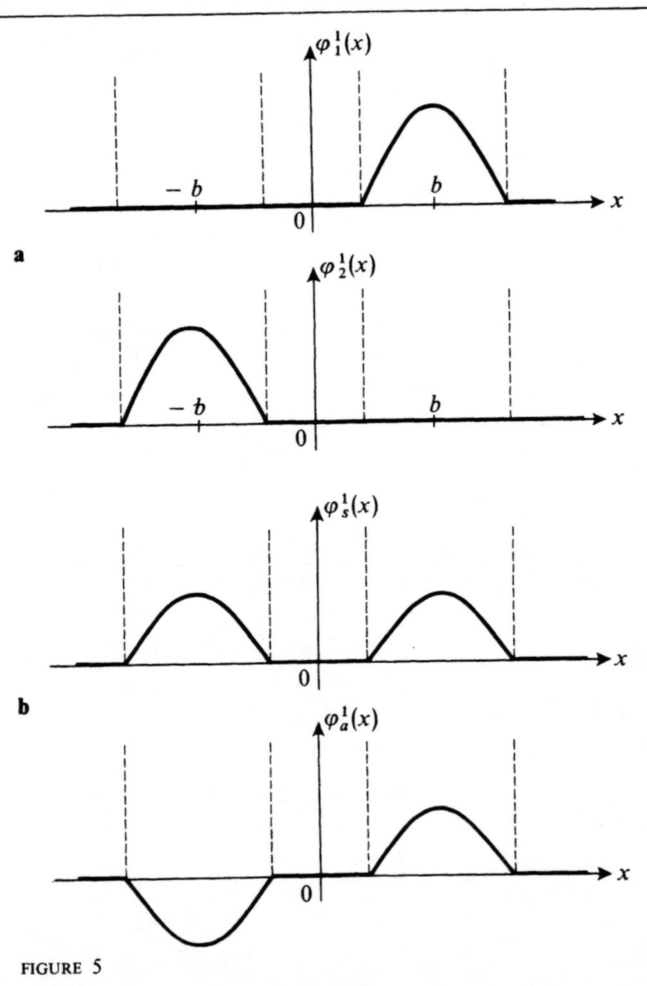


FIGURE 5

**The states $\varphi_1^1(x)$ and $\varphi_2^1(x)$, shown in figure a, are stationary states with the same energy, respectively localized in the right-hand well and the left-hand well of figure 3. To use the symmetry of the problem, it is more convenient to choose as stationary states the symmetrical state $\varphi_s^1(x)$ and the antisymmetrical state $\varphi_a^1(x)$, linear combinations of $\varphi_1^1(x)$ and $\varphi_2^1(x)$ (figure b).**

### b. FINITE POTENTIAL BARRIER

Let us try to find the shape of the eigenfunctions of the first energy levels when $V_0$ has a finite value (assumed, nevertheless, to be greater than the energy of these levels).

Inside the two "square" potential wells (dashed lines in figure 2), $V(x) = 0$. The wave function is therefore of the form:

$$\begin{aligned} \chi(x) &= A \sin\left[k\left(b + \frac{a}{2} - x\right)\right] \quad &\text{if} \quad b - \frac{a}{2} \leqslant x \leqslant b + \frac{a}{2} \\ \chi(x) &= A' \sin\left[k\left(b + \frac{a}{2} + x\right)\right] \quad &\text{if} \quad b - \frac{a}{2} \leqslant -x \leqslant b + \frac{a}{2} \end{aligned} \tag{5}$$

where $k$ is related to the energy $E$ of the level by the relation:

$$E = \frac{\hbar^2 k^2}{2m} \tag{6}$$

As in the preceding paragraph, $\chi(x)$ always goes to zero at $x = \pm (b + a/2)$, since $V(x)$ becomes infinite at these two points. On the other hand, since $V_0$ is finite, $\chi(x)$ no longer goes to zero at $x = \pm (b - a/2)$; consequently, $k$ no longer satisfies relation (2).

Once again, since $V(x)$ is even, we can look for eigenfunctions of the Hamiltonian, $\chi_s(x)$ and $\chi_a(x)$, which are respectively even and odd. Let us denote by $A_s$ and $A'_s$, $A_a$ and $A'_a$ the values of the coefficients $A$ and $A'$, introduced in (5), which correspond to $\chi_s(x)$ and $\chi_a(x)$. We have, obviously:

$$\begin{aligned} A'_s &= A_s \\ A'_a &= -A_a \end{aligned} \tag{7}$$

The eigenvalues associated with $\chi_s$ and $\chi_a$ will be denoted by $E_s$ and $E_a$, which enables us, using (6), to define the corresponding values $k_s$ and $k_a$ of the parameter $k$.

In the interval $-(b - a/2) \leqslant x \leqslant (b - a/2)$, the wave function is no longer zero, as it was before, since $V_0$ is finite. It must be a linear combination, even or odd depending on whether we are considering $\chi_s$ or $\chi_a$, of exponentials $e^{q_{s,a}x}$ and $e^{-q_{s,a}x}$; $q_s$ and $q_a$ are defined in terms of $E_{s,a}$ and $V_0$ by:

$$q_{s,a} = \sqrt{\frac{2m}{\hbar^2}(V_0 - E_{s,a})} = \sqrt{\alpha^2 - k_{s,a}^2} \tag{8}$$

with:

$$V_0 = \frac{\hbar^2 \alpha^2}{2m} \tag{9}$$

Therefore, for $-(b - a/2) \leqslant x \leqslant (b - a/2)$, the functions $\chi_s$ and $\chi_a$ are written:

$$\begin{aligned} \chi_s(x) &= B_s \cosh(q_s x) \\ \chi_a(x) &= B_a \sinh(q_a x) \end{aligned} \tag{10}$$

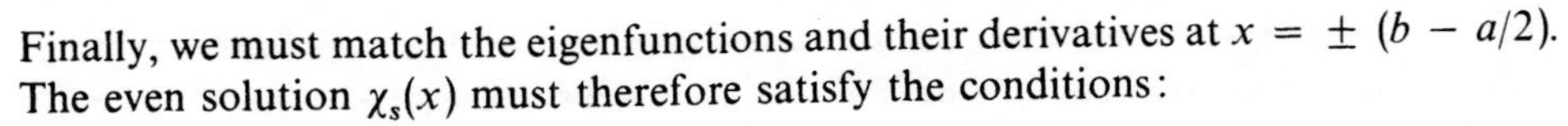



Finally, we must match the eigenfunctions and their derivatives at $x = \pm (b - a/2)$. The even solution $\chi_s(x)$ must therefore satisfy the conditions:

$$A_s \sin (k_s a) = B_s \cosh \left[ q_s \left( b - \frac{a}{2} \right) \right]$$

$$- A_s k_s \cos (k_s a) = B_s q_s \sinh \left[ q_s \left( b - \frac{a}{2} \right) \right] \tag{11}$$

Since $A_s$ and $B_s$ cannot be zero simultaneously, we can take the ratio of equations (11) to obtain:

$$\tan (k_s a) = - \frac{k_s}{q_s} \coth \left[ q_s \left( b - \frac{a}{2} \right) \right] \tag{12}$$

For the odd solution $\chi_a(x)$, we obtain in the same way:

$$\tan (k_a a) = - \frac{k_a}{q_a} \tanh \left[ q_a \left( b - \frac{a}{2} \right) \right] \tag{13}$$

If $q_s$ and $q_a$ are replaced by their values in terms of $k_s$ and $k_a$, relations (12) and (13) can be written:

$$\tan (k_s a) = - \frac{k_s}{\sqrt{\alpha^2 - k_s^2}} \coth \left[ \sqrt{\alpha^2 - k_s^2} \left( b - \frac{a}{2} \right) \right] \tag{14}$$

and:

$$\tan (k_a a) = - \frac{k_a}{\sqrt{\alpha^2 - k_a^2}} \tanh \left[ \sqrt{\alpha^2 - k_a^2} \left( b - \frac{a}{2} \right) \right] \tag{15}$$

In theory, therefore, the problem is solved. Relations (14) and (15) express the energy quantization since they give the possible values of $k_s$ and $k_a$ and therefore, thanks to relation (6), the energies $E_s$ and $E_a$ (with the condition that they be less than $V_0$). The transcendental equations (14) and (15) can be solved graphically. A certain number of roots are found : $k_s^1$, $k_s^2$, ..., $k_a^1$, $k_a^2$, ... The root $k_s^n$ is different from $k_a^n$, since equations (14) and (15) are not the same : *the energies* $E_s^n$ and $E_a^n$ *are therefore different*. Of course, when $V_0$ becomes very large, $k_s^n$ and $k_a^n$ both approach the value $n\pi/a$ found in the preceding section; this can be seen by letting $\alpha$ approach infinity in equations (14) and (15), which yields $\tan(k_{s,a}a) = 0$, an equation equivalent to (2). The energies $E_s^n$ and $E_a^n$ therefore approach the value $E_n = \hbar^2 n^2 \pi^2 / 2ma^2$ calculated in the preceding section for $V_0$ approaching infinity. Finally, it is easy to see that, the more $V_0$ exceeds $E_n$, the closer together the two energies $E_s^n$ and $E_a^n$ will be.

The exact values of $E_s^n$ and $E_a^n$ are of little importance to us here. We shall content ourselves with sketching the shape of the energy spectrum in figure 6, which shows what happens to the energies of levels $E_1$ and $E_2$ of figure 4 when the finite height $V_0$ of the potential barrier is taken into account. We see that the tunnel effect across this barrier removes the degeneracy of $E_1$ and $E_2$, giving rise to doublets, $(E_s^1, E_a^1)$ and $(E_s^2, E_a^2)$ (assuming, of course, that all these energies are less than $V_0$). Since the $(E_s^1, E_a^1)$ doublet is the deeper one, it is clear that $|E_s^1 - E_a^1| < |E_s^2 - E_a^2|$. Finally, the distance between the doublets is much greater than the spacing within each doublet (experimentally, their ratio is of the order of a thousand). These spacings enable us, moreover, to define new Bohr frequencies:

$$\Omega_1 = \frac{E_a^1 - E_s^1}{\hbar} \quad , \quad \Omega_2 = \frac{E_a^2 - E_s^2}{\hbar}, \ldots$$

whose physical significance we shall study in the next paragraph (the corresponding transitions are represented by arrows in figure 6).

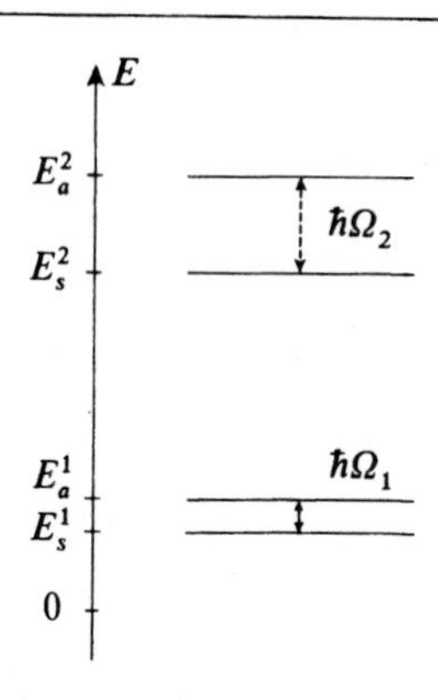


FIGURE 6

**When one takes the finite height $V_0$ of the barrier into account, one finds that the energy spectrum of figure 4 is modified: each level splits into two distinct ones. The Bohr frequencies $\Omega_1/2\pi$ and $\Omega_2/2\pi$ corresponding to tunnelling from one well to the other are the inversion frequencies of the ammonia molecule for the first two vibration levels. The tunnel effect is more important in the higher vibration level, so $\Omega_2 > \Omega_1$.**

Finally, in figure 7, we have shown the shape of the eigenfunctions $\chi_s^1(x)$ and $\chi_a^1(x)$ which are given by equations (5), (7) and (10), once $k_s^1$ and $k_a^1$ have been determined from (14) and (15). We see that they greatly resemble the func-

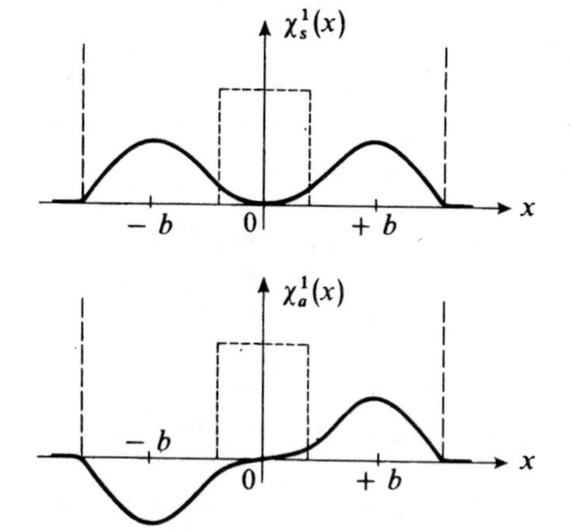


FIGURE 7

**Wave functions associated with the levels $E_s'$ and $E_a'$ in figure 6. Note the analogy with the functions in figure 5-b; however, these new functions do not vanish on the interval $-b + a/2 \leqslant x \leqslant b - a/2$.**

tions $\varphi_s^1(x)$ and $\varphi_a^1(x)$ of figure 5, the essential difference being that the wave function is no longer zero in the interval $-(b - a/2) \leqslant x \leqslant (b - a/2)$. The reason for introducing the $\varphi_s^1$ and $\varphi_a^1$ basis in the preceding paragraph can now be understood : the eigenfunctions $\chi_s^1$ and $\chi_a^1$, in the presence of the tunnel effect, resemble $\varphi_s^1$ and $\varphi_a^1$ much more than $\varphi_1^1$ and $\varphi_2^1$.

c. EVOLUTION OF THE MOLECULE. INVERSION FREQUENCY

Assume that at time $t = 0$, the molecule is in the state :

$$| \psi(t = 0) \rangle = \frac{1}{\sqrt{2}} [ | \chi_s^1 \rangle + | \chi_a^1 \rangle ] \tag{16}$$

The state vector $| \psi(t) \rangle$ at time $t$ can be obtained by using the general formula (D-54) of chapter III; we obtain :

$$| \psi(t) \rangle = \frac{1}{\sqrt{2}} e^{-i\frac{E_s^1 + E_a^1}{2\hbar}t} [e^{+i\Omega_1 t/2} | \chi_s^1 \rangle + e^{-i\Omega_1 t/2} | \chi_a^1 \rangle] \tag{17}$$

From this we deduce the probability density:

$$|\psi(x, t)|^2 = \frac{1}{2} [\chi_s^1(x)]^2 + \frac{1}{2} [\chi_a^1(x)]^2 + \cos (\Omega_1 t) \chi_s^1(x) \chi_a^1(x) \tag{18}$$

The variation with respect to time of this probability density is simple to obtain graphically from the curves of figure 7. They are shown in figure 8. For $t = 0$ (fig. 8-a), we see that the initial state chosen in (16) corresponds to a probability density which is concentrated in the right-hand well (in the left-hand well, the functions $\chi_s^1$ and $\chi_a^1$ are of opposite sign and very close in absolute value, so their sum is practically zero). It can therefore be said that the particle, initially, is practically in the right-hand well. At time $t = \pi/2\Omega_1$ (fig. 8-b), it has moved

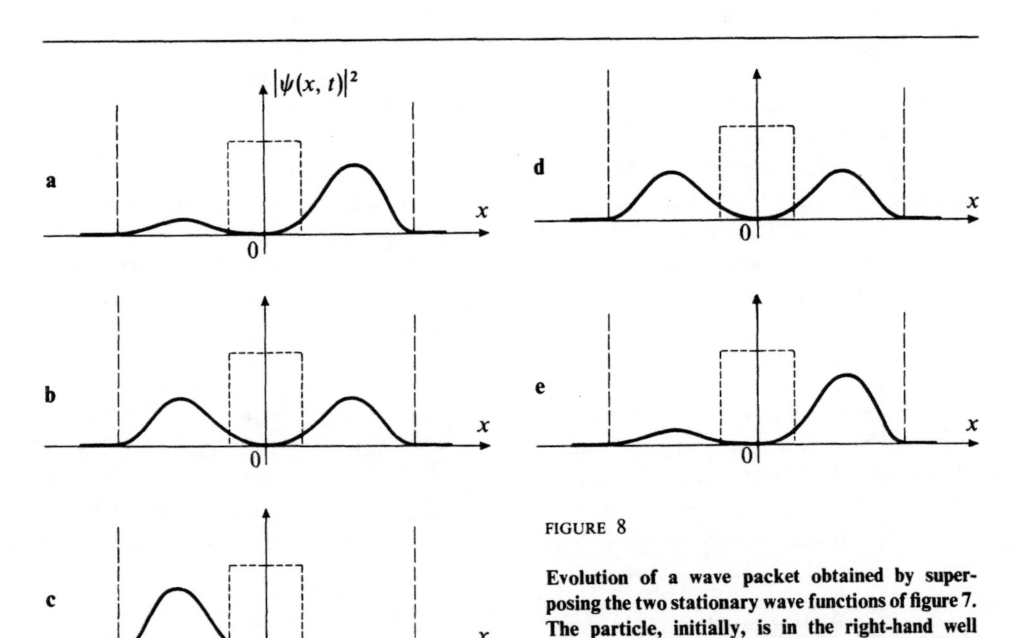


FIGURE 8

**Evolution of a wave packet obtained by superposing the two stationary wave functions of figure 7. The particle, initially, is in the right-hand well (fig. a), tunnels into the left-hand well (fig. b) and, after a certain time, becomes localized there (fig. c); then it returns to the right-hand well (fig. d) and the initial state (fig. e), and so on.**

appreciably, through the tunnel effect, into the left-hand well, is practically there at time $t = \pi/\Omega_1$ (fig. 8-c), and then performs the process in reverse (figures 8-d and 8-e).

The fictitious particle therefore moves from one side of the potential barrier to the other with the frequency $\Omega_1/2\pi$, which means that the plane of the hydrogen atoms continually passes from one side of the nitrogen atom to the other. This is why the frequency $\Omega_1/2\pi$ is called the *inversion frequency* of the molecule. Note that this inversion frequency has no classical analogue, since its existence is related to the tunnel effect of the fictitious particle across the potential barrier.

Since the nitrogen atom tends to attract the electrons of the three hydrogen atoms, the ammonia molecule possesses an electric dipole moment which is proportional to the mean value $\langle X \rangle$ of the position of the fictitious particle we have studied; we see in figure 8 that this dipole moment is an oscillating function of time. Under these conditions, the ammonia molecule is capable of emitting or absorbing electromagnetic radiation of frequency $\Omega_1/2\pi$.

Experimentally, this is indeed observed; the value of $\Omega_1$ falls in the domain of centimeter waves. In radioastronomy, ammonia molecules in interstellar space have been shown to emit and absorb electromagnetic waves of this frequency. Let us also point out that the principle of the ammonia maser is based on the stimulated emission of these waves by the $NH_3$ molecule.

## 3. The ammonia molecule considered as a two-level system

We see in figure 6 that we have a situation which is analogous to the one mentioned in the introduction of § C of chapter IV. The system under study possesses two levels, $E_s^1$ and $E_a^1$, which are very close to each other and very far from all other levels $E_s^2$, $E_a^2$, ... If we are interested only in the two levels $E_s^1$ and $E_a^1$, we can "forget" all the others (the exact justification for such an approximation will be given in the framework of perturbation theory in chapter XI).

We are going to return to the preceding discussion with a slightly different point of view and show that the general considerations of chapter IV concerning two-level systems can be applied to the ammonia molecule. This point of view will also enable us to study very simply the effect of a static external electric field on this molecule.

### a. THE STATE SPACE

The state space we are going to consider is spanned by the two orthogonal vectors $|\varphi_1^1\rangle$ and $|\varphi_2^1\rangle$, whose wave functions are given by (3). As we explained above, we shall ignore the other states $|\varphi_1^n\rangle$ and $|\varphi_2^n\rangle$ for which $n > 1$. In the states $|\varphi_1^1\rangle$ and $|\varphi_2^1\rangle$, the nitrogen atom is either above or below the plane of the hydrogen atoms. We introduced in (4) a second orthonormal basis of the state space, composed of the even and odd vectors:

$$
\begin{aligned}
|\varphi_s^1\rangle &= \frac{1}{\sqrt{2}}[|\varphi_1^1\rangle + |\varphi_2^1\rangle] \\
|\varphi_a^1\rangle &= \frac{1}{\sqrt{2}}[|\varphi_1^1\rangle - |\varphi_2^1\rangle]
\end{aligned}
\tag{19}
$$

There is the same probability in these two states of finding the nitrogen atom above or below the plane of the hydrogen atoms.

### b. ENERGY LEVELS. REMOVAL OF THE DEGENERACY DUE TO THE TRANSPARENCY OF THE POTENTIAL BARRIER

When the height $V_0$ of the potential barrier is infinite, the states $|\varphi_1^1\rangle$ and $|\varphi_2^1\rangle$ have the same energy (as do the states $|\varphi_s^1\rangle$ and $|\varphi_a^1\rangle$), so that $H_0$, the Hamiltonian of the system, is written:

$$H_0 = E_1 \times \mathbb{1} \tag{20}$$

(where $\mathbb{1}$ is the identity operator in the two-dimensional state space).

To take into account phenomenologically the fact that the barrier is not infinite, we add to $H_0$ a perturbation $W$ which is non-diagonal in the $\{ |\varphi_1^1\rangle, |\varphi_2^1\rangle \}$ basis and is represented by the matrix:

$$W = -A \begin{pmatrix} 0 & 1 \\ 1 & 0 \end{pmatrix} \tag{21}$$

where $A$ is a real positive coefficient*.

If we want to find the stationary states of the molecule, we must now diagonalize the total Hamiltonian operator $H = H_0 + W$, whose matrix is written:

$$H = \begin{pmatrix} E_1 & -A \\ -A & E_1 \end{pmatrix} \tag{22}$$

A simple calculation gives the eigenvalues and eigenvectors of $H$:

$$\begin{array}{ll} E_1 + A \text{ corresponding to the eigenket } & |\varphi_a^1\rangle \\ E_1 - A \quad \text{"} \quad \text{"} \quad \text{"} & |\varphi_s^1\rangle \end{array} \tag{23}$$

We see that, under the effect of the perturbation $W$, the two levels, which were degenerate when $A$ was zero, now split; an energy difference, equal to $2A$, appears, and the new eigenstates are the states $|\varphi_s^1\rangle$ and $|\varphi_a^1\rangle$. We again find the results of § 2.

If, at time $t = 0$, the molecule is in the state $|\varphi_1^1\rangle$:

$$|\psi(t=0)\rangle = |\varphi_1^1\rangle = \frac{1}{\sqrt{2}}[|\varphi_s^1\rangle + |\varphi_a^1\rangle] \tag{24}$$

the state vector at time $t$ will be:

$$\begin{aligned} |\psi(t)\rangle &= \frac{1}{\sqrt{2}} e^{-iE_1t/\hbar} [e^{iAt/\hbar} |\varphi_s^1\rangle + e^{-iAt/\hbar} |\varphi_a^1\rangle] \\ &= e^{-iE_1t/\hbar} \left[ \cos\left(\frac{At}{\hbar}\right) |\varphi_1^1\rangle + i \sin\left(\frac{At}{\hbar}\right) |\varphi_2^1\rangle \right] \end{aligned} \tag{25}$$

* We are forced to assume $A > 0$ in order to obtain the relative disposition of the $E_s^1$ and $E_a^1$ levels of figure 6 [see eigenvalues (23)].

In a measurement performed at time $t$, we therefore have a probability $\cos^2 (At/\hbar)$ of finding the molecule in the state $|\varphi_1^1\rangle$ (the nitrogen atom above the plane of the hydrogen atoms) and $\sin^2 (At/\hbar)$ of finding it in the state $|\varphi_2^1\rangle$ (the nitrogen atom below). Thus we again find that, under the effect of the coupling $W$, the ammonia molecule inverts periodically.

COMMENT:

The perturbation $W$ [given in (21)] describes (phenomenologically) the fact that the potential barrier is finite. This approach is less precise than the discussion above, since we obtain here eigenfunctions $\varphi_s^1(x)$ and $\varphi_a^1(x)$ which, unlike $\chi_s^1$ and $\chi_a^1$, go to zero in the region $(-b + a/2) \leqslant x \leqslant (b - a/2)$. This much more simple description nevertheless explains two fundamental physical effects: the removal of the degeneracy of $E_1$ and the periodic oscillation of the molecule between the states $|\varphi_1^1\rangle$ and $|\varphi_2^1\rangle$ (inversion).

### c. INFLUENCE OF A STATIC ELECTRIC FIELD

We saw above that, in the states $|\varphi_1^1\rangle$ and $|\varphi_2^1\rangle$, the electric dipole moment of the molecule takes on two opposite values which we shall denote by $+\eta$ and $-\eta$. If we call $D$ the observable associated with this physical quantity, we can therefore assume that $D$ is represented in the $\{|\varphi_1^1\rangle, |\varphi_2^1\rangle\}$ basis by a diagonal matrix whose eigenvalues are $+\eta$ and $-\eta$:

$$D = \begin{pmatrix} \eta & 0 \\ 0 & -\eta \end{pmatrix} \tag{26}$$

When the molecule is placed in a static electric field $\mathscr{E}$*, the interaction energy with this field is:

$$W'(\mathscr{E}) = -\mathscr{E}D \tag{27}$$

This term of the Hamiltonian** is represented in the $\{|\varphi_1^1\rangle, |\varphi_2^1\rangle\}$ basis by the matrix:

$$W'(\mathscr{E}) = -\eta\mathscr{E}\begin{pmatrix} 1 & 0 \\ 0 & -1 \end{pmatrix} \tag{28}$$

Let us then write the matrix which represents, in the $\{|\varphi_1^1\rangle, |\varphi_2^1\rangle\}$ basis, the total Hamiltonian operator of the molecule, $H_0 + W + W'(\mathscr{E})$:

$$H_0 + W + W'(\mathscr{E}) = \begin{pmatrix} E_1 - \eta\mathscr{E} & -A \\ -A & E_1 + \eta\mathscr{E} \end{pmatrix} \tag{29}$$

* For the sake of simplicity, we assume here that this field is parallel to the $Ox$ axis of figure 1 (one-dimensional model).

** In $W'(\mathscr{E})$, $D$ is an observable, while $\mathscr{E}$ is a classical quantity which is externally imposed (*cf.* note, page 446).

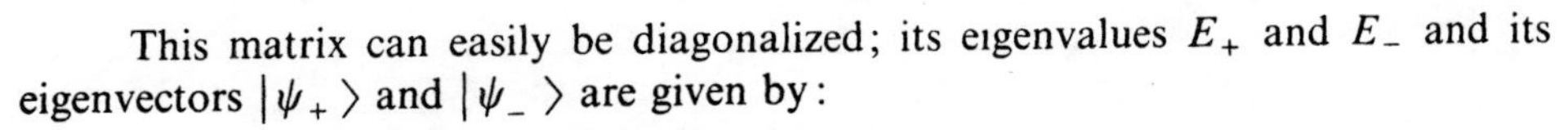


This matrix can easily be diagonalized; its eigenvalues $E_+$ and $E_-$ and its eigenvectors $|\psi_+\rangle$ and $|\psi_-\rangle$ are given by:

$$E_+ = E_1 + \sqrt{A^2 + \eta^2\mathscr{E}^2}$$
$$E_- = E_1 - \sqrt{A^2 + \eta^2\mathscr{E}^2} \tag{30}$$

and:

$$|\psi_+\rangle = \cos\frac{\theta}{2}|\varphi_1^1\rangle - \sin\frac{\theta}{2}|\varphi_2^1\rangle$$
$$|\psi_-\rangle = \sin\frac{\theta}{2}|\varphi_1^1\rangle + \cos\frac{\theta}{2}|\varphi_2^1\rangle \tag{31}$$

where we have set:

$$\tan\theta = -\frac{A}{\eta\mathscr{E}} \qquad 0 \leqslant \theta < \pi \tag{32}$$

[*cf.* complement $B_{IV}$, relations (9), (10), (22) and (23); since $A$ is real and negative, the angle $\varphi$ introduced in that complement is here equal to $\pi$].

When $\mathscr{E}$ is zero, $\theta = \pi/2$, and we again obtain the results of §3-b, since:

$$E_+(\mathscr{E}=0) = E_1 + A$$
$$E_-(\mathscr{E}=0) = E_1 - A \tag{33}$$

with:

$$|\psi_+(\mathscr{E}=0)\rangle = |\varphi_a^1\rangle$$
$$|\psi_-(\mathscr{E}=0)\rangle = |\varphi_s^1\rangle \tag{34}$$

When, for arbitrary $\mathscr{E}$, $A$ is zero (a perfectly opaque potential barrier), we obtain:

$$E_+(A=0) = E_1 + \eta|\mathscr{E}|$$
$$E_-(A=0) = E_1 - \eta|\mathscr{E}| \tag{35}$$

with, if $\mathscr{E}$ is positive*:

$$|\psi_-(A=0)\rangle = |\varphi_1^1\rangle$$
$$|\psi_+(A=0)\rangle = -|\varphi_2^1\rangle \tag{36}$$

In this case, the energies therefore vary linearly with $\mathscr{E}$ (dashed straight lines in figure 9). Physically, results (35) and (36) are easy to understand: when the electric field alone acts on the molecule, it "pulls" the positively charged hydrogen atoms above or below the nitrogen atom; this is why the stationary states are $|\varphi_1^1\rangle$ and $|\varphi_2^1\rangle$.

When the electric field $\mathscr{E}$ and the coupling constant $A$ are both arbitrary, the states $|\psi_+\rangle$ and $|\psi_-\rangle$ are linear superpositions of the states $|\varphi_1^1\rangle$ and $|\varphi_2^1\rangle$ (and of the states $|\varphi_s^1\rangle$ and $|\varphi_a^1\rangle$ as well), and result from a compromise between the action of the electric field, which tends to pull the hydrogen atoms to one side

* If $\mathscr{E}$ is negative, the roles of $|\varphi_1^1\rangle$ and $|\varphi_2^1\rangle$ are inverted in (36).

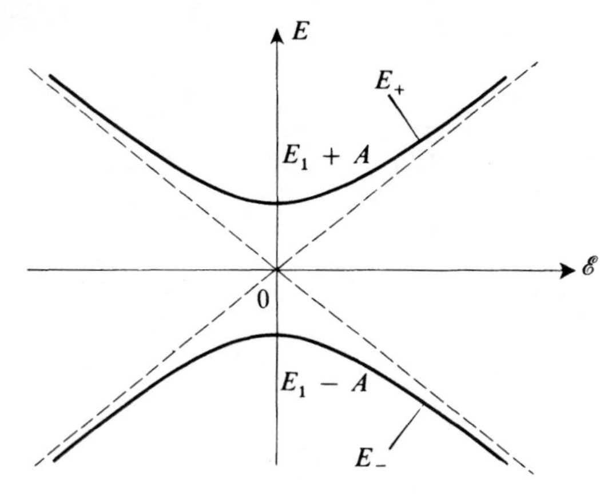


FIGURE 9

**Influence of an electric field $\mathscr{E}$ on the first two levels of the ammonia molecule (their spacing $2A$ in a zero field is due to the tunnel effect coupling). For weak $\mathscr{E}$, the molecule acquires a dipole moment proportional to $\mathscr{E}$, and the corresponding energy varies with $\mathscr{E}^2$. For large $\mathscr{E}$, the dipole moment approaches a limit (corresponding to the nitrogen atom either above or below the plane of the hydrogen atoms), and the energy becomes a linear function of $\mathscr{E}$.**

of the nitrogen atom, and that of the coupling $W$, which tends to draw the nitrogen atom accross the potential barrier. The variation of the energies $E_+$ and $E_-$ is shown graphically in figure 9, in which we see the phenomenon of anti-crossing (*cf.* chap. IV, §C-2-b) due to the coupling $W$. $E_+$ and $E_-$ correspond to the two branches of a hyperbola whose asymptotes are the dashed lines associated with the energies in the absence of coupling.

Finally, we can calculate the mean values of the electric dipole moment $D$ in each of the two stationary states $|\psi_+\rangle$ and $|\psi_-\rangle$. Using (26) and (31), we find :

$$\langle \psi_+ | D | \psi_+ \rangle = - \langle \psi_- | D | \psi_- \rangle = \eta \cos \theta \tag{37}$$

which, according to (32), yields:

$$\langle \psi_+ | D | \psi_+ \rangle = - \langle \psi_- | D | \psi_- \rangle = - \frac{\eta^2 \mathscr{E}}{\sqrt{A^2 + \eta^2 \mathscr{E}^2}} \tag{38}$$

For $\mathscr{E} = 0$, these two mean values are zero. This corresponds to the fact that, in the two states $|\varphi^1_{s,a}\rangle$, the particle has an equal probability of being in one or the other of the two wells. On the other hand, when $\eta\mathscr{E} \gg A$, we again find the dipole moment $+\eta$ (or $-\eta$) corresponding to the state $|\varphi^1_1\rangle$ (or $|\varphi^1_2\rangle$).

When the electric field is weak ($\eta\mathscr{E} \ll A$), formulas (38) can be written in the form:

$$\langle \psi_+ | D | \psi_+ \rangle = - \langle \psi_- | D | \psi_- \rangle = - \frac{\eta^2}{A} \mathscr{E} \tag{39}$$

We see that the molecule in the stationary state $|\psi_+\rangle$ (or $|\psi_-\rangle$) acquires an electric dipole moment proportional to the external field $\mathscr{E}$. If we define an electric susceptibility of the molecule in the state $|\psi_-\rangle$ by the relation:

$$\langle \psi_- | D | \psi_- \rangle = \varepsilon_- \mathscr{E} \tag{40}$$

we find, according to (39), that:

$$\varepsilon_- = \frac{\eta^2}{A} \tag{41}$$

(the same calculations are valid for $|\psi_+\rangle$ and yield $\varepsilon_+ = -\varepsilon_-$).

COMMENT:

In a weak field, formulas (30) can be expanded in power series of $\eta\mathscr{E}/A$:

$$E_- = E_1 - A - \frac{1}{2}\frac{\eta^2\mathscr{E}^2}{A} + \dots \tag{42-a}$$

$$E_+ = E_1 + A + \frac{1}{2}\frac{\eta^2\mathscr{E}^2}{A} + \dots \tag{42-b}$$

Let us now consider ammonia molecules moving in a region where $\mathscr{E}$ is weak but where $\mathscr{E}^2$ has a strong gradient in the $Ox$ direction (i.e. along the axis of the molecules):

$$\frac{\mathrm{d}}{\mathrm{d}x}(\mathscr{E}^2) = \lambda \tag{43}$$

According to (42-a), the molecules in the state $|\psi_-\rangle$ are subjected to a force parallel to $Ox$ which is equal to:

$$F_- = -\frac{\mathrm{d}E_-}{\mathrm{d}x} = \frac{1}{2}\lambda\frac{\eta^2}{A} \tag{44}$$

Relation (42-b) indicates that the molecules in the state $|\psi_+\rangle$ are subjected to an opposite force:

$$F_+ = -\frac{\mathrm{d}E_+}{\mathrm{d}x} = -F_- \tag{45}$$

This result is the basis of the method which is used in the ammonia maser to sort the molecules and select those in the higher energy state. The device used is analogous to the Stern-Gerlach apparatus: a beam of ammonia molecules crosses a region where there is a strong electric field gradient; the molecules follow different trajectories depending on whether they are in one state or the other; one can, using a suitable diaphragm, isolate either one of the two states.

**References and suggestions for further reading:**

Feynman III (1.2), §8-6 and chap. 9; Alonso and Finn III (1.4), §2-8; article by Vuylsteke (1.34); Townes and Schawlow (12.10), chap. 12; see (15.11) for references to original articles on masers; articles by Lyons (15.14), Gordon (15.15), and Turner (12.14).

See also Encrenaz (12.11), chap. VI.

# Complement $H_{IV}$

## EFFECTS OF A COUPLING BETWEEN A STABLE STATE AND AN UNSTABLE STATE



### 1. Introduction. Notation

The effects of a coupling $W$ between two states $|\varphi_1\rangle$ and $|\varphi_2\rangle$ of energies $E_1$ and $E_2$ were discussed in detail in § C of chapter IV. What modifications appear when one of the two states ($|\varphi_1\rangle$, for example) is unstable?

The concepts of an unstable state and a lifetime were introduced in complement $K_{III}$. We shall assume, for example, that $|\varphi_1\rangle$ is an excited atomic state. When the atom is in this state, it can fall back to a lower energy level through spontaneous emission of one or several photons, with a probability $1/\tau_1$ per unit time : $\tau_1$ is the lifetime of the unstable state $|\varphi_1\rangle$. On the other hand, we assume that in the absence of the coupling $W$, the state $|\varphi_2\rangle$ is stable ($\tau_2$ is infinite).

We saw in complement $K_{III}$ that a simple way of taking the instability of a state into account consists of adding an imaginary term to the corresponding energy. We shall therefore replace the energy $E_1$ of the state $|\varphi_1\rangle$ by:

$$E'_1 = E_1 - i\frac{\hbar}{2}\gamma_1 \tag{1}$$

with:

$$\gamma_1 = \frac{1}{\tau_1} \tag{2}$$

(since $\tau_2$ is infinite, $\gamma_2$ is zero and $E'_2 = E_2$). In the absence of coupling, the matrix which represents the "Hamiltonian" $H_0$ of the system can therefore be written in the $\{ |\varphi_1\rangle, |\varphi_2\rangle \}$ basis*:

$$H_0 = \begin{pmatrix} E'_1 & 0 \\ 0 & E'_2 \end{pmatrix} = \begin{pmatrix} E_1 - i\frac{\hbar}{2}\gamma_1 & 0 \\ 0 & E_2 \end{pmatrix} \tag{3}$$

* The operator $H_0$ is not Hermitian and is therefore not really a Hamiltonian (see the comment at the end of complement $K_{III}$).

## 2. Influence of a weak coupling on states of different energies

Now let us assume, as in § C of chapter IV, that we add to $H_0$ a perturbation $W$, whose matrix in the $\{ |\varphi_1\rangle, |\varphi_2\rangle \}$ basis is:

$$W = \begin{pmatrix} 0 & W_{12} \\ W_{21} & 0 \end{pmatrix} \tag{4}$$

What now happens to the energies and lifetimes of the levels?

Let us calculate the eigenvalues $\varepsilon_1'$ and $\varepsilon_2'$ of the matrix:

$$H = H_0 + W = \begin{pmatrix} E_1 - i\frac{\hbar}{2}\gamma_1 & W_{12} \\ W_{21} & E_2 \end{pmatrix} \tag{5}$$

$\varepsilon_1'$ and $\varepsilon_2'$ are the solutions of the equation in $\varepsilon$:

$$\varepsilon^2 - \varepsilon\left(E_1 + E_2 - i\frac{\hbar}{2}\gamma_1\right) + E_1E_2 - i\frac{\hbar}{2}\gamma_1E_2 - |W_{12}|^2 = 0 \tag{6}$$

To simplify the calculation, we shall confine ourselves to the case where the coupling is weak

$$\left(|W_{12}| \ll \sqrt{(E_1 - E_2)^2 + \frac{\hbar^2}{4}\gamma_1^2}\right);$$

we then find:

$$\begin{aligned} \varepsilon_1' &\simeq E_1 - i\frac{\hbar}{2}\gamma_1 + \frac{|W_{12}|^2}{E_1 - E_2 - i\frac{\hbar}{2}\gamma_1} \\ \varepsilon_2' &\simeq E_2 + \frac{|W_{12}|^2}{E_2 - E_1 + i\frac{\hbar}{2}\gamma_1} \end{aligned} \tag{7}$$

The energies of the eigenstates in the presence of the coupling are the real parts of $\varepsilon_1'$ and $\varepsilon_2'$; the lifetimes are inversely proportional to their imaginary parts. We see from (7) that the coupling changes, to second order in $|W_{12}|$, both the energies and the lifetimes. In particular, we observe that $\varepsilon_1'$ and $\varepsilon_2'$ are both complex when $|W_{12}|$ is not zero: in the presence of the coupling, there is no longer any stable state. We can write $\varepsilon_2'$ in the form:

$$\varepsilon_2' = \Delta_2 - i\frac{\hbar}{2}\Gamma_2 \tag{8}$$

with:

$$\Delta_2 = E_2 + \frac{(E_2 - E_1)|W_{12}|^2}{(E_2 - E_1)^2 + \frac{\hbar^2}{4}\gamma_1^2} \tag{9-a}$$

$$\Gamma_2 = \gamma_1 \frac{|W_{12}|^2}{(E_2 - E_1)^2 + \frac{\hbar^2}{4}\gamma_1^2} \tag{9-b}$$

The state $|\varphi_2\rangle$ therefore acquires, under the effect of the coupling, a finite lifetime whose inverse is given in (9-b) (Bethe's formula). This result is easy to

understand physically: if the system at $t = 0$ is in the stable state $| \varphi_2 \rangle$, there is a non-zero probability at a subsequent time $t$ of finding it in the state $| \varphi_1 \rangle$, in which the system has a finite lifetime. It is sometimes said figuratively that "the coupling brings into the stable state part of the instability of the other state". Moreover, it can be seen from expressions (7) that, as in the case studied in §C of chapter IV, the smaller the difference between the unperturbed energies $E_1$ and $E_2$, the more effectively the perturbation acts on the energies and lifetimes. We shall therefore study in the next section the case where this difference is zero.

## 3. Influence of an arbitrary coupling on states of the same energy

When the energies $E_1$ and $E_2$ are equal, the operator $H$ is written, if we make its trace appear explicitly, as in § 2 of complement $\mathbf{B}_{IV}$:

$$H = \left( E_1 - i\frac{\hbar}{4}\gamma_1 \right) \mathbb{1} + K \tag{10}$$

where $\mathbb{1}$ is the identity operator and $K$ is the operator which, in the $\{ | \varphi_1 \rangle, | \varphi_2 \rangle \}$ basis, has for its matrix:

$$(K) = \begin{pmatrix} -i\frac{\hbar}{4}\gamma_1 & W_{12} \\ W_{12}^* & i\frac{\hbar}{4}\gamma_1 \end{pmatrix} \tag{11}$$

The eigenvalues $k_1$ and $k_2$ of $K$ are the two solutions of the characteristic equation:

$$k^2 = |W_{12}|^2 - \frac{\hbar^2}{16}\gamma_1^2 \tag{12}$$

They therefore have opposite values:

$$k_1 = - k_2 \tag{13}$$

which yields for the eigenvalues of $H$:

$$\begin{aligned} \varepsilon_1' &= E_1 - i\frac{\hbar}{4}\gamma_1 + k_1 \\ \varepsilon_2' &= E_1 - i\frac{\hbar}{4}\gamma_1 - k_1 \end{aligned} \tag{14}$$

The eigenvectors of $H$ and $K$ are the same; a simple calculation enables us to obtain these vectors $| \psi_1' \rangle$ and $| \psi_2' \rangle$*:

$$\begin{aligned} | \psi_1' \rangle &= W_{12} | \varphi_1 \rangle + \left( k_1 + i\frac{\hbar}{4}\gamma_1 \right) | \varphi_2 \rangle \\ | \psi_2' \rangle &= W_{12} | \varphi_1 \rangle + \left( - k_1 + i\frac{\hbar}{4}\gamma_1 \right) | \varphi_2 \rangle \end{aligned} \tag{15}$$

* For the calculation performed here, it is not indispensable to normalize $| \psi_1' \rangle$ and $| \psi_2' \rangle$. Note also that since $H$ is not Hermitian, $| \psi_1' \rangle$ and $| \psi_2' \rangle$ are not orthogonal.

Assume that the system at time $t = 0$ is in the state $| \varphi_2 \rangle$ (which would be stable in the absence of the coupling):

$$| \psi(t = 0) \rangle = | \varphi_2 \rangle = \frac{1}{2k_1} [| \psi'_1 \rangle - | \psi'_2 \rangle] \tag{16}$$

Using (14), we see that, at time $t$, the state vector is :

$$| \psi(t) \rangle = \frac{1}{2k_1} e^{-iE_1 \frac{t}{\hbar}} e^{-\frac{1}{4}\gamma_1 t} [e^{-ik_1 t/\hbar} | \psi'_1 \rangle - e^{ik_1 t/\hbar} | \psi'_2 \rangle] \tag{17}$$

The probability $\mathscr{P}_{21}(t)$ of finding the system at time $t$ in the state $| \varphi_1 \rangle$ is :

$$\begin{aligned} \mathscr{P}_{21}(t) &= |\langle \varphi_1 | \psi(t) \rangle|^2 \\ &= \frac{1}{4\,|k_1|^2} e^{-\gamma_1 t/2} |e^{-ik_1 t/\hbar} \langle \varphi_1 | \psi'_1 \rangle - e^{ik_1 t/\hbar} \langle \varphi_1 | \psi'_2 \rangle|^2 \\ &= \frac{1}{4\,|k_1|^2} e^{-\gamma_1 t/2} |W_{12}|^2 |e^{-ik_1 t/\hbar} - e^{ik_1 t/\hbar}|^2 \end{aligned} \tag{18}$$

We shall distinguish between several cases:

($i$) When the condition:

$$|W_{12}| > \frac{\hbar}{4}\gamma_1 \tag{19}$$

is satisfied, we obtain directly, using (12):

$$k_1 = -k_2 = \sqrt{|W_{12}|^2 - \left(\frac{\hbar}{4}\gamma_1\right)^2} \tag{20}$$

and the eigenvalues $\varepsilon'_1$ and $\varepsilon'_2$ are given by:

$$\begin{aligned} \varepsilon'_1 &= E_1 + \sqrt{|W_{12}|^2 - \left(\frac{\hbar}{4}\gamma_1\right)^2} - i\frac{\hbar}{4}\gamma_1 \\ \varepsilon'_2 &= E_1 - \sqrt{|W_{12}|^2 - \left(\frac{\hbar}{4}\gamma_1\right)^2} - i\frac{\hbar}{4}\gamma_1 \end{aligned} \tag{21}$$

$\varepsilon'_1$ and $\varepsilon'_2$ have the same imaginary part, but different real parts. The states $| \psi'_1 \rangle$ and $| \psi'_2 \rangle$ therefore have the same lifetime, $2\tau_1$, but different energies.

Substituting (20) into (18), we obtain:

$$\mathscr{P}_{21}(t) = \frac{|W_{12}|^2}{|W_{12}|^2 - \left(\frac{\hbar}{4}\gamma_1\right)^2} e^{-\gamma_1 t/2} \sin^2\left(\sqrt{|W_{12}|^2 - \left(\frac{\hbar}{4}\gamma_1\right)^2}\,\frac{t}{\hbar}\right) \tag{22}$$

The form of this result recalls Rabi's formula [*cf.* chap. IV, equation (C-32)]. The function $\mathscr{P}_{21}(t)$ is represented by a damped sinusoid with time constant $2\tau_1$ (fig. 1). Condition (19) thus expresses the fact that the coupling is sufficiently strong to make the system oscillate between the states $| \varphi_1 \rangle$ and $| \varphi_2 \rangle$ before the instability of the state $| \varphi_1 \rangle$ can make itself felt.

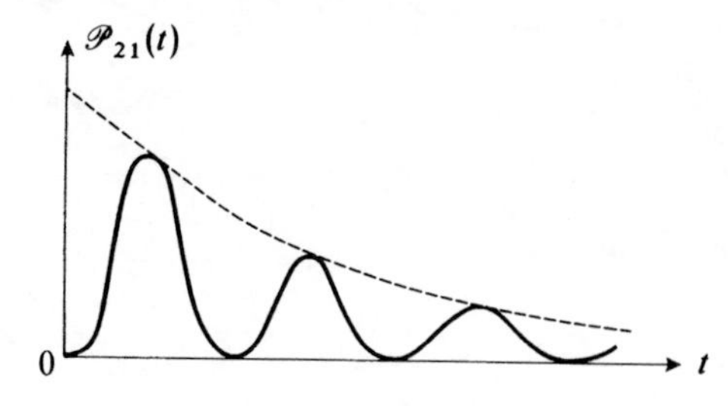


FIGURE 1

**Effect of a strong coupling between a stable state $|\varphi_2\rangle$ and an unstable state $|\varphi_1\rangle$. If the system is initially in the state $|\varphi_2\rangle$, the probability $\mathscr{P}_{21}(t)$ of finding it in the state $|\varphi_1\rangle$ at time $t$ presents damped oscillations.**

(*ii*) If, on the other hand, the condition:

$$|W_{12}| < \frac{\hbar}{4}\gamma_1 \tag{23}$$

is satisfied, we then have:

$$k_1 = -k_2 = i\sqrt{\left(\frac{\hbar}{4}\gamma_1\right)^2 - |W_{12}|^2} \tag{24}$$

and:

$$\begin{aligned} \varepsilon_1' &= E_1 - i\left[\frac{\hbar}{4}\gamma_1 - \sqrt{\left(\frac{\hbar}{4}\gamma_1\right)^2 - |W_{12}|^2}\right] \\ \varepsilon_2' &= E_1 - i\left[\frac{\hbar}{4}\gamma_1 + \sqrt{\left(\frac{\hbar}{4}\gamma_1\right)^2 - |W_{12}|^2}\right] \end{aligned} \tag{25}$$

The states $|\psi_1'\rangle$ and $|\psi_2'\rangle$ then have the same energy and different lifetimes. Formula (18) becomes:

$$\mathscr{P}_{21}(t) = \frac{|W_{12}|^2}{\left(\frac{\hbar}{4}\gamma_1\right)^2 - |W_{12}|^2}\, e^{-\gamma_1 t/2} \sinh^2\left(\sqrt{\left(\frac{\hbar}{4}\gamma_1\right)^2 - |W_{12}|^2}\,\frac{t}{\hbar}\right) \tag{26}$$

This time, $\mathscr{P}_{21}(t)$ is a sum of damped exponentials (fig. 2).

This result has a simple physical interpretation: condition (23) expresses the fact that the lifetime $\tau_1$ is so short that the system is completely damped before the coupling $W$ has had the time to make it oscillate between the states $|\varphi_1\rangle$ and $|\varphi_2\rangle$.

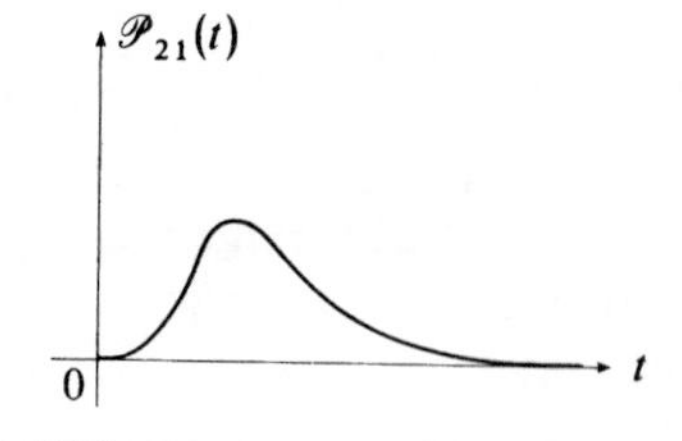


FIGURE 2

**When the coupling is weak, oscillations between the states $|\varphi_1\rangle$ and $|\varphi_2\rangle$ do not have time to occur.**

($iii$) Finally, let us examine the case where we have exactly:

$$|W_{12}| = \frac{\hbar}{4}\gamma: \tag{27}$$

We see then from (14) that the states $| \psi'_1 \rangle$ and $| \psi'_2 \rangle$ both have the same energy $E_1$ and the same lifetime $2\tau_1$.

Equations (22) and (26), in this case, take on indeterminate forms, which both yield:

$$\mathscr{P}_{21}(t) = \frac{|W_{12}|^2}{\hbar^2}\, t^2\, \mathrm{e}^{-\gamma_1 t/2} \tag{28}$$

COMMENT:

The preceding discussion is very analogous to that of the classical motion of a damped harmonic oscillator. Conditions (19), (23) and (27) correspond respectively to weak, strong and critical damping.

**References and suggestions for further reading:**

An important application of the phenomenon discussed in this complement is the shortening of the lifetime of a metastable state due to an electric field. See: Lamb and Retherford (3.11), App. II; Sobel'man (11.12), chap. 8, § 28-5.

## Complement $J_{IV}$

## EXERCISES

**1.** Consider a spin 1/2 particle of magnetic moment $\mathbf{M} = \gamma \mathbf{S}$. The spin state space is spanned by the basis of the $| + \rangle$ and $| - \rangle$ vectors, eigenvectors of $S_z$ with eigenvalues $+\hbar/2$ and $-\hbar/2$. At time $t = 0$, the state of the system is:

$$| \psi(t = 0) \rangle = | + \rangle$$

*a.* If the observable $S_x$ is measured at time $t = 0$, what results can be found, and with what probabilities?

*b.* Instead of performing the preceding measurement, we let the system evolve under the influence of a magnetic field parallel to $Oy$, of modulus $B_0$. Calculate, in the $\{ | + \rangle, | - \rangle \}$ basis, the state of the system at time $t$.

*c.* At this time $t$, we measure the observables $S_x$, $S_y$, $S_z$. What values can we find, and with what probabilities? What relation must exist between $B_0$ and $t$ for the result of one of the measurements to be certain? Give a physical interpretation of this condition.

**2.** Consider a spin 1/2 particle, as in the previous exercise (using the same notation).

*a.* At time $t = 0$, we measure $S_y$ and find $+\hbar/2$. What is the state vector $| \psi(0) \rangle$ immediately after the measurement?

*b.* Immediately after this measurement, we apply a uniform time-dependent field parallel to $Oz$. The Hamiltonian operator of the spin $H(t)$ is then written:

$$H(t) = \omega_0(t)\, S_z$$

Assume that $\omega_0(t)$ is zero for $t < 0$ and $t > T$ and increases linearly from 0 to $\omega_0$ when $0 \leqslant t \leqslant T$ ($T$ is a given parameter whose dimensions are those of time). Show that at time $t$ the state vector can be written:

$$| \psi(t) \rangle = \frac{1}{\sqrt{2}} \left[ e^{i\theta(t)} | + \rangle + i\, e^{-i\theta(t)} | - \rangle \right]$$

where $\theta(t)$ is a real function of $t$ (to be calculated by the student).

*c.* At a time $t = \tau > T$, we measure $S_y$. What results can we find, and with what probabilities? Determine the relation which must exist between $\omega_0$ and $T$ in order for us to be sure of the result. Give the physical interpretation.

**3.** Consider a spin 1/2 particle placed in a magnetic field $\mathbf{B}_0$ with components:

$$\begin{cases} B_x = \dfrac{1}{\sqrt{2}} B_0 \\ B_y = 0 \\ B_z = \dfrac{1}{\sqrt{2}} B_0 \end{cases}$$

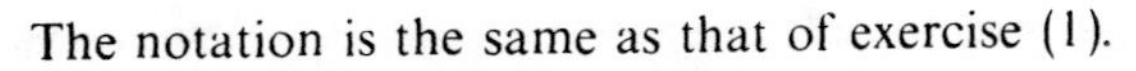

The notation is the same as that of exercise (1).

*a*. Calculate the matrix representing, in the $\{ | + \rangle, | - \rangle \}$ basis, the operator $H$, the Hamiltonian of the system.

*b*. Calculate the eigenvalues and the eigenvectors of $H$.

*c*. The system at time $t = 0$ is in the state $| - \rangle$. What values can be found if the energy is measured, and with what probabilities?

*d*. Calculate the state vector $| \psi(t) \rangle$ at time $t$. At this instant, $S_x$ is measured; what is the mean value of the results that can be obtained? Give a geometrical interpretation.

**4.** Consider the experimental device described in § B-2-b of chapter IV (*cf.* fig. 8) : a beam of atoms of spin 1/2 passes through one apparatus, which serves as a "polarizer" in a direction which makes an angle $\theta$ with $Oz$ in the $xOz$ plane, and then through another apparatus, the "analyzer", which measures the $S_z$ component of the spin. We assume in this exercise that between the polarizer and the analyzer, over a length $L$ of the atomic beam, a magnetic field $\mathbf{B}_0$ is applied which is uniform and parallel to $Ox$. We call $v$ the speed of the atoms and $T = L/v$ the time during which they are submitted to the field $\mathbf{B}_0$. We set $\omega_0 = - \gamma B_0$.

*a*. What is the state vector $| \psi_1 \rangle$ of a spin at the moment it enters the analyzer?

*b*. Show that when the measurement is performed in the analyzer, there is a probability equal to $\frac{1}{2}(1 + \cos\theta \cos\omega_0 T)$ of finding $+ \hbar/2$ and $\frac{1}{2}(1 - \cos\theta\cos\omega_0 T)$ of finding $- \hbar/2$. Give a physical interpretation.

*c*. (This question and the following one involve the concept of a density operator, defined in complement $E_{III}$. The reader is also advised to refer to complement $E_{IV}$.) Show that the density matrix $\rho_1$ of a particle which enters the analyzer is written, in the $\{ | + \rangle, | - \rangle \}$ basis:

$$\rho_1 = \frac{1}{2}\begin{pmatrix} 1 + \cos\theta\cos\omega_0 T & \sin\theta + i\cos\theta\sin\omega_0 T \\ \sin\theta - i\cos\theta\sin\omega_0 T & 1 - \cos\theta\cos\omega_0 T \end{pmatrix}$$

Calculate $\text{Tr}\{ \rho_1 S_x \}$, $\text{Tr}\{ \rho_1 S_y \}$ and $\text{Tr}\{ \rho_1 S_z \}$. Give an interpretation. Does the density operator $\rho_1$ describe a pure state?

*d*. Now assume that the speed of an atom is a random variable, and hence the time $T$ is known only to within a certain uncertainty $\Delta T$. In addition, the field $B_0$ is assumed to be sufficiently strong that $\omega_0 \Delta T \gg 1$. The possible values of the product $\omega_0 T$ are then (modulus $2\pi$) all values included between 0 and $2\pi$, all of which are equally probable.

In this case, what is the density operator $\rho_2$ of an atom at the moment it enters the analyzer? Does $\rho_2$ correspond to a pure case? Calculate the quantities $\text{Tr}\{ \rho_2 S_x \}$, $\text{Tr}\{ \rho_2 S_y \}$ and $\text{Tr}\{ \rho_2 S_z \}$. What is your interpretation? In which

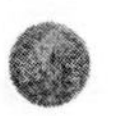


case does the density operator describe a completely polarized spin? A completely unpolarized spin?

Describe qualitatively the phenomena observed at the analyzer exit when $\omega_0$ varies from zero to a value where the condition $\omega_0 \Delta T \gg 1$ is satisfied.

### 5. **Evolution operator of a spin 1/2** (*cf.* complement $F_{III}$)

Consider a spin 1/2, of magnetic moment $\mathbf{M} = \gamma \mathbf{S}$, placed in a magnetic field $\mathbf{B}_0$ of components $B_x = -\omega_x/\gamma$, $B_y = -\omega_y/\gamma$, $B_z = -\omega_z/\gamma$.
We set:

$$\omega_0 = -\gamma \,|\mathbf{B}_0|$$

*a.* Show that the evolution operator of this spin is:

$$U(t, 0) = e^{-iMt}$$

where $M$ is the operator:

$$M = \frac{1}{\hbar}[\omega_x S_x + \omega_y S_y + \omega_z S_z] = \frac{1}{2}[\omega_x \sigma_x + \omega_y \sigma_y + \omega_z \sigma_z]$$

where $\sigma_x$, $\sigma_y$ and $\sigma_z$ are the three Pauli matrices (*cf.* complement $A_{IV}$).

Calculate the matrix which represents $M$ in the $\{ |+\rangle, |-\rangle \}$ basis of eigenvectors of $S_z$. Show that:

$$M^2 = \frac{1}{4}[\omega_x^2 + \omega_y^2 + \omega_z^2] = \left(\frac{\omega_0}{2}\right)^2$$

*b.* Put the evolution operator into the form:

$$U(t, 0) = \cos\left(\frac{\omega_0 t}{2}\right) - \frac{2i}{\omega_0} M \sin\left(\frac{\omega_0 t}{2}\right)$$

*c.* Consider a spin which at time $t = 0$ is in the state $|\psi(0)\rangle = |+\rangle$. Show that the probability $\mathscr{P}_{++}(t)$ of finding it in the state $|+\rangle$ at time $t$ is:

$$\mathscr{P}_{++}(t) = |\langle + | U(t, 0) | + \rangle|^2$$

and derive the relation:

$$\mathscr{P}_{++}(t) = 1 - \frac{\omega_x^2 + \omega_y^2}{\omega_0^2} \sin^2\left(\frac{\omega_0 t}{2}\right)$$

Give a geometrical interpretation.

**6.** Consider the system composed of two spin 1/2's, $\mathbf{S}_1$ and $\mathbf{S}_2$, and the basis of four vectors $|\pm, \pm\rangle$ defined in complement $D_{IV}$. The system at time $t = 0$ is in the state

$$|\psi(0)\rangle = \frac{1}{2}|++\rangle + \frac{1}{2}|+-\rangle + \frac{1}{\sqrt{2}}|--\rangle$$

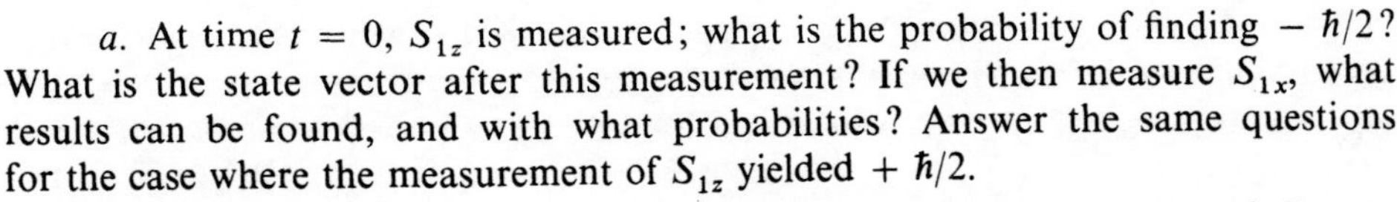

*a.* At time $t = 0$, $S_{1z}$ is measured; what is the probability of finding $-\hbar/2$? What is the state vector after this measurement? If we then measure $S_{1x}$, what results can be found, and with what probabilities? Answer the same questions for the case where the measurement of $S_{1z}$ yielded $+\hbar/2$.

*b.* When the system is in the state $|\psi(0)\rangle$ written above, $S_{1z}$ and $S_{2z}$ are measured simultaneously. What is the probability of finding opposite results? Identical results?

*c.* Instead of performing the preceding measurements, we let the system evolve under the influence of the Hamiltonian:

$$H = \omega_1 S_{1z} + \omega_2 S_{2z}$$

What is the state vector $|\psi(t)\rangle$ at time $t$? Calculate at time $t$ the mean values $\langle \mathbf{S}_1 \rangle$ and $\langle \mathbf{S}_2 \rangle$. Give a physical interpretation.

*d.* Show that the lengths of the vectors $\langle \mathbf{S}_1 \rangle$ and $\langle \mathbf{S}_2 \rangle$ are less than $\hbar/2$. What must be the form of $|\psi(0)\rangle$ for each of these lengths to be equal to $+\hbar/2$?

**7.** Consider the same system of two spin 1/2's as in the preceding exercise; the state space is spanned by the basis of four states $|\pm, \pm\rangle$.

*a.* Write the $4 \times 4$ matrix representing, in this basis, the $S_{1y}$ operator. What are the eigenvalues and eigenvectors of this matrix?

*b.* The normalized state of the system is:

$$|\psi\rangle = \alpha|++\rangle + \beta|+-\rangle + \gamma|-+\rangle + \delta|--\rangle$$

where $\alpha$, $\beta$, $\gamma$ and $\delta$ are given complex coefficients. $S_{1x}$ and $S_{2y}$ are measured simultaneously; what results can be found, and with what probabilities? What happens to these probabilities if $|\psi\rangle$ is a tensor product of a vector of the state space of the first spin and a vector of the state space of the second spin?

*c.* Same questions for a measurement of $S_{1y}$ and $S_{2y}$.

*d.* Instead of performing the preceding measurements, we measure only $S_{2y}$. Calculate, first from the results of *b* and then from those of *c*, the probability of finding $-\hbar/2$.

**8.** Consider an electron of a linear triatomic molecule formed by three equidistant atoms. We use $|\varphi_A\rangle$, $|\varphi_B\rangle$, $|\varphi_C\rangle$ to denote three orthonormal states of this electron, corresponding respectively to three wave functions localized about the nuclei of atoms $A$, $B$, $C$. We shall confine ourselves to the subspace of the state space spanned by $|\varphi_A\rangle$, $|\varphi_B\rangle$ and $|\varphi_C\rangle$.

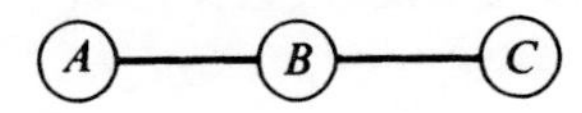


When we neglect the possibility of the electron jumping from one nucleus to another, its energy is described by the Hamiltonian $H_0$ whose eigenstates are the three states $|\varphi_A\rangle$, $|\varphi_B\rangle$, $|\varphi_C\rangle$ with the same eigenvalue $E_0$. The coupling between the states $| \varphi_A \rangle$, $| \varphi_B \rangle$, $| \varphi_C \rangle$ is described by an additional Hamiltonian $W$ defined by:

$$W | \varphi_A \rangle = - a | \varphi_B \rangle$$
$$W | \varphi_B \rangle = - a | \varphi_A \rangle - a | \varphi_C \rangle$$
$$W | \varphi_C \rangle = - a | \varphi_B \rangle$$

where $a$ is a real positive constant.

1. Calculate the energies and stationary states of the Hamiltonian $H = H_0 + W$.

2. The electron at time $t = 0$ is in the state $| \varphi_A \rangle$. Discuss qualitatively the localization of the electron at subsequent times $t$. Are there any values of $t$ for which it is perfectly localized about atom $A$, $B$ or $C$?

3. Let $D$ be the observable whose eigenstates are $| \varphi_A \rangle$, $| \varphi_B \rangle$, $| \varphi_C \rangle$ with respective eigenvalues $-d$, $0$, $d$. $D$ is measured at time $t$; what values can be found, and with what probabilities?

4. When the initial state of the electron is arbitrary, what are the Bohr frequencies that can appear in the evolution of $\langle D \rangle$? Give a physical interpretation of $D$. What are the frequencies of the electromagnetic waves that can be absorbed or emitted by the molecule?

**9.** A molecule is composed of six identical atoms $A_1$, $A_2$, ... $A_6$ which form a regular hexagon. Consider an electron which can be localized on each of the atoms. Call $| \varphi_n \rangle$ the state in which it is localized on the $n$th atom ($n = 1, 2, \ldots 6$). The electron states will be confined to the space spanned by the $| \varphi_n \rangle$, assumed to be orthonormal.

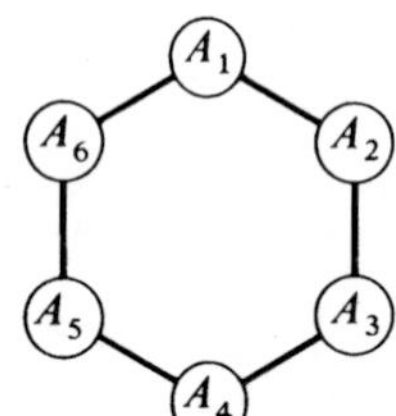


*a.* Define an operator $R$ by the following relations:

$$R | \varphi_1 \rangle = | \varphi_2 \rangle; \; R | \varphi_2 \rangle = | \varphi_3 \rangle; \ldots; \; R | \varphi_6 \rangle = | \varphi_1 \rangle$$

Find the eigenvalues and eigenstates of $R$. Show that the eigenvectors of $R$ form a basis of the state space.

*b.* When the possibility of the electron passing from one site to another is neglected, its energy is described by a Hamiltonian $H_0$ whose eigenstates are the six states $| \varphi_n \rangle$, with the same eigenvalue $E_0$. As in the previous exercise, we describe the possibility of the electron jumping from one atom to another by adding a perturbation $W$ to the Hamiltonian $H_0$; $W$ is defined by:

$$W | \varphi_1 \rangle = - a | \varphi_6 \rangle - a | \varphi_2 \rangle; \qquad W | \varphi_2 \rangle = - a | \varphi_1 \rangle - a | \varphi_3 \rangle;$$
$$\ldots \ldots; \qquad W | \varphi_6 \rangle = - a | \varphi_5 \rangle - a | \varphi_1 \rangle$$

Show that $R$ commutes with the total Hamiltonian $H = H_0 + W$. From this deduce the eigenstates and eigenvalues of $H$. In these eigenstates, is the electron localized? Apply these considerations to the benzene molecule.

**Exercise 9**

Reference: Feynman III (1.2), § 15-4.

CHAPTER V

# The one-dimensional harmonic oscillator

## OUTLINE OF CHAPTER V

# A. INTRODUCTION

## 1. Importance of the harmonic oscillator in physics

This chapter is devoted to the study of a particularly important physical system : the one-dimensional harmonic oscillator.

The simplest example of such a system is that of a particle of mass $m$ moving in a potential which depends only on $x$ and has the form:

$$V(x) = \frac{1}{2} k x^2 \tag{A-1}$$

($k$ is a real positive constant). The particle is attracted towards the $x = 0$ plane [the minimum of $V(x)$, corresponding to positions of stable equilibrium] by a restoring force:

$$F_x = - \frac{dV}{dx} = - kx \tag{A-2}$$

which is proportional to the distance $x$ between the particle and the $x = 0$ plane. We know that in classical mechanics, the projection onto $Ox$ of the particle's motion is a sinusoidal oscillation about $x = 0$, of angular frequency:

$$\omega = \sqrt{\frac{k}{m}} \tag{A-3}$$

Actually, a large number of systems are governed (at least approximately) by the harmonic oscillator equations. Whenever one studies the behavior of a physical system in the neighborhood of a stable equilibrium position, one arrives at equations which, in the limit of small oscillations, are those of a harmonic oscillator (see § A-2). The results we shall derive in this chapter are applicable, therefore, to a whole series of important physical phenomena – for example, the vibrations of the atoms of a molecule about their equilibrium position, the oscillations of atoms or ions of a crystalline lattice (phonons)★.

The harmonic oscillator is also involved in the study of the electromagnetic field. We know that in a cavity, there exist an infinite number of possible stationary waves (normal modes of the cavity). The electromagnetic field can be expanded in terms of these modes and it can been shown, using Maxwell's equations, that each of the coefficients of this expansion (which describe the state of the field at each instant) obeys a differential equation which is identical to that of a harmonic oscillator whose angular frequency $\omega$ is that of the associated normal mode. In other words, the electromagnetic field is formally equivalent to a set of independent harmonic oscillators (*cf.* complement $K_V$). The quantization of the field is obtained by quantizing these oscillators associated with the various normal modes of the cavity. Recall, moreover, that it was the study of the behavior of these oscillators at thermal equilibrium (blackbody radiation) which, historically, led Planck to

★ Complement $A_V$ is devoted to a qualitative study of some physical examples of harmonic oscillators.

introduce, for the first time in physics, the constant $h$ which bears his name. We shall see (*cf.* complement $L_V$) that the mean energy of a harmonic oscillator in thermodynamic equilibrium at the temperature $T$ is different for classical and quantum mechanical oscillators.

The harmonic oscillator also plays an important role in the description of a set of identical particles which are all in the same quantum mechanical state (they must obviously be bosons, *cf.* chap. XIV). As we shall see later, this is because the energy levels of a harmonic oscillator are equidistant, the spacing between two adjacent levels being equal to $\hbar\omega$. With the energy level labelled by the integer $n$ (situated at a distance $n\hbar\omega$ above the ground state) can then be associated a set of $n$ identical particles (or quanta), each possessing an energy $\hbar\omega$. The transition of the oscillator from level $n$ to level $n + 1$ or $n - 1$ corresponds to the creation or destruction of a quantum of energy $\hbar\omega$. In this chapter, we shall introduce the operators $a^\dagger$ and $a$, which enable us to describe this transition from level $n$ to level $n + 1$ or $n - 1$. These operators, called creation or destruction operators, are used throughout quantum statistical mechanics and quantum field theory*.

The detailed study of the harmonic oscillator in quantum mechanics is therefore extremely important from a physical point of view. Moreover, we are dealing with a quantum mechanical system for which the Schrödinger equation can be solved rigorously. Having studied spin 1/2 and two-level systems in chapter IV, we shall therefore now consider another simple example which illustrates the general formalism of quantum mechanics. We shall show in particular how to solve an eigenvalue equation by dealing only with the operators and the commutation relations (this technique will also be applied to angular momentum). We shall also study in a detailed way the motion of wave packets, particularly at the classical limit (*cf.* complement $G_V$ on quasi-classical states).

In § A-2, we shall review some results related to the classical oscillator before stating (§ A-3) certain general properties of the eigenvalues of the Hamiltonian $H$. Then, in §§ B and C, we shall determine these eigenvalues and eigenvectors by introducing creation and destruction operators and using only the consequences of the canonical commutation relation $[X, P] = i\hbar$, as well as the particular form of $H$. § D is devoted to a physical study of the stationary states of the oscillator and wave packets formed by linear superpositions of these stationary states.

## 2. The harmonic oscillator in classical mechanics

The potential energy $V(x)$ [formula (A-1)] is shown in figure 1. The motion of the particle is governed by the dynamical equation:

$$m\frac{\mathrm{d}^2x}{\mathrm{d}t^2} = -\frac{\mathrm{d}V}{\mathrm{d}x} = -kx \tag{A-4}$$

* The aim of quantum field theory is to describe interactions between particles in the relativistic domain, especially the interactions between electrons, positrons and photons. It is clear that creation and destruction operators should play an important role, since such processes are indeed observed experimentally (absorption or emission of photons, pair creation...).

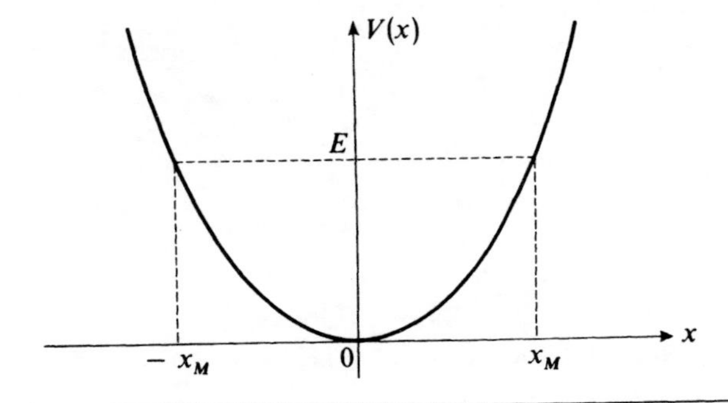


FIGURE 1

**The potential energy $V(x)$ of a one-dimensional harmonic oscillator. The amplitude of the classical motion of energy $E$ is $x_M$.**

The general solution of this equation is of the form :

$$x = x_M \cos(\omega t - \varphi) \tag{A-5}$$

where $\omega$ is defined by (A-3), and the constants of integration $x_M$ and $\varphi$ are determined by the initial conditions of the motion. The particle therefore *oscillates sinusoidally* about the point $O$, with an amplitude $x_M$ and an angular frequency $\omega$.

The kinetic energy of the particle is :

$$T = \frac{1}{2} m \left( \frac{\mathrm{d}x}{\mathrm{d}t} \right)^2 = \frac{p^2}{2m} \tag{A-6}$$

where $p = m \dfrac{\mathrm{d}x}{\mathrm{d}t}$ is the momentum of the particle. The total energy is:

$$E = T + V = \frac{p^2}{2m} + \frac{1}{2} m\omega^2 x^2 \tag{A-7}$$

Substituting solution (A-5) into this equation, we find :

$$E = \frac{1}{2} m\omega^2 x_M^2 \tag{A-8}$$

The energy of the particle is therefore time-independent (this is a general property of conservative systems) and can take on any positive (or zero) value, since $x_M$ is *a priori* arbitrary.

If we fix the total energy $E$, the limits of the classical motion $x = \pm x_M$ can be determined from figure 1 by taking the intersection of the parabola with the line parallel to $Ox$ of ordinate $E$. At these points $x = \pm x_M$, the potential energy is at a maximum and equal to $E$, and the kinetic energy is zero. On the other hand, at $x = 0$, the potential energy is zero and the kinetic energy is maximum.

COMMENT :

Consider an arbitrary potential $V(x)$ which has a minimum at $x = x_0$ (fig. 2). Expanding the function $V(x)$ in a Taylor's series in the neighborhood of $x_0$, we obtain :

$$V(x) = a + b(x - x_0)^2 + c(x - x_0)^3 + \ldots \tag{A-9}$$

The coefficients of this expansion are given by:

$$a = V(x_0)$$
$$b = \frac{1}{2}\left(\frac{\mathrm{d}^2 V}{\mathrm{d}x^2}\right)_{x=x_0}$$
$$c = \frac{1}{3!}\left(\frac{\mathrm{d}^3 V}{\mathrm{d}x^3}\right)_{x=x_0} \tag{A-10}$$

and the linear term in $(x - x_0)$ is zero since $x_0$ corresponds to a minimum of $V(x)$. The force derived from the potential $V(x)$ is, in the neighborhood of $x_0$:

$$F_x = -\frac{\mathrm{d}V}{\mathrm{d}x} = -2b(x - x_0) - 3c(x - x_0)^2 + \dots \tag{A-11}$$

Since $x = x_0$ represents a minimum, the coefficient $b$ is positive.

The point $x = x_0$ corresponds to a stable equilibrium position for the particle: $F_x$ is zero for $x = x_0$. Moreover, for $(x - x_0)$ sufficiently small, $F_x$ and $(x - x_0)$ have opposite signs since $b$ is positive.

If the amplitude of the motion of the particle about $x_0$ is sufficiently small for the term in $(x - x_0)^3$ of (A-9) [and therefore, the corresponding term in $(x - x_0)^2$ of (A-11)] to be negligible compared to the preceding ones, we have a harmonic oscillator since the dynamical equation can then be approximated by:

$$m\frac{\mathrm{d}^2 x}{\mathrm{d}t^2} \simeq -2b\,(x - x_0) \tag{A-12}$$

The corresponding angular frequency $\omega$ is related to the second derivative of $V(x)$ at $x = x_0$ by the formula:

$$\omega = \sqrt{\frac{2b}{m}} = \sqrt{\frac{1}{m}\left(\frac{\mathrm{d}^2 V}{\mathrm{d}x^2}\right)_{x=x_0}} \tag{A-13}$$

Since the amplitude of the motion must remain small, the energy of the harmonic oscillator will be low.

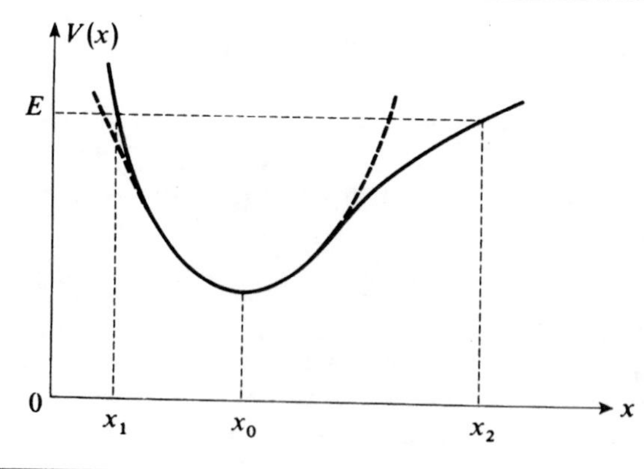


FIGURE 2

**In the neighborhood of a minimum, any potential $V(x)$ can be approximated by a parabolic potential (dashed line). In the potential $V(x)$, a classical particle of energy $E$ oscillates between $x_1$ and $x_2$.**

For higher energies $E$, the particle will be in *periodic but not sinusoidal motion* between the limits $x_1$ and $x_2$ (fig. 2). If we expand the function $x(t)$ in a Fourier series which gives the position of the particle, we shall find, not one, but several, sinusoidal terms, whose frequencies are integral multiples of the lowest frequency. We then say that we are dealing with an *anharmonic* oscillator. Note also that in this case, the period of the motion is not generally $2\pi/\omega$, where $\omega$ is given by formula (A-13).

## 3. General properties of the quantum mechanical Hamiltonian

In quantum mechanics, the classical quantities $x$ and $p$ are replaced respectively by the observables $X$ and $P$, which satisfy:

$$[X, P] = i\hbar \tag{A-14}$$

It is then easy to obtain the Hamiltonian operator of the system from (A-7):

$$H = \frac{P^2}{2m} + \frac{1}{2} m\omega^2 X^2 \tag{A-15}$$

Since $H$ is time-independent (conservative system), the quantum mechanical study of the harmonic oscillator reduces to the solution of the eigenvalue equation:

$$H \,|\, \varphi \rangle = E \,|\, \varphi \rangle \tag{A-16}$$

which is written, in the $\{ \,|\, x \rangle \}$ representation:

$$\left[ -\frac{\hbar^2}{2m}\frac{\mathrm{d}^2}{\mathrm{d}x^2} + \frac{1}{2} m\omega^2 x^2 \right] \varphi(x) = E \; \varphi(x) \tag{A-17}$$

Before undertaking the detailed study of equation (A-16), let us indicate some important properties that can be deduced from the form (A-1) of the potential function:

(*i*) *The eigenvalues of the Hamiltonian are positive.* It can be shown that, in general (complement $\mathrm{M_{III}}$), if the potential function $V(x)$ has a lower bound, the eigenvalues $E$ of the Hamiltonian $H = \dfrac{P^2}{2m} + V(X)$ are greater than the minimum of $V(x)$:

$$V(x) \geqslant V_m \quad \text{requires} \quad E > V_m \tag{A-18}$$

For the harmonic oscillator which we are studying here, we have chosen the energy origin such that $V_m$ is zero.

(*ii*) *The eigenfunctions of H have a definite parity.* This is due to the fact that the potential $V(x)$ is an even function:

$$V(-\,x) = V(x) \tag{A-19}$$

We can then (*cf.* complements $\mathrm{F_{II}}$ and $\mathrm{C_V}$) look for eigenfunctions of $H$, in the $\{ \,|\, x \rangle \}$ representation, amongst the functions which have a definite parity (in fact, we shall see that the eigenvalues of $H$ are not degenerate; consequently, the wave functions associated with the stationary states are necessarily either even or odd).

(*iii*) *The energy spectrum is discrete*. Whatever the value of the total energy, the classical motion is limited to a bounded region of the $Ox$ axis (fig. 1), and it can be shown (complement $M_{III}$) that in this case, the eigenvalues of the Hamiltonian form a discrete set.

We shall derive these properties (in a more precise form) in the following sections. However, it is interesting to note that they can be obtained simply by applying to the harmonic oscillator some general theorems concerning one-dimensional problems.

## B. EIGENVALUES OF THE HAMILTONIAN

We are now going to study the eigenvalue equation (A-16). First of all, using only the canonical commutation relation (A-14), we shall find the spectrum of the Hamiltonian $H$ written in (A-15).

### 1. Notation

We shall begin by introducing some useful notations.

#### a. THE $\hat{X}$ AND $\hat{P}$ OPERATORS

The observables $X$ and $P$ obviously have dimensions (those of a length and a momentum, respectively). Since $\omega$ has the dimensions of the inverse of a time and $\hbar$, of an action (product of an energy and a time), it is easy to see that the observables $\hat{X}$ and $\hat{P}$ defined by:

$$\hat{X} = \sqrt{\frac{m\omega}{\hbar}}\, X$$

$$\hat{P} = \frac{1}{\sqrt{m\hbar\omega}}\, P \tag{B-1}$$

are dimensionless.

If we use these new operators, the canonical commutation relation will be written:

$$[\hat{X}, \hat{P}] = i \tag{B-2}$$

and the Hamiltonian can be put in the form:

$$H = \hbar\omega\, \hat{H} \tag{B-3}$$

with:

$$\hat{H} = \frac{1}{2}(\hat{X}^2 + \hat{P}^2) \tag{B-4}$$

We shall therefore seek the solutions of the eigenvalue equation:

$$\hat{H} \mid \varphi_v^i \rangle = \varepsilon_v \mid \varphi_v^i \rangle \tag{B-5}$$

where the operator $\hat{H}$ and the eigenvalues $\varepsilon_v$ are dimensionless. The index $v$ can belong to either a discrete or a continuous set, and the additional index $i$ enables us to distinguish between the various possible orthogonal eigenvectors associated with the same eigenvalue $\varepsilon_v$.

### b. THE $a$, $a^\dagger$ AND $N$ OPERATORS

If $\hat{X}$ and $P$ were numbers and not operators, we could write the sum $\hat{X}^2 + \hat{P}^2$ appearing in expression (B-4) for $\hat{H}$ in the form of a product of linear terms : $(\hat{X} - i\hat{P})(\hat{X} + i\hat{P})$. In fact, since $\hat{X}$ and $\hat{P}$ are non-commuting operators, $\hat{X}^2 + \hat{P}^2$ is not equal to $(\hat{X} - i\hat{P})(\hat{X} + i\hat{P})$. We shall show, however, that the introduction of operators proportional to $\hat{X} + i\hat{P}$ and $\hat{X} - i\hat{P}$ enables us to simplify considerably our search for eigenvalues and eigenvectors of $\hat{H}$.

We therefore set*:

$$a = \frac{1}{\sqrt{2}}(\hat{X} + i\hat{P}) \tag{B-6-a}$$

$$a^\dagger = \frac{1}{\sqrt{2}}(\hat{X} - i\hat{P}) \tag{B-6-b}$$

These formulas can immediately be inverted to yield:

$$\hat{X} = \frac{1}{\sqrt{2}}(a^\dagger + a) \tag{B-7-a}$$

$$\hat{P} = \frac{i}{\sqrt{2}}(a^\dagger - a) \tag{B-7-b}$$

Since $\hat{X}$ and $\hat{P}$ are Hermitian, $a$ and $a^\dagger$ are not (because of the factor $i$), but are adjoints of each other.

The commutator of $a$ and $a^\dagger$ is easy to calculate from (B-6) and (B-2):

$$\begin{aligned}[a, a^\dagger] &= \frac{1}{2}[\hat{X} + i\hat{P}, \hat{X} - i\hat{P}] \\ &= \frac{i}{2}[\hat{P}, \hat{X}] - \frac{i}{2}[\hat{X}, \hat{P}]\end{aligned} \tag{B-8}$$

that is:

$$[a, a^\dagger] = 1 \tag{B-9}$$

This relation is completely equivalent to the canonical commutation relation (A-14).

* Until now, we have designated operators by capital letters. However, to conform to standard usage, we shall use the small letters $a$ and $a^\dagger$ for the operators (B-6).

Finally, let us derive some simple formulas which will be useful to us in the rest of this chapter. Let us first calculate $a^\dagger a$:

$$\begin{aligned} a^\dagger a &= \frac{1}{2}(\hat{X} - i\hat{P})(\hat{X} + i\hat{P}) \\ &= \frac{1}{2}(\hat{X}^2 + \hat{P}^2 + i\hat{X}\hat{P} - i\hat{P}\hat{X}) \\ &= \frac{1}{2}(\hat{X}^2 + \hat{P}^2 - 1) \end{aligned} \tag{B-10}$$

Comparing this with expression (B-4), we see that:

$$\hat{H} = a^\dagger a + \frac{1}{2} = \frac{1}{2}(\hat{X} - i\hat{P})(\hat{X} + i\hat{P}) + \frac{1}{2} \tag{B-11}$$

Unlike the situation in the classical case, $\hat{H}$ cannot be put in the form of a product of linear terms. The non-commutativity of $\hat{X}$ and $\hat{P}$ is at the origin of the additional term 1/2 which appears on the right-hand side of (B-11). Similarly, it can be shown that :

$$\hat{H} = aa^\dagger - \frac{1}{2} \tag{B-12}$$

Let us now introduce the operator $N$ defined by:

$$N = a^\dagger a \tag{B-13}$$

This operator is Hermitian since:

$$N^\dagger = a^\dagger (a^\dagger)^\dagger = a^\dagger a = N \tag{B-14}$$

Moreover, according to (B-11):

$$\hat{H} = N + \frac{1}{2} \tag{B-15}$$

so that *the eigenvectors of $\hat{H}$ are eigenvectors of $N$, and vice versa.*

Finally, let us calculate the commutators of $N$ with $a$ and $a^\dagger$:

$$\begin{aligned} [N, a] &= [a^\dagger a, a] = a^\dagger [a, a] + [a^\dagger, a]a = -a \\ [N, a^\dagger] &= [a^\dagger a, a^\dagger] = a^\dagger [a, a^\dagger] + [a^\dagger, a^\dagger]\, a = a^\dagger \end{aligned} \tag{B-16}$$

that is:

$$[N, a] = -a \tag{B-17-a}$$

$$[N, a^\dagger] = a^\dagger \tag{B-17-b}$$

Our study of the harmonic oscillator will be based on the use of the $a$, $a^\dagger$ and $N$ operators. We have replaced the eigenvalue equation of $H$, which we first wrote in the form (B-5), by that of $N$:

$$N \,|\, \varphi_\nu^i \rangle = \nu \,|\, \varphi_\nu^i \rangle \tag{B-18}$$

When this equation is solved, we shall know that the eigenvector $|\varphi_\nu^i\rangle$ of $N$ is also an eigenvector of $H$ with the eigenvalue $E_\nu = (\nu + 1/2)\hbar\omega$ [formulas (B-3) and (B-15)]:

$$H|\varphi_\nu^i\rangle = (\nu + 1/2)\hbar\omega|\varphi_\nu^i\rangle \tag{B-19}$$

The solution of equation (B-18) will be based on the commutation relation (B-9), which is equivalent to the initial relation (A-14), and on formulas (B-17), which are consequences of it.

## 2. Determination of the spectrum

### a. LEMMAS

α. *Lemma I (property of the eigenvalues of N)*

The eigenvalues $\nu$ of the operator $N$ are positive or zero.

Consider an arbitrary eigenvector $|\varphi_\nu^i\rangle$ of $N$. The square of the norm of the vector $a|\varphi_\nu^i\rangle$ is positive or zero:

$$\| a|\varphi_\nu^i\rangle \|^2 = \langle \varphi_\nu^i | a^\dagger a | \varphi_\nu^i \rangle \geqslant 0 \tag{B-20}$$

Let us then use definition (B-13) of $N$:

$$\langle \varphi_\nu^i | a^\dagger a | \varphi_\nu^i \rangle = \langle \varphi_\nu^i | N | \varphi_\nu^i \rangle = \nu \langle \varphi_\nu^i | \varphi_\nu^i \rangle \tag{B-21}$$

Since $\langle \varphi_\nu^i | \varphi_\nu^i \rangle$ is positive, comparison of (B-20) and (B-21) shows that:

$$\nu \geqslant 0 \tag{B-22}$$

β. *Lemma II (properties of the vector $a|\varphi_\nu^i\rangle$)*

Let $|\varphi_\nu^i\rangle$ be a (non-zero) eigenvector of $N$ with the eigenvalue $\nu$.

We shall prove the following:

(*i*) If $\nu = 0$, the ket $a|\varphi_{\nu=0}^i\rangle$ is zero.

(*ii*) If $\nu > 0$, the ket $a|\varphi_\nu^i\rangle$ is a non-zero eigenvector of $N$ with the eigenvalue $\nu - 1$.

(*i*) According to (B-21), the square of the norm of $a|\varphi_\nu^i\rangle$ is zero if $\nu = 0$; now, the norm of a vector is zero if and only if this vector is zero. Consequently, if $\nu = 0$ is an eigenvalue of $N$, all eigenvectors $|\varphi_0^i\rangle$ associated with this eigenvalue satisfy the relation:

$$a|\varphi_0^i\rangle = 0 \tag{B-23}$$

Let us now show that relation (B-23) is characteristic of these eigenvectors. Consider a vector $|\varphi\rangle$ which satisfies:

$$a|\varphi\rangle = 0 \tag{B-24}$$

Multiply both sides of this equation from the left by $a^\dagger$:

$$a^\dagger a|\varphi\rangle = N|\varphi\rangle = 0 \tag{B-25}$$

Any vector which satisfies (B-24) is therefore an eigenvector of $N$ with the eigenvalue $\nu = 0$.

(*ii*) Now let us assume that $\nu$ is strictly positive. According to (B-21), the vector $a \mid \varphi_\nu^i \rangle$ is then non-zero, since the square of its norm is not equal to zero.

Let us show that $a \mid \varphi_\nu^i \rangle$ is an eigenvector of $N$. To do this, let us apply the operator relation (B-17-a) to the vector $\mid \varphi_\nu^i \rangle$:

$$\begin{aligned} [N, a] \mid \varphi_\nu^i \rangle &= - a \mid \varphi_\nu^i \rangle \\ Na \mid \varphi_\nu^i \rangle &= aN \mid \varphi_\nu^i \rangle - a \mid \varphi_\nu^i \rangle \\ &= a\nu \mid \varphi_\nu^i \rangle - a \mid \varphi_\nu^i \rangle \end{aligned} \tag{B-26}$$

Therefore:

$$N[a \mid \varphi_\nu^i \rangle] = (\nu - 1)[a \mid \varphi_\nu^i \rangle] \tag{B-27}$$

which shows that $a \mid \varphi_\nu^i \rangle$ is an eigenvector of $N$ with the eigenvalue $\nu - 1$.

γ. *Lemma III (properties of the vector $a^\dagger \mid \varphi_\nu^i \rangle$)*

Let $\mid \varphi_\nu^i \rangle$ be a (non-zero) eigenvector of $N$ of eigenvalue $\nu$.

We shall prove the following:

(*i*) $a^\dagger \mid \varphi_\nu^i \rangle$ is always non-zero.

(*ii*) $a^\dagger \mid \varphi_\nu^i \rangle$ is an eigenvector of $N$ with the eigenvalue $\nu + 1$.

(*i*) It is easy to calculate the norm of the vector $a^\dagger \mid \varphi_\nu^i \rangle$, using formulas (B-9) and (B-13):

$$\begin{aligned} \| a^\dagger \mid \varphi_\nu^i \rangle \|^2 &= \langle \varphi_\nu^i \mid aa^\dagger \mid \varphi_\nu^i \rangle \\ &= \langle \varphi_\nu^i \mid (N + 1) \mid \varphi_\nu^i \rangle \\ &= (\nu + 1) \langle \varphi_\nu^i \mid \varphi_\nu^i \rangle \end{aligned} \tag{B-28}$$

Since, according to lemma I, $\nu$ is positive or zero, the ket $a^\dagger \mid \varphi_\nu^i \rangle$ always has a non-zero norm and, consequently, is never zero.

(*ii*) The proof of the fact that $a^\dagger \mid \varphi_\nu^i \rangle$ is an eigenvector of $N$ is analogous to that of lemma II; starting from relation (B-17-b) between operators, we obtain:

$$\begin{aligned} [N, a^\dagger] \mid \varphi_\nu^i \rangle &= a^\dagger \mid \varphi_\nu^i \rangle \\ Na^\dagger \mid \varphi_\nu^i \rangle &= a^\dagger N \mid \varphi_\nu^i \rangle + a^\dagger \mid \varphi_\nu^i \rangle = (\nu + 1) a^\dagger \mid \varphi_\nu^i \rangle \end{aligned} \tag{B-29}$$

#### b. THE SPECTRUM OF $N$ IS COMPOSED OF NON-NEGATIVE INTEGERS

Consider an arbitrary eigenvalue $\nu$ of $N$ and a non-zero eigenvector $\mid \varphi_\nu^i \rangle$ associated with this eigenvalue.

According to lemma I, $\nu$ is necessarily positive or zero. First, let us assume $\nu$ to be non-integral. We are now going to show that such a hypothesis contradicts lemma I and must consequently be excluded. If $\nu$ is non-integral, we can always find an integer $n \geqslant 0$ such that:

$$n < \nu < n + 1 \tag{B-30}$$

Now let us consider the series of vectors:

$$|\varphi_\nu^i\rangle,\ a\,|\varphi_\nu^i\rangle\ \dots\ a^n\,|\varphi_\nu^i\rangle \tag{B-31}$$

According to lemma II, each of the vectors $a^p\,|\varphi_\nu^i\rangle$ of this series (with $0 \leqslant p \leqslant n$) is non-zero and an eigenvector of $N$ with the eigenvalue $\nu - p$ (*cf.* fig. 3). The proof is by iteration : $|\varphi_\nu^i\rangle$ is non-zero by hypothesis; $a\,|\varphi_\nu^i\rangle$ is non-zero (since $\nu > 0$) and corresponds to the eigenvalue $\nu - 1$ of $N$ ...; $a^p\,|\varphi_\nu^i\rangle$ is obtained when $a$ acts on $a^{p-1}\,|\varphi_\nu^i\rangle$, an eigenvector of $N$ with the strictly positive eigenvalue $\nu - p + 1$, since $p \leqslant n$ and $\nu > n$ [*cf.* (B-30)].

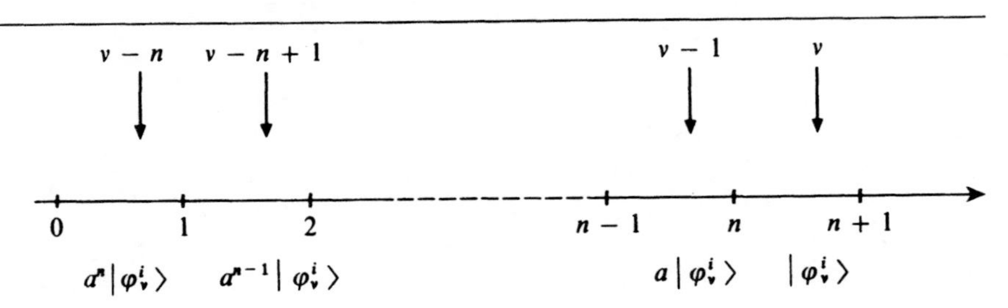


FIGURE 3

**Letting $a$ act several times on the ket $|\varphi_\nu^i\rangle$, we can construct eigenvectors of $N$ with eigenvalues $\nu - 1$, $\nu - 2$, etc...**

Now let $a$ act on the ket $a^n\,|\varphi_\nu^i\rangle$. Since $\nu - n > 0$ according to (B-30), the action of $a$ on $a^n\,|\varphi_\nu^i\rangle$ (an eigenvector of $N$ with the eigenvalue $\nu - n > 0$) yields a non-zero vector (lemma II). Moreover, again according to lemma II, $a^{n+1}\,|\varphi_\nu^i\rangle$ is an eigenvector of $N$ with the eigenvalue $\nu - n - 1$, which is strictly negative according to (B-30). If $\nu$ is non-integral, we can therefore construct a non-zero eigenvector of $N$ with a strictly negative eigenvalue. Since this is impossible, according to lemma I, the hypothesis of non-integral $\nu$ must be rejected.

What now happens if:

$$\nu = n \tag{B-32}$$

with $n$ a positive integer or zero? In the series of vectors (B-31), $a^n\,|\varphi_n^i\rangle$ is non-zero and an eigenvector of $N$ with the eigenvalue 0. According to lemma II (§ *i*), we therefore have:

$$a^{n+1}\,|\varphi_n^i\rangle = 0 \tag{B-33}$$

The series of vectors obtained by repeated action of the operator $a$ on $|\varphi_n^i\rangle$ is therefore limited when $n$ is integral. It is then never possible to obtain a non-zero eigenvector of $N$ which corresponds to a negative eigenvalue.

In conclusion, $\nu$ can only be a non-negative integer.

Lemma III can then be used to show that the spectrum of $N$ indeed includes all positive or zero integers. We have already constructed an eigenvector of $N$ with an eigenvalue of zero ($a^n\,|\varphi_n^i\rangle$). All we must do is let $(a^\dagger)^k$ act on such a vector in order to obtain an eigenvector of $N$ of eigenvalue $k$, where $k$ is an arbitrary positive integer.

If we then refer to formula (B-19), we conclude that the eigenvalues of $H$ are of the form:

$$E_n = \left(n + \frac{1}{2}\right)\hbar\omega \tag{B-34}$$

with $n = 0, 1, 2, \ldots$ Therefore, in quantum mechanics, *the energy of the harmonic oscillator is quantized* and cannot take on any arbitrary value. Note also that the smallest value (the ground state) is not zero, but $\hbar\omega/2$ (see § D-2 below).

#### c. INTERPRETATION OF THE $a$ AND $a^\dagger$ OPERATORS

If we start with an eigenstate $|\varphi_n^i\rangle$ of $H$ corresponding to the eigenvalue $E_n = (n + 1/2)\hbar\omega$, application of the operator $a$ yields an eigenvector associated with the eigenvalue $E_{n-1} = (n + 1/2)\hbar\omega - \hbar\omega$, and application of $a^\dagger$ yields, in the same way, the energy $E_{n+1} = (n + 1/2)\hbar\omega + \hbar\omega$.

For this reason, $a$ is said to be a *destruction operator* and $a^\dagger$, a *creation operator*: their action on an eigenvector of $N$ makes an energy quantum $\hbar\omega$ appear or disappear.

### 3. Degeneracy of the eigenvalues

We shall now show that the energy levels of the one-dimensional harmonic oscillator, given by equation (B-34), are not degenerate.

#### a. THE GROUND STATE IS NON-DEGENERATE

The eigenstates of $H$ associated with the eigenvalue $E_0 = \hbar\omega/2$, that is, the eigenstates of $N$ associated with the eigenvalue $n = 0$, according to lemma II of § 2-a-$\beta$, must all satisfy the equation:

$$a\,|\varphi_0^i\rangle = 0 \tag{B-35}$$

To find the degeneracy of the $E_0$ level, all we must do is see how many linearly independent kets satisfy (B-35).

Using definition (B-6-a) of $a$ and relations (B-1), we can write (B-35) in the form:

$$\frac{1}{\sqrt{2}}\left[\sqrt{\frac{m\omega}{\hbar}}\,X + \frac{i}{\sqrt{m\hbar\omega}}\,P\right]|\varphi_0^i\rangle = 0 \tag{B-36}$$

In the $\{\,|x\rangle\,\}$ representation, this relation becomes:

$$\left(\frac{m\omega}{\hbar}\,x + \frac{\mathrm{d}}{\mathrm{d}x}\right)\varphi_0^i(x) = 0 \tag{B-37}$$

where:

$$\varphi_0^i(x) = \langle x \mid \varphi_0^i \rangle \tag{B-38}$$

Therefore we must solve a first-order differential equation. Its general solution is:

$$\varphi_0^i(x) = c\, e^{-\frac{1}{2}\frac{m\omega}{\hbar}x^2} \tag{B-39}$$

where $c$ is the constant of integration. The various solutions of (B-37) are all proportional to each other. Consequently, to within a multiplicative factor, there exists only one ket $|\varphi_0\rangle$ which satisfies (B-35): the ground state $E_0 = \hbar\omega/2$ is not degenerate.

### b. ALL THE STATES ARE NON-DEGENERATE

We have just seen that the ground state is not degenerate. Let us show by recurrence that this is also the case for all the other states.

To do this, all we need prove is that, if the level $E_n = (n + 1/2)\hbar\omega$ is not degenerate, the level $E_{n+1} = (n + 1 + 1/2)\hbar\omega$ is not either. Let us therefore assume that there exists, to within a constant factor, only one vector $|\varphi_n\rangle$ such that:

$$N \mid \varphi_n \rangle = n \mid \varphi_n \rangle \tag{B-40}$$

Then consider an eigenvector $|\varphi_{n+1}^i\rangle$ corresponding to the eigenvalue $n + 1$:

$$N \mid \varphi_{n+1}^i \rangle = (n + 1) \mid \varphi_{n+1}^i \rangle \tag{B-41}$$

We know that the ket $a|\varphi_{n+1}^i\rangle$ is not zero and that it is an eigenvector of $N$ with the eigenvalue $n$ (*cf.* lemma II). Since this ket is not degenerate by hypothesis, there exists a number $c^i$ such that:

$$a \mid \varphi_{n+1}^i \rangle = c^i \mid \varphi_n \rangle \tag{B-42}$$

It is simple to invert this equation by applying $a^\dagger$ to both sides:

$$a^\dagger a \mid \varphi_{n+1}^i \rangle = c^i a^\dagger \mid \varphi_n \rangle \tag{B-43}$$

that is, taking (B-13) and (B-41) into account:

$$\mid \varphi_{n+1}^i \rangle = \frac{c^i}{n + 1} a^\dagger \mid \varphi_n \rangle \tag{B-44}$$

We already knew that $a^\dagger|\varphi_n\rangle$ was an eigenvector of $N$ with the eigenvalue $(n + 1)$; we see here that all kets $|\varphi_{n+1}^i\rangle$ associated with the eigenvalue $(n + 1)$ are proportional to $a^\dagger|\varphi_n\rangle$. They are therefore proportional to each other: the eigenvalue $(n + 1)$ is not degenerate.

Thus, since the eigenvalue $n = 0$ is not degenerate (see § a), the eigenvalue $n = 1$ is not either, nor is $n = 2$, etc... : all the eigenvalues of $N$ and, consequently, all those of $H$, are non-degenerate. This enables us to write simply $|\varphi_n\rangle$ for the eigenvector of $H$ associated with the eigenvalue $E_n = (n + 1/2)\hbar\omega$.

# C. EIGENSTATES OF THE HAMILTONIAN

In this section, we are going to study the principal properties of the eigenstates of the operator $N$ and of the Hamiltonian $H$.

## 1. The $\{ | \varphi_n \rangle \}$ representation

We shall assume that $N$ and $H$ are observables, that is, their eigenvectors constitute a basis in the space $\mathscr{E}_x$, the state space of a particle in a one-dimensional problem (this could be proved by considering the wave functions associated with the eigenstates of $N$, which we shall calculate in § 2 below). Since none of the eigenvalues of $N$ (or of $H$) is degenerate (see § B-3), $N$ (or $H$) alone constitutes a C.S.C.O. in $\mathscr{E}_x$.

### a. THE BASIS VECTORS IN TERMS OF $| \varphi_0 \rangle$

The vector $| \varphi_0 \rangle$ associated with $n = 0$ is the vector of $\mathscr{E}_x$ which satisfies:

$$a \, | \varphi_0 \rangle = 0 \tag{C-1}$$

It is defined to within a constant factor; we shall assume $| \varphi_0 \rangle$ to be normalized, so the indeterminacy is reduced to a global phase factor of the form $e^{i\theta}$, with $\theta$ real.

According to lemma III of § B-2-a, the vector $| \varphi_1 \rangle$ which corresponds to $n = 1$ is proportional to $a^\dagger | \varphi_0 \rangle$:

$$| \varphi_1 \rangle = c_1 \, a^\dagger \, | \varphi_0 \rangle \tag{C-2}$$

We shall determine $c_1$ by requiring $| \varphi_1 \rangle$ to be normalized and choosing the phase of $| \varphi_1 \rangle$ (relative to $| \varphi_0 \rangle$) such that $c_1$ is real and positive. The square of the norm of $| \varphi_1 \rangle$, according to (C-2), is equal to:

$$\begin{aligned} \langle \varphi_1 | \varphi_1 \rangle &= |c_1|^2 \langle \varphi_0 | a a^\dagger | \varphi_0 \rangle \\ &= |c_1|^2 \langle \varphi_0 | (a^\dagger a + 1) | \varphi_0 \rangle \end{aligned} \tag{C-3}$$

where (B-9) has been used. Since $| \varphi_0 \rangle$ is a normalized eigenstate of $N = a^\dagger a$ with the eigenvalue zero, we find:

$$\langle \varphi_1 | \varphi_1 \rangle = |c_1|^2 = 1 \tag{C-4}$$

With the preceding phase convention, we have $c_1 = 1$ and, consequently:

$$| \varphi_1 \rangle = a^\dagger \, | \varphi_0 \rangle \tag{C-5}$$

Similarly, we can construct $| \varphi_2 \rangle$ from $| \varphi_1 \rangle$:

$$| \varphi_2 \rangle = c_2 \, a^\dagger \, | \varphi_1 \rangle \tag{C-6}$$

We require $| \varphi_2 \rangle$ to be normalized and choose its phase such that $c_2$ is real and positive:

$$\begin{aligned} \langle \varphi_2 | \varphi_2 \rangle &= |c_2|^2 \langle \varphi_1 | aa^\dagger | \varphi_1 \rangle \\ &= |c_2|^2 \langle \varphi_1 | (a^\dagger a + 1) | \varphi_1 \rangle \\ &= 2 |c_2|^2 = 1 \end{aligned} \tag{C-7}$$

Therefore:

$$| \varphi_2 \rangle = \frac{1}{\sqrt{2}} a^\dagger | \varphi_1 \rangle = \frac{1}{\sqrt{2}} (a^\dagger)^2 | \varphi_0 \rangle \tag{C-8}$$

if we take (C-5) into account.

This procedure can easily be generalized. If we know $| \varphi_{n-1} \rangle$ (which is normalized), then the normalized vector $| \varphi_n \rangle$ is written:

$$| \varphi_n \rangle = c_n a^\dagger | \varphi_{n-1} \rangle \tag{C-9}$$

Since:

$$\begin{aligned} \langle \varphi_n | \varphi_n \rangle &= |c_n|^2 \langle \varphi_{n-1} | aa^\dagger | \varphi_{n-1} \rangle \\ &= n |c_n|^2 = 1 \end{aligned} \tag{C-10}$$

we choose, with the same phase conventions as above:

$$c_n = \frac{1}{\sqrt{n}} \tag{C-11}$$

With these successive phase choices, we can obtain all the $| \varphi_n \rangle$ from $| \varphi_0 \rangle$:

$$\begin{aligned} | \varphi_n \rangle &= \frac{1}{\sqrt{n}} a^\dagger | \varphi_{n-1} \rangle = \frac{1}{\sqrt{n}} \frac{1}{\sqrt{n-1}} (a^\dagger)^2 | \varphi_{n-2} \rangle = \dots \\ &= \frac{1}{\sqrt{n}} \frac{1}{\sqrt{n-1}} \cdots \frac{1}{\sqrt{2}} (a^\dagger)^n | \varphi_0 \rangle \end{aligned} \tag{C-12}$$

that is:

$$| \varphi_n \rangle = \frac{1}{\sqrt{n!}} (a^\dagger)^n | \varphi_0 \rangle \tag{C-13}$$

### b. ORTHONORMALIZATION AND CLOSURE RELATIONS

Since $H$ is Hermitian, the kets $| \varphi_n \rangle$ corresponding to different values of $n$ are orthogonal. Since each of them is also normalized, they satisfy the orthonormalization relation:

$$\langle \varphi_{n'} | \varphi_n \rangle = \delta_{nn'} \tag{C-14}$$

In addition, $H$ is an observable (we shall assume this here without proof); the set of the $| \varphi_n \rangle$ therefore constitutes a basis in $\mathscr{E}_x$. This is shown in the closure relation:

$$\sum_n | \varphi_n \rangle \langle \varphi_n | = 1 \tag{C-15}$$

COMMENT:

It can be verified directly from expression (C-13) that the kets $| \varphi_n \rangle$ are orthonormal:

$$\langle \varphi_{n'} | \varphi_n \rangle = \frac{1}{\sqrt{n!\, n'!}} \langle \varphi_0 | a^{n'} a^{\dagger n} | \varphi_0 \rangle \tag{C-16}$$

But:

$$\begin{aligned} a^{n'} a^{\dagger n} | \varphi_0 \rangle &= a^{n'-1}(aa^\dagger)\, a^{\dagger n-1} | \varphi_0 \rangle \\ &= a^{n'-1}(a^\dagger a + 1)\, a^{\dagger n-1} | \varphi_0 \rangle \\ &= n\, a^{n'-1} a^{\dagger n-1} | \varphi_0 \rangle \end{aligned} \tag{C-17}$$

(using the fact that $a^{\dagger n-1} | \varphi_0 \rangle$ is an eigenstate of $N = a^\dagger a$ with the eigenvalue $n - 1$). Thus can we reduce the exponents of $a$ and $a^\dagger$ by iteration. We obtain, finally:

$$\text{if } n < n' : \quad \langle \varphi_0 | a^{n'} a^{\dagger n} | \varphi_0 \rangle = n \times (n-1) \times \ldots 2 \times 1 \langle \varphi_0 | a^{n'-n} | \varphi_0 \rangle \tag{C-18-a}$$

$$\text{if } n > n' : \quad \langle \varphi_0 | a^{n'} a^{\dagger n} | \varphi_0 \rangle = n \times (n-1) \ldots (n - n' + 1) \langle \varphi_0 | (a^\dagger)^{n-n'} | \varphi_0 \rangle \tag{C-18-b}$$

$$\text{if } n = n' : \quad \langle \varphi_0 | a^{n'} a^{\dagger n} | \varphi_0 \rangle = n \times (n-1) \times \ldots 2 \times 1 \langle \varphi_0 | \varphi_0 \rangle \tag{C-18-c}$$

The expression (C-18-a) is zero because $a | \varphi_0 \rangle = 0$. Similarly, (C-18-b) is equal to zero because $\langle \varphi_0 | (a^\dagger)^{n-n'} | \varphi_0 \rangle$ can be considered to be the scalar product of $| \varphi_0 \rangle$ and the bra associated with $a^{n-n'} | \varphi_0 \rangle$, which is zero if $n > n'$. Finally, if we substitute (C-18-c) into (C-16), we see that $\langle \varphi_n | \varphi_n \rangle$ is equal to 1.

### c. ACTION OF THE VARIOUS OPERATORS

The observables $X$ and $P$ are linear combinations of the operators $a$ and $a^\dagger$ [formulas (B-1) and (B-7)]. Consequently, all physical quantities can be expressed in terms of $a$ and $a^\dagger$. Now, the action of $a$ and $a^\dagger$ on the $| \varphi_n \rangle$ vectors is especially simple [see equations (C-19) below]. In most cases, it is therefore desirable to use the $\{ | \varphi_n \rangle \}$ representation to calculate the matrix elements and mean values of the various observables.

With the phase conventions introduced in §a above, the action of the $a$ and $a^\dagger$ operators on the vectors of the $\{ | \varphi_n \rangle \}$ basis is given by:

$$a^\dagger | \varphi_n \rangle = \sqrt{n+1}\, | \varphi_{n+1} \rangle \tag{C-19-a}$$

$$a | \varphi_n \rangle = \sqrt{n}\, | \varphi_{n-1} \rangle \tag{C-19-b}$$

We have already proved (C-19-a): it suffices to replace $n$ by $n + 1$ in equations (C-9) and (C-11). To obtain (C-19-b), multiply both sides of (C-9) on the left by the operator $a$ and use (C-11):

$$a | \varphi_n \rangle = \frac{1}{\sqrt{n}} aa^\dagger | \varphi_{n-1} \rangle = \frac{1}{\sqrt{n}} (a^\dagger a + 1) | \varphi_{n-1} \rangle = \sqrt{n}\, | \varphi_{n-1} \rangle \tag{C-20}$$

COMMENT:

The adjoint equations of (C-19-a) and (C-19-b) are:

$$\langle \varphi_n | a = \sqrt{n+1} \langle \varphi_{n+1} | \tag{C-21-a}$$

$$\langle \varphi_n | a^\dagger = \sqrt{n} \langle \varphi_{n-1} | \tag{C-21-b}$$

Note that $a$ decreases or increases $n$ by one unit depending on whether it acts on the ket $| \varphi_n \rangle$ or on the bra $\langle \varphi_n |$. Similarly, $a^\dagger$ increases or decreases $n$ by one unit, depending on whether it acts on the ket $| \varphi_n \rangle$ or on the bra $\langle \varphi_n |$.

Starting with (C-19) and using (B-1) and (B-7), we immediately find the expressions for the kets $X | \varphi_n \rangle$ and $P | \varphi_n \rangle$:

$$X | \varphi_n \rangle = \sqrt{\frac{\hbar}{m\omega}} \frac{1}{\sqrt{2}} (a^\dagger + a) | \varphi_n \rangle$$

$$= \sqrt{\frac{\hbar}{2m\omega}} \left[ \sqrt{n+1} | \varphi_{n+1} \rangle + \sqrt{n} | \varphi_{n-1} \rangle \right] \tag{C-22-a}$$

$$P | \varphi_n \rangle = \sqrt{m\hbar\omega} \frac{i}{\sqrt{2}} (a^\dagger - a) | \varphi_n \rangle$$

$$= i \sqrt{\frac{m\hbar\omega}{2}} \left[ \sqrt{n+1} | \varphi_{n+1} \rangle - \sqrt{n} | \varphi_{n-1} \rangle \right] \tag{C-22-b}$$

The matrix elements of the $a$, $a^\dagger$, $X$ and $P$ operators in the $\{ | \varphi_n \rangle \}$ representation are therefore:

$$\langle \varphi_{n'} | a | \varphi_n \rangle = \sqrt{n} \, \delta_{n',n-1} \tag{C-23-a}$$

$$\langle \varphi_{n'} | a^\dagger | \varphi_n \rangle = \sqrt{n+1} \, \delta_{n',n+1} \tag{C-23-b}$$

$$\langle \varphi_{n'} | X | \varphi_n \rangle = \sqrt{\frac{\hbar}{2m\omega}} \left[ \sqrt{n+1} \, \delta_{n',n+1} + \sqrt{n} \, \delta_{n',n-1} \right] \tag{C-23-c}$$

$$\langle \varphi_{n'} | P | \varphi_n \rangle = i \sqrt{\frac{m\hbar\omega}{2}} \left[ \sqrt{n+1} \, \delta_{n',n+1} - \sqrt{n} \, \delta_{n',n-1} \right] \tag{C-23-d}$$

The matrices representing $a$ and $a^\dagger$ are indeed Hermitian conjugates of each other, as can be seen from their explicit expressions:

$$(a) = \begin{pmatrix} 0 & \sqrt{1} & 0 & 0 & \cdots\cdots\cdots & \\ 0 & 0 & \sqrt{2} & 0 & \cdots\cdots\cdots & \\ 0 & 0 & 0 & \sqrt{3} & \cdots\cdots\cdots & \\ \vdots & \vdots & \vdots & \vdots & & \\ 0 & 0 & 0 & 0 & 0 & \sqrt{n} \ldots \\ \vdots & \vdots & \vdots & \vdots & \vdots & \vdots \end{pmatrix} \tag{C-24-a}$$

$$(a^\dagger) = \begin{pmatrix} 0 & 0 & 0 & \cdots\cdots\cdots & \\ \sqrt{1} & 0 & 0 & \cdots\cdots\cdots & \\ 0 & \sqrt{2} & 0 & \cdots\cdots\cdots & \\ 0 & 0 & \sqrt{3} & \cdots\cdots\cdots & \\ \vdots & \vdots & \vdots & & \\ 0 & 0 & 0 & \sqrt{n+1} & 0 \ldots \\ \vdots & \vdots & \vdots & \vdots & \vdots \end{pmatrix} \tag{C-24-b}$$

As for the matrices representing $X$ and $P$, they are both Hermitian : the matrix associated with $X$ is, to within a constant factor, the sum of the two preceding ones; the matrix associated with $P$ is proportional to their difference, but the presence of the factor $i$ in (C-22-b) re-establishes its Hermiticity.

## 2. Wave functions associated with the stationary states

We shall now use the $\{ | x \rangle \}$ representation and write the functions $\varphi_n(x) = \langle x | \varphi_n \rangle$ which then represent the eigenstates of the Hamiltonian.

We have already determined the function $\varphi_0(x)$ which represents the ground state $| \varphi_0 \rangle$ (*cf.* § B-3-a) :

$$\varphi_0(x) = \langle x | \varphi_0 \rangle = \left( \frac{m\omega}{\pi\hbar} \right)^{1/4} e^{-\frac{1}{2}\frac{m\omega}{\hbar}x^2} \tag{C-25}$$

The constant which appears before the exponential insures the normalization of $\varphi_0(x)$.

To obtain the functions $\varphi_n(x)$ associated with the other stationary states of the harmonic oscillator, all we need to do is use expression (C-13) for the ket $| \varphi_n \rangle$ and the fact that, in the $\{ | x \rangle \}$ representation, $a^\dagger$ is represented by :

$$\frac{1}{\sqrt{2}} \left[ \sqrt{\frac{m\omega}{\hbar}}\, x - \sqrt{\frac{\hbar}{m\omega}} \frac{d}{dx} \right],$$

since $X$ is represented by multiplication by $x$ and $P$, by $\frac{\hbar}{i}\frac{d}{dx}$ [formula (B-6-b)].

Thus we obtain :

$$\begin{aligned} \varphi_n(x) = \langle x | \varphi_n \rangle &= \frac{1}{\sqrt{n!}} \langle x | (a^\dagger)^n | \varphi_0 \rangle \\ &= \frac{1}{\sqrt{n!}} \frac{1}{\sqrt{2^n}} \left[ \sqrt{\frac{m\omega}{\hbar}}\, x - \sqrt{\frac{\hbar}{m\omega}} \frac{d}{dx} \right]^n \varphi_0(x) \end{aligned} \tag{C-26}$$

that is:

$$\varphi_n(x) = \left[\frac{1}{2^n n!}\left(\frac{\hbar}{m\omega}\right)^n\right]^{1/2} \left(\frac{m\omega}{\pi\hbar}\right)^{1/4} \left[\frac{m\omega}{\hbar}x - \frac{\mathrm{d}}{\mathrm{d}x}\right]^n \mathrm{e}^{-\frac{1}{2}\frac{m\omega}{\hbar}x^2} \tag{C-27}$$

It is easy to see from this expression that $\varphi_n(x)$ is the product of $\mathrm{e}^{-\frac{1}{2}\frac{m\omega}{\hbar}x^2}$ and a polynomial of degree $n$ and parity $(-1)^n$ called a *Hermite polynomial* (*cf.* complements $B_V$ and $C_V$).

A simple calculation gives the first several functions $\varphi_n(x)$:

$$\varphi_1(x) = \left[\frac{4}{\pi}\left(\frac{m\omega}{\hbar}\right)^3\right]^{1/4} x\,\mathrm{e}^{-\frac{1}{2}\frac{m\omega}{\hbar}x^2}$$

$$\varphi_2(x) = \left(\frac{m\omega}{4\pi\hbar}\right)^{1/4} \left[2\frac{m\omega}{\hbar}x^2 - 1\right] \mathrm{e}^{-\frac{1}{2}\frac{m\omega}{\hbar}x^2} \tag{C-28}$$

These functions are shown in figure 4 and the corresponding probability densities, in figure 5. Figure 6 gives the shape of the wave function $\varphi_n(x)$ and that of the probability density $|\varphi_n(x)|^2$ for $n = 10$.

We see from these figures that when $n$ increases, the region of the $Ox$ axis in which $\varphi_n(x)$ takes on non-negligible values becomes larger. This corresponds to the fact, in classical mechanics, that the amplitude of the particle's motion increases with the energy [*cf.* fig. 1 and relation (A-8)]. It follows that the mean value of the potential energy grows with $n$ [*cf.* comment (*ii*) of §D-1], since $\varphi_n(x)$, when $n$ is large, takes on non-negligible values in regions of the $x$-axis where $V(x)$ is large. Moreover, we see in these figures that the number of zeros of $\varphi_n(x)$ is $n$ (*cf.* complement $B_V$, where this property is derived). This implies that the mean kinetic energy of the particle increases with $n$ [*cf.* comment (*ii*) of §D-1], since this energy is given by:

$$\frac{1}{2m}\langle P^2 \rangle = -\frac{\hbar^2}{2m}\int_{-\infty}^{+\infty} \varphi_n^*(x)\frac{\mathrm{d}^2}{\mathrm{d}x^2}\varphi_n(x)\,\mathrm{d}x \tag{C-29}$$

When the number of zeros of $\varphi_n(x)$ increases, the curvature of the wave function increases, and, in (C-29), the second derivative $\frac{\mathrm{d}^2}{\mathrm{d}x^2}\varphi_n(x)$ takes on larger and larger values.

Finally, when $n$ is large, we observe (see, for example, figure 6) that the probability density $|\varphi_n(x)|^2$ is large for $x \simeq \pm x_M$ [where $x_M$ is the amplitude of the classical motion of energy $E_n$; *cf.* (A-8)]. This result is related to a feature of the motion predicted by classical mechanics: the classical particle has a zero velocity at $x = \pm x_M$; therefore, on the average, it spends more time in the neighborhood of these two points than in the center of the interval $-x_M \leqslant x \leqslant x_M$.

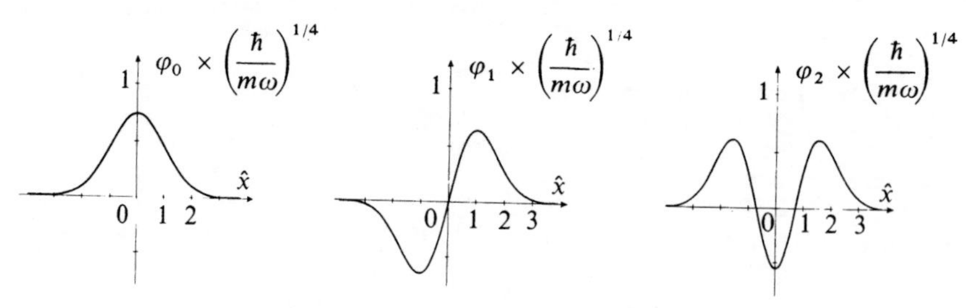


FIGURE 4

**Wave functions associated with the first three levels of a harmonic oscillator.**

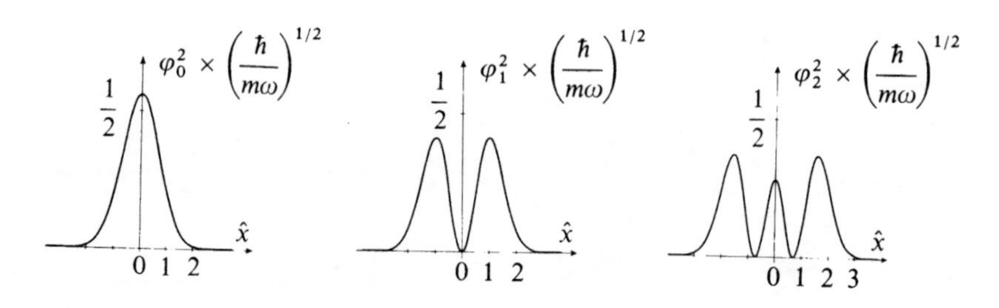


FIGURE 5

**Probability densities associated with the first three levels of a harmonic oscillator.**

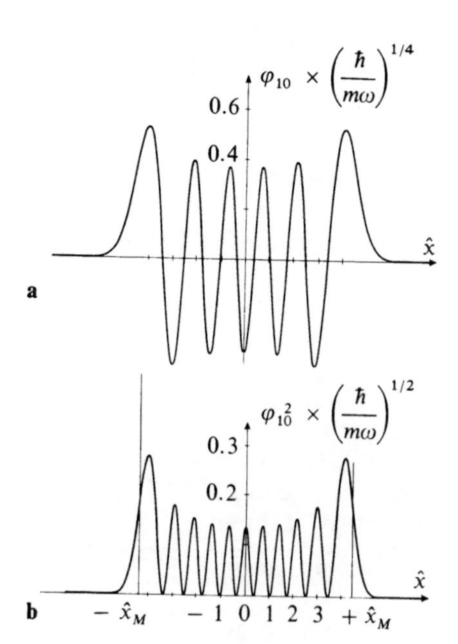


FIGURE 6

**Shape of the wave function (fig. a) and of the probability density (fig. b) for the $n = 10$ level of a harmonic oscillator.**

# D. DISCUSSION

## 1. Mean values and root-mean-square deviations of *X* and *P* in a state $|\varphi_n\rangle$

Neither $X$ nor $P$ commutes with $H$, and the eigenstates $|\varphi_n\rangle$ of $H$ are not eigenstates of $X$ or $P$. Consequently, if the harmonic oscillator is in a stationary state $|\varphi_n\rangle$, a measurement of the observable $X$ or the observable $P$ can, *a priori*, yield any result (since the spectra of $X$ and $P$ include all real numbers). We shall now calculate the mean values of $X$ and $P$ in such a stationary state and then their root-mean-square deviations $\Delta X$ and $\Delta P$, which will enable us to verify the uncertainty relation.

As we indicated in §C-1-c, we shall perform these calculations with the help of the operators $a$ and $a^\dagger$. As far as the mean values of $X$ and $P$ are concerned, the result follows directly from formulas (C-22), which show that neither $X$ nor $P$ has diagonal matrix elements:

$$\begin{aligned}\langle \varphi_n | X | \varphi_n \rangle &= 0\\ \langle \varphi_n | P | \varphi_n \rangle &= 0\end{aligned} \qquad \text{(D-1)}$$

To obtain the root-mean-square deviations $\Delta X$ and $\Delta P$, we must calculate the mean values of $X^2$ and $P^2$:

$$\begin{aligned}(\Delta X)^2 &= \langle \varphi_n |X^2| \varphi_n \rangle - (\langle \varphi_n | X | \varphi_n \rangle)^2 = \langle \varphi_n | X^2 | \varphi_n \rangle\\ (\Delta P)^2 &= \langle \varphi_n |P^2| \varphi_n \rangle - (\langle \varphi_n | P | \varphi_n \rangle)^2 = \langle \varphi_n | P^2 | \varphi_n \rangle\end{aligned} \qquad \text{(D-2)}$$

But, according to (B-1) and (B-7):

$$\begin{aligned}X^2 &= \frac{\hbar}{2m\omega}(a^\dagger + a)(a^\dagger + a)\\ &= \frac{\hbar}{2m\omega}(a^{\dagger 2} + aa^\dagger + a^\dagger a + a^2)\\ P^2 &= -\frac{m\hbar\omega}{2}(a^\dagger - a)(a^\dagger - a)\\ &= -\frac{m\hbar\omega}{2}(a^{\dagger 2} - aa^\dagger - a^\dagger a + a^2)\end{aligned} \qquad \text{(D-3)}$$

The terms in $a^2$ and $a^{\dagger 2}$ do not contribute to the diagonal matrix elements, since $a^2|\varphi_n\rangle$ is proportional to $|\varphi_{n-2}\rangle$ and $a^{\dagger 2}|\varphi_n\rangle$ to $|\varphi_{n+2}\rangle$ both of which are orthogonal to $|\varphi_n\rangle$. On the other hand:

$$\begin{aligned}\langle \varphi_n |(a^\dagger a + aa^\dagger) | \varphi_n \rangle &= \langle \varphi_n | (2a^\dagger a + 1) | \varphi_n \rangle\\ &= 2n + 1\end{aligned} \qquad \text{(D-4)}$$

Consequently:

$$(\Delta X)^2 = \langle \varphi_n | X^2 | \varphi_n \rangle = \left(n + \frac{1}{2}\right)\frac{\hbar}{m\omega} \qquad \text{(D-5-a)}$$

$$(\Delta P)^2 = \langle \varphi_n | P^2 | \varphi_n \rangle = \left(n + \frac{1}{2}\right)m\hbar\omega \qquad \text{(D-5-b)}$$

The product $\Delta X \,.\, \Delta P$ is therefore equal to:

$$\Delta X \,.\, \Delta P = \left(n + \frac{1}{2}\right)\hbar \tag{D-6}$$

We again find (*cf.* complement $C_{III}$) that it is greater than or equal to $\hbar/2$. In fact, this lower bound is attained for $n = 0$, that is, for the ground state (§ 2 below).

COMMENTS:

(*i*) If $x_M$ denotes the amplitude of the classical motion whose energy is $E_n = (n + 1/2)\hbar\omega$, it is easy to see, using (A-8) and (D-5-a), that:

$$\Delta X = \frac{1}{\sqrt{2}} x_M \tag{D-7}$$

Similarly, if $p_M$ denotes the oscillation amplitude of the corresponding classical momentum:

$$p_M = m\omega \; x_M \tag{D-8}$$

we obtain:

$$\Delta P = \frac{1}{\sqrt{2}} p_M \tag{D-9}$$

It is not surprising that $\Delta X$ is of the order of the interval $[- x_M, + x_M]$ over which the classical motion occurs (*cf.* fig. 1): we saw at the end of § C, that it is approximately inside this interval that $\varphi_n(x)$ takes on non-negligible values. Also, it is easy to understand why, when $n$ increases, $\Delta X$ does also. For large $n$, the probability density $|\varphi_n(x)|^2$ has two symmetric peaks situated approximately at $x = \pm x_M$. The root-mean-square deviation cannot then be much smaller than the distance between the peaks, even if each of them is very sharp (*cf.* chap. III, § C-5 and the discussion of § 1-b of complement $A_{III}$). An analogous argument can be set forth for $\Delta P$ (*cf.* complement $D_V$).

(*ii*) The mean potential energy of a particle in the state $|\varphi_n\rangle$ is:

$$\langle V(X) \rangle = \frac{1}{2} m\omega^2 \langle X^2 \rangle \tag{D-10}$$

that is, since $\langle X \rangle$ is zero [*cf.* (D-1)]:

$$\langle V(X) \rangle = \frac{1}{2} m\omega^2(\Delta X)^2 \tag{D-11}$$

Similarly, we could find the mean kinetic energy of this particle:

$$\langle \frac{P^2}{2m} \rangle = \frac{1}{2m} (\Delta P)^2 \tag{D-12}$$

Substituting relations (D-5) into (D-11) and (D-12), we obtain:

$$\begin{aligned} \langle V(X) \rangle &= \frac{1}{2}\left(n + \frac{1}{2}\right)\hbar\omega = \frac{E_n}{2} \\ \langle \frac{P^2}{2m} \rangle &= \frac{1}{2}\left(n + \frac{1}{2}\right)\hbar\omega = \frac{E_n}{2} \end{aligned} \tag{D-13}$$

The mean potential and kinetic energies are therefore equal. This is an illustration of the virial theorem (*cf.* exercise 10 of complement $L_{III}$).

(*iii*) A stationary state $|\varphi_n\rangle$ has no equivalent in classical mechanics; its energy is not zero although the mean values $\langle X \rangle$ and $\langle P \rangle$ are. Nevertheless, there is a certain analogy between the state $|\varphi_n\rangle$ and that of a classical particle whose position is given by (A-5) [where $x_M$ is related to the energy $E_n$ by relation (A-8)], but for which the initial phase $\varphi$ of the motion is chosen at random (all values included between 0 and $2\pi$ have the same probability). The mean values of $x$ and $p$ are then zero, since:

$$\begin{cases} \overline{x}_{cl} = x_M \dfrac{1}{2\pi}\displaystyle\int_0^{2\pi} \cos(\omega t - \varphi)\, d\varphi = 0 \\ \overline{p}_{cl} = -\, p_M \dfrac{1}{2\pi}\displaystyle\int_0^{2\pi} \sin(\omega t - \varphi)\, d\varphi = 0 \end{cases} \tag{D-14}$$

Moreover, we find, for the root-mean-square deviations of the position and the momentum, values identical to those of the state $|\varphi_n\rangle$ [formulas (D-7) and (D-9)]:

$$\begin{aligned} \overline{x_{cl}^2} &= x_M^2 \frac{1}{2\pi}\int_0^{2\pi} \cos^2(\omega t - \varphi)\, d\varphi = \frac{x_M^2}{2} \\ \overline{p_{cl}^2} &= p_M^2 \frac{1}{2\pi}\int_0^{2\pi} \sin^2(\omega t - \varphi)\, d\varphi = \frac{p_M^2}{2} \end{aligned} \tag{D-15}$$

that is:

$$\begin{aligned} \delta x_{cl} &= \sqrt{\overline{x_{cl}^2} - (\overline{x}_{cl})^2} = \frac{x_M}{\sqrt{2}} \\ \delta p_{cl} &= \sqrt{\overline{p_{cl}^2} - (\overline{p}_{cl})^2} = \frac{p_M}{\sqrt{2}} \end{aligned} \tag{D-16}$$

## 2. Properties of the ground state

In classical mechanics, the lowest energy of the harmonic oscillator is obtained when the particle is at rest (zero momentum and kinetic energy) at the $x$-origin ($x = 0$ and therefore zero potential energy). The situation is completely different in quantum mechanics: the minimum energy state is $|\varphi_0\rangle$, whose *energy is not zero*, and the associated wave function has a certain *spatial extension*, characterized by the root-mean-square deviation $\Delta X = \sqrt{\hbar/2m\omega}$.

This essential difference between the quantum and classical results can be seen to have its source in the uncertainty relations, which forbid the simultaneous minimization of the kinetic energy and the potential energy. As we pointed out in complements $C_I$ and $M_{III}$, the ground state corresponds to a compromise in which the sum of these two energies is as small as possible.

In the special case of a harmonic oscillator, it is possible to state these qualitative considerations semi-quantitatively, and thus find the order of magnitude of the energy and the spatial extension of the ground state. If the distance $\xi$ characterizes this spatial extension, the mean potential energy will be of the order of:

$$\overline{V} \simeq \frac{1}{2} m\omega^2 \xi^2 \tag{D-17}$$

But $\Delta P$ is then equal to about $\hbar/\xi$, so the mean kinetic energy is approximately:

$$\overline{T} = \frac{\overline{p^2}}{2m} \simeq \frac{\hbar^2}{2m\xi^2} \tag{D-18}$$

The order of magnitude of the total energy is therefore:

$$\overline{E} = \overline{T} + \overline{V} \simeq \frac{\hbar^2}{2m\xi^2} + \frac{1}{2} m\omega^2 \xi^2 \tag{D-19}$$

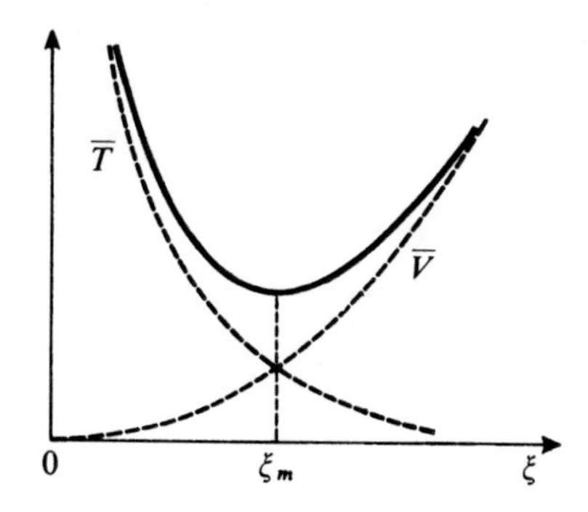


FIGURE 7

**Variation of the potential energy $\overline{V}$ and of the kinetic energy $\overline{T}$ with respect to a parameter $\xi$ which characterizes the spatial extension of the wave function about $x = 0$. Since the potential energy is at a minimum at $x = 0$, $\overline{V}$ is a function which increases with $\xi(\overline{V} \propto \xi^2)$. On the other hand, according to Heisenberg's uncertainty relation, the kinetic energy $\overline{T}$ is a decreasing function of $\xi$. The lowest possible total energy, obtained for $\xi = \xi_m$, results from a compromise in which the sum $\overline{T} + \overline{V}$ is at a minimum.**

The variation of $\overline{T}$, $\overline{V}$ and $\overline{E}$ with respect to $\xi$ is shown in figure 7. For small values of $\xi$, $\overline{T}$ prevails over $\overline{V}$; the opposite occurs for large values of $\xi$. The ground state therefore corresponds approximately to the minimum of the function (D-19); it is easy to see that this minimum occurs at:

$$\xi_m \simeq \sqrt{\frac{\hbar}{m\omega}} \tag{D-20}$$

and is equal to:

$$\overline{E}_m \simeq \hbar\omega \tag{D-21}$$

We again find the correct orders of magnitude of $E_0$ and $\Delta X$ in the state $| \varphi_0 \rangle$.

The harmonic oscillator possesses the pecularity that, because of the form of the potential $V(x)$, the product $\Delta X \,.\, \Delta P$ actually attains its lower bound, $\hbar/2$, in the ground state $| \varphi_0 \rangle$ [formula (D-6)]. This is related to the fact (*cf.* complement $C_{III}$) that the wave function of the ground state is Gaussian.

### 3. Time evolution of the mean values

Consider a harmonic oscillator whose state at $t = 0$ is:

$$| \psi(0) \rangle = \sum_{n=0}^{\infty} c_n(0) \, | \varphi_n \rangle \tag{D-22}$$

($| \psi(0) \rangle$ is assumed to be normalized). Its state $| \psi(t) \rangle$ at $t$ can be obtained by using rule (D-54) of chapter III :

$$\begin{aligned} | \psi(t) \rangle &= \sum_{n=0}^{\infty} c_n(0) \, e^{-i E_n t/\hbar} \, | \varphi_n \rangle \\ &= \sum_{n=0}^{\infty} c_n(0) \, e^{-i\left(n+\frac{1}{2}\right)\omega t} \, | \varphi_n \rangle \end{aligned} \tag{D-23}$$

The mean value of any physical quantity $A$ is therefore given as a function of time by:

$$\langle \psi(t) | A | \psi(t) \rangle = \sum_{m=0}^{\infty} \sum_{n=0}^{\infty} c_m^*(0) \, c_n(0) \, A_{mn} \, e^{i(m-n)\omega t} \tag{D-24}$$

with:

$$A_{mn} = \langle \varphi_m | A | \varphi_n \rangle \tag{D-25}$$

Since $m$ and $n$ are integers, the time evolution of the mean values involves only the frequency $\omega/2\pi$ and its various harmonics, which constitute the Bohr frequencies of the harmonic oscillator.

Let us consider, in particular, the mean values of the observables $X$ and $P$. According to formulas (C-22), the only non-zero matrix elements $X_{mn}$ and $P_{mn}$ are those for which $m = n \pm 1$. Consequently, the mean values of $X$ and $P$ include only terms in $e^{\pm i\omega t}$; they are sinusoidal functions of time with angular frequency $\omega$. This obviously relates to the classical solution of the harmonic oscillator problem. Moreover, as we pointed out in the discussion of Ehrenfest's theorem (chap. III, § D-1-d-$\gamma$), the form of the harmonic oscillator potential implies that for all $| \psi \rangle$ the mean values of $X$ and $P$ rigorously satisfy the classical equations of motion. Thus, according to general formulas (D-34) and (D-35) of chapter III :

$$\frac{d}{dt} \langle X \rangle = \frac{1}{i\hbar} \langle [X, H] \rangle = \frac{\langle P \rangle}{m} \tag{D-26-a}$$

$$\frac{d}{dt} \langle P \rangle = \frac{1}{i\hbar} \langle [P, H] \rangle = - m\omega^2 \langle X \rangle \tag{D-26-b}$$

If we integrate these equations, we obtain:

$$\langle X \rangle(t) = \langle X \rangle(0) \cos \omega t + \frac{1}{m\omega} \langle P \rangle(0) \sin \omega t$$

$$\langle P \rangle(t) = \langle P \rangle(0) \cos \omega t - m\omega \langle X \rangle(0) \sin \omega t \tag{D-27}$$

We again find the sinusoidal form indicated by formula (D-24).

COMMENT:

It is important to note that this analogy with the classical situation appears only when $| \psi(0) \rangle$ is a superposition of states $| \varphi_n \rangle$ of the type of (D-22), where several coefficients $c_n(0)$ are non-zero. If all these coefficients except one are equal to zero, the oscillator is in a stationary state and the mean values of all the observables are constant over time.

It follows that, in a stationary state $| \varphi_n \rangle$, the behavior of a harmonic oscillator is totally different from that predicted by classical mechanics, even if $n$ is very large (the limit of large quantum numbers). If we want to construct a wave packet whose average position oscillates over time, we must superpose different states $| \varphi_n \rangle$ (see complement $G_V$).

**References and suggestions for further reading :**

Dirac (1.13), §34; Messiah (1.17), chap. XII.

COMPLEMENTS OF CHAPTER V

| | |
|---|---|
| $A_V$: SOME EXAMPLES OF HARMONIC OSCILLATORS | $A_V$: demonstrates, in some examples chosen from various fields, the importance of the quantum mechanical harmonic oscillator in physics. Semi-quantitative and rather simple; recommended for a first reading. |
| $B_V$: STUDY OF THE STATIONARY STATES IN THE $\{\lvert \mathbf{r} \rangle\}$ REPRESENTATION. HERMITE POLYNOMIALS | $B_V$: technical study of the stationary wave functions of the harmonic oscillator. Intended to serve as a reference. |
| $C_V$: SOLVING THE EIGENVALUE EQUATION OF THE HARMONIC OSCILLATOR BY THE POLYNOMIAL METHOD | $C_V$: another method which yields the results of chapter V. Reveals the relation between energy quantization and the behavior of the wave functions at infinity. Moderately difficult. |
| $D_V$: STUDY OF THE STATIONARY STATES IN THE $\{\lvert \mathbf{p} \rangle\}$ REPRESENTATION. | $D_V$: shows that, in a stationary state of the harmonic oscillator, the momentum probability distribution has the same form as that of position. Fairly simple. |
| $E_V$: THE ISOTROPIC THREE-DIMENSIONAL HARMONIC OSCILLATOR | $E_V$: generalization of the results of chapter V to three dimensions. Recommended for a first reading, since it is simple and important. |
| $F_V$: A CHARGED HARMONIC OSCILLATOR PLACED IN A UNIFORM ELECTRIC FIELD | $F_V$: a direct and simple application of the results of chapter V (except for §3, which uses the translation operator introduced in complement $E_{II}$). Recommended for a first reading. |
| $G_V$: COHERENT "QUASI-CLASSICAL" STATES OF THE HARMONIC OSCILLATOR | $G_V$: detailed study of the "quasi-classical" states of the harmonic oscillator, which illustrates the relation between quantum and classical mechanics. Important because of its applications to the quantum theory of radiation. Moderately difficult; can be omitted in a first reading. |
| $H_V$: NORMAL VIBRATIONAL MODES OF TWO COUPLED HARMONIC OSCILLATORS | $H_V$: study, in the very simple case of two coupled harmonic oscillators, of normal vibrational modes of a system. Recommended, since it is simple and physically important. |

$J_V$: **VIBRATIONAL MODES OF AN INFINITE LINEAR CHAIN OF COUPLED HARMONIC OSCILLATORS; PHONONS**

$K_V$: **VIBRATIONAL MODES OF A CONTINUOUS PHYSICAL SYSTEM. APPLICATION TO RADIATION; PHOTONS**

$J_V$, $K_V$: introduction, using simplified models, of concepts which are particularly important in physics. Rather difficult (graduate level); can be reserved for later study.

$J_V$: determination of the normal vibrational modes of a linear chain of coupled oscillators, leading to the concept of the phonon, fundamental to solid state physics.

$K_V$: normal vibrational modes of a continuous system. A simple way to introduce photons in the quantum mechanical study of the electromagnetic field.

---

$L_V$: **THE ONE-DIMENSIONAL HARMONIC OSCILLATOR IN THERMODYNAMIC EQUILIBRIUM AT A TEMPERATURE *T***

$L_V$: application of the density operator (introduced in complement $E_{III}$) to a harmonic oscillator in thermodynamic equilibrium. Important from a physical point of view, but requires knowledge of $E_{III}$.

---

$M_V$: **EXERCISES**

# Complement $A_V$

# SOME EXAMPLES OF HARMONIC OSCILLATORS



We mentioned in the introduction to chapter V that the results obtained in the study of the harmonic oscillator are applicable to numerous cases in physics, especially those concerning small oscillations of a system about a position of stable equilibrium (where the potential energy is at a minimum). The aim of this complement is to describe some examples of such oscillations and to point out their physical importance: vibration of the nuclei in a diatomic molecule or a crystalline lattice, torsional oscillations in a molecule, motion of a muon $\mu^-$ inside a heavy nucleus. We do not intend to discuss these phenomena in great detail here. We shall confine ourselves to a simple, qualitative discussion.

## 1. Vibration of the nuclei of a diatomic molecule

### a. INTERACTION ENERGY OF TWO ATOMS

The formation of a molecule from two neutral atoms occurs because the interaction energy $V(r)$ of these two atoms has a minimum ($r$ is the distance between them). The form of $V(r)$ is shown in figure 1. When $r$ is very large, the two atoms do not interact and $V(r)$ approaches a constant which we shall choose as the energy origin. Then, as $r$ decreases, $V(r)$ varies approximately like $-1/r^6$: the corresponding attractive forces are the Van der Waals forces (which we shall study in complement $C_{XI}$). When $r$ becomes so small that the electronic wave functions overlap, $V(r)$ decreases faster and passes through a minimum at $r = r_e$; it then increases and becomes very large as $r$ approaches zero.

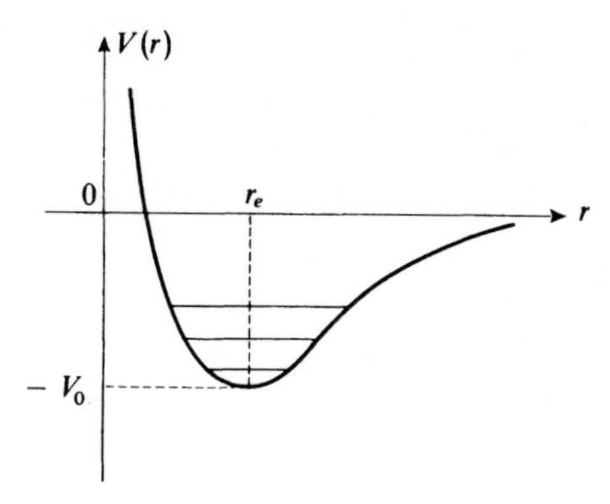


FIGURE 1

**Form of the interaction potential between two atoms which can form a stable molecule. Classically, $V_0$ is the dissociation energy of the molecule and $r_e$, the distance between the two nuclei in the equilibrium position. In quantum mechanics, one obtains vibrational states (the horizontal lines inside the well) whose energies are all greater than $-V_0$.**

The minimum of $V(r)$ is responsible for the phenomenon of the chemical bond which can form between the two atoms. We have already pointed out, in §C-2-c of chapter IV (taking the molecule $H_2^+$ as an example), that the cause of this lowering of the energy is a delocalization phenomenon of the electronic states (quantum resonance) which allows the electrons to profit from the attraction of the two nuclei. The rapid rise of $V(r)$ at small distances is due to the repulsion of the nuclei.

If the nuclei were classical particles, they would have stable equilibrium positions separated by $r = r_e$. The depth $V_0$ of the potential well at $r = r_e$ is called, classically, the dissociation energy of the molecule : it is, in fact, the energy that must be furnished to the two atoms in order to separate them. The larger $V_0$, the more stable the molecule.

The theoretical and experimental determination of the curve of figure 1 is a very important problem in atomic and molecular physics. We shall see that by studying the vibrations of the nuclei we get a certain amount of information about this curve.

COMMENT (the Born-Oppenheimer approximation) :

The quantum mechanical description of a diatomic molecule is actually a very complex problem; it involves finding the stationary states of a system of particles, the nuclei and the electrons, all interacting with each other. In general, it is impossible to solve the Schrödinger equation for such a system exactly. A significant simplication arises from the fact that the mass of the electrons is much smaller than that of the nuclei. It follows from this that one can, in a first approximation, study the two motions separately. One begins by determining the motion of the electrons for a fixed value of the distance $r$ between the two nuclei ; thus one obtains a series of stationary states for the electronic system, of energies $E_1(r)$, $E_2(r)$... Then one considers the ground state, of energy $E_1(r)$, of the electronic system; when $r$ varies because of the motion of the nuclei, the electronic system always remains in the ground state, for all $r$. This means that the system's wave function adapts itself instantaneously to any change in $r$ : the electrons, which are very mobile, are said to follow "adiabatically" the motion of the nuclei. In the study of this motion, the electronic energy $E_1(r)$ then plays the role of a potential energy of interaction between these two nuclei. This interaction potential depends on the distance between the nuclei, $r$, and adds to their electrostatic repulsion $Z_1Z_2\ e^2/r$

(where $Z_1$ and $Z_2$ are the atomic numbers of the two nuclei; we have set $e^2 = q^2/4\pi\varepsilon_0$, where $q$ is the charge of the electron). The total potential energy $V(r)$ of the system of the two nuclei, which enables us to determine their motion, is then :

$$V(r) = E_1(r) + \frac{Z_1 Z_2 e^2}{r} \tag{1}$$

It is this function that is shown in figure 1.

### b. MOTION OF THE NUCLEI

#### α. *Separation of the rotational and vibrational motions*

We are thus faced with a problem which involves the motion of two particles of masses $m_1$ and $m_2$, whose interaction is described by the potential $V(r)$ of figure 1, which depends only on the distance between them. The problem is complicated by the existence of several degrees of freedom : vibrational (variation of $r$) and rotational (variation of the polar angles $\theta$ and $\varphi$ which give the direction of the axis of the molecule). In addition, these degrees of freedom are coupled : when the molecule vibrates, its moment of inertia changes because of the variation of $r$, and the rotational energy is modified.

If we confine ourselves to small amplitude vibrations, it can be shown that the coupling between vibrational and rotational degrees of freedom is negligible since the relative variation of the moment of inertia is very small during the vibration. The problem is then reduced (as we shall see in detail in complement $F_{VII}$) to two independent problems : in the first place, the study of the rotation of a "dumbbell" composed of two masses $m_1$ and $m_2$ separated by a fixed distance $r_e$*; plus a one-dimensional problem (in which $r$ is the only variable) involving a fictitious particle whose mass $m$ is equal to the reduced mass of $m_1$ and $m_2$ (*cf.* chap. VII, § B):

$$m = \frac{m_1 m_2}{m_1 + m_2} \tag{2}$$

moving in the potential $V(r)$ of figure 1. We must then solve the eigenvalue equation:

$$\left[ -\frac{\hbar^2}{2m}\frac{\mathrm{d}^2}{\mathrm{d}r^2} + V(r) \right] \varphi(r) = E\,\varphi(r) \tag{3}$$

We shall concentrate on the latter problem here.

#### β. *Vibrational states*

If we confine ourselves to small amplitude oscillations, we can make a limited expansion of $V(r)$ in the neighborhood of its minimum, at $r = r_e$:

$$V(r) = -V_0 + \frac{1}{2} V''(r_e)(r - r_e)^2 + \frac{1}{6} V'''(r_e)(r - r_e)^3 + \dots \tag{4}$$

* We shall study this system (also called a "rigid rotator") quantum mechanically in complement $C_{VI}$, once we have introduced angular momentum.

The discussion in § A-2 of chapter V shows that if we neglect higher than second-order terms in expression (4), we are left with the equation of a one-dimensional harmonic oscillator centered at $r = r_e$, of angular frequency:

$$\omega = \sqrt{\frac{V''(r_e)}{m}} \tag{5}$$

The vibrational states $| \varphi_v \rangle$, shown by the horizontal lines in figure 1, therefore have energies given by:

$$E_v = \left(v + \frac{1}{2}\right)\hbar\omega - V_0 \tag{6}$$

where $v = 0, 1, 2, ...$ ($v$ is used instead of $n$ in the notation of molecular vibrations).

According to the discussion in § D-3 of chapter V, the mean value $\langle R \rangle$ of the distance between the two nuclei oscillates about $r_e$ with a frequency of $\omega/2\pi$ which can thus be seen to be the vibrational frequency of the molecule.

COMMENTS:

(*i*) Even in the ground state, the wave function of a harmonic oscillator has a finite spread, of the order of $\sqrt{\hbar/2m\omega}$ (*cf.* § D-2 of chapter V). The distance between the two nuclei of the molecule in the vibrational ground state is therefore defined only to within $\sqrt{\hbar/2m\omega}$. An important condition for the decoupling of the vibrational and rotational degrees of freedom is therefore that:

$$\sqrt{\frac{\hbar}{2m\omega}} \ll r_e \tag{7}$$

(*ii*) When the reduced mass $m$ is known, the measurement of $\omega$ yields, according to (5), the second derivative $V''(r_e)$. When the quantum number $v$ increases, it is no longer possible to neglect terms in $(r - r_e)^3$ in expression (4) (which indicate the deviation of the potential well from a parabolic form). The oscillator then becomes anharmonic. Studying the effects of the term in $(r - r_e)^3$ of (4) by perturbation theory (as we shall do in complement $A_{XI}$), one finds that the separation $E_{v+1} - E_v$ of two neighboring states is not the same for large and small values of $v$. Studying the variation of $E_{v+1} - E_v$ with respect to $v$ thus enables us to obtain the coefficient $V'''(r_e)$ of the term in $(r - r_e)^3$. Thus we see how the study of the frequencies of molecular vibration enables us to define more precisely the form of the curve $V(r)$ in the neighborhood of its minimum.

γ. *Order of magnitude of the vibrational frequencies*

Molecular vibrational frequencies are commonly expressed in $cm^{-1}$, that is, by giving the inverse of the wavelength $\lambda$ (expressed in cm) of an electromagnetic wave of the same frequency $\nu$. Note that 1 $cm^{-1}$ corresponds to a frequency of $3 \times 10^{10}$ Hertz and to an energy of $1.24 \times 10^{-4}$ eV.

The vibrational frequencies of diatomic molecules fall between several tens and several thousands of $cm^{-1}$. The corresponding wavelengths therefore go from a few microns to a few hundred microns, consequently falling in the infrared.

Formula (5) shows that as $m$ decreases, $\omega$ increases. $\omega$ also increases with $V''(r_e)$, that is, with a greater curvature of the potential well at $r = r_e$. Since $r_e$ is always of the same order of magnitude (a few Å), $V''(r_e)$ increases with the depth $V_0$ of the well: $\omega$ therefore increases with the chemical stability We shall consider some concrete illustrations of the preceding observations.

The vibrational frequencies of the hydrogen and deuterium molecules ($H_2$ and $D_2$) are, respectively (not taking into account anharmonicity corrections):

$$\begin{aligned} \nu_{H_2} &= 4\,401 \text{ cm}^{-1} \\ \nu_{D_2} &= 3\,112 \text{ cm}^{-1} \end{aligned} \tag{8}$$

The curve $V(r)$ is the same in these two cases : the chemical bond between the two atoms depends only on the electronic atmosphere. However, the reduced mass of $H_2$ is half as large as that of $D_2$. We must therefore have, according to (5), $\nu_{H_2} = \sqrt{2}\,\nu_{D_2}$. This is in agreement with the experimental values (8).

Now let us consider an example of two molecules which have about the same reduced mass but very different chemical stabilities. The molecule $^{79}Br^{85}Rb$ is chemically stable (halogen-alkaline bond); its vibrational frequency is 181 $\text{cm}^{-1}$. Molecules of $^{84}Kr^{85}Rb$ have been observed recently in optical pumping experiments. Their chemical stability is much lower, because krypton, which is a rare gas, is practically inert from a chemical point of view (in fact, the cohesion of the molecule is due only to Van der Waals forces). These molecules have been found to have a vibrational frequency of the order of 13 $\text{cm}^{-1}$. The considerable difference between this figure and the preceding one is due solely to the difference in chemical stability of the two types of molecules since the reduced masses are, to within a few per cent, practically the same.

### c. EXPERIMENTAL OBSERVATIONS OF NUCLEAR VIBRATION

We shall now explain how nuclear vibration can be detected experimentally. In particular, we shall consider the interaction of the molecule with an electromagnetic wave.

#### α. *Infrared absorption and emission*

First, let us assume the molecule to be *heteropolar* (composed of two different atoms). Since the electrons are attracted towards the more electronegative atom, the molecule generally has a permanent dipole moment $D(r)$ which depends on the distance $r$ between the two nuclei. Expanding $D(r)$ in the neighborhood of the equilibrium position $r = r_e$ :

$$D(r) = d_0 + d_1(r - r_e) + \dots \tag{9}$$

where $d_0$ and $d_1$ are real constants.

When the molecule is in a linear superposition $| \psi(t) \rangle$ of several stationary vibrational states $| \varphi_v \rangle$, the mean value $\langle \psi(t) | D(R) | \psi(t) \rangle$ of its electric dipole moment oscillates about the value $d_0$ with a frequency of $\omega/2\pi$. The oscillatory term arises from the mean value of the term $d_1(R - r_e)$ of (9) ($R - r_e$ plays the same role in our problem as the observable $X$ of the harmonic oscillator studied in §D-3 of chapter V). Now $(R - r_e)$ has a non-zero matrix element between two states $| \varphi_v \rangle$ and $| \varphi_{v'} \rangle$ only when $v - v' = \pm 1$. This selection rule enables us to understand why only one Bohr frequency $\omega/2\pi$, appears in the motion of $\langle D(R) \rangle(t)$ [the harmonic frequencies evidently appear when one takes into account the anharmonicity of the potential and terms of higher order in expansion (9); their intensity is however much weaker].

This vibration of the electric dipole moment results in a coupling between the molecule and the electromagnetic field; the molecule can consequently absorb or emit radiation of frequency $\nu$. In terms of photons, the molecule can absorb a photon of energy $h\nu$ and move from the state $| \varphi_v \rangle$ to the state $| \varphi_{v+1} \rangle$ (fig. 2-a) or emit a photon $h\nu$ by going from $| \varphi_v \rangle$ to $| \varphi_{v-1} \rangle$ (fig. 2-b).

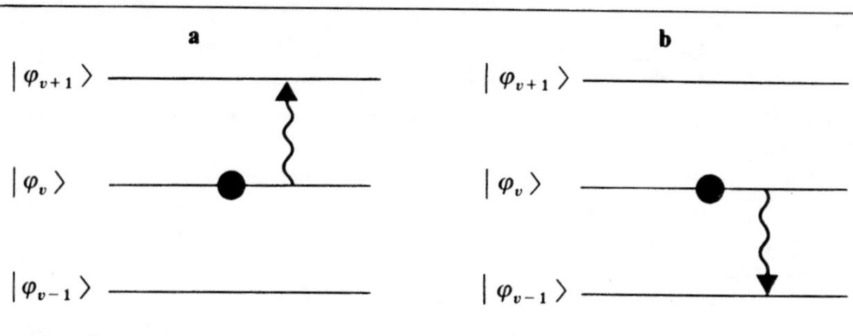


FIGURE 2

**Absorption (fig. a) or emission (fig. b) of a photon of energy $h\nu$ by a heteropolar molecule going from the vibrational state $| \varphi_v \rangle$ to the state $| \varphi_{v+1} \rangle$ or $| \varphi_{v-1} \rangle$.**

β. *The Raman effect*

Now let us consider a *homopolar* molecule (consisting of two identical atoms). Because of symmetry, the permanent electric dipole moment is then zero for all $r$, and the molecule is "inactive" in the infrared.

Imagine that an optical wave of frequency $\Omega/2\pi$ strikes this molecule. This frequency, much higher than those considered previously, is able to excite the electrons of the molecule; under the effect of the optical wave, the electrons will undergo forced oscillation and re-emit radiation of the same frequency in all directions. This is the well-known phenomenon of the molecular scattering of light (Rayleigh scattering)*. What new phenomena are produced by the vibration of the molecule?

What happens can be explained qualitatively in the following way. The electronic susceptibility** of the molecule is generally a function of the distance $r$ between the two nuclei. When $r$ varies (recall that this variation is slow compared to the motion of the electrons), the amplitude of the *induced* electric dipole moment, which vibrates at a frequency of $\Omega/2\pi$, varies. The time dependence of the dipole moment is therefore that of a sinusoid of frequency $\Omega/2\pi$ whose amplitude is modulated at the frequency of the molecular vibration $\omega/2\pi$, which

* In complement $A_{XIII}$, we shall use quantum mechanics to study the forced motion of the electrons of an atom under the effect of incident light waves.

** Under the effect of the field $\mathbf{E}_0\, e^{i\Omega t}$ of the incident optical wave, the electronic cloud of the molecule acquires an induced dipole moment $\mathbf{D}$ given by:

$$\mathbf{D} = \chi(\Omega)\, \mathbf{E}_0\, e^{i\Omega t}$$

$\chi(\Omega)$ is, by definition, the electronic susceptibility of the molecule. The important point here is that $\chi$ depends on $r$.

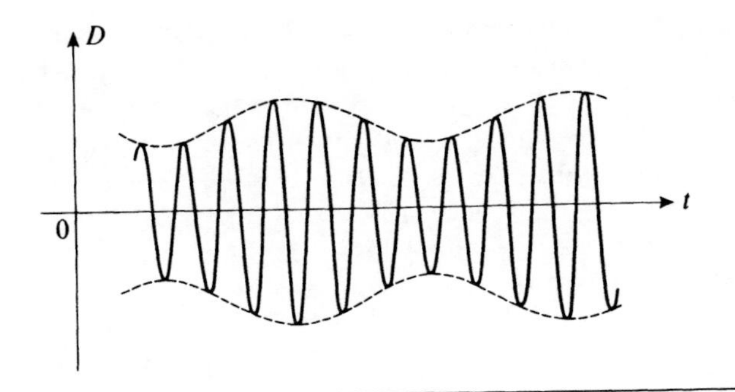


FIGURE 3

**The vibration of a molecule modulates the amplitude of the oscillating electric dipole induced by an incident light wave.**

is much smaller (fig. 3). The frequency distribution of the light emitted by the molecule is given by the Fourier transform of the motion of the electric dipole shown in figure 3. It is easy to see (fig. 4) that there exists a central line of frequency $\Omega/2\pi$ (Rayleigh scattering) and two shifted lines, of frequency $(\Omega - \omega)/2\pi$ (Raman-Stokes scattering) and frequency $(\Omega + \omega)/2\pi$ (Raman-anti-Stokes scattering).

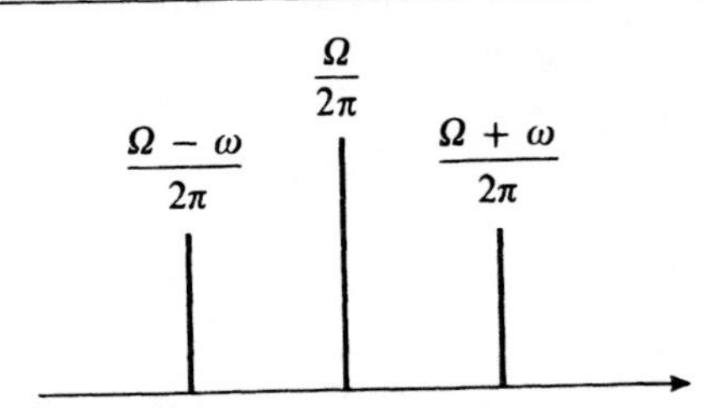


FIGURE 4

**Spectrum of the oscillations shown in figure 3. In addition to the central line, whose frequency is the same as that of the incident light wave (Rayleigh line), two shifted lines appear (the Raman-Stokes and Raman-anti-Stokes lines). The frequency shift is equal to the vibrational frequency of the molecule.**

It is very simple to interpret these lines in terms of photons. Consider an optical photon of energy $\hbar\Omega$ which strikes the molecule when it is in the state $| \varphi_v \rangle$ (fig. 5-a). If the molecule does not change vibrational states during the

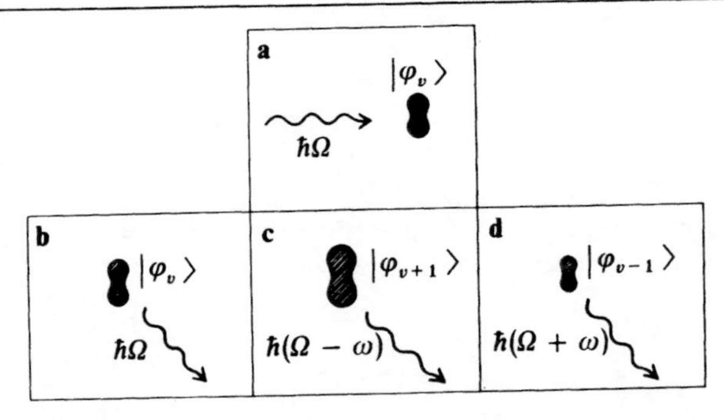


FIGURE 5

**Schematic representation of the scattering of a photon of energy $\hbar\Omega$ by a molecule which is initially in the vibrational state $| \varphi_v \rangle$ (fig. a) : Rayleigh scattering without a change in the vibrational state (fig. b); Raman-Stokes or Raman-anti-Stokes scattering with a change in the molecule's state from $| \varphi_v \rangle$ to $| \varphi_{v+1} \rangle$ (fig. c) or to $| \varphi_{v-1} \rangle$ (fig. d).**

scattering process, the scattering is elastic. Because of conservation of energy, the scattered photon has the same energy as the incident photon (fig. 5-b; Rayleigh line). However, the molecule, during the scattering process, can make a transition from the state $| \varphi_v \rangle$ to the state $| \varphi_{v+1} \rangle$. The molecule acquires an energy $\hbar\omega$ at the expense of the scattered photon, whose energy therefore is $\hbar(\Omega - \omega)$ (fig. 5-c): the scattering is *inelastic* (Raman-Stokes line). Finally, the molecule may move from the state $| \varphi_v \rangle$ to the state $| \varphi_{v-1} \rangle$, in which case the scattered photon will have an energy of $\hbar(\Omega + \omega)$ (fig. 5-d; Raman-anti-Stokes line).

COMMENTS:

(*i*) The Raman effect can also be observed with heteropolar molecules.

(*ii*) The Raman effect has recently enjoyed a revival of interest because of the development of lasers. If, in the cavity of a laser oscillating at a frequency of $\Omega/2\pi$, one places a cell filled with a substance which exhibits the Raman effect, one can, in certain cases, obtain an amplification (stimulated Raman effect) and hence a laser oscillation at the frequency $(\Omega - \omega)/2\pi$, where $\omega$ is the vibrational frequency of the molecules in the cell (Raman laser). Thus, by varying this substance, one can vary the oscillation frequency of the laser.

(*iii*) The study of Raman and infrared spectra of molecules is useful in chemistry because it permits the identification of the various bonds which exist in a complex molecule. For example, the vibration frequency of a group of two carbon atoms depends on whether the bond between them is single, double or triple.

## 2. Vibration of the nuclei in a crystal

### a. THE EINSTEIN MODEL

A crystal consists of a system of atoms (or ions) which are regularly distributed in space, forming a periodic lattice. For simplicity, let us choose a one-dimensional model in which we consider a linear chain of atoms. The average position of the nucleus of the $q$th atom is:

$$x_q^0 = qd \tag{10}$$

where $d$ is the distance between adjacent atoms (on the order of a few Å).

Let $U(x_1, x_2, ..., x_q, ...)$ be the total potential energy of the crystal nuclei, which depends on their positions $x_1, x_2, ..., x_q, ...$ If $x_q - x_q^0$ is not too large, that is, if each nucleus is not too far from its equilibrium position, $U(x_1, x_2, ..., x_q, ...)$ has, in certain cases, the following simple form:

$$U(x_1, x_2, ... x_q, ...) \simeq U_0 + \sum_q \frac{1}{2}(x_q - x_q^0)^2 U_0'' + ... \tag{11}$$

where $U_0$ and $U_0''$ are real constants (with $U_0'' > 0$). The absence of terms linear in $x_q - x_q^0$ shows that $x_q^0$ is a stable equilibrium position for the nucleus ($q$) (a minimum of $U$). We add to (11) the total kinetic energy:

$$T = \sum_q \frac{p_q^2}{2m} \tag{12}$$

where $p_q$ is the momentum of the nucleus ($q$) of mass $m$. The total Hamiltonian $H$ of the system is, to within the constant $U_0$, a sum of Hamiltonians of one-dimensional harmonic oscillators centered at each nucleus ($q$):

$$H = U_0 + \sum_q \left[ \frac{p_q^2}{2m} + \frac{1}{2}(x_q - x_q^0)^2 U_0'' \right] \tag{13}$$

Consequently, in this simplified model, each nucleus vibrates about its equilibrium position independently of its neighbors, with an angular frequency:

$$\omega = \sqrt{\frac{U_0''}{m}} \tag{14}$$

As in the case of the diatomic molecule, $\omega$ increases when $m$ decreases and when the curvature of the potential attracting the nucleus towards its equilibrium position increases.

COMMENT:

In the simple model which we have just presented, each nucleus vibrates independently of the others. This is because the proposed potential $U$ does not contain any terms which are simultaneously dependent on more than one of the variables $x_q$, as it would if it described internuclear interactions. This model is not realistic since such interactions do, in fact, exist. In complement $J_V$, we shall present a more elaborate model which takes into account the coupling between each nucleus and its two nearest neighbors. We shall see that it is still possible, in this model, to put the total Hamiltonian of the system in the form of a sum of Hamiltonians of independent harmonic oscillators.

### b. THE QUANTUM MECHANICAL NATURE OF CRYSTALLINE VIBRATIONS

Despite its very schematic character, the Einstein model enables us to understand a certain number of phenomena related to the quantum mechanical nature of crystalline vibrations. The low temperature behavior of the constant volume specific heat, which cannot be explained using classical mechanics, will be described in complement $L_V$ in connection with the study of the properties of a harmonic oscillator in thermodynamic equilibrium. Here, we shall discuss a spectacular effect related to the finite spread of the wave functions associated with the position of each atom in the ground state.

At absolute zero, under a pressure of one atmosphere, all substances except helium are solids. To solidify helium, it is necessary to apply a pressure of at least 25 atmospheres. Can this peculiarity be explained qualitatively?

First let us try to understand the phenomenon of the melting of an ordinary substance. At absolute zero, the atoms are practically localized at their equilibrium positions; the spread of their wave functions about the $x_q^0$ is given by [*cf.* formula (D-5-a) of chapter V]:

$$\Delta X \simeq \sqrt{\frac{\hbar}{2m\omega}} = \left[ \frac{\hbar^2}{4mU_0''} \right]^{1/4} \tag{15}$$

[where we have used expression (14) for $\omega$]. $\Delta X$ is, in general, very small. When the crystal is heated, the nuclei move into higher and higher vibrational states: in classical language, they vibrate with a larger and larger amplitude; in quantum mechanical language, the spread of their wave functions increases [with the square root of the vibrational quantum number; see formula (D-5-a) of chapter V]. When this spread is no longer negligible with respect to the interatomic distance $d$, the crystal melts (see §4-c of complement $L_V$, in which this phenomenon is studied more quantitatively).

It is impossible to solidify helium at ordinary pressures. This corresponds to the fact that, even at absolute zero, the spread of the wave function given by (15) is not negligible compared to $d$. This results from the fact that the mass of helium is very small and its chemical affinity, very weak (the curvature $U_0''$ of the potential in the neighborhood of each minimum is very small, since the potential wells are very shallow). The effect of both of these factors, in formula (15), is the same: a large spread $\Delta X$. An increase in the pressure results in an increase in $U_0''$ and therefore, in $\omega$; consequently, $\Delta X$ decreases. This is due to the fact that, at high pressures, each helium atom is "wedged" between its neighbors: the smaller the average distance between these neighbors (the higher the pressure), the sharper the potential minimum (the greater $U_0''$). Thus we see how an increase in the pressure makes the solidification of helium possible.

## 3. Torsional oscillations of a molecule: ethylene

### a. STRUCTURE OF THE ETHYLENE MOLECULE $C_2H_4$

The structure of the molecule $C_2H_4$ is well-known: the six atoms of the molecule are in the same plane (fig. 6) and the angles between the various $C-H$ and $C-C$ bonds are close to 120°.

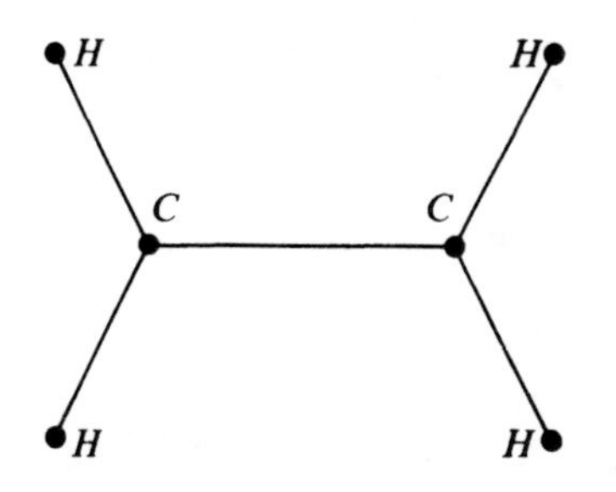


FIGURE 6

**Plane structure of the ethylene molecule.**

Now imagine that, without changing the relative positions of the bonds of each carbon atom, we rotate one of the $CH_2$ groups, about the $C-C$ axis, through an angle $\alpha$ with respect to the other one. Figure 7 represents the molecule as seen along the $C-C$ axis: the $C-H$ bonds of one $CH_2$ group are shown in solid lines and those of the other one, in dashed lines. How does the potential energy $V(\alpha)$ of the molecule vary with respect to $\alpha$?

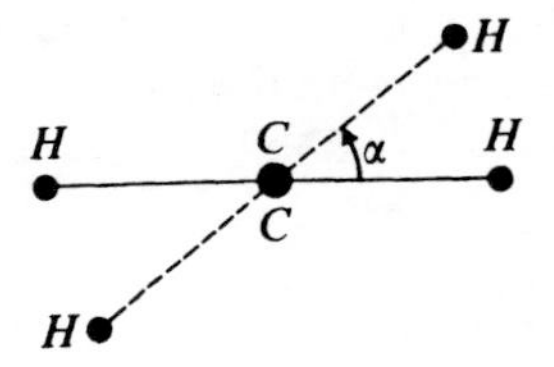


FIGURE 7

**Torsion of the ethylene molecule (seen along the C-C axis) : one of the $CH_2$ groups has rotated with respect to the other one through an angle $\alpha$ about the C-C axis.**

Since the stable structure of the molecule is planar, the angle $\alpha = 0$ must correspond to a minimum of $V(\alpha)$. It is also clear that $\alpha = \pi$ corresponds to another minimum of $V(\alpha)$, since the two structures associated with $\alpha = 0$ and $\alpha = \pi$ are undistinguishable. $V(\alpha)$ therefore has the form shown in figure 8 [$\alpha$ varies from $-\pi/2$ to $3\pi/2$; $V(0)$ is chosen as the energy origin].

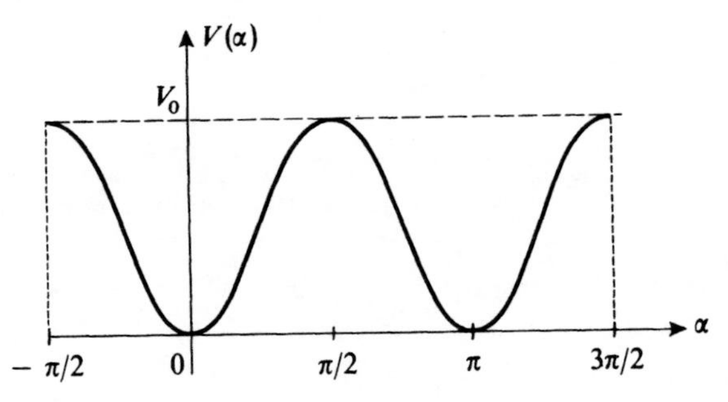


FIGURE 8

**The potential energy of the molecule depends on the torsion angle $\alpha$; $V(\alpha)$ is minimal for $\alpha = 0$ and $\alpha = \pi$ (planar structures).**

The two stable positions $\alpha = 0$ and $\alpha = \pi$ are separated by a potential barrier of height $V_0$. The potential of figure 8 is often approximated by the simple formula:

$$V(\alpha) = \frac{V_0}{2}(1 - \cos 2\alpha) \tag{16}$$

COMMENT:

Quantum mechanics enables us to interpret all the features of the $C_2H_4$ molecule which we have just described. In this molecule, each carbon atom has four valence electrons. Three of these electrons ($\sigma$ electrons) are found to have wave functions which are symmetrical about three coplanar lines making angles of 120° with each other, and defining the directions of the chemical bonds (fig. 6). These wave functions overlap those of the electrons of the neighboring atoms to a considerable extent, and it is this overlap that insures the stability of the C−H bonds and of part of the C−C bond (this phenomenon is called "$sp^2$ hybridation" and will be studied in greater detail in complement $E_{VII}$). The last valence electron of each carbon atom ($\pi$ electron) has a wave function which is symmetrical about a line passing through C and perpendicular to the plane defined by C and its three neighbors. The overlap of the wave functions of the two $\pi$ electrons is maximum and, consequently, the chemical stability of the double bond is greatest when the two lines associated with the $\pi$ electrons are parallel, that is, when the six atoms of the molecules are in the same plane. The structure of figure 6 is thus entirely explained.

Since $V(\alpha)$ can be approximated by a parabola in the neighborhood of its two minima, the molecule performs, about its two stable equilibrium positions, torsional oscillations which we shall now examine. First, we shall review rapidly the corresponding classical equations.

### b. CLASSICAL EQUATIONS OF MOTION

We denote by $\alpha_1$ and $\alpha_2$ the angles formed by the planes of the two $CH_2$ groups with a fixed plane passing through the C–C axis (fig. 9). The angle in figure 7 is obviously:

$$\alpha = \alpha_1 - \alpha_2 \tag{17}$$

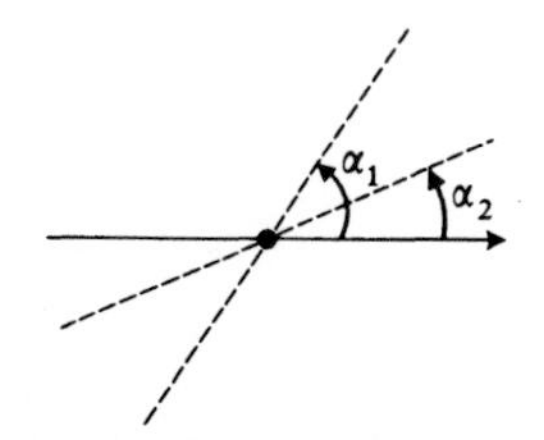


FIGURE 9

**To write the classical equations of motion, we denote by $\alpha_1$ and $\alpha_2$ the angles formed by the planes of the two $CH_2$ groups with a fixed plane.**

Let $I$ be the moment of inertia of one of the $CH_2$ groups with respect to the C–C axis. Since the potential energy depends only on $\alpha = \alpha_1 - \alpha_2$, the dynamical equations describing the rotation of each group are written:

$$\begin{cases} I \dfrac{\mathrm{d}^2\alpha_1}{\mathrm{d}t^2} = - \dfrac{\partial}{\partial \alpha_1} V(\alpha_1 - \alpha_2) = - \dfrac{\mathrm{d}}{\mathrm{d}\alpha} V(\alpha) \\ I \dfrac{\mathrm{d}^2\alpha_2}{\mathrm{d}t^2} = - \dfrac{\partial}{\partial \alpha_2} V(\alpha_1 - \alpha_2) = + \dfrac{\mathrm{d}}{\mathrm{d}\alpha} V(\alpha) \end{cases} \tag{18}$$

Adding and subtracting these two equations we obtain:

$$\frac{\mathrm{d}^2}{\mathrm{d}t^2}(\alpha_1 + \alpha_2) = 0 \tag{19-a}$$

$$I \frac{\mathrm{d}^2\alpha}{\mathrm{d}t^2} = - 2 \frac{\mathrm{d}}{\mathrm{d}\alpha} V(\alpha) \tag{19-b}$$

Equation (19-a) indicates that the entire molecule can rotate freely about the C–C axis independently of the torsional motion: the angle $(\alpha_1 + \alpha_2)/2$ of the plane bisecting the planes of the two $CH_2$ groups is a linear function of time. Equation (19-b) describes the torsional motion (rotation of one group with respect to the other). Let us consider this motion in the immediate neighborhood of one of the stable equilibrium positions, $\alpha = 0$. Expanding expression (16) in the neighborhood of $\alpha = 0$:

$$V(\alpha) \simeq V_0 \, \alpha^2 \tag{20}$$

Substituting (20) into (19-b), we obtain:

$$\frac{d^2\alpha}{dt^2} + \frac{4V_0}{I}\alpha = 0 \tag{21}$$

We recognize (21) as the equation of a one-dimensional harmonic oscillator ($\alpha$ is the only variable) of angular frequency:

$$\omega_t = 2\sqrt{\frac{V_0}{I}} \tag{22}$$

For the $C_2H_4$ molecule, $\omega_t$ is of the order of 825 $cm^{-1}$.

### c. QUANTUM MECHANICAL TREATMENT

In the neighborhood of its two equilibrium positions $\alpha = 0$ and $\alpha = \pi$, the molecule possesses "torsional states" of quantized energy $E_n = (n + 1/2)\hbar\omega_t$, with $n = 0, 1, 2, ...$ In a first approximation, each energy level $E_n = (n + 1/2)\hbar\omega_t$ is therefore doubly degenerate, since for each one there are two states $|\varphi_n\rangle$ and $|\varphi'_n\rangle$ whose wave functions $\varphi_n(\alpha)$ and $\varphi'_n(\alpha)$ differ only in that one is centered at $\alpha = 0$ and the other, at $\alpha = \pi$ (fig. 10-a and 10-b).

In fact, we must also take into account a typically quantum mechanical effect: the tunnel effect across the potential barrier separating the two minima (fig. 8).

a

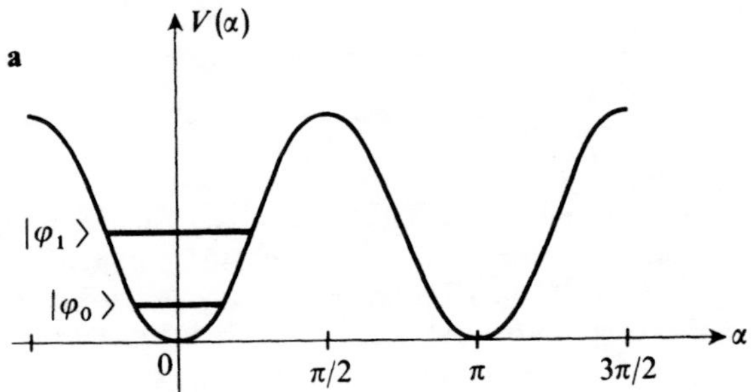



b

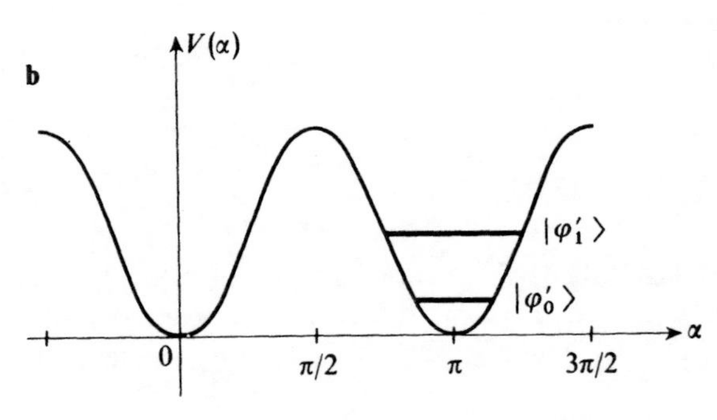



FIGURE 10

**When one neglects the tunnel effect across the potential barriers at $\alpha = \pi/2$ and $\alpha = 3\pi/2$, one can find torsional states of the molecule localized in the wells centered at $\alpha = 0$ (fig. a) and $\alpha = \pi$ (fig. b).**

We have already encountered a situation of this type, in complement $G_{IV}$, in connection with the inversion of the $NH_3$ molecule. Calculations analogous to those in that complement would enable us to show here that the degeneracy between the two states $|\varphi_n\rangle$ and $|\varphi'_n\rangle$ is removed by the tunnel effect. Thus, for each value of $n$, two stationary states, $|\psi^n_+\rangle$ and $|\psi^n_-\rangle$, appear (to a first

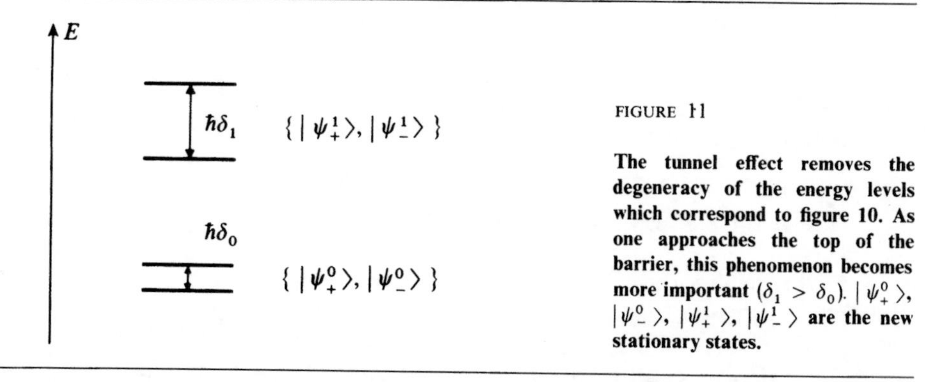


FIGURE 11

**The tunnel effect removes the degeneracy of the energy levels which correspond to figure 10. As one approaches the top of the barrier, this phenomenon becomes more important ($\delta_1 > \delta_0$). $|\psi^0_+\rangle$, $|\psi^0_-\rangle$, $|\psi^1_+\rangle$, $|\psi^1_-\rangle$ are the new stationary states.**

approximation, they are symmetrical and antisymmetrical linear combinations of $|\varphi_n\rangle$ and $|\varphi'_n\rangle$). The larger $n$ (that is, the closer the initial energy $E_n$ is to $V_0$, and hence, the more important the tunnel effect), the greater their energy difference $\hbar\delta_n$. However, $\hbar\delta_n$ is always much smaller than the distance $\hbar\omega_t$ between adjacent levels $n$ and $n \pm 1$ (fig. 11).

For the mean value of the angle $\alpha$, quantum mechanics therefore predicts the following motion : rapid oscillations of frequency $\omega_t$ about one of the two values $\alpha = 0$ and $\alpha = \pi$, upon which are superposed much slower oscillations between $\alpha = 0$ and $\alpha = \pi$, at the Bohr frequencies $\delta_0/2\pi$, $\delta_1/2\pi$, $\delta_2/2\pi$...

COMMENT :

States of course exist for which the energy is greater than the maximum height $V_0$ of the potential barrier of figure 8. These states correspond to a rotational kinetic energy which is large enough for one of the $CH_2$ groups to be considered as rotating almost freely with respect to the other one (while being, nevertheless, periodically slowed down and accelerated by the potential of figure 8).

The ethane molecule $C_2H_6$ behaves in this way. The absence of $\pi$ electrons in this molecule permits a much freer rotation of one of the $CH_3$ groups with respect to the other (the potential barrier $V_0$ is much lower). In this case, the potential $V(\alpha)$, which tends to oppose the free rotation of one of the $CH_3$ groups with respect to the other, has a period of $2\pi/3$ because of symmetry.

## 4. Heavy muonic atoms

The muon $\mu^-$ (sometimes called, for historical reasons, the "$\mu$ meson") is a particle which has the same properties as the electron except that its mass is 207 times greater*. In particular, it is not sensitive to strong interactions, and

* The muon is unstable : it decays into an electron and two neutrinos.

its coupling with nuclei is essentially electromagnetic. A muon $\mu^-$ which has been slowed down in matter can be attracted by the Coulomb field of an atomic nucleus and can form a bound state with the nucleus. The system thus constituted is called a muonic atom.

### a. COMPARISON WITH THE HYDROGEN ATOM

In chapter VII (§C), we shall study the bound states of two particles of opposite charge and, in particular, those of the hydrogen atom. We shall see that the results of quantum mechanics concerning the energies of bound states are the same as those of the Bohr model (chap. VII, §C-2). Similarly, the spread of the wave functions which describe these bound states is of the order of the Bohr orbital radius. Let us therefore begin by using this simple model to calculate the energies and spreads of the first bound states of a muon $\mu^-$ in the Coulomb field of a heavy atom such as lead ($Z = 82$, $A = 207$).

If we consider the nucleus to be infinitely heavy, the $n$th Bohr orbital has an energy of:

$$E_n = -\frac{Z^2 m e^4}{2\hbar^2}\frac{1}{n^2} \tag{23}$$

where $Z$ is the atomic number of the nucleus, $e^2 = q^2/4\pi\varepsilon_0$ (where $q$ is the electron charge), and $m$ represents the mass of the electron or of the muon, depending on the case. When one goes from hydrogen to the muonic atom under study here, $E_n$ is multiplied by a factor of $Z^2 m_\mu/m_e = (82)^2 \times (207) = 1.4 \times 10^6$. From this we deduce that, for the muonic atom :

$$\begin{cases} E_1 = -\ 19\ \text{MeV} \\ E_2 = -\ 4.7\ \text{MeV} \end{cases} \tag{24}$$

The radius of the $n$th Bohr orbital is given by:

$$r_n = \frac{n^2\hbar^2}{Zme^2} \tag{25}$$

For hydrogen, $r_1 \simeq 0.5$ Å. Here, this number must be divided by $Zm_\mu/m_e$, which gives:

$$\begin{cases} r_1 = 3 \times 10^{-13}\ \text{cm} \\ r_2 = 12 \times 10^{-13}\ \text{cm} \end{cases} \tag{26}$$

In the preceding calculations, we have implicitly assumed the nucleus to be pointlike (in the Bohr model and in the theory presented in chapter VII, §C, the potential energy is taken equal to $-Ze^2/r$). The small values found for $r_1$ and $r_2$ [formulas (26)] show us that this viewpoint is not at all valid for a heavy muonic atom. The lead nucleus has a non-negligible radius $\rho_0$, on the order of $8.5 \times 10^{-13}$ cm (recall that the radius of a nucleus increases with $A^{1/3}$). The preceding qualitative calculation therefore leaves us with the impression that the

spread of the wave functions of the muon may be smaller than the nucleus★. Consequently, we must reconsider the problem completely and first calculate the potential "seen" by the muon on the *inside* as well as on the outside of the nuclear charge distribution.

### b. THE HEAVY MUONIC ATOM TREATED AS A HARMONIC OSCILLATOR

We shall use a rough model of the lead nucleus: we shall assume its charge to be evenly distributed throughout a sphere of radius $\rho_0 = 8.5 \times 10^{-13}$ cm.

When the distance $r$ of the muon from the center of this sphere is greater than $\rho_0$, its potential energy is given by:

$$V(r) = -\frac{Ze^2}{r} \qquad \text{for} \qquad r \geqslant \rho_0 \tag{27}$$

For $r < \rho_0$, one can calculate the electrostatic force acting on the muon, using Gauss' theorem; it is directed towards the center of the sphere and its absolute value is:

$$Ze^2\left(\frac{r}{\rho_0}\right)^3 \frac{1}{r^2} = \frac{Ze^2}{\rho_0^3} r \tag{28}$$

This force is derived from the potential energy:

$$V(r) = \frac{1}{2}\frac{Ze^2}{\rho_0^3} r^2 + C \qquad \text{for} \qquad r \leqslant \rho_0 \tag{29}$$

The constant $C$ is determined by the condition that expressions (27) and (29) be identical for $r = \rho_0$:

$$C = -\frac{3}{2}\frac{Ze^2}{\rho_0} \tag{30}$$

Figure 12 represents the potential energy of the muon, plotted with respect to $r$.

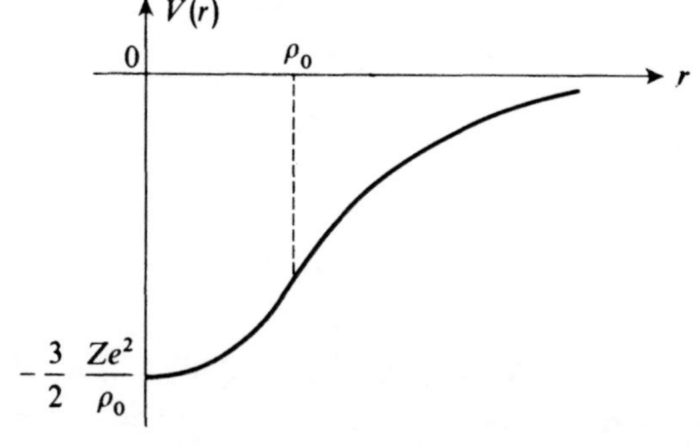


FIGURE 12

**Form of the potential $V(r)$ seen by a muon $\mu^-$ attracted by a nucleus of radius $\rho_0$ situated at $r = 0$. When $r < \rho_0$, the variation of the potential is parabolic (if the charge density of the nucleus is uniform); when $r > \rho_0$, $V(r)$ varies like $1/r$ (Coulomb's law).**

★ For hydrogen, the spread of the wave functions, on the order of an Angström, is about $10^5$ times larger than the dimensions of the proton, which can therefore be treated like a point. The new situation encountered here results from several factors which reinforce each other: increased $m$ and increased $Z$, which results in a greater electrostatic force and a larger nuclear radius.

Inside the nucleus, the potential is parabolic. The orders of magnitude we calculated in §a indicate that it would not be realistic to choose a pure Coulomb potential for the ground state of the muonic lead atom since the wave function is actually concentrated in the region where the potential is parabolic. It is therefore certainly preferable to consider the muon to be "elastically bound" to the nucleus in this case. We then have a three-dimensional harmonic oscillator (complement $E_V$) whose angular frequency is:

$$\omega = \sqrt{\frac{Ze^2}{m_\mu \rho_0^3}} \tag{31}$$

In fact, we shall see that the wave function of the ground state of this harmonic oscillator is not zero outside the nucleus, so the harmonic approximation is not perfect either.

COMMENT:

It is interesting that the physical system studied here presents many analogies with the first atomic model, proposed by J. J. Thomson. This physicist assumed the positive charge of the atom to be distributed in a sphere whose radius was of the order of a few Angströms, with the electrons moving in the parabolic potential existing inside this charge distribution (model of the elastically bound electron). We know from Rutherford's experiments that the nucleus is much smaller and that such a model does not correspond to reality for atoms.

### c. ORDER OF MAGNITUDE OF THE ENERGIES AND SPREAD OF THE WAVE FUNCTIONS

If we substitute into expression (31) the numerical values:

$Z = 82$ $\qquad c = 3 \times 10^8$ m/sec

$\frac{e^2}{\hbar c} \simeq \frac{1}{137}$ $\qquad m_\mu = 207\, m_e = 1.86 \times 10^{-28}$ kg

$\hbar \simeq 1.05 \times 10^{-34}$ Joule . sec $\qquad \rho_0 = 8.5 \times 10^{-15}$ m

we find:

$$\omega \simeq 1.3 \times 10^{22} \text{ rad. sec}^{-1} \tag{32}$$

which corresponds to an energy $\hbar\omega$ on the order of:

$$\hbar\omega \simeq 8.4 \text{ MeV} \tag{33}$$

We can compare $\hbar\omega$ to the total depth of the well $\frac{3}{2}\frac{Ze^2}{\rho_0}$, which is equal to:

$$\frac{3}{2}\frac{Ze^2}{\rho_0} \simeq 21 \text{ MeV} \tag{34}$$

We see that $\hbar\omega$ is smaller than this depth, but not small enough for us to be able to neglect completely the non-parabolic part of $V(r)$.

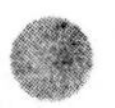



Similarly, the spread of the ground state, if the well were perfectly parabolic, would be on the order of:

$$\sqrt{\frac{\hbar}{2m_\mu\omega}} \simeq 4.7 \times 10^{-13} \text{ cm} \tag{35}$$

The qualitative predictions of § 4-a are therefore confirmed : a large part of the wave function of the muon is inside the nucleus. Nevertheless, what happens outside cannot be completely neglected.

The exact calculation of the energies and the wave functions is therefore more complicated than it would be for a simple harmonic oscillator. The Schrödinger equation corresponding to the potential of figure 12 must be solved (taking into account, in addition, spin, relativistic corrections, etc...). Such a calculation is important : the study of the energy of photons emitted by a heavy muonic atom contributes information about the structure of the nucleus, for example concerning the real charge distribution inside the nuclear volume.

COMMENT :

In the case of ordinary atoms (with an electron instead of a muon), it is valid to neglect the effects of the deviation of the potential from the $-Ze^2/r$ law. However, one can take this deviation into account by using perturbation theory (*cf.* chap. XI). In complement $D_{XI}$, we shall study this "volume effect" of the nucleus on the atomic energy levels.

**References and suggestions for further reading :**

Molecular vibrations: Karplus and Porter (12.1), chap. 7; Pauling and Wilson (1.9), chap. X; Herzberg (12.4), Vol. I, chap. III, § 1; Landau and Lifshitz (1.19), chaps. XI and XIII.

Stimulated Raman effect: Baldwin (15.19), § 5.2; see also Schawlow's article (15.17).

Torsion oscillations: Herzberg (12.4), Vol. II, chap. II, § 5 d; Kondratiev (11.6), § 37

The Einstein model: Kittel (13.2), chap. 6; Seitz (13.4), chap. III; Ziman (13.3), chap. 2; see also Bertman and Guyer's article (13.20).

Muonic atoms: Cagnac and Pebay-Peyroula (11.2), § XIX-7; Weissenberg (16.19), § 4-2; see also De Benedetti's article (11.21).

## Complement $B_V$

# STUDY OF THE STATIONARY STATES IN THE $\{ | x \rangle \}$ REPRESENTATION. HERMITE POLYNOMIALS



We now intend to study, in a little more detail than in §C-2 of chapter V, the wave functions $\varphi_n(x) = \langle x | \varphi_n \rangle$ associated with the stationary states $| \varphi_n \rangle$ of the harmonic oscillator. Before undertaking this study, we shall define the Hermite polynomials and mention their principal properties.

## 1. Hermite polynomials

### a. DEFINITION AND SIMPLE PROPERTIES

Consider the Gaussian function:

$$F(z) = e^{-z^2} \tag{1}$$

represented by the bell-shaped curve in figure 1. The successive derivatives of $F$ are given by:

$$F'(z) = -2z\, e^{-z^2} \tag{2}$$

$$F''(z) = (4z^2 - 2)\, e^{-z^2} \tag{3}$$

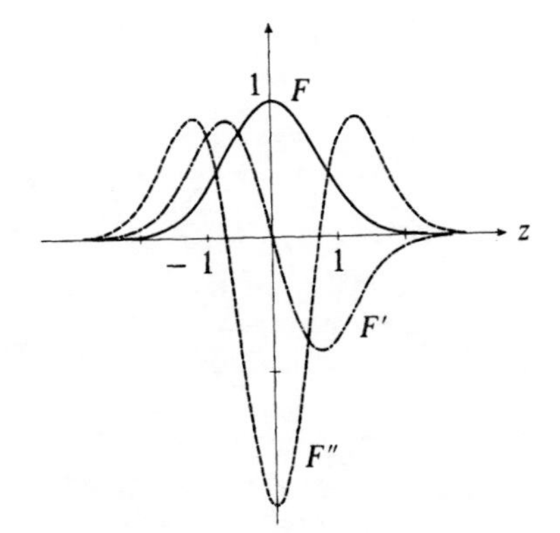


FIGURE 1

**Shape of the Gaussian function $F(z)$ and of its first and second derivatives $F'(z)$ and $F''(z)$.**

The $n$th-order derivative, $F^{(n)}(z)$, can be written :

$$F^{(n)}(z) = (-1)^n H_n(z) \, e^{-z^2} \tag{4}$$

where $H_n(z)$ is an $n$th-degree polynomial in $z$. The proof is by recurrence. This relation is valid for $n = 1, 2$ [*cf.* equations (2) and (3)]. Assume it is true for $n - 1$:

$$F^{(n-1)}(z) = (-1)^{n-1} H_{n-1}(z) \, e^{-z^2} \tag{5}$$

where $H_{n-1}(z)$ is a polynomial of degree $n - 1$. We then obtain relation (4) directly by differentiation, if we set:

$$H_n(z) = \left(2z - \frac{d}{dz}\right) H_{n-1}(z) \tag{6}$$

Since $H_{n-1}(z)$ is a polynomial of degree $n - 1$ in $z$, we see from this last relation that $H_n(z)$ is indeed an $n$th-degree polynomial. The polynomial $H_n(z)$ is called the *nth-degree Hermite polynomial*. Its definition is therefore:

$$H_n(z) = (-1)^n \, e^{z^2} \frac{d^n}{dz^n} e^{-z^2} \tag{7}$$

We see from (2) and (3) that $H_1(z)$ and $H_2(z)$ are, respectively, even and odd. Moreover, relation (6) shows that if $H_{n-1}(z)$ has a definite parity, $H_n(z)$ has the opposite parity. From this, we deduce that the parity of $H_n(z)$ is $(-1)^n$.

The zeros of $H_n(z)$ correspond to those of the $n$th-order derivative of the function $F(z)$. We are going to show that $H_n(z)$ has $n$ real zeros, between which one finds those of $H_{n-1}$. It can be seen from figure 1 and from relations (1), (2) and (3) that this is true for $n = 0, 1, 2$. Arguing by recurrence, we can generalize this result: assume that $H_{n-1}(z)$ has $n - 1$ real zeros; if $z_1$ and $z_2$ are two consecutive zeros of $H_{n-1}(z)$ and therefore of $F^{(n-1)}(z)$, Rolle's theorem shows that the derivative $F^{(n)}(z)$ of $F^{(n-1)}(z)$ goes to zero at a point $z_3$ between $z_1$ and $z_2$; therefore, $H_n(z_3) = 0$. Since, in addition, $F^{(n-1)}(z)$ goes to zero when $z \longrightarrow -\infty$ and when $z \longrightarrow +\infty$, $F^{(n)}(z)$ and $H_n(z)$ have at least $n$ real zeros [and not more, because $H_n(z)$ is $n$th-degree] between which are interposed those of $H_{n-1}(z)$.

### b. GENERATING FUNCTION

Consider the function of $z$ and $\lambda$:

$$F(z + \lambda) = e^{-(z+\lambda)^2} \tag{8}$$

Taylor's formula enables us to write:

$$\begin{aligned} F(z + \lambda) &= \sum_{n=0}^{\infty} \frac{\lambda^n}{n!} F^{(n)}(z) \\ &= \sum_{n=0}^{\infty} \frac{\lambda^n}{n!} (-1)^n H_n(z) \, e^{-z^2} \end{aligned} \tag{9}$$

Multiplying this relation by $e^{z^2}$ and replacing $\lambda$ by $-\lambda$ we obtain:

$$e^{z^2} F(z - \lambda) = \sum_{n=0}^{\infty} \frac{\lambda^n}{n!} H_n(z) \tag{10}$$

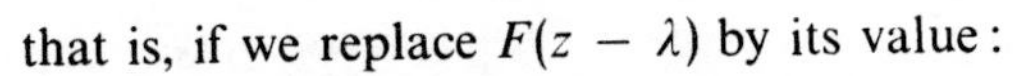

that is, if we replace $F(z - \lambda)$ by its value:

$$\mathrm{e}^{-\lambda^2 + 2\lambda z} = \sum_{n=0}^{\infty} \frac{\lambda^n}{n!} H_n(z) \tag{11}$$

The Hermite polynomials can therefore be obtained from the series expansion in $\lambda$ of the function $\mathrm{e}^{-\lambda^2 + 2\lambda z}$, which for this reason, is called the generating function of the Hermite polynomials.

Relation (11) gives us another definition of the Hermite polynomials $H_n(z)$:

$$H_n(z) = \left\{ \frac{\partial^n}{\partial \lambda^n} \mathrm{e}^{-\lambda^2 + 2\lambda z} \right\}_{\lambda = 0} \tag{12}$$

## c. RECURRENCE RELATIONS; DIFFERENTIAL EQUATION

We have already obtained, in (6), one recurrence relation. It is easy to obtain others by differentiating relation (11). A differentiation with respect to $z$ yields:

$$2\lambda\, \mathrm{e}^{-\lambda^2 + 2\lambda z} = \sum_{n=0}^{\infty} \frac{\lambda^n}{n!} \frac{\mathrm{d}}{\mathrm{d}z} H_n(z) \tag{13}$$

that is, replacing $\mathrm{e}^{-\lambda^2 + 2\lambda z}$ by the expansion (11) and setting equal terms of the same power in $\lambda$:

$$\frac{\mathrm{d}}{\mathrm{d}z} H_n(z) = 2n\, H_{n-1}(z) \tag{14}$$

Similarly, if we differentiate (11) with respect to $\lambda$, an analogous argument yields:

$$H_n(z) = 2z\, H_{n-1}(z) - 2(n-1)\, H_{n-2}(z) \tag{15}$$

Finally, it is not difficult to obtain a differential equation satisfied by the polynomials $H_n(z)$. Differentiating (14) and using (6), we obtain:

$$\begin{aligned} \frac{\mathrm{d}^2}{\mathrm{d}z^2} H_n(z) &= 2n \frac{\mathrm{d}}{\mathrm{d}z} H_{n-1}(z) \\ &= 2n\,[2zH_{n-1}(z) - H_n(z)] \end{aligned} \tag{16}$$

that is, replacing $H_{n-1}(z)$ by its value as given in (14):

$$\left[ \frac{\mathrm{d}^2}{\mathrm{d}z^2} - 2z \frac{\mathrm{d}}{\mathrm{d}z} + 2n \right] H_n(z) = 0 \tag{17}$$

## d. EXAMPLES

Definition (7) or the recurrence relation (6) (which amounts to the same thing) enables us to calculate the first Hermite polynomials easily:

$$\begin{aligned} H_0(z) &= 1 \\ H_1(z) &= 2z \\ H_2(z) &= 4z^2 - 2 \\ H_3(z) &= 8z^3 - 12z \end{aligned} \tag{18}$$

In general:

$$H_n(z) = \left(2z - \frac{d}{dz}\right)^n 1 \tag{19}$$

## 2. The eigenfunctions of the harmonic oscillator Hamiltonian

### a. GENERATING FUNCTION

Consider the function:

$$K(\lambda, x) = \sum_{n=0}^{\infty} \frac{1}{\sqrt{n!}} \lambda^n \langle x \mid \varphi_n \rangle \tag{20}$$

Using the relation [*cf.* chap. V, relation (C-13)]:

$$\mid \varphi_n \rangle = \frac{1}{\sqrt{n!}} (a^\dagger)^n \mid \varphi_0 \rangle \tag{21}$$

we obtain:

$$\begin{aligned} K(\lambda, x) &= \sum_{n=0}^{\infty} \langle x \mid \frac{(\lambda a^\dagger)^n}{n!} \mid \varphi_0 \rangle \\ &= \langle x \mid e^{\lambda a^\dagger} \mid \varphi_0 \rangle \end{aligned} \tag{22}$$

Introducing, as in chapter V, the dimensionless operators $\hat{X}$ and $\hat{P}$:

$$\begin{cases} \hat{X} = \beta X \\ \hat{P} = \dfrac{P}{\beta \hbar} \end{cases} \tag{23}$$

where the parameter $\beta$, which has the dimensions of an inverse length, is defined by:

$$\beta = \sqrt{\frac{m\omega}{\hbar}} \tag{24}$$

The operator:

$$e^{\lambda a^\dagger} = e^{\frac{\lambda}{\sqrt{2}}(\hat{X} - i\hat{P})} \tag{25}$$

can be calculated by using formula (63) of complement $\mathbf{B_{II}}$, where we set:

$$\begin{cases} A = \dfrac{\lambda}{\sqrt{2}} \hat{X} \\ B = -\dfrac{i\lambda}{\sqrt{2}} \hat{P} \end{cases} \tag{26}$$

We obtain:

$$\begin{aligned} e^{\lambda a^\dagger} &= e^{\frac{\lambda}{\sqrt{2}}\hat{X}} \, e^{-\frac{i\lambda}{\sqrt{2}}\hat{P}} \, e^{\frac{i}{4}\lambda^2 [\hat{X}, \hat{P}]} \\ &= e^{\frac{\lambda}{\sqrt{2}}\hat{X}} \, e^{-\frac{i\lambda}{\sqrt{2}}\hat{P}} \, e^{-\lambda^2/4} \end{aligned} \tag{27}$$

Substituting this result into (22), we obtain :

$$K(\lambda, x) = e^{-\lambda^2/4} \langle x \mid e^{(\lambda/\sqrt{2})\hat{X}} e^{(-i\lambda/\sqrt{2})\hat{P}} \mid \varphi_0 \rangle$$

$$= e^{-\lambda^2/4} e^{\beta\lambda x/\sqrt{2}} \langle x \mid e^{(-i\lambda/\sqrt{2}) P/\beta\hbar} \mid \varphi_0 \rangle \tag{28}$$

Now, we have [*cf.* complement $E_{II}$, formula (15)] :

$$\langle x \mid e^{-i\frac{\lambda P}{\beta\hbar\sqrt{2}}} = \langle x - \lambda/\beta\sqrt{2} \mid \tag{29}$$

and (28) can be written :

$$K(\lambda, x) = e^{-\lambda^2/4} e^{\beta\lambda x/\sqrt{2}} \langle x - \lambda/\beta\sqrt{2} \mid \varphi_0 \rangle$$

$$= e^{-\lambda^2/4} e^{\beta\lambda x/\sqrt{2}} \varphi_0(x - \lambda/\beta\sqrt{2}) \tag{30}$$

Using formula (C-25) of chapter V, we obtain, finally :

$$K(\lambda, x) = \left(\frac{\beta^2}{\pi}\right)^{1/4} \exp\left\{-\frac{\beta^2 x^2}{2} + \beta\lambda x\sqrt{2} - \frac{\lambda^2}{2}\right\} \tag{31}$$

According to definition (20), all we must do to find the wave functions $\varphi_n(x) = \langle x \mid \varphi_n \rangle$ is expand this expression in powers of $\lambda$ :

$$K(\lambda, x) = \sum_{n=0}^{\infty} \frac{\lambda^n}{\sqrt{n!}} \varphi_n(x) \tag{32}$$

$K(\lambda, x)$ is called the generating function of the $\varphi_n(x)$.

#### b. $\varphi_n(x)$ IN TERMS OF THE HERMITE POLYNOMIALS

Replacing, in formula (11), $\lambda$ by $\lambda/\sqrt{2}$ and $z$ by $\beta x$, we obtain :

$$\exp\left\{-\frac{\lambda^2}{2} + \beta\lambda x\sqrt{2}\right\} = \sum_{n=0}^{\infty} \left(\frac{\lambda}{\sqrt{2}}\right)^n \frac{1}{n!} H_n(\beta x) \tag{33}$$

Substituting this expression into (31) :

$$K(\lambda, x) = \left(\frac{\beta^2}{\pi}\right)^{1/4} \sum_{n=0}^{\infty} \left(\frac{\lambda}{\sqrt{2}}\right)^n \frac{1}{n!} e^{-\beta^2 x^2/2} H_n(\beta x) \tag{34}$$

Setting equal the coefficients of the different powers of $\lambda$ in (32) and (34), we obtain :

$$\varphi_n(x) = \left(\frac{\beta^2}{\pi}\right)^{1/4} \frac{1}{\sqrt{2^n \; n!}} e^{-\beta^2 x^2/2} H_n(\beta x) \tag{35}$$

The shape of the function $\varphi_n(x)$ is therefore analogous to that of the $n$th-order derivative of the Gaussian function $F(x)$ considered in §1 above; $\varphi_n(x)$ is of parity $(-1)^n$ and possesses $n$ zeros interposed between those of $\varphi_{n+1}(x)$. We mentioned in §C-2 of chapter V that this property is related to the increase in the average kinetic energy of the states $\mid \varphi_n \rangle$ when $n$ increases.

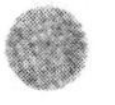



### c. RECURRENCE RELATIONS

Let us write the equations:

$$\begin{cases} a \mid \varphi_n \rangle = \sqrt{n} \mid \varphi_{n-1} \rangle \\ a^\dagger \mid \varphi_n \rangle = \sqrt{n+1} \mid \varphi_{n+1} \rangle \end{cases} \tag{36}$$

in the $\{ \mid x \rangle \}$ representation. Using the definitions of $a$ and $a^\dagger$ [*cf.* chap. V, relations (B-6)], we see that in the $\{ \mid x \rangle \}$ representation, the action of these operators is given by:

$$\begin{aligned} a &\Longrightarrow \frac{\beta}{\sqrt{2}}\left[x + \frac{1}{\beta^2}\frac{\mathrm{d}}{\mathrm{d}x}\right] \\ a^\dagger &\Longrightarrow \frac{\beta}{\sqrt{2}}\left[x - \frac{1}{\beta^2}\frac{\mathrm{d}}{\mathrm{d}x}\right] \end{aligned} \tag{37}$$

Equations (36) therefore become:

$$\begin{cases} \dfrac{\beta}{\sqrt{2}}\left[x + \dfrac{1}{\beta^2}\dfrac{\mathrm{d}}{\mathrm{d}x}\right]\varphi_n(x) = \sqrt{n}\,\varphi_{n-1}(x) \\ \dfrac{\beta}{\sqrt{2}}\left[x - \dfrac{1}{\beta^2}\dfrac{\mathrm{d}}{\mathrm{d}x}\right]\varphi_n(x) = \sqrt{n+1}\,\varphi_{n+1}(x) \end{cases} \tag{38}$$

Let us take the sum and difference of these equations:

$$x\,\beta\sqrt{2}\,\varphi_n(x) = \sqrt{n}\,\varphi_{n-1}(x) + \sqrt{n+1}\,\varphi_{n+1}(x) \tag{39}$$

$$\frac{\sqrt{2}}{\beta}\frac{\mathrm{d}}{\mathrm{d}x}\varphi_n(x) = \sqrt{n}\,\varphi_{n-1}(x) - \sqrt{n+1}\,\varphi_{n+1}(x) \tag{40}$$

COMMENT:

If we replace the functions $\varphi_n(x)$ in (39) and (40) by their expressions given in (35), we obtain, after simplification (setting $\hat{x} = \beta x$):

$$2\hat{x}\,H_n(\hat{x}) = 2n\,H_{n-1}(\hat{x}) + H_{n+1}(\hat{x}) \tag{41}$$

$$2\left[-\,\hat{x}\,H_n(\hat{x}) + \frac{\mathrm{d}}{\mathrm{d}\hat{x}}H_n(\hat{x})\right] = 2n\,H_{n-1}(\hat{x}) - H_{n+1}(\hat{x}) \tag{42}$$

By taking the sum and the difference of these equations, we obtain relations (6) and (14) of §1.

#### References

Messiah (1.17), App. B, §III; Arfken (10.4), chap. 13, §1; Angot (10.2), §7.8.

## Complement $C_V$

# SOLVING THE EIGENVALUE EQUATION OF THE HARMONIC OSCILLATOR BY THE POLYNOMIAL METHOD



In §B of chapter V, the method used to calculate the energies of the harmonic oscillator stationary states $|\varphi_n\rangle$ is based on the use of the operators $a$, $a^\dagger$ and $N$ and their commutation relations. It is possible to obtain the same results in a completely different way, by solving the eigenvalue equation of the Hamiltonian $H$ in the $\{|x\rangle\}$ representation. This is what we are going to do in this complement.

## 1. Changing the function and the variable

In the $\{|x\rangle\}$ representation, the eigenvalue equation of $H$ is written :

$$\left[-\frac{\hbar^2}{2m}\frac{d^2}{dx^2} + \frac{1}{2}m\omega^2x^2\right]\varphi(x) = E\,\varphi(x) \tag{1}$$

As in chapter V, let us introduce the dimensionless operators $\hat{X}$ and $\hat{P}$ :

$$\begin{cases} \hat{X} = \beta X \\ \hat{P} = \dfrac{P}{\beta\hbar} \end{cases} \tag{2}$$

where the parameter $\beta$, which has the dimensions of an inverse length, is defined by:

$$\beta = \sqrt{\frac{m\omega}{\hbar}} \tag{3}$$

Let us denote by $|\xi_{\hat{x}}\rangle$ the eigenvector of $\hat{X}$, whose eigenvalue is $\hat{x}$ :

$$\hat{X}\,|\,\xi_{\hat{x}}\rangle = \hat{x}\,|\,\xi_{\hat{x}}\rangle \tag{4}$$

The orthonormalization and closure relations of the kets $|\xi_{\hat{x}}\rangle$ are written :

$$\langle\,\xi_{\hat{x}}\,|\,\xi_{\hat{x}'}\,\rangle = \delta(\hat{x} - \hat{x}') \tag{5}$$

$$\int_{-\infty}^{+\infty} d\hat{x}\,|\,\xi_{\hat{x}}\,\rangle\langle\,\xi_{\hat{x}}\,| = 1 \tag{6}$$

The ket $| \xi_{\hat{x}} \rangle$ is obviously an eigenvector of $X$, with the eigenvalue $\hat{x}/\beta$. Therefore, when :

$$\hat{x} = \beta x \tag{7}$$

the two kets $| x \rangle$ and $| \xi_{\hat{x}} \rangle$ are proportional. However, they are not equal. Writing the closure relation for the kets $| x \rangle$ :

$$\int_{-\infty}^{+\infty} \mathrm{d}x \, | x \rangle \langle x | = 1 \tag{8}$$

and making the change of variables given in (7), we obtain :

$$\int_{-\infty}^{+\infty} \frac{\mathrm{d}\hat{x}}{\beta} | x = \hat{x}/\beta \rangle \langle x = \hat{x}/\beta | = 1 \tag{9}$$

Comparison with (6) shows that we can, for example, set :

$$| x = \hat{x}/\beta \rangle = \sqrt{\beta} \, | \xi_{\hat{x}} \rangle \tag{10}$$

to orthonormalize the kets $| \xi_{\hat{x}} \rangle$ with respect to $\hat{x}$, since the kets $| x \rangle$ are orthonormal with respect to $x$.

Let $| \varphi \rangle$ be an arbitrary ket, $\varphi(x) = \langle x | \varphi \rangle$ its wave function in the $\{ | x \rangle \}$ representation, and $\hat{\varphi}(\hat{x}) = \langle \hat{x} | \varphi \rangle$ its wave function in the $\{ | \xi_{\hat{x}} \rangle \}$ representation. According to (10):

$$\hat{\varphi}(\hat{x}) = \langle \xi_{\hat{x}} | \varphi \rangle = \frac{1}{\sqrt{\beta}} \langle x = \hat{x}/\beta | \varphi \rangle \tag{11}$$

that is:

$$\hat{\varphi}(\hat{x}) = \frac{1}{\sqrt{\beta}} \varphi \, (x = \hat{x}/\beta) \tag{12}$$

If $| \varphi \rangle$ is normalized, relation (8) yields :

$$\langle \varphi | \varphi \rangle = \langle \varphi | \left( \int_{-\infty}^{+\infty} \mathrm{d}x \, | x \rangle \langle x | \right) | \varphi \rangle = \int_{-\infty}^{+\infty} \varphi^*(x) \, \varphi(x) \, \mathrm{d}x = 1 \tag{13}$$

and relation (6) gives :

$$\langle \varphi | \varphi \rangle = \langle \varphi | \left( \int_{-\infty}^{+\infty} \mathrm{d}\hat{x} \, | \xi_{\hat{x}} \rangle \langle \xi_{\hat{x}} | \right) | \varphi \rangle = \int_{-\infty}^{+\infty} \hat{\varphi}^*(\hat{x}) \, \hat{\varphi}(\hat{x}) \, \mathrm{d}\hat{x} = 1 \tag{14}$$

The wave function $\varphi(x)$ is therefore normalized with respect to the variable $x$, as is $\hat{\varphi}(\hat{x})$ with respect to the variable $\hat{x}$. [This can be shown by using the integral in (13), in which we make the change of variables indicated in (7)].

Now, substituting (7) and (12) into (1), we obtain :

$$\frac{1}{2} \left[ - \frac{\mathrm{d}^2}{\mathrm{d}\hat{x}^2} + \hat{x}^2 \right] \hat{\varphi}(\hat{x}) = \varepsilon \, \hat{\varphi}(\hat{x}) \tag{15}$$

setting:

$$\varepsilon = \frac{E}{\hbar\omega} \tag{16}$$

Equation (15) is more convenient than equation (1) since all the quantities appearing in it are dimensionless.

## 2. The polynomial method

### a. THE ASYMPTOTIC FORM OF $\hat{\varphi}(\hat{x})$

Equation (15) can be written:

$$\left[\frac{d^2}{d\hat{x}^2} - (\hat{x}^2 - 2\varepsilon)\right]\hat{\varphi}(\hat{x}) = 0 \tag{17}$$

Let us try to predict intuitively the behavior of $\hat{\varphi}(\hat{x})$ for very large $\hat{x}$. To do this, consider the functions:

$$G_{\pm}(\hat{x}) = e^{\pm \hat{x}^2/2} \tag{18}$$

They are solutions of the differential equations:

$$\left[\frac{d^2}{d\hat{x}^2} - (\hat{x}^2 \pm 1)\right]G_{\pm}(\hat{x}) = 0 \tag{19}$$

When $\hat{x}$ approaches infinity:

$$\hat{x}^2 \pm 1 \sim \hat{x}^2 \sim \hat{x}^2 - 2\varepsilon \tag{20}$$

and equations (17) and (19) take on the same form asymptotically. We should therefore expect the solutions of equation (17) to behave, for large $\hat{x}$*, either like $e^{\hat{x}^2/2}$ or like $e^{-\hat{x}^2/2}$. From a physical point of view, the only functions $\hat{\varphi}(\hat{x})$ which are of interest to us are those which are bounded everywhere. This restricts us to solutions of (17) which behave like $e^{-\hat{x}^2/2}$ (if they exist). This is why we shall set:

$$\hat{\varphi}(\hat{x}) = e^{-\hat{x}^2/2}\, h(\hat{x}) \tag{21}$$

Substituting (21) into (17), we obtain:

$$\frac{d^2}{d\hat{x}^2}h(\hat{x}) - 2\hat{x}\frac{d}{d\hat{x}}h(\hat{x}) + (2\varepsilon - 1)\,h(\hat{x}) = 0 \tag{22}$$

We are going to show how this equation can be solved by expanding $h(\hat{x})$ in a power series. Then we shall impose the condition that its solutions be physically acceptable.

* The solutions of equation (17) are not necessarily equivalent to $e^{\hat{x}^2/2}$ or $e^{-\hat{x}^2/2}$ when $\hat{x} \to \infty$ : the intuitive arguments which we have given do not exclude, for example, the possibility that $\hat{\varphi}(\hat{x})$ may behave like the product of $e^{\hat{x}^2/2}$ or $e^{-\hat{x}^2/2}$ and a power of $\hat{x}$.

### b. THE CALCULATION OF $h(\hat{x})$ IN THE FORM OF A SERIES EXPANSION

As we pointed out in §A-3 of chapter V, the solutions of equation (1) [or, which amounts to the same thing, of (17)] can be sought amongst either even or odd functions. Since the function $e^{-\hat{x}^2/2}$ is even, we can therefore set:

$$h(\hat{x}) = \hat{x}^p (a_0 + a_2\hat{x}^2 + a_4\hat{x}^4 + ... + a_{2m}\,\hat{x}^{2m} + ...) \tag{23}$$

with $a_0 \neq 0$ (where $a_0\hat{x}^p$ is, by definition, the first non-zero term of the expansion) Writing (23) in the form:

$$h(\hat{x}) = \sum_{m=0}^{\infty} a_{2m}\,\hat{x}^{2m+p} \tag{24}$$

we easily obtain:

$$\frac{d}{d\hat{x}}\,h(\hat{x}) = \sum_{m=0}^{\infty} (2m + p)\,a_{2m}\,\hat{x}^{(2m+p-1)} \tag{25}$$

and:

$$\frac{d^2}{d\hat{x}^2}\,h(\hat{x}) = \sum_{m=0}^{\infty} (2m + p)(2m + p - 1)\,a_{2m}\,\hat{x}^{(2m+p-2)} \tag{26}$$

Let us substitute (24), (25) and (26) into (22). For this equation to be satisfied, each term of the series expansion of the left-hand side must be equal to zero. For the general term in $\hat{x}^{2m+p}$, this condition is written:

$$(2m + p + 2)(2m + p + 1)\,a_{2m+2} = (4m + 2p - 2\varepsilon + 1)\,a_{2m} \tag{27}$$

The term of lowest degree is in $\hat{x}^{p-2}$; its coefficient will be zero if:

$$p(p - 1)\,a_0 = 0 \tag{28}$$

Since $a_0$ is not zero, we therefore have either $p = 0$ [the function $\varphi(x)$ is then even] or $p = 1$ [$\varphi(x)$ is then odd].

Relation (27) can be written:

$$a_{2m+2} = \frac{4m + 2p + 1 - 2\varepsilon}{(2m + p + 2)(2m + p + 1)}\,a_{2m} \tag{29}$$

which is a recurrence relation between the coefficients $a_{2m}$. Since $a_0$ is not zero, (29) enables us to calculate $a_2$ in terms of $a_0$, $a_4$ in terms of $a_2$, and so on.

For arbitrary $\varepsilon$, we therefore have the series expansion of two linearly independent solutions of equation (22), corresponding respectively to $p = 0$ and $p = 1$.

### c. QUANTIZATION OF THE ENERGY

We most now choose, from amongst all the solutions found in the preceding section, those which satisfy the physical conditions that $\hat{\varphi}(\hat{x})$ be bounded everywhere.

For most values of $\varepsilon$, the numerator of (29) does not go to zero for any positive or zero integer $m$. Since none of the coefficients $a_{2m}$ is then zero, the series has an infinite number of terms.

It can be shown that the asymptotic behavior of such a series makes it physically unacceptable. We see from (29) that:

$$\frac{a_{2m+2}}{a_{2m}} \underset{m \to \infty}{\sim} \frac{1}{m} \tag{30}$$

Now consider the power series expansion of the function $e^{\lambda \hat{x}^2}$ (where $\lambda$ is a real parameter):

$$e^{\lambda \hat{x}^2} = \sum_{m=0}^{\infty} b_{2m} \, \hat{x}^{2m} \tag{31}$$

with:

$$b_{2m} = \frac{\lambda^m}{m!} \tag{32}$$

For this second series, we therefore have:

$$\frac{b_{2m+2}}{b_{2m}} = \frac{m!}{(m+1)!} \frac{\lambda^{m+1}}{\lambda^m} = \frac{\lambda}{m+1} \underset{m \to \infty}{\sim} \frac{\lambda}{m} \tag{33}$$

If we choose the value of the parameter $\lambda$ such that:

$$0 < \lambda < 1 \tag{34}$$

we see from (30) and (33) that there exists an integer $M$ such that the condition $m > M$ implies:

$$\frac{a_{2m+2}}{a_{2m}} > \frac{b_{2m+2}}{b_{2m}} > 0 \tag{35}$$

We can deduce from this that, when condition (34) is satisfied, we have:

$$|\hat{x}^{-p} h(\hat{x}) - P(\hat{x})| \geqslant \left| \frac{a_{2M}}{b_{2M}} \right| |e^{\lambda \hat{x}^2} - Q(\hat{x})| \tag{36}$$

where $P(\hat{x})$ and $Q(\hat{x})$ are polynomials of degree $2M$ given by the first $M + 1$ terms of series (23) and (31). When $\hat{x}$ approaches infinity, (36) gives:

$$|h(\hat{x})| \underset{\hat{x} \to \infty}{\geqslant} \left| \frac{a_{2M}}{b_{2M}} \right| \hat{x}^p \, e^{\lambda \hat{x}^2} \tag{37}$$

and therefore:

$$|\hat{\varphi}(\hat{x})| \underset{\hat{x} \to \infty}{\geqslant} \left| \frac{a_{2M}}{b_{2M}} \right| \hat{x}^p \, e^{(\lambda - 1/2)\hat{x}^2} \tag{38}$$

Since we can choose $\lambda$ such that:

$$1/2 < \lambda < 1 \tag{39}$$

$|\hat{\varphi}(\hat{x})|$ is not bounded when $\hat{x} \longrightarrow \infty$. We must therefore reject this solution, which makes no sense physically.

There is only one possibility left: that the numerator of (29) goes to zero for a value $m_0$ of $m$. We then have:

$$\begin{cases} a_{2m} \neq 0 & \text{if} \quad m \leqslant m_0 \\ a_{2m} = 0 & \text{if} \quad m > m_0 \end{cases} \tag{40}$$

and the power series expansion of $h(\hat{x})$ reduces to a polynomial of degree $2m_0 + p$. The behavior at infinity of $\hat{\varphi}(\hat{x})$ is then determined by that of the exponential $e^{-\hat{x}^2/2}$, and $\hat{\varphi}(\hat{x})$ is physically acceptable (since it is square-integrable).

The fact that the numerator of (29) goes to zero at $m = m_0$ imposes the condition:

$$2\varepsilon = 2(2m_0 + p) + 1 \tag{41}$$

If we set:

$$2m_0 + p = n \tag{42}$$

equation (41) can be written:

$$\varepsilon = \varepsilon_n = n + 1/2 \tag{43}$$

where $n$ is an arbitrary positive integer or zero (since $m$ is an arbitrary positive integer or zero and since $p$ is equal to 0 or 1). Condition (43) introduces the quantization of the harmonic oscillator energy, since it implies [*cf.* (16)]:

$$E_n = \left(n + \frac{1}{2}\right)\hbar\omega \tag{44}$$

We have thus obtained relation (B-34) of chapter V.

#### d. STATIONARY WAVE FUNCTIONS

The polynomial method also yields the eigenfunctions associated with the various energies $E_n$, in the form:

$$\hat{\varphi}_n(\hat{x}) = e^{-\hat{x}^2/2}\, h_n(\hat{x}) \tag{45}$$

where $h_n(\hat{x})$ is an $n$th degree polynomial. According to (23) and (42), $h_n(\hat{x})$ is an even function if $n$ is even and an odd function if $n$ is odd.

The ground state is obtained for $n = 0$, that is, for $m_0 = p = 0$; $h_0(\hat{x})$ is then a constant, and:

$$\hat{\varphi}_0(\hat{x}) = a_0\, e^{-\hat{x}^2/2} \tag{46}$$

A simple calculation shows that, to normalize $\hat{\varphi}_0(\hat{x})$ with respect to the variable $\hat{x}$, it suffices to choose:

$$a_0 = \pi^{-1/4} \tag{47}$$

Then, using (12), we find:

$$\varphi_0(x) = \left(\frac{\beta^2}{\pi}\right)^{1/4} e^{-\beta^2 x^2/2} \tag{48}$$

which is indeed the expression given in chapter V [formula (C-25)].

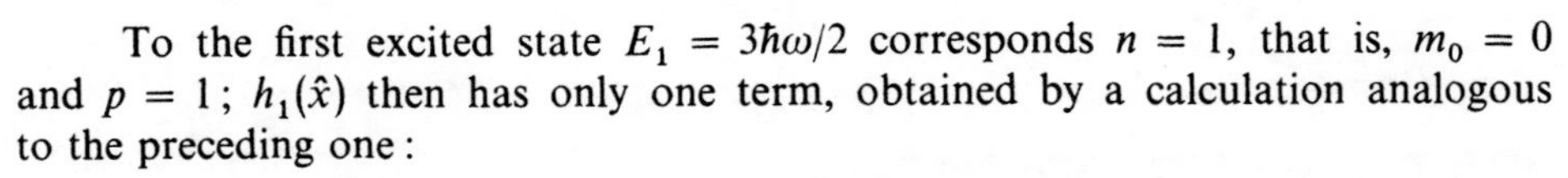



To the first excited state $E_1 = 3\hbar\omega/2$ corresponds $n = 1$, that is, $m_0 = 0$ and $p = 1$; $h_1(\hat{x})$ then has only one term, obtained by a calculation analogous to the preceding one:

$$\begin{cases} \hat{\varphi}_1(\hat{x}) = \left(\dfrac{4}{\pi}\right)^{1/4} \hat{x}\, \mathrm{e}^{-\hat{x}^2/2} \\ \varphi_1(x) = \left(\dfrac{4\beta^6}{\pi}\right)^{1/4} x\, \mathrm{e}^{-\beta^2 x^2/2} \end{cases} \tag{49}$$

For $n = 2$, we have $m_0 = 1$ and $p = 0$. Relation (29) then yields:

$$a_2 = -2a_0 \tag{50}$$

which finally leads to:

$$\begin{cases} \hat{\varphi}_2(\hat{x}) = \left(\dfrac{1}{4\pi}\right)^{1/4} (2\hat{x}^2 - 1)\, \mathrm{e}^{-\hat{x}^2/2} \\ \varphi_2(x) = \left(\dfrac{\beta^2}{4\pi}\right)^{1/4} (2\beta^2 x^2 - 1)\, \mathrm{e}^{-\beta^2 x^2/2} \end{cases} \tag{51}$$

For arbitrary $n$, $h_n(\hat{x})$ is the polynomial solution of equation (22), which can be written, taking the quantization condition (43) into account:

$$\left[\frac{\mathrm{d}^2}{\mathrm{d}\hat{x}^2} - 2\hat{x}\frac{\mathrm{d}}{\mathrm{d}\hat{x}} + 2n\right] h(\hat{x}) = 0 \tag{52}$$

We recognize (52) to be the differential equation satisfied by the Hermite polynomial $H_n(\hat{x})$ [see equation (17) of complement $B_V$]. The polynomial $h_n(\hat{x})$ is therefore proportional to $H_n(\hat{x})$, where the proportionality factor is determined by normalization of $\hat{\varphi}(\hat{x})$. This is in agreement with formula (35) of complement $B_V$.

**References**

Mathematical treatment of differential equations: Morse and Feshbach (10.13), chaps. 5 and 6; Courant and Hilbert (10.11), § V-11.

# Complement $D_V$

## STUDY OF THE STATIONARY STATES IN THE $\{ | p \rangle \}$ REPRESENTATION



The distribution of the possible momenta of a particle in the state $| \varphi_n \rangle$ is given by the wave function $\bar{\varphi}_n(p)$ in the $\{ | p \rangle \}$ representation, the Fourier transform of the wave function $\varphi_n(x)$ in the $\{ | x \rangle \}$ representation. We shall show in this complement that in the case of the harmonic oscillator, the functions $\varphi_n$ and $\bar{\varphi}_n$ are the same (to within multiplicative factors). Thus, in a stationary state, the probability distributions of the momentum and the position have similar forms.

### 1. Wave functions in momentum space

#### a. CHANGING THE VARIABLE AND THE FUNCTION

In complement $C_V$, we introduced, for simplicity, the operator:

$$\hat{X} = \beta X \tag{1}$$

where:

$$\beta = \sqrt{\frac{m\omega}{\hbar}} \tag{2}$$

as well as the eigenkets $| \xi_{\hat{x}} \rangle$ of $\hat{X}$ and the wave function $\hat{\varphi}(\hat{x})$ in the $\{ | \xi_{\hat{x}} \rangle \}$ representation. We shall follow a similar procedure for the operator:

$$\hat{P} = \frac{P}{\beta\hbar} \tag{3}$$

We shall therefore call $| \pi_{\hat{p}} \rangle$ the eigenkets of $\hat{P}$:

$$\hat{P} \, | \pi_{\hat{p}} \rangle = \hat{p} \, | \pi_{\hat{p}} \rangle \tag{4}$$

and denote by $\hat{\bar{\varphi}}(\hat{p})$ the wave function in the $\{ | \pi_{\hat{p}} \rangle \}$ representation:

$$\hat{\bar{\varphi}}(\hat{p}) = \langle \pi_{\hat{p}} | \varphi \rangle \tag{5}$$

Just as the ket $| \xi_{\hat{x}} \rangle$ is proportional to the ket $| x = \hat{x}/\beta \rangle$, the ket $| \pi_{\hat{p}} \rangle$ is proportional to the ket $| p = \beta\hbar\hat{p} \rangle$. If we change $\beta$ to $1/\beta\hbar$ [*cf.* (1) and (3)], equation (10) of complement $C_V$ shows that:

$$| \pi_{\hat{p}} \rangle = \sqrt{\beta\hbar} \, | p = \beta\hbar\hat{p} \rangle \tag{6}$$

The wave function $\hat{\bar{\varphi}}(\hat{p})$ in the $\{ | \pi_{\hat{p}} \rangle \}$ representation is therefore related to the wave function $\bar{\varphi}(p)$ in the $\{ | p \rangle \}$ representation by :

$$\hat{\bar{\varphi}}(\hat{p}) = \sqrt{\beta\hbar}\, \bar{\varphi}(p = \beta\hbar\hat{p}) \tag{7}$$

Furthermore, we can use (6) and relation (10) of complement $C_V$ to obtain :

$$\langle \xi_{\hat{x}} | \pi_{\hat{p}} \rangle = \frac{e^{i\hat{p}\hat{x}}}{\sqrt{2\pi}} \tag{8}$$

We therefore have, using definition (5) and the closure relation for the $\{ | \xi_{\hat{x}} \rangle \}$ basis:

$$\begin{aligned} \hat{\bar{\varphi}}(\hat{p}) &= \int_{-\infty}^{+\infty} \langle \pi_{\hat{p}} | \xi_{\hat{x}} \rangle \langle \xi_{\hat{x}} | \varphi \rangle \, d\hat{x} \\ &= \frac{1}{\sqrt{2\pi}} \int_{-\infty}^{+\infty} e^{-i\hat{p}\hat{x}} \, \hat{\varphi}(\hat{x}) \, d\hat{x} \end{aligned} \tag{9}$$

The function $\hat{\bar{\varphi}}$ is therefore the Fourier transform of $\hat{\varphi}$.

#### b. DETERMINATION OF $\hat{\bar{\varphi}}_n(\hat{p})$

We have seen [*cf.* equation (15) of complement $C_V$] that the stationary wave functions $\hat{\varphi}(\hat{x})$ of the harmonic oscillator satisfy the equation :

$$\frac{1}{2}\left[ -\frac{d^2}{d\hat{x}^2} + \hat{x}^2 \right] \hat{\varphi}(\hat{x}) = \varepsilon \, \hat{\varphi}(\hat{x}) \tag{10}$$

Now, the Fourier transform of $\frac{d^2}{d\hat{x}^2}\hat{\varphi}(\hat{x})$ is $-\hat{p}^2\hat{\bar{\varphi}}(\hat{p})$ and that of $\hat{x}^2\hat{\varphi}(\hat{x})$ is $-\frac{d^2}{d\hat{p}^2}\hat{\bar{\varphi}}(\hat{p})$; The Fourier transform of equation (10) is therefore :

$$\frac{1}{2}\left[ \hat{p}^2 - \frac{d^2}{d\hat{p}^2} \right] \hat{\bar{\varphi}}(\hat{p}) = \varepsilon \, \hat{\bar{\varphi}}(\hat{p}) \tag{11}$$

If we compare (10) and (11), we see that the functions $\hat{\varphi}_n$ and $\hat{\bar{\varphi}}_n$ satisfy the same differential equation. Moreover, we know that this equation, when $\varepsilon = n + 1/2$ (where $n$ is a positive integer or zero), has only one square-integrable solution (the eigenvalues $\varepsilon_n$ are non-degenerate; *cf.* chapter V, § B-3). We can conclude that $\hat{\varphi}_n$ and $\hat{\bar{\varphi}}_n$ are proportional. Since these two functions are normalized, the proportionality factor is a complex number of modulus 1, so that :

$$\hat{\bar{\varphi}}_n(\hat{p}) = e^{i\theta_n} \, \hat{\varphi}_n \, (\hat{x} = \hat{p}) \tag{12}$$

where $e^{i\theta_n}$ is a phase factor which we shall now find.

#### c. CALCULATION OF THE PHASE FACTOR

The wave function of the ground state is given by [*cf.* complement $C_V$, formulas (46) and (47)]:

$$\hat{\varphi}_0(\hat{x}) = \pi^{-1/4} \, e^{-\hat{x}^2/2} \tag{13}$$

This is a Gaussian function; its Fourier transform is therefore [*cf.* appendix I, relation (50)]:

$$\hat{\bar{\varphi}}_0(\hat{p}) = \pi^{-1/4}\, e^{-\hat{p}^2/2} \tag{14}$$

This implies that $\theta_0$ is zero.

To find $\theta_n$, let us write, in the $\{ |\xi_{\hat{x}}\rangle \}$ and $\{ |\pi_{\hat{p}}\rangle \}$ representations, the relation:

$$a^\dagger |\varphi_n\rangle = \sqrt{n+1}\, |\varphi_{n+1}\rangle \tag{15}$$

In the $\{ |\xi_{\hat{x}}\rangle \}$ representation, $\hat{X}$ and $\hat{P}$ act like $\hat{x}$ and $\frac{1}{i}\frac{d}{d\hat{x}}$; $a^\dagger$ therefore acts like $\frac{1}{\sqrt{2}}\left(\hat{x} - \frac{d}{d\hat{x}}\right)$. In the $\{ |\hat{p}\rangle \}$ representation, $\hat{X}$ acts like $i\frac{d}{d\hat{p}}$ and $\hat{P}$ like $\hat{p}$; $a^\dagger$ therefore acts like $\frac{i}{\sqrt{2}}\left(\frac{d}{d\hat{p}} - \hat{p}\right)$.

In the $\{ |\xi_{\hat{x}}\rangle \}$ representation, relation (15) therefore becomes:

$$\hat{\varphi}_{n+1}(\hat{x}) = \frac{1}{\sqrt{2(n+1)}}\left(\hat{x} - \frac{d}{d\hat{x}}\right)\hat{\varphi}_n(\hat{x}) \tag{16}$$

while in the $\{ |\pi_{\hat{p}}\rangle \}$ representation, it becomes:

$$\hat{\bar{\varphi}}_{n+1}(\hat{p}) = \frac{i}{\sqrt{2(n+1)}}\left(\frac{d}{d\hat{p}} - \hat{p}\right)\hat{\bar{\varphi}}_n(\hat{p}) \tag{17}$$

We therefore have:

$$e^{i\theta_{n+i}} = -i\, e^{i\theta_n} \tag{18}$$

that is, knowing that $\theta_0 = 0$:

$$e^{i\theta_n} = (-i)^n \tag{19}$$

Thus we obtain:

$$\hat{\bar{\varphi}}_n(\hat{p}) = (-i)^n\, \hat{\varphi}_n(\hat{x} = \hat{p}) \tag{20}$$

or, returning to the functions $\varphi_n$ and $\bar{\varphi}_n$:

$$\bar{\varphi}_n(p) = (-i)^n \frac{1}{\beta\sqrt{\hbar}}\, \varphi_n\!\left(x = \frac{p}{\beta^2\hbar}\right) \tag{21}$$

## 2. Discussion

Consider a particle in the state $|\varphi_n\rangle$. When the position of the particle is measured, one has a probability $\rho_n(x)\,dx$ of finding a result between $x$ and $x + dx$, where $\rho_n(x)$ is given by:

$$\rho_n(x) = |\varphi_n(x)|^2 \tag{22}$$

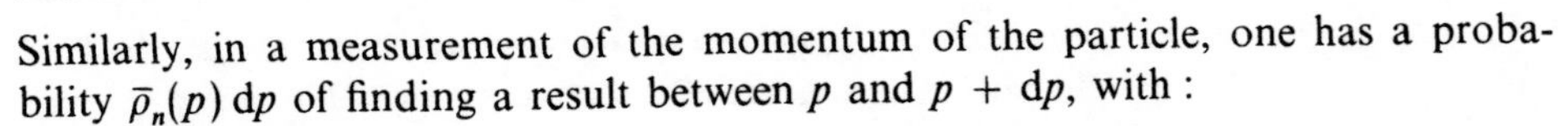



Similarly, in a measurement of the momentum of the particle, one has a probability $\bar{\rho}_n(p)\,\mathrm{d}p$ of finding a result between $p$ and $p + \mathrm{d}p$, with :

$$\bar{\rho}_n(p) = |\bar{\varphi}_n(p)|^2 \tag{23}$$

Relation (21) then yields :

$$\bar{\rho}_n(p) = \frac{1}{m\omega}\,\rho_n\!\left(x = \frac{p}{m\omega}\right) \tag{24}$$

which shows that the momentum distribution in a stationary state has the same form as the position distribution.

We see, for example (*cf.* fig. 6 of chapter V), that if $n$ is large, $\bar{\rho}_n(p)$ has a peak at each of the two values :

$$p = \pm\, m\omega x_M = \pm\, p_M \tag{25}$$

where $p_M$ is the maximum momentum of the classical particle moving in the potential well with an energy $E_n$. An argument analogous to the one set forth at the end of § C-2 of chapter V enables us to understand this result. When the momentum of the classical particle is equal to $\pm\, p_M$, its acceleration is zero (its velocity is stationary), and the values $\pm\, p_M$ of the momentum are, averaging over time, the most probable ones. Comment (*i*) of §D-1 of chapter V concerning the probability density $\rho_n(x)$ can easily be transposed to this context; for example, when $n$ is large, the root-mean-square deviation $\Delta P$ can be interpreted as being the order of magnitude of the distance between the peaks of $\bar{\rho}_n(p)$ situated at $p = \pm\, p_M$.

It is also possible to understand directly from figure 6-a of chapter V why these values of the momentum are highly probable when $n$ is large. The wave function then performs a large number of oscillations between the two peaks, analogous to those of a sinusoid. This happens because the differential equation for the wave function [*cf.* formula (A-17) of chapter V] when $E \gg m\omega^2x^2/2$ becomes :

$$\frac{\mathrm{d}^2}{\mathrm{d}x^2}\,\varphi(x) + \frac{2mE}{\hbar^2}\,\varphi(x) \simeq 0 \tag{26}$$

which yields, according to the definition of $p_M$ :

$$\varphi(x) \simeq A\,\mathrm{e}^{ip_M \cdot x/\hbar} + A'\,\mathrm{e}^{-ip_M \cdot x/\hbar} \tag{27}$$

The wave function (when $n$ is large) therefore looks like a sinusoid of wavelength $h/p_M$ over a relatively large region of the $Ox$ axis. This sinusoid can be considered to be the sum of two progressive waves [*cf.* (27)] associated with the two opposite momenta $\pm\, p_M$ (corresponding to the to-and-fro motion of the particle in the well). It is not surprising, therefore, that the probability density $\bar{\rho}_n(p)$ should be large in the neighborhood of the values $p = \pm\, p_M$.

An analogous argument also enables us to understand the order of magnitude of the product $\Delta X\,.\,\Delta P$. This product is equal to [*cf.* chap. V, relations (D-6), (D-7) and (D-9)] :

$$\Delta X\,.\,\Delta P = \left(n + \frac{1}{2}\right)\hbar = \frac{x_M \,.\, p_M}{2}. \tag{28}$$

When $n$ increases, the amplitudes $x_M$ and $p_M$ of the oscillations increase, and the product $\Delta X \,.\, \Delta P$ takes on values much greater than its minimum value $\hbar/2$. We might wonder why this is the case, since we have seen in several examples that when the width $\Delta X$ of a function increases, the width $\Delta P$ of its Fourier transform decreases. This is indeed what would happen for the functions $\varphi_n(x)$ if, in the interval $-x_M \leqslant x \leqslant +x_M$ where they take on non-negligible values, they varied slowly, reaching, for example, a single maximum or minimum. This is in fact the case for small values of $n$, for which the value of the product $\Delta X \,.\, \Delta P$ is indeed near its minimum. However, when $n$ is large, the functions $\varphi_n(x)$ perform numerous oscillations in the interval $-x_M \leqslant x \leqslant +x_M$, where they have $n$ zeros. One can therefore associate with them wavelengths of the order of $\lambda \simeq x_M/n \simeq \Delta X/n$, corresponding to momenta of the particle situated in a domain of dimension $\Delta P$ given by:

$$\Delta P \simeq \frac{h}{\lambda} \simeq \frac{nh}{\Delta X} \tag{29}$$

We thus find again that:

$$\Delta X \,.\, \Delta P \simeq nh \tag{30}$$

This situation is somewhat analogous to the one studied in §1 of complement $A_{III}$, in connection with the infinite one-dimensional well.

## Complement $E_V$

# THE ISOTROPIC THREE-DIMENSIONAL HARMONIC OSCILLATOR



In chapter V, we studied the one-dimensional harmonic oscillator. We shall show here how we can use the results of this study to treat the three-dimensional harmonic oscillator.

## 1. The Hamiltonian operator

Consider a spinless particle of mass $m$ which can move in three-dimensional space. The particle is subject to a central force (that is, a force which is constantly directed towards the coordinate origin $O$) whose absolute value is proportional to the distance of the particle from the point $O$:

$$\mathbf{F} = -k\,\mathbf{r} \tag{1}$$

($k$ is a positive constant).

This force field is derived from the potential energy:

$$V(\mathbf{r}) = \frac{1}{2}k\,\mathbf{r}^2 = \frac{1}{2}m\omega^2\mathbf{r}^2 \tag{2}$$

where the angular frequency $\omega$ is defined as for the one-dimensional harmonic oscillator:

$$\omega = \sqrt{\frac{k}{m}} \tag{3}$$

The classical Hamiltonian is therefore:

$$\mathscr{H}(\mathbf{r}, \mathbf{p}) = \frac{\mathbf{p}^2}{2m} + \frac{1}{2}m\omega^2\mathbf{r}^2 \tag{4}$$

From this we immediately deduce the Hamiltonian operator, using the quantization rules (chap. III, §B-5):

$$H = \frac{\mathbf{P}^2}{2m} + \frac{1}{2}m\omega^2\mathbf{R}^2 \tag{5}$$

Since the Hamiltonian $H$ is time-independent, we shall solve its eigenvalue equation:

$$H\,|\,\psi\,\rangle = E\,|\,\psi\,\rangle \tag{6}$$

where $|\psi\rangle$ belongs to the state space $\mathscr{E}_\mathbf{r}$ of a particle in three-dimensional space.

COMMENT:

Since $V(\mathbf{r})$ depends only on the distance $r = |\mathbf{r}|$ of the particle from the origin [$V(\mathbf{r})$ is consequently invariant under an arbitrary rotation], this harmonic oscillator is said to be *isotropic*. Nevertheless, the calculations which follow can easily be generalized to the case of an anisotropic harmonic oscillator, for which:

$$V(\mathbf{r}) = \frac{m}{2}(\omega_x^2 x^2 + \omega_y^2 y^2 + \omega_z^2 z^2) \tag{7}$$

where the three constants $\omega_x$, $\omega_y$ and $\omega_z$ are different.

## 2. Separation of the variables in Cartesian coordinates

Recall that the state space $\mathscr{E}_\mathbf{r}$ can be considered (*cf.* chap. II, §F) to be the tensor product:

$$\mathscr{E}_\mathbf{r} = \mathscr{E}_x \otimes \mathscr{E}_y \otimes \mathscr{E}_z \tag{8}$$

where $\mathscr{E}_x$ is the state space of a particle moving along $Ox$, that is, the space associated with the wave functions $\varphi(x)$. $\mathscr{E}_y$ and $\mathscr{E}_z$ are defined analogously.

Now, expression (5) for the Hamiltonian $H$ can be written in the form:

$$\begin{aligned} H &= \frac{1}{2m}(P_x^2 + P_y^2 + P_z^2) + \frac{1}{2}m\omega^2(X^2 + Y^2 + Z^2) \\ &= H_x + H_y + H_z \end{aligned} \tag{9}$$

with:

$$H_x = \frac{P_x^2}{2m} + \frac{1}{2}m\omega^2 X^2 \tag{10}$$

and similar definitions for $H_y$ and $H_z$. $H_x$ is a function only of $X$ and $P_x$: $H_x$ is therefore the extension into $\mathscr{E}_\mathbf{r}$ of an operator which actually acts in $\mathscr{E}_x$. Similarly, $H_y$ and $H_z$ act only in $\mathscr{E}_y$ and $\mathscr{E}_z$ respectively. In $\mathscr{E}_x$, $H_x$ is a one-dimensional harmonic oscillator Hamiltonian. The same is true for $H_y$ and $H_z$ in $\mathscr{E}_y$ and $\mathscr{E}_z$.

$H_x$, $H_y$ and $H_z$ commute. Each of them therefore commutes with their sum $H$. Consequently, the eigenvalue equation (6) can be solved by seeking the eigenvectors of $H$ which are also eigenvectors of $H_x$, $H_y$ and $H_z$. Now, we already know the eigenvectors and eigenvalues of $H_x$ in $\mathscr{E}_x$, as well as those of $H_y$ in $\mathscr{E}_y$ and of $H_z$ in $\mathscr{E}_z$:

$$H_x|\varphi_{n_x}\rangle = \left(n_x + \frac{1}{2}\right)\hbar\omega|\varphi_{n_x}\rangle \; ; \; |\varphi_{n_x}\rangle \in \mathscr{E}_x \tag{11-a}$$

$$H_y|\varphi_{n_y}\rangle = \left(n_y + \frac{1}{2}\right)\hbar\omega|\varphi_{n_y}\rangle \; ; \; |\varphi_{n_y}\rangle \in \mathscr{E}_y \tag{11-b}$$

$$H_z|\varphi_{n_z}\rangle = \left(n_z + \frac{1}{2}\right)\hbar\omega|\varphi_{n_z}\rangle \; ; \; |\varphi_{n_z}\rangle \in \mathscr{E}_z \tag{11-c}$$

($n_x$, $n_y$ and $n_z$ are positive integers or zero). From this we deduce (*cf.* chap. II, §F) that the eigenstates common to $H$, $H_x$, $H_y$ and $H_z$ are of the form:

$$|\psi_{n_x,n_y,n_z}\rangle = |\varphi_{n_x}\rangle|\varphi_{n_y}\rangle|\varphi_{n_z}\rangle \tag{12}$$

 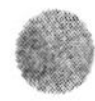

According to equations (9) and (11):

$$H \mid \psi_{n_x,n_y,n_z} \rangle = \left(n_x + n_y + n_z + \frac{3}{2}\right)\hbar\omega \mid \psi_{n_x,n_y,n_z} \rangle \tag{13}$$

that is, the eigenvectors of $H$ are seen to be *tensor products of eigenvectors* of $H_x$, $H_y$ and $H_z$ respectively, and the eigenvalues of $H$, to be *sums of eigenvalues* of these three operators.

According to equation (13), the energy levels $E_n$ of the isotropic three-dimensional harmonic oscillator are of the form:

$$E_n = \left(n + \frac{3}{2}\right)\hbar\omega \tag{14}$$

with:

$$n \text{ a positive integer or zero} \tag{15}$$

since $n$ is the sum $n_x + n_y + n_z$ of three numbers, each of which can take on any non-negative integral value.

Furthermore, formula (12) enables us to deduce the properties of the vectors $\mid \psi_{n_x,n_y,n_z} \rangle$, common eigenstates of $H$, $H_x$, $H_y$ and $H_z$, from those derived in §C-1 of chapter V for $\mid \varphi_{n_x} \rangle$ (which are also valid for $\mid \varphi_{n_y} \rangle$ and $\mid \varphi_{n_z} \rangle$).

Let us introduce three pairs of creation and annihilation operators:

$$a_x = \sqrt{\frac{m\omega}{2\hbar}}\, X + \frac{i}{\sqrt{2m\hbar\omega}}\, P_x \qquad a_x^\dagger = \sqrt{\frac{m\omega}{2\hbar}}\, X - \frac{i}{\sqrt{2m\hbar\omega}}\, P_x \tag{16-a}$$

$$a_y = \sqrt{\frac{m\omega}{2\hbar}}\, Y + \frac{i}{\sqrt{2m\hbar\omega}}\, P_y \qquad a_y^\dagger = \sqrt{\frac{m\omega}{2\hbar}}\, Y - \frac{i}{\sqrt{2m\hbar\omega}}\, P_y \tag{16-b}$$

$$a_z = \sqrt{\frac{m\omega}{2\hbar}}\, Z + \frac{i}{\sqrt{2m\hbar\omega}}\, P_z \qquad a_z^\dagger = \sqrt{\frac{m\omega}{2\hbar}}\, Z - \frac{i}{\sqrt{2m\hbar\omega}}\, P_z \tag{16-c}$$

These operators are the extensions into $\mathscr{E}_\mathbf{r}$ of operators acting in $\mathscr{E}_x$, $\mathscr{E}_y$ and $\mathscr{E}_z$. The canonical commutation relations between the components of **R** and **P** imply that the only non-zero commutators of the six operators defined in (16) are:

$$[a_x, a_x^\dagger] = [a_y, a_y^\dagger] = [a_z, a_z^\dagger] = 1 \tag{17}$$

Note that two operators with different indices always commute, as is logical because they act in different spaces. The action of the operators $a_x$ and $a_x^\dagger$ on the states $\mid \psi_{n_x,n_y,n_z} \rangle$ is given by the formulas:

$$\begin{aligned} a_x \mid \psi_{n_x,n_y,n_z} \rangle &= (a_x \mid \varphi_{n_x} \rangle) \mid \varphi_{n_y} \rangle \mid \varphi_{n_z} \rangle \\ &= \sqrt{n_x} \mid \varphi_{n_x - 1} \rangle \mid \varphi_{n_y} \rangle \mid \varphi_{n_z} \rangle \\ &= \sqrt{n_x} \mid \psi_{n_x - 1,n_y,n_z} \rangle \end{aligned} \tag{18-a}$$

$$\begin{aligned} a_x^\dagger \mid \psi_{n_x,n_y,n_z} \rangle &= (a_x^\dagger \mid \varphi_{n_x} \rangle) \mid \varphi_{n_y} \rangle \mid \varphi_{n_z} \rangle \\ &= \sqrt{n_x + 1} \mid \varphi_{n_x+1} \rangle \mid \varphi_{n_y} \rangle \mid \varphi_{n_z} \rangle \\ &= \sqrt{n_x + 1} \mid \psi_{n_x+1,n_y,n_z} \rangle \end{aligned} \tag{18-b}$$

For $a_y$, $a_y^\dagger$ and $a_z$, $a_z^\dagger$, we have analogous relations.

We also know [*cf.* equation (C-13) of chapter V] that:

$$| \varphi_{n_x} \rangle = \frac{1}{\sqrt{n_x!}} (a_x^\dagger)^{n_x} | \varphi_0 \rangle \tag{19}$$

where $| \varphi_0 \rangle$ is the vector of $\mathscr{E}_x$ which satisfies the condition:

$$a_x | \varphi_0 \rangle = 0 \tag{20}$$

In $\mathscr{E}_y$ and $\mathscr{E}_z$, there are analogous expressions for $| \varphi_{n_y} \rangle$ and $| \varphi_{n_z} \rangle$. Consequently, according to (12):

$$| \psi_{n_x,n_y,n_z} \rangle = \frac{1}{\sqrt{n_x!\, n_y!\, n_z!}} (a_x^\dagger)^{n_x} (a_y^\dagger)^{n_y} (a_z^\dagger)^{n_z} | \psi_{0,0,0} \rangle \tag{21}$$

where $| \psi_{0,0,0} \rangle$ is the tensor product of the ground states of the three one-dimensional oscillators, so that:

$$a_x | \psi_{0,0,0} \rangle = a_y | \psi_{0,0,0} \rangle = a_z | \psi_{0,0,0} \rangle = 0 \tag{22}$$

Finally, recall that, since $| \psi_{n_x,n_y,n_z} \rangle$ is a tensor product, the associated wave function is of the form:

$$\langle \mathbf{r} | \psi_{n_x,n_y,n_z} \rangle = \varphi_{n_x}(x) \quad \varphi_{n_y}(y) \quad \varphi_{n_z}(z) \tag{23}$$

where $\varphi_{n_x}$, $\varphi_{n_y}$ and $\varphi_{n_z}$ are stationary wave functions of the one-dimensional harmonic oscillator (chap. V, §C-2). For example:

$$\langle \mathbf{r} | \psi_{0,0,0} \rangle = \left( \frac{m\omega}{\pi\hbar} \right)^{3/4} \mathrm{e}^{-\frac{m\omega}{2\hbar}(x^2 + y^2 + z^2)} \tag{24}$$

## 3. Degeneracy of the energy levels

We showed in § B-3 of chapter V that $H_x$ constitutes a C.S.C.O. in $\mathscr{E}_x$; the same is true for $H_y$ in $\mathscr{E}_y$ and for $H_z$ in $\mathscr{E}_z$. According to § F of chapter II, $\{ H_x, H_y, H_z \}$ is thus a C.S.C.O. in $\mathscr{E}_\mathbf{r}$. Therefore, there exists (to within a constant factor) a unique ket $| \psi_{n_x,n_y,n_z} \rangle$ of $\mathscr{E}_\mathbf{r}$ corresponding to a given set of eigenvalues of $H_x$, $H_y$ and $H_z$, that is, to fixed non-negative integers $n_x$, $n_y$ and $n_z$.

However, $H$ alone does not form a C.S.C.O. since *the energy levels $E_n$ are degenerate*. If we choose an eigenvalue of $H$, $E_n = (n + 3/2)\hbar\omega$ (which amounts to fixing a non-negative integer $n$), all the kets of the $\{ | \psi_{n_x,n_y,n_z} \rangle \}$ basis which satisfy:

$$n_x + n_y + n_z = n \tag{25}$$

are eigenvectors of $H$ with the eigenvalue $E_n$.

The degree of degeneracy $g_n$ of $E_n$ is therefore equal to the number of different sets $\{ n_x, n_y, n_z \}$ which satisfy condition (25). To find $g_n$, we can proceed as follows: with $n$ fixed, choose $n_x$ first, giving it one of the values:

$$n_x = 0, 1, 2, \ldots n \tag{26}$$

With $n_x$ thus fixed, we must have:

$$n_y + n_z = n - n_x \tag{27}$$

There are then $(n - n_x + 1)$ possibilities for the pair $\{ n_y, n_z \}$:

$$\{ n_y, n_z \} = \{ 0, n - n_x \}, \{ 1, n - n_x - 1 \}, \dots \{ n - n_x, 0 \} \tag{28}$$

The degree of degeneracy $g_n$ of $E_n$ is therefore equal to:

$$g_n = \sum_{n_x=0}^{n} (n - n_x + 1) \tag{29}$$

This sum is easy to calculate:

$$g_n = (n + 1) \sum_{n_x=0}^{n} 1 - \sum_{n_x=0}^{n} n_x = \frac{(n + 1)(n + 2)}{2} \tag{30}$$

Consequently, only the ground state $E_0 = \frac{3}{2}\hbar\omega$ is non-degenerate.

COMMENT:

The kets $| \psi_{n_x,n_y,n_z} \rangle$ constitute an orthonormal system of eigenvectors of $H$, which forms a basis in $\mathscr{E}_{\mathbf{r}}$. Since the eigenvalues $E_n$ of $H$ are degenerate, this system is not unique. We shall see in particular in complement $B_{VII}$ that, in order to solve equation (6), it is possible to use a set of constants of the motion other than $\{ H_x, H_y, H_z \}$: thus we obtain a basis of $\mathscr{E}_{\mathbf{r}}$ which is different from the preceding one, although still consisting of eigenvectors of $H$. The kets of this new basis are orthonormal linear combinations of the $| \psi_{n_x,n_y,n_z} \rangle$ belonging to each of the eigensubspaces of $H$, that is, corresponding to a fixed value of the sum $n_x + n_y + n_z$.

## Complement $F_V$

## A CHARGED HARMONIC OSCILLATOR IN A UNIFORM ELECTRIC FIELD



The one-dimensional harmonic oscillator studied in chapter V consists of a particle of mass $m$ having a potential energy:

$$V(X) = \frac{1}{2} m\omega^2 X^2 \tag{1}$$

Assume, in addition, that this particle has a charge $q$ and that it is placed in a uniform electric field $\mathscr{E}$ parallel to $Ox$. What are its new stationary states and the corresponding energies?

The classical potential energy of a particle placed in a uniform field $\mathscr{E}$ is equal to*:

$$w(\mathscr{E}) = -q\mathscr{E}x \tag{2}$$

To obtain the quantum mechanical Hamiltonian operator $H'(\mathscr{E})$ in the presence of the field $\mathscr{E}$, we must therefore add to the potential energy (1) of the harmonic oscillator the term:

$$W(\mathscr{E}) = -q\mathscr{E}X \tag{3}$$

which gives:

$$H'(\mathscr{E}) = \frac{P^2}{2m} + \frac{1}{2} m\omega^2 X^2 - q\mathscr{E}X \tag{4}$$

We must now find the eigenvalues and eigenvectors of this operator. To this end, we shall use two different methods. First, we shall consider directly the eigenvalue equation of $H'(\mathscr{E})$ in the $\{ | x \rangle \}$ representation; it is very simple to interpret the results obtained. Then we shall show how the problem can be solved in a purely operator formalism.

### 1. Eigenvalue equation of $H'(\mathscr{E})$ in the $\{ | x \rangle \}$ representation

Let $| \varphi' \rangle$ be an eigenvector of $H'(\mathscr{E})$:

$$H'(\mathscr{E}) | \varphi' \rangle = E' | \varphi' \rangle \tag{5}$$

* We use the convention of zero potential energy for the particle at $x = 0$.

Using (4), we can write this equation in the $\{ | x \rangle \}$ representation:

$$\left[ -\frac{\hbar^2}{2m} \frac{d^2}{dx^2} + \frac{1}{2} m\omega^2 x^2 - q\mathscr{E} x \right] \varphi'(x) = E' \varphi'(x) \tag{6}$$

Completing the square with respect to $x$ on the left-hand side of (6):

$$\left[ -\frac{\hbar^2}{2m} \frac{d^2}{dx^2} + \frac{1}{2} m\omega^2 \left( x - \frac{q\mathscr{E}}{m\omega^2} \right)^2 - \frac{q^2 \mathscr{E}^2}{2m\omega^2} \right] \varphi'(x) = E' \varphi'(x) \tag{7}$$

Let us now replace the variable $x$ by a new variable $u$, setting:

$$u = x - \frac{q\mathscr{E}}{m\omega^2} \tag{8}$$

Through the intermediary of $x$, $\varphi'$ is then a function of $u$, and equation (7) becomes:

$$\left[ -\frac{\hbar^2}{2m} \frac{d^2}{du^2} + \frac{1}{2} m\omega^2 u^2 \right] \varphi'(u) = E'' \varphi'(u) \tag{9}$$

with:

$$E'' = E' + \frac{q^2 \mathscr{E}^2}{2m\omega^2} \tag{10}$$

Thus we see that equation (9) is the same as the one which enabled us to obtain the stationary states of the harmonic oscillator in the absence of an electric field in the $\{ | x \rangle \}$ representation [*cf.* chap. V, relation (A-17)]. Therefore, we have already solved this equation, and we know that the acceptable values of $E''$ are given by:

$$E''_n = \left( n + \frac{1}{2} \right) \hbar\omega \tag{11}$$

(where $n$ is a positive integer or zero).

Relations (10) and (11) show that in the presence of the electric field, the energies $E'$ of the stationary states of the harmonic oscillator are modified:

$$E'_n(\mathscr{E}) = \left( n + \frac{1}{2} \right) \hbar\omega - \frac{q^2 \mathscr{E}^2}{2m\omega^2} \tag{12}$$

The entire spectrum of the harmonic oscillator is therefore shifted by the quantity $q^2 \mathscr{E}^2 / 2m\omega^2$.

Now, let us show that the eigenfunctions $\varphi'_n(x)$ associated with the energies (12) can all be obtained from the $\varphi_n(x)$ by a translation along $Ox$. The solution of (9) corresponding to a given value of $n$ is $\varphi_n(u)$ [where the function $\varphi_n$ is given, for example, by formula (35) of complement $B_V$]. According to (8), we have:

$$\varphi'_n(x) = \varphi_n \left( x - \frac{q\mathscr{E}}{m\omega^2} \right) \tag{13}$$

This translation comes from the fact that the electric field exerts a force on the particle★.

★ It can be seen from (13) that the function $\varphi'_n(x)$ is obtained from $\varphi_n(x)$ by a translation of $q\mathscr{E}/m\omega^2$; if the product $q\mathscr{E}$ is positive, the translation is performed in the positive $x$-direction, which is indeed the direction of the force exerted by $\mathscr{E}$.

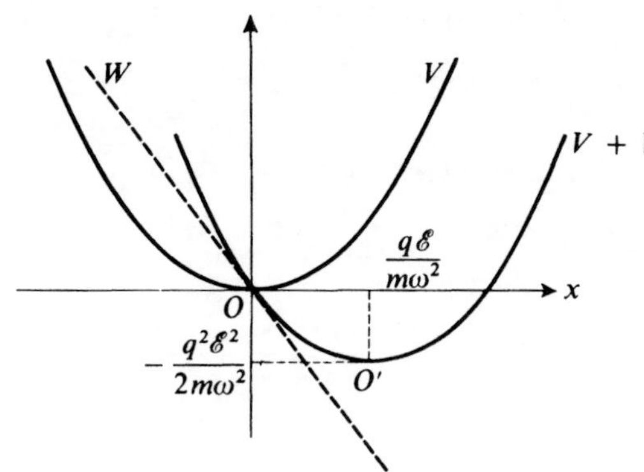


FIGURE 1

**The presence of a uniform electric field has the effect of adding a linear term $W$ to the potential energy $V$ of the harmonic oscillator; the total potential $V + W$ is then represented by a displaced parabola.**

COMMENT:

The change of variable given in (8) allows us to reduce the case of an arbitrary electric field to an already solved problem, the one in which $\mathscr{E}$ was zero. The only effect of the electric field is to change the $x$-origin [*cf.* (13)] and the energy origin [*cf.* (12)]. This result can easily be understood graphically (*cf.* fig. 1). When $\mathscr{E}$ is zero, the potential energy $V(x)$ is represented by a parabola centered at $O$. When $\mathscr{E}$ is not zero, it is necessary to add to this potential energy the quantity $-q\mathscr{E}x$, which corresponds to the dashed line in this figure; the curve representing $V + W$ is again a parabola. Thus, in the presence of the field $\mathscr{E}$ we still have a harmonic oscillator. Since the two parabolas are superposable, they correspond to the same value of $\omega$ and therefore to the same energy difference between the levels. However, their minima $O$ and $O'$ are different, as is consistent with formulas (12) and (13).

## 2. Discussion

### a. ELECTRICAL SUSCEPTIBILITY OF AN ELASTICALLY BOUND ELECTRON

In certain situations, the electrons of an atom or a molecule behave, to a good approximation, as if they were "elastically bound", that is, as if each of them were a harmonic oscillator. We shall prove this for atoms in complement $A_{XIII}$, using time-dependent perturbation theory.

The contribution of each electron to the electric dipole moment of the atom is described by the operator:

$$D = qX \tag{14}$$

where $q$ is the charge of the electron ($q < 0$) and $X$ the corresponding position observable. We are going to examine the mean value of $D$ in the model of the elastically bound electron.

In the absence of an electric field, the mean value of the electric dipole moment in a stationary state of the oscillator is zero:

$$\langle D \rangle = q \langle \varphi_n | X | \varphi_n \rangle = 0 \tag{15}$$

[see formulas (D-1) of chapter V].

Now, let us assume that the field $\mathscr{E}$ is turned on so slowly that the state of the electron changes gradually from $| \varphi_n \rangle$ to $| \varphi'_n \rangle$ ($n$ remaining the same). The mean dipole moment is now different from zero, since:

$$\langle D \rangle' = q \langle \varphi'_n | X | \varphi'_n \rangle = q \int_{-\infty}^{+\infty} \mathrm{d}x \, x \, |\varphi'_n(x)|^2 \tag{16}$$

Using (8) and (13), we obtain:

$$\langle D \rangle' = q \int_{-\infty}^{+\infty} u \, |\varphi_n(u)|^2 \, \mathrm{d}u + \frac{q^2 \mathscr{E}}{m\omega^2} \int_{-\infty}^{+\infty} |\varphi_n(u)|^2 \, \mathrm{d}u = \frac{q^2 \mathscr{E}}{m\omega^2} \tag{17}$$

since the first integral is zero by symmetry. $\langle D \rangle'$ is therefore proportional to $\mathscr{E}$. In this model, the electrical susceptibility of the atomic electron under consideration is equal to:

$$\chi = \frac{\langle D \rangle'}{\mathscr{E}} = \frac{q^2}{m\omega^2} \tag{18}$$

It is positive, whatever the sign of $q$.

It is simple to interpret result (18) physically. The effect of the electric field is to shift the classical equilibrium position of the electron, that is, the mean value of its position in quantum mechanics [see formula (13)]. This results in the appearance of an induced dipole moment. $\chi$ decreases when $\omega$ increases because the oscillator is less easily deformable when the restoring force (which is proportional to $\omega^2$) is larger.

### b. INTERPRETATION OF THE ENERGY SHIFT

Using the model just described, we can interprêt formula (12) by calculating the variation in the mean kinetic and potential energies of the electron when it passes from the state $| \varphi_n \rangle$ to the state $| \varphi'_n \rangle$.

The variation in the kinetic energy is, in fact, zero (as can be understood intuitively from figure 1, for example):

$$\left\langle \frac{P^2}{2m} \right\rangle' - \left\langle \frac{P^2}{2m} \right\rangle = - \frac{\hbar^2}{2m} \left[ \int_{-\infty}^{+\infty} \varphi'^*_n(x) \frac{\mathrm{d}^2}{\mathrm{d}x^2} \varphi'_n(x) \, \mathrm{d}x - \int_{-\infty}^{+\infty} \varphi^*_n(x) \frac{\mathrm{d}^2}{\mathrm{d}x^2} \varphi_n(x) \, \mathrm{d}x \right] = 0 \tag{19}$$

according to formula (13).

The variation in the potential energy can be treated in two terms:

– the first term, $\langle W(\mathscr{E}) \rangle'$, corresponds to the electrical potential energy of the dipole in the field $\mathscr{E}$. Since the dipole is parallel to the field, we have, according to (17):

$$\langle W(\mathscr{E}) \rangle' = - \mathscr{E} \langle D \rangle' = - \frac{q^2 \mathscr{E}^2}{m\omega^2} \tag{20}$$

– the second term, $\langle V(X) \rangle' - \langle V(X) \rangle$, arises from the electric field modification of the wave function of the level labeled by the quantum number $n$. The "elastic" potential energy of the particle therefore changes by a quantity:

$$\langle V(X) \rangle' - \langle V(X) \rangle = \frac{1}{2} m\omega^2 \left[ \int_{-\infty}^{+\infty} x^2 \, |\varphi_n'(x)|^2 \, \mathrm{d}x - \int_{-\infty}^{+\infty} x^2 \, |\varphi_n(x)|^2 \, \mathrm{d}x \right] \tag{21}$$

The first integral can be calculated by using (13) and the change of variable (8):

$$\int_{-\infty}^{+\infty} x^2 \, |\varphi_n'(x)|^2 \, \mathrm{d}x = \int_{-\infty}^{+\infty} u^2 \, |\varphi_n(u)|^2 \, \mathrm{d}u + \frac{2q\mathscr{E}}{m\omega^2} \int_{-\infty}^{+\infty} u \, |\varphi_n(u)|^2 \, \mathrm{d}u + \left( \frac{q\mathscr{E}}{m\omega^2} \right)^2 \int_{-\infty}^{+\infty} |\varphi_n(u)|^2 \, \mathrm{d}u \tag{22}$$

Since $\varphi_n(u)$ is normalized, and since the integral of $u \, |\varphi_n(u)|^2$ is zero by symmetry, we obtain, finally:

$$\langle V(X) \rangle' - \langle V(X) \rangle = \frac{q^2 \mathscr{E}^2}{2m\omega^2} \tag{23}$$

We see why this result should be positive, since the electric field moves the particle away from the point $O$ and attracts it into a region where the "elastic" potential energy $V(x)$ is larger.

Adding (20) to (23), we again find that the energy of the state $| \varphi_n' \rangle$ is less than that of the state $| \varphi_n \rangle$ by $q^2 \mathscr{E}^2 / 2m\omega^2$.

## 3. Use of the translation operator

We shall see in this section that, instead of using the $\{ | x \rangle \}$ representation, as we have done until now, we can argue directly in terms of the operator $H'(\mathscr{E})$ given in (4). More precisely, we are going to show that a unitary transformation (which corresponds to a translation of the wave function along the $x$-axis) transforms the operator $H = H'(\mathscr{E} = 0)$ into the operator $H'(\mathscr{E})$ (to within an additive constant which does not change the eigenvectors). Since the eigenvectors and eigenvalues of $H$ were determined in chapter V, this approach enables us to solve our problem.

Therefore, consider the operator:

$$U(\lambda) = \mathrm{e}^{-\lambda(a - a^\dagger)} \tag{24}$$

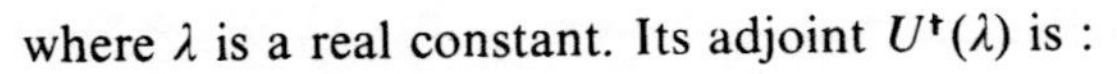



where $\lambda$ is a real constant. Its adjoint $U^\dagger(\lambda)$ is :

$$U^\dagger(\lambda) = e^{\lambda(a - a^\dagger)} \tag{25}$$

It is clear that:

$$U(\lambda)\, U^\dagger(\lambda) = U^\dagger(\lambda)\, U(\lambda) = 1 \tag{26}$$

$U(\lambda)$ is therefore a unitary operator. Under the corresponding unitary transformation, $H$ becomes:

$$\begin{aligned} \tilde{H} &= U(\lambda)\, H\, U^\dagger(\lambda) \\ &= \hbar\omega \left[ \frac{1}{2} + U(\lambda)\, a^\dagger a\, U^\dagger(\lambda) \right] \end{aligned} \tag{27}$$

We must now calculate the operator:

$$U(\lambda)\, a^\dagger a\, U^\dagger(\lambda) = \tilde{a}^\dagger \tilde{a} \tag{28}$$

with:

$$\begin{aligned} \tilde{a} &= U(\lambda)\; a\; U^\dagger(\lambda) \\ \tilde{a}^\dagger &= U(\lambda)\, a^\dagger\, U^\dagger(\lambda) \end{aligned} \tag{29}$$

To obtain $\tilde{a}$ and $\tilde{a}^\dagger$, we use formula (63) of complement $B_{II}$ (which can be applied here since the commutator of $a$ and $a^\dagger$ is equal to 1), which yields:

$$\begin{aligned} U(\lambda) &= e^{-\lambda a + \lambda a^\dagger} = e^{-\lambda a}\, e^{\lambda a^\dagger}\, e^{\lambda^2/2} \\ U^\dagger(\lambda) &= e^{-\lambda a^\dagger + \lambda a} = e^{-\lambda a^\dagger}\, e^{\lambda a}\, e^{-\lambda^2/2} \end{aligned} \tag{30}$$

Also, formula (51) of complement $B_{II}$ enables us to write:

$$\begin{cases} [e^{-\lambda a}, a^\dagger] = -\lambda\, e^{-\lambda a} \\ [e^{\lambda a^\dagger}, a] \;\; = -\lambda\, e^{\lambda a^\dagger} \end{cases} \tag{31}$$

that is:

$$\begin{aligned} e^{-\lambda a}\, a^\dagger\, e^{\lambda a} &= a^\dagger - \lambda \\ e^{\lambda a^\dagger}\, a\, e^{-\lambda a^\dagger} &= a - \lambda \end{aligned} \tag{32}$$

Thus it follows that:

$$\begin{aligned} \tilde{a} &= e^{-\lambda a}\, e^{\lambda a^\dagger}\, a\, e^{-\lambda a^\dagger}\, e^{\lambda a} \\ &= e^{-\lambda a}(a - \lambda)\, e^{\lambda a} = a - \lambda \end{aligned} \tag{33}$$

and, similarly:

$$\tilde{a}^\dagger = a^\dagger - \lambda \tag{34}$$

$\tilde{H}$ is therefore given by:

$$\begin{aligned} \tilde{H} &= \hbar\omega \left[ \frac{1}{2} + (a^\dagger - \lambda)\,(a - \lambda) \right] \\ &= \hbar\omega \left[ \frac{1}{2} + a^\dagger a - \lambda(a + a^\dagger) + \lambda^2 \right] \\ &= H - \lambda\hbar\omega\,(a + a^\dagger) + \lambda^2 \hbar\omega \end{aligned} \tag{35}$$

Since $(a + a^\dagger)$ is proportional to the operator $X$ [formulas (B-1) and (B-7) of chapter V], it suffices to set:

$$\lambda = \frac{q\mathscr{E}}{\omega}\sqrt{\frac{1}{2m\hbar\omega}} \tag{36}$$

to obtain:

$$\tilde{H} = H - q\mathscr{E}X + \frac{q^2\mathscr{E}^2}{2m\omega^2} = H'(\mathscr{E}) + \frac{q^2\mathscr{E}^2}{2m\omega^2} \tag{37}$$

The two operators $\tilde{H}$ and $H'(\mathscr{E})$ therefore have the same eigenvectors, and their eigenvalues differ by $q^2\mathscr{E}^2/2m\omega^2$. Now, we know (*cf.* complement $C_{II}$, § 2) that if the eigenvectors of $H$ are the kets $| \varphi_n \rangle$, those of $\tilde{H}$ are the kets:

$$| \tilde{\varphi}_n \rangle = U(\lambda) | \varphi_n \rangle \tag{38}$$

and the corresponding eigenvalues of $H$ and $\tilde{H}$ are the same. The stationary states $| \varphi'_n \rangle$ of the harmonic oscillator in the presence of the field $\mathscr{E}$ are therefore the state $| \tilde{\varphi}_n \rangle$ given by (38). The associated eigenvalue of $H'(\mathscr{E})$ is, according to (37):

$$E'_n(\mathscr{E}) = \left(n + \frac{1}{2}\right)\hbar\omega - \frac{q^2\mathscr{E}^2}{2m\omega^2} \tag{39}$$

which is the same as formula (12) of the preceding section. Expression (38) for the eigenvectors can be put into the form:

$$| \varphi'_n \rangle = | \tilde{\varphi}_n \rangle = e^{-i\frac{q\mathscr{E}}{m\hbar\omega^2}P} | \varphi_n \rangle \tag{40}$$

using (24) and (36), as well as formulas (B-1) and (B-7) of chapter V. We interpreted, in complement $E_{II}$, the operator $e^{-iaP/\hbar}$ as being the translation operator over an algebraic distance $a$ along $Ox$. $| \varphi'_n \rangle$ is therefore the state obtained from $| \varphi_n \rangle$ by a translation $q\mathscr{E}/m\omega^2$, just as is indicated by formula (13).

**References**

The elastically bound electron : see references of complement $A_{XIII}$.

## Complement $G_V$

# COHERENT "QUASI-CLASSICAL" STATES OF THE HARMONIC OSCILLATOR



The properties of the stationary states $| \varphi_n \rangle$ of the harmonic oscillator were studied in chapter V; for example, in § D, we saw that the mean values $\langle X \rangle$ and $\langle P \rangle$ of the position and the momentum of the oscillator are zero in such a state. Now, in classical mechanics, it is well-known that the position $x$ and the momentum $p$ are oscillating functions of time, which always remain zero only if the energy of the motion is also zero [*cf.* chap. V, relations (A-5) and (A-8)]. Furthermore, we know that quantum mechanics must yield the same results as classical mechanics in the limiting case where the harmonic oscillator has an energy much greater than the quantum $\hbar\omega$ (limit of large quantum numbers).

Thus, we may ask the following question : is it possible to construct quantum mechanical states leading to physical predictions which are almost identical to the classical ones, at least for a macroscopic oscillator? We shall see in this complement that such quantum states exist: they are coherent linear superpositions of all the states $| \varphi_n \rangle$. We shall call them "quasi-classical states" or "coherent states of the harmonic oscillator".

The problem we are considering here is of great general interest in quantum mechanics. As we saw in the introduction to chapter V and in complement $A_V$, many physical systems can be likened to a harmonic oscillator, at least to a first approximation. For all these systems, it is important to understand, in the framework of quantum mechanics, how to move gradually from the case in which the results given by the classical approximation are sufficient to the case in which quantum effects are preponderant. Electromagnetic radiation is a very important example of such a system. Depending on the experiment, it either reveals its quantum mechanical nature (as is the case in the experiment discussed

in § A-2-a of chapter I, in which the light intensity is very low) or else can be treated classically. "Coherent states" of electromagnetic radiation were recently introduced by Glauber and are now in current use in the domain of quantum optics.

The position, the momentum, and the energy of a harmonic oscillator are described in quantum mechanics by operators which do not commute; they are incompatible physical quantities. It is not possible, therefore, to construct a state in which they are all perfectly well-defined. We shall confine ourselves to looking for a state vector such that, for all $t$, the mean values $\langle X \rangle$, $\langle P \rangle$ and $\langle H \rangle$ are as close as possible to the corresponding classical values. This will lead us to a compromise in which none of these three observables is perfectly known. We shall see, nevertheless, that the root-mean-square deviations $\Delta X$, $\Delta P$ and $\Delta H$ are, in the macroscopic limit, completely negligible.

## 1. Quasi-classical states

### a. INTRODUCING THE PARAMETER $\alpha_0$ TO CHARACTERIZE A CLASSICAL MOTION

The classical equations of motion of a one-dimensional harmonic oscillator, of mass $m$ and angular frequency $\omega$, are written:

$$\begin{cases} \dfrac{\mathrm{d}}{\mathrm{d}t} x(t) = \dfrac{1}{m} p(t) & \text{(1-a)} \\[2ex] \dfrac{\mathrm{d}}{\mathrm{d}t} p(t) = -\, m\omega^2 \, x(t) & \text{(1-b)} \end{cases}$$

The quantum mechanical calculations we shall perform later will be simplified by the introduction of the dimensionless quantities:

$$\begin{cases} \hat{x}(t) = \beta \, x(t) \\[1ex] \hat{p}(t) = \dfrac{1}{\hbar\beta} \, p(t) \end{cases} \tag{2}$$

where:

$$\beta = \sqrt{\frac{m\omega}{\hbar}} \tag{3}$$

Equations (1) can then be written:

$$\begin{cases} \dfrac{\mathrm{d}}{\mathrm{d}t} \hat{x}(t) = \omega \, \hat{p}(t) & \text{(4-a)} \\[2ex] \dfrac{\mathrm{d}}{\mathrm{d}t} \hat{p}(t) = -\, \omega \, \hat{x}(t) & \text{(4-b)} \end{cases}$$

The classical state of the harmonic oscillator is determined at the time $t$ when we know its position $x(t)$ and its momentum $p(t)$, that is, $\hat{x}(t)$ and $\hat{p}(t)$. We shall therefore combine these two real numbers into a single dimensionless complex number $\alpha(t)$ defined by:

$$\alpha(t) = \frac{1}{\sqrt{2}} \left[ \hat{x}(t) + i \, \hat{p}(t) \right] \tag{5}$$

The set of two equations (4) is equivalent to the single equation :

$$\frac{d}{dt}\alpha(t) = -i\omega\,\alpha(t) \tag{6}$$

whose solution is :

$$\alpha(t) = \alpha_0\,e^{-i\omega t} \tag{7}$$

where we have set :

$$\alpha_0 = \alpha(0) = \frac{1}{\sqrt{2}}[\,\hat{x}(0) + i\,\hat{p}(0)\,] \tag{8}$$

Now consider the points $M_0$ and $M$ in the complex plane that correspond to the complex numbers $\alpha_0$ and $\alpha(t)$ [fig. 1]. $M$ is at $M_0$ at $t = 0$ and describes, with an angular velocity $-\omega$, a circle centered at $O$ of radius $OM_0$.

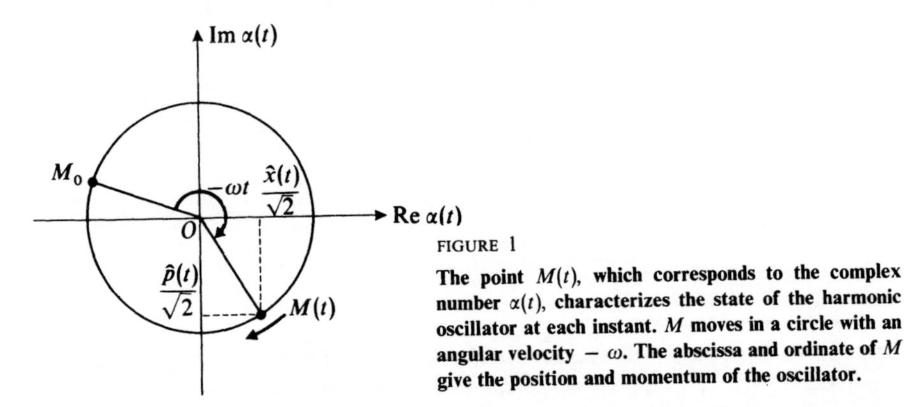


FIGURE 1

**The point $M(t)$, which corresponds to the complex number $\alpha(t)$, characterizes the state of the harmonic oscillator at each instant. $M$ moves in a circle with an angular velocity $-\omega$. The abscissa and ordinate of $M$ give the position and momentum of the oscillator.**

Since, according to (5), the coordinates of $M$ are equal to $\hat{x}(t)/\sqrt{2}$ and $\hat{p}(t)/\sqrt{2}$, we thus obtain a very simple geometrical representation of the time evolution of the state of the system. Every possible motion corresponding to given initial conditions is entirely characterized by the point $M_0$, that is, by the complex number $\alpha_0$ (the modulus of $\alpha_0$ gives the amplitude of the oscillation and the argument of $\alpha_0$, its phase). According to (5) and (7), we have :

$$\begin{cases} \hat{x}(t) = \dfrac{1}{\sqrt{2}}[\,\alpha_0\,e^{-i\omega t} + \alpha_0^*\,e^{i\omega t}\,] & \text{(9-a)} \\[2ex] \hat{p}(t) = -\dfrac{i}{\sqrt{2}}[\,\alpha_0\,e^{-i\omega t} - \alpha_0^*\,e^{i\omega t}\,] & \text{(9-b)} \end{cases}$$

As for the classical energy $\mathscr{H}$ of the system, it is constant in time and equal to :

$$\begin{aligned} \mathscr{H} &= \frac{1}{2m}[p(0)]^2 + \frac{1}{2}m\omega^2[x(0)]^2 \\ &= \frac{\hbar\omega}{2}\{\,[\hat{x}(0)]^2 + [\hat{p}(0)]^2\,\} \end{aligned} \tag{10}$$

which yields, taking (8) into account:

$$\mathscr{H} = \hbar\omega \, |\alpha_0|^2 \tag{11}$$

For a macroscopic oscillator, the energy $\mathscr{H}$ is much greater than the quantum $\hbar\omega$, so:

$$|\alpha_0| \gg 1 \tag{12}$$

### b. CONDITIONS DEFINING QUASI-CLASSICAL STATES

We are looking for a quantum mechanical state for which at every instant the mean values $\langle X \rangle$, $\langle P \rangle$ and $\langle H \rangle$ are practically equal to the values $x$, $p$ and $\mathscr{H}$ which correspond to a given classical motion.

To calculate $\langle X \rangle$, $\langle P \rangle$ and $\langle H \rangle$, we use the expressions:

$$\hat{X} = \beta X = \frac{1}{\sqrt{2}}(a + a^\dagger)$$

$$\hat{P} = \frac{1}{\hbar\beta} P = -\frac{i}{\sqrt{2}}(a - a^\dagger) \tag{13}$$

and:

$$H = \hbar\omega\left(a^\dagger a + \frac{1}{2}\right) \tag{14}$$

For an arbitrary state $|\psi(t)\rangle$, the time evolution of the matrix element $\langle a \rangle(t) = \langle \psi(t) | a | \psi(t) \rangle$ is given by (*cf.* §D-1-d of chapter III):

$$i\hbar \frac{\mathrm{d}}{\mathrm{d}t} \langle a \rangle(t) = \langle [a, H] \rangle(t) \tag{15}$$

Now:

$$[a, H] = \hbar\omega \, [a, a^\dagger a] = \hbar\omega \, a \tag{16}$$

which implies:

$$i \frac{\mathrm{d}}{\mathrm{d}t} \langle a \rangle(t) = \omega \langle a \rangle(t) \tag{17}$$

that is:

$$\langle a \rangle(t) = \langle a \rangle(0) \, \mathrm{e}^{-i\omega t} \tag{18}$$

The evolution of $\langle a^\dagger \rangle(t) = \langle \psi(t) | a^\dagger | \psi(t) \rangle$ obeys the complex conjugate equation:

$$\begin{aligned} \langle a^\dagger \rangle(t) &= \langle a^\dagger \rangle(0) \, \mathrm{e}^{i\omega t} \\ &= \langle a \rangle^*(0) \, \mathrm{e}^{i\omega t} \end{aligned} \tag{19}$$

(18) and (19) are analogous to the classical equation (7).

Substituting (18) and (19) into (13), we obtain:

$$\begin{cases} \langle \hat{X} \rangle(t) = \dfrac{1}{\sqrt{2}} [\langle a \rangle(0) \, e^{-i\omega t} + \langle a \rangle^*(0) \, e^{i\omega t}] \\ \langle \hat{P} \rangle(t) = -\dfrac{i}{\sqrt{2}} [\langle a \rangle(0) \, e^{-i\omega t} - \langle a \rangle^*(0) \, e^{i\omega t}] \end{cases} \quad (20)$$

Comparing these results with (9), we see that, in order to have at all times $t$:

$$\begin{cases} \langle \hat{X} \rangle(t) = \hat{x}(t) \\ \langle \hat{P} \rangle(t) = \hat{p}(t) \end{cases} \quad (21)$$

it is necessary and sufficient to set, at the instant $t = 0$, the condition:

$$\langle a \rangle(0) = \alpha_0 \quad (22)$$

where $\alpha_0$ is the complex parameter characterizing the classical motion which we are trying to reproduce quantum mechanically. The normalized state vector $| \psi(t) \rangle$ of the oscillator must therefore satisfy the condition:

$$\langle \psi(0) | a | \psi(0) \rangle = \alpha_0 \quad (23)$$

We must now require its mean value:

$$\langle H \rangle = \hbar\omega \langle a^\dagger a \rangle(0) + \frac{\hbar\omega}{2} \quad (24)$$

to be equal to the classical energy $\mathscr{H}$ given by (11). Since, for a classical oscillator, $|\alpha_0|$ is much greater than 1 [*cf.* (12)], we shall neglect the term $\hbar\omega/2$ (of purely quantum mechanical origin; see §D-2 of chapter V) with respect to $\hbar\omega \, |\alpha_0|^2$. The second condition on the state vector can now be written:

$$\langle a^\dagger a \rangle(0) = |\alpha_0|^2 \quad (25)$$

that is:

$$\langle \psi(0) | a^\dagger a | \psi(0) \rangle = |\alpha_0|^2 \quad (26)$$

We shall see that conditions (23) and (26) are sufficient to determine the normalized state vector $| \psi(0) \rangle$ (to within a constant phase factor).

### c. QUASI-CLASSICAL STATES ARE EIGENVECTORS OF THE OPERATOR $a$

We introduce the operator $b(\alpha_0)$ defined by:

$$b(\alpha_0) = a - \alpha_0 \quad (27)$$

We then have:

$$b^\dagger(\alpha_0) \, b(\alpha_0) = a^\dagger a - \alpha_0 a^\dagger - \alpha_0^* a + \alpha_0^* \alpha_0 \quad (28)$$

and the square of the norm of the ket $b(\alpha_0) \, | \psi(0) \rangle$ is:

$$\begin{aligned} \langle \psi(0) | b^\dagger(\alpha_0) \, b(\alpha_0) | \psi(0) \rangle = \\ \langle \psi(0) | a^\dagger a | \psi(0) \rangle - \alpha_0 \langle \psi(0) | a^\dagger | \psi(0) \rangle \\ - \alpha_0^* \langle \psi(0) | a | \psi(0) \rangle + \alpha_0^* \alpha_0 \end{aligned} \quad (29)$$

Substituting into this relation conditions (23) and (26), we obtain :

$$\langle \psi(0) \mid b^{\dagger}(\alpha_0)\, b(\alpha_0) \mid \psi(0) \rangle = \alpha_0^* \alpha_0 - \alpha_0 \alpha_0^* - \alpha_0^* \alpha_0 + \alpha_0^* \alpha_0 = 0 \tag{30}$$

The ket $b(\alpha_0) \mid \psi(0) \rangle$, whose norm is zero, is therefore zero :

$$b(\alpha_0) \mid \psi(0) \rangle = 0 \tag{31}$$

that is :

$$a \mid \psi(0) \rangle = \alpha_0 \mid \psi(0) \rangle \tag{32}$$

Conversely, if the normalized vector $\mid \psi(0) \rangle$ satisfies this relation, it is obvious that conditions (23) and (26) are satisfied.

We therefore arrive at the following result: the quasi-classical state, associated with a classical motion characterized by the parameter $\alpha_0$, is such that $\mid \psi(0) \rangle$ is an eigenvector of the operator $a$ with the eigenvalue $\alpha_0$.

In what follows, we shall denote the eigenvector of $a$ with eigenvalue $\alpha$ by $\mid \alpha \rangle$:

$$a \mid \alpha \rangle = \alpha \mid \alpha \rangle \tag{33}$$

[we shall show later that the solution of (33) is unique to within a constant factor].

## 2. Properties of the $\mid \alpha \rangle$ states

### a. EXPANSION OF $\mid \alpha \rangle$ ON THE BASIS OF THE STATIONARY STATES $\mid \varphi_n \rangle$

Let us determine the ket $\mid \alpha \rangle$ which is a solution of (33) by using an expansion on the states $\mid \varphi_n \rangle$ :

$$\mid \alpha \rangle = \sum_n c_n(\alpha) \mid \varphi_n \rangle \tag{34}$$

We then have :

$$a \mid \alpha \rangle = \sum_n c_n(\alpha) \sqrt{n} \mid \varphi_{n-1} \rangle \tag{35}$$

and, substituting this relation into (33), we obtain :

$$c_{n+1}(\alpha) = \frac{\alpha}{\sqrt{n+1}}\, c_n(\alpha) \tag{36}$$

This relation enables us to determine by recurrence all the coefficients $c_n(\alpha)$ in terms of $c_0(\alpha)$:

$$c_n(\alpha) = \frac{\alpha^n}{\sqrt{n!}}\, c_0(\alpha) \tag{37}$$

It follows that, when $c_0(\alpha)$ is fixed, all the $c_n(\alpha)$ are also fixed. The vector $\mid \alpha \rangle$ is therefore unique to within a multiplicative factor. We shall choose $c_0(\alpha)$ real and

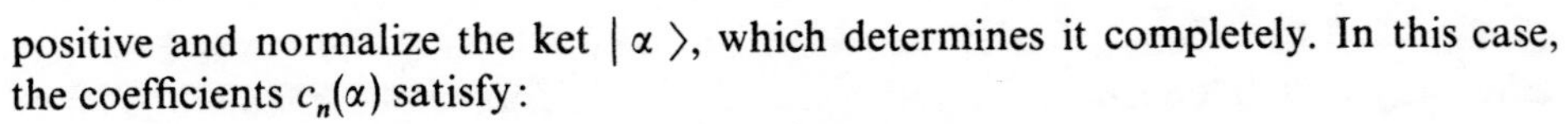


positive and normalize the ket $| \alpha \rangle$, which determines it completely. In this case, the coefficients $c_n(\alpha)$ satisfy:

$$\sum_n |c_n(\alpha)|^2 = 1 \tag{38}$$

that is:

$$|c_0(\alpha)|^2 \sum_n \frac{|\alpha|^{2n}}{n!} = |c_0(\alpha)|^2 \, e^{|\alpha|^2} = 1 \tag{39}$$

With the convention we have chosen:

$$c_0(\alpha) = e^{-|\alpha|^2/2} \tag{40}$$

and, finally:

$$| \alpha \rangle = e^{-|\alpha|^2/2} \sum_n \frac{\alpha^n}{\sqrt{n!}} | \varphi_n \rangle \tag{41}$$

#### b. POSSIBLE VALUES OF THE ENERGY IN AN $| \alpha \rangle$ STATE

Let us consider an oscillator in the state $| \alpha \rangle$. We see from (41) that a measurement of the energy can yield the result $E_n = (n + 1/2)\hbar\omega$ with the probability:

$$\mathscr{P}_n(\alpha) = |c_n(\alpha)|^2 = \frac{|\alpha|^{2n}}{n!} \, e^{-|\alpha|^2} \tag{42}$$

The probability distribution obtained, $\mathscr{P}_n(\alpha)$, is therefore a Poisson distribution. Since:

$$\mathscr{P}_n(\alpha) = \frac{|\alpha|^2}{n} \mathscr{P}_{n-1}(\alpha) \tag{43}$$

it is easy to verify that $\mathscr{P}_n(\alpha)$ reaches its maximum value when:

$$n = \text{the integral part of } |\alpha|^2 \tag{44}$$

To calculate the mean value $\langle H \rangle_\alpha$ of the energy, we can use (42) and the expression:

$$\langle H \rangle_\alpha = \sum_n \mathscr{P}_n(\alpha) \left[ n + \frac{1}{2} \right] \hbar\omega \tag{45}$$

Nevertheless, it is quicker to notice that, since the adjoint relation of (33) is:

$$\langle \alpha | a^\dagger = \alpha^* \langle \alpha | \tag{46}$$

we have:

$$\langle \alpha | a^\dagger a | \alpha \rangle = \alpha^* \alpha \tag{47}$$

and therefore:

$$\langle H \rangle_\alpha = \hbar\omega \langle \alpha | \left[ a^\dagger a + \frac{1}{2} \right] | \alpha \rangle = \hbar\omega \left[ |\alpha|^2 + \frac{1}{2} \right] \tag{48}$$

Comparing this result to (44), we see that, when $|\alpha| \gg 1$, $\langle H \rangle_\alpha$ is not very different, in relative value, from the energy $E_n$ which corresponds to the maximum value of $\mathscr{P}_n(\alpha)$.

Let us calculate the mean value $\langle H^2 \rangle_\alpha$:

$$\langle H^2 \rangle_\alpha = \hbar^2\omega^2 \langle \alpha \left| \left( a^\dagger a + \frac{1}{2} \right)^2 \right| \alpha \rangle \tag{49}$$

Using (33) and the fact that $[a, a^\dagger] = 1$, we easily obtain:

$$\langle H^2 \rangle_\alpha = \hbar^2\omega^2 \left[ |\alpha|^4 + 2\,|\alpha|^2 + \frac{1}{4} \right] \tag{50}$$

from which we get:

$$\Delta H_\alpha = \hbar\omega\,|\alpha| \tag{51}$$

If we compare (48) and (51), we see that, if $|\alpha|$ is very large, we have:

$$\frac{\Delta H_\alpha}{\langle H \rangle_\alpha} \simeq \frac{1}{|\alpha|} \ll 1 \tag{52}$$

The relative value of the energy of the state $|\,\alpha \rangle$ is very well-defined.

COMMENT:

Since:

$$H = \left( N + \frac{1}{2} \right) \hbar\omega \tag{53}$$

we immediately obtain from (48) and (51):

$$\begin{cases} \langle N \rangle_\alpha = |\alpha|^2 \\ \Delta N_\alpha = |\alpha| \end{cases} \tag{54}$$

Thus we see that, to obtain a quasi-classical state, we must linearly superpose a very large number of states $|\,\varphi_n \rangle$ since $\Delta N_\alpha \gg 1$. However, the relative value of the dispersion over $N$ is very small:

$$\frac{\Delta N_\alpha}{\langle N \rangle_\alpha} = \frac{1}{|\alpha|} \ll 1 \tag{55}$$

### c. CALCULATION OF $\langle X \rangle$, $\langle P \rangle$, $\Delta X$ AND $\Delta P$ IN AN $|\,\alpha \rangle$ STATE

The mean values $\langle X \rangle$ and $\langle P \rangle$ can be obtained by expressing $X$ and $P$ in terms of $a$ and $a^\dagger$ [formula (13)] and using (33) and (46). We obtain:

$$\begin{aligned} \langle X \rangle_\alpha &= \langle \alpha \,|\, X \,|\, \alpha \rangle = \sqrt{\frac{2\hbar}{m\omega}}\ \mathrm{Re}\,(\alpha) \\ \langle P \rangle_\alpha &= \langle \alpha \,|\, P \,|\, \alpha \rangle = \sqrt{2m\hbar\omega}\ \mathrm{Im}\,(\alpha) \end{aligned} \tag{56}$$

An analogous calculation yields:

$$\langle X^2 \rangle_\alpha = \frac{\hbar}{2m\omega}\left[ (\alpha + \alpha^*)^2 + 1 \right]$$
$$\langle P^2 \rangle_\alpha = \frac{m\hbar\omega}{2}\left[ 1 - (\alpha - \alpha^*)^2 \right] \tag{57}$$

and therefore:

$$\Delta X_\alpha = \sqrt{\frac{\hbar}{2m\omega}}$$
$$\Delta P_\alpha = \sqrt{\frac{m\hbar\omega}{2}} \tag{58}$$

Neither $\Delta X_\alpha$ nor $\Delta P_\alpha$ depends on $\alpha$. Note also that $\Delta X \,.\, \Delta P$ takes on its minimum value:

$$\Delta X_\alpha \,.\, \Delta P_\alpha = \hbar/2 \tag{59}$$

#### d. THE OPERATOR $D(\alpha)$; THE WAVE FUNCTIONS $\psi_\alpha(x)$

Consider the operator $D(\alpha)$ defined by:

$$D(\alpha) = e^{\alpha a^\dagger - \alpha^* a} \tag{60}$$

This operator is unitary, since:

$$D^\dagger(\alpha) = e^{\alpha^* a - \alpha a^\dagger} \tag{61}$$

immediately implies:

$$D(\alpha) D^\dagger(\alpha) = D^\dagger(\alpha) D(\alpha) = 1 \tag{62}$$

Since the commutator of the operators $\alpha a^\dagger$ and $-\alpha^* a$ is equal to $\alpha^*\alpha$, which is a number, we can use relation (63) of complement $B_{II}$ to write:

$$D(\alpha) = e^{-|\alpha|^2/2}\, e^{\alpha a^\dagger}\, e^{-\alpha^* a} \tag{63}$$

Now let us calculate the ket $D(\alpha) | \varphi_0 \rangle$; since:

$$e^{-\alpha^* a} | \varphi_0 \rangle = \left[ 1 - \alpha^* a + \frac{\alpha^{*2}}{2!} a^2 + \ldots \right] | \varphi_0 \rangle$$
$$= | \varphi_0 \rangle \tag{64}$$

then:

$$D(\alpha) | \varphi_0 \rangle = e^{-|\alpha|^2/2}\, e^{\alpha a^\dagger} | \varphi_0 \rangle$$
$$= e^{-|\alpha|^2/2} \sum_n \frac{(\alpha a^\dagger)^n}{n!} | \varphi_0 \rangle$$
$$= e^{-|\alpha|^2/2} \sum_n \frac{\alpha^n}{\sqrt{n!}} | \varphi_n \rangle \tag{65}$$

Comparing (41) and (65), we see that:

$$| \alpha \rangle = D(\alpha) | \varphi_0 \rangle \tag{66}$$

$D(\alpha)$ is therefore the unitary transformation which transforms the ground state $| \varphi_0 \rangle$ into the quasi-classical state $| \alpha \rangle$.

Formula (66) will enable us to obtain the wave function:

$$\psi_\alpha(x) = \langle x | \alpha \rangle \tag{67}$$

which characterizes the quasi-classical state $| \alpha \rangle$ in the $\{ | x \rangle \}$ representation. To calculate:

$$\psi_\alpha(x) = \langle x | D(\alpha) | \varphi_0 \rangle \tag{68}$$

we shall write the operator $\alpha a^\dagger - \alpha^* a$ in terms of $X$ and $P$:

$$\alpha a^\dagger - \alpha^* a = \sqrt{\frac{m\omega}{\hbar}} \left( \frac{\alpha - \alpha^*}{\sqrt{2}} \right) X - \frac{i}{\sqrt{m\hbar\omega}} \left( \frac{\alpha + \alpha^*}{\sqrt{2}} \right) P \tag{69}$$

Using formula (63) of complement $B_{II}$ again, we obtain:

$$D(\alpha) = e^{\alpha a^\dagger - \alpha^* a} = e^{\sqrt{\frac{m\omega}{\hbar}} \frac{\alpha - \alpha^*}{\sqrt{2}} X} \, e^{-\frac{i}{\sqrt{m\hbar\omega}} \frac{\alpha + \alpha^*}{\sqrt{2}} P} \, e^{\frac{\alpha^{*2} - \alpha^2}{4}} \tag{70}$$

Substituting this result into (68), we find:

$$\begin{aligned} \psi_\alpha(x) &= e^{\frac{\alpha^{*2} - \alpha^2}{4}} \langle x | e^{\sqrt{\frac{m\omega}{\hbar}} \frac{\alpha - \alpha^*}{\sqrt{2}} X} \, e^{-\frac{i}{\sqrt{m\hbar\omega}} \frac{\alpha + \alpha^*}{\sqrt{2}} P} | \varphi_0 \rangle \\ &= e^{\frac{\alpha^{*2} - \alpha^2}{4}} \, e^{\sqrt{\frac{m\omega}{\hbar}} \frac{\alpha - \alpha^*}{\sqrt{2}} x} \langle x | e^{-\frac{i}{\sqrt{m\hbar\omega}} \frac{\alpha + \alpha^*}{\sqrt{2}} P} | \varphi_0 \rangle \end{aligned} \tag{71}$$

Now, the operator $e^{-i\lambda P/\hbar}$ is the translation operator of $\lambda$ along $Ox$ (*cf.* complement $E_{II}$):

$$\langle x | e^{-\frac{i}{\sqrt{m\hbar\omega}} \frac{\alpha + \alpha^*}{\sqrt{2}} P} = \langle x - \sqrt{\frac{\hbar}{2m\omega}} (\alpha + \alpha^*) | \tag{72}$$

Relation (71) therefore yields:

$$\psi_\alpha(x) = e^{\frac{\alpha^{*2} - \alpha^2}{4}} \, e^{\sqrt{\frac{m\omega}{\hbar}} \frac{\alpha - \alpha^*}{\sqrt{2}} x} \varphi_0\left( x - \sqrt{\frac{\hbar}{2m\omega}} (\alpha + \alpha^*) \right) \tag{73}$$

If we write $\alpha$ and $\alpha^*$ in terms of $\langle X \rangle_\alpha$ and $\langle P \rangle_\alpha$ [formulas (56)], $\psi_\alpha(x)$ becomes:

$$\psi_\alpha(x) = e^{i\theta_\alpha} \, e^{i\langle P \rangle_\alpha x/\hbar} \varphi_0(x - \langle X \rangle_\alpha) \tag{74}$$

where the global phase factor $e^{i\theta_\alpha}$ is defined by:

$$e^{i\theta_\alpha} = e^{\frac{\alpha^{*2} - \alpha^2}{4}} \tag{75}$$

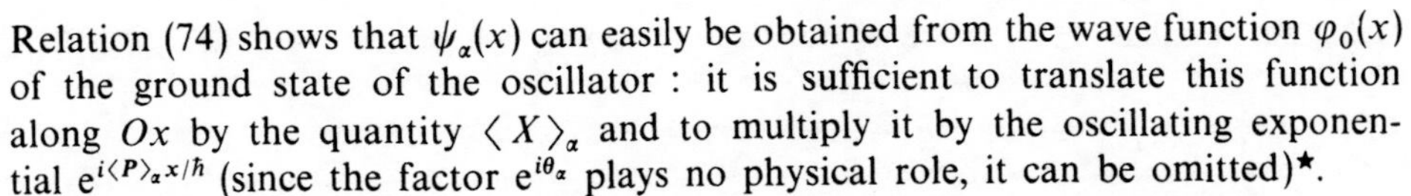

Relation (74) shows that $\psi_\alpha(x)$ can easily be obtained from the wave function $\varphi_0(x)$ of the ground state of the oscillator : it is sufficient to translate this function along $Ox$ by the quantity $\langle X \rangle_\alpha$ and to multiply it by the oscillating exponential $e^{i\langle P \rangle_\alpha x/\hbar}$ (since the factor $e^{i\theta_\alpha}$ plays no physical role, it can be omitted)★.

If we replace $\varphi_0$ in (74) by its explicit expression, we obtain, finally :

$$\psi_\alpha(x) = e^{i\theta_\alpha}\left(\frac{m\omega}{\pi\hbar}\right)^{1/4} \exp\left\{-\left[\frac{x - \langle X \rangle_\alpha}{2\,\Delta X_\alpha}\right]^2 + i\langle P \rangle_\alpha \frac{x}{\hbar}\right\} \tag{76}$$

The form of the wave packet associated with the $|\alpha\rangle$ state is therefore given by :

$$|\psi_\alpha(x)|^2 = \sqrt{\frac{m\omega}{\pi\hbar}}\exp\left\{-\frac{1}{2}\left[\frac{x - \langle X \rangle_\alpha}{\Delta X_\alpha}\right]^2\right\} \tag{77}$$

For any $|\alpha\rangle$ state, we obtain a Gaussian wave packet. This result should be compared with the fact that the product $\Delta X_\alpha \,.\, \Delta P_\alpha$ is always minimal (*cf.* complement $C_{III}$).

### e. THE SCALAR PRODUCT OF TWO $|\alpha\rangle$ STATES. CLOSURE RELATION

The $|\alpha\rangle$ states are eigenvectors of the non-Hermitian operator $a$. There is therefore no obvious reason for these states to satisfy orthogonality and closure relations. In this section, we shall investigate this question.

First, we shall consider two eigenkets $|\alpha\rangle$ and $|\alpha'\rangle$ of the operator $a$. Relation (41) gives their scalar product, since :

$$\langle \alpha | \alpha' \rangle = \sum_n c_n^*(\alpha)\, c_n(\alpha') \tag{78}$$

We therefore have :

$$\begin{aligned} \langle \alpha | \alpha' \rangle &= e^{-|\alpha|^2/2}\, e^{-|\alpha'|^2/2} \sum_n \frac{(\alpha^*\alpha')^n}{n!} \\ &= e^{-|\alpha|^2/2}\, e^{-|\alpha'|^2/2}\, e^{\alpha^*\alpha'} \end{aligned} \tag{79}$$

from which we conclude :

$$|\langle \alpha | \alpha' \rangle|^2 = e^{-|\alpha - \alpha'|^2} \tag{80}$$

This scalar product is therefore never zero.

However, we shall show that the $|\alpha\rangle$ states do satisfy a closure relation, which is written :

$$\frac{1}{\pi}\iint |\alpha\rangle\langle\alpha|\, \mathrm{d}\{\mathrm{Re}\,\alpha\}\, \mathrm{d}\{\mathrm{Im}\,\alpha\} = 1 \tag{81}$$

★ The exponential $e^{i\langle P \rangle_\alpha x/\hbar}$ is obviously not a global phase factor since its value depends on $x$. The presence of this exponential in (74) insures that the mean value of $P$ in the state described by $\psi_\alpha(x)$ be equal to $\langle P \rangle_\alpha$.

To do so, we replace $| \alpha \rangle$, on the left-hand side of (81), by its expression (41). This yields :

$$\frac{1}{\pi}\iint e^{-|\alpha|^2} \sum_n \frac{\alpha^n}{\sqrt{n!}} | \varphi_n \rangle \sum_m \frac{\alpha^{*m}}{\sqrt{m!}} \langle \varphi_m | \, d\{ \text{Re}\, \alpha \} \, d\{ \text{Im}\, \alpha \} \tag{82}$$

that is, going into polar coordinates in the complex $\alpha$ plane (setting $\alpha = \rho \, e^{i\varphi}$) :

$$\frac{1}{\pi}\int_0^\infty \rho \, d\rho \int_0^{2\pi} d\varphi \, e^{-\rho^2} \sum_{nm} e^{i(n-m)\varphi} \frac{\rho^{n+m}}{\sqrt{n!\,m!}} | \varphi_n \rangle \langle \varphi_m | \tag{83}$$

The integral over $\varphi$ is easily calculated :

$$\int_0^{2\pi} e^{i(n-m)\varphi} \, d\varphi = 2\pi \, \delta_{nm} \tag{84}$$

which yields for (83) :

$$\sum_n I_n \frac{1}{n!} | \varphi_n \rangle \langle \varphi_n | \tag{85}$$

with :

$$I_n = 2 \int_0^\infty \rho \, d\rho \, e^{-\rho^2} \rho^{2n} = \int_0^\infty du \, e^{-u} \, u^n \tag{86}$$

Integrating by parts, we find a recurrence relation for the $I_n$ :

$$I_n = n \, I_{n-1} \tag{87}$$

whose solution is :

$$I_n = n! \, I_0 = n! \tag{88}$$

Substituting this result into (85), we see that the left-hand side of formula (81) can finally be written :

$$\sum_n | \varphi_n \rangle \langle \varphi_n | \tag{89}$$

which proves that formula.

## 3. Time evolution of a quasi-classical state

Consider a harmonic oscillator which, at the instant $t = 0$, is in a particular $|\alpha\rangle$ state :

$$| \psi(0) \rangle = | \alpha_0 \rangle \tag{90}$$

How do its physical properties evolve over time? We already know (*cf.* § 1-b) that the mean values $\langle X \rangle(t)$ and $\langle P \rangle(t)$ always remain equal to the corresponding classical values. We shall now study other interesting properties of the state vector $| \psi(t) \rangle$.

### a. A QUASI-CLASSICAL STATE ALWAYS REMAINS AN EIGENVECTOR OF *a*

Starting with (41), we can use the general rule to obtain $| \psi(t) \rangle$ when the Hamiltonian is not time-dependent (*cf.* chap. III, § D-2-a):

$$\begin{aligned} | \psi(t) \rangle &= e^{-|\alpha_0|^2/2} \sum_n \frac{\alpha_0^n}{\sqrt{n!}} e^{-iE_n t/\hbar} | \varphi_n \rangle \\ &= e^{-i\omega t/2} e^{-|\alpha_0|^2/2} \sum_n \frac{\alpha_0^n e^{-in\omega t}}{\sqrt{n!}} | \varphi_n \rangle \end{aligned} \tag{91}$$

If we compare this result with (41), we see that, to go from $| \psi(0) \rangle = | \alpha_0 \rangle$ to $| \psi(t) \rangle$, all we must do is change $\alpha_0$ to $\alpha_0 e^{-i\omega t}$ and multiply the ket obtained by $e^{-i\omega t/2}$ (which is a global phase factor with no physical consequences):

$$| \psi(t) \rangle = e^{-i\omega t/2} | \alpha = \alpha_0 e^{-i\omega t} \rangle \tag{92}$$

In other words, we see that a quasi-classical state remains an eigenvector of $a$ for all time, with an eigenvalue $\alpha_0 e^{-i\omega t}$ which is nothing more than the parameter $\alpha(t)$ of figure 1 (corresponding to the point $M$), which characterizes at all times the classical oscillator whose motion is reproduced by the state $| \psi(t) \rangle$.

### b. EVOLUTION OF PHYSICAL PROPERTIES

Using (92) and changing $\alpha$ to $\alpha_0 e^{-i\omega t}$ in (56), we immediately obtain:

$$\begin{cases} \langle X \rangle(t) = \sqrt{\dfrac{2\hbar}{m\omega}} \text{Re} \left[\alpha_0 e^{-i\omega t}\right] \\ \langle P \rangle(t) = \sqrt{2m\hbar\omega} \text{ Im} \left[\alpha_0 e^{-i\omega t}\right] \end{cases} \tag{93}$$

As predicted, these equations are similar to the classical equations (9).

The average energy of the oscillator is time-independent:

$$\langle H \rangle = \hbar\omega \left[ |\alpha_0|^2 + \frac{1}{2} \right] \tag{94}$$

According to (51) and (58), the root-mean-square deviations $\Delta H$, $\Delta X$ and $\Delta P$ are equal to:

$$\Delta H = \hbar\omega |\alpha_0| \tag{95}$$

and:

$$\begin{cases} \Delta X = \sqrt{\dfrac{\hbar}{2m\omega}} \\ \Delta P = \sqrt{\dfrac{m\hbar\omega}{2}} \end{cases} \tag{96}$$

$\Delta X$ and $\Delta P$ are not time-dependent; the wave packet remains a minimum wave packet for all time.

### c. MOTION OF THE WAVE PACKET

Let us calculate the wave function at time $t$:

$$\psi(x, t) = \langle x \mid \psi(t) \rangle \tag{97}$$

where $|\psi(t)\rangle$ is given by (92). From (76), we obtain:

$$\psi(x, t) = e^{i\theta_\alpha} \left(\frac{m\omega}{\pi\hbar}\right)^{1/4} e^{-i\omega t/2} e^{i\frac{x\langle P\rangle(t)}{\hbar}} e^{-\left[\frac{x - \langle X\rangle(t)}{2\Delta X}\right]^2} \tag{98}$$

At $t$, the wave packet is still Gaussian. Its form does not vary with time, since:

$$|\psi(x, t)|^2 = |\varphi_0[x - \langle X \rangle(t)]|^2 \tag{99}$$

Thus, it remains "minimum" for all time [*cf.* (96)].

Figure 2 shows the motion of the wave packet, which performs a periodic oscillation ($T = 2\pi/\omega$) along the $Ox$ axis, without becoming distorted. We saw in complement $G_I$ that a Gaussian wave packet, when it is free, becomes distorted as it propagates, since its width varies ("spreading" of the wave packet). We see here that nothing of the sort occurs when a wave packet is subject to the influence of a parabolic potential. Physically, this result arises from the fact that the tendency of the wave packet to spread is compensated by the potential, whose effect is to push the wave packet towards the origin from regions where $V(x)$ is large.

What happens to these results when $|\alpha|$ is very large? The root-mean-square deviations $\Delta X$ and $\Delta P$ do not change, as is shown by (96). On the other hand, the

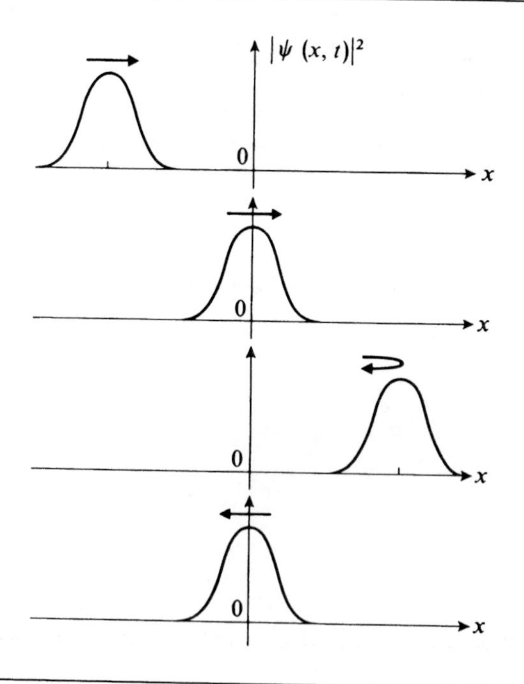


FIGURE 2

**Motion of the Gaussian wave packet associated with an $|\alpha\rangle$ state: under the effect of the parabolic potential $V(x)$, the wave packet oscillates without becoming distorted.**

oscillation amplitudes of $\langle X \rangle(t)$ and $\langle P \rangle(t)$ become much larger than $\Delta X$ and $\Delta P$. By choosing a sufficiently large value of $|\alpha|$, one can obtain a quantum mechanical motion for which the position and momentum of the oscillator are, in relative value, as well-defined as might be desired. Therefore when $|\alpha| \gg 1$, an $|\alpha\rangle$ state describes very well the motion of a macroscopic harmonic oscillator, for which the position, the momentum and the energy can be considered to be classical quantities.

## 4. Example: quantum mechanical treatment of a macroscopic oscillator

Let us consider a concrete example: a macroscopic body of mass $m = 1$ kg suspended from a rope of length $l = 0.1$ m and placed in the gravitational field ($g \simeq 10$ m/s$^2$). We know that, for small oscillations, the period $T$ of the motion is given by:

$$T = 2\pi\sqrt{\frac{l}{g}} \tag{100}$$

In our case, we obtain:

$$\begin{aligned} T &\simeq 0.63 \text{ s} \\ \omega &= 10 \text{ rd/s} \end{aligned} \tag{101}$$

Let us now assume that this oscillator performs a periodic motion of amplitude $x_M = 1$ cm. What is the quantum mechanical state which best represents its oscillation?

We have seen that this state is an $|\alpha\rangle$ state in which, according to (93), $\alpha$ satisfies the relation:

$$|\alpha| = \sqrt{\frac{m\omega}{2\hbar}}\, x_M \tag{102}$$

that is, in our case:

$$|\alpha| \simeq \sqrt{5} \times 10^{15} \simeq 2.2 \times 10^{15} \gg 1 \tag{103}$$

(the argument of $\alpha$ is determined by the initial phase of the motion).

The root-mean-square deviations $\Delta X$ and $\Delta P$ are then:

$$\begin{cases} \Delta X = \sqrt{\dfrac{\hbar}{2m\omega}} \simeq 2.2 \times 10^{-18} \text{m} \ll x_M \\ \Delta P = \sqrt{\dfrac{m\hbar\omega}{2}} \simeq 2.2 \times 10^{-17} \text{ kg m/s} \end{cases} \tag{104}$$

The root-mean-square deviation $\Delta V$ of the velocity is equal to:

$$\Delta V \simeq 2.2 \times 10^{-17} \text{ m/s} \tag{105}$$

Since the maximum velocity of the oscillator is 0.1 m/s, we see that the uncertainties of its position and its velocity are completely negligible compared to the quantities involved in the problem. For example, $\Delta X$ is less than a Fermi ($10^{-15}$ m), that is, the approximate size of a nucleus. Measuring a macroscopic length with this accuracy is obviously out of the question.

Finally, the energy of the oscillator is known with an excellent relative accuracy, since, according to (52):

$$\frac{\Delta H}{\langle H \rangle} \simeq \frac{1}{|\alpha|} \simeq 0.4 \times 10^{-15} \ll 1 \qquad (106)$$

The laws of classical mechanics are therefore quite adequate for the study of the evolution of the macroscopic oscillator.

**References and suggestions for further reading:**

Glauber's lectures in (15.2).

## Complement $H_V$

# NORMAL VIBRATIONAL MODES OF TWO COUPLED HARMONIC OSCILLATORS



This complement is devoted to the study of the motion of two coupled (one-dimensional) harmonic oscillators. Such a study is of interest because it permits the introduction, in a very simple case, of a physically important concept : that of *normal vibrational modes.* This concept, encountered in quantum mechanics as well as in classical mechanics, appears in numerous problems: for example, in the study of atomic vibrations of a crystal (*cf.* complement $J_V$) and of the vibrations of electromagnetic radiation (*cf.* complement $K_V$).

Let us therefore consider two particles (1) and (2), of the same mass $m$, moving along the $Ox$ axis, with abscissas $x_1$ and $x_2$. To begin, we assume their potential energy to be:

$$U_0(x_1, x_2) = \frac{1}{2} m\omega^2(x_1 - a)^2 + \frac{1}{2} m\omega^2(x_2 + a)^2 \tag{1}$$

When $x_1 = a$ and $x_2 = -a$, the potential energy $U_0(x_1, x_2)$ is minimal, and the two particles are in stable equilibrium. If the particles move from these equilibrium positions, they are subject to the forces $F_1$ and $F_2$, respectively:

$$\begin{cases} F_1 = -\dfrac{\partial}{\partial x_1} U_0(x_1, x_2) = -m\omega^2(x_1 - a) \\ F_2 = -\dfrac{\partial}{\partial x_2} U_0(x_1, x_2) = -m\omega^2(x_2 + a) \end{cases} \tag{2}$$

and their motion is given by:

$$\begin{cases} m\dfrac{d^2}{dt^2} x_1(t) = -m\omega^2(x_1 - a) \\ m\dfrac{d^2}{dt^2} x_2(t) = -m\omega^2(x_2 + a) \end{cases} \tag{3}$$

Each particle is therefore in independent sinusoidal motion, centered at its equilibrium position. The amplitude of the motion of each particle is arbitrary★ and can be fixed by a suitable choice of the initial conditions.

Now let us assume the potential energy $U(x_1, x_2)$ of the two particles to be:

$$U(x_1, x_2) = U_0(x_1, x_2) + V(x_1, x_2) \tag{4}$$

with:

$$V(x_1, x_2) = \lambda m\omega^2(x_1 - x_2)^2 \tag{5}$$

(where $\lambda$ is a dimensionless positive constant which we shall call the coupling constant). To the forces $F_1$ and $F_2$ written in (2), we must add, respectively, the forces $F_1'$ and $F_2'$ given by:

$$\begin{cases} F_1' = -\dfrac{\partial}{\partial x_1} V(x_1, x_2) = 2\lambda m\omega^2(x_2 - x_1) \\ F_2' = -\dfrac{\partial}{\partial x_2} V(x_1, x_2) = 2\lambda m\omega^2(x_1 - x_2) \end{cases} \tag{6}$$

We see that the introduction of $V(x_1, x_2)$ takes into account an attractive force between the particles which is proportional to the distance between them. The two particles (1) and (2) are therefore no longer independent; what is their motion now? Before attacking this problem from a quantum mechanical point of view, we shall recall the results given by classical mechanics.

## 1. Vibration of the two particles in classical mechanics

### a. SOLVING THE EQUATIONS OF MOTION

In the presence of the coupling $V(x_1, x_2)$, we must replace (3) by the system of coupled differential equations:

$$\begin{cases} m\dfrac{d^2}{dt^2} x_1(t) = -m\omega^2(x_1 - a) + 2\lambda m\omega^2(x_2 - x_1) \\ m\dfrac{d^2}{dt^2} x_2(t) = -m\omega^2(x_2 + a) + 2\lambda m\omega^2(x_1 - x_2) \end{cases} \tag{7}$$

We know how to solve such a system (*cf.*, for example, chapter IV §C-3-a). We diagonalize the matrix $K$ of the coefficients appearing on the right-hand side of (7):

$$K = -m\omega^2 \begin{pmatrix} 1 + 2\lambda & -2\lambda \\ -2\lambda & 1 + 2\lambda \end{pmatrix} \tag{8}$$

We then replace $x_1(t)$ and $x_2(t)$ by linear combinations of these two functions (given by the eigenvectors of $K$) whose time dependence obeys uncoupled linear differential equations (with coefficients which are the eigenvalues of $K$).

★ Of course, the choice of the potential (1) implies that we are not taking into account the collisions that could occur if sufficiently large amplitudes were chosen.

In this case, these linear combinations are:

$$x_G(t) = \frac{1}{2}\left[x_1(t) + x_2(t)\right] \tag{9}$$

(the position of the center of mass of the two particles) and:

$$x_R(t) = x_1(t) - x_2(t) \tag{10}$$

(the abscissa of the "relative particle"). Substituting (9) and (10) into (7), we obtain (taking the sum and the difference):

$$\frac{d^2}{dt^2} x_G(t) = -\omega^2 x_G(t)$$

$$\frac{d^2}{dt^2} x_R(t) = -\omega^2 \left[x_R(t) - 2a\right] - 4\lambda\omega^2 x_R(t) \tag{11}$$

These equations can be integrated immediately:

$$\begin{cases} x_G(t) = x_G^0 \cos(\omega_G t + \theta_G) \\ x_R(t) = \dfrac{2a}{1 + 4\lambda} + x_R^0 \cos(\omega_R t + \theta_R) \end{cases} \tag{12}$$

with:

$$\begin{cases} \omega_G = \omega \\ \omega_R = \omega\sqrt{1 + 4\lambda} \end{cases} \tag{13}$$

$x_G^0$, $x_R^0$, $\theta_G$ and $\theta_R$ are integration constants fixed by the initial conditions. To obtain the motion of particles (1) and (2), all we must do is invert formulas (9) and (10):

$$\begin{cases} x_1(t) = x_G(t) + \dfrac{1}{2} x_R(t) \\ x_2(t) = x_G(t) - \dfrac{1}{2} x_R(t) \end{cases} \tag{14}$$

and substitute (12) into these equations.

### b. THE PHYSICAL MEANING OF EACH OF THE MODES

Through the change of functions performed in (9) and (10), we have been able to find the motion of the two interacting particles by associating with them two *fictitious particles* (*G*) and (*R*), of abscissas $x_G(t)$ and $x_R(t)$. These fictitious particles do not interact; their motions are independent, so their amplitudes and phases can be fixed arbitrarily by a suitable choice of the initial conditions. For example, it is possible to require one of the two fictitious particles to be motionless without this being the case for the other one: we then say that a *vibrational mode* of the system is excited. It must be understood that, in a vibrational mode, the real particles (1) and (2) are both in motion with the same angular frequency ($\omega_R$ or $\omega_G$, depending on the mode). No solution of the equations of motion exists for which one of the two real particles (1) or (2) remains motionless while the other one vibrates. If, at the instant $t = 0$, one were to give an initial velocity to only one of the two particles, (1) or (2), the coupling force would set the other one in motion (*cf.* discussion of §c below).

The simplest case is of course the one in which neither of the two modes is excited. In formulas (12) such a situation corresponds to $x_G^0 = x_R^0 = 0$; $x_G(t)$ and $x_R(t)$ then always remain equal to zero and $2a/(1 + 4\lambda)$ respectively, which, according to (14), yields:

$$x_1 = -x_2 = \frac{a}{1 + 4\lambda} \tag{15}$$

The system does not oscillate and the two particles (1) and (2) remain motionless in their new equilibrium positions given by (15) (it can be verified that, for these values of $x_1$ and $x_2$, the forces exerted on the particles are zero; the fact that these equilibrium positions are closer in the presence of the coupling than when $\lambda = 0$ is due to their mutual attraction).

To excite only the mode corresponding to $x_G(t)$, one places the two particles (1) and (2) at the initial instant at the same distance $2a/(1 + 4\lambda)$ as in the preceding case, and one gives them equal velocities. One then finds that $x_R(t)$ remains equal to $2a/(1 + 4\lambda)$ (the initial conditions require $x_R^0$ to be zero). The two particles move "in unison", performing the same motion without the distance between them varying. For this mode, the two-particle system can be treated like a single undeformable particle of mass $2m$ on which is exerted the force $F_1 + F_2 = -2m\omega^2 x_G(t)$. We then see why the angular frequency of this mode is $\omega_G = \omega$ [*cf.* formula (A-3) of chapter V].

To excite only the mode corresponding to $x_R(t)$, one chooses an initial state in which the positions and initial velocities of the two particles are opposite. One then finds that, at every subsequent instant, $x_G(t) = 0$, and the two particles move symmetrically with respect to the origin $O$. For this mode, the distance $(x_2 - x_1)$ varies and the attractive force between the two particles enters into the equations of motion; this is the reason why the angular frequency of this mode is not $\omega$ but $\omega_R = \omega\sqrt{1 + 4\lambda}$.

The dynamical variables $x_G(t)$ and $x_R(t)$ associated with the independent modes, that is, with the fictitious particles ($G$) and ($R$), are called *normal variables.*

## c. MOTION OF THE SYSTEM IN THE GENERAL CASE

In the general case, both modes are excited and the positions $x_1(t)$ and $x_2(t)$ are both given by the superposition of two oscillations of different frequencies $\omega_G$ and $\omega_R$ [*cf.* formulas (14)]. The motion of the system is not periodic, except in the case in which the ratio $\omega_G/\omega_R$ is rational★.

Let us investigate, for example, what happens if, at the initial time $t_0$, particle (1) is motionless at its equilibrium position $x_1 = a/(1 + 4\lambda)$, while particle (2) has a non-zero velocity (this is, in classical mechanics, the analogue of the problem studied in §C-3-b of chapter IV). In the absence of coupling,

★ If $\omega_G/\omega_R = 1/\sqrt{1 + 4\lambda}$ is equal to an irreducible rational fraction $p_1/p_2$, the period of the motion is : $T = 2\pi p_1/\omega_G = 2\pi p_2/\omega_R$.

particle (2) would oscillate alone and particle (1) would remain motionless. We shall show that the coupling sets particle (1) in motion. Two different angular frequencies $\omega_G$ and $\omega_R$ appear in the time evolution of $x_1(t)$ and $x_2(t)$. The two corresponding oscillations give rise to a beat phenomenon (fig. 1), whose frequency is:

$$\nu = \frac{\omega_R - \omega_G}{2\pi} = \frac{\omega}{2\pi}[\sqrt{1 + 4\lambda} - 1] \tag{16}$$

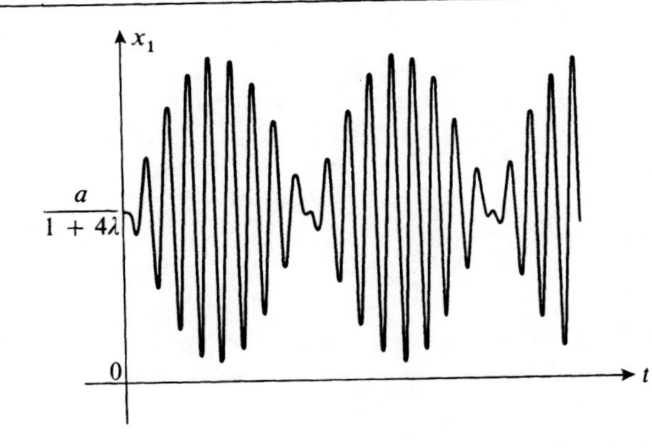


FIGURE 1

**Oscillations of the position of particle (1), assumed to be motionless at its equilibrium position at $t = 0$, particle (2) having an initial velocity. A beat phenomenon is produced between the two modes, and the amplitude of the oscillations of particle (1) varies over time.**

If the coupling is weak ($\lambda \ll 1$), this frequency $\nu \simeq \lambda\omega/\pi$ is negligible with respect to $\omega_R$ and $\omega_G$. In this case, as long as $(t - t_0) \ll 1/\nu$, particle (2) is practically the only one to oscillate; the vibrational energy is then slowly transferred to particle (1), whose amplitude of oscillation increases, while that of (2) decreases. After a certain time, the original situation is inverted: particle (1) oscillates strongly while particle (2) is practically motionless. Then the amplitude of (1) slowly decreases and that of (2) increases until the energy is again almost entirely localized in oscillator (2). The same process is repeated indefinitely. The effect of a weak coupling is to cause the energy of the oscillator associated with particle (1) to be constantly transferred to the one associated with particle (2) and vice versa, with a frequency proportional to the intensity of the coupling.

COMMENTS:

(*i*) If $p_1$ and $p_2$ are the respective momenta of particles (1) and (2), the classical Hamiltonian of the system under study can be written:

$$\mathscr{H}(x_1, x_2, p_1, p_2) = \frac{p_1^2}{2m} + \frac{p_2^2}{2m} + U_0(x_1, x_2) + V(x_1, x_2) \tag{17}$$

If we set:

$$\begin{cases} p_G(t) = p_1(t) + p_2(t) \\ p_R(t) = \frac{1}{2}[\, p_1(t) - p_2(t)\,] \end{cases} \tag{18}$$

and:

$$\begin{cases} \mu_G = 2m \\ \mu_R = \dfrac{m}{2} \end{cases} \tag{19}$$

it can be verified that $\mathscr{H}$ becomes:

$$\mathscr{H} = \frac{p_G^2}{2\mu_G} + \frac{1}{2}\mu_G\omega_G^2 x_G^2 + \frac{p_R^2}{2\mu_R} + \frac{1}{2}\mu_R\omega_R^2\left(x_R - \frac{2a}{1+4\lambda}\right)^2 + m\omega^2 a^2 \frac{4\lambda}{1+4\lambda} \tag{20}$$

By a suitable change in the energy origin, one can eliminate the last term of this expression, which is constant. $\mathscr{H}$ can then be seen to be the sum of two energies, each of which corresponds to one of the modes. Unlike the situation in (17), in which the terms in $x_1 x_2$ of $V(x_1, x_2)$ are responsible for a coupling between the particles, there is no coupling term in (20) between the modes: they are indeed independent.

(*ii*) We have assumed, for simplicity, the masses $m_1$ and $m_2$ of particles (1) and (2) to be equal. It is easy to eliminate this hypothesis by replacing (9), (10), (18) and (19) by:

$$\begin{cases} x_G(t) = \dfrac{m_1 x_1(t) + m_2 x_2(t)}{m_1 + m_2} \\ p_G(t) = p_1(t) + p_2(t) \\ \mu_G = m_1 + m_2 \end{cases} \tag{21}$$

(the position, momentum and mass associated with the center of mass) and:

$$\begin{cases} x_R(t) = x_1(t) - x_2(t) \\ p_R(t) = \dfrac{m_2 p_1(t) - m_1 p_2(t)}{m_1 + m_2} \\ \mu_R = \dfrac{m_1 m_2}{m_1 + m_2}. \end{cases} \tag{22}$$

(the position, momentum and mass of the "relative particle"). One then finds a result analogous to (20).

(*iii*) In the absence of coupling, the two modes have the same angular frequency $\omega$; in the presence of coupling, two different angular frequencies $\omega_G$ and $\omega_R$ appear. This is an example of a result which is often found in physics: the effect of a coupling between two oscillations is, in most cases, to separate their normal frequencies (the same phenomenon would occur here if the two oscillators originally had different angular frequencies). If, instead of two, we have an infinite number of oscillators (which, if isolated, would have the same frequency), we shall see in complement $J_V$ that the effect of the coupling is to create an infinite number of different frequencies for the modes.

## 2. Vibrational states of the system in quantum mechanics

Let us now reconsider the problem from a quantum mechanical point of view. We must now replace the positions $x_1(t)$, $x_2(t)$ and the momenta $p_1(t)$, $p_2(t)$ of the particles by operators, which we shall denote, respectively, by $X_1$, $X_2$, $P_1$, $P_2$. We then introduce, by analogy with (9), (10) and (18), the observables:

$$\begin{cases} X_G = \frac{1}{2}(X_1 + X_2) \\ P_G = P_1 + P_2 \end{cases} \tag{23}$$

$$\begin{cases} X_R = X_1 - X_2 \\ P_R = \frac{1}{2}(P_1 - P_2) \end{cases} \tag{24}$$

To see if the operator $H$, the Hamiltonian of the system, can be put into a form analogous to (20), we shall begin by examining the commutation relations of $X_G$, $P_G$, $X_R$ and $P_R$.

### a. COMMUTATION RELATIONS

Since all the observables concerning only particle (1) commute with those concerning particle (2), the only non-zero commutators involving $X_1$, $X_2$, $P_1$ and $P_2$ are:

$$\begin{aligned} [X_1, P_1] &= i\hbar \\ [X_2, P_2] &= i\hbar \end{aligned} \tag{25}$$

In particular, $X_1$ commutes with $X_2$, and we see immediately that:

$$[X_G, X_R] = 0 \tag{26}$$

Similarly:

$$[P_G, P_R] = 0 \tag{27}$$

Calculating the commutator $[X_G, P_G]$, we obtain:

$$\begin{aligned} [X_G, P_G] &= \frac{1}{2}\{ [X_1, P_1] + [X_1, P_2] + [X_2, P_1] + [X_2, P_2] \} \\ &= \frac{1}{2}\{ i\hbar + i\hbar \} = i\hbar \end{aligned} \tag{28}$$

Similarly, one finds:

$$[X_R, P_R] = i\hbar \tag{29}$$

The two commutators $[X_G, P_R]$ and $[X_R, P_G]$ remain to be examined; they are equal to:

$$\begin{aligned} [X_G, P_R] &= \frac{1}{4}\{ [X_1, P_1] - [X_1, P_2] + [X_2, P_1] - [X_2, P_2] \} \\ &= \frac{1}{4}\{ i\hbar - i\hbar \} = 0 \end{aligned} \tag{30}$$

and, similarly :

$$[X_R, P_G] = 0 \tag{31}$$

We can thus consider $X_G$, $P_G$ and $X_R$, $P_R$ to be the position and momentum operators of two distinct particles. Formulas (28) and (29) are the canonical commutation relations for each of these particles. Moreover, relations (26), (27), (30) and (31) express the fact that all the observables concerning one of them commute with all those which concern the other one.

### b. TRANSFORMATION OF THE HAMILTONIAN OPERATOR

In the presence of the coupling $V(X_1, X_2)$, we have :

$$H = T + U \tag{32}$$

with :

$$T = \frac{1}{2m}(P_1^2 + P_2^2) \tag{33}$$

(the kinetic energy operator) and :

$$U = \frac{1}{2} m\omega^2 [ (X_1 - a)^2 + (X_2 + a)^2 + 2\lambda(X_1 - X_2)^2 ] \tag{34}$$

(the potential energy operator). Since $P_1$ and $P_2$ commute, (33) can be transformed as if these operators were numbers; we find :

$$T = \frac{1}{2\mu_G} P_G^2 + \frac{1}{2\mu_R} P_R^2 \tag{35}$$

where $\mu_G$ and $\mu_R$ are defined in (19). Similarly, since $X_1$ and $X_2$ commute, we have, as above [formula (20)] :

$$U = \frac{1}{2} \mu_G\omega_G^2 X_G^2 + \frac{1}{2} \mu_R\omega_R^2 \left( X_R - \frac{2a}{1 + 4\lambda} \right)^2 + m\omega^2 a^2 \frac{4\lambda}{1 + 4\lambda} \tag{36}$$

where $\omega_G$ and $\omega_R$ are given by (13).

Thus we see that $H$ can be put into a form which is analogous to (20), in which there is no coupling term :

$$H = H_G + H_R + m\omega^2 a^2 \frac{4\lambda}{1 + 4\lambda} \tag{37}$$

with :

$$\begin{cases} H_G = \dfrac{P_G^2}{2\mu_G} + \dfrac{1}{2} \mu_G\omega_G^2 X_G^2 \\ H_R = \dfrac{P_R^2}{2\mu_R} + \dfrac{1}{2} \mu_R\omega_R^2 \left[ X_R - \dfrac{2a}{1 + 4\lambda} \right]^2 \end{cases} \tag{38}$$

### c. STATIONARY STATES OF THE SYSTEM

The state space of the system is the tensor product $\mathscr{E}(1) \otimes \mathscr{E}(2)$ of the state spaces of particles (1) and (2); it is also the tensor product $\mathscr{E}(G) \otimes \mathscr{E}(R)$ of the state spaces of the fictitious particles, the "center of mass" and the "relative particle" associated with each of the two modes. Since $H$ is the sum of two operators $H_G$ and $H_R$ which act only in $\mathscr{E}(G)$ and $\mathscr{E}(R)$ respectively (the constant $m\omega^2 a^2 \dfrac{4\lambda}{1 + 4\lambda}$ merely introduces a shift in the energy origin), we know (chap. II, §F) that we can find a basis of eigenvectors of $H$ in the form :

$$| \varphi \rangle = | \varphi^G \rangle | \varphi^R \rangle \tag{39}$$

where $| \varphi^G \rangle$ and $| \varphi^R \rangle$ are, respectively, eigenvectors of $H_G$ and $H_R$ in $\mathscr{E}(G)$ and $\mathscr{E}(R)$. Now, $H_G$ and $H_R$ are Hamiltonians of one-dimensional harmonic oscillators, whose eigenvectors and eigenvalues we know. If the operators $a_G^\dagger$ and $a_R^\dagger$ are defined by :

$$\begin{cases} a_G^\dagger = \dfrac{1}{\sqrt{2}}\left[\sqrt{\dfrac{\mu_G \omega_G}{\hbar}}\, X_G - i \dfrac{P_G}{\sqrt{\mu_G \hbar \omega_G}}\right] \\[2ex] a_R^\dagger = \dfrac{1}{\sqrt{2}}\left[\sqrt{\dfrac{\mu_R \omega_R}{\hbar}}\, X'_R - i \dfrac{P_R}{\sqrt{\mu_R \hbar \omega_R}}\right] \end{cases} \tag{40-a}$$

with :

$$X'_R = X_R - \frac{2a}{1 + 4\lambda} \tag{40-b}$$

and if $| \varphi_0^G \rangle$ and $| \varphi_0^R \rangle$ denote respectively the ground states of $H_G$ and $H_R$, the eigenvectors of $H_G$ are the vectors :

$$| \varphi_n^G \rangle = \frac{1}{\sqrt{n!}} (a_G^\dagger)^n \, | \varphi_0^G \rangle \tag{41}$$

whose eigenvalues are :

$$E_n^G = \left(n + \frac{1}{2}\right) \hbar \omega_G \tag{42}$$

those of $H_R$ being :

$$| \varphi_p^R \rangle = \frac{1}{\sqrt{p!}} (a_R^\dagger)^p \, | \varphi_0^R \rangle \tag{43}$$

with the eigenvalues :

$$E_p^R = \left(p + \frac{1}{2}\right) \hbar \omega_R \tag{44}$$

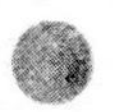


Thus we have here a situation which is analogous to the one encountered in the study of a two-dimensional anisotropic (since $\omega_G \neq \omega_R$) harmonic oscillator. The stationary states of the system are given by:

$$| \varphi_{n,p} \rangle = | \varphi_n^G \rangle | \varphi_p^R \rangle = \frac{(a_G^\dagger)^n (a_R^\dagger)^p}{\sqrt{n!\,p!}} | \varphi_{0,0} \rangle \tag{45}$$

and their energies are:

$$\begin{aligned} E_{n,p} &= E_n^G + E_p^R + m\omega^2 a^2 \frac{4\lambda}{1 + 4\lambda} \\ &= \left(n + \frac{1}{2}\right)\hbar\omega_G + \left(p + \frac{1}{2}\right)\hbar\omega_R + m\omega^2 a^2 \frac{4\lambda}{1 + 4\lambda} \end{aligned} \tag{46}$$

The operators $a_G$ and $a_G^\dagger$ or ($a_R$ and $a_R^\dagger$) can thus be seen to be destruction or creation operators of an energy quantum in the mode corresponding to ($G$) [or ($R$)]. We see from (45) that, through the repeated action of $a_G^\dagger$ and $a_R^\dagger$, we can obtain stationary states of the system in which the number of quanta in each mode is arbitrary. The action of $a_G^\dagger$, $a_G$, $a_R^\dagger$ or $a_R$ on the stationary states $| \varphi_{n,p} \rangle$ is very simple:

$$\begin{aligned} a_G^\dagger | \varphi_{n,p} \rangle &= \sqrt{n + 1} | \varphi_{n+1,p} \rangle \\ a_G | \varphi_{n,p} \rangle &= \sqrt{n} | \varphi_{n-1,p} \rangle \\ a_R^\dagger | \varphi_{n,p} \rangle &= \sqrt{p + 1} | \varphi_{n,p+1} \rangle \\ a_R | \varphi_{n,p} \rangle &= \sqrt{p} | \varphi_{n,p-1} \rangle \end{aligned} \tag{47}$$

In general, there are no degenerate levels since there do not exist two different pairs of integers $\{ n, p \}$ and $\{ n', p' \}$ such that:

$$n\omega_G + p\omega_R = n'\omega_G + p'\omega_R \tag{48}$$

(except when the ratio $\omega_R/\omega_G = \sqrt{1 + 4\lambda}$ is rational).

#### d. EVOLUTION OF THE MEAN VALUES

The most general state of the system is a linear superposition of stationary states $| \varphi_{n,p} \rangle$:

$$| \varphi(t) \rangle = \sum_{n,p} c_{n,p}(t) | \varphi_{n,p} \rangle \tag{49}$$

with:

$$c_{n,p}(t) = c_{n,p}(0)\, e^{-iE_{n,p}t/\hbar} \tag{50}$$

According to relations (40) and their adjoints, $X_G(X_R)$ is a linear combination of $a_G$ and $a_G^\dagger$ (of $a_R$ and $a_R^\dagger$). We then see, by using (47), that $X_G$ has non-zero matrix elements between two states $| \varphi_{n,p} \rangle$ and $| \varphi_{n',p'} \rangle$ only when $n - n' = \pm 1$, $p = p'$ (for $X'_R$, we would have $n = n'$, $p - p' = \pm 1$). From this we deduce that

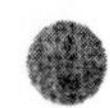

the only Bohr frequencies which can appear in the time evolution of $\langle X_G \rangle(t)$ and $\langle X_R \rangle(t)$ are, respectively*:

$$\frac{E_{n\pm 1,p} - E_{n,p}}{\hbar} = \pm\, \omega_G$$

$$\frac{E_{n,p\pm 1} - E_{n,p}}{\hbar} = \pm\, \omega_R \tag{51}$$

Thus we again find that $\langle X_G \rangle(t)$ and $\langle X_R \rangle(t)$ oscillate at angular frequencies of $\omega_G$ and $\omega_R$, which recalls the classical result obtained in §1-a.

**References and suggestions for further reading:**

Coupling between two classical oscillators: Berkeley 3 (7.1), §§1.4 and 3.3; Alonso and Finn (6.1), Vol. I, §12.10.

* For these frequencies actually to appear, at least one of the products $c^*_{n\pm 1,p}c_{n,p}$ or $c^*_{n,p\pm 1}c_{n,p}$ must be different from zero.

## Complement $J_V$

# VIBRATIONAL MODES OF AN INFINITE LINEAR CHAIN OF COUPLED HARMONIC OSCILLATORS; PHONONS



In complement $H_V$, we studied the motion of a system of two coupled harmonic oscillators. We concluded, in essence, that, while the individual dynamical variables of each oscillator do not evolve independently, it is possible to introduce linear combinations of them (normal variables) which possess the important property of being uncoupled. Such variables describe vibrational normal modes of well-defined frequencies. Expressed in terms of these normal variables, the Hamiltonian of the system appears in the form of a sum of Hamiltonians of independent harmonic oscillators, thus making quantization simple.

In this complement we shall show that these ideas are also applicable to a system formed by an infinite series of identical harmonic oscillators, regularly spaced along an axis, each one coupled to its neighbors.

To do this, we shall determine the various vibrational normal modes of the system and show that each one corresponds to a collective vibration of the system of particles characterized by an angular frequency $\Omega$ and a wave vector $k$. The process of finding the eigenstates and eigenvalues of the quantum mechanical Hamiltonian is then greatly simplified by the fact that the total energy of the system is the sum of the energies associated with each vibrational normal mode.

The results obtained will enable us to indicate how vibrations propagate in a crystal and to introduce the concept of a phonon, a central idea in solid state physics. Of course, in this complement, we shall emphasize the introduction and quantization of the normal modes, and not the detailed properties of phonons, which would be treated in a solid state physics course.

## 1. Classical treatment

### a. EQUATIONS OF MOTION

Let us consider an infinite chain of identical one-dimensional harmonic oscillators, each one labeled by an integer $q$ (positive, negative or zero). The particle $M_q$ of mass $m$ which constitutes oscillator $(q)$ has its equilibrium position at the point whose abscissa is $ql$ (fig. 1), where $l$ is the unit distance of the oscillator chain. We denote by $x_q$ the (algebraic) displacement of oscillator $(q)$ with respect to its equilibrium position. The state of the system at the instant $t$ is defined by specifying the dynamical variables $x_q(t)$ and their time derivatives $\dot{x}_q(t)$ at this instant.

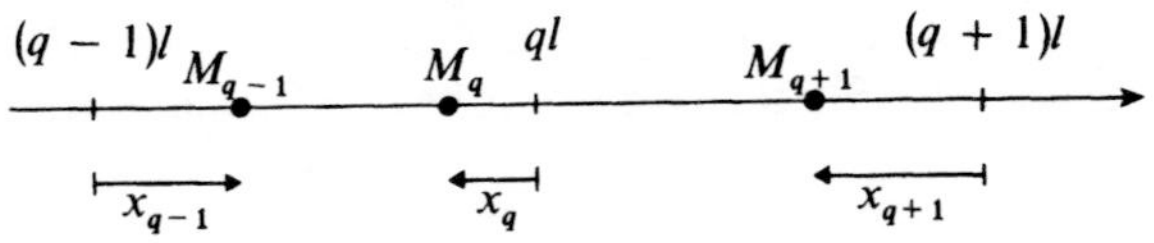

FIGURE 1

**Infinite chain of oscillators; the displacement of the** $q$th **particle with respect to its equilibrium position** $ql$ **is denoted by** $x_q$.

In the absence of interactions between the various particles, the potential energy of the system is:

$$U(..., x_{-1}, x_0, x_{+1}, ...) = \sum_{q=-\infty}^{+\infty} \frac{1}{2} m\omega^2 x_q^2 \tag{1}$$

where $\omega$ is the angular frequency of each oscillator. The evolution of the system is then given by the equations:

$$m \frac{\mathrm{d}^2}{\mathrm{d}t^2} x_q(t) = -m\omega^2 x_q(t) \tag{2}$$

whose solutions are:

$$x_q(t) = x_q^M \cos(\omega t - \varphi_q) \tag{3}$$

where the integration constants $x_q^M$ and $\varphi_q$ are fixed by the initial conditions of the motion. The oscillators therefore vibrate independently.

Now imagine that these particles are interacting. For simplicity, we shall assume that one need take into account only the forces exerted on a particle by its two nearest neighbors and that these forces are attractive and proportional to the distance. Thus, particle $(q)$ is subjected to two new attractive forces exerted by particles $(q + 1)$ and $(q - 1)$. These forces are proportional to $|M_q M_{q+1}|$ and $|M_q M_{q-1}|$ (the coefficient of proportionality being the same in both cases). The total force $F_q$ to which particle $(q)$ is subjected can therefore be written:

$$\begin{aligned} F_q &= -m\omega^2 x_q - m\omega_1^2[ql + x_q - (q+1)l - x_{q+1}] \\ &\qquad\qquad - m\omega_1^2[ql + x_q - (q-1)l - x_{q-1}] \\ &= -m\omega^2 x_q - m\omega_1^2(x_q - x_{q+1}) - m\omega_1^2(x_q - x_{q-1}) \end{aligned} \tag{4}$$

where $\omega_1$ is a constant [having the dimensions of an inverse time] which characterizes the intensity of the coupling. Equations (2) must now be replaced by:

$$m \frac{d^2}{dt^2} x_q(t) = - m\omega^2 x_q(t) - m\omega_1^2 [2x_q(t) - x_{q+1}(t) - x_{q-1}(t)] \tag{5}$$

It can easily be verified that the interaction forces [terms in $\omega_1^2$ of (4)] are derived from the potential energy of the coupling $V$ given by:

$$V(..., x_{-1}, x_0, x_{+1}, ...) = \frac{1}{2} m\omega_1^2 \sum_{q=-\infty}^{+\infty} (x_q - x_{q+1})^2 \tag{6}$$

According to (5), the evolution of $x_q$ depends on $x_{q+1}$ and $x_{q-1}$. Therefore, we must solve an infinite system of coupled differential equations. Before we introduce new variables which allow these equations to be uncoupled, it is interesting to try to find simple solutions of equations (5) and investigate their physical significance.

#### b. SIMPLE SOLUTIONS OF THE EQUATIONS OF MOTION

##### α. *Existence of simple solutions*

The infinite chain of coupled oscillators which we are studying is analogous to an infinite macroscopic spring. Now, we know that progressive longitudinal waves (corresponding to expansions and compressions) can propagate along this spring. Under the influence of a sinusoidal wave of this type, of wave vector $k$ and angular frequency $\Omega$, the point of the spring whose abscissa is $x$ at equilibrium is found at time $t$, at $x + u(x, t)$, with:

$$u(x, t) = \mu\, e^{i(kx - \Omega t)} + \mu^* e^{-i(kx - \Omega t)} \tag{7}$$

Such solutions of the equations of motion (5) do indeed exist. However, since the oscillator chain is not a continuous medium, the effects of the wave are observed only at a series of points, corresponding to the abscissas $x = ql$; $u(ql, t)$ thus represents the displacement of oscillator ($q$) at time $t$:

$$x_q(t) = u(ql, t) = \mu\, e^{i(kql - \Omega t)} + \mu^* e^{-i(kql - \Omega t)} \tag{8}$$

It is easy to verify that this expression is a solution of equations (5) if $\Omega$ and $k$ satisfy:

$$- m\Omega^2 = - m\omega^2 - m\omega_1^2 [2 - e^{ikl} - e^{-ikl}] \tag{9}$$

$\Omega$ is therefore related to $k$ by the "dispersion relation":

$$\Omega(k) = \sqrt{\omega^2 + 4\omega_1^2 \sin^2 \left(\frac{kl}{2}\right)} \tag{10}$$

which we shall discuss in detail later (§ 1-b-δ).

##### β. *Physical interpretation*

In the solution (8) of the equations of motion, all the oscillators are vibrating at the same frequency $\Omega/2\pi$, with the same amplitude $|2\mu|$, but with a phase which depends periodically on their rest positions. It is as if the displacements of the various oscillators were determined by a progressive sinusoidal wave of wave vector $k$ and phase velocity:

$$v_\varphi(k) = \frac{\Omega(k)}{k} \tag{11}$$

This is easy to show. Using (8), we see that:

$$x_{[q_1 + q_2]}(t) = x_{q_1}\left(t - \frac{q_2 l}{v_\varphi}\right) \tag{12}$$

Thus, oscillator $(q_1 + q_2)$ performs the same motion as oscillator $(q_1)$, shifted by the time taken by the wave to travel, at a velocity $v_\varphi$, the distance $q_2 l$ separating the two oscillators. Since all the oscillators are then in motion, solutions (8) are called "collective modes" of vibration of the system.

### γ. *Possible values of the wave vector k*

Consider two values of the wave vector, $k$ and $k'$, which differ by an integral number of $2\pi/l$:

$$k' = k + \frac{2n\pi}{l} \quad \text{with } n \text{ an integer (positive or negative)} \tag{13}$$

We have, obviously:

$$\begin{cases} e^{ik'ql} = e^{ikql} \\ \Omega(k') = \Omega(k) \end{cases} \tag{14}$$

where the second relation follows directly from (10).

We see from (8) that the two progressive waves $k$ and $k'$ lead to the same motion for the oscillators and are, consequently, *physically indistinguishable*. Therefore in the problem we are studying here, it suffices to let $k$ vary over an interval of $2\pi/l$. For reasons of symmetry, we choose:

$$-\frac{\pi}{l} < k \leqslant \frac{\pi}{l} \tag{15}$$

The corresponding interval is often called the "first Brillouin zone".

### δ. *Dispersion relation*

The dispersion relation (10) which gives the angular frequency $\Omega(k)$ associated with each value of $k$ enables us to study the propagation of vibrations in the system. If, for example, we form a "wave packet" by superposing waves with different wave vectors, we know that it has a group velocity given by:

$$v_G = \frac{d\Omega(k)}{dk} \tag{16}$$

which is different from $v_\varphi$.

Figure 2 shows the form of the variation of $\Omega(k)$ with respect to $k$, where $k$ varies within the first Brillouin zone.

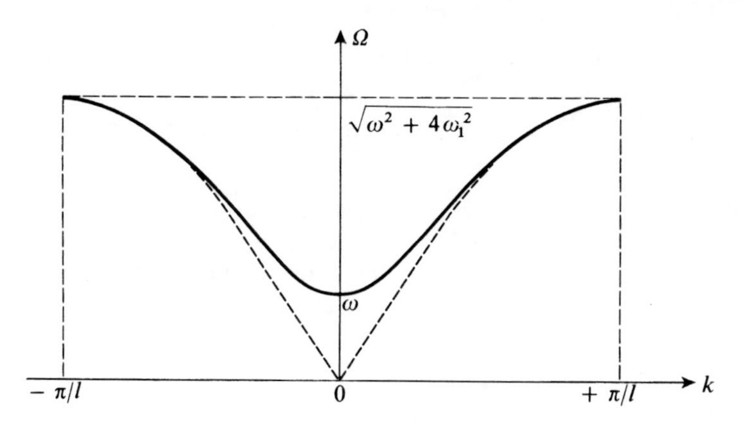


FIGURE 2

**Dispersion relation giving the variation of the angular frequency of the vibrational normal modes with respect to the wave number $k$ in the first Brillouin zone $[-\pi/l, +\pi/l]$. The dashed line corresponds to the case $\omega = 0$.**

It is clear from this figure that $\Omega(k)$ cannot take on arbitrary values: a vibration of frequency $\nu$ can propagate freely in the medium only if $\nu$ falls within the "allowed band":

$$\frac{\omega}{2\pi} \leqslant \nu \leqslant \frac{\sqrt{\omega^2 + 4\omega_1^2}}{2\pi} \tag{17}$$

The other values of $\nu$ correspond to "forbidden bands". The two limiting frequencies of interval (17) are often called "cut-off frequencies".

The mode of lowest angular frequency, $\Omega(0) = \omega$, has a zero wave vector $k$; it corresponds to an in-phase vibration of all the oscillators, whose particles are moving "together" without changing their relative distances (fig. 3). This explains why the angular frequency of this mode is the same as in the absence of coupling (*cf.* complement $H_V$, §1-b).

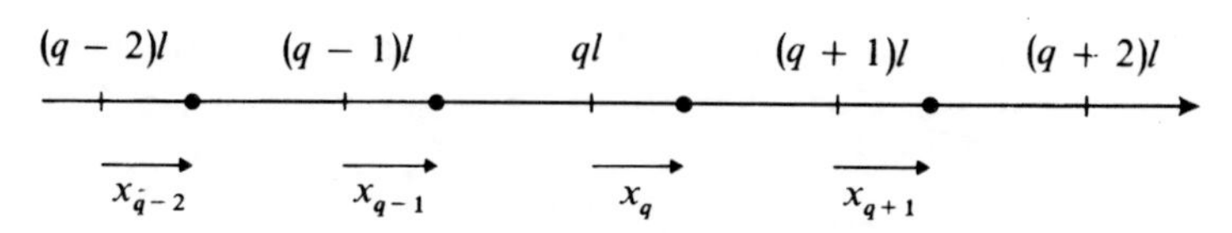


FIGURE 3

**The lowest frequency mode ($k = 0, \Omega = \omega$) corresponds to a displacement of the system of oscillators "as a whole". This is why its frequency does not depend on the coupling $V$.**

As for the mode of highest angular frequency, $\Omega(\pm \pi/l) = \sqrt{\omega^2 + 4\omega_1^2}$, in the corresponding vibration of the system, two adjacent oscillators are completely out of phase (fig. 4); the effect of the attractive forces due to the coupling $V$ is then maximal.

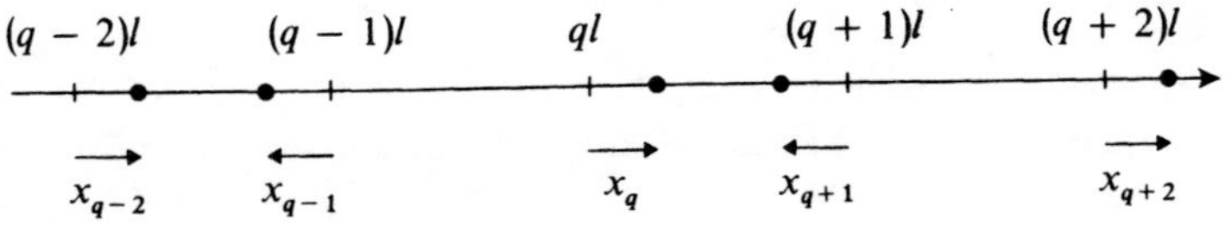

$(q-2)l$ $(q-1)l$ $ql$ $(q+1)l$ $(q+2)l$

$x_{q-2}$ $x_{q-1}$ $x_q$ $x_{q+1}$ $x_{q+2}$

FIGURE 4

**The modes $k = \pm \pi/l$ are those in which two neighboring oscillators are completely out of phase; the coupling $V$ strongly modifies their frequency.**

## c. NORMAL VARIABLES

### α. *Obtaining uncoupled equations*

Returning to the equations of motion (5), let us introduce new dynamical variables (linear combinations of the $x_q$) which evolve independently.

To do so, we multiply both sides of equation (5) by the quantity $e^{-ikql}$ and sum over $q$. If we notice that:

$$\begin{aligned}\sum_{q=-\infty}^{+\infty} x_{q\pm 1}\, e^{-iqkl} &= e^{\pm ikl} \sum_{q=-\infty}^{+\infty} x_{q\pm 1}\, e^{-i(q\pm 1)kl} \\ &= e^{\pm ikl} \sum_{q=-\infty}^{+\infty} x_q\, e^{-iqkl}\end{aligned} \tag{18}$$

and if we set:

$$\sum_{q=-\infty}^{+\infty} x_q(t)\, e^{-iqkl} = \xi(k, t) \tag{19}$$

we see that (5) becomes:

$$\frac{\partial^2}{\partial t^2} \xi(k, t) = -\left[\omega^2 + \omega_1^2(2 - e^{ikl} - e^{-ikl})\right] \xi(k, t) \tag{20}$$

that is, taking (9) into account:

$$\frac{\partial^2}{\partial t^2} \xi(k, t) = -\,\Omega^2(k)\, \xi(k, t) \tag{21}$$

This relation shows that the time evolution of $\xi(k, t)$ is independent of that of $\xi(k', t)$ for $k'$ different from $k$. The quantities $\xi(k, t)$ introduced in (19) are therefore completely uncoupled and have a remarkably simple equation of motion.

COMMENTS:

(*i*) Equations (5) are easy to uncouple because the problem is invariant under a translation of the system of oscillators by a quantity $\pm\, l$ (replacement of $q$ by $q \pm 1$). This invariance is itself due to the fact that the chain is regular and infinite.

(*ii*) In reality, every chain is of course finite, even if it contains a very large number $\mathcal{N}$ of oscillators. To find its vibrational normal modes, one must thus take into account the boundary conditions at the two ends of the chain, and the problem becomes much more complicated (edge effects). Instead of obtaining, as we do here, a continuous infinity of vibrational normal modes which correspond to the various values of $k$ in the first Brillouin zone, one finds a finite number of eigenmodes, equal to the number $\mathcal{N}$ of oscillators. When one is concerned only with the behavior of the chain far from the ends, one often introduces artificial boundary conditions which are different from the real boundary conditions but which have the advantage of simplifying the calculations while conserving the essential physical properties. Thus, one requires the two end oscillators to have the same motion ("periodic" boundary conditions, also called "Born-Von Karman conditions"). Since we shall have the opportunity to return to this question in connection with the study of other periodic structures (*cf.* complement $F_{XI}$; see also § 1-c of complement $C_{XIV}$), we shall not dwell any further on periodic boundary conditions, but shall continue our discussion, confining ourselves to the simple case of an infinite chain.

The function $\xi(k, t)$ introduced in (19) is, by definition, the sum of a Fourier series whose coefficients are the displacements $x_q(t)$. It is a periodic function, of period $2\pi/l$, which is therefore perfectly well-defined when its values in the interval $-\pi/l \leqslant k \leqslant \pi/l$ are specified [this is the first Brillouin zone, defined in (15)]. $\xi(k, t)$ depends on the positions of all the oscillators at time $t$. Conversely, these positions are unambiguously defined when the values of $\xi$ in the interval (15) are given for time $t$. This is true because it is possible to invert relation (19) since, using:

$$\int_{-\pi/l}^{+\pi/l} \mathrm{d}k \; \mathrm{e}^{i(q-q')kl} = \frac{2\pi}{l}\,\delta_{qq'} \tag{22}$$

we obtain:

$$x_q(t) = \frac{l}{2\pi}\int_{-\pi/l}^{+\pi/l} \mathrm{d}k \; \xi(k, t)\, \mathrm{e}^{iqkl} \tag{23}$$

Note also that, since the displacements $x_q(t)$ are real, the function $\xi(k, t)$ satisfies:

$$\xi(-k, t) = \xi^*(k, t) \tag{24}$$

Similarly, one can, using the momenta $p_q(t) = m\dot{x}_q(t)$, define the function:

$$\pi(k, t) = \sum_q p_q(t)\, \mathrm{e}^{-ikql} \tag{25}$$

and the $p_q(t)$ can be expressed as:

$$p_q(t) = \frac{l}{2\pi}\int_{-\pi/l}^{+\pi/l} \pi(k, t)\, \mathrm{e}^{ikql}\, \mathrm{d}k \tag{26}$$

The fact that the $p_q(t)$ are real implies that:

$$\pi(-k, t) = \pi^*(k, t) \tag{27}$$

Differentiating both sides of (19) term by term, and using (25) and then (21), we finally obtain:

$$\begin{cases} m \dfrac{\partial}{\partial t} \xi(k, t) = \pi(k, t) & \text{(28-a)} \\ \dfrac{\partial}{\partial t} \pi(k, t) = - m\Omega^2(k)\, \xi(k, t) & \text{(28-b)} \end{cases}$$

At time $t$, the dynamical state of the system is characterized by the specification of the $x_q(t)$ and $p_q(t)$ for all integers (positive, negative or zero) as well as by that of the "normal variables" $\xi(k, t)$ and $\pi(k, t)$ (where $k$ can take on any value in the first Brillouin zone). The equations of motion (28) of the normal variables corresponding to each value of $k$ describe the evolution of the position and momentum of a harmonic oscillator of mass $m$ and angular frequency $\Omega(k)$; however, $\xi$ and $\pi$ are complex. *Thus we have reduced the study of an infinite but discrete chain of coupled harmonic oscillators to that of a continuous system of fictitious independent oscillators* (labeled by the index $k$).

COMMENT:

Rigorously, these fictitious oscillators are not completely independent since, according to conditions (24) and (27), the initial values $\xi(k, 0)$ and $\pi(k, 0)$ must satisfy:

$$\begin{aligned} \xi(k, 0) &= \xi^*(-k, 0) \\ \pi(k, 0) &= \pi^*(-k, 0) \end{aligned} \tag{29}$$

β. *Normal variables $\alpha(k, t)$ associated with the progressive waves*

It is convenient (see also § 1-a of complement $G_V$) to condense the two normal variables $\xi(k, t)$ and $\pi(k, t)$ into one, $\alpha(k, t)$, defined by:

$$\alpha(k, t) = \frac{1}{\sqrt{2}} [\, \hat{\xi}(k, t) + i\hat{\pi}(k, t)\, ] \tag{30}$$

where $\hat{\xi}(k, t)$ and $\hat{\pi}(k, t)$ are dimensionless quantities proportional to $\xi(k, t)$ and $\pi(k, t)$:

$$\begin{cases} \hat{\xi}(k, t) = \beta(k)\, \xi(k, t) \\ \hat{\pi}(k, t) = \dfrac{1}{\hbar\beta(k)} \pi(k, t) \end{cases} \tag{31}$$

To simplify the quantum mechanical calculations presented later, we shall set:

$$\beta(k) = \sqrt{\frac{m\Omega(k)}{\hbar}} \tag{32}$$

It is easy to show, using (30), that the two equations (28) are equivalent to the single equation:

$$i\frac{\partial}{\partial t}\alpha(k, t) = \Omega(k)\,\alpha(k, t) \tag{33}$$

which is of first order in $t$ [$\alpha(k, t)$ is completely defined by the specification of $\alpha(k, 0)$, while $\xi(k, t)$ depends on $\xi(k, 0)$ and $\pi(k, 0)$]. The general solution of (33) is:

$$\alpha(k, t) = \alpha(k, 0)\,\mathrm{e}^{-i\Omega(k)t} \tag{34}$$

Using (19) and (25), we easily obtain the expression for $\alpha(k, t)$ in terms of the $x_q(t)$ and $p_q(t)$:

$$\alpha(k, t) = \frac{1}{\sqrt{2}}\beta(k)\sum_q \mathrm{e}^{-iqkl}\left[x_q(t) + i\frac{p_q(t)}{m\Omega(k)}\right] \tag{35}$$

Let us show, conversely, that the $x_q(t)$ and $p_q(t)$ can be simply expressed in terms of the $\alpha(k, t)$. According to (24) and (27):

$$\begin{aligned}\alpha^*(-k, t) &= \frac{1}{\sqrt{2}}[\,\hat{\xi}^*(-k, t) - i\,\hat{\pi}^*(-k, t)\,]\\ &= \frac{1}{\sqrt{2}}[\,\hat{\xi}(k, t) - i\,\hat{\pi}(k, t)\,]\end{aligned} \tag{36}$$

From this we deduce:

$$\begin{cases}\hat{\xi}(k, t) = \dfrac{1}{\sqrt{2}}[\,\alpha(k, t) + \alpha^*(-k, t)\,] & \text{(37-a)}\\[2ex] \hat{\pi}(k, t) = -\dfrac{i}{\sqrt{2}}[\,\alpha(k, t) - \alpha^*(-k, t)\,] & \text{(37-b)}\end{cases}$$

which allows us to write formula (23) in the form:

$$x_q(t) = \frac{l}{2\pi\sqrt{2}}\left\{\int_{-\frac{\pi}{l}}^{+\frac{\pi}{l}}\frac{\alpha(k, t)}{\beta(k)}\mathrm{e}^{iqkl}\,\mathrm{d}k + \int_{-\frac{\pi}{l}}^{+\frac{\pi}{l}}\frac{\alpha^*(-k, t)}{\beta(k)}\mathrm{e}^{iqkl}\,\mathrm{d}k\right\} \tag{38}$$

Changing $k$ to $-k$ in the second integral, we finally obtain [$\beta(k)$ is an even function of $k$]:

$$x_q(t) = \frac{l}{2\pi\sqrt{2}}\left\{\int_{-\frac{\pi}{l}}^{+\frac{\pi}{l}}\frac{\alpha(k, t)}{\beta(k)}\mathrm{e}^{iqkl}\,\mathrm{d}k + \int_{-\frac{\pi}{l}}^{+\frac{\pi}{l}}\frac{\alpha^*(k, t)}{\beta(k)}\mathrm{e}^{-iqkl}\,\mathrm{d}k\right\} \tag{39}$$

An analogous calculation, starting with (26), yields:

$$p_q(t) = \frac{l}{2\pi\sqrt{2}}\frac{\hbar}{i}\left\{\int_{-\frac{\pi}{l}}^{+\frac{\pi}{l}}\beta(k)\,\alpha(k, t)\,\mathrm{e}^{iqkl}\,\mathrm{d}k - \int_{-\frac{\pi}{l}}^{+\frac{\pi}{l}}\beta(k)\,\alpha^*(k, t)\,\mathrm{e}^{-iqkl}\,\mathrm{d}k\right\} \tag{40}$$

The state of the system is therefore described by the $\alpha(k, t)$ as well as by the set of $x_q(t)$ and $p_q(t)$.

If we replace, in (39), $\alpha(k, t)$ by its general expression (34), $x_q(t)$ takes on the form:

$$x_q(t) = \frac{l}{2\pi\sqrt{2}} \left\{ \int_{-\frac{\pi}{l}}^{+\frac{\pi}{l}} dk \frac{\alpha(k, 0)}{\beta(k)} e^{i[qkl - \Omega(k)t]} + c.c. \right\} \tag{41}$$

The most general solution of the problem of the chain of coupled oscillators is therefore a linear superposition of the progressive waves introduced in § 1-b $\left(\text{where the coefficients of this linear combination are } \frac{l}{2\pi\sqrt{2}} \frac{\alpha(k, 0)}{\beta(k)}\right)$. These progressive waves constitute the vibrational normal modes of the system★.

COMMENT:

For each value of $k$, the two terms appearing on the right-hand sides of (39) and (40) are complex conjugates of each other. This insures the reality of the $x_q(t)$ and $p_q(t)$ without the necessity of imposing an arbitrary condition on the $\alpha(k, t)$. The $\alpha(k, t)$, consequently, are truly independent variables.

## d. TOTAL ENERGY AND ENERGY OF EACH OF THE MODES

The total energy of the system under consideration is the sum of the kinetic energies of each particle ($q$) and the potential energies (1) and (6):

$$\mathscr{H}(\ldots x_{-1}, x_0, x_{+1}, \ldots p_{-1}, p_0, p_{+1} \ldots) =$$

$$\sum_{q=-\infty}^{\infty} \left[ \frac{1}{2m} p_q^2 + \frac{1}{2} m\omega^2 x_q^2 + \frac{1}{2} m\omega_1^2 (x_q - x_{q+1})^2 \right] \tag{42}$$

We shall see in this section that this energy can be expressed very simply in terms of the energies which can be associated with each of the modes.

Let us therefore calculate the various sums involved in (42). Since the displacements $x_q$ are the coefficients of the Fourier series which defines the function $\xi(k, t)$, Parseval's relation [appendix I, relation (18)] immediately yields:

$$\sum_{q=-\infty}^{+\infty} (x_q)^2 = \frac{l}{2\pi} \int_{-\frac{\pi}{l}}^{+\frac{\pi}{l}} |\xi(k, t)|^2 \, dk \tag{43}$$

$$\sum_{q=-\infty}^{+\infty} (p_q)^2 = \frac{l}{2\pi} \int_{-\pi/l}^{+\pi/l} |\pi(k, t)|^2 \, dk \tag{44}$$

★ We could also have introduced the modes corresponding to stationary waves in the system (sum of two progressive waves of the same frequency and opposite velocities). We would then have obtained equivalent results but the motion of the system would have been expanded on a different "basis". An expansion of this type is used in complement $K_V$.

The sum which, in (42), corresponds to the coupling, now remains to be calculated. To do so, notice as in (18) that, if the displacements $x_q$ are the coefficients of the Fourier series of $\xi(k, t)$, the $x_{q+1}$ are those of $e^{ikl}\,\xi(k, t)$. The quantities $(x_q - x_{q+1})$ are therefore the coefficients of the Fourier series of $(1 - e^{ikl})\xi(k, t)$, and Parseval's relation yields:

$$\sum_{q=-\infty}^{+\infty} (x_q - x_{q+1})^2 = \frac{l}{2\pi}\int_{-\frac{\pi}{l}}^{+\frac{\pi}{l}} |(1 - e^{ikl})\xi(k, t)|^2 \, dk$$
$$= \frac{l}{2\pi}\int_{-\frac{\pi}{l}}^{+\frac{\pi}{l}} 4\sin^2\left(\frac{kl}{2}\right) |\xi(k, t)|^2 \, dk \tag{45}$$

Substituting (43), (44) and (45) into (42), we finally obtain:

$$\mathscr{H} = \frac{l}{2\pi}\int_{-\frac{\pi}{l}}^{+\frac{\pi}{l}} \left\{ \frac{m}{2}\left[\omega^2 + 4\omega_1^2 \sin^2\left(\frac{kl}{2}\right)\right] |\xi(k, t)|^2 + \frac{1}{2m}|\pi(k, t)|^2 \right\} dk \tag{46}$$

We shall write this result in the form:

$$\mathscr{H} = \frac{l}{2\pi}\int_{-\frac{\pi}{l}}^{+\frac{\pi}{l}} h(k)\, dk \tag{47}$$

with:

$$h(k) = \frac{1}{2} m\Omega^2(k)\, |\xi(k, t)|^2 + \frac{1}{2m}|\pi(k, t)|^2 \tag{48}$$

where $\Omega(k)$ is given by (10). $\mathscr{H}$ is thus the sum (in fact, the integral) of the energies associated with the fictitious uncoupled harmonic oscillators for which $\xi(k, t)$ gives the position and $\pi(k, t)$, the momentum.

We can also express $h(k)$ in terms of the variables $\alpha(k, t)$ associated with each normal mode. Using (37), we transform expression (48) into:

$$h(k) = \frac{1}{2}\hbar\Omega(k)\left[\, \alpha(k, t)\alpha^*(k, t) + \alpha^*(-k, t)\alpha(-k, t)\,\right] \tag{49}$$

that is, taking (34) into account:

$$h(k) = \frac{1}{2}\hbar\Omega(k)\left[\, \alpha(k, 0)\alpha^*(k, 0) + \alpha^*(-k, 0)\alpha(-k, 0)\,\right] \tag{50}$$

$h(k)$ is therefore time-independent, which is not surprising since $h(k)$ is the energy of a harmonic oscillator. Furthermore, we again find in (47) that the fictitious oscillators are independent, since the total energy $\mathscr{H}$ is simply the sum of the energies associated with each of them.

Substituting expression (49) into (47), we obtain:

$$\mathscr{H} = \frac{l}{2\pi}\int_{-\frac{\pi}{l}}^{+\frac{\pi}{l}} dk\, \frac{1}{2}\hbar\Omega(k)\left[\, \alpha(k, t)\alpha^*(k, t) + \alpha(-k, t)\alpha^*(-k, t)\,\right] \tag{51}$$

We can then change $k$ to $-k$ in the integral of the second term and consider $\mathscr{H}$ to be the sum of the energies $h'(k)$ associated with the normal modes characterized by the $\alpha(k, t)$:

$$\mathscr{H} = \frac{l}{2\pi} \int_{-\frac{\pi}{l}}^{+\frac{\pi}{l}} \mathrm{d}k \; h'(k) \tag{52}$$

with:

$$\begin{aligned} h'(k) &= \hbar\,\Omega(k)\,\alpha^*(k, t)\,\alpha(k, t) \\ &= \hbar\,\Omega(k)\,\alpha^*(k, 0)\,\alpha(k, 0) \end{aligned} \tag{53}$$

## 2. Quantum mechanical treatment

The quantum mechanical treatment of the problem of the infinite chain of coupled oscillators is based, in accordance with the general quantization rules, on the replacement of the classical quantities $x_q(t)$ and $p_q(t)$ by the observables $X_q$ and $P_q$ which satisfy the canonical commutation relations:

$$[X_{q_1}, P_{q_2}] = i\hbar\,\delta_{q_1 q_2} \tag{54}$$

### a. STATIONARY STATES IN THE ABSENCE OF COUPLING

In the absence of coupling ($\omega_1 = 0$), the Hamiltonian $H$ of the system can be written:

$$\begin{aligned} H(\omega_1 = 0) &= \sum_q \left[\frac{1}{2} m\omega^2 X_q^2 + \frac{1}{2m} P_q^2\right] \\ &= \sum_q H_q \end{aligned} \tag{55}$$

where $H_q$ is the Hamiltonian of a one-dimensional harmonic oscillator acting in the state space of particle ($q$).

We introduce the operator $a_q$ defined by:

$$a_q = \frac{1}{\sqrt{2}}\left[\sqrt{\frac{m\omega}{\hbar}}\, X_q + \frac{i}{\sqrt{m\hbar\omega}}\, P_q\right] \tag{56}$$

$H_q$ can then be written:

$$H_q = \frac{1}{2}(a_q a_q^\dagger + a_q^\dagger a_q)\,\hbar\omega = \left(a_q^\dagger a_q + \frac{1}{2}\right)\hbar\omega \tag{57}$$

$a_q^\dagger$ and $a_q$ are the creation and annihilation operators of an energy quantum for oscillator ($q$). We know (chap. V, §C-1-a) that the eigenstates of $H_q$ are given by:

$$|\,\varphi_{n_q}^q\,\rangle = \frac{1}{\sqrt{(n_q)!}}(a_q^\dagger)^{n_q}\,|\,\varphi_0^q\,\rangle \tag{58}$$

where $| \varphi_0^q \rangle$ is the ground state of oscillator $(q)$ and $n_q$ is a positive integer or zero. If we choose as the energy origin the energy of the ground state [which amounts to omitting, in (57), the term 1/2], we obtain for the energy $E_{n_q}^q$ of the state $| \varphi_{n_q}^q \rangle$:

$$E_{n_q}^q = n_q \hbar\omega \tag{59}$$

In the absence of coupling, the stationary states of the global system are tensor products of the form:

$$... \otimes | \varphi_{n_{-1}}^{-1} \rangle \otimes | \varphi_{n_0}^0 \rangle \otimes | \varphi_{n_1}^1 \rangle \otimes ... \tag{60}$$

and their energies are*:

$$E = \sum_q E_{n_q}^q = [... + n_{-1} + n_0 + n_1 + ...] \hbar\omega \tag{61}$$

The ground state, whose energy was chosen for the origin, is not degenerate since $E = 0$ is obtained, in (61), only for:

$$n_q = 0 \qquad \text{for all } q \tag{62}$$

The corresponding state (60) is therefore unique. On the other hand, all the other levels are infinitely degenerate. For example, to the first level, of energy $\hbar\omega$, correspond all states (60) for which all the numbers $n_q$ are zero except for one, which is itself equal to one. All but one of the oscillators are then in their ground states. It is because the excitation can be localized in any one of the oscillators that the level $E = \hbar\omega$ is infinitely degenerate.

#### b. EFFECTS OF THE COUPLING

When the coupling is not zero, the Hamiltonian operator becomes:

$$H = H(\omega_1 = 0) + V \tag{63}$$

with:

$$V = \frac{1}{2} m\omega_1^2 \sum_q (X_q - X_{q+1})^2 \tag{64}$$

The states (60) are no longer, in this case, the stationary states of the system as they are eigenstates of $H(\omega_1 = 0)$, but not of $V$. To see this, we write $V$ in terms of the operators $a_q$ and $a_q^\dagger$:

$$V = \frac{1}{4} \hbar\omega_1 \frac{\omega_1}{\omega} \sum_q (a_q + a_q^\dagger - a_{q+1} - a_{q+1}^\dagger)^2 \tag{65}$$

Now, it is clear that the action of $V$ on a state of type (60) changes the state: the numbers $n_q$ are no longer "good quantum numbers" since, for example, $V$ can transfer an excitation from site $(q)$ to site $(q + 1)$ (term in $a_{q+1}^\dagger a_q$).

* If we had not changed the energy origin of each oscillator by omitting the term 1/2 in (57), we would have found an infinite energy for the system, whatever the quantum number $n_q$. This difficulty does not arise if, instead of an infinite chain, one considers a chain formed by a very large but finite number of oscillators. However, problems related to "edge effects" then appear.

To find the stationary states of the system in the presence of the coupling, it is useful, as in classical mechanics, to introduce "normal variables", that is, operators associated with the normal modes of the system.

### c. NORMAL OPERATORS. COMMUTATION RELATIONS

To the normal variables $\xi(k, t)$ and $\pi(k, t)$ correspond the operators $\Xi(k)$ and $\Pi(k)$ defined by:

$$\Xi(k) = \sum_q X_q \, e^{-iqkl} \tag{66-a}$$

$$\Pi(k) = \sum_q P_q \, e^{-iqkl} \tag{66-b}$$

The domain of variation of the continuous parameter $k$ is again limited to the first Brillouin zone (15). Note that since the normal variables $\xi(k, t)$ and $\pi(k, t)$ are complex, the associated operators $\Xi(k)$ and $\Pi(k)$ are not Hermitian, unlike $X_q$ and $P_q$. The relations which correspond to (24) and (27) are here:

$$\Xi(-k) = \Xi^\dagger(k) \tag{67-a}$$

$$\Pi(-k) = \Pi^\dagger(k) \tag{67-b}$$

The canonical commutation relations (54) enable us to calculate the commutators of $\Xi(k)$ and $\Pi(k)$. We immediately see that $\Xi(k)$ and $\Xi(k')$ commute, as do $\Pi(k)$ and $\Pi(k')$. As for the commutator $[\Xi(k), \Pi^\dagger(k')]$, it can be written :

$$\begin{aligned} [\Xi(k), \Pi^\dagger(k')] &= \sum_q \sum_{q'} [X_q, P_{q'}] \, e^{-iqkl} \, e^{+iq'k'l} \\ &= i\hbar \sum_q e^{-iq(k-k')l} \end{aligned} \tag{68}$$

Using formula (31) of appendix II and the fact that $k$ and $k'$ both belong to interval (15), we obtain:

$$[\Xi(k), \Pi^\dagger(k')] = i\hbar \frac{2\pi}{l} \delta(k - k') \tag{69}$$

We saw in §1-c-β that it is convenient to condense the two normal variables $\xi(k, t)$ and $\pi(k, t)$ into one, $\alpha(k, t)$ [formula (30)]. The operator associated with $\alpha(k, t)$ will be:

$$a(k) = \frac{1}{\sqrt{2}} \left[ \beta(k) \, \Xi(k) + \frac{i}{\hbar \beta(k)} \Pi(k) \right] \tag{70}$$

where $\beta(k)$ is defined in (32). Note that the adjoint of $a(k)$ can be written:

$$a^\dagger(k) = \frac{1}{\sqrt{2}} \left[ \beta(k) \, \Xi^\dagger(k) - \frac{i}{\hbar \beta(k)} \Pi^\dagger(k) \right] \tag{71}$$

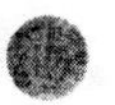



Using (69) and (67), we find without difficulty that:

$$[a(k), a(k')] = [a^\dagger(k), a^\dagger(k')] = 0 \tag{72-a}$$

$$[a(k), a^\dagger(k')] = \frac{2\pi}{l}\,\delta(k - k') \tag{72-b}$$

To the classical quantity $h(k)$ defined in (48) corresponds the operator:

$$H(k) = \frac{1}{2m}\,\Pi(k)\,\Pi^\dagger(k) + \frac{1}{2}\,m\Omega^2(k)\,\Xi(k)\,\Xi^\dagger(k) \tag{73}$$

since $\Xi(k)$ and $\Xi^\dagger(k)$ commute, as do $\Pi(k)$ and $\Pi^\dagger(k)$. To obtain the equivalent of the classical formula (49), one must take into consideration the fact that $a(k)$ and $a^\dagger(k)$ do not commute; the order in which these operators appear must therefore be retained throughout the calculation. Relations (37), taking (67) into account, can be written here:

$$\beta(k)\,\Xi(k) = \frac{1}{\sqrt{2}}\left[\,a(k) + a^\dagger(-\,k)\,\right] \tag{74-a}$$

$$\frac{1}{\hbar\beta(k)}\,\Pi(k) = -\frac{i}{\sqrt{2}}\left[\,a(k) - a^\dagger(-\,k)\,\right] \tag{74-b}$$

Substituting these expressions into (73), we find:

$$H(k) = \frac{1}{2}\,\hbar\Omega(k)\left[\,a(k)a^\dagger(k) + a^\dagger(-\,k)a(-\,k)\,\right] \tag{75}$$

As in (52), we can put the total Hamiltonian $H$ of the system in the form:

$$H = \frac{l}{2\pi}\int_{-\frac{\pi}{l}}^{+\frac{\pi}{l}} \mathrm{d}k\; H'(k) \tag{76}$$

with:

$$H'(k) = \frac{1}{2}\,\hbar\Omega(k)\left[\,a(k)a^\dagger(k) + a^\dagger(k)a(k)\,\right] \tag{77}$$

$a(k)$ and $a^\dagger(k)$ thus can be seen to be annihilation and creation operators analogous to those of a harmonic oscillator. However, since $k$ is a continuous index, the commutation relations (72) involve $\delta(k - k')$ instead of a Kronecker delta, so $H'(k)$ must remain in the symmetrical form (77). It can easily be shown that the various operators $H'(k)$ commute:

$$[H'(k), H'(k')] = 0 \tag{78}$$

#### d. STATIONARY STATES IN THE PRESENCE OF COUPLING

According to formulas (76) and (77), the ground state $|\,0\,\rangle$ of the system of coupled oscillators is defined by the condition:

$$a(k)\,|\,0\,\rangle = 0 \tag{79}$$

for all values of $k$. The other stationary states can be obtained from the state $|\,0\,\rangle$ by

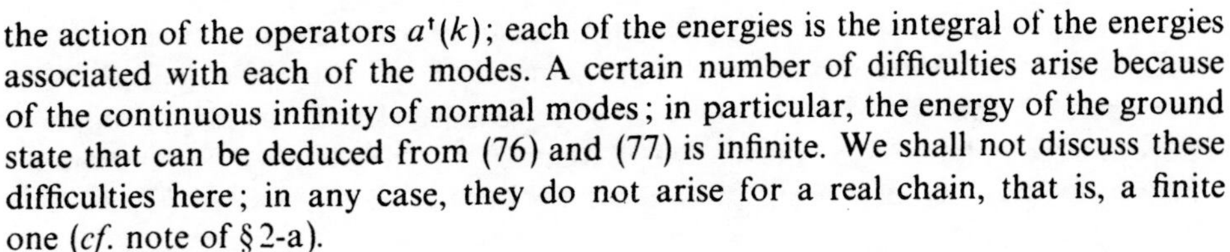

the action of the operators $a^\dagger(k)$; each of the energies is the integral of the energies associated with each of the modes. A certain number of difficulties arise because of the continuous infinity of normal modes; in particular, the energy of the ground state that can be deduced from (76) and (77) is infinite. We shall not discuss these difficulties here; in any case, they do not arise for a real chain, that is, a finite one (*cf.* note of § 2-a).

Formula (10) gives the value of the energy quantum $\hbar\Omega(k)$ associated with each of the modes. It therefore indicates what energy quanta the system can absorb or emit: they must correspond to frequencies situated within the allowed band (17).

## 3. Application to the study of vibrations in a crystal: phonons

### a. OUTLINE OF THE PROBLEM

Consider a solid body, composed of a large number of atoms (or ions) whose equilibrium positions are regularly arranged in space, forming a crystalline lattice. For simplicity, we shall assume that this lattice is one-dimensional and can be treated like an infinite chain of atoms. We intend to use here the results of the preceding section to study the motion of the nuclei of these atoms about their equilibrium positions.

With this object in view, we shall make use of the same approximation as in the study of molecular vibrations (Born-Oppenheimer approximation; *cf.* complement $A_V$, comment of § 1-a). We shall assume that the motion of the electrons can be calculated by considering the positions of the nuclei to be fixed parameters $x_q$. Thus, we shall solve the corresponding Schrödinger equation (actually, this equation is too complex to be solved exactly; in practice, one must again settle for approximations). We shall then denote the energy of the electronic system in its ground state by $E_{el}(..., x_{-1}, x_0, x_1, ...)$ where $x_q$ is the displacement of nucleus ($q$) from its equilibrium position. It can be shown that it is then possible to calculate the motion of the nuclei, to a good approximation, by assuming that they possess a total potential energy $U_N(..., x_{-1}, x_0, x_1, ...)$ equal to the sum of their electrostatic interaction energy and $E_{el}(..., x_{-1}, x_0, x_1, ...)$.

In fact, we shall further simplify the problem by making some reasonable hypotheses about $U_N$ (this is indispensable, since we do not know $E_{el}$). We shall assume that $U_N$ describes essentially the interactions of each of the nuclei with its nearest neighbors (in an infinite linear chain, each nucleus has two such neighbors), that is, that the forces exerted between non-adjacent nuclei can be neglected. In addition, we shall grant that, in the range of values that the displacements $x_q$ can attain, $U_N$ is well represented by an expression of the form:

$$U_N \simeq \frac{1}{2} m\omega_1^2 \sum_q (x_q - x_{q+1})^2 \tag{80}$$

where $m$ is the mass of a nucleus and $\omega_1$ characterizes the intensity of its interaction with its neighbors. We shall not, therefore, take into account terms of higher order in $(x_q - x_{q+1})$, that is, the anharmonicity of the potential.

Since expression (80) is identical to (6), we can apply the results of the preceding sections to the simple model of a solid body that we have just defined. Note, nevertheless, that we must choose $\omega = 0$ because $U_N$ is the total potential energy of the system of the nuclei, which interact with their neighbors, but are not elastically bound to their equilibrium positions*.

### b. NORMAL MODES. SPEED OF SOUND IN THE CRYSTAL

Each of the vibrational normal modes of the crystal is characterized by a wave vector $k$ and an angular frequency $\Omega(k)$. In solid state physics, the energy quantum associated with a mode is called a "phonon". The phonons can be considered to be particles of energy $\hbar\Omega(k)$ and momentum $\hbar k$. Actually, a phonon is not a true particle, since its existence involves a state of collective vibration of the real particles which constitute the crystal. The phonons are sometimes said to be "quasi-particles": they are entirely analogous to the fictitious particles, of position $x_G(t)$ and $x_R(t)$, introduced in complement $H_V$. In addition, a phonon can be created or destroyed by giving to or taking from the crystal the corresponding vibrational energy, while (at least in the non-relativistic domain to which we are restricting ourselves), a particle such as an electron cannot be created or destroyed. In this connection, note that, since the number of phonons in a given mode is arbitrary, phonons are bosons (chap. XIV).

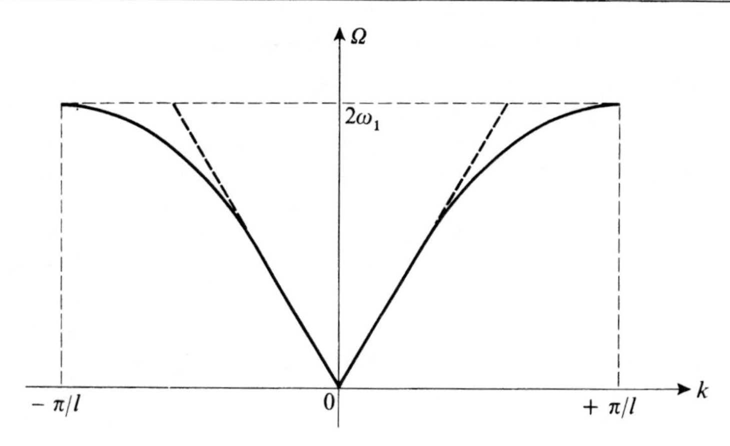


FIGURE 5

**Dispersion relation for phonons (curve of figure 2 for $\omega = 0$); the slope of the curve at the origin gives the speed of sound in the crystal.**

* The Einstein model, which we described in complement $A_V$, is based on a different hypothesis: each nucleus is assumed to "see" an average potential, due to its interactions with the other nuclei, but practically independent of the exact positions of these other nuclei. To a first approximation, this average potential is assumed to be parabolic, and one has a system of independent harmonic oscillators. Here, on the other hand, we are studying a somewhat more elaborate model, in which we explicitly (although approximately) take into account interactions between the nuclei.

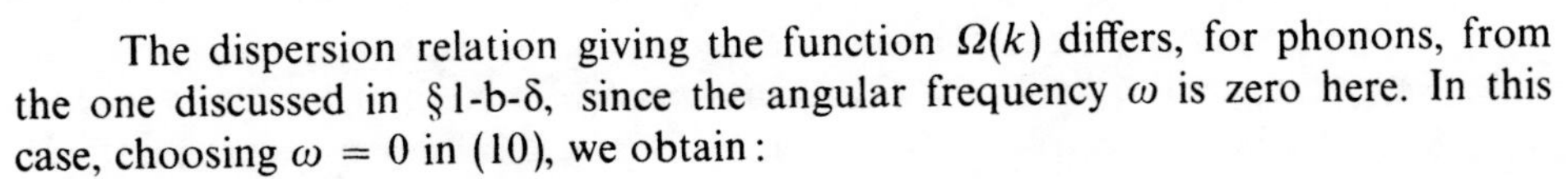

The dispersion relation giving the function $\Omega(k)$ differs, for phonons, from the one discussed in § 1-b-δ, since the angular frequency $\omega$ is zero here. In this case, choosing $\omega = 0$ in (10), we obtain:

$$\Omega(k) = 2\omega_1 \left| \sin\left(\frac{kl}{2}\right) \right| \tag{81}$$

The curve which represents $\Omega(k)$ is given in figure 5; it is composed of two half-arcs of a sinusoid. Unlike what happens for $\omega$ not equal to zero, $\Omega(k)$ now goes to zero for $k = 0$ and varies linearly when $k$ is very small, since, as long as:

$$|k| \ll \frac{1}{l} \tag{82}$$

we have:

$$\Omega(k) \simeq \omega_1 \, |kl| = v_s \, |k| \tag{83}$$

where:

$$v_s = \omega_1 l \tag{84}$$

Condition (82) means that the wavelength $2\pi/|k|$ associated with the mode being considered must be much greater than the separation between nuclei. For such wavelengths, the discontinuous structure of the chain is negligible, and the medium is not dispersive: the phase velocity $\Omega(k)/|k| \simeq v_s$ is independent of $k$, which implies that a wave packet involving only small values of $k$ (of the same sign) propagates without being deformed at the velocity $v_s$. Since acoustical wavelengths satisfy (82), $v_s$ is the speed of sound in the crystal.

When $|k|$ is of the order of $1/l$, the discontinuous structure of the chain becomes important, and the angular frequency $\Omega(k)$ increases less rapidly with $|k|$ than formula (83) would indicate (in figure 5, the curve deviates from the straight dashed lines which are its tangents at the origin). The medium is then dispersive, and a wave packet moves with a group velocity:

$$v_G = \frac{\mathrm{d}\Omega(k)}{\mathrm{d}k} \neq \frac{\Omega(k)}{k} \tag{85}$$

Finally, when $k$ approaches the edges of the first Brillouin zone ($k \longrightarrow \pm \pi/l$), we see from figure 5 that the group velocity approaches zero. As in an electromagnetic waveguide, the propagation velocity goes to zero when the cut-off frequency (here $\omega_1/2\pi$) is attained.

Figure 5 can also be seen as giving the spectrum of possible phonon energies $\hbar\Omega(k)$ in terms of their momentum $\hbar k$. Knowledge of such a spectrum, for a real crystal, is very important. It gives the precise energies and momenta that the crystal can supply or absorb when it interacts with another system. For example, the inelastic scattering of light by a crystal (the Brillouin effect) can be interpreted as the result of the destruction or creation of a phonon, with a change in the energy and momentum of the incident photon (total energy and momentum being conserved throughout the process).

COMMENT:

The simple one-dimensional model described here has allowed us to present some important physical concepts, which remain valid for a real crystal: energy quanta associated with the normal modes, dispersion of the medium, allowed and forbidden frequency (and, therefore, energy) bands. In reality, the crystalline lattice is three-dimensional, and a normal mode is characterized by a true wave vector $\mathbf{k}$; $\Omega$ then depends, in general, not only on the absolute value of $\mathbf{k}$, but also on its direction. Also, the situation may arise (as is the case for an ionic crystal) in which the vertices of the lattice are not all occupied by identical particles, but rather, for example, by two different types of particles in regular alternation*. Then, for each wave vector $\mathbf{k}$, several angular frequencies $\Omega(\mathbf{k})$ appear. Some of them, which go to zero when $|\mathbf{k}| \longrightarrow 0$, constitute "acoustic branches" like the one encountered above; the others belong to what are called "optical branches"**, in which a phonon of zero momentum has a non-zero energy. It would be out of the question to study all these problems here, although they are of primordial importance in solid state physics.

**References and suggestions for further reading:**

Chains of coupled classical oscillators: Berkeley 3 (7.1), §§2.4 and 3.5.
See section 13 of the bibliography, particularly Kittel (13.2), chap. 5.
Other examples of collective oscillations: Feynman III (1.2), chap. 15.

* A real crystal also contains impurities and imperfections distributed at random. Here we are speaking only about perfect crystals.

** This name arises from the fact that, in an ionic crystal, the "optical" phonons are coupled to electromagnetic waves like those in the visible domain, whose wavelengths are much larger than the atomic separation.

## Complement $K_V$

# VIBRATIONAL MODES OF A CONTINUOUS PHYSICAL SYSTEM. APPLICATION TO RADIATION ; PHOTONS



## 1. Outline of the problem

In complements $H_V$ and $J_V$, we introduced the idea of normal variables for a system of two or a countable infinity of coupled harmonic oscillators. The aim of this complement is to show that the same ideas can also be applied to the electromagnetic field, which is a *continuous* physical system (there is no natural lower bound for the wavelength of radiation).

This study raises a certain number of delicate problems. Therefore, before beginning, and in order to make a smooth transition between this complement and the preceding ones, $H_V$ and $J_V$, we shall start by studying in §2 the vibrational modes of a continuous *mechanical* system : a vibrating string. It is obvious that, on the atomic scale, such a system is not continuous : the string is composed of a very large number of atoms. However, we shall ignore this atomic structure and treat the string as if it were really continuous, since the fundamental aim of the calculation is to show how normal variables can be introduced for a continuous system. Since, moreover, we are dealing with a mechanical system, we can, without difficulty, define the conjugate momenta of the normal variables, calculate the Hamiltonian of the system, and show that it indeed appears in the form of a sum of Hamiltonians of independent one-dimensional harmonic oscillators. We shall also discuss in detail the quantization of such a system.

The results obtained in §2 will enable us to attack, in §3, the problem of the vibrational modes of radiation. We shall show that the study of radiation confined to a parallelepiped cavity leads to equations which are very similar to those of the vibrating string. The same transformations allow the introduction of completely uncoupled normal variables for the radiation (associated with the standing waves which can exist inside the cavity). Then we shall generalize the results obtained in §2 to derive simply the concept of a photon (as it is not feasible here to show rigorously how one can introduce, for a non-mechanical system like the electromagnetic field, conjugate momenta, a Lagrangian and a Hamiltonian).

## 2. Vibrational modes of a continuous mechanical system: example of a vibrating string

### a. NOTATION. DYNAMICAL VARIABLES OF THE SYSTEM

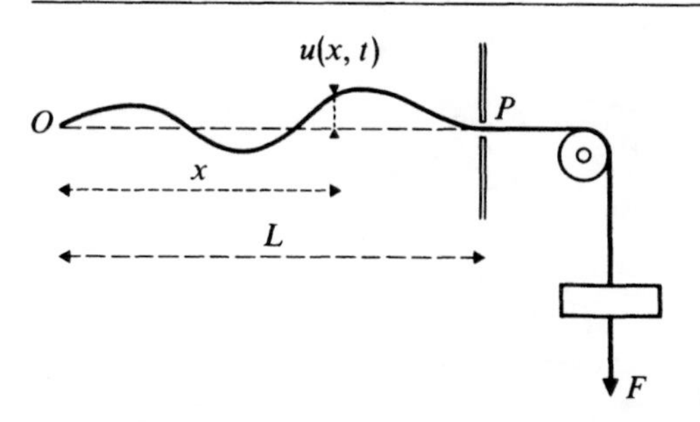


FIGURE 1

**Vibrating string passing through two fixed points $O$ and $P$ and having tension $F$; $u(x, t)$ denotes the deviation with respect to the equilibrium position of the point of the string situated at a distance $x$ from $O$.**

The string is fixed at a point $O$ (fig. 1). It passes through a very small hole $P$ pierced in a plate, and a weight exerts a tension $F$ on it. For simplicity, we assume that the string always remains in the same plane, passing through $O$ and $P$. Its state is defined at time $t$ when we know at this time the displacement $u(x, t)$ of the various points (labeled by their abscissas $x$ on $OP$), as well as the corresponding velocities $\dfrac{\partial u(x, t)}{\partial t}$. The constraints imposed at $O$ and $P$ are expressed by the boundary conditions:

$$u(0, t) = u(L, t) = 0 \tag{1}$$

where 0 and $L$ are the abscissas of the points $O$ and $P$.

It is important to remember that, in this problem, the dynamical variables are the displacements $u(x, t)$ at each point of abscissa $x$: there is a continuous infinity of dynamical variables. Consequently, $x$ is not a dynamical variable, but rather a continuous index which labels the dynamical variable with which we are concerned ($x$ plays the same role as indices 1 and 2 of complement $H_V$ or as the index $q$ of complement $J_V$).

### b. CLASSICAL EQUATIONS OF MOTION

Let $\mu$ be the mass per unit length of the string. If we assume the string to be perfectly flexible and if we confine ourselves to small displacements, a classical calculation enables us to obtain the partial differential equation satisfied by $u$. We find:

$$\left(\frac{1}{v^2}\frac{\partial^2}{\partial t^2} - \frac{\partial^2}{\partial x^2}\right) u(x, t) = 0 \tag{2}$$

where:

$$v = \sqrt{F/\mu} \tag{3}$$

is the propagation velocity of a perturbation along the string.

Such an equation expresses the fact that the evolution of the variable $u$ corresponding to the point $x$ depends on the variables $u$ at infinitely near points (via $\partial^2 u/\partial x^2$). Thus the variables $u(x, t)$ are all coupled to each other. We can then ask the following question: is it possible, as in complements $H_V$ and $J_V$, to introduce new, uncoupled variables which are linear combinations of the variables $u(x, t)$ associated with the various points $x$?

### c. INTRODUCTION OF THE NORMAL VARIABLES

Consider the set of functions of $x$:

$$f_k(x) = \sqrt{\frac{2}{L}} \sin\left(k\frac{\pi x}{L}\right) \tag{4}$$

where $k$ is a positive integer : $k = 1, 2, 3, ...$ The $f_k(x)$ satisfy the same boundary conditions as $u(x, t)$:

$$f_k(0) = f_k(L) = 0 \tag{5}$$

In addition, it is easy to verify the relations:

$$\int_0^L f_k(x)\, f_{k'}(x)\, \mathrm{d}x = \delta_{kk'} \tag{6}$$

(orthonormalization relation) and:

$$\left(\frac{\mathrm{d}^2}{\mathrm{d}x^2} + \frac{k^2\pi^2}{L^2}\right) f_k(x) = 0 \tag{7}$$

It can be shown that any function which goes to zero at $x = 0$ and $x = L$ [as is the case for $u(x, t)$] can be expanded in one and only one way in terms of the $f_k(x)$. We can therefore write:

$$u(x, t) = \sum_{k=1}^{\infty} q_k(t)\, f_k(x) \tag{8}$$

$q_k(t)$ can be obtained easily, using (6):

$$q_k(t) = \int_0^L u(x, t)\, f_k(x)\, \mathrm{d}x \tag{9}$$

The state of the string at the instant $t$ can be defined by the set of values $\left\{u(x, t), \frac{\partial}{\partial t} u(x, t)\right\}$ corresponding to the various points $x$ as well as by the set of numbers $\{ q_k(t), \dot{q}_k(t) \}$. The new variables $q_k(t)$ just introduced are linear combinations of the old $u(x, t)$, as can be see from (9). The converse is obviously also true [*cf.* formula (8)].

To obtain the equation satisfied by the $q_k(t)$, we substitute expansion (8) into the equation of motion (2). Using (7), we obtain, after a simple calculation:

$$\sum_{k=1}^{\infty} f_k(x)\left[\frac{1}{v^2}\frac{\mathrm{d}^2}{\mathrm{d}t^2}q_k(t) + \frac{k^2\pi^2}{L^2}q_k(t)\right] = 0 \tag{10}$$

that is, since the $f_k(x)$ are linearly independent:

$$\left[\frac{\mathrm{d}^2}{\mathrm{d}t^2} + \omega_k^2\right]q_k(t) = 0 \tag{11}$$

with:

$$\omega_k = \frac{k\pi v}{L} \tag{12}$$

Thus we see that the new variables $q_k(t)$, also called normal variables, evolve independently: they are *uncoupled.* Moreover, equation (11) is identical to that of a one-dimensional harmonic oscillator of angular frequency $\omega_k$, so that:

$$q_k(t) = A_k \cos(\omega_k t - \varphi_k) \tag{13}$$

Each of the terms $q_k(t)f_k(x)$ appearing on the right-hand side of (8) consequently represents a standing wave of frequency $\omega_k/2\pi$ and half-wavelength $L/k$. Each normal variable $q_k$ is therefore associated with a vibrational normal mode of the string, the most general motion of the string being a linear superposition of these normal modes.

COMMENT:

In complement $J_V$, we started with an infinite discrete set of harmonic oscillators and introduced a continuous infinity of normal variables. Here we find ourselves in the opposite situation: the $u(x, t)$ form a continuous set with respect to the index $x$, while, because of the boundary conditions, the normal variables $q_k(t)$ are labeled by a discrete index $k$.

## d. CLASSICAL HAMILTONIAN

### α. *Kinetic energy*

The kinetic energy of the segment of string included between $x$ and $x + \mathrm{d}x$ is $\frac{1}{2}\mu\,\mathrm{d}x\left[\frac{\partial u(x,t)}{\partial t}\right]^2$. It follows that the total kinetic energy $\mathscr{T}$ of the string is equal to:

$$\mathscr{T} = \frac{\mu}{2}\int_0^L \left[\frac{\partial u(x,t)}{\partial t}\right]^2 \mathrm{d}x \tag{14}$$

$\mathscr{T}$ can be expressed simply in terms of the $q_k$, using (8):

$$\mathscr{T} = \frac{\mu}{2}\sum_k\sum_{k'}\frac{\mathrm{d}q_k(t)}{\mathrm{d}t}\frac{\mathrm{d}q_{k'}(t)}{\mathrm{d}t}\int_0^L f_k(x)\,f_{k'}(x)\,\mathrm{d}x \tag{15}$$

which can also be written, taking (6) into account:

$$\mathscr{T} = \frac{\mu}{2} \sum_k \left( \frac{\mathrm{d}q_k}{\mathrm{d}t} \right)^2 \tag{16}$$

β. *Potential energy*

Consider the segment of string included between the abscissas $x$ and $x + \mathrm{d}x$. It makes an angle $\theta$ with the $Ox$ axis such that:

$$\tan \theta = \frac{\partial u(x, t)}{\partial x} \tag{17}$$

Its length is therefore equal to:

$$\frac{\mathrm{d}x}{\cos \theta} = \mathrm{d}x \left[ 1 + \tan^2 \theta \right]^{\frac{1}{2}} \tag{18}$$

Since the displacements are small, $\theta$ is very small, and we can write:

$$\frac{\mathrm{d}x}{\cos \theta} \simeq \mathrm{d}x \left[ 1 + \frac{1}{2} \left( \frac{\partial u(x, t)}{\partial x} \right)^2 \right] \tag{19}$$

From this we deduce that the total increase in the length of the string with respect to its equilibrium position (which corresponds to $u \equiv 0$ for all $x$) is equal to:

$$\Delta L = \frac{1}{2} \int_0^L \left( \frac{\partial u(x, t)}{\partial x} \right)^2 \mathrm{d}x \tag{20}$$

Now, $\Delta L$ represents the distance over which the weight stretching the rope has risen. The potential energy $\mathscr{V}$ of the string, with respect to the value corresponding to the equilibrium position, is therefore equal to:

$$\mathscr{V} = F \Delta L = \frac{1}{2} F \int_0^L \left( \frac{\partial u(x, t)}{\partial x} \right)^2 \mathrm{d}x \tag{21}$$

$\mathscr{V}$ can also be expressed in terms of the normal variables $q_k$. A simple calculation, using (8) and (4), yields:

$$\mathscr{V} = \frac{F}{2} \sum_k \frac{k^2 \pi^2}{L^2} q_k^2 \tag{22}$$

γ. *Conjugate momenta of the $q_k$; classical Hamiltonian*

The Lagrangian $\mathscr{L}$ of the system (*cf.* appendix III) can be written:

$$\mathscr{L} = \mathscr{T} - \mathscr{V} = \frac{\mu}{2} \sum_k [\dot{q}_k^2 - \omega_k^2 q_k^2] \tag{23}$$

From this, we deduce the expression for the conjugate momentum $p_k$ of $q_k$:

$$p_k = \frac{\partial \mathscr{L}}{\partial \dot{q}_k} = \mu \dot{q}_k \tag{24}$$

so that we finally obtain, for the Hamiltonian $\mathscr{H}(q_k, p_k)$ of the system, the expression:

$$\mathscr{H} = \mathscr{T} + \mathscr{V} = \sum_k \left[ \frac{p_k^2}{2\mu} + \frac{1}{2} \mu \, \omega_k^2 q_k^2 \right] \tag{25}$$

that is:

$$\mathscr{H} = \sum_k h_k \tag{26}$$

with:

$$h_k = \frac{p_k^2}{2\mu} + \frac{1}{2} \mu \omega_k^2 q_k^2 \tag{27}$$

Since $p_k$ and $q_k$ are conjugate variables, we recognize $h_k$ to be the Hamiltonian of a one-dimensional harmonic oscillator of angular frequency $\omega_k$. $\mathscr{H}$ is therefore a sum of Hamiltonians of independent one-dimensional harmonic oscillators (independent, since the normal variables are uncoupled).

It is useful to introduce, as in complements $B_V$ and $C_V$, the dimensionless variables:

$$\hat{q}_k = \beta_k q_k \tag{28-a}$$

$$\hat{p}_k = \frac{1}{\beta_k \hbar} p_k \tag{28-b}$$

where:

$$\beta_k = \sqrt{\frac{\mu \omega_k}{\hbar}} \tag{29}$$

is a (dimensional) constant. $\mathscr{H}$ can then be written:

$$\mathscr{H} = \sum_k \frac{1}{2} \hbar \omega_k \left[ \hat{q}_k^2 + \hat{p}_k^2 \right] \tag{30}$$

#### e. QUANTIZATION

##### α. *Preliminary comment*

The calculations performed in this section are, of course, not intended to reveal typically quantum effects in the motion of a macroscopic vibrating string. The vibrational frequencies $\omega_k/2\pi$ which can be excited on such a string are so low (of the order of a kilohertz at most) and the elementary energies $\hbar\omega_k$ so much smaller than the macroscopic energy of the string, that a classical description is quite sufficient. It might be thought that $\omega_k$ could be as large as we like because $k$ has no upper bound in formula (12). In fact, for sufficiently small wavelengths $2L/k$, the rigidity of the string can no longer be neglected, and equation (2) is no longer valid. Furthermore, as we pointed out in the introduction, the string is not really a continuous system, and it would make no sense to consider wavelengths smaller than the interatomic distance.

The calculations which will be presented here must be considered as a simple first approach to problems posed by the quantum mechanical description of radiation. Now, radiation constitutes a truly continuous system (no natural lower bound exists for the wavelength) and satisfies an equation which is analogous to (2), whatever frequencies and wavelengths may be involved★.

β. *Eigenstates and eigenvalues of the quantum mechanical Hamiltonian H*

We quantize each oscillator by associating with $\hat{q}_k$ and $\hat{p}_k$ [see formula (28)] observables $\hat{Q}_k$ and $\hat{P}_k$ such that:

$$[\hat{Q}_k, \hat{P}_k] = i \tag{31}$$

Since the normal variables are uncoupled, we also assume that the operators relating to two different oscillators commute. We therefore have:

$$[\hat{Q}_k, \hat{P}_{k'}] = i\delta_{kk'} \tag{32}$$

Let:

$$H_k = \frac{1}{2}\hbar\omega_k\,(\hat{Q}_k^2 + \hat{P}_k^2) \tag{33}$$

be the quantum mechanical Hamiltonian of oscillator $k$. From the results of chapter V, we know its eigenstates and eigenvalues:

$$H_k \,|\, n_k \rangle = \left(n_k + \frac{1}{2}\right)\hbar\omega_k \,|\, n_k \rangle \tag{34}$$

where $n_k$ is a non-negative integer (to simplify the notation, we shall write $|\, n_k \rangle$ instead of $|\, \varphi_{n_k} \rangle$).

Since the $H_k$ commute, we can choose the eigenstates of $H$ to be in the form of tensor products of the $|\, n_k \rangle$:

$$|\, n_1 \rangle \,|\, n_2 \rangle \ldots |\, n_k \rangle \ldots = |\, n_1, n_2, \ldots, n_k, \ldots \rangle \tag{35}$$

The ground or "vacuum" state corresponds to all the $n_k$ equal to zero:

$$|\, 0, 0, \ldots 0 \ldots \rangle = |\, 0 \rangle \tag{36}$$

When we choose the energy origin to be the energy of the state $|\, 0 \rangle$, the energy of state (35) is equal to:

$$E_{n_1\, n_2 \ldots n_k \ldots} = \sum_k n_k\, \hbar\omega_k \tag{37}$$

A state such as (35) can be considered to represent a set of $n_1$ energy quanta $\hbar\omega_1$, ..., $n_k$ energy quanta $\hbar\omega_k$, ... These vibrational quanta are analogous to the phonons studied in complement $J_V$.

Finally, using $\hat{P}_k$ and $\hat{Q}_k$, we can introduce, as in §B of chapter V, creation and annihilation operators for an energy quantum $\hbar\omega_k$:

$$a_k = \frac{1}{\sqrt{2}}(\hat{Q}_k + i\hat{P}_k) \tag{38}$$

★ If we were really interested in a microscopic "vibrating string" (for example, a linear macromolecule), it would be more realistic to consider, as in complement $J_V$, a chain of atoms and to study not only their longitudinal displacements but also their transversal ones (transverse phonons).

where $a_k^\dagger$ is the adjoint of $a_k$. We then have:

$$[a_k, a_{k'}^\dagger] = \delta_{kk'} \tag{39}$$

and:

$$\begin{cases} a_k \mid n_1, n_2, ..., n_k, ... \rangle = \sqrt{n_k} \mid n_1, n_2, ..., n_k - 1, ... \rangle \\ a_k^\dagger \mid n_1, n_2, ..., n_k, ... \rangle = \sqrt{n_k + 1} \mid n_1, n_2, ..., n_k + 1, ... \rangle \end{cases} \tag{40}$$

All the states (35) can be expressed in terms of the vacuum state $| 0 \rangle$:

$$| n_1, n_2, ..., n_k, ... \rangle = \frac{(a_1^\dagger)^{n_1}}{\sqrt{n_1!}} \frac{(a_2^\dagger)^{n_2}}{\sqrt{n_2!}} ... \frac{(a_k^\dagger)^{n_k}}{\sqrt{n_k!}} ... | 0 \rangle \tag{41}$$

γ. *Quantum mechanical state of the system*

The most general quantum mechanical state of the system is a linear superposition of the states $| n_1, n_2, ... n_k, ... \rangle$:

$$| \psi(t) \rangle = \sum_{n_1, n_2, ..., n_k ...} c_{n_1, n_2, ..., n_k ...}(t) \mid n_1, n_2, ..., n_k ... \rangle \tag{42}$$

The equation of motion of $| \psi(t) \rangle$ is the Schrödinger equation:

$$i\hbar \frac{\mathrm{d}}{\mathrm{d}t} | \psi(t) \rangle = H \mid \psi(t) \rangle \tag{43}$$

Using (37) and (43), we easily obtain:

$$c_{n_1, n_2, ..., n_k ...}(t) = c_{n_1, n_2, ..., n_k ...}(0) \, \mathrm{e}^{-i \sum_k n_k \omega_k t} \tag{44}$$

δ. *Observables associated with the dynamical variables $u(x, t)$*

When the system is quantized, $u(x, t)$ becomes an observable $U(x)$ which does not depend on $t$* and which is obtained by replacing in (8) $q_k(t)$ by the observable $Q_k$:

$$\begin{aligned} U(x) &= \sum_k f_k(x) \, Q_k \\ &= \sum_k \frac{1}{\beta_k \sqrt{2}} f_k(x) \, [a_k + a_k^\dagger] \end{aligned} \tag{45}$$

Thus we see that a displacement observable $U(x)$ can be defined for each value of $x$, and that it depends linearly on the creation and annihilation operators $a_k^\dagger$ and $a_k$.

It is interesting to compare the mean value of $U(x)$, $\langle \psi(t) \mid U(x) \mid \psi(t) \rangle$, with the classical quantity $u(x, t)$. Since, according to (40), $a_k$ and $a_k^\dagger$ can link only states whose energy difference is $\mp \hbar\omega_k$, we deduce from (45) that the only Bohr frequencies which can appear in the evolution of $\langle U(x) \rangle(t)$ are the frequencies $\omega_1/2\pi$, $\omega_2/2\pi$, ... $\omega_k/2\pi$, ... associated, respectively, with the spatial

* Recall that, in quantum mechanics, the time dependence is usually contained in the state vector and not in the observables (*cf.* discussion of § D-1-d of chapter III).

functions $f_1(x), f_2(x), ..., f_k(x), ...$ Thus we find for $\langle U(x) \rangle(t)$ a linear superposition of the standing waves which can exist on the string. This analogy can, moreover, be pursued further. Let us calculate the derivative $\frac{\partial^2}{\partial t^2} \langle U(x) \rangle(t)$; using the fact that [*cf.* complement $G_V$, equation (17)]:

$$\frac{d}{dt} \langle a_k \rangle = - i\omega_k \langle a_k \rangle \tag{46}$$

and relations (7) and (12), we easily find that the mean value of $U(x)$ given by (45) satisfies the differential equation:

$$\left( \frac{1}{v^2} \frac{\partial^2}{\partial t^2} - \frac{\partial^2}{\partial x^2} \right) \langle U(x) \rangle(t) = 0 \tag{47}$$

which is identical to (2).

Finally, note that since $Q_k$ does not commute with $H_k$, $U(x)$ does not commute with $H$. The displacement and the total energy are therefore, in quantum mechanics, incompatible physical quantities.

## 3. Vibrational modes of radiation: photons

### a. NOTATION. EQUATIONS OF MOTION

The classical state of the electromagnetic field at a given $t$ is defined when we know, for this time, the value of the components of the electric field $\mathbf{E}$ and the magnetic field $\mathbf{B}$ at each point $\mathbf{r}$ of space. As in §2 above, we therefore have a continuous infinity of dynamical variables: the six components $E_x$, $E_y$, $E_z$ and $B_x$, $B_y$, $B_z$ at each point $\mathbf{r}$.

In order to stress the importance of the idea of normal variables (or normal modes) of a field, we shall introduce a simplification which consists of forgetting the vector nature of the fields $\mathbf{E}$ and $\mathbf{B}$: we shall base our arguments on a scalar field $\mathscr{S}(\mathbf{r}, t)$ which obeys (like each of the components of $\mathbf{E}$ and $\mathbf{B}$) the equation:

$$\left( \frac{1}{c^2} \frac{\partial^2}{\partial t^2} - \Delta \right) \mathscr{S}(\mathbf{r}, t) = 0 \tag{48}$$

where $c$ is the speed of light.

We shall assume the field to be confined to a parallelepiped cavity whose inside walls are perfectly conducting and whose edges, parallel to $Ox$, $Oy$, $Oz$, have, respectively, the lengths $L_1$, $L_2$, $L_3$. As boundary conditions, we require $\mathscr{S}(\mathbf{r}, t)$ to be zero on the walls of the cavity (in the real problem, it is, for example, the tangential components of the electric field $\mathbf{E}$ that must go to zero on these walls). We can therefore write:

$$\begin{aligned} \mathscr{S}(x = 0, y, z, t) &= \mathscr{S}(x = L_1, y, z, t) = \mathscr{S}(x, y = 0, z, t) = ... \\ &= \mathscr{S}(x, y, z = L_3, t) = 0 \end{aligned} \tag{49}$$

### b. INTRODUCTION OF THE NORMAL VARIABLES

Consider the set of functions of $x, y, z$:

$$f_{klm}(x, y, z) = \sqrt{\frac{8}{L_1 L_2 L_3}} \sin\left(\frac{k\pi x}{L_1}\right) \sin\left(\frac{l\pi y}{L_2}\right) \sin\left(\frac{m\pi z}{L_3}\right) \tag{50}$$

where $k$, $l$, $m$ are positive integers ($k, l, m = 1, 2, 3, \ldots$). The $f_{klm}(x, y, z)$ go to zero on the walls of the cavity and therefore satisfy the same boundary conditions as $\mathscr{S}(x, y, z, t)$:

$$f_{klm}(x = 0, y, z) = f_{klm}(x = L_1, y, z) = \ldots = f_{klm}(x, y, z = L_3) = 0 \tag{51}$$

In addition, the following relations are simple to verify:

$$\int_0^{L_1} \mathrm{d}x \int_0^{L_2} \mathrm{d}y \int_0^{L_3} \mathrm{d}z \, f_{klm}(x, y, z) \, f_{k'l'm'}(x, y, z) = \delta_{kk'} \, \delta_{ll'} \, \delta_{mm'} \tag{52}$$

and:

$$\left[\Delta + \left(\frac{k^2}{L_1^2} + \frac{l^2}{L_2^2} + \frac{m^2}{L_3^2}\right)\pi^2\right] f_{klm}(x, y, z) = 0 \tag{53}$$

Any function which goes to zero on the walls of the cavity, $\mathscr{S}(\mathbf{r}, t)$ in particular, can be expanded in one and only one way in terms of the $f_{klm}(x, y, z)$. We therefore have:

$$\mathscr{S}(x, y, z, t) = \sum_{k,l,m} q_{klm}(t) \, f_{klm}(x, y, z) \tag{54}$$

Formula (54) can easily be inverted, with the help of (52):

$$q_{klm}(t) = \int_0^{L_1} \mathrm{d}x \int_0^{L_2} \mathrm{d}y \int_0^{L_3} \mathrm{d}z \, f_{klm}(x, y, z) \, \mathscr{S}(x, y, z, t) \tag{55}$$

Thus we see that the field at time $t$ is described by the set of variables $q_{klm}(t)$ as well as by the set of variables $\mathscr{S}(x, y, z, t)$. Formulas (54) and (55) enable us to go from one set to the other one.

Substituting (54) into (48) and using (53), we obtain, after a simple calculation:

$$\left[\frac{\mathrm{d}^2}{\mathrm{d}t^2} + \omega_{klm}^2\right] q_{klm}(t) = 0 \tag{56}$$

where:

$$\omega_{klm}^2 = c^2 \pi^2 \left[\frac{k^2}{L_1^2} + \frac{l^2}{L_2^2} + \frac{m^2}{L_3^2}\right] \tag{57}$$

The normal variables $q_{klm}(t)$ are therefore *uncoupled*. According to (56), $q_{klm}(t)$ varies like $A \cos(\omega_{klm} t - \varphi)$. Each of the terms $q_{klm}(t) f_{klm}(x, y, z)$ of the sum (54) therefore represents a standing wave (a vibrational normal mode of the field in the cavity) characterized by its frequency $\omega_{klm}/2\pi$ and its spatial dependence in the three directions $Ox$, $Oy$, $Oz$ (half-wavelengths $L_1/k$, $L_2/l$ and $L_3/m$ respectively).

Thus we have been able to generalize the results of §2-c without difficulty. Note, nevertheless, that when the vectorial nature of the electromagnetic field is taken into account, the structure of the modes is more complex. However, the general idea is the same, and one reaches similar conclusions.

### c. CLASSICAL HAMILTONIAN

Basing our discussion on the very close analogy between the results of §2-c and those of §3-b, we shall assume without proof that one can associate with the field $\mathscr{S}(\mathbf{r}, t)$ a Lagrangian $\mathscr{L}$, from which can be deduced the equation of motion (48), the conjugate momenta $p_{klm}(t)$ of the normal variables, and finally, the expression for the Hamiltonian $\mathscr{H}$ of the system. The only point which is important here is that this Hamiltonian is analogous to (30):

$$\mathscr{H} = \sum_{k,l,m} \frac{1}{2} \hbar\omega_{klm} \left[(\hat{q}_{klm})^2 + (\hat{p}_{klm})^2\right] \tag{58}$$

where $\hat{q}_{klm}$ and $\hat{p}_{klm}$ are dimensionless variables proportional to the $q_{klm}$ and $p_{klm}$:

$$\hat{q}_{klm} = \beta_{klm} q_{klm} \qquad \hat{p}_{klm} = \frac{1}{\hbar\beta_{klm}} p_{klm} \tag{59}$$

$\beta_{klm}$ is a (dimensional) constant which is analogous to the one introduced in (29).

COMMENTS:

(*i*) The equation of motion of each normal variable $q_{klm}$ [established in (56)] is analogous to that of a one-dimensional harmonic oscillator of angular frequency $\omega_{klm}$. Thus we see why we obtain for $\mathscr{H}$ a sum of Hamiltonians of independent one-dimensional harmonic oscillators. It is possible, moreover, to obtain (56) from (58). The Hamilton-Jacobi equations (*cf.* appendix III) can, in fact, be written, taking (59) into account :

$$\begin{cases} \dfrac{\mathrm{d}\hat{q}_{klm}}{\mathrm{d}t} = \dfrac{1}{\hbar}\dfrac{\partial\mathscr{H}}{\partial\hat{p}_{klm}} \\[2ex] \dfrac{\mathrm{d}\hat{p}_{klm}}{\mathrm{d}t} = -\dfrac{1}{\hbar}\dfrac{\partial\mathscr{H}}{\partial\hat{q}_{klm}} \end{cases} \tag{60}$$

that is, with the form (58) of $\mathscr{H}$:

$$\frac{\mathrm{d}\hat{q}_{klm}}{\mathrm{d}t} = \omega_{klm}\,\hat{p}_{klm} \tag{61-a}$$

$$\frac{\mathrm{d}\hat{p}_{klm}}{\mathrm{d}t} = -\,\omega_{klm}\,\hat{q}_{klm} \tag{61-b}$$

Eliminating $\hat{p}_{klm}$ between these two equations, we indeed find (56).

(*ii*) For the real electromagnetic field, composed of two fields **E** and **B**, one can also obtain expression (58) for $\mathscr{H}$ directly without using the Lagrangian. One simply takes the total

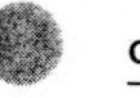



energy $\mathscr{H}$ of the field as the sum of the electrical and magnetic energies contained in the cavity:

$$\mathscr{H} = \frac{\varepsilon_0}{2}\int_0^{L_1} dx \int_0^{L_2} dy \int_0^{L_3} dz \,[\mathbf{E}^2 + c^2\mathbf{B}^2] \tag{62}$$

and uses for **E** and **B** expansions analogous to (54). Thus one finds that the terms in $\hat{p}_{klm}^2$ and $\hat{q}_{klm}^2$ of (58) correspond respectively to the electrical and magnetic energies.

### d. QUANTIZATION

Now, starting with equation (58), we can carry out the same operations as in §2-e.

#### α. *Eigenstates and eigenvalues of H*

We associate with $\hat{q}_{klm}$ and $\hat{p}_{klm}$ two observables $\hat{Q}_{klm}$ and $\hat{P}_{klm}$ whose commutator is equal to $i$. Since observables relating to two different modes commute, we have, in general:

$$[\hat{Q}_{klm}, \hat{P}_{k'l'm'}] = i\delta_{kk'}\,\delta_{ll'}\,\delta_{mm'} \tag{63}$$

Let $H_{klm}$ be the Hamiltonian associated with the mode $(klm)$:

$$H_{klm} = \frac{\hbar\omega_{klm}}{2}[(\hat{Q}_{klm})^2 + (\hat{P}_{klm})^2] \tag{64}$$

We know its eigenstates and eigenvalues:

$$H_{klm}\,|\,n_{klm}\rangle = \left(n_{klm} + \frac{1}{2}\right)\hbar\omega_{klm}\,|\,n_{klm}\rangle \tag{65}$$

where $n_{klm}$ is a non-negative integer.

Since the $H_{klm}$ commute, we can choose the eigenstates of $H = \sum_{klm} H_{klm}$ to be in the form of tensor products of the $|\,n_{klm}\rangle$:

$$|\,n_{111}, n_{211}, n_{121}, n_{112}, \ldots, n_{klm}, \ldots\rangle \tag{66}$$

The ground state, called the "vacuum", corresponds to all the $n_{klm}$ equal to zero :

$$|\,0, 0, 0, 0, \ldots, 0, \ldots\rangle = |\,0\rangle \tag{67}$$

With respect to the vacuum state energy, the energy of state (66) is equal to:

$$E_{n_{111},\ldots n_{klm}\ldots} = \sum_{klm} n_{klm}\,\hbar\omega_{klm} \tag{68}$$

A state such as (66) can be considered to represent a set of $n_{111}$ energy quanta $\hbar\omega_{111}$, ... $n_{klm}$ energy quanta $\hbar\omega_{klm}$ ... These quanta are none other than the *photons*. Thus we see that a certain type of photon is associated with each normal mode of the cavity.

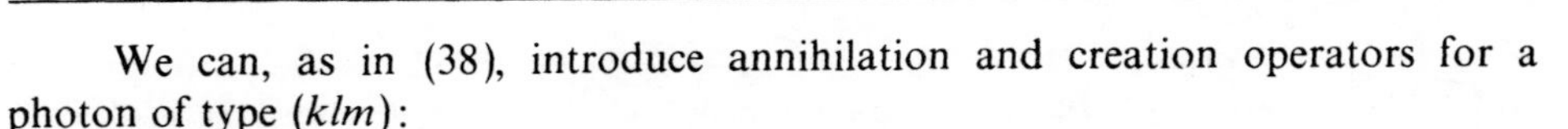



We can, as in (38), introduce annihilation and creation operators for a photon of type $(klm)$:

$$\begin{cases} a_{klm} = \dfrac{1}{\sqrt{2}}(\hat{Q}_{klm} + i\hat{P}_{klm}) \\ a^{\dagger}_{klm} = \dfrac{1}{\sqrt{2}}(\hat{Q}_{klm} - i\hat{P}_{klm}) \end{cases} \tag{69}$$

and establish formulas identical to (39), (40) and (41):

$$[a_{klm}, a^{\dagger}_{k'l'm'}] = \delta_{kk'}\,\delta_{ll'}\,\delta_{mm'} \tag{70}$$

$$\begin{cases} a_{klm} \mid n_{111}, ..., n_{klm}, ... \rangle = \sqrt{n_{klm}} \mid n_{111}, ..., n_{klm} - 1, ... \rangle \\ a^{\dagger}_{klm} \mid n_{111}, ..., n_{klm}, ... \rangle = \sqrt{n_{klm} + 1} \mid n_{111}, ..., n_{klm} + 1, ... \rangle \end{cases} \tag{71}$$

$$\mid n_{111}, ..., n_{klm}, ... \rangle = \frac{(a^{\dagger}_{111})^{n_{111}}}{\sqrt{n_{111}!}} \cdots \frac{(a^{\dagger}_{klm})^{n_{klm}}}{\sqrt{n_{klm}!}} \cdots \mid 0 \rangle \tag{72}$$

## β. *Quantum state of the field*

The most general state of the field is a linear superposition of states (66):

$$\mid \psi(t) \rangle = \sum_{n_{111}, ..., n_{klm} ...} c_{n_{111}, ..., n_{klm}, ...}(t) \mid n_{111}, ..., n_{klm}, ... \rangle \tag{73}$$

The Schrödinger equation:

$$i\hbar \frac{\mathrm{d}}{\mathrm{d}t} \mid \psi(t) \rangle = H \mid \psi(t) \rangle \tag{74}$$

enables us to obtain the coefficients $c_{n_{111}, ..., n_{klm}, ...}(t)$ in the form:

$$c_{n_{111}, ..., n_{klm}, ...}(t) = c_{n_{111}, ..., n_{klm}, ...}(0)\, \mathrm{e}^{-i \sum_{klm} n_{klm}\, \omega_{klm}\, t} \tag{75}$$

## γ. *Field operator*

Upon quantization, $\mathscr{S}(\mathbf{r}, t)$ becomes an observable $S(\mathbf{r})$ which no longer depends on $t$ and is obtained by replacing in (54) $q_{klm}(t)$ by $Q_{klm}$:

$$S(\mathbf{r}) = \sum_{klm} \frac{1}{\beta_{klm}} f_{klm}(\mathbf{r})\, \hat{Q}_{klm} \tag{76}$$

One can also, with the help of (69), express $S(\mathbf{r})$ in terms of the creation and annihilation operators:

$$S(\mathbf{r}) = \frac{1}{\sqrt{2}} \sum_{klm} \frac{1}{\beta_{klm}} f_{klm}(\mathbf{r})\, [a_{klm} + a^{\dagger}_{klm}] \tag{77}$$

The same arguments as in §2-e-δ enable us to show, using formulas (71) and (75), that the only Bohr frequencies which can appear in the time evolution of the mean value of the field:

$$\langle S(\mathbf{r}) \rangle(t) = \langle \psi(t) | S(\mathbf{r}) | \psi(t) \rangle$$

are the frequencies $\omega_{111}/2\pi$, $\omega_{211}/2\pi$, ..., $\omega_{klm}/2\pi$, ... associated, respectively, with the spatial functions $f_{111}(\mathbf{r})$, $f_{211}(\mathbf{r})$, ... $f_{klm}(\mathbf{r})$... Thus we find for $\langle S(\mathbf{r}) \rangle(t)$ a linear superposition of the classical standing waves which can exist in the cavity. A calculation identical to the one in §2-e-δ would enable us to show that $\langle S(\mathbf{r}) \rangle$ satisfies equation (48).

Finally, we find that $S(\mathbf{r})$ and $H$ do not commute. It is therefore impossible, in quantum theory, to know simultaneously and with certainty both the number of photons and the value of the electromagnetic field at a point in space.

COMMENT:

For the electromagnetic field, coherent states can be constructed which are analogous to the ones introduced in complement $G_V$ and which represent the best possible compromise between the incompatible quantities, field and energy.

δ. *Vacuum fluctuations*

We saw in §D-1 of chapter V that, in the ground state of a harmonic oscillator, $\langle X \rangle$ is zero while $\langle X^2 \rangle$ is not, and we discussed the physical meaning of this typically quantum mechanical effect.

In the problem we are studying here, $S(\mathbf{r})$ presents many analogies with the $X$ operator of chapter V; we see from (77) that $S(\mathbf{r})$ is a *linear* combination of creation and annihilation operators. Consider the mean value of $S(\mathbf{r})$ in the ground state $|0\rangle$ of the field, that is, the "vacuum" state of photons. Since the diagonal elements of $a$ and $a^\dagger$ are zero according to (71), we see that:

$$\langle 0 | S(\mathbf{r}) | 0 \rangle = 0 \tag{78}$$

On the other hand, the corresponding matrix element of $[S(\mathbf{r})]^2$ is not zero. According to (71):

$$\begin{cases} a_{klm} | 0 \rangle = 0 \\ \langle 0 | a^\dagger_{k'l'm'} = 0 \\ \langle 0 | a_{k'l'm'} a^\dagger_{klm} | 0 \rangle = \delta_{kk'} \, \delta_{ll'} \, \delta_{mm'} \end{cases} \tag{79}$$

Therefore, a simple calculation enables us to establish, using (77), that:

$$\langle 0 | [S(\mathbf{r})]^2 | 0 \rangle = \frac{1}{2} \sum_{klm} \frac{1}{\beta^2_{klm}} [f_{klm}(\mathbf{r})]^2 \tag{80}$$

From this we see that in a vacuum, that is, in the absence of photons, the electromagnetic field $S(\mathbf{r})$ at a point of space has a zero mean value but a *non-zero root-mean-square deviation*. This means, for example, that if we perform *one* measurement of $S(\mathbf{r})$, we can find a non-zero result (varying, of course, from one measurement to another), even if there is no photon present in space. This effect has no equivalent in classical theory, in which, when the energy is zero, the field is rigorously zero. The preceding result is often expressed by saying that the "vacuum state" of photons is subject to fluctuations of the field, characterized by (78) and (80) and called *vacuum fluctuations*.

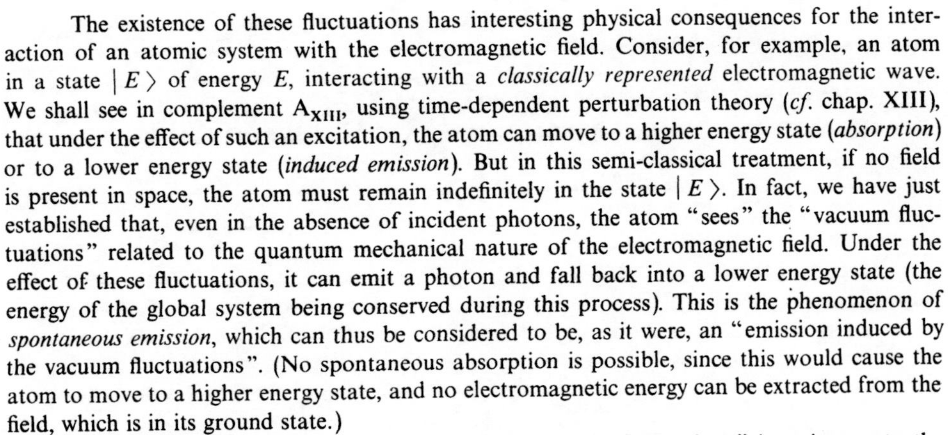

The existence of these fluctuations has interesting physical consequences for the interaction of an atomic system with the electromagnetic field. Consider, for example, an atom in a state $| E \rangle$ of energy $E$, interacting with a *classically represented* electromagnetic wave. We shall see in complement $A_{XIII}$, using time-dependent perturbation theory (*cf.* chap. XIII), that under the effect of such an excitation, the atom can move to a higher energy state (*absorption*) or to a lower energy state (*induced emission*). But in this semi-classical treatment, if no field is present in space, the atom must remain indefinitely in the state $| E \rangle$. In fact, we have just established that, even in the absence of incident photons, the atom "sees" the "vacuum fluctuations" related to the quantum mechanical nature of the electromagnetic field. Under the effect of these fluctuations, it can emit a photon and fall back into a lower energy state (the energy of the global system being conserved during this process). This is the phenomenon of *spontaneous emission*, which can thus be considered to be, as it were, an "emission induced by the vacuum fluctuations". (No spontaneous absorption is possible, since this would cause the atom to move to a higher energy state, and no electromagnetic energy can be extracted from the field, which is in its ground state.)

It can also be shown that another effect of "vacuum fluctuations" is to impart to the atomic electrons an erratic motion which slightly modifies the energies of the levels. The observation of this effect in the hydrogen atom spectrum (the "Lamb shift") constituted the point of departure for the development of modern quantum electrodynamics.

COMMENT:

In the preceding discussions, we have always chosen the energy of the vacuum state as the energy origin. In fact, harmonic oscillator theory gives us the absolute value of the energy of the vacuum state :

$$E_0 = \sum_{klm} \frac{1}{2} \hbar \omega_{klm} \tag{81}$$

There is obviously a close relationship between $E_0$ and the electrical and magnetic energy associated with "vacuum fluctuations". One of the difficulties of quantum electrodynamics, of which we have just given a brief sketch, is that the sum (81) is in fact infinite, as is, moreover, (80)! Nevertheless, it is possible to surmount this difficulty : using the procedure called "renormalization", one manages to bypass infinite quantities and calculate the physical effects which are actually observable, such as the "Lamb shift", with remarkable accuracy. It is obviously out of the question to consider these vast problems here.

**References and suggestions for further reading:**

Modes of a continuous string in classical mechanics: Berkeley 3 (1.1), §§2.1, 2.2 and 2.3.

Quantization of the electromagnetic field: Mandl (2.9); Schiff (1.18), chap. 14; Messiah (1.17), chap. XXI; Bjorken and Drell (2.10), chap. 11; Power (2.11); Heitler (2.13).

The Lamb shift : Lamb and Retherford (3.11); Frisch (3.13); Kuhn (11.1), chap. III, §A 5 e; Series (11.7), chaps. VIII, IX and X.

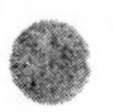



## Complement $L_V$

# ONE-DIMENSIONAL HARMONIC OSCILLATOR IN THERMODYNAMIC EQUILIBRIUM AT A TEMPERATURE $T$



This complement is devoted to the study of the physical properties of a one-dimensional harmonic oscillator in thermodynamic equilibrium with a reservoir at temperature $T$. We know (*cf.* complement $E_{III}$) that such an oscillator is not in a pure state (it is impossible to describe its state by a ket $| \psi \rangle$). The partial information which we possess about it and the results of statistical mechanics enable us to characterize it by a statistical mixture of stationary states $| \varphi_n \rangle$ with weights proportional to $e^{-E_n/kT}$ ($k$ : Boltzmann constant; $E_n$: energy of the state $| \varphi_n \rangle$). We saw in complement $E_{III}$ (§5-a) that the corresponding density operator is then written:

$$\rho = Z^{-1}\, e^{-H/kT} \tag{1}$$

where $H$ is the Hamiltonian operator, and:

$$Z = \text{Tr}\, e^{-H/kT} \tag{2}$$

is a normalization factor which insures that:

$$\text{Tr}\, \rho = 1 \tag{3}$$

($Z$ is the partition function).

We shall calculate the mean value $\langle H \rangle$ of the oscillator's energy, interpret the result obtained, and show how it enters into numerous problems in physics (blackbody radiation, specific heat of solids, ...). Finally, we shall establish and discuss the expression for the probability density of the particle's position (the observable $X$).

## 1. Mean value of the energy

### a. PARTITION FUNCTION

The energies $E_n$ of the states $| \varphi_n \rangle$ are, according to the results of §B of chapter V, equal to $(n + 1/2)\hbar\omega$. Since the energy levels are not degenerate, we have, according to (2):

$$Z = \sum_{n=0}^{\infty} \langle \varphi_n | e^{-H/kT} | \varphi_n \rangle$$

$$= \sum_{n=0}^{\infty} e^{-(n+1/2)\hbar\omega/kT}$$

$$= e^{-\hbar\omega/2kT}[1 + e^{-\hbar\omega/kT} + e^{-2\hbar\omega/kT} + ...] \tag{4}$$

Inside the brackets, we recognize a geometrical progression of $e^{-\hbar\omega/kT}$. Therefore:

$$Z = \frac{e^{-\hbar\omega/2kT}}{1 - e^{-\hbar\omega/kT}} \tag{5}$$

### b. CALCULATION OF $\langle H \rangle$

According to formula (31) of complement $E_{III}$ and expression (1) for $\rho$:

$$\langle H \rangle = \mathrm{Tr}\,(H\rho) = Z^{-1}\,\mathrm{Tr}\,(H\,e^{-H/kT}) \tag{6}$$

Writing the trace explicitly in the $\{ | \varphi_n \rangle \}$ basis, we obtain:

$$\langle H \rangle = Z^{-1} \sum_{n=0}^{\infty} (n + 1/2)\ \hbar\omega\, e^{-(n+1/2)\hbar\omega/kT} \tag{7}$$

To calculate this quantity, we differentiate both sides of (4) with respect to $T$:

$$\frac{dZ}{dT} = \frac{1}{kT^2} \sum_{n=0}^{\infty} (n + 1/2)\ \hbar\omega\, e^{-(n+1/2)\hbar\omega/kT} \tag{8}$$

We see that:

$$\langle H \rangle = kT^2 \frac{1}{Z}\frac{dZ}{dT} \tag{9}$$

Using (5), we then find, after a simple calculation:

$$\boxed{\langle H \rangle = \frac{\hbar\omega}{2} + \frac{\hbar\omega}{e^{\hbar\omega/kT} - 1}} \tag{10}$$

COMMENTS:

(*i*) *Isotropic three-dimensional oscillator*

Using the results and notation of complement $E_V$, we can write:

$$\langle H \rangle = \langle H_x \rangle + \langle H_y \rangle + \langle H_z \rangle \tag{11}$$

where $\langle H_x \rangle$ is given by:

$$\langle H_x \rangle = Z^{-1}\,\text{Tr}\,(H_x\,\text{e}^{-H/kT})$$

$$= \frac{\displaystyle\sum_{n_x=0}^{\infty}\sum_{n_y=0}^{\infty}\sum_{n_z=0}^{\infty}(n_x + 1/2)\,\hbar\omega\,\text{e}^{-[(n_x+1/2)+(n_y+1/2)+(n_z+1/2)]\hbar\omega/kT}}{\displaystyle\sum_{n_x=0}^{\infty}\sum_{n_y=0}^{\infty}\sum_{n_z=0}^{\infty}\text{e}^{-[(n_x+1/2)+(n_y+1/2)+(n_z+1/2)]\hbar\omega/kT}} \tag{12}$$

The sums over $n_y$ and $n_z$ can be separated out and are identical in the numerator and denominator, so that:

$$\langle H_x \rangle = \frac{\displaystyle\sum_{n_x=0}^{\infty}(n_x + 1/2)\,\hbar\omega\,\text{e}^{-(n_x+1/2)\hbar\omega/kT}}{\displaystyle\sum_{n_x=0}^{\infty}\text{e}^{-(n_x+1/2)\hbar\omega/kT}} \tag{13}$$

Aside from the replacement of $n$ by $n_x$, the final expression is identical to the one calculated in the preceding section; $\langle H_x \rangle$ is therefore equal to the value given in (10). It is easy to show that the same is true for $\langle H_y \rangle$ and $\langle H_z \rangle$. Therefore, we have established the following result: at thermodynamic equilibrium, the mean energy of an isotropic three-dimensional oscillator is equal to three times that of a one-dimensional oscillator of the same angular frequency.

(*ii*) *Classical oscillator*

The energy $\mathscr{H}(x, p)$ of a classical one-dimensional oscillator is equal to:

$$\mathscr{H}(x, p) = \frac{p^2}{2m} + \frac{1}{2}m\omega^2 x^2 \tag{14}$$

In expression (14), $x$ and $p$ can take on any values between $-\infty$ and $+\infty$. According to the results of classical statistical mechanics, the mean energy of this classical oscillator is given by:

$$\langle \mathscr{H} \rangle = \frac{\displaystyle\int_{-\infty}^{+\infty}\int_{-\infty}^{+\infty}\mathscr{H}(x, p)\,\text{e}^{-\mathscr{H}(x,p)/kT}\,\text{d}x\,\text{d}p}{\displaystyle\int_{-\infty}^{+\infty}\int_{-\infty}^{+\infty}\text{e}^{-\mathscr{H}(x,p)/kT}\,\text{d}x\,\text{d}p} \tag{15}$$

Substituting (14) into (15), we find, after a simple calculation:

$$\langle \mathscr{H} \rangle = kT \tag{16}$$

An argument analogous to that of comment (*i*) shows that result (16) must be multiplied by 3 when we go from one to three dimensions.

## 2. Discussion

### a. COMPARISON WITH THE CLASSICAL OSCILLATOR

In figure 1, the solid line gives the mean energy $\langle H \rangle$ of the one-dimensional quantum mechanical oscillator as a function of $T$. The dashed line corresponds to the mean energy $\langle \mathscr{H} \rangle$ of the classical oscillator.

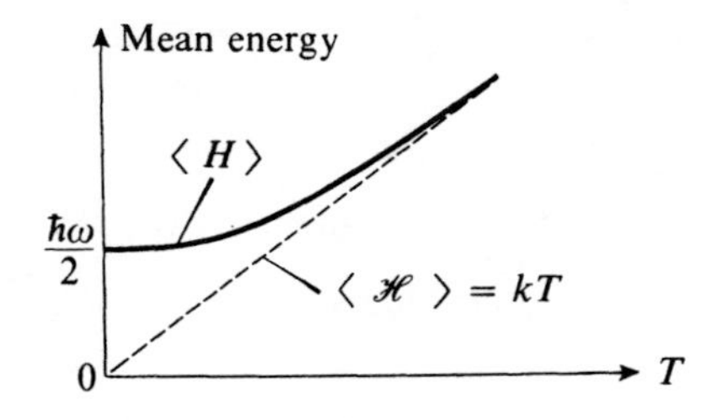


FIGURE 1

**As a function of the temperature, variation of the mean energy of a quantum mechanical oscillator (solid line) compared with that of a classical oscillator (straight dashed line).**

For $T = 0$, $\langle H \rangle = \hbar\omega/2$. This result corresponds to the fact that at absolute zero, one is sure that the oscillator is in the ground state $|\varphi_0\rangle$, with energy $\hbar\omega/2$ ($\hbar\omega/2$ is, for this reason, sometimes called the "zero point energy"). As for the classical oscillator, it is motionless ($p = 0$) at its stable equilibrium position ($x = 0$), and its energy is zero: $\langle \mathscr{H} \rangle = 0$.

As long as $T$ remains small – more precisely, as long as $kT \ll \hbar\omega$ – only the population of the ground state is appreciable, and $\langle H \rangle$ remains practically equal to $\hbar\omega/2$ : in this region, the solid-line curve of figure 1 has a horizontal tangent. We can see this directly from expression (10), which can be written, for small $T$:

$$\langle H \rangle \simeq \frac{\hbar\omega}{2} + \hbar\omega\, e^{-\hbar\omega/kT} \tag{17}$$

On the other hand, for large $T$, that is, for $kT \gg \hbar\omega$, the same formula yields:

$$\langle H \rangle = \frac{\hbar\omega}{2} + kT\left(1 - \frac{1}{2}\frac{\hbar\omega}{kT} + \ldots\right) \tag{18}$$

or:

$$\langle H \rangle \simeq kT \tag{19}$$

to within an infinitesimal of the order of $kT(\hbar\omega/kT)^2$. The asymptote of the curve giving $\langle H \rangle$ as a function of $T$ is therefore the straight line $\langle \mathscr{H} \rangle = kT$.

In conclusion, the quantum mechanical and classical oscillators have the same mean energy, $kT$, at high temperatures ($kT \gg \hbar\omega$). Striking differences appear at low temperatures ($kT \lesssim \hbar\omega$): it is no longer possible to ignore the quantization of the oscillator's energy once the energy $kT$ which characterizes the reservoir falls to the order of the distance $\hbar\omega$ separating two adjacent levels of the oscillator.

#### b. COMPARISON WITH A TWO-LEVEL SYSTEM

It is interesting to compare the preceding results with those obtained for a two-level system. Let $|\psi_1\rangle$ and $|\psi_2\rangle$ be the corresponding states, with energies $E_1$ and $E_2$ (with $E_1 < E_2$). For such a system, the general equation (6) yields:

$$\langle H \rangle = \frac{E_1 \, e^{-E_1/kT} + E_2 \, e^{-E_2/kT}}{e^{-E_1/kT} + e^{-E_2/kT}} \tag{20}$$

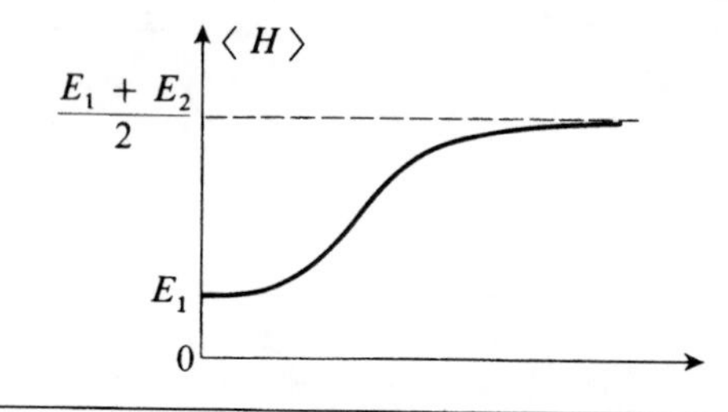


FIGURE 2

**Mean energy of a quantum mechanical system with two energy levels $E_1$ and $E_2$, in thermodynamic equilibrium at a temperature $T$.**

The mean energy of a two-level system, given by (20), is shown in figure 2. For small $T$ ($kT \ll E_2 - E_1$), the terms in $e^{-E_1/kT}$ are preponderant in both the numerator and the denominator of (20) (since $E_1 < E_2$) and we obtain:

$$\langle H \rangle \underset{T \to 0}{\longrightarrow} E_1 \tag{21}$$

It can be verified that the curve starts with a horizontal tangent. For large $T$ ($kT \gg E_2 - E_1$), the asymptote of the curve is the straight line parallel to the $T$-axis of ordinate $(E_1 + E_2)/2$. The preceding results are easy to understand: for $T = 0$, the system is in its ground state $|\psi_1\rangle$, of energy $E_1$; at high temperatures, the populations of the two levels are practically equal, and $\langle H \rangle$ approaches half the sum of the two energies $E_1$ and $E_2$.

Although the solid-line curves of figures 1 and 2 have the same shape at low temperatures, we see that this is not at all true at high temperatures. For the harmonic oscillator, $\langle H \rangle$ is not bounded and increases linearly with $T$, while, for a two-level system, $\langle H \rangle$ cannot exceed a certain value. This difference is due to the fact that the energy spectrum of the harmonic oscillator extends upward indefinitely: when $T$ increases, levels of higher and higher $n$ are occupied, and this causes $\langle H \rangle$ to increase. On the other hand, for a two-level system, once the populations of the two levels are equalized, an additional increase in the temperature does not change the mean energy.

## 3. Applications

#### a. BLACKBODY RADIATION

We have already pointed out, in the introduction to chapter V (and in complement $K_V$, where we justified this result more precisely), that the electromagnetic field in a cavity is equivalent to a set of independent one-dimensional harmonic oscillators. Each of these oscillators is associated with one of the standing waves which can exist inside the cavity (normal modes) and has the same angular frequency as this wave. Let us show that this result, combined with those obtained above for $\langle \mathscr{H} \rangle$ and $\langle H \rangle$, leads very simply to the Rayleigh-Jeans law and the Planck law for blackbody radiation.

Let $\mathscr{V}$ be the volume of the cavity, whose walls are assumed to be perfectly reflecting. The first modes of the cavity (those of lowest frequency) depend strongly on the form of the cavity. On the other hand, for the high-frequency modes (those whose wavelength $\lambda = c/\nu$ is much smaller than the dimensions of the cavity), a classical electromagnetic calculation shows that, if $N(\nu)\,\mathrm{d}\nu$ denotes the number of modes whose frequency is between $\nu$ and $\nu + \mathrm{d}\nu$, $N(\nu)$ is practically independent of the form of the cavity and equal to:

$$N(\nu) = \frac{8\pi\nu^2}{c^3}\mathscr{V} \tag{22}$$

Let $u(\nu)\,\mathrm{d}\nu$ be the electromagnetic energy per unit volume of the cavity contained in the frequency band $(\nu, \nu + \mathrm{d}\nu)$ when the cavity is in thermodynamic equilibrium at a temperature $T$. To obtain the energy $\mathscr{V}\,u(\nu)\,\mathrm{d}\nu$, one must multiply the number of modes whose frequency is between $\nu$ and $\nu + \mathrm{d}\nu$ by the mean energy of the corresponding harmonic oscillators. We calculated this energy above; it is equal to $\langle \mathscr{H} \rangle$ or $\langle H \rangle - \hbar\omega/2$*, depending on whether the problem is treated classically or quantum mechanically. We then obtain, using (10), (16) and (22):

$$u_{cl}(\nu) = \frac{8\pi\nu^2}{c^3}kT \tag{23}$$

in a classical treatment, and:

$$u_Q(\nu) = \frac{8\pi h\nu^3}{c^3}\frac{1}{e^{h\nu/kT} - 1} \tag{24}$$

in a quantum mechanical treatment.

We recognize (23) to be Rayleigh-Jeans' law and (24) to be Planck's law, which reduces to the preceding one in the limit of low frequencies or high temperatures ($h\nu/kT \ll 1$). The differences between these two laws reflect those which exist between the two curves of figure 1. At high frequencies, difficulties arise in Rayleigh-Jeans' law: the quantity $u_{cl}(\nu)$ given in (23) approaches infinity when $\nu \longrightarrow \infty$, which is physically absurd. In order to remedy this defect, Planck was led to postulate that the energy of each oscillator varied discontinuously, by jumps proportional to $\nu$ (energy quantization); thus he obtained formula (24), which accounts perfectly for the experimental results.

### b. BOSE-EINSTEIN DISTRIBUTION LAW

Instead of calculating the mean value $\langle H \rangle$ of the energy, as we did in §1, let us calculate that of the operator $N$. Since, according to formula (B-15) of chapter V:

$$H = \left(N + \frac{1}{2}\right)\hbar\omega \tag{25}$$

* We use $\langle H \rangle - \hbar\omega/2$ and not $\langle H \rangle$ for the following reason: $u(\nu)$ represents an electromagnetic energy which can be extracted from the cavity. At absolute zero, all the oscillators are in their ground states and no energy can be radiated outward because the system is in its lowest energy state; $u(\nu)$ must therefore be zero at absolute zero, as experimental observations indeed show it to be. This requires us to define the mean energy of the field in the cavity with respect to the value corresponding to $T = 0$.

we deduce from result (10) that:

$$\langle N \rangle = \frac{1}{e^{h\nu/kT} - 1} \tag{26}$$

The fact that the levels of a one-dimensional harmonic oscillator are equidistant enables us to associate with the oscillator in the state $| \varphi_n \rangle$ a set of $n$ identical particles (quanta) of the same energy $h\nu$. In this interpretation, the operators $a^\dagger$ and $a$, which take $| \varphi_n \rangle$ into $| \varphi_{n+1} \rangle$ or $| \varphi_{n-1} \rangle$, create or destroy a particle. $N$ is thus the operator associated with the number of particles ($| \varphi_n \rangle$ is the eigenstate of $N$ with the eigenvalue $n$).

In the special case of the electromagnetic field, the quanta associated with each harmonic oscillator are none other than *photons*. To each mode of the cavity considered in the preceding paragraph correspond photons of a certain type, characterized by the frequency, polarization, and spatial distribution of the mode. Expression (26) gives the mean number of photons associated with a mode of frequency $\nu$ at thermodynamic equilibrium. We recognize (26) to be the Bose-Einstein distribution law, which can be derived in a more general way; here, we have established it very simply by studying the harmonic oscillator and interpreting the states $| \varphi_n \rangle$.

COMMENT:

To be rigorous, we should write the Bose-Einstein distribution law for bosons of energy $\varepsilon$:

$$\langle N \rangle = \frac{1}{e^{(\varepsilon - \mu)/kT} - 1} \tag{27}$$

where $\mu$ is the chemical potential. In the case of photons, $\mu = 0$. This is due to the fact that the total number of photons in the global system radiation-reservoir is not fixed, because of the possibility of absorption or emission of photons by the walls.

### c. SPECIFIC HEATS OF SOLIDS AT CONSTANT VOLUME

We shall confine ourselves here to the Einstein model (*cf.* complement $A_V$), in which a solid is considered to be composed of $\mathcal{N}$ atoms vibrating independently about their equilibrium positions with the same angular frequency $\omega_E$. The internal energy $U$ of the solid at the temperature $T$ is therefore equal to the sum of the mean energies of the $\mathcal{N}$ isotropic three-dimensional oscillators in thermodynamic equilibrium at this temperature. Using comment ($i$) of § 1, we see that:

$$U = 3\mathcal{N} \langle H \rangle \tag{28}$$

where $\langle H \rangle$ is the mean energy of a one-dimensional harmonic oscillator of angular frequency $\omega_E$. We know that the constant volume specific heat $c_V$ is the derivative of the internal energy $U$ with respect to the temperature:

$$c_V = \frac{dU}{dT} = 3\mathcal{N} \frac{d}{dT} \langle H \rangle \tag{29}$$

which, taking (10) into account, yields:

$$c_V = 3\mathcal{N}k \frac{\left(\dfrac{\hbar\omega_E}{kT}\right)^2 e^{\hbar\omega_E/kT}}{[e^{\hbar\omega_E/kT} - 1]^2} \tag{30}$$

The variation of $c_V$ with $T$ is shown in figure 3. According to (29), $c_V$ is proportional to the derivative of the solid-line curve of figure 1. It is therefore very simple to describe the behavior of the specific heat $c_V$ as a function of the temperature. In figure 1, we see that $\langle H \rangle$ has

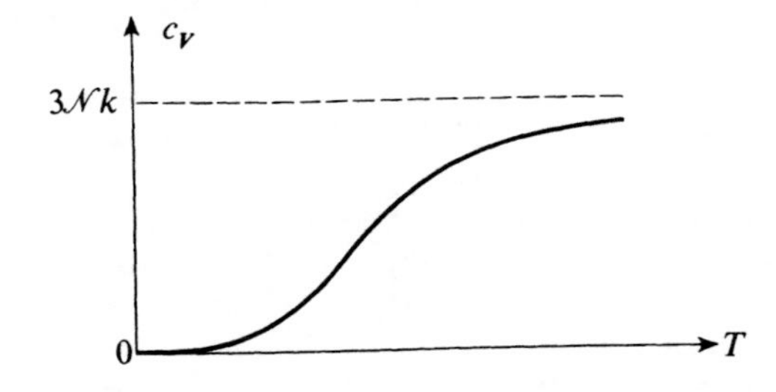


FIGURE 3

**(Constant volume) specific heat $c_V$ of a solid in Einstein's model. The high-temperature limit corresponds to the classical Dulong-Petit law.**

a horizontal tangent at the origin and increases very slowly; $c_V$ is therefore zero for $T = 0$ and also increases very slowly. On the other hand, for large $T$ ($kT \gg \hbar\omega_E$), $\langle H \rangle$ approaches $kT$; we deduce that $c_V$ approaches a constant, $3\mathcal{N}k$, independent of $\omega_E$. The transition region corresponds to $\hbar\omega_E/kT \simeq 1$.

The asymptote of figure 3 corresponds to the Dulong-Petit law: if one takes an atom-gram of any solid, $\mathcal{N}$ is equal to Avogadro's number and the limiting value of $c_V$ is equal to 3 $R$ (where $R$ is the ideal gas constant), that is, to about 6 cal. degree$^{-1}$ mole$^{-1}$.

As we pointed out above, the quantum mechanical nature of crystalline vibrations manifests itself at low temperatures when $kT$ falls to the order of $\hbar\omega_E$ or less. Insofar as $c_V$ is concerned, this means that the specific heat approaches zero when $T$ approaches zero. It is as if the degrees of freedom corresponding to crystalline vibrations were "frozen" beneath a certain temperature and no longer entered into the specific heat. This can be understood physically: at absolute zero, each oscillator is in its ground state $|\varphi_0\rangle$; as long as $kT \ll \hbar\omega_E$, it cannot absorb any thermal energy, since its first excited state has an energy far greater than $kT$.

COMMENTS:

(*i*) *Comparison with the specific heat of a two-level system*

We can apply an analogous argument to a sample composed of a set of two-level systems (for example, a paramagnetic sample composed of $\mathcal{N}$ spin 1/2 particles): its specific heat $c_V$ is given, to within a coefficient, by the derivative of the curve of figure 2. For such a system, the variation of $c_V$ with $T$ is shown in figure 4.

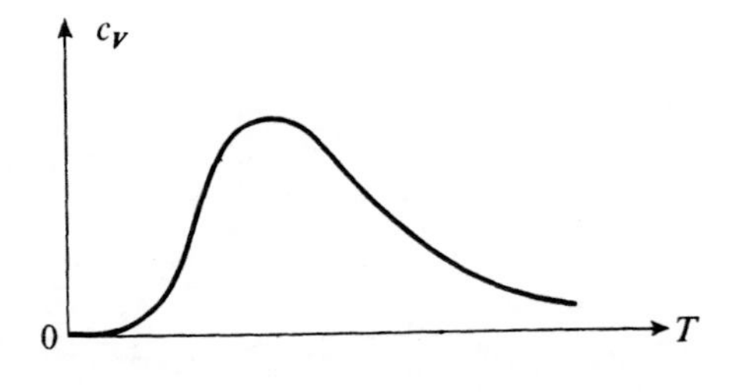


FIGURE 4

**Specific heat $c_V$ for a set of two-level systems. $c_V$ approaches zero at high temperatures because the energy spectrum has an upper bound.**

The behavior for $T \simeq 0$ is the same as in the case of figure 3. However, we see that $c_V$ approaches zero when $kT \gg E_2 - E_1$, since the mean energy then becomes independent of $T$ and is equal to $(E_1 + E_2)/2$ (*cf.* fig. 2). For a two-level system, $c_V$ therefore has a maximum (Schottky anomaly) whose physical interpretation is the following : like the harmonic oscillator, the two-level system cannot absorb any thermal energy at very low temperatures, as long as $E_2 - E_1 \gg kT$; $c_V$ is therefore zero at the origin. Then, as $T$ increases, $E_2$ becomes populated, and $c_V$ increases. When the temperature is high enough for the two populations to be practically equal, the system cannot absorb any more thermal energy, since the populations can no longer change : $c_V$ therefore approaches zero when $T \longrightarrow \infty$.

(*ii*) Einstein's model enables us to understand simply why the specific heat $c_V$ approaches zero when the temperature $T$ approaches zero (a classically inexplicable result). However, it is too schematic to describe the exact dependence of $c_V$ at low temperatures.

In a real crystal, the various oscillators are coupled. This gives rise to a set of vibrational normal modes (phonons) whose frequencies go from zero to a certain cutoff frequency (*cf.* complement $J_V$). (30) must then be summed over the different possible frequencies $\nu$ (taking into account the fact that the number of modes whose frequencies are included between $\nu$ and $\nu + d\nu$ depends on $\nu$). Thus one finds an expression for the specific heat which, at low temperatures, varies like $T^3$ (this is confirmed experimentally).

## 4. Probability distribution of the observable $X$

### a. DEFINITION OF THE PROBABILITY DENSITY $\rho(x)$

Let us return to the one-dimensional harmonic oscillator in thermodynamic equilibrium. We seek the probability $\rho(x)\,dx$ of finding, in a measurement of the position $X$ of the particle, a result included between $x$ and $x + dx$. It is clear that $\rho(x)$ plays an important role in a large number of physical problems. For example, for a solid described by Einstein's model, the width of $\rho(x)$ gives an idea of the amplitude of atomic vibrations; the study of the variation of this width with respect to $T$ makes it possible to understand the phenomenon of melting [which occurs when the width of $\rho(x)$ is no longer negligible compared to the interatomic distance].

When the oscillator is in the stationary state $|\varphi_n\rangle$, the corresponding probability density $\rho_n(x)$ is:

$$\rho_n(x) = |\varphi_n(x)|^2 = \langle x \mid \varphi_n \rangle \langle \varphi_n \mid x \rangle \tag{31}$$

At thermodynamic equilibrium, the oscillator is described by a statistical mixture of the states $|\varphi_n\rangle$ with the weights: $Z^{-1}\,e^{-E_n/kT}$. The probability density $\rho(x)$ is then:

$$\rho(x) = Z^{-1} \sum_n \rho_n(x)\, e^{-E_n/kT} \tag{32}$$

$\rho(x)$ is the weighted sum of the probability densities $\rho_n(x)$ associated with the various states $|\varphi_n\rangle$. Some of the $\rho_n(x)$ are shown in figures 5 and 6 of chapter V. We shall see later that the oscillations of the functions $\rho_n(x)$ which are visible in these figures disappear in the summation over $n$: we shall show that $\rho(x)$ is simply a Gaussian function.

The probability density $\rho(x)$ defined in (32) is related very simply to the density operator $\rho$ of the harmonic oscillator in thermodynamic equilibrium. Using (31) and (32), we obtain:

$$\rho(x) = Z^{-1} \sum_n e^{-E_n/kT} \langle x \mid \varphi_n \rangle \langle \varphi_n \mid x \rangle \tag{33}$$

On the right-hand side, we can bring in the operator $e^{-H/kT}$ which, taking into account the closure relation for the states $|\varphi_n\rangle$, can be written:

$$e^{-H/kT} = e^{-H/kT} \sum_n | \varphi_n \rangle \langle \varphi_n | = \sum_n e^{-E_n/kT} | \varphi_n \rangle \langle \varphi_n | \tag{34}$$

We then see that:

$$\rho(x) = Z^{-1} \langle x \mid e^{-H/kT} \mid x \rangle = \langle x | \rho | x \rangle \tag{35}$$

where the density operator $\rho$ is given by formula (1). $\rho(x)$ can then be seen to be the diagonal element of $\rho$ which corresponds to the ket $|x\rangle$.

### b. CALCULATION OF $\rho(x)$

We know that:

$$H = \hbar\omega\left(a^\dagger a + \frac{1}{2}\right) \tag{36}$$

so that $\rho(x)$ can be written in the form:

$$\rho(x) = Z^{-1} e^{-\lambda/2} F_\lambda(x) \tag{37}$$

with:

$$\lambda = \frac{\hbar\omega}{kT} \tag{38}$$

and:

$$F_\lambda(x) = \langle x \mid e^{-\lambda a^\dagger a} \mid x \rangle \tag{39}$$

In order to know $\rho(x)$, all we need to do, therefore, is evaluate this diagonal matrix element.

To do this, let us calculate the variation in $F_\lambda(x)$ when $x$ is changed to $x + \mathrm{d}x$. Since the ket $| x + \mathrm{d}x \rangle$ is given by [*cf.* complement $E_{II}$, relation (20)]:

$$| x + \mathrm{d}x \rangle = \left(1 - i\frac{\mathrm{d}x}{\hbar} P\right) | x \rangle \tag{40}$$

we obtain, substituting this relation and the adjoint relation into (39) (neglecting infinitesimals of second order in $\mathrm{d}x$):

$$F_\lambda(x + \mathrm{d}x) = F_\lambda(x) + i\frac{\mathrm{d}x}{\hbar} \langle x \mid [P, e^{-\lambda a^\dagger a}] \mid x \rangle \tag{41}$$

The matrix element appearing on the right-hand side of (41) involves the operator $P$, which is proportional to $(a - a^\dagger)$. Now, it is the $X$ operator, proportional to $(a + a^\dagger)$, which acts in a simple way on the kets $| x \rangle$. We shall therefore transform $[P, e^{-\lambda a^\dagger a}]$ so as to make $X$

appear. We shall begin by seeking the relation between $a\,e^{-\lambda a^\dagger a}$ and $e^{-\lambda a^\dagger a}\,a$. This can be obtained very simply in the $\{ | \varphi_n \rangle \}$ representation :

$$a\,e^{-\lambda a^\dagger a} | \varphi_n \rangle = \sqrt{n}\,e^{-\lambda n} | \varphi_{n-1} \rangle \tag{42-a}$$

$$e^{-\lambda a^\dagger a}\,a | \varphi_n \rangle = \sqrt{n}\,e^{-\lambda(n-1)} | \varphi_{n-1} \rangle \tag{42-b}$$

that is:

$$e^{-\lambda a^\dagger a}\,a = e^{\lambda}\,a\,e^{-\lambda a^\dagger a} \tag{43}$$

which can also be written:

$$\left(1 - \tanh\frac{\lambda}{2}\right) e^{-\lambda a^\dagger a}\,a = \left(1 + \tanh\frac{\lambda}{2}\right) a\,e^{-\lambda a^\dagger a} \tag{44}$$

Similarly, it can be shown that:

$$e^{-\lambda a^\dagger a}\,a^\dagger = e^{-\lambda}\,a^\dagger\,e^{-\lambda a^\dagger a} \tag{45}$$

that is:

$$\left(1 + \tanh\frac{\lambda}{2}\right) e^{-\lambda a^\dagger a}\,a^\dagger = \left(1 - \tanh\frac{\lambda}{2}\right) a^\dagger\,e^{-\lambda a^\dagger a} \tag{46}$$

We now subtract, term by term, relations (44) and (46); we obtain:

$$[a - a^\dagger, e^{-\lambda a^\dagger a}] = -\tanh\frac{\lambda}{2}\,[a + a^\dagger, e^{-\lambda a^\dagger a}]_+ \tag{47}$$

where the symbol $[A, B]_+$ denotes the anticommutator:

$$[A, B]_+ = AB + BA \tag{48}$$

If we take into account the numerical factors which result from formulas (B-1) and (B-7) in chapter V, (47) finally becomes:

$$[P, e^{-\lambda a^\dagger a}] = im\omega\,\tanh\frac{\lambda}{2}\,[X, e^{-\lambda a^\dagger a}]_+ \tag{49}$$

Substituting this result into relation (41):

$$\begin{aligned} F_\lambda(x + \mathrm{d}x) - F_\lambda(x) &= -\frac{m\omega}{\hbar}\,\mathrm{d}x\,\tanh\frac{\lambda}{2} \langle x | [X, e^{-\lambda a^\dagger a}]_+ | x \rangle \\ &= -2x\frac{m\omega}{\hbar}\,\tanh\frac{\lambda}{2}\,F_\lambda(x)\,\mathrm{d}x \end{aligned} \tag{50}$$

$F_\lambda(x)$ therefore satisfies the differential equation:

$$\frac{\mathrm{d}}{\mathrm{d}x} F_\lambda(x) + \frac{2x}{\xi^2} F_\lambda(x) = 0 \tag{51}$$

where $\xi$, which has the dimensions of a length, is defined by:

$$\xi = \sqrt{\frac{\hbar}{m\omega}\coth\frac{\lambda}{2}} = \sqrt{\frac{\hbar}{m\omega}\coth\left(\frac{\hbar\omega}{2kT}\right)} \tag{52}$$

Equation (51) can be integrated directly:

$$F_\lambda(x) = F_\lambda(0)\,e^{-x^2/\xi^2} \tag{53}$$

Therefore, we know $\rho(x)$ to within a constant factor, since, according to (37):

$$\rho(x) = Z^{-1}\, e^{-\lambda/2}\, F_\lambda(0)\, e^{-x^2/\xi^2} \tag{54}$$

Since we know that the integral of $\rho(x)$ over the whole $x$-axis must be equal to 1, we obtain finally:

$$\rho(x) = \frac{1}{\xi\sqrt{\pi}}\, e^{-x^2/\xi^2} \tag{55}$$

The function $\rho(x)$ is thus a Gaussian, whose width is characterized by the length $\xi$ defined in (52).

### c. DISCUSSION

Starting from the probability density (55), it is easy to calculate:

$$\begin{aligned} \langle X \rangle &= 0 \\ \langle X^2 \rangle &= (\Delta X)^2 = \frac{\xi^2}{2} \end{aligned} \tag{56}$$

Figure 5 shows the variation of $(\Delta X)^2$ with respect to $T$. We see from (52) that $(\Delta X)^2$ is equal to $\hbar/2m\omega$ when $T = 0$. This result is not surprising: at $T = 0$, the oscillator is in its ground state, and $\rho(x)$ is equal to $|\varphi_0(x)|^2$; $\Delta X$ is found to be the root-mean-square deviation of $X$ in the ground state [*cf.* formula (D-5-a) of chapter V]. Then, $(\Delta X)^2$ increases, and, when $kT \gg \hbar\omega$:

$$(\Delta X)^2 \underset{T\to\infty}{\simeq} \frac{kT}{m\omega^2} \tag{57}$$

In this case, furthermore, $\rho(x)$ becomes identical to the probability density of a classical oscillator in thermodynamic equilibrium at the temperature $T$:

$$\rho_{cl}(x) = \frac{e^{-V(x)/kT}}{\int_{-\infty}^{+\infty} e^{-V(x)/kT}\, dx} = \frac{1}{\sqrt{\frac{2\pi kT}{m\omega^2}}}\, e^{-\frac{m\omega^2 x^2}{2kT}} \tag{58}$$

which leads to $(\Delta x_{cl})^2 = kT/m\omega^2$ (the straight dashed line in figure 5). Here again, classical and quantum mechanical predictions meet for $kT \gg \hbar\omega$.

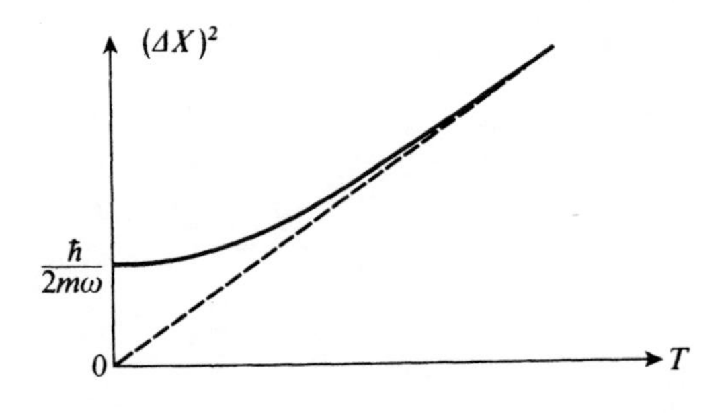


FIGURE 5

**Variation with respect to the temperature $T$ of $(\Delta X)^2$, for a harmonic oscillator in thermodynamic equilibrium. When $T \longrightarrow \infty$, $\Delta X$ is identical to the classical value, shown by the dashed line; at low temperatures, quantum mechanical effects (Heisenberg uncertainty relation) prevent $\Delta X$ from approaching zero.**

Now, let us apply the preceding results to the problem of melting of a solid body (for simplicity, we shall choose the one-dimensional Einstein model; see complement $A_V$). Experiments show that the solid melts when $\Delta X$ is of the order of an appreciable fraction of the interatomic distance $d$. Consequently, the melting point temperature $T_f$ is given approximately by:

$$\frac{\xi^2}{2d^2} \simeq \eta^2 \tag{59}$$

where $\xi$ can be replaced by its expression (52), with $T = T_f$. Assuming that $T_f$ is large enough that $kT_f \gg \hbar\omega_E$, we can use in (59) the asymptotic form (57)★, and we obtain the law for $T_f$:

$$\frac{kT_f}{m\omega_E^2} \simeq \eta^2 d^2 \tag{60}$$

If we set:

$$\hbar\omega_E = k\Theta_E \tag{61}$$

($\Theta_E$ is called the "Einstein temperature") and if we note that $d$ does not vary very much from one substance to another (anyway, much less than $\omega_E$, that is, $\Theta_E$), we find the approximate law:

$$\frac{T_f}{m\Theta_E^2} \simeq C^{te} \tag{62}$$

The melting point temperature of a crystal is therefore approximately proportional to the square of a vibrational frequency which is characteristic of the crystal.

### d. BLOCH'S THEOREM

Consider the operator $e^{-iqX}$, where $q$ is a real variable. Its mean value:

$$\langle e^{-iqX} \rangle = \text{Tr}\,[\rho\, e^{-iqX}] \tag{63}$$

[where $\rho$ is given by (1)] is a function of $q$, which we shall denote by $f(q)$:

$$f(q) = \langle e^{-iqX} \rangle \tag{64}$$

In probability theory, $f(q)$ is called the characteristic function of the random variable $x$.

It is easy to calculate $f(q)$ if we place ourselves in the $\{ | x \rangle \}$ representation:

$$\begin{aligned} f(q) &= \int_{-\infty}^{+\infty} dx\, \langle x | \rho\, e^{-iqX} | x \rangle \\ &= \int_{-\infty}^{+\infty} dx\, \langle x | \rho | x \rangle\, e^{-iqx} \\ &= \int_{-\infty}^{+\infty} dx\, \rho(x)\, e^{-iqx} \end{aligned} \tag{65}$$

★ This is not always possible. Recall that helium remains liquid at atmospheric pressure, even at $T = 0$; $\xi$ is never negligible compared to $d$, whatever $T$ may be (*cf.* complement $A_V$).

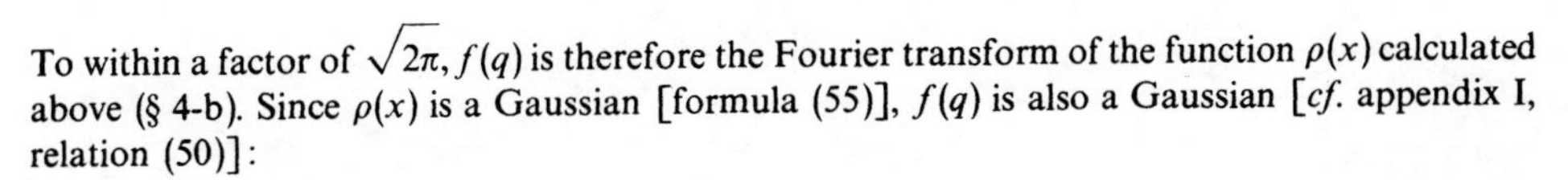


To within a factor of $\sqrt{2\pi}$, $f(q)$ is therefore the Fourier transform of the function $\rho(x)$ calculated above (§ 4-b). Since $\rho(x)$ is a Gaussian [formula (55)], $f(q)$ is also a Gaussian [*cf.* appendix I, relation (50)]:

$$f(q) = e^{-\xi^2 q^2/4} \tag{66}$$

which, according to formula (56), can be written:

$$\langle e^{-iqX} \rangle = e^{-\frac{q^2}{2}\langle X^2 \rangle} \tag{67}$$

We could perform calculations analogous to those of §§ 4-a and b above for the observable $P$ instead of $X$. We would then define the probability density $\bar{\rho}(p)$ by:

$$\bar{\rho}(p) = Z^{-1} \sum_n e^{-E_n/kT} |\bar{\varphi}_n(p)|^2 \tag{68}$$

Formula (24) of complement $D_V$ shows that:

$$\bar{\rho}(p) = \frac{1}{m\omega} \rho\left(x = \frac{p}{m\omega}\right) \tag{69}$$

Therefore:

$$\bar{\rho}(p) = \frac{1}{m\omega\xi\sqrt{\pi}} e^{-\frac{p^2}{m^2\omega^2\xi^2}} \tag{70}$$

Consequently, the study of $\langle e^{-iqP} \rangle$ would lead to the same result as in (67):

$$\langle e^{-iqP} \rangle = e^{-\frac{q^2}{2}\langle P^2 \rangle} \tag{71}$$

The generalization of formulas (67) and (71) is known as Bloch's theorem : if $G(X, P)$ is an arbitrary linear combination of the position $X$ and the momentum $P$ of a one-dimensional harmonic oscillator in thermodynamic equilibrium at the temperature $T$, then:

$$\langle e^{-iqG} \rangle = e^{-\frac{q^2}{2}\langle G^2 \rangle} \tag{72}$$

This theorem is used in solid state physics, for example in the theory of emission without recoil by the nuclei of a crystalline lattice (the Mössbauer effect).

**References and suggestions for further reading:**

Specific heats: Kittel (8.2), chap. 6, p. 91 and 100; Kittel (13.2), chap. 6; Seitz (13.4), chap. III; Ziman (13.3), chap. 2.

Blackbody radiation: Eisberg and Resnick (1.3), chap. 1; Kittel (8.2), chap. 15; Reif (8.4) §§ 9-13 to 9-15; Bruhat (8.3), chap. XXII.

Bloch's theorem: Messiah (1.17), chap. XII, § II-12.

## Complement $M_V$

## EXERCISES

**1.** Consider a harmonic oscillator of mass $m$ and angular frequency $\omega$. At time $t = 0$, the state of this oscillator is given by:

$$|\psi(0)\rangle = \sum_n c_n |\varphi_n\rangle$$

where the states $|\varphi_n\rangle$ are stationary states with energies $(n + 1/2)\hbar\omega$.

*a.* What is the probability $\mathscr{P}$ that a measurement of the oscillator's energy performed at an arbitrary time $t > 0$, will yield a result greater than $2\hbar\omega$? When $\mathscr{P} = 0$, what are the non-zero coefficients $c_n$?

*b.* From now on, assume that only $c_0$ and $c_1$ are different from zero. Write the normalization condition for $|\psi(0)\rangle$ and the mean value $\langle H\rangle$ of the energy in terms of $c_0$ and $c_1$. With the additional requirement $\langle H \rangle = \hbar\omega$, calculate $|c_0|^2$ and $|c_1|^2$.

*c.* As the normalized state vector $|\psi(0)\rangle$ is defined only to within a global phase factor, we fix this factor by choosing $c_0$ real and positive. We set: $c_1 = |c_1|\, e^{i\theta_1}$. We assume that $\langle H \rangle = \hbar\omega$ and that:

$$\langle X \rangle = \frac{1}{2}\sqrt{\frac{\hbar}{m\omega}}$$

Calculate $\theta_1$.

*d.* With $|\psi(0)\rangle$ so determined, write $|\psi(t)\rangle$ for $t > 0$ and calculate the value of $\theta_1$ at $t$. Deduce the mean value $\langle X\rangle(t)$ of the position at $t$.

**2. Anisotropic three-dimensional harmonic oscillator**

In a three-dimensional problem, consider a particle of mass $m$ and of potential energy:

$$V(X, Y, Z) = \frac{m\omega^2}{2}\left[\left(1 + \frac{2\lambda}{3}\right)(X^2 + Y^2) + \left(1 - \frac{4\lambda}{3}\right)Z^2\right]$$

where $\omega$ and $\lambda$ are constants which satisfy:

$$\omega \geqslant 0 \quad , \quad 0 \leqslant \lambda < \frac{3}{4}$$

*a.* What are the eigenstates of the Hamiltonian and the corresponding energies?

*b.* Calculate and discuss, as functions of $\lambda$, the variation of the energy, the parity and the degree of degeneracy of the ground state and the first two excited states.

## 3. Harmonic oscillator: two particles

Two particles of the same mass $m$, with positions $X_1$ and $X_2$ and momenta $P_1$ and $P_2$, are subject to the same potential:

$$V(X) = \frac{1}{2} m\omega^2 X^2$$

The two particles do not interact.

*a.* Write the operator $H$, the Hamiltonian of the two-particle system. Show that $H$ can be written:

$$H = H_1 + H_2$$

where $H_1$ and $H_2$ act respectively only in the state space of particle (1) and in that of particle (2). Calculate the energies of the two-particle system, their degrees of degeneracy, and the corresponding wave functions.

*b.* Does $H$ form a C.S.C.O.? Same question for the set $\{ H_1, H_2 \}$. We denote by $| \Phi_{n_1,n_2} \rangle$ the eigenvectors common to $H_1$ and $H_2$. Write the orthonormalization and closure relations for the states $| \Phi_{n_1,n_2} \rangle$.

*c.* Consider a system which, at $t = 0$, is in the state:

$$| \psi(0) \rangle = \frac{1}{2} (| \Phi_{0,0} \rangle + | \Phi_{1,0} \rangle + | \Phi_{0,1} \rangle + | \Phi_{1,1} \rangle)$$

What results can be found, and with what probabilities, if at this time one measures:

- the total energy of the system?
- the energy of particle (1)?
- the position or velocity of this particle?

**4.** (This exercise is a continuation of the preceding one and uses the same notation.)

The two-particle system, at $t = 0$, is in the state $| \psi(0) \rangle$ given in exercise 3.

*a.* At $t = 0$, one measures the total energy $H$ and one finds the result $2\hbar\omega$.

$\alpha$. Calculate the mean values of the position, the momentum, and the energy of particle (1) at an arbitrary positive $t$. Same question for particle (2).

$\beta$. At $t > 0$, one measures the energy of particle (1). What results can be found, and with what probabilities? Same question for a measurement of the position of particle (1); trace the curve for the corresponding probability density.

*b.* Instead of measuring the total energy $H$, at $t = 0$, one measures the energy $H_2$ of particle (2); the result obtained is $\hbar\omega/2$. What happens to the answers to questions $\alpha$ and $\beta$ of *a*?

**5.** (This exercise is a continuation of exercise 3 and uses the same notation.)
We denote by $| \Phi_{n_1,n_2} \rangle$ the eigenstates common to $H_1$ and $H_2$, of eigenvalues $(n_1 + 1/2)\hbar\omega$ and $(n_2 + 1/2)\hbar\omega$. The "two particle exchange" operator $P_e$ is defined by:

$$P_e | \Phi_{n_1,n_2} \rangle = | \Phi_{n_2,n_1} \rangle$$

*a.* Prove that $P_e^{-1} = P_e$ and that $P_e$ is unitary. What are the eigenvalues of $P_e$? Let $B' = P_e B P_e^\dagger$ be the observable resulting from the transformation by $P_e$ of an arbitrary observable $B$. Show that the condition $B' = B$ ($B$ invariant under exchange of the two particles) is equivalent to $[B, P_e] = 0$.

*b.* Show that:

$$P_e H_1 P_e^\dagger = H_2$$
$$P_e H_2 P_e^\dagger = H_1$$

Does $H$ commute with $P_e$? Calculate the action of $P_e$ on the observables $X_1$, $P_1$, $X_2$, $P_2$.

*c.* Construct a basis of eigenvectors common to $H$ and $P_e$. Do these two operators form a C.S.C.O.? What happens to the spectrum of $H$ and the degeneracy of its eigenvalues if one retains only the eigenvectors $| \Phi \rangle$ of $H$ for which $P_e | \Phi \rangle = - | \Phi \rangle$?

**6. Charged harmonic oscillator in a variable electric field**

A one-dimensional harmonic oscillator is composed of a particle of mass $m$, charge $q$ and potential energy $V(X) = \frac{1}{2} m\omega^2 X^2$. We assume in this exercise that the particle is placed in an electric field $\mathscr{E}(t)$ parallel to $Ox$ and time-dependent, so that to $V(x)$ must be added the potential energy:

$$W(t) = - q\mathscr{E}(t)X$$

*a.* Write the Hamiltonian $H(t)$ of the particle in terms of the operators $a$ and $a^\dagger$. Calculate the commutators of $a$ and $a^\dagger$ with $H(t)$.

*b.* Let $\alpha(t)$ be the number defined by:

$$\alpha(t) = \langle \psi(t) | a | \psi(t) \rangle$$

where $| \psi(t) \rangle$ is the normalized state vector of the particle under study. Show from the results of the preceding question that $\alpha(t)$ satisfies the differential equation:

$$\frac{\mathrm{d}}{\mathrm{d}t} \alpha(t) = - i\omega\, \alpha(t) + i\lambda(t)$$

where $\lambda(t)$ is defined by:

$$\lambda(t) = \frac{q}{\sqrt{2m\hbar\omega}} \mathscr{E}(t)$$

Integrate this differential equation. At time $t$, what are the mean values of the position and momentum of the particle?

*c.* The ket $|\varphi(t)\rangle$ is defined by:

$$|\varphi(t)\rangle = [a - \alpha(t)] \, |\psi(t)\rangle$$

where $\alpha(t)$ has the value calculated in *b*. Using the results of questions *a* and *b*, show that the evolution of $|\varphi(t)\rangle$ is given by:

$$i\hbar \frac{\mathrm{d}}{\mathrm{d}t} |\varphi(t)\rangle = [H(t) + \hbar\omega] \, |\varphi(t)\rangle$$

How does the norm of $|\varphi(t)\rangle$ vary with time?

*d.* Assuming that $|\psi(0)\rangle$ is an eigenvector of $a$ with the eigenvalue $\alpha(0)$, show that $|\psi(t)\rangle$ is also an eigenvector of $a$, and calculate its eigenvalue.

Find at time $t$ the mean value of the unperturbed Hamiltonian

$$H_0 = H(t) - W(t)$$

as a function of $\alpha(0)$. Give the root-mean-square deviations $\Delta X$, $\Delta P$ and $\Delta H_0$; how do they vary with time?

*e.* Assume that at $t = 0$, the oscillator is in the ground state $|\varphi_0\rangle$. The electric field acts between times 0 and $T$ and then falls to zero. When $t > T$, what is the evolution of the mean values $\langle X \rangle(t)$ and $\langle P \rangle(t)$? Application : assume that between 0 and $T$, the field $\mathscr{E}(t)$ is given by $\mathscr{E}(t) = \mathscr{E}_0 \cos(\omega' t)$; discuss the phenomena observed (resonance) in terms of $\Delta\omega = \omega' - \omega$. If, at $t > T$, the energy is measured, what results can be found, and with what probabilities?

**7.** Consider a one-dimensional harmonic oscillator of Hamiltonian $H$ and stationary states $|\varphi_n\rangle$:

$$H \, |\varphi_n\rangle = (n + 1/2)\hbar\omega \, |\varphi_n\rangle$$

The operator $U(k)$ is defined by:

$$U(k) = \mathrm{e}^{ikX}$$

where $k$ is real.

*a.* Is $U(k)$ unitary? Show that, for all $n$, its matrix elements satisfy the relation:

$$\sum_{n'} |\langle \varphi_n | U(k) | \varphi_{n'} \rangle|^2 = 1$$

*b.* Express $U(k)$ in terms of the operators $a$ and $a^\dagger$. Use Glauber's formula [formula (63) of complement $B_{II}$] to put $U(k)$ in the form of a product of exponential operators.

*c.* Establish the relations:

$$e^{\lambda a} | \varphi_0 \rangle = | \varphi_0 \rangle$$

$$\langle \varphi_n | e^{\lambda a^\dagger} | \varphi_0 \rangle = \frac{\lambda^n}{\sqrt{n!}}$$

where $\lambda$ is an arbitrary complex parameter.

*d.* Find the expression, in terms of $E_k = \hbar^2 k^2/2m$ and $E_\omega = \hbar\omega$, for the matrix element:

$$\langle \varphi_0 | U(k) | \varphi_n \rangle$$

What happens when $k$ approaches zero? Could this result have been predicted directly?

**8.** The evolution operator $U(t, 0)$ of a one-dimensional harmonic oscillator is written:

$$U(t, 0) = e^{-iHt/\hbar}$$

with:

$$H = \hbar\omega\left(a^\dagger a + \frac{1}{2}\right)$$

*a.* Consider the operators:

$$\tilde{a}(t) = U^\dagger(t, 0)\, a\, U(t, 0)$$

$$\tilde{a}^\dagger(t) = U^\dagger(t, 0)\, a^\dagger\, U(t, 0)$$

By calculating their action on the eigenkets $| \varphi_n \rangle$ of $H$, find the expression for $\tilde{a}(t)$ and $\tilde{a}^\dagger(t)$ in terms of $a$ and $a^\dagger$.

*b.* Calculate the operators $\tilde{X}(t)$ and $\tilde{P}(t)$ obtained from $X$ and $P$ by the unitary transformation:

$$\tilde{X}(t) = U^\dagger(t, 0)\, X\, U(t, 0)$$

$$\tilde{P}(t) = U^\dagger(t, 0)\, P\, U(t, 0)$$

How can the relations so obtained be interpreted?

*c.* Show that $U^\dagger\left(\frac{\pi}{2\omega}, 0\right) | x \rangle$ is an eigenvector of $P$ and specify its eigenvalue.

Similarly, establish that $U^\dagger\left(\frac{\pi}{2\omega}, 0\right) | p \rangle$ is an eigenvector of $X$.

*d.* At $t = 0$, the wave function of the oscillator is $\psi(x, 0)$. How can one obtain from $\psi(x, 0)$ the wave function of the oscillator at all subsequent times $t_q = q\pi/2\omega$ (where $q$ is a positive integer)?

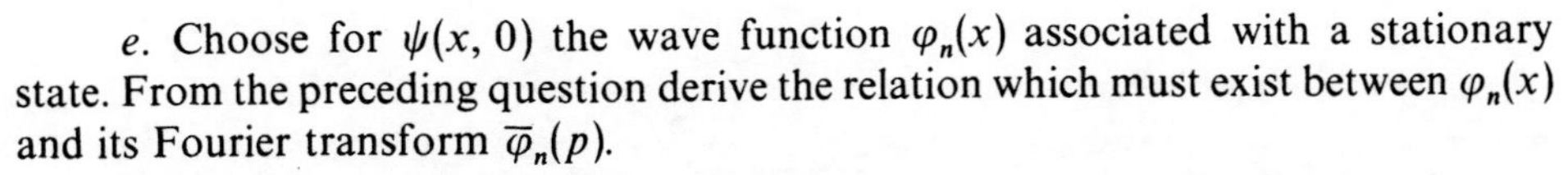

*e*. Choose for $\psi(x, 0)$ the wave function $\varphi_n(x)$ associated with a stationary state. From the preceding question derive the relation which must exist between $\varphi_n(x)$ and its Fourier transform $\overline{\varphi}_n(p)$.

*f*. Describe qualitatively the evolution of the wave function in the following cases:

(*i*) $\psi(x, 0) = \mathrm{e}^{ikx}$ where $k$, real, is given.

(*ii*) $\psi(x, 0) = \mathrm{e}^{-\rho x}$ where $\rho$ is real and positive.

(*iii*)

$$\psi(x, 0) \begin{cases} = \dfrac{1}{\sqrt{a}} & \text{if } -\dfrac{a}{2} \leqslant x \leqslant \dfrac{a}{2} \\ = 0 & \text{everywhere else} \end{cases}$$

(*iv*) $\psi(x, 0) = \mathrm{e}^{-\rho^2 x^2}$ where $\rho$ is real.

CHAPTER VI

# General properties of angular momentum in quantum mechanics

## OUTLINE OF CHAPTER VI

**A. INTRODUCTION: THE IMPORTANCE OF ANGULAR MOMENTUM**

---

**B. COMMUTATION RELATIONS CHARACTERISTIC OF ANGULAR MOMENTUM**

1. Orbital angular momentum
2. Generalization: definition of an angular momentum
3. Statement of the problem

---

**C. GENERAL THEORY OF ANGULAR MOMENTUM**

1. Definitions and notation
   *a.* The $J_+$ and $J_-$ operators
   *b.* Notation for the eigenvalues of $\mathbf{J}^2$ and $J_z$
   *c.* Eigenvalue equations for $\mathbf{J}^2$ and $J_z$
2. Eigenvalue of $\mathbf{J}^2$ and $J_z$
   *a.* Lemmas
   *b.* Determination of the spectrum of $\mathbf{J}^2$ and $J_z$
3. " Standard " $\{ |k, j, m\rangle \}$ representations
   *a.* The basis states
   *b.* The $\mathscr{E}(k,j)$ spaces
   *c.* Matrices representing the angular momentum operators

---

**D. APPLICATION TO ORBITAL ANGULAR MOMENTUM**

1. Eigenvalues and eigenfunctions of $\mathbf{L}^2$ and $L_z$
   *a.* Eigenvalue equations in the $\{ |\mathbf{r}\rangle \}$ representation
   *b.* Values of $l$ and $m$
   *c.* Fundamental properties of the spherical harmonics
   *d.* " Standard " bases of the wave function space of a spinless particle
2. Physical considerations
   *a.* Study of a $|k, l, m\rangle$ state
   *b.* Calculation of the physical predictions concerning measurements of $\mathbf{L}^2$ and $L_z$

---

# A. INTRODUCTION: THE IMPORTANCE OF ANGULAR MOMENTUM

The present chapter is the first in a series of four chapters (VI, VII, IX and X) devoted to the study of angular momenta in quantum mechanics. This is an extremely important problem, and the results we are going to establish are used in many domains of physics : the classification of atomic, molecular and nuclear spectra, the spin of elementary particles, magnetism, etc...

We already know the important role played by angular momentum in classical mechanics; the total angular momentum of an isolated physical system is a *constant of the motion*. Furthermore, this is also true in certain cases in which the system is not isolated. For example, if a point particle $P$, of mass $m$, is moving in a central potential (one which depends only on the distance between $P$ and a fixed point $O$), the force to which $P$ is subjected is always directed towards $O$. Its moment with respect to $O$ is consequently zero, and the angular momentum theorem implies that:

$$\frac{\mathrm{d}}{\mathrm{d}t} \boldsymbol{\mathscr{L}} = \mathbf{0} \tag{A-1}$$

where $\boldsymbol{\mathscr{L}}$ is the angular momentum of $P$ with respect to $O$. This fact has important consequences: the motion of the particle $P$ is limited to a fixed plane (the plane passing through $O$ and perpendicular to the angular momentum $\boldsymbol{\mathscr{L}}$); moreover, this motion obeys the law of constant areal velocity (Kepler's second law).

All these properties have their equivalents in quantum mechanics. With the angular momentum $\boldsymbol{\mathscr{L}}$ of a classical system is associated an observable $\mathbf{L}$, actually a set of three observables, $L_x$, $L_y$ and $L_z$, which correspond to the three components of $\boldsymbol{\mathscr{L}}$ in a Cartesian frame. If the physical system under study is a point moving in a central potential, we shall see in chapter VII that $L_x$, $L_y$ and $L_z$ are constants of the motion in a quantum mechanical sense, that is, they commute with the Hamiltonian $H$ describing the particle in the central potential $V(r)$. This important property considerably simplifies the determination and classification of eigenstates of $H$.

Also, we described the Stern-Gerlach experiment in chapter IV and this revealed the *quantization of angular momentum* : the component, along a fixed axis, of the intrinsic angular momentum of an atom can take on only certain discrete values. We shall see that all angular momenta are quantized in this way. This enables us to understand atomic magnetism, the Zeeman effect, etc... Furthermore, to analyze all these phenomena, we must introduce *typically quantum mechanical angular momenta, which have no classical equivalents* (intrinsic angular momenta of elementary particles; chap. IX).

From now on, we shall denote by *orbital angular momentum* any angular momentum which has a classical equivalent (and by $\mathbf{L}$, the corresponding observables), and by *spin angular momentum* any intrinsic angular momentum of an elementary particle (for which we reserve the letter $\mathbf{S}$). In a complex system, such as an atom, a nucleus or a molecule, the orbital angular momenta $\mathbf{L}_i$ of the various elementary particles which constitute the system (electrons, protons, neutrons,...)

combine with each other and with the spin angular momenta $\mathbf{S}_i$ of these same particles to form the *total angular momentum* $\mathbf{J}$ of the system. The way in which angular momenta are combined in quantum mechanics (coupling of angular momenta) will be studied in chapter X. Finally, let us add that $\mathbf{J}$ will also be used to denote an arbitrary angular momentum when it is not necessary to specify whether we are dealing with an orbital angular momentum, a spin, or a combination of several angular momenta.

Before beginning the study of the physical problems just mentioned (energy levels of a particle in a central potential, spin, the Zeeman effect, addition of angular momenta,...), we shall establish, in this chapter, the *general* quantum mechanical properties associated with all angular momenta, whatever their nature.

These properties follow from commutation relations satisfied by the three observables $J_x$, $J_y$ and $J_z$, the components of an arbitrary angular momentum $\mathbf{J}$. The origin of these commutation relations is discussed in §B : for an orbital angular momentum, they are simply consequences of the quantization rules (§B-5 of chapter III) and the canonical commutation relations [formulas (E-30) of chapter II]; for spin angular momenta, which have no classical equivalents, they actually serve as definitions of the corresponding observables*. In §C, we study the consequences of these commutation relations which are characteristic of angular momenta. In particular, we discuss space quantization, that is, the fact that any component of an angular momentum possesses a discrete spectrum. Finally, the general results so obtained are applied, in §D, to the orbital angular momentum of a particle.

## B. COMMUTATION RELATIONS CHARACTERISTIC OF ANGULAR MOMENTUM

### 1. Orbital angular momentum

To obtain the observables $L_x$, $L_y$, $L_z$ associated in quantum mechanics with the three components of the angular momentum $\boldsymbol{\mathscr{L}}$ of a spinless particle, we simply apply the quantization rules stated in §B-5 of chapter III. If we consider the component $\mathscr{L}_x$ of the classical angular momentum:

$$\mathscr{L}_x = yp_z - zp_y \tag{B-1}$$

We associate with the position variables $y$ and $z$, the observables $Y$ and $Z$, and with the momentum variables $p_y$ and $p_z$, the observables $P_y$ and $P_z$. Although formula (B-1) involves products of two classical variables, no precautions need be taken in replacing them by the corresponding observables, since $Y$ and $P_z$ commute, as do $Z$ and $P_y$ [see the canonical commutation relations (E-30) of chapter II]. We therefore do not need to symmetrize expression (B-1) in order to obtain the operator $L_x$:

$$L_x = YP_z - ZP_y \tag{B-2}$$

* The fundamental origin of these commutation relations is purely geometrical. We shall discuss this point in detail in complement $B_{VI}$, in which we demonstrate the intimate relation between the angular momentum of a system with respect to a point $O$ and the *geometrical rotations* of this system about $O$.

For the same reason ($Y$ and $P_z$ commute, as do $Z$ and $P_y$), the operator thus obtained is Hermitian.

In the same way, we find the operators $L_y$ and $L_z$ corresponding to the components $\mathscr{L}_y$ and $\mathscr{L}_z$ of the classical angular momentum. This allows us to write:

$$\mathbf{L} = \mathbf{R} \times \mathbf{P} \tag{B-3}$$

Since we know the canonical commutation relations of the position $\mathbf{R}$ and momentum $\mathbf{P}$ observables, we can easily calculate the commutators of the operators $L_x$, $L_y$ and $L_z$.

For example, let us evaluate $[L_x, L_y]$:

$$\begin{aligned}[L_x, L_y] &= [YP_z - ZP_y, ZP_x - XP_z] \\ &= [YP_z, ZP_x] + [ZP_y, XP_z]\end{aligned} \tag{B-4}$$

since $YP_z$ commutes with $XP_z$, and $ZP_y$, with $ZP_x$. We then have:

$$\begin{aligned}[L_x, L_y] &= Y[P_z, Z]P_x + X[Z, P_z]P_y \\ &= -i\hbar\, YP_x + i\hbar\, XP_y \\ &= i\hbar\, L_z\end{aligned} \tag{B-5}$$

Analogous calculations yield the other two commutators, and we obtain, finally:

$$\begin{aligned}[L_x, L_y] &= i\hbar\, L_z \\ [L_y, L_z] &= i\hbar\, L_x \\ [L_z, L_x] &= i\hbar\, L_y\end{aligned} \tag{B-6}$$

Thus we have established the commutation relations for the components of the angular momentum of a spinless particle.

This result can immediately be generalized to a system of $N$ spinless particles. The total angular momentum of such a system is, in quantum mechanics:

$$\mathbf{L} = \sum_{i=1}^{N} \mathbf{L}_i \tag{B-7}$$

with:

$$\mathbf{L}_i = \mathbf{R}_i \times \mathbf{P}_i \tag{B-8}$$

Now, each of the individual angular momenta $\mathbf{L}_i$ satisfies the commutation relations (B-6) and commutes with $\mathbf{L}_j$ when $j$ is not equal to $i$ (operators acting in state spaces of different particles). Thus we see that relations (B-6) remain valid for the total angular momentum $\mathbf{L}$.

## 2. Generalization: definition of an angular momentum

The three operators associated with the components of an arbitrary classical angular momentum therefore satisfy the commutation relations (B-6). It can be shown, moreover (*cf.* complement $B_{VI}$), that the origin of these relations lies in the geometric properties of rotations in three-dimensional space. This is why we shall adopt a more general point of view and define an angular momentum **J** as any set of three observables $J_x$, $J_y$, $J_z$ which satisfies:

$$\boxed{\begin{aligned}[J_x, J_y] &= i\hbar J_z \\ [J_y, J_z] &= i\hbar J_x \\ [J_z, J_x] &= i\hbar J_y\end{aligned}} \tag{B-9}$$

We then introduce the operator:

$$\mathbf{J}^2 = J_x^2 + J_y^2 + J_z^2 \tag{B-10}$$

the (scalar) square of the angular momentum **J**. This operator is Hermitian, since $J_x$, $J_y$ and $J_z$ are Hermitian. We shall assume that it is an observable. Let us show that $\mathbf{J}^2$ commutes with the three components of **J**:

$$\boxed{[\mathbf{J}^2, \mathbf{J}] = \mathbf{0}} \tag{B-11}$$

We perform the calculation for $J_x$, for example:

$$\begin{aligned}[\mathbf{J}^2, J_x] &= [J_x^2 + J_y^2 + J_z^2, J_x] \\ &= [J_y^2, J_x] + [J_z^2, J_x]\end{aligned} \tag{B-12}$$

since $J_x$ obviously commutes with itself and, therefore, with its square. The other two commutators can be obtained easily from (B-9).

$$\begin{aligned}[J_y^2, J_x] &= J_y[J_y, J_x] + [J_y, J_x]J_y \\ &= -\, i\hbar\, J_yJ_z - i\hbar\, J_zJ_y\end{aligned} \tag{B-13-a}$$

$$\begin{aligned}[J_z^2, J_x] &= J_z[J_z, J_x] + [J_z, J_x]J_z \\ &= i\hbar\, J_zJ_y + i\hbar\, J_yJ_z\end{aligned} \tag{B-13-b}$$

The sum of these two commutators, which enters into (B-12), is indeed zero.

Angular momentum theory in quantum mechanics is founded entirely on the commutation relations (B-9). Note that these relations imply that it is impossible to measure simultaneously the three components of an angular momentum; however, $\mathbf{J}^2$ and any component of **J** are compatible.

## 3. Statement of the problem

Let us return to the example of a spinless particle in a central potential, mentioned in the introduction. We shall see in chapter VII that, in this case, the three components of the angular momentum **L** of the particle commute with the Hamiltonian $H$; thus, this is also true for the operator $\mathbf{L}^2$. We then have at our disposal four constants of the motion: $\mathbf{L}^2$, $L_x$, $L_y$, $L_z$. But these four operators do not all commute; to form a complete set of commuting observables with $H$, we must pick only $\mathbf{L}^2$ and one of the three other operators, $L_z$ for example. For a particle subject to a central potential, we can then look for eigenstates of the Hamiltonian $H$ which are also eigenvectors of $\mathbf{L}^2$ and $L_z$, without restricting the generality of the problem. However, it is impossible to obtain a basis of the state space composed of eigenvectors common to the three components of $\mathbf{L}$, as these three observables do not commute.

The situation is the same in the general case: since the components of an arbitrary angular momentum $\mathbf{J}$ do not commute, they are not simultaneously diagonalizable. We shall therefore seek the system of eigenvectors common to $\mathbf{J}^2$ and $J_z$, observables corresponding to the square of the absolute value of the angular momentum and to its component along the $Oz$ axis.

# C. GENERAL THEORY OF ANGULAR MOMENTUM

In this section, we shall determine the spectrum of $\mathbf{J}^2$ and $J_z$ for the general case and study their common eigenvectors. The reasoning will be analogous to the one used in chapter V for the harmonic oscillator.

## 1. Definitions and notation

### a. THE $J_+$ AND $J_-$ OPERATORS

Instead of using the components $J_x$ and $J_y$ of the angular momentum $\mathbf{J}$, it is more convenient to introduce the following linear combinations:

$$\boxed{\begin{aligned} J_+ &= J_x + iJ_y \\ J_- &= J_x - iJ_y \end{aligned}} \tag{C-1}$$

Like the operators $a$ and $a^\dagger$ of the harmonic oscillator, $J_+$ and $J_-$ are not Hermitian: they are adjoints of each other.

In the rest of this section, we shall use only the operators $J_+$, $J_-$, $J_z$ and $\mathbf{J}^2$. It is straightforward, using (B-9) and (B-11), to show that these operators satisfy the commutation relations:

$$[J_z, J_+] = \hbar J_+ \tag{C-2}$$

$$[J_z, J_-] = -\hbar J_- \tag{C-3}$$

$$[J_+, J_-] = 2\hbar J_z \tag{C-4}$$

$$[\mathbf{J}^2, J_+] = [\mathbf{J}^2, J_-] = [\mathbf{J}^2, J_z] = 0 \tag{C-5}$$

Let us calculate the products $J_+J_-$ and $J_-J_+$. We find:

$$\begin{aligned} J_+J_- &= (J_x + iJ_y)(J_x - iJ_y) \\ &= J_x^2 + J_y^2 - i[J_x, J_y] \\ &= J_x^2 + J_y^2 + \hbar J_z \end{aligned} \tag{C-6-a}$$

$$\begin{aligned} J_-J_+ &= (J_x - iJ_y)(J_x + iJ_y) \\ &= J_x^2 + J_y^2 + i[J_x, J_y] \\ &= J_x^2 + J_y^2 - \hbar J_z \end{aligned} \tag{C-6-b}$$

Using definition (B-10) of the operator $\mathbf{J}^2$, we can write these expressions in the form:

$$J_+J_- = \mathbf{J}^2 - J_z^2 + \hbar J_z \tag{C-7-a}$$

$$J_-J_+ = \mathbf{J}^2 - J_z^2 - \hbar J_z \tag{C-7-b}$$

Adding relations (C-7) term by term, we obtain:

$$\mathbf{J}^2 = \frac{1}{2}(J_+J_- + J_-J_+) + J_z^2 \tag{C-8}$$

### b. NOTATION FOR THE EIGENVALUES OF $\mathbf{J}^2$ AND $J_z$

According to (B-10), $\mathbf{J}^2$ is the sum of the squares of three Hermitian operators. Consequently, for any ket $|\psi\rangle$, the matrix element $\langle\psi|\mathbf{J}^2|\psi\rangle$ is positive or zero:

$$\begin{aligned}\langle\psi|\mathbf{J}^2|\psi\rangle &= \langle\psi|J_x^2|\psi\rangle + \langle\psi|J_y^2|\psi\rangle + \langle\psi|J_z^2|\psi\rangle \\ &= \|J_x|\psi\rangle\|^2 + \|J_y|\psi\rangle\|^2 + \|J_z|\psi\rangle\|^2 \geqslant 0\end{aligned} \tag{C-9}$$

Note that this could have been expected, since $\mathbf{J}^2$ corresponds to the square of the absolute value of the angular momentum $\mathbf{J}$. From this we see, in particular, that *all the eigenvalues of* $\mathbf{J}^2$ *are positive or zero*, since if $|\psi\rangle$ is an eigenvector of $\mathbf{J}^2$, $\langle\psi|\mathbf{J}^2|\psi\rangle$ is the product of the corresponding eigenvalue and the square of the norm of $|\psi\rangle$, which is always positive.

We shall write the eigenvalues of $\mathbf{J}^2$ in the form $j(j+1)\hbar^2$, with the *convention*:

$$j \geqslant 0 \tag{C-10}$$

This notation is intended to simplify the arguments which follow; it does not influence the result. Since $\mathbf{J}$ has the dimensions of $\hbar$, an eigenvalue of $\mathbf{J}^2$ is necessarily of the form $\lambda\hbar^2$, where $\lambda$ is a real dimensionless number. We have just seen that $\lambda$ must be positive or zero; it can then easily be shown that the second-degree equation in $j$:

$$j(j+1) = \lambda \tag{C-11}$$

always has one and only one positive or zero root. Therefore, if we use (C-10), the specification of $\lambda$ determines $j$ uniquely; any eigenvalue of $\mathbf{J}^2$ can thus be written $j(j+1)\hbar^2$, with $j$ positive or zero.

As for the eigenvalues of $J_z$, which have the same dimensions as $\hbar$, they are traditionally written $m\hbar$, where $m$ is a dimensionless number.

### c. EIGENVALUE EQUATIONS FOR $\mathbf{J}^2$ AND $J_z$

We shall label the eigenvectors common to $\mathbf{J}^2$ and $J_z$ by the indices $j$ and $m$ which characterize the associated eigenvalues. However, $\mathbf{J}^2$ and $J_z$ do not in general constitute a C.S.C.O. (see, for example, § A of chapter VII), and it is necessary to introduce a third index in order to distinguish between the different eigenvectors corresponding to the same eigenvalues $j(j+1)\hbar^2$ and $m\hbar$ of $\mathbf{J}^2$ and $J_z$ (this point will be expanded in § 3-a below). We shall call this index $k$ (which does not necessarily imply that it is always a discrete index).

We shall therefore try to solve the simultaneous eigenvalue equations:

$$\begin{aligned} \mathbf{J}^2 \,|\, k, j, m \rangle &= j\,(j+1)\hbar^2 \,|\, k, j, m \rangle \\ J_z \,|\, k, j, m \rangle &= m\hbar \,|\, k, j, m \rangle \end{aligned} \tag{C-12}$$

## 2. Eigenvalues of $\mathbf{J}^2$ and $J_z$

As in § B-2 of chapter V, we shall begin by proving three lemmas which will then enable us to determine the spectrum of $\mathbf{J}^2$ and $J_z$.

### a. LEMMAS

#### α. *Lemma I (Properties of the eigenvalues of $\mathbf{J}^2$ and $J_z$)*

If $j(j+1)\hbar^2$ and $m\hbar$ are the eigenvalues of $\mathbf{J}^2$ and $J_z$ associated with the same eigenvector $|k, j, m\rangle$, then $j$ and $m$ satisfy the inequality:

$$-j \leqslant m \leqslant j \tag{C-13}$$

To prove this, consider the vectors $J_+ |k, j, m\rangle$ and $J_- |k, j, m\rangle$, and note that the square of their norms is positive or zero:

$$\| J_+ \,|\, k, j, m \rangle \|^2 = \langle k, j, m \,|\, J_- J_+ \,|\, k, j, m \rangle \qquad \geqslant 0 \tag{C-14-a}$$

$$\| J_- \,|\, k, j, m \rangle \|^2 = \langle k, j, m \,|\, J_+ J_- \,|\, k, j, m \rangle \qquad \geqslant 0 \tag{C-14-b}$$

To calculate the left-hand sides of these inequalities, we can use formulas (C-7). We find (if we assume $|k, j, m\rangle$ to be normalized):

$$\begin{aligned} \langle k, j, m \,|\, J_- J_+ \,|\, k, j, m \rangle &= \langle k, j, m \,|\, (\mathbf{J}^2 - J_z^2 - \hbar J_z) \,|\, k, j, m \rangle \\ &= j(j+1)\hbar^2 - m^2\hbar^2 - m\hbar^2 \end{aligned} \tag{C-15-a}$$

$$\begin{aligned} \langle k, j, m \,|\, J_+ J_- \,|\, k, j, m \rangle &= \langle k, j, m \,|\, (\mathbf{J}^2 - J_z^2 + \hbar J_z) \,|\, k, j, m \rangle \\ &= j(j+1)\hbar^2 - m^2\hbar^2 + m\hbar^2 \end{aligned} \tag{C-15-b}$$

Substituting these expressions into inequalities (C-14), we obtain:

$$j(j+1) - m(m+1) = (j-m)(j+m+1) \geqslant 0 \tag{C-16-a}$$

$$j(j+1) - m(m-1) = (j-m+1)(j+m) \geqslant 0 \tag{C-16-b}$$

that is:

$$-(j+1) \leqslant m \leqslant j \tag{C-17-a}$$

$$-j \leqslant m \leqslant j+1 \tag{C-17-b}$$

These two conditions are satisfied simultaneously only if $m$ satisfies inequality (C-13).

#### β. *Lemma II (Properties of the vector $J_- |k, j, m\rangle$)*

Let $|k, j, m\rangle$ be an eigenvector of $\mathbf{J}^2$ and $J_z$ with the eigenvalues $j(j+1)\hbar^2$ and $m\hbar$.

(*i*) If $m = -j$, $J_- |k, j, -j\rangle = 0$.

(*ii*) If $m > -j$, $J_- |k, j, m\rangle$ is a non-null eigenvector of $\mathbf{J}^2$ and $J_z$ with the eigenvalues $j(j+1)\hbar^2$ and $(m-1)\hbar$.

(*i*) According to (C-15-b), the square of the norm of $J_- |k, j, m\rangle$ is equal to $\hbar^2[j(j+1) - m(m-1)]$ and therefore goes to zero for $m = -j$. Since the norm of a vector goes to zero if and only if this vector is a null vector, we conclude that all vectors $J_- |k, j, -j\rangle$ are null:

$$m = -j \Longrightarrow J_- |k, j, -j\rangle = 0 \tag{C-18}$$

It is easy to establish the converse of (C-18):

$$J_- |k, j, m\rangle = 0 \Longrightarrow m = -j \tag{C-19}$$

Letting $J_+$ act on both sides of the equation appearing in (C-19), and using (C-7-a), we obtain:

$$\begin{aligned} &\hbar^2[j(j+1) - m^2 + m] \,|k, j, m\rangle \\ &\qquad = \hbar^2(j+m)(j-m+1) \,|k, j, m\rangle = 0 \end{aligned} \tag{C-20}$$

Using (C-13), (C-20) has only one solution, $m = -j$.

(*ii*) Now assume $m$ to be greater than $-j$. According to (C-15-b), $J_- |k, j, m\rangle$ is then a non-null vector since the square of its norm is different from zero.

Let us show that it is an eigenvector of $\mathbf{J}^2$ and $J_z$. The operators $J_-$ and $\mathbf{J}^2$ commute; consequently:

$$[\mathbf{J}^2, J_-] \,|k, j, m\rangle = 0 \tag{C-21}$$

which can be written:

$$\begin{aligned} \mathbf{J}^2 J_- |k, j, m\rangle &= J_- \mathbf{J}^2 |k, j, m\rangle \\ &= j(j+1)\hbar^2 J_- |k, j, m\rangle \end{aligned} \tag{C-22}$$

This relation expresses the fact that $J_- |k, j, m\rangle$ is an eigenvector of $\mathbf{J}^2$ with the eigenvalue $j(j+1)\hbar^2$.

Moreover, if we apply operator equation (C-3) to $|k, j, m\rangle$:

$$[J_z, J_-] \,|k, j, m\rangle = -\hbar J_- |k, j, m\rangle \tag{C-23}$$

that is:

$$\begin{aligned} J_z J_- |k, j, m\rangle &= J_- J_z |k, j, m\rangle - \hbar J_- |k, j, m\rangle \\ &= m\hbar J_- |k, j, m\rangle - \hbar J_- |k, j, m\rangle \\ &= (m-1)\hbar \, J_- |k, j, m\rangle \end{aligned} \tag{C-24}$$

$J_- |k, j, m\rangle$ is therefore an eigenvector of $J_z$ with the eigenvalue $(m-1)\hbar$.

γ. *Lemma III (Properties of the vector $J_+ |k, j, m\rangle$)*

Let $|k, j, m\rangle$ be an eigenvector of $\mathbf{J}^2$ and $J_z$ with the eigenvalues $j(j+1)\hbar^2$ and $m\hbar$.

(*i*) If $m = j$, $J_+ |k, j, j\rangle = 0$.

(*ii*) If $m < j$, $J_+ |k, j, m\rangle$ is a non-null eigenvector of $\mathbf{J}^2$ and $J_z$ with the eigenvalues $j(j+1)\hbar^2$ and $(m+1)\hbar$.

($i$) The argument is similar to that of §C-2-a-β. According to (C-14-a), the square of the norm of $J_+ | k, j, m \rangle$ is zero if $m = j$. Therefore:

$$m = j \Longrightarrow J_+ | k, j, j \rangle = 0 \tag{C-25}$$

The converse can be proved in the same way:

$$J_+ | k, j, m \rangle = 0 \Longleftrightarrow m = j \tag{C-26}$$

($ii$) If $m < j$, an argument analogous to that of § β-($ii$) yields, using formulas (C-5) and (C-2):

$$\mathbf{J}^2 J_+ | k, j, m \rangle = j(j+1)\hbar^2 J_+ | k, j, m \rangle \tag{C-27}$$

$$J_z J_+ | k, j, m \rangle = (m+1)\hbar \, J_+ | k, j, m \rangle \tag{C-28}$$

#### b. DETERMINATION OF THE SPECTRUM OF $\mathbf{J}^2$ AND $J_z$

We shall now show that the three lemmas above enable us to determine the possible values of $j$ and $m$.

Let $|k, j, m\rangle$ be a non-null eigenvector of $\mathbf{J}^2$ and $J_z$ with the eigenvalues $j(j+1)\hbar^2$ and $m\hbar$. According to lemma I, $-j \leqslant m \leqslant j$. It is therefore certain that a positive or zero integer $p$ exists such that:

$$-j \leqslant m - p < -j + 1 \tag{C-29}$$

Now consider the series of vectors:

$$| k, j, m \rangle, J_- | k, j, m \rangle, ..., (J_-)^p | k, j, m \rangle \tag{C-30}$$

According to lemma II, each of the vectors $(J_-)^n |k, j, m\rangle$ of this series ($n = 0, 1, ..., p$) is a non-null eigenvector of $\mathbf{J}^2$ and $J_z$ with the eigenvalues $j(j+1)\hbar^2$ and $(m-n)\hbar$.

The proof is by iteration. By hypothesis, $|k, j, m\rangle$ is non-null and corresponds to the eigenvalues $j(j+1)\hbar^2$ and $m\hbar$. $(J_-)^n |k, j, m\rangle$ is obtained by the action of $J_-$ on $(J_-)^{n-1} |k, j, m\rangle$, which is an eigenvector of $\mathbf{J}^2$ and $J_z$ with the eigenvalues $j(j+1)\hbar^2$ and $(m-n+1)\hbar$, the latter eigenvalue being necessarily greater than $-j$, since, according to (C-29):

$$m - n + 1 \geqslant m - p + 1 \geqslant -j + 1 \tag{C-31}$$

It follows, according to point ($ii$) of lemma II, that $(J_-)^n |k, j, m\rangle$ is a non-null eigenvector of $\mathbf{J}^2$ and $J_z$, the corresponding eigenvalues being $j(j+1)\hbar^2$ and $(m-n)\hbar$.

Now let $J_-$ act on $(J_-)^p |k, j, m\rangle$. Let us first assume that the eigenvalue $(m-p)\hbar$ of $J_z$ associated with $(J_-)^p |k, j, m\rangle$ is greater than $-j\hbar$, that is, that:

$$m - p > -j \tag{C-32}$$

By point ($ii$) of lemma II, $J_-(J_-)^p |k, j, m\rangle$ is then non-null and corresponds to the eigenvalues $j(j+1)\hbar^2$ and $(m-p-1)\hbar$. This is in contradiction with lemma I since, according to (C-29):

$$m - p - 1 < -j \tag{C-33}$$

We must therefore have $m - p$ equal to $-j$. In this case, $(J_-)^p | k, j, m \rangle$ corresponds to the eigenvalue $-j$ of $J_z$, and, according to point (*i*) of lemma II, $J_-(J_-)^p | k, j, m \rangle$ is zero. The vector series (C-30) obtained by the repeated action of $J_-$ on $| k, j, m \rangle$ is therefore limited and the contradiction with lemma I is removed.

We have now shown that there exists a positive or zero integer $p$ such that:

$$m - p = -j \tag{C-34}$$

A completely analogous argument, based on lemma III, would show that there exists a positive or zero integer $q$ such that:

$$m + q = j \tag{C-35}$$

since the vector series:

$$| k, j, m \rangle, J_+ | k, j, m \rangle, ..., (J_+)^q | k, j, m \rangle \tag{C-36}$$

must be limited if there is to be no contradiction with lemma I.

Combining (C-34) and (C-35), we obtain:

$$p + q = 2j \tag{C-37}$$

$j$ is therefore equal to a positive or zero integer divided by 2. It follows that $j$ is necessarily integral or half-integral*. Furthermore, if there exists a non-null vector $| k, j\, m \rangle$, all the vectors of series (C-30) and (C-36) are also non-null and eigenvectors of $\mathbf{J}^2$ with the eigenvalue $j(j+1)\hbar^2$, as well as of $J_z$ with the eigenvalues:

$$-j\hbar, (-j+1)\hbar, (-j+2)\hbar, ..., (j-2)\hbar, (j-1)\hbar, j\hbar \tag{C-38}$$

We summarize the results obtained above as follows:

Let $\mathbf{J}$ be an arbitrary angular momentum, obeying the commutation relations (B-9). If $j(j+1)\hbar^2$ and $m\hbar$ denote the eigenvalues of $\mathbf{J}^2$ and $J_z$, then:

– the only values possible for $j$ are positive integers or half-integers or zero, that is: 0, 1/2, 1, 3/2, 2, ... (these values are the only ones possible, but they are not all necessarily realized for all angular momenta).

– for a fixed value of $j$, the only values possible for $m$ are the $(2j + 1)$ numbers: $-j, -j+1, ..., j-1, j$; $m$ is therefore integral if $j$ is integral and half-integral if $j$ is half-integral. All these values of $m$ are realized if one of them is.

### 3. « Standard » $\{ | k, j, m \rangle \}$ representations

We shall now study the eigenvectors common to $\mathbf{J}^2$ and $J_z$, which form a basis of the state space since $\mathbf{J}^2$ and $J_z$ are, by hypothesis, observables.

* A number is said to be "half-integral" if it is equal to an odd number divided by 2.

### a. THE BASIS STATES

Consider an angular momentum $\mathbf{J}$ acting in a state space $\mathscr{E}$. We shall show how to construct an orthonormal basis in $\mathscr{E}$ composed of eigenvectors common to $\mathbf{J}^2$ and $J_z$.

Take a pair of eigenvalues, $j(j+1)\hbar^2$ and $m\hbar$, that are actually found in the case we are considering. The set of eigenvectors associated with this pair of eigenvalues forms a vector subspace of $\mathscr{E}$ which we shall denote by $\mathscr{E}(j, m)$; the dimension $g(j, m)$ of this subspace may well be greater than 1, since $\mathbf{J}^2$ and $J_z$ do not generally constitute a C.S.C.O. We choose in $\mathscr{E}(j, m)$ an arbitrary orthonormal basis, $\{ |k, j, m\rangle ; k = 1, 2, ..., g(j, m) \}$.

If $m$ is not equal to $j$, there must exist another subspace $\mathscr{E}(j, m+1)$ in $\mathscr{E}$ composed of eigenvectors of $\mathbf{J}^2$ and $J_z$ associated with the eigenvalues $j(j+1)\hbar^2$ and $(m+1)\hbar$. Similarly, if $m$ is not equal to $-j$, there exists a subspace $\mathscr{E}(j, m-1)$. In the case where $m$ is not equal to $j$ or $-j$, we shall construct orthonormal bases in $\mathscr{E}(j, m+1)$ and in $\mathscr{E}(j, m-1)$, starting with the one chosen in $\mathscr{E}(j, m)$.

First, let us show that, if $k_1$ is not equal to $k_2$, $J_+ |k_1, j, m\rangle$ and $J_+ |k_2, j, m\rangle$ are orthogonal, as are $J_- |k_1, j, m\rangle$ and $J_- |k_2, j, m\rangle$. We can find the scalar product of $J_\pm |k_1, j, m\rangle$ and $J_\pm |k_2, j, m\rangle$ by using formulas (C-7):

$$\begin{aligned}\langle k_2, j, m | J_\mp J_\pm | k_1, j, m\rangle &= \langle k_2, j, m | (\mathbf{J}^2 - J_z^2 \mp \hbar J_z) | k_1, j, m\rangle \\ &= [j(j+1) - m(m \pm 1)]\hbar^2 \langle k_2, j, m | k_1, j, m\rangle\end{aligned} \tag{C-39}$$

These scalar products are therefore zero if $k_1 \neq k_2$ since the basis of $\mathscr{E}(j, m)$ is orthonormal; if $k_1 = k_2$, the square of the norm of $J_\pm |k_1, j, m\rangle$ is equal to:

$$[j(j+1) - m(m \pm 1)]\hbar^2$$

Now let us consider the set of the $g(j, m)$ vectors defined by:

$$|k, j, m+1\rangle = \frac{1}{\hbar\sqrt{j(j+1) - m(m+1)}} J_+ |k, j, m\rangle \tag{C-40}$$

Because of what we have just shown, these vectors are orthonormal. We shall show that they constitute a basis in $\mathscr{E}(j, m+1)$. Assume that there exists, in $\mathscr{E}(j, m+1)$, a vector $|\alpha, j, m+1\rangle$ orthogonal to all the $|k, j, m+1\rangle$ obtained from (C-40). The vector $J_- |\alpha, j, m+1\rangle$ would not be null since $(m+1)$ cannot be equal to $-j$; it would belong to $\mathscr{E}(j, m)$ and would be orthogonal to all vectors $J_- |k, j, m+1\rangle$. Now, according to (C-40), $J_- |k, j, m+1\rangle$ is proportional to $J_- J_+ |k, j, m\rangle$, that is, to $|k, j, m\rangle$ [formula (C-7-b)]. Therefore, $J_- |\alpha, j, m+1\rangle$ would be a non-null vector of $\mathscr{E}(j, m)$ which would be orthogonal to all vectors of the $\{ |k, j, m\rangle \}$ basis. But this is impossible. Consequently, the set of vectors (C-40) constitutes a basis in $\mathscr{E}(j, m+1)$.

It can be shown, using a completely analogous argument, that the vectors $|k, j, m-1\rangle$ defined by:

$$|k, j, m-1\rangle = \frac{1}{\hbar\sqrt{j(j+1) - m(m-1)}} J_- |k, j, m\rangle \tag{C-41}$$

form an orthonormal basis in $\mathscr{E}(j, m-1)$.

We see, in particular, that the dimension of subspaces $\mathscr{E}(j, m+1)$ and $\mathscr{E}(j, m-1)$ is equal to that of $\mathscr{E}(j, m)$. In other words, this dimension is independent of $m$*:

$$g(j, m+1) = g(j, m-1) = g(j, m) = g(j) \tag{C-42}$$

We then proceed as follows. For each value of $j$ actually found in the problem under consideration, we choose one of the subspaces associated with this value of $j$, for example, $\mathscr{E}(j, j)$, which corresponds to $m = j$. In this subspace, we choose an arbitrary orthonormal basis, $\{ |k, j, j\rangle ; k = 1, 2, ..., g(j) \}$. Then, using formula (C-41), we construct, by iteration, the basis to which each of the other $2j$ subspaces $\mathscr{E}(j, m)$ will be related: the arrows of table (VI-1) indicate the method used. By treating all the values of $j$ found in the problem in this way, we

TABLE VI.1

**Schematic representation of the construction of the $(2j+1)g(j)$ vectors of a "standard basis" associated with a fixed value of $j$. Starting with each of the $g(j)$ vectors $|k, j, j\rangle$ of the first line, one uses the action of $J_-$ to construct the $(2j+1)$ vectors of the corresponding column.**

**Each subspace $\mathscr{E}(j, m)$ is spanned by the $g(j)$ vectors situated in the same row. Each subspace $\mathscr{E}(k,j)$ is spanned by the $(2j+1)$ vectors of the corresponding column.**

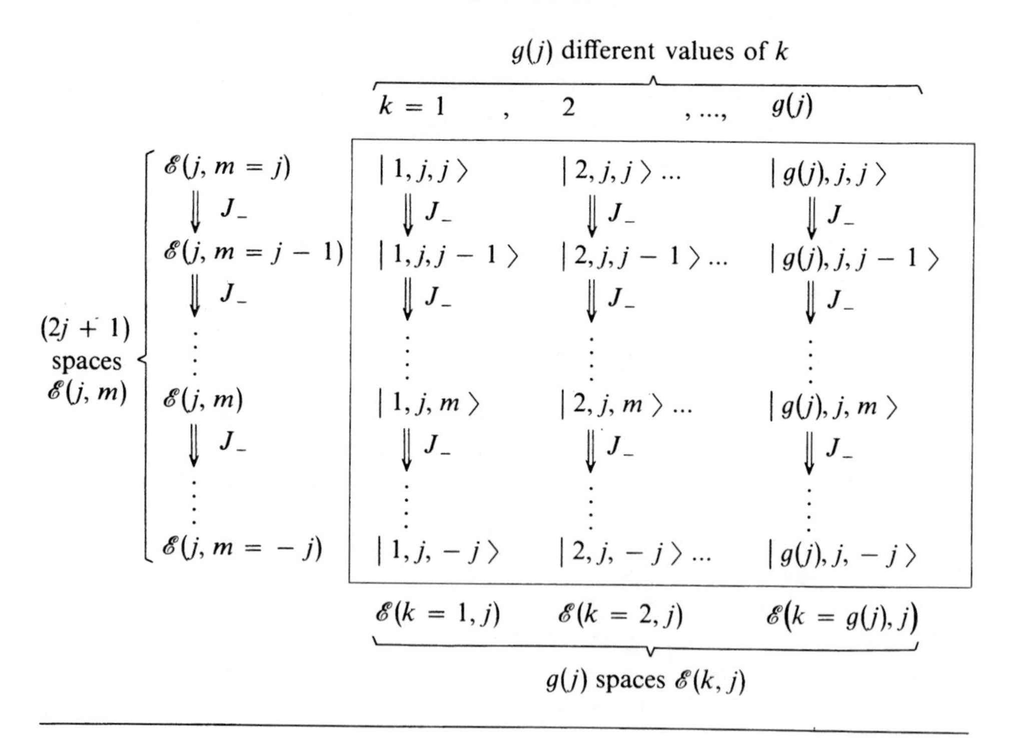


* If this dimension is infinite, the result must be interpreted as follows : there is a one-to-one correspondence between the basis vectors of two subspaces corresponding to the same value of $j$.

arrive at what is called a *standard basis* of the state space $\mathscr{E}$. The orthonormalization and closure relations for such a basis are:

$$\langle k, j, m \mid k', j', m' \rangle = \delta_{kk'} \delta_{jj'} \delta_{mm'} \tag{C-43-a}$$

$$\sum_j \sum_{m=-j}^{+j} \sum_{k=1}^{g(j)} \mid k, j, m \rangle \langle k, j, m \mid = 1 \tag{C-43-b}$$

COMMENTS:

(*i*) The use of formula (C-41) implies a choice of phases: the basis vectors in $\mathscr{E}(j, m - 1)$ are chosen to be proportional, with a real and positive coefficient, to the vectors obtained by application of $J_-$ to the basis of $\mathscr{E}(j, m)$.

(*ii*) Formulas (C-40) and (C-41) are compatible, since, if we apply $J_+$ to both sides of (C-41) and take (C-7-a) into account, we find (C-40) [with $m$ replaced by $(m - 1)$]. This means that one is not obliged to start, as we did, with the maximum value $m = j$ and (C-41) in order to construct bases of the subspaces $\mathscr{E}(j, m)$ corresponding to a given value of $j$.

In most cases, in order to define a standard basis, one uses observables $A$, $B$, ... which commute with the three components of $\mathbf{J}$* and form a C.S.C.O. with $\mathbf{J}^2$ and $J_z$ (we shall see a concrete example of this in §A of chapter VII):

$$[A, \mathbf{J}] = [B, \mathbf{J}] = \ldots = \mathbf{0} \tag{C-44}$$

For the sake of simplicity, we shall assume that only one of these observables $A$ is required to make a C.S.C.O. with $\mathbf{J}^2$ and $J_z$. Under these conditions, each of the subspaces $\mathscr{E}(j, m)$ defined above is globally invariant under the action of $A$: if $\mid \psi_{j,m} \rangle$ is an arbitrary vector of $\mathscr{E}(j, m)$, $A \mid \psi_{j,m} \rangle$ is still, according to (C-44), an eigenvector of $\mathbf{J}^2$ and $J_z$:

$$\mathbf{J}^2 A \mid \psi_{j,m} \rangle = A\,\mathbf{J}^2 \mid \psi_{j,m} \rangle = j(j + 1)\hbar^2 A \mid \psi_{j,m} \rangle$$

$$J_z A \mid \psi_{j,m} \rangle = A\,J_z \mid \psi_{j,m} \rangle = m\hbar A \mid \psi_{j,m} \rangle \tag{C-45}$$

with the same eigenvalues as $\mid \psi_{j,m} \rangle$. Thus $A \mid \psi_{j,m} \rangle$ also belongs to $\mathscr{E}(j, m)$. If we then choose a value of $j$, we can diagonalize $A$ inside the corresponding subspace $\mathscr{E}(j, j)$. We denote by $a_{k,j}$ the various eigenvalues found in this way: the index $j$ indicates in which space $\mathscr{E}(j, j)$ they were found, and the index $k$ (assumed to be discrete, for simplicity) distinguishes between them. A single vector (written $\mid k, j, j \rangle$) of $\mathscr{E}(j, j)$ is associated with each eigenvalue $a_{k,j}$, since $A$, $\mathbf{J}^2$ and $J_z$ form, by hypothesis, a C.S.C.O.:

$$A \mid k, j, j \rangle = a_{k,j} \mid k, j, j \rangle \tag{C-46}$$

The set $\{ \mid k, j, j \rangle;\ j \text{ fixed};\ k = 1, 2, \ldots, g(j) \}$ constitutes an orthonormal basis in $\mathscr{E}(j, j)$, from which we construct, using the method described above, a basis in the other subspaces $\mathscr{E}(j, m)$ related to the value of $j$ chosen. By applying this procedure successively for all values of $j$, we arrive at a "standard" basis, $\{ \mid k, j, m \rangle \}$ of the state space, all of whose vectors are eigenstates, not only of $\mathbf{J}^2$ and $J_z$, but also of $A$:

$$A \mid k, j, m \rangle = a_{k,j} \mid k, j, m \rangle \tag{C-47}$$

* An operator which commutes with the three components of the total angular momentum of a physical system is said to be "scalar" (*cf.* complement $\mathrm{B_{VI}}$).

This can be shown as follows. If hypothesis (C-44) is satisfied, $A$ commutes with $J_-$, which means that $J_- \mid k, j, j \rangle$, that is, $\mid k, j, j - 1 \rangle$, is an eigenvector of $A$ with the same eigenvalue as $\mid k, j, j \rangle$:

$$A J_- \mid k,j,j \rangle = J_- A \mid k,j,j \rangle = a_{k,j} J_- \mid k,j,j \rangle \tag{C-48}$$

By repeating this process, it is easy to prove relation (C-47).

COMMENTS:

(*i*) An observable which commutes with $\mathbf{J}^2$ and $J_z$ does not necessarily commute with $J_x$ and $J_y$ ($J_z$ is itself an example). Consequently, it should not be necessary, in order to form a C.S.C.O. with $\mathbf{J}^2$ and $J_z$, to choose observables which commute with the three components of $\mathbf{J}$ as in (C-44). However, if $A$ did not commute with $J_+$ and $J_-$ (that is, with $J_x$ and $J_y$), $J_\pm \mid k, j, m \rangle$ would not necessarily be an eigenvector of $A$ with the same eigenvalue as $\mid k, j, m \rangle$.

(*ii*) The spectrum of $A$ is the same in all the subspaces $\mathscr{E}(j, m)$ associated with the same value of $j$. However, the eigenvalues $a_{k,j}$ generally depend on $j$ (this point will be illustrated by concrete examples in §§ A and C of chapter VII).

### b. THE SPACES $\mathscr{E}(k, j)$

In the preceding section, we introduced a "standard basis" of the state space by starting with a basis chosen in the subspace $\mathscr{E}(j, m = j)$ and constructing a basis of $\mathscr{E}(j, m = j - 1)$, then one of $\mathscr{E}(j, m = j - 2)$, ..., $\mathscr{E}(j, m)$, etc... The state space can be considered to be the direct sum of all the orthogonal subspaces $\mathscr{E}(j, m)$, where $m$ varies by integral jumps from $-j$ to $+j$ and $j$ takes on all the values actually found in the problem. This means that any vector of $\mathscr{E}$ can be written in one and only one way as a sum of vectors, each belonging to a particular subspace $\mathscr{E}(j, m)$.

Nevertheless, the use of the subspaces $\mathscr{E}(j, m)$ presents certain disadvantages. First of all, their dimension $g(j)$ depends on the physical system being considered and is not necessarily known. In addition, the subspaces $\mathscr{E}(j, m)$ are not invariant under the action of $\mathbf{J}$ since, by the very means of construction of the vectors $\mid k, j, m \rangle$, $J_+$ and $J_-$ have non-zero matrix elements between vectors of $\mathscr{E}(j, m)$ and those of $\mathscr{E}(j, m \pm 1)$.

We shall therefore introduce other subspaces of $\mathscr{E}$, the spaces $\mathscr{E}(k, j)$. Instead of grouping together the kets $\mid k, j, m \rangle$ with fixed indices $j$ and $m$ [which span $\mathscr{E}(j, m)$], we shall now group together those for which $k$ and $j$ have given values, and we shall call $\mathscr{E}(k, j)$ the subspace which they span. This amounts to associating, in table (VI-1), the $(2j + 1)$ vectors of one column [instead of the $g(j)$ vectors of one row].

$\mathscr{E}$ can then be seen to be the direct sum of the orthogonal subspaces $\mathscr{E}(k, j)$, which have the simpler properties :

– the dimension of $\mathscr{E}(k, j)$ is $(2j + 1)$, whatever the value of $k$ and whatever the physical system under consideration.

– $\mathscr{E}(k, j)$ is globally invariant under the action of $\mathbf{J}$: any component $J_u$ of $\mathbf{J}$ [or a function $F(\mathbf{J})$ of $\mathbf{J}$], acting on a ket of $\mathscr{E}(k, j)$, yields another ket also belonging to $\mathscr{E}(k, j)$★. This result is not difficult to establish, since $J_u$ [or $F(\mathbf{J})$] can always be expressed in terms of $J_z$, $J_+$ and $J_-$. Now, $J_z$, acting on $|k, j, m\rangle$, yields a ket proportional to $|k, j, m\rangle$; $J_+$, a ket proportional to $|k, j, m+1\rangle$; and $J_-$, a ket proportional to $|k, j, m-1\rangle$. The existence of the property in question therefore follows from the very means of construction of the "standard basis" $\{ |k, j, m\rangle \}$.

### c. MATRICES REPRESENTING THE ANGULAR MOMENTUM OPERATORS

Using the subspaces $\mathscr{E}(k, j)$ considerably simplifies the search for the matrix which represents, in a "standard" basis, a component $J_u$ of $\mathbf{J}$ [or an arbitrary function $F(\mathbf{J})$]. The matrix elements between two basis kets belonging to two different subspaces $\mathscr{E}(k, j)$ are zero. The matrix therefore has the following form:

| | $\mathscr{E}(k, j)$ | $\mathscr{E}(k', j)$ | $\mathscr{E}(k', j')$ | ... |
|---|---|---|---|---|
| $\mathscr{E}(k, j)$ | matrix $(2j+1) \times (2j+1)$ | 0 | 0 | 0 |
| $\mathscr{E}(k', j)$ | 0 | matrix $(2j+1) \times (2j+1)$ | 0 | 0 |
| $\mathscr{E}(k', j')$ | 0 | 0 | matrix $(2j'+1) \times (2j'+1)$ | 0 |
| ⋮ | 0 | 0 | 0 | 0 |

(C-49)

All we must then do is calculate the finite-dimensional matrices which represent the operator under consideration inside each of the subspaces $\mathscr{E}(k, j)$.

Another very important simplification arises from the fact that each of these finite submatrices is independent of $k$ and of the physical system under study; it depends only on $j$ and, of course, on the operator which we want to represent. To see this, note that the definition of the $|k, j, m\rangle$ [*cf.* (C-12), (C-40) and (C-41)] implies that:

$$\begin{aligned} J_z |k, j, m\rangle &= m\hbar |k, j, m\rangle \\ J_+ |k, j, m\rangle &= \hbar\sqrt{j(j+1) - m(m+1)}\, |k, j, m+1\rangle \\ J_- |k, j, m\rangle &= \hbar\sqrt{j(j+1) - m(m-1)}\, |k, j, m-1\rangle \end{aligned} \qquad \text{(C-50)}$$

★ It can also easily be shown that $\mathscr{E}(k, j)$ is "irreducible" with respect to $\mathbf{J}$: there exists no subspace of $\mathscr{E}(k, j)$ other than $\mathscr{E}(k, j)$ itself which is globally invariant under the action of the various components of $\mathbf{J}$.

that is:

$$\begin{aligned} \langle k, j, m \mid J_z \mid k', j', m' \rangle &= m\hbar\, \delta_{kk'}\, \delta_{jj'}\, \delta_{mm'} \\ \langle k, j, m \mid J_{\pm} \mid k', j', m' \rangle &= \hbar\sqrt{j(j+1) - m'(m' \pm 1)}\, \delta_{kk'}\, \delta_{jj'}\, \delta_{m,m'\pm 1} \end{aligned} \qquad \text{(C-51)}$$

These relations show that the matrix elements representing the components of **J** depend only on $j$ and $m$ and not on $k$.

In order to know, in all cases, the matrix associated with an arbitrary component $J_u$ in a standard basis, all we need to do, therefore, is calculate, once and for all, the "universal" matrices $(J_u)^{(j)}$ which represent $J_u$ inside the subspaces $\mathscr{E}(k, j)$ for all possible values of $j$ ($j = 0, 1/2, 1, 3/2, ...$). When we study a particular physical system and its angular momentum **J**, we shall determine the values of $j$ actually found in the problem, as well as the number of subspaces $\mathscr{E}(k, j)$ associated with each of them [that is, its degree of degeneracy $(2j + 1)g(j)$]. We know that the matrix representing $J_u$ in this particular case has the "block-diagonal" form (C-49), and we can therefore construct it from the "universal" matrices which we have just defined: for each value of $j$, we shall have $g(j)$ "blocks" identical to $(J_u)^{(j)}$.

Let us give some examples of $(J_u)^{(j)}$ matrices:

($i$) $j = 0$

The subspaces $(k, j = 0)$ are one-dimensional, since zero is the only possible value for $m$. The $(J_u)^{(0)}$ matrices therefore reduce to numbers, which, according to (C-51), are zero.

($ii$) $j = 1/2$

The subspaces $(k, j = 1/2)$ are two-dimensional ($m = 1/2$ or $-1/2$). If we choose the basis vectors in this order ($m = 1/2, m = -1/2$), we find, using (C-51):

$$(J_z)^{(1/2)} = \frac{\hbar}{2}\begin{pmatrix} 1 & 0 \\ 0 & -1 \end{pmatrix} \qquad \text{(C-52)}$$

and:

$$(J_+)^{(1/2)} = \hbar\begin{pmatrix} 0 & 1 \\ 0 & 0 \end{pmatrix} \qquad (J_-)^{(1/2)} = \hbar\begin{pmatrix} 0 & 0 \\ 1 & 0 \end{pmatrix} \qquad \text{(C-53)}$$

that is, using (C-1):

$$(J_x)^{(1/2)} = \frac{\hbar}{2}\begin{pmatrix} 0 & 1 \\ 1 & 0 \end{pmatrix} \qquad (J_y)^{(1/2)} = \frac{\hbar}{2}\begin{pmatrix} 0 & -i \\ i & 0 \end{pmatrix} \qquad \text{(C-54)}$$

The matrix representing $\mathbf{J}^2$ is therefore:

$$(\mathbf{J}^2)^{(1/2)} = \frac{3}{4}\hbar^2\begin{pmatrix} 1 & 0 \\ 0 & 1 \end{pmatrix} \qquad \text{(C-55)}$$

We thus find the matrices which were introduced without justification in chapter IV, §A-2.

(*iii*) $j = 1$

We now have (order of the basis vectors : $m = 1, m = 0, m = -1$):

$$(J_z)^{(1)} = \hbar \begin{pmatrix} 1 & 0 & 0 \\ 0 & 0 & 0 \\ 0 & 0 & -1 \end{pmatrix} \tag{C-56}$$

$$(J_+)^{(1)} = \hbar \begin{pmatrix} 0 & \sqrt{2} & 0 \\ 0 & 0 & \sqrt{2} \\ 0 & 0 & 0 \end{pmatrix} \qquad (J_-)^{(1)} = \hbar \begin{pmatrix} 0 & 0 & 0 \\ \sqrt{2} & 0 & 0 \\ 0 & \sqrt{2} & 0 \end{pmatrix} \tag{C-57}$$

and therefore:

$$(J_x)^{(1)} = \frac{\hbar}{\sqrt{2}} \begin{pmatrix} 0 & 1 & 0 \\ 1 & 0 & 1 \\ 0 & 1 & 0 \end{pmatrix} \qquad (J_y)^{(1)} = \frac{\hbar}{\sqrt{2}} \begin{pmatrix} 0 & -i & 0 \\ i & 0 & -i \\ 0 & i & 0 \end{pmatrix} \tag{C-58}$$

and:

$$(\mathbf{J}^2)^{(1)} = 2\hbar^2 \begin{pmatrix} 1 & 0 & 0 \\ 0 & 1 & 0 \\ 0 & 0 & 1 \end{pmatrix} \tag{C-59}$$

COMMENT:

It can be verified that matrices (C-56) and (C-58) satisfy the commutation relations (B-9).

(*iv*) *j arbitrary*

We use relations (C-51), which, according to (C-1), can also be written:

$$\langle k, j, m \mid J_x \mid k', j', m' \rangle = \frac{\hbar}{2} \delta_{kk'} \delta_{jj'} \times \left[\sqrt{j(j+1) - m'(m'+1)}\, \delta_{m,m'+1} + \sqrt{j(j+1) - m'(m'-1)}\, \delta_{m,m'-1}\right] \tag{C-60}$$

and:

$$\langle k, j, m \mid J_y \mid k', j', m' \rangle = \frac{\hbar}{2i} \delta_{kk'} \delta_{jj'} \times \left[\sqrt{j(j+1) - m'(m'+1)}\, \delta_{m,m'+1} - \sqrt{j(j+1) - m'(m'-1)}\, \delta_{m,m'-1}\right] \tag{C-61}$$

The matrix $(J_z)^{(j)}$ is therefore diagonal and its elements are the $(2j + 1)$ values of $m\hbar$. The only non-zero matrix elements of $(J_x)^{(j)}$ and $(J_y)^{(j)}$ are those directly above and directly below the diagonal: $(J_x)^{(j)}$ is symmetrical and real, and $(J_y)^{(j)}$ is antisymmetrical and pure imaginary.

Since the kets $|k, j, m\rangle$ are, by construction, eigenvectors of $\mathbf{J}^2$, we have:

$$\langle k, j, m \mid \mathbf{J}^2 \mid k', j', m' \rangle = j(j+1)\hbar^2 \delta_{kk'} \delta_{jj'} \delta_{mm'} \tag{C-62}$$

The matrix $(\mathbf{J}^2)^{(j)}$ is therefore proportional to the $(2j + 1) \times (2j \times 1)$ unit matrix: its diagonal elements are all equal to $j(j + 1)\hbar^2$.

COMMENT:

The $Oz$ axis which we have chosen as the "quantization axis" is entirely arbitrary. All directions in space are physically equivalent, and we should expect the eigenvalues of $J_x$ or $J_y$ to be the same as those of $J_z$ (their eigenvectors, however, are different, since $J_x$ and $J_y$ do not commute with $J_z$). It can indeed be verified that the eigenvalues of the $(J_x)^{(1/2)}$ and $(J_y)^{(1/2)}$ matrices [formulas (C-54)] are $\pm \frac{\hbar}{2}$, and that those of the $(J_x)^{(1)}$ and $(J_y)^{(1)}$ matrices [formulas (C-58)] are $+ \hbar$, $0$, $- \hbar$. This result is general: inside a given subspace $\mathscr{E}(k, j)$, the eigenvalues of $J_x$ and $J_y$ (like those of the component $J_u = \mathbf{J} \cdot \mathbf{u}$ of $\mathbf{J}$ along an arbitrary unit vector $\mathbf{u}$) are $j\hbar$, $(j - 1)\hbar$, ..., $(-j + 1)\hbar$, $-j\hbar$. The corresponding eigenvectors (eigenvectors common to $\mathbf{J}^2$ and $J_x$, $\mathbf{J}^2$ and $J_y$, or $\mathbf{J}^2$ and $J_u$) are linear combinations of the $|k, j, m\rangle$ with $k$ and $j$ fixed.

To conclude this section devoted to "standard" representations, we summarize:

An orthonormal basis $\{ |k, j, m\rangle \}$ of the state space, composed of eigenvectors common to $\mathbf{J}^2$ and $J_z$:

$$\mathbf{J}^2 |k, j, m\rangle = j(j + 1)\hbar^2 |k, j, m\rangle$$
$$J_z |k, j, m\rangle = m\hbar |k, j, m\rangle$$

is called a "standard basis" if the action of the operators $J_+$ and $J_-$ on the basis vectors is given by:

$$J_+ |k, j, m\rangle = \hbar\sqrt{j(j + 1) - m(m + 1)}\,|k, j, m + 1\rangle$$
$$J_- |k, j, m\rangle = \hbar\sqrt{j(j + 1) - m(m - 1)}\,|k, j, m - 1\rangle$$

## D. APPLICATION TO ORBITAL ANGULAR MOMENTUM

In §C, we studied the general properties of angular momenta, derived uniquely from the commutation relations (B-9). We shall now return to the orbital angular momentum $\mathbf{L}$ of a spinless particle [formula (B-3)] and see how the general theory just developed applies to this particular case. Using the $\{ |\mathbf{r}\rangle \}$ representation, we shall show that the eigenvalues of the operator $\mathbf{L}^2$ are the numbers $l(l + 1)\hbar^2$ corresponding to *all positive integral or zero* $l$: of the possible values for $j$ found in §C-2-b, the only ones allowed in this case are the integral values, all of which are present. Then we shall indicate the eigenfunctions common to $\mathbf{L}^2$ and $L_z$ and their principal properties. Finally, we shall study these eigenstates from a physical point of view.

## 1. Eigenvalues and eigenfunctions of $\mathbf{L}^2$ and $L_z$

### a. EIGENVALUE EQUATION IN THE $\{ | \mathbf{r} \rangle \}$ REPRESENTATION

In the $\{ | \mathbf{r} \rangle \}$ representation, the observables $\mathbf{R}$ and $\mathbf{P}$ correspond respectively to multiplication by $\mathbf{r}$ and to the differential operator $\frac{\hbar}{i}\boldsymbol{\nabla}$. The three components of the angular momentum $\mathbf{L}$ can then be written:

$$L_x = \frac{\hbar}{i}\left( y \frac{\partial}{\partial z} - z \frac{\partial}{\partial y} \right) \tag{D-1-a}$$

$$L_y = \frac{\hbar}{i}\left( z \frac{\partial}{\partial x} - x \frac{\partial}{\partial z} \right) \tag{D-1-b}$$

$$L_z = \frac{\hbar}{i}\left( x \frac{\partial}{\partial y} - y \frac{\partial}{\partial x} \right) \tag{D-1-c}$$

It is more convenient to work in spherical (or polar) coordinates, since, as we shall see, the various angular momentum operators act only on the angular variables $\theta$ and $\varphi$, and not on the variable $r$. Instead of characterizing the vector $\mathbf{r}$ by its Cartesian components $x$, $y$, $z$, we label the corresponding point $M$ in space ($\mathbf{OM} = \mathbf{r}$) by its spherical coordinates $r$, $\theta$, $\varphi$ (fig. 1):

$$\begin{cases} x = r \sin\theta \cos\varphi \\ y = r \sin\theta \sin\varphi \\ z = r\cos\theta \end{cases}$$

with:

$$\begin{cases} r \geqslant 0 \\ 0 \leqslant \theta \leqslant \pi \\ 0 \leqslant \varphi < 2\pi \end{cases} \tag{D-2}$$

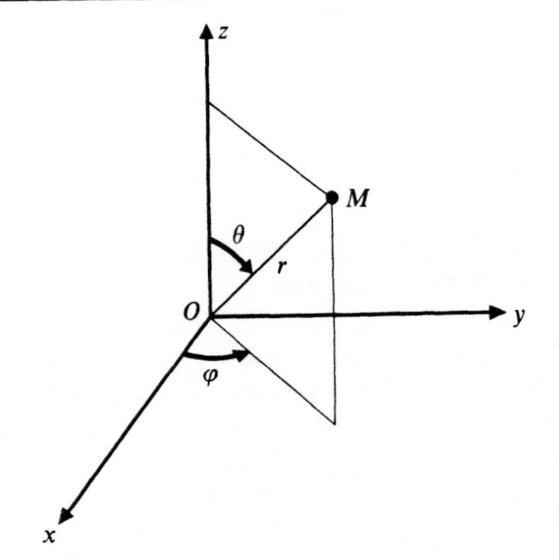


FIGURE 1

**Definition of the spherical coordinates $r$, $\theta$, $\varphi$ of an arbitrary point in space.**

The volume element $d^3r = \mathrm{d}x\,\mathrm{d}y\,\mathrm{d}z$ is written in spherical coordinates:

$$\begin{aligned} d^3r &= r^2 \sin\theta \,\mathrm{d}r\,\mathrm{d}\theta\,\mathrm{d}\varphi \\ &= r^2\,\mathrm{d}r\,\mathrm{d}\Omega \end{aligned} \tag{D-3}$$

where:

$$\mathrm{d}\Omega = \sin\theta\,\mathrm{d}\theta\,\mathrm{d}\varphi \tag{D-4}$$

is the solid angle element about the direction of polar angles $\theta$ and $\varphi$.

Applying the classical technique of changing variables, we obtain, from formulas (D-1) and (D-2), the following expressions (the calculations are rather time-consuming but pose no great problems):

$$L_x = i\hbar\left(\sin\varphi\frac{\partial}{\partial\theta} + \frac{\cos\varphi}{\tan\theta}\frac{\partial}{\partial\varphi}\right) \tag{D-5-a}$$

$$L_y = i\hbar\left(-\cos\varphi\frac{\partial}{\partial\theta} + \frac{\sin\varphi}{\tan\theta}\frac{\partial}{\partial\varphi}\right) \tag{D-5-b}$$

$$L_z = \frac{\hbar}{i}\frac{\partial}{\partial\varphi} \tag{D-5-c}$$

which yield:

$$\mathbf{L}^2 = -\hbar^2\left(\frac{\partial^2}{\partial\theta^2} + \frac{1}{\tan\theta}\frac{\partial}{\partial\theta} + \frac{1}{\sin^2\theta}\frac{\partial^2}{\partial\varphi^2}\right) \tag{D-6-a}$$

$$L_+ = \hbar\,\mathrm{e}^{i\varphi}\left(\frac{\partial}{\partial\theta} + i\cot\theta\frac{\partial}{\partial\varphi}\right) \tag{D-6-b}$$

$$L_- = \hbar\,\mathrm{e}^{-i\varphi}\left(-\frac{\partial}{\partial\theta} + i\cot\theta\frac{\partial}{\partial\varphi}\right) \tag{D-6-c}$$

In the $\{\,|\,\mathbf{r}\,\rangle\,\}$ representation, the eigenfunctions associated with the eigenvalues $l(l+1)\hbar^2$ of $\mathbf{L}^2$ and $m\hbar$ of $L_z$ are the solutions of the partial differential equations:

$$\left\{\begin{aligned} &-\left\{\frac{\partial^2}{\partial\theta^2} + \frac{1}{\tan\theta}\frac{\partial}{\partial\theta} + \frac{1}{\sin^2\theta}\frac{\partial^2}{\partial\varphi^2}\right\}\psi(r,\theta,\varphi) = l\,(l+1)\,\psi(r,\theta,\varphi) && \text{(D-7-a)} \\ &-i\frac{\partial}{\partial\varphi}\psi(r,\theta,\varphi) = m\,\psi(r,\theta,\varphi) && \text{(D-7-b)} \end{aligned}\right.$$

Since the general results of §C are applicable to the orbital angular momentum, we already know that $l$ is integral or half-integral and that, for fixed $l$, $m$ can take on only the values $-l, -l+1, ..., l-1, l$.

In equations (D-7), $r$ does not appear in any differential operator, so we can consider it to be a parameter and *take into account only the $\theta$- and $\varphi$-dependence of $\psi$*. Thus, we denote by $Y_l^m(\theta, \varphi)$ a common eigenfunction of $\mathbf{L}^2$ and $L_z$ which corresponds to the eigenvalues $l(l+1)\hbar^2$ and $m\hbar$:

$$\mathbf{L}^2\,Y_l^m(\theta,\varphi) = l(l+1)\hbar^2\,Y_l^m(\theta,\varphi) \tag{D-8-a}$$

$$L_z\,Y_l^m(\theta,\varphi) = m\hbar\,Y_l^m(\theta,\varphi) \tag{D-8-b}$$

To be completely rigorous, we would have to introduce an additional index in order to distinguish between the various solutions of (D-8) which correspond to the same pair of values of $l$ and $m$. In fact, as we shall see further on, these equations have only one solution (to within a constant factor) for each pair of allowed values of $l$ and $m$; this is why the indices $l$ and $m$ are sufficient.

COMMENTS:

(i) Equations (D-8) give the $\theta$-and $\varphi$-dependence of the eigenfunctions of $\mathbf{L}^2$ and $L_z$. Once the solution $Y_l^m(\theta, \varphi)$ of these equations has been found, these eigenfunctions will be obtained in the form:

$$\psi_{l,m}(r, \theta, \varphi) = f(r)Y_l^m(\theta, \varphi) \tag{D-9}$$

where $f(r)$ is a function of $r$* which appears as an integration constant for the partial differential equations (D-7). The fact that $f(r)$ is arbitrary shows that $\mathbf{L}^2$ and $L_z$ do not form a C.S.C.O. in the space $\mathscr{E}_\mathbf{r}$ of functions of $\mathbf{r}$ (or of $r, \theta, \varphi$).

(ii) In order to normalize $\psi_{l,m}(r, \theta, \varphi)$, it is convenient to normalize $Y_l^m(\theta, \varphi)$ and $f(r)$ separately (as we shall do here). We then have, taking (D-4) into account:

$$\int_0^{2\pi} \mathrm{d}\varphi \int_0^{\pi} \sin\theta \, |Y_l^m(\theta, \varphi)|^2 \, \mathrm{d}\theta = 1 \tag{D-10}$$

and:

$$\int_0^{\infty} r^2 \, |f(r)|^2 \, \mathrm{d}r = 1 \tag{D-11}$$

### b. VALUES OF *l* AND *m*

#### α. *l and m must be integral*

Using expression (D-5-c) for $L_z$, we can write (D-8-b) in the form:

$$\frac{\hbar}{i}\frac{\partial}{\partial\varphi} Y_l^m(\theta, \varphi) = m\hbar \, Y_l^m(\theta, \varphi) \tag{D-12}$$

which shows that $Y_l^m(\theta, \varphi)$ is equal to:

$$Y_l^m(\theta, \varphi) = F_l^m(\theta) \, \mathrm{e}^{im\varphi} \tag{D-13}$$

We can cover all space by letting $\varphi$ vary between 0 and $2\pi$. Since a wave function must be continuous at all points in space**, we must have, in particular:

$$Y_l^m(\theta, \varphi = 0) = Y_l^m(\theta, \varphi = 2\pi) \tag{D-14}$$

which implies that:

$$\mathrm{e}^{2im\pi} = 1 \tag{D-15}$$

* $f(r)$ must be such that $\psi_{l,m}(r, \theta, \varphi)$ is square-integrable.

** If $Y_l^m(\theta, \varphi)$ were not continuous for $\varphi = 0$, it would not be differentiable and could not be an eigenfunction of the differential operators (D-5-c) and (D-6-a). For example, $\frac{\partial}{\partial\varphi} Y_l^m(\theta, \varphi)$ would produce a function $\delta(\varphi)$, which is incompatible with (D-12).

According to the results of § C, $m$ is integral or half-integral. Relation (D-15) shows that, *in the case of an orbital angular momentum, $m$ must be an integer* ($e^{2im\pi}$ would be equal to $-1$ if $m$ were half-integral). But we know that $m$ and $l$ are either both integral or both half-integral: it follows that *$l$ must also be an integer.*

β. *All integral values (positive or zero) of $l$ can be found*

Choose an integral value of $l$ (positive or zero). We know from the general theory of §C that $Y_l^l(\theta, \varphi)$ must satisfy:

$$L_+ Y_l^l(\theta, \varphi) = 0 \tag{D-16}$$

which yields, taking (D-6-b) and (D-13) into account:

$$\left\{ \frac{d}{d\theta} - l \cot \theta \right\} F_l^l(\theta) = 0 \tag{D-17}$$

This first order equation can be integrated immediately if we notice that:

$$\cot \theta \, d\theta = \frac{d(\sin \theta)}{\sin \theta} \tag{D-18}$$

Its general solution is:

$$F_l^l(\theta) = c_l(\sin \theta)^l \tag{D-19}$$

where $c_l$ is a normalization constant*

Consequently, for each positive or zero integral value of $l$, there exists a function $Y_l^l(\theta, \varphi)$ which is unique (to within a constant factor):

$$Y_l^l(\theta, \varphi) = c_l(\sin \theta)^l \, e^{il\varphi} \tag{D-20}$$

Through the repeated action of $L_-$, we construct $Y_l^{l-1}$, ..., $Y_l^m$, ... $Y_l^{-l}$. Thus we see that there corresponds, to the pair of eigenvalues $l(l+1)\hbar^2$ and $m\hbar$ (where $l$ is an arbitrary positive integer or zero and $m$ is another integer such that $-l \leqslant m \leqslant l$), *one and only one eigenfunction*: $Y_l^m(\theta, \varphi)$, which can be unambiguously calculated from (D-20). The eigenfunctions $Y_l^m(\theta, \varphi)$ are called *spherical harmonics.*

### c. FUNDAMENTAL PROPERTIES OF THE SPHERICAL HARMONICS

The spherical harmonics $Y_l^m(\theta, \varphi)$ will be studied in greater detail in complement $A_{VI}$. Here we shall confine ourselves to summarizing this study by stating without proof its principal results.

α. *Recurrence relations*

According to the general results of §C, we have:

$$L_\pm Y_l^m(\theta, \varphi) = \hbar \sqrt{l(l+1) - m(m \pm 1)} \; Y_l^{m \pm 1}(\theta, \varphi) \tag{D-21}$$

* Inversely, one can easily show that the function obtained in this way is actually an eigenfunction of $\mathbf{L}^2$ and $L_z$ with eigenvalues $l(l+1)\hbar^2$ and $l\hbar$. According to (D-5-c) and (D-13), it is immediately seen that $L_z \, Y_l^l(\theta, \varphi) = l\hbar \, Y_l^l(\theta, \varphi)$. Then, using this equation and (D-16), as well as (C-7-b), one can show that $Y_l^l(\theta, \varphi)$ is also an eigenfunction of $\mathbf{L}^2$ with the expected eigenvalue.

Using expressions (D-6-b) and (D-6-c) for the operators $L_+$ and $L_-$ and the fact that $Y_l^m(\theta, \varphi)$ is the product of a function of $\theta$ alone and $e^{im\varphi}$, we obtain:

$$e^{i\varphi}\left(\frac{\partial}{\partial\theta} - m \cot\theta\right)Y_l^m(\theta, \varphi) = \sqrt{l(l+1) - m(m+1)}\; Y_l^{m+1}(\theta, \varphi) \quad \text{(D-22-a)}$$

$$e^{-i\varphi}\left(-\frac{\partial}{\partial\theta} - m \cot\theta\right)Y_l^m(\theta, \varphi) = \sqrt{l(l+1) - m(m-1)}\; Y_l^{m-1}(\theta, \varphi) \quad \text{(D-22-b)}$$

β. *Orthonormalization and closure relations*

Equations (D-7) determine the spherical harmonics only to within a constant factor. We now choose this factor so as to orthonormalize the $Y_l^m(\theta, \varphi)$ (as functions of the angular variables $\theta$ and $\varphi$):

$$\int_0^{2\pi} d\varphi \int_0^{\pi} \sin\theta \, d\theta \; Y_{l'}^{m'*}(\theta, \varphi) Y_l^m(\theta, \varphi) = \delta_{l'l}\,\delta_{m'm} \quad \text{(D-23)}$$

Furthermore, any function of $\theta$ and $\varphi$, $f(\theta, \varphi)$, can be expanded in terms of the spherical harmonics:

$$f(\theta, \varphi) = \sum_{l=0}^{\infty} \sum_{m=-l}^{+l} c_{l,m} Y_l^m(\theta, \varphi) \quad \text{(D-24)}$$

with:

$$c_{l,m} = \int_0^{2\pi} d\varphi \int_0^{\pi} \sin\theta \, d\theta \; Y_l^{m*}(\theta, \varphi) f(\theta, \varphi) \quad \text{(D-25)}$$

The spherical harmonics therefore constitute an orthonormal basis in the space $\mathscr{E}_\Omega$ of functions of $\theta$ and $\varphi$. This fact is expressed by the closure relation:

$$\sum_{l=0}^{\infty} \sum_{m=-l}^{l} Y_l^m(\theta, \varphi) Y_l^{m*}(\theta', \varphi') = \delta(\cos\theta - \cos\theta')\,\delta(\varphi - \varphi')$$

$$= \frac{1}{\sin\theta}\,\delta(\theta - \theta')\,\delta(\varphi - \varphi') \quad \text{(D-26)}$$

[it is $\delta(\cos\theta - \cos\theta')$, and not $\delta(\theta - \theta')$ which enters into the right-hand side of the closure relation because the integrations over the variable $\theta$ are performed using the differential element $\sin\theta \, d\theta = -\, d(\cos\theta)$].

γ. *Parity and complex conjugation*

First of all, recall that the change from $\mathbf{r}$ to $-\mathbf{r}$ (reflection through the coordinate origin) is expressed in spherical coordinates by (fig. 2):

$$\begin{aligned} r &\Longrightarrow r \\ \theta &\Longrightarrow \pi - \theta \\ \varphi &\Longrightarrow \pi + \varphi \end{aligned} \quad \text{(D-27)}$$

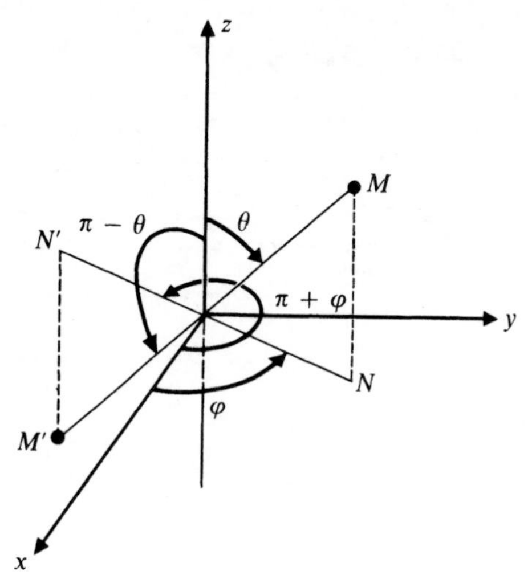


FIGURE 2

**Transformation in spherical coordinates of an arbitrary point by reflection through the origin; $r$ is not changed, $\theta$ becomes $\pi - \theta$, and $\varphi$ becomes $\pi + \varphi$.**

It is simple (see complement $A_{VI}$) to show that:

$$Y_l^m(\pi - \theta, \pi + \varphi) = (-1)^l Y_l^m(\theta, \varphi) \tag{D-28}$$

The spherical harmonics are therefore functions with a definite parity, which is independent of $m$; they are even if $l$ is even and odd if $l$ is odd.

Also, it can easily be seen that:

$$[Y_l^m(\theta, \varphi)]^* = (-1)^m Y_l^{-m}(\theta, \varphi) \tag{D-29}$$

### d. "STANDARD" BASES OF THE WAVE FUNCTION SPACE OF A SPINLESS PARTICLE

As we have already noted [comment ($i$) of §D-1-a], $\mathbf{L}^2$ and $L_z$ do not constitute a C.S.C.O. in the wave function space of a spinless particle. We shall now indicate, relying on the reasoning and results of §C-3, the form of the "standard" bases of this space.

Let $\mathscr{E}(l, m = l)$ be the subspace of eigenfunctions common to $\mathbf{L}^2$ and $L_z$, of eigenvalues $l(l+1)\hbar^2$ and $l\hbar$, where $l$ is a fixed positive integer or zero. The first step in the construction of a "standard" basis (*cf.* §C-3) consists of choosing an arbitrary orthonormal basis in each of the $\mathscr{E}(l, m = l)$. We shall denote by $\psi_{k,l,l}(\mathbf{r})$ the functions which constitute the basis chosen in $\mathscr{E}(l, m = l)$, the index $k$ (assumed to be discrete for simplicity) serving to distinguish between the various functions of this basis. By repeated application of the operator $L_-$ on the $\psi_{k,l,l}(\mathbf{r})$, we then construct the functions $\psi_{k,l,m}(\mathbf{r})$ which complete the "standard" basis for $m \neq l$; they satisfy equations (C-12) and (C-50), which become here:

$$\begin{aligned} \mathbf{L}^2\, \psi_{k,l,m}(\mathbf{r}) &= l(l+1)\hbar^2\, \psi_{k,l,m}(\mathbf{r}) \\ L_z\, \psi_{k,l,m}(\mathbf{r}) &= m\hbar\, \psi_{k,l,m}(\mathbf{r}) \end{aligned} \tag{D-30}$$

and:

$$L_{\pm}\psi_{k,l,m}(\mathbf{r}) = \hbar\sqrt{l(l+1) - m(m \pm 1)}\,\psi_{k,l,m\pm 1}(\mathbf{r}) \tag{D-31}$$

But we saw in §D-1-a that all eigenfunctions common to $\mathbf{L}^2$ and $L_z$ which correspond to given eigenvalues $l(l+1)\hbar^2$ and $m\hbar$ have the same angular dependence, that of $Y_l^m(\theta, \varphi)$; only their radial dependence differs. From equations (D-30), we therefore deduce that $\psi_{k,l,m}(\mathbf{r})$ has the form:

$$\psi_{k,l,m}(\mathbf{r}) = R_{k,l,m}(r)Y_l^m(\theta, \varphi) \tag{D-32}$$

Let us now show that, if the $\psi_{k,l,m}(\mathbf{r})$ constitute a "standard" basis, the radial functions $R_{k,l,m}(r)$ are independent of $m$. Since the differential operators $L_{\pm}$ do not act on the $r$-dependence, we have, according to (D-21):

$$\begin{aligned} L_{\pm}\psi_{k,l,m}(\mathbf{r}) &= R_{k,l,m}(r)\, L_{\pm}\, Y_l^m(\theta, \varphi) \\ &= \hbar\sqrt{l(l+1) - m(m \pm 1)}\, R_{k,l,m}(r)\, Y_l^{m\pm 1}(\theta, \varphi) \end{aligned} \tag{D-33}$$

Comparison with (D-31) shows that the radial functions must satisfy, for all $r$:

$$R_{k,l,m\pm 1}(r) = R_{k,l,m}(r) \tag{D-34}$$

and are consequently independent of $m$. The functions $\psi_{k,l,m}(\mathbf{r})$ of a "standard" basis of the wave function space of a (spinless) particle are therefore necessarily of the form:

$$\psi_{k,l,m}(\mathbf{r}) = R_{k,l}(r)Y_l^m(\theta, \varphi) \tag{D-35}$$

The orthonormalization relation for such a basis is:

$$\begin{aligned} \int d^3r\, \psi_{k,l,m}^*(\mathbf{r})\, \psi_{k',l',m'}(\mathbf{r}) &= \int_0^\infty r^2\, \mathrm{d}r\, R_{k,l}^*(r)R_{k',l'}(r) \\ &\times \int_0^{2\pi} \mathrm{d}\varphi \int_0^\pi \sin\theta\, \mathrm{d}\theta\, Y_l^{m*}(\theta, \varphi)Y_{l'}^{m'}(\theta, \varphi) = \delta_{kk'}\,\delta_{ll'}\,\delta_{mm'} \end{aligned} \tag{D-36}$$

Since the spherical harmonics are orthonormal [formula (D-23)], we obtain, finally:

$$\int_0^\infty r^2\, \mathrm{d}r\; R_{k,l}^*(r)\; R_{k',l}(r) = \delta_{kk'} \tag{D-37}$$

The radial functions $R_{k,l}(r)$ are therefore normalized with respect to the variable $r$; moreover, two radial functions corresponding to the same value of $l$ but to different indices $k$ are orthogonal.

COMMENTS:

(*i*) Formula (D-37) is simply a consequence of the fact that the functions $\psi_{k,l,l}(\mathbf{r}) = R_{k,l}(r)Y_l^l(\theta, \varphi)$ chosen as a basis in the subspace $\mathscr{E}(l, m = l)$ are orthonormal. It is therefore essential that the index $l$ be the same for the two functions $R_{k,l}$ appearing on the left-hand side. For $l \neq l'$, $\psi_{k,l,m}(\mathbf{r})$

and $\psi_{k',l',m'}(\mathbf{r})$ are orthogonal anyway because of their angular dependence (they are eigenfunctions of the Hermitian operator $\mathbf{L}^2$ with different eigenvalues). The integral:

$$\int_0^\infty r^2 \, \mathrm{d}r \, R_{k,l}^*(r) \, R_{k',l'}(r) \tag{D-38}$$

may therefore take on any value *a priori* if $l$ and $l'$ are different.

(*ii*) In general, the radial functions $R_{k,l}(r)$ depend on $l$, for the following reason. A function of the form $f(r)g(\theta, \varphi)$ can be continuous at the coordinate origin ($r = 0$, $\theta$ and $\varphi$ arbitrary) only if $g(\theta, \varphi)$ reduces to a constant or if $f(r)$ goes to zero at $r = 0$ [since if $g(\theta, \varphi)$ depends on $\theta$ and $\varphi$, the limit of $f(r)g(\theta, \varphi)$ when $r \longrightarrow 0$ depends on the direction along which one approaches the origin if $f(0)$ is not zero]. Consequently, if we want the basis functions $\psi_{k,l,m}(\mathbf{r})$ to be continuous, only the radial functions corresponding to $l = 0$ can be non-zero at $r = 0$ [$Y_0^0(\theta, \varphi)$ is indeed a constant]. Similarly, if we require the $\psi_{k,l,m}(\mathbf{r})$ to be differentiable (once or several times) at the origin, we obtain conditions for the $R_{k,l}(r)$ which depend on the value of $l$.

## 2. Physical considerations

### a. STUDY OF A $|k, l, m\rangle$ STATE

Consider a (spinless) particle in an eigenstate $|k, l, m\rangle$ of $\mathbf{L}^2$ and $L_z$ [whose associated wave function is $\psi_{k,l,m}(\mathbf{r})$], that is, a state in which the square of its angular momentum and the projection of this angular momentum along the $Oz$ axis have well-defined values [$l(l+1)\hbar^2$ and $m\hbar$ respectively].

Suppose that we want to measure the component along the $Ox$ or $Oy$ axis of the angular momentum of this particle. Since $L_x$ and $L_y$ do not commute with $L_z$, $|k, l, m\rangle$ is an eigenstate neither of $L_x$ nor of $L_y$; we cannot, therefore, predict with certainty the result of such a measurement. Let us calculate the mean values and root-mean-square deviations of $L_x$ and $L_y$ in the state $|k, l, m\rangle$.

These calculations can be performed very simply if we express $L_x$ and $L_y$ in terms of $L_+$ and $L_-$ by inverting formulas (C-1):

$$L_x = \frac{1}{2}(L_+ + L_-)$$

$$L_y = \frac{1}{2i}(L_+ - L_-) \tag{D-39}$$

Thus we see that $L_x|k, l, m\rangle$ and $L_y|k, l, m\rangle$ are linear combinations of $|k, l, m+1\rangle$ and $|k, l, m-1\rangle$; this leads to:

$$\langle k, l, m | L_x | k, l, m \rangle = \langle k, l, m | L_y | k, l, m \rangle = 0 \tag{D-40}$$

Furthermore:

$$\langle k, l, m \mid L_x^2 \mid k, l, m \rangle = \frac{1}{4} \langle k, l, m \mid (L_+^2 + L_-^2 + L_+L_- + L_-L_+) \mid k, l, m \rangle$$

$$\langle k, l, m \mid L_y^2 \mid k, l, m \rangle = \qquad \text{(D-41)}$$

$$- \frac{1}{4} \langle k, l, m \mid (L_+^2 + L_-^2 - L_+L_- - L_-L_+) \mid k, l, m \rangle$$

The terms in $L_+^2$ and $L_-^2$ do not contribute to the result, since $L^2{}_{\pm} \mid k, l, m \rangle$ is proportional to $\mid k, l, m \pm 2 \rangle$. In addition, formula (C-8) yields:

$$L_+L_- + L_-L_+ = 2(\mathbf{L}^2 - L_z^2) \qquad \text{(D-42)}$$

We therefore obtain :

$$\begin{aligned} \langle k, l, m \mid L_x^2 \mid k, l, m \rangle &= \langle k, l, m \mid L_y^2 \mid k, l, m \rangle \\ &= \frac{1}{2} \langle k, l, m \mid (\mathbf{L}^2 - L_z^2) \mid k, l, m \rangle \\ &= \frac{\hbar^2}{2} [l(l + 1) - m^2] \end{aligned} \qquad \text{(D-43)}$$

Thus, in the state $\mid k, l, m \rangle$:

$$\langle L_x \rangle = \langle L_y \rangle = 0 \qquad \text{(D-44-a)}$$

$$\Delta L_x = \Delta L_y = \hbar \sqrt{\frac{1}{2} [l(l + 1) - m^2]} \qquad \text{(D-44-b)}$$

These results suggest the following picture. Consider a *classical* angular momentum, whose modulus is equal to $\hbar \sqrt{l(l + 1)}$ and whose projection along $Oz$ is $m\hbar$ (fig. 3):

$$\begin{aligned} |\mathbf{OL}| &= \hbar \sqrt{l(l + 1)} \\ \overline{OH} &= m\hbar \end{aligned} \qquad \text{(D-45)}$$

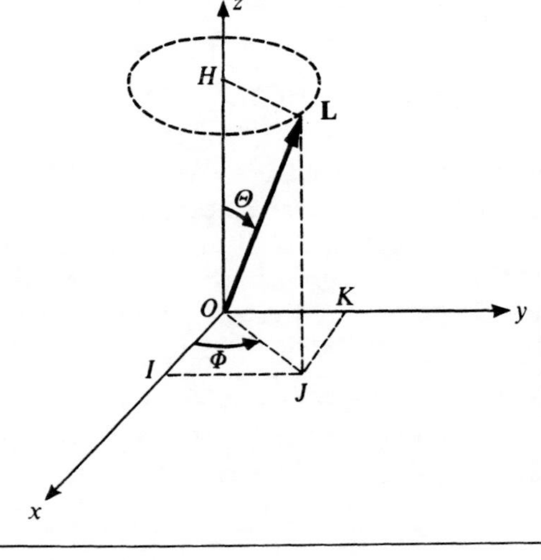


FIGURE 3

**A classical model which can be applied to the orbital angular momentum of a particle in a state $\mid l, m \rangle$. We assume that the distance $|\mathbf{OL}|$ and the angle $\Theta$ are known, but that $\Phi$ is a random variable whose probability density is constant inside the interval $[0, 2\pi]$. The classical mean values of the components of OL, as well as those of the squares of these components, are then equal to the corresponding quantum mechanical mean values.**

We denote by $\Theta$ and $\Phi$ the polar angles which characterize its direction. Since the triangle $OLJ$ has a right angle at $J$, and $OH = JL$, we have:

$$OJ = \sqrt{OL^2 - OH^2} = \hbar\sqrt{l(l+1) - m^2} \tag{D-46}$$

Consequently, the components of such a classical angular momentum would be:

$$\begin{aligned} \overline{OI} &= \hbar\sqrt{l(l+1) - m^2}\cos\Phi \\ \overline{OK} &= \hbar\sqrt{l(l+1) - m^2}\sin\Phi \\ \overline{OH} &= \hbar\sqrt{l(l+1)}\cos\Theta = m\hbar \end{aligned} \tag{D-47}$$

Now let us assume that $|\mathbf{OL}|$ and $\Theta$ are known and that $\Phi$ is a random variable which can take on any value in the interval $[0, 2\pi]$, all these values being equally probable (an evenly distributed random variable). We then have, averaging over $\Phi$:

$$\langle \overline{OI} \rangle \propto \int_0^{2\pi} \cos\Phi \, \mathrm{d}\Phi = 0 \tag{D-48-a}$$

$$\langle \overline{OK} \rangle \propto \int_0^{2\pi} \sin\Phi \, \mathrm{d}\Phi = 0 \tag{D-48-b}$$

which corresponds to (D-44-a). In addition:

$$\langle \overline{OI}^2 \rangle = \frac{1}{2\pi}\hbar^2[l(l+1) - m^2]\int_0^{2\pi} \cos^2\Phi \, \mathrm{d}\Phi = \frac{\hbar^2}{2}[l(l+1) - m^2] \tag{D-49}$$

and, similarly:

$$\langle \overline{OK}^2 \rangle = \frac{\hbar^2}{2}[l(l+1) - m^2] \tag{D-50}$$

These mean values are identical to the ones we found in (D-44). Consequently, the angular momentum of a particle in the state $| k, l, m \rangle$ behaves, insofar as the mean values of its components and their squares are concerned, like a classical angular momentum of magnitude $\hbar\sqrt{l(l+1)}$ having a projection $m\hbar$ along $Oz$, but for which $\Phi$ is a random variable evenly distributed between 0 and $2\pi$.

Of course, this picture must be used carefully: we have shown throughout this chapter how much the quantum mechanical properties of angular momenta differ from their classical properties. In particular, we must stress the fact that an individual measurement of $L_x$ or $L_y$ on a particle in the state $| k, l, m \rangle$ cannot yield an arbitrary value between $-\hbar\sqrt{l(l+1) - m^2}$ and $+\hbar\sqrt{l(l+1) - m^2}$, as the preceding model might lead us to believe. The only possible results are the eigenvalues of $L_x$ or $L_y$ (we saw at the end of §C that these are the same as those of $L_z$), that is, since $l$ is fixed here, one of the $(2l+1)$ values $l\hbar$, $(l-1)\hbar$, ..., $(-l+1)\hbar$, $-l\hbar$.

### b. CALCULATION OF THE PHYSICAL PREDICTIONS CONCERNING MEASUREMENTS OF $\mathbf{L}^2$ AND $L_z$

Consider a particle whose state is described by the (normalized) wave function:

$$\langle \mathbf{r} \mid \psi \rangle = \psi(\mathbf{r}) = \psi(r, \theta, \varphi) \tag{D-51}$$

We know that a measurement of $\mathbf{L}^2$ can yield only the results $0, 2\hbar^2, 6\hbar^2, ..., l(l+1)\hbar^2, ...$, and a measurement of $L_z$, only the results $0, \pm\hbar, \pm 2\hbar, ..., m\hbar, ...$ How can we calculate the probabilities of these different results from the wave function $\psi(r, \theta, \varphi)$?

#### α. *General formulas*

Let us denote by $\mathscr{P}_{\mathbf{L}^2, L_z}(l, m)$ the probability of finding, in a simultaneous measurement of $\mathbf{L}^2$ and $L_z$, the results $l(l+1)\hbar^2$ and $m\hbar$. This probability can be obtained by expanding $\psi(\mathbf{r})$ on a basis composed of eigenfunctions of $\mathbf{L}^2$ and $L_z$; we shall choose a "standard" basis of the type introduced in §D-1-d:

$$\psi_{k,l,m}(\mathbf{r}) = R_{k,l}(r) Y_l^m(\theta, \varphi) \tag{D-52}$$

$\psi(\mathbf{r})$ can then be written:

$$\psi(\mathbf{r}) = \sum_k \sum_l \sum_m c_{k,l,m} \, R_{k,l}(r) Y_l^m(\theta, \varphi) \tag{D-53}$$

where the coefficients $c_{k,l,m}$ can be calculated by using the usual formula:

$$\begin{aligned} c_{k,l,m} &= \int d^3r \; \psi_{k,l,m}^*(\mathbf{r}) \, \psi(\mathbf{r}) \\ &= \int_0^\infty r^2 \, \mathrm{d}r \, R_{k,l}^*(r) \int_0^{2\pi} \mathrm{d}\varphi \int_0^\pi \sin\theta \, \mathrm{d}\theta \, Y_l^{m*}(\theta, \varphi) \, \psi(r, \theta, \varphi) \end{aligned} \tag{D-54}$$

According to the postulates of chapter III, the probability $\mathscr{P}_{\mathbf{L}^2, L_z}(l, m)$ is given, under these conditions, by:

$$\mathscr{P}_{\mathbf{L}^2, L_z}(l, m) = \sum_k |c_{k,l,m}|^2 \tag{D-55}$$

If we measure only $\mathbf{L}^2$, the probability of finding the result $l(l+1)\hbar^2$ is equal to:

$$\mathscr{P}_{\mathbf{L}^2}(l) = \sum_{m=-l}^{+l} \mathscr{P}_{\mathbf{L}^2, L_z}(l, m) = \sum_k \sum_{m=-l}^{+l} |c_{k,l,m}|^2 \tag{D-56}$$

Similarly, if it is only $L_z$ that we wish to measure, the probability of obtaining $m\hbar$ is:

$$\mathscr{P}_{L_z}(m) = \sum_{l \geqslant |m|} \mathscr{P}_{\mathbf{L}^2, L_z}(l, m) = \sum_k \sum_{l \geqslant |m|} |c_{k,l,m}|^2 \tag{D-57}$$

(the restriction $l \geqslant |m|$ is automatically satisfied, since there are no coefficients $c_{k,l,m}$ for which $|m|$ would be greater than $l$).

Actually, since $\mathbf{L}^2$ and $L_z$ act only on $\theta$ and $\varphi$, we see that it is the $\theta$- and $\varphi$-dependence of the wave function $\psi(\mathbf{r})$ that count in the preceding probability calculations. To be more precise, consider $\psi(r, \theta, \varphi)$ as a function of $\theta$ and $\varphi$ depending on the parameter $r$. Like any other function of $\theta$ and $\varphi$, $\psi$ can then be expanded in terms of the spherical harmonics:

$$\psi(r, \theta, \varphi) = \sum_l \sum_m a_{l,m}(r)\, Y_l^m(\theta, \varphi) \tag{D-58}$$

The coefficients $a_{l,m}$ of this expansion depend on the "parameter" $r$ and are given by:

$$a_{l,m}(r) = \int_0^{2\pi} \mathrm{d}\varphi \int_0^{\pi} \sin\theta \,\mathrm{d}\theta\, Y_l^{m*}(\theta, \varphi)\, \psi(r, \theta, \varphi) \tag{D-59}$$

If we compare expressions (D-58) and (D-53), we see that the $c_{k,l,m}$ are the coefficients of the expansion of $a_{l,m}(r)$ on the functions $R_{k,l}(r)$:

$$a_{l,m}(r) = \sum_k c_{k,l,m}\, R_{k,l}(r) \tag{D-60}$$

with, taking (D-54) and (D-59) into account:

$$c_{k,l,m} = \int_0^{\infty} r^2 \,\mathrm{d}r\, R_{k,l}^*(r)\, a_{l,m}(r) \tag{D-61}$$

Using (D-37) and (D-60), we also obtain:

$$\int_0^{\infty} r^2 \,\mathrm{d}r\, |a_{l,m}(r)|^2 = \sum_k |c_{k,l,m}|^2 \tag{D-62}$$

The probability $\mathscr{P}_{\mathbf{L}^2,L_z}(l, m)$ [formula (D-55)] can therefore also be written in the form:

$$\mathscr{P}_{\mathbf{L}^2,L_z}(l, m) = \int_0^{\infty} r^2 \,\mathrm{d}r\, |a_{l,m}(r)|^2 \tag{D-63}$$

From this, we can deduce, as in (D-56) and (D-57):

$$\mathscr{P}_{\mathbf{L}^2}(l) = \sum_{m=-l}^{+l} \int_0^{\infty} r^2 \,\mathrm{d}r\, |a_{l,m}(r)|^2 \tag{D-64}$$

and:

$$\mathscr{P}_{L_z}(m) = \sum_{l \geqslant |m|} \int_0^{\infty} r^2 \,\mathrm{d}r\, |a_{l,m}(r)|^2 \tag{D-65}$$

[here again, $a_{l,m}(r)$ exists only for $l \geqslant |m|$]. Consequently, to obtain the physical predictions concerning measurements of $\mathbf{L}^2$ and $L_z$, we may consider the wave function as depending only on $\theta$ and $\varphi$. We then expand it in terms of the spherical harmonics as in (D-58) and apply formulas (D-63), (D-64) and (D-65).

Similarly, since $L_z$ acts only on $\varphi$, it is the $\varphi$-dependence of the wave function $\psi(\mathbf{r})$ which counts in the calculation of $\mathscr{P}_{L_z}(m)$. To see this, we shall use the fact that the spherical harmonics are products of a function of $\theta$ alone and a function of $\varphi$ alone. We shall write them in the form:

$$Y_l^m(\theta, \varphi) = Z_l^m(\theta) \frac{e^{im\varphi}}{\sqrt{2\pi}} \tag{D-66}$$

so that each of the functions of the product is normalized, since we have:

$$\int_0^{2\pi} d\varphi \frac{e^{-im\varphi}}{\sqrt{2\pi}} \frac{e^{im'\varphi}}{\sqrt{2\pi}} = \delta_{mm'} \tag{D-67}$$

Substituting this formula into the orthonormalization relation (D-23) for the spherical harmonics, we find:

$$\int_0^{\pi} \sin\theta \, d\theta \, Z_l^{m*}(\theta) \, Z_{l'}^m(\theta) = \delta_{ll'} \tag{D-68}$$

[for reasons analogous to those indicated in comment ($i$) of §D-1-d, the same value of $m$ is involved in both functions $Z_l^m$ of the left-hand side].

If we consider $\psi(r, \theta, \varphi)$ to be a function of $\varphi$ defined in the interval $[0, 2\pi]$ and depending on the "parameters" $r$ and $\theta$, we can expand it in a Fourier series:

$$\psi(r, \theta, \varphi) = \sum_m b_m(r, \theta) \frac{e^{im\varphi}}{\sqrt{2\pi}} \tag{D-69}$$

where the coefficients $b_m(r, \theta)$ can be calculated from the formula:

$$b_m(r, \theta) = \frac{1}{\sqrt{2\pi}} \int_0^{2\pi} d\varphi \, e^{-im\varphi} \, \psi(r, \theta, \varphi) \tag{D-70}$$

If we compare formulas (D-69) and (D-70) with (D-58) and (D-59), we see that the $a_{l,m}(r)$ for fixed $m$ are the coefficients of the expansion of $b_m(r, \theta)$ on the functions $Z_l^m$ corresponding to the same value of $m$:

$$b_m(r, \theta) = \sum_l a_{l,m}(r) \, Z_l^m(\theta) \tag{D-71}$$

with:

$$a_{l,m}(r) = \int_0^{\pi} \sin\theta \, d\theta \, Z_l^{m*}(\theta) \, b_m(r, \theta) \tag{D-72}$$

With (D-68) taken into account, expansion (D-71) requires that:

$$\int_0^{\pi} \sin\theta \, d\theta \, |b_m(r, \theta)|^2 = \sum_l |a_{l,m}(r)|^2 \tag{D-73}$$

Substituting this formula into (D-65), we obtain $\mathscr{P}_{L_z}(m)$ in the form:

$$\mathscr{P}_{L_z}(m) = \int_0^{\infty} r^2 \, dr \int_0^{\pi} \sin\theta \, d\theta \, |b_m(r, \theta)|^2 \tag{D-74}$$

Therefore, as far as measurements of $L_z$ alone are concerned, all we need to do is consider the wave function as depending solely on $\varphi$ and expand it in a Fourier series as in (D-69) in order to calculate the probabilities of the various possible results.

We might be tempted to think that an argument analogous to the preceding ones would give $\mathscr{P}_{\mathbf{L}^2}(l)$ in terms of the expansion of $\psi(r, \theta, \varphi)$ with respect to the variable $\theta$ alone. In fact, this is not the case: predictions concerning a measurement of $\mathbf{L}^2$ alone involve both the $\theta$- and the $\varphi$-dependence of the wave function; this is related to the fact that $\mathbf{L}^2$ acts on both $\theta$ and $\varphi$. We must therefore use formula (D-64).

β. *Special cases and examples*

Suppose that the wave function $\psi(\mathbf{r})$ which represents the state of the particle appears in the form of a product of a function of $r$ alone and a function of $\theta$ and $\varphi$:

$$\psi(r, \theta, \varphi) = f(r)\, g(\theta, \varphi) \tag{D-75}$$

We can always assume $f(r)$ and $g(\theta, \varphi)$ to be separately normalized:

$$\int_0^\infty r^2 \,\mathrm{d}r\, |f(r)|^2 = 1 \tag{D-76-a}$$

$$\int_0^{2\pi} \mathrm{d}\varphi \int_0^\pi \sin\theta \,\mathrm{d}\theta\, |g(\theta, \varphi)|^2 = 1 \tag{D-76-b}$$

To obtain the expansion (D-58) of such a wave function, all we must do is expand $g(\theta, \varphi)$ in terms of the spherical harmonics:

$$g(\theta, \varphi) = \sum_l \sum_m d_{l,m}\, Y_l^m(\theta, \varphi) \tag{D-77}$$

with:

$$d_{l,m} = \int_0^{2\pi} \mathrm{d}\varphi \int_0^\pi \sin\theta \,\mathrm{d}\theta\, Y_l^{m*}(\theta, \varphi)\, g(\theta, \varphi) \tag{D-78}$$

In this case, therefore, the coefficients $a_{l,m}(r)$ of formula (D-58) are all proportional to $f(r)$:

$$a_{l,m}(r) = d_{l,m}\, f(r) \tag{D-79}$$

With (D-76-a) taken into account, expression (D-63) for the probability $\mathscr{P}_{\mathbf{L}^2, L_z}(l, m)$ here becomes simply:

$$\mathscr{P}_{\mathbf{L}^2, L_z}(l, m) = |d_{l,m}|^2 \tag{D-80}$$

This probability is totally *independent of the radial part $f(r)$ of the wave function.*

Similarly, let us consider the case in which the wave function $\psi(r, \theta, \varphi)$ is the product of three functions of a single variable:

$$\psi(r, \theta, \varphi) = f(r)\, h(\theta)\, k(\varphi) \tag{D-81}$$

which we shall assume to be separately normalized:

$$\int_0^\infty r^2 \, dr \, |f(r)|^2 = \int_0^\pi \sin\theta \, d\theta \, |h(\theta)|^2 = \int_0^{2\pi} d\varphi \, |k(\varphi)|^2 = 1 \tag{D-82}$$

Of course, (D-81) is a special case of (D-75), and the results we have just established apply here. But, in addition, if we are interested only in a measurement of $L_z$, all we must do is expand $k(\varphi)$ in the form:

$$k(\varphi) = \sum_m e_m \frac{e^{im\varphi}}{\sqrt{2\pi}} \tag{D-83}$$

where:

$$e_m = \frac{1}{\sqrt{2\pi}} \int_0^{2\pi} d\varphi \, e^{-im\varphi} \, k(\varphi) \tag{D-84}$$

in order to obtain the equivalent of formula (D-69), with:

$$b_m(r, \theta) = e_m \, f(r) \, h(\theta) \tag{D-85}$$

According to (D-82), $\mathscr{P}_{L_z}(m)$ is then given by (D-74) as:

$$\mathscr{P}_{L_z}(m) = |e_m|^2 \tag{D-86}$$

The preceding considerations can be illustrated by some very simple examples. First, let us assume that the wave function $\psi(\mathbf{r})$ is in fact independent of $\theta$ and $\varphi$, so that:

$$\begin{cases} h(\theta) = \dfrac{1}{\sqrt{2}} \\[2mm] k(\varphi) = \dfrac{1}{\sqrt{2\pi}} \end{cases} \tag{D-87}$$

We then have:

$$g(\theta, \varphi) = \frac{1}{\sqrt{4\pi}} = Y_0^0(\theta, \varphi) \tag{D-88}$$

Thus a measurement of $\mathbf{L}^2$ or of $L_z$ must yield zero.

Now, let us modify only the $\theta$-dependence, choosing:

$$\begin{cases} h(\theta) = \sqrt{\dfrac{3}{2}} \cos\theta \\[2mm] k(\varphi) = \dfrac{1}{\sqrt{2\pi}} \end{cases} \tag{D-89}$$

In this case:

$$g(\theta, \varphi) = \sqrt{\frac{3}{4\pi}} \cos\theta = Y_1^0(\theta, \varphi) \tag{D-90}$$

We are again sure of the results of measuring $\mathbf{L}^2$ or $L_z$. For $\mathbf{L}^2$, we can only obtain $2\hbar^2$; for $L_z$, 0. It can be verified that this modification of the $\theta$-dependence has not changed the physical predictions concerning the measurement of $L_z$.

On the other hand, if we modify the $\varphi$-dependence by setting, for example:

$$\begin{cases} h(\theta) = \dfrac{1}{\sqrt{2}} \\ g(\varphi) = \dfrac{e^{i\varphi}}{\sqrt{2\pi}} \end{cases} \tag{D-91}$$

$g(\theta, \varphi)$ is no longer equal to a single spherical harmonic. According to (D-86), all the probabilities $\mathscr{P}_{L_z}(m)$ are zero except for:

$$\mathscr{P}_{L_z}(m = 1) = 1 \tag{D-92}$$

But the predictions concerning a measurement of $\mathbf{L}^2$ are also changed with respect to the case (D-87). In order to calculate these predictions, we must expand the function:

$$g(\theta, \varphi) = \frac{1}{\sqrt{4\pi}} e^{i\varphi} \tag{D-93}$$

on the spherical harmonics. It can be verified that all the $Y_l^m(\theta, \varphi)$, with odd $l$ and $m = 1$, actually appear in the expansion of the function (D-93). Consequently, we are no longer sure of the result of a measurement of $\mathbf{L}^2$ (the probabilities of the various possible results can be calculated from the expression for the spherical harmonics). We therefore conclude from this example that, as pointed out at the end of §α, the $\varphi$-dependence of the wave function also enters into the calculation of predictions concerning measurements of $\mathbf{L}^2$.

**References and suggestions for further reading:**

Dirac (1.13), §§35 and 36; Messiah (1.17), chap. XIII; Rose (2.19); Edmonds (2.21).

COMPLEMENTS OF CHAPTER VI

| | |
|---|---|
| $A_{VI}$: SPHERICAL HARMONICS | $A_{VI}$: detailed study of the spherical harmonics $Y_l^m(\theta, \varphi)$; establishes certain properties used in chapter VI, as well as in certain subsequent complements. |
| $B_{VI}$: ANGULAR MOMENTUM AND ROTATIONS | $B_{VI}$: brings out the close relation which exists between the angular momentum **J** of a quantum mechanical system and the spatial rotations that can be performed on it. Shows that the commutation relations between the components of **J** express purely geometrical properties of these rotations; introduces the concept of a scalar or vector observable, which will reappear in other complements (especially $D_X$). Important theoretically; however, sometimes difficult; can be reserved for later study. |
| $C_{VI}$: ROTATION OF DIATOMIC MOLECULES | $C_{VI}$: a simple and direct application of quantum mechanical properties of angular momentum: pure rotational spectra of heteropolar diatomic molecules and Raman rotational spectra. Elementary level. Because of the importance of the phenomena studied in physics and chemistry, can be recommended for a first reading. |
| $D_{VI}$: ANGULAR MOMENTUM OF STATIONARY STATES OF A TWO-DIMENSIONAL HARMONIC OSCILLATOR | $D_{VI}$: can be considered as a worked example. Studies the stationary states of the two-dimensional harmonic oscillator; in order to class these states by angular momentum, introduces the concept of " circular quanta ". Not theoretically difficult. Some results will be used in complement $E_{VI}$. |
| $E_{VI}$: A CHARGED PARTICLE IN A MAGNETIC FIELD; LANDAU LEVELS | $E_{VI}$: general study of the quantum mechanical properties of a charged particle in a magnetic field, followed by a study of the important special case in which the magnetic field is uniform (Landau levels). Not theoretically difficult. Recommended for a first reading, which can, however, be confined to §§ 1-a and b, 2-a and b, 3-a. |
| $F_{VI}$: EXERCISES | |

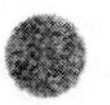



# Complement $A_{VI}$

## SPHERICAL HARMONICS



This complement is devoted to the study of the form and principal properties of spherical harmonics. It includes the proofs of certain results which were stated without proof in § D-1-c of chapter VI.

### 1. Calculation of spherical harmonics

In order to calculate the various spherical harmonics $Y_l^m(\theta, \varphi)$, we shall use the method indicated in chapter VI (§D-1-c): starting with the expression for $Y_l^l(\theta, \varphi)$, we shall use the operator $L_-$ to obtain by iteration the spherical harmonics corresponding to the same value of $l$ and the $(2l + 1)$ values of $m$ associated with it. Recall that the operators $L_+$ and $L_-$ act only on the angular dependence of a wave function and can be written:

$$L_\pm = \hbar\, e^{\pm i\varphi}\left[\pm \frac{\partial}{\partial\theta} + i \cot\theta \frac{\partial}{\partial\varphi}\right] \tag{1}$$

#### a. DETERMINATION OF $Y_l^l(\theta, \varphi)$

We have seen (§D-1-b of chapter VI) that $Y_l^l(\theta, \varphi)$ can be calculated from the equation:

$$L_+ Y_l^l(\theta, \varphi) = 0 \tag{2}$$

and from the fact that:

$$Y_l^l(\theta, \varphi) = F_l^l(\theta)\, e^{il\varphi} \tag{3}$$

Thus we obtained:

$$Y_l^l(\theta, \varphi) = c_l (\sin\theta)^l\, e^{il\varphi} \tag{4}$$

where $c_l$ is an arbitrary constant.

First, let us determine the absolute value of $c_l$ by requiring $Y_l^l(\theta, \varphi)$ to be normalized with respect to the angular variables $\theta$ and $\varphi$:

$$\int_0^{2\pi} d\varphi \int_0^{\pi} \sin\theta \, d\theta \, |Y_l^l(\theta, \varphi)|^2 = |c_l|^2 \int_0^{2\pi} d\varphi \int_0^{\pi} \sin\theta \, d\theta \, (\sin\theta)^{2l} = 1 \tag{5}$$

We obtain:

$$|c_l|^2 = 1/2\pi I_l \tag{6}$$

where $I_l$ is given by:

$$I_l = \int_0^{\pi} \sin\theta \, d\theta \, (\sin\theta)^{2l} = \int_{-1}^{+1} du \, (1 - u^2)^l \tag{7}$$

(setting $u = \cos\theta$). $I_l$ can easily be calculated by recurrence since:

$$I_l = \int_{-1}^{+1} du \, (1 - u^2)(1 - u^2)^{l-1} = I_{l-1} - \int_{-1}^{+1} du \, u^2(1 - u^2)^{l-1} \tag{8}$$

An integration by parts of the last integral yields:

$$I_l = I_{l-1} - \frac{1}{2l} I_l \tag{9}$$

We therefore have:

$$I_l = \frac{2l}{2l + 1} I_{l-1} \tag{10}$$

with:

$$I_0 = \int_{-1}^{+1} du = 2 \tag{11}$$

From this, we can immediately derive the value of $I_l$:

$$I_l = \frac{(2l)!!}{(2l + 1)!!} I_0 = \frac{2^{2l+1}(l!)^2}{(2l + 1)!} \tag{12}$$

$Y_l^l(\theta, \varphi)$ is then normalized if:

$$|c_l| = \frac{1}{2^l l!} \sqrt{\frac{(2l + 1)!}{4\pi}} \tag{13}$$

In order to define $c_l$ completely, we must choose its phase. It is customary to choose:

$$c_l = \frac{(-1)^l}{2^l \, l!} \sqrt{\frac{(2l + 1)!}{4\pi}} \tag{14}$$

We shall see later that, with this convention, $Y_l^0(\theta, \varphi)$ (which is independent of $\varphi$) has a real positive value for $\theta = 0$.

### b. GENERAL EXPRESSION FOR $Y_l^m(\theta, \varphi)$

We shall obtain the other spherical harmonics $Y_l^m(\theta, \varphi)$ by successive application of the operator $L_-$ to the $Y_l^l(\theta, \varphi)$ which we have just determined. First, we shall prove a convenient formula which will enable us to simplify the calculations.

#### α. *The action of $(L_\pm)^p$ on a function of the form $e^{in\varphi}F(\theta)$*

The action of the operators $L_+$ and $L_-$ on a function of the form $e^{in\varphi}F(\theta)$ (where $n$ is any integer) is given by:

$$L_\pm[e^{in\varphi}F(\theta)] = \mp \hbar\, e^{i(n\pm 1)\varphi} (\sin\theta)^{1\pm n} \frac{d}{d(\cos\theta)} [(\sin\theta)^{\mp n}F(\theta)] \tag{15}$$

More generally:

$$(L_\pm)^p[e^{in\varphi}F(\theta)] = (\mp \hbar)^p\, e^{i(n\pm p)\varphi} (\sin\theta)^{p\pm n} \frac{d^p}{d(\cos\theta)^p} [(\sin\theta)^{\mp n}F(\theta)] \tag{16}$$

First, let us prove formula (15). We know that:

$$\frac{d}{d(\cos\theta)} = \frac{d\theta}{d(\cos\theta)}\frac{d}{d\theta} = -\frac{1}{\sin\theta}\frac{d}{d\theta} \tag{17}$$

and therefore:

$$\begin{aligned}(\sin\theta)^{1\pm n}\frac{d}{d(\cos\theta)}[(\sin\theta)^{\mp n}F(\theta)] &= \\ &= (\sin\theta)^{1\pm n}\left(-\frac{1}{\sin\theta}\right)\left[\mp n(\sin\theta)^{\mp n-1}\cos\theta\, F(\theta) + (\sin\theta)^{\mp n}\frac{dF(\theta)}{d\theta}\right] \\ &= -\left[\mp n\cot\theta\, F(\theta) + \frac{dF(\theta)}{d\theta}\right]\end{aligned} \tag{18}$$

Consequently:

$$\begin{aligned}\mp e^{i(n\pm 1)\varphi}(\sin\theta)^{1\pm n}\frac{d}{d(\cos\theta)}[(\sin\theta)^{\mp n}F(\theta)] &= \left[-n\cot\theta \pm \frac{\partial}{\partial\theta}\right] e^{i(n\pm 1)\varphi}F(\theta) \\ &= e^{\pm i\varphi}\left[\pm\frac{\partial}{\partial\theta} + i\cot\theta\frac{\partial}{\partial\varphi}\right] e^{in\varphi}F(\theta)\end{aligned} \tag{19}$$

We recognize expression (1) for the operators $L_+$ and $L_-$; relation (19) is therefore identical to (15).

Now, to establish formula (16), we can reason by recurrence, since, for $p = 1$, (16) reduces to (15), which we have just proved. Let us therefore assume that relation (16) is true for $(p - 1)$:

$$(L_\pm)^{p-1}[e^{in\varphi}F(\theta)] = (\mp\hbar)^{p-1} e^{i(n\pm p\mp 1)\varphi}(\sin\theta)^{p-1\pm n} \times \frac{d^{p-1}}{d(\cos\theta)^{p-1}}[(\sin\theta)^{\mp n}F(\theta)] \tag{20}$$

and let us show that it is then also valid for $p$. To do so, we apply $L_\pm$ to both sides of (20); for the right-hand side, we can use formula (15), making the substitutions:

$$\begin{aligned} n &\Longrightarrow n \pm p \mp 1 \\ F(\theta) &\Longrightarrow (\sin\theta)^{p-1\pm n}\frac{d^{p-1}}{d(\cos\theta)^{p-1}}[(\sin\theta)^{\mp n}F(\theta)]\end{aligned} \tag{21}$$

We then obtain:

$$
\begin{aligned}
(L_\pm)^p[e^{in\varphi}F(\theta)] &= (\mp \hbar)^p\, e^{i(n\pm p)\varphi}(\sin\theta)^{\pm n+p} \times \\
&\times \frac{d}{d(\cos\theta)}\left\{(\sin\theta)^{\mp n-p+1}(\sin\theta)^{p-1\pm n}\frac{d^{p-1}}{d(\cos\theta)^{p-1}}[(\sin\theta)^{\mp n}F(\theta)]\right\} \\
&= (\mp \hbar)^p\, e^{i(n\pm p)\varphi}(\sin\theta)^{p\pm n}\frac{d^p}{d(\cos\theta)^p}[(\sin\theta)^{\mp n}F(\theta)]
\end{aligned} \tag{22}
$$

Formula (16) is therefore proven by recurrence.

β. *Calculation of $Y_l^m(\theta, \varphi)$ from $Y_l^l(\theta, \varphi)$*

As we have already indicated (chapter VI, §D-1-c-α), the spherical harmonics $Y_l^m(\theta, \varphi)$ must satisfy:

$$
\begin{aligned}
L_\pm Y_l^m(\theta, \varphi) &= \hbar\sqrt{l(l+1) - m(m\pm 1)}\, Y_l^{m\pm 1}(\theta, \varphi) \\
&= \hbar\sqrt{(l\mp m)(l\pm m+1)}\, Y_l^{m\pm 1}(\theta, \varphi)
\end{aligned} \tag{23}
$$

These relations automatically insure that $Y_l^{m\pm 1}$ is normalized if $Y_l^m$ is. Also, they fix the relative phases of spherical harmonics corresponding to the same value of $l$ and different values of $m$.

In particular, we can calculate $Y_l^m(\theta, \varphi)$ from $Y_l^l(\theta, \varphi)$ by using the operator $L_-$ given by (1) and formula (23). Thus, we shall obtain directly a normalized function $Y_l^m(\theta, \varphi)$ whose phase will be determined by the convention used for $Y_l^l(\theta, \varphi)$ [formula (14)]. To go from $Y_l^l(\theta, \varphi)$ to $Y_l^m(\theta, \varphi)$, we must apply $(l-m)$ times the operator $L_-$; according to (23), we thus obtain:

$$
\begin{aligned}
&(L_-)^{l-m} Y_l^l(\theta, \varphi) \\
&= (\hbar)^{l-m}\sqrt{(2l)(1)\times(2l-1)(2)\times \dots \times (l+m+1)(l-m)}\, Y_l^m(\theta, \varphi)
\end{aligned} \tag{24}
$$

that is:

$$
Y_l^m(\theta, \varphi) = \sqrt{\frac{(l+m)!}{(2l)!(l-m)!}}\left(\frac{L_-}{\hbar}\right)^{l-m} Y_l^l(\theta, \varphi) \tag{25}
$$

Finally, using expression (4) for $Y_l^l(\theta, \varphi)$ [where the coefficient $c_l$ is given by (14)] and formula (16) (with $n = l$ and $p = l - m$), we can write (25) explicitly in the form:

$$
Y_l^m(\theta, \varphi) = \frac{(-1)^l}{2^l\, l!}\sqrt{\frac{2l+1}{4\pi}\frac{(l+m)!}{(l-m)!}}\, e^{im\varphi}(\sin\theta)^{-m}\frac{d^{l-m}}{d(\cos\theta)^{l-m}}(\sin\theta)^{2l} \tag{26}
$$

γ. *Calculation of $Y_l^m(\theta, \varphi)$ from $Y_l^{-l}(\theta, \varphi)$*

In order to obtain expression (26), we started with the result of § 1-a. It is, of course, just as easy to calculate $Y_l^{-l}(\theta, \varphi)$ first and then use the operator $L_+$. The expression thus obtained for $Y_l^m$ is different from (26), although the two are completely equivalent.

Let us therefore calculate $Y_l^{-l}(\theta, \varphi)$ from (26)★. Since:

$$(\sin \theta)^{2l} = (1 - \cos^2 \theta)^l \tag{27}$$

is a polynomial of degree $2l$ in $\cos \theta$, only its highest-order term contributes to $Y_l^{-l}(\theta, \varphi)$:

$$\frac{d^{2l}}{d(\cos \theta)^{2l}} (\sin \theta)^{2l} = (-1)^l (2l)! \tag{28}$$

We therefore immediately find that:

$$Y_l^{-l}(\theta, \varphi) = \frac{1}{2^l \, l!} \sqrt{\frac{(2l+1)!}{4\pi}} \, e^{-il\varphi} (\sin \theta)^l \tag{29}$$

$Y_l^m(\theta, \varphi)$ can then be obtained by applying $(l + m)$ times the operator $L_+$ to $Y_l^{-l}(\theta, \varphi)$. Using (23) and (16), we finally arrive at:

$$Y_l^m(\theta, \varphi) = \frac{(-1)^{l+m}}{2^l \, l!} \sqrt{\frac{2l+1}{4\pi} \frac{(l-m)!}{(l+m)!}} \, e^{im\varphi} (\sin \theta)^m \frac{d^{l+m}}{d(\cos \theta)^{l+m}} (\sin \theta)^{2l} \tag{30}$$

### c. EXPLICIT EXPRESSIONS FOR $l = 0$, 1 AND 2

General formulas (26) and (30) yield the spherical harmonics for the first values of $l$:

$$Y_0^0 = \frac{1}{\sqrt{4\pi}} \tag{31}$$

$$\begin{cases} Y_1^{\pm 1}(\theta, \varphi) = \mp \sqrt{\dfrac{3}{8\pi}} \sin \theta \, e^{\pm i\varphi} \\ Y_1^0(\theta, \varphi) = \sqrt{\dfrac{3}{4\pi}} \cos \theta \end{cases} \tag{32}$$

$$\begin{cases} Y_2^{\pm 2}(\theta, \varphi) = \sqrt{\dfrac{15}{32\pi}} \sin^2 \theta \, e^{\pm 2i\varphi} \\ Y_2^{\pm 1}(\theta, \varphi) = \mp \sqrt{\dfrac{15}{8\pi}} \sin \theta \cos \theta \, e^{\pm i\varphi} \\ Y_2^0(\theta, \varphi) = \sqrt{\dfrac{5}{16\pi}} (3 \cos^2 \theta - 1) \end{cases} \tag{33}$$

★ We could obviously calculate $Y_l^{-l}$ from the equation:

$$L_- Y_l^{-l}(\theta, \varphi) = 0$$

However, its phase would then remain arbitrary. By using (26), we shall determine $Y_l^{-l}(\theta, \varphi)$ completely, and its phase will be a consequence of the convention chosen in § 1-a.

## 2. Properties of spherical harmonics

### a. RECURRENCE RELATIONS

By their very construction, the spherical harmonics satisfy relations (23); that is, using (1):

$$e^{\pm i\varphi}\left[\pm\frac{\partial}{\partial\theta} - m\cot\theta\right] Y_l^m(\theta,\varphi) = \sqrt{l(l+1) - m(m\pm 1)}\, Y_l^{m\pm 1}(\theta,\varphi) \tag{34}$$

Also note the following formula, which is often useful:

$$\cos\theta\, Y_l^m(\theta,\varphi) = \sqrt{\frac{(l+m+1)(l-m+1)}{(2l+1)(2l+3)}}\, Y_{l+1}^m(\theta,\varphi) + \sqrt{\frac{(l+m)(l-m)}{(2l+1)(2l-1)}}\, Y_{l-1}^m(\theta,\varphi) \tag{35}$$

Here is an outline of its proof. According to (25):

$$\cos\theta\, Y_l^m = \sqrt{\frac{(l+m)!}{(2l)!\,(l-m)!}}\cos\theta\left(\frac{L_-}{\hbar}\right)^{l-m} Y_l^l(\theta,\varphi) \tag{36}$$

Now, using expression (1) for $L_-$, it is easy to verify that:

$$[L_-, \cos\theta] = \hbar\, e^{-i\varphi}\sin\theta \tag{37}$$

and:

$$[L_-, e^{-i\varphi}\sin\theta] = 0 \tag{38}$$

Using a recurrence argument, we can then calculate the commutator of $\left(\frac{L_-}{\hbar}\right)^k$ and $\cos\theta$, since, if we assume that:

$$\left[\left(\frac{L_-}{\hbar}\right)^{k-1}, \cos\theta\right] = (k-1)\, e^{-i\varphi}\sin\theta\left(\frac{L_-}{\hbar}\right)^{k-2} \tag{39}$$

we obtain:

$$\left[\left(\frac{L_-}{\hbar}\right)^{k}, \cos\theta\right] = \left(\frac{L_-}{\hbar}\right)^{k-1}\left[\frac{L_-}{\hbar}, \cos\theta\right] + \left[\left(\frac{L_-}{\hbar}\right)^{k-1}, \cos\theta\right]\frac{L_-}{\hbar}$$

$$= \left(\frac{L_-}{\hbar}\right)^{k-1} e^{-i\varphi}\sin\theta + (k-1)\, e^{-i\varphi}\sin\theta\left(\frac{L_-}{\hbar}\right)^{k-1} \tag{40}$$

that is:

$$\left[\left(\frac{L_-}{\hbar}\right)^{k}, \cos\theta\right] = k\, e^{-i\varphi}\sin\theta\left(\frac{L_-}{\hbar}\right)^{k-1} = k\left(\frac{L_-}{\hbar}\right)^{k-1} e^{-i\varphi}\sin\theta \tag{41}$$

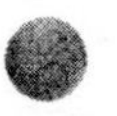



This relation has therefore been established by recurrence. We can use it to write (36) in the form:

$$\cos\theta\, Y_l^m = \sqrt{\frac{(l+m)!}{(2l)!\,(l-m)!}}\left[\left(\frac{L_-}{\hbar}\right)^{l-m}\cos\theta\, Y_l^l - (l-m)\left(\frac{L_-}{\hbar}\right)^{l-m-1} e^{-i\varphi}\sin\theta\, Y_l^l\right] \tag{42}$$

Using (4) and (14), we can easily show that:

$$e^{-i\varphi}\sin\theta\, Y_l^l = -\sqrt{\frac{2l+1}{2l}}\,(1-\cos^2\theta)\, Y_{l-1}^{l-1} \tag{43}$$

If we then calculate the explicit expressions for $Y_{l+1}^l$ and $Y_{l+1}^{l-1}$ from the general expression (26), we find that:

$$\cos\theta\, Y_l^l = \frac{1}{\sqrt{2l+3}}\, Y_{l+1}^l \tag{44-a}$$

$$\cos^2\theta\, Y_{l-1}^{l-1} = \frac{2}{2l+1}\sqrt{\frac{l}{2l+3}}\, Y_{l+1}^{l-1} + \frac{1}{2l+1}\, Y_{l-1}^{l-1} \tag{44-b}$$

Substituting relations (43) and (44) into (42) and using (23), we obtain (35).

### b. ORTHONORMALIZATION AND CLOSURE RELATIONS

Because of the way we constructed them, the spherical harmonics constitute a set of normalized functions; they are also orthogonal, since they are eigenfunctions of the Hermitian operators $\mathbf{L}^2$ and $L_z$ with different eigenvalues. The corresponding orthonormalization relation is:

$$\int_0^{2\pi} d\varphi \int_0^{\pi} \sin\theta\, d\theta\, Y_l^{m*}(\theta,\varphi)\, Y_{l'}^{m'}(\theta,\varphi) = \delta_{ll'}\,\delta_{mm'} \tag{45}$$

It can be shown (here, we shall assume) that any square-integrable function of $\theta$ and $\varphi$ can be expanded in one and only one way on the spherical harmonics:

$$f(\theta,\varphi) = \sum_l \sum_m c_{l,m} Y_l^m(\theta,\varphi) \tag{46}$$

with:

$$c_{l,m} = \int_0^{2\pi} d\varphi \int_0^{\pi} \sin\theta\, d\theta\, Y_l^{m*}(\theta,\varphi)\, f(\theta,\varphi) \tag{47}$$

The set of spherical harmonics therefore constitutes an orthonormal basis of the space of square-integrable functions of $\theta$ and $\varphi$. This can be expressed by the closure relation:

$$\sum_l \sum_m Y_l^m(\theta,\varphi)\, Y_l^{m*}(\theta',\varphi') = \delta(\cos\theta - \cos\theta')\,\delta(\varphi-\varphi') \tag{48}$$

### c. PARITY

The parity operation on a function defined in ordinary space (*cf.* complement $F_{II}$) consists of replacing in this function the coordinates of any point in space by those of the point symmetrical to it with respect to the origin of the reference frame:

$$\mathbf{r} \Longrightarrow -\mathbf{r} \tag{49}$$

In spherical coordinates, this operation is expressed by the substitutions (fig. 2 of chapter VI):

$$\begin{aligned} r &\Longrightarrow r \\ \theta &\Longrightarrow \pi - \theta \\ \varphi &\Longrightarrow \pi + \varphi \end{aligned} \tag{50}$$

Consequently, if we are using a standard basis for the wave function space of a spinless particle (§ D-1-d of chapter VI), the radial part of the basis functions $\psi_{k,l,m}(\mathbf{r})$ is unchanged by the parity operation. The only transformation is that of the spherical harmonics, which we shall now describe.

First, note that in the substitution of (50):

$$\begin{aligned} \sin\theta &\Longrightarrow \sin\theta \\ \cos\theta &\Longrightarrow -\cos\theta \\ e^{im\varphi} &\Longrightarrow (-1)^m e^{im\varphi} \end{aligned} \tag{51}$$

Under these conditions, the function $Y_l^l(\theta, \varphi)$ which we calculated in § 1-a is transformed into:

$$Y_l^l(\pi - \theta, \pi + \varphi) = (-1)^l Y_l^l(\theta, \varphi) \tag{52}$$

Moreover:

$$\begin{aligned} \frac{\partial}{\partial\theta} &\Longrightarrow -\frac{\partial}{\partial\theta} \\ \frac{\partial}{\partial\varphi} &\Longrightarrow \frac{\partial}{\partial\varphi} \end{aligned} \tag{53}$$

Relations (51) and (53) show that the operators $L_+$ and $L_-$ [formulas (1)] remain unchanged [which means that $L_+$ and $L_-$ are even operators, in the sense defined in complement $F_{II}$ (§ 2-a)]. Consequently, according to result (52) and formula (25), which enables us to calculate $Y_l^m(\theta, \varphi)$:

$$Y_l^m(\pi - \theta, \pi + \varphi) = (-1)^l Y_l^m(\theta, \varphi) \tag{54}$$

The spherical harmonics are therefore functions whose parity is well-defined and independent of $m$: they are even for $l$ even and odd for $l$ odd.

### d. COMPLEX CONJUGATION

Because of their $\varphi$-dependence, the spherical harmonics are complex-valued functions. It can be seen directly, by comparing (26) and (30), that:

$$[Y_l^m(\theta, \varphi)]^* = (-1)^m Y_l^{-m}(\theta, \varphi) \tag{55}$$

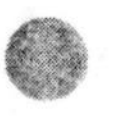



## e. RELATION BETWEEN THE SPHERICAL HARMONICS AND THE LEGENDRE POLYNOMIALS AND ASSOCIATED LEGENDRE FUNCTIONS

The $\theta$-dependence of the spherical harmonics resides in functions known as Legendre polynomials and associated Legendre functions. We shall neither prove nor even enumerate all the properties of these functions, but shall simply indicate their relation with the spherical harmonics.

### α. *$Y_l^0(\theta)$ is proportional to a Legendre polynomial*

For $m = 0$, formulas (26) and (30) yield:

$$Y_l^0(\theta) = \frac{(-1)^l}{2^l\, l!}\sqrt{\frac{2l+1}{4\pi}}\frac{\mathrm{d}^l}{\mathrm{d}(\cos\theta)^l}(\sin\theta)^{2l} \tag{56}$$

which can be written in the form:

$$Y_l^0(\theta) = \sqrt{\frac{2l+1}{4\pi}}\, P_l(\cos\theta) \tag{57}$$

setting:

$$P_l(u) = \frac{(-1)^l}{2^l\, l!}\frac{\mathrm{d}^l}{\mathrm{d}u^l}(1-u^2)^l \tag{58}$$

According to its definition (58), $P_l(u)$ is an $l$th order polynomial in $u$ of parity $(-1)^l$*:

$$P_l(-u) = (-1)^l P_l(u) \tag{59}$$

$P_l(u)$ is the $l$th order Legendre polynomial. It is easy to show that it has $l$ zeros in the interval $[-1, +1]$, and that the numerical coefficient in (58) insures that:

$$P_l(1) = 1 \tag{60}$$

It can also be proven that the Legendre polynomials form a set of orthogonal functions:

$$\int_{-1}^{+1} \mathrm{d}u\, P_l(u)\, P_{l'}(u) = \int_0^\pi \sin\theta\, \mathrm{d}\theta\, P_l(\cos\theta)\, P_{l'}(\cos\theta) = \frac{2}{2l+1}\delta_{ll'} \tag{61}$$

on which can be expanded functions of $\theta$ alone:

$$f(\theta) = \sum_l c_l\, P_l(\cos\theta) \tag{62}$$

* Parity with respect to the variable $u$. Note, however, that the parity operation in space [formulas (50)] amounts to changing $\cos\theta$ to $-\cos\theta$; property (59) can be expressed by:

$$Y_l^0(\pi - \theta) = (-1)^l Y_l^0(\theta)$$

which is a special case of (54).

with:

$$c_l = \frac{2l + 1}{2} \int_0^\pi \sin\theta \, d\theta \, P_l(\cos\theta) \, f(\theta) \tag{63}$$

COMMENT:

According to (57) and (60):

$$Y_l^0(0) = \sqrt{\frac{2l + 1}{4\pi}} \tag{64}$$

As we pointed out in §1-a, the phase convention chosen for $Y_l^l(\theta, \varphi)$ gives a real positive value to $Y_l^0(0)$.

β. *$Y_l^m(\theta, \varphi)$ is proportional to an associated Legendre function*

For $m$ positive, $Y_l^m(\theta, \varphi)$ can be obtained by applying $L_+$ to $Y_l^0(\theta)$; using (23):

$$Y_l^m(\theta, \varphi) = \sqrt{\frac{(l - m)!}{(l + m)!}} \left(\frac{L_+}{\hbar}\right)^m Y_l^0(\theta) \qquad (m \geqslant 0) \tag{65}$$

Using formulas (1) and (16), we then find:

$$Y_l^m(\theta, \varphi) = (-1)^m \sqrt{\frac{2l + 1}{4\pi} \frac{(l - m)!}{(l + m)!}} \, P_l^m(\cos\theta) \, e^{im\varphi} \qquad (m \geqslant 0) \tag{66}$$

where $P_l^m$ is an associated Legendre function, defined by:

$$P_l^m(u) = \sqrt{(1 - u^2)^m} \, \frac{d^m}{du^m} P_l(u) \qquad (-1 \leqslant u \leqslant +1) \tag{67}$$

$P_l^m(u)$ is the product of $\sqrt{(1 - u^2)^m}$ and a polynomial of degree $(l - m)$ and parity $(-1)^{l-m}$; $P_l^0(u)$ is the Legendre polynomial $P_l(u)$. The set of $P_l^m$ for fixed $m$ constitutes an orthogonal system of functions:

$$\int_{-1}^{+1} du \, P_l^m(u) \, P_{l'}^m(u) = \int_0^\pi \sin\theta \, d\theta \, P_l^m(\cos\theta) \, P_{l'}^m(\cos\theta) = \frac{2}{2l + 1} \frac{(l + m)!}{(l - m)!} \delta_{ll'} \tag{68}$$

on which can be expanded functions of $\theta$ alone.

Formula (66) is valid for $m$ positive (or zero); for negative $m$, it suffices to use relation (55) to obtain:

$$Y_l^m(\theta, \varphi) = \sqrt{\frac{2l + 1}{4\pi} \frac{(l + m)!}{(l - m)!}} \, P_l^{-m}(\cos\theta) \, e^{im\varphi} \qquad (m < 0) \tag{69}$$

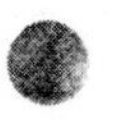



γ. *Spherical harmonic addition theorem*

Consider two arbitrary directions in space, $Ou'$ and $Ou''$, defined respectively by the polar angles $(\theta', \varphi')$ and $(\theta'', \varphi'')$, and call the angle between them $\alpha$. The following relation can be proven:

$$\frac{2l+1}{4\pi} P_l(\cos\alpha) = \sum_{m=-l}^{+l} (-1)^m \, Y_l^m(\theta', \varphi') \, Y_l^{-m}(\theta'', \varphi'') \tag{70}$$

(where $P_l$ is the $l$th-order Legendre polynomial). It is known as the "spherical harmonics addition theorem".

We shall indicate the main steps of an elementary proof of relation (70). First of all, note that, if $\cos\alpha$ is expressed in terms of the polar angles $(\theta', \varphi')$ and $(\theta'', \varphi'')$, the left-hand side of (70) can be considered to be a function of $\theta'$ and $\varphi'$; it can therefore be expanded on the spherical harmonics $Y_{l'}^{m'}(\theta', \varphi')$. The coefficients of this expansion, which are, of course, functions of the other two variables, $\theta''$ and $\varphi''$, can also be expanded on the spherical harmonics $Y_{l''}^{m''}(\theta'', \varphi'')$. We must therefore have:

$$\frac{2l+1}{4\pi} P_l(\cos\alpha) = \sum_{l',m'} \sum_{l'',m''} c_{l',m';l'',m''} Y_{l'}^{m'}(\theta', \varphi') \, Y_{l''}^{m''}(\theta'', \varphi'') \tag{71}$$

where the problem is to calculate the coefficients $c_{l',m';l'',m''}$. They can be obtained by the following process:

(*i*) in the first place, these coefficients are different from zero only for:

$$l' = l'' = l \tag{72}$$

To show this, first fix the direction $Ou''$; $P_l(\cos\alpha)$ then depends only on $\theta'$ and $\varphi'$. If the $Oz$ axis is chosen along $Ou''$, $\cos\alpha = \cos\theta'$ and $P_l(\cos\alpha)$ is proportional to $Y_l^0(\theta')$ [*cf.* relation (57)]. To generalize to the case in which the direction of $Ou''$ is arbitrary, we perform a rotation which takes $Oz$ onto this direction: $\cos\alpha$ remains unchanged, as does $P_l(\cos\alpha)$. Since the rotation operators (complement $B_{VI}$, §3-c-γ) commute with $\mathbf{L}^2$, the transform of $Y_l^0(\theta')$ remains an eigenfunction of $\mathbf{L}^2$ with the eigenvalue $l(l+1)\hbar^2$, that is, a linear combination of the spherical harmonics $Y_l^{m'}(\theta', \varphi')$; we therefore have $l' = l$. Similarly, it can be established that $l'' = l$.

(*ii*) under a rotation of both directions $Ou'$ and $Ou''$ through an angle $\beta$ about $Oz$, the angle $\alpha$ is not changed, and neither are $\theta'$ and $\theta''$, while $\varphi'$ and $\varphi''$ become $\varphi' + \beta$ and $\varphi'' + \beta$. The left-hand side of (71) therefore does not change in value, and each term of the right-hand side is multiplied by $e^{i(m'+m'')\beta}$. Consequently, the only non-zero coefficients in the sum of the right-hand side are those which satisfy:

$$m' + m'' = 0 \tag{73}$$

(*iii*) combining results (72) and (73), we see that formula (71) can be written in the form:

$$\frac{2l+1}{4\pi} P_l(\cos\alpha) = \sum_{m=-l}^{+l} (-1)^m \, c_m \, Y_l^m(\theta', \varphi') \, Y_l^{-m}(\theta'', \varphi'') \tag{74}$$

If we set $\theta' = \theta''$ and $\varphi' = \varphi''$, we obtain, according to (60):

$$\frac{2l+1}{4\pi} = \sum_{m=-l}^{+l} (-1)^m \, c_m \, Y_l^m(\theta', \varphi') \, Y_l^{-m}(\theta', \varphi') \tag{75}$$

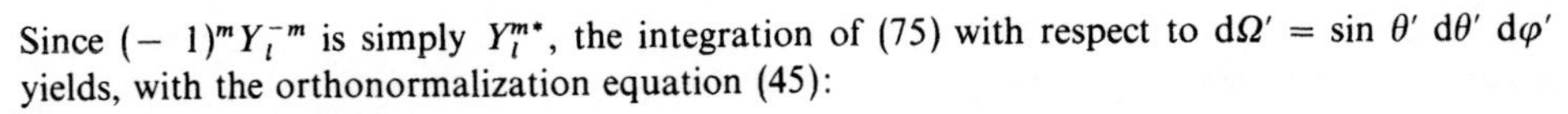



Since $(-1)^m Y_l^{-m}$ is simply $Y_l^{m*}$, the integration of (75) with respect to $d\Omega' = \sin\theta'\, d\theta'\, d\varphi'$ yields, with the orthonormalization equation (45):

$$2l + 1 = \sum_{m=-l}^{+l} c_m \tag{76}$$

We now take the square of the modulus of both sides of (74) and integrate over $d\Omega'$ and $d\Omega''$. Using relation (45), it is easy to see that the right-hand side yields $\sum_{m=-l}^{+l} |c_m|^2$. As far as the left-hand side is concerned, we can again take advantage of the invariance of the angle $\alpha$ witl respect to a rotation in order to show that $\int d\Omega'\, |P_l(\cos\alpha)|^2$ is actually independent of $(\theta'', \varphi'')$. If we then choose $Ou''$ along $Oz$ to evaluate this integral, we find, according to relation (61):

$$\int d\Omega'\, |P_l(\cos\alpha)|^2 = \int d\Omega\, |P_l(\cos\theta)|^2 = 2\pi \times \frac{2}{2l+1} \tag{77}$$

Integrating over $d\Omega''$, we find a second relation between the coefficients $c_m$:

$$2l + 1 = \sum_{m=-l}^{+l} |c_m|^2 \tag{78}$$

(*iv*) equations (76) and (78) suffice for the determination of the $(2l + 1)$ coefficients $c_m$: they are all equal to 1. To show this, consider, in a normed $(2l + 1)$-dimensional vector space, the vector **X** of components $x_m = \dfrac{c_m}{\sqrt{2l+1}}$ and the vector **Y** of components $y_m = \dfrac{1}{\sqrt{2l+1}}$. The Schwartz inequality indicates that:

$$(\mathbf{X}^* . \mathbf{X})(\mathbf{Y}^* . \mathbf{Y}) \geqslant |\mathbf{Y}^* . \mathbf{X}|^2 \tag{79}$$

where there is equality if and only if **X** and **Y** are proportional. Now, (76) and (78) show that this is the case : $x_m$ and $c_m$ are therefore independent of $m$, as is $y_m$, and we have, necessarily, $c_m = 1$. This concludes the proof of formula (70).

**References :**

Messiah (1.17), App. B, §IV; Arfken (10.4), chap. 12; Edmonds (2.21), Table 1; Butkov (10.8), chap. 9, §§5 and 8; Whittaker and Watson (10.12), chap. XV; Bateman (10.39), chap. III; Bass (10.1), vol. I, § 17-7.

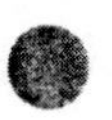



# Complement $B_{VI}$

## ANGULAR MOMENTUM AND ROTATIONS



## 1. Introduction

We indicated in chapter VI (§B-2) that the commutation relations between the components of an angular momentum are actually the expression of the geometrical properties of rotations in ordinary three-dimensional space. This is what we intend to show in this complement, where we investigate the relation between rotations and angular momentum operators.

Consider a physical system $(S)$ whose quantum mechanical state, at a given time, is characterized by the ket $|\psi\rangle$ of the state space $\mathscr{E}$. We perform a rotation $\mathscr{R}$ on this system; in this new position, the state of the system is described by a ket $|\psi'\rangle$ which is different from $|\psi\rangle$. Given the geometrical transformation $\mathscr{R}$, the problem is to determine $|\psi'\rangle$ from $|\psi\rangle$. We shall see that it has the following solution: with every geometrical rotation $\mathscr{R}$ can be associated a linear operator $R$ acting in the state space $\mathscr{E}$ such that:

$$|\psi'\rangle = R|\psi\rangle \tag{1}$$

Let us immediately stress the necessity of distinguishing between the geometrical rotation $\mathscr{R}$, which operates in ordinary space, and its "image" $R$, which acts in the state space:

$$\mathscr{R} \Longrightarrow R \tag{2}$$

We shall begin (§2) by reviewing the principal properties of geometrical rotations $\mathscr{R}$. We shall not embark upon a detailed study of them; rather, we shall simply note some results which will be useful to us later. Then, in §3, we shall use the example of a spinless particle to define the rotation operators $R$ precisely, to study their most important properties, and to determine their relation to the angular momentum operators **L**. We shall then be able to interpret the commutation relations amongst the components of the angular momentum **L** as the image, in the space $\mathscr{E}_{\mathbf{r}}$, of purely geometrical characteristics of rotations $\mathscr{R}$. We shall then generalize (§ 4) these concepts to arbitrary quantum mechanical systems. In §5, we shall examine the behavior of the observables describing the physical quantities which are measurable in this system under a rotation of the system. This will lead us to classify observables according to how they transform under a rotation (scalar, vector, tensor observables). Finally, in §6, we shall briefly consider the problem of rotation invariance and indicate some important consequences of this invariance.

## 2. Brief study of geometrical rotations $\mathscr{R}$

### a. DEFINITION. PARAMETRIZATION

A rotation $\mathscr{R}$ is a one-to-one transformation of three-dimensional space which conserves a point of this space, the angles and the distances, as well as the handedness of the reference frames*. We shall be concerned here with the set of rotations which conserve a given point $O$, which we shall choose as the origin of the reference frame. A rotation can then be characterized by the *axis of rotation* (given by its unit vector **u** or its polar angles $\theta$ and $\varphi$) and the angle of rotation $\alpha$ ($0 \leqslant \alpha < 2\pi$). To determine a rotation, three parameters are required; they can be chosen to be the components of the vector:

$$\boldsymbol{\alpha} = \alpha \mathbf{u} \tag{3}$$

whose absolute value is equal to the angle of rotation and whose direction defines the axis of rotation. Note that a rotation can also be characterized by three angles, called Euler angles. We shall denote by $\mathscr{R}_{\mathbf{u}}(\alpha)$ the geometrical rotation through an angle $\alpha$ about the axis defined by the unit vector **u**.

The set of rotations $\mathscr{R}$ constitutes a group: the product of two rotations (that is, the transformation resulting from the successive application of these two rotations) is also a rotation; there exists an identity rotation (rotation through a zero angle about an arbitrary axis); for every rotation $\mathscr{R}_{\mathbf{u}}(\alpha)$ there is an inverse rotation, $\mathscr{R}_{-\mathbf{u}}(\alpha)$. The group of rotations is not commutative: in general, the product of two rotations depends on the order in which they are performed**:

$$\mathscr{R}_{\mathbf{u}}(\alpha)\mathscr{R}_{\mathbf{u}'}(\alpha') \neq \mathscr{R}_{\mathbf{u}'}(\alpha')\mathscr{R}_{\mathbf{u}}(\alpha) \tag{4}$$

* This last property is imposed in order to exclude reflections with respect to a point or a plane.

** When one writes $\mathscr{R}_2\mathscr{R}_1$, this means that rotation $\mathscr{R}_1$ must be performed first, $\mathscr{R}_2$ being applied subsequently to the result obtained.

Recall, however, that two rotations performed about the same axis always commute:

$$\mathscr{R}_{\mathbf{u}}(\alpha)\,\mathscr{R}_{\mathbf{u}}(\alpha') = \mathscr{R}_{\mathbf{u}}(\alpha')\,\mathscr{R}_{\mathbf{u}}(\alpha) = \mathscr{R}_{\mathbf{u}}(\alpha + \alpha') \tag{5}$$

(if necessary, $2\pi$ is subtracted from $\alpha + \alpha'$, to keep it within the interval $[0,2\pi]$).

### b. INFINITESIMAL ROTATIONS

An infinitesimal rotation is defined as a rotation which is infinitesimally close to the identity rotation, that is, a rotation $\mathscr{R}_{\mathbf{u}}(\mathrm{d}\alpha)$ through an infinitesimal angle $\mathrm{d}\alpha$ about an arbitrary axis $\mathbf{u}$. It is easy to see that the transform of a vector $\mathbf{OM}$ under the infinitesimal rotation $\mathscr{R}_{\mathbf{u}}(\mathrm{d}\alpha)$ can be written, to first order in $\mathrm{d}\alpha$:

$$\mathscr{R}_{\mathbf{u}}(\mathrm{d}\alpha)\,\mathbf{OM} = \mathbf{OM} + \mathrm{d}\alpha\,\mathbf{u} \times \mathbf{OM} \tag{6}$$

Every finite rotation can be decomposed into an infinite number of infinitesimal rotations, since the angle of rotation can vary continuously and since, according to (5):

$$\mathscr{R}_{\mathbf{u}}(\alpha + \mathrm{d}\alpha) = \mathscr{R}_{\mathbf{u}}(\alpha)\,\mathscr{R}_{\mathbf{u}}(\mathrm{d}\alpha) = \mathscr{R}_{\mathbf{u}}(\mathrm{d}\alpha)\,\mathscr{R}_{\mathbf{u}}(\alpha) \tag{7}$$

where $\mathscr{R}_{\mathbf{u}}(\mathrm{d}\alpha)$ is an infinitesimal rotation. Thus, the study of the rotation group can be reduced to an examination of infinitesimal rotations*.

Before ending this rapid survey of the properties of geometrical rotations, we note the following relation which will be useful to us later:

$$\mathscr{R}_{\mathbf{e}_y}(-\,\mathrm{d}\alpha')\,\mathscr{R}_{\mathbf{e}_x}(\mathrm{d}\alpha)\,\mathscr{R}_{\mathbf{e}_y}(\mathrm{d}\alpha')\,\mathscr{R}_{\mathbf{e}_x}(-\,\mathrm{d}\alpha) = \mathscr{R}_{\mathbf{e}_z}(\mathrm{d}\alpha\,\mathrm{d}\alpha') \tag{8}$$

where $\mathbf{e}_x$, $\mathbf{e}_y$ and $\mathbf{e}_z$ denote the unit vectors of the three coordinate axes $Ox$, $Oy$ and $Oz$ respectively. If $\mathrm{d}\alpha$ and $\mathrm{d}\alpha'$ are first-order infinitesimal angles, this relation is correct to the second order. It describes, in a special case, the non-commutative structure of the rotation group.

To prove relation (8), let us apply its left-hand side to an arbitrary vector $\mathbf{OM}$. We use formula (6) to find the vector $\mathbf{OM}'$, the transform of $\mathbf{OM}$ under the succesive action of the four infinitesimal rotations. It can be seen immediately that if $\mathrm{d}\alpha$ is zero, the left-hand side of (8) reduces to the product $\mathscr{R}_{\mathbf{e}_y}(-\,\mathrm{d}\alpha')\,\mathscr{R}_{\mathbf{e}_y}(\mathrm{d}\alpha')$, which is equal to the identity rotation [see (5)]; the vector $\mathbf{OM}' - \mathbf{OM}$ must therefore be proportional to $\mathrm{d}\alpha$. For an analogous reason, it must also be proportional to $\mathrm{d}\alpha'$. Consequently, the difference $\mathbf{OM}' - \mathbf{OM}$ is proportional to $\mathrm{d}\alpha\mathrm{d}\alpha'$.

Therefore, to calculate $\mathbf{OM}'$ to second order, we may restrict ourselves to first order in each of the two infinitesimal angles $\mathrm{d}\alpha$ and $\mathrm{d}\alpha'$. First of all, according to (6):

$$\mathscr{R}_{\mathbf{e}_x}(-\,\mathrm{d}\alpha)\,\mathbf{OM} = \mathbf{OM} - \mathrm{d}\alpha\,\mathbf{e}_x \times \mathbf{OM} \tag{9}$$

We must then apply $\mathscr{R}_{\mathbf{e}_y}(\mathrm{d}\alpha')$ to this vector; this can be done by again using (6):

$$\begin{aligned}\mathscr{R}_{\mathbf{e}_y}(\mathrm{d}\alpha')\,\mathscr{R}_{\mathbf{e}_x}(-\,\mathrm{d}\alpha)\,\mathbf{OM} &\\ &= (\mathbf{OM} - \mathrm{d}\alpha\,\mathbf{e}_x \times \mathbf{OM}) + \mathrm{d}\alpha'\,\mathbf{e}_y \times (\mathbf{OM} - \mathrm{d}\alpha\,\mathbf{e}_x \times \mathbf{OM})\\ &= \mathbf{OM} - \mathrm{d}\alpha\,\mathbf{e}_x \times \mathbf{OM} + \mathrm{d}\alpha'\,\mathbf{e}_y \times \mathbf{OM} - \mathrm{d}\alpha\,\mathrm{d}\alpha'\,\mathbf{e}_y \times (\mathbf{e}_x \times \mathbf{OM})\end{aligned} \tag{10}$$

* However, limiting ourselves to infinitesimal rotations, we lose sight of a "global" property of the finite rotation group: the fact that a rotation through an angle of $2\pi$ is the identity transformation. The rotation operators (see § 3) constructed from infinitesimal operators do not always have this global property. In certain cases (see complement $A_{IX}$), the operator associated with a $2\pi$ rotation is not the unit operator but its opposite.

The action of $\mathscr{R}_{\mathbf{e}_x}(\mathrm{d}\alpha)$ on the vector appearing on the right-hand side of (10) results in the addition of the following infinitesimal terms to this vector:

$$\mathrm{d}\alpha\, \mathbf{e}_x \times \mathbf{OM} + \mathrm{d}\alpha\, \mathrm{d}\alpha'\, \mathbf{e}_x \times (\mathbf{e}_y \times \mathbf{OM}) \tag{11}$$

obtained by the vector multiplication of the right-hand side of (10) by $\mathrm{d}\alpha\mathbf{e}_x$, with only the first order terms in $\mathrm{d}\alpha$ being retained. Therefore:

$$\begin{aligned}\mathscr{R}_{\mathbf{e}_x}(\mathrm{d}\alpha)\, \mathscr{R}_{\mathbf{e}_y}(\mathrm{d}\alpha')\, \mathscr{R}_{\mathbf{e}_x}(-\,\mathrm{d}\alpha)\, \mathbf{OM} \\ = \mathbf{OM} + \mathrm{d}\alpha'\, \mathbf{e}_y \times \mathbf{OM} + \mathrm{d}\alpha\, \mathrm{d}\alpha'[\mathbf{e}_x \times (\mathbf{e}_y \times \mathbf{OM}) - \mathbf{e}_y \times (\mathbf{e}_x \times \mathbf{OM})]\end{aligned} \tag{12}$$

Finally, $\mathbf{OM}'$ is equal to the sum of the vector just obtained and its vector product by $-\,\mathrm{d}\alpha'\,\mathbf{e}_y$. To first order in $\mathrm{d}\alpha'$, this vector product can be written simply :

$$-\,\mathrm{d}\alpha'\, \mathbf{e}_y \times \mathbf{OM} \tag{13}$$

which means that:

$$\begin{aligned}\mathscr{R}_{\mathbf{e}_y}(-\,\mathrm{d}\alpha')\, \mathscr{R}_{\mathbf{e}_x}(\mathrm{d}\alpha)\, \mathscr{R}_{\mathbf{e}_y}(\mathrm{d}\alpha')\, \mathscr{R}_{\mathbf{e}_x}(-\,\mathrm{d}\alpha)\, \mathbf{OM} \\ = \mathbf{OM} + \mathrm{d}\alpha\, \mathrm{d}\alpha'[\mathbf{e}_x \times (\mathbf{e}_y \times \mathbf{OM}) - \mathbf{e}_y \times (\mathbf{e}_x \times \mathbf{OM})]\end{aligned} \tag{14}$$

It is then easy to transform the double vector products; we find:

$$\begin{aligned}\mathscr{R}_{\mathbf{e}_y}(-\,\mathrm{d}\alpha')\, \mathscr{R}_{\mathbf{e}_x}(\mathrm{d}\alpha)\, \mathscr{R}_{\mathbf{e}_y}(\mathrm{d}\alpha')\, \mathscr{R}_{\mathbf{e}_x}(-\,\mathrm{d}\alpha)\mathbf{OM} &= \mathbf{OM} + \mathrm{d}\alpha\, \mathrm{d}\alpha'\, \mathbf{e}_z \times \mathbf{OM} \\ &= \mathscr{R}_{\mathbf{e}_z}(\mathrm{d}\alpha\, \mathrm{d}\alpha')\, \mathbf{OM}\end{aligned} \tag{15}$$

Since this relation is true for any vector $\mathbf{OM}$, expression (8) is verified.

## 3. Rotation operators in state space. Example: a spinless particle

In this section, we consider a physical system composed of a single (spinless) particle in three-dimensional space.

### a. EXISTENCE AND DEFINITION OF ROTATION OPERATORS

At a given time, the quantum mechanical state of the particle is characterized, in the state space $\mathscr{E}_\mathbf{r}$, by the ket $|\psi\rangle$ with which is associated the wave function $\psi(\mathbf{r}) = \langle \mathbf{r} | \psi \rangle$. Let us perform a rotation $\mathscr{R}$ on this system which associates with the point $\mathbf{r}_0(x_0, y_0, z_0)$ of space the point $\mathbf{r}'_0(x'_0, y'_0, z'_0)$ such that:

$$\mathbf{r}'_0 = \mathscr{R}\,\mathbf{r}_0 \tag{16}$$

Let $|\psi'\rangle$ be the state vector of the system after rotation, and $\psi'(\mathbf{r}) = \langle \mathbf{r} | \psi' \rangle$, the corresponding wave function. It is natural to assume that the value of the initial wave function $\psi(\mathbf{r})$ at the point $\mathbf{r}_0$ will be found, after rotation, to be the value of the final wave function $\psi'(\mathbf{r})$ at the point $\mathbf{r}'_0$ given by (16):

$$\psi'(\mathbf{r}'_0) = \psi(\mathbf{r}_0) \tag{17}$$

that is:

$$\psi'(\mathbf{r}'_0) = \psi(\mathscr{R}^{-1}\mathbf{r}'_0) \tag{18}$$

Since this equation is valid for any point $\mathbf{r}_0'$ in space, it can be written in the form:

$$\psi'(\mathbf{r}) = \psi(\mathscr{R}^{-1}\mathbf{r}) \tag{19}$$

By definition, the operator $R$ in the state space $\mathscr{E}_{\mathbf{r}}$ associated with the geometrical rotation $\mathscr{R}$ being treated is the one which acts on the state $|\psi\rangle$ before rotation to yield the state $|\psi'\rangle$ after the rotation $\mathscr{R}$:

$$|\psi'\rangle = R|\psi\rangle \tag{20}$$

$R$ is called a "rotation operator". Relation (19) characterizes its action in the $\{|\mathbf{r}\rangle\}$ representation:

$$\langle \mathbf{r}|R|\psi\rangle = \langle \mathscr{R}^{-1}\mathbf{r}|\psi\rangle \tag{21}$$

where $|\mathscr{R}^{-1}\mathbf{r}\rangle$ is the basis ket of this representation determined by the components of the vector $\mathscr{R}^{-1}\mathbf{r}$.

COMMENT:

If the state of the particle after rotation were $e^{i\theta}|\psi'\rangle$ (where $\theta$ is an arbitrary real number) instead of $|\psi'\rangle$, its physical properties would not be modified. In other words, relation (17) could be replaced by:

$$\psi'(\mathbf{r}_0') = e^{i\theta}\,\psi(\mathbf{r}_0) \tag{22}$$

$\theta$ would obviously be independent of $\mathbf{r}_0$ but could depend on the rotation $\mathscr{R}$. We shall not treat this difficulty here.

### b. PROPERTIES OF ROTATION OPERATORS $R$

#### α. *$R$ is a linear operator*

This essential property of rotation operators follows from their very definition. If the state $|\psi\rangle$ before the rotation is a linear superposition of states, for example:

$$|\psi\rangle = \lambda_1|\psi_1\rangle + \lambda_2|\psi_2\rangle \tag{23}$$

formula (21) indicates that:

$$\begin{aligned}\langle \mathbf{r}|R|\psi\rangle &= \lambda_1\langle \mathscr{R}^{-1}\mathbf{r}|\psi_1\rangle + \lambda_2\langle \mathscr{R}^{-1}\mathbf{r}|\psi_2\rangle \\ &= \lambda_1\langle \mathbf{r}|R|\psi_1\rangle + \lambda_2\langle \mathbf{r}|R|\psi_2\rangle\end{aligned} \tag{24}$$

Since this relation is true for any ket of the $\{|\mathbf{r}\rangle\}$ basis, we deduce that $R$ is a linear operator:

$$R|\psi\rangle = R[\lambda_1|\psi_1\rangle + \lambda_2|\psi_2\rangle] = \lambda_1 R|\psi_1\rangle + \lambda_2 R|\psi_2\rangle \tag{25}$$

#### β. *$R$ is unitary*

In formula (21), the ket $|\psi\rangle$ can be arbitrary. The action of the operator $R$ on the bra $\langle \mathbf{r}|$ is therefore given by:

$$\langle \mathbf{r}|R = \langle \mathscr{R}^{-1}\mathbf{r}| \tag{26}$$

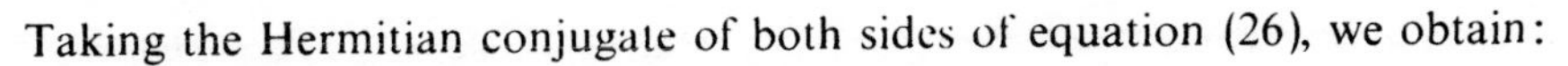

Taking the Hermitian conjugate of both sides of equation (26), we obtain:

$$R^{\dagger} \mid \mathbf{r} \rangle = \mid \mathscr{R}^{-1}\mathbf{r} \rangle \tag{27}$$

Moreover, if we recall that the ket $|\mathbf{r}\rangle$ represents a state in which the particle is perfectly localized at the point $\mathbf{r}$, we see that:

$$R \mid \mathbf{r} \rangle = \mid \mathscr{R}\,\mathbf{r} \rangle \tag{28}$$

This equation simply expresses the fact that if the particle was localized at the point $\mathbf{r}$ before the rotation, it will be localized at the point $\mathbf{r}' = \mathscr{R}\mathbf{r}$ after the rotation. To get (28) from (21), we choose a basis state $|\mathbf{r}_0\rangle$ for $|\psi\rangle$:

$$\langle \mathbf{r} \mid R \mid \mathbf{r}_0 \rangle = \langle \mathscr{R}^{-1}\mathbf{r} \mid \mathbf{r}_0 \rangle = \delta[(\mathscr{R}^{-1}\mathbf{r}) - \mathbf{r}_0] \tag{29}$$

where we have used the orthonormalization relation of the $\{ |\mathbf{r}\rangle \}$ basis. Furthermore*:

$$\delta[(\mathscr{R}^{-1}\mathbf{r}) - \mathbf{r}_0] = \delta[\mathbf{r} - (\mathscr{R}\,\mathbf{r}_0)] \tag{30}$$

Substituting (30) into (29), we indeed find that:

$$\langle \mathbf{r} \mid R \mid \mathbf{r}_0 \rangle = \delta[\mathbf{r} - (\mathscr{R}\,\mathbf{r}_0)] = \langle \mathbf{r} \mid \mathscr{R}\,\mathbf{r}_0 \rangle \tag{31}$$

that is, since $\{ |\mathbf{r}\rangle \}$ is a basis of $\mathscr{E}_{\mathbf{r}}$:

$$R \mid \mathbf{r}_0 \rangle = \mid \mathscr{R}\,\mathbf{r}_0 \rangle \tag{32}$$

Starting with formulas (27) and (28), it is easy to show that:

$$RR^{\dagger} = R^{\dagger}R = 1 \tag{33}$$

since the action of $RR^{\dagger}$ or $R^{\dagger}R$ on any vector of the $\{ |\mathbf{r}\rangle \}$ basis yields the same vector; for example:

$$RR^{\dagger} \mid \mathbf{r} \rangle = R \mid \mathscr{R}^{-1}\mathbf{r} \rangle = \mid \mathscr{R}\mathscr{R}^{-1}\mathbf{r} \rangle = \mid \mathbf{r} \rangle \tag{34}$$

The operator $R$ is therefore unitary.

COMMENT:

The operator $R$ therefore conserves the scalar product and the norm of vectors that it transforms:

$$\left.\begin{array}{l} |\psi'\rangle = R\,|\psi\rangle \\ |\varphi'\rangle = R\,|\varphi\rangle \end{array}\right\} \Longrightarrow \langle \varphi' \mid \psi' \rangle = \langle \varphi \mid \psi \rangle \tag{35}$$

This property is very important from the physical point of view since the probability amplitudes which yield physical predictions appear in the form of scalar products of two kets.

* Relation (30) can easily be established by using the definition of delta "functions" and the fact that a rotation conserves the infinitesimal volume element.

γ. *The set of operators R constitutes a representation of the rotation group*

We have pointed out (§ 2) that the geometrical rotations form a group; in particular, the product of two rotations $\mathscr{R}_1$ and $\mathscr{R}_2$ is always a rotation:

$$\mathscr{R}_2\mathscr{R}_1 = \mathscr{R}_3 \tag{36}$$

With the three geometrical rotations $\mathscr{R}_1$, $\mathscr{R}_2$ and $\mathscr{R}_3$ are associated, in the state space $\mathscr{E}_{\mathbf{r}}$, three rotation operators $R_1$, $R_2$ and $R_3$, respectively. If the three geometrical rotations satisfy relation (36), we shall show that the corresponding rotation operators are such that:

$$R_2R_1 = R_3 \tag{37}$$

($R_2R_1$ is a product of operators of $\mathscr{E}_{\mathbf{r}}$ as defined in chapter II, §B-3-a).

Consider a particle whose state is described by an arbitrary ket $|\mathbf{r}\rangle$ of the basis characterizing the $\{|\mathbf{r}\rangle\}$ representation. If we perform the rotation $\mathscr{R}_1$ on this particle, its state becomes:

$$R_1 \,|\,\mathbf{r}\,\rangle = |\mathscr{R}_1\mathbf{r}\,\rangle \tag{38}$$

by definition of $R_1$. Now, we perform the rotation $\mathscr{R}_2$ on the new state we have just obtained; the state of the particle after this second rotation is, according to (38) and the definition of $R_2$:

$$R_2R_1 \,|\,\mathbf{r}\,\rangle = R_2 \,|\,\mathscr{R}_1\mathbf{r}\,\rangle = |\,\mathscr{R}_2\mathscr{R}_1\mathbf{r}\,\rangle \tag{39}$$

If we take (36) into account, we see that relation (39) is equivalent to:

$$R_2R_1 \,|\,\mathbf{r}\,\rangle = |\,\mathscr{R}_3\mathbf{r}\,\rangle \tag{40}$$

Now, the operator $R_3$, associated with the rotation $\mathscr{R}_3$, is such that:

$$R_3 \,|\,\mathbf{r}\,\rangle = |\,\mathscr{R}_3\mathbf{r}\,\rangle \tag{41}$$

Relation (37) is therefore proven, since the ket $|\mathbf{r}\rangle$ under consideration can be chosen in an arbitrary way from the kets of the $\{|\mathbf{r}\rangle\}$ basis.

To express the important result which we have just established, one says that the correspondence $\mathscr{R} \Longrightarrow R$ between geometrical rotations and rotation operators conserves the group law, or that the set of operators $R$ constitutes a "representation" of the rotation group. Of course, with the identity rotation, we associate the identity operator in $\mathscr{E}_{\mathbf{r}}$, and with the rotation $\mathscr{R}^{-1}$ (the inverse of a rotation $\mathscr{R}$), the operator $R^{-1}$, which is the inverse of the one which corresponds to $\mathscr{R}$ (we showed in §3-b-β that $R^{-1} = R^\dagger$).

## c. EXPRESSION FOR ROTATION OPERATORS IN TERMS OF ANGULAR MOMENTUM OBSERVABLES

α. *Infinitesimal rotation operators*

First of all, let us consider an infinitesimal rotation about the $Oz$ axis, $\mathscr{R}_{\mathbf{e}_z}(\mathrm{d}\alpha)$. If we apply it to a particle whose state is described by the wave function $\psi(\mathbf{r})$,

we know from (19) that the wave function $\psi'(\mathbf{r})$ associated with the state of the particle after rotation satisfies:

$$\psi'(\mathbf{r}) = \psi[\mathscr{R}_{\mathbf{e}_z}^{-1}(\mathrm{d}\alpha)\,\mathbf{r}] \tag{42}$$

But if $(x, y, z)$ are the components of $\mathbf{r}$, those of $\mathscr{R}_{\mathbf{e}_z}^{-1}(\mathrm{d}\alpha)\,\mathbf{r}$ can easily be calculated from (6):

$$\mathscr{R}_{\mathbf{e}_z}^{-1}(\mathrm{d}\alpha)\,\mathbf{r} = \mathscr{R}_{-\mathbf{e}_z}(\mathrm{d}\alpha)\,\mathbf{r} = (\,\mathbf{r} - \mathrm{d}\alpha\,\mathbf{e}_z \times \mathbf{r}\,) \qquad \begin{cases} x + y\,\mathrm{d}\alpha \\ y - x\,\mathrm{d}\alpha \\ z \end{cases} \tag{43}$$

Equation (42) can then be written in the form:

$$\psi'(x, y, z) = \psi(x + y\,\mathrm{d}\alpha, y - x\,\mathrm{d}\alpha, z) \tag{44}$$

which yields, to first order in $\mathrm{d}\alpha$:

$$\begin{aligned} \psi'(x, y, z) &= \psi(x, y, z) + \mathrm{d}\alpha\left[y\frac{\partial\psi}{\partial x} - x\frac{\partial\psi}{\partial y}\right] \\ &= \psi(x, y, z) - \mathrm{d}\alpha\left[x\frac{\partial}{\partial y} - y\frac{\partial}{\partial x}\right]\psi(x, y, z) \end{aligned} \tag{45}$$

Within the brackets, we recognize, to within a factor of $\frac{\hbar}{i}$, the expression in the $\{\,|\,\mathbf{r}\,\rangle\,\}$ representation for the operator $L_z = XP_y - YP_x$. We therefore obtain the result:

$$\psi'(\mathbf{r}) = \langle\,\mathbf{r}\,|\,\psi'\,\rangle = \langle\,\mathbf{r}\,|\left(1 - \frac{i}{\hbar}\mathrm{d}\alpha\,L_z\right)|\,\psi\,\rangle \tag{46}$$

Now, by definition of the operator $R_{\mathbf{e}_z}(\mathrm{d}\alpha)$ associated with the rotation $\mathscr{R}_{\mathbf{e}_z}(\mathrm{d}\alpha)$:

$$|\,\psi'\,\rangle = R_{\mathbf{e}_z}(\mathrm{d}\alpha)\,|\,\psi\,\rangle \tag{47}$$

Therefore, since the original state $|\,\psi\,\rangle$ is arbitrary, we find, finally, that:

$$\boxed{R_{\mathbf{e}_z}(\mathrm{d}\alpha) = 1 - \frac{i}{\hbar}\mathrm{d}\alpha\,L_z} \tag{48}$$

The preceding argument can easily be generalized to an infinitesimal rotation about an arbitrary axis. We therefore have, in general:

$$\boxed{R_{\mathbf{u}}(\mathrm{d}\alpha) = 1 - \frac{i}{\hbar}\mathrm{d}\alpha\,\mathbf{L}\,.\,\mathbf{u}} \tag{49}$$

COMMENT:

(46) can also be quickly established by using the spherical coordinates $(r, \theta, \varphi)$, since $L_z$ then corresponds to the differential operator $\frac{\hbar}{i}\frac{\partial}{\partial\varphi}$.

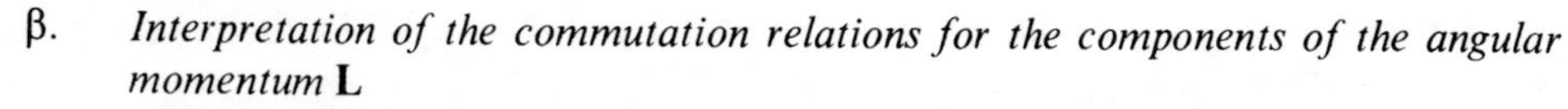

β. *Interpretation of the commutation relations for the components of the angular momentum* **L**

What then is the "image" in the state space $\mathscr{E}_{\mathbf{r}}$ of relation (8)? According to the results of § 3-b-γ and the expressions we have just obtained, this relation implies that, to first order with respect to each of the angles dα and dα′:

$$\left[1+\frac{i}{\hbar}\mathrm{d}\alpha' L_y\right]\left[1-\frac{i}{\hbar}\mathrm{d}\alpha L_x\right]\left[1-\frac{i}{\hbar}\mathrm{d}\alpha' L_y\right]\left[1+\frac{i}{\hbar}\mathrm{d}\alpha L_x\right]=1-\frac{i}{\hbar}\mathrm{d}\alpha\,\mathrm{d}\alpha' L_z \tag{50}$$

Expanding the left-hand side and setting the coefficients of dα dα′ equal, we easily find that relation (50) reduces to:

$$[L_x, L_y] = i\hbar L_z \tag{51}$$

Of course, the two other commutation relations for the components of **L** can be found, by an analogous argument, from the formulas obtained from (8) by cyclic permutation of the vectors $\mathbf{e}_x$, $\mathbf{e}_y$ and $\mathbf{e}_z$.

Thus, *the commutation relations of the orbital angular momentum of a particle can be seen to be consequences of the non-commutative structure of the geometrical rotation group.*

γ. *Finite rotation operators*

Now, consider a rotation $\mathscr{R}_{\mathbf{e}_z}(\alpha)$ through an arbitrary angle α about the *Oz* axis. According to formula (7), the operator $R_{\mathbf{e}_z}(\alpha)$ associated with such a rotation must satisfy (again using the results of § 3-b-γ):

$$R_{\mathbf{e}_z}(\alpha + \mathrm{d}\alpha) = R_{\mathbf{e}_z}(\alpha)\, R_{\mathbf{e}_z}(\mathrm{d}\alpha) \tag{52}$$

where the two operators of the right-hand side commute. But we know the expression for $R_{\mathbf{e}_z}(\mathrm{d}\alpha)$, so we have:

$$R_{\mathbf{e}_z}(\alpha + \mathrm{d}\alpha) = R_{\mathbf{e}_z}(\alpha)\left[1 - \frac{i}{\hbar}\mathrm{d}\alpha\; L_z\right] \tag{53}$$

that is:

$$R_{\mathbf{e}_z}(\alpha + \mathrm{d}\alpha) - R_{\mathbf{e}_z}(\alpha) = -\frac{i}{\hbar}\mathrm{d}\alpha\; R_{\mathbf{e}_z}(\alpha)\, L_z \tag{54}$$

Here again, $R_{\mathbf{e}_z}(\alpha)$ and $L_z$ must commute. Although we are dealing with operators, the solution of equation (54) is formally the same as it would be if we were considering an ordinary function of the variable α:

$$\boxed{R_{\mathbf{e}_z}(\alpha) = \mathrm{e}^{-\frac{i}{\hbar}\alpha L_z}} \tag{55}$$

Indeed, if we recall (*cf.* complement $B_{II}$, § 4) that the exponential of an operator is defined by the corresponding power series expansion, it is easy to verify that expression (55) is the solution of equation (54). Moreover, the "integration constant" is equal to 1, since we know that:

$$R_{\mathbf{e}_z}(0) = 1 \tag{56}$$

As in § 3-c-α above, it is easy to generalize this result to a finite rotation about an arbitrary axis:

$$R_{\mathbf{u}}(\alpha) = e^{-\frac{i}{\hbar}\alpha \mathbf{L}.\mathbf{u}} \tag{57}$$

COMMENTS:

(*i*) Formula (57) can be written explicitly in the form:

$$R_{\mathbf{u}}(\alpha) = e^{-\frac{i}{\hbar}\alpha(L_x u_x + L_y u_y + L_z u_z)} \tag{58}$$

where $u_x$, $u_y$ and $u_z$ are the components of the unit vector **u**. Recall, however, that, since $L_x$, $L_y$ and $L_z$ do not commute:

$$R_{\mathbf{u}}(\alpha) \neq e^{-\frac{i}{\hbar}\alpha L_x u_x} \, e^{-\frac{i}{\hbar}\alpha L_y u_y} \, e^{-\frac{i}{\hbar}\alpha L_z u_z} \tag{59}$$

(*ii*) It can be seen from expression (57) that the operator $R_{\mathbf{u}}(\alpha)$ is unitary. Since the components of **L** are Hermitian:

$$[R_{\mathbf{u}}(\alpha)]^\dagger = e^{\frac{i}{\hbar}\alpha \mathbf{L}.\mathbf{u}} \tag{60}$$

we have (as **L** . **u** obviously commutes with itself):

$$[R_{\mathbf{u}}(\alpha)]^\dagger R_{\mathbf{u}}(\alpha) = R_{\mathbf{u}}(\alpha)[R_{\mathbf{u}}(\alpha)]^\dagger = 1 \tag{61}$$

(*iii*) In the special case envisaged in this section, we find that:

$$R_{\mathbf{u}}(2\pi) = 1 \tag{62}$$

We shall confine ourselves to proving this result for the rotation through $2\pi$ about the $Oz$ axis (the generalization of this proof involves no difficulties). To this end, consider an arbitrary ket $|\psi\rangle$, and expand it on a basis composed of eigenvectors of the observable $L_z$:

$$|\psi\rangle = \sum_{m,\tau} c_{m,\tau} |m, \tau\rangle \tag{63}$$

with:

$$L_z |m, \tau\rangle = m\hbar |m, \tau\rangle \tag{64}$$

($\tau$ symbolizes the indices other than $m$ which are necessary in order to specify the vectors of the basis used, which may, for example, be a "standard" basis $\{ |k, l, m\rangle \}$ such as those introduced in §C-3 of chapter VI). The action of $R_{\mathbf{e}_z}(\alpha)$ on $|\psi\rangle$ is then easy to obtain:

$$\begin{aligned} R_{\mathbf{e}_z}(\alpha)|\psi\rangle &= \sum_{m,\tau} c_{m,\tau} \, e^{-\frac{i}{\hbar}\alpha L_z} |m, \tau\rangle \\ &= \sum_{m,\tau} c_{m,\tau} \, e^{-i\alpha m} |m, \tau\rangle \end{aligned} \tag{65}$$

But we know that, for the orbital angular momentum of a particle, *m is always integral*. Consequently, when $\alpha$ attains the value $2\pi$, all the factors $e^{-i\alpha m}$ become equal to 1, and :

$$R_{\mathbf{e}_z}(2\pi) \mid \psi \rangle = \sum_{m,\tau} c_{m,\tau} \mid m, \tau \rangle = \mid \psi \rangle \tag{66}$$

Since this relation is satisfied for all $|\psi\rangle$, we deduce that $R_{\mathbf{e}_z}(2\pi)$ is the identity operator.

The preceding argument clearly indicates that *formula* (62) *would not be valid if half-integral values of m were not excluded.* Indeed, we shall see in complement $A_{IX}$ that, for a spin 1/2, the operator associated with a rotation of $2\pi$ is equal to $-1$ and not 1 ; this result is related to the fact that we constructed the finite rotations from infinitesimal rotations (*cf.* note, page 692).

## 4. Rotation operators in the state space of an arbitrary system

We shall now generalize the concepts we introduced and the results we obtained for a special case (in §3).

### a. SYSTEM OF SEVERAL SPINLESS PARTICLES

First of all, the arguments of §3 can be extended without difficulty to systems composed of several spinless particles. We shall quickly demonstrate this, choosing as an example a system of two spinless particles, (1) and (2).

The state space $\mathscr{E}$ of such a system is the tensor product of the state spaces $\mathscr{E}_{\mathbf{r}_1}$ and $\mathscr{E}_{\mathbf{r}_2}$ of the two particles:

$$\mathscr{E} = \mathscr{E}_{\mathbf{r}_1} \otimes \mathscr{E}_{\mathbf{r}_2} \tag{67}$$

We shall use the same notation as in §F-4-b of chapter II. Starting from the position and momentum observables ($\mathbf{R}_1$ and $\mathbf{P}_1$ on the one hand, $\mathbf{R}_2$ and $\mathbf{P}_2$ on the other), we can define an orbital angular momentum for each of the particles :

$$\begin{aligned} \mathbf{L}_1 &= \mathbf{R}_1 \times \mathbf{P}_1 \\ \mathbf{L}_2 &= \mathbf{R}_2 \times \mathbf{P}_2 \end{aligned} \tag{68}$$

The components of $\mathbf{L}_1$, as well as those of $\mathbf{L}_2$, satisfy the commutation relations characteristic of angular momenta.

Consider a vector which is a tensor product of a vector of $\mathscr{E}_{\mathbf{r}_1}$ and a vector of $\mathscr{E}_{\mathbf{r}_2}$:

$$| \psi \rangle = | \varphi(1) \rangle \otimes | \chi(2) \rangle \tag{69}$$

$|\psi\rangle$ represents the state of the system formed by particle (1) in the state $|\varphi(1)\rangle$ and particle (2) in the state $|\chi(2)\rangle$. If we perform a rotation through an angle $\alpha$ about $\mathbf{u}$ on the two-particle system, the state of the system after the rotation

corresponds to the two particles in the "rotated states" $|\varphi'(1)\rangle$ and $|\chi'(2)\rangle$ respectively:

$$|\psi'\rangle = |\varphi'(1)\rangle \otimes |\chi'(2)\rangle = [R_{\mathbf{u}}^1(\alpha)\,|\varphi(1)\rangle] \otimes [R_{\mathbf{u}}^2(\alpha)\,|\chi(2)\rangle] \tag{70}$$

where $R_{\mathbf{u}}^1(\alpha)$ and $R_{\mathbf{u}}^2(\alpha)$ are the rotation operators in $\mathscr{E}_{\mathbf{r}_1}$ and $\mathscr{E}_{\mathbf{r}_2}$ :

$$R_{\mathbf{u}}^1(\alpha) = e^{-\frac{i}{\hbar}\alpha\,\mathbf{L}_1.\mathbf{u}} \tag{71-a}$$

$$R_{\mathbf{u}}^2(\alpha) = e^{-\frac{i}{\hbar}\alpha\,\mathbf{L}_2.\mathbf{u}} \tag{71-b}$$

Relation (70) can also be written, by definition of the tensor product of two operators (chap. II, §F-2-b):

$$|\psi'\rangle = [R_{\mathbf{u}}^1(\alpha) \otimes R_{\mathbf{u}}^2(\alpha)]\,|\varphi(1)\rangle \otimes |\chi(2)\rangle \tag{72}$$

Since every vector of $\mathscr{E}$ is a linear combination of vectors analogous to (69), the rotation transform $|\psi'\rangle$ of an arbitrary vector $|\psi\rangle$ of $\mathscr{E}$ is:

$$|\psi'\rangle = [R_{\mathbf{u}}^1(\alpha) \otimes R_{\mathbf{u}}^2(\alpha)]\,|\psi\rangle \tag{73}$$

Using formula (F-14) of chapter II and the fact that $\mathbf{L}_1$ and $\mathbf{L}_2$ commute (they are operators which relate to different particles), we obtain for the rotation operators in $\mathscr{E}$:

$$R_{\mathbf{u}}^1(\alpha) \otimes R_{\mathbf{u}}^2(\alpha) = e^{-\frac{i}{\hbar}\alpha\,\mathbf{L}_1.\mathbf{u}}\,e^{-\frac{i}{\hbar}\alpha\,\mathbf{L}_2.\mathbf{u}} = e^{-\frac{i}{\hbar}\alpha\,\mathbf{L}.\mathbf{u}} \tag{74}$$

where:

$$\mathbf{L} = \mathbf{L}_1 + \mathbf{L}_2 \tag{75}$$

is the total angular momentum of the two-particle system. All the formulas of the preceding section therefore remain valid as long as $\mathbf{L}$ represents the *total angular momentum*.

COMMENTS:

(*i*) $\mathbf{L}$ is an operator which acts in $\mathscr{E}$. In (75), $\mathbf{L}_1$ is, rigorously, the extension of the operator $\mathbf{L}_1$ acting in $\mathscr{E}_{\mathbf{r}_1}$ into $\mathscr{E}$ (an analogous comment could be made for $\mathbf{L}_2$). To simplify the notation, we shall not use different symbols for $\mathbf{L}_1$ and its extension into $\mathscr{E}$ (*cf.* chap. II, §F-2-c).

(*ii*) We might consider performing a rotation on only one of the two particles, for example, particle (1). In the course of such a "partial rotation", a vector such as (69) transforms into:

$$[R_{\mathbf{u}}^1(\alpha)|\varphi(1)\rangle] \otimes |\chi(2)\rangle \tag{76}$$

where only the state of particle (1) is modified. As above, it can be shown that the effect of a rotation performed only on particle (1) on an arbitrary state $|\psi\rangle$ of $\mathscr{E}$ is described by the operator:

$$R_{\mathbf{u}}^1(\alpha) \otimes \mathbb{1}(2) = e^{-\frac{i}{\hbar}\alpha\mathbf{L}_1.\mathbf{u}} \tag{77}$$

where $\mathbb{1}$ is the unit operator in $\mathscr{E}_{\mathbf{r}_2}$ [in (77), $\mathbf{L}_1$ acts in $\mathscr{E}$].

### b. AN ARBITRARY SYSTEM

The starting point of the arguments elaborated thus far is equation (19), which gives the transformation law of the state vector of the system in terms of that of its wave function. In the case of an arbitrary quantum mechanical system (which does not necessarily have a classical analogue), one cannot use the same method. For example, for a particle with spin, the operators $X$, $Y$ and $Z$ no longer form a C.S.C.O., and the state of the particle can no longer be defined by a wave function $\psi(x, y, z)$ (*cf.* chap. IX). One must reason directly in the state space $\mathscr{E}$ of the system. Without going into detail, we shall assume here that an operator $R$ acting in $\mathscr{E}$ can be associated with any geometrical rotation $\mathscr{R}$; if the system is initially in the state $|\psi\rangle$, the rotation $\mathscr{R}$ takes it into the state:

$$|\psi'\rangle = R|\psi\rangle \tag{78}$$

where the operator $R$ is linear and unitary (*cf.* comment of §3-b-β).

As far as the group law of the rotations $\mathscr{R}$ is concerned, it is conserved by the operators $R$, but only locally: the product of two geometrical rotations, at least one of which is infinitesimal, is represented in the state space $\mathscr{E}$ by the product of the corresponding operators $R$ (which implies, in particular, that the "image" of a rotation through an angle of zero is the identity operator). However, the operator associated with a geometrical rotation through an angle $2\pi$ is not necessarily the identity operator [*cf.* comment (*iii*) of § 3-c-γ and complement $A_{IX}$].

Now, let us consider an infinitesimal rotation $\mathscr{R}_{\mathbf{e}_z}(\mathrm{d}\alpha)$ about the $Oz$ axis. Since the group law is conserved for infinitesimal rotations, the operator $R_{\mathbf{e}_z}(\mathrm{d}\alpha)$ is necessarily of the form:

$$R_{\mathbf{e}_z}(\mathrm{d}\alpha) = 1 - \frac{i}{\hbar}\,\mathrm{d}\alpha\, J_z \tag{79}$$

where $J_z$ is a Hermitian operator since $R_{\mathbf{e}_z}(\mathrm{d}\alpha)$ is unitary (*cf.* complement $C_{II}$, §3). This relation is the *definition* of $J_z$. Similarly, the Hermitian operators $J_x$ and $J_y$ can be introduced by starting with infinitesimal rotations about the $Ox$ and $Oy$ axes. The total angular momentum $\mathbf{J}$ of the system is then defined in terms of its three components $J_x$, $J_y$ and $J_z$.

Now we can use the reasoning of §3-c-β: the geometrical relation (8) implies that the components of $\mathbf{J}$ satisfy commutation relations which are identical to those of orbital angular momenta. Thus, *the total angular momentum of any quantum mechanical system is related to the corresponding rotation operators; the commutation relations amongst its components follow directly*, which enables us to use them, as in chapter VI (§ B-2), to characterize any angular momentum.

Finally, let us show that, with $J_x$, $J_y$ and $J_z$ defined as we have just indicated, the operator $R_{\mathbf{u}}(\mathrm{d}\alpha)$ associated with an arbitrary infinitesimal rotation is written ($u_x$, $u_y$ and $u_z$ being the components of the unit vector $\mathbf{u}$):

$$R_{\mathbf{u}}(\mathrm{d}\alpha) = 1 - \frac{i}{\hbar}\,\mathrm{d}\alpha\,(J_x u_x + J_y u_y + J_z u_z) \tag{80}$$

which can then be condensed into the form:

$$R_{\mathbf{u}}(\mathrm{d}\alpha) = 1 - \frac{i}{\hbar}\,\mathrm{d}\alpha\,\mathbf{J}\cdot\mathbf{u} \tag{81}$$

Formula (80) is simply a consequence of the geometrical relation:

$$\mathscr{R}_{\mathbf{u}}(\mathrm{d}\alpha) = \mathscr{R}_{\mathbf{e}_x}(u_x\,\mathrm{d}\alpha)\,\mathscr{R}_{\mathbf{e}_y}(u_y\,\mathrm{d}\alpha)\,\mathscr{R}_{\mathbf{e}_z}(u_z\,\mathrm{d}\alpha) \tag{82}$$

valid to first order in $\mathrm{d}\alpha$, and which can be obtained directly from formula (6).

We have thus generalized expressions (48) and (49) for infinitesimal rotation operators. Since the group law is conserved locally (see above), relation (52) and the argument following it remain valid. Consequently, the finite rotation operators have expressions analogous to (55) and (57):

$$R_{\mathbf{u}}(\alpha) = \mathrm{e}^{-\frac{i}{\hbar}\alpha\,\mathbf{J}.\mathbf{u}} \tag{83}$$

## 5. Rotation of observables

We now know how the vector representing the state of a quantum mechanical system transforms under rotation. But in quantum mechanics, the state of a system and the physical quantities are described independently. Therefore, we shall now indicate what happens to observables upon rotation.

### a. GENERAL TRANSFORMATION LAW

Consider an observable $A$, relating to a given physical system; we shall assume, to simplify the notation, the spectrum of $A$ to be discrete and non-degenerate:

$$A\,|\,u_n\rangle = a_n\,|\,u_n\rangle \tag{84}$$

In order to understand how this observable is affected by a rotation, we shall imagine that we have a device which can measure $A$ in the physical system under consideration. Now, the observable $A'$, the transform of $A$ with respect to the geometrical rotation $\mathscr{R}$, is by definition what is measured by the device when it has been subjected to the rotation $\mathscr{R}$.

Let us assume the system to be in the eigenstate $|\,u_n\rangle$ of $A$: the device for measuring $A$ in this system will give the result $a_n$ without fail. But just before performing the measurement, we apply a rotation $\mathscr{R}$ to the physical system and, simultaneously, to the measurement device; their relative positions are unchanged. Consequently, if the observable $A$ which we are considering describes a physical quantity attached only to the system which we have rotated (that is, independent of other systems or devices which we have not rotated), then, in its new position, the measurement device will still give the *same* result $a_n$ without fail. Now, after rotation, the device, by definition, measures $A'$, and the system is in the state:

$$|\,u_n'\rangle = R\,|\,u_n\rangle \tag{85}$$

We must therefore have:

$$A\,|\,u_n\rangle = a_n\,|\,u_n\rangle \Longrightarrow A'\,|\,u_n'\rangle = a_n\,|\,u_n'\rangle \tag{86}$$

Combining (85) and (86), we find:

$$A'R\,|\,u_n\rangle = a_nR\,|\,u_n\rangle \tag{87}$$

that is:

$$R^{\dagger}A'R \mid u_n \rangle = a_n \mid u_n \rangle \tag{88}$$

since the inverse of $R$ is $R^{\dagger}$. The set of vectors $\mid u_n \rangle$ constitutes a basis in the state space ($A$ is an observable), so we have:

$$R^{\dagger}A'R = A \tag{89}$$

that is:

$$\boxed{A' = RAR^{\dagger}} \tag{90}$$

In the special case of an infinitesimal rotation $\mathscr{R}_{\mathbf{u}}(\mathrm{d}\alpha)$, the general expression (81), substituted into (90), gives, to first order in $\mathrm{d}\alpha$:

$$\begin{aligned} A' &= \left(1 - \frac{i}{\hbar}\mathrm{d}\alpha\, \mathbf{J}.\mathbf{u}\right)A\left(1 + \frac{i}{\hbar}\mathrm{d}\alpha\, \mathbf{J}.\mathbf{u}\right) \\ &= A - \frac{i}{\hbar}\mathrm{d}\alpha\, [\mathbf{J}.\mathbf{u}, A] \end{aligned} \tag{91}$$

COMMENTS:

(*i*) In the case of a spinless particle, relation (90) implies that:

$$\langle \mathbf{r} \mid A' \mid \mathbf{r}' \rangle = \langle \mathbf{r} \mid RAR^{\dagger} \mid \mathbf{r}' \rangle \tag{92}$$

Using (26) and (27), we therefore obtain:

$$\langle \mathbf{r} \mid A' \mid \mathbf{r}' \rangle = \langle \mathscr{R}^{-1}\mathbf{r} \mid A \mid \mathscr{R}^{-1}\mathbf{r}' \rangle \tag{93}$$

The transformation which enables us to obtain $A'$ from $A$ is therefore completely analogous to the one which gives $\mid \psi' \rangle$ in terms of $\mid \psi \rangle$ [*cf.* (19)].

(*ii*) Consider the case in which the observable $A$ is associated with a classical quantity $\mathscr{A}$. $\mathscr{A}$ is then a function of the positions $r_i$ and momenta $p_i$ of the particles which constitute the system; the operator $A$ is obtained from this function by applying the quantization rules given in chapter III. We know how to find the quantity $\mathscr{A}'$ associated with $\mathscr{A}$ by a rotation $\mathscr{R}$ in classical mechanics: for example, if $\mathscr{A}$ is a scalar, $\mathscr{A}'$ is the same as $\mathscr{A}$; if $\mathscr{A}$ is the component along an axis $Ou$ of a vectorial quantity, $\mathscr{A}'$ is the component of this same vectorial quantity along the axis which is the result of the transformation of $Ou$ by the rotation $\mathscr{R}$. We can also construct the quantum mechanical operator corresponding to $\mathscr{A}'$, by applying the same quantization rules as above. It can be shown that this operator is the same as the operator $A'$ given in (90); this is what is shown in figure 1.

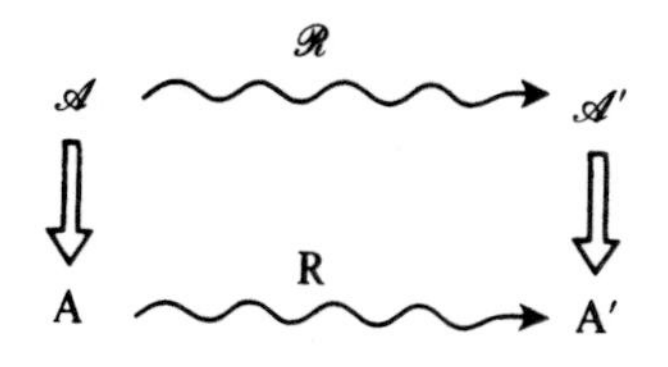


FIGURE 1

**Behavior, under a rotation $\mathscr{R}$, of a classical physical quantity $\mathscr{A}$ and of the associated observable $A$.**

#### b. SCALAR OBSERVABLES

An observable $A$ is said to be scalar if:

$$A' = A \tag{94}$$

for all $R$. According to (91), this implies that:

$$[A, \mathbf{J}] = \mathbf{0} \tag{95}$$

*A scalar observable commutes with the three components of the total angular momentum.*

There are numerous examples of scalar observables. $\mathbf{J}^2$ is always a scalar (this results, as we saw in § B-2 of chapter VI, from the commutation relations which characterize an angular momentum). For a spinless particle, $\mathbf{R}^2$, $\mathbf{P}^2$ and $\mathbf{R} . \mathbf{P}$, which correspond to classical scalar quantities, are scalars. It is easy to show, moreover (*cf.* § 5-c below) that $\mathbf{R}^2$, $\mathbf{P}^2$ and $\mathbf{R} . \mathbf{P}$ satisfy (95). We shall also see later (§6) that the Hamiltonian of an isolated physical system is a scalar.

#### c. VECTOR OBSERVABLES

A vector observable $\mathbf{V}$ is a set of three observables $V_x$, $V_y$, $V_z$ (its Cartesian components) which is transformed by rotation according to the characteristic law of vectors. The transform, under a rotation $\mathscr{R}$, of the component $V_u = \mathbf{V} . \mathbf{u}$ of $\mathbf{V}$ along a given axis $Ou$ (of unit vector $\mathbf{u}$) must be the component $V_{u'} = \mathbf{V} . \mathbf{u}'$ of $\mathbf{V}$ along the axis $Ou'$ derived from $Ou$ by the rotation $\mathscr{R}$.

Consider, for example, the component $V_x$ of such an observable. We shall examine its behavior under infinitesimal rotations about each of the coordinate axes. $V_x$ is obviously unchanged by a rotation about $Ox$; according to (91), this can be expressed in the form:

$$[J_x, V_x] = 0 \tag{96}$$

If we perform a rotation $\mathscr{R}_{\mathbf{e}_y}(\mathrm{d}\alpha)$ about the $Oy$ axis, the transform of $V_x$ is the observable $(V_x)'$ given by (91) as:

$$(V_x)' = V_x - \frac{i}{\hbar}\,\mathrm{d}\alpha\,[J_y, V_x] \tag{97}$$

But $V_x$ is the component of $\mathbf{V}$ along the $Ox$ axis, of unit vector $\mathbf{e}_x$. The rotation $\mathscr{R}_{\mathbf{e}_y}(\mathrm{d}\alpha)$ takes $\mathbf{e}_x$ onto $\mathbf{e}'_x$ such that [formula (6)]:

$$\begin{aligned}\mathbf{e}'_x &= \mathbf{e}_x + \mathrm{d}\alpha\,\mathbf{e}_y \times \mathbf{e}_x \\ &= \mathbf{e}_x - \mathrm{d}\alpha\,\mathbf{e}_z\end{aligned} \tag{98}$$

Consequently, if $\mathbf{V}$ is a vector observable, $(V_x)'$ must be the same as $\mathbf{V} . \mathbf{e}'_x$:

$$\begin{aligned}(V_x)' &= \mathbf{V} . \mathbf{e}_x - \mathrm{d}\alpha\,\mathbf{V} . \mathbf{e}_z \\ &= V_x - \mathrm{d}\alpha V_z\end{aligned} \tag{99}$$

Comparing (97) and (99), we see that:

$$[V_x, J_y] = i\hbar V_z \tag{100}$$

For an infinitesimal rotation $\mathscr{R}_{\mathbf{e}_z}(\mathrm{d}\alpha)$ about the $Oz$ axis, an argument analogous to the one above leads to the relation :

$$[J_z, V_x] = i\hbar V_y \tag{101}$$

By studying the behavior of $V_y$ and $V_z$ under infinitesimal rotations, one can prove the formulas which can be derived from (96), (100) and (101) by cyclic permutation of the indices $x$, $y$, $z$. The set of relations obtained in this way is characteristic of a vector observable : they imply that an arbitrary infinitesimal rotation transforms $\mathbf{V} \cdot \mathbf{u}$ into $\mathbf{V} \cdot \mathbf{u}'$, where $\mathbf{u}'$ is the transform of $\mathbf{u}$ with respect to the rotation under consideration.

It is clear that the angular momentum $\mathbf{J}$ itself is a vector observable ; (96), (100) and (101) then follow from the commutation relations characterizing angular momenta. For a system composed of a single spinless particle, $\mathbf{R}$ and $\mathbf{P}$ are vector observables, as can easily be verified from the canonical commutation relations. Thus the vector notation we use for $\mathbf{R}$, $\mathbf{P}$, $\mathbf{L}$ and $\mathbf{J}$ is justified.

COMMENTS:

(*i*) The scalar product $\mathbf{V} \cdot \mathbf{W}$ of two vector observables, defined by the customary formula:

$$\mathbf{V} \cdot \mathbf{W} = V_x W_x + V_y W_y + V_z W_z \tag{102}$$

is a scalar operator. To verify this, we can calculate, for example, the commutator of $\mathbf{V} \cdot \mathbf{W}$ with $J_x$:

$$\begin{aligned} [\mathbf{V} \cdot \mathbf{W}, J_x] &= [V_y W_y, J_x] + [V_z W_z, J_x] \\ &= V_y [W_y, J_x] + [V_y, J_x] W_y + V_z [W_z, J_x] + [V_z, J_x] W_z \\ &= - i\hbar V_y W_z - i\hbar V_z W_y + i\hbar V_z W_y + i\hbar V_y W_z \\ &= 0 \end{aligned} \tag{103}$$

We have already pointed out that $\mathbf{J}^2$, $\mathbf{R}^2$, $\mathbf{P}^2$ and $\mathbf{R} \cdot \mathbf{P}$ are scalar observables.

(*ii*) It is the *total angular momentum of the system* under study which appears in relations (96), (100) and (101). The following example illustrates the importance of this fact : if, for a two-particle system, we were to use $\mathbf{L}_1$ instead of $\mathbf{L} = \mathbf{L}_1 + \mathbf{L}_2$, $\mathbf{R}_2$ would appear to be a set of three scalar observables and not a vector observable.

## 6. Rotation invariance

The discussion presented in the preceding sections does not have as its sole purpose the justification of the definition of angular momenta in terms of the commutation relations. The importance of rotations in physics is essentially related to the fact that physical laws are rotation invariant. We are going to explain in this section exactly what this means, and we shall indicate some of the consequences of this fundamental property.

### a. INVARIANCE OF PHYSICAL LAWS

Consider a physical system $(S)$, classical or quantum mechanical, which we subject to a rotation $\mathscr{R}$ at some given time. If we take the precaution of rotating, at the same time as the system $(S)$ under consideration, all other systems or devices which can influence it, *the physical properties and behavior of $(S)$ are not modified.* This means that the physical laws governing the system have remained the same : the physical laws are said to be rotation invariant. Note that this property is not at all obvious : there exist transformations – those of similarity*, for example – with respect to which the physical laws are not invariant**. It is therefore advisable to consider rotation invariance to be a postulate which is justified by the experimental verification of its consequences.

When we say that the physical properties and behavior of a system are unchanged by a rotation performed at the time $t_0$, this statement covers two observations :

($i$) the properties of the system at this time are not modified (although the description of the state of the system and the physical quantities are; see preceding sections). In quantum mechanics, this implies that the transform $A'$ of an arbitrary observable $A$ has the same spectrum, and that the probability of finding one of the eigenvalues of this spectrum in a measurement of $A'$ on the system after rotation is the same as it was for the measurement of $A$ on the system before rotation. From this, it can be deduced that the operators $R$ describing rotations in state space are linear and unitary, or antilinear and unitary (that is, anti-unitary***).

($ii$) the time evolution of the system is not affected. To state this point more precisely, let us denote the state of the system by $|\psi(t_0)\rangle$; under a rotation performed at $t_0$, this state becomes :

$$|\psi'(t_0)\rangle = R|\psi(t_0)\rangle \tag{104}$$

We now let the system evolve freely and compare its state $|\psi'(t)\rangle$ at a subsequent time $t$ to the state $|\psi(t)\rangle$ which it would have attained if it had been allowed to evolve freely from $|\psi(t_0)\rangle$. If the behavior of the system is not modified, we must have :

$$|\psi'(t)\rangle = R|\psi(t)\rangle \tag{105}$$

that is, for all $t$, the state $|\psi'(t)\rangle$ must be obtained from $|\psi(t)\rangle$ by the same rotation as in (104). Therefore, if $|\psi(t)\rangle$ is a solution of the Schrödinger equation, $R|\psi(t)\rangle$ is also a solution of this equation : *the transform of a possible motion of the system is also a possible motion.* We shall see in §b that this implies that the Hamiltonian $H$ of the system is a scalar observable.

* Consider, for example, a hydrogen atom. If we multiply the distance between the proton and the electron by a constant $\lambda \neq 1$ (without modifying the charges and masses of the particles), we obtain a system whose evolution no longer obeys either classical or quantum mechanical physical laws.

** Let us also point out that experiments have shown that the laws governing the $\beta$-decay of nuclei are not invariant under reflection with respect to a plane (non-conservation of parity).

*** All transformations which leave the physical laws invariant are described by unitary operators, except for time reversal, with which is associated an anti-unitary operator.

The invariance of physical laws under rotation is related to the *symmetry properties* of the equations which state these laws mathematically. To understand the origin of these symmetries, consider, for example, a system composed of a single (spinless) particle. The expression of the physical laws governing such a system explicitly involves the parameters $\mathbf{r}(x, y, z)$ and $\mathbf{p}(p_x, p_y, p_z)$ which characterize the position of the particle and its momentum : in classical mechanics, **r** and **p** define at each instant the state of the particle; in quantum mechanics, although the meaning of these parameters is a little less simple, they appear in the wave function $\psi(\mathbf{r})$ and its Fourier transform $\bar{\psi}(\mathbf{p})$. When the particle is subjected to an instantaneous rotation $\mathscr{R}$, **r** and **p** are transformed into $\mathbf{r}'$ and $\mathbf{p}'$ such that:

$$\begin{aligned} \mathbf{r}' &= \mathscr{R}\mathbf{r} \\ \mathbf{p}' &= \mathscr{R}\mathbf{p} \end{aligned} \tag{106}$$

If we replace **r** by $\mathscr{R}^{-1}\mathbf{r}'$ and **p** by $\mathscr{R}^{-1}\mathbf{p}'$ in the equations which express the physical laws, we obtain relations which now involve $\mathbf{r}'$ and $\mathbf{p}'$. The invariance of physical laws under the rotation $\mathscr{R}$ thus implies that *the form of the equations for* $\mathbf{r}'$ *and* $\mathbf{p}'$ *is the same as that of the equations for* **r** *and* **p** : simply omitting the primes labeling the new parameters must give us back the initial equations. It is clear that this considerably restrains the number of possible forms of these equations.

COMMENTS:

(*i*) What happens when we perform a rotation on a system which is not isolated? Consider, for example, a particle in an external potential. If we rotate this system, *without simultaneously rotating the sources of the external potential*, the subsequent evolution of the system is, in general, modified*. In classical mechanics, the forces exerted on the particle are not the same in its new position. In quantum mechanics, $\psi'(\mathbf{r}, t) = \psi(\mathscr{R}^{-1}\mathbf{r}, t)$ is a solution of a Schrödinger equation in which the potential $V(\mathbf{r})$ is replaced by $V(\mathscr{R}^{-1}\mathbf{r})$, which is, in general, different from $V(\mathbf{r})$. Therefore, the transform of a possible motion is no longer a possible motion. The presence of the external potential destroys, so to speak, the homogeneity of the space in which the system under study evolves.

However, the external potential may present certain symmetries which allow certain rotations to be performed on the physical system without its behavior being modified. If there exist such rotations $\mathscr{R}_0$ such that $V(\mathscr{R}_0^{-1}\mathbf{r})$ is the same as $V(\mathbf{r})$, the properties of the system are unchanged by one of these rotations $\mathscr{R}_0$. This is the case, for example, for central potentials, that is, those which depend only on the distance to a fixed point $O$ : the rotations $\mathscr{R}_0$ are then all rotations which conserve the point $O$ (*cf.* chap. VII).

(*ii*) Let us return to the case of isolated physical systems. Thus far, we have adopted an "*active*" *viewpoint* : the observer remains fixed, and the physical system is rotated.

* If the particle is placed in a vector potential, its properties immediately after the rotation may be profoundly modified. Consider, for example, a spinless particle in an external magnetic field. According to the transformation law (19), the probability current given by formula (D-20) of chapter III cannot in general be derived by rotation of the initial current, as it depends on the vector potential describing the magnetic field.

The physical interpretation of this phenomenon is as follows. We can imagine that, instead of rotating the particle, we rotate the magnetic field rapidly and in the opposite direction. The wave function has not had the time to change : this corresponds to formula (19). If the physical properties are modified, it is because an induced electromotive field has appeared and acted on the particle. This action does not depend on the exact way in which we rotate the magnetic field, provided that we do so quickly enough.

We can also define a "*passive*" *viewpoint* : the observer rotates, and, without touching the system being studied, uses a new coordinate frame, derived from the initial frame by the given rotation. Rotation invariance is then expressed in the following way : in his new position (that is, using his new coordinate axes), the observer describes physical phenomena by laws which have the same form as in the old frame. Nothing allows him to assert that one of his positions is more fundamental than the other : it is impossible to define an absolute orientation in space by the study of any physical phenomenon. It is clear that, for an isolated system, a "passive" rotation is equivalent to the "active" rotation through an equal angle about the opposite axis.

### b. CONSEQUENCE: CONSERVATION OF ANGULAR MOMENTUM

We indicated in §6-a that rotation invariance is related to symmetry properties of the equations which express the physical laws. Here, we shall study the case of the Schrödinger equation, and we shall show that *the Hamiltonian of an isolated physical system is a scalar observable.*

Consider an isolated system in the state $|\psi(t_0)\rangle$. We perform an arbitrary rotation $\mathcal{R}$ at time $t_0$ ; the state of the system becomes:

$$|\psi'(t_0)\rangle = R\,|\psi(t_0)\rangle \tag{107}$$

where $R$ is the "image" of the rotation $\mathcal{R}$. If we now let the system evolve freely from $|\psi'(t_0)\rangle$, its state at the instant $t_0 + \mathrm{d}t$, according to the Schrödinger equation, will be :

$$|\psi'(t_0 + \mathrm{d}t)\rangle = |\psi'(t_0)\rangle + \frac{\mathrm{d}t}{i\hbar} H\,|\psi'(t_0)\rangle \tag{108}$$

Now, if we had not performed the rotation, the state of the system at time $t_0 + \mathrm{d}t$ would have been:

$$|\psi(t_0 + \mathrm{d}t)\rangle = |\psi(t_0)\rangle + \frac{\mathrm{d}t}{i\hbar} H\,|\psi(t_0)\rangle \tag{109}$$

Rotation invariance implies (*cf.* §6-a) that :

$$|\psi'(t_0 + \mathrm{d}t)\rangle = R\,|\psi(t_0 + \mathrm{d}t)\rangle \tag{110}$$

where $R$ is the same as in (107). According to the two preceding equations, this implies that :

$$RH\,|\psi(t_0)\rangle = H\,|\psi'(t_0)\rangle \tag{111}$$

that is:

$$RH|\psi(t_0)\rangle = HR|\psi(t_0)\rangle \tag{112}$$

Since $|\psi(t_0)\rangle$ is arbitrary, it follows that $H$ commutes with all rotation operators. For this to be so, it is necessary and sufficient that $H$ commute with the infinitesimal rotation operators, that is, with the three components of the total angular momentum $\mathbf{J}$ of the system:

$$[H, \mathbf{J}] = 0 \tag{113}$$

$H$ is therefore a scalar observable.

Rotation invariance is therefore related to the fact that *the total angular momentum of an isolated system is a constant of the motion*: conservation of angular momentum can be seen to be a consequence of rotation invariance.

COMMENTS:

(*i*) The Hamiltonian of a non-isolated system is not, in general, a scalar. However, if certain rotations exist which leave the system invariant [comment (*i*) of § 6-a], the Hamiltonian commutes with the corresponding operators. Thus, the Hamiltonian of a particle in a central potential commutes with the operator **L** associated with the angular momentum of the particle with respect to the center of forces.

(*ii*) For an isolated system composed of several interacting particles, the Hamiltonian commutes with the *total* angular momentum. However, it does not generally commute with the individual angular momentum of each particle. For the transform of a possible motion to remain a possible motion, the rotation must be performed on the whole system, not on only some of the particles.

## c. APPLICATIONS

We have just shown that rotation invariance means that the total angular momentum **J** of an isolated system is a constant of the motion in the quantum mechanical sense. It is therefore useful to determine the stationary states of such a system (eigenstates of the Hamiltonian) which are also eigenstates of $\mathbf{J}^2$ and $J_z$. We can then choose for the state space a standard basis, $\{ | k, j, m \rangle \}$, composed of eigenstates common to $H$, $\mathbf{J}^2$ and $J_z$:

$$\begin{aligned} H \, | k, j, m \rangle &= E \, | k, j, m \rangle \\ \mathbf{J}^2 | k, j, m \rangle &= j(j+1)\hbar^2 \, | k, j, m \rangle \\ J_z \, | k, j, m \rangle &= m\hbar \, | k, j, m \rangle \end{aligned} \tag{114}$$

### α. *Essential rotational degeneracy*

Since the Hamiltonian $H$ is a scalar observable, it commutes with $J_+$ and $J_-$. From this fact, we deduce that $| k, j, m+1 \rangle$ and $| k, j, m-1 \rangle$, which are respectively proportional to $J_+ \, | k, j, m \rangle$ and $J_- \, | k, j, m \rangle$, are eigenstates of $H$ with the same eigenvalue as $| k, j, m \rangle$ [the argument is the same as for formula (C-48) of chapter VI]. Thus it can be shown by iteration that the $(2j + 1)$ vectors of the standard basis characterized by the given values of $k$ and $j$ have the same energy. The corresponding degeneracy of the eigenvalues of $H$ is called "essential" because it arises from rotation invariance and occurs for any form of the Hamiltonian $H$. Of course, in certain cases, the energy levels can present additional degeneracies, which are called "accidental". We shall see an example of this in chapter VII, § C.

### β. *Matrix elements of observables in a standard basis*

When we study a physical quantity in an isolated system, the knowledge of the behavior of the associated observable under a rotation enables us to establish some of its properties, without having to consider its precise form. We can predict that only some of its matrix elements in a standard basis such as $\{ | k, j, m \rangle \}$ will be different from zero, and we can give the relations between them. Thus, a scalar observable has matrix elements only between two basis vectors whose values of $j$ are equal, as are their values of $m$ [this results from the fact that this observable commutes with $\mathbf{J}^2$ and $J_z$; *cf.* theorems on commuting observables, §D-3-a of chapter II]. Moreover, these non-zero elements are independent of $m$ (since the scalar observable also commutes with $J_+$ and $J_-$). For vector or tensor observables, these properties are contained in the *Wigner-Eckart theorem*, which we shall prove later in a special case (*cf.* complement $D_X$), and which is frequently used in the areas of physics in which phenomena are treated by quantum mechanics (atomic, molecular, and nuclear physics, elementary particle physics, etc.).

**References and suggestions for further reading:**

Symmetry and conservation laws : Feynman III (1.2), chap. 17; Schiff (1.18), chap. 7; Messiah (1.17), chap. XV; see also the articles by Morrisson (2.28), Feinberg and Goldhaber (2.29), Wigner (2.30).

Relation with group therory : Messiah (1.17), appendix D; Meijer and Bauer (2.18), chaps. 5 and 6; Bacry (10.31), chap. 6; Wigner (2.23), chaps. 14 and 15.

# Complement $C_{VI}$

## ROTATION OF DIATOMIC MOLECULES



### 1. Introduction

In §1 of complement $A_V$, we studied the vibrations of the two nuclei of a diatomic molecule about their equilibrium position, neglecting the rotation of these two nuclei about their center of mass. Thus we obtained stationary vibrational states of energies $E_v$ whose wave functions $\varphi_v(r)$ depended only on the distance $r$ between the nuclei.

Here, we shall adopt a complementary point of view: we shall study the rotation of the two nuclei about their center of mass, neglecting their vibrations. That is, we shall assume that the distance $r$ between them remains fixed and equal to $r_e$ (where $r_e$ represents the distance between the two nuclei in the stable equilibrium position of the molecule; see figure 1 of complement $A_V$). The wave functions of the stationary rotational states then can depend only on the polar angles $\theta$ and $\varphi$ which define the direction of the molecular axis. We shall see that these wave functions are the spherical harmonics $Y_l^m(\theta, \varphi)$ [studied in chapter VI (§ D-1) and in complement $A_{VI}$], and that they correspond to a rotational energy $E_l$ which depends only on $l$.

Actually, in the center of mass frame, the molecule both rotates and vibrates, and the wave functions of its stationary states must be functions of the three variables $r$, $\theta$ and $\varphi$. We shall show in complement $F_{VII}$ that, to a first approximation, these wave functions are of the form $\frac{1}{r}\varphi_v(r)Y_l^m(\theta, \varphi)$ and correspond to the energy $E_v + E_l$. This result justifies the approach adopted here, which consists of considering only one degree of freedom – rotational or vibrational – at a time*. We shall begin in § 2 by presenting the classical study of a system

* In complement $F_{VII}$, we shall also study the corrections introduced by the coupling between the vibrational and rotational degrees of freedom.

of two masses separated by a fixed distance (rigid rotator). The quantum mechanical treatment of this problem will then be taken up in §3, where we shall use the results of chapter VI concerning orbital angular momentum. Finally, in §4, we shall describe some experimental manifestations of the rotation of diatomic molecules (pure and Raman rotational spectra).

## 2. Rigid rotator. Classical study

### a. NOTATION

Two particles, of mass $m_1$ and $m_2$, are separated by a fixed distance $r_e$. Their center of mass $O$ is chosen as the origin of a coordinate frame $Oxyz$ with respect to which the direction of the axis connecting them is defined by means of the polar angles $\theta$ and $\varphi$ (fig. 1). The distances $OM_1$ and $OM_2$ are denoted respectively by $r_1$ and $r_2$; by definition of the center of mass:

$$m_1 r_1 = m_2 r_2 \tag{1}$$

which allows us to write:

$$\frac{r_1}{m_2} = \frac{r_2}{m_1} = \frac{r_e}{m_1 + m_2} \tag{2}$$

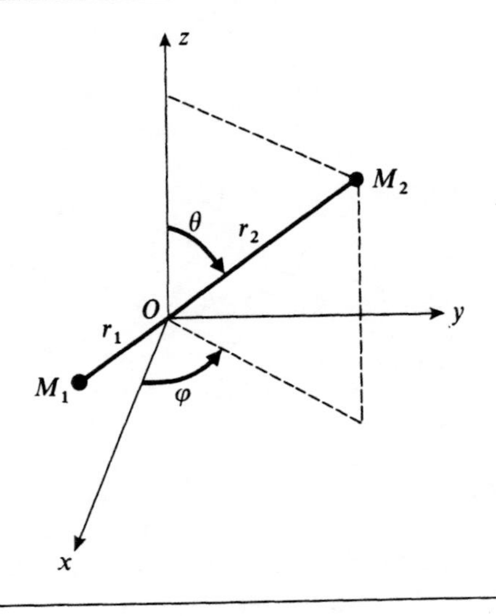


FIGURE 1

**Parameters defining the position of the rigid rotator $M_1M_2$ whose center of mass is at the origin $O$ of the reference frame; the distances $r_1$ and $r_2$ are fixed; only the polar angles $\theta$ and $\varphi$ can vary.**

The moment of inertia $I$ of the system with respect to $O$ is equal to:

$$I = m_1 r_1^2 + m_2 r_2^2 \tag{3}$$

Introducing the reduced mass:

$$\mu = \frac{m_1 m_2}{m_1 + m_2} \tag{4}$$

and using (2), we can put $I$ in the form:

$$I = \mu r_e^2 \tag{5}$$

#### b. MOTION OF THE ROTATOR. ANGULAR MOMENTUM AND ENERGY

If no external force acts on the rotator, the total angular momentum $\mathscr{L}$ of the system with respect to the point $O$ is a constant of the motion. The rotator therefore rotates about $O$ in a plane perpendicular to the fixed vector $\mathscr{L}$, with a constant angular velocity $\omega_R$. The modulus of $\mathscr{L}$ is related to $\omega_R$ by:

$$|\mathscr{L}| = m_1 r_1 r_1 \omega_R + m_2 r_2 r_2 \omega_R = I\,\omega_R \tag{6}$$

that is, using (5):

$$|\mathscr{L}| = \mu r_e^2 \omega_R \tag{7}$$

The rotational frequency of the system $\nu_R = \omega_R/2\pi$ is proportional to the angular momentum $|\mathscr{L}|$ and inversely proportional to the moment of inertia $I$.

In the center of mass frame, the total energy $\mathscr{H}$ of the system reduces to the rotational kinetic energy:

$$\mathscr{H} = \frac{1}{2} I \omega_R^2 \tag{8}$$

which can also be written, using (6) and (5):

$$\mathscr{H} = \frac{\mathscr{L}^2}{2I} = \frac{\mathscr{L}^2}{2\mu r_e^2} \tag{9}$$

#### c. THE FICTITIOUS PARTICLE ASSOCIATED WITH THE ROTATOR

Formulas (5), (7) and (9) show that the problem we are studying here is formally equivalent to that of a fictitious particle of mass $\mu$ forced to remain at a fixed distance $r_e$ from the point $O$, about which it rotates with the angular velocity $\omega_R$. $\mathscr{L}$ is the angular momentum of this fictitious particle with respect to $O$.

### 3. Quantization of the rigid rotator

#### a. THE QUANTUM MECHANICAL STATE AND OBSERVABLES OF THE ROTATOR

Since $r_e$ is fixed, the parameters defining the position of the rotator (or that of the associated fictitious particle) are the polar angles $\theta$ and $\varphi$ of figure 1. The quantum mechanical state of the rotator will then be described by a wave function $\psi(\theta, \varphi)$ which depends only on these two parameters. $\psi(\theta, \varphi)$ is square-integrable; we shall assume it to be normalized:

$$\int_0^{2\pi} d\varphi \int_0^{\pi} \sin\theta\, d\theta\, |\psi(\theta, \varphi)|^2 = 1 \tag{10}$$

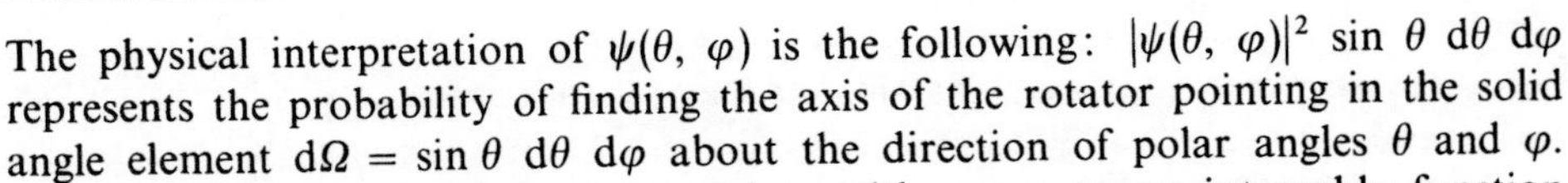

The physical interpretation of $\psi(\theta, \varphi)$ is the following: $|\psi(\theta, \varphi)|^2 \sin\theta \, d\theta \, d\varphi$ represents the probability of finding the axis of the rotator pointing in the solid angle element $d\Omega = \sin\theta \, d\theta \, d\varphi$ about the direction of polar angles $\theta$ and $\varphi$.

Using Dirac notation, we associate with every square-integrable function $\psi(\theta, \varphi)$, a ket $| \psi \rangle$ of the state space $\mathscr{E}_\Omega$:

$$\psi(\theta, \varphi) \longleftrightarrow | \psi \rangle \in \mathscr{E}_\Omega \tag{11}$$

The scalar product of $| \psi \rangle$ and $| \chi \rangle$ is, by definition:

$$\langle \chi | \psi \rangle = \int d\Omega \, \chi^*(\theta, \varphi) \, \psi(\theta, \varphi) \tag{12}$$

where $\chi(\theta, \varphi)$ and $\psi(\theta, \varphi)$ are the wave functions associated with $| \chi \rangle$ and $| \psi \rangle$.

The quantum mechanical Hamiltonian $H$ of the rotator (or of the associated fictitious particle) can be obtained by replacing $\mathscr{L}^2$ in expression (9) for the classical energy by the operator $\mathbf{L}^2$ studied in § D of chapter VI:

$$H = \frac{\mathbf{L}^2}{2\mu r_e^2} \tag{13}$$

$H$ is an operator acting in $\mathscr{E}_\Omega$. According to formula (D-6-a) of chapter VI, if $| \psi \rangle$ is represented by the wave function $\psi(\theta, \varphi)$, $H | \psi \rangle$ is represented by:

$$H | \psi \rangle \longleftrightarrow -\frac{\hbar^2}{2\mu r_e^2} \left[ \frac{\partial^2}{\partial\theta^2} + \frac{1}{\tan\theta} \frac{\partial}{\partial\theta} + \frac{1}{\sin^2\theta} \frac{\partial^2}{\partial\varphi^2} \right] \psi(\theta, \varphi) \tag{14}$$

Other observables of interest, which we shall study later, are those which correspond to the three algebraic projections $x$, $y$, $z$ of the segment $M_1 M_2$ ($x$, $y$, $z$ are also the coordinates of the fictitious particle):

$$\begin{aligned} x &= r_e \sin\theta \cos\varphi \\ y &= r_e \sin\theta \sin\varphi \\ z &= r_e \cos\theta \end{aligned} \tag{15}$$

The importance of these variables will be seen in § 4-a. The observables $X$, $Y$, $Z$ corresponding to $x$, $y$, $z$ act in $\mathscr{E}_\Omega$. With the kets $X | \psi \rangle$, $Y | \psi \rangle$, $Z | \psi \rangle$, are associated the functions:

$$\begin{aligned} X | \psi \rangle &\longleftrightarrow r_e \sin\theta \cos\varphi \, \psi(\theta, \varphi) \\ Y | \psi \rangle &\longleftrightarrow r_e \sin\theta \sin\varphi \, \psi(\theta, \varphi) \\ Z | \psi \rangle &\longleftrightarrow r_e \cos\theta \, \psi(\theta, \varphi) \end{aligned} \tag{16}$$

COMMENT:

As we have already pointed out in the introduction, the true wave functions of the molecule depend on $r$, $\theta$, $\varphi$. Similarly, the observables of this molecule, obtained from the corresponding classical quantities by the quantization rules of chapter III, act on these functions of three variables and not solely on the functions of $\theta$ and $\varphi$. In complement $F_{VII}$, we shall justify the point of view we are adopting here, namely, ignoring the radial part of the wave functions and considering $r$ to be a fixed parameter which is equal to $r_e$ [*cf.* formulas (14) and (16)].

### b. EIGENSTATES AND EIGENVALUES OF THE HAMILTONIAN

We determined the eigenvalues of the operator $\mathbf{L}^2$ in §D of chapter VI: they are of the form $l(l+1)\hbar^2$, where $l$ is any non-negative integer. Furthermore, we know an orthonormal system of eigenfunctions of $\mathbf{L}^2$: the spherical harmonics $Y_l^m(\theta, \varphi)$, which constitute a basis in the space of functions which are square-integrable in $\theta$ and $\varphi$ (§D-1-c-β of chapter VI). We shall denote by $| l, m \rangle$ the ket of $\mathscr{E}_\Omega$ associated with $Y_l^m(\theta, \varphi)$:

$$Y_l^m(\theta, \varphi) \longleftrightarrow | l, m \rangle \tag{17}$$

We see from (13) that:

$$H | l, m \rangle = \frac{l(l+1)\hbar^2}{2\mu r_e^2} | l, m \rangle \tag{18}$$

It is customary to set:

$$B = \frac{\hbar}{4\pi I} = \frac{\hbar}{4\pi\mu r_e^2} \tag{19}$$

$B$ is called the "rotational constant" and has the dimensions of a frequency*. The eigenvalues of $H$ are thus of the form:

$$E_l = Bh\,l(l+1) \tag{20}$$

Since, for a given value of $l$, there exist $(2l+1)$ spherical harmonics $Y_l^m(\theta, \varphi)$ ($m = -l, -l+1, ..., l$), we see that each eigenvalue $E_l$ is $(2l+1)$-fold degenerate. Figure 2 represents the first energy levels of the rotator. The separation of two adjacent levels, $l$ and $l-1$, is equal to:

$$E_l - E_{l-1} = Bh\left[l(l+1) - l(l-1)\right] = 2Bhl \tag{21}$$

and increases linearly with $l$.

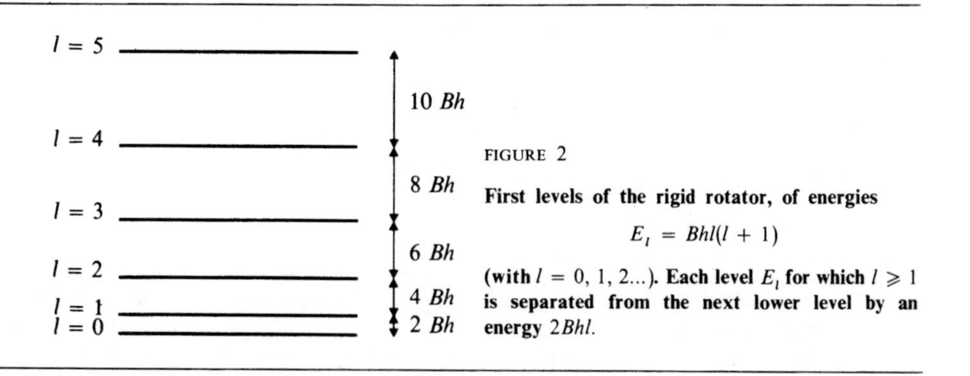


FIGURE 2

**First levels of the rigid rotator, of energies**

$$E_l = Bhl(l+1)$$

**(with $l = 0, 1, 2...$). Each level $E_l$ for which $l \geqslant 1$ is separated from the next lower level by an energy $2Bhl$.**

* The speed of light $c$ is sometimes placed in the denominator of the right-hand side of (19). $B$ then has the dimensions of an inverse length and is expressed in $cm^{-1}$ (in the $CGS$ system).

The eigenstates of $H$ satisfy the following orthogonality and closure relations (deduced from those satisfied by the spherical harmonics, §D-1-c-β of chapter VI):

$$\langle l, m \mid l', m' \rangle = \delta_{ll'}\,\delta_{mm'}$$
$$\sum_{l=0}^{\infty}\sum_{m=-l}^{+l} \mid l, m \rangle \langle l, m \mid = 1 \tag{22}$$

The most general quantum state of the rotator can be expanded on the states $\mid l, m \rangle$:

$$\mid \psi(t) \rangle = \sum_{l=0}^{\infty}\sum_{m=-l}^{+l} c_{l,m}(t) \mid l, m \rangle \tag{23}$$

The component:

$$c_{l,m}(t) = \langle l, m \mid \psi(t) \rangle = \int \mathrm{d}\Omega\, Y_l^{m*}(\theta, \varphi)\, \psi(\theta, \varphi; t) \tag{24}$$

evolves in time in accordance with the equation:

$$c_{l,m}(t) = c_{l,m}(0)\,\mathrm{e}^{-iE_l t/\hbar} \tag{25}$$

### c. STUDY OF THE OBSERVABLE $Z$

Earlier, we introduced the observables $X$, $Y$, $Z$ which correspond to the projections onto the three axes of the segment $M_1M_2$. In this section, we shall study the evolution of the mean values of these observables and compare the results obtained with those predicted by classical mechanics. We shall confine ourselves to the calculation of $\langle Z \rangle(t)$ since $\langle X \rangle(t)$ and $\langle Y \rangle(t)$ have analogous properties.

A Bohr frequency $\frac{E_l - E_{l'}}{h}$ can appear in the function $\langle Z \rangle(t)$ if $Z$ has a non-zero matrix element between a state $\mid l, m \rangle$ of energy $E_l$ and a state $\mid l', m' \rangle$ of energy $E_{l'}$. The first problem is therefore to find the non-zero matrix elements of $Z$. To solve this problem, we shall use the following relation, which can be established by using the mathematical properties of spherical harmonics [complement $A_{VI}$, formula (35)]:

$$\cos\theta\, Y_l^m(\theta, \varphi) = \sqrt{\frac{l^2 - m^2}{4l^2 - 1}}\, Y_{l-1}^m(\theta, \varphi) + \sqrt{\frac{(l+1)^2 - m^2}{4(l+1)^2 - 1}}\, Y_{l+1}^m(\theta, \varphi) \tag{26}$$

From this, we deduce, using (16), (17) and (22):

$$\langle l', m' \mid Z \mid l, m \rangle = r_e\,\delta_{mm'}\left[\delta_{l',l-1}\sqrt{\frac{l^2 - m^2}{4l^2 - 1}} + \delta_{l',l+1}\sqrt{\frac{(l+1)^2 - m^2}{4(l+1)^2 - m^2}}\right] \tag{27}$$

COMMENT:

According to (27), the selection rules satisfied by $Z$ are: $\Delta l = \pm 1$, $\Delta m = 0$. It can be shown that for $X$ and $Y$ we have: $\Delta l = \pm 1$, $\Delta m = \pm 1$. Since the energies depend only on $l$, the Bohr frequencies are the same for $\langle X \rangle$, $\langle Y \rangle$ and $\langle Z \rangle$.

The operator $Z$ can therefore connect only states belonging to two adjacent levels of figure 2 (the corresponding transitions are represented by vertical arrows in figure 2). The only Bohr frequencies which appear in the evolution of $\langle Z \rangle(t)$ are thus of the form:

$$\nu_{l,l-1} = \frac{E_l - E_{l-1}}{h} = 2Bl \tag{28}$$

They form a series of equidistant frequencies, separated by the interval $2B$ (fig. 3).

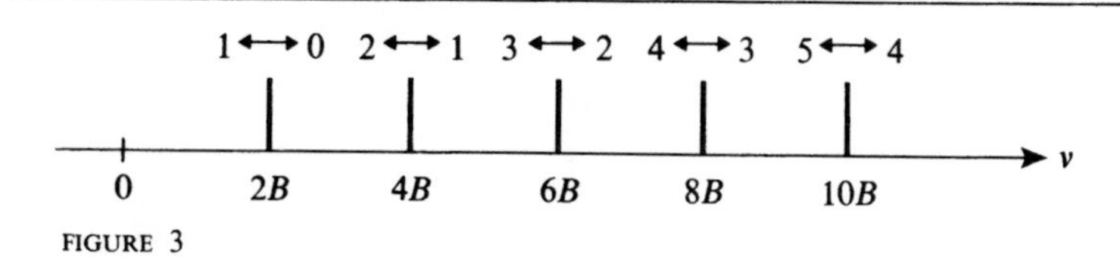


FIGURE 3

**Frequencies appearing in the evolution of the mean value of the observable $Z$. Because of the selection rule $\Delta l = \pm 1$, only the Bohr frequencies $2Bl$ (with $l \geqslant 1$), associated with two adjacent levels $E_l$ and $E_{l-1}$ in figure 2, are observed.**

The mean value $\langle Z \rangle(t)$ can evolve only at a well-defined series of frequencies. This is unlike the classical case, in which the frequency of rotation $\nu_R$ of the rotator can take on any value.

According to (27), if the system is in a stationary state $| l, m \rangle$, $\langle Z \rangle(t)$ is always zero, even for large $l$. To obtain a quantum mechanical state in which $\langle Z \rangle$ behaves like the corresponding classical variable $z$, one must superpose a large number of states $| l, m \rangle$. If we assume that the state of the system is given by formula (23), and that the numbers $|c_{l,m}(0)|^2$ have values which vary with $l$ as is shown in figure 4, the most probable value of $l$, $l_M$, is very large; the spread $\Delta l$ of the values of $l$ is also very large in absolute value, but very small in relative value:

$$l_M, \Delta l \gg 1 \tag{29-a}$$

$$\frac{\Delta l}{l_M} \ll 1 \tag{29-b}$$

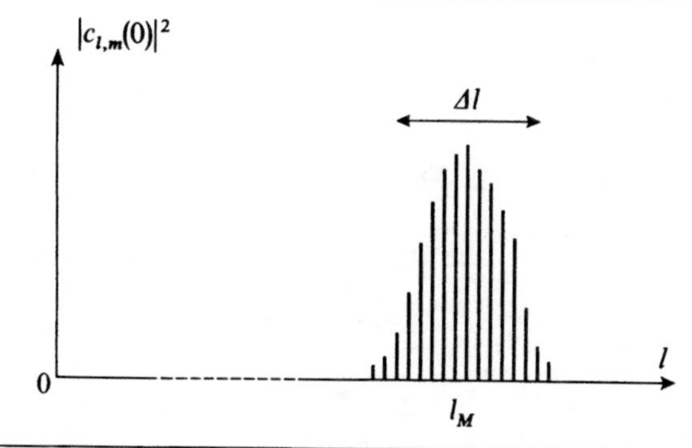


FIGURE 4

**Square of the moduli of the expansion coefficients of a «quasi-classical» state on the stationary states $| l, m \rangle$ of the rigid rotator. The spread $\Delta l$ is large; however, since the most probable value of $l$, $l_M$, is very large, we have $\Delta l / l_M \ll 1$, and the relative accuracy with respect to $l$ is very good.**

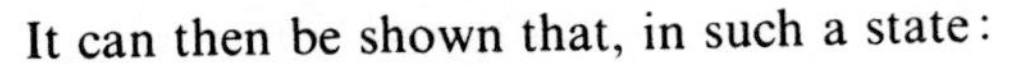

It can then be shown that, in such a state:

$$\langle \mathbf{L} \rangle^2 \simeq \langle \mathbf{L}^2 \rangle \simeq l_M(l_M + 1)\hbar^2 \simeq l_M^2 \hbar^2 \tag{30}$$

In addition, the Bohr frequencies which appear in $\langle Z \rangle(t)$ are then all very close (in relative value) to:

$$\nu_M = 2Bl_M \tag{31}$$

Eliminating $l_M$ between (30) and (31), we obtain, according to (19):

$$\nu_M \simeq \frac{2B|\langle \mathbf{L} \rangle|}{\hbar} = \frac{|\langle \mathbf{L} \rangle|}{2\pi I} \tag{32}$$

which is the equivalent of the classical relation (6).

COMMENT:

It is interesting to study in greater detail the motion of the wave packet corresponding to the state of figure 4. It is represented by a function of $\theta$ and $\varphi$ and can be considered to evolve on the sphere of unit radius. The preceding discussion shows that this wave packet rotates on the sphere with the average frequency $\nu_M$. Because of the spread $\Delta l$ of $l$ and the corresponding spread $2B\Delta l$ of the Bohr frequencies which enter into $\langle X \rangle$, $\langle Y \rangle$ and $\langle Z \rangle$, the wave packet becomes distorted over time. This distortion becomes appreciable after a time of the order of:

$$\tau \simeq \frac{1}{2B\,\Delta l} \tag{33}$$

Since the spread of $l$ is small in relative value, we have:

$$\nu_M \tau \simeq \frac{l_M}{\Delta l} \gg 1 \tag{34}$$

The distortion of the wave packet is therefore slow, relative to its rotation.

In fact, the Bohr frequencies of the system form a *discrete* series of *equidistant* frequencies, separated by the interval $2B$. The motion which results from the superposition of these frequencies is therefore periodic, of period:

$$T = \frac{1}{2B} \tag{35}$$

with, according to (29-a):

$$T \gg \tau \gg \frac{1}{\nu_M} \tag{36}$$

The distortion of the wave packet is therefore not irreversible; it follows a cycle which is repeated periodically. This is related to the fact that the wave packet evolves on the unit sphere, which is a bounded surface. This behavior should be compared with that of free wave packets (irreversible spreading; complement $G_I$) and that of the quasi-classical states of the harmonic oscillator (oscillation without distortion; complement $G_V$).

## 4. Experimental evidence for the rotation of molecules

### a. HETEROPOLAR MOLECULES. PURE ROTATIONAL SPECTRUM

#### α. *Description of the spectrum*

If the molecule is composed of two different atoms, the electrons are attracted by the more electronegative atom, and the molecule generally has a *permanent* electric dipole moment $d_0$, directed along the molecular axis. The projection of the electric dipole moment onto $Oz$ becomes, in quantum mechanics, an observable proportional to $Z$. We have seen that $\langle Z \rangle(t)$ evolves at all the Bohr frequencies $2Bl$ ($l = 1, 2, 3, ...$) shown in figure 3. Thus we see how the molecule is coupled to the electromagnetic field and can absorb or emit radiation polarized parallel to $Oz$*, on the condition that the frequency of this radiation is equal to one of the Bohr frequencies $2Bl$.

The corresponding absorption or emission spectrum of the molecule is called the "pure rotational spectrum". It is composed of a series of equidistant lines, the frequency separation between two successive lines being equal to $2B$, as in figure 3. The absorption (or emission) of the line of frequency $2Bl$ corresponds to the passage of the molecule from the level $l - 1$ to the level $l$ (or from the level $l$ to the level $l - 1$), at the same time that a photon of frequency $2Bl$ is absorbed (or emitted). Figure 5 represents this process schematically [(5-a) represents the absorption and (5-b), the emission of a photon of frequency $2Bl$].

The pure rotational spectra of diatomic molecules therefore provide direct experimental proof of the quantization of the observable $\mathbf{L}^2$.

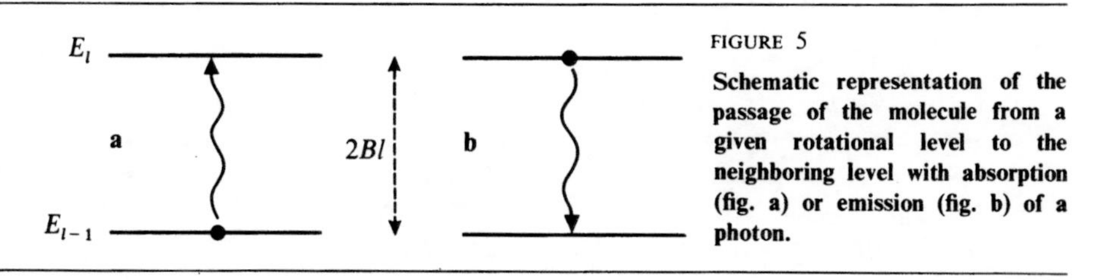


FIGURE 5

**Schematic representation of the passage of the molecule from a given rotational level to the neighboring level with absorption (fig. a) or emission (fig. b) of a photon.**

#### β. *Comparison with the "pure vibrational" spectrum*

In § 1-c-α of complement $A_V$, we studied the "pure vibrational" spectrum of a heteropolar diatomic molecule. It is interesting to compare this spectrum with the pure rotational spectrum which we are studying here.

(*i*) The rotational frequencies of a diatomic molecule are generally much lower than the vibrational frequencies. The separation $2B/c$ of two rotational lines varies between a few tenths of a $cm^{-1}$ and a few dozen $cm^{-1}$'s. For small values of $l$, the rotational frequencies $2Bl$ therefore correspond to wavelengths of the order of a centimeter or a millimeter. Thus for HCl, the separation $2B/c$ is equal to

* If we study the motion of $\langle X \rangle(t)$ and $\langle Y \rangle(t)$, we see that the molecule can also absorb or emit radiation polarized parallel to $Ox$ or $Oy$.

20.8 $cm^{-1}$, while the vibration frequency, which corresponds to 2 886 $cm^{-1}$, is more than a hundred times greater.

Pure rotational spectra therefore fall in the very far infrared or the microwave domain.

COMMENT:

As we shall show in complement $F_{VII}$, the rotation of molecules is also responsible for a fine structure of vibrational spectra (vibration-rotation spectra). $2B$ can then be measured in a domain of wavelengths which is no longer that of microwaves. The same comment applies to the Raman rotational effect (§ 4-b below), which appears as a rotational structure of an optical line.

(*ii*) The "pure vibrational" spectrum studied in complement $A_V$ has only one *vibrational line*. This is due to the fact that the various vibrational levels are equidistant (if the anharmonicity of the potential is neglected) and, consequently, only one Bohr frequency appears in the dipole motion (selection rule $\Delta v = \pm 1$). On the other hand, the pure rotational spectrum consists of a *series of equidistant lines*.

(*iii*) We indicated in complement $A_V$ that the permanent electric dipole moment of the heteropolar molecule can be expanded in powers of $r - r_e$ in the neighborhood of the stable equilibrium position of the molecule:

$$D(r) = d_0 + d_1(r - r_e) + \dots \tag{37}$$

For the pure vibrational spectrum to appear, $D(r)$ must vary with $r$: $d_1$ must therefore be different from zero. On the other hand, even if $r$ remains fixed and equal to $r_e$, the rotation of the molecule modulates the projection of the electric dipole onto one of the axes, provided that $d_0$ is different from zero. Thus we see that the study of the intensity of vibrational and rotational lines permits the separate measurement of the coefficients $d_1$ and $d_0$ of (37).

$\gamma$. *Applications*

The study of pure rotational spectra has some interesting applications; we shall mention three examples.

(*i*) Measurement of the separation $2B$ of two neighboring lines yields the moment of inertia $I$ of the molecule, according to (19). If we know $m_1$ and $m_2$, we can deduce $r_e$, the separation of the two nuclei in the stable equilibrium position of the molecule [$r_e$ is the abscissa of the minimum of the curve $V(r)$ of figure 1 in complement $A_V$]. Recall that measurement of the vibrational frequency yields the curvature of $V(r)$ at $r = r_e$.

(*ii*) Consider two diatomic molecules $N-M$ and $N-M'$, in which two isotopes $M$ and $M'$ of the same element are bound to the same atom $N$. Since the distances $r_e$ between the nuclei are equal in the two molecules, measurement of the ratio of the corresponding coefficients $B$, which can be performed with great accuracy, yields the ratio of the masses of the two isotopes $M$ and $M'$.

One could also compare the vibrational frequencies of the two molecules, but it is preferable to use the rotational spectrum, since the rotational frequencies vary with $1/\mu$ [formula (19)], while the vibrational frequencies vary with $1/\sqrt{\mu}$ [formula (5) of $A_V$].

(*iii*) In the study of a sample containing a great number of identical molecules, the relative intensities of the lines (in absorption or emission) of the pure rotational spectrum yields information about the distribution of the molecules among the various levels $E_l$. Unlike what happens in the case of the vibrational spectrum, transitions between two given adjacent levels (arrows of figure 2) occur at a particular frequency, which is characteristic of these two levels. Thus, the intensity of the corresponding line depends on the number of molecules which are found in the two levels.

This information can be used to determine the temperature of a medium*. If thermodynamic equilibrium has been attained, we know that the probability that a given molecule is in a particular state of energy $E_l$ is proportional to $e^{-E_l/kT}$; since the degeneracy of the rotational level $E_l$ is $(2l + 1)$, the total probability $\mathscr{P}_l$ of finding the molecule being considered in one of the states of the level $E_l$ (the "population" of the level $E_l$) is:

$$\begin{aligned} \mathscr{P}_l &= \frac{1}{Z}(2l + 1)\, e^{-E_l/kT} \\ &= \frac{1}{Z}(2l + 1)\, e^{-l(l+1)hB/kT} \end{aligned} \tag{38}$$

where:

$$Z = \sum_{l=0}^{\infty} (2l + 1)\, e^{-l(l+1)hB/kT} \tag{39}$$

is the partition function. If we are studying a system containing a large number of molecules whose interactions can be neglected, $\mathscr{P}_l$ gives the fraction of them whose energy is $E_l$.

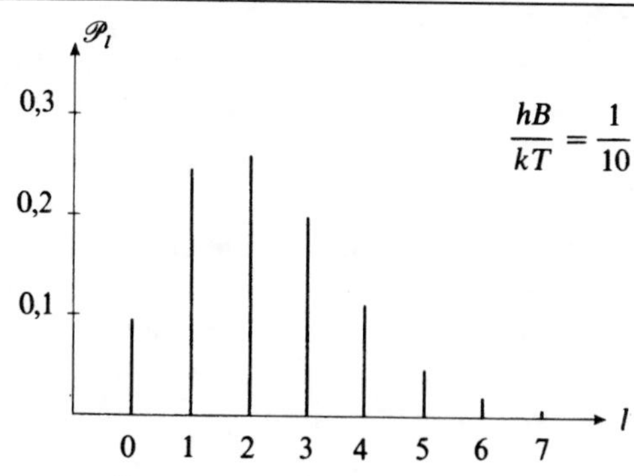


FIGURE 6

**Population $\mathscr{P}_l$ of the various rotational levels $E_l$ at thermodynamic equilibrium. The fact that $\mathscr{P}_l$ begins by increasing with $l$ arises from the $(2l + 1)$-fold degeneracy of the levels $E_l$. When $l$ becomes sufficiently large, the Boltzmann factor $e^{-E_l/kT}$ prevails and is responsible for the decrease in $\mathscr{P}_l$.**

* Actually, the vibration-rotation or Raman rotational spectra are more often used, since they fall into more convenient frequency ranges than does the pure rotational spectrum.

At ordinary temperatures, $hB$ is much smaller than $kT$, so that several rotational levels are populated. Note that the presence of the factor $(2l + 1)$ means that it is not the lowest levels which are the most populated : figure 6 indicates the shape of $\mathcal{P}_l$ as a function of $l$ for a temperature $T$ such that $hB/kT$ is of the order of 1/10. Recall that the vibrational levels, on the other hand, are non-degenerate, and that their separation is much greater than $hB$; consequently, when the distribution of the molecules between the two rotational levels is that of figure 6, they are practically all in the vibrational ground state ($v = 0$).

### b. HOMOPOLAR MOLECULES. RAMAN ROTATIONAL SPECTRA

As we pointed out in §1-c-β of complement $A_V$, a homopolar molecule (that is, a molecule composed of two identical atoms) has no permanent electric dipole moment: in formula (37), we have $d_0 = d_1 = ... = 0$. The vibration and rotation of the molecule induce no coupling with the electromagnetic field, and the molecule is consequently "inactive" in the near infrared (vibration) and the microwave domain (rotation). Like the vibration (*cf.* § 1-c-β of $A_V$), the rotation of the molecule can, however, be observed via the inelastic scattering of light (the Raman effect).

#### α. *The Raman rotational effect. Classical treatment*

We have already introduced, in complement $A_V$, the susceptibility $\chi$ of a molecule in the optical domain: an incident light wave, whose electric field is $\mathbf{E}\, e^{i\Omega t}$, sets the electrons of the molecule in forced motion and causes an electric dipole $\mathbf{D}\, e^{i\Omega t}$, oscillating at the same frequency as the incident wave, to appear. $\chi$ is the coefficient of proportionality between $\mathbf{D}$ and $\mathbf{E}$. If $\mathbf{E}$ is parallel to the axis of the molecule, $\chi$ depends on the distance $r$ between the two nuclei : when the molecule vibrates, $\chi$ vibrates at the same frequency. This is the origin of the Raman vibrational effect described in §1-c-β of $A_V$.

Actually, a diatomic molecule is an anisotropic system. When the angle between the molecular axis and $\mathbf{E}$ is arbitrary, $\mathbf{D}$ is not generally parallel to $\mathbf{E}$: the relation between $\mathbf{D}$ and $\mathbf{E}$ is tensorial ($\chi$ is the "susceptibility tensor"). $\mathbf{D}$ is parallel to $\mathbf{E}$ in the two following simple cases: $\mathbf{E}$ parallel to the molecular

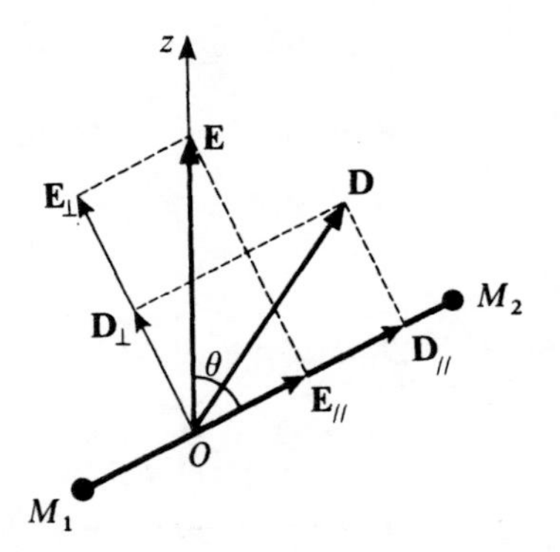


FIGURE 7

**Decomposition of the electric field E into a component $\mathbf{E}_{//}$ parallel to the molecular axis and a component $\mathbf{E}_\perp$ perpendicular to this axis. These fields induce electric dipoles $\chi_{//}\mathbf{E}_{//}$ and $\chi_\perp\mathbf{E}_\perp$ collinear with the corresponding fields. However, since $\chi_{//}$ and $\chi_\perp$ have different values (the molecule is anisotropic), the induced dipole $\mathbf{D} = \chi_{//}\mathbf{E}_{//} + \chi_\perp\mathbf{E}_\perp$ is not collinear with E.**

axis (we then have $\chi = \chi_{//}$), and $\mathbf{E}$ perpendicular to this axis ($\chi = \chi_{\perp}$). In the general case, we choose the $Oz$ axis along the electric field $\mathbf{E}$ of the light wave (assumed to be polarized); we consider a molecule whose axis points in the direction of polar angles $\theta$ and $\varphi$ and calculate the component along $Oz$ of the dipole induced by $\mathbf{E}$ on this molecule. $\mathbf{E}$ can be decomposed into a component $\mathbf{E}_{//}$, parallel to the molecular axis, and a component $\mathbf{E}_{\perp}$, perpendicular to $M_1M_2$ and contained in the plane formed by $Oz$ and $M_1M_2$ (fig. 7). The dipole induced on the molecule by the field $\mathbf{E}\cos\Omega t$ is then equal to:

$$\mathbf{D} = (\chi_{//}\mathbf{E}_{//} + \chi_{\perp}\mathbf{E}_{\perp})\cos\Omega t \tag{40}$$

Its projection onto $Oz$ can be calculated immediately:

$$\begin{aligned} D_z &= (\cos\theta\,\chi_{//}|\mathbf{E}_{//}| + \sin\theta\,\chi_{\perp}|\mathbf{E}_{\perp}|)\cos\Omega t \\ &= (\cos^2\theta\,\chi_{//} + \sin^2\theta\,\chi_{\perp})E\cos\Omega t \\ &= [\chi_{\perp} + (\chi_{//} - \chi_{\perp})\cos^2\theta]E\cos\Omega t \end{aligned} \tag{41}$$

We see that $D_z$ depends on $\theta$, since $\chi_{//}$ and $\chi_{\perp}$ are not equal (the molecule is anisotropic).

To see what happens when the molecule rotates, we shall begin by reasoning classically. The fact that the molecule rotates at the frequency $\frac{\omega_R}{2\pi}$ means that $\cos\theta$ oscillates at the same frequency:

$$\cos\theta = \alpha\cos(\omega_R t - \beta) \tag{42}$$

where $\alpha$ and $\beta$ depend on the initial conditions and the orientation of the angular momentum $\mathscr{L}$ (which is fixed). Thus we see that the term in $\cos^2\theta$ of (41) gives rise to components of $D_z$ which oscillate at frequencies of $\frac{\Omega \pm 2\omega_R}{2\pi}$, in addition to the component which varies at the frequency $\frac{\Omega}{2\pi}$. The fact that the rotation of the molecule at the frequency $\frac{\omega_R}{2\pi}$ modulates its polarizability at twice its frequency is easy to understand: after half a rotation, performed in half a period, the molecule returns to the same geometrical position with respect to the incident light wave. The light re-emitted with a polarization parallel to $Oz$ is that which is radiated by $D_z$. It has an unshifted line of frequency $\frac{\Omega}{2\pi}$ (Rayleigh line), as well as two shifted lines, one on each side of the Rayleigh line, of frequencies $\frac{\Omega - 2\omega_R}{2\pi}$ (Raman-Stokes line) and $\frac{\Omega + 2\omega_R}{2\pi}$ (Raman-anti-Stokes line).

### β. *Quantum mechanical selection rules. Form of the Raman spectrum*

Quantum mechanically, Raman scattering corresponds to an inelastic scattering process in which the molecule goes from level $E_l$ to level $E_{l'}$, while the energy $\hbar\Omega$ of the photon becomes $\hbar\Omega + E_l - E_{l'}$ (the total energy of the system is conserved during this process).

The quantum theory of the Raman effect (which we shall not discuss here) indicates that the probability of such a process involves the matrix elements of $(\chi_{//} - \chi_{\perp}) \cos^2 \theta + \chi_{\perp}$ between the initial state $Y_l^m(\theta, \varphi)$ and the final state $Y_{l'}^{m'}(\theta, \varphi)$ of the molecule:

$$\int \mathrm{d}\Omega \, Y_{l'}^{m'*}(\theta, \varphi) \, [(\chi_{//} - \chi_{\perp}) \cos^2 \theta + \chi_{\perp}] \, Y_l^m(\theta, \varphi) \tag{43}$$

It can be shown, using the properties of the spherical harmonics, that such a matrix element is different from zero only if*:

$$l' - l = 0, + 2, - 2 \tag{44}$$

There is only one Rayleigh line (which corresponds to $l = l'$). Since the rotational levels are not equidistant, there are several Raman-anti-Stokes lines (which correspond to $l' = l - 2$), of frequencies:

$$\begin{cases} \dfrac{\Omega}{2\pi} + \dfrac{E_{l'+2} - E_{l'}}{h} = \dfrac{\Omega}{2\pi} + 4B\left(l' + \dfrac{3}{2}\right) \\ \text{with } l' = 0, 1, 2, \ldots \end{cases} \tag{45}$$

and several Raman-Stokes lines (which correspond to $l' = l + 2$), of frequencies:

$$\begin{cases} \dfrac{\Omega}{2\pi} + \dfrac{E_l - E_{l+2}}{h} = \dfrac{\Omega}{2\pi} - 4B\left(l + \dfrac{3}{2}\right) \\ \text{with } l = 0, 1, 2, \ldots \end{cases} \tag{46}$$

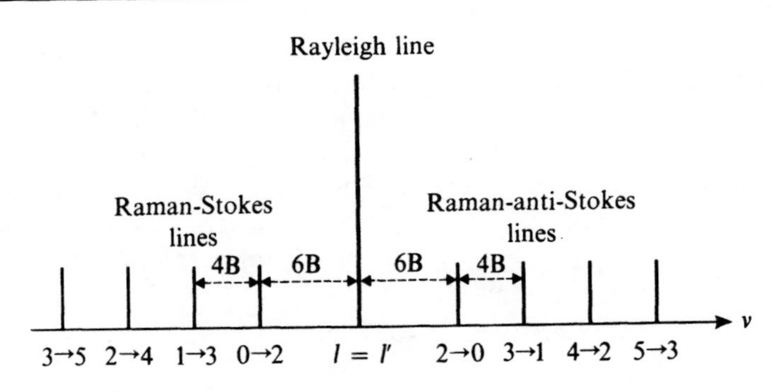


FIGURE 8

**Raman rotational spectrum of a molecule. This molecule, initially in the rotational level $E_l$, inelastically scatters an incident photon of energy $\hbar\Omega$. After the scattering, the molecule has moved to the rotational level $E_{l'}$, and the energy of the photon is $\hbar\Omega + E_l - E_{l'}$ (conservation of energy).**

**If $l = l'$, the scattered photon has the same frequency $\nu = \Omega/2\pi$ as the incident photon; this yields the Rayleigh line. But it is also possible to have $l' - l = \pm 2$; if $l' = l + 2$, the frequency of the scattered photon is lower (Stokes scattering; if $l' = l - 2$, it is higher (anti-Stokes scattering). Since the rotational levels $E_l$ are not equidistant (cf. fig. 2), there are as many Stokes or anti-Stokes lines as there are values of $l$. These lines are labeled in the figure by $l \longrightarrow l'$ (with $l' = l \pm 2$).**

* The integral (43) is also zero if $m \neq m'$. If we consider light re-emitted with a different polarization state from that of the incident light wave, we obtain the following selection rules for $m$: $\Delta m = 0, \pm 1, \pm 2$.

The form of the Raman rotational spectrum is shown in figure 8. The Stokes and anti-Stokes lines occur symmetrically with respect to the Rayleigh line. The separation between two adjacent Stokes (or anti-Stokes) lines is equal to $4B$, that is, to *twice* the separation which would be found between two adjacent lines of the pure rotational spectrum if it existed. Moreover, since the vibrational frequency is much larger than $B$, the Raman-Stokes and anti-Stokes vibrational lines are situated much further to the left and to the right of the Rayleigh line than the rotational Raman lines and hence do not appear in the figure (these lines also have rotational structures analogous to that of figure 8).

COMMENTS:

(*i*) Consider a wave packet like those studied in §3-c, that is, one for which the values of $l$ are grouped about a very large integer $l_M$ (fig. 4). According to (45) and (46), the frequencies of the various Stokes and anti-Stokes lines will be very close (in relative value) to:

$$\frac{\Omega}{2\pi} \pm 4Bl_M \tag{47}$$

that is, according to (31):

$$\frac{\Omega}{2\pi} \pm 2\nu_M \tag{48}$$

where $\nu_M$ is the average rotational frequency of the molecule. Thus, the quantum mechanical treatment, at the classical limit, yields the results of §4-b-α.

(*ii*) In Raman rotational spectra, the Stokes and anti-Stokes lines appear with comparable intensities since levels of large $l$ have large populations, as $hB$ is much smaller than $kT$. This is necessary for the observation of anti-Stokes lines, for which the initial state of the molecule must be at least $l = 2$. However, the anti-Stokes vibrational line has a much smaller intensity than the Stokes line. The vibrational energy is much larger than $kT$; the population of the vibrational ground state $v = 0$ is much larger than the others, and Stokes processes $v = 0 \longrightarrow v = 1$ are much more frequent than anti-Stokes processes $v = 1 \longrightarrow v = 0$.

(*iii*) The Raman rotational effect also exists for heteropolar molecules.

**References and suggestions for further reading:**

Karplus and Porter (12.1), §7.4; Herzberg (12.4), Vol. I, chap. III, §§1 and 2; Landau and Lifshitz (1.19), chaps. XI and XIII; Townes and Schawlow (12.10), chaps. 1 to 4.

## Complement $D_{VI}$

# ANGULAR MOMENTUM OF STATIONARY STATES OF A TWO-DIMENSIONAL HARMONIC OSCILLATOR



In this complement, we shall be concerned with the quantum mechanical properties of a two-dimensional harmonic oscillator. The quantum mechanical problem is exactly soluble and does not involve complicated calculations. Furthermore, this subject provides an opportunity to study a simple application of the properties of the orbital angular momentum **L**, since, as we shall see, the stationary states of such an oscillator can be classified with respect to the possible values of the observable $L_z$. In addition, the results obtained will be useful in the next complement, $E_{VI}$.

## 1. Introduction

### a. REVIEW OF THE CLASSICAL PROBLEM

A physical particle always moves in three-dimensional space. However, if its potential energy depends only on $x$ and $y$, the problem can be treated in two dimensions. We shall assume here that this potential energy can be written:

$$V(x, y) = \frac{\mu}{2} \omega^2(x^2 + y^2) \tag{1}$$

where $\mu$ is the mass of the particle and $\omega$ is a constant. The classical Hamiltonian of the system is then:

$$\mathscr{H} = \mathscr{H}_{xy} + \mathscr{H}_z \tag{2}$$

with:

$$\mathscr{H}_{xy} = \frac{1}{2\mu}(p_x^2 + p_y^2) + \frac{1}{2}\mu\omega^2(x^2 + y^2)$$

$$\mathscr{H}_z = \frac{1}{2\mu}p_z^2 \tag{3}$$

where $p_x$, $p_y$, $p_z$ are the three components of the momentum $\mathbf{p}$ of the particle. $\mathscr{H}_{xy}$ is a two-dimensional harmonic oscillator Hamiltonian.

The equations of motion can easily be integrated to yield:

$$\begin{cases} p_z(t) = p_0 \\ z(t) = \dfrac{p_0}{\mu}t + z_0 \end{cases} \tag{4}$$

$$\begin{cases} x(t) = x_M \cos(\omega t - \varphi_x) \\ p_x(t) = -\mu\omega x_M \sin(\omega t - \varphi_x) \end{cases} \tag{5}$$

$$\begin{cases} y(t) = y_M \cos(\omega t - \varphi_y) \\ p_y(t) = -\mu\omega y_M \sin(\omega t - \varphi_y) \end{cases} \tag{6}$$

where $p_0$, $z_0$, $x_M$, $\varphi_x$, $y_M$, $\varphi_y$ are constants which depend on the initial conditions (we assume $x_M$ and $y_M$ to be positive).

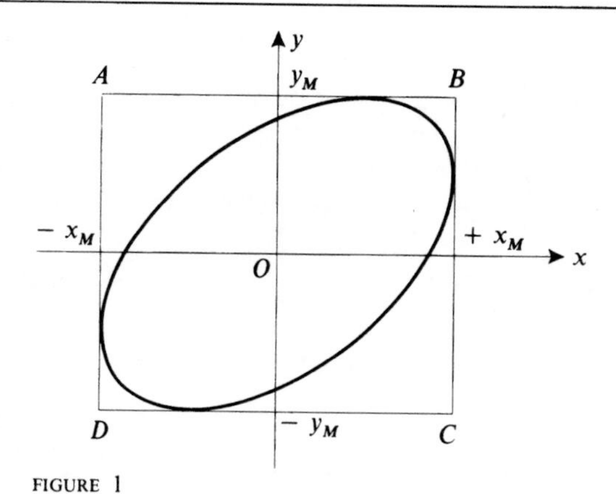


FIGURE 1

**Projection of the classical trajectory of a particle in a two-dimensional harmonic potential onto the $xOy$ plane; we obtain an ellipse inscribed in the rectangle $ABCD$.**

We see that the projection of the particle onto $Oz$ describes a uniform motion with a velocity of $p_0/\mu$. The projection onto the $xOy$ plane describes an ellipse inscribed in the rectangle $ABCD$ of figure 1. The direction the particle takes on this ellipse depends on the phase difference $\varphi_y - \varphi_x$. When $\varphi_y - \varphi_x = \pm\pi$, the ellipse reduces to the line $AC$. When $\varphi_y - \varphi_x$ is between $-\pi$ and 0, the particle moves clockwise on the ellipse ("left-handed" motion), with the axes of the ellipse parallel to $Ox$ and $Oy$ for $\varphi_y - \varphi_x = -\pi/2$. When $\varphi_y - \varphi_x = 0$, the ellipse reduces to the line $BD$. Finally, when $\varphi_y - \varphi_x$ is between 0 and $\pi$, the particle moves counterclockwise on the ellipse ("right-handed" motion), with the axes parallel to $Ox$ and $Oy$ for $\varphi_y - \varphi_x = +\pi/2$. Note that the ellipse reduces to a circle if $\varphi_y - \varphi_x = \pm\pi/2$ and $x_M = y_M$.

It is easy to determine several constants of the motion related to the projection of the motion onto the $xOy$ plane:

– the total energy $\mathscr{H}_{xy}$, which, according to (3), (5), (6), is equal to:

$$\mathscr{H}_{xy} = \frac{1}{2}\mu\omega^2(x_M^2 + y_M^2) \tag{7}$$

– the energies:

$$\mathscr{H}_x = \frac{1}{2}\mu\omega^2 x_M^2 \tag{8-a}$$

$$\mathscr{H}_y = \frac{1}{2}\mu\omega^2 y_M^2 \tag{8-b}$$

of the projections of the motion onto $Ox$ and $Oy$;

– the component of the orbital angular momentum $\mathscr{L}$ of the particle along $Oz$:

$$\mathscr{L}_z = xp_y - yp_x \tag{9}$$

which, according to (5) and (6), is equal to:

$$\mathscr{L}_z = \mu\omega x_M y_M \sin(\varphi_y - \varphi_x) \tag{10}$$

We see that $\mathscr{L}_z$ is positive or negative depending on whether the motion is counterclockwise ($0 < \varphi_y - \varphi_x < \pi$) or clockwise ($-\pi < \varphi_y - \varphi_x < 0$). $\mathscr{L}_z$ is zero for the two rectilinear motions ($\varphi_y - \varphi_x = \pm\pi$ and $\varphi_y - \varphi_x = 0$). Finally, for a motion at a given energy, that is, according to (7), for a fixed value of $x_M^2 + y_M^2$, $|\mathscr{L}_z|$ is maximal when $\varphi_y - \varphi_x = \pm\pi/2$ and the product $x_M y_M$ is maximal, which implies $x_M = y_M$. Of all motions at a given energy, it is the counterclockwise (clockwise) motion which corresponds to the maximal (minimal) algebraic value of $\mathscr{L}_z$.

### b. THE PROBLEM IN QUANTUM MECHANICS

The quantization rules of chapter III enable us to obtain $H$, $H_{xy}$, $H_z$ from $\mathscr{H}$, $\mathscr{H}_{xy}$, $\mathscr{H}_z$. The stationary states $|\varphi\rangle$ of the particle are given by:

$$H|\varphi\rangle = (H_{xy} + H_z)|\varphi\rangle = E|\varphi\rangle \tag{11}$$

with:

$$H_{xy} = \frac{P_x^2 + P_y^2}{2\mu} + \frac{1}{2}\mu\omega^2(X^2 + Y^2) \tag{12-a}$$

$$H_z = \frac{P_z^2}{2\mu} \tag{12-b}$$

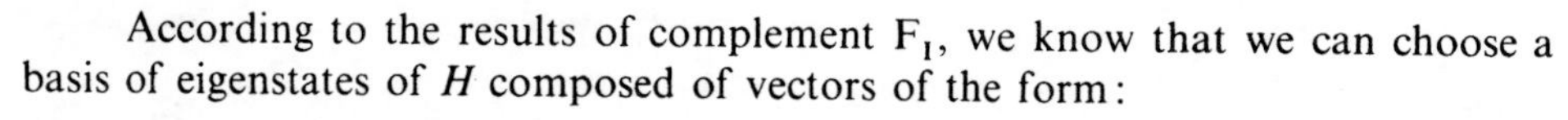



According to the results of complement $F_I$, we know that we can choose a basis of eigenstates of $H$ composed of vectors of the form:

$$| \varphi \rangle = | \varphi_{xy} \rangle \otimes | \varphi_z \rangle \tag{13}$$

where $| \varphi_{xy} \rangle$ is an eigenvector of $H_{xy}$ in the state space $\mathscr{E}_{xy}$ associated with the variables $x$ and $y$:

$$H_{xy} | \varphi_{xy} \rangle = E_{xy} | \varphi_{xy} \rangle \tag{14}$$

and $| \varphi_z \rangle$ is an eigenvector of $H_z$ in the space $\mathscr{E}_z$ associated with the variable $z$:

$$H_z | \varphi_z \rangle = E_z | \varphi_z \rangle \tag{15}$$

The total energy associated with the state (13) is then:

$$E = E_{xy} + E_z \tag{16}$$

Now, equation (15), which in fact describes the stationary states of a free particle in a one-dimensional problem, can be solved immediately; it yields:

$$\langle z | \varphi_z \rangle = \frac{1}{\sqrt{2\pi\hbar}} e^{ip_z z/\hbar} \tag{17}$$

(where $p_z$ is an arbitrary real constant), with:

$$E_z = \frac{p_z^2}{2\mu} \tag{18}$$

The problem therefore reduces to the determination of the solutions of equation (14), that is, the energies and stationary states of a two-dimensional harmonic oscillator. This is the problem we shall now try to solve.

We shall see that the eigenvalues $E_{xy}$ of $H_{xy}$ are degenerate : $H_{xy}$ alone does not constitute a C.S.C.O. in $\mathscr{E}_{xy}$. We must therefore add one or several other observables to $H_{xy}$ in order to construct a C.S.C.O. In fact, we find in quantum mechanics the same constants of the motion as in classical mechanics: $H_x$ and $H_y$, the energies of the projection of the motion onto $Ox$ and $Oy$; and $L_z$, the component along $Oz$ of the orbital angular momentum $\mathbf{L}$. Since $L_z$ commutes with neither $H_x$ nor $H_y$, we shall see that a C.S.C.O. can be formed of $H_{xy}$, $H_x$ and $H_y$ (§2) or of $H_{xy}$ and $L_z$ (§3).

COMMENTS:

(*i*) Formula (18) indicates that the eigenvalues $E_z$ of $H_z$ are all two-fold degenerate in the space $\mathscr{E}_z$. Furthermore, the degeneracy in $\mathscr{E} = \mathscr{E}_{xy} \otimes \mathscr{E}_z$ of the eigenvalues (16) of the total Hamiltonian $H$ is not due solely to the degeneracy of $E_{xy}$ in $\mathscr{E}_{xy}$ and of $E_z$ in $\mathscr{E}_z$: two eigenvectors of $H$ of the form (13) can have the same total energy $E$ without their corresponding values of $E_{xy}$ (and of $E_z$) being equal.

(*ii*) $H$ commutes with the component $L_z$ of $\mathbf{L}$, but not with $L_x$ and $L_y$. This results from the fact that the potential energy written in (1) is rotation-invariant only about $Oz$. Moreover, of the three operators $L_x$, $L_y$ and $L_z$, only one, $L_z$, acts only in $\mathscr{E}_{xy}$. In the study of the two-dimensional harmonic oscillator, therefore, we shall use only the observable $L_z$. In complement $B_{VII}$, we shall study the isotropic three-dimensional harmonic oscillator, whose potential energy is invariant with respect to any rotation about an axis which passes through the origin; we shall see that all the components of **L** then commute with the Hamiltonian.

## 2. Classification of the stationary states by the quantum numbers $n_x$ and $n_y$

### a. ENERGIES; STATIONARY STATES

To obtain the solutions of the eigenvalue equation (14), note that $H_{xy}$ can be written:

$$H_{xy} = H_x + H_y \tag{19}$$

where $H_x$ and $H_y$ are both Hamiltonians of one-dimensional harmonic oscillators:

$$\begin{aligned} H_x &= \frac{P_x^2}{2\mu} + \frac{1}{2}\mu\omega^2 X^2 \\ H_y &= \frac{P_y^2}{2\mu} + \frac{1}{2}\mu\omega^2 Y^2 \end{aligned} \tag{20}$$

We know the eigenstates $|\varphi_{n_x}\rangle$ of $H_x$ in $\mathscr{E}_x$ and the eigenstates $|\varphi_{n_y}\rangle$ of $H_y$ in $\mathscr{E}_y$. Their energies are, respectively, $E_x = (n_x + 1/2)\hbar\omega$ and $E_y = (n_y + 1/2)\hbar\omega$ (where $n_x$ and $n_y$ are positive integers or zero). The eigenstates of $H_{xy}$ can thus be chosen in the form:

$$|\varphi_{n_x,n_y}\rangle = |\varphi_{n_x}\rangle \otimes |\varphi_{n_y}\rangle \tag{21}$$

where the corresponding energy $E_{xy}$ is given by:

$$\begin{aligned} E_{xy} &= \left(n_x + \frac{1}{2}\right)\hbar\omega + \left(n_y + \frac{1}{2}\right)\hbar\omega \\ &= (n_x + n_y + 1)\hbar\omega \end{aligned} \tag{22}$$

According to the properties of the one-dimensional harmonic oscillator, $E_x$ is non-degenerate in $\mathscr{E}_x$, and $E_y$ in $\mathscr{E}_y$. Consequently, a vector $|\varphi_{n_x,n_y}\rangle$ of $\mathscr{E}_{xy}$ which is unique to within a constant factor corresponds to a pair $\{n_x, n_y\}$ : $H_x$ and $H_y$ form a C.S.C.O. in $\mathscr{E}_{xy}$.

It will prove convenient to use the operators $a_x$ and $a_y$ (destruction operators of a quantum, relative to $Ox$ and $Oy$ respectively), defined by:

$$\begin{aligned} a_x &= \frac{1}{\sqrt{2}}\left(\beta X + i\frac{P_x}{\beta\hbar}\right) \\ a_y &= \frac{1}{\sqrt{2}}\left(\beta Y + i\frac{P_y}{\beta\hbar}\right) \end{aligned} \tag{23}$$

with:

$$\beta = \sqrt{\frac{\mu\omega}{\hbar}} \tag{24}$$

Since $a_x$ and $a_y$ act in different spaces, $\mathscr{E}_x$ and $\mathscr{E}_y$, the only non-zero commutators between the four operators $a_x$, $a_y$, $a_x^\dagger$, $a_y^\dagger$, are:

$$[a_x, a_x^\dagger] = [a_y, a_y^\dagger] = 1 \tag{25}$$

The operators $N_x$ (the number of quanta relative to the $Ox$ axis) and $N_y$ (the number of quanta relative to the $Oy$ axis) are given by:

$$\begin{aligned} N_x &= a_x^\dagger a_x \\ N_y &= a_y^\dagger a_y \end{aligned} \tag{26}$$

which enables us to write $H_{xy}$ in the form:

$$H_{xy} = H_x + H_y = (N_x + N_y + 1)\,\hbar\omega \tag{27}$$

We have, obviously:

$$\begin{aligned} N_x \,|\, \varphi_{n_x,n_y} \rangle &= n_x \,|\, \varphi_{n_x,n_y} \rangle \\ N_y \,|\, \varphi_{n_x,n_y} \rangle &= n_y \,|\, \varphi_{n_x,n_y} \rangle \end{aligned} \tag{28}$$

The ground state $|\,\varphi_{0,0}\rangle$ is given by:

$$|\,\varphi_{0,0}\rangle = |\,\varphi_{n_x=0}\rangle \otimes |\,\varphi_{n_y=0}\rangle \tag{29}$$

The state $|\,\varphi_{n_x,n_y}\rangle$ defined by (21) can be obtained from $|\,\varphi_{0,0}\rangle$ by the successive application of the operators $a_x^\dagger$ and $a_y^\dagger$:

$$|\,\varphi_{n_x,n_y}\rangle = \frac{1}{\sqrt{n_x!\,n_y!}}\,(a_x^\dagger)^{n_x}\,(a_y^\dagger)^{n_y}\,|\,\varphi_{0,0}\rangle \tag{30}$$

The corresponding wave function is the product of $\varphi_{n_x}(x)$ and $\varphi_{n_y}(y)$ [*cf.* complement $B_V$, formula (35)]:

$$\varphi_{n_x,n_y}(x, y) = \frac{\beta}{\sqrt{\pi (2)^{n_x+n_y}(n_x)!\,(n_y)!}}\,e^{-\beta^2(x^2+y^2)/2}\,H_{n_x}(\beta x)\,H_{n_y}(\beta y) \tag{31}$$

#### b. $H_{xy}$ DOES NOT CONSTITUE A C.S.C.O. IN $\mathscr{E}_{xy}$

We see from (22) that the eigenvalues of $H_{xy}$ are of the form:

$$E_{xy} = E_n = (n + 1)\,\hbar\omega \tag{32}$$

where:

$$n = n_x + n_y \tag{33}$$

is a positive integer or zero. To each value of the energy correspond the various orthogonal eigenvectors:

$$|\,\varphi_{n_x=n,n_y=0}\rangle, \; |\,\varphi_{n_x=n-1,n_y=1}\rangle, \; \ldots, \; |\,\varphi_{n_x=0,n_y=n}\rangle \tag{34}$$

Since there are $(n + 1)$ of these vectors, the eigenvalue $E_n$ is $(n + 1)$-fold degenerate in $\mathscr{E}_{xy}$. $H_{xy}$ alone does not, therefore, constitute a C.S.C.O. On the other hand, we have seen that $\{ H_x, H_y \}$ is a C.S.C.O.; this is also, obviously, true of $\{ H_{xy}, H_x \}$ and $\{ H_{xy}, H_y \}$.

## 3. Classification of the stationary states in terms of their angular momenta

### a. SIGNIFICANCE AND PROPERTIES OF THE OPERATOR $L_z$

In the preceding section, we identified the stationary states by the quantum numbers $n_x$ and $n_y$. But the $Ox$ and $Oy$ axes do not enjoy a privileged position in this problem. Since the potential energy is invariant under rotation about $Oz$, we could just as well have chosen another system of orthogonal axes $Ox'$ and $Oy'$ in the $xOy$ plane; we would have then obtained stationary states different from the preceding ones.

Therefore, in order to take better advantage of the symmetry of the problem, we shall now consider the component $L_z$ of the angular momentum, defined by:

$$L_z = XP_y - YP_x \tag{35}$$

Expressing $X$ and $P_x$ in terms of $a_x$ and $a_x^\dagger$, and $Y$ and $P_y$ in terms of $a_y$ and $a_y^\dagger$, we get:

$$L_z = i\hbar(a_x a_y^\dagger - a_x^\dagger a_y) \tag{36}$$

Now, the expression for $H_{xy}$ in terms of the same operators is:

$$H_{xy} = (a_x^\dagger a_x + a_y^\dagger a_y + 1)\, \hbar\omega \tag{37}$$

Since:

$$\begin{aligned} \left[a_x a_y^\dagger, a_x^\dagger a_x + a_y^\dagger a_y\right] &= a_x a_y^\dagger - a_x a_y^\dagger = 0 \\ \left[a_x^\dagger a_y, a_x^\dagger a_x + a_y^\dagger a_y\right] &= -\, a_x^\dagger a_y + a_x^\dagger a_y = 0 \end{aligned} \tag{38}$$

we find that:

$$\left[H_{xy}, L_z\right] = 0 \tag{39}$$

We shall therefore look for a basis of eigenvectors common to $H_{xy}$ and $L_z$.

### b. RIGHT AND LEFT CIRCULAR QUANTA

We introduce the operators $a_d$ and $a_g$ defined by:

$$\begin{aligned} a_d &= \frac{1}{\sqrt{2}}(a_x - ia_y) \\ a_g &= \frac{1}{\sqrt{2}}(a_x + ia_y) \end{aligned} \tag{40}$$

We see from this definition that the action of $a_d$ (or $a_g$) on $| \varphi_{n_x, n_y} \rangle$ yields a state which is a linear combination of $| \varphi_{n_x - 1, n_y} \rangle$ and $| \varphi_{n_x, n_y - 1} \rangle$, that is, a stationary state which has one less energy quantum $\hbar\omega$. Similarly, the action of $a_d^\dagger$ (or $a_g^\dagger$) on $| \varphi_{n_x, n_y} \rangle$ yields another stationary state which has one more energy quantum. In fact, we shall see that $a_d$ (or $a_g$) is analogous to $a_x$ (or $a_y$), and that $a_d$ and $a_g$ can be interpreted as being destruction operators of a right and left "circular quantum" respectively.

First of all, using (40) and (25), it is simple to verify that the only non-zero commutators between the four operators $a_d$, $a_g$, $a_d^\dagger$, $a_g^\dagger$ are:

$$[a_d, a_d^\dagger] = [a_g, a_g^\dagger] = 1 \tag{41}$$

These relations are indeed analogous to (25). Moreover, $H_{xy}$ can be written, in terms of these operators, in a way that is similar to (37); since:

$$\begin{aligned} a_d^\dagger a_d &= \frac{1}{2}(a_x^\dagger a_x + a_y^\dagger a_y - i a_x^\dagger a_y + i a_x a_y^\dagger) \\ a_g^\dagger a_g &= \frac{1}{2}(a_x^\dagger a_x + a_y^\dagger a_y + i a_x^\dagger a_y - i a_x a_y^\dagger) \end{aligned} \tag{42}$$

we have:

$$H_{xy} = (a_d^\dagger a_d + a_g^\dagger a_g + 1)\hbar\omega \tag{43}$$

In addition, using (36), we see that:

$$L_z = \hbar(a_d^\dagger a_d - a_g^\dagger a_g) \tag{44}$$

If we introduce the operators $N_d$ and $N_g$ (the number of right and left "circular quanta"):

$$\begin{aligned} N_d &= a_d^\dagger a_d \\ N_g &= a_g^\dagger a_g \end{aligned} \tag{45}$$

formulas (43) and (44) become:

$$\begin{aligned} H_{xy} &= (N_d + N_g + 1)\hbar\omega \\ L_z &= \hbar(N_d - N_g) \end{aligned} \tag{46}$$

Thus, while maintaining $H$ in a form as simple as (27), we have simplified that of $L_z$.

### c. STATIONARY STATES OF WELL-DEFINED ANGULAR MOMENTUM

Using the operators $a_d$ and $a_g$, we can now go through the same arguments we used for $a_x$ and $a_y$. It follows that the spectra of $N_d$ and $N_g$ are composed of all positive integers and zero. In addition, specifying a pair $\{ n_d, n_g \}$ of such integers determines uniquely (to within a constant factor) the eigenvector common to $N_d$ and $N_g$, associated with these eigenvalues, which is written:

$$| \chi_{n_d, n_g} \rangle = \frac{1}{\sqrt{(n_d)!\,(n_g)!}} (a_d^\dagger)^{n_d} (a_g^\dagger)^{n_g} | \varphi_{0,0} \rangle \tag{47}$$

$N_d$ and $N_g$ therefore form a C.S.C.O. in $\mathscr{E}_{xy}$. Thus we see, by using (46), that $|\chi_{n_d,n_g}\rangle$ is also an eigenvector of $H_{xy}$ and of $L_z$, with the eigenvalues $(n+1)\hbar\omega$ and $m\hbar$, where $n$ and $m$ are given by:

$$\begin{aligned} n &= n_d + n_g \\ m &= n_d - n_g \end{aligned} \tag{48}$$

Equations (48) enable us to understand the origin of the name of right or left "circular quanta". The action of the operator $a_d^\dagger$ on $|\chi_{n_d,n_g}\rangle$ yields a state with one more quantum, to which, since $m$ has increased by one, an additional angular momentum $+\hbar$ must be attributed (this corresponds to a counterclockwise rotation about $Oz$). Similarly, $a_g^\dagger$ yields a state with one more quantum, of angular momentum $-\hbar$ (clockwise rotation).

Since $n_d$ and $n_g$ are positive integers (or zero), our results are in agreement with those of the preceding section: the eigenvalues of $H_{xy}$ are of the form $(n+1)\hbar\omega$, where $n$ is a positive integer or zero; their degree of degeneracy is $(n+1)$ since, for fixed $n$, we can have:

$$\begin{aligned} n_d &= n \quad ; n_g = 0 \\ n_d &= n-1 ; n_g = 1 \\ &\vdots \\ n_d &= 0 \quad ; n_g = n \end{aligned} \tag{49}$$

Furthermore, we see that the eigenvalues of $L_z$ are of the form $m\hbar$, where $m$ is a positive or negative integer or zero, which is the result that was established for the general case in chapter VI. In addition, table (49) tells us which values of $m$ are associated with a given value of $n$. For example, for the ground state, we have $n_d = n_g = 0$, and therefore, necessarily, $m = 0$; for the first excited state, we can have $n_d = 1$ and $n_g = 0$, or $n_d = 0$ and $n_g = 1$, which yields either $m = +1$ or $m = -1$. In general, formulas (48) and (49) show that, for a given energy level $(n+1)\hbar\omega$, the possible values of $m$ are:

$$m = n, n-2, n-4, \ldots, -n+2, -n \tag{50}$$

It follows that, to a pair of values of $n$ and $m$, there corresponds a single vector (to within a constant factor):

$$|\chi_{n_d=\frac{n+m}{2},\, n_g=\frac{n-m}{2}}\rangle$$

$H$ and $L_z$ therefore form a C.S.C.O. in $\mathscr{E}_{xy}$.

COMMENT:

For a given value of the total energy (labeled by $n$), the states $|\chi_{n_d=n,n_g=0}\rangle$ and $|\chi_{n_d=0,n_g=n}\rangle$ therefore correspond to the maximal ($n\hbar$) and minimal ($-n\hbar$) values of $L_z$. These states therefore recall the classical right and left circular motions associated with a given value of the total energy, for which $\mathscr{L}_z$ takes on its maximal and minimal values (see §1-a).

### d. WAVE FUNCTIONS ASSOCIATED WITH THE EIGENSTATES COMMON TO $H_{xy}$ AND $L_z$

To conserve the symmetry of the problem with respect to rotation about $Oz$, we shall use polar coordinates, setting:

$$\begin{aligned} x &= \rho \cos \varphi \qquad \rho \geqslant 0 \\ y &= \rho \sin \varphi \qquad 0 \leqslant \varphi < 2\pi \end{aligned} \tag{51}$$

Now, what is the action of the operators $a_d$ and $a_g$ on a function of $\rho$ and $\varphi$? We shall begin by determining their action on a function of $x$ and $y$. Knowing that of $X$ and $P_x$ and therefore that of $a_x$ (and, by analogy, that of $a_y$), we can use (40), which yields:

$$a_d \Longrightarrow \frac{1}{2}\left[\beta(x - iy) + \frac{1}{\beta}\left(\frac{\partial}{\partial x} - i\frac{\partial}{\partial y}\right)\right] \tag{52}$$

According to the rules for differentiating functions of several variables, we then obtain:

$$a_d \Longrightarrow \frac{e^{-i\varphi}}{2}\left[\beta\rho + \frac{1}{\beta}\frac{\partial}{\partial\rho} - \frac{i}{\beta\rho}\frac{\partial}{\partial\varphi}\right] \tag{53}$$

Similarly:

$$a_d^\dagger \Longrightarrow \frac{e^{i\varphi}}{2}\left[\beta\rho - \frac{1}{\beta}\frac{\partial}{\partial\rho} - \frac{i}{\beta\rho}\frac{\partial}{\partial\varphi}\right] \tag{54}$$

and:

$$\begin{aligned} a_g &\Longrightarrow \frac{e^{i\varphi}}{2}\left[\beta\rho + \frac{1}{\beta}\frac{\partial}{\partial\rho} + \frac{i}{\beta\rho}\frac{\partial}{\partial\varphi}\right] \\ a_g^\dagger &\Longrightarrow \frac{e^{-i\varphi}}{2}\left[\beta\rho - \frac{1}{\beta}\frac{\partial}{\partial\rho} + \frac{i}{\beta\rho}\frac{\partial}{\partial\varphi}\right] \end{aligned} \tag{55}$$

To calculate the wave functions $\chi_{n_d,n_g}(\rho, \varphi)$, simply apply the differential operators which represent $a_d^\dagger$ and $a_g^\dagger$ to the function $\chi_{0,0}(\rho, \varphi)$, which is, according to (31):

$$\chi_{0,0}(\rho, \varphi) = \frac{\beta}{\sqrt{\pi}} e^{-\beta^2\rho^2/2} \tag{56}$$

Now it can be seen from (54) and (55) that the action of $a_d^\dagger$ (or of $a_g^\dagger$) on a function of the form $e^{im\varphi}F(\rho)$ is given by:

$$\begin{aligned} a_d^\dagger[e^{im\varphi}F(\rho)] &= \frac{e^{i(m+1)\varphi}}{2}\left[\left(\beta\rho + \frac{m}{\beta\rho}\right)F(\rho) - \frac{1}{\beta}\frac{dF}{d\rho}\right] \\ a_g^\dagger[e^{im\varphi}F(\rho)] &= \frac{e^{i(m-1)\varphi}}{2}\left[\left(\beta\rho - \frac{m}{\beta\rho}\right)F(\rho) - \frac{1}{\beta}\frac{dF}{d\rho}\right] \end{aligned} \tag{57}$$

Through the repeated application of these relations to (56), we see that the $\varphi$-dependence of $\chi_{n_d,n_g}(\rho, \varphi)$ is simply given by: $e^{i(n_d-n_g)\varphi}$. This is a general result, established in chapter VI: the $\varphi$-dependence of an eigenfunction of $L_z$ of eigenvalue $m\hbar$ is $e^{im\varphi}$.

If, in (57), we choose $F(\rho) = \rho^m \, e^{-\beta^2\rho^2/2}$, then:

$$a_d^\dagger[e^{im\varphi}\rho^m \, e^{-\beta^2\rho^2/2}] = \beta e^{i(m+1)\varphi}\rho^{m+1} \, e^{-\beta^2\rho^2/2} \tag{58}$$

Applying the operator $a_d^\dagger$ to the function $\chi_{0,0}(\rho)$ $n_d$ times, we obtain:

$$\chi_{n_d,0}(\rho, \varphi) = \frac{\beta}{\sqrt{\pi(n_d)!}} \, e^{in_d\varphi}(\beta\rho)^{n_d} \, e^{-\beta^2\rho^2/2} \tag{59}$$

An analogous calculation yields:

$$\chi_{0,n_g}(\rho, \varphi) = \frac{\beta}{\sqrt{\pi(n_g)!}} \, e^{-in_g\varphi}(\beta\rho)^{n_g} \, e^{-\beta^2\rho^2/2} \tag{60}$$

These wave functions are normalized. For a given energy level $(n + 1)\hbar\omega$, the wave functions (59) and (60) correspond to the limiting values $+ n$ and $- n$ of the quantum number $m$. Their $\rho$-dependence is particularly simple: their modulus reaches a maximum for $\rho = \sqrt{n}/\beta$. Therefore (as in the case of a one-dimensional harmonic oscillator), the spatial spread of these wave functions increases with the energy $(n + 1)\hbar\omega$ with which they are associated.

In the same way, application of the operators $a_d^\dagger$ (or $a_g^\dagger$) to (59) and (60) permits the construction of the functions $\chi_{n_d,n_g}(\rho, \varphi)$ for any $n_d$ and $n_g$. The results obtained for the first excited levels are given in table I.

| | | |
|---|---|---|
| $n = 0$ | $m = 0$ | $\chi_{0,0}(\rho) = \frac{\beta}{\sqrt{\pi}} \, e^{-\beta^2\rho^2/2}$ |
| $n = 1$ | $m = 1$ | $\chi_{1,0}(\rho, \varphi) = \frac{\beta}{\sqrt{\pi}} \, \beta\rho \, e^{-\beta^2\rho^2/2} \, e^{i\varphi}$ |
| | $m = -1$ | $\chi_{0,1}(\rho, \varphi) = \frac{\beta}{\sqrt{\pi}} \, \beta\rho \, e^{-\beta^2\rho^2/2} \, e^{-i\varphi}$ |
| $n = 2$ | $m = 2$ | $\chi_{2,0}(\rho, \varphi) = \frac{\beta}{\sqrt{2\pi}} (\beta\rho)^2 \, e^{-\beta^2\rho^2/2} \, e^{2i\varphi}$ |
| | $m = 0$ | $\chi_{1,1}(\rho, \varphi) = \frac{\beta}{\sqrt{\pi}} [(\beta\rho)^2 - 1] \, e^{-\beta^2\rho^2/2}$ |
| | $m = -2$ | $\chi_{0,2}(\rho, \varphi) = \frac{\beta}{\sqrt{2\pi}} (\beta\rho)^2 \, e^{-\beta^2\rho^2/2} \, e^{-2i\varphi}$ |

TABLE I

**Eigenfunctions common to the Hamiltonian $H_{xy}$ and the observable $L_z$, for the first levels of the two-dimensional harmonic oscillator.**

COMMENT:

The functions $\chi_{n_d,0}(\rho, \varphi)$ given in (59) are proportional to $e^{-\beta^2\rho^2/2}(\beta\rho\, e^{i\varphi})^{n_d}$. More generally, all their linear combinations are of the form:

$$F(\rho, \varphi) = e^{-\beta^2\rho^2/2} f(\beta\rho\, e^{i\varphi}) \tag{61}$$

(where $f$ is an arbitrary function of one variable) and are eigenfunctions of $N_g$ with the eigenvalue zero. It can easily be shown from (55) that:

$$a_g F(\rho, \varphi) = 0 \tag{62}$$

Similarly, the subspace of eigenfunctions of $N_d$ of eigenvalue zero is composed of functions of the form:

$$G(\rho, \varphi) = e^{-\beta^2\rho^2/2} g(\beta\rho\, e^{-i\varphi}) \tag{63}$$

## 4. Quasi-classical states

Using the properties of the one-dimensional harmonic oscillator, we can easily calculate the time evolution of the state vector and the mean values of the various observables of the two-dimensional oscillator. For example, it is not difficult to show that in the mean values $\langle X \rangle(t)$ and $\langle Y \rangle(t)$, as well as $\langle P_x \rangle(t)$ and $\langle P_y \rangle(t)$, only the Bohr frequency $\omega$ appears. Moreover, it can be shown that these mean values exactly obey the classical equations of motion. In this section, we shall be concerned with the properties and evolution of the quasi-classical states of the two-dimensional harmonic oscillator.

### a. DEFINITION OF THE STATES $|\alpha_x, \alpha_y\rangle$ AND $|\alpha_d, \alpha_g\rangle$

To construct a quasi-classical state of the two-dimensional harmonic oscillator, we can base our reasoning on the one-dimensional oscillator (*cf.* complement $G_V$). Recall that, in a quasi-classical state associated with a given classical motion, the mean values $\langle X \rangle(t)$ and $\langle P \rangle(t)$ coincide at each instant with $x(t)$ and $p(t)$. Similarly, the mean value of the Hamiltonian $H$ is equal (to within a half-quantum $\hbar\omega/2$) to the classical energy. We showed in complement $G_V$ that, at any time, the quasi-classical states are eigenstates of the destruction operator $a$ and can be written:

$$|\alpha\rangle = \sum_n c_n(\alpha)\,|\varphi_n\rangle \tag{64}$$

where $\alpha$ is the eigenvalue of $a$, and:

$$c_n(\alpha) = \frac{\alpha^n}{\sqrt{n!}} e^{-|\alpha|^2/2} \tag{65}$$

In the case which concerns us here, we can use the rules of the tensor product to obtain the quasi-classical states in the form:

$$|\alpha_x, \alpha_y\rangle = |\alpha_x\rangle \otimes |\alpha_y\rangle = \sum_{n_x=0}^{\infty} \sum_{n_y=0}^{\infty} c_{n_x}(\alpha_x)\, c_{n_y}(\alpha_y)\,|\varphi_{n_x,n_y}\rangle \tag{66}$$

with:

$$\begin{aligned} a_x \mid \alpha_x, \alpha_y \rangle &= \alpha_x \mid \alpha_x, \alpha_y \rangle \\ a_y \mid \alpha_x, \alpha_y \rangle &= \alpha_y \mid \alpha_x, \alpha_y \rangle \end{aligned} \tag{67}$$

We are then sure that $\langle X \rangle$, $\langle P_x \rangle$, $\langle H_x \rangle$, $\langle Y \rangle$, $\langle P_y \rangle$, $\langle H_y \rangle$ are the same as the corresponding classical quantities. Now, returning to definition (40) and using (67), we see that:

$$\begin{aligned} a_d \mid \alpha_x, \alpha_y \rangle &= \alpha_d \mid \alpha_x, \alpha_y \rangle \\ a_g \mid \alpha_x, \alpha_y \rangle &= \alpha_g \mid \alpha_x, \alpha_y \rangle \end{aligned} \tag{68}$$

with:

$$\begin{aligned} \alpha_d &= \frac{1}{\sqrt{2}}(\alpha_x - i\alpha_y) \\ \alpha_g &= \frac{1}{\sqrt{2}}(\alpha_x + i\alpha_y) \end{aligned} \tag{69}$$

Therefore, the state $\mid \alpha_x, \alpha_y \rangle$ is also an eigenvector of $a_d$ and $a_g$ with the eigenvalues given in (69). We shall denote by $|\alpha_d, \alpha_g\rangle$ the eigenvector common to $a_d$ and $a_g$ associated with the eigenvalues $\alpha_d$ and $\alpha_g$. It is easy to show that the expansion of $\mid \alpha_d, \alpha_g \rangle$ on the $\{ \mid \chi_{n_d,n_g} \rangle \}$ basis has the same form as that of $\mid \alpha_x, \alpha_y \rangle$ on the $\{ \mid \varphi_{n_x,n_y} \rangle \}$ basis:

$$\mid \alpha_d, \alpha_g \rangle = \sum_{n_d=0}^{\infty} \sum_{n_g=0}^{\infty} c_{n_d}(\alpha_d)\, c_{n_g}(\alpha_g) \mid \chi_{n_d,n_g} \rangle \tag{70}$$

where the coefficients $c_n$ are given by (65). It follows from (68) and (69) that:

$$\mid \alpha_x, \alpha_y \rangle = \mid \alpha_d = \frac{\alpha_x - i\alpha_y}{\sqrt{2}}, \alpha_g = \frac{\alpha_x + i\alpha_y}{\sqrt{2}} \rangle \tag{71}$$

Because of the properties of the states $|\alpha\rangle$ (*cf.* complement $G_V$, § 3-a), we see that if:

$$\mid \psi(0) \rangle = \mid \alpha_x, \alpha_y \rangle = \mid \alpha_d, \alpha_g \rangle \tag{72}$$

the state vector at the instant $t$ will be:

$$\begin{aligned} \mid \psi(t) \rangle &= \mathrm{e}^{-i\omega t} \mid \mathrm{e}^{-i\omega t}\alpha_x, \mathrm{e}^{-i\omega t}\alpha_y \rangle \\ &= \mathrm{e}^{-i\omega t} \mid \mathrm{e}^{-i\omega t}\alpha_d, \mathrm{e}^{-i\omega t}\alpha_g \rangle \end{aligned} \tag{73}$$

### b. MEAN VALUES AND ROOT-MEAN-SQUARE DEVIATIONS OF THE VARIOUS OBSERVABLES

We set:

$$\begin{aligned} \alpha_x &= |\alpha_x|\, \mathrm{e}^{i\varphi_x} \\ \alpha_y &= |\alpha_y|\, \mathrm{e}^{i\varphi_y} \end{aligned} \tag{74}$$

Using formulas (93) of complement $G_V$, we obtain:

$$\begin{cases} \langle X \rangle(t) = \dfrac{\sqrt{2}}{\beta} |\alpha_x| \cos(\omega t - \varphi_x) \\ \langle Y \rangle(t) = \dfrac{\sqrt{2}}{\beta} |\alpha_y| \cos(\omega t - \varphi_y) \end{cases} \tag{75}$$

$$\begin{cases} \langle P_x \rangle(t) = -\mu\omega \dfrac{\sqrt{2}}{\beta} |\alpha_x| \sin(\omega t - \varphi_x) \\ \langle P_y \rangle(t) = -\mu\omega \dfrac{\sqrt{2}}{\beta} |\alpha_y| \sin(\omega t - \varphi_y) \end{cases} \tag{76}$$

Comparing (75) and (76) with (5) and (6), we see that:

$$\begin{aligned} \alpha_x &= \frac{\beta x_M}{\sqrt{2}} e^{i\varphi_x} \\ \alpha_y &= \frac{\beta y_M}{\sqrt{2}} e^{i\varphi_y} \end{aligned} \tag{77}$$

where $x_M$, $\varphi_x$, $y_M$, $\varphi_y$ are the parameters defining the classical motion which the state $|\alpha_x, \alpha_y\rangle$ best reproduces.

Also:

$$\begin{aligned} \langle N_x \rangle &= |\alpha_x|^2 \\ \langle N_y \rangle &= |\alpha_y|^2 \end{aligned} \tag{78}$$

and:

$$\begin{aligned} \langle N_d \rangle &= |\alpha_d|^2 = \frac{1}{2}\left[|\alpha_x|^2 + |\alpha_y|^2 + i(\alpha_x\alpha_y^* - \alpha_x^*\alpha_y)\right] \\ \langle N_g \rangle &= |\alpha_g|^2 = \frac{1}{2}\left[|\alpha_x|^2 + |\alpha_y|^2 - i(\alpha_x\alpha_y^* - \alpha_x^*\alpha_y)\right] \end{aligned} \tag{79}$$

that is, according to (46):

$$\langle H_{xy} \rangle = \hbar\omega\left(|\alpha_x|^2 + |\alpha_y|^2 + 1\right) = \hbar\omega\left(|\alpha_d|^2 + |\alpha_g|^2 + 1\right) \tag{80}$$

and:

$$\langle L_z \rangle = 2\hbar |\alpha_x| |\alpha_y| \sin(\varphi_y - \varphi_x) = \hbar\left(|\alpha_d|^2 - |\alpha_g|^2\right) \tag{81}$$

According to (77), $\langle L_z \rangle$ is the same as the classical value of $\mathscr{L}_z$ [formula (10)].

Now let us consider the root-mean-square deviations of the position and momentum and then of the energy and angular momentum in a state $|\alpha_x, \alpha_y\rangle$. Directly applying the results of complement $G_V$, we obtain:

$$\begin{aligned} \Delta X &= \Delta Y = \frac{1}{\beta\sqrt{2}} \\ \Delta P_x &= \Delta P_y = \frac{\mu\omega}{\beta\sqrt{2}} \end{aligned} \tag{82}$$

The root-mean-square deviations of the position and momentum are independent of $\alpha_x$ and $\alpha_y$; if $|\alpha_x|$ and $|\alpha_y|$ are much greater than 1, the position and momentum of the oscillator have a very small spread about $\langle X \rangle$, $\langle Y \rangle$ and $\langle P_x \rangle$, $\langle P_y \rangle$.

Finally, let us calculate the root-mean-square deviations $\Delta H_{xy}$ for the energy and $\Delta L_z$ for the angular momentum. As in complement $G_V$:

$$\begin{aligned} \Delta N_x &= |\alpha_x| \\ \Delta N_y &= |\alpha_y| \\ \Delta N_d &= |\alpha_d| \\ \Delta N_g &= |\alpha_g| \end{aligned} \tag{83}$$

But the Hamiltonian $H_{xy}$ involves $N = N_x + N_y$, and $L_z$ is proportional to $N_d - N_g$. We must now calculate, for example:

$$\begin{aligned} (\Delta N)^2 &= \langle (N_x + N_y)^2 \rangle - \langle (N_x + N_y) \rangle^2 \\ &= (\Delta N_x)^2 + (\Delta N_y)^2 + 2[\langle N_x N_y \rangle - \langle N_x \rangle \langle N_y \rangle] \end{aligned} \tag{84}$$

According to (66), the state of the system is a tensor product, which means that the observables $N_x$ and $N_y$ are not correlated:

$$\langle N_x N_y \rangle = \langle N_x \rangle \langle N_y \rangle \tag{85}$$

It follows that:

$$(\Delta N)^2 = (\Delta N_x)^2 + (\Delta N_y)^2 \tag{86}$$

that is:

$$\Delta H_{xy} = \hbar\omega \sqrt{|\alpha_x|^2 + |\alpha_y|^2} = \hbar\omega \sqrt{|\alpha_d|^2 + |\alpha_g|^2} \tag{87}$$

Similarly:

$$\Delta L_z = \hbar \sqrt{|\alpha_d|^2 + |\alpha_g|^2} = \hbar \sqrt{|\alpha_x|^2 + |\alpha_y|^2} \tag{88}$$

## Complement $E_{VI}$

# A CHARGED PARTICLE IN A MAGNETIC FIELD: LANDAU LEVELS



Thus far, we have been considering, for various special cases, the properties of a particle subjected to a scalar potential $V(\mathbf{r})$ (representing, for example, the effect of an electric field on a charged particle). Chapter V (the harmonic oscillator) and chapter VII (particle subjected to a central potential) treat other examples of scalar potentials. Here we shall be concerned with a complementary problem, that of the properties of a particle subjected to a vector potential $\mathbf{A}(\mathbf{r})$ (a charged particle placed in a magnetic field). We shall encounter a number of purely quantum mechanical effects, such as equally spaced energy levels in a uniform magnetic field (Landau levels)*. Before studying the problem from a quantum mechanical point of view, we shall rapidly review some classical results.

## 1. Review of the classical problem

### a. MOTION OF THE PARTICLE

When a particle of position $\mathbf{r}$ and charge $q$ is subject to a magnetic field $\mathbf{B}(\mathbf{r})$, the force $\mathbf{f}$ exerted on it is given by the Lorentz force:

$$\mathbf{f} = q\,\mathbf{v} \times \mathbf{B}(\mathbf{r}) \tag{1}$$

where:

$$\mathbf{v} = \frac{\mathrm{d}\mathbf{r}}{\mathrm{d}t} \tag{2}$$

* This equal spacing is, as we shall show, a consequence of the properties of the harmonic oscillator, and it could have been treated in chapter V. However, we shall also see that the properties of angular momentum are useful in the study and classification of the stationary states of the particle. This is why this complement follows chapter VI.

is the velocity of the particle. Its motion obeys the fundamental law of dynamics:

$$\mu \frac{\mathrm{d}\mathbf{v}}{\mathrm{d}t} = \mathbf{f} \tag{3}$$

(where $\mu$ is the mass of the particle).

In the rest of this complement, we shall often be considering the case in which the magnetic field is uniform; we shall choose its direction along the $Oz$ axis. By solving the equation of motion (3), one can show that in this case the three coordinates $x(t)$, $y(t)$ and $z(t)$ of the particle are given by:

$$\begin{aligned} x(t) &= x_0 + \sigma \cos(\omega_c t - \varphi_0) \\ y(t) &= y_0 + \sigma \sin(\omega_c t - \varphi_0) \\ z(t) &= v_{0z} t + z_0 \end{aligned} \tag{4}$$

where $x_0$, $y_0$, $z_0$, $\sigma$, $\varphi_0$ and $v_{0z}$ are six constant parameters which depend on the initial conditions; the "cyclotron frequency" $\omega_c$ is given by:

$$\omega_c = -q \frac{B}{\mu} \tag{5}$$

Equations (4) show that the projection of the position $M$ of the particle onto the $xOy$ plane performs a uniform circular motion, of angular velocity $\omega_c$ and initial phase $\varphi_0$, on a circle of radius $\sigma$ whose center is the point $C_0$, with coordinates $x_0$ and $y_0$. The motion of the projection of $M$ onto $Oz$ is simply rectilinear and uniform. It follows that the particle moves in space along a circular helix (*cf.* fig. 1), whose axis is parallel to $Oz$ and passes through $C_0$.

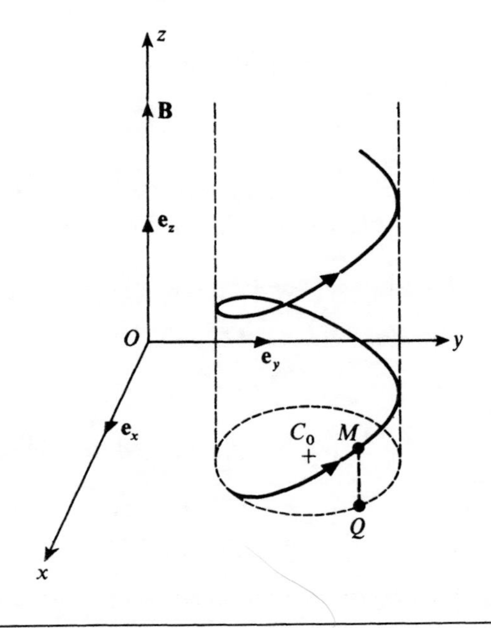


FIGURE 1

**Classical trajectory of a charged particle in a uniform magnetic field parallel to $Oz$. The particle moves at constant velocity along a circular helix whose axis, parallel to $Oz$, passes through the point $C_0$. The figure is drawn for $q < 0$ (the case of the electron), that is, $\omega_c > 0$.**

If we are concerned only with the motion of the point $Q$, the projection of $M$ onto the $xOy$ plane, we study the behavior of the vector:

$$\boldsymbol{\rho} = x\,\mathbf{e}_x + y\,\mathbf{e}_y \tag{6}$$

(where $\mathbf{e}_x$ and $\mathbf{e}_y$ are the unit vectors of the $Ox$ and $Oy$ axes). The velocity of $Q$ is :

$$\mathbf{v}_\perp = \frac{\mathrm{d}\boldsymbol{\rho}}{\mathrm{d}t} \tag{7}$$

It is therefore convenient to introduce the components $x'$ and $y'$ of the vector $\mathbf{C}_0\mathbf{Q}$:

$$\begin{aligned} x' &= x - x_0 \\ y' &= y - y_0 \end{aligned} \tag{8}$$

Since $Q$ performs a uniform circular motion about $C_0$, we have:

$$\mathbf{v}_\perp = \omega_c\,\mathbf{e}_z \times \mathbf{C}_0\mathbf{Q} \tag{9}$$

(where $\mathbf{e}_z$ is the unit vector of $Oz$). This implies that the coordinates $x_0$ and $y_0$ of $C_0$ are related to the coordinates of $Q$ and to the components of $\mathbf{v}_\perp$ by:

$$\begin{aligned} x_0 &= x - \frac{1}{\omega_c} v_y \\ y_0 &= y + \frac{1}{\omega_c} v_x \end{aligned} \tag{10}$$

#### b. THE VECTOR POTENTIAL. THE CLASSICAL LAGRANGIAN AND HAMILTONIAN

To describe the magnetic field $\mathbf{B}(\mathbf{r})$, one can use a vector potential $\mathbf{A}(\mathbf{r})$ which is, by definition, related to $\mathbf{B}(\mathbf{r})$ by:

$$\mathbf{B}(\mathbf{r}) = \boldsymbol{\nabla} \times \mathbf{A}(\mathbf{r}) \tag{11}$$

For example, if the field $\mathbf{B}$ is uniform, one can choose:

$$\mathbf{A}(\mathbf{r}) = -\frac{1}{2}\mathbf{r} \times \mathbf{B} \tag{12}$$

We know, furthermore, that when $\mathbf{B}(\mathbf{r})$ is given, condition (11) does not determine $\mathbf{A}(\mathbf{r})$ uniquely : a gradient of an arbitrary function of $\mathbf{r}$ can be added to $\mathbf{A}(\mathbf{r})$ without changing $\mathbf{B}(\mathbf{r})$*.

It can be shown (*cf.* appendix III, §4-b) that the Lagrange function $\mathscr{L}(\mathbf{r}, \mathbf{v})$ of the particle is given by:

$$\mathscr{L}(\mathbf{r}, \mathbf{v}) = \frac{1}{2}\mu\,\mathbf{v}^2 + q\,\mathbf{v}.\,\mathbf{A}(\mathbf{r}) \tag{13}$$

It follows that $\mathbf{p}$, the conjugate momentum of the position $\mathbf{r}$, is related to $\mathbf{v}$ and $\mathbf{A}(\mathbf{r})$ by:

$$\mathbf{p} = \boldsymbol{\nabla}_{\mathbf{v}}\mathscr{L}(\mathbf{r}, \mathbf{v}) = \mu\,\mathbf{v} + q\mathbf{A}(\mathbf{r}) \tag{14}$$

* For example, for a uniform field parallel to $Oz$, one could choose, instead of the vector $\mathbf{A}(\mathbf{r})$ given by (12), the vector whose components are $A_x = 0$, $A_y = xB$, $A_z = 0$.

The classical Hamiltonian $\mathscr{H}(\mathbf{r}, \mathbf{p})$ is then:

$$\mathscr{H}(\mathbf{r}, \mathbf{p}) = \frac{1}{2\mu}[\mathbf{p} - q\mathbf{A}(\mathbf{r})]^2 \tag{15}$$

It will prove convenient to set:

$$\mathscr{H}(\mathbf{r}, \mathbf{p}) = \mathscr{H}_{\perp}(\mathbf{r}, \mathbf{p}) + \mathscr{H}_{/\!/}(\mathbf{r}, \mathbf{p}) \tag{16}$$

with:

$$\begin{aligned} \mathscr{H}_{\perp}(\mathbf{r}, \mathbf{p}) &= \frac{1}{2\mu}\{[p_x - qA_x(\mathbf{r})]^2 + [p_y - qA_y(\mathbf{r})]^2\} \\ \mathscr{H}_{/\!/}(\mathbf{r}, \mathbf{p}) &= \frac{1}{2\mu}[p_z - qA_z(\mathbf{r})]^2 \end{aligned} \tag{17}$$

COMMENT:

In this case, unlike that of a scalar potential $V(\mathbf{r})$, relation (14) shows that the momentum $\mathbf{p}$ is not equal to the mechanical momentum $\mu\mathbf{v}$. Also, comparing (14) with (15), we see that $\mathscr{H}$ is equal to the kinetic energy $\mu\mathbf{v}^2/2$ of the particle; this results from the fact that since the Lorentz force written in (1) is always perpendicular to $\mathbf{v}$, it does no work during the motion. Similarly, it must be noted that the angular momentum:

$$\mathscr{L} = \mathbf{r} \times \mathbf{p} \tag{18}$$

is different from the moment of the mechanical momentum $\mu\mathbf{v}$:

$$\boldsymbol{\lambda} = \mathbf{r} \times \mu\mathbf{v} \tag{19}$$

### c. CONSTANTS OF THE MOTION IN A UNIFORM FIELD

Consider the special case in which the field $\mathbf{B}$ is uniform. The motion of the particle (§1-a) is such that $\mathscr{H}_{/\!/}$ and $\mathscr{H}_{\perp}$, defined in (17), are constants of the motion*.

If we substitute (14) into (10), we obtain:

$$\begin{aligned} x_0 &= x - \frac{1}{\mu\omega_c}[p_y - qA_y(\mathbf{r})] \\ y_0 &= y + \frac{1}{\mu\omega_c}[p_x - qA_x(\mathbf{r})] \end{aligned} \tag{20}$$

It follows that the radius $\sigma$ of the helical trajectory satisfies:

$$\begin{aligned} \sigma^2 = (x - x_0)^2 + (y - y_0)^2 &= \left(\frac{1}{\mu\omega_c}\right)^2 \{[p_y - qA_y(\mathbf{r})]^2 + [p_x - qA_x(\mathbf{r})]^2\} \\ &= \frac{2}{\mu\omega_c^2}\mathscr{H}_{\perp} \end{aligned} \tag{21}$$

$\sigma^2$ is therefore proportional to the Hamiltonian $\mathscr{H}_{\perp}$.

* This follows from the fact that, according to (14) and (17), $\mathscr{H}_{\perp}$ and $\mathscr{H}_{/\!/}$ are equal, respectively, to the kinetic energies $\mu\mathbf{v}_{\perp}^2/2$ and $\mu v_z^2/2$ associated with the motions which are perpendicular and parallel to $Oz$.

Similarly, let $\boldsymbol{\theta}$ be the moment of the mechanical momentum $\mu\mathbf{v}$ with respect to the center $C_0$ of the circle:

$$\boldsymbol{\theta} = \mathbf{C_0M} \times \mu\,\mathbf{v} \tag{22}$$

The component $\theta_z$ of this moment can then be written, with (20) taken into account:

$$\begin{aligned}\theta_z &= \mu\,[(x - x_0)v_y - (y - y_0)v_x] \\ &= \frac{1}{\mu\omega_c}\{[p_y - qA_y(\mathbf{r})]^2 + [p_x - qA_x(\mathbf{r})]^2\} = \frac{2}{\omega_c}\mathscr{H}_\perp\end{aligned} \tag{23}$$

$\theta_z$ is therefore a constant of the motion, as might have been expected. On the other hand, the component $\lambda_z$ of the moment of the mechanical momentum $\mu\mathbf{v}$ with respect to $O$ is not, in general, constant, since:

$$\lambda_z = \theta_z + \mu[x_0v_y(t) - y_0v_x(t)] \tag{24}$$

Therefore, according to (4), $\lambda_z$ varies sinusoidally in time.

Finally, consider the projection $\mathscr{L}_z$ onto $Oz$ of the angular momentum $\boldsymbol{\mathscr{L}}$:

$$\mathscr{L}_z = xp_y - yp_x \tag{25}$$

According to (14), it can be written:

$$\mathscr{L}_z = x[\mu\,v_y + qA_y(\mathbf{r})] - y[\mu\,v_x + qA_x(\mathbf{r})] \tag{26}$$

It therefore depends explicitly on the gauge chosen, that is, on the vector potential $\mathbf{A}$ picked to describe the magnetic field. In most cases, $\mathscr{L}_z$ is not a constant of the motion. Nevertheless, if one chooses the gauge given in (12), one obtains from (4):

$$\mathscr{L}_z = \frac{qB}{2}(x_0^2 + y_0^2 - \sigma^2) \tag{27}$$

$\mathscr{L}_z$ is then a constant of the motion.

Relation (27) does not have a simple physical interpretation, since it is valid only in a particular gauge. However, it will prove useful to us in the following sections for the quantum mechanical study of the problem.

## 2. General quantum mechanical properties of a particle in a magnetic field

### a. QUANTIZATION. HAMILTONIAN

Consider a particle placed in an arbitrary magnetic field described by the vector potential $\mathbf{A}(x, y, z)$. In quantum mechanics, the vector potential becomes an operator, a function of three observables, $X$, $Y$ and $Z$. The operator $H$, the Hamiltonian of the particle, can be obtained from (15):

$$H = \frac{1}{2\mu}[\mathbf{P} - q\mathbf{A}(X, Y, Z)]^2 \tag{28}$$

According to (14), the operator $\mathbf{V}$ associated with the velocity of the particle is given by:

$$\mathbf{V} = \frac{1}{\mu}[\mathbf{P} - q\mathbf{A}(X, Y, Z)] \tag{29}$$

which enables us to write $H$ in the form:

$$H = \frac{\mu}{2}\mathbf{V}^2 \tag{30}$$

### b. COMMUTATION RELATIONS

The observables **R** and **P** satisfy the canonical commutation relations:

$$[X, P_x] = [Y, P_y] = [Z, P_z] = i\hbar \tag{31}$$

The other commutators between components of **R** and **P** are zero. Two components of **P** therefore commute. However, we see from (29) that the same is not true for **V**; for example:

$$[V_x, V_y] = -\frac{q}{\mu^2}\{[P_x, A_y(\mathbf{R})] + [A_x(\mathbf{R}), P_y]\} \tag{32}$$

This expression is easy to calculate, using the rule given in complement $B_{II}$ [*cf.* formula (48)]:

$$[V_x, V_y] = \frac{iq\hbar}{\mu^2}\left\{\frac{\partial A_y}{\partial X} - \frac{\partial A_x}{\partial Y}\right\} = \frac{iq\hbar}{\mu^2} B_z(\mathbf{R}) \tag{33-a}$$

Similarly, it can be shown that:

$$[V_y, V_z] = \frac{iq\hbar}{\mu^2} B_x(\mathbf{R}) \tag{33-b}$$

$$[V_z, V_x] = \frac{iq\hbar}{\mu^2} B_y(\mathbf{R}) \tag{33-c}$$

The magnetic field therefore enters explicitly into the commutation relations for the velocity.

However, since $\mathbf{A}(\mathbf{R})$ commutes with $X$, $Y$ and $Z$, relation (29) implies that:

$$[X, V_x] = \frac{1}{\mu}[X, P_x] = \frac{i\hbar}{\mu} \tag{34-a}$$

and, similarly:

$$[Y, V_y] = [Z, V_z] = \frac{i\hbar}{\mu} \tag{34-b}$$

(the other commutators between a component of **R** and a component of **V** are zero). From these relations, it can be deduced (*cf.* complement $C_{III}$) that:

$$\Delta X \,.\, \Delta V_x \geqslant \frac{\hbar}{2\mu} \tag{35}$$

(with analogous inequalities for the components along $Oy$ and $Oz$). The physical consequences of the Heisenberg uncertainty relations are therefore not modified by the presence of a magnetic field.

Finally, let us calculate the commutation relations between the components of the operator:

$$\mathbf{\Lambda} = \mu \ \mathbf{R} \times \mathbf{V} \tag{36}$$

associated with the moment with respect to $O$ of the mechanical momentum*. We obtain:

$$\begin{aligned}[\Lambda_x, \Lambda_y] &= \mu^2[YV_z - ZV_y, ZV_x - XV_z] \\ &= \mu^2 Y \{ [V_z, Z]V_x + Z[V_z, V_x] \} - \mu^2 Z^2 [V_y, V_x] \\ &\qquad + \mu^2 X \{ Z[V_y, V_z] + [Z, V_z]V_y \}\end{aligned} \tag{37}$$

that is, with (33) and (34) taken into account:

$$[\Lambda_x, \Lambda_y] = i\hbar \{ -\mu YV_x + qYZB_y + qZ^2B_z + qXZB_x + \mu XV_y \} \tag{38}$$

It follows that:

$$[\Lambda_x, \Lambda_y] = i\hbar \{ \Lambda_z + qZ \ \mathbf{R} . \mathbf{B}(\mathbf{R}) \} \tag{39}$$

(the other commutators can be obtained by cyclic permutation of the indices $x$, $y$ and $z$). When the field $\mathbf{B}$ is not zero, the commutation relations of $\mathbf{\Lambda}$ are completely different from those of $\mathbf{L}$. The operator $\mathbf{\Lambda}$ therefore does not, *a priori*, possess the properties of angular momenta proved in chapter VI.

#### c. PHYSICAL CONSEQUENCES

##### α. *Evolution of* $\langle \mathbf{R} \rangle$

The time variation of the mean position of the particle is given by Ehrenfest's theorem:

$$i\hbar \frac{\mathrm{d}}{\mathrm{d}t} \langle \mathbf{R} \rangle = \langle [\mathbf{R}, H] \rangle = \langle \left[ \mathbf{R}, \frac{\mu}{2} \mathbf{V}^2 \right] \rangle \tag{40}$$

[according to formula (30)]. Equations (34) are not difficult to interpret, since, substituted into (40), they yield:

$$\frac{\mathrm{d}}{\mathrm{d}t} \langle \mathbf{R} \rangle = \langle \mathbf{V} \rangle \tag{41}$$

As in the case in which the magnetic field is zero, the mean velocity is therefore equal to the derivative of $\langle \mathbf{R} \rangle$. Equation (41) is the quantum mechanical analogue of (2).

##### β. *Evolution of* $\langle \mathbf{V} \rangle$. *The Lorentz force*

Let us calculate the time derivative of the mean value $\langle \mathbf{V} \rangle$ of the velocity:

$$i\hbar \frac{\mathrm{d}}{\mathrm{d}t} \langle \mathbf{V} \rangle = \langle \left[ \mathbf{V}, \frac{\mu}{2} \mathbf{V}^2 \right] \rangle \tag{42}$$

* Of course, the components of the angular momentum $\mathbf{L} = \mathbf{R} \times \mathbf{P}$ always satisfy the usual commutation relations.

Since, according to relations (33):

$$[\mathbf{V}^2, V_x] = [V_x^2 + V_y^2 + V_z^2, V_x]$$
$$= V_y[V_y, V_x] + [V_y, V_x]V_y + V_z[V_z, V_x] + [V_z, V_x]V_z$$
$$= \frac{iq\hbar}{\mu^2}\{ - V_yB_z(\mathbf{R}) - B_z(\mathbf{R})V_y + V_zB_y(\mathbf{R}) + B_y(\mathbf{R})V_z \} \tag{43}$$

it is easy to see that:

$$\mu\frac{\mathrm{d}}{\mathrm{d}t}\langle \mathbf{V} \rangle = \langle \mathbf{F}(\mathbf{R},\mathbf{V}) \rangle \tag{44}$$

where the operator $\mathbf{F}(\mathbf{R}, \mathbf{V})$ is defined by:

$$\mathbf{F}(\mathbf{R},\mathbf{V}) = \frac{q}{2}\{ \mathbf{V} \times \mathbf{B}(\mathbf{R}) - \mathbf{B}(\mathbf{R}) \times \mathbf{V} \} \tag{45}$$

The last two relations are simply the analogues of the classical relations (1) and (3). Here, we obtain a symmetrized expression for $\mathbf{F}(\mathbf{R}, \mathbf{V})$ (*cf.* chap. III, § B-5), since $\mathbf{R}$ and $\mathbf{V}$ do not commute.

γ. *Evolution of* $\langle \mathbf{\Lambda} \rangle$

Now let us evaluate:

$$i\hbar\frac{\mathrm{d}}{\mathrm{d}t}\langle \mathbf{\Lambda} \rangle = \langle [\mathbf{\Lambda}, H] \rangle \tag{46}$$

To do so, let us calculate, for example, the commutator $[XV_y - YV_x, H]$:

$$[XV_y - YV_x, H] = X[V_y, H] + [X, H]V_y - Y[V_x, H] - [Y, H]V_x$$
$$= \frac{i\hbar}{\mu}(XF_y - YF_x) + i\hbar(V_xV_y - V_yV_x) \tag{47}$$

But $X$ and $V_y$ commute, as do $Y$ and $V_x$. The commutator we are calculating is therefore equal to:

$$[V_yX - V_xY, H] = V_y[X, H] + [V_y, H]X - V_x[Y, H] - [V_x, H]Y$$
$$= \frac{i\hbar}{\mu}(F_yX - F_xY) + i\hbar(V_yV_x - V_xV_y) \tag{48}$$

Taking half the sum of these two expressions, we find $\frac{\mathrm{d}}{\mathrm{d}t}\langle \Lambda_z \rangle$ in the form:

$$\frac{\mathrm{d}}{\mathrm{d}t}\langle \Lambda_z \rangle = \frac{1}{2}\langle XF_y - YF_x - F_xY + F_yX \rangle \tag{49}$$

Analogous arguments give the derivative of $\langle \Lambda_x \rangle$ and $\langle \Lambda_y \rangle$; finally:

$$\frac{\mathrm{d}}{\mathrm{d}t}\langle \mathbf{\Lambda} \rangle = \frac{1}{2}\langle \mathbf{R} \times \mathbf{F}(\mathbf{R},\mathbf{V}) - \mathbf{F}(\mathbf{R},\mathbf{V}) \times \mathbf{R} \rangle \tag{50}$$

The classical analogue of this relation is:

$$\frac{d}{dt}\boldsymbol{\lambda} = \mathbf{r} \times \mathbf{f}(\mathbf{r}, \mathbf{v}) \tag{51}$$

which expresses a well-known theorem: the time derivative of the moment of the mechanical momentum with respect to a fixed point $O$ is equal to the moment with respect to $O$ of the force exerted on the particle.

## 3. Case of a uniform magnetic field

When the magnetic field is uniform, the preceding general study can easily be pursued further. We choose the direction of the field $\mathbf{B}$ as the $Oz$ axis. The commutation relations (33) then become, using definition (5):

$$[V_x, V_y] = -i\frac{\hbar\omega_c}{\mu} \tag{52-a}$$

$$[V_y, V_z] = [V_z, V_x] = 0 \tag{52-b}$$

COMMENT:

Applying the results of complement $C_{III}$ to $V_x$ and $V_y$, we can see from (52-a) that their root-mean-square deviations satisfy:

$$\Delta V_x \cdot \Delta V_y \geqslant \frac{\hbar\,|\omega_c|}{2\mu} \tag{53}$$

The components of the velocity $\mathbf{V}_\perp$ are therefore incompatible physical quantities.

### a. EIGENVALUES OF THE HAMILTONIAN

By analogy with (16), $H$ can be written in the form:

$$H = H_\perp + H_{/\!/} \tag{54}$$

with:

$$H_\perp = \frac{\mu}{2}(V_x^2 + V_y^2) \tag{55-a}$$

$$H_{/\!/} = \frac{\mu}{2}V_z^2 \tag{55-b}$$

According to (52-b):

$$[H_\perp; H_{/\!/}] = 0 \tag{56}$$

We can now look for a basis of eigenvectors common to $H_\perp$ (eigenvalues $E_\perp$) and $H_{/\!/}$ (eigenvalues $E_{/\!/}$); they will automatically be eigenvectors of $H$, with the eigenvalues:

$$E = E_\perp + E_{/\!/} \tag{57}$$

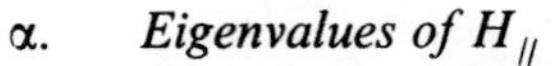

α. *Eigenvalues of $H_{//}$*

The eigenvectors of the operator $V_z$ are also eigenvectors of $H_{//}$. Now, $Z$ and $V_z$ are two Hermitian operators which satisfy the relation:

$$[Z, V_z] = \frac{i\hbar}{\mu} \tag{58}$$

We can therefore apply to them the results of complement $E_{II}$; in particular, the spectrum of $V_z$ includes all the real numbers.

Consequently, the eigenvalues of $H_{//}$ are of the form:

$$E_{//} = \frac{\mu}{2} v_z^2 \tag{59}$$

where $v_z$ is a real arbitrary constant. The spectrum of $H_{//}$ is therefore continuous: the energy $E_{//}$ can take on any positive value or zero.

The interpretation of this result is obvious : $H_{//}$ describes the kinetic energy of a free particle moving along $Oz$ (as in classical mechanics; § 1-a).

β. *Eigenvalues of $H_\perp$*

We shall assume, for example, that the particle under consideration has a negative charge $q$; the cyclotron frequency $\omega_c$ is then positive [formula (5)]*.

We set:

$$\begin{aligned} \hat{Q} &= \sqrt{\frac{\mu}{\hbar\omega_c}}\, V_y \\ \hat{S} &= \sqrt{\frac{\mu}{\hbar\omega_c}}\, V_x \end{aligned} \tag{60}$$

Relation (52-a) can then be written:

$$[\hat{Q}, \hat{S}] = i \tag{61}$$

and $H_\perp$ becomes:

$$H_\perp = \frac{\hbar\omega_c}{2}(\hat{Q}^2 + \hat{S}^2) \tag{62}$$

$H_\perp$ then takes on the form of the Hamiltonian of a one-dimensional harmonic oscillator [*cf.* chap. V, relation (B-4)]. $\hat{Q}$ and $\hat{S}$, which satisfy (61), play the roles of the position $\hat{X}$ and momentum $\hat{P}$ of this oscillator.

The arguments set forth in § B-2 of chapter V for the operators $\hat{X}$ and $\hat{P}$ can be repeated here for $\hat{Q}$ and $\hat{S}$. For example, it can easily be shown that if $|\varphi_\perp\rangle$ is an eigenvector of $H_\perp$ :

$$H_\perp |\varphi_\perp\rangle = E_\perp |\varphi_\perp\rangle \tag{63}$$

* For a positive charge $q$, one can keep the convention of positive $\omega_c$ by choosing the direction of the $Oz$ axis opposite to the magnetic field.

the kets:

$$| \varphi'_\perp \rangle = \frac{1}{\sqrt{2}}(\hat{Q} + i\hat{S}) | \varphi_\perp \rangle \tag{64-a}$$

$$| \varphi''_\perp \rangle = \frac{1}{\sqrt{2}}(\hat{Q} - i\hat{S}) | \varphi_\perp \rangle \tag{64-b}$$

are also eigenvectors of $H_\perp$:

$$H_\perp | \varphi'_\perp \rangle = (E_\perp - \hbar\omega_c) | \varphi'_\perp \rangle \tag{65-a}$$

$$H_\perp | \varphi''_\perp \rangle = (E_\perp + \hbar\omega_c) | \varphi''_\perp \rangle \tag{65-b}$$

From this we deduce that the possible values of $E_\perp$ are given by:

$$E_\perp = \left(n + \frac{1}{2}\right)\hbar\omega_c \tag{66}$$

where $n$ is a positive integer or zero.

γ. *Eigenvalues of H*

According to the preceding results, the eigenvalues of the total Hamiltonian $H$ are of the form:

$$E(n, v_z) = \left(n + \frac{1}{2}\right)\hbar\omega_c + \frac{1}{2}\mu v_z^2 \tag{67}$$

The corresponding levels are called *Landau levels.*

For a given value of $v_z$, all possible values of $n$ (positive integers or zero) actually found. By the repeated action of the operators $\frac{1}{\sqrt{2}}(\hat{Q} \pm i\hat{S})$on an eigenvector of $H$ of eigenvalue $E(n, v_z)$, one can obtain an energy state $E(n', v_z)$ from (65). where $n'$ is any integer but where $v_z$ has not changed (since $\hat{Q}$ and $\hat{S}$ commute with $H_{//}$). Therefore, although the energy of the motion along $Oz$ is not quantized, that of the motion projected onto $xOy$ is.

COMMENT:

We showed in chapter V (§ B-3) that the energy levels of the one-dimensional harmonic oscillator are non-degenerate in $\mathscr{E}_x$. The situation is different here, since the particle under study is moving in three-dimensional space. Since $\frac{1}{\sqrt{2}}(\hat{Q} + i\hat{S}) = \sqrt{\frac{\mu}{2\hbar\omega_c}}(V_y + iV_x)$ is the destruction operator of a quantum $\hbar\omega_c$, the eigenvectors of $H_\perp$ corresponding to $n = 0$ are solutions of the equation:

$$(V_y + iV_x) | \varphi \rangle = 0 \tag{68}$$

On the one hand, vectors which are solutions of (68) can be eigenvectors of $H_{//}$ with an arbitrary (positive) eigenvalue. On the other, even for a fixed value of $v_z$, equation (68) is a partial differential equation with respect to $x$ and $y$ which has an infinite number of solutions. The energies $E(n = 0, v_z)$ are therefore infinitely degenerate. By using the creation operator for a quantum, it can easily be shown that this is true for all the levels $E(n, v_z)$, for any $n$ (non-negative integer).

### b. THE OBSERVABLES IN A PARTICULAR GAUGE

In order to state the above results more precisely, we shall calculate the stationary states of the system. This will enable us to study their physical properties. It is now necessary to choose a gauge; we shall choose the one given by (12). The components of the velocity are then:

$$
\begin{aligned}
V_x &= \frac{P_x}{\mu} - \frac{\omega_c}{2} Y \\
V_y &= \frac{P_y}{\mu} + \frac{\omega_c}{2} X \\
V_z &= \frac{P_z}{\mu}
\end{aligned}
\tag{69}
$$

#### α. *The Hamiltonians $H_\perp$ and $H_{//}$. Relation with the two-dimensional harmonic oscillator*

Substituting (69) into (55), we obtain:

$$
H_\perp = \frac{P_x^2 + P_y^2}{2\mu} + \frac{\omega_c}{2} L_z + \frac{\mu\,\omega_c^2}{8}(X^2 + Y^2) \tag{70-a}
$$

$$
H_{//} = \frac{P_z^2}{2\mu} \tag{70-b}
$$

where $L_z$ is the component along $Oz$ of the angular momentum $\mathbf{L} = \mathbf{R} \times \mathbf{P}$.

In the $\{ | \mathbf{r} \rangle \}$ representation, $H_{//}$ is an operator which acts only on the variable $z$, while $H_\perp$ acts only on the variables $x$ and $y$. We can therefore find a basis of eigenvectors of $H$ by solving in $\mathscr{E}_z$ the eigenvalue equation of $H_{//}$, and then, in $\mathscr{E}_{xy}$, that of $H_\perp$. All we must then do is take the tensor products of the vectors obtained.

Actually, the eigenvalue equation of $H_{//}$ simply leads to the wave functions:

$$
\varphi(z) = \frac{1}{\sqrt{2\pi\hbar}} e^{ip_z z/\hbar} \tag{71}
$$

with:

$$
E_{//} = \frac{p_z^2}{2\mu} \tag{72}
$$

[we again find (59)]. Therefore, we shall concentrate on solving the eigenvalue equation of $H_\perp$ in $\mathscr{E}_{xy}$; the wave functions we shall be considering now depend on $x$ and $y$, and not on $z$.

Comparing (70-a) with expression (12-a) of complement $D_{VI}$, we see that $H_\perp$ can be expressed simply in terms of the Hamiltonian $H_{xy}$ of a two-dimensional harmonic oscillator:

$$H_\perp = H_{xy} + \frac{\omega_c}{2} L_z \tag{73}$$

if we choose for the value of the constant which enters into $H_{xy}$:

$$\omega = \frac{\omega_c}{2} \tag{74}$$

Now, in complement $D_{VI}$, we saw that $H_{xy}$ and $L_z$ form a C.S.C.O. in $\mathscr{E}_{xy}$, and we constructed a basis of eigenvectors $| \chi_{n_d, n_g} \rangle$ common to these two observables [*cf.* formula (47) of $D_{VI}$]. The $| \chi_{n_d, n_g} \rangle$ are also eigenvectors of $H_\perp$; complement $D_{VI}$ therefore gives the solutions to the eigenvalue equation of $H_\perp$.

COMMENTS:

(*i*) In § 3-a, we saw that $H_\perp$ can be written in a form which is analogous to that of a Hamiltonian of a one-dimensional harmonic oscillator. Here, we find that, in a particular gauge, this same operator $H_\perp$ is also simply related to the Hamiltonian $H_{xy}$ of a two-dimensional harmonic oscillator. These two results are not contradictory; they simply correspond to two different decompositions of the same Hamiltonian, which must obviously lead to the same physical conclusions.

(*ii*) One must not lose sight of the fact that the Hamiltonian $H_\perp$ involves a physical problem which is completely different from that of the two-dimensional harmonic oscillator; the charged particle is subjected to a vector potential (describing a uniform magnetic field) and not a harmonic scalar potential (which would describe, for example, a non-uniform electric field). It so happens that, in the gauge chosen, the effects of the magnetic field can be likened to those of a fictitious harmonic scalar potential.

β. *Expression for the observables in terms of the creation and destruction operators of circular quanta*

First of all, we shall express the observables describing the quantities associated with the particle in terms of the operators $a_d$ and $a_g$ defined by equations (40) of complement $D_{VI}$ and their adjoints $a_d^\dagger$ and $a_g^\dagger$ (we shall also use the operators $N_d = a_d^\dagger a_d$ and $N_g = a_g^\dagger a_g$).

Substituting relations (46) of $D_{VI}$ into (73), we obtain*:

$$H_\perp = \left(N_d + \frac{1}{2}\right)\hbar\omega_c \tag{75}$$

* Recall that we have assumed $\omega_c$ to be positive. If $\omega_c$ were negative, the indices $d$ and $g$ would have to be inverted in a certain number of the following formulas; for example, (75) would become:

$H_\perp = (N_g + 1/2)\hbar\,|\omega_c|$.

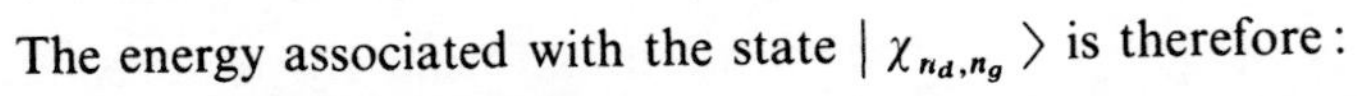

The energy associated with the state $| \chi_{n_d, n_g} \rangle$ is therefore:

$$E_\perp = \left( n_d + \frac{1}{2} \right) \hbar \omega_c \tag{76}$$

as we found in (66). Moreover, since $E_\perp$ is independent of $n_g$, we see that all the eigenvalues of $H_\perp$ are infinitely degenerate.

Using relations (23) and (40) of $D_{VI}$, we can see that:

$$\begin{aligned} X &= \frac{1}{2\beta} (a_d + a_d^\dagger + a_g + a_g^\dagger) \\ Y &= \frac{i}{2\beta} (a_d - a_d^\dagger - a_g + a_g^\dagger) \end{aligned} \tag{77}$$

where, using (74), $\beta$ is defined by:

$$\beta = \sqrt{\frac{\mu \omega_c}{2\hbar}} \tag{78}$$

Similarly:

$$\begin{aligned} P_x &= \frac{i\hbar\beta}{2} (- a_d + a_d^\dagger - a_g + a_g^\dagger) \\ P_y &= \frac{\hbar\beta}{2} (a_d + a_d^\dagger - a_g - a_g^\dagger) \end{aligned} \tag{79}$$

These expressions, substituted into (69), yield:

$$\begin{aligned} V_x &= - \frac{i\omega_c}{2\beta} (a_d - a_d^\dagger) \\ V_y &= \frac{\omega_c}{2\beta} (a_d + a_d^\dagger) \end{aligned} \tag{80}$$

Since $a_d$ and $a_d^\dagger$ do not commute with $N_d$, it can be seen by using (75) that, as in classical mechanics, $V_x$ and $V_y$ are not constants of the motion; in addition, using the commutation relations of $a_d$ and $a_d^\dagger$, we indeed obtain (52-a).

It is also interesting to study the quantum mechanical operators associated with the various variables introduced in the description of the classical motion (§ 1): the coordinates $(x_0, y_0)$ of the center $C_0$ of the classical trajectory, the components $(x', y')$ of the vector $\mathbf{C_0Q}$, etc. As above, we shall denote each of these operators by the capital letter corresponding to the small letter which designates the corresponding classical variable. By analogy with (10), we therefore set:

$$X_0 = X - \frac{1}{\omega_c} V_y = \frac{1}{2\beta} (a_g + a_g^\dagger) \tag{81-a}$$

$$Y_0 = Y + \frac{1}{\omega_c} V_x = \frac{i}{2\beta} (a_g^\dagger - a_g) \tag{81-b}$$

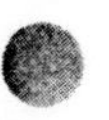



The operators $a_g$ and $a_g^\dagger$ commute with $N_d$; it follows that $X_0$ and $Y_0$ are constants of the motion. Formulas (81) also imply that:

$$[X_0, Y_0] = \frac{i}{2\beta^2} = \frac{i\hbar}{\mu\omega_c} \tag{82}$$

Consequently, $X_0$ and $Y_0$ are incompatible physical quantities, their root-mean-square deviations being related by:

$$\Delta X_0 \,.\, \Delta Y_0 \geqslant \frac{\hbar}{2\mu\omega_c} \tag{83}$$

We also define:

$$\begin{aligned} X' &= X - X_0 = \frac{1}{2\beta}(a_d + a_d^\dagger) \\ Y' &= Y - Y_0 = \frac{i}{2\beta}(a_d - a_d^\dagger) \end{aligned} \tag{84}$$

We immediately see that $X'$ and $Y'$, as in classical mechanics, are not constants of the motion; moreover, $X'$ and $Y'$ are simply proportional to $V_y$ and $V_x$ respectively:

$$\begin{aligned} V_x &= -\,\omega_c Y' \\ V_y &= \omega_c X' \end{aligned} \tag{85}$$

like the corresponding classical variables [formula (9)]. According to (53), equations (85) imply:

$$\Delta X' \,.\, \Delta Y' \geqslant \frac{\hbar}{2\mu\omega_c} \tag{86}$$

Let $\Sigma^2$ be the operator corresponding to $\sigma^2$ (square of the radius of the classical trajectory):

$$\Sigma^2 = (X - X_0)^2 + (Y - Y_0)^2 \tag{87}$$

According to (81), we have:

$$\Sigma^2 = \left(\frac{1}{\omega_c}\right)^2 (V_x^2 + V_y^2) = \frac{2}{\mu\omega_c^2} H_\perp \tag{88}$$

$\Sigma^2$ is therefore a constant of the motion, as is $\sigma^2$ in classical mechanics.

Finally, the operator associated with the moment of the mechanical momentum $\mu\mathbf{v}$ with respect to $O$ is:

$$\Theta_z = \mu[(X - X_0)V_y - (Y - Y_0)V_x] \tag{89}$$

and formulas (81) indicate that:

$$\Theta_z = \frac{2}{\omega_c} H_\perp \tag{90}$$

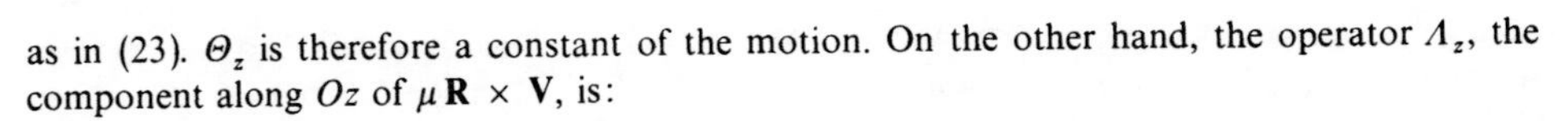



as in (23). $\Theta_z$ is therefore a constant of the motion. On the other hand, the operator $\Lambda_z$, the component along $Oz$ of $\mu \mathbf{R} \times \mathbf{V}$, is:

$$\Lambda_z = \frac{2}{\omega_c} H_\perp + \hbar(a_d a_g + a_d^\dagger a_g^\dagger) \tag{91}$$

and therefore does not commute with $H_\perp$.

## c. THE STATIONARY STATES

We indicated above that the eigenvalues of the Hamiltonian $H_\perp$ are all infinitely degenerate in $\mathscr{E}_{xy}$. For each positive or zero integer $n$, there exists an infinite-dimensional subspace $\mathscr{E}_{xy}^{(n)}$ of $\mathscr{E}_{xy}$, all of whose kets are eigenvectors of $H_\perp$ with the same eigenvalues $(n + 1/2)\hbar\omega_c$. In this section, we shall study different bases which can be chosen in each of these subspaces. First, we shall indicate the general properties of the stationary states, valid for any basis of eigenstates of $H_\perp$.

### α. *General properties*

Relations (88) and (90) show that an arbitrary stationary state is necessarily an eigenvector of $\Sigma^2$ and $\Theta_z$; the corresponding physical quantities are therefore always well-defined in such a state and are equal to:

$$\begin{aligned} &(2n + 1)\frac{\hbar}{\mu\omega_c} \qquad \text{for} \quad \Sigma^2 \\ &(2n + 1)\,\hbar \qquad\quad \text{for} \quad \Theta_z \end{aligned} \tag{92}$$

The values of $\Sigma^2$ and $\Theta_z$ are proportional to the energy; this corresponds to the classical description of the motion (*cf.* § 1).

It follows from (80) and (84) that $X'$, $Y'$, $V_x$ and $V_y$ have no matrix elements inside a given subspace $\mathscr{E}_{xy}^{(n)}$; it follows, for a stationary state, that:

$$\begin{aligned} &\langle V_x \rangle = \langle V_y \rangle = 0 \\ &\langle X' \rangle = \langle Y' \rangle = 0 \end{aligned} \tag{93}$$

Nevertheless, since $V_x$ and $V_y$ (and therefore $X'$ and $Y'$) are not constants of the motion, the corresponding physical quantities do not have perfectly well-defined values in a stationary state. In fact, by using (80), (84) and the properties of the one-dimensional harmonic oscillator [*cf.* chap. V, relation (D-5)], it can be shown that:

$$\begin{aligned} \Delta V_x &= \omega_c \Delta Y' = \sqrt{\left(n + \frac{1}{2}\right)\frac{\hbar\omega_c}{\mu}} \\ \Delta V_y &= \omega_c \Delta X' = \sqrt{\left(n + \frac{1}{2}\right)\frac{\hbar\omega_c}{\mu}} \end{aligned} \tag{94}$$

in agreement with (53). Moreover, we see that the only stationary states in which the product $\Delta V_x \,.\, \Delta V_y$ (or $\Delta X' \,.\, \Delta Y'$) takes on its minimal value are the ground states ($n = 0$).

COMMENT:

The various ground states are solutions of the equation:

$$a_d \,|\, \varphi \rangle = 0 \tag{95-a}$$

that is, using (80):

$$(V_y + iV_x) \,|\, \varphi \rangle = 0 \tag{95-b}$$

as we found in (68).

β. *The states* $|\chi_{n_d,n_g}\rangle$

As we saw in complement $D_{VI}$, the fact that $H_\perp$ and $L_z$ form a C.S.C.O. in $\mathscr{E}_{xy}$ can be used to construct a basis of eigenvectors common to these two observables. This basis is composed of the vectors $|\chi_{n_d,n_g}\rangle$, since, according to (75) and formula (46) of complement $D_{VI}$:

$$H_\perp \,|\, \chi_{n_d,n_g} \rangle = \left(n_d + \frac{1}{2}\right)\hbar\omega_c \,|\, \chi_{n_d,n_g} \rangle \tag{96-a}$$

$$L_z \,|\, \chi_{n_d,n_g} \rangle = (n_d - n_g)\,\hbar \,|\, \chi_{n_d,n_g} \rangle \tag{96-b}$$

The subspace $\mathscr{E}_{xy}^{(n)}$ defined by specifying the (non-negative) integer $n$ is therefore spanned by the set of vectors $|\chi_{n_d,n_g}\rangle$ such that $n_d = n$. The eigenvalues of $L_z$ associated with these different vectors are of the form $m\hbar$, and, for fixed $n$, $m$ is an integer which can vary between $-\infty$ and $n$ (for example, all the ground states correspond to negative values of $m$; this is related to the hypothesis $\omega_c > 0$ posed above).

The wave functions associated with the states $|\chi_{n_d,n_g}\rangle$ were calculated in complement $D_{VI}$ (§ 3-d).

Note that the states $|\chi_{n_d,n_g}\rangle$ are eigenstates of the operator $L_z$, but not of the operator $\Lambda_z$ associated with the moment of the mechanical momentum. This can be seen directly from formula (91).

In a state $|\chi_{n_d,n_g}\rangle$, the mean values $\langle X_0 \rangle$ and $\langle Y_0 \rangle$ are zero, according to (81). However, neither $X_0$ nor $Y_0$ corresponds to perfectly well-defined physical quantities, since, by using the properties of the one-dimensional harmonic oscillator, it can easily be shown that, in a state $|\chi_{n_d,n_g}\rangle$:

$$\begin{aligned} \Delta X_0 &= \sqrt{\left(n_g + \frac{1}{2}\right)\frac{\hbar}{\mu\omega_c}} \\ \Delta Y_0 &= \sqrt{\left(n_g + \frac{1}{2}\right)\frac{\hbar}{\mu\omega_c}} \end{aligned} \tag{97}$$

The minimal value of the product $\Delta X_0 \,.\, \Delta Y_0$ is therefore attained for the states $|\chi_{n_d,n_g=0}\rangle$, that is, the states of each energy level $E_\perp = (n + 1/2)\hbar\omega_c$ for which $L_z$ takes on its maximal value $n\hbar$ [*cf.* (96)].

However, let us define the operator:

$$\Gamma^2 = X_0^2 + Y_0^2 \tag{98}$$

It corresponds to the square of the distance from the center $C_0$ of the trajectory to the origin. Using (81), we easily find:

$$\begin{aligned} \Gamma^2 &= \frac{\hbar}{\mu\omega_c}(a_g a_g^\dagger + a_g^\dagger a_g) \\ &= \frac{\hbar}{\mu\omega_c}(2N_g + 1) \end{aligned} \tag{99}$$

The state $| \chi_{n_d,n_g} \rangle$ is therefore an eigenstate of $\Gamma^2$ with the eigenvalue $\frac{\hbar}{\mu\omega_c}(2n_g + 1)$; the fact that this value can never go to zero is related to the non-commutativity of the operators $X_0$ and $Y_0$.

COMMENT:

The operator $L_z$, according to (75) and (99), is given by:

$$L_z = \hbar(N_d - N_g) = \hbar\left[\frac{H_\perp}{\hbar\omega_c} - \frac{1}{2} - \frac{\mu\omega_c}{2\hbar}\Gamma^2 + \frac{1}{2}\right] \tag{100}$$

that is, according to (88):

$$L_z = \frac{\mu\omega_c}{2}(\Sigma^2 - \Gamma^2) = \frac{qB}{2}(\Gamma^2 - \Sigma^2) \tag{101}$$

which is the equivalent of the classical relation (27).

γ. *Other types of stationary states*

Any linear combination of vectors $| \chi_{n_d,n_g} \rangle$ associated with the same value of $n_d$ is an eigenstate of $H_\perp$ and therefore possesses the properties stated in § 3-c-α. By a suitable choice of the coefficients of the linear combination, one can obtain stationary states which possess other interesting properties as well.

We know, for example (§ 3-b-β). that $X_0$ and $Y_0$ are constants of the motion. However, since $X_0$ and $Y_0$ do not commute, there are no eigenstates common to these two operators. This means that, in quantum mechanics, it is not possible to obtain a state in which the two coordinates of the point $C_0$ are known.

To construct the eigenstates common to $H_\perp$ and $X_0$, we can use the properties of the one-dimensional harmonic oscillator; formula (81-a) shows that $X_0$ has the same expression, to within a constant factor, as the position operator $X_g$ of a one-dimensional oscillator whose destruction operator is $a_g$:

$$X_0 = \frac{1}{\beta\sqrt{2}}\hat{X}_g \tag{102}$$

Since we know the wave functions $\hat{\varphi}_k(\hat{x})$ associated with the stationary states $| \hat{\varphi}_k \rangle$ of a one-dimensional harmonic oscillator (*cf.* complement $B_V$, § 2-b), we know how to write the eigenvectors $| \hat{x} \rangle$ of the position operator as linear combinations of the states $| \hat{\varphi}_k \rangle$:

$$\begin{aligned} | \hat{x} \rangle &= \sum_{k=0}^{\infty} | \hat{\varphi}_k \rangle \langle \hat{\varphi}_k | \hat{x} \rangle \\ &= \sum_{k=0}^{\infty} \hat{\varphi}_k^*(\hat{x}) | \hat{\varphi}_k \rangle \end{aligned} \tag{103}$$

In order to obtain the eigenstates common to $H_\perp$ and $X_0$ it suffices to apply this result to the states $| \chi_{n_d,n_g=k} \rangle$; the vector:

$$| \xi_{n,x_0} \rangle = \sum_{k=0}^{\infty} \hat{\varphi}_k^*(\beta\sqrt{2}\,x_0) | \chi_{n_d=n,n_g=k} \rangle \tag{104}$$

is a common eigenvector of $H_\perp$ and $X_0$ with the eigenvalues $(n + 1/2)\hbar\omega_c$ and $x_0$.

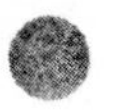


The eigenstates $|\eta_{n,y_0}\rangle$ common to $H_\perp$ and $Y_0$ can be found in an analogous fashion. Relation (81-b) indicates that $Y_0$ is proportional to the momentum operator $P_g$ of the ficticious one-dimensional oscillator just used:

$$Y_0 = \frac{1}{\beta\sqrt{2}}\hat{P}_g \tag{105}$$

Consequently [see formula (20) of complement $D_V$]:

$$|\eta_{n,y_0}\rangle = \sum_{k=0}^{\infty} i^k \hat{\varphi}_k^*(\beta\sqrt{2}y_0)\,|\chi_{n_d=n,n_g=k}\rangle \tag{106}$$

We have just constructed the states in which either $X_0$ or $Y_0$ is perfectly well-defined. We can also determine the stationary states in which the product $\Delta X_0 . \Delta Y_0$ reaches its minimal value, given by (83). For a one-dimensional harmonic oscillator, we studied in complement $G_V$ the states in which the product $\Delta\hat{X} . \Delta\hat{P}$ is minimal; these are the quasi-classical states, given by:

$$|\alpha\rangle = \sum_{k=0}^{\infty} c_k(\alpha)\,|\varphi_k\rangle \tag{107}$$

with:

$$c_k(\alpha) = \frac{\alpha^k}{\sqrt{k!}}\,e^{-|\alpha|^2/2} \tag{108}$$

In these states:

$$\Delta\hat{X} = \Delta\hat{P} = \frac{1}{\sqrt{2}} \tag{109}$$

It follows that, in the case which interests us here, the state:

$$|\theta_{n,\alpha_0}\rangle = \sum_{k=0}^{\infty} c_k(\alpha_0)\,|\chi_{n_d=n,n_g=k}\rangle \tag{110}$$

yields, for $X_0$ and $Y_0$, root-mean-square deviations:

$$\Delta X_0 = \Delta Y_0 = \frac{1}{2\beta} \tag{111}$$

The product $\Delta X_0 . \Delta Y_0$ is therefore minimal.

COMMENT:

Since the magnetic field is uniform, the physical problem we are considering is invariant with respect to translation. Thus far, this symmetry has been masked by the choice of the particular gauge (12), which gives the origin $O$ a privileged position with respect to all other points in space. Consequently, neither the Hamiltonian $H$ nor its eigenstates are invariant with respect to translation. We know, however (*cf.* complement $H_{III}$) that the physical predictions of quantum mechanics are gauge-invariant. These predictions must remain the same if, by a change of gauge, we give a point other than $O$ a privileged position. Consequently, the translation symmetry must reappear when we study the physical properties of a given state.

To show this more precisely, let us assume that, at a given instant, the state of the particle is characterized in the gauge (12) by the ket $|\psi\rangle$ with which the wave function $\langle \mathbf{r} \mid \psi \rangle = \psi(\mathbf{r})$ is associated. We then perform a translation $\mathcal{T}$ defined by the vector $\mathbf{a}$, and consider the ket $| \psi_T \rangle$ defined by:

$$| \psi_T \rangle = e^{-\frac{i}{\hbar}\mathbf{P}.\mathbf{a}} | \psi \rangle \tag{112}$$

with which, according to the results of complement $E_{II}$, is associated the wave function:

$$\psi_T(\mathbf{r}) = \langle \mathbf{r} \mid \psi_T \rangle = \psi(\mathbf{r} - \mathbf{a}) \tag{113}$$

The same translation can be applied to the vector potential, which becomes:

$$\mathbf{A}_T(\mathbf{r}) = \mathbf{A}(\mathbf{r} - \mathbf{a}) = -\frac{1}{2}(\mathbf{r} - \mathbf{a}) \times \mathbf{B} \tag{114}$$

$\mathbf{A}_T(\mathbf{r})$ clearly describes the same magnetic field as $\mathbf{A}(\mathbf{r})$. Since the physical properties attached to a given state vector depend only on this state vector and the potential $\mathbf{A}$ chosen, they must undergo the translation $\mathcal{T}$ when $\psi(\mathbf{r})$ and $\mathbf{A}(\mathbf{r})$ are replaced by expressions (113) and (114). It is simple to use these relations to obtain the expression for the probability density associated with $| \psi_T \rangle$:

$$\rho_T(\mathbf{r}) = |\psi_T(\mathbf{r})|^2 = |\psi(\mathbf{r} - \mathbf{a})|^2 = \rho(\mathbf{r} - \mathbf{a}) \tag{115}$$

and that for the current $\mathbf{J}_T(\mathbf{r})$, calculated with the vector potential $\mathbf{A}_T(\mathbf{r})$:

$$\begin{aligned} \mathbf{J}_T(\mathbf{r}) &= \frac{1}{2\mu}\left\{ \psi_T^*(\mathbf{r})\left[\frac{\hbar}{i}\,\nabla + \frac{q}{2}(\mathbf{r} - \mathbf{a}) \times \mathbf{B}\right]\psi_T(\mathbf{r}) + \text{c.c.} \right\} \\ &= \frac{1}{2\mu}\left\{ \psi^*(\mathbf{r} - \mathbf{a})\left[\frac{\hbar}{i}\,\nabla + \frac{q}{2}(\mathbf{r} - \mathbf{a}) \times \mathbf{B}\right]\psi(\mathbf{r} - \mathbf{a}) + \text{c.c.} \right\} \\ &= \mathbf{J}(\mathbf{r} - \mathbf{a}) \end{aligned} \tag{116}$$

[where $\mathbf{J}(\mathbf{r})$ is the probability current associated with $\psi(\mathbf{r})$ in the gauge (12)]. The ket $| \psi_T \rangle$ therefore describes, *in the new gauge* $\mathbf{A}_T(\mathbf{r})$, a state whose physical properties are related by the translation $\mathcal{T}$ to those corresponding to the ket $| \psi \rangle$ in the gauge $\mathbf{A}(\mathbf{r})$.

Let us show, moreover, that the translation of a possible motion yields another possible motion; this will conclude the proof of the translation invariance of the problem. To do so, consider the Schrödinger equation in the $\{ | \mathbf{r} \rangle \}$ representation, in the gauge $\mathbf{A}(\mathbf{r})$.:

$$i\hbar \frac{\partial}{\partial t}\psi(\mathbf{r}, t) = \frac{1}{2\mu}\left[\frac{\hbar}{i}\,\nabla - q\mathbf{A}(\mathbf{r})\right]^2 \psi(\mathbf{r}, t) \tag{117}$$

Changing $\mathbf{r}$ to $\mathbf{r} - \mathbf{a}$ in this equation, we obtain, using (113) and (114):

$$i\hbar \frac{\partial}{\partial t}\psi_T(\mathbf{r}, t) = \frac{1}{2\mu}\left[\frac{\hbar}{i}\,\nabla - q\mathbf{A}_T(\mathbf{r})\right]^2 \psi_T(\mathbf{r}, t) \tag{118}$$

The operator which appears on the right-hand side of (118) is none other than the Hamiltonian in the gauge $\mathbf{A}_T(\mathbf{r})$. Consequently, if $\psi(\mathbf{r}, t)$ describes, in the gauge $\mathbf{A}(\mathbf{r})$, a possible motion of the system, $\psi_T(\mathbf{r}, t)$ describes, in the equivalent gauge $\mathbf{A}_T(\mathbf{r})$, another possible motion, which, according to what we have just shown, is nothing more than the result of a translation of the first motion. In particular, if

$$\psi(\mathbf{r}, t) = \varphi(\mathbf{r})\, e^{-iEt/\hbar}$$

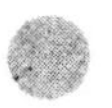



is a stationary state [in the gauge $\mathbf{A}(\mathbf{r})$],

$$\psi_T(\mathbf{r}, t) = \varphi_T(\mathbf{r})\, e^{-iEt/\hbar}$$

is another stationnary state of the same energy [in the gauge $\mathbf{A}_T(\mathbf{r})$].

If we want to continue to use the gauge (12) after having performed the translation $\mathcal{T}$ on the physical state of the particle, we must describe the translated state by a mathematical ket $|\psi'_T\rangle$ which is different from $|\psi_T\rangle$. According to § 3-b-α of complement $H_{III}$, the ket $|\psi'_T\rangle$ can be obtained from $|\psi_T\rangle$ by a unitary transformation:

$$|\psi'_T\rangle = T_\chi\,|\psi_T\rangle \tag{119}$$

The operator $T_\chi$ is given by:

$$T_\chi = e^{\frac{i}{\hbar} q\chi(\mathbf{R})} \tag{120}$$

where $\chi(\mathbf{r})$ is the function characterizing the gauge transformation performed. Here, the potential after the gauge change is:

$$\mathbf{A}(\mathbf{r}) = -\frac{1}{2}\mathbf{r}\times\mathbf{B} = \mathbf{A}_T(\mathbf{r}) - \frac{1}{2}\mathbf{a}\times\mathbf{B} \tag{121}$$

so that:

$$\chi(\mathbf{r}) = -\frac{1}{2}\mathbf{r}\,.\,(\mathbf{a}\times\mathbf{B}) \tag{122}$$

Substituting (112), (120) and (122) into (119), we finally obtain:

$$|\psi'_T\rangle = U(\mathbf{a})\,|\psi\rangle \tag{123}$$

with:

$$U(\mathbf{a}) = e^{-i\frac{q}{2\hbar}\mathbf{R}.(\mathbf{a}\times\mathbf{B})}\, e^{-\frac{i}{\hbar}\mathbf{P}.\mathbf{a}} \tag{124}$$

Therefore, if we remain in the gauge $\mathbf{A}(\mathbf{r})$, the translation operator is $U(\mathbf{a})$, given by (124).

The components of $\mathbf{R}$ and $\mathbf{P}$ along two perpendicular axes enter into formula (124); they therefore commute, so we can write:

$$U(\mathbf{a}) = e^{-i\frac{q}{2\hbar}\mathbf{R}.(\mathbf{a}\times\mathbf{B}) - \frac{i}{\hbar}\mathbf{P}.\mathbf{a}} \tag{125}$$

When $\mathbf{a}$ is a vector of the $xOy$ plane, a simple calculation, using formulas (10) and (69), yields:

$$U(\mathbf{a}) = e^{i\frac{q}{\hbar}(\mathbf{a}\times\mathbf{R}_0).\mathbf{B}} \tag{126}$$

with:

$$\mathbf{R}_0 = X_0\,\mathbf{e}_x + Y_0\,\mathbf{e}_y \tag{127}$$

The operators $X_0$ and $Y_0$ (coordinates of the center of the circle) are therefore associated with infinitesimal translations along $Oy$ and $Ox$ respectively.

## d. TIME EVOLUTION

### α. *Mean values of the observables*

We have already encountered a certain number of physical quantities which are constants of the motion: $X_0$, $Y_0$, $\Theta_z$, $\Sigma^2$. Whatever the state of the system, their mean values are time-independent.

Let us examine the time evolution of the mean values $\langle X \rangle$, $\langle Y \rangle$, $\langle V_x \rangle$, $\langle V_y \rangle$ and $\langle X' \rangle$, $\langle Y' \rangle$. We immediately see from the expressions given in § 3-b-β that the corresponding operators have matrix elements only between states $| \chi_{n_d n_g} \rangle$ whose values of $n_d$ differ by $\pm 1$ (or 0). The evolution of these mean values therefore involves only one Bohr frequency, which is none other than the cyclotron frequency $\omega_c/2\pi$ defined in (5).

This result is completely analogous to the one given by classical mechanics.

### β. *Quasi-classical states*

Assume that at $t = 0$ the state of the particle is:

$$| \psi_\perp(0) \rangle = | \alpha_d, \alpha_g \rangle \tag{128}$$

where the ket $| \alpha_d, \alpha_g \rangle$ is defined by expression (70) of complement $D_{VI}$. Since expression (75) for $H_\perp$ involves $N_d$ but not $N_g$, the state vector $| \psi_\perp(t) \rangle$ at the instant $t$ is obtained by changing $\alpha_d$ to $\alpha_d \, e^{-i\omega_c t}$:

$$| \psi_\perp(t) \rangle = e^{-i\omega_c t/2} \, | \alpha_d \, e^{-i\omega_c t}, \alpha_g \rangle \tag{129}$$

[*cf.* expression (92) of complement $G_V$].

We set:

$$\begin{aligned} \alpha_d &= |\alpha_d| \, e^{i\varphi_d} \\ \alpha_g &= |\alpha_g| \, e^{i\varphi_g} \end{aligned} \tag{130}$$

Relations (80), (81) and (84) then show that:

$$\begin{cases} \langle X_0 \rangle = \dfrac{1}{2\beta}(\alpha_g + \alpha_g^*) = \dfrac{|\alpha_g|}{\beta} \cos \varphi_g \\[2ex] \langle Y_0 \rangle = \dfrac{i}{2\beta}(\alpha_g^* - \alpha_g) = \dfrac{|\alpha_g|}{\beta} \sin \varphi_g \end{cases} \tag{131}$$

$$\begin{cases} \langle X' \rangle(t) = \dfrac{1}{2\beta}(\alpha_d \, e^{-i\omega_c t} + \alpha_d^* \, e^{i\omega_c t}) = \dfrac{|\alpha_d|}{\beta} \cos(\omega_c t - \varphi_d) \\[2ex] \langle Y' \rangle(t) = \dfrac{i}{2\beta}(\alpha_d \, e^{-i\omega_c t} - \alpha_d^* \, e^{i\omega_c t}) = \dfrac{|\alpha_d|}{\beta} \sin(\omega_c t - \varphi_d) \end{cases} \tag{132}$$

and:

$$\begin{cases} \langle V_x \rangle(t) = -\dfrac{|\alpha_d|}{\beta} \omega_c \sin(\omega_c t - \varphi_d) \\[2ex] \langle V_y \rangle(t) = \dfrac{|\alpha_d|}{\beta} \omega_c \cos(\omega_c t - \varphi_d) \end{cases} \tag{133}$$

Moreover, the properties of the states $| \alpha \rangle$ imply that:

$$\begin{aligned} \langle H_\perp \rangle &= \hbar\omega_c \left( |\alpha_d|^2 + \frac{1}{2} \right) \\ \langle \Theta_z \rangle &= 2\hbar \left( |\alpha_d|^2 + \frac{1}{2} \right) \\ \langle \Sigma^2 \rangle &= \frac{1}{\beta^2} \left( |\alpha_d|^2 + \frac{1}{2} \right) \end{aligned} \tag{134}$$

All these results are extremely close to those given by classical mechanics [*cf.* (4)]. We see that $|\alpha_d|$ is related to the radius $\sigma$ of the classical trajectory and $\varphi_d$, to the initial phase $\varphi_0$, while $|\alpha_g|$ is related to the distance $OC_0$, and $\varphi_g$ corresponds to the polar angle of the vector $\mathbf{OC}_0$.

Furthermore, the properties of the states $| \alpha \rangle$ can be used to show that:

$$\Delta X_0 = \Delta Y_0 = \Delta X' = \Delta Y' = \frac{1}{2\beta} \tag{135-a}$$

$$\Delta V_x = \Delta V_y = \frac{\omega_c}{2\beta} \tag{135-b}$$

(the products $\Delta X_0 \, . \, \Delta Y_0$, $\Delta X' \, . \, \Delta Y'$ and $\Delta V_x \, . \, \Delta V_y$ therefore take on their minimal values), and :

$$\begin{aligned} \Delta H_\perp &= \hbar\omega_c |\alpha_d| \\ \Delta \Theta_z &= 2\hbar |\alpha_d| \\ \Delta \Sigma^2 &= \frac{1}{\beta^2} |\alpha_d| \end{aligned} \tag{136}$$

As for the deviations $\Delta X$ and $\Delta Y$, they can be calculated by using the fact that:

$$| \psi_\perp(t) \rangle = e^{-i\omega_c t/2} \, | \alpha_x = \frac{\alpha_d \, e^{-i\omega_c t} + \alpha_g}{\sqrt{2}}, \alpha_y = \frac{i\alpha_d \, e^{-i\omega_c t} - i\alpha_g}{\sqrt{2}} \rangle \tag{137}$$

[where $| \alpha_x, \alpha_y \rangle$ is defined by relation (66) of $D_{VI}$], which yields:

$$\Delta X = \Delta Y = \sqrt{\frac{\hbar}{\mu\omega_c}} = \frac{1}{\beta\sqrt{2}} \tag{138}$$

($\Delta P_x$ and $\Delta P_y$ can easily be obtained in the same way).

If the conditions:

$$|\alpha_d| \gg 1, \; |\alpha_g| \gg 1 \tag{139}$$

are satisfied, we see, therefore, that the various physical quantities (position, velocity, energy, ...) are, in relative value, very well defined. The states (129) therefore represent "quasi-classical" states of the charged particle placed in a uniform magnetic field.

COMMENT:

If $\alpha_d = 0$, we obtain:

$$\begin{cases} \langle H_\perp \rangle = \dfrac{1}{2}\hbar\omega_c \\ \Delta H_\perp = 0 \end{cases} \tag{140}$$

The states:

$$| \alpha_x, \alpha_y = - i\alpha_x \rangle \tag{141}$$

therefore correspond to the ground state.

**References and suggestions for further reading:**

Landau and Lifshitz (1.19), chap. XVI, §§ 124 and 125; Ter Haar (1.23), chap. 6.

Application to solid state physics : Mott and Jones (13.7), chap. VI, § 6 ; Kittel (13.2), chap. 8, p. 239 and chap. 9, p. 290.

## Complement $F_{VI}$

## EXERCISES

**1.** Consider a system of angular momentum $j = 1$, whose state space is spanned by the basis $\{ | +1 \rangle, | 0 \rangle, | -1 \rangle \}$ of three eigenvectors common to $\mathbf{J}^2$ (eigenvalue $2\hbar^2$) and $J_z$ (respective eigenvalues $+\hbar$, 0 and $-\hbar$). The state of the system is:

$$| \psi \rangle = \alpha | +1 \rangle + \beta | 0 \rangle + \gamma | -1 \rangle$$

where $\alpha$, $\beta$, $\gamma$ are three given complex parameters.

*a.* Calculate the mean value $\langle \mathbf{J} \rangle$ of the angular momentum in terms of $\alpha$, $\beta$ and $\gamma$.

*b.* Give the expression for the three mean values $\langle J_x^2 \rangle$, $\langle J_y^2 \rangle$ and $\langle J_z^2 \rangle$ in terms of the same quantities.

**2.** Consider an arbitrary physical system whose four-dimensional state space is spanned by a basis of four eigenvectors $| j, m_z \rangle$ common to $\mathbf{J}^2$ and $J_z$ ($j = 0$ or $1$; $-j \leqslant m_z \leqslant +j$), of eigenvalues $j(j+1)\hbar^2$ and $m_z\hbar$, such that:

$$J_\pm | j, m_z \rangle = \hbar \sqrt{j(j+1) - m_z(m_z \pm 1)} \, | j, m_z \pm 1 \rangle$$
$$J_+ | j, j \rangle = J_- | j, -j \rangle = 0$$

*a.* Express in terms of the kets $| j, m_z \rangle$, the eigenstates common to $\mathbf{J}^2$ and $J_x$, to be denoted by $| j, m_x \rangle$.

*b.* Consider a system in the normalized state:

$$| \psi \rangle = \alpha | j = 1, m_z = 1 \rangle + \beta | j = 1, m_z = 0 \rangle + \gamma | j = 1, m_z = -1 \rangle + \delta | j = 0, m_z = 0 \rangle$$

(*i*) What is the probability of finding $2\hbar^2$ and $\hbar$ if $\mathbf{J}^2$ and $J_x$ are measured simultaneously?

(*ii*) Calculate the mean value of $J_z$ when the system is in the state $| \psi \rangle$, and the probabilities of the various possible results of a measurement bearing only on this observable.

(*iii*) Same questions for the observable $\mathbf{J}^2$ and for $J_x$.

(*iv*) $J_z^2$ is now measured; what are the possible results, their probabilities, and their mean value?

**3.** Let $\mathbf{L} = \mathbf{R} \times \mathbf{P}$ be the angular momentum of a system whose state is $\mathscr{E}_\mathbf{r}$. Prove the commutation relations:

$$[L_i, R_j] = i\hbar \, \varepsilon_{ijk} \, R_k$$
$$[L_i, P_j] = i\hbar \, \varepsilon_{ijk} \, P_k$$
$$[L_i, \mathbf{P}^2] = [L_i, \mathbf{R}^2] = [L_i, \mathbf{R} \cdot \mathbf{P}] = 0$$

where $L_i$, $R_j$, $P_j$ denote arbitrary components of **L**, **R**, **P** in an orthonormal system, and $\varepsilon_{ijk}$ is defined by :

$$\varepsilon_{ijk} \begin{cases} = 0 & \text{if two (or three) of the indices } i, j, k \text{ are equal} \\ = 1 & \text{if these indices are an even permutation of } x, y, z \\ = -1 & \text{if the permutation is odd.} \end{cases}$$

## 4. Rotation of a polyatomic molecule

Consider a system composed of $N$ different particles, of positions $\mathbf{R}_1, ..., \mathbf{R}_m, ..., \mathbf{R}_N$, and momenta $\mathbf{P}_1, ..., \mathbf{P}_m, ..., \mathbf{P}_N$. We set:

$$\mathbf{J} = \sum_m \mathbf{L}_m$$

with:

$$\mathbf{L}_m = \mathbf{R}_m \times \mathbf{P}_m$$

*a.* Show that the operator **J** satisfies the commutation relations which define an angular momentum, and deduce from this that if **V** and **V**′ denote two ordinary vectors of three-dimensional space, then:

$$[\mathbf{J} \cdot \mathbf{V}, \mathbf{J} \cdot \mathbf{V}'] = i\hbar(\mathbf{V} \times \mathbf{V}') \cdot \mathbf{J}$$

*b.* Calculate the commutators of **J** with the three components of $\mathbf{R}_m$ and with those of $\mathbf{P}_m$. Show that:

$$[\mathbf{J}, \mathbf{R}_m \cdot \mathbf{R}_p] = 0$$

*c.* Prove that:

$$[\mathbf{J}, \mathbf{J} \cdot \mathbf{R}_m] = 0$$

and deduce from this the relation:

$$[\mathbf{J} \cdot \mathbf{R}_m, \mathbf{J} \cdot \mathbf{R}_{m'}] = i\hbar(\mathbf{R}_{m'} \times \mathbf{R}_m) \cdot \mathbf{J} = i\hbar \; \mathbf{J} \cdot (\mathbf{R}_{m'} \times \mathbf{R}_m)$$

We set:

$$\mathbf{W} = \sum_m a_m \mathbf{R}_m$$

$$\mathbf{W}' = \sum_m a'_m \mathbf{R}_m$$

where the coefficients $a_m$ and $a'_m$ are given. Show that:

$$[\mathbf{J} \cdot \mathbf{W}, \mathbf{J} \cdot \mathbf{W}'] = -i\hbar(\mathbf{W} \times \mathbf{W}') \cdot \mathbf{J}$$

Conclusion: what is the difference between the commutation relations of the components of **J** along fixed axes and those of the components of **J** along the moving axes of the system being studied?

*d.* Consider a molecule which is formed by $N$ unaligned atoms whose relative distances are assumed to be invariant (a rigid rotator). **J** is the sum of the angular momenta of the atoms with respect to the center of mass of the molecule, situated at a fixed point $O$; the $Oxyz$ axes constitute a fixed orthonormal frame. The three principal inertial axes of the system are denoted by $O\alpha$, $O\beta$ and $O\gamma$, with the ellipsoid of inertia assumed to be an ellipsoid of revolution about $O\gamma$ (a symmetrical rotator). The rotational energy of the molecule is then :

$$H = \frac{1}{2}\left[\frac{J_\gamma^2}{I_{//}} + \frac{J_\alpha^2 + J_\beta^2}{I_\perp}\right]$$

where $J_\alpha$, $J_\beta$ and $J_\gamma$ are the components of $\mathbf{J}$ along the unit vectors $\mathbf{w}_\alpha$, $\mathbf{w}_\beta$ and $\mathbf{w}_\gamma$ of the moving axes $O\alpha$, $O\beta$, $O\gamma$ attached to the molecule, and $I_{//}$ and $I_\perp$ are the corresponding moments of inertia. We grant that :

$$J_\alpha^2 + J_\beta^2 + J_\gamma^2 = J_x^2 + J_y^2 + J_z^2 = \mathbf{J}^2$$

(*i*) Derive the commutation relations of $J_\alpha$, $J_\beta$, $J_\gamma$ from the results of *c*.

(*ii*) We introduce the operators $N_\pm = J_\alpha \pm iJ_\beta$. Using the general arguments of chapter VI, show that one can find eigenvectors common to $\mathbf{J}^2$ and $J_\gamma$, of eigenvalues $J(J+1)\hbar^2$ and $K\hbar$, with $K = -J, -J+1, ..., J-1, J$.

(*iii*) Express the Hamiltonian $H$ of the rotator in terms of $\mathbf{J}^2$ and $J_\gamma^2$. Find its eigenvalues.

(*iv*) Show that one can find eigenstates common to $\mathbf{J}^2$, $J_z$ and $J_\gamma$, to be denoted by $|J, M, K\rangle$ [the respective eigenvalues are $J(J+1)\hbar^2$, $M\hbar$, $K\hbar$]. Show that these states are also eigenstates of $H$.

(*v*) Calculate the commutators of $J_\pm$ and $N_\pm$ with $\mathbf{J}^2$, $J_z$, $J_\gamma$. Derive from them the action of $J_\pm$ and $N_\pm$ on $|J, M, K\rangle$. Show that the eigenvalues of $H$ are at least $2(2J+1)$-fold degenerate if $K \neq 0$, and $(2J+1)$-fold degenerate if $K = 0$.

(*vi*) Draw the energy diagram of the rigid rotator ($J$ is an integer since $\mathbf{J}$ is a sum of orbital angular momenta; *cf.* chapter X). What happens to this diagram when $I_{//} = I_\perp$ (spherical rotator)?

**5.** A system whose state space is $\mathscr{E}_\mathbf{r}$ has for its wave function:

$$\psi(x, y, z) = N(x + y + z)\,\mathrm{e}^{-r^2/\alpha^2}$$

where $\alpha$, which is real, is given and $N$ is a normalization constant.

*a.* The observables $L_z$ and $\mathbf{L}^2$ are measured; what are the probabilities of finding 0 and $2\hbar^2$? Recall that:

$$Y_1^0(\theta, \varphi) = \sqrt{\frac{3}{4\pi}}\cos\theta$$

*b.* If one also uses the fact that:

$$Y_1^{\pm 1}(\theta, \varphi) = \mp\sqrt{\frac{3}{8\pi}}\sin\theta\,\mathrm{e}^{\pm i\varphi}$$

is it possible to predict directly the probabilities of all possible results of measurements of $\mathbf{L}^2$ and $L_z$ in the system of wave function $\psi(x, y, z)$?

**6.** Consider a system of angular momentum $l = 1$. A basis of its state space is formed by the three eigenvectors of $L_z$ : $|+1\rangle$, $|0\rangle$, $|-1\rangle$, whose eigenvalues are, respectively, $+\hbar$, 0, and $-\hbar$, and which satisfy:

$$L_{\pm} | m \rangle = \hbar\sqrt{2} | m \pm 1 \rangle$$
$$L_{+} | 1 \rangle = L_{-} | -1 \rangle = 0$$

This system, which possesses an electric quadrupole moment, is placed in an electric field gradient, so that its Hamiltonian can be written:

$$H = \frac{\omega_0}{\hbar}(L_u^2 - L_v^2)$$

where $L_u$ and $L_v$ are the components of **L** along the two directions $Ou$ and $Ov$ of the $xOz$ plane which form angles of 45° with $Ox$ and $Oz$; $\omega_0$ is a real constant.

*a.* Write the matrix which represents $H$ in the $\{ |+1\rangle, |0\rangle, |-1\rangle \}$ basis. What are the stationary states of the system, and what are their energies? (These states are to be written $|E_1\rangle$, $|E_2\rangle$, $|E_3\rangle$, in order of decreasing energies.)

*b.* At time $t = 0$, the system is in the state:

$$|\psi(0)\rangle = \frac{1}{\sqrt{2}}[|+1\rangle - |-1\rangle]$$

What is the state vector $|\psi(t)\rangle$ at time $t$? At $t$, $L_z$ is measured; what are the probabilities of the various possible results?

*c.* Calculate the mean values $\langle L_x \rangle(t)$, $\langle L_y \rangle(t)$ and $\langle L_z \rangle(t)$ at $t$. What is the motion performed by the vector $\langle \mathbf{L} \rangle$?

*d.* At $t$, a measurement of $L_z^2$ is performed.

(*i*) Do times exist when only one result is possible?

(*ii*) Assume that this measurement has yielded the result $\hbar^2$. What is the state of the system immediately after the measurement? Indicate, without calculation, its subsequent evolution.

**7.** Consider rotations in ordinary three-dimensional space, to be denoted by $\mathcal{R}_{\mathbf{u}}(\alpha)$, where **u** is the unit vector which defines the axis of rotation and $\alpha$ is the angle of rotation.

*a.* Show that, if $M'$ is the transform of $M$ under an infinitesimal rotation of angle $\varepsilon$, then:

$$\mathbf{OM}' = \mathbf{OM} + \varepsilon\mathbf{u} \times \mathbf{OM}$$

*b.* If **OM** is represented by the column vector $\begin{pmatrix} x \\ y \\ z \end{pmatrix}$, what is the matrix associated with $\mathcal{R}_{\mathbf{u}}(\varepsilon)$? Derive from it the matrices which represent the components of the operator $\boldsymbol{\mathcal{M}}$ defined by:

$$\mathcal{R}_{\mathbf{u}}(\varepsilon) = 1 + \varepsilon\, \boldsymbol{\mathcal{M}} . \mathbf{u}$$

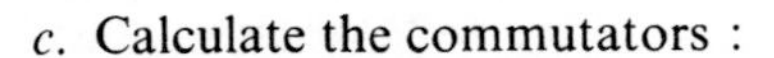

*c.* Calculate the commutators :

$$[\mathcal{M}_x, \mathcal{M}_y]; \quad [\mathcal{M}_y, \mathcal{M}_z]; \quad [\mathcal{M}_z, \mathcal{M}_x]$$

What are the quantum mechanical analogues of the purely geometrical relations obtained?

*d.* Starting with the matrix which represents $\mathcal{M}_z$, calculate the one which represents $e^{\alpha \mathcal{M}_z}$; show that $\mathcal{R}_z(\alpha) = e^{\alpha \mathcal{M}_z}$; what is the analogue of this relation in quantum mechanics?

**8.** Consider a particle in three-dimensional space, whose state vector is $|\psi\rangle$, and whose wave function is $\psi(\mathbf{r}) = \langle \mathbf{r} | \psi \rangle$. Let $A$ be an observable which commutes with $\mathbf{L} = \mathbf{R} \times \mathbf{P}$, the orbital angular momentum of the particle. Assuming that $A$, $\mathbf{L}^2$ and $L_z$ form a C.S.C.O. in $\mathscr{E}_\mathbf{r}$, call $|n, l, m\rangle$ their common eigenkets, whose eigenvalues are, respectively, $a_n$ (the index $n$ is assumed to be discrete), $l(l+1)\hbar^2$ and $m\hbar$.

Let $U(\varphi)$ be the unitary operator defined by:

$$U(\varphi) = e^{-i\varphi L_z/\hbar}$$

where $\varphi$ is a real dimensionless parameter. For an arbitrary operator $K$, we call $\tilde{K}$ the transform of $K$ by the unitary operator $U(\varphi)$:

$$\tilde{K} = U(\varphi) K U^\dagger(\varphi)$$

*a.* We set $L_+ = L_x + iL_y$, $L_- = L_x - iL_y$. Calculate $\tilde{L}_+ |n, l, m\rangle$ and show that $L_+$ and $\tilde{L}_+$ are proportional; calculate the proportionality constant. Same question for $L_-$ and $\tilde{L}_-$.

*b.* Express $\tilde{L}_x$, $\tilde{L}_y$ and $\tilde{L}_z$ in terms of $L_x$, $L_y$ and $L_z$. What geometrical transformation can be associated with the transformation of $\mathbf{L}$ into $\tilde{\mathbf{L}}$?

*c.* Calculate the commutators $[X \pm iY, L_z]$ and $[Z, L_z]$. Show that the kets $(X \pm iY)|n, l, m\rangle$ and $Z|n, l, m\rangle$ are eigenvectors of $L_z$ and calculate their eigenvalues. What relation must exist between $m$ and $m'$ for the matrix element $\langle n', l', m' | X \pm iY | n, l, m\rangle$ to be non-zero? Same question for:

$$\langle n', l', m' | Z | n, l, m\rangle.$$

*d.* By comparing the matrix elements of $\widetilde{X \pm iY}$ and $\tilde{Z}$ with those of $X \pm iY$ and $Z$, calculate $\tilde{X}$, $\tilde{Y}$, $\tilde{Z}$ in terms of $X$, $Y$, $Z$. Give a geometrical interpretation.

**9.** Consider a physical system of fixed angular momentum $l$, whose state space is $\mathscr{E}_l$, and whose state vector is $|\psi\rangle$; its orbital angular momentum operator is denoted by $\mathbf{L}$. We assume that a basis of $\mathscr{E}_l$ is composed of $2l + 1$ eigenvectors $|l, m\rangle$ of $L_z$ $(-l \leqslant m \leqslant +l)$, associated with the wave functions $f(r) Y_l^m(\theta, \varphi)$. We call $\langle \mathbf{L} \rangle = \langle \psi | \mathbf{L} | \psi \rangle$ the mean value of $\mathbf{L}$.

*a.* We begin by assuming that:

$$\langle L_x \rangle = \langle L_y \rangle = 0$$

Out of all the possible states of the system, what are those for which the sum $(\Delta L_x)^2 + (\Delta L_y)^2 + (\Delta L_z)^2$ is minimal? Show that, for these states, the root-mean-square deviation $\Delta L_\alpha$ of the component of $\mathbf{L}$ along an axis which is at an angle $\alpha$ with $Oz$ is given by:

$$\Delta L_\alpha = \hbar \sqrt{\frac{l}{2}} \sin \alpha$$

*b.* We now assume that $\langle \mathbf{L} \rangle$ has an arbitrary direction with respect to the $Oxyz$ axes. We denote by $OXYZ$ a frame whose $OZ$ axis is directed along $\langle \mathbf{L} \rangle$, with the $OY$ axis in the $xOy$ plane.

(*i*) Show that the state $| \psi_0 \rangle$ of the system for which:

$$(\Delta L_x)^2 + (\Delta L_y)^2 + (\Delta L_z)^2$$

is minimal is such that:

$$\begin{aligned} &(L_x + iL_y) | \psi_0 \rangle = 0 \\ &L_z | \psi_0 \rangle = l\hbar | \psi_0 \rangle \end{aligned}$$

(*ii*) Let $\theta_0$ be the angle between $Oz$ and $OZ$, and $\varphi_0$, the angle between $Oy$ and $OY$; prove the relations:

$$L_X + iL_Y = \cos^2 \frac{\theta_0}{2} e^{-i\varphi_0} L_+ - \sin^2 \frac{\theta_0}{2} e^{i\varphi_0} L_- - \sin \theta_0 L_z$$

$$L_Z = \sin \frac{\theta_0}{2} \cos \frac{\theta_0}{2} e^{-i\varphi_0} L_+ + \sin \frac{\theta_0}{2} \cos \frac{\theta_0}{2} e^{i\varphi_0} L_- + \cos \theta_0 L_z$$

If we set:

$$| \psi_0 \rangle = \sum_m d_m | l, m \rangle$$

show that:

$$d_m = \tan \frac{\theta_0}{2} e^{i\varphi_0} \sqrt{\frac{l + m + 1}{l - m}} d_{m+1}$$

Express $d_m$ in terms of $d_l$, $\theta_0$, $\varphi_0$ and $l$.

(*iii*) To calculate $d_l$, show that the wave function associated with $| \psi_0 \rangle$ is $\psi_0(X, Y, Z) = c_l \dfrac{(X + iY)^l}{r^l} f(r)$ [where $c_l$ is defined by equation (D-20) of chapter VI], the one associated with $| l, l \rangle$ being $c_l \dfrac{(x + iy)^l}{r^l} f(r)$. By replacing $X$, $Y$ and $Z$ in this expression for $\psi_0(X, Y, Z)$ by their values in terms of $x$, $y$, $z$, find the value of $d_l$ and the relation:

$$d_m = \left( \sin \frac{\theta_0}{2} \right)^{l-m} \left( \cos \frac{\theta_0}{2} \right)^{l+m} e^{-im\varphi_0} \sqrt{\frac{(2l)!}{(l + m)! \, (l - m)!}}$$

(*iv*) With the system in the state $|\psi_0\rangle$, $L_z$ is measured. What are the probabilities of the various possible results? What is the most probable result? Show that, if $l$ is much greater than 1, the results correspond to the classical limit.

**10.** Let **J** be the angular momentum operator of an arbitrary physical system whose state vector is $|\psi\rangle$.

*a.* Can states of the system be found for which the root-mean-square deviations $\Delta J_x$, $\Delta J_y$ and $\Delta J_z$ are simultaneously zero?

*b.* Prove the relation:

$$\Delta J_x \, . \, \Delta J_y \geqslant \frac{\hbar}{2} |\langle J_z \rangle|$$

and those obtained by cyclic permutation of $x$, $y$, $z$.

Let $\langle \mathbf{J} \rangle$ be the mean value of the angular momentum of the system. The $Oxyz$ axes are assumed to be chosen in such a way that $\langle J_x \rangle = \langle J_y \rangle = 0$. Show that:

$$(\Delta J_x)^2 + (\Delta J_y)^2 \geqslant \hbar \, |\langle J_z \rangle|$$

*c.* Show that the two inequalities proven in question *b.* both become equalities if and only if $J_+ |\psi\rangle = 0$ or $J_- |\psi\rangle = 0$.

*d.* The system under consideration is a spinless particle for which $\mathbf{J} = \mathbf{L} = \mathbf{R} \times \mathbf{P}$. Show that it is not possible to have both $\Delta L_x \, . \, \Delta L_y = \frac{\hbar}{2} |\langle L_z \rangle|$ and $(\Delta L_x)^2 + (\Delta L_y)^2 = \hbar \, |\langle L_z \rangle|$ unless the wave function of the system is of the form:

$$\psi(r, \theta, \varphi) = F(r, \sin\theta \, e^{\pm i\varphi})$$

**11.** Consider a three-dimensional harmonic oscillator, whose state vector $|\psi\rangle$ is:

$$|\psi\rangle = |\alpha_x\rangle \otimes |\alpha_y\rangle \otimes |\alpha_z\rangle$$

where $|\alpha_x\rangle$, $|\alpha_y\rangle$ and $|\alpha_z\rangle$ are quasi-classical states (*cf.* complement $G_V$) for one-dimensional harmonic oscillators moving along $Ox$, $Oy$ and $Oz$, respectively. Let $\mathbf{L} = \mathbf{R} \times \mathbf{P}$ be the orbital angular momentum of the three-dimensional oscillator.

*a.* Prove:

$$\langle L_z \rangle = i\hbar \, (\alpha_x \alpha_y^* - \alpha_x^* \alpha_y)$$
$$\Delta L_z = \hbar \sqrt{|\alpha_x|^2 + |\alpha_y|^2}$$

and the analogous expressions for the components of **L** along $Ox$ and $Oy$.

*b.* We now assume that:

$$\langle L_x \rangle = \langle L_y \rangle = 0 \quad , \quad \langle L_z \rangle = \lambda\hbar > 0$$

Show that $\alpha_z$ must be zero. We then fix the value of $\lambda$. Show that, in order to minimize $\Delta L_x + \Delta L_y$, we must choose:

$$\alpha_x = -\,i\alpha_y = \sqrt{\frac{\lambda}{2}}\,e^{i\varphi_0}$$

(where $\varphi_0$ is an arbitrary real number). Do the expressions $\Delta L_x \,.\, \Delta L_y$ and $(\Delta L_x)^2 + (\Delta L_y)^2$ in this case have minimum values which are compatible with the inequalities obtained in question *b.* of the preceding exercise?

*c.* Show that the state of a system for which the preceding conditions are satisfied is necessarily of the form:

$$|\,\psi\,\rangle = \sum_k c_k(\alpha_d)\,|\,\chi_{n_d = k, n_g = 0, n_z = 0}\,\rangle$$

with:

$$|\,\chi_{n_d = k, n_g = 0, n_z = 0}\,\rangle = \frac{(a_x^\dagger + ia_y^\dagger)^k}{\sqrt{2^k k!}}\,|\,\varphi_{n_x = 0, n_y = 0, n_z = 0}\,\rangle$$

$$c_k(\alpha) = \frac{\alpha^k}{\sqrt{k!}}\,e^{-|\alpha|^2/2} \quad ; \quad \alpha_d = e^{i\varphi_0}\sqrt{\lambda}$$

(the results of complement $G_V$ and of § 4 of complement $D_{VI}$ can be used). Show that the angular dependence of $|\,\chi_{n_d = k, n_g = 0, n_z = 0}\,\rangle$ is $(\sin\theta\, e^{i\varphi})^k$.

$\mathbf{L}^2$ is measured on a system in the state $|\,\psi\,\rangle$. Show that the probabilities of the various possible results are given by a Poisson distribution. What results can be obtained in a measurement of $L_z$ which follows a measurement of $\mathbf{L}^2$ whose result was $l(l + 1)\hbar^2$?

**Reference:**

Exercise 4 : Landau and Lifshitz (1.19), § 101 ; Ter Haar (1.23), §§ 8.13 and 8.14.

CHAPTER VII

# Particle in a central potential. The hydrogen atom

## OUTLINE OF CHAPTER VII

In this chapter, we shall consider the quantum mechanical properties of a particle placed in a central potential [that is, a potential $V(r)$ which depends only on the distance $r$ from the origin]. This problem is closely related to the study of angular momentum presented in the preceding chapter. As we shall see in § A, the fact that $V(r)$ is invariant under any rotation about the origin means that the Hamiltonian $H$ of the particle commutes with the three components of the orbital angular momentum operator $\mathbf{L}$. This considerably simplifies the determination of the eigenfunctions and eigenvalues of $H$, since these functions can be required to be eigenfunctions of $\mathbf{L}^2$ and $L_z$ as well. This immediately defines their angular dependence and the eigenvalue equation of $H$ can be replaced by a differential equation involving only the variable $r$.

The importance of this problem derives from a property which will be established in § B : a two-particle system in which the interaction is described by a potential energy which depends only on the relative positions of the particles can be reduced to a simpler problem involving only one fictitious particle. In addition, when the interaction potential of the two particles depends only on the distance between them, the fictitious particle's motion is governed by a central potential. This explains why the problem considered in this chapter is of such general interest: it is encountered in quantum mechanics whenever we investigate the behavior of an isolated system composed of two interacting particles.

In § C, we shall apply the general methods already described to a special case: that in which $V(r)$ is a Coulomb potential. The hydrogen atom, composed of an electron and a proton which electrostatically attract each other, supplies the simplest example of a system of this type. It is not the only one : in addition to hydrogen isotopes (deuterium, tritium), there are the hydrogenoid ions, which are systems composed of a single electron and a nucleus, such as the ions $He^+$, $Li^{++}$, etc... (other examples will be given in complement $A_{VII}$). For these systems, we shall explicitly calculate the energies of the bound states and the corresponding wave functions. We also recall the fact that, historically, quantum mechanics was introduced in order to explain atomic properties (in particular, those of the simplest atom, hydrogen), which could not be accounted for by classical mechanics. The remarkable agreement between the theoretical predictions and the experimental observations constitutes one of the most spectacular successes of this branch of physics. Finally, it should be noted that the exact results concerning the hydrogen atom serve as the basis of all approximate calculations relating to more complex atoms (having several electrons).

# A. STATIONARY STATES OF A PARTICLE IN A CENTRAL POTENTIAL

In this section, we consider a (spinless) particle of mass $\mu$, subject to a central force derived from the potential $V(r)$ (the center of force is chosen as the origin).

## 1. Outline of the problem

### a. REVIEW OF SOME CLASSICAL RESULTS

The force acting on the classical particle situated at the point $M$ (with $\mathbf{OM} = \mathbf{r}$) is equal to:

$$\mathbf{F} = - \boldsymbol{\nabla} V(r) = - \frac{\mathrm{d}V}{\mathrm{d}r} \frac{\mathbf{r}}{r} \tag{A-1}$$

$\mathbf{F}$ is therefore always directed towards $O$, and its moment with respect to this point is therefore always zero. If:

$$\boldsymbol{\mathscr{L}} = \mathbf{r} \times \mathbf{p} \tag{A-2}$$

is the angular momentum of the particle with respect to $O$, the angular momentum theorem then implies that:

$$\frac{\mathrm{d}\boldsymbol{\mathscr{L}}}{\mathrm{d}t} = \mathbf{0} \tag{A-3}$$

$\boldsymbol{\mathscr{L}}$ is therefore a *constant of the motion*, and the particle's trajectory is therefore necessarily situated in the plane passing through $O$ and perpendicular to $\boldsymbol{\mathscr{L}}$.

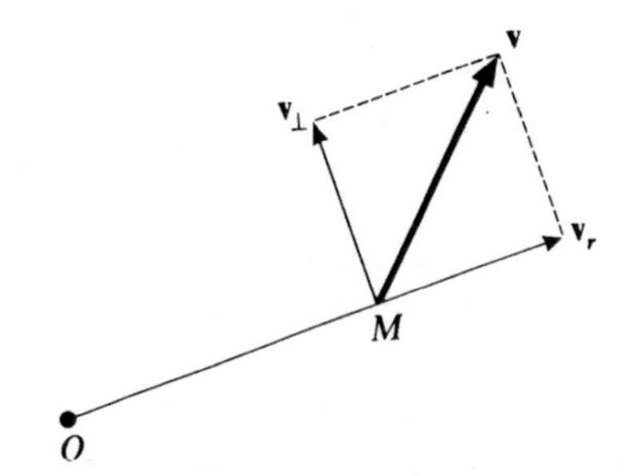


FIGURE 1

**Radial component $\mathbf{v}_r$ and tangential component $\mathbf{v}_\perp$ of a particle's velocity.**

Now let us consider (fig. 1) the position (denoted by $\mathbf{OM} = \mathbf{r}$) and velocity $\mathbf{v}$ of the particle at the instant $t$. The two vectors $\mathbf{r}$ and $\mathbf{v}$ lie in the plane of the trajectory, and the velocity $\mathbf{v}$ can be decomposed into the radial component $\mathbf{v}_r$ (along the axis defined by $\mathbf{r}$) and the tangential component $\mathbf{v}_\perp$ (along the axis perpendicular to $\mathbf{r}$).

The radial velocity, the algebraic value of $\mathbf{v}_r$, is the time derivative of the distance of the particle from the point $O$:

$$v_r = \frac{\mathrm{d}r}{\mathrm{d}t} \tag{A-4}$$

The tangential velocity can be expressed in terms of $r$ and the angular momentum $\mathscr{L}$, since:

$$|\mathbf{r} \times \mathbf{v}| = r\,|\mathbf{v}_\perp| \tag{A-5}$$

so that the modulus of the angular momentum $\mathscr{L}$ is equal to:

$$|\mathscr{L}| = |\mathbf{r} \times \mu\,\mathbf{v}| = \mu\, r\,|\mathbf{v}_\perp| \tag{A-6}$$

The total energy of the particle:

$$E = \frac{1}{2}\mu\,\mathbf{v}^2 + V(r) = \frac{1}{2}\mu\,\mathbf{v}_r^2 + \frac{1}{2}\mu\,\mathbf{v}_\perp^2 + V(r) \tag{A-7}$$

can then be written:

$$E = \frac{1}{2}\mu\,v_r^2 + \frac{\mathscr{L}^2}{2\mu\,r^2} + V(r) \tag{A-8}$$

The classical Hamiltonian of the system is then:

$$\mathscr{H} = \frac{p_r^2}{2\mu} + \frac{\mathscr{L}^2}{2\mu\,r^2} + V(r) \tag{A-9}$$

where:

$$p_r = \mu\frac{\mathrm{d}r}{\mathrm{d}t} \tag{A-10}$$

is the conjugate momentum of $r$, and $\mathscr{L}^2$ must be expressed in terms of the variables $r$, $\theta$, $\varphi$ and their conjugate momenta $p_r$, $p_\theta$, $p_\varphi$. One finds (*cf.* appendix III, § 4-a):

$$\mathscr{L}^2 = p_\theta^2 + \frac{1}{\sin^2\theta}\,p_\varphi^2 \tag{A-11}$$

In expression (A-9), the kinetic energy is broken into two terms: the radial kinetic energy and the kinetic energy of rotation about $O$. The reason is that, since $V(r)$ is independent of $\theta$ and $\varphi$ in this case, the angular variables and their conjugate momenta appear only in the $\mathscr{L}^2$ term. In fact, if we are interested in the evolution of $r$, we can use the fact that $\mathscr{L}$ is a constant of the motion, and replace $\mathscr{L}^2$ by a constant in expression (A-9). The Hamiltonian $\mathscr{H}$ then appears as a function only of the radial variables $r$ and $p_r$ ($\mathscr{L}^2$ plays the role of a parameter), and the result is a differential equation involving only one variable, $r$:

$$\frac{\mathrm{d}p_r}{\mathrm{d}t} = \mu\frac{\mathrm{d}^2 r}{\mathrm{d}t^2} = -\frac{\partial\mathscr{H}}{\partial r} = \frac{\mathscr{L}^2}{\mu\,r^3} - \frac{\mathrm{d}V}{\mathrm{d}r} \tag{A-12}$$

It is just as if we had a one-dimensional problem (with $r$ varying only between 0 and $+\infty$), with a particle of mass $\mu$ subjected to the "effective potential":

$$V_{\text{eff}}(r) = V(r) + \frac{\mathscr{L}^2}{2\mu r^2} \tag{A-13}$$

We shall see that the situation is analogous in quantum mechanics.

### b. THE QUANTUM MECHANICAL HAMILTONIAN

In quantum mechanics, we want to solve the eigenvalue equation of the Hamiltonian $H$, the observable associated with the total energy. This equation is written, in the $\{ |\mathbf{r}\rangle \}$ representation:

$$\left[ -\frac{\hbar^2}{2\mu} \Delta + V(r) \right] \varphi(\mathbf{r}) = E\, \varphi(\mathbf{r}) \tag{A-14}$$

Since the potential $V$ depends only on the distance $r$ of the particle from the origin, spherical coordinates (*cf.* § D-1-a of chapter VI) are best adapted to the problem. We therefore express the Laplacian $\Delta$ in spherical coordinates*:

$$\Delta = \frac{1}{r}\frac{\partial^2}{\partial r^2} r + \frac{1}{r^2}\left( \frac{\partial^2}{\partial \theta^2} + \frac{1}{\tan\theta}\frac{\partial}{\partial\theta} + \frac{1}{\sin^2\theta}\frac{\partial^2}{\partial\varphi^2} \right) \tag{A-15}$$

and look for eigenfunctions $\varphi(\mathbf{r})$ which are functions of the variables $r$, $\theta$, $\varphi$.

If we compare expression (A-15) with the one for the operator $\mathbf{L}^2$ [formula (D-6-a) of chapter VI], we see that the quantum mechanical Hamiltonian $H$ can be put in a form which is completely analogous to (A-9):

$$\boxed{H = -\frac{\hbar^2}{2\mu}\frac{1}{r}\frac{\partial^2}{\partial r^2} r + \frac{1}{2\mu r^2}\mathbf{L}^2 + V(r)} \tag{A-16}$$

The angular dependence of the Hamiltonian is contained entirely in the $\mathbf{L}^2$ term, which is an operator here. We could, in fact, perfect the analogy by defining an operator $P_r$, which would allow us to write the first term of (A-16) like the one in (A-9).

We shall now show how one can solve the eigenvalue equation:

$$\left[ -\frac{\hbar^2}{2\mu}\frac{1}{r}\frac{\partial^2}{\partial r^2} r + \frac{1}{2\mu r^2}\mathbf{L}^2 + V(r) \right] \varphi(r, \theta, \varphi) = E\, \varphi(r, \theta, \varphi) \tag{A-17}$$

* Expression (A-15) gives the Laplacian only for non-zero $r$. This is because of the privileged position of the origin in spherical coordinates; it can be seen, moreover, that expression (A-15) is not defined for $r = 0$.

## 2. Separation of variables

### a. ANGULAR DEPENDENCE OF THE EIGENFUNCTIONS

We know [*cf.* formulas (D-5) of chapter VI] that the three components of the angular momentum operator **L** act only on the angular variables $\theta$ and $\varphi$; consequently, they commute with all operators which act only on the $r$-dependence. In addition, they commute with $\mathbf{L}^2$. Therefore, according to expression (A-16) for the Hamiltonian, *the three components of* **L** *are constants of the motion** in the quantum mechanical sense:

$$[H, \mathbf{L}] = \mathbf{0} \tag{A-18}$$

Obviously, $H$ also commutes with $\mathbf{L}^2$.

Although we have at our disposition four constants of the motion ($L_x$, $L_y$, $L_z$ and $\mathbf{L}^2$), we cannot use all four of them to solve equation (A-17) because they do not commute with each other; we shall use only $\mathbf{L}^2$ and $L_z$. Since the three observables $H$, $\mathbf{L}^2$ and $L_z$ commute, we can find a basis of the state space $\mathscr{E}_\mathbf{r}$ of the particle composed of eigenfunctions common to these three observables. We can, therefore, without restricting the generality of the problem posed in § 1 above, require the functions $\varphi(r, \theta, \varphi)$, solutions of equation (A-17), to be eigenfunctions of $\mathbf{L}^2$ and $L_z$ as well. We must then solve the system of differential equations:

$$H\,\varphi(\mathbf{r}) = E\,\varphi(\mathbf{r}) \tag{A-19-a}$$

$$\mathbf{L}^2\varphi(\mathbf{r}) = l(l+1)\hbar^2\varphi(\mathbf{r}) \tag{A-19-b}$$

$$L_z\varphi(\mathbf{r}) = m\hbar\,\varphi(\mathbf{r}) \tag{A-19-c}$$

But we already know the general form of the common eigenfunctions of $\mathbf{L}^2$ and $L_z$ (chap. VI, § D-1-b-β): the solutions $\varphi(\mathbf{r})$ of equations (A-19), corresponding to fixed values of $l$ and $m$, are necessarily products of a function of $r$ alone and the spherical harmonic $Y_l^m(\theta, \varphi)$:

$$\varphi(\mathbf{r}) = R(r)\,Y_l^m(\theta, \varphi) \tag{A-20}$$

Whatever the radial function $R(r)$, $\varphi(\mathbf{r})$ is a solution of equations (A-19-b) and (A-19-c). The only problem which remains to be solved is therefore how to determine $R(r)$ such that $\varphi(\mathbf{r})$ is also an eigenfunction of $H$ [equation (A-19-a)].

### b. THE RADIAL EQUATION

We shall now substitute expressions (A-16) and (A-20) into equation (A-19-a). Since $\varphi(\mathbf{r})$ is an eigenfunction of $\mathbf{L}^2$ with the eigenvalue $l(l+1)\hbar^2$, we see that $Y_l^m(\theta, \varphi)$ is a common factor on both sides. After simplifying, we obtain the radial equation:

$$\left[-\frac{\hbar^2}{2\mu}\frac{1}{r}\frac{\mathrm{d}^2}{\mathrm{d}r^2}r + \frac{l(l+1)\hbar^2}{2\mu r^2} + V(r)\right]R(r) = E\,R(r) \tag{A-21}$$

* Equation (A-18) expresses the fact that $H$ is a scalar operator with respect to rotations about the point $O$ (see complement $B_{VI}$). This is true because the potential energy is invariant under rotations about $O$.

Actually, a solution of (A-21), substituted into (A-20), does not necessarily yield a solution of the eigenvalue equation (A-14) of the Hamiltonian. As we have already pointed out (*cf.* note on page 778), expression (A-15) for the Laplacian is not necessarily valid at $r = 0$. We must therefore make sure that the behavior of the solutions $R(r)$ of (A-21) at the origin is sufficiently regular for (A-20) to be in fact a solution of (A-14).

Instead of solving the partial differential equation (A-17) in the three variables $r$, $\theta$, $\varphi$, we must now solve a differential equation involving only the variable $r$, but dependent on a parameter $l$: we are looking for eigenvalues and eigenfunctions of an operator $H_l$ which is different for each value of $l$.

In other words, we consider separately, in the state space $\mathscr{E}_{\mathbf{r}}$, the subspaces $\mathscr{E}(l, m)$ corresponding to fixed values of $l$ and $m$ (*cf.* chap. VI, § C-3-a), studying the eigenvalue equation of $H$ in each of these subspaces (which is possible because $H$ commutes with $\mathbf{L}^2$ and $L_z$). The equation to be solved depends on $l$, but not on $m$; it is therefore the same in the $(2l + 1)$ subspaces $\mathscr{E}(l, m)$ associated with a given value of $l$. We shall denote by $E_{k,l}$ the eigenvalues of $H_l$, that is, the eigenvalues of the Hamiltonian $H$ inside a given subspace $\mathscr{E}(l, m)$. The index $k$, which can be discrete or continuous, represents the various eigenvalues associated with the same value of $l$. As for the eigenfunctions of $H_l$, we shall label them with the same two indices as the eigenvalues: $R_{k,l}(r)$. It is not obvious that this is sufficient: several radial functions might exist and be eigenfunctions of the same operator $H_l$ with the same eigenvalue $E_{k,l}$; we shall see in § 3-b that this is not the case and that, consequently, the two indices $k$ and $l$ are sufficient to characterize the different radial functions. We shall therefore rewrite equation (A-21) in the form:

$$\left[ -\frac{\hbar^2}{2\mu}\frac{1}{r}\frac{\mathrm{d}^2}{\mathrm{d}r^2}r + \frac{l(l+1)\hbar^2}{2\mu r^2} + V(r) \right] R_{k,l}(r) = E_{k,l}\, R_{k,l}(r) \tag{A-22}$$

We can simplify the differential operator to be studied by a change in functions. We set:

$$R_{k,l}(r) = \frac{1}{r}\, u_{k,l}(r) \tag{A-23}$$

Multiplying both sides of (A-22) by $r$, we obtain for $u_{k,l}(r)$ the following differential equation:

$$\boxed{\left[ -\frac{\hbar^2}{2\mu}\frac{\mathrm{d}^2}{\mathrm{d}r^2} + \frac{l(l+1)\hbar^2}{2\mu r^2} + V(r) \right] u_{k,l}(r) = E_{k,l}\, u_{k,l}(r)} \tag{A-24}$$

This equation is analogous to the one which we would have to solve if, in a *one-dimensional* problem, a particle of mass $\mu$ were moving in an *effective potential* $V_{\text{eff}}(r)$ such that:

$$V_{\text{eff}}(r) = V(r) + \frac{l(l+1)\hbar^2}{2\mu r^2} \tag{A-25}$$

Nevertheless, we must not lose sight of the fact that the variable $r$ can take on only *non-negative* real values. The term $l(l+1)\hbar^2/2\mu r^2$ which is added to the potential $V(r)$ is always positive or zero; the corresponding force (equal to minus the gradient of this term) always tends to repel the particle from the force center $O$; this is why this term is called the *centrifugal potential* (or centrifugal barrier). Figure 2 represents the shape of the effective potential $V_{\text{eff}}(r)$ for various values of $l$ in the case in which $V(r)$ is an attractive Coulomb potential $[V(r) = -\, e^2/r]$: for $l \geqslant 1$, the presence of the centrifugal term, which predominates for small $r$ values, causes $V_{\text{eff}}$ to be repulsive for short distances.

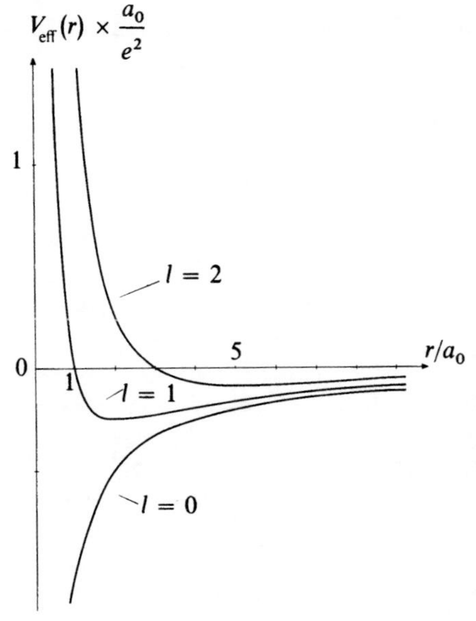


FIGURE 2

**Shape of the effective potential $V_{\text{eff}}(r)$ for the first values of $l$ in the case where $V(r) = -\frac{e^2}{r}$. When $l = 0$, $V_{\text{eff}}(r)$ is simply equal to $V(r)$. When $l$ takes on the values 1, 2, etc., $V_{\text{eff}}(r)$ is obtained by adding to $V(r)$ the centrifugal potential $\frac{l(l+1)\hbar^2}{2\mu r^2}$, which approaches $+\infty$ when $r$ approaches zero.**

### c. BEHAVIOR OF THE SOLUTIONS OF THE RADIAL EQUATION AT THE ORIGIN

We have already pointed out that it is necessary to examine the behavior of the solutions $R(r)$ of the radial equation (A-21) at the origin in order to know if they are really solutions of (A-14).

We shall assume that when $r$ approaches zero, the potential $V(r)$ remains finite, or at least approaches infinity less rapidly than $1/r$ (this hypothesis is true in most cases encountered in physics and, in particular, in the case of the Coulomb potential, to be studied in § C). We shall consider a solution of (A-22) and assume that it behaves at the origin like $r^s$:

$$R_{k,l}(r) \underset{r \to 0}{\simeq} C\, r^s \tag{A-26}$$

Substituting (A-26) into (A-22), and setting the coefficient of the dominant term equal to zero, we obtain the equation:

$$-\, s(s+1) + l(l+1) = 0 \tag{A-27}$$

and, consequently:

$$\begin{cases} \text{either} \quad s = l & \text{(A-28-a)} \\ \text{or} \quad s = -(l+1) & \text{(A-28-b)} \end{cases}$$

For a given value of $E_{k,l}$, there are therefore two linearly independent solutions of the second-order equation (A-22), behaving at the origin like $r^l$ and $1/r^{l+1}$, respectively. But those which behave like $1/r^{l+1}$ must be rejected, since it can be shown that $\frac{1}{r^{l+1}} Y_l^m(\theta, \varphi)$ is not a solution of the eigenvalue equation (A-14) for $r = 0$*. From this, we see that acceptable solutions of (A-24) go to zero at the origin for all $l$, since:

$$u_{k,l}(r) \underset{r \to 0}{\sim} C\, r^{l+1} \tag{A-29}$$

Consequently, to equation (A-24) must be added the condition:

$$\boxed{u_{k,l}(0) = 0} \tag{A-30}$$

COMMENT:

In equation (A-24), $r$, the distance of the particle from the origin, varies only between 0 and $+\infty$. However, thanks to condition (A-30), we can assume that we are actually dealing with a one-dimensional problem, in which the particle can theoretically move along the entire axis, but in which the effective potential is infinite for all negative values of the variable. We know that, in such a case, the wave function must be identically zero on the negative half-axis; condition (A-30) insures the continuity of the wave function at $r = 0$.

## 3. Stationary states of a particle in a central potential

### a. QUANTUM NUMBERS

We can summarize the results of §2 as follows: the fact that the potential $V(r)$ is independent of $\theta$ and $\varphi$ makes it possible:

($i$) to require the eigenfunctions of $H$ to be *simultaneous eigenfunctions of* $\mathbf{L}^2$ *and* $L_z$, *which determines their angular dependence:*

$$\varphi_{k,l,m}(\mathbf{r}) = R_{k,l}(r) Y_l^m(\theta, \varphi) = \frac{1}{r} u_{k,l}(r) Y_l^m(\theta, \varphi) \tag{A-31}$$

($ii$) *to replace the eigenvalue equation of* $H$, an equation involving partial derivatives with respect to $r$, $\theta$, $\varphi$, *by a differential equation involving only the variable* $r$ and depending on a parameter $l$ [equation (A-24)], with condition (A-30) imposed.

* This is because the Laplacian of $\frac{1}{r^{l+1}} Y_l^m(\theta, \varphi)$ involves the $l$th derivatives of $\delta(\mathbf{r})$ (*cf.* appendix II, end of §4).

These results can be compared with those recalled in § 1-a, of which they are the quantum mechanical analogues.

In principle, the functions $\varphi_{k,l,m}(r, \theta, \varphi)$ must be square-integrable, that is, normalizable:

$$\int |\varphi_{k,l,m}(r, \theta, \varphi)|^2 r^2 \, dr \, d\Omega = 1 \tag{A-32}$$

Their form (A-31) allows us to separate radial and angular integrations:

$$\int |\varphi_{k,l,m}(r, \theta, \varphi)|^2 r^2 \, dr \, d\Omega = \int_0^\infty r^2 \, dr \, |R_{k,l}(r)|^2 \int d\Omega \, |Y_l^m(\theta, \varphi)|^2 \tag{A-33}$$

But the spherical harmonics $Y_l^m(\theta, \varphi)$ are normalized with respect to $\theta$ and $\varphi$; condition (A-32) therefore reduces to:

$$\int_0^\infty r^2 \, dr \, |R_{k,l}(r)|^2 = \int_0^\infty dr \, |u_{k,l}(r)|^2 = 1 \tag{A-34}$$

Actually, we know that it is often convenient to accept eigenfunctions of the Hamiltonian which are not square-integrable. If the spectrum of $H$ has a continuous part, we shall require only that the corresponding eigenfunctions be orthonormalized in the extended sense, that is, that they satisfy a condition of the form:

$$\int_0^\infty r^2 \, dr \, R_{k',l}^*(r) \, R_{k,l}(r) = \int_0^\infty dr \, u_{k',l}^*(r) \, u_{k,l}(r) = \delta(k' - k) \tag{A-35}$$

where $k$ is a continuous index.

In (A-34) and (A-35), the integrals converge at their lower limit, $r = 0$ [condition (A-30)]. This is physically satisfying since the probability of finding the particle in any volume of finite dimensions is then always finite. It is therefore only because of the behavior of the wave functions for $r \longrightarrow \infty$ that, in the case of a continuous spectrum, the normalization integrals (A-35) diverge if $k = k'$.

Finally, the eigenfunctions of the Hamiltonian $H$ of a particle placed in a central potential $V(r)$ depend on at least three indices [formula (A-31)]: $\varphi_{k,l,m}(r, \theta, \varphi) = R_{k,l}(r) Y_l^m(\theta, \varphi)$ is a simultaneous eigenfunction of $H$, $\mathbf{L}^2$ and $L_z$ with the respective eigenvalues $E_{k,l}$, $l(l+1)\hbar^2$ and $m\hbar$. $k$ is called the *radial* quantum number; $l$, the *azimuthal* quantum number; and $m$, the *magnetic* quantum number. The radial part $R_{k,l}(r) = \frac{1}{r} u_{k,l}(r)$ of the eigenfunction and the eigenvalue $E_{k,l}$ of $H$ are independent of the magnetic quantum number and are given by the radial equation (A-24). The angular part of the eigenfunction depends only on $l$ and $m$ and not on $k$; it does not depend on the form of the potential $V(r)$.

### b. DEGENERACY OF THE ENERGY LEVELS

Finally, we shall consider the degeneracy of the energy levels, that is, of the eigenvalues of the Hamiltonian $H$. The $(2l + 1)$ functions $\varphi_{k,l,m}(r, \theta, \varphi)$ with $k$ and $l$ fixed and $m$ varying from $-l$ to $+l$ are eigenfunctions of $H$ with the same eigenvalue $E_{k,l}$ [these $(2l + 1)$ functions are clearly orthogonal, since they correspond to different eigenvalues of $L_z$]. The level $E_{k,l}$ is therefore at least $(2l + 1)$-fold degenerate. This degeneracy, which exists for all potentials $V(r)$, is called an *essential degeneracy*: it is due to the fact that the Hamiltonian $H$ contains $\mathbf{L}^2$, but not $L_z$★, which means that $m$ does not appear in the radial equation. It is also possible for one of the eigenvalues $E_{k,l}$ of the radial equation corresponding to a given value of $l$ to be the same as an eigenvalue $E_{k',l'}$ associated with another radial equation, characterized by $l' \neq l$. This occurs only for certain potentials $V(r)$. The resulting degeneracies are called *accidental* (we shall see in §C that the energy states of the hydrogen atom present accidental degeneracies).

We must now show that, for a fixed value of $l$, the radial equation has at most one physically acceptable solution for each eigenvalue $E_{k,l}$. This actually results from condition (A-30). The radial equation, since it is a second-order differential equation, has *a priori* two linearly independent solutions for each value of $E_{k,l}$. Condition (A-30) eliminates one of them, so there is at most one acceptable solution for each value of $E_{k,l}$. We must also consider the behavior of the solutions for $r$ approaching infinity; if $V(r) \longrightarrow 0$ when $r \longrightarrow \infty$, the negative values of $E_{k,l}$ for which the solution we have just chosen is also acceptable at infinity (that is, bounded) form a discrete set (see example of §C below and complement $B_{VII}$).

It follows from the preceding considerations that $H$, $\mathbf{L}^2$ *and* $L_z$ *constitute a C.S.C.O.*★★. If we fix three eigenvalues $E_{k,l}$, $l(l + 1)\hbar^2$ and $m\hbar$, there corresponds to them a single function $\varphi_{k,l,m}(\mathbf{r})$. The eigenvalue of $\mathbf{L}^2$ indicates which equation yields the radial function; the eigenvalue of $H$ determines this radial function $R_{k,l}(r)$ uniquely, as we have just seen; finally, there exists only one spherical harmonic $Y_l^m(\theta, \varphi)$ for a given $l$ and $m$.

## B. MOTION OF THE CENTER OF MASS AND RELATIVE MOTION FOR A SYSTEM OF TWO INTERACTING PARTICLES

Consider a system of two spinless particles, of masses $m_1$ and $m_2$ and positions $\mathbf{r}_1$ and $\mathbf{r}_2$. We assume that the forces exerted on these particles are derived from a *potential energy* $V(\mathbf{r}_1 - \mathbf{r}_2)$ *which depends only on* $\mathbf{r}_1 - \mathbf{r}_2$. This is true if there are no forces originating outside the system (that is, the system is isolated), and if the interactions between the two particles are derived from a potential. This potential must depend only on $\mathbf{r}_1 - \mathbf{r}_2$, since only the relative positions of the two particles are involved. We shall show that the study of such a system can be reduced to that of a single particle placed in the potential $V(\mathbf{r})$.

★ This essential degeneracy appears whenever the Hamiltonian is rotation-invariant (*cf.* complement $B_{VI}$). This is why it is encountered in numerous physical problems.

★★ Actually, we have not proven that these operators are observables, that is, that the set of $\varphi_{k,l,m}(\mathbf{r})$ form a basis in the state space $\mathscr{E}_r$.

## 1. Motion of the center of mass and relative motion in classical mechanics

In classical mechanics, the two-particle system is described by the Lagrangian (*cf.* appendix III):

$$\mathscr{L}(\mathbf{r}_1, \dot{\mathbf{r}}_1 ; \mathbf{r}_2, \dot{\mathbf{r}}_2) = T - V = \frac{1}{2} m_1 \dot{\mathbf{r}}_1^2 + \frac{1}{2} m_2 \dot{\mathbf{r}}_2^2 - V(\mathbf{r}_1 - \mathbf{r}_2) \tag{B-1}$$

and the conjugate momenta of the six coordinates of the two particles are the components of the mechanical momenta:

$$\begin{aligned} \mathbf{p}_1 &= m_1 \dot{\mathbf{r}}_1 \\ \mathbf{p}_2 &= m_2 \dot{\mathbf{r}}_2 \end{aligned} \tag{B-2}$$

The study of the motion of the two particles is simplified by replacing the positions $\mathbf{r}_i$ by the three *coordinates of the center of mass* (or center of gravity):

$$\mathbf{r}_G = \frac{m_1 \mathbf{r}_1 + m_2 \mathbf{r}_2}{m_1 + m_2} \tag{B-3}$$

and the three *relative coordinates*★:

$$\mathbf{r} = \mathbf{r}_1 - \mathbf{r}_2 \tag{B-4}$$

Formulas (B-3) and (B-4) can be inverted to yield:

$$\begin{aligned} \mathbf{r}_1 &= \mathbf{r}_G + \frac{m_2}{m_1 + m_2} \mathbf{r} \\ \mathbf{r}_2 &= \mathbf{r}_G - \frac{m_1}{m_1 + m_2} \mathbf{r} \end{aligned} \tag{B-5}$$

The Lagrangian can then be written, in terms of the new variables $\mathbf{r}_G$ and $\mathbf{r}$:

$$\begin{aligned} \mathscr{L}(\mathbf{r}_G, \dot{\mathbf{r}}_G ; \mathbf{r}, \dot{\mathbf{r}}) &= \frac{1}{2} m_1 \left[ \dot{\mathbf{r}}_G + \frac{m_2}{m_1 + m_2} \dot{\mathbf{r}} \right]^2 + \frac{1}{2} m_2 \left[ \dot{\mathbf{r}}_G - \frac{m_1}{m_1 + m_2} \dot{\mathbf{r}} \right]^2 - V(\mathbf{r}) \\ &= \frac{1}{2} M \dot{\mathbf{r}}_G^2 + \frac{1}{2} \mu \, \dot{\mathbf{r}}^2 - V(\mathbf{r}) \end{aligned} \tag{B-6}$$

where:

$$M = m_1 + m_2 \tag{B-7}$$

is the *total mass* of the system, and:

$$\mu = \frac{m_1 m_2}{m_1 + m_2} \tag{B-8-a}$$

★ Definition (B-4) introduces a slight asymmetry between the two particles.

is its *reduced mass* (the geometrical mean of the two masses $m_1$ and $m_2$), which is also given by:

$$\frac{1}{\mu} = \frac{1}{m_1} + \frac{1}{m_2} \tag{B-8-b}$$

The conjugate momenta of the variables $\mathbf{r}_G$ and $\mathbf{r}$ are obtained by differentiating expression (B-6) with respect to the components of $\dot{\mathbf{r}}_G$ and $\dot{\mathbf{r}}$. Using (B-3), (B-4) and (B-2), we find:

$$\mathbf{p}_G = M\,\dot{\mathbf{r}}_G = m_1\dot{\mathbf{r}}_1 + m_2\dot{\mathbf{r}}_2 = \mathbf{p}_1 + \mathbf{p}_2 \tag{B-9-a}$$

$$\mathbf{p} = \mu\,\dot{\mathbf{r}} = \frac{m_2\mathbf{p}_1 - m_1\mathbf{p}_2}{m_1 + m_2} \tag{B-9-b}$$

or:

$$\frac{\mathbf{p}}{\mu} = \frac{\mathbf{p}_1}{m_1} - \frac{\mathbf{p}_2}{m_2} \tag{B-9-c}$$

$\mathbf{p}_G$ is the *total momentum* of the system, and $\mathbf{p}$ is called the *relative momentum* of the two particles.

We can then express the classical Hamiltonian of the system in terms of the new dynamical variables which we have just introduced:

$$\mathscr{H}(\mathbf{r}_G, \mathbf{p}_G; \mathbf{r}, \mathbf{p}) = \frac{\mathbf{p}_G^2}{2M} + \frac{\mathbf{p}^2}{2\mu} + V(\mathbf{r}) \tag{B-10}$$

From this, the equations of motion can immediately be derived [formulas (27) of appendix III]:

$$\dot{\mathbf{p}}_G = \mathbf{0} \tag{B-11}$$

$$\dot{\mathbf{p}} = -\nabla V(\mathbf{r}) \tag{B-12}$$

The first term of expression (B-10) represents the kinetic energy of a fictitious particle whose mass $M$ would be the sum $m_1 + m_2$ of the masses of the two real particles, whose position would be that of the center of mass of the system [formula (B-3)], and whose momentum $\mathbf{p}_G$ would be the total momentum $\mathbf{p}_1 + \mathbf{p}_2$ of the system. Equation (B-11) indicates that this fictitious particle is in uniform rectilinear motion (free particle). This result is well known in classical mechanics: the center of mass of a system of particles moves like a single particle whose mass is the total mass of this system, subjected to the resultant of all the forces exerted on the various particles. Here, this resultant is zero since the only forces present are internal ones obeying the principle of action and reaction.

Since the center of mass is in uniform rectilinear motion with respect to the initially chosen frame, the frame in which it is at rest ($\mathbf{p}_G = \mathbf{0}$) is also an inertial frame. In this *center of mass frame*, the first term of (B-10) is zero. The classical Hamiltonian, that is, the total energy of the system, then reduces to :

$$\mathscr{H}_r = \frac{\mathbf{p}^2}{2\mu} + V(\mathbf{r}) \tag{B-13}$$

$\mathscr{H}_r$ is the energy associated with the *relative motion* of the two particles. It is obviously this relative motion that is the most interesting in the study of the two interacting particles. It can be described by introducing a fictitious particle, called the *relative particle*: its mass is the reduced mass $\mu$ of the two real particles, its position is characterized by the relative coordinates $\mathbf{r}$, and its momentum is the relative momentum $\mathbf{p}$. Since its motion obeys equation (B-12), it behaves as if it were subject to a potential $V(\mathbf{r})$ equal to the potential energy of interaction between the two real particles.

The study of the relative motion of two interacting particles therefore reduces to that of the motion of a single fictitious particle, characterized by formulas (B-4), (B-8) and (B-9-c). This last equation expresses the fact that the velocity $\mathbf{p}/\mu$ of the relative particle is indeed the difference between the velocities of the two particles, that is, their relative velocity.

## 2. Separation of variables in quantum mechanics

The considerations of the preceding section can easily be transposed to quantum mechanics, as we shall now show.

### a. OBSERVABLES ASSOCIATED WITH THE CENTER OF MASS AND THE RELATIVE PARTICLE

The operators $\mathbf{R}_1$, $\mathbf{P}_1$ and $\mathbf{R}_2$, $\mathbf{P}_2$, which describe the positions and momenta of the two particles of the system, satisfy the canonical commutation relations:

$$\begin{aligned}[][X_1, P_{1x}] &= i\hbar \\ [X_2, P_{2x}] &= i\hbar\end{aligned} \qquad \text{(B-14)}$$

with analogous expressions for the components along $Oy$ and $Oz$. All the observables labeled by the index 1 commute with all those of index 2, and all the observables relating to one of the axes $Ox$, $Oy$ or $Oz$ commute with those corresponding to another one of these axes.

Now let us define the observables $\mathbf{R}_G$ and $\mathbf{R}$ by formulas similar to (B-3) and (B-4):

$$\mathbf{R}_G = \frac{m_1\mathbf{R}_1 + m_2\mathbf{R}_2}{m_1 + m_2} \qquad \text{(B-15-a)}$$

$$\mathbf{R} = \mathbf{R}_1 - \mathbf{R}_2 \qquad \text{(B-15-b)}$$

and the observables $\mathbf{P}_G$ and $\mathbf{P}$ by formulas similar to (B-9):

$$\mathbf{P}_G = \mathbf{P}_1 + \mathbf{P}_2 \qquad \text{(B-16-a)}$$

$$\mathbf{P} = \frac{m_2\mathbf{P}_1 - m_1\mathbf{P}_2}{m_1 + m_2} \qquad \text{(B-16-b)}$$

It is easy to calculate the various commutators of these new observables. The results are as follows:

$$[X_G, P_{Gx}] = i\hbar \qquad \text{(B-17-a)}$$

$$[X, P_x] = i\hbar \qquad \text{(B-17-b)}$$

with analogous expressions for the components along $Oy$ and $Oz$; all the other commutators are zero. Consequently, **R** and **P**, like $\mathbf{R}_G$ and $\mathbf{P}_G$, satisfy canonical commutation relations. Moreover, every observable of the set $\{ \mathbf{R}, \mathbf{P} \}$ commutes with every observable of the set $\{ \mathbf{R}_G, \mathbf{P}_G \}$.

*We can also interpret* **R** *and* **P**, *on the one hand, and* $\mathbf{R}_G$ *and* $\mathbf{P}_G$, *on the other, as being the position and momentum observables of two distinct fictitious particles.*

### b. EIGENVALUES AND EIGENFUNCTIONS OF THE HAMILTONIAN

The Hamiltonian operator of the system is obtained from formulas (B-1) and (B-2) and the quantization rules of chapter III:

$$H = \frac{\mathbf{P}_1^2}{2m_1} + \frac{\mathbf{P}_2^2}{2m_2} + V(\mathbf{R}_1 - \mathbf{R}_2) \tag{B-18}$$

Since definitions (B-15) and (B-16) are formally identical to (B-3), (B-4) and (B-9), and since all the momentum operators commute, a simple algebraic calculation yields the equivalent of expression (B-10).

$$H = \frac{\mathbf{P}_G^2}{2M} + \frac{\mathbf{P}^2}{2\mu} + V(\mathbf{R}) \tag{B-19}$$

The Hamiltonian $H$ then appears as the sum of two terms:

$$H = H_G + H_r \tag{B-20}$$

with:

$$H_G = \frac{\mathbf{P}_G^2}{2M} \tag{B-21-a}$$

$$H_r = \frac{\mathbf{P}^2}{2\mu} + V(\mathbf{R}) \tag{B-21-b}$$

which commute, according to the results of §a:

$$[H_G, H_r] = 0 \tag{B-22}$$

$H_G$ and $H_r$ therefore commute with $H$. It follows that there exists a basis of eigenvectors of $H$ which are also eigenvectors of $H_G$ and $H_r$; we shall therefore look for solutions of the system:

$$\begin{aligned} H_G \,|\, \varphi \rangle &= E_G \,|\, \varphi \rangle \\ H_r \,|\, \varphi \rangle &= E_r \,|\, \varphi \rangle \end{aligned} \tag{B-23}$$

which immediately implies, according to (B-20):

$$H \,|\, \varphi \rangle = E \,|\, \varphi \rangle \tag{B-24}$$

with:

$$E = E_G + E_r \tag{B-25}$$

Consider the $\{ |\mathbf{r}_G, \mathbf{r}\rangle \}$ representation, whose basis vectors are the eigenvectors common to the observables $\mathbf{R}_G$ and $\mathbf{R}$. In this representation, a state is characterized by a wave function $\varphi(\mathbf{r}_G, \mathbf{r})$ which is a function of six variables. The action of the operators $\mathbf{R}_G$ and $\mathbf{R}$ is expressed by the multiplication of the wave functions by the variables $\mathbf{r}_G$ and $\mathbf{r}$ respectively. $\mathbf{P}_G$ and $\mathbf{P}$ become the differential operators $\frac{\hbar}{i}\nabla_G$ and $\frac{\hbar}{i}\nabla$ (where $\nabla_G$ denotes the set of three operators $\partial/\partial x_G$, $\partial/\partial y_G$ and $\partial/\partial z_G$). The state space $\mathscr{E}$ of the system can then be considered to be the tensor product $\mathscr{E}_{\mathbf{r}_G} \otimes \mathscr{E}_{\mathbf{r}}$ of the state space $\mathscr{E}_{\mathbf{r}_G}$ associated with the observable $\mathbf{R}_G$ and the space $\mathscr{E}_{\mathbf{r}}$ associated with $\mathbf{R}$. $H_G$ and $H_r$ then appear as the extensions into $\mathscr{E}$ of operators actually acting only in $\mathscr{E}_{\mathbf{r}_G}$ and $\mathscr{E}_{\mathbf{r}}$ respectively. We can therefore, as we saw in §F of chapter II, find a basis of eigenvectors $|\varphi\rangle$ satisfying (B-23), in the form:

$$|\varphi\rangle = |\chi_G\rangle \otimes |\omega_r\rangle \tag{B-26}$$

with:

$$\begin{cases} H_G|\chi_G\rangle = E_G|\chi_G\rangle \\ |\chi_G\rangle \in \mathscr{E}_{\mathbf{r}_G} \end{cases} \tag{B-27-a}$$

$$\begin{cases} H_r|\omega_r\rangle = E_r|\omega_r\rangle \\ |\omega_r\rangle \in \mathscr{E}_{\mathbf{r}} \end{cases} \tag{B-27-b}$$

Writing these equations in the $\{|\mathbf{r}_G\rangle\}$ and $\{|\mathbf{r}\rangle\}$ representations respectively, we obtain:

$$-\frac{\hbar^2}{2M}\Delta_G\,\chi_G(\mathbf{r}_G) = E_G\,\chi_G(\mathbf{r}_G) \tag{B-28-a}$$

$$\left[-\frac{\hbar^2}{2\mu}\Delta + V(\mathbf{r})\right]\omega_r(\mathbf{r}) = E_r\,\omega_r(\mathbf{r}) \tag{B-28-b}$$

The first of these equations, (B-28-a), shows that the particle associated with the center of mass of the system is free, as in classical mechanics. We know its solutions: they are, for example, the plane waves:

$$\chi_G(\mathbf{r}_G) = \frac{1}{(2\pi\hbar)^{3/2}}\,e^{\frac{i}{\hbar}\mathbf{p}_G\cdot\mathbf{r}_G} \tag{B-29}$$

whose energy is equal to:

$$E_G = \frac{\mathbf{p}_G^2}{2M} \tag{B-30}$$

$E_G$ can take on any positive value or zero; it is the kinetic energy corresponding to a translation of the system as a whole.

The more interesting equation from a physical point of view is the second one, (B-28-b), which concerns the relative particle. It describes the behavior of the system of the two interacting particles in the center of mass frame. If the interaction potential of the two real particles depends only on the distance between them, $|\mathbf{r}_1 - \mathbf{r}_2|$, and not on the direction of the vector $\mathbf{r}_1 - \mathbf{r}_2$, the relative particle is subject to a central potential $V(r)$; the problem is then reduced to the one treated in §A.

COMMENT:

The total angular momentum of the system of the two real particles is :

$$\mathbf{J} = \mathbf{L}_1 + \mathbf{L}_2 \tag{B-31}$$

with:

$$\begin{aligned} \mathbf{L}_1 &= \mathbf{R}_1 \times \mathbf{P}_1 \\ \mathbf{L}_2 &= \mathbf{R}_2 \times \mathbf{P}_2 \end{aligned} \tag{B-32}$$

It can easily be shown that it can also be written:

$$\mathbf{J} = \mathbf{L}_G + \mathbf{L} \tag{B-33}$$

where:

$$\begin{aligned} \mathbf{L}_G &= \mathbf{R}_G \times \mathbf{P}_G \\ \mathbf{L} &= \mathbf{R} \times \mathbf{P} \end{aligned} \tag{B-34}$$

are the angular momenta of the fictitious particles (according to the results of § a, $\mathbf{L}_G$ and $\mathbf{L}$ satisfy the commutation relations which characterize angular momenta, and the components of $\mathbf{L}$ commute with those of $\mathbf{L}_G$).

## C. THE HYDROGEN ATOM

### 1. Introduction

The hydrogen atom consists of a proton, of mass:

$$m_p = 1.7 \times 10^{-27}\ \text{kg} \tag{C-1}$$

and charge:

$$q = 1.6 \times 10^{-19}\ \text{Coulomb} \tag{C-2}$$

and of an electron, of mass:

$$m_e = 0.91 \times 10^{-30}\ \text{kg} \tag{C-3}$$

and charge $-q$. The interaction between these two particles is essentially electrostatic. The corresponding potential energy is :

$$V(r) = -\frac{q^2}{4\pi\varepsilon_0}\frac{1}{r} = -\frac{e^2}{r} \tag{C-4}$$

where $r$ denotes the distance between the two particles, and:

$$\frac{q^2}{4\pi\varepsilon_0} = e^2 \tag{C-5}$$

Using the results of § B, we confine ourselves to the study of this system in the center of mass frame. The classical Hamiltonian which describes the relative motion of the two particles is then★ :

$$\mathscr{H}(\mathbf{r}, \mathbf{p}) = \frac{\mathbf{p}^2}{2\mu} - \frac{e^2}{r} \tag{C-6}$$

Since $m_p \gg m_e$ [formulas (C-1) and (C-3)], the reduced mass $\mu$ of the system is very close to $m_e$:

$$\mu = \frac{m_e m_p}{m_e + m_p} \simeq m_e\left(1 - \frac{m_e}{m_p}\right) \tag{C-7}$$

(the correction term $m_e/m_p$ is on the order of 1/1 800). This means that the center of mass of the system is practically in the same place as the proton, and that the relative particle can be identified, to a very good approximation, with the electron. This is why we shall adopt the slightly inaccurate convention of calling the relative particle the electron and the center of mass the proton.

## 2. The Bohr model

We shall briefly review the results of the Bohr model which relate to the hydrogen atom. This model, which is based on the concept of a trajectory, is incompatible with the ideas of quantum mechanics. However, it allows us to introduce, in a very simple way, fundamental quantities such as the ionization energy $E_I$ of the hydrogen atom and a parameter which characterizes atomic dimensions (the Bohr radius $a_0$). In addition, it so happens that the energies $E_n$ given by the Bohr theory are the same as the eigenvalues of the Hamiltonian which we shall calculate in § 3. Finally, quantum mechanical theory is in agreement with some of the intuitive images of the Bohr model (*cf.* § 4-c-β below).

This semi-classical model is based on the hypothesis that the electron describes a *circular orbit* of radius $r$ about the proton obeying the following equations :

$$E = \frac{1}{2}\mu v^2 - \frac{e^2}{r} \tag{C-8}$$

$$\frac{\mu v^2}{r} = \frac{e^2}{r^2} \tag{C-9}$$

$$\mu v r = n\hbar \; ; \quad n \text{ a positive integer} \tag{C-10}$$

The first two equations are classical ones. (C-8) expresses the fact that the total energy $E$ of the electron is the sum of its kinetic energy $\mu v^2/2$ and its potential energy $- e^2/r$. (C-9) is none other than the fundamental equation of Newtonian dynamics ($e^2/r^2$ is the Coulomb force exerted on the electron, and $v^2/r$ is the acceleration of its uniform circular motion). The third equation expresses the quantization condition, introduced empirically by Bohr in order to explain the existence of

★ Henceforth, we shall omit the index $r$ which was used to label in § B the quantities corresponding to the relative motion.

discrete energy levels: he postulated that only circular orbits satisfying this condition are possible trajectories for the electron. The different orbits, as well as the corresponding values of the various physical quantities, are labeled by the integer $n$ which is associated with them.

A very simple algebraic calculation then yields the expressions for $E_n$, $r_n$ and $v_n$:

$$E_n = -\frac{1}{n^2} E_I \tag{C-11-a}$$

$$r_n = n^2 a_0 \tag{C-11-b}$$

$$v_n = \frac{1}{n} v_0 \tag{C-11-c}$$

with:

$$E_I = \frac{\mu e^4}{2\hbar^2} \tag{C-12-a}$$

$$a_0 = \frac{\hbar^2}{\mu e^2} \tag{C-12-b}$$

$$v_0 = \frac{e^2}{\hbar} \tag{C-12-c}$$

When this model was proposed by Bohr, it marked an important step towards the understanding of atomic phenomena, since it yielded the correct values for the energy levels of the hydrogen atom. These values indeed follow the $1/n^2$ (the Balmer formula) law indicated by expression (C-11-a). Moreover, the experimentally measured *ionization energy* (the energy which must be supplied to the hydrogen atom in its ground state in order to remove the electron) is equal to the numerical value of $E_I$:

$$E_I \simeq 13.6 \text{ eV} \tag{C-13}$$

Finally, the *Bohr radius* $a_0$ indeed characterizes atomic dimensions:

$$a_0 \simeq 0.52 \text{ Å} \tag{C-14}$$

COMMENT:

Complement $C_I$ shows how the uncertainty principle, applied to the hydrogen atom, explains the existence of a stable ground state and permits the evaluation of the order of magnitude of its energy and its spatial extension.

### 3. Quantum mechanical theory of the hydrogen atom

We shall now take up the question of the determination of the eigenvalues and eigenfunctions of the Hamiltonian $H$ which describes the relative motion of the proton and the electron in the center of mass frame [formula (C-6)]. In the $\{ |\mathbf{r}\rangle \}$ representation, the eigenvalue equation of the Hamiltonian $H$ is written:

$$\left[ -\frac{\hbar^2}{2\mu}\Delta - \frac{e^2}{r} \right] \varphi(\mathbf{r}) = E\, \varphi(\mathbf{r}) \tag{C-15}$$

Since the potential $-e^2/r$ is central, we can apply the results of § A : the eigenfunctions $\varphi(\mathbf{r})$ are of the form:

$$\varphi_{k,l,m}(\mathbf{r}) = \frac{1}{r}\, u_{k,l}(r)\, Y_l^m(\theta, \varphi) \tag{C-16}$$

$u_{k,l}(r)$ is given by the radial equation, (A-24), that is:

$$\left[ -\frac{\hbar^2}{2\mu}\frac{\mathrm{d}^2}{\mathrm{d}r^2} + \frac{l(l+1)\hbar^2}{2\mu\, r^2} - \frac{e^2}{r} \right] u_{k,l}(r) = E_{k,l}\, u_{k,l}(r) \tag{C-17}$$

We add to this equation condition (A-30):

$$u_{k,l}(0) = 0 \tag{C-18}$$

It can be shown that the spectrum of $H$ includes a discrete part (negative eigenvalues) and a continuous part (positive eigenvalues). Consider figure 3, which shows the effective potential for a given value of $l$ (the figure is drawn for $l \neq 0$, but the reasoning remains valid for $l = 0$).

For a positive value of $E$, the classical motion is not bounded in space: for the value $E > 0$ chosen in figure 3, it is limited on the left by the abscissa of point $A$, but it is not limited on the right*. As a result (*cf.* complement $M_{III}$) equation (C-17) has acceptable solutions for any $E > 0$. The spectrum of $H$ is therefore continuous for $E > 0$, and the corresponding eigenfunctions are not square-integrable.

On the other hand, for $E < 0$, the classical motion is bounded: it is confined to the region between the abscissas of the two points $B$ and $C$**. We shall see later that equation (C-17) has acceptable solutions only for certain discrete values of $E$. The spectrum of $H$ is therefore discrete for $E < 0$, and the corresponding eigenfunctions are square-integrable.

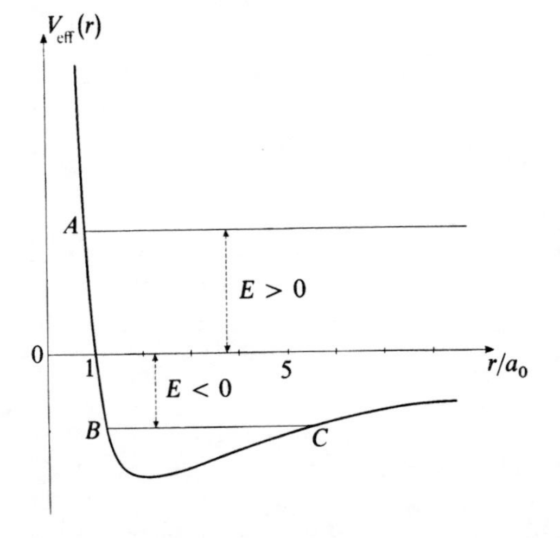


FIGURE 3

**For a positive value of the energy $E$, the classical motion is not bounded. The spectrum of the quantum mechanical Hamiltonian $H$ is therefore continuous for $E > 0$, and the corresponding eigenfunctions are not normalizable. On the other hand, for negative $E$, the classical motion is limited to the interval $BC$. The spectrum of $H$ is therefore discrete for $E < 0$, and the corresponding eigenfunctions are normalizable.**

* For a $-1/r$ potential, the classical trajectories are conic sections; unbounded motion follows a hyperbola or a parabola.

** The classical trajectory is then an ellipse or a circle.

### a. CHANGE OF VARIABLES

To simplify the reasoning, we shall choose $a_0$ and $E_I$ [formulas (C-12)] as the units of length and energy. That is, we shall introduce the dimensionless quantities:

$$\rho = r/a_0 \tag{C-19}$$

$$\lambda_{k,l} = \sqrt{-E_{k,l}/E_I} \tag{C-20}$$

(the quantity under the radical sign is positive, since we are looking for the bound states).

With expressions (C-12-a) and (C-12-b) for $E_I$ and $a_0$ taken into account, the radial equation (C-17) becomes simply:

$$\left[\frac{d^2}{d\rho^2} - \frac{l(l+1)}{\rho^2} + \frac{2}{\rho} - \lambda_{k,l}^2\right] u_{k,l}(\rho) = 0 \tag{C-21}$$

### b. SOLVING THE RADIAL EQUATION

In order to solve equation (C-21), we shall use the method illustrated in complement $C_V$, expanding $u_{k,l}(\rho)$ in a power series.

#### α. *Asymptotic behavior*

Let us determine the asymptotic behavior of $u_{k,l}(\rho)$ qualitatively. When $\rho$ approaches infinity, the terms in $1/\rho$ and $1/\rho^2$ become negligible compared to the constant term $\lambda_{k,l}^2$, so that equation (C-21) practically reduces to:

$$\left[\frac{d^2}{d\rho^2} - \lambda_{k,l}^2\right] u_{k,l}(\rho) = 0 \tag{C-22}$$

whose solutions are $e^{\pm\rho\lambda_{k,l}}$. This argument is not rigorous, since we have completely neglected the terms in $1/\rho$ and $1/\rho^2$; actually, it can be shown that $u_{k,l}(\rho)$ is equal to $e^{\pm\rho\lambda_{k,l}}$ multiplied by a power of $\rho$.

We shall later be led by physical considerations to require the function $u_{k,l}(\rho)$ to be bounded at infinity, and hence to reject the solutions of (C-21) whose asymptotic behavior is governed by $e^{+\rho\lambda_{k,l}}$. This is why we perform the change of function:

$$u_{k,l}(\rho) = e^{-\rho\lambda_{k,l}} y_{k,l}(\rho) \tag{C-23}$$

Although this change of function singles out $e^{-\rho\lambda_{k,l}}$, it clearly does not eliminate solutions in $e^{+\rho\lambda_{k,l}}$, which must be identified and then rejected at the end of the calculation. The differential equation which $y_{k,l}(\rho)$ must satisfy can easily be derived from (C-21):

$$\left\{\frac{d^2}{d\rho^2} - 2\lambda_{k,l}\frac{d}{d\rho} + \left[\frac{2}{\rho} - \frac{l(l+1)}{\rho^2}\right]\right\} y_{k,l}(\rho) = 0 \tag{C-24}$$

Condition (C-18) must be associated with this equation, that is:

$$y_{k,l}(0) = 0 \tag{C-25}$$

β. *Solutions in the form of power series*

Consider the expansion of $y_{k,l}(\rho)$ in powers of $\rho$:

$$y_{k,l}(\rho) = \rho^s \sum_{q=0}^{\infty} c_q \, \rho^q \tag{C-26}$$

By definition, $c_0$ is the first non-zero coefficient of this expansion:

$$c_0 \neq 0 \tag{C-27}$$

Condition (C-25) implies that $s$ is strictly positive.

We calculate $\frac{\mathrm{d}}{\mathrm{d}\rho} y_{k,l}(\rho)$ and $\frac{\mathrm{d}^2}{\mathrm{d}\rho^2} y_{k,l}(\rho)$ from (C-26):

$$\frac{\mathrm{d}}{\mathrm{d}\rho} y_{k,l}(\rho) = \sum_{q=0}^{\infty} (q + s) \, c_q \, \rho^{q+s-1} \tag{C-28-a}$$

$$\frac{\mathrm{d}^2}{\mathrm{d}\rho^2} y_{k,l}(\rho) = \sum_{q=0}^{\infty} (q + s)(q + s - 1) \, c_q \, \rho^{q+s-2} \tag{C-28-b}$$

To obtain the left-hand side of (C-24), we multiply expressions (C-26), (C-28-a) and (C-28-b) respectively by the factors $\left[\frac{2}{\rho} - \frac{l(l+1)}{\rho^2}\right]$, $-2\lambda_{k,l}$ and 1. According to (C-24), the series so determined must be identically zero, that is, all its coefficients must be zero.

The lowest order term is in $\rho^{s-2}$. Taking its coefficient as zero, we obtain:

$$[-l(l+1) + s(s-1)]c_0 = 0 \tag{C-29}$$

If we take (C-27) into account, we see that $s$ can take on one of two values:

$$\begin{cases} s = l + 1 & \text{(C-30-a)} \\ s = -l & \text{(C-30-b)} \end{cases}$$

(in agreement with the general result of § A-2-c). We have seen that only (C-30-a) gives a behavior at the origin which can lead to an acceptable solution [condition (C-25)]. Setting the coefficient of the general term in $\rho^{q+s-2}$ equal to zero, we obtain (with $s = l + 1$) the following recurrence relation :

$$q(q + 2l + 1) \, c_q = 2[(q + l)\lambda_{k,l} - 1]c_{q-1} \tag{C-31}$$

If we fix $c_0$, this relation enables us to calculate $c_1$, then $c_2$, and thus by recurrence all the coefficients $c_q$. Since $c_q/c_{q-1}$ approaches zero when $q \longrightarrow \infty$, the corresponding series is convergent for all $\rho$. Thus we have determined, for any value of $\lambda_{k,l}$, the solution of (C-24) which satisfies condition (C-25).

### c. ENERGY QUANTIZATION. RADIAL FUNCTIONS

We are now going to require the preceding solution to have a physically acceptable asymptotic behavior (*cf.* § b-α below). This will involve quantization of the possible values of $\lambda_{k,l}$.

If the term in brackets on the right-hand side of (C-31) does not go to zero for any integer $q$, expansion (C-26) is a true infinite series, for which:

$$\frac{c_q}{c_{q-1}} \underset{q \to \infty}{\sim} \frac{2\lambda_{k,l}}{q} \tag{C-32}$$

Now, the power series expansion of the function $e^{2\rho\lambda_{k,l}}$ is written:

$$\begin{cases} e^{2\rho\lambda_{k,l}} = \sum_{q=0}^{\infty} d_q \, \rho^q \\ d_q = \dfrac{(2\lambda_{k,l})^q}{q!} \end{cases} \tag{C-33}$$

which implies:

$$\frac{d_q}{d_{q-1}} = \frac{2\lambda_{k,l}}{q} \tag{C-34}$$

If we compare (C-32) and (C-34), it is easy to see* that, for large values of $\rho$, the series being considered behaves like $e^{2\rho\lambda_{k,l}}$. The corresponding function $u_{k,l}$ [formula (C-23)] is then proportional to $e^{+\rho\lambda_{k,l}}$, which is not physically acceptable.

Consequently, we must reject all cases in which expansion (C-26) is an infinite series. The only possible values of $\lambda_{k,l}$ are those for which (C-26) has only a finite number of terms, that is, those for which $y_{k,l}$ reduces to a polynomial. The corresponding function $u_{k,l}$ is then physically acceptable, since its asymptotic behavior is dominated by $e^{-\rho\lambda_{k,l}}$. Therefore, all we need is an integer $k$ such that the term in brackets of the right-hand side of (C-31) goes to zero for $q = k$: the corresponding coefficient $c_k$ is then zero, as are all those of higher order, since the fact that $c_k$ is zero means that $c_{k+1}$ is as well, and so on. For fixed $l$, we label the corresponding values of $\lambda_{k,l}$ by this integer $k$ (note that $k$ is greater than or equal to 1, since $c_0$ never goes to zero). We then have, according to (C-31):

$$\lambda_{k,l} = \frac{1}{k + l} \tag{C-35}$$

For a given $l$, the only negative energies possible are therefore [formula (C-20)] :

$$E_{k,l} = \frac{-E_I}{(k + l)^2} \quad ; \quad k = 1, 2, 3, \ldots \tag{C-36}$$

We shall discuss this result in § 4.

* The reader can find a fuller discussion relating to an analogous problem in complement $C_V$.

$y_{k,l}$ is therefore a polynomial, whose term of lowest order is in $\rho^{l+1}$ and whose term of highest order is in $\rho^{k+l}$. Its various coefficients can be calculated in terms of $c_0$ by solving recurrence relation (C-31), which can be written, using (C-35):

$$c_q = -\frac{2(k-q)}{q(q+2l+1)(k+l)}c_{q-1} \tag{C-37}$$

It is easy to show that:

$$c_q = (-1)^q\left(\frac{2}{k+l}\right)^q \frac{(k-1)!}{(k-q-1)!}\,\frac{(2l+1)!}{q!\,(q+2l+1)!}c_0 \tag{C-38}$$

$u_{k,l}(\rho)$ is then given by formula (C-23), and $c_0$ is determined (to within a phase factor) by normalization condition (A-34) [we must first, of course, return to the variable $r$ by using (C-19)]. Finally, we obtain the true function $R_{k,l}(r)$ by dividing $u_{k,l}(r)$ by $r$. The following three examples give an idea of the form of these radial functions:

$$R_{k=1,l=0}(r) = 2(a_0)^{-3/2}\,\mathrm{e}^{-r/a_0} \tag{C-39-a}$$

$$R_{k=2,l=0}(r) = 2(2a_0)^{-3/2}\left(1-\frac{r}{2a_0}\right)\mathrm{e}^{-r/2a_0} \tag{C-39-b}$$

$$R_{k=1,l=1}(r) = (2a_0)^{-3/2}\frac{1}{\sqrt{3}}\frac{r}{a_0}\,\mathrm{e}^{-r/2a_0} \tag{C-39-c}$$

## 4. Discussion of the results

### a. ORDER OF MAGNITUDE OF ATOMIC PARAMETERS

Formulas (C-36) and (C-39) show that, for the hydrogen atom, the ionization energy $E_I$, defined by (C-12-a), and the Bohr radius, given by (C-12-b), play an important role. These quantities give an order of magnitude of the energies and spatial extensions of the wave functions associated with the bound states of the hydrogen atom.

(C-12-a) and (C-12-b) can be written in the form:

$$E_I = \frac{1}{2}\alpha^2\mu c^2 \tag{C-40-a}$$

$$a_0 = \frac{1}{\alpha}\lambda\!\!\!^{-}_c \tag{C-40-b}$$

where $\alpha$ is the *fine structure constant*, which is a dimensionless constant which plays a very important role in physics:

$$\alpha = \frac{e^2}{\hbar c} = \frac{q^2}{4\pi\varepsilon_0\hbar c} \simeq \frac{1}{137} \tag{C-41}$$

and where $\lambda\!\!\!^{-}_c$ is defined by:

$$\lambda\!\!\!^{-}_c = \frac{\hbar}{\mu c} \tag{C-42}$$

Since $\mu$ is almost the same as $m_e$, the rest mass of the electron, $\lambda_c$ is practically equal to the *Compton wavelength* of the electron, which is given by :

$$\frac{\hbar}{m_e c} \simeq 3.8 \times 10^{-3} \text{Å} \tag{C-43}$$

Relation (C-40-b) therefore indicates that $a_0$ is on the order of one hundred times the Compton wavelength of the electron. Relation (C-40-a) shows that the order of magnitude of the binding energy of the electron is between $10^{-4}\mu c^2$ and $10^{-5}\mu c^2$, where $\mu c^2$ is practically equal to the rest energy of the electron :

$$m_e c^2 \simeq 0.51 \times 10^6 \text{ eV} \tag{C-44}$$

It follows that :

$$E_I \ll m_e c^2 \tag{C-45}$$

This justifies our choice of the non-relativistic Schrödinger equation to describe the hydrogen atom. Of course, relativistic effects, although small, do exist; nevertheless, their smallness allows them to be studied by perturbation theory (*cf.* chap. XI and XII).

### b. ENERGY LEVELS

#### α. *Possible values of the quantum numbers; degeneracies*

For fixed $l$, there exists an infinite number of possible energy values [formula (C-36)], corresponding to $k = 1, 2, 3, ...$ Each of them is at least $(2l + 1)$-fold degenerate : this is an *essential degeneracy* related to the fact that the

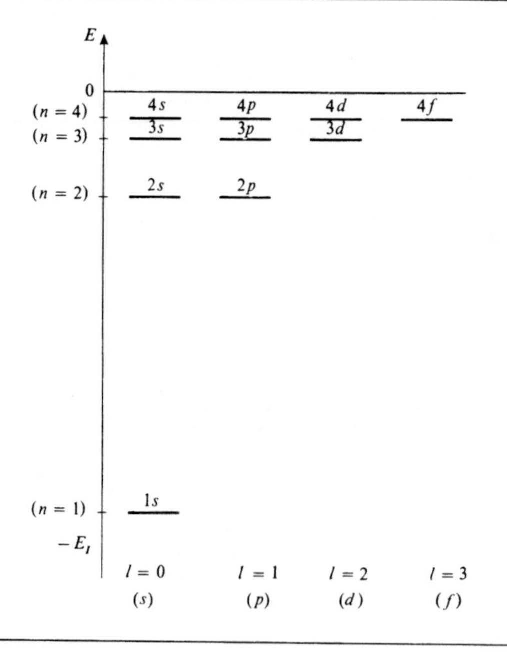


FIGURE 4

**Energy levels of the hydrogen atom. The energy $E_n$ of each level depends only on $n$. If $n$ is fixed, several values of $l$ are possible : $l = 0, 1, 2, ..., n - 1$. To each of these values of $l$ correspond $(2l + 1)$ possible values for $m$ :**

$$m = -l, -l + 1, ..., l.$$

**Consequently, the level $E_n$ is $n^2$-fold degenerate.**

radial equation depends only on the quantum number $l$ and not on $m$ (§ A-3). But, in addition, there exist *accidental degeneracies*: equation (C-36) indicates that two eigenvalues $E_{k,l}$ and $E_{k',l'}$ corresponding to different radial equations ($l' \neq l$) are equal if $k + l = k' + l'$. Figure 4, in which the first eigenvalues associated with $l = 0$, 1, 2 and 3 are shown on a common energy scale, clearly reveals several accidental degeneracies.

In the special case of the hydrogen atom, $E_{k,l}$ does not depend on $k$ and $l$ separately, but only on their sum. We set:

$$n = k + l \tag{C-46}$$

The various energy states are labeled by the integer $n$ (greater than or equal to 1), and (C-36) becomes:

$$E_n = -\frac{1}{n^2} E_I \tag{C-47}$$

According to (C-46), it is equivalent to specify $k$ and $l$ or $n$ and $l$ to determine the eigenfunctions. Following convention, from now on we shall use the quantum numbers $n$ and $l$. The energy is fixed by $n$, which is called the *principal quantum number*; a given value of $n$ characterizes what is called an *electron shell*.

Since $k$ is necessarily an integer which is greater than or equal to 1 (§ 3-c above), there is only a finite number of values of $l$ associated with the same value of $n$. According to (C-46), if $n$ is fixed, one can have:

$$l = 0, 1, 2, ..., n - 1 \tag{C-48}$$

The shell characterized by $n$ is said to contain $n$ *sub-shells**, each one corresponding to one of the values of $l$ given in (C-48). Finally, each sub-shell contains $(2l + 1)$ distinct states, associated with the $(2l + 1)$ possible values of $m$ for fixed $l$.

The total degeneracy of the energy level $E_n$ is therefore:

$$g_n = \sum_{l=0}^{n-1} (2l + 1) = 2\frac{(n - 1)n}{2} + n = n^2 \tag{C-49}$$

We shall see in chapter IX that the existence of electron spin multiplies this number by 2 (if we also take into account the proton spin, which is equal to that of the electron, we obtain another factor of 2).

#### β. *Spectroscopic notation*

For historical reasons (dating from the period, before the development of quantum mechanics, in which the study of spectra resulted in an empirical

* The concept of a sub-shell can even be found in the semi-classical model of Sommerfeld. This model assigns to each value $n$ of Bohr's quantum number $n$ elliptical orbits of the same energy and different angular momenta. One of these orbits is circular; it is the one which corresponds to the maximum value of the angular momentum.

classification of the numerous lines observed), letters of the alphabet are associated with the various values of $l$. The correspondence is as follows:

$$\begin{aligned} l = 0 &\longleftrightarrow s \\ l = 1 &\longleftrightarrow p \\ l = 2 &\longleftrightarrow d \\ l = 3 &\longleftrightarrow f \\ l = 4 &\longleftrightarrow g \\ \vdots \quad & \quad \vdots \end{aligned} \tag{C-50}$$

alphabetical order

Therefore, *spectroscopic notation* labels a sub-shell by the corresponding number $n$ followed by the letter which characterizes the value of $l$. Thus, the ground level [which is non-degenerate, according to (C-49)], sometimes called the "$K$ shell", includes only the $1s$ sub-shell; the first excited level, or "$L$ shell", includes the $2s$ and $2p$ sub-shells; the second excited level ("$M$ shell") includes the $3s$, $3p$ and $3d$ sub-shells, etc. (The capital letters sometimes associated with the successive shells follow an alphabetical order, starting with the letter $K$.)

### c. WAVE FUNCTIONS

The wave functions associated with the eigenstates common to $\mathbf{L}^2$, $L_z$ and the Hamiltonian $H$ of the hydrogen atom are generally labeled, not by the three quantum numbers $k$, $l$, $m$, as we have done until now, but by $n$, $l$ and $m$ [passage from one set to the other simply involves use of relation (C-46)]. Since the operators $H$, $\mathbf{L}^2$ and $L_z$ constitute a C.S.C.O. (*cf.* §A-3), specification of the three integers $n$, $l$ and $m$, which is equivalent to that of the eigenvalues of $H$, $\mathbf{L}^2$ and $L_z$, unambiguously determines the corresponding eigenfunction $\varphi_{n,l,m}(\mathbf{r})$.

#### α. *Angular dependence*

As is the case for any central potential, the functions $\varphi_{n,l,m}(\mathbf{r})$ are products of a radial function and a spherical harmonic $Y_l^m(\theta, \varphi)$. To visualize their angular dependence, we can measure off a distance on the axis characterized by the polar angles $\theta$ and $\varphi$ which is proportional to $|\varphi_{n,l,m}(r, \theta, \varphi)|^2$ for any *fixed* $r$, that is, proportional to $|Y_l^m(\theta, \varphi)|^2$. Thus, we obtain a surface of revolution about the $Oz$ axis, since we know that $Y_l^m(\theta, \varphi)$ depends on $\varphi$ only through the factor $e^{im\varphi}$ (§D-1-b of chapter VI); consequently, $|Y_l^m(\theta, \varphi)|^2$ is independent of $\varphi$. We can therefore represent its cross-section by a plane containing $Oz$. This is what is done in figure 5, for $m = 0$ and $l = 0$, 1 and 2 [the corresponding spherical harmonics are given in complement $A_{VI}$, formulas (31), (32) and (33)] : $Y_0^0$ is a constant, and is therefore spherically symmetric; $|Y_1^0|^2$ is proportional to $\cos^2\theta$, and $|Y_2^0|^2$, to $(3\cos^2\theta - 1)^2$.

#### β. *Radial dependence*

The radial functions $R_{n,l}(r)$, each of which characterizes a sub-shell, can be calculated from the results of §3-c, paying attention, however, to the change in

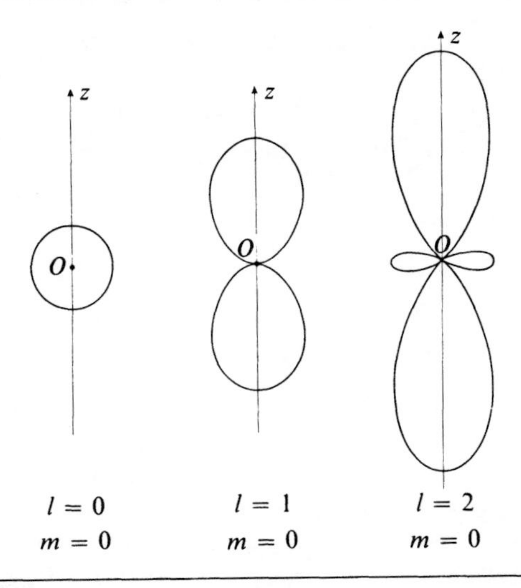


FIGURE 5

**Angular dependence, $Y_l^m(\theta, \varphi)$, of some stationary wave functions of the hydrogen atom, corresponding to well-defined values of $l$ and $m$. For each direction of polar angles $\theta$, $\varphi$, the value of $|Y_l^m(\theta, \varphi)|^2$ is recorded; a surface of revolution about the $Oz$ axis is thus obtained. When $l = 0$, this surface is a sphere centered at $O$; it becomes more complicated for higher values of $l$.**

notation introduced by formula (C-46). Figure 6 represents the variation with respect to $r$ of the three radial functions given in (C-39):

$$R_{k=1,l=0} \equiv R_{n=1,l=0} \ ; \ R_{k=2,l=0} \equiv R_{n=2,l=0} \ ; \ R_{k=1,l=1} \equiv R_{n=2,l=1} \qquad \text{(C-51)}$$

The behavior of $R_{n,l}(r)$ in the neighborhood of $r = 0$ is that of $r^l$ (see discussion of § A-2-c). Consequently, *only states belonging to s sub-shells* ($l = 0$) *give a non-zero probability of presence at the origin.* The greater $l$, the larger the region around the proton in which the position probability of the electron is negligible. This has a certain number of physical consequences, particularly in the phenomenon of electron capture by certain nuclei, and in the hyperfine structure of lines (*cf.* chap. XII, § B-2).

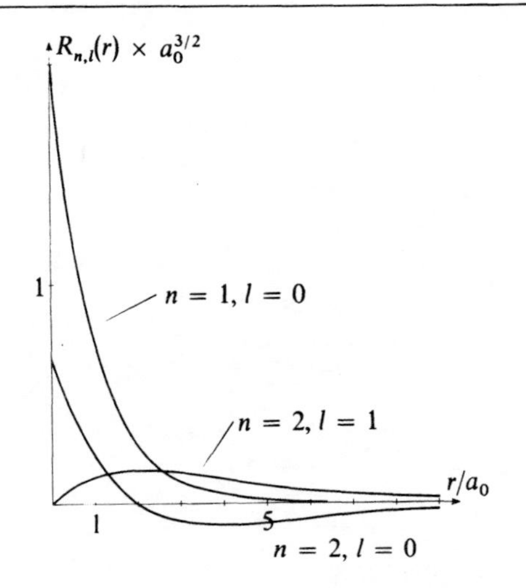


FIGURE 6

**Radial dependence $R_{n,l}(r)$ of wave functions associated with the first few levels of the hydrogen atom. When $r \longrightarrow 0$, $R_{n,l}(r)$ behaves like $r^l$; only the $s$ states (for which $l = 0$) have a non-zero position probability at the origin.**

Finally, we can derive formula (C-11-b) for the successive Bohr radii. To do so, consider the various states for which $l = n - 1$★. We calculate the variation with $r$ of the probability density for each of the preceding levels in an infinitesimal solid angle $\mathrm{d}\Omega$ about a fixed direction of polar angles $\theta$ and $\varphi$. In general, the position probability for the electron in the volume element $d^3r = r^2\,\mathrm{d}r\,\mathrm{d}\Omega$ situated at the point $(r, \theta, \varphi)$ is given by:

$$\begin{aligned} d^3\mathscr{P}_{n,l,m}(r, \theta, \varphi) &= |\varphi_{n,l,m}(r, \theta, \varphi)|^2\, r^2\,\mathrm{d}r\,\mathrm{d}\Omega \\ &= |R_{n,l}(r)|^2\, r^2\,\mathrm{d}r \times |Y_l^m(\theta, \varphi)|^2\,\mathrm{d}\Omega \end{aligned} \tag{C-52}$$

Here, we have fixed $\theta$, $\varphi$ and $\mathrm{d}\Omega$. The probability of finding the electron between $r$ and $r + \mathrm{d}r$, inside the solid angle under consideration, is then proportional to $r^2\,|R_{n,l}(r)|^2\,\mathrm{d}r$. The corresponding density is therefore, to within a constant factor, $r^2\,|R_{n,l}(r)|^2$ (the factor $r^2$ arises from the expression for the volume element in spherical coordinates). The cases which interest us here are those for which $l = n - 1$, that is, $k = n - l = 1$; § 3-c then indicates that the polynomial which enters into $R_{n,l}(r)$ contains only one term in $(r/a_0)^{n-1}$. The desired probability density is therefore proportional to:

$$\begin{aligned} f_n(r) &= \frac{r^2}{a_0^2}\left[\left(\frac{r}{a_0}\right)^{n-1} \mathrm{e}^{-r/na_0}\right]^2 \\ &= \left(\frac{r}{a_0}\right)^{2n} \mathrm{e}^{-2r/na_0} \end{aligned} \tag{C-53}$$

This function has a maximum for:

$$r = r_n = n^2 a_0 \tag{C-54}$$

which is the radius of the Bohr orbit corresponding to the energy $E_n$.

Finally, the table below gives the expressions for the wave functions of the first energy levels:

| | |
|---|---|
| 1s level | $\varphi_{n=1,l=0,m=0} = \dfrac{1}{\sqrt{\pi a_0^3}}\,\mathrm{e}^{-r/a_0}$ |
| 2s level | $\varphi_{n=2,l=0,m=0} = \dfrac{1}{\sqrt{8\pi a_0^3}}\left(1 - \dfrac{r}{2a_0}\right)\mathrm{e}^{-r/2a_0}$ |
| 2p level | $\varphi_{n=2,l=1,m=1} = -\dfrac{1}{8\sqrt{\pi a_0^3}}\dfrac{r}{a_0}\,\mathrm{e}^{-r/2a_0}\sin\theta\,\mathrm{e}^{i\varphi}$ <br> $\varphi_{n=2,l=1,m=0} = \dfrac{1}{4\sqrt{2\pi a_0^3}}\dfrac{r}{a_0}\,\mathrm{e}^{-r/2a_0}\cos\theta$ <br> $\varphi_{n=2,l=1,m=-1} = \dfrac{1}{8\sqrt{\pi a_0^3}}\dfrac{r}{a_0}\,\mathrm{e}^{-r/2a_0}\sin\theta\,\mathrm{e}^{-i\varphi}$ |

★ These states correspond to the circular orbits of the Sommerfeld theory (*cf.* note on page 799).

**References and suggestions for further reading:**

Particle in a central potential: Messiah (1.17), chap. IX; Schiff (1.18), §16.

The Bohr-Sommerfeld atom and the old quantum therory: Cagnac and Pebay-Peyroula (11.2), chaps. V, VI and XIII; Born (11.4), chap. V, §§1 and 2; Pauling and Wilson (1.9), chap. II; Tomonaga (1.8), Vol. I; Eisberg and Resnick (1.3), chap. 4.

Hydrogen-like wave functions: Levine (12.3), §6.5; Karplus and Porter (12.1), §§3.8 and 3.10; Eisberg and Resnick (1.3), §§7.6 and 7.7.

Degeneracy related to a $1/r$ potential (dynamical group): Borowitz (1.7), §13.7; Schiff (1.18), §30; Bacry (10.31), §6.11.

Mathematical treatment of differential equations: Morse and Feshbach (10.13), chaps. 5 and 6; Courant and Hilbert (10.11), vol. I, §V-11.

COMPLEMENTS OF CHAPTER VII

$A_{VII}$: HYDROGEN-LIKE SYSTEMS

$A_{VII}$: presentation of various hydrogen-like systems to which the calculations of chapter VII can be applied directly. The accent is placed on physical discussions and on the influence of the masses of the particles involved in the system. Simple, advised for a first reading.

---

$B_{VII}$: A SOLUBLE EXAMPLE OF A CENTRAL POTENTIAL: THE ISOTROPIC THREE-DIMENSIONAL HARMONIC OSCILLATOR

$B_{VII}$: study of another case (a three-dimensional harmonic oscillator) in which it is possible to calculate exactly the energy levels of a particle in a central potential by the method of chapter VII (solution of the radial equation). Not theoretically difficult; may be considered as a worked example.

---

$C_{VII}$: PROBABILITY CURRENTS ASSOCIATED WITH THE STATIONARY STATES OF THE HYDROGEN ATOM

$C_{VII}$: completes the results of §C-4-c of chapter VII concerning the properties of the stationary states of the hydrogen atom by calculating their probability currents. Short and simple, useful for complement $D_{VII}$.

---

$D_{VII}$: THE HYDROGEN ATOM PLACED IN A UNIFORM MAGNETIC FIELD. PARAMAGNETISM AND DIAMAGNETISM. THE ZEEMAN EFFECT

$E_{VII}$: STUDY OF SOME ATOMIC ORBITALS. HYBRID ORBITALS

$F_{VII}$: VIBRATIONAL-ROTATIONAL LEVELS OF DIATOMIC MOLECULES

$D_{VII}$, $E_{VII}$, $F_{VII}$: discussion of a certain number of physical phenomena, using the results of chapter VII.

$D_{VII}$: the properties of an atom in a magnetic field (diamagnetism, paramagnetism, Zeeman effect). Moderately difficult, important because of its numerous applications.

$E_{VII}$: complement intended to introduce the concept of an atomic hybrid orbital, essential for the understanding of certain properties of the chemical bond. No theoretical difficulty. Stresses the geometrical aspect of wave functions.

$F_{VII}$: direct application of the theory of chapter VII to the study of the vibrational-rotational spectrum of heteropolar diatomic molecules. Sequel to complements $A_V$ (§1) and $C_{VI}$; moderately difficult.

---

$G_{VII}$: EXERCISES

$G_{VII}$: exercise 2 studies the influence of a uniform magnetic field on the levels of a simple physical system, in an exactly soluble case. It thus provides a concrete illustration of the general considerations of complements $C_{VII}$ and $D_{VII}$ concerning the influence of the paramagnetic and diamagnetic terms of the Hamiltonian.

## Complement $A_{VII}$

# HYDROGEN-LIKE SYSTEMS



The calculations of chapter VII, which enabled us to determine numerous physical properties of the hydrogen atom (energy levels, spatial distribution of the wave functions, etc.) are based on the fact that the system under study is composed of two particles (an electron and a proton) whose mutual attraction energy is inversely proportional to the distance between them. There exist numerous other systems in physics which fulfil this condition: deuterium or tritium, muonium, positronium, muonic atoms, etc. The results obtained in chapter VII are therefore directly applicable to these examples. All we need to do is change the constants introduced in the calculations (masses and charges of the two particles). This is what we shall do in this complement, in which we shall study, in particular, how the Bohr radius and ionization energy $E_I$ are modified in each of the systems to be considered. The wave functions associated with their stationary states and the corresponding energies will then be obtained by replacing, in formulas (C-39) and (C-47) of chapter VII, $a_0$ and $E_I$ by their new values, which give the order of magnitude of the spatial extension of the wave functions and the binding energies of these systems.

We recall the expressions for $a_0$ and $E_I$:

$$a_0 = \lambda_c \frac{1}{\alpha} = \frac{\hbar^2}{\mu e^2} \tag{1}$$

$$E_I = \frac{1}{2}\mu c^2 \alpha^2 = \frac{\mu e^4}{2\hbar^2} \tag{2}$$

where $\mu$ is the reduced mass of the electron-proton system:

$$\mu = \mu(\text{H}) = \frac{m_e m_p}{m_e + m_p} \simeq m_e\left(1 - \frac{m_e}{m_p}\right) \tag{3}$$

and $e^2$ characterizes the intensity of the attractive potential $V(r)$:

$$V(r) = -\frac{e^2}{|\mathbf{r}_1 - \mathbf{r}_2|} \tag{4}$$

In the case of hydrogen, we have seen that:

$$a_0(\mathrm{H}) \simeq 0.52\ \text{Å} \tag{5-a}$$

$$E_I(\mathrm{H}) = 13.6\ \text{eV} \simeq 2.2\ 10^{-18}\ \text{J} \tag{5-b}$$

How can we obtain the corresponding values for a system of two arbitrary particles, of masses $m_1$ and $m_2$, whose attraction energy is:

$$V'(r) = -\frac{Ze^2}{|\mathbf{r}_1 - \mathbf{r}_2|} \tag{6}$$

(where $Z$ is a dimensionless parameter)? All we must do is calculate the reduced mass $\mu$ of the system by replacing $m_e$ and $m_p$ by $m_1$ and $m_2$ in (3):

$$\mu = \frac{m_1 m_2}{m_1 + m_2} \tag{7-a}$$

and substitute the result obtained into (1) and (2), being careful to perform the substitution:

$$e^2 \Longrightarrow Z\,e^2 \tag{7-b}$$

This is what we shall do in a certain number of physical examples.

## 1. Hydrogen-like systems with one electron

### a. ELECTRICALLY NEUTRAL SYSTEMS

#### α. *Heavy isotopes of hydrogen*

The physical systems closest to the hydrogen atom are its two isotopes, deuterium and tritium. In these atoms, the proton is replaced by a nucleus having the same charge but possessing either one or two neutrons in addition to the proton. The mass of the deuterium nucleus is approximately $2m_p$, and that of the tritium nucleus, $3m_p$. The reduced masses therefore become, in these two cases:

$$\mu(\text{deuterium}) \simeq m_e\left(1 - \frac{m_e}{2m_p}\right) \tag{8-a}$$

$$\mu(\text{tritium}) \simeq m_e\left(1 - \frac{m_e}{3m_p}\right) \tag{8-b}$$

Since:

$$\frac{m_e}{m_p} \simeq \frac{1}{1\,836} \ll 1 \tag{9}$$

it is clear that the reduced masses of deuterium and tritium are very close to that of hydrogen, and that they can be replaced by $m_e$ without great inaccuracy.

If we substitute either (3), or (8-a), or (8-b) into formulas (1) and (2), we see that the Bohr radii and energies of the hydrogen, deuterium and tritium atoms are practically the same. Nevertheless, there are slight differences, of the order of a thousandth in relative value. These differences can be detected experimentally. For example, with an optical spectrograph of sufficient resolution, it can be

observed that the wavelengths of the lines emitted by hydrogen atoms are slightly greater than those emitted by deuterium, which are in turn greater than those emitted by tritium. This slight shift in the emitted wavelengths is related to the fact that the nucleus is not infinitely heavy, and does not remain fixed while the electron moves; this is called the "nuclear finite mass effect". Experiments have verified that formulas (7-a), (1) and (2) account very precisely for this effect.

### β. *Muonium*

The muon is a particle whose fundamental properties are the same as those of the electron, except for a difference in mass (the mass $m_\mu$ of the muon is equal to 207 $m_e$). In particular, the muon is not sensitive to nuclear forces (strong interactions). There are two types of muons, the $\mu^-$ and the $\mu^+$, whose charges are respectively equal to those of the electron $e^-$ and the positron $e^+$*. Like all other charged particles, the muon is sensitive to electromagnetic interactions.

We can therefore consider a physical system formed by a $\mu^+$ muon and an electron $e^-$, in which the electrostatic attraction is the same as for a proton and an electron. Bound states therefore exist. This is, so to speak, a light isotope of hydrogen, in which the $\mu^+$ muon replaces the proton. This "isotope" is called muonium (its atomic mass is on the order of $m_\mu/m_p \simeq 0.1$).

It is not difficult to use the results in chapter VII to find the ionization energy and Bohr radius associated with muonium; (1), (2) and (7) yield:

$$a_0(\text{muonium}) = a_0(\text{H})\frac{1 + m_e/m_\mu}{1 + m_e/m_p} \simeq a_0(\text{H})\left(1 + \frac{1}{200}\right) \tag{10-a}$$

$$E_I(\text{muonium}) = E_I(\text{H})\frac{1 + m_e/m_p}{1 + m_e/m_\mu} \simeq E_I(\text{H})\left(1 - \frac{1}{200}\right) \tag{10-b}$$

Since the muon is approximately ten times lighter than the proton, the nuclear finite mass effect is about ten times greater for muonium than for hydrogen; however, since the electron is distinctly lighter than the muon, this effect remains small (on the order of 0.5 %). For example, the wavelengths of the optical lines emitted by muonium should be close to those of the corresponding lines for hydrogen. Actually, the optical emission spectrum of muonium has not yet been observed.

Experimentally, the existence of muonium was revealed by its instability: the $\mu^+$ muon decays, emitting a positron and two neutrinos, and the lifetime of muonium is $2.2 \times 10^{-6}$ seconds. The positron resulting from this decay can be detected. It is emitted preferentially in the direction of the $\mu^+$ muon spin** (non-conservation of parity in weak interactions). Detection of the positrons then leads to the experimental determination of this direction. Since, in addition, the spin of the $\mu^+$ muon of a muonium atom is coupled to that of the electron (hyperfine structure coupling; *cf.* chap. XII and complements), its precession frequency in a magnetic field is different from that of a free muon. Measurement of this frequency thus reveals the existence of muonium atoms.

* In addition, like $e^-$ and $e^+$, $\mu^-$ and $\mu^+$ are antiparticles of each other.

** Like the electron, the muon has a spin of $\frac{1}{2}$ with which is associated a magnetic moment $\mathbf{M}_\mu = \frac{q_\mu}{m_\mu}\mathbf{S}$.

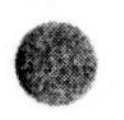



The study of muonium is of very great interest, theoretically as well as experimentally. The two particles which constitute this system are not subject to strong interactions, so that its energy levels (in particular, the hyperfine structure of the ground state $1s$) can be calculated with great precision, without bringing in any "nuclear correction" (for the hydrogen atom, on the other hand, one must take into account the internal structure and polarizability of the proton, which are due to strong interactions). Comparison between theoretical predictions and experimental results provides a very severe test of quantum electrodynamics. Recent measurement of the hyperfine structure of muonium led to one of the best determinations to date of the fine structure constant $\alpha = e^2/\hbar c$.

γ. *Positronium*

Positronium is a bound system composed of an electron $e^-$ and a positron $e^+$. Like muonium, it can be said by extension to be an isotope of hydrogen, with the proton being replaced by a positron. However, it must be noted that the situation is not quite the same: in the hydrogen atom, the proton (which is much heavier than the electron) remains almost motionless, while in positronium, the positron, the antiparticle of the electron, has the same mass and consequently the same velocity as the electron when the center of mass of positronium is fixed (*cf.* fig. 1-b).

According to (7-a), the reduced mass associated with positronium is:

$$\mu(\text{positronium}) = \frac{m_e}{2} \tag{11}$$

Therefore:

$$a_0\ (\text{positronium}) \simeq 2\, a_0(\text{H}) \tag{12-a}$$

$$\text{E}_I(\text{positronium}) \simeq \frac{1}{2} E_I(\text{H}) \tag{12-b}$$

Thus, for a given state of positronium, the average electron-positron distance is twice the electron-proton distance for the corresponding state of the hydrogen atom (*cf.* fig. 1). The differences between the energies of the stationary states, however, are twice as small, and the optical line spectrum emitted by positronium is obtained by doubling all the wavelengths of that of hydrogen.

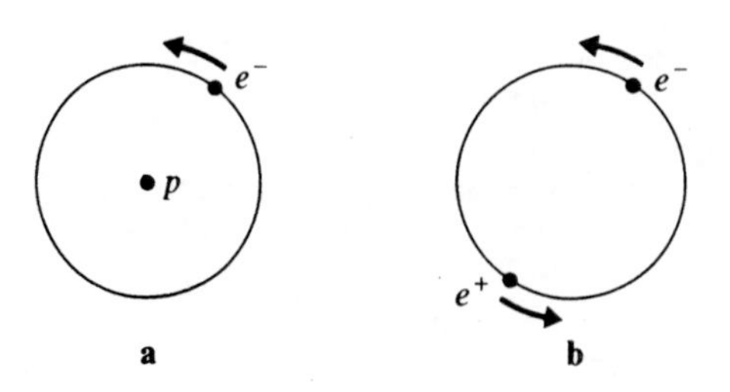


FIGURE 1

**Schematic representation of the hydrogen (electron + proton system) and positronium (electron + positron system) atoms. Since the proton is much heavier than the electron, it is located practically at the center of mass of the hydrogen atom; the electron "revolves" about the proton at a distance $a_0(H)$. On the other hand, the positron is equal in mass to the electron; these particles therefore both revolve about their center of mass, the distance between them being $a_0$ (positronium) = $2a_0(H)$.**

COMMENT:

One should not conclude from formula (12-a) that the radius of positronium is twice that of the hydrogen atom. The Bohr radius gives an idea of the extension of the wave functions associated with the "relative particle" (*cf.* chap. VII, § B), whose position $\mathbf{r}_1 - \mathbf{r}_2$ is related to the distance between the two particles and not to the distance between them and the center of mass $G$. Figure 1 clearly shows, moreover, that the hydrogen and positronium atoms are of equal size. In general, all hydrogen-like systems for which the attractive potential is given by (6) with $Z = 1$ have exactly the same radius, since formula (B-5) of chapter VII shows that:

$$\mathbf{r}_1 - \mathbf{r}_G = \frac{m_2}{m_1 + m_2}\mathbf{r} = \frac{\mu}{m_1}\mathbf{r} \tag{13}$$

Using (1), which gives the order of magnitude of the spatial extension of the wave function $\varphi_{100}(\mathbf{r})$ of the ground state, we see that the "radius" $\rho$ of the atom can be defined by:

$$\rho = \frac{\hbar^2}{m_1 Z e^2} \tag{14}$$

where $m_1$ is the mass of the lighter particle (the heavier particle is found closer to the center of mass). In all the systems considered until now, $Z = 1$ and $m_1 = m_e$; their radii are therefore the same. We shall see cases later in which the radius $\rho$ is smaller, either because $m_1 \neq m_e$ or because $Z \neq 1$.

The optical spectrum of positronium has only been observed very recently [*cf.* ref. (11-23)]. As for the (hyperfine) structure of the ground state (due to the interaction between the magnetic moments of the electron and the positron), it has been accurately determined (*cf.* complement $C_{XII}$).

The fact that positronium, like muonium, is a purely electrodynamic system (neither the electron nor the positron is sensitive to strong interactions) explains the importance attached to its theoretical and experimental study.

Let us also point out that positronium is an unstable system. Since the ground state is a $1s$ state, the electron and positron come into contact and annihilate, yielding two or three photons, depending on the hyperfine structure level they started from. The study of the corresponding decay rate is also of great interest in quantum electrodynamics.

### δ. *Hydrogen-like systems in solid state physics*

Atomic physics is not the only domain of application of the theory presented in chapter VII. For example, the "donor atoms" localized in semiconductors constitute approximately hydrogen-like systems in solid state physics. Consider a silicon crystal; in the silicon lattice, each atom uses its four valence electrons to form four tetrahedral bonds with its neighbors. If a pentavalent atom like phosphorus (a donor atom) is introduced into the lattice in place of a silicon atom, it must lose a valence electron, and its overall charge becomes positive. It then behaves like a center which can retain the electron and form a hydrogen-

like system with it. Actually, the force acting on the electron cannot be calculated directly from Coulomb's law in a vacuum since silicon has a large dielectric constant $\varepsilon \simeq 12$, so that (4) must be replaced by:

$$V(r) = - \frac{e^2}{\varepsilon \left| \mathbf{r}_1 - \mathbf{r}_2 \right|} \tag{15}$$

To be completely rigorous, we should have to replace the electron mass by the "effective mass" $m^*$ of the electron in silicon, which is different from the free electron mass because of interactions with the charges of the nuclei in the crystal. Nevertheless, we shall confine ourselves to a qualitative discussion, noting that the effect of the high value of $\varepsilon$ is to decrease $e^2$ in (15), that is, according to (1), to increase the Bohr radius by a factor on the order of 10. The donor atom impurity is therefore similar to a very large hydrogen atom, whose wave functions are spread over distances much greater than the unit cell length of the silicon lattice.

Let us briefly describe another hydrogen-like system in solid state physics: the exciton. Consider a semi-conductor crystal. In the absence of external perturbations, the outside electrons of the atoms forming the crystal are all in states belonging to the "valence band" (the temperature is assumed to be sufficiently low; *cf.* complement $C_{XIV}$). By suitably illuminating the crystal, we can, through the absorption of a photon, cause an electron to go into the "conduction band" (which contains a set of energy levels which are higher than those of the valence band). There is then one electron missing from the valence band. We can treat this band as if it contained a particle of charge opposite to that of an electron, called a "hole". The hole can attract an electron of the valence band and form a bound system with it : the exciton. The exciton, like hydrogen, has a series of energy levels between which it can undergo transitions. Its existence can be demonstrated by a measurement of light absorption by the crystal.

#### b. HYDROGEN-LIKE IONS

The neutral helium atom is composed of two electrons and a positively charged nucleus of charge $-2q_e$. Such a system, which consists of three particles, cannot be studied with the theory of chapter VII. However, if an electron is somehow removed from the helium atom, the $He^+$ ion is similar to a hydrogen atom; the only differences are in the nuclear charge, which is twice that of the proton (the total charge of the ion is positive and equal to $-q_e$) and its mass (which, for $^4He$, is approximately four times that of the proton). Of course, there are other hydrogen-like ions : the $Li^{++}$ ion (the lithium atom, when it is not ionized, has $Z = 3$ electrons), the $Be^{+++}$ ion ($Z = 4$), etc...

Let us then consider a system formed by a nucleus, of mass $M$ and positive charge $-Zq_e$, and an electron. If we substitute (7-b) in (1) and (2), we obtain:

$$a_0(Z) \simeq \frac{a_0(H)}{Z} \tag{16}$$

$$E_I(Z) \simeq Z^2 E_I(H) \tag{17}$$

(since $M \gg m_e$, we have neglected the difference between the reduced mass of hydrogen and that of the hydrogen-like ion under study; the consequences of the nuclear finite mass effect on $a_0$ and $E_I$ are, in effect, negligible compared to those due to the charge variation). Hydrogen-like ions are therefore all smaller than the hydrogen atom, as one would expect since the nucleus and the electron are more strongly bound. Moreover, their energy increases rapidly with $Z$ (quadratically): for example, the energy which must be supplied to a $Li^{++}$ ion to remove its last electron is greater than 100 eV. This is why the electromagnetic frequencies which can be emitted or absorbed by a hydrogen-like ion fall into the ultraviolet, and even, when $Z$ is large enough, into the domain of X-rays.

## 2. Hydrogen-like systems without an electron

Thus far, the systems we have considered all include an electron. Nevertheless, there exist numerous other particles having the same charge $q_e$, able to form a hydrogen-like system with a nucleus of charge $-Zq_e$. We shall give a few examples. The "atoms" which we are going to describe here are of course less common than the "usual" atoms which appear in Mendeleev's classification. They are unstable, and, in order to study them, it is necessary to use high-energy particle accelerators to produce the particles needed for their formation. This is why they are sometimes called "exotic atoms".

### a. MUONIC ATOMS

We have already mentioned some essential features of the muon and pointed out the existence of the $\mu^-$ muon. When this particle is attracted by a positively charged atomic nucleus, it can form a bound system with it which is called a "muonic atom"★.

Consider, for example, the simplest muonic atom, which is composed of a $\mu^-$ muon and a proton; this is a neutral system whose Bohr radius is:

$$a_0(\mu^-, p^+) \simeq \frac{\hbar^2}{m_\mu e^2} \simeq \frac{a_0(H)}{200} \tag{18}$$

and whose ionization energy is:

$$E_I(\mu^-, p^+) \simeq \frac{m_\mu e^4}{2\hbar^2} \simeq 200\, E_I(H) \tag{19}$$

The size of this muonic atom is therefore on the order of several thousandths of an Angström. Its spectrum is obtained from that of hydrogen by dividing the wavelengths by 200; it therefore falls into the domain of soft X-rays.

★ We could also imagine a bound system composed of a $\mu^+$ muon and a $\mu$  muon. However, given the low intensity of muon beams which we know how to produce, such an atom is difficult to create and has never been observed.

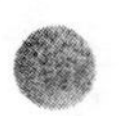



What happens if, instead of revolving about a proton, the $\mu^-$ is captured by a nucleus $N$ whose charge is $Z$ times greater, like lead, for example, for which $Z = 82$*? Formulas (1) and (2) then yield:

$$a_0(\mu^-, N) \simeq \frac{a_0(H)}{200\,Z} \tag{20}$$

$$E_I(\mu^-, N) \simeq 200\,Z^2 E_I(H) \tag{21}$$

Setting $Z = 82$ in these formulas, we find for the transitions of the lead muonic atom energies equal to several MeV (1 MeV = $10^6$ electron volts). However, it must be noted that formulas (1) and (2) are no longer valid in this case, since equation (20) would yield:

$$a_0(\mu^-, \text{Pb}) \simeq 3 \times 10^{-5}\,\text{Å} = 3\ \text{Fermi} \tag{22}$$

that is, a radius slightly smaller than the radius of the lead nucleus. The calculations of chapter VII are therefore no longer valid. This is because they are based on form (6) of the potential $V(r)$, which is correct** only when the particles under study, separated by distances much greater than their dimensions, can be considered to be point particles. This hypothesis, very well satisfied for hydrogen, is not valid in the case studied here.

However, (20) and (21) give the correct order of magnitude of the energies and radius of the lead muonic atom. The physical consequences of the existence of a non-zero spatial extension of the nucleus ("volume effect") will be studied in greater detail in complement $D_{XI}$. We take this occasion to note, however, that one of the reasons for interest in muonic atoms is precisely related to this type of effect: the $\mu^-$ muon "explores", as it were, the internal structure of the nucleus***, and the energy levels of muonic atoms depend on the electrical charge distribution and magnetism inside the nucleus (recall that the muon is not sensitive to nuclear forces). Thus, the study of these states can furnish information which is very useful in nuclear physics.

#### b. HADRONIC ATOMS

"Hadrons" are those particles which are subject to strong interactions, as opposed to "leptons", which are not. Electrons and muons, whose bound states in a Coulomb potential we have studied thus far, are leptons. Protons, neutrons and mesons such as the $\pi$ meson, etc... are hadrons. Of the latter, those which are negatively charged can form a hydrogen-like bound system with an atomic nucleus,

* Such a system can be formed by directing a $\mu^-$ beam onto a target of lead atoms. When a $\mu$ is captured by a lead nucleus, it revolves about it at a distance which is about 200 times smaller than the distance to the electrons of the innermost shell of the atom. The nuclear charge is therefore practically the only one to affect the muon. Thus, in studying the states of the muonic atom, we can simply ignore the electrons.

** Inside the nucleus, the potential is approximately parabolic (*cf.* complements $A_V$, § 4 and $D_{XI}$).

*** The concept of the impenetrability of two solid bodies is macroscopic. In quantum mechanics, nothing prevents the wave functions of two particles of different nature from overlapping.

called a "hadronic atom". For example, the nucleus-$\pi^-$ meson system yields a "pionic atom"; the nucleus-$\Sigma^-$ particle system, a "sigmaonic atom"*; the nucleus-$K^-$ meson system, a "kaonic atom"; the nucleus-antiproton system, an "antiprotonic atom", etc... All of the systems just cited have actually been observed and studied. They are all unstable, but their lifetimes are long enough for us to observe some of their spectral lines. Hydrogen atom theory, which takes into account only the electrostatic interaction between the two particles under consideration, does not, of course, apply to such systems, in which strong interactions play an important role. However, since they are of very short range, strong interactions can be neglected in the study of excited states of the hadronic atom (other than the $s$ states), in which the two particles are far apart. The theory of chapter VII is then applicable, as are formulas (1) and (2), which, in all these cases, lead to much smaller Bohr radii and much greater energies than for hydrogen. Thus, measurement of the spectral frequencies emitted by pionic atoms gives a very precise determination of the mass of the $\pi^-$ meson.

**References and suggestions for further reading:**

Exotic atoms: see the subsection "Exotic atoms" of section 11 of the bibliography; see also Cagnac and Pebay-Peyroula (11.2), chap. XIX, §7; Weissenberg (16.19), chap. 4, §2 and chap. 6.

Excitons: Kittel (13.2), chap. 17, p. 538; Ziman (13.3), §6.7.

* The term "mesic atom" is sometimes used to denote a system involving a meson. Similarly, since the $\Sigma^-$ is a hyperon (a particle heavier than the proton), the sigmaonic atom is sometimes called a "hyperonic atom".

# Complement $B_{VII}$

## A SOLUBLE EXAMPLE OF A CENTRAL POTENTIAL: THE ISOTROPIC THREE-DIMENSIONAL HARMONIC OSCILLATOR



In this complement we shall examine a special case of a central potential for which the radial equation is exactly soluble: the isotropic three-dimensional harmonic oscillator. We have already treated this problem (complement $E_V$) by considering the state space $\mathscr{E}_\mathbf{r}$ as the tensor product $\mathscr{E}_x \otimes \mathscr{E}_y \otimes \mathscr{E}_z$; this amounts, in the $\{ | \mathbf{r} \rangle \}$ representation, to separating the variables in Cartesian coordinates. We thus obtained three differential equations, one in the $x$-variable, one in $y$, and the third in $z$. Here we intend to seek the stationary states which are also eigenstates of $\mathbf{L}^2$ and $L_z$ by separating the variables in polar coordinates. We shall then indicate how the two bases of $\mathscr{E}_\mathbf{r}$ obtained by these two different methods are related to each other.

We shall also study, in complement $A_{VIII}$, the stationary states of well-defined angular momentum of a free particle. This can be considered to be another special case of a central potential $[V(r) \equiv 0]$ which leads to an exactly soluble radial equation.

A three-dimensional harmonic oscillator is composed of a (spinless) particle of mass $\mu$ subjected to the potential:

$$V(x, y, z) = \frac{1}{2} \mu [\omega_x^2 x^2 + \omega_y^2 y^2 + \omega_z^2 z^2] \tag{1}$$

where $\omega_x$, $\omega_y$ and $\omega_z$ are real positive constants. The oscillator is said to be isotropic if:

$$\omega_x = \omega_y = \omega_z = \omega \tag{2}$$

Since the potential (1) is the sum of a function of $x$ alone, a function of $y$ alone and a function of $z$ alone, we can solve the eigenvalue equation of the Hamiltonian:

$$H = \frac{\mathbf{P}^2}{2\mu} + V(\mathbf{R}) \tag{3}$$

by separating the variables $x$, $y$ and $z$ in the $\{ | \mathbf{r} \rangle \}$ representation. This is what was done in complement $E_V$. The energy levels, for an isotropic oscillator, are then found to be of the form:

$$E_n = \left(n + \frac{3}{2}\right)\hbar\omega \tag{4}$$

where $n$ is any positive integer or zero. The degree of degeneracy $g_n$ of the level $E_n$ is equal to:

$$g_n = \frac{1}{2}(n + 1)(n + 2) \tag{5}$$

and the associated eigenfunctions are:

$$\varphi_{n_x,n_y,n_z}(x, y, z) = \left(\frac{\beta^2}{\pi}\right)^{3/4} \frac{1}{\sqrt{2^{n_x+n_y+n_z}\, n_x!\, n_y!\, n_z!}}\, e^{-\frac{\beta^2}{2}(x^2+y^2+z^2)} \times H_{n_x}(\beta x)\, H_{n_y}(\beta y)\, H_{n_z}(\beta z) \tag{6}$$

with:

$$\beta = \sqrt{\frac{\mu\omega}{\hbar}} \tag{7}$$

[$H_p(u)$ denotes the Hermite polynomial of degree $p$; *cf.* complement $B_V$]. $\varphi_{n_x,n_y,n_z}$ is an eigenfunction of the Hamiltonian $H$ with the eigenvalue $E_n$ such that:

$$n = n_x + n_y + n_z \tag{8}$$

If the oscillator under consideration is isotropic*, the potential (1) is a function only of the distance $r$ between the particle and the origin:

$$V(r) = \frac{1}{2}\mu\omega^2 r^2 \tag{9}$$

Consequently, the three components of the orbital angular momentum **L** are constants of the motion. We want to find the common eigenstates of $H$, $\mathbf{L}^2$ and $L_z$. To do so, we could proceed, as in complement $D_{VI}$, by introducing operators related to right and left circular quanta and to "longitudinal" quanta corresponding to the third degree of freedom along $Oz$ (an outline of this method is given at the end of this complement). However, we prefer to use this example to illustrate the method elaborated in chapter VII (§A) and solve the radial equation by the polynomial method.

## 1. Solving the radial equation

For a fixed value of the quantum number $l$, the radial functions $R_{k,l}(r)$ and energies $E_{k,l}$ are given by the equation:

$$\left[-\frac{\hbar^2}{2\mu}\frac{1}{r}\frac{d^2}{dr^2}r + \frac{1}{2}\mu\omega^2 r^2 + \frac{l(l+1)\hbar^2}{2\mu r^2}\right]R_{k,l}(r) = E_{k,l}\, R_{k,l}(r) \tag{10}$$

* Separation of the polar variables $r$, $\theta$, $\varphi$ is possible only for an isotropic oscillator.

We set:

$$R_{k,l}(r) = \frac{1}{r} u_{k,l}(r) \tag{11-a}$$

$$\varepsilon_{k,l} = \frac{2\mu E_{k,l}}{\hbar^2} \tag{11-b}$$

Equation (10) then becomes:

$$\left[\frac{\mathrm{d}^2}{\mathrm{d}r^2} - \beta^4 r^2 - \frac{l(l+1)}{r^2} + \varepsilon_{k,l}\right] u_{k,l}(r) = 0 \tag{12}$$

[where $\beta$ is the constant defined in (7)]. We must add the condition at the origin:

$$u_{k,l}(0) = 0 \tag{13}$$

For large $r$, (12) virtually reduces to:

$$\left[\frac{\mathrm{d}^2}{\mathrm{d}r^2} - \beta^4 r^2\right] u_{k,l}(r) \underset{r \to \infty}{\simeq} 0 \tag{14}$$

The asymptotic behavior of the solutions of equation (12) is therefore dominated by $\mathrm{e}^{\beta^2 r^2/2}$ or $\mathrm{e}^{-\beta^2 r^2/2}$. Only the second possibility is physically acceptable. This leads us to the change of functions:

$$u_{k,l}(r) = \mathrm{e}^{-\beta^2 r^2/2}\, y_{k,l}(r) \tag{15}$$

It is easy to find that $y_{k,l}(r)$ must satisfy:

$$\frac{\mathrm{d}^2}{\mathrm{d}r^2} y_{k,l} - 2\beta^2 r \frac{\mathrm{d}}{\mathrm{d}r} y_{k,l} + \left[\varepsilon_{k,l} - \beta^2 - \frac{l(l+1)}{r^2}\right] y_{k,l} = 0 \tag{16-a}$$

$$y_{k,l}(0) = 0 \tag{16-b}$$

Now we shall seek $y_{k,l}(r)$ in the form of a power series in $r$:

$$y_{k,l}(r) = r^s \sum_{q=0}^{\infty} a_q r^q \tag{17}$$

where, by definition, $a_0$ is the coefficient of the first non-zero term:

$$a_0 \neq 0 \tag{18}$$

When we substitute expansion (17) into equation (16-a), the term of lowest order is in $r^{s-2}$; its coefficient is zero if:

$$[s(s-1) - l(l+1)]a_0 = 0 \tag{19}$$

With conditions (18) and (16-b) taken into account, the only way to satisfy relation (19) is to choose:

$$s = l + 1 \tag{20}$$

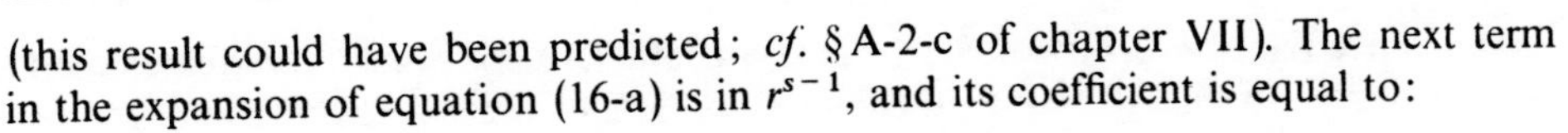



(this result could have been predicted; *cf.* § A-2-c of chapter VII). The next term in the expansion of equation (16-a) is in $r^{s-1}$, and its coefficient is equal to:

$$[s(s+1) - l(l+1)]a_1 \tag{21}$$

Since $s$ is already fixed by (20), this coefficient can go to zero only if:

$$a_1 = 0 \tag{22}$$

Finally, let us set the coefficient of the general term in $r^{q+s}$ equal to zero:

$$[(q+s+2)(q+s+1) - l(l+1)]\, a_{q+2} + [\varepsilon_{k,l} - \beta^2 - 2\beta^2(q+s)]\, a_q = 0 \tag{23}$$

that is, using (20):

$$(q+2)(q+2l+3)\, a_{q+2} = [(2q+2l+3)\beta^2 - \varepsilon_{k,l}]\, a_q \tag{24}$$

We therefore obtain a recurrence relation for the coefficients $a_q$ of expansion (17).

Note, first of all, that this recurrence relation, combined with result (22), implies that *all coefficients $a_q$ of odd rank $q$ are zero.* As for the coefficients of even rank, they must all be proportional to $a_0$. If the value of $\varepsilon_{k,l}$ is such that no integer $q$ makes the term in brackets on the right-hand side of (24) go to zero, we find the solution $y_{k,l}$ of (16) in the form of an infinite power series, for which:

$$\frac{a_{q+2}}{a_q} \underset{q \to \infty}{\sim} \frac{2\beta^2}{q} \tag{25}$$

This behavior is the same as that of the coefficients appearing in the expansion of the function $e^{\beta^2 r^2}$, since:

$$e^{\beta^2 r^2} = \sum_{p=0}^{\infty} c_{2p}\, r^{2p} \tag{26}$$

with:

$$c_{2p} = \frac{\beta^{2p}}{p!} \tag{27}$$

and, consequently:

$$\frac{c_{2p+2}}{c_{2p}} \underset{p \to \infty}{\sim} \frac{\beta^2}{p} \tag{28}$$

Since it is $2p$ which corresponds to the even integer $q$ of the expansion of $y_{k,l}$, (28) is indeed identical to (25). From this, we can see that if (17) really contains an infinite number of terms, the asymptotic behavior of $y_{k,l}$ is dominated by $e^{\beta^2 r^2}$, which renders this function physically unacceptable [*cf.* relation (15)].

The only cases which are interesting from a physical point of view are therefore those in which there exists an even integer $k$, positive or zero, such that:

$$\varepsilon_{k,l} = (2k + 2l + 3)\beta^2 \tag{29}$$

Recurrence relation (24) indicates that the coefficients of even rank greater than $k$ are then zero. Since all the coefficients of odd rank are also zero, expansion (17) reduces to a polynomial, and the radial function $u_{k,l}(r)$ given by (15) decreases exponentially as $r$ goes to infinity.

## 2. Energy levels and stationary wave functions

Using definitions (7) and (11-b), relation (29) gives the energies $E_{k,l}$ associated with a given value of $l$:

$$E_{k,l} = \hbar\omega\left(k + l + \frac{3}{2}\right) \tag{30}$$

where $k$ is any even positive integer or zero. Since $E_{k,l}$ actually depends only on the sum:

$$n = k + l \tag{31}$$

accidental degeneracies appear: the energy levels of the isotropic three-dimensional harmonic oscillator are of the form:

$$E_n = \left(n + \frac{3}{2}\right)\hbar\omega \tag{32}$$

$l$ is any positive integer or zero, and $k$ is any even positive integer or zero; $n$ can therefore take on all integral values, positive or zero. This is in agreement with result (4).

We shall fix an energy $E_n$, that is, an integer $n$, positive or zero. The values of $k$ and $l$ which can be associated with it according to (31) are the following:

$$(k, l) = (0, n), \quad (2, n - 2), \dots (n - 2, 2), \quad (n, 0) \qquad \text{if } n \text{ is even} \tag{33-a}$$
$$(k, l) = (0, n), \quad (2, n - 2), \dots (n - 3, 3), \quad (n - 1, 1) \qquad \text{if } n \text{ is odd} \tag{33-b}$$

From this, we can immediately get the values of $l$ associated with the first values of $n$:

$$\begin{aligned} n &= 0 : \quad l = 0 \\ n &= 1 : \quad l = 1 \\ n &= 2 : \quad l = 0, 2 \\ n &= 3 : \quad l = 1, 3 \\ n &= 4 : \quad l = 0, 2, 4 \end{aligned} \tag{34}$$

Figure 1 represents, with the same conventions as for the hydrogen atom (*cf.* figure 4 of chapter VII), the lowest energy levels of an isotropic three-dimensional harmonic oscillator.

For each pair $(k, l)$, there exists one and only one radial function $u_{k,l}(r)$, that is, $(2l + 1)$ common eigenfunctions of $H$, $\mathbf{L}^2$ and $L_z$:

$$\varphi_{k,l,m}(\mathbf{r}) = \frac{1}{r}\, u_{k,l}(r)\, Y_l^m(\theta, \varphi) \tag{35}$$

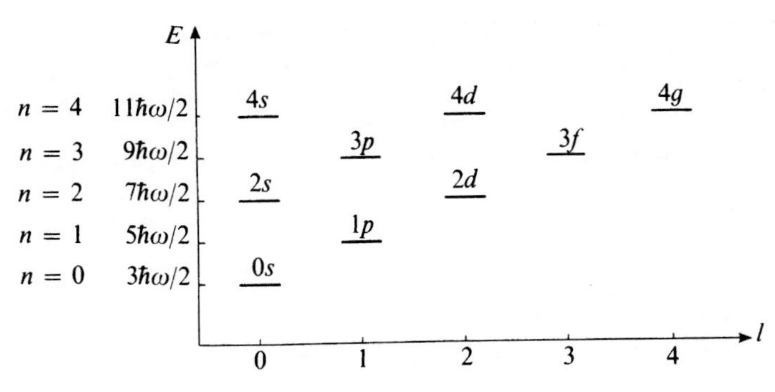


FIGURE 1

**Lowest energy levels of the three-dimensional harmonic oscillator. When $n$ is even, $l$ can take on $\frac{n}{2} + 1$ values: $l = n, n - 2, \dots 0$. When $n$ is odd, $l$ can take on $\frac{n+1}{2}$ values: $l = n, n - 2, \dots 1$. With the possible values of $m(-l \leqslant m \leqslant l)$ taken into account, the degree of degeneracy of the level $E_n$ is $\frac{(n+1)(n+2)}{2}$.**

Consequently, the degree of degeneracy of the energy $E_n$ under consideration is equal to:

$$g_n = \sum_{l=0,2,\cdots,n} (2l + 1) \qquad \text{if } n \text{ is even} \tag{36-a}$$

$$g_n = \sum_{l=1,3,\cdots,n} (2l + 1) \qquad \text{if } n \text{ is odd} \tag{36-b}$$

These sums are simple to calculate, and we again obtain result (5):

$$\text{for even } n: \quad g_n = \sum_{p=0}^{n/2} (4p + 1) = \frac{1}{2}(n + 1)(n + 2) \tag{37-a}$$

$$\text{for odd n}: \quad g_n = \sum_{p=0}^{(n-1)/2} (4p + 3) = \frac{1}{2}(n + 1)(n + 2) \tag{37-b}$$

For each of the pairs $(k, l)$ given in (33), the results of §1 enable us to determine the corresponding radial function $u_{k,l}(r)$ (to within the factor $a_0$) and, therefore, the $(2l + 1)$ common eigenfunctions of $H$ and $\mathbf{L}^2$, of eigenvalues $E_n$ and $l(l + 1)\hbar^2$. We shall calculate, for example, the wave functions associated in this way with the three lowest energy levels.

For the ground state $E_0 = \frac{3}{2}\hbar\omega$, we must have:

$$k = l = 0 \tag{38}$$

$y_{0,0}(r)$ then reduces to $a_0 r$. If we choose $a_0$ to be real and positive, the normalized function $\varphi_{k=l=m=0}$ can be written:

$$\varphi_{0,0,0}(\mathbf{r}) = \left(\frac{\beta^2}{\pi}\right)^{3/4} \mathrm{e}^{-\beta^2 r^2/2} \tag{39}$$

Since the ground state is not degenerate ($g_0 = 1$), $\varphi_{0,0,0}$ is the same as the function $\varphi_{n_x = n_y = n_z = 0}$ which is found by separating the Cartesian variables $x$, $y$, and $z$ [*cf.* formula (6)].

With the first excited state $\left(E_1 = \frac{5}{2}\hbar\omega\right)$, which is three-fold degenerate, is again associated a single pair $(k, l)$ :

$$\begin{cases} k = 0 \\ l = 1 \end{cases} \tag{40}$$

and $y_{0,1} = a_0 r^2$. The three functions of the basis defined by $\mathbf{L}^2$ and $L_z$ are therefore :

$$\varphi_{0,1,m}(\mathbf{r}) = \sqrt{\frac{8}{3}\frac{\beta^{3/2}}{\pi^{1/4}}}\,\beta r\, e^{-\beta^2 r^2/2}\, Y_1^m(\theta, \varphi) \qquad m = 1, 0, -1 \tag{41}$$

We know [*cf.* complement $A_{VI}$, formulas (32)] that the spherical harmonics $Y_1^m$ are such that:

$$\begin{aligned} r Y_1^0(\theta, \varphi) &= \sqrt{\frac{3}{4\pi}}\, z \\ \frac{r}{\sqrt{2}}\left[Y_1^{-1} - Y_1^1\right] &= \sqrt{\frac{3}{4\pi}}\, x \\ \frac{r}{\sqrt{2}}\left[Y_1^{-1} + Y_1^1\right] &= -\, i\sqrt{\frac{3}{4\pi}}\, y \end{aligned} \tag{42}$$

and that the Hermite polynomial of order 1 is [*cf.* complement $B_V$, formulas (18)] :

$$H_1(u) = 2u \tag{43}$$

Consequently, it is clear that the three functions $\varphi_{0,1,m}$ are related to the functions $\varphi_{n_x, n_y, n_z}$ of basis (6) by the equations:

$$\begin{aligned} \varphi_{n_x = 0, n_y = 0, n_z = 1} &= \varphi_{k = 0, l = 1, m = 0} \\ \varphi_{n_x = 1, n_y = 0, n_z = 0} &= \frac{1}{\sqrt{2}}\left[\varphi_{k=0,l=1,m=-1} - \varphi_{k=0,l=1,m=1}\right] \\ \varphi_{n_x = 0, n_y = 1, n_z = 0} &= \frac{i}{\sqrt{2}}\left[\varphi_{k=0,l=1,m=-1} + \varphi_{k=0,l=1,m=1}\right] \end{aligned} \tag{44}$$

Finally, consider the second excited state, of energy $E_2 = \frac{7}{2}\hbar\omega$. It is six-fold degenerate, and the quantum numbers $k$ and $l$ can take on the values:

$$k = 0, \quad l = 2 \tag{45-a}$$

$$k = 2, \quad l = 0 \tag{45-b}$$

The function $y_{0,2}(r)$ which corresponds to the values (45-a) is simply $a_0 r^3$. For the values (45-b), $y_{2,0}$ contains two terms; using (24) and (29), we easily find :

$$y_{2,0}(r) = a_0 r\left[1 - \frac{2}{3}\beta^2 r^2\right] \tag{46}$$

The six basis functions in the eigensubspace associated with $E_2$ are thus of the form:

$$\varphi_{0,2,m}(\mathbf{r}) = \sqrt{\frac{16}{15}\frac{\beta^{3/2}}{\pi^{1/4}}}\beta^2 r^2 \, \mathrm{e}^{-\beta^2 r^2/2} \, Y_2^m(\theta, \varphi) \qquad m = 2, 1, 0, -1, -2 \quad \text{(47-a)}$$

$$\varphi_{2,0,0}(\mathbf{r}) = \sqrt{\frac{3}{2}\frac{\beta^{3/2}}{\pi^{3/4}}}\left(1 - \frac{2}{3}\beta^2 r^2\right) \mathrm{e}^{-\beta^2 r^2/2} \quad \text{(47-b)}$$

Knowing the explicit expressions for the spherical harmonics [formulas (33) of complement $A_{VI}$] and the Hermite polynomials [formulas (18) of complement $B_V$], we can easily prove the following relations:

$$\varphi_{k=2,l=0,m=0} = -\frac{1}{\sqrt{3}}\left[\varphi_{n_x=2,n_y=0,n_z=0} + \varphi_{n_x=0,n_y=2,n_z=0} + \varphi_{n_x=0,n_y=0,n_z=2}\right]$$

$$\frac{1}{\sqrt{2}}\left[\varphi_{k=0,l=2,m=2} + \varphi_{k=0,l=2,m=-2}\right] = \frac{1}{\sqrt{2}}\left[\varphi_{n_x=2,n_y=0,n_z=0} - \varphi_{n_x=0,n_y=2,n_z=0}\right]$$

$$\frac{1}{\sqrt{2}}\left[\varphi_{k=0,l=2,m=2} - \varphi_{k=0,l=2,m=-2}\right] = i\,\varphi_{n_x=1,n_y=1,n_z=0}$$

$$\frac{1}{\sqrt{2}}\left[\varphi_{k=0,l=2,m=1} - \varphi_{k=0,l=2,m=-1}\right] = -\,\varphi_{n_x=1,n_y=0,n_z=1}$$

$$\frac{1}{\sqrt{2}}\left[\varphi_{k=0,l=2,m=1} + \varphi_{k=0,l=2,m=-1}\right] = -\,i\,\varphi_{n_x=0,n_y=1,n_z=1}$$

$$\varphi_{k=0,l=2,m=0} = \sqrt{\frac{2}{3}}\left[\varphi_{n_x=0,n_y=0,n_z=2} - \frac{1}{2}\varphi_{n_x=2,n_y=0,n_z=0} - \frac{1}{2}\varphi_{n_x=0,n_y=2,n_z=0}\right] \quad (48)$$

COMMENT:

As we pointed out in the beginning of this complement, we can apply a method analogous to the one presented in complement $D_{VI}$ to the isotropic three-dimensional harmonic oscillator. If $a_x$, $a_y$ and $a_z$ are the annihilation operators which act in the state spaces $\mathscr{E}_x$, $\mathscr{E}_y$ and $\mathscr{E}_z$ respectively, we define:

$$a_d = \frac{1}{\sqrt{2}}(a_x - ia_y) \quad \text{(49-a)}$$

$$a_g = \frac{1}{\sqrt{2}}(a_x + ia_y) \quad \text{(49-b)}$$

It can be shown that $a_d$ and $a_g$ behave like independent annihilation operators (complement $D_{VI}$, §3-b). The Hamiltonian $H$ and the angular momentum operators can then be expressed in terms of $a_d$, $a_g$, $a_z$ and their adjoints:

$$H = \hbar\omega\left(N_d + N_g + N_z + \frac{3}{2}\right) \tag{50-a}$$

$$L_z = \hbar(N_d - N_g) \tag{50-b}$$

$$L_+ = \hbar\sqrt{2}\,(a_z^\dagger a_g - a_d^\dagger a_z) \tag{50-c}$$

$$L_- = \hbar\sqrt{2}\,(a_g^\dagger a_z - a_z^\dagger a_d) \tag{50-d}$$

The common eigenvectors $|\chi_{n_d,n_g,n_z}\rangle$ of the observables $N_d$, $N_g$ and $N_z$ can be obtained through the action of the creation operators $a_d^\dagger$, $a_g^\dagger$ and $a_z^\dagger$ on the ground state $|0, 0, 0\rangle$ of the Hamiltonian $H$ [this state is unique to within a constant factor; *cf.* formulas (6) and (39)]:

$$|\chi_{n_d,n_g,n_z}\rangle = \frac{1}{\sqrt{n_d!\,n_g!\,n_z!}}(a_d^\dagger)^{n_d}(a_g^\dagger)^{n_g}(a_z^\dagger)^{n_z}\,|0, 0, 0\rangle \tag{51}$$

According to (50-a) and (50-b), $|\chi_{n_d,n_g,n_z}\rangle$ is an eigenvector of $H$ and $L_z$ with the eigenvalues $(n_d + n_g + n_z + 3/2)\hbar\omega$ and $(n_d - n_g)\hbar$. The eigensubspace $\mathscr{E}_n$ associated with a given energy $E_n$ can therefore be spanned by the set of vectors $|\chi_{n_d,n_g,n_z}\rangle$ such that:

$$n_d + n_g + n_z = n \tag{52}$$

Of these, the eigenvector of $L_z$ with the largest eigenvalue compatible with $E_n$ is $|\chi_{n,0,0}\rangle$, whose eigenvalue is $n\hbar$. This ket, according to (50-c), satisfies:

$$L_+\,|\chi_{n,0,0}\rangle = 0 \tag{53}$$

Consequently*, it is an eigenvector of $\mathbf{L}^2$ with the eigenvalue $n(n + 1)\hbar^2$, and it can be identified with the ket of the $\{\,|\varphi_{k,l,m}\rangle\,\}$ basis such that:

$$\begin{aligned} k + l &= n \\ l = m &= n \end{aligned} \tag{54}$$

Therefore:

$$|\varphi_{k=0,l=n,m=n}\rangle = |\chi_{n_d=n,n_g=0,n_z=0}\rangle \tag{55}$$

Application of the operator $L_-$ [formula (50-d)] to both sides of relation (55) yields:

$$|\varphi_{0,n,n-1}\rangle = -\,|\chi_{n-1,0,1}\rangle \tag{56}$$

* This result follows directly from relation (C-7-b) of chapter VI, which, applied to $|\chi_{n,0,0}\rangle$, yields:

$$\mathbf{L}^2\,|\chi_{n,0,0}\rangle = \hbar^2(n^2 + n)\,|\chi_{n,0,0}\rangle$$

The eigenvalue $(n-2)\hbar$ of $L_z$, unlike the two preceding ones, is two-fold degenerate in $\mathscr{E}_n$: two orthogonal vectors, $|\chi_{n-2,0,2}\rangle$ and $|\chi_{n-1,1,0}\rangle$ correspond to it. Using (50-d) again in order to apply $L_-$ to (56), we find that:

$$|\varphi_{0,n,n-2}\rangle = \sqrt{\frac{2(n-1)}{2n-1}}\,|\chi_{n-2,0,2}\rangle - \frac{1}{\sqrt{2n-1}}\,|\chi_{n-1,1,0}\rangle \tag{57}$$

It can be shown that the action of $L_+$ on the linear combination orthogonal to (57) yields the null vector. This linear combination must therefore be an eigenvector of $\mathbf{L}^2$ with the eigenvalue $(n-2)(n-1)\hbar^2$. This gives, to within a phase factor:

$$|\varphi_{2,n-2,n-2}\rangle = \frac{1}{\sqrt{2n-1}}\,|\chi_{n-2,0,2}\rangle + \sqrt{\frac{2(n-1)}{2n-1}}\,|\chi_{n-1,1,0}\rangle \tag{58}$$

We can thus relate, by iteration*, the two bases, $\{|\chi_{n_d,n_g,n_z}\rangle\}$ and $\{|\varphi_{k,l,m}\rangle\}$. Of course, replacing $a_d^\dagger$ and $a_g^\dagger$ by functions of $a_x^\dagger$ and $a_y^\dagger$ in (51), we can express $|\chi_{n_d,n_g,n_z}\rangle$ as a linear combination of the vectors $|\varphi_{n_x,n_y,n_z}\rangle$ whose wave functions are given by (6).

**References and suggestions for further reading:**

Other soluble examples (spherical square well, etc.): Messiah (1.17), chap. IX, § 10; Schiff (1.18), § 15; see also Flügge (1.24), §§ 58 to 79.

* An argument analogous to the one just outlined will be used in chapter X to add two angular momenta.

# Complement $C_{VII}$

## PROBABILITY CURRENTS ASSOCIATED WITH THE STATIONARY STATES OF THE HYDROGEN ATOM



The normalized wave functions $\varphi_{n,l,m}(\mathbf{r})$ associated with the stationary states of the hydrogen atom were determined in chapter VII. $\varphi_{n,l,m}(\mathbf{r})$ is the product of the spherical harmonic $Y_l^m(\theta, \varphi)$ and the function $R_{n,l}(r)$ calculated in § C-3 of this chapter:

$$\varphi_{n,l,m}(\mathbf{r}) = R_{n,l}(r) Y_l^m(\theta, \varphi) \tag{1}$$

The spatial variation of the probability density:

$$\rho_{n,l,m}(\mathbf{r}) = |\varphi_{n,l,m}(\mathbf{r})|^2 \tag{2}$$

was then studied, at least for the lowest energy states.

It is important, however, to understand that a stationary state is not characterized only by the value of the probability density $\rho_{n,l,m}(\mathbf{r})$ at all points of space. We must also associate with it a probability current which can be expressed as:

$$\mathbf{J}_{n,l,m}(\mathbf{r}) = \frac{\hbar}{2\mu i} \varphi^*_{n,l,m}(\mathbf{r}) \nabla \varphi_{n,l,m}(\mathbf{r}) + \text{c.c.} \tag{3}$$

[we assume here that the vector potential $\mathbf{A}(\mathbf{r}, t)$ is zero; $\mu$ denotes the mass of the particle].

Thus, we can associate with the quantum state of a particle, a "fluid" (called the "probability fluid") whose density at each point of space is $\rho(\mathbf{r})$. This fluid is not motionless; it is in a state of flow characterized by the current density $\mathbf{J}$. In a stationary state, $\rho$ and $\mathbf{J}$ are not time-dependent : the fluid is in a steady state of flow.

To complete the results of chapter VII concerning the physical properties of the stationary states of the hydrogen atom, we shall now study the probability currents $\mathbf{J}_{n,l,m}(\mathbf{r})$.

### 1. General expression for the probability current

Consider an arbitrary normalized wave function $\psi(\mathbf{r})$. We introduce the real quantities $\alpha(\mathbf{r})$ [the modulus of $\psi(\mathbf{r})$] and $\xi(\mathbf{r})$ [the argument of $\psi(\mathbf{r})$] by setting:

$$\psi(\mathbf{r}) = \alpha(\mathbf{r})\, e^{i\xi(\mathbf{r})} \tag{4}$$

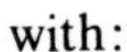

with:

$$\alpha(\mathbf{r}) \geqslant 0 \; ; \quad 0 \leqslant \xi(\mathbf{r}) < 2\pi \tag{5}$$

If we substitute (4) into the expressions for the probability density $\rho(\mathbf{r})$ and the current $\mathbf{J}(\mathbf{r})$, we obtain [still assuming the vector potential $\mathbf{A}(\mathbf{r})$ to be zero]:

$$\rho(\mathbf{r}) = \alpha^2(\mathbf{r}) \tag{6}$$

$$\mathbf{J}(\mathbf{r}) = \frac{\hbar}{\mu} \alpha^2(\mathbf{r}) \, \nabla\xi(\mathbf{r}) \tag{7}$$

Therefore, $\rho(\mathbf{r})$ depends only on the modulus of the wave function, while $\mathbf{J}(\mathbf{r})$ brings in its phase [for example, $\mathbf{J}(\mathbf{r})$ is zero if this phase is constant throughout all space].

COMMENT:

If the wave function $\psi(\mathbf{r})$ is given, it is obvious that $\rho(\mathbf{r})$ and $\mathbf{J}(\mathbf{r})$ are perfectly well-defined. Conversely, is there always one and only one function $\psi(\mathbf{r})$ corresponding to given values of $\rho(\mathbf{r})$ and $\mathbf{J}(\mathbf{r})$?

According to (6), the modulus $\alpha(\mathbf{r})$ of the wave function can be obtained directly from $\rho(\mathbf{r})$*; the argument $\xi(\mathbf{r})$ must satisfy the equation:

$$\nabla\xi(\mathbf{r}) = \frac{\mu}{\hbar} \frac{\mathbf{J}(\mathbf{r})}{\rho(\mathbf{r})} \tag{8}$$

We know that such an equation has a solution only if:

$$\nabla \times \frac{\mathbf{J}(\mathbf{r})}{\rho(\mathbf{r})} = \mathbf{0} \tag{9}$$

It then has an infinite number of solutions, which differ from each other by a constant. Since this constant corresponds to a global phase factor, it follows, if condition (9) is satisfied, that the wave function of the particle is perfectly well-defined by the specification of $\rho(\mathbf{r})$ and $\mathbf{J}(\mathbf{r})$. If condition (9) is not fulfilled, no wave function exists which corresponds to the values of $\rho(\mathbf{r})$ and $\mathbf{J}(\mathbf{r})$ under consideration.

## 2. Application to the stationary states of the hydrogen atom

### a. STRUCTURE OF THE PROBABILITY CURRENT

When the wave function has the form of (1), where $R_{n,l}(r)$ is a real function and $Y_l^m(\theta, \varphi)$ is the product of $e^{im\varphi}$ and a real function, we have:

$$\begin{aligned} \alpha_{n,l,m}(\mathbf{r}) &= |R_{n,l}(r)| \, |Y_l^m(\theta, \varphi)| \\ \xi_{n,l,m}(\mathbf{r}) &= m\varphi \end{aligned} \tag{10}$$

* Of course, in order to be a probability density, $\rho(\mathbf{r})$ must be positive everywhere.

Applying formula (7) and using the expression for the gradient in polar coordinates, we obtain:

$$\mathbf{J}_{n,l,m}(\mathbf{r}) = \frac{\hbar}{\mu} m \frac{\rho_{n,l,m}(\mathbf{r})}{r \sin\theta} \mathbf{e}_\varphi(\mathbf{r}) \tag{11}$$

where $\mathbf{e}_\varphi(\mathbf{r})$ is the unit vector which forms with $Oz$ and $\mathbf{r}$ a right-handed Cartesian coordinate frame.

Probability current variations in a plane perpendicular to $Oz$ are shown in figure 1.

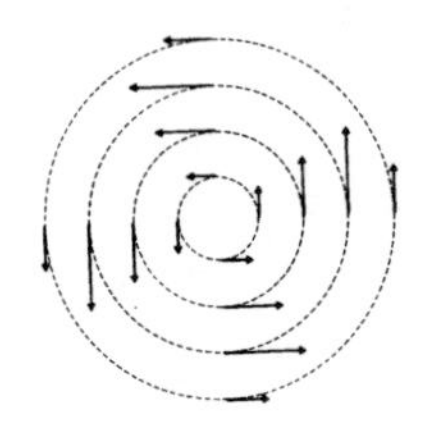

FIGURE 1

**Structure of the probability current associated with a stationary state $|\varphi_{n,l,m}\rangle$ of the hydrogen atom (shown in a plane perpendicular to $Oz$). The index $m$ refers to the eigenvalue $m\hbar$ of $L_z$. If $m > 0$, the probability fluid rotates counterclockwise about $Oz$; if $m < 0$, clockwise. If $m = 0$, the probability current is zero at all points of space.**

According to (11), the current at each point $M$ is perpendicular to the plane defined by $M$ and the $Oz$ axis: the probability fluid rotates about $Oz$. Since $|\mathbf{J}|$ is not proportional to $r \sin\theta\, \rho(\mathbf{r})$, the system does not rotate as a whole. The eigenvalue $m\hbar$ of the observable $L_z$ can be interpreted as the classical angular momentum associated with this rotational motion of the probability fluid. The contribution of the volume element $d^3r$, situated at the point $\mathbf{r}$, to the angular momentum with respect to the origin can be written:

$$d\mathscr{L} = \mu\, \mathbf{r} \times \mathbf{J}_{n,l,m}(\mathbf{r})\, d^3r \tag{12}$$

By symmetry, the resultant of all these elementary momenta is directed along $Oz$; it is equal to:

$$\mathscr{L}_z = \mu \int d^3r\, \mathbf{e}_z \cdot [\mathbf{r} \times \mathbf{J}_{n,l,m}(\mathbf{r})] \tag{13}$$

Using expression (11) for $\mathbf{J}_{n,l,m}(\mathbf{r})$, we easily obtain:

$$\begin{aligned} \mathscr{L}_z &= \mu \int d^3r\; r\, |\mathbf{J}_{n,l,m}(\mathbf{r})| \sin\theta \\ &= m\hbar \int d^3r\; \rho_{n,l,m}(\mathbf{r}) \\ &= m\hbar \end{aligned} \tag{14}$$

### b. EFFECT OF A MAGNETIC FIELD

The results obtained thus far are valid only if the vector potential $\mathbf{A}(\mathbf{r})$ is zero; let us examine what happens when this is not the case. Assume, for example, that we place the hydrogen atom in a uniform magnetic field $\mathbf{B}$. This field can be described by the vector potential:

$$\mathbf{A}(\mathbf{r}) = -\frac{1}{2}\mathbf{r} \times \mathbf{B} \tag{15}$$

What is the probability current associated with the ground state?

For simplicity, we shall also assume that the magnetic field $\mathbf{B}$ does not modify the wave function of the ground state*. The probability current can then be calculated from the general expression for $\mathbf{J}$ [*cf.* relation (D-20) of chapter III]. This yields:

$$\begin{aligned} \mathbf{J}_{n,l,m}(\mathbf{r}) &= \frac{1}{2\mu}\left\{\varphi^*_{n,l,m}(\mathbf{r})\left[\frac{\hbar}{i}\boldsymbol{\nabla} - q\mathbf{A}(\mathbf{r})\right]\varphi_{n,l,m}(\mathbf{r}) + \text{c.c.}\right\} \\ &= \frac{1}{\mu}\,\rho_{n,l,m}(\mathbf{r})\left[\hbar\boldsymbol{\nabla}\xi_{n,l,m}(\mathbf{r}) - q\mathbf{A}(\mathbf{r})\right] \end{aligned} \tag{16}$$

For the ground state and a field $\mathbf{B}$ directed along $Oz$, we obtain, using (15):

$$\mathbf{J}_{1,0,0}(\mathbf{r}) = \frac{\omega_c}{2}\,\rho_{1,0,0}(\mathbf{r})\,\mathbf{e}_z \times \mathbf{r} \tag{17}$$

where the cyclotron frequency $\omega_c$ is defined by:

$$\omega_c = -\frac{qB}{\mu} \tag{18}$$

Therefore, the probability current of the ground state is not zero in the presence of a magnetic field as it is when $\mathbf{B} = \mathbf{0}$. Expression (17) indicates that the probability fluid rotates as a whole about $\mathbf{B}$ with the angular velocity $\omega_c/2$. Physically, this result arises from the fact that when the magnetic field $\mathbf{B}$ is turned on, a transient electric field $\mathbf{E}(t)$ must exist. Under its influence, the electron, while remaining in the ground state, goes into rotation about the proton, with an angular velocity depending only on the value of $\mathbf{B}$ (and not on the precise way in which the field was turned on during the transition period).

COMMENT:

The particular choice of gauge made in (15) allowed us to retain the same wave functions we used with only a negligible error in the absence of a field (*cf.* note at the bottom of this page). With another gauge, the wave functions would have been different (*cf.* complement $H_{III}$) and, in (16), the term containing $\mathbf{A}(\mathbf{r})$ explicitly would not have been the only one to contribute to the value of $\mathbf{J}(\mathbf{r})$, to first order in $\mathbf{B}$. Nevertheless, we would have found (17) at the end of the calculation, since the physical result must not be gauge-dependent.

* Since the Hamiltonian $H$ depends on $\mathbf{B}$, this is obviously not rigorously correct. Nevertheless, by considering the expression for $H$ [*cf.* formulas (6) and (7) of complement $D_{VII}$], we can show that, for the gauge chosen in (15) and $\mathbf{B}$ directed along $Oz$, the functions $\varphi_{n,l,m}(\mathbf{r})$ are eigenfunctions of $H$ to within a term of second order in $\mathbf{B}$. By using the perturbation theory of chapter XI, one can show that, for magnetic fields normally produced in the laboratory, this second-order term is negligible.

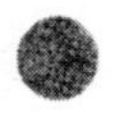



# Complement $D_{VII}$

## THE HYDROGEN ATOM PLACED IN A UNIFORM MAGNETIC FIELD. PARAMAGNETISM AND DIAMAGNETISM. THE ZEEMAN EFFECT



In chapter VII, we studied the quantum mechanical properties of a free hydrogen atom, that is, of the system formed by an electron and a proton exerting an electrostatic attraction on each other but not interacting with any external field. This complement is devoted to the study of the new effects which appear when this atom is placed in a static magnetic field. We shall confine ourselves to the case in which this field is uniform, as it always is, moreover, in practice, since the magnetic fields that can be produced in the laboratory vary very little in relative value over distances comparable to atomic dimensions.

We have already studied the behavior of an electron subjected either to an electric field alone (*cf.* for example, chapter VII) or to a magnetic field alone (*cf.* complement $E_{VI}$). Here, we shall generalize these discussions by calculating the energy levels of an electron subjected both to the influence of the internal electric field of the atom and to an external magnetic field. Under these conditions, the exact solution of the Schrödinger equation may seem to be a very complicated problem. However, we shall see that this problem can be simplified considerably by means of certain approximations. In the first place, we shall totally neglect the nuclear finite mass effect*. Then we shall use the fact that, in practice, the effect of the external magnetic field is much smaller than that of the internal electric field of the atom : the atomic level shifts due to the magnetic field are much smaller than the energy separations in a zero field.

The discussion presented in this complement will enable us to introduce and explain certain effects which are important in atomic physics. We shall see, in

* For the hydrogen atom, such an approximation is justified by the fact that the proton is considerably heavier than the electron. For muonium (*cf.* complement $A_{VII}$), the approximation is not as good and it becomes totally inapplicable for the case of positronium. We note, moreover, that, in the presence of a magnetic field, it is not rigorously possible to separate the motion of the center of mass. If one wished to take the nuclear finite mass effect into account in this complement, it would not suffice to replace the mass $m_e$ of the electron by the reduced mass $\mu$ of the electron-proton system.

particular, how atomic paramagnetism and diamagnetism appear in the quantum mechanical formalism. In addition, we shall be able to predict the modifications occurring in the optical spectrum emitted by hydrogen atoms when they are placed in a static magnetic field (the Zeeman effect).

## 1. The Hamiltonian of the problem. The paramagnetic term and the diamagnetic term

### a. EXPRESSION FOR THE HAMILTONIAN

Consider a spinless particle, of mass $m_e$ and charge $q$, subjected simultaneously to a scalar central potential $V(r)$ and a vector potential $\mathbf{A}(\mathbf{r})$. Its Hamiltonian is :

$$H = \frac{1}{2m_e}[\mathbf{P} - q\mathbf{A}(\mathbf{R})]^2 + V(\mathbf{R}) \tag{1}$$

When the magnetic field $\mathbf{B} = \boldsymbol{\nabla} \times \mathbf{A}(\mathbf{r})$ is uniform, the vector potential $\mathbf{A}$ can be put into the form:

$$\mathbf{A}(\mathbf{r}) = -\frac{1}{2}\mathbf{r} \times \mathbf{B} \tag{2}$$

To substitute this expression into (1), we shall calculate the quantity:

$$[\mathbf{P} - q\mathbf{A}(\mathbf{R})]^2 = \mathbf{P}^2 + \frac{q}{2}[\mathbf{P} . (\mathbf{R} \times \mathbf{B}) + (\mathbf{R} \times \mathbf{B}) . \mathbf{P}] + \frac{q^2}{4}(\mathbf{R} \times \mathbf{B})^2 \tag{3}$$

Now, $\mathbf{B}$ is actually a constant and not an operator. All observables therefore commute with $\mathbf{B}$, so we can write, using the rules of vector calculus:

$$[\mathbf{P} - q\mathbf{A}(\mathbf{R})]^2 = \mathbf{P}^2 + \frac{q}{2}[\mathbf{B} . (\mathbf{P} \times \mathbf{R}) - (\mathbf{R} \times \mathbf{P}) . \mathbf{B}] + \frac{q^2}{4}[\mathbf{R}^2\mathbf{B}^2 - (\mathbf{R} . \mathbf{B})^2] \tag{4}$$

On the right-hand side of this expression, the angular momentum $\mathbf{L}$ of the particle appears :

$$\mathbf{L} = \mathbf{R} \times \mathbf{P} = -\mathbf{P} \times \mathbf{R} \tag{5}$$

We can therefore write $H$ in the form:

$$H = H_0 + H_1 + H_2 \tag{6}$$

where $H_0$, $H_1$ and $H_2$ are defined by:

$$H_0 = \frac{\mathbf{P}^2}{2m_e} + V(\mathbf{R}) \tag{7-a}$$

$$H_1 = -\frac{\mu_B}{\hbar}\mathbf{L} . \mathbf{B} \tag{7-b}$$

$$H_2 = \frac{q^2\mathbf{B}^2}{8m_e}\mathbf{R}_\perp^2 \tag{7-c}$$

In these relations, $\mu_B$ denotes the Bohr magneton (whose dimensions are those of a magnetic moment):

$$\mu_B = \frac{q\hbar}{2m_e} \tag{8}$$

and the operator $\mathbf{R}_\perp$ is the projection of $\mathbf{R}$ onto a plane perpendicular to $\mathbf{B}$:

$$\mathbf{R}_\perp^2 = \mathbf{R}^2 - \frac{(\mathbf{R} \cdot \mathbf{B})^2}{\mathbf{B}^2} \tag{9}$$

If we choose a system of orthonormal axes $Oxyz$ such that $\mathbf{B}$ is parallel to $Oz$, we have:

$$\mathbf{R}_\perp^2 = X^2 + Y^2 \tag{10}$$

COMMENT:

When the field $\mathbf{B}$ is zero, $H$ becomes equal to $H_0$, which is the sum of the kinetic energy $\mathbf{P}^2/2m_e$ and the potential energy $V(\mathbf{R})$. Nevertheless, we must not conclude from this that when $\mathbf{B}$ is not zero, $\mathbf{P}^2/2m_e$ still represents the kinetic energy of the electron. We have seen (*cf.* complement $H_{III}$) that the physical meaning of operators acting in the state space changes when the vector potential is not zero. For example, the momentum $\mathbf{P}$ no longer represents the mechanical momentum $\mathbf{\Pi} = m_e\mathbf{V}$, and the kinetic energy is then equal to :

$$\frac{\mathbf{\Pi}^2}{2m_e} = \frac{1}{2m_e}[\mathbf{P} - q\mathbf{A}(\mathbf{R})]^2 \tag{11}$$

The meaning of the term $\mathbf{P}^2/2m_e$, taken alone, depends on the gauge chosen. With the one defined by (2), it can easily be shown to correspond to the "relative" kinetic energy $\mathbf{\Pi}_R^2/2m_e$, where $\mathbf{\Pi}_R$ is the mechanical momentum of the particle with respect to the "Larmor frame" rotating about $\mathbf{B}$ with angular velocity $\omega_L = -qB/2m_e$. The term $H_2$ then describes the kinetic energy $\mathbf{\Pi}_E^2/2m_e$ related to the drag velocity of the frame. As for $H_1$, it corresponds to the cross term $\mathbf{\Pi}_E \cdot \mathbf{\Pi}_R/m_e$.

## b. ORDER OF MAGNITUDE OF THE VARIOUS TERMS

In the presence of the magnetic field $\mathbf{B}$, two new terms, $H_1$ and $H_2$, therefore appear in $H$. Before examining their physical meaning in greater detail, we shall calculate the order of magnitude of the energy differences $\Delta E$ (or the frequency differences $\Delta E/h$) associated with them.

As far as $H_0$ is concerned, we already know the corresponding energy differences $\Delta E_0$ (*cf.* chap. VII). The associated frequencies are of the order of:

$$\frac{\Delta E_0}{h} \simeq 10^{14} \text{ to } 10^{15} \text{ Hz} \tag{12}$$

Also, by using (7-b), we see that $\Delta E_1$ is approximately given by:

$$\frac{\Delta E_1}{h} \simeq \frac{1}{h}\left(\frac{\mu_B}{\hbar}\hbar B\right) = \frac{\omega_L}{2\pi} \tag{13}$$

where $\omega_L$ is the Larmor angular velocity$^\star$:

$$\omega_L = -\frac{qB}{2\mu} \tag{14}$$

A simple numerical calculation shows that, for an electron, the Larmor frequency is such that:

$$\frac{\nu_L}{B} = \frac{\omega_L}{2\pi B} \simeq 1.40 \times 10^{10} \text{ Hz/tesla} = 1.40 \text{ MHz/gauss} \tag{15}$$

Now, with the fields usually produced in the laboratory (which rarely exceed 100,000 gauss), we have:

$$\frac{\omega_L}{2\pi} \lesssim 10^{11} \text{ Hz} \tag{16}$$

Comparing (12) and (16), we see that:

$$\Delta E_1 \ll \Delta E_0 \tag{17}$$

Let us show, similarly, that:

$$\Delta E_2 \ll \Delta E_1 \tag{18}$$

To do so, we shall evaluate the order of magnitude $\Delta E_2$ of the energies associated with $H_2$. The matrix elements of the operator $\mathbf{R}_\perp^2 = X^2 + Y^2$ are of the same order of magnitude as $a_0^2$, where $a_0 = \hbar^2/m_e e^2$ characterizes atomic dimensions. Thus we obtain:

$$\Delta E_2 \simeq \frac{q^2 B^2}{m_e} a_0^2 \tag{19}$$

We then find the ratio:

$$\frac{\Delta E_2}{\Delta E_1} \simeq \frac{q^2 B^2}{m_e} a_0^2 \frac{1}{\hbar \omega_L} = 2\hbar \frac{qB}{m_e} \frac{m_e a_0^2}{\hbar^2} \tag{20}$$

Now, according to formulas (C-12-a) and (C-12-b) of chapter VII:

$$\Delta E_0 \simeq \frac{\hbar^2}{m_e a_0^2} \tag{21}$$

Relation (20) therefore yields, with (13) taken into account:

$$\frac{\Delta E_2}{\Delta E_1} \simeq \frac{\Delta E_1}{\Delta E_0} \tag{22}$$

which, according to (17), proves (18).

$\star$ Note that the Larmor frequency $\frac{\omega_L}{2\pi}$ is half the cyclotron frequency.

Therefore, the effects of the magnetic field always remain, in practice, much smaller than those due to the internal field of the atom. Moreover, it is generally sufficient, when we study them, to retain only the term $H_1$, compared to which $H_2$ is negligible ($H_2$ will be taken into account only in the special cases in which the contribution of $H_1$ is zero)*.

#### c. INTERPRETATION OF THE PARAMAGNETIC TERM

Consider, first of all, the term $H_1$ given by (7-b). We shall see that it can be interpreted to be the coupling energy $-\mathbf{M}_1 \cdot \mathbf{B}$ of the field $\mathbf{B}$ and the magnetic moment $\mathbf{M}_1$ related to the revolution of the electron in its orbit.

For this purpose, we shall begin by calculating the magnetic moment $\mathcal{M}$ classically associated with a charge $q$ in a circular orbit of radius $r$ (fig. 1). If the speed of the particle is $v$, its motion is equivalent to a current:

$$i = q\frac{v}{2\pi r} \tag{23}$$

Since the surface $S$ defined by this current is:

$$S = \pi r^2 \tag{24}$$

the magnetic moment $\mathcal{M}$ is given by:

$$|\mathcal{M}| = i \times S = \frac{q}{2} rv \tag{25}$$

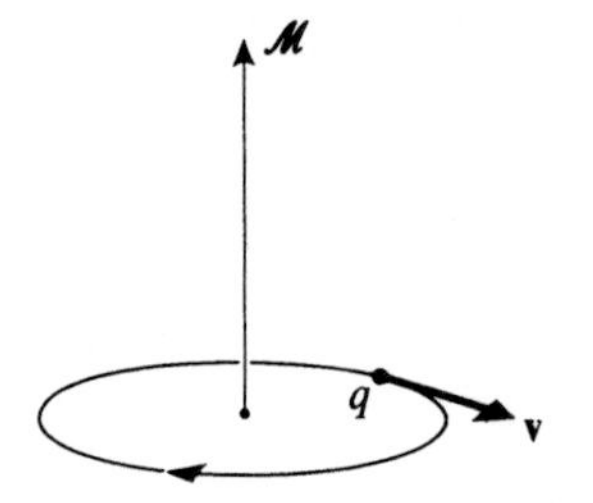


FIGURE 1

**Classically, the motion of an electron in its orbit can be regarded as a current loop of magnetic moment $\mathcal{M}$.**

Introducing the angular momentum $\mathcal{L}$, which, since the velocity is tangential, has a modulus of:

$$|\mathcal{L}| = m_e rv \tag{26}$$

we can write (25) in the form:

$$\mathcal{M} = \frac{q}{2m_e}\mathcal{L} \tag{27}$$

* The Zeeman effect of a three-dimensional harmonic oscillator can be calculated without approximations (*cf.* problem 2 of complement $G_{VII}$). This is true because $V(\mathbf{R})$ and $H_2$ then have analogous forms. This example is interesting since it enables us to analyze the contributions of $H_1$ and $H_2$ in a soluble case.

(this is a vector relation since $\mathscr{L}$ and $\mathscr{M}$ are parallel, as both are perpendicular to the plane of the classical orbit).

The quantum mechanical analogue of (27) is the operator relation:

$$\mathbf{M}_1 = \frac{q}{2m_e}\mathbf{L} \tag{28}$$

We can therefore write $H_1$ in the form:

$$H_1 = -\mathbf{M}_1 \cdot \mathbf{B} \tag{29}$$

This confirms the interpretation given above: $H_1$ corresponds to the coupling between the magnetic field $\mathbf{B}$ and the permanent atomic magnetic moment ($\mathbf{M}_1$ is independent of $\mathbf{B}$). $H_1$ is called the paramagnetic coupling term.

COMMENTS:

(*i*) According to (28), the eigenvalues of any component of the magnetic moment $\mathbf{M}_1$ are of the form:

$$\left(\frac{q}{2m_e}\right) \times (m\hbar) = m\mu_B \tag{30}$$

where $m$ is an integer. $\mu_B$ therefore gives the order of magnitude of the magnetic moment associated with the orbital moment of the electron. This is why definition (8) is useful. In the MKSA system:

$$\mu_B \simeq -9.27 \times 10^{-24} \text{ Joule/tesla} \tag{31}$$

(*ii*) As we shall see in chapter IX, the electron possesses, in addition to the orbital angular momentum $\mathbf{L}$, a spin angular momentum $\mathbf{S}$. With this observable is associated a magnetic moment, $\mathbf{M}_S$, proportional to $\mathbf{S}$:

$$\mathbf{M}_S = 2\frac{\mu_B}{\hbar}\mathbf{S} \tag{32}$$

Although the magnetic effects due to the spin are important, we shall ignore them for now (we shall return to them in complement $\text{D}_{\text{XII}}$).

(*iii*) The classical argument presented above is not completely correct. We have confused the angular momentum:

$$\mathscr{L} = \mathbf{r} \times \mathbf{p} \tag{33}$$

with the moment of the mechanical momentum:

$$\boldsymbol{\lambda} = \mathbf{r} \times m_e\mathbf{v} = \mathscr{L} - q\,\mathbf{r} \times \mathbf{A}(\mathbf{r}) \tag{34}$$

In fact, the error is small. As we shall see in the next section it simply amounts to neglecting $H_2$ relative to $H_1$.

### d. INTERPRETATION OF THE DIAMAGNETIC TERM

Consider a zero angular momentum state of the hydrogen atom (for example, the ground state). The correction supplied by $H_1$ to the energy of this state is also zero. Thus, to determine the effect of the field **B**, we must now take into account the presence of $H_2$. How should the corresponding energy be interpreted?

We have seen (*cf.* complement $C_{VII}$, §2-b) that, in the presence of a uniform magnetic field, the probability current associated with the electron is modified. This current is cylindrically symmetrical with respect to **B**. It corresponds to a uniform rotation of the probability fluid, clockwise if $q$ is positive and counter-clockwise if $q$ is negative. With the corresponding electric current is then associated a magnetic moment $\langle \mathbf{M}_2 \rangle$ antiparallel to **B**, and, therefore, a positive coupling energy, which explains the physical origin of the term $H_2$.

To see this more precisely, we shall return to the classical argument of the preceding section, taking into account the fact [*cf.* comment (*iii*) of §c] that the magnetic moment $\mathcal{M}$ is in fact proportional to $\boldsymbol{\lambda} = \mathbf{r} \times m_e \mathbf{v}$ (and not to $\mathcal{L} = \mathbf{r} \times \mathbf{p}$):

$$\mathcal{M} = \frac{q}{2m_e} \boldsymbol{\lambda} = \frac{q}{2m_e} [\mathcal{L} - q\,\mathbf{r} \times \mathbf{A}(\mathbf{r})] \tag{35}$$

When $\mathcal{L}$ is zero, $\mathcal{M}$ reduces, in gauge (2), to:

$$\mathcal{M}_2 = \frac{q^2}{4m_e} \mathbf{r} \times (\mathbf{r} \times \mathbf{B}) = \frac{q^2}{4m_e} [(\mathbf{r} \cdot \mathbf{B})\mathbf{r} - \mathbf{r}^2\mathbf{B}] \tag{36}$$

$\mathcal{M}_2$ is proportional to the value of the magnetic field*. It therefore represents the *moment induced* by **B** in the atom. Its coupling energy with **B** is:

$$\begin{aligned} W_2 &= -\int_0^{\mathbf{B}} \mathcal{M}_2(\mathbf{B}') \cdot \mathrm{d}\mathbf{B}' = -\frac{1}{2} \mathcal{M}_2(\mathbf{B}) \cdot \mathbf{B} \\ &= \frac{q^2}{8m_e} [\mathbf{r}^2\mathbf{B}^2 - (\mathbf{r} \cdot \mathbf{B})^2] \\ &= \frac{q^2}{8m_e} \mathbf{r}_\perp^2 \mathbf{B}^2 \end{aligned} \tag{37}$$

as we found in (7-c). Therefore, the interpretation given above has been confirmed : $H_2$ describes the coupling between the field **B** and the magnetic moment $\mathbf{M}_2$ induced in the atom. Since the induced moment, according to Lenz' law, opposes the applied field, the coupling energy is positive. $H_2$ is called the diamagnetic term of the Hamiltonian.

* $\mathcal{M}_2$ is not collinear with **B**. However, it can be shown that, in the ground state of the hydrogen atom, the mean value $\langle \mathbf{M}_2 \rangle$ of the operator associated with $\mathcal{M}_2$ is antiparallel to **B**. This is in agreement with the result obtained above from the structure of the probability current.

COMMENT:

As we have already pointed out [*cf.* (18)], atomic diamagnetism is a weak phenomenon which is concealed by paramagnetism when both are present. As is shown by (37) (and the calculations of § 1-b), this result is related to the small size of the atomic radius : for magnetic fields of the type usually produced, the magnetic flux intercepted by an atom is very small. It must not be concluded that we can always neglect $H_2$ relative to $H_1$, whatever the physical problem. For example, in the case of a free electron (for which the radius of the classical orbit would be infinite in a zero magnetic field), we saw in complement $E_{VI}$ that the contribution of the diamagnetic term is as important as that of the paramagnetic term.

## 2. The Zeeman effect

Now that we have explained the physical significance of the various terms appearing in the Hamiltonian, we shall look more closely at their effects on the spectrum of the hydrogen atom. More precisely, we shall examine the way in which the emission of the optical line called the "resonance line" ($\lambda \simeq 1\,200$ Å) is modified when the hydrogen atom is placed in a static magnetic field. We shall see that this changes not only the frequency, but also the polarization, of the atomic lines : this is what is usually called the "Zeeman effect".

*Important comment:* In reality, because of the existence of electron and proton spins, the resonance line of hydrogen includes several neighboring components (fine and hyperfine structure; *cf.* chap. XII). Moreover, the spin degrees of freedom profoundly modify the effect of a magnetic field on the various components of the resonance line (the Zeeman effect of the hydrogen atom is sometimes called "anomalous"). Since we are ignoring the effects of spin here, the following calculations do not truly correspond to the real physical situation. However, they can easily be generalized to take spins into account (*cf.* complement $D_{XII}$). Moreover, the results we shall obtain (the appearance of several Zeeman components of different frequencies and polarizations) remain qualitatively valid.

### a. ENERGY LEVELS OF THE ATOM IN THE PRESENCE OF THE MAGNETIC FIELD

The resonance line of hydrogen corresponds to an atomic transition between the ground state $1s$ ($n = 1$ ; $l = m = 0$) and the excited state $2p$ ($n = 2$ ; $l = 1$ ; $m = +1, 0, -1$). While the angular momentum is zero in the ground state, it is not so in the excited state; in calculating optical line modifications in the presence of the magnetic field **B**, we therefore make a small error by neglecting the effects of the diamagnetic term $H_2$, which amounts to taking $H_0 + H_1$ for the Hamiltonian.

We denote by $|\varphi_{n,l,m}\rangle$ the common eigenstates of $H_0$ (eigenvalue $E_n = -E_I/n^2$), $\mathbf{L}^2$ [eigenvalue $l(l+1)\hbar^2$] and $L_z$ (eigenvalue $m\hbar$). The wave functions of these states are those which were calculated in chapter VII:

$$\varphi_{n,l,m}(r, \theta, \varphi) = R_{n,l}(r)\, Y_l^m(\theta, \varphi) \tag{38}$$

We choose the $Oz$ axis parallel to $\mathbf{B}$; it is not difficult to see that the states $|\varphi_{n,l,m}\rangle$ are then also eigenvectors of $H_0 + H_1$:

$$\begin{aligned}(H_0 + H_1)\,|\varphi_{n,l,m}\rangle &= \left(H_0 - \frac{\mu_B}{\hbar} B\, L_z\right) |\varphi_{n,l,m}\rangle \\ &= (E_n - m\mu_B B)\,|\varphi_{n,l,m}\rangle\end{aligned} \tag{39}$$

If we neglect the diamagnetic term, the stationary states of the atom placed in the field $\mathbf{B}$ are therefore still the $|\varphi_{n,l,m}\rangle$; only the corresponding energies are modified.

In particular, for the states involved in the resonance line, we see that:

$$(H_0 + H_1)\,|\varphi_{1,0,0}\rangle = -E_I\,|\varphi_{1,0,0}\rangle \tag{40-a}$$

$$(H_0 + H_1)\,|\varphi_{2,1,m}\rangle = [-E_I + \hbar(\Omega + m\omega_L)]\,|\varphi_{2,1,m}\rangle \tag{40-b}$$

where:

$$\Omega = \frac{E_2 - E_1}{\hbar} = \frac{3E_I}{4\hbar} \tag{41}$$

is the angular frequency of the resonance line in a zero field.

### b. ELECTRIC DIPOLE OSCILLATIONS

#### α. *Matrix elements of the operator associated with the dipole*

Let:

$$\mathbf{D} = q\mathbf{R} \tag{42}$$

be the electric dipole operator of the atom. To calculate the mean value $\langle \mathbf{D} \rangle$ of this dipole, we begin by evaluating the matrix elements of $\mathbf{D}$.

Under reflection through the origin, $\mathbf{D}$ is changed into $-\mathbf{D}$: the electric dipole is therefore an odd operator (*cf.* complement $F_{II}$). Now, the states $|\varphi_{n,l,m}\rangle$ also have a well-defined parity: since their angular dependence is given by $Y_l^m(\theta, \varphi)$, their parity is $+1$ if $l$ is even and $-1$ if $l$ is odd (*cf.* complement $A_{VI}$). It follows, in particular, that:

$$\begin{cases} \langle \varphi_{1,0,0} | \mathbf{D} | \varphi_{1,0,0} \rangle = \mathbf{0} \\ \langle \varphi_{2,1,m'} | \mathbf{D} | \varphi_{2,1,m} \rangle = \mathbf{0} \end{cases} \tag{43}$$

for all $m$ and $m'$.

The non-zero matrix elements of $\mathbf{D}$ are therefore necessarily non-diagonal elements. To calculate the matrix elements $\langle \varphi_{2\,1\,m} | \mathbf{D} | \varphi_{1\,0\,0} \rangle$, it is convenient to note that $x$, $y$ and $z$ can easily be expressed in terms of the spherical harmonics:

$$\begin{cases} x = \sqrt{\frac{2\pi}{3}}\, r[Y_1^{-1}(\theta, \varphi) - Y_1^1(\theta, \varphi)] \\ y = i\sqrt{\frac{2\pi}{3}}\, r[Y_1^{-1}(\theta, \varphi) + Y_1^1(\theta, \varphi)] \\ z = \sqrt{\frac{4\pi}{3}}\, r Y_1^0(\theta, \varphi) \end{cases} \tag{44}$$

In the expressions for the desired matrix elements, we therefore have:

– on the one hand, a radial integral, which we shall set equal to $\chi$:

$$\chi = \int_0^\infty R_{2,1}(r)\, R_{1,0}(r)\, r^3\, \mathrm{d}r \tag{45}$$

– and on the other hand, an angular integral which, thanks to relations (44), reduces to a scalar product of spherical harmonics, which can be calculated directly from their orthogonality relations. We obtain, finally:

$$\begin{cases} \langle \varphi_{2,1,1} | D_x | \varphi_{1,0,0} \rangle = - \langle \varphi_{2,1,-1} | D_x | \varphi_{1,0,0} \rangle = - \frac{q\chi}{\sqrt{6}} \\ \langle \varphi_{2,1,0} | D_x | \varphi_{1,0,0} \rangle = 0 \end{cases} \tag{46-a}$$

$$\begin{cases} \langle \varphi_{2,1,1} | D_y | \varphi_{1,0,0} \rangle = \langle \varphi_{2,1,-1} | D_y | \varphi_{1,0,0} \rangle = \frac{iq\chi}{\sqrt{6}} \\ \langle \varphi_{2,1,0} | D_y | \varphi_{1,0,0} \rangle = 0 \end{cases} \tag{46-b}$$

$$\begin{cases} \langle \varphi_{2,1,1} | D_z | \varphi_{1,0,0} \rangle = \langle \varphi_{2,1,-1} | D_z | \varphi_{1,0,0} \rangle = 0 \\ \langle \varphi_{2,1,0} | D_z | \varphi_{1,0,0} \rangle = \frac{q\chi}{\sqrt{3}} \end{cases} \tag{46-c}$$

#### β. *Calculation of the mean value of the dipole*

The results of §α indicate that, if the system is in a stationary state, the mean value of the operator **D** is zero. Let us assume, rather, that the state vector of the system is initially a linear superposition of the ground state $1s$ and one of the $2p$ states:

$$| \psi_m(0) \rangle = \cos\alpha \, | \varphi_{1,0,0} \rangle + \sin\alpha \, | \varphi_{2,1,m} \rangle \tag{47}$$

with $m = +1, 0$ or $-1$ ($\alpha$ is a real parameter). We then immediately obtain the state vector at time $t$:

$$| \psi_m(t) \rangle = \cos\alpha \, | \varphi_{1,0,0} \rangle + \sin\alpha \, \mathrm{e}^{-i(\Omega + m\omega_L)t} \, | \varphi_{2,1,m} \rangle \tag{48}$$

(we have omitted the global phase factor $\mathrm{e}^{iE_I t/\hbar}$, which is of no physical consequence).

To calculate the mean value of the electric dipole:

$$\langle \mathbf{D} \rangle_m(t) = \langle \psi_m(t) | \mathbf{D} | \psi_m(t) \rangle \tag{49}$$

we shall use results (46) and (48), and cite three cases:

(*i*) if $m = 1$, we obtain:

$$\begin{cases} \langle D_x \rangle_1 = -\dfrac{q\chi}{\sqrt{6}} \sin 2\alpha \cos [(\Omega + \omega_L)t] \\ \langle D_y \rangle_1 = -\dfrac{q\chi}{\sqrt{6}} \sin 2\alpha \sin [(\Omega + \omega_L)t] \\ \langle D_z \rangle_1 = 0 \end{cases} \tag{50}$$

The vector $\langle \mathbf{D} \rangle_1(t)$ therefore rotates in the $xOy$ plane about the $Oz$ axis, in the counterclockwise direction and with the angular velocity $\Omega + \omega_L$.

(*ii*) if $m = 0$:

$$\begin{cases} \langle D_x \rangle_0 = \langle D_y \rangle_0 = 0 \\ \langle D_z \rangle_0 = \dfrac{q\chi}{\sqrt{3}} \sin 2\alpha \cos \Omega t \end{cases} \tag{51}$$

The motion of $\langle \mathbf{D} \rangle_0(t)$ is now a linear oscillation along the $Oz$ axis, of angular frequency $\Omega$.

(*iii*) if $m = -1$:

$$\begin{cases} \langle D_x \rangle_{-1} = \dfrac{q\chi}{\sqrt{6}} \sin 2\alpha \cos [(\Omega - \omega_L)t] \\ \langle D_y \rangle_{-1} = -\dfrac{q\chi}{\sqrt{6}} \sin 2\alpha \sin [(\Omega - \omega_L)t] \\ \langle D_z \rangle_{-1} = 0 \end{cases} \tag{52}$$

The vector $\langle \mathbf{D} \rangle_{-1}(t)$ again rotates in the $xOy$ plane about $Oz$, but this time in the clockwise direction and with the angular velocity $\Omega - \omega_L$.

### c. FREQUENCY AND POLARIZATION OF EMITTED RADIATION

In the three cases ($m = +1, 0$ and $-1$), the mean value of the electric dipole is an oscillating function of time. It is clear that such a dipole radiates electromagnetic energy.

Since the atomic dimensions are negligible compared to the optical wavelength, the atom's radiation at great distances can be treated like that of a dipole. We shall assume that the characteristics of the light emitted (or absorbed) by the atom during transition between a state $|\varphi_{2,1,m}\rangle$ and the ground state are correctly given

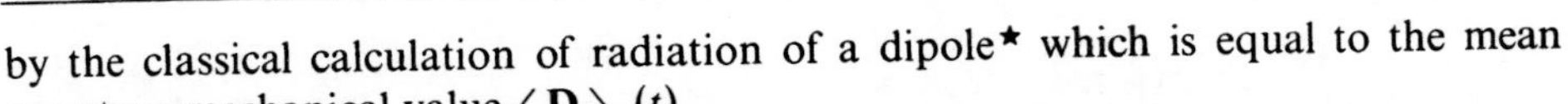

by the classical calculation of radiation of a dipole★ which is equal to the mean quantum mechanical value $\langle \mathbf{D} \rangle_m(t)$.

To state the problem precisely, we shall assume that we want to study the radiation emitted by a sample containing a great number of hydrogen atoms, which have somehow been excited into a $2p$ state. In most experiments actually performed, the excitation of the atoms is isotropic, and the three states $|\varphi_{2,1,1}\rangle$, $|\varphi_{2,1,0}\rangle$ and $|\varphi_{2,1,-1}\rangle$ occur with equal probability. Therefore, we shall begin by calculating the radiation diagram for each of the cases of the preceding sections. Then we shall obtain the radiation actually emitted by the atomic system by taking, for each spatial direction, the sum of the light intensities emitted in each case.

(*i*) If $m = 1$, the angular frequency of the emitted radiation is $(\Omega + \omega_L)$. The optical line frequency is therefore slightly shifted by the magnetic field. In accordance with the laws of classical electromagnetism applied to a rotating dipole such as $\langle \mathbf{D} \rangle_1(t)$, the radiation emitted in the $Oz$ direction is circularly polarized (the corresponding polarization is called $\sigma_+$). However, the radiation emitted in a direction of the $xOy$ plane is linearly polarized (parallel to this plane). In other directions, the polarization is elliptical.

(*ii*) If $m = 0$, we must consider a dipole oscillating linearly along $Oz$, with angular frequency $\Omega$, that is, the same as in a zero field. The wavelength of the radiation is therefore not changed by the field $\mathbf{B}$. Its polarization is always linear, whatever the propagation direction being considered. For example, for a propagation direction situated in the $xOy$ plane, this polarization is parallel to $Oz$ ($\pi$ polarization). No radiation is emitted in the $Oz$ direction (an oscillating linear dipole does not radiate along its axis).

(*iii*) If $m = -1$, the results are analogous to those for $m = 1$. The only difference is that the angular frequency of the radiation is $(\Omega - \omega_L)$ instead of $(\Omega + \omega_L)$, and the dipole rotates in the opposite direction; this changes, for example, the direction of the circular polarization ($\sigma_-$ polarization).

If we now assume that there are equal numbers of excited atoms in the three states $m = +1, 0$ and $-1$, we see that:

– in an arbitrary spatial direction, three optical frequencies are emitted: $\Omega/2\pi$, $(\Omega \pm \omega_L)/2\pi$. The polarization associated with the first one is linear, and that associated with the others is, in general, elliptical;

– in a direction perpendicular to the field $\mathbf{B}$, the three polarizations are linear (*cf.* fig. 2). The first one is parallel to $\mathbf{B}$, and the other two are perpendicular. The intensity of the central line is twice that of each of the shifted lines [*cf.* formulas (50), (51) and (52)]. In a direction parallel to $\mathbf{B}$, only the two shifted frequencies $(\Omega \pm \omega_L)/2\pi$ are emitted, and the associated light polarizations are both circular but opposite in direction (*cf.* fig. 3).

★ If we wanted to treat the problem entirely quantum mechanically, we should have to use the quantum mechanical theory of radiation. In particular, the return of the atom to the ground state by spontaneous emission of a photon could only be understood in the framework of this theory. However, the results we shall obtain here semi-classically would remain essentially valid as far as radiation is concerned.

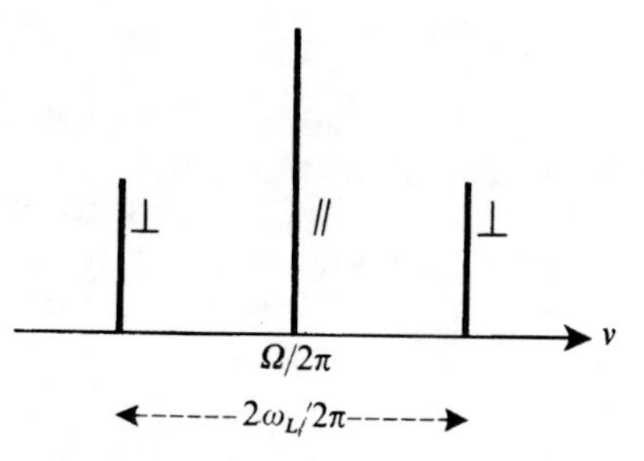



FIGURE 2

**The Zeeman components of the resonance line of hydrogen observed in a direction perpendicular to the magnetic field B (ignoring electron spin). We obtain a component of unshifted frequency $\nu$, polarized parallel to B, and two shifted components $\pm\, \omega_L/2\pi$, polarized perpendicularly to B.**

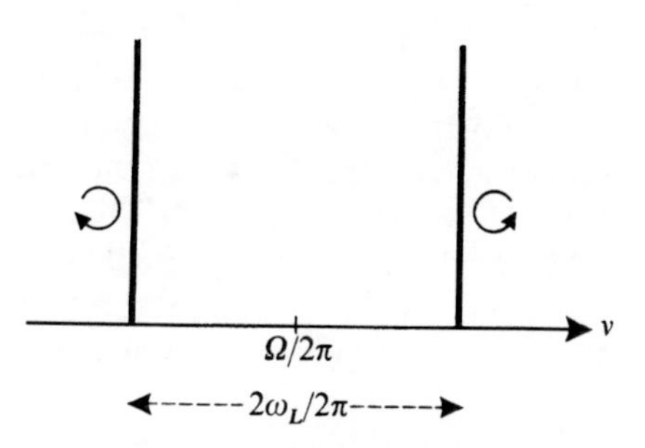



FIGURE 3

**When the observation is carried out along the direction of the field B, only two Zeeman components are obtained, circularly polarized in opposite directions and shifted by $\pm\, \omega_L/2\pi$.**

COMMENT:

The atom therefore emits $\sigma_+$-polarized radiation in going from the state $|\varphi_{2,1,1}\rangle$ to the state $|\varphi_{1,0,0}\rangle$, $\sigma_-$ in going from $|\varphi_{2,1,-1}\rangle$ to $|\varphi_{1,0,0}\rangle$, and $\pi$ in going from $|\varphi_{2,1,0}\rangle$ to $|\varphi_{1,0,0}\rangle$. Formulas (46) furnish a simple rule for finding these polarizations. Consider the operators $D_x + iD_y$, $D_x - iD_y$ and $D_z$; their only non-zero matrix elements between the $2p$ and $1s$ states taken in this order are:

$$\langle \varphi_{2,1,1} | D_x + iD_y | \varphi_{1,0,0} \rangle, \quad \langle \varphi_{2,1,-1} | D_x - iD_y | \varphi_{1,0,0} \rangle$$

and $\langle \varphi_{2,1,0} | D_z | \varphi_{1,0,0} \rangle$.

To the $\sigma_+$, $\sigma_-$ and $\pi$ polarizations, therefore, correspond the operators $D_x + iD_y$, $D_x - iD_y$ and $D_z$, respectively. This is a general rule: there is emission of electric dipole radiation when the operator **D** has a non-zero matrix element between the atom's initial state and its final state. The polarization of this radiation is $\sigma_+$, $\sigma_-$ or $\pi$ depending on whether the non-zero matrix element* is that of $D_x + iD_y$, $D_x - iD_y$ or $D_z$.

**References and suggestions for further reading:**

Paramagnetism and diamagnetism: Feynman II (7.2), chaps. 34 and 35; Cagnac and Pebay-Peyroula (11.2), chaps. VIII and IX; Kittel (13.2), chap. 14; Slater (1.6), chap. 14; Flügge (1.24), §§128 and 160.

Dipole radiation: Cagnac and Pebay-Peyroula (11.2), Annex III; Panofsky and Phillips (7.6), §14-7; Jackson (7.5), §9-2.

Angular momentum of radiation and selection rules: Cagnac and Pebay-Peyroula (11.2), chap. XI.

* The order of the states in the matrix element must be respected in order not to confuse $\sigma_+$ with $\sigma_-$.

# Complement $E_{VII}$

## SOME ATOMIC ORBITALS. HYBRID ORBITALS



### 1. Introduction

In §C of chapter VII, we determined an orthonormal basis of stationary states for the electron of the hydrogen atom. The corresponding wave functions are:

$$\varphi_{n,l,m}(\mathbf{r}) = R_{n,l}(r)Y_l^m(\theta, \varphi) \tag{1}$$

and the quantum numbers $n$, $l$, $m$ refer, respectively, to the energy $E_n = -E_I/n^2$, the square of the angular momentum $l(l+1)\hbar^2$, and the $Oz$-component of the angular momentum $m\hbar$.

By linearly superposing stationary states of the same energy, that is, of the same quantum number $n$, we can construct new stationary states which no longer necessarily correspond to well-defined values of $l$ and $m$. In this complement, we intend to study the properties of some of these new stationary states – in particular, the angular dependence of the associated wave functions.

The wave functions (1) are often called *atomic orbitals.* A linear superposition of orbitals of the same $n$ but different $l$ and $m$ is called a *hybrid orbital.* We shall see that a hybrid orbital can extend further in certain spatial directions than the (pure) orbitals from which it is constructed. It is this property, important in the formation of chemical bonds, which justifies the introduction of hybrid orbitals.

Although the calculations presented in this complement are rigorously valid only for the hydrogen atom, we shall also indicate qualitatively how they explain the geometrical structure of the various bonds formed by an atom with several valence electrons.

## 2. Atomic orbitals associated with real wave functions

In expression (1), the radial function $R_{n,l}(r)$ is real. However, $Y_l^m(\theta, \varphi)$, except for $m = 0$, is a complex function of $\varphi$, since:

$$Y_l^m(\theta, \varphi) = F_l^m(\theta)\ \mathrm{e}^{im\varphi} \tag{2}$$

where $F_l^m(\theta)$ is a real function of $\theta$.

Atomic orbitals are therefore generally complex functions. By superposing the $\varphi_{n,l,m}(\mathbf{r})$ and $\varphi_{n,l,-m}(\mathbf{r})$ orbitals, one can, however, construct real orbitals whose advantage is their simple angular dependence, which can be represented graphically without having to take the square of the modulus of the wave function (as we did in § C-4-c-α of chapter VII).

### a. *s* ORBITALS ($l = 0$)

When $l = m = 0$, the wave function $\varphi_{n,0,0}(\mathbf{r})$ is real, and we say we are dealing with an "*s* orbital". We shall denote the corresponding stationary state by $|\varphi_{ns}\rangle$. To represent the angular dependence of the *ns* orbital, we fix $r$ and measure off in each direction of polar angles $\theta$ and $\varphi$ a line segment of length $\varphi_{ns}(r, \theta, \varphi)$. The surface obtained by varying $\theta$ and $\varphi$ is a sphere centered at $O$ (fig. 1).

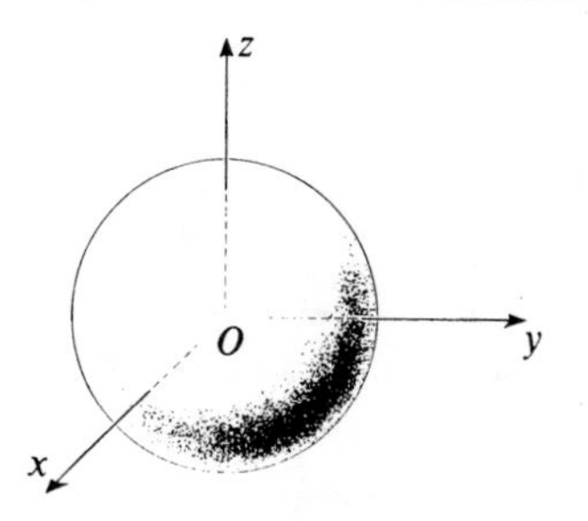


FIGURE 1

**An *s* orbital is spherically symmetrical: the wave function depends on neither $\theta$ nor $\varphi$.**

### b. *p* ORBITALS ($l = 1$)

#### α. $p_z$, $p_x$, $p_y$ *orbitals*

If we use the expression for the three spherical harmonics $Y_1^m(\theta, \varphi)$ given in complement $A_{VI}$ [formulas (32)], we obtain, for the three atomic orbitals $\varphi_{n,1,m}(\mathbf{r})$ corresponding to $l = 1$:

$$\begin{cases} \varphi_{n,1,1}(\mathbf{r}) = -\sqrt{\dfrac{3}{8\pi}}\, R_{n,1}(r) \sin\theta\ \mathrm{e}^{i\varphi} & \text{(3-a)} \\ \varphi_{n,1,0}(\mathbf{r}) = \sqrt{\dfrac{3}{4\pi}}\, R_{n,1}(r) \cos\theta & \text{(3-b)} \\ \varphi_{n,1,-1}(\mathbf{r}) = \sqrt{\dfrac{3}{8\pi}}\, R_{n,1}(r) \sin\theta\ \mathrm{e}^{-i\varphi} & \text{(3-c)} \end{cases}$$

We now form the three linear superpositions:

$$\left\{\begin{array}{ll}\varphi_{n,1,0}(\mathbf{r}) & \text{(4-a)}\\ -\dfrac{1}{\sqrt{2}}\left[\varphi_{n,1,1}(\mathbf{r}) - \varphi_{n,1,-1}(\mathbf{r})\right] & \text{(4-b)}\\ \dfrac{i}{\sqrt{2}}\left[\varphi_{n,1,1}(\mathbf{r}) + \varphi_{n,1,-1}(\mathbf{r})\right] & \text{(4-c)}\end{array}\right.$$

It is easy to see that the three preceding wave functions can also be written:

$$\left\{\begin{array}{ll}\sqrt{\dfrac{3}{4\pi}}\,R_{n,1}(r)\,\dfrac{z}{r} & \text{(5-a)}\\ \sqrt{\dfrac{3}{4\pi}}\,R_{n,1}(r)\,\dfrac{x}{r} & \text{(5-b)}\\ \sqrt{\dfrac{3}{4\pi}}\,R_{n,1}(r)\,\dfrac{y}{r} & \text{(5-c)}\end{array}\right.$$

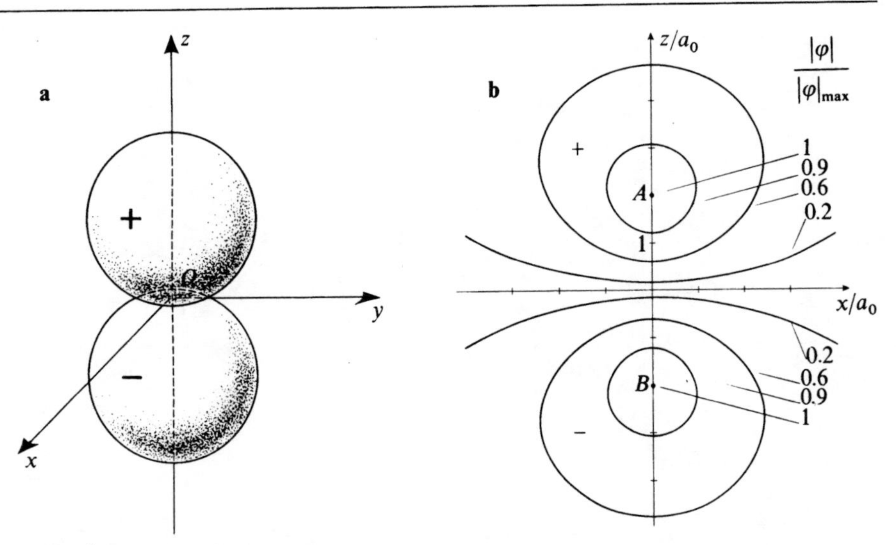


FIGURE 2

**Two possible representations of a $p_z$ orbital ($l = 1$, $m = 0$)**
**fig. a: angular dependence of this orbital. With $r$ fixed, we display $|\varphi_{n,l=1,m=0}(r, \theta, \varphi)|$ for each direction $\theta$, $\varphi$. Thus we obtain two spheres, tangential at $O$ to the $xOy$ plane. The sign indicated on each of them is that of the wave function (which is real).**
**fig. b: cross sections in the $xOz$ plane of a family of surfaces, each one corresponding to a given value for $|\varphi_{n,l=1,m=0}(r, \theta, \varphi)|$ [we have chosen values equal to 0.2, 0.6, 0.9 times the maximum value of $|\varphi|$ at points $A$ and $B$]. These are surfaces of revolution about $Oz$. The sign indicated is that of the wave function (which is real). Unlike the one in figure a, the representation in figure b depends on the radial part of the wave function (the one chosen here corresponds to the $n = 2$ state of the hydrogen atom).**

These are real functions of $r$, $\theta$, $\varphi$ which, like the $\varphi_{n,1,m}(\mathbf{r})$, are orthonormal and form a basis in the subspace $\mathscr{E}_{n,l=1}$. They are called, respectively, "$p_z$, $p_x$, $p_y$ orbitals". We shall write the wave functions (5) $\varphi_{np_z}(\mathbf{r})$, $\varphi_{np_x}(\mathbf{r})$ and $\varphi_{np_y}(\mathbf{r})$.

Two distinct geometrical representations enable us to visualize the form of an orbital $\psi(r, \theta, \varphi)$. First of all, if we are interested in the angular dependence of the orbital, we fix $r$ and measure off a line segment of length $|\psi(r, \theta, \varphi)|$ along each direction of polar angles $\theta$ and $\varphi$. Thus, the angular dependence of the $2p_z$ orbital is that of $z/r = \cos\theta$. As $\varphi$ varies between 0 and $2\pi$, and $\theta$ varies between 0 and $\pi$, the end of the line segment of length $|\cos\theta|$ drawn in the direction of polar angles $\theta$ and $\varphi$ describes two spheres centered on the $Oz$ axis, tangential at $O$ to the $xOy$ plane and mirror images with respect to the $xOy$ plane (fig. 2-a). The sign indicated in the figure is that of the wave function, which is real. Another possible representation of the orbital $\psi(r, \theta, \varphi)$ is obtained by tracing a family of surfaces, each one corresponding to a given value of $|\psi(r, \theta, \varphi)|$ (surfaces of equal probability density). This is what is done for the $2p_z$ orbital in figure 2-b (here again, the sign indicated is that of the wave function, which is real). In the rest of this complement, we shall use one or the other of these two representations.

The $p_x$ and $p_y$ orbitals can be obtained respectively from the $p_z$ orbitals by rotations through angles of $+\pi/2$ and $-\pi/2$ about $Oy$ and $Ox$ (*cf.* figures 3 and 4, which use a geometric representation identical to the one in figure 2-a).

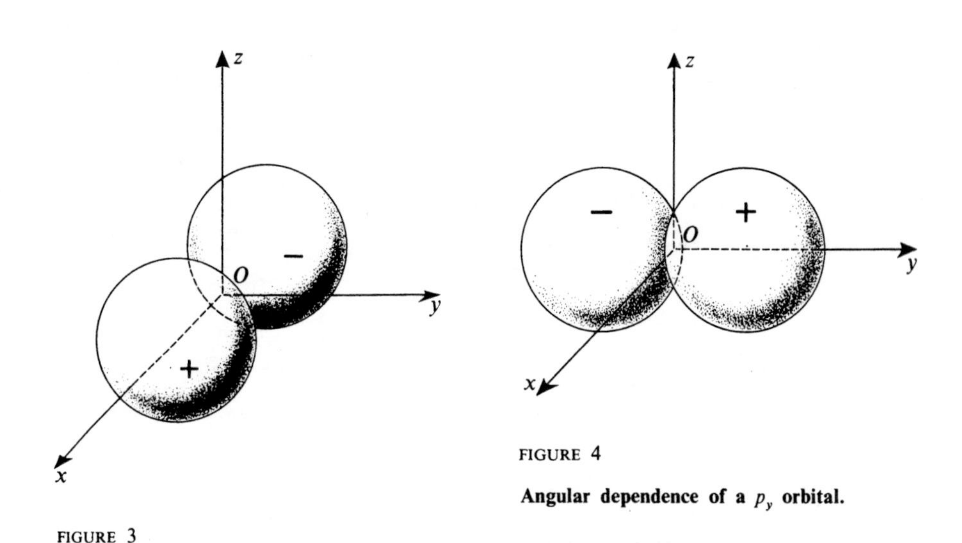


FIGURE 3

**Angular dependence of a $p_x$ orbital (the representation adopted is that of figure 2-a).**

FIGURE 4

**Angular dependence of a $p_y$ orbital.**

Unlike an $s$ orbital, which is spherically symmetrical, the $p_z$, $p_x$, $p_y$ orbitals therefore point along the $Oz$, $Ox$, $Oy$ axes, respectively.

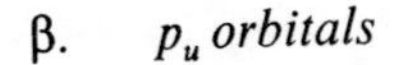

### β. $p_u$ *orbitals*

The choice of the $Ox$, $Oy$, $Oz$ axes is obviously arbitrary. By linearly superposing the $p_x$, $p_y$ and $p_z$ orbitals, we should therefore be able to construct a $p_u$ orbital having the same form but directed along an arbitrary $Ou$ axis.

Let $Ou$ be such an axis, forming angles $\alpha$, $\beta$, $\gamma$ with $Ox$, $Oy$, $Oz$. We have, obviously:

$$\cos^2 \alpha + \cos^2 \beta + \cos^2 \gamma = 1 \tag{6}$$

Consider the state:

$$\cos \alpha \mid np_x \rangle + \cos \beta \mid np_y \rangle + \cos \gamma \mid np_z \rangle \tag{7}$$

which, according to (6), is normalized. We can, using formulas (5), put the corresponding wave function in the form:

$$\sqrt{\frac{3}{4\pi}}\, R_{n,1}(r) \frac{x \cos \alpha + y \cos \beta + z \cos \gamma}{r} = \sqrt{\frac{3}{4\pi}}\, R_{n,1}(r) \frac{u}{r} \tag{8}$$

where:

$$u = x \cos \alpha + y \cos \beta + z \cos \gamma \tag{9}$$

is the projection of $\mathbf{r}$ onto the $Ou$ axis. Comparison with (5) indicates that the orbital so constructed is indeed a $p_u$ orbital.

Therefore, any real and normalized linear superposition of $p_x$, $p_y$ and $p_z$ orbitals:

$$\lambda\, \varphi_{np_x}(\mathbf{r}) + \mu\, \varphi_{np_y}(\mathbf{r}) + \nu\, \varphi_{np_z}(\mathbf{r}) \tag{10}$$

can be considered to be a $p_u$ orbital directed along the $Ou$ direction, defined by:

$$\begin{cases} \cos \alpha = \lambda \\ \cos \beta = \mu \\ \cos \gamma = \nu \end{cases} \tag{11}$$

### γ. *Example: structure of the* $H_2O$ *and* $H_3N$ *molecules*

To a first approximation (*cf.* complement $A_{XIV}$), in a many-electron atom, each electron can be considered to move independently of the others in a central potential $V_c(r)$ which is the sum of the electrostatic attractive potential of the nucleus and a "mean potential" due to the repulsion of the other electrons. Each electron can therefore be found in a state characterized by three quantum numbers, $n$, $l$, $m$. However, since the potential $V_c(r)$ no longer varies exactly like $1/r$, the energy no longer depends only on $n$, but also on $l$. We shall see in complement $A_{XIV}$ that the energy of the $2s$ state is slightly lower than that of the $2p$ state; the $3s$ state is also lower than the $3p$ state, which is, in turn, lower than the $3d$ state, etc.

The existence of spin and Pauli's principle (which we shall study in chapters IX and XIV) imply that the $1s$, $2s$, ... sub-shells can contain only two electrons; the $2p$, $3p$, ... sub-shells, six electrons, ...; the $nl$ sub-shells, $2(2l + 1)$ electrons (the factor $2l + 1$ arises from the degeneracy related to $L_z$, and the factor 2, from the electron spin).

Thus, for the oxygen atom, which has eight electrons, the $1s$ and $2s$ sub-shells are filled and contain four electrons in all. The four remaining electrons are in the $2p$ sub-shell: two of them (with opposite spins) can fill one of the three $2p$ orbitals, for example, $2p_z$; the other two are then distributed in the remaining $2p_x$ and $2p_y$ orbitals. These last two electrons are the valence electrons: they are "unpaired", which means that the orbitals in which they can be found can accept another electron. The $2p_x$ and $2p_y$ wave functions of the valence electrons of oxygen are therefore directed along two perpendicular axes. Now, it can be shown that, the greater the overlapping of the wave functions of the two electrons participating in a chemical bond, the greater the stability of this bond. The two hydrogen atoms which will bind with the oxygen atom to form a water molecule must therefore be centered respectively on the $Ox$ and $Oy$ axes. Then the spherical $1s$ orbital of the valence electron of each hydrogen will maximally overlap one of the $2p_x$ and $2p_y$ orbitals of the valence electrons of oxygen. Figure 5 represents the shape of the probability clouds associated with the valence electrons of the oxygen and hydrogen atoms in the water molecule. The graphical representation used is analogous to the one in figure 2-b. We have drawn, for each electron, a surface defined as follows: the probability density has the same value at all points of this surface; this value is chosen in such a way that the total probability contained inside the surface has a fixed value close to 1 (0.9, for example).

The preceding argument enables us to understand the form of the $H_2O$ molecule. The angle between the two OH bonds should be close to 90°. Actually, the angle found experimentally is 104°. The deviation from the value 90° arises, in part, from the electrostatic repulsion between the two protons of the hydrogen atoms, which tends to open up the angle between the two OH bonds*.

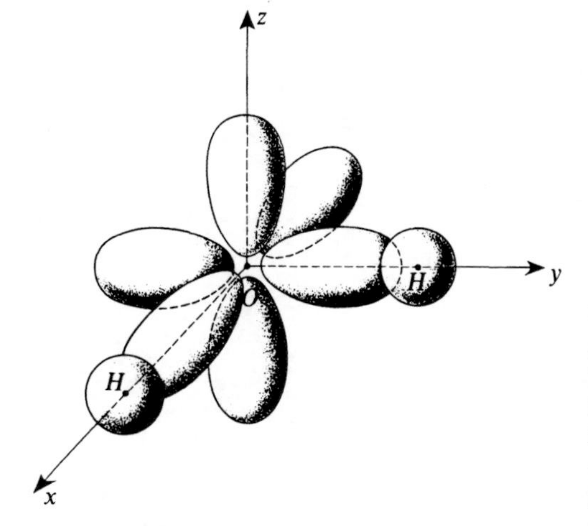


FIGURE 5

**Schematic structure of the water molecule $H_2O$. The $2p_x$ and $2p_y$ orbitals yields bonds making an angle of approximately 90° (the real angle is 104° because of the electrostatic repulsion between the two protons).**

An analogous argument explains the pyramidal form of the $NH_3$ molecule. The three valence electrons of nitrogen occupy the $2p_x$, $2p_y$, $2p_z$ orbitals, directed at right angles to each other. Here again, the electrostatic repulsion between the protons of the three hydrogen atoms causes the bond angle to go from 90° to 108° (through a slight hybridization of the $2s$ and $2p$ orbitals).

* The opening up of the angle between the two OH bonds can be described as the result of a slight $sp^3$ hybridization of the $2p$ and $2s$ orbitals (*cf.* §5).

### c. OTHER VALUES OF *l*

We have confined ourselves thus far to the *s* and *p* orbitals. Actually, an orthonormal basis of real orbitals can be constructed for every value of *l*. If we note that [*cf.* relation (D-29) of chapter VI]:

$$[Y_l^m(\theta, \varphi)]^* = (-1)^m Y_l^{-m}(\theta, \varphi) \tag{12}$$

we immediately see that (for $m \neq 0$) the two complex functions $\varphi_{n,l,m}(\mathbf{r})$ and $\varphi_{n,l,-m}(\mathbf{r})$ can be replaced by the two functions:

$$\frac{1}{\sqrt{2}}[\varphi_{n,l,m}(\mathbf{r}) + (-1)^m \varphi_{n,l,-m}(\mathbf{r})] \tag{13-a}$$

$$\frac{i}{\sqrt{2}}[\varphi_{n,l,m}(\mathbf{r}) - (-1)^m \varphi_{n,l,-m}(\mathbf{r})] \tag{13-b}$$

which are real and orthonormal.

Thus, for $l = 2$ ("*d* orbitals"), we can construct five real orbitals, for which the angular dependence is given by:

$$\sqrt{\frac{1}{2}}(3\cos^2\theta - 1), \sqrt{6}\sin\theta\cos\theta\cos\varphi, \sqrt{6}\sin\theta\cos\theta\sin\varphi,$$

$$\sqrt{\frac{3}{2}}\sin^2\theta\cos 2\varphi, \sqrt{\frac{3}{2}}\sin^2\theta\sin 2\varphi$$

($d_{3z^2-r^2}$, $d_{zx}$, $d_{zy}$, $d_{x^2-y^2}$, $d_{xy}$ orbitals).
The form of these orbitals is a little more complicated than that of the *s* and *p* orbitals to which we shall confine ourselves here. However, it is possible to apply to them arguments of the same type as those below.

## 3. *sp* hybridization

### a. INTRODUCTION OF *sp* HYBRID ORBITALS

Returning to the hydrogen atom, we shall consider the subspace $\mathscr{E}_{ns} \oplus \mathscr{E}_{np}$, subtended by the four real orbitals $\varphi_{ns}(\mathbf{r})$, $\varphi_{np_x}(\mathbf{r})$, $\varphi_{np_y}(\mathbf{r})$ and $\varphi_{np_z}(\mathbf{r})$ (which correspond to the same energy). We shall show that, by linearly superposing *ns* and *np* orbitals, we can construct other real orbitals, which form an orthonormal basis in $\mathscr{E}_{ns} \oplus \mathscr{E}_{np}$ and possess some interesting properties.

We shall begin by linearly superposing the $\varphi_{ns}(\mathbf{r})$ and $\varphi_{np_z}(\mathbf{r})$ orbitals alone, without using $\varphi_{np_x}(\mathbf{r})$ and $\varphi_{np_y}(\mathbf{r})$. We shall therefore replace the two functions $\varphi_{ns}(\mathbf{r})$ and $\varphi_{np_z}(\mathbf{r})$ by the two real orthonormal linear combination:

$$\begin{cases} \cos\alpha\, \varphi_{ns}(\mathbf{r}) + \sin\alpha\, \varphi_{np_z}(\mathbf{r}) & \text{(14-a)} \\ \sin\alpha\, \varphi_{ns}(\mathbf{r}) - \cos\alpha\, \varphi_{np_z}(\mathbf{r}) & \text{(14-b)} \end{cases}$$

In addition, we shall require the two orbitals (14-a) and (14-b) to have the same geometrical form. Since this form depends only on the relative amounts of the $s$ and $p$ orbitals in the linear superposition, we see immediately that we must have $\sin\alpha = \cos\alpha$, that is, $\alpha = \pi/4$. The two new orbitals we are introducing are therefore of the form:

$$\begin{cases} \varphi_{n,s,p_z}(\mathbf{r}) = \dfrac{1}{\sqrt{2}}\left[\varphi_{ns}(\mathbf{r}) + \varphi_{np_z}(\mathbf{r})\right] & \text{(15-a)} \\ \varphi'_{n,s,p_z}(\mathbf{r}) = \dfrac{1}{\sqrt{2}}\left[\varphi_{ns}(\mathbf{r}) - \varphi_{np_z}(\mathbf{r})\right] & \text{(15-b)} \end{cases}$$

and correspond to what is called "*sp* hybridization". Thus we have constructed a new orthonormal basis of $\mathscr{E}_{ns} \oplus \mathscr{E}_{np}$, composed of $\varphi_{n,s,p_z}(\mathbf{r})$, $\varphi'_{n,s,p_z}(\mathbf{r})$, $\varphi_{np_x}(\mathbf{r})$ and $\varphi_{np_y}(\mathbf{r})$.

#### b. PROPERTIES OF *sp* HYBRID ORBITALS

To study the angular dependence of the $\varphi_{n,s,p_z}(\mathbf{r})$ and $\varphi'_{n,s,p_z}(\mathbf{r})$ hybrid orbitals, we shall choose a given value $r_0$ of $r$ and set:

$$\begin{aligned} \lambda &= \sqrt{\frac{1}{4\pi}}\, R_{n,0}(r_0) \\ \mu &= \sqrt{\frac{3}{4\pi}}\, R_{n,1}(r_0) \end{aligned} \qquad (16)$$

Thus we obtain, using (5) and (15), the angular functions:

$$\begin{cases} \dfrac{1}{\sqrt{2}}(\lambda + \mu\cos\theta) \\ \dfrac{1}{\sqrt{2}}(\lambda - \mu\cos\theta) \end{cases} \qquad (17)$$

which we shall represent, using the same method as in §2 (*cf.* fig. 2-a), by measuring off, along each direction of polar angles $\theta$ and $\varphi$, a line segment of length $\dfrac{1}{\sqrt{2}}|\lambda + \mu\cos\theta|$ or $\dfrac{1}{\sqrt{2}}|\lambda - \mu\cos\theta|$ and indicating by a plus or minus sign whether the wave function is positive or negative. Figure 6 represents the cross sections in the $xOz$ plane of the surfaces so obtained, which have cylindrical symmetry with respect to $Oz$ (we have assumed $\mu > \lambda > 0$). The $\varphi_{n,s,p_z}(\mathbf{r})$ orbital can be transformed into the $\varphi'_{n,s,p_z}(\mathbf{r})$ by a reflection through the point $O$. It can be seen that the $\varphi_{n,s,p_z}(\mathbf{r})$ orbital has no simple symmetry with respect to the point $O$. This asymmetry is due to the fact that the $\varphi_{np_z}(\mathbf{r})$ and $\varphi_{ns}(\mathbf{r})$ orbitals of which it is formed (and which are shown in figure 6-c) have opposite parities. In the region where $z > 0$, $\varphi_{ns}(\mathbf{r})$ and $\varphi_{np_z}(\mathbf{r})$ have the same sign and add, while in the region where $z < 0$, $\varphi_{ns}(\mathbf{r})$ and $\varphi_{np_z}(\mathbf{r})$ have opposite signs and subtract. The conclusions are reversed for $\varphi'_{n,s,p_z}(\mathbf{r})$.

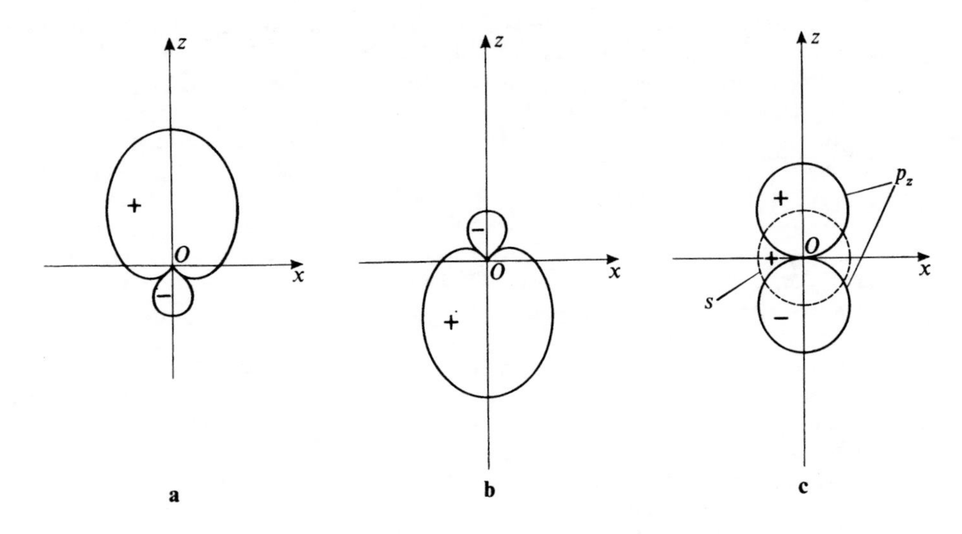


FIGURE 6

**Angular dependence of the $\varphi_{n,s,p_z}(\mathbf{r})$ (fig. a) and $\varphi'_{n,s,p_z}(\mathbf{r})$ (fig. b), hybrid orbitals obtained from the $\varphi_{n\,s}(\mathbf{r})$ and $\varphi_{n\,p_z}(\mathbf{r})$ orbitals, which have opposite parities (fig. c). A hybrid orbital can extend further in certain directions than the pure orbitals from which it is obtained.**

The $\varphi_{n,s,p_z}(\mathbf{r})$ orbital therefore extends further in the positive direction of the $Oz$ axis than in the negative direction since, for fixed $r$, the values it takes on are greater (in absolute value) for $\theta = 0$ than for $\theta = \pi$. In general, for large values of $r$, $\lambda$ and $\mu$ are such that the values of the $\varphi_{n,s,p_z}(\mathbf{r})$ orbital in the positive direction of the $Oz$ axis are larger than those taken on separately by the $\varphi_{ns}(\mathbf{r})$ and $\varphi_{np_z}(\mathbf{r})$ orbitals [the same conclusions are valid for the $\varphi'_{n,s,p_z}(\mathbf{r})$ orbital and the negative direction of the $Oz$ axis].

This property plays an important role in the study of the chemical bond. To understand this qualitatively, assume that, in a particular atom $A$, one of the valence electrons can be either in the $ns$ orbital or in one of the $np$ orbitals. Then suppose that another atom $B$ is in the neighborhood of the first one, and we call $Oz$ the axis joining $A$ and $B$. The $\varphi_{n,s,p_z}(\mathbf{r})$ orbital of $A$ will overlap more of the orbitals of the valence electrons of $B$ than the $\varphi_{ns}(\mathbf{r})$ or $\varphi_{np_z}(\mathbf{r})$ orbitals. Thus we see that hybridization of the orbitals of $A$ can lead to a greater stability of the chemical bond, since this stability increases, as we have already pointed out, with the overlapping of the electronic orbitals of $A$ and $B$ involved in the bond.

### c. EXAMPLE: THE STRUCTURE OF ACETYLENE

The carbon atom has six electrons. When this atom is free, two of these electrons are in the $1s$ sub-shell, two in the $2s$ sub-shell, and two in the $2p$ sub-shell. Only these last two are unpaired, and we should therefore expect carbon to be bivalent. This is indeed what is observed in some of its compounds. However, carbon is usually present in a quadrivalent form. This

arises from the fact that when a carbon atom is bound to other atoms, one of its $2s$ electrons can leave this sub-shell and place itself in the third $2p$ orbital, which is unoccupied in the free carbon atom. There are then four unpaired electrons, whose wave functions are the result of a hybridization of the four orbitals, $2s$, $2p_x$, $2p_y$ and $2p_z$.

Thus, in the acetylene molecule $C_2H_2$, the four valence electrons of each carbon atom are distributed as follows: two electrons are in the $\varphi_{2,s,p_z}(\mathbf{r})$ and $\varphi'_{2,s,p_z}(\mathbf{r})$ hybrid orbitals we have just introduced, and the other two are in the $\varphi_{2p_x}(\mathbf{r})$ and $\varphi_{2p_y}(\mathbf{r})$ orbitals studied in §2-b. According to figures 6-a and 6-b, the two electrons of each carbon atom which occupy the $\varphi_{2,s,p_z}(\mathbf{r})$ and $\varphi'_{2,s,p_z}(\mathbf{r})$ hybrid orbitals participate in bonds separated by an angle of 180° : the first one with the other carbon atom, and the second one with one of the two hydrogen atoms (whose valence electrons occupy $1s$ orbitals). Thus we understand why the $C_2H_2$ molecule is linear (*cf.* fig. 7, where we have used the same type of graphical representation as in figure 5).

As for the $2p_x$ orbitals centered on each of the carbon atoms, they present a partial lateral overlapping, as do the two $2p_y$ orbitals, as is shown by the solid lines in figure 7. They contribute to the reinforcement of the chemical stability of the molecule. The two carbon atoms thus form a *triple bond* between them. One bond is produced by the $\varphi_{2,s,p_z}(\mathbf{r})$ and $\varphi'_{2,s,p_z}(\mathbf{r})$ hybrid orbitals, each centered on one of the two atoms and cylindrically symmetrical with respect to the $Oz$ axis ($\sigma$ bond). Two bonds are produced by the $\varphi_{2p_x}(\mathbf{r})$ and $\varphi_{2p_y}(\mathbf{r})$ orbitals, which are symmetrical with respect to the $xOz$ and $yOz$ planes ($\pi$ bonds).

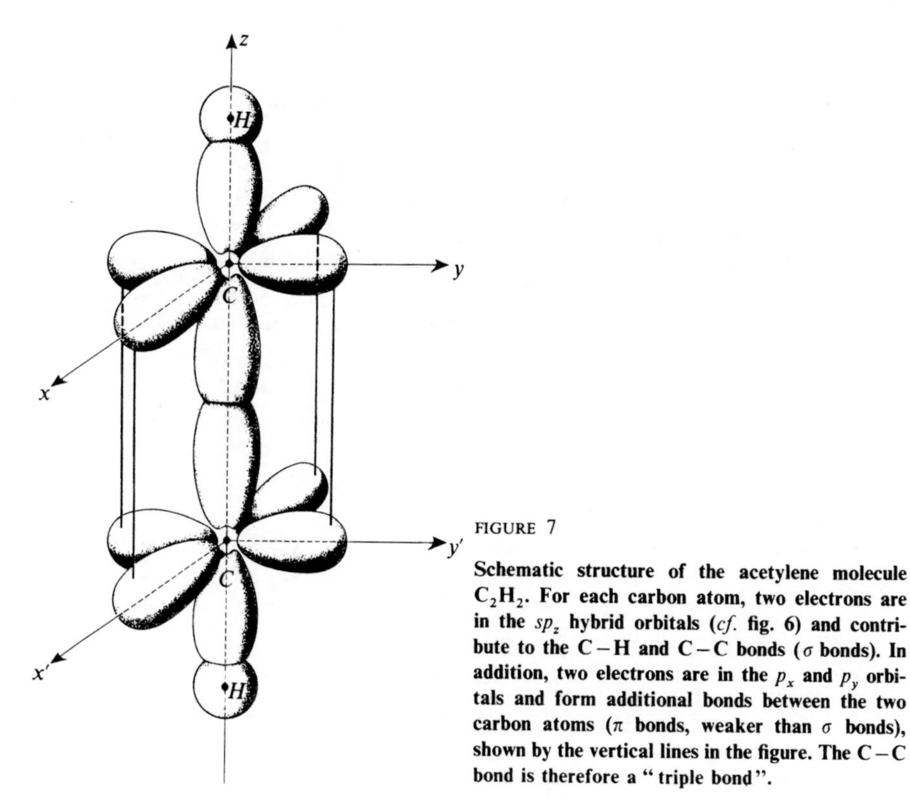


FIGURE 7

**Schematic structure of the acetylene molecule $C_2H_2$. For each carbon atom, two electrons are in the $sp_z$ hybrid orbitals (*cf.* fig. 6) and contribute to the C – H and C – C bonds ($\sigma$ bonds). In addition, two electrons are in the $p_x$ and $p_y$ orbitals and form additional bonds between the two carbon atoms ($\pi$ bonds, weaker than $\sigma$ bonds), shown by the vertical lines in the figure. The C – C bond is therefore a "triple bond".**

COMMENT:

As we have already pointed out, the $2p$ sub-shell, in a many-electron atom, has an energy greater than that of the $2s$ sub-shell. The movement of an electron from the $2s$ sub-shell to the $2p$ sub-shell is therefore not energetically favorable. However, the energy needed for this excitation is fully compensated by the increase in stability due to the hybrid orbitals involved in the C–H and C–C bonds.

## 4. $sp^2$ hybridization

### a. INTRODUCTION OF $sp^2$ HYBRID ORBITALS

We shall now return to the four orbitals $\varphi_{ns}(\mathbf{r})$, $\varphi_{np_x}(\mathbf{r})$, $\varphi_{np_y}(\mathbf{r})$ and $\varphi_{np_z}(\mathbf{r})$, and replace the first three by the three following real combinations:

$$\begin{cases} \varphi_{n,s,p_x,p_y}(\mathbf{r}) = a\,\varphi_{ns}(\mathbf{r}) + b\,\varphi_{np_x}(\mathbf{r}) + c\,\varphi_{np_y}(\mathbf{r}) & \text{(18-a)} \\ \varphi'_{n,s,p_x,p_y}(\mathbf{r}) = a'\varphi_{ns}(\mathbf{r}) + b'\varphi_{np_x}(\mathbf{r}) + c'\varphi_{np_y}(\mathbf{r}) & \text{(18-b)} \\ \varphi''_{n,s,p_x,p_y}(\mathbf{r}) = a''\varphi_{ns}(\mathbf{r}) + b''\varphi_{np_x}(\mathbf{r}) + c''\varphi_{np_y}(\mathbf{r}) & \text{(18-c)} \end{cases}$$

We require the three wave functions (18) to be equivalent, that is, to be transformable into each other under rotation about $Oz$. Consequently, the proportion of the $\varphi_{ns}(\mathbf{r})$ orbital must be the same in each of them:

$$a = a' = a'' \tag{19}$$

It is always possible to choose the axes so as to make the first orbital (18-a) symmetrical about the $xOz$ plane. We can therefore choose:

$$c = 0 \tag{20}$$

By taking the three orbitals (18) to be normalized and orthogonal, we obtain six relations which enable us to determine* the six coefficients $a$, $b$, $b'$, $b''$, $c'$, $c''$. A simple calculation yields:

$$\begin{cases} \varphi_{n,s,p_x,p_y}(\mathbf{r}) = \dfrac{1}{\sqrt{3}}\,\varphi_{ns}(\mathbf{r}) + \sqrt{\dfrac{2}{3}}\,\varphi_{np_x}(r) & \text{(21-a)} \\ \varphi'_{n,s,p_x,p_y}(\mathbf{r}) = \dfrac{1}{\sqrt{3}}\,\varphi_{ns}(\mathbf{r}) - \dfrac{1}{\sqrt{6}}\,\varphi_{np_x}(\mathbf{r}) + \dfrac{1}{\sqrt{2}}\,\varphi_{np_y}(\mathbf{r}) & \text{(21-b)} \\ \varphi''_{n,s,p_x,p_y}(\mathbf{r}) = \dfrac{1}{\sqrt{3}}\,\varphi_{ns}(\mathbf{r}) - \dfrac{1}{\sqrt{6}}\,\varphi_{np_x}(\mathbf{r}) - \dfrac{1}{\sqrt{2}}\,\varphi_{np_y}(\mathbf{r}) & \text{(21-c)} \end{cases}$$

We have thus produced what is called "$sp^2$ hybridization". The three hybrid orbitals (21) and the $\varphi_{np_z}(\mathbf{r})$ orbital form a new orthonormal basis in the space $\mathscr{E}_{ns} \oplus \mathscr{E}_{np}$.

* Actually, the signs of $a$, $b$ and $c'$ can be chosen arbitrarily.

### b. PROPERTIES OF THE $sp^2$ HYBRID ORBITALS

We shall use the same graphical representation as in figure 6.

The $\varphi_{n,s,p_x,p_y}(\mathbf{r})$ orbital has cylindrical symmetry with respect to $Ox$. Figure 8-a represents the cross section in the $xOy$ plane of the surface which describes its angular dependence for fixed $r$. The form of the curve obtained is completely analogous to that of figure 6-a : the orbital points along the positive direction of the $Ox$ axis.

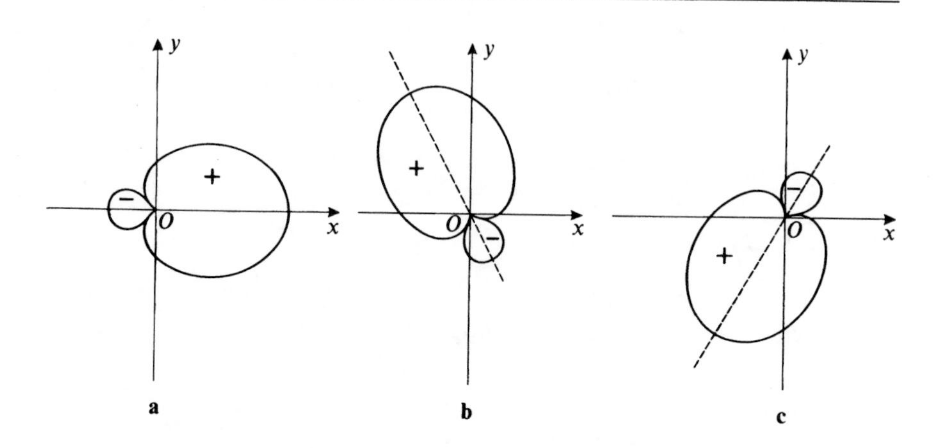


FIGURE 8

**Angular dependence of the three orthogonal $sp^2$ orbitals. The $\varphi_{n,s,p_x,p_y}$, $\varphi'_{n,s,p_x,p_y}$ and $\varphi''_{n,s,p_x,p_y}$ orbitals can be transformed into each other by rotations through 120° about $Oz$.**

By using expression (4-b) for $\varphi_{np_x}(\mathbf{r})$, we can easily obtain the action on $|\varphi_{np_x}\rangle$ of the operator which performs a rotation through an angle $\alpha$ about $Oz$, $e^{-i\alpha L_z/\hbar}$:

$$e^{-i\alpha L_z/\hbar}|\varphi_{np_x}\rangle = \cos\alpha\,|\varphi_{np_x}\rangle + \sin\alpha\,|\varphi_{np_y}\rangle \tag{22}$$

Also, we obviously have:

$$e^{-i\alpha L_z/\hbar}|\varphi_{ns}\rangle = |\varphi_{ns}\rangle \tag{23}$$

Formulas (21) then indicate that:

$$|\varphi'_{n,s,p_x,p_y}\rangle = e^{-2i\frac{\pi}{3}L_z/\hbar}|\varphi_{n,s,p_x,p_y}\rangle \tag{24-a}$$

$$|\varphi''_{n,s,p_x,p_y}\rangle = e^{2i\frac{\pi}{3}L_z/\hbar}|\varphi_{n,s,p_x,p_y}\rangle \tag{24-b}$$

The two orbitals (21-b) and (21-c) can therefore be obtained from the orbital (21-a) by rotations through angles $2\pi/3$ and $-2\pi/3$ about $Oz$. Figures (8-b) and (8-c) give the cross sections in the $xOy$ plane of the surfaces which describe their angular dependence.

### c. EXAMPLE: THE STRUCTURE OF ETHYLENE

As in the acetylene molecule, each of the two carbon atoms of the ethylene molecule $C_2H_4$ has four valence electrons (one electron in the $2s$ sub-shell and three electrons in the $2p$ sub-shell).

Three of these four electrons occupy $sp^2$ hybrid orbitals of the type of those just considered. They are the ones which, for each carbon atom, form the bonds with the neighboring carbon atom and the two hydrogen atoms of the $CH_2$ group. Thus we see why the three bonds, C – C, C – H, C – H, originating from one carbon atom are coplanar and form angles of 120° with each other (*cf.* fig. 9, in which we have used the same graphical representation as in figures 5 and 7). The remaining electron of each carbon atom occupies the $2p_z$ orbital. The $2p_z$ orbitals of the two carbons present a partial lateral overlapping, shown by the solid lines in figure 9.

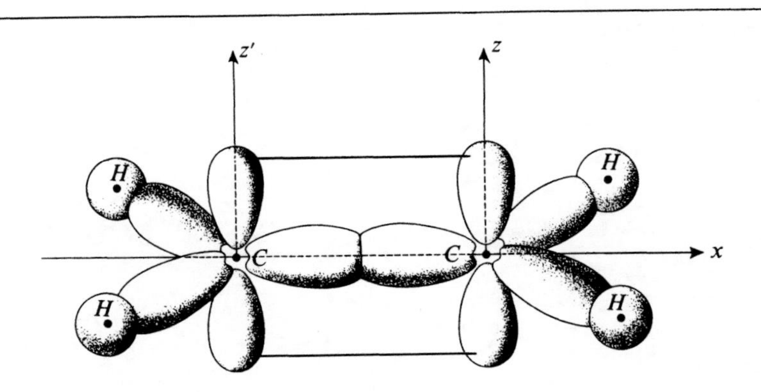


FIGURE 9

**Schematic structure of the ethylene molecule $C_2H_4$. The two carbon atoms form a double bond with each other : one $\sigma$ bond due to $sp^2$ orbitals of the type of those shown in figure 8 (the other two $sp^2$ hybrid orbitals at 120° with this one form the C – H bonds), and one $\pi$ bond, due to the overlapping of the $p_z$ orbitals.**

The two carbon atoms of the ethylene molecule are therefore connected by a *double bond* : one bond formed by two hybrid orbitals of the $sp^2$ type, cylindrically symmetrical with respect to the $Ox$ axis joining the two carbon atoms ($\sigma$ bond), and one bond formed by two $2p_z$ orbitals, symmetrical about the $xOz$ plane ($\pi$ bond). It is the latter bond which blocks the rotation of one $CH_2$ group with respect to the other one. If one of the $CH_2$ groups were to rotate with respect to the other one about the axis joining the two carbons, the axes of the two orbitals $2p_z$ and $2p_{z'}$ (fig. 9) would no longer be parallel. This would diminish their lateral overlapping and, consequently, the stability of the system. Thus we see why the six atoms of the ethylene molecule are in the same plane.

## 5. $sp^3$ hybridization

### a. INTRODUCTION OF $sp^3$ HYBRID ORBITALS

We shall now superpose the four orbitals, $\varphi_{ns}(\mathbf{r})$, $\varphi_{np_x}(\mathbf{r})$, $\varphi_{np_y}(\mathbf{r})$, $\varphi_{np_z}(\mathbf{r})$, to form the four hybrid orbitals:

$$\begin{cases} \varphi_{n,s,p_x,p_y,p_z}(\mathbf{r}) = a\,\varphi_{ns}(\mathbf{r}) + b\,\varphi_{np_x}(\mathbf{r}) + c\,\varphi_{np_y}(\mathbf{r}) + d\,\varphi_{np_z}(\mathbf{r}) & \text{(25-a)} \\ \varphi'_{n,s,p_x,p_y,p_z}(\mathbf{r}) = a'\varphi_{ns}(\mathbf{r}) + b'\,\varphi_{np_x}(\mathbf{r}) + c'\,\varphi_{np_y}(\mathbf{r}) + d'\,\varphi_{np_z}(\mathbf{r}) & \text{(25-b)} \\ \varphi''_{n,s,p_x,p_y,p_z}(\mathbf{r}) = a''\varphi_{ns}(\mathbf{r}) + b''\,\varphi_{np_x}(\mathbf{r}) + c''\,\varphi_{np_y}(\mathbf{r}) + d''\,\varphi_{np_z}(\mathbf{r}) & \text{(25-c)} \\ \varphi'''_{n,s,p_x,p_y,p_z}(\mathbf{r}) = a'''\varphi_{ns}(\mathbf{r}) + b'''\varphi_{np_x}(\mathbf{r}) + c'''\varphi_{np_y}(\mathbf{r}) + d'''\varphi_{np_z}(\mathbf{r}) & \text{(25-d)} \end{cases}$$

We again require the four orbitals to have the same geometrical form, which means that:

$$a = a' = a'' = a''' \tag{26}$$

We can arbitrarily choose the symmetry axis of one of the orbitals, then the plane containing this axis and that of a second orbital. This reduces the number of free parameters to 10; we can find them by taking the four orbitals (25) to be orthonormal.

We shall content ourselves here with giving a possible set of such hybrid orbitals, defined by:

$$\begin{cases} a = b = c = d = \dfrac{1}{2} \\ a' = -\,b' = -\,c' = d' = \dfrac{1}{2} \\ a'' = -\,b'' = c'' = -\,d'' = \dfrac{1}{2} \\ a''' = b''' = -\,c''' = -\,d''' = \dfrac{1}{2} \end{cases} \tag{27}$$

and which can immediately be shown to be orthonormal and of the same geometrical form. All the other possible sets can be obtained from this one by rotation.

We have thus produced what is called a "$sp^3$ hybridization". The four orbitals (25) corresponding to the coefficients (27) form a new orthonormal basis in the space $\mathscr{E}_{ns} \oplus \mathscr{E}_{np}$.

### b. PROPERTIES OF $sp^3$ HYBRID ORBITALS

The four orbitals constructed in § 5-a are analogous in form to those studied in §§ 3 and 4. They point respectively in the directions of the vectors whose components are:

$$\begin{cases} (1,\,1,\,1) \\ (-\,1,\,-\,1,\,1) \\ (-\,1,\,1,\,-\,1) \\ (1,\,-\,1,\,-\,1) \end{cases} \tag{28}$$

The axes of the four $sp^3$ orbitals are therefore arranged like the straight lines joining the center of a regular tetrahedron to the four corners of this tetrahedron. The angle between any two of these straight lines is equal to 109°28′.

### c. EXAMPLE: THE STRUCTURE OF METHANE

In the methane molecule $CH_4$, the four valence electrons of the carbon atom each occupy one of the four $sp^3$ hybrid orbitals studied above. This immediately explains why the four hydrogen atoms form the corners of a regular tetrahedron centered on the carbon atom (fig. 10).

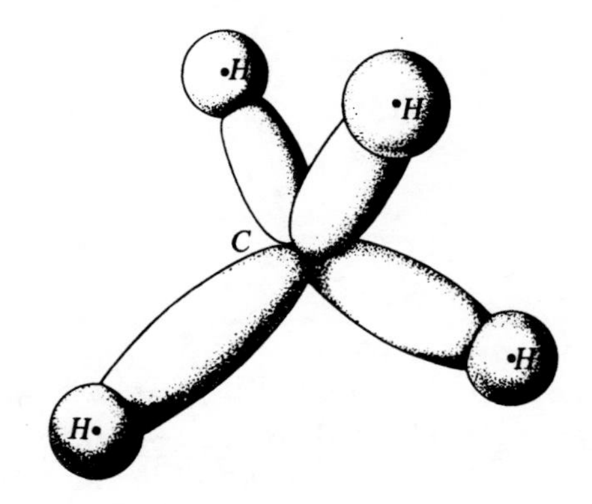


FIGURE 10

**Schematic structure of the methane molecule. The $sp^3$ orbitals produce bonds arranged like the straight lines joining the center of a tetrahedron to its four corners (angles of 109°28′).**

In the ethane molecule $C_2H_6$, one of the hydrogen atoms of methane is replaced by a $CH_3$ group. The two carbon atoms are then connected by a single bond, formed by two $sp^3$ hybrid orbitals, cylindrically symmetrical with respect to the straight line joining the two carbon atoms. The absence of a double bond permits the practically free rotation of one $CH_3$ group with respect to the other one.

**References and suggestions for further reading:**

Various geometrical representations of orbitals: Levine (12.3), §6.6; Karplus and Porter (12.1), §3.10.

Hybrid orbitals : Karplus and Porter (12.1), §6.3; Alonso and Finn III (1.4), §5-5; Eyring et al (12.5), chap. XII, §12 b; Coulson (12.6), chap. VIII; Pauling (12.2), chap. III, §§13 and 14.

## Complement $F_{VII}$

# VIBRATIONAL-ROTATIONAL LEVELS OF DIATOMIC MOLECULES



## 1. Introduction

In this complement, we shall use the results of chapter VII to study quantum mechanically the stationary states of the system formed by the two nuclei of a diatomic molecule. We shall simultaneously take into account all the degrees of freedom of the system: vibration of the two nuclei about their equilibrium position and rotation of the system about the center of mass. We shall show that the results obtained in complements $A_V$ and $C_{VI}$, in which one degree of freedom at a time was considered, are valid to a first approximation. In addition, a certain number of corrections due to the "centrifugal distortion" of the molecule and to the vibration-rotation coupling will be calculated and interpreted.

We saw in § 1-a of complement $A_V$ (the Born-Oppenheimer approximation) that the potential energy $V(r)$ of interaction between the two nuclei depends only on the distance $r$ between them and has the form shown in figure 1: $V(r)$ is

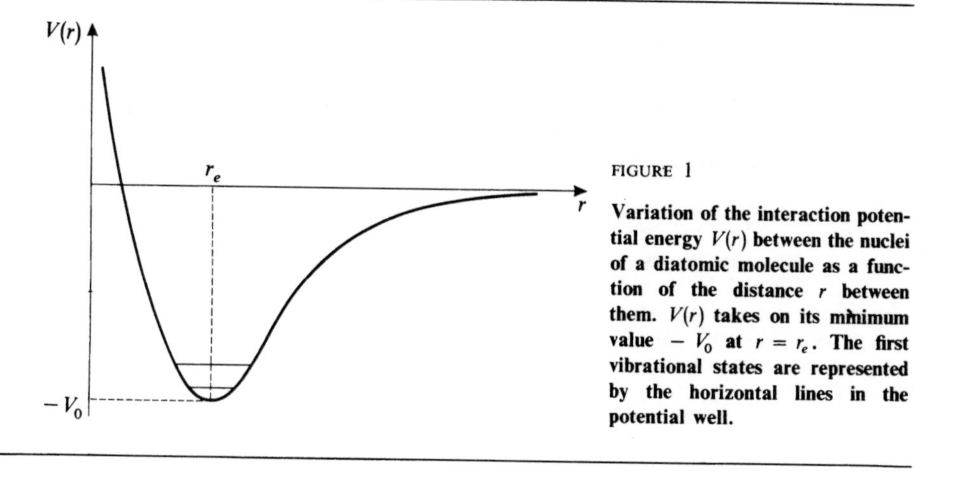


FIGURE 1

**Variation of the interaction potential energy $V(r)$ between the nuclei of a diatomic molecule as a function of the distance $r$ between them. $V(r)$ takes on its minimum value $-V_0$ at $r = r_e$. The first vibrational states are represented by the horizontal lines in the potential well.**

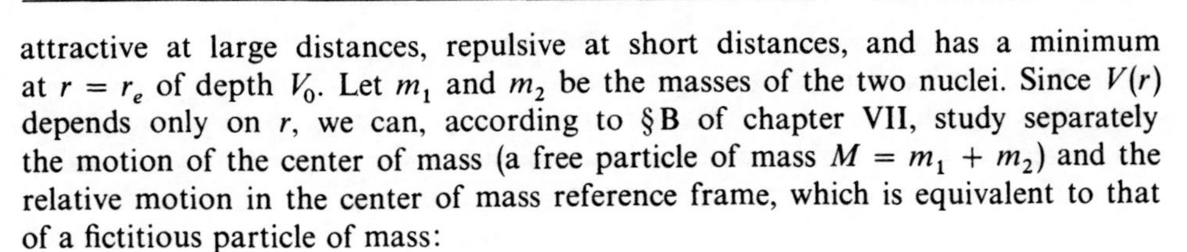

attractive at large distances, repulsive at short distances, and has a minimum at $r = r_e$ of depth $V_0$. Let $m_1$ and $m_2$ be the masses of the two nuclei. Since $V(r)$ depends only on $r$, we can, according to §B of chapter VII, study separately the motion of the center of mass (a free particle of mass $M = m_1 + m_2$) and the relative motion in the center of mass reference frame, which is equivalent to that of a fictitious particle of mass:

$$\mu = \frac{m_1 m_2}{m_1 + m_2} \tag{1}$$

placed in the potential $V(r)$ of figure 1.

If we are interested only in the relative motion, the stationary states of the system, according to the results of §A of chapter VII, are described by the wave functions:

$$\varphi_{v,l,m}(r, \theta, \varphi) = \frac{1}{r}\, u_{v,l}(r)\, Y_l^m(\theta, \varphi) \tag{2}$$

where the corresponding energies $E_{v,l}$ and the radial functions $u_{v,l}(r)$ are given by the equation:

$$\left[ -\frac{\hbar^2}{2\mu}\frac{\mathrm{d}^2}{\mathrm{d}r^2} + V(r) + \frac{l(l+1)\hbar^2}{2\mu r^2} \right] u_{v,l}(r) = E_{v,l}\, u_{v,l}(r) \tag{3}$$

COMMENT:

Rigorously speaking, we implicitly assume in all of this complement (as in $A_V$ and $C_{VI}$) that the projection of the total orbital angular momentum of the electrons onto the internuclear axis is zero, as is their total spin. The total angular momentum of the molecule then arises only from the rotation of the two nuclei. Such a situation is found in virtually all diatomic molecules in their ground states. In the general case, terms will also appear in the nuclear interaction energy which do not depend exclusively on the distance $r$.

## 2. Approximate solution of the radial equation

The radial equation has the same form as the eigenvalue equation of the Hamiltonian of a one-dimensional problem in which a particle of mass $\mu$ is placed in the effective potential:

$$V_{\text{eff}}(r) = V(r) + \frac{l(l+1)\hbar^2}{2\mu r^2} \tag{4}$$

### a. THE ZERO ANGULAR MOMENTUM STATES ($l = 0$)

For $l = 0$, the "centrifugal potential" $l(l+1)\hbar^2/2\mu r^2$ is zero, and $V_{\text{eff}}(r)$ is then the same as $V(r)$. In the neighborhood of the minimum at $r = r_e$, $V(r)$ can be expanded in powers of $r - r_e$:

$$V(r) = -V_0 + f(r - r_e)^2 - g(r - r_e)^3 + \ldots \tag{5}$$

$f$ and $g$ are positive since $r = r_e$ is a minimum and the potential increases faster for $r < r_e$ than for $r > r_e$.

We begin by neglecting the term in $(r - r_e)^3$ and terms of higher order. The potential is then purely parabolic, and we know the eigenstates and eigenvalues of the Hamiltonian. If we set:

$$\omega = \sqrt{\frac{2f}{\mu}} \tag{6}$$

we obtain levels whose energy is:

$$E_{v,0} = -V_0 + \left(v + \frac{1}{2}\right)\hbar\omega \tag{7}$$

$$(v = 0, 1, 2, ...)$$

with the associated wave functions (*cf.* chap. V and complement $B_V$):

$$u_v(r) = \left(\frac{\beta^2}{\pi}\right)^{1/4} \frac{1}{\sqrt{2^v v!}} e^{-\beta^2(r-r_e)^2/2} H_v[\beta(r - r_e)] \tag{8}$$

with:

$$\beta = \sqrt{\frac{\mu\omega}{\hbar}} \tag{9}$$

($H_v$ is a Hermite polynomial). In figure 1, we have represented the first two energy levels by horizontal lines. The length of the lines gives an idea of the extension $(\Delta r)_v$ of the wave functions corresponding to these states. Recall [chap. V, formula (D-5-a)] that:

$$(\Delta r)_v \simeq \sqrt{\left(v + \frac{1}{2}\right)\frac{\hbar}{\mu\omega}} \tag{10}$$

For the preceding calculation to be valid, it is necessary, in a region of width $(\Delta r)$ about $r = r_e$, for the term in $(r - r_e)^3$ of (5) to be always negligible compared to the term in $(r - r_e)^2$. We must therefore have:

$$f \gg g\,(\Delta r)_v = g\,(\Delta r)_0 \sqrt{v + \frac{1}{2}} \tag{11}$$

where $(\Delta r)_0$ is the extension of the ground state:

$$(\Delta r)_0 = \sqrt{\frac{\hbar}{\mu\omega}} \tag{12}$$

This implies, in particular that:

$$f \gg g\,(\Delta r)_0 \tag{13}$$

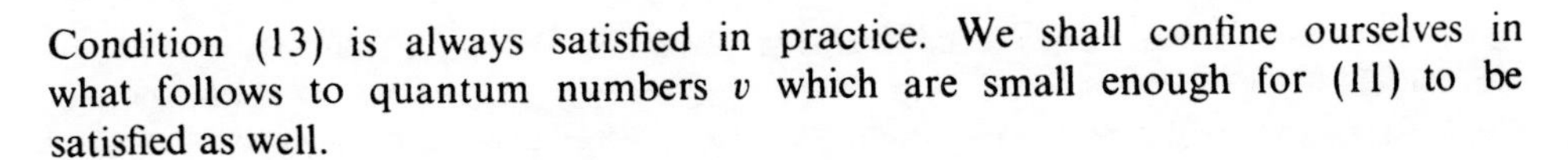


Condition (13) is always satisfied in practice. We shall confine ourselves in what follows to quantum numbers $v$ which are small enough for (11) to be satisfied as well.

COMMENT:

Expansion (5) is obviously not valid at $r = 0$, where $V(r)$ is infinite. The preceding argument implicitly assumes that:

$$(\Delta r)_v \ll r_e \tag{14}$$

In this case, the wave functions (8) are practically zero at the origin, and almost identical to the exact solutions of the radial equation (3), which must be rigorously zero at $r = 0$ (*cf.* §A-2-c of chapter VII).

**b. GENERAL CASE ($l$ ANY POSITIVE INTEGER)**

α. *Evaluation of the effect of the centrifugal potential*

At $r = r_e$, the centrifugal potential is equal to:

$$\frac{l(l+1)\hbar^2}{2\mu r_e^2} = Bh\,l(l+1) \tag{15}$$

where:

$$B = \frac{\hbar}{4\pi\mu r_e^2} \tag{16}$$

is the rotational constant introduced in complement $C_{VI}$. We have already pointed out in that complement (§4-a-β) that the energy $2Bh$ (the distance between two adjacent lines of the pure rotational spectrum) is always very much smaller than $\hbar\omega$ (the vibrational quantum):

$$2Bh \ll \hbar\omega \tag{17}$$

We shall confine ourselves here to rotational quantum numbers $l$ sufficiently small that:

$$Bh\,l(l+1) \ll \hbar\omega \tag{18}$$

In a domain of small width $\Delta r$ about $r = r_e$, the variation of the centrifugal potential is of the order of:

$$\frac{l(l+1)\hbar^2}{\mu r_e^3}\Delta r = 2Bh\,l(l+1)\frac{\Delta r}{r_e} \tag{19}$$

That of the potential $V(r)$ is approximately:

$$f(\Delta r)^2 = \frac{1}{2}\mu\omega^2(\Delta r)^2 = \frac{1}{2}\hbar\omega\frac{(\Delta r)^2}{(\Delta r)_0^2} \tag{20}$$

where we have used (12). We know from §2-a that the extension $\Delta r$ of the wave functions we shall be considering is negligible with respect to $r_e$, but certainly at least of the order of $(\Delta r)_0$. Consequently, in the region of space in which the wave functions have significant amplitudes, the variation (19) of the centrifugal potential is, according to (18), much smaller than that of $V(r)$ found in (20). We can then, to a first approximation, replace the centrifugal potential in equation (4), by its value (15) at $r = r_e$, which gives for the effective potential:

$$V_{eff}(r) \simeq V(r) + Bh\, l(l+1) \tag{21}$$

β. *Energy levels and stationary wave functions*

By using (21) and neglecting terms of order greater than two in expansion (5), we can put the radial equation (3) in the form:

$$\left[-\frac{\hbar^2}{2\mu}\frac{d^2}{dr^2} + \frac{1}{2}\mu\omega^2(r - r_e)^2\right]u_{v,l}(r) = \left[E_{v,l} + V_0 - Bh\, l(l+1)\right]u_{v,l}(r) \tag{22}$$

which is completely analogous to the eigenvalue equation of a one-dimensional harmonic oscillator.

From this, we deduce immediately that the term in brackets on the right-hand side must be equal to $(v + 1/2)\hbar\omega$, where $v = 0, 1, 2, \ldots$; this yields the possible energies $E_{v,l}$ of the molecule:

$$E_{v,l} = -V_0 + \left(v + \frac{1}{2}\right)\hbar\omega + Bh\, l(l+1) \tag{23}$$

with:

$$\begin{cases} v = 0, 1, 2 \ldots \\ l = 0, 1, 2 \ldots \end{cases}$$

As for the radial functions, they do not depend on $l$, since the differential operator appearing on the left-hand side of (22) does not depend on $l$. Consequently, we have:

$$u_{v,l}(r) = u_v(r) \tag{24}$$

where $u_v(r)$ was given in (8). Expression (2) for the wave functions of the stationary states can then be written, in this approximation:

$$\varphi_{v,l,m}(r, \theta, \varphi) = \frac{1}{r}\, u_v(r)\, Y_l^m(\theta, \varphi) \tag{25}$$

Thus we see that the energies of the stationary states are the sums of the energies calculated in complements $A_V$ and $C_{VI}$, in which only one degree of freedom at a time (vibration or rotation) was taken into account. In addition, the wave functions are the products of the wave functions found in these two complements, to within a factor of $1/r$.

Figure 2 shows the first two vibrational levels $v = 0$ and $v = 1$, with their rotational structure due to the term $Bh\, l(l+1)$.

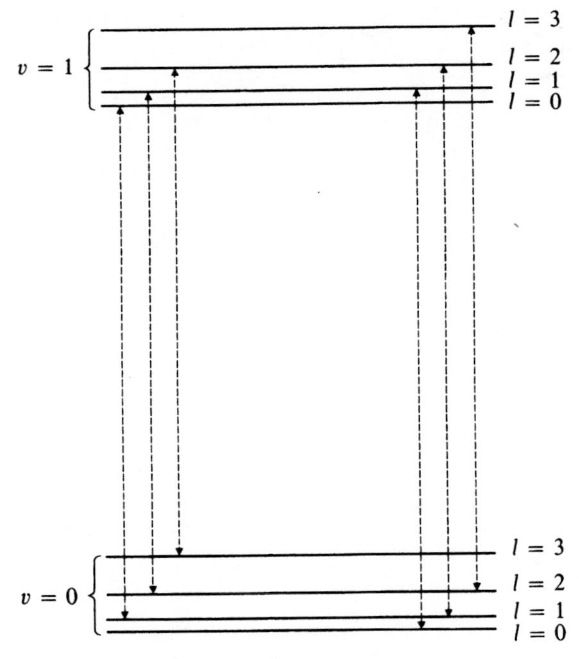


FIGURE 2

**Diagram showing the first two vibrational levels ($v = 0$ and $v = 1$) of a diatomic molecule and their rotational structure ($l = 0, 1, 2, \ldots$). Within the limits of the approximations, this rotational structure is the same for the various vibrational levels. For a heteropolar molecule, the transitions represented by the vertical arrows in the figure yield the lines of the vibrational-rotational spectrum of the molecule. These lines fall in the infrared. These transitions obey the selection rule $\Delta l = l' - l = \pm 1$.**

### c. THE VIBRATIONAL-ROTATIONAL SPECTRUM

We shall confine ourselves to the study of the infrared absorption or emission spectrum, thereby assuming the molecule to be heteropolar (calculations analogous to those presented in § 1-c-β of $A_V$ and § 4-b of $C_{VI}$ could be performed when dealing with homopolar molecules and the Raman effect).

α. *Selection rules*

Recall that the dipole moment $D(r)$ of the molecule is directed along the straight line joining the two nuclei and can be expanded in powers of $r - r_e$ about $r_e$:

$$D(r) = d_0 + d_1(r - r_e) + \ldots \tag{26}$$

The projection of this dipole moment onto $Oz$ is equal to $D(r) \cos \theta$ (where $\theta$ is the angle between the axis of the molecule and $Oz$).

We want to determine the frequency spectrum of electromagnetic waves polarized along $Oz$ that the molecule can absorb or emit as a consequence of the variation of its electric dipole. As we have done several times before, we shall look for the Bohr frequencies that can appear in the time evolution of the mean value of $D(r) \cos \theta$. All we must do, then, is find for what values of $v'$, $l'$, $m'$ and $v$, $l$, $m$ the matrix element:

$$\langle \varphi_{v',l',m'} \mid D(r) \cos \theta \mid \varphi_{v,l,m} \rangle$$

$$= \int r^2 \, \mathrm{d}r \, \mathrm{d}\Omega \; \varphi^*_{v',l',m'}(r, \theta, \varphi) \, D(r) \cos \theta \; \varphi_{v,l,m}(r, \theta, \varphi) \tag{27}$$

is different from zero. Using expression (25) for the wave functions, we put this matrix element in the form:

$$\left[\int_0^\infty \mathrm{d}r\; u_{v'}^*(r)\, D(r)\, u_v(r)\right] \times \left[\int \mathrm{d}\Omega\; Y_{l'}^{m'*}(\theta, \varphi) \cos\theta\, Y_l^m(\theta, \varphi)\right] \tag{28}$$

We thus obtain a product of two integrals which have already been treated in complements $A_V$ and $C_{VI}$. The second integral is different from zero only if:

$$l' - l = +1, -1 \tag{29}$$

As for the first one, if we confine ourselves to the terms in $d_0$ and $d_1$ of (26), it is different from zero only if:

$$v' - v = 0, +1, -1 \tag{30}$$

The set of lines corresponding to $v - v' = 0$ constitutes the pure rotational spectrum studied in complement $C_{VI}$ (its intensity is proportional to $d_0^2$). As for the lines $v' - v = \pm 1$, $l' - l = \pm 1$, of intensity proportional to $d_1^2$, they constitute the vibrational-rotational spectrum which we shall now briefly describe.

COMMENT:

The selection rule $l' - l = \pm 1$ arises from the angular dependence of the wave functions. It is therefore independent of the approximation used to solve the radial equation (3), while (30) is valid only in the harmonic approximation.

### β. *Form of the spectrum*

Let $v'$ be the larger of the two vibrational quantum numbers under consideration ($v' = v + 1$). The vibrational-rotational lines can be separated into two groups:

– the lines $v' = v + 1$, $l' = l + 1 \longleftrightarrow v, l$, of frequencies:

$$\frac{\omega}{2\pi} + B(l+1)(l+2) - Bl(l+1) = \frac{\omega}{2\pi} + 2B(l+1) \tag{31}$$

with $l = 0, 1, 2, \ldots$
(these lines correspond to the transitions indicated by the arrows on the right-hand side of figure 2).

– the lines $v' = v + 1$, $l' = l - 1 \longleftrightarrow v, l$, of frequencies:

$$\frac{\omega}{2\pi} + Bl'(l'+1) - B(l'+1)(l'+2) = \frac{\omega}{2\pi} - 2B(l'+1) \tag{32}$$

with $l' = 0, 1, 2, \ldots$
(transitions indicated by the arrows on the left-hand side of figure 2).

The vibrational-rotational spectrum therefore has the form shown in figure 3. It contains two groups of equidistant lines, symmetrical with respect to the vibrational frequency $\omega/2\pi$. All these lines together constitute a "band". The group of lines corresponding to frequencies (31) is called the "$R$ branch", and the one corresponding to frequencies (32), the "$P$ branch". In each branch, the distance between two adjacent lines is $2B$. The central interval separating the two branches is of width $4B$: there is no line at the pure vibrational frequency $\omega/2\pi$ (there is often said to be a "missing line" in the spectrum).

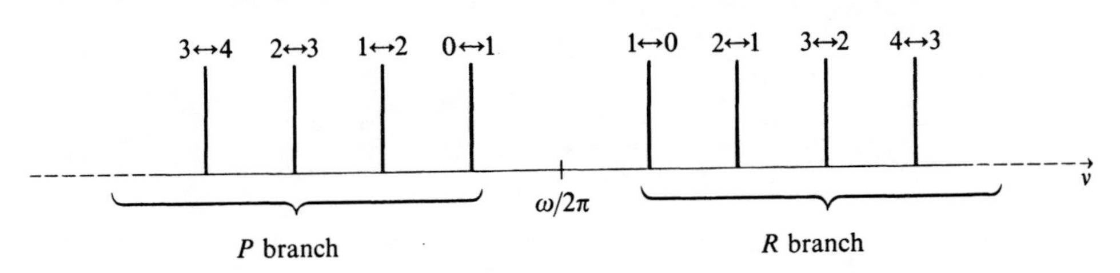


FIGURE 3

**The vibrational-rotational spectrum for a heteropolar molecule. Since transitions between levels of figure 2 with the same value of $l$ are forbidden by the selection rules, no line has the pure vibrational frequency $\frac{\omega}{2\pi}$. Transitions in which the molecule passes from the level $(v', l')$ to the level $(v = v' - 1, l = l' - 1)$ correspond to frequencies $\frac{\omega}{2\pi} + 2B(l + 1)$ (lines of the "$R$ branch"). Transitions in which the molecule passes from the level $(v', l')$ to the level $(v = v' - 1, l = l' + 1)$ correspond to frequencies $\frac{\omega}{2\pi} - 2B(l' + 1)$ (lines of the "$P$ branch"). The different lines are labeled $l' \longleftrightarrow l$ in the figure.**

COMMENT:

The "pure vibrational" spectrum, studied in $A_V$ and composed of a single line at $\omega/2\pi$, therefore does not exist in practice. It is only when one uses a spectroscopic device with low resolution that one can ignore the rotational structure of the vibrational-rotational line and treat the band of figure 3 like a single line centered at $\omega/2\pi$ (recall that $\omega/2\pi \gg 2B$).

## 3. Evaluation of some corrections

The calculations of the preceding section are based on the approximation which consists of replacing the centrifugal potential by its value at $r = r_e$ in the radial equation. The effective potential $V_{\text{eff}}(r)$ can then be obtained from $V(r)$ by a simple vertical translation [formula (21)].

In this section, we shall study the corrections which must be performed on the results of § 2 in order to take into account the slow variation of the centrifugal potential about $r = r_e$. To do so, we shall use its expansion in powers of $(r - r_e)$:

$$\frac{l(l+1)\hbar^2}{2\mu r^2} = \frac{l(l+1)\hbar^2}{2\mu r_e^2} - \frac{l(l+1)\hbar^2}{\mu r_e^3}(r - r_e) + \frac{3l(l+1)\hbar^2}{2\mu r_e^4}(r - r_e)^2 + \dots \quad (33)$$

### a. MORE PRECISE STUDY OF THE FORM OF THE EFFECTIVE POTENTIAL $V_{eff}(r)$

If we use (5) and (33), the expansion of the effective potential (4) in the neighborhood of $r = r_e$ can be written :

$$V_{\text{eff}}(r) = -V_0 + f(r - r_e)^2 - g(r - r_e)^3 + \ldots + \frac{l(l+1)\hbar^2}{2\mu r_e^2} - \frac{l(l+1)\hbar^2}{\mu r_e^3}(r - r_e) + \frac{3l(l+1)\hbar^2}{2\mu r_e^4}(r - r_e)^2 + \ldots \quad (34)$$

We shall see that the variation of the centrifugal potential in the neighborhood of $r = r_e$ produces, for $l$ different from zero, the following effects:

(*i*) The position $\tilde{r}_e$ of the minimum of $V_{\text{eff}}(r)$ does not coincide exactly with $r_e$.

(*ii*) The value $V_{\text{eff}}(\tilde{r}_e)$ of this minimum is slightly different from $-V_0 + Bh\, l(l+1)$.

(*iii*) The curvature of $V_{\text{eff}}(r)$ at $r = \tilde{r}_e$ [which fixes, as in formula (6), the angular frequency of the equivalent harmonic oscillator] is no longer strictly given by the coefficient $f$.

We shall evaluate these various effects by using expansion (34). As far as the first two are concerned, we can neglect terms of order higher than 2 in $V(r)$, and those of order higher than 1 in the centrifugal potential, since the distance $\tilde{r}_e - r_e$ which we shall find is very small [it will even be small relative to $(\Delta r)_0$]. In fact, we shall be able to verify *a posteriori* that :

$$g(\tilde{r}_e - r_e) \ll f \quad (35\text{-a})$$

$$\frac{3l(l+1)\hbar^2}{2\mu r_e^4}(\tilde{r}_e - r_e) \ll \frac{l(l+1)\hbar^2}{\mu r_e^3} \quad (35\text{-b})$$

#### α. *Position and value of the minimum of $V_{\text{eff}}(r)$*

If, in expansion (34), we keep only the first two terms of $V(r)$ and the first two terms of the centrifugal potential, $\tilde{r}_e$ is given by:

$$2f(\tilde{r}_e - r_e) \simeq \frac{l(l+1)\hbar^2}{\mu r_e^3} \quad (36)$$

that is:

$$\tilde{r}_e - r_e \simeq \frac{l(l+1)\hbar^2}{2\mu f r_e^3} = \frac{Bh\, l(l+1)}{f r_e} \quad (37)$$

According to (6) and (12), we have:

$$\frac{\tilde{r}_e - r_e}{(\Delta r)_0} \simeq \frac{2Bh\, l(l+1)}{\hbar\omega}\frac{(\Delta r)_0}{r_e} \ll 1 \quad (38)$$

which, with (13) and (14) taken into account, proves (35-a) and (35-b).

Substituting this value of $\tilde{r}_e$ into the expansion of $V_{\text{eff}}(r)$, we find:

$$V_{\text{eff}}(\tilde{r}_e) \simeq -V_0 + Bh\, l(l+1) - Gh[l(l+1)]^2 \tag{39}$$

with:

$$G = \frac{\hbar^3}{8\pi\mu^2 r_e^6 f} \tag{40}$$

β. *Curvature of $V_{\text{eff}}(r)$ at its minimum*

In the neighborhood of $r = \tilde{r}_e$, we can therefore write $V_{\text{eff}}(r)$ in the form:

$$V_{\text{eff}}(r) = V_{\text{eff}}(\tilde{r}_e) + f'(r - \tilde{r}_e)^2 - g'(r - \tilde{r}_e)^3 + \ldots \tag{41}$$

The coefficient $f'$ is related to the curvature of $V_{\text{eff}}(r)$ at $r = \tilde{r}_e$:

$$f' = \frac{1}{2}\left[\frac{d^2}{dr^2} V_{\text{eff}}(r)\right]_{r=\tilde{r}_e} \tag{42}$$

To evaluate the difference between $f'$ and $f$, we must take into account the term in $(r - r_e)^3$ of $V(r)$ in expansion (34) and, consequently, also the term in $(r - r_e)^2$ of the centrifugal potential. A simple calculation then yields, using (37):

$$2f' \simeq 2f + \frac{3l(l+1)\hbar^2}{\mu r_e^4} - \frac{3g\, l(l+1)\hbar^2}{\mu r_e^3 f} \tag{43}$$

The angular frequency $\omega$ defined by (6) must therefore be replaced by:

$$\omega' = \sqrt{\frac{2f'}{\mu}} \tag{44}$$

Expanding the square root, we easily find:

$$\omega' = \omega - 2\pi\alpha_e\, l(l+1) \tag{45}$$

with:

$$\alpha_e = \frac{3\hbar^2\omega}{8\pi\mu r_e^3 f}\left[\frac{g}{f} - \frac{1}{r_e}\right] \tag{46}$$

We could carry out an analogous calculation to determine $g'$. Actually, since the term in $(r - \tilde{r}_e)^3$ of (41) adds only a small correction to the results obtained using the first two terms, we shall neglect the variation of $\frac{d^3}{dr^3} V_{\text{eff}}(r)$ when we go from $r_e$ to $\tilde{r}_e$, and take $g' \simeq g$.

In conclusion, in the neighborhood of its minimum, we can write $V_{\text{eff}}(r)$ in the form:

$$V_{\text{eff}}(r) \simeq V_{\text{eff}}(\tilde{r}_e) + \frac{1}{2}\mu\omega'^2(r - \tilde{r}_e)^2 - g(r - \tilde{r}_e)^3 \tag{47}$$

where $\tilde{r}_e$, $V_{\text{eff}}(\tilde{r}_e)$, $\omega'$ are given by (37), (39) and (45).

### b. ENERGY LEVELS AND WAVE FUNCTIONS OF THE STATIONARY STATES

With expression (47) for $V_{eff}(r)$, the radial equation becomes:

$$\left[-\frac{\hbar^2}{2\mu}\frac{d^2}{dr^2} + \frac{1}{2}\mu\omega'^2(r - \tilde{r}_e)^2 - g(r - \tilde{r}_e)^3\right]u_{v,l}(r) = [E_{v,l} - V_{eff}(\tilde{r}_e)]u_{v,l}(r) \quad (48)$$

If, as in §2, we neglect the term in $g(r - \tilde{r}_e)^3$, we recognize the eigenvalue equation of a one-dimensional harmonic oscillator of angular frequency $\omega'$ whose equilibrium position is $r = \tilde{r}_e$. From this, we deduce that the only possible values of the term in brackets on the right-hand side are $(v + 1/2)\hbar\omega'$, with $v = 0, 1, 2, ...$ According to (39), we therefore have:

$$E_{v,l} = -V_0 + \left(v + \frac{1}{2}\right)\hbar\omega' + Bh\,l(l + 1) - Gh[l(l + 1)]^2 \quad (49)$$

As for the wave functions of the stationary states, they have the same form as in (25). All we need to do in expression (8) for the radial function is replace $r_e$ by $\tilde{r}_e$ and $\beta$ by:

$$\beta' = \sqrt{\frac{\mu\omega'}{\hbar}} \quad (50)$$

We have taken into account the term in $g(r - r_e)^3$ in the calculation of the new angular frequency $\omega'$. For the calculation to be consistent, it is then necessary to evaluate the corrections in the eigenvalues and eigenfunctions of the radial equation due to the presence of this term on the left-hand side of (48). We shall do this in complement $A_{XI}$, using perturbation theory. Here, we shall content ourselves with stating the result concerning the eigenvalues: we must add to expression (49) for the energy the term:

$$\xi\hbar\omega'\left(v + \frac{1}{2}\right)^2 + \frac{7}{60}\xi\hbar\omega' \quad (51)$$

where:

$$\xi = -\frac{15}{4}\frac{g^2\hbar}{\mu^3\omega'^5} \quad (52)$$

is a dimensionless quantity much smaller than 1 (hence $\omega'$ may be replaced by $\omega$ in this corrective term).

### c. INTERPRETATION OF THE VARIOUS CORRECTIONS

#### α. *Centrifugal distortion of the molecule*

The discussion of §3-a-α shows that the distance between the two nuclei increases when the molecule rotates. According to (37), this increase in distance becomes larger when $l(l + 1)$ becomes larger, that is, when the molecule rotates faster.

This is quite comprehensible : in classical terms, one would say that the "centrifugal force" tends to separate the two nuclei until it is balanced by the restoring force $2f(\tilde{r}_e - r_e)$ due to the potential $V(r)$.

The molecule is therefore not really a "rigid rotator". The variation $\tilde{r}_e - r_e$ of the average distance between the nuclei produces an increase in the moment of inertia of the molecule and, consequently, a decrease (at constant angular momentum) in the rotational energy. This decrease is only partially compensated by the increase in the potential energy $V(\tilde{r}_e) - V(r_e)$. This is the physical origin of the energy correction $-\ Ghl^2(l + 1)^2$ which appears in (49). This correction, whose sign is negative, increases much faster with $l$ than the rotational energy $Bh\, l(l + 1)$. This can be seen experimentally: the lines of the pure rotational spectrum are not rigorously equidistant; the separation of the lines decreases when $l$ increases.

### β. *Vibrational-rotational coupling*

We shall group the second and third terms of (49) and replace $\omega'$ by its expression (45). We obtain:

$$\left(v + \frac{1}{2}\right)\hbar\omega' + Bh\, l(l + 1) = \left(v + \frac{1}{2}\right)\hbar\omega + Bh\, l(l + 1) - \alpha_e h\, l(l + 1)\left(v + \frac{1}{2}\right) \tag{53}$$

The first two terms on the right-hand side of (53) are the vibrational and rotational energies calculated in complements $A_V$ and $C_{VI}$. The third term, which depends on the two quantum numbers $v$ and $l$, represents the effects of the coupling of the vibrational and rotational degrees of freedom.

We can rewrite (53) in the form:

$$\left(v + \frac{1}{2}\right)\hbar\omega + B_v h\, l(l + 1) \tag{54}$$

with:

$$B_v = B - \alpha_e\left(v + \frac{1}{2}\right) \tag{55}$$

It looks as if each vibrational level had an effective rotational constant $B_v$ depending on $v$ associated with it.

To explain this coupling of the vibration and rotation of the molecule, we shall argue in classical terms. The rotational constant $B$ is proportional to $1/r^2$ [formula (16)]. When the molecule vibrates, $r$ varies, and, consequently, so does $B$. Since the vibrational frequencies are much higher than the rotational ones, we can define an effective rotational constant of the molecule in a given vibrational state : this will be the average of $B$ taken over a time interval which is much longer than the vibrational period. We must therefore take the time average of $1/r^2$ in the vibrational state under consideration.

In this way, we can interpret the two terms of opposite sign which appear in expression (46) for $\alpha_e$. The first of these terms, which is proportional to $g$, is due to the anharmonicity of the potential $V(r)$, which increases with the amplitude of vibration (that is, in fact, with $v$). Given the asymmetric form of $V(r)$ (fig. 1), the molecule "spends more time" in the region $r > r_e$ than in the region $r < r_e$. It follows that the average value of $1/r^2$ is less than $1/r_e^2$ : the anharmonicity decreases the effective rotational constant. This can be seen in formulas (55) and (46). Actually, even if the vibrational motion were perfectly symmetrical with respect to $r_e$ (that is, if $g$ were zero), the average value of $1/r^2$ would not be equal to $1/r_e^2$, since:

$$\langle \frac{1}{r^2} \rangle \neq \frac{1}{\langle r \rangle^2} \tag{56}$$

This is the origin of the second term of expression (46) : when the average of $1/r^2$ is taken, small values of $r$ are favorized, so that $\langle 1/r^2 \rangle$ is greater than $1/\langle r \rangle^2$; hence the sign of this second correction.

The overall sign of $\alpha_e$ results from the competition between the two preceding effects. In general, it is the anharmonicity term which dominates, so that $\alpha_e$ is positive and $B_v$ is less than $B$.

COMMENTS:

(*i*) Vibrational-rotational coupling exists even in the vibrational ground state $v = 0$:

$$B_0 = B - \frac{1}{2}\alpha_e \tag{57}$$

This is another manifestation of the finite extension $(\Delta r)_0$ of the wave function of the $v = 0$ state.

(*ii*) Experimentally, vibrational-rotational coupling appears in the following way: if $\alpha_e$ is positive, the rotational structure is slightly more compact in the higher vibrational state $v'$ than in the lower vibrational state $v = v' - 1$. It is easy to show that the $P$ and $R$ branches of figure 3 are affected differently. Adjacent lines are no longer completely equidistant and are, on the average, closer together in the $R$ branch than in the $P$ branch.

To sum up, the energy of a vibrational-rotational level of a diatomic molecule, labeled by the quantum numbers $v$ and $l$, is given by:

$$E_{v,l} = -V_0 + \left(v + \frac{1}{2}\right)\hbar\omega + \left[B - \alpha_e\left(v + \frac{1}{2}\right)\right] h\, l(l+1) - Gh\, l^2(l+1)^2 + \xi\left(v + \frac{1}{2}\right)^2 \hbar\omega + \frac{7}{60}\xi\hbar\omega \tag{58}$$

$V_0$: dissociation energy of the molecule;
$\omega/2\pi$: vibrational frequency;
$B$: rotational constant given by (16);
$G$, $\alpha_e$, $\xi$: dimensionless constants given by (40), (46) and (52).

**References and suggestions for further reading:**

Molecular spectra: Eisberg and Resnick (1.3), chap. 12; Pauling and Wilson (1.9), chap. X; Karplus and Porter (12.1), chap. 7; Herzberg (12.4), Vol. I, chap. III, §§2 b and 2 c; Landau and Lifshitz (1.19), chaps. XI and XIII.

Nuclear vibration and rotation: Valentin (16.1), §VII-2.

# Complement $G_{VII}$

## EXERCISES

### 1. Particle in a cylindrically symmetrical potential

Let $\rho$, $\varphi$, $z$ be the cylindrical coordinates of a spinless particle ($x = \rho \cos \varphi$, $y = \rho \sin \varphi$; $\rho \geqslant 0$, $0 \leqslant \varphi < 2\pi$). Assume that the potential energy of this particle depends only on $\rho$, and not on $\varphi$ and $z$. Recall that:

$$\frac{\partial^2}{\partial x^2} + \frac{\partial^2}{\partial y^2} = \frac{\partial^2}{\partial \rho^2} + \frac{1}{\rho}\frac{\partial}{\partial \rho} + \frac{1}{\rho^2}\frac{\partial^2}{\partial \varphi^2}$$

*a.* Write, in cylindrical coordinates, the differential operator associated with the Hamiltonian. Show that $H$ commutes with $L_z$ and $P_z$. Show from this that the wave functions associated with the stationary states of the particle can be chosen in the form:

$$\varphi_{n,m,k}(\rho, \varphi, z) = f_{n,m}(\rho)\, e^{im\varphi}\, e^{ikz}$$

where the values that can be taken on by the indices $m$ and $k$ are to be specified.

*b.* Write, in cylindrical coordinates, the eigenvalue equation of the Hamiltonian $H$ of the particle. Derive from it the differential equation which yields $f_{n,m}(\rho)$.

*c.* Let $\Sigma_y$ be the operator whose action, in the $\{ | \mathbf{r} \rangle \}$ representation, is to change $y$ to $-y$ (reflection with respect to the $xOz$ plane). Does $\Sigma_y$ commute with $H$? Show that $\Sigma_y$ anticommutes with $L_z$, and show from this that $\Sigma_y | \varphi_{n,m,k} \rangle$ is an eigenvector of $L_z$. What is the corresponding eigenvalue? What can be concluded concerning the degeneracy of the energy levels of the particle? Could this result be predicted directly from the differential equation established in (*b*)?

### 2. Three-dimensional harmonic oscillator in a uniform magnetic field

N.B. The object of this exercise is to study a simple physical system for which the effect of a uniform magnetic field can be calculated exactly. Thus, it is possible in this case to compare precisely the relative importance of the "paramagnetic" and "diamagnetic" terms, and to study in detail the modification of the wave function of the ground state due to the effect of the diamagnetic term. (The reader may wish to refer to complements $D_{VI}$ and $B_{VII}$.)

Consider a particle of mass $\mu$, whose Hamiltonian is:

$$H_0 = \frac{\mathbf{P}^2}{2\mu} + \frac{1}{2}\mu\omega_0^2\,\mathbf{R}^2$$

(an isotropic three-dimensional harmonic oscillator), where $\omega_0$ is a given positive constant.

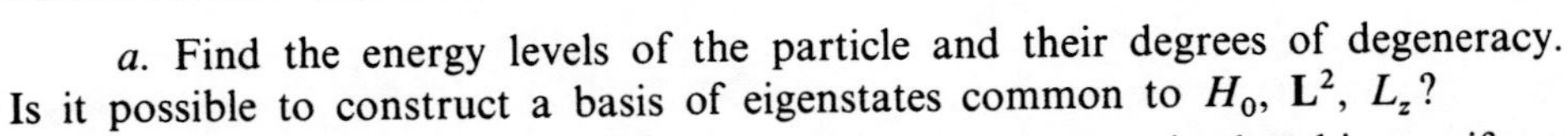



*a*. Find the energy levels of the particle and their degrees of degeneracy. Is it possible to construct a basis of eigenstates common to $H_0$, $\mathbf{L}^2$, $L_z$?

*b*. Now, assume that the particle, which has a charge $q$, is placed in a uniform magnetic field $\mathbf{B}$ parallel to $Oz$. We set $\omega_L = -qB/2\mu$. The Hamiltonian $H$ of the particle is then, if we choose the gauge $\mathbf{A} = -\frac{1}{2}\mathbf{r} \times \mathbf{B}$:

$$H = H_0 + H_1(\omega_L)$$

where $H_1$ is the sum of an operator which is linearly dependent on $\omega_L$ (the paramagnetic term) and an operator which is quadratically dependent on $\omega_L$ (the diamagnetic term). Show that the new stationary states of the system and their degrees of degeneracy can be determined exactly.

*c*. Show that if $\omega_L$ is much smaller than $\omega_0$, the effect of the diamagnetic term is negligible compared to that of the paramagnetic term.

*d*. We now consider the first excited state of the oscillator, that is, the states whose energies approach $5\hbar\omega_0/2$ when $\omega_L \longrightarrow 0$. To first order in $\omega_L/\omega_0$, what are the energy levels in the presence of the field $\mathbf{B}$ and their degrees of degeneracy (the Zeeman effect for a three-dimensional harmonic oscillator)? Same questions for the second excited state.

*e*. Now consider the ground state. How does its energy vary as a function of $\omega_L$ (the diamagnetic effect on the ground state)? Calculate the magnetic susceptibility $\chi$ of this state. Is the ground state, in the presence of the field $\mathbf{B}$, an eigenvector of $\mathbf{L}^2$? of $L_z$? of $L_x$? Give the form of its wave function and the corresponding probability current. Show that the effect of the field $\mathbf{B}$ is to compress the wave function about $Oz$ (in a ratio $[1 + (\omega_L/\omega_0)^2]^{1/4}$) and to induce a current.

# Bibliography

## 1. QUANTUM MECHANICS: GENERAL REFERENCES


### INTRODUCTORY TEXTS

#### Quantum physics

(1.1) E. H. WICHMANN, *Berkeley Physics Course, Vol. 4: Quantum Physics*, McGraw-Hill, New York (1971).

(1.2) R. P. FEYNMAN, R. B. LEIGHTON and M. SANDS, *The Feynman Lectures on Physics, Vol. III: Quantum Mechanics*, Addison-Wesley, Reading, Mass. (1965).

(1.3) R. EISBERG and R. RESNICK, *Quantum Physics of Atoms, Molecules, Solids, Nuclei and Particules*, Wiley, New York (1974).

(1.4) M. ALONSO and E. J. FINN, *Fundamental University Physics, Vol. III: Quantum and Statistical Physics*, Addison-Wesley, Reading, Mass. (1968).

(1.5) U. FANO and L. FANO, *Basic Physics of Atoms and Molecules*, Wiley, New York (1959).

(1.6) J. C. SLATER, *Quantum Theory of Matter*, McGraw-Hill, New York (1968).

#### Quantum mechanics

(1.7) S. BOROWITZ, *Fundamentals of Quantum Mechanics*, Benjamin, New York (1967).

(1.8) S. I. TOMONAGA, *Quantum Mechanics, Vol. 1: Old Quantum Theory*, North Holland, Amsterdam (1962).

(1.9) L. PAULING and E. B. WILSON JR., *Introduction to Quantum Mechanics*, McGraw-Hill, New York (1935).

(1.10) Y. AYANT et E. BELORIZKY, *Cours de Mécanique Quantique*, Dunod, Paris (1969).

(1.11) P. T. MATTHEWS, *Introduction to Quantum Mechanics*, McGraw-Hill, New York (1963).

(1.12) J. AVERY, *The Quantum Theory of Atoms, Molecules and Photons*, McGraw-Hill, London (1972).

### MORE ADVANCED TEXTS:

(1.13) P. A. M. DIRAC, *The Principles of Quantum Mechanics*, Oxford University Press (1958).

(1.14) R. H. DICKE and J. P. WITTKE, *Introduction to Quantum Mechanics*, Addison-Wesley, Reading, Mass. (1966).

(1.15) D. I. BLOKHINTSEV, *Quantum Mechanics*, D. Reidel, Dordrecht (1964).

(1.16) E. MERZBACHER, *Quantum Mechanics*, Wiley, New York (1970).

(1.17) A. MESSIAH, *Mécanique Quantique*, Vols 1 and 2, Dunod, Paris (1964). English translation : *Quantum Mechanics*, North Holland, Amsterdam (1961).

(1.18) L. I. SCHIFF, *Quantum Mechanics*, McGraw-Hill, New York (1968).

(1.19) L. D. LANDAU and E. M. LIFSHITZ, *Quantum Mechanics, Nonrelativistic Theory*, Pergamon Press, Oxford (1965).

(1.20) A. S. DAVYDOV, *Quantum Mechanics*, Translated, edited and with additions by D. Ter HAAR, Pergamon Press, Oxford (1965).

(1.21) H. A. BETHE and R. W. JACKIW, *Intermediate Quantum Mechanics*, Benjamin, New York (1968).

(1.22) H. A. KRAMERS, *Quantum Mechanics*, North Holland, Amsterdam (1958).

### PROBLEMS IN QUANTUM MECHANICS

(1.23) *Selected Problems in Quantum Mechanics*, Collected and edited by D. Ter HAAR, Infosearch, London (1964).

(1.24) S. FLÜGGE, *Practical Quantum Mechanics*, I and II, Springer-Verlag, Berlin (1971).

### ARTICLES

(1.25) E. SCHRÖDINGER, "What is Matter?", *Scientific American*, **189**, 52 (Sept. 1953).

(1.26) G. GAMOW, "The Principle of Uncertainty", *Scientific American*, **198**, 51 (Jan. 1958).

(1.27) G. GAMOW, "The Exclusion Principle", *Scientific American*, **201**, 74 (July 1959).

(1.28) M. BORN and W. BIEM, "Dualism in Quantum Theory", *Physics Today*, **21**, p. 51 (Aug. 1968).

(1.29) W. E. LAMB JR., "An Operational Interpretation of Nonrelativistic Quantum Mechanics", *Physics Today*, **22**, 23 (April 1969).

(1.30) M. O. SCULLY and M. SARGENT III, "The Concept of the Photon", *Physics Today*, **25**, 38 (March 1972).

(1.31) A. EINSTEIN, "Zur Quantentheorie der Strahlung", *Physik. Z.*, **18**, 121 (1917).

(1.32) A. GOLDBERG, H. M. SCHEY and J. L. SCHWARTZ, "Computer-Generated Motion Pictures of One-Dimensional Quantum-Mechanical Transmission and Reflection Phenomena", *Am. J. Phys.*, **35**, 177 (1967).

(1.33) R. P. FEYNMAN, F. L. VERNON JR. and R. W. HELLWARTH, "Geometrical Representation of the Schrödinger Equation for Solving Maser Problems", *J. Appl. Phys.*, **28**, 49 (1957).

(1.34) A. A. VUYLSTEKE, "Maser States in Ammonia-Inversion", *Am. J. Phys.*, **27**, 554 (1959).

## 2. QUANTUM MECHANICS: MORE SPECIALIZED REFERENCES

### COLLISIONS

(2.1) T. Y. WU and T. OHMURA, *Quantum Theory of Scattering*, Prentice Hall, Englewood Cliffs (1962).

(2.2) R. G. NEWTON, *Scattering Theory of Waves and Particles*, McGraw-Hill, New York (1966).

(2.3) P. ROMAN, *Advanced Quantum Theory*, Addison-Wesley, Reading, Mass. (1965).

(2.4) M. L. GOLDBERGER and K. M. WATSON, *Collision Theory*, Wiley, New York (1964).

(2.5) N. F. MOTT and H. S. W. MASSEY, *The Theory of Atomic Collisions*, Oxford University Press (1965).

## RELATIVISTIC QUANTUM MECHANICS

(2.6) J. D. BJORKEN and S. D. DRELL, *Relativistic Quantum Mechanics*, McGraw-Hill, New York (1964).

(2.7) J. J. SAKURAI, *Advanced Quantum Mechanics*, Addison-Wesley, Reading, Mass. (1967).

(2.8) V. B. BERESTETSKII, E. M. LIFSHITZ and L. P. PITAEVSKII, *Relativistic Quantum Theory*, Pergamon Press, Oxford (1971).

## FIELD THEORY. QUANTUM ELECTRODYNAMICS

(2.9) F. MANDL, *Introduction to Quantum Field Theory*, Wiley Interscience, New York (1959).

(2.10) J. D. BJORKEN and S. D. DRELL, *Relativistic Quantum Fields*, McGraw-Hill, New York (1965).

(2.11) E. A. POWER, *Introductory Quantum Electrodynamics*, Longmans, London (1964).

(2.12) R. P. FEYNMAN, *Quantum Electrodynamics*, Benjamin, New York (1961).

(2.13) W. HEITLER, *The Quantum Theory of Radiation*, Clarendon Press, Oxford (1954).

(2.14) A. I. AKHIEZER and V. B. BERESTETSKII, *Quantum Electrodynamics*, Wiley Interscience, New York (1965).

(2.15) N. N. BOGOLIUBOV and D. V. SHIRKOV, *Introduction to the Theory of Quantized Fields*, Interscience Publishers, New York (1959); *Introduction à la Théorie des Champs*, Dunod, Paris (1960).

(2.16) S. S. SCHWEBER, *An Introduction to Relativistic Quantum Field Theory*, Harper and Row, New York (1961).

(2.17) M. M. STERNHEIM, "Resource Letter TQE-1 : Tests of Quantum Electrodynamics", *Am. J. Phys.*, **40**, 1363 (1972).

## ROTATIONS AND GROUP THEORY

(2.18) P. H. E. MEIJER and E. BAUER, *Group Theory*, North Holland, Amsterdam (1962).

(2.19) M. E. ROSE, *Elementary Theory of Angular Momentum*, Wiley, New York (1957).

(2.20) M. E. ROSE, *Multipole Fields*, Wiley, New York (1955).

(2.21) A. R. EDMONDS, *Angular Momentum in Quantum Mechanics*, Princeton University Press (1957).

(2.22) M. TINKHAM, *Group Theory and Quantum Mechanics*, McGraw-Hill, New York (1964).

(2.23) E. P. WIGNER, *Group Theory and its Application to the Quantum Mechanics of Atomic Spectra*, Academic Press, New York (1959).

(2.24) D. PARK, "Resource Letter SP-I on Symmetry in Physics", *Am. J. Phys.*, **36**, 577 (1968).

## MISCELLANEOUS

(2.25) R. P. FEYNMAN and A. R. HIBBS, *Quantum Mechanics and Path Integrals*, McGraw-Hill, New York (1965).

(2.26) J. M. ZIMAN, *Elements of Advanced Quantum Theory*, Cambridge University Press (1969).

(2.27) F. A. KAEMPFFER, *Concepts in Quantum Mechanics*, Academic Press, New York (1965).

## ARTICLES

(2.28) P. MORRISON, "The Overthrow of Parity", *Scientific American*, **196**, 45 (April 1957).

(2.29) G. FEINBERG and M. GOLDHABER, "The Conservation Laws of Physics", *Scientific American*, **209**, 36 (Oct. 1963).

(2.30) E. P. WIGNER, "Violations of Symmetry in Physics", *Scientific American*, **213**, 28 (Dec. 1965).

(2.31) U. FANO, "Description of States in Quantum Mechanics by Density Matrix and Operator Techniques", *Rev. Mod. Phys.*, **29**, 74 (1957).

(2.32) D. Ter HAAR, "Theory and Applications of the Density Matrix", *Rept. Progr. Phys.*, **24**, 304 (1961).

(2.33) V. F. WEISSKOPF and E. WIGNER, "Berechnung der Natürlichen Linienbreite auf Grund der Diracschen Lichttheorie", *Z. Physik*, **63**, 54 (1930).

(2.34) A. DALGARNO and J. T. LEWIS, "The Exact Calculation of Long-Range Forces between Atoms by Perturbation Theory", *Proc. Roy. Soc.*, **A 233**, 70 (1955).

(2.35) A. DALGARNO and A. L. STEWART, "On the Perturbation Theory of Small Disturbances", *Proc. Roy. Soc.*, **A 238**, 269 (1957).

(2.36) C. SCHWARTZ, "Calculations in Schrödinger Perturbation Theory", *Annals of Physics* (New York), **6**, 156 (1959).

(2.37) J. O. HIRSCHFELDER, W. BYERS BROWN and S. T. EPSTEIN, "Recent Developments in Perturbation Theory", in *Advances in Quantum Chemistry*, P. O. LOWDIN ed., Vol. I, Academic Press, New York (1964).

(2.38) R. P. FEYNMAN, "Space Time Approach to Nonrelativistic Quantum Mechanics", *Rev. Mod. Phys.*, **20**, 367 (1948).

(2.39) L. VAN HOVE, "Correlations in Space and Time and Born Approximation Scattering in Systems of Interacting Particles", *Phys. Rev.*, **95**, 249 (1954).

## 3. QUANTUM MECHANICS: FUNDAMENTAL EXPERIMENTS

*Interference effects with weak light:*

(3.1) G. I. TAYLOR, "Interference Fringes with Feeble Light", *Proc. Camb. Phil. Soc.*, **15**, 114 (1909).

(3.2) G. T. REYNOLDS, K. SPARTALIAN and D. B. SCARL, "Interference Effects Produced by Single Photons", *Nuovo Cimento*, **61 B**, 355 (1969).

*Experimental verification of Einstein's law for the photoelectric effect; measurement of h :*

(3.3) A. L. HUGHES, "On the Emission Velocities of Photoelectrons", *Phil. Trans. Roy. Soc.*, **212**, 205 (1912).

(3.4) R. A. MILLIKAN, "A Direct Photoelectric Determination of Planck's h", *Phys. Rev.* 7, **355** (1916).

*The Franck-Hertz experiment :*

(3.5) J. FRANCK und G. HERTZ, "Über Zusammenstöße Zwischen Elecktronen und den Molekülen des Quecksilberdampfes und die Ionisierungsspannung desselben", *Verhandlungen der Deutschen Physikalischen Gesellschaft*, **16**, 457 (1914).
"Über Kinetik von Elektronen und Ionen in Gasen", *Physikalische Zeitschrift*, **17**, 409 (1916).

*The proportionality between the magnetic moment and the angular momentum :*

(3.6) A. EINSTEIN und J. W. DE HAAS, "Experimenteller Nachweis der Ampereschen Molekularströme", *Verhandlungen der Deutschen Physikalischen Gesellschaft*, **17**, 152 (1915).

(3.7) E. BECK, "Zum Experimentellen Nachweis der Ampereschen Molekularströme", *Annalen der Physik* (Leipzig), **60**, 109 (1919).

*The Stern-Gerlach experiment:*

(3.8) W. GERLACH und O. STERN, "Der Experimentelle Nachweis der Richtungsquantelung im Magnetfeld", *Zeitschrift für Physik*, **9**, 349 (1922).

*The Compton effect:*

(3.9) A. H. COMPTON, "A Quantum Theory of the Scattering of X-Rays by Light Elements", *Phys. Rev.*, **21**, 483 (1923).
"Wavelength Measurements of Scattered X-Rays", *Phys. Rev.*, **21**, 715 (1923).

*Electron diffraction:*

(3.10) C. DAVISSON and L. H. GERMER, "Diffraction of Electrons by a Crystal of Nickel", *Phys. Rev.*, **30**, 705 (1927).

*The Lamb shift:*

(3.11) W. E. LAMB JR. and R. C. RETHERFORD, "Fine Structure of the Hydrogen Atom",
I – *Phys. Rev.*, **79**, 549 (1950),
II – *Phys. Rev.*, **81**, 222 (1951).

*Hyperfine structure of the hydrogen ground state:*

(3.12) S. B. CRAMPTON, D. KLEPPNER and N. F. RAMSEY, "Hyperfine Separation of Ground State Atomic Hydrogen", *Phys. Rev. Letters*, **11**, 338 (1963).

*Several fundamental experiments are described in:*

(3.13) O. R. FRISCH, "Molecular Beams", *Scientific American*, **212**, 58 (May 1965).

## 4. QUANTUM MECHANICS: HISTORY

(4.1) L. DE BROGLIE, "Recherches sur la Théorie des Quanta", *Annales de Physique*, **3**, 22, Paris (1925).

(4.2) N. BOHR, "The Solvay Meetings and the Development of Quantum Mechanics", *Essays 1958-1962 on Atomic Physics and Human Knowledge*, Vintage, New York (1966).

(4.3) W. HEISENBERG, *Physics and Beyond: Encounters and Conversations*, Harper and Row, New York (1971).
*La Partie et le Tout*, Albin Michel, Paris (1972).

(4.4) *Niels Bohr, His life and work as seen by his friends and colleagues*, S. ROZENTAL, ed., North Holland, Amsterdam (1967).

(4.5) A. EINSTEIN, M. and H. BORN, *Correspondance 1916-1955*, Editions du Seuil, Paris (1972). See also *La Recherche*, **3**, 137 (Feb. 1972).

(4.6) *Theoretical Physics in the Twentieth Century*, M. FIERZ and V. F. WEISSKOPF eds., Wiley Interscience, New York (1960).

(4.7) *Sources of Quantum Mechanics*, B. L. VAN DER WAERDEN ed., North Holland, Amsterdam (1967); Dover, New York (1968).

(4.8) M. JAMMER, *The Conceptual Development of Quantum Mechanics*, McGraw-Hill, New York (1966). This book traces the historical development of quantum mechanics. Its very numerous footnotes provide a multitude of references. See also (5.12).

ARTICLES

(4.9) K. K. DARROW, "The Quantum Theory", *Scientific American*, **186**, 47 (March 1952).

(4.10) M. J. KLEIN, "Thermodynamics and Quanta in Planck's work", *Physics Today*, **19**, 23 (Nov. 1966).

(4.11) H. A. MEDICUS, "Fifty years of Matter Waves", *Physics Today*, **27**, 38 (Feb. 1974).

Reference (5.11) contains a large number of references to the original texts.

## 5. QUANTUM MECHANICS: DISCUSSION OF ITS FOUNDATIONS

GENERAL PROBLEMS:

(5.1) D. BOHM, *Quantum Theory*, Constable, London (1954).

(5.2) J. M. JAUCH, *Foundations of Quantum Mechanics*, Addison-Wesley, Reading, Mass. (1968).

(5.3) B. D'ESPAGNAT, *Conceptual Foundations of Quantum Mechanics*, Benjamin, New York (1971).

(5.4) Proceedings of the International School of Physics "Enrico Fermi" (Varenna), Course IL; *Foundations of Quantum Mechanics*, B. D'ESPAGNAT ed., Academic Press, New York (1971).

(5.5) B. S. DEWITT, "Quantum Mechanics and Reality", *Physics Today*, **23**, 30, (Sept. 1970).

(5.6) "Quantum Mechanics debate", *Physics Today*, **24**, 36 (April 1971).

See also (1.28).

MISCELLANEOUS INTERPRETATIONS

(5.7) N. BOHR, "Discussion with Einstein on Epistemological Problems in Atomic Physics", in *A. Einstein: Philosopher-Scientist*, P. A. SCHILPP ed., Harper and Row, New York (1959).

(5.8) M. BORN, *Natural Philosophy of Cause and Chance*, Oxford University Press, London (1951); Clarendon Press, Oxford (1949).

(5.9) L. DE BROGLIE, *Une Tentative d'Interprétation Causale et Non Linéaire de la Mécanique Ondulatoire: la Théorie de la Double Solution*, Gauthier-Villars, Paris (1956); *Etude Critique des Bases de l'Interprétation Actuelle de la Mécanique Ondulatoire*, Gauthier-Villars, Paris (1963).

(5.10) *The Many-Worlds Interpretation of Quantum Mechanics*, B. S. DEWITT and N. GRAHAM eds., Princeton University Press (1973).

A very complete set of references, classified and annotated, can be found in:

(5.11) B. S. DEWITT and R. N. GRAHAM, "Resource Letter IQM-1 on the Interpretation of Quantum Mechanics", *Am. J. Phys.* **39**, 724 (1971).

(5.12) M. JAMMER, *The Philosophy of Quantum Mechanics*, Wiley-Interscience, New York (1974). A general presentation of the different interpretations of the Quantum Mechanics formalism. Gives numerous references.

MEASUREMENT THEORY

(5.13) K. GOTTFRIED, *Quantum Mechanics*, Vol. I, Benjamin, New York (1966).

(5.14) D. I. BLOKHINTSEV, *Principes Essentiels de la Mécanique Quantique*, Dunod, Paris (1968).

(5.15) A. SHIMONY, "Role of the Observer in Quantum Theory", *Am. J. Phys.*, **31**, 755 (1963).

HIDDEN VARIABLES AND "PARADOXES":

(5.16) A. EINSTEIN, B. PODOLSKY and N. ROSEN, "Can Quantum-Mechanical Description of Physical Reality Be Considered Complete?", *Phys. Rev.* **47**, 777 (1935).
N. BOHR, "Can Quantum Mechanical Description of Physical Reality Be Considered Complete?", *Phys. Rev.* **48**, 696 (1935).

(5.17) *Paradigms and Paradoxes, the Philosophical Challenge of the Quantum Domain*, R. G. COLODNY ed., University of Pittsburg Press (1972).

(5.18) J. S. BELL, "On the Problem of Hidden Variables in Quantum Mechanics", *Rev. Mod. Phys.* **38**, 447 (1966).

See also reference (4.8), as well as (5.11) and chap. 7 of (5.12).

## 6. CLASSICAL MECHANICS

INTRODUCTORY LEVEL

(6.1) M. ALONSO and E. J. FINN, *Fundamental University Physics, Vol. I: Mechanics*, Addison-Wesley, Reading, Mass. (1967).

(6.2) C. KITTEL, W. D. KNIGHT and M. A. RUDERMAN, *Berkeley Physics Course, Vol. 1: Mechanics*, McGraw-Hill, New York (1962).

(6.3) R. P. FEYNMAN, R. B. LEIGHTON and M. SANDS, *The Feynman Lectures on Physics, Vol. I: Mechanics, Radiation, and Heat*, Addison-Wesley, Reading, Mass. (1966).

(6.4) J. B. MARION, *Classical Dynamics of Particles and Systems*, Academic Press, New York (1965).

MORE ADVANCED LEVEL:

(6.5) A. SOMMERFELD, *Lectures on Theoretical Physics, Vol. I: Mechanics*, Academic Press, New York (1964).

(6.6) H. GOLDSTEIN, *Classical Mechanics*, Addison-Wesley, Reading, Mass. (1959).

(6.7) L. D. LANDAU and E. M. LIFSHITZ, *Mechanics*, Pergamon Press, Oxford (1960).

## 7. ELECTROMAGNETISM AND OPTICS

INTRODUCTORY LEVEL

(7.1) E. M. PURCELL, *Berkeley Physics Course, Vol. 2: Electricity and Magnetism*, McGraw-Hill, New York (1965).
F. S. CRAWFORD JR., *Berkeley Physics Course, Vol. 3: Waves*, McGraw-Hill, New York (1968).

(7.2) R. P. FEYNMAN, R. B. LEIGHTON and M. SANDS, *The Feynman Lectures on Physics, Vol. II: Electromagnetism and Matter*, Addison-Wesley, Reading Mass. (1966).

(7.3) M. ALONSO and E. J. FINN, *Fundamental University Physics, Vol. II: Fields and Waves*, Addison-Wesley, Reading, Mass. (1967).

(7.4) E. HECHT and A. ZAJAC, *Optics*, Addison-Wesley, Reading, Mass. (1974).

### MORE ADVANCED LEVEL

(7.5) J. D. JACKSON, *Classical Electrodynamics*, 2ᵈ ed. Wiley, New York (1975).

(7.6) W. K. H. PANOFSKY and M. PHILLIPS, *Classical Electricity and Magnetism*, Addison-Wesley, Reading, Mass. (1964).

(7.7) J. A. STRATTON, *Electromagnetic Theory*, McGraw-Hill, New York (1941).

(7.8) M. BORN and E. WOLF, *Principles of Optics*, Pergamon Press, London (1964).

(7.9) A. SOMMERFELD, *Lectures on Theoretical Physics, Vol. IV: Optics*, Academic Press, New York (1964).

(7.10) G. BRUHAT, *Optique*, 5ᵉ Edition revised and completed by A. KASTLER, Masson, Paris (1954).

(7.11) L. LANDAU and E. LIFSHITZ, *The Classical Theory of Fields*, Addison-Wesley, Reading, Mass. (1951); Pergamon Press, London (1951).

(7.12) L. D. LANDAU and E. M. LIFSHITZ, *Electrodynamics of Condinuous Media*, Pergamon Press, Oxford (1960).

(7.13) L. BRILLOUIN, *Wave Propagation and Group Velocity*, Academic Press, New York (1960).

## 8. THERMODYNAMICS. STATISTICAL MECHANICS

### INTRODUCTORY LEVEL

(8.1) F. REIF, *Berkeley Physics Course, Vol. 5: Statistical Physics*, McGraw-Hill, New York (1967).

(8.2) C. KITTEL, *Thermal Physics*, Wiley, New York (1969).

(8.3) G. BRUHAT, *Thermodynamique*, 5ᵉ Edition revised by A. KASTLER, Masson, Paris (1962).

See also references (1.4), part. 2, and (6.3).

### MORE ADVANCED LEVEL

(8.4) F. REIF, *Fundamentals of Statistical and Thermal Physics*, McGraw-Hill, New York (1965).

(8.5) R. CASTAING, *Thermodynamique Statistique*, Masson, Paris (1970).

(8.6) P. M. MORSE, *Thermal Physics*, Benjamin, New York (1964).

(8.7) R. KUBO, *Statistical Mechanics*, North Holland, Amsterdam and Wiley, New York (1965).

(8.8) L. D. LANDAU and E. M. LIFSHITZ, *Course of Theoretical Physics, Vol. 5: Statistical Physics*, Pergamon Press, London (1963).

(8.9) H. B. CALLEN, *Thermodynamics*, Wiley, New York (1961).

(8.10) A. B. PIPPARD, *The Elements of Classical Thermodynamics*, Cambridge University Press (1957).

(8.11) R. C. TOLMAN, *The Principles of Statistical Mechanics*, Oxford University Press (1950).

## 9. RELATIVITY

### INTRODUCTORY LEVEL

(9.1) J. H. SMITH, *Introduction to Special Relativity*, Benjamin, New York (1965).

See also references (6.2) and (6.3).

### MORE ADVANCED LEVEL

(9.2) J. L. SYNGE, *Relativity: The Special Theory*, North Holland, Amsterdam (1965).
(9.3) R. D. SARD, *Relativistic Mechanics*, Benjamin, New York (1970).
(9.4) J. AHARONI, *The Special Theory of Relativity*, Oxford University Press, London (1959).
(9.5) C. MØLLER, *The Theory of Relativity*, Oxford University Press, London (1972).
(9.6) P. G. BERGMANN, *Introduction to the Theory of Relativity*, Prentice Hall, Englewood Cliffs (1960).
(9.7) C. W. MISNER, K. S. THORNE and J. A. WHEELER, *Gravitation*, Freeman, San Francisco (1973).

See also the electromagnetism references, in particular (7.5) and (7.11).

Also valuable:

(9.8) A. EINSTEIN, *Quatre Conférences sur la Théorie de la Relativité*, Gauthier-Villars, Paris (1971).
(9.9) A. EINSTEIN, *La Théorie de la Relativité Restreinte et Générale. La Relativité et le Problème de l'Espace*, Gauthier-Villars, Paris (1971).
9.10) A. EINSTEIN, *The Meaning of Relativity*, Methuen, London (1950).
9.11) A. EINSTEIN, *Relativity, the Special and General Theory, a Popular Exposition*, Methuen, London (1920); H. Holt, New York (1967).

A much more complete list of references can be found in :

(9.12) G. HOLTON, Resource Letter SRT-1 on Special Relativity Theory, *Am. J. Phys.* **30**, 462 (1962).


## 10. MATHEMATICAL METHODS


### ELEMENTARY GENERAL TEXTS

(10.1) J. BASS, *Cours de Mathématiques*, Vols. I, II and III, Masson, Paris (1961).
(10.2) A. ANGOT, *Compléments de Mathématiques*, Revue d'Optique, Paris (1961).
(10.3) T. A. BAK and J. LICHTENBERG, *Mathematics for Scientists*, Benjamin, New York (1966).
(10.4) G. ARFKEN, *Mathematical Methods for Physicists*, Academic Press, New York (1966).
(10.5) J. D. JACKSON, *Mathematics for Quantum Mechanics*, Benjamin, New York (1962).

### MORE ADVANCED GENERAL TEXTS

(10.6) J. MATHEWS and R. L. WALKER, *Mathematical Methods of Physics*, Benjamin, New York (1970).
(10.7) L. SCHWARTZ, *Mathematics for the Physical Sciences*, Hermann, Paris (1968). *Méthodes mathématiques pour les sciences physiques*, Hermann, Paris (1965).
(10.8) E. BUTKOV, *Mathematical Physics*, Addison-Wesley, Reading, Mass. (1968).
(10.9) H. CARTAN, *Elementary Theory of Analytic Functions of One or Several Complex Variables*, Addison-Wesley, Reading, Mass. (1966). *Théorie élémentaire des fonctions analytiques d'une ou plusieurs variables complexes*, Hermann, Paris (1961).
(10.10) J. VON NEUMANN, *Mathematical Foundations of Quantum Mechanics*, Princeton University Press (1955).
(10.11) R. COURANT and D. HILBERT, *Methods of Mathematical Physics*, Vols. I and II, Wiley, Interscience, New York (1966).

(10.12) E. T. WHITTAKER and G. N. WATSON, *A Course of Modern Analysis*, Cambridge University Press (1965).

(10.13) P. M. MORSE and H. FESHBACH, *Methods of Theoretical Physics*, McGraw-Hill, New York (1953).

## LINEAR ALGEBRA. HILBERT SPACES

(10.14) A. C. AITKEN, *Determinants and Matrices*, Oliver and Boyd, Edinburgh (1956).

(10.15) R. K. EISENSCHITZ, *Matrix Algebra for Physicists*, Plenum Press, New York (1966).

(10.16) M. C. PEASE III, *Methods of Matrix Algebra*, Academic Press, New York (1965).

(10.17) J. L. SOULE, *Linear Operators in Hilbert Space*, Gordon and Breach, New York (1967).

(10.18) W. SCHMEIDLER, *Linear Operators in Hilbert Space*, Academic Press, New York (1965).

(10.19) N. I. AKHIEZER and I. M. GLAZMAN, *Theory of Linear Operators in Hilbert Space*, Ungar, New York (1961).

## FOURIER TRANSFORMS; DISTRIBUTIONS

(10.20) R. STUART, *Introduction to Fourier Analysis*, Chapman and Hall, London (1969).

(10.21) M. J. LIGHTHILL, *Introduction to Fourier Analysis and Generalized Functions*, Cambridge University Press (1964).

(10.22) L. SCHWARTZ, *Théorie des Distributions*, Hermann, Paris (1967).

(10.23) I. M. GEL'FAND and G. E. SHILOV, *Generalized Functions*, Academic Press, New York (1964).

(10.24) F. OBERHETTINGER, *Tabellen zur Fourier Transformation*, Springer-Verlag, Berlin (1957).

## PROBABILITY AND STATISTICS

(10.25) J. BASS, *Elements of Probability Theory*, Academic Press, New York (1966). *Éléments* de *Calcul des Probabilités*, Masson, Paris (1974).

(10.26) P. G. HOEL, S. C. PORT and C. J. STONE, *Introduction to Probability Theory*, Houghton-Mifflin, Boston (1971).

(10.27) H. G. TUCKER, *An Introduction to Probability and Mathematical Statistics*, Academic Press, New York (1965).

(10.28) J. LAMPERTI, *Probability*, Benjamin, New York (1966).

(10.29) W. FELLER, *An Introduction to Probability Theory and its Applications*, Wiley, New York (1968).

(10.30) L. BREIMAN, *Probability*, Addison-Wesley, Reading, Mass. (1968).

## GROUP THEORY

### Applied to physics:

(10.31) H. BACRY, *Lectures on Group Theory*, Gordon and Breach, New York (1967).

(10.32) M. HAMERMESH, *Group Theory and its Application to Physical Problems*, Addison-Wesley, Reading, Mass. (1962).

See also (2.18), (2.22) and (2.23) or reference (16.13) which gives a simple introduction to continuous groups in physics.

### More mathematical:

(10.33) G. PAPY, *Groups*, Macmillan, New York (1964).

(10.34) A. G. KUROSH, *The Theory of Groups*, Chelsea, New York (1960).

(10.35) L. S. PONTRYAGIN, *Topological Groups*, Gordon and Breach, New York (1966).

### SPECIAL FUNCTIONS AND TABLES

(10.36) A. GRAY and G. B. MATHEWS, *A Treatise on Bessel Functions and their Applications to Physics*, Dover, New York (1966).

(10.37) E. D. RAINVILLE, *Special Functions*, Macmillan, New York (1965).

(10.38) W. MAGNUS, F. OBERHETTINGER and R. P. SONI, *Formulas and Theorems for the Special Functions of Mathematical Physics*, Springer-Verlag, Berlin (1966).

(10.39) BATEMAN MANUSCRIPT PROJECT, *Higher Transcendental Functions*, Vols. I, II and III, A. ERDELYI ed., McGraw-Hill, New York (1953).

(10.40) M. ABRAMOWITZ and I. A. STEGUN, *Handbook of Mathematical Functions*, Dover, New York (1965).

(10.41) L. J. COMRIE, *Chambers's Shorter Six-Figure Mathematical Tables*, Chambers, London (1966).

(10.42) E. JAHNKE and F. EMDE, *Tables of Functions*, Dover, New York (1945).

(10.43) V. S. AIZENSHTADT, V. I. KRYLOV and A. S. METEL'SKII, *Tables of Laguerre Polynomials and Functions*, Pergamon Press, Oxford (1966).

(10.44) H. B. DWIGHT, *Tables of Integrals and Other Mathematical Data*, Macmillan, New York (1965).

(10.45) D. BIERENS DE HAAN, *Nouvelles Tables d'Intégrales Définies*, Hafner, New York (1957).

(10.46) F. OBERHETTINGER and L. BADII, *Tables of Laplace Transforms*, Springer-Verlag, Berlin (1973).

(10.47) BATEMAN MANUSCRIPT PROJECT, *Tables of Integral Transforms*, Vols. I and II, A. ERDELYI ed., McGraw-Hill, New York (1954).

(10.48) M. ROTENBERG, R. BIVINS, N. METROPOLIS and J. K. WOOTEN JR., *The 3-j and 6-j symbols*, M.I.T. Technology Press (1959); Crosby Lockwood and Sons, London.

## 11. ATOMIC PHYSICS

### INTRODUCTORY LEVEL

(11.1) H. G. KUHN, *Atomic Spectra*, Longman, London (1969).

(11.2) B. CAGNAC and J. C. PEBAY-PEYROULA, *Physique Atomique*, Vols. 1 and 2, Dunod, Paris (1971).
English translation : *Modern Atomic Physics*, Vol. 1 : *Fundamental Principles*, and 2: *Quantum Theory and its Application*, Macmillan, London (1975).

(11.3) A. G. MITCHELL and M. W. ZEMANSKY, *Resonance Radiation and Excited Atoms*, Cambridge University Press, London (1961).

(11.4) M. BORN, *Atomic Physics*, Blackie and Son, London (1951).

(11.5) H. E. WHITE, *Introduction to Atomic Spectra*, McGraw-Hill, New York (1934).

(11.6) V. N. KONDRATIEV, *La Structure des Atomes et des Molécules*, Masson, Paris (1964).

See also (1.3) and (12.1).

### MORE ADVANCED LEVEL

(11.7) G. W. SERIES, *The Spectrum of Atomic Hydrogen*, Oxford University Press, London (1957).

(11.8) J. C. SLATER, *Quantum Theory of Atomic Structure*, Vols. I and II, McGraw-Hill, New York (1960).

(11.9) A. E. RUARK and H. C. UREY, *Atoms, Molecules and Quanta*, Vols. I and II, Dover, New York (1964).

(11.10) *Handbuch der Physik, Vols. XXXV and XXXVI, Atoms*, S. FLÜGGE ed., Springer-Verlag Berlin (1956 and 1957).

(11.11) N. F. RAMSEY, *Molecular Beams*, Oxford University Press, London (1956).

(11.12) I. I. SOBEL'MAN, *Introduction to the Theory of Atomic Spectra*, Pergamon Press, Oxford (1972).

(11.13) E. U. CONDON and G. H. SHORTLEY, *The Theory of Atomic Spectra*, Cambridge University Press (1953).

### ARTICLES

Numerous references to articles and books, classified and discussed, can be found in:

(11.14) J. C. ZORN, "Resource Letter MB-1 on Experiments with Molecular Beams, *Am. J. Phys.* **32**, 721 (1964).

See also: (3.13).

(11.15) V. F. WEISSKOPF, "How Light Interacts with Matter", *Scientific American*, **219**, 60 (Sept. 1968).

(11.16) H. R. CRANE, "The g Factor of the Electron", *Scientific American*, **218**, 72 (Jan. 1968).

(11.17) M. S. ROBERTS, "Hydrogen in Galaxies", *Scientific American*, **208**, 94 (June 1963).

(11.18) S. A. WERNER, R. COLELLA, A. W. OVERHAUSER and C. F. EAGEN, "Observation of the Phase Shift of a Neutron due to Precession in a Magnetic Field", *Phys. Rev. Letters*, **35**, 1053 (1975).

See also H. RAUCH, A. ZEILINGER, G. BADUREK, W. BAUSPIESS, U. BONSE, *Phys. Letters* **54**A, 425 (1975).

### EXOTIC ATOMS

(11.19) H. C. CORBEN and S. DE BENEDETTI, "The Ultimate Atom", *Scientific American*, **191**, 88 (Dec. 1954).

(11.20) V. W. HUGHES, "The Muonium Atom", *Scientific American*, **214**, 93, (April 1966). "Muonium", *Physics Today*, **20**, 29 (Dec. 1967).

(11.21) S. DE BENEDETTI, "Mesonic Atoms", *Scientific American*, **195**, 93, (Oct. 1956).

(11.22) C. E. WIEGAND, "Exotic Atoms", *Scientific American*, **227**, 102, (Nov. 1972).

(11.23) V. W. HUGHES, "Quantum Electrodynamics: experiment", in *Atomic Physics*, B. Bederson, V. W. Cohen and F. M. Pichanick eds., Plenum Press, New York (1969).

(11.24) R. DE VOE, P. M. MC INTYRE, A. MAGNON, D, Y. STOWELL, R. A. SWANSON and V. L. TELEGDI, "Measurement of the muonium Hfs Splitting and of the muon moment by double resonance, and new value of α", *Phys. Rev. Letters*, **25**, 1779 (1970).

(11.25) K. F. CANTER, A. P. MILLS JR. and S. BERKO, "Observations of Positronium Lyman-Radiation", *Phys. Rev. Letters*, **34**, 177 (1975). "Fine-Structure Measurement in the First Excited State of Positronium" *Phys. Rev. Letters*, **34**, 1541 (1975).

## 12. MOLECULAR PHYSICS

### INTRODUCTORY LEVEL

(12.1) M. KARPLUS and R. N. PORTER, *Atoms and Molecules*, Benjamin, New York (1970).

(12.2) L. PAULING, *The Nature of the Chemical Bond*, Cornell University Press (1948).

See also (1.3), chap. 12; (1.5) and (11.6).

**MORE ADVANCED LEVEL**

(12.3) I. N. LEVINE, *Quantum Chemistry*, Allyn and Bacon, Boston (1970).

(12.4) G. HERZBERG, *Molecular Spectra and Molecular Structure*, Vol. I: *Spectra of Diatomic Molecules*, and Vol. II: *Infrared and Raman Spectra of Polyatomic Molecules*, D. Van Nostrand Company, Princeton (1963 and 1964).

(12.5) H. EYRING, J. WALTER and G. E. KIMBALL, *Quantum Chemistry*, Wiley, New York (1963).

(12.6) C. A. COULSON, *Valence*, Oxford at the Clarendon Press (1952).

(12.7). J. C. SLATER, *Quantum Theory of Molecules and Solids*, Vol. 1: *Electronic Structure of Molecules*, McGraw-Hill, New York (1963).

(12.8) *Handbuch der Physik, Vol. XXXVII, 1 and 2, Molecules*, S. FLÜGGE, ed., Springer Verlag, Berlin (1961).

(12.9) D. LANGBEIN, *Theory of Van der Waals Attraction*, Springer Tracts in Modern Physics, Vol. 72, Springer Verlag, Berlin (1974).

(12.10) C. H. TOWNES and A. L. SCHAWLOW, *Microwave Spectroscopy*, McGraw-Hill, New York (1955).

(12.11) P. ENCRENAZ, *Les Molécules interstellaires*, Delachaux et Niestlé, Neuchâtel (1974).

See also (11.9), (11.11) and (11.14).

**ARTICLES**

(12.12) B. V. DERJAGUIN, "The Force Between Molecules", *Scientific American*, **203**, 47 (July 1960).

(12.13) A. C. WAHL, "Chemistry by Computer", *Scientific American*, **222**, 54 (April 1970).

(12.14) B. E. TURNER, "Interstellar Molecules", *Scientific American*, **228**, 51 (March 1973).

(12.15) P. M. SOLOMON, "Interstellar Molecules", *Physics Today*, **26**, 32 (March 1973).

See also (16.25).

## 13. SOLID STATE PHYSICS

**INTRODUCTORY LEVEL**

(13.1) C. KITTEL, *Elementary Solid State Physics*, Wiley, New York (1962).

(13.2) C. KITTEL, *Introduction to Solid State Physics*, 3ᵉ ed., Wiley, New York (1966).

(13.3) J. M. ZIMAN, *Principles of the Theory of Solids*, Cambridge University Press, London (1972).

(13.4) F. SEITZ, *Modern Theory of Solids*, McGraw-Hill, New York (1940).

**MORE ADVANCED LEVEL**

**General texts:**

(13.5) C. KITTEL, *Quantum Theory of Solids*, Wiley, New York (1963).

(13.6) R. E. PEIERLS, *Quantum Theory of Solids*, Oxford University Press, London (1964).

(13.7) N. F. MOTT and H. JONES, *The Theory of the Properties of Metals and Alloys*, Clarendon Press, Oxford (1936); Dover, New York (1958).

**More specialized texts:**

(13.8) M. BORN and K. HUANG, *Dynamical Theory of Crystal Lattices*, Oxford University Press, London (1954).

(13.9) J. M. ZIMAN, *Electrons and Phonons*, Oxford University Press, London (1960).

(13.10) H. JONES, *The Theory of Brillouin Zones and Electronic States in Crystals*, North Holland, Amsterdam (1962).

(13.11) J. CALLAWAY, *Energy Band Theory*, Academic Press, New York (1964).

(13.12) R. A. SMITH, *Wave Mechanics of Crystalline Solids*, Chapman and Hall, London (1967).

(13.13) D. PINES and P. NOZIERES, *The Theory of Quantum Liquids*, Benjamin, New York (1966).

(13.14) D. A. WRIGHT, *Semiconductors*, Associated Book Publishers, London (1966).

(13.15) R. A. SMITH, *Semiconductors*, Cambridge University Press, London (1964).

ARTICLES

(13.16) R. L. SPROULL, "The Conduction of Heat in Solids", *Scientific American*, **207**, 92 (Dec. 1962).

(13.17) A. R. MACKINTOSH, "The Fermi Surface of Metals", *Scientific American*, **209**, 110 (July 1963).

(13.18) D. N. LANGENBERG, D. J. SCALAPINO and B. N. TAYLOR, "The Josephson Effects", *Scientific American* **214**, 30 (May 1966).

(13.19) G. L. POLLACK, "Solid Noble Gases", *Scientific American*, **215**, 64 (Oct. 1966).

(13.20) B. BERTMAN and R. A. GUYER, "Solid Helium", *Scientific American*, **217**, 85 (Aug. 1967).

(13.21) N. MOTT, "The Solid State", *Scientific American*, **217**, 80 (Sept. 1967).

(13.22) M. Ya. AZBEL', M. I. KAGANOV and I. M. LIFSHITZ, "Conduction Electrons in Metals", *Scientific American*, **228**, 88 (Jan. 1973).

(13.23) W. A. HARRISON, "Electrons in Metals", *Physics Today*, **22**, 23 (Oct. 1969).

## 14. MAGNETIC RESONANCE

(14.1) A. ABRAGAM, *The Principles of Nuclear Magnetism*, Clarendon Press, Oxford (1961).

(14.2) C. P. SLICHTER, *Principles of Magnetic Resonance*, Harper and Row, New York (1963).

(14.3) G. E. PAKE, *Paramagnetic Resonance*, Benjamin, New York (1962).

See also Ramsey (11.11), Chaps. V, VI and VII.

ARTICLES

(14.4) G. E. PAKE, "Fundamentals of Nuclear Magnetic Resonance Absorption, I and II, *Am. J. Phys.*, **18**, 438 and 473 (1950).

(14.5) E. M. PURCELL, "Nuclear Magnetism", *Am. J. Phys.*, **22**, 1 (1954).

(14.6) G. E. PAKE, "Magnetic Resonance", *Scientific American*, **199**, 58 (Aug. 1958).

(14.7) K. WÜTHRICH and R. C. SHULMAN, "Magnetic Resonance in Biology", *Physics Today*, **23**, 43 (April 1970).

(14.8) F. BLOCH, "Nuclear Induction", *Phys. Rev.* **70**, 460 (1946).

Numerous other references, in particular to original articles, can be found in:

(14.9) R. E. NORBERG, "Resource Letter NMR-EPR-1 on Nuclear Magnetic Resonance and Electron Paramagnetic Resonance", *Am. J. Phys.*, **33**, 71 (1965).

## 15. QUANTUM OPTICS; MASERS AND LASERS

### OPTICAL PUMPING; MASERS AND LASERS


(15.1) R. A. BERNHEIM, *Optical Pumping: An Introduction*, Benjamin, New York (1965). This book contains many references. In addition, several important original papers are reprinted.

(15.2) *Quantum Optics and Electronics, Les Houches Lectures 1964*, C. DE WITT, A. BLANDIN and C. COHEN-TANNOUDJI eds., Gordon and Breach, New York (1965).

(15.3) *Quantum Optics, Proceedings of the Scottish Universities Summer School 1969*, S. M. KAY and A. MAITLAND eds., Academic Press, London (1970). These two summer-school books contain several useful texts related to optical pumping and quantum electronics.

(15.4) W. E. LAMB JR., *Quantum Mechanical Amplifiers*, in *Lectures in Theoretical Physics*, Vol. II, W. BRITTIN and D. DOWNS eds., Interscience Publishers, New York (1960).

(15.5) M. SARGENT III, M. O. SCULLY and W. E. LAMB JR., *Laser Physics*, Addison-Wesley, New York (1974).

(15.6) A. E. SIEGMAN, *An Introduction to Lasers and Masers*, McGraw-Hill, New York (1971).

(15.7) L. ALLEN, *Essentials of Lasers*, Pergamon Press, Oxford (1969). This small book contains several reprints of original papers on lasers.

(15.8) L. ALLEN and J. H. EBERLY, *Optical Resonance and Two-Level Atoms*, Wiley Interscience, New York (1975).

(15.9) A. YARIV, *Quantum Electronics*, Wiley, New York (1967).

(15.10) H. M. NUSSENZVEIG, *Introduction to Quantum Optics*, Gordon and Breach, London (1973).


### ARTICLES

Two "Resource Letters" give, discuss and classify many useful references:


(15.11) H. W. MOOS, "Resource Letter MOP-1 on Masers (Microwave through Optical) and on Optical Pumping", *Am. J. Phys.*, **32**, 589 (1964).

(15.12) P. CARRUTHERS, "Resource Letter QSL-1 on Quantum and Statistical Aspects of Light", *Am. J. Phys.*, **31**, 321 (1963).


Reprints of many important papers on Lasers have been collected in:


(15.13) *Laser Theory*, F. S. BARNES ed., I.E.E.E. Press, New York (1972).

(15.14) H. LYONS, "Atomic Clocks", *Scientific American*, **196**, 71 (Feb. 1957).

(15.15) J. P. GORDON, "The Maser", *Scientific American*, **199**, 42 (Dec. 1958).

(15.16) A. L. BLOOM, "Optical Pumping", *Scientific American*, **203**, 72 (Oct. 1960).

(15.17) A. L. SCHAWLOW, "Optical Masers", *Scientific American*, **204**, 52 (June 1961). "Advances in Optical Masers", *Scientific American*, **209**, 34 (July 1963). "Laser Light", *Scientific American*, **219**, 120 (Sept. 1968).

(15.18) M. S. FELD and V. S. LETOKHOV, "Laser Spectroscopy", *Scientific American*, **229**, 69 (Dec. 1973).


### NON-LINEAR OPTICS


(15.19) G. C. BALDWIN, *An Introduction to Non-Linear Optics*, Plenum Press, New York (1969).

(15.20) F. ZERNIKE and J. E. MIDWINTER, *Applied Non-Linear Optics*, Wiley Interscience, New York (1973).

(15.21) N. BLOEMBERGEN, *Non-Linear Optics*, Benjamin, New York (1965).
See also this author's lectures in references (15.2) and (15.3).

ARTICLES

(15.22) J. A. GIORDMAINE, "The Interaction of Light with Light", *Scientific American*, **210**, 38 (Apr. 1964).
"Non-Linear Optics", *Physics Today*, **22**, 39 (Jan. 1969).

## 16. NUCLEAR PHYSICS AND PARTICLE PHYSICS

INTRODUCTION TO NUCLEAR PHYSICS

(16.1) L. VALENTIN, *Physique Subatomique: Noyaux et Particules*, Hermann, Paris (1975).
(16.2) D. HALLIDAY, *Introductory Nuclear Physics*, Wiley, New York (1960).
(16.3) R. D. EVANS, *The Atomic Nucleus*, McGraw-Hill, New York (1955).
(16.4) M. A. PRESTON, *Physics of the Nucleus*, Addison-Wesley, Reading, Mass. (1962).
(16.5) E. SEGRE, *Nuclei and Particles*, Benjamin, New York (1965).

MORE ADVANCED NUCLEAR PHYSICS TEXTS

(16.6) A. DESHALIT and H. FESHBACH, *Theoretical Nuclear Physics, Vol. 1: Nuclear Structure*, Wiley, New York (1974).
(16.7) J. M. BLATT and V. F. WEISSKOPF, *Theoretical Nuclear Physics*, Wiley, New York (1963).
(16.8) E. FEENBERG, *Shell Theory of the Nucleus*, Princeton University Press (1955).
(16.9) A. BOHR and B. R. MOTTELSON, *Nuclear Structure*, Benjamin, New York (1969).

INTRODUCTION TO PARTICLE PHYSICS

(16.10) D. H. FRISCH and A. M. THORNDIKE, *Elementary Particles*, Van Nostrand, Princeton (1964).
(16.11) C. E. SWARTZ, *The Fundamental Particles*, Addison-Wesley, Reading, Mass. (1965).
(16.12) R. P. FEYNMAN, *Theory of Fundamental Processes*, Benjamin, New York (1962).
(16.13) R. OMNES, *Introduction à l'Etude des Particules Elémentaires*, Ediscience, Paris (1970).
(16.14) K. NISHIJIMA, *Fundamental Particles*, Benjamin, New York (1964).

MORE ADVANCED PARTICLE PHYSICS TEXTS

(16.15) B. DIU, *Qu'est-ce qu'une Particule Elémentaire?* Masson, Paris (1965).
(16.16) J. J. SAKURAI, *Invariance Principles and Elementary Particles*, Princeton University Press (1964).
(16.17) G. KÄLLEN, *Elementary Particle Physics*, Addison-Wesley, Reading, Mass. (1964).
(16.18) A. D. MARTIN and T. D. SPEARMAN, *Elementary Particle Theory*, North Holland, Amsterdam (1970).
(16.19) A. O. WEISSENBERG, *Muons*, North Holland, Amsterdam (1967).

ARTICLES

(16.20) M. G. MAYER, "The Structure of the Nucleus", *Scientific American*, **184**, 22 (March 1951).
(16.21) R. E. PEIERLS, "The Atomic Nucleus", *Scientific American*, **200**, 75 (Jan. 1959).
(16.22) E. U. BARANGER, "The present status of the nuclear shell model", *Physics Today*, **26**, 34 (June 1973).

(16.23) S. DE BENEDETTI, "Mesonic Atoms", *Scientific American*, **195**, 93 (Oct. 1956).

(16.24) S. DE BENEDETTI, "The Mössbauer Effect", *Scientific American*, **202**, 72 (April 1960).

(16.25) R. H. HERBER, "Mössbauer Spectroscopy", *Scientific American*, **225**, 86 (Oct. 1971).

(16.26) S. PENMAN, "The Muon", *Scientific American*, **205**, 46 (July 1961).

(16.27) R. E. MARSHAK, "The Nuclear Force", *Scientific American*, **202**, 98 (March 1960).

(16.28) M. GELL-MANN and E. P. ROSENBAUM, "Elementary Particles", *Scientific American*, **197**, 72 (July 1957).

(16.29) G. F. CHEW, M. GELL-MANN and A. H. ROSENFELD, "Strongly Interacting Particles", *Scientific American*, **210**, 74 (Feb. 1964).

(16.30) V. F. WEISSKOPF, "The Three Spectroscopies", *Scientific American*, **218**, 15 (May 1968).

(16.31) U. AMALDI, "Proton Interactions at High Energies", *Scientific American*, **229**, 36 (Nov. 1973).

(16.32) S. WEINBERG, "Unified Theories of Elementary-Particle Interaction", *Scientific American*, **231**, 50 (July 1974).

(16.33) S. D. DRELL, "Electron-Positron Annihilation and the New Particles", *Scientific American*, **232**, 50 (June 1975).

(16.34) R. WILSON, "Form Factors of Elementary Particles", *Physics Today*, **22**, 47 (Jan. 1969).

(16.35) E. S. ABERS and B. W. LEE, "Gauge Theories", *Physics Reports*, **9C**, 1, Amsterdam (1973).

# Index

N.B. *References followed by* (e) *refer to an exercise*

Achevé d'imprimer en décembre 2011
par la Sté ACORT Europe
www.cogetefi.com

Dépôt légal à parution

*Imprimé en France*

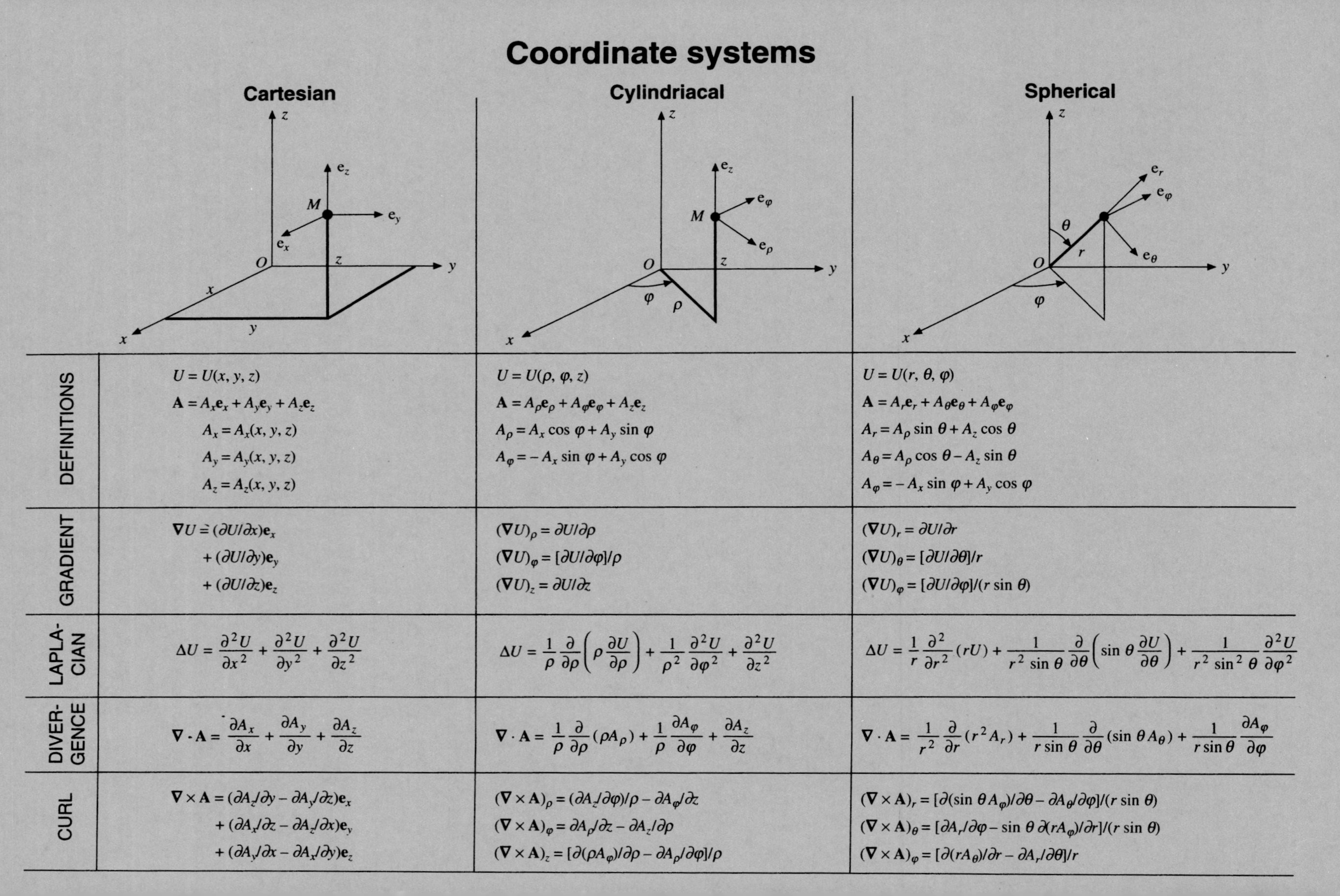


# Coordinate systems

| | Cartesian | Cylindriacal | Spherical |
|---|---|---|---|
| DEFINITIONS | $U = U(x, y, z)$<br>$\mathbf{A} = A_x\mathbf{e}_x + A_y\mathbf{e}_y + A_z\mathbf{e}_z$<br>$A_x = A_x(x, y, z)$<br>$A_y = A_y(x, y, z)$<br>$A_z = A_z(x, y, z)$ | $U = U(\rho, \varphi, z)$<br>$\mathbf{A} = A_\rho\mathbf{e}_\rho + A_\varphi\mathbf{e}_\varphi + A_z\mathbf{e}_z$<br>$A_\rho = A_x \cos\varphi + A_y \sin\varphi$<br>$A_\varphi = -A_x \sin\varphi + A_y \cos\varphi$ | $U = U(r, \theta, \varphi)$<br>$\mathbf{A} = A_r\mathbf{e}_r + A_\theta\mathbf{e}_\theta + A_\varphi\mathbf{e}_\varphi$<br>$A_r = A_\rho \sin\theta + A_z \cos\theta$<br>$A_\theta = A_\rho \cos\theta - A_z \sin\theta$<br>$A_\varphi = -A_x \sin\varphi + A_y \cos\varphi$ |
| GRADIENT | $\nabla U = (\partial U/\partial x)\mathbf{e}_x$<br>$+ (\partial U/\partial y)\mathbf{e}_y$<br>$+ (\partial U/\partial z)\mathbf{e}_z$ | $(\nabla U)_\rho = \partial U/\partial\rho$<br>$(\nabla U)_\varphi = [\partial U/\partial\varphi]/\rho$<br>$(\nabla U)_z = \partial U/\partial z$ | $(\nabla U)_r = \partial U/\partial r$<br>$(\nabla U)_\theta = [\partial U/\partial\theta]/r$<br>$(\nabla U)_\varphi = [\partial U/\partial\varphi]/(r \sin\theta)$ |
| LAPLA-CIAN | $\Delta U = \frac{\partial^2 U}{\partial x^2} + \frac{\partial^2 U}{\partial y^2} + \frac{\partial^2 U}{\partial z^2}$ | $\Delta U = \frac{1}{\rho}\frac{\partial}{\partial\rho}\left(\rho\frac{\partial U}{\partial\rho}\right) + \frac{1}{\rho^2}\frac{\partial^2 U}{\partial\varphi^2} + \frac{\partial^2 U}{\partial z^2}$ | $\Delta U = \frac{1}{r}\frac{\partial^2}{\partial r^2}(rU) + \frac{1}{r^2\sin\theta}\frac{\partial}{\partial\theta}\left(\sin\theta\frac{\partial U}{\partial\theta}\right) + \frac{1}{r^2\sin^2\theta}\frac{\partial^2 U}{\partial\varphi^2}$ |
| DIVER-GENCE | $\nabla\cdot\mathbf{A} = \frac{\partial A_x}{\partial x} + \frac{\partial A_y}{\partial y} + \frac{\partial A_z}{\partial z}$ | $\nabla\cdot\mathbf{A} = \frac{1}{\rho}\frac{\partial}{\partial\rho}(\rho A_\rho) + \frac{1}{\rho}\frac{\partial A_\varphi}{\partial\varphi} + \frac{\partial A_z}{\partial z}$ | $\nabla\cdot\mathbf{A} = \frac{1}{r^2}\frac{\partial}{\partial r}(r^2 A_r) + \frac{1}{r\sin\theta}\frac{\partial}{\partial\theta}(\sin\theta A_\theta) + \frac{1}{r\sin\theta}\frac{\partial A_\varphi}{\partial\varphi}$ |
| CURL | $\nabla\times\mathbf{A} = (\partial A_z/\partial y - \partial A_y/\partial z)\mathbf{e}_x$<br>$+ (\partial A_x/\partial z - \partial A_z/\partial x)\mathbf{e}_y$<br>$+ (\partial A_y/\partial x - \partial A_x/\partial y)\mathbf{e}_z$ | $(\nabla\times\mathbf{A})_\rho = (\partial A_z/\partial\varphi)/\rho - \partial A_\varphi/\partial z$<br>$(\nabla\times\mathbf{A})_\varphi = \partial A_\rho/\partial z - \partial A_z/\partial\rho$<br>$(\nabla\times\mathbf{A})_z = [\partial(\rho A_\varphi)/\partial\rho - \partial A_\rho/\partial\varphi]/\rho$ | $(\nabla\times\mathbf{A})_r = [\partial(\sin\theta A_\varphi)/\partial\theta - \partial A_\theta/\partial\varphi]/(r\sin\theta)$<br>$(\nabla\times\mathbf{A})_\theta = [\partial A_r/\partial\varphi - \sin\theta\,\partial(rA_\varphi)/\partial r]/(r\sin\theta)$<br>$(\nabla\times\mathbf{A})_\varphi = [\partial(rA_\theta)/\partial r - \partial A_r/\partial\theta]/r$ |

## Useful Identities

$U$ : scalar field; $\mathbf{A}$, $\mathbf{B}$, . . . : vector fields.

$$\nabla \times (\nabla U) = 0 \qquad \nabla \cdot (\nabla U) = \Delta U$$

$$\nabla \cdot (\nabla \times \mathbf{A}) = 0 \qquad \nabla \times (\nabla \times \mathbf{A}) = \nabla (\nabla \cdot \mathbf{A}) - \nabla \mathbf{A}$$

$$\mathbf{L} = \frac{\hbar}{i} \mathbf{r} \times \nabla$$

$$\nabla = \frac{\mathbf{r}}{r} \frac{\partial}{\partial r} - \frac{i}{\hbar r^2} \mathbf{r} \times \mathbf{L}$$

$$\Delta = \frac{1}{r} \frac{\partial^2}{\partial r^2} r - \frac{\mathbf{L}^2}{\hbar^2 r^2}$$

$$\mathbf{A} \times (\mathbf{B} \times \mathbf{C}) = (\mathbf{A} \cdot \mathbf{C})\mathbf{B} - (\mathbf{A} \cdot \mathbf{B})\mathbf{C}$$

$$\mathbf{A} \times (\mathbf{B} \times \mathbf{C}) + \mathbf{B} \times (\mathbf{C} \times \mathbf{A}) + \mathbf{C} \times (\mathbf{A} \times \mathbf{B}) = 0$$

$$(\mathbf{A} \times \mathbf{B}) \cdot (\mathbf{C} \times \mathbf{D}) = (\mathbf{A} \cdot \mathbf{C})(\mathbf{B} \cdot \mathbf{D}) - (\mathbf{A} \cdot \mathbf{D})(\mathbf{B} \cdot \mathbf{C})$$

$$\begin{aligned}(\mathbf{A} \times \mathbf{B}) \times (\mathbf{C} \times \mathbf{D}) &= [(\mathbf{A} \times \mathbf{B}) \cdot \mathbf{D}]\mathbf{C} - [(\mathbf{A} \times \mathbf{B}) \cdot \mathbf{C}]\mathbf{D} \\ &= [(\mathbf{C} \times \mathbf{D}) \cdot \mathbf{A}]\mathbf{B} - [(\mathbf{C} \times \mathbf{D}) \cdot \mathbf{B}]\mathbf{A}\end{aligned}$$

$$\begin{aligned}
\nabla (UV) &= U \nabla V + V \nabla U \\
\Delta (UV) &= U \Delta V + 2(\nabla U) \cdot (\nabla V) + V \Delta U \\
\nabla \cdot (U\mathbf{A}) &= U \nabla \cdot \mathbf{A} + \mathbf{A} \cdot \nabla U \\
\nabla \times (U\mathbf{A}) &= U \nabla \times \mathbf{A} + (\nabla U) \times \mathbf{A} \\
\nabla \cdot (\mathbf{A} \times \mathbf{B}) &= \mathbf{B} \cdot (\nabla \times \mathbf{A}) - \mathbf{A} \cdot (\nabla \times \mathbf{B}) \\
\nabla (\mathbf{A} \cdot \mathbf{B}) &= \mathbf{A} \times (\nabla \times \mathbf{B}) + \mathbf{B} \times (\nabla \times \mathbf{A}) + \mathbf{B} \cdot \nabla \mathbf{A} + \mathbf{A} \cdot \nabla \mathbf{B} \\
\nabla \times (\mathbf{A} \times \mathbf{B}) &= \mathbf{A} (\nabla \cdot \mathbf{B}) - \mathbf{B} (\nabla \cdot \mathbf{A}) + \mathbf{B} \cdot \nabla \mathbf{A} - \mathbf{A} \cdot \nabla \mathbf{B}
\end{aligned}$$

N.B. : $\mathbf{B} \cdot \nabla \mathbf{A}$ vector field whose components are:

$$(\mathbf{B} \cdot \nabla \ \mathbf{A})_i = B_j \partial_j A_i = \sum_j B_j \frac{\partial}{\partial x_j} A_i$$

$$(i = x, y, z)$$